KIRK-OTHMER ENCYCLOPEDIA OF

CHEMICAL TECHNOLOGY

Fifth Edition

VOLUME 3

KIRK-OTHMER ENCYCLOPEDIA OF CHEMICAL TECHNOLOGY, FIFTH EDITION
EDITORIAL STAFF

Vice President, STM Books: **Janet Bailey**

Executive Editor: **Jacqueline I. Kroschwitz**

Editor: **Arza Seidel**

Managing Editor: **Michalina Bickford**

Director, Book Production and Manufacturing: **Camille P. Carter**

Production Manager: **Shirley Thomas**

Senior Production Editor: **Kellsee Chu**

Illustration Manager: **Dean Gonzalez**

Editorial Assistant: **Liam Kuhn**

KIRK-OTHMER ENCYCLOPEDIA OF

CHEMICAL TECHNOLOGY

Fifth Edition

VOLUME 3

Kirk-Othmer Encyclopedia of Chemical Technology
is available Online in full color and with additional content at
http://www3.interscience.wiley.com/cgi-bin/mrwhome/104554789/HOME.

Ⓦ **WILEY-INTERSCIENCE**

A John Wiley & Sons, Inc., Publication

Library of Congress Cataloging-in-Publication Data:

Kirk-Othmer encyclopedia of chemical technology. – 5th ed.
    p. cm.
Editor-in-chief, Arza Seidel.
"A Wiley-Interscience publication."
Includes index.
    ISBN 0-471-48494-6 (set) – ISBN 0-471-48520-9 (v. 3)
    1. Chemistry, Technical–Encyclopedias. I. Title: Encyclopedia of
chemical technology. II. Kroschwitz, Jacqueline I.
    TP9.K54 2004
    660'.03–dc22   2003021960

CONTENTS

CONTRIBUTORS

Elizabeth Aguinaldo, *FMC Corporation, Princeton, NJ,* Barium Compounds

Jerry Andrews, *Noveon Kalama, Inc., Kalama, WA,* Benzaldehyde, Benzoic Acid

George Blomgren, *Eveready Battery Co., Inc. West Lake, Ohio,* Batteries, Primary Cells

Claudio Boffito, *Saes Getters SpA, Milan, Italy,* Barium

Ralph Brodd, *Broddarp of Nevada, Inc., Henderson, NV,* Batteries, Introduction

Edward Brown, *Noveon Kalama, Inc., Kalama, WA,* Benzaldehyde, Benzoic Acid

Kathryn R. Bullock, *Johnson Controls, Inc., Milwaukee, WI,* Batteries, Secondary Cells

Karen Bush, *The R.W. Johnson Pharmaceutical Research Institute, Raritan, NJ,* Antibacterial Agents, Overview

S. C. Carapella, Jr., *Consultant,* Arsenic and Arsenic Alloys

Phillip DeLassus, *The University of Texas, Edinburg, TX,* Barrier Polymers

Martin Dexter, *Consultant,* Antioxidants, Polymers

Rocco DiSanto, *Cambrex Technical Center, North Brunswick, NJ,* Biocatalysis

Patrick Dibello, *FMC Corporation, Princeton, NJ,* Barium Compounds

G. O. Doak, *North Carolina State University, Raleigh, NC,* Antimony Compounds

Leon D. Freedman, *North Carolina State University, Raleigh, NC,* Antimony Compounds

William Fruscella, *Unocal Corporation El Segundo, CA,* Benzene

Paul Gifford, *Ovonic Battery Company, Troy, MI,* Batteries, Secondary Cells

Dasantila Golemi-Kotra, *Wayne State University, Detroit, MI,* Antibiotic Resistance

Martha Hesser, *Noveon Kalama, Inc., Kalama, WA,* Benzaldehyde, Benzoic Acid

Ramachandra S. Hosmane, *University of Maryland, Baltimore County, Baltimore, MD,* Antiviral Agents

James Hunter, *Eveready Battery Co., Inc. West Lake, OH,* Batteries, Primary Cells

Donald J. Kaczynski, *Brush Wellman, Inc. Oak Harbor, OH,* Beryllium, Beryllium Alloys and Composites; Beryllium Compounds

Sanjay Kamat, *Cambrex Technical Center, North Brunswick, NJ,* Biocatalysis

R. E. King III, *Ciba Specialty Chemicals, Tarrytown, NY,* Antioxidants, Polymers

Michael Ladisch, *Purdue University, West Lafayette, IN,* Bioseparations

Serena Laschi, *University of Florence, Florence, Italy,* Biosensors

Michael Lewis, *University of California, Davis, CA,* Beer and Brewing

T. Li, *ASARCO Inc., Phoenix, AZ,* Antimony and Antimony Alloys

Charles Lindahl, *Elf Atochem North America, Philadelphia, PA,* Antimony Compounds; Barium Compounds

G. Gilbert Long, *North Carolina State University, Raleigh, NC,* Antimony Compounds

Mark Macielag, *The R.W. Johnson Pharmaceutical Research Institute, Raritan, NJ,* Antibacterial Agents, Overview

Tariq Mahmood, *Elf Atochem North America, Philadelphia, PA,* Antimony Compounds; Barium Compounds

James Manganaro, *FMC Corporation, Princeton, NJ,* Barium Compounds

Marco Mascini, *University of Florence, Florence, Italy,* Biosensors

Shahriar Mobashery, *Wayne State University, Detroit, MI,* Antibiotic Resistance

Jarl Opgrande, *Noveon Kalama, Inc., Kalama, WA,* Benzaldehyde, Benzoic Acid

Mark Paisley, *Future Energy Resources Corporation, Columbus, OH,* Biomass Energy

Sachin Pannuri, *Cambrex Technical Center, North Brunswick, NJ,* Biocatalysis

Charles Anthony Peterson, *Intel Corporation,* Atomic Force Microscopy—AFM

John R. Pierson, *Johnson Controls, Inc., Milwaukee, WI,* Batteries, Secondary Cells

Roger C. Prince, *Exxon Research and Engineering Company,* Bioremediation

Jinhao Qiu, *Tohoku University, Sendai, Japan,* Biomaterials, Prosthetics, and Biomedical Devices

Alvin Salkind, *Rutgers University, New Brunswick, NJ,* Batteries, Secondary Cells

Dror Sarid, *University of Arizona, Tucson, AZ,* Atomic Force Microscopy—AFM

Rosalie A. Schnick, *Consultant, Lacrosse, WI,* Aquaculture Chemicals

Robert R. Stickney, *Texas A & M University, Bryan, TX,* Aquaculture

Michael Szycher, *CardioTech International, Inc., Woburn, MA,* Biomaterials, Prosthetics, and Biomedical Devices

Mami Tanaka, *Tohoku University, Sendai, Japan,* Biomaterials, Prosthetics, and Biomedical Devices

O. S. Thirunavukkarasu, *University of Regina, Regina, Saskatchewan, Canada,* Arsenic, Environmental Impact, Health Effects, and Treatment Methods

Richard Thomas, *Ciba Specialty Chemicals, Tarrytown, NY,* Antioxidants, Polymers

Jefferson Tilley, *Hoffmann LaRoche, Inc., Nutley, NJ,* Antiobesity Drugs

Anthony P. F. Turner, *Cranfield University at Silsoe, Bedfordshire, United Kingdom,* Biosensors

Robert Virta, *U.S. Geological Survey,* Asbestos

T. Viraraghavan, *University of Regina, Regina, Saskatchewan, Canada,* Arsenic, Environmental Impact, Health Effects, and Treatment Methods

Michele A. Weidner-Wells, *The R.W. Johnson Pharmaceutical Research Institute, Raritan, NJ,* Antibacterial Agents, Overview

Michael Zviely, *Frutarom, Ltd. Haifa, Israel,* Aroma Chemicals

CONVERSION FACTORS, ABBREVIATIONS, AND UNIT SYMBOLS

SI Units (Adopted 1960)

The International System of Units (abbreviated SI), is implemented throughout the world. This measurement system is a modernized version of the MKSA (meter, kilogram, second, ampere) system, and its details are published and controlled by an international treaty organization (The International Bureau of Weights and Measures) (1).

SI units are divided into three classes:

BASE UNITS

length	meter[†] (m)
mass	kilogram (kg)
time	second (s)
electric current	ampere (A)
thermodynamic temperature[‡]	kelvin (K)
amount of substance	mole (mol)
luminous intensity	candela (cd)

SUPPLEMENTARY UNITS

plane angle	radian (rad)
solid angle	steradian (sr)

DERIVED UNITS AND OTHER ACCEPTABLE UNITS

These units are formed by combining base units, suplementary units, and other derived units (2–4). Those derived units having special names and symbols are marked with an asterisk in the list below.

[†] The spellings "metre" and "litre" are preferred by ASTM; however, "-er" is used in the *Encyclopedia*.

[‡] Wide use is made of Celsius temperature (t) defined by

$$t = T - T_0$$

where T is the thermodynamic temperature, expressed in kelvin, and $T_0 = 273.15$ K by definition. A temperature interval may be expressed in degrees Celsius as well as in kelvin.

Quantity	Unit	Symbol	Acceptable equivalent
*absorbed dose	gray	Gy	J/Kg
acceleration	meter per second squared	m/s^2	
*activity (of a radionuclide)	becquerel	Bq	1/s
area	square kilometer	km^2	
	square hectometer	hm^2	ha (hectare)
	square meter	m^2	
concentration (of amount of substance)	mole per cubic meter	mol/m^3	
current density	ampere per square meter	A/m^2	
density, mass density	kilogram per cubic meter	kg/m^3	g/L; mg/cm^3
dipole moment (quantity)	coulomb meter	$C \cdot m$	
*dose equivalent	sievert	Sv	J/kg
*electric capacitance	farad	F	C/V
*electric charge, quantity of electricity	coulomb	C	$A \cdot s$
electric charge density	coulomb per cubic meter	C/m^3	
*electric conductance	siemens	S	A/V
electric field strength	volt per meter	V/m	
electric flux density	coulomb per square meter	C/m^2	
*electric potential, potential difference, electromotive force	volt	V	W/A
*electric resistance	ohm	Ω	V/A
*energy, work, quantity of heat	megajoule	MJ	
	kilojoule	kJ	
	joule	J	$N \cdot m$
	electronvolt[†]	eV[†]	
	kilowatt-hour[†]	$kW \cdot h$[†]	
energy density	joule per cubic meter	J/m^3	
*force	kilonewton	kN	
	newton	N	$kg \cdot m/s^2$

[†]This non-SI unit is recognized by the CIPM as having to be retained because of practical importance or use in specialized fields (1).

Quantity	Unit	Symbol	Acceptable equivalent
*frequency	megahertz	MHz	
	hertz	Hz	1/s
heat capacity, entropy	joule per kelvin	J/K	
heat capacity (specific), specific entropy	joule per kilogram kelvin	J/(kg·K)	
heat-transfer coefficient	watt per square meter kelvin	W/(m²·K)	
*illuminance	lux	lx	lm/m²
*inductance	henry	H	Wb/A
linear density	kilogram per meter	kg/m	
luminance	candela per square meter	cd/m²	
*luminous flux	lumen	lm	cd·sr
magnetic field strength	ampere per meter	A/m	
*magnetic flux	weber	Wb	V·s
*magnetic flux density	tesla	T	Wb/m²
molar energy	joule per mole	J/mol	
molar entropy, molar heat capacity	joule per mole kelvin	J/(mol·K)	
moment of force, torque	newton meter	N·m	
momentum	kilogram meter per second	kg·m/s	
permeability	henry per meter	H/m	
permittivity	farad per meter	F/m	
*power, heat flow rate, radiant flux	kilowatt	kW	
	watt	W	J/s
power density, heat flux density, irradiance	watt per square meter	W/m²	
*pressure, stress	megapascal	MPa	
	kilopascal	kPa	
	pascal	Pa	N/m²
sound level	decibel	dB	
specific energy	joule per kilogram	J/kg	
specific volume	cubic meter per kilogram	m³/kg	
surface tension	newton per meter	N/m	
thermal conductivity	watt per meter kelvin	W/(m·K)	
velocity	meter per second	m/s	
	kilometer per hour	km/h	
viscosity, dynamic	pascal second	Pa·s	
	millipascal second	mPa·s	
viscosity, kinematic	square meter per second	m²/s	
	square millimeter per second	mm²/s	

Quantity	Unit	Symbol	Acceptable equivalent
volume	cubic meter	m^3	
	cubic diameter	dm^3	L (liter) (5)
	cubic centimeter	cm^3	mL
wave number	1 per meter	m^{-1}	
	1 per centimeter	cm^{-1}	

In addition, there are 16 prefixes used to indicate order of magnitude, as follows

Multiplication factor	Prefix	symbol	Note
10^{18}	exa	E	
10^{15}	peta	P	
10^{12}	tera	T	
10^9	giga	G	
10^6	mega	M	
10^3	kilo	k	
10^2	hecto	h^a	[a]Although hecto, deka, deci, and
10	deka	da^a	centi are SI prefixes, their use
10^{-1}	deci	d^a	should be avoided except for SI
10^{-2}	centi	c^a	unit-multiples for area and
10^{-3}	milli	m	volume and nontechnical use of
10^{-6}	micro	μ	centimeter, as for body and
10^{-9}	nano	n	clothing measurement.
10^{-12}	pico	p	
10^{-15}	femto	f	
10^{-18}	atto	a	

For a complete description of SI and its use the reader is referred to ASTM E380 (4) and the article UNITS AND CONVERSION FACTORS which appears in Vol. 24.

A representative list of conversion factors from non-SI to SI units is presented herewith. Factors are given to four significant figures. Exact relationships are followed by a dagger. A more complete list is given in the latest editions of ASTM E380 (4) and ANSI Z210.1 (6).

Conversion Factors to SI Units

To convert from	To	Multiply by
acre	square meter (m^2)	4.047×10^3
angstrom	meter (m)	1.0×10^{-10}[†]
are	square meter (m^2)	1.0×10^{2}[†]
astronomical unit	meter (m)	1.496×10^{11}

[†]Exact.

To convert from	To	Multiply by
atmosphere, standard	pascal (Pa)	1.013×10^5
bar	pascal (Pa)	$1.0 \times 10^{5\dagger}$
barn	square meter (m²)	$1.0 \times 10^{-28\dagger}$
barrel (42 U.S. liquid gallons)	cubic meter (m³)	0.1590
Bohr magneton (μ_B)	J/T	9.274×10^{-24}
Btu (International Table)	joule (J)	1.055×10^3
Btu (mean)	joule (J)	1.056×10^3
Btu (thermochemical)	joule (J)	1.054×10^3
bushel	cubic meter(m³)	3.524×10^{-2}
calorie (International Table)	joule (J)	4.187
calorie (mean)	joule (J)	4.190
calorie (thermochemical)	joule (J)	$4.184^\dagger$
centipoise	pascal second (Pa·s)	$1.0 \times 10^{-3\dagger}$
centistokes	square millimeter per second (mm²/s)	$1.0^\dagger$
cfm (cubic foot per minute)	cubic meter per second (m³s)	4.72×10^{-4}
cubic inch	cubic meter (m³)	1.639×10^{-5}
cubic foot	cubic meter (m³)	2.832×10^{-2}
cubic yard	cubic meter (m³)	0.7646
curie	becquerel (Bq)	$3.70 \times 10^{10\dagger}$
debye	coulomb meter (C·m)	3.336×10^{-30}
degree (angle)	radian (rad)	1.745×10^{-2}
denier (international)	kilogram per meter (kg/m)	1.111×10^{-7}
	tex[‡]	0.1111
dram (apothecaries')	kilogram (kg)	3.888×10^{-3}
dram (avoirdupois)	kilogram (kg)	1.772×10^{-3}
dram (U.S. fluid)	cubic meter (m³)	3.697×10^{-6}
dyne	newton (N)	$1.0 \times 10^{-5\dagger}$
dyne/cm	newton per meter (N/m)	$1.0 \times 10^{-3\dagger}$
electronvolt	joule (J)	1.602×10^{-19}
erg	joule (J)	$1.0 \times 10^{-7\dagger}$
fathom	meter (m)	1.829
fluid ounce (U.S.)	cubic meter (m³)	2.957×10^{-5}
foot	meter (m)	$0.3048^\dagger$
footcandle	lux (lx)	10.76
furlong	meter (m)	2.012×10^{-2}
gal	meter per second squared (m/s²)	$1.0 \times 10^{-2\dagger}$
gallon (U.S. dry)	cubic meter (m³)	4.405×10^{-3}
gallon (U.S. liquid)	cubic meter (m³)	3.785×10^{-3}
gallon per minute (gpm)	cubic meter per second (m³/s)	6.309×10^{-5}
	cubic meter per hour (m³/h)	0.2271

[†]Exact.
[‡]See footnote on p. ix.

To convert from	To	Multiply by
gauss	tesla (T)	1.0×10^{-4}
gilbert	ampere (A)	0.7958
gill (U.S.)	cubic meter (m^3)	1.183×10^{-4}
grade	radian	1.571×10^{-2}
grain	kilogram (kg)	6.480×10^{-5}
gram force per denier	newton per tex (N/tex)	8.826×10^{-2}
hectare	square meter (m^2)	$1.0 \times 10^{4\dagger}$
horsepower (550 ft · lbf/s)	watt (W)	7.457×10^2
horsepower (boiler)	watt (W)	9.810×10^3
horsepower (electric)	watt (W)	$7.46 \times 10^{2\dagger}$
hundredweight (long)	kilogram (kg)	50.80
hundredweight (short)	kilogram (kg)	45.36
inch	meter (m)	$2.54 \times 10^{-2\dagger}$
inch of mercury (32°F)	pascal (Pa)	3.386×10^3
inch of water (39.2°F)	pascal (Pa)	2.491×10^2
kilogram-force	newton (N)	9.807
kilowatt hour	megajoule (MJ)	$3.6^\dagger$
kip	newton (N)	4.448×10^3
knot (international)	meter per second (m/S)	0.5144
lambert	candela per square meter (cd/m^3)	3.183×10^3
league (British nautical)	meter (m)	5.559×10^3
league (statute)	meter (m)	4.828×10^3
light year	meter (m)	9.461×10^{15}
liter (for fluids only)	cubic meter (m^3)	$1.0 \times 10^{-3\dagger}$
maxwell	weber (Wb)	$1.0 \times 10^{-8\dagger}$
micron	meter (m)	$1.0 \times 10^{-6\dagger}$
mil	meter (m)	$2.54 \times 10^{-5\dagger}$
mile (statue)	meter (m)	1.609×10^3
mile (U.S. nautical)	meter (m)	$1.852 \times 10^{3\dagger}$
mile per hour	meter per second (m/s)	0.4470
millibar	pascal (Pa)	1.0×10^2
millimeter of mercury (0°C)	pascal (Pa)	$1.333 \times 10^{2\dagger}$
minute (angular)	radian	2.909×10^{-4}
myriagram	kilogram (Kg)	10
myriameter	kilometer (Km)	10
oersted	ampere per meter (A/m)	79.58
ounce (avoirdupois)	kilogram (kg)	2.835×10^{-2}
ounce (troy)	kilogram (kg)	3.110×10^{-2}
ounce (U.S. fluid)	cubic meter (m^3)	2.957×10^{-5}
ounce-force	newton (N)	0.2780
peck (U.S.)	cubic meter (m^3)	8.810×10^{-3}
pennyweight	kilogram (kg)	1.555×10^{-3}
pint (U.S. dry)	cubic meter (m^3)	5.506×10^{-4}

†Exact.

To convert from	To	Multiply by
pint (U.S. liquid)	cubic meter (m^3)	4.732×10^{-4}
poise (absolute viscosity)	pascal second (Pa·s)	$0.10^{\dagger}$
pound (avoirdupois)	kilogram (kg)	0.4536
pound (troy)	kilogram (kg)	0.3732
poundal	newton (N)	0.1383
pound-force	newton (N)	4.448
pound force per square inch (psi)	pascal (Pa)	6.895×10^3
quart (U.S. dry)	cubic meter (m^3)	1.101×10^{-3}
quart (U.S. liquid)	cubic meter (m^3)	9.464×10^{-4}
quintal	kilogram (kg)	$1.0 \times 10^{-2\dagger}$
rad	gray (Gy)	$1.0 \times 10^{-2\dagger}$
rod	meter (m)	5.029
roentgen	coulomb per kilogram (C/kg)	2.58×10^{-4}
second (angle)	radian (rad)	$4.848 \times 10^{-6\dagger}$
section	square meter (m^2)	2.590×10^6
slug	kilogram (kg)	14.59
spherical candle power	lumen (lm)	12.57
square inch	square meter (m^2)	6.452×10^{-4}
square foot	square meter (m^2)	9.290×10^{-2}
square mile	square meter (m^2)	2.590×10^6
square yard	square meter (m^2)	0.8361
stere	cubic meter (m^3)	$1.0^{\dagger}$
stokes (kinematic viscosity)	square meter per second (m^2/s)	$1.0 \times 10^{-4\dagger}$
tex	kilogram per meter (kg/m)	$1.0 \times 10^{-6\dagger}$
ton (long, 2240 pounds)	kilogram (kg)	1.016×10^3
ton (metric) (tonne)	kilogram (kg)	$1.0 \times 10^{3\dagger}$
ton (short, 2000 pounds)	kilogram (kg)	9.072×10^2
torr	pascal (Pa)	1.333×10^2
unit pole	weber (Wb)	1.257×10^{-7}
yard	meter (m)	$0.9144^{\dagger}$

†Exact.

Abbreviations and Unit Symbols

Following is a list of common abbreviations and unit symbnols used in the Encyclopedia. In general they agree with those listed in *American National Standard Abbreviations for Use on Drawings and in Text* (*ANSI Y1.1*) (6) and *American National Standard Letter Symbols for Units in Science and Technology* (*ANSI Y10*) (6). Also included is a list of acronyms for a number of private and government organizations as well as common industrial solvents, polymers, and other chemicals.

Rules for Writing Unit Symbols (4):

1. Unit symbols are printed in upright letters (roman) regardless of the type style used in the surrounding text.
2. Unit symbols are unaltered in the plural.
3. Unit symbols are not followed by a period except when used at the end of a sentence.
4. Letter unit symbols are generally printed lower-case (for example, cd for candela) unless the unit name has been derived from a proper name, in which case the first letter of the symbol is capitalized (W, Pa). Prefixes and unit symbols retain their prescribed form regardless of the surrounding typography.
5. In the complete expression for a quantity, a space should be left between the numerical value and the unit symbol. For example, write 2.37 lm, *not* 2.37 lm, and 35 mm, *not* 35 mm. When the quantity is used in an adjectival sense, a hyphen is often used, for example, 35-mm film. *Exception:* No space is left between the numerical value and the symbols of degree, minute, and second of plane angle, degree Celsius, and the percent sign.
6. No space is used between the prefix and unit symbol (for example, kg).
7. Symbols, not abbreviations, should be used for units. For example, use "A," not "amp," for ampere.
8. When multiplying unit symbols, use a raised dot:

$$N \cdot m \text{ for newton meter}$$

In the case of $W \cdot h$, the dot may be omitted, thus:

$$Wh$$

An exception to this practice is made for computer printouts, automatic typewriter work, etc, where the raised dot is not possible, and a dot on the line may be used.

9. When dividing unit symbols, use one of the following forms:

$$m/s \quad or \quad m \cdot s^{-1} \quad or \quad \frac{m}{s}$$

In no case should more than one slash be used in the same expression unless parentheses are inserted to avoid ambiguity. For example, write:

$$J/(mol \cdot K) \quad or \quad J \cdot mol^{-1} \cdot K^{-1} \quad or \quad (J/mol)/K$$

but *not*

$$J/mol/K$$

10. Do not mix symbols and unit names in the same expression. Write:

$$\text{joules per kilogram} \quad or \quad \text{J/kg} \quad or \quad \text{J} \cdot \text{kg}^{-1}$$

but *not*

$$\text{joules/kilogram} \quad nor \quad \text{Joules/kg} \quad nor \quad \text{Joules} \cdot \text{kg}^{-1}$$

ABBREVIATIONS AND UNITS

A	ampere	AOAC	Association of Official Analytical Chemists
A	anion (eg, HA)		
A	mass number	AOCS	American Oil Chemists' Society
a	atto (prefix for 10^{-18})		
AATCC	American Association of Textile Chemists and Colorists	APHA	American Public Health Association
		API	American Petroleum Institute
ABS	acrylonitrile–butadiene–styrene		
		aq	aqueous
abs	absolute	Ar	aryl
ac	alternating current, *n.*	*ar-*	aromatic
a-c	alternating current, *adj.*	*as-*	Asymmetric(al)
ac-	alicyclic	ASHRAE	American Society of Heating, Refrigerating, and Air Conditioning Engineers
acac	acetylacetonate		
ACGIH	American Conference of Governmental Industrial Hygienists		
		ASM	American Society for Metals
ACS	American Chemical Society	ASME	American Society of Mechanical Engineers
AGA	American Gas Association		
Ah	ampere hour	ASTM	American Society for Testing and Materials
AIChE	American Institute of Chemical Engineers		
		at no.	atomic number
AIME	American Institute of Mining, metallurgical, and Petroleum Engineers	at wt	atomic weight
		av(g)	average
		AWS	American Welding Society
		b	bonding orbital
AIP	American Institute of Physics	bbl	barrel
		bcc	body-centered cubic
AISI	American Iron and Steel Institute	BCT	body-centered tetragonal
		Bé	Baumé
alc	alcohol(ic)	BET	Brunauer-Emmett-Teller (adsorption equation)
Alk	alkyl		
alk	alkaline (not alkali)	bid	twice daily
amt	amount	Boc	*t*-butyloxycarbonyl
amu	atomic mass unit	BOD	biochemical (biological) oxygen demand
ANSI	American National Standards Institute		
		bp	boiling point
AO	atomic orbital	Bq	becquerel

C	coulomb	dil	dilute
°C	degree Celsius	DIN	Deutsche Industrie
C-	denoting attachment to		Normen
	carbon	*dl-*; DL-	racemic
c	centi (prefix for 10^{-2})	DMA	dimethylacetamide
c	critical	DMF	dimethylformamide
ca	circa (Approximately)	DMG	dimethyl glyoxime
cd	candela; current density;	DMSO	dimethyl sulfoxide
	circular dichroism	DOD	Department of Defense
CFR	Code of Federal	DOE	Department of Energy
	Regulations	DOT	Department of
cgs	centimeter-gram-second		Transportation
CI	Color Index	DP	degree of polymerization
cis-	isomer in which	dp	dew point
	substituted groups are	DPH	diamond pyramid
	on some side of double		hardness
	bond between C atoms	dstl(d)	distill(ed)
cl	carload	dta	differential thermal
cm	centimeter		analysis
cmil	circular mil	*(E)-*	entgegen; opposed
cmpd	compound	ϵ	dielectric constant
CNS	central nervous system		(unitless number)
CoA	coenzyme A	*e*	electron
COD	chemical oxygen demand	ECU	electrochemical unit
coml	commerical(ly)	ed.	edited, edition, editor
cp	chemically pure	ED	effective dose
cph	close-packed hexagonal	EDTA	ethylenediaminetetra-
CPSC	Consumer Product Safety		acetic acid
	Commission	emf	electromotive force
cryst	crystalline	emu	electromagnetic unit
cub	cubic	en	ethylene diamine
D	debye	eng	engineering
D-	denoting configurational	EPA	Environmental Protection
	relationship		Agency
d	differential operator	epr	electron paramagnetic
d	day; deci (prefix for 10^{-1})		resonance
d	density	eq.	equation
d-	*dextro-*, dextrorotatory	esca	electron spectroscopy for
da	deka (prefix for 10^{-1})		chemical analysis
dB	decibel	esp	especially
dc	direct current, *n.*	esr	electron-spin resonance
d-c	direct current, *adj.*	est(d)	estimate(d)
dec	decompose	estn	estimation
detd	determined	esu	electrostatic unit
detn	determination	exp	experiment, experimental
Di	didymium, a mixture of all	ext(d)	extract(ed)
	lanthanons	F	farad (capacitance)
dia	diameter	*F*	fraday (96,487 C)

f	femto (prefix for 10^{-15})	hyd	hydrated, hydrous
FAO	Food and Agriculture Organization (United Nations)	hyg	hygroscopic
		Hz	hertz
		^{i}(eg, Pr^i)	iso (eg, isopropyl)
fcc	face-centered cubic	i-	inactive (eg, i-methionine)
FDA	Food and Drug Administration	IACS	international Annealed Copper Standard
FEA	Federal Energy Administration	ibp	initial boiling point
		IC	integrated circuit
FHSA	Federal Hazardous Substances Act	ICC	Interstate Commerce Commission
fob	free on board	ICT	International Critical Table
fp	freezing point		
FPC	Federal Power Commission	ID	inside diameter; infective dose
FRB	Federal Reserve Board		
frz	freezing	ip	intraperitoneal
G	giga (prefix for 10^9)	IPS	iron pipe size
G	gravitational constant $= 6.67 \times 10^{11} \mathrm{N} \cdot \mathrm{m}^2/\mathrm{kg}^2$	ir	infrared
		IRLG	Interagency Regulatory Liaison Group
g	gram		
(g)	gas, only as in H_2O(g)	ISO	International Organization Standardization
g	gravitatonal acceleration		
gc	gas chromatography		
gem-	geminal	ITS-90	International Temperature Scale (NIST)
glc	gas–liquid chromatography		
g-mol wt; gmw	gram-molecular weight	IU	International Unit
		IUPAC	International Union of Pure and Applied Chemistry
GNP	gross national product		
gpc	gel-permeation chromatography		
		IV	iodine value
GRAS	Generally Recognized as Safe	iv	intravenous
		J	joule
grd	ground	K	kelvin
Gy	gray	k	kilo (prefix for 10^3)
H	henry	kg	kilogram
h	hour; hecto (prefix for 10^2)	L	denoting configurational relationship
ha	hectare		
HB	Brinell hardness number	L	liter (for fluids only) (5)
Hb	hemoglobin	l-	*levo*-, levorotatory
hcp	hexagonal close-packed	(l)	liquid, only as in NH_3(l)
hex	hexagonal	LC$_{50}$	conc lethal to 50% of the animals tested
HK	Knoop hardness number		
hplc	high performance liquid chromatography	LCAO	linear combnination of atomic orbitals
HRC	Rockwell hardness (C scale)	lc	liquid chromatography
		LCD	liquid crystal display
HV	Vickers hardness number	lcl	less than carload lots

LD_{50}	dose lethal to 50% of the animals tested	N	newton (force)
		N	normal (concentration); neutron number
LED	light-emitting diode		
liq	liquid	N-	denoting attachment to nitrogen
lm	lumen		
ln	logarithm (natural)	n (as n_D^{20})	index of refraction (for 20°C and sodium light)
LNG	liquefied natural gas		
log	logarithm (common)		
LOI	limiting oxygen index	n (as Bu^n),	normal (straight-chain structure)
LPG	liquefied petroleum gas	n-	
ltl	less than truckload lots	n	neutron
lx	lux	n	nano (prefix for 10^9)
M	mega (prefix for 10^6); metal (as in MA)	na	not available
		NAS	National Academy of Sciences
M	molar; actual mass		
$\overline{M}_w$	weight-average mol wt	NASA	National Aeronautics and Space Administration
$\overline{M}_n$	number-average mol wt		
m	meter; milli (prefix for 10^{-3})	nat	natural
		ndt	nondestructive testing
m	molal	neg	negative
m-	meta	NF	*National Formulary*
max	maximum	NIH	National Institutes of Health
MCA	Chemical Manufacturers' Association (was Manufacturing Chemists Association)		
		NIOSH	National Institute of Occupational Safety and Health
MEK	methyl ethyl ketone	NIST	National Institute of Standards and Technology (formerly National Bureau of Standards)
meq	milliequivalent		
mfd	manufactured		
mfg	manufacturing		
mfr	manufacturer		
MIBC	methyl isobutyl carbinol	nmr	nuclear magnetic resonance
MIBK	methyl isobutyl ketone		
MIC	minimum inhibiting concentration	NND	New and Nonofficial Drugs (AMA)
min	minute; minimum	no.	number
mL	milliliter	NOI-(BN)	not otherwise indexed (by name)
MLD	minimum lethal dose		
MO	molecular orbital	NOS	not otherwise specified
mo	month	nqr	nuclear quadruple resonance
mol	mole		
mol wt	molecular weight	NRC	Nuclear Regulatory Commission; National Research Council
mp	melting point		
MR	molar refraction		
ms	mass spectrometry	NRI	New Ring Index
MSDS	material safety data sheet	NSF	National Science Foundation
mxt	mixture		
μ	micro (prefix for 10^{-6})	NTA	nitrilotriacetic acid

NTP	normal temperature and pressure (25°C and 101.3 kPa or 1 atm)	pwd	powder
		py	pyridine
		qv	quod vide (which see)
NTSB	National Transportation Safety Board	R	univalent hydrocarbon radical
O-	denoting attachment to oxygen	(*R*)-	rectus (clockwise configuration)
o-	ortho	*r*	precision of data
OD	outside diameter	rad	radian; radius
OPEC	Organization of Petroleum Exporting Countries	RCRA	Resource Conservation and Recovery Act
o-phen	*o*-phenanthridine	rds	rate-determining step
OSHA	Occupational Safety and Health Administration	ref.	reference
		rf	radio frequency, *n.*
owf	on weight of fiber	r-f	radio frequency, *adj.*
Ω	ohm	rh	relative humidity
P	peta (prefix for 10^{15})	RI	Ring Index
p	pico (prefix for 10^{-12})	rms	root-mean square
p-	para	rpm	rotations per minute
p	proton	rps	revolutions per second
p.	page	RT	room temperature
Pa	Pascal (pressure)	RTECS	Registry of Toxic Effects of Chemical Substances
PEL	personal exposure limit based on an 8-h exposure	^{s}(eg, Bu^s); *sec*-	secondary (eg, secondary butyl)
pd	potential difference	S	siemens
pH	negative logarithm of the effective hydrogen ion concentration	(*S*)-	sinister (counterclockwise configuration)
		S-	denoting attachment to sulfur
phr	parts per hundred of resin (rubber)	*s*-	symmetric(al)
p-i-n	positive-intrinsic-negative	S	second
pmr	proton magnetic resonance	(s)	solid, only as in $H_2O(s)$
p-n	positive-negative	SAE	Society of Automotive Engineers
po	per os (oral)		
POP	polyoxypropylene	SAN	styrene-acrylonitrile
pos	positive	sat(d)	saturate(d)
pp.	pages	satn	saturation
ppb	parts per billion (10^9)	SBS	styrene–butadiene–styrene
ppm	parts per milion (10^6)	sc	subcutaneous
ppmv	parts per million by volume	SCF	self-consistent field; standard cubic feet
ppmwt	parts per million by weight		
PPO	poly(phenyl oxide)	Sch	Schultz number
ppt(d)	precipitate(d)	sem	scanning electron microscope(y)
pptn	precipitation		
Pr (no.)	foreign prototype (number)	SFs	Saybolt Furol seconds
pt	point; part	sl sol	slightly soluble
PVC	poly(vinyl chloride)	sol	soluble

soln	solution	*trans-*	isomer in which substituted groups are on opposite sides of double bond between C atoms
soly	solubility		
sp	specific; species		
sp gr	specific gravity		
sr	steradian		
std	standard	TSCA	Toxic Substances Control Act
STP	standard temperature and pressure (0°C and 101.3 kPa)		
		TWA	time-weighted average
		Twad	Twaddell
sub	sublime(s)	UL	Underwriters' Laboratory
SUs	Saybolt Universal seconds	USDA	United States Department of Agriculture
syn	synthetic		
t (eg, Bu^t), t-, tert-	tertiary (eg, tertiary butyl)	USP	*United States Pharmacopeia*
T	tera (prefix for 10^{12}); tesla (magnetic flux density)	uv	ultraviolet
		V	volt (emf)
t	metric to (tonne)	var	variable
t	temperature	*vic-*	vicinal
TAPPI	Technical Association of the Pulp and Paper Industry	vol	volume (not volatile)
		vs	versus
		v sol	very soluble
TCC	Tagliabue closed cup	W	watt
tex	tex (linear density)	Wb	weber
T_g	glass-transition temperature	Wh	watt hour
		WHO	World Health Organization (United Nations)
tga	thermogravimetric analysis		
		wk	week
THF	tetrahydrofuran	yr	year
tlc	thin layer chromatography	(Z)-	zusammen; together; atomic number
TLV	threshold limit value		

Non-SI (Unacceptable and Obsolete) Units		Use
Å	angstrom	nm
at	atmosphere, technical	Pa
atm	atmosphere, standard	Pa
b	barn	cm^2
bar†	bar	Pa
bbl	barrel	m^3
bhp	brake horsepower	W
Btu	British thermal unit	J
bu	bushel	m^3; L
cal	calorie	J
cfm	cubic foot per minute	m^3/s
Ci	curie	Bq
cSt	centistokes	mm^2/s
c/s	cycle per second	Hz
cu	cubic	exponential form

†Do not use bar (10^5 Pa) or millibar (10^2 Pa) because they are not SI units, and are accepted internationally only in special fields because of existing usage.

Non-SI (Unacceptable and Obsolete) Units		Use
D	debye	$C \cdot m$
den	denier	tex
dr	dram	kg
dyn	dyne	N
dyn/cm	dyne per centimeter	mN/m
erg	erg	J
eu	entropy unit	J/K
°F	degree Fahrenheit	°C; K
fc	footcandle	lx
fl	footlambert	lx
fl oz	fluid ounce	m^3; L
ft	foot	m
ft $\cdot$ lbf	foot pound-force	J
gf den	gram-force per denier	N/tex
G	gauss	T
Gal	gal	m/s^2
gal	gallon	m^3; L
Gb	gilbert	A
gpm	gallon per minute	(m^3/s); (m^3/h)
gr	grain	kg
hp	horsepower	W
ihp	indicated horsepower	W
in.	inch	m
in. Hg	inch of mercury	Pa
in. H_2O	inch of water	Pa
in.-lbf	inch pound-force	J
kcal	kilo-calorie	J
kgf	kilogram-force	N
kilo	for kilogram	kg
L	lambert	lx
lb	pound	kg
lbf	pound-force	N
mho	mho	S
mi	mile	m
MM	million	M
mm Hg	millimeter of mercury	Pa
mμ	millimicron	nm
mph	miles per hour	km/h
μ	micron	μm
Oe	oersted	A/m
oz	ounce	kg
ozf	ounce-force	N
η	poise	$Pa \cdot s$
P	poise	$Pa \cdot s$
ph	phot	lx
psi	pounds-force per square inch	Pa
psia	pounds-force per square inch absolute	Pa
psig	pounds-force per square inch gage	Pa
qt	quart	m^3; L
°R	degree Rankine	K
rd	rad	Gy
sb	stilb	lx
SCF	standard cubic foot	m^3
sq	square	exponential form
thm	therm	J
yd	yard	m

BIBLIOGRAPHY

1. The International Bureau of Weights and Measures, BIPM (Parc Saint-Cloud, France) is described in Ref. 4. This bureau operates under the exclusive supervision of the International Committee for Weights and Measures (CIPM).

2. *Metric Editorial Guide (ANMC-78-1)*, latest ed., American National Metric Council, 900 Mix Avenue, Suite 1 Hamden CT 06514-5106, 1981.

3. *SI Units and Recommendations for the Use of Their Multiples and of Certain Other Units (ISO 1000-1992)*, American National Standards Institute, 25 W 43rd St., New York, 10036, 1992.

4. Based on IEEE/ASTM-SI-10 *Standard for use of the International System of Units (SI): The Modern Metric System* (Replaces ASTM380 and ANSI/IEEE Std 268-1992), ASTM International, West Conshohocken, PA., 2002. See also www.astm.org

5. *Fed. Reg.*, Dec. 10, 1976 (41 FR 36414).

6. For ANSI address, see Ref. 3. See also www.ansi.org

A

Continued

ANTIBACTERIAL AGENTS, OVERVIEW

1. Introduction

Antibacterial agents are synthetic compounds derived from petrochemical sources and other small chemical building blocks that either kill or prevent the growth of bacteria. For the purposes of this survey, antibacterial agents are distinguished from antibiotics, antiseptics, disinfectants, and preservatives. Antibiotics are chemical substances isolated from natural sources, or their semi-synthetic derivatives, that kill microorganisms or inhibit their growth (see also ANTIBIOTICS). Antiseptics are chemical substances with antimicrobial properties that are used on the surface of living tissues, such as the skin or mucous membranes. In contrast to antibacterial agents and antibiotics, antiseptics do not necessarily exhibit selective toxicity for the microbial cell relative to the host cell. Disinfectants are chemical substances that kill microorganisms when applied to inanimate objects. Preservatives are generally static agents that slow the decomposition of organic substances by inhibiting the growth of microorganisms.

Antibacterial agents are commonly used to treat and/or prevent infections due to pathogenic bacteria in humans and animals. Although injectable dosage forms of some antibacterial agents have been developed, most of the drugs used in modern antibacterial chemotherapy were designed to achieve high systemic blood levels following oral administration. Consequently, antibacterial agents, with molecular weights in the range of 135–400 amu, tend to be less complex molecules than antibiotics. Given their synthetic origin, the antibacterial agents also contain few or no chiral centers, in contrast to antibiotics derived from natural sources, which frequently contain multiple contiguous stereocenters.

1

Thousands of analogues of antibacterial agents have been prepared in an effort to identify compounds with an enhanced spectrum of activity, improved pharmacokinetics, or a greater safety margin. Nevertheless, a relatively small number (~100) of antibacterial agents have been marketed for clinical or veterinary use.

The mechanism of action of antibacterial agents varies depending on the structural class. Some agents interfere with bacterial deoxyribonucleic acid (DNA) or protein synthesis (quinolones, oxazolidinones); others inhibit the activity of an enzyme or enzymes involved in bacterial cell metabolism (sulfonamides, diaminopyrimidines, nitrofurans, isoniazid, ethionamide). In the case of the sulfonamides and the antitubercular agents, the molecular target(s) are unique to the bacterial cell. The quinolones, oxazolidinones, and diaminopyrimidines are selectively toxic to bacteria owing to their greater affinity for the bacterial target than the mammalian counterpart. Some classes of antibacterial agents, such as the sulfonamides, nitrofurans, and oxazolidinones are bacteriostatic (ie, bacterial cell growth is inhibited). Others, such as the quinolones (against gram-positive and gram-negative bacteria) and isoniazid (against mycobacteria), are bactericidal (ie, bacteria are killed).

2. History

A systematic search for synthetic antiinfective agents began in the early 1900s with Ehrlich's pioneering research as Director of the Institute for Experimental Therapy in Frankfurt. In 1904, Ehrlich demonstrated the curative effect of the azo-dye trypan red (**1**) [574-64-1] in mice infected with trypanosomiasis (in humans, diseases such as sleeping sickness and Chagas' disease) (1). The drug was ineffective in humans, but this discovery proved that small molecular weight compounds were of value in the treatment of infectious diseases. Ehrlich's interest in organoarsenical compounds led to the subsequent discovery in 1909 of arsphenamine (Salvarsan) (**2**) [139-93-5] and neoarsphenamine (Neosalvarsan) (**3**) [457-60-3] as treatments for syphilis in humans (2), validating the scientific principles of chemotherapy he first enunciated in the late 1890s. During the next 20 years, the vast majority of research in antiinfective chemotherapy was directed toward the discovery of agents for the treatment of protozoal and parasitic diseases.

(**1**)

HO—⟨benzene⟩—As=As—⟨benzene⟩—OH with H$_2$N and NH$_2$ substituents

(2)

HO—⟨benzene⟩—As=As—⟨benzene⟩—OH with HN–CH$_2$–O–S–ONa and NH$_2$ substituents

(3)

Application of Ehrlich's principles to antibacterial research began in the mid-1930s with the work of Domagk and colleagues at the German chemical firm, I.G. Farbenindustrie. Domagk led the group investigating the effects of various chemical dyes in experimental animal models of infection. In 1932, he discovered that the azo-dye, sulfamidochrysoidine (Prontosil) (**4**) [103-12-8], was effective in preventing death in mice infected with hemolytic streptococci and in rabbits infected with staphylococci (3). Subsequent clinical studies demonstrated the remarkable properties of this drug in the treatment of puerperal sepsis, meningitis and pneumonia (4). Later research by a French group at the Institute Pasteur showed that sulfamidochrysoidine was actually a prodrug, which is metabolized in the body to the bioactive compound sulfanilamide (**5**) [63-74-1] (5). The identification of the active metabolite facilitated the rapid development of the antibacterial sulfonamides (the sulfa drugs) derived from the sulfanilamide molecular template (6). One of these analogues, sulfamethoxazole (**6**), is still widely used in the clinic. Domagk's discovery proved that it was possible to discover small synthetic agents effective against bacterial infection, and stimulated interest in the identification of additional substances with broad-spectrum antibacterial activity.

(4)   (5)   (6)

The work of Dodd, Stillman, and others on nitroheterocycle derivatives in the mid-1940s led to a number of nitrofuran drugs, among which nitrofurantoin (**7**) [67-20-9] and nitrofurazone (**8**) [59-87-0] are still used in clinical practice (7–9), in spite of multiple laboratory findings of mutagenicity (10,11). Nitrofurantoin, in particular, is active against a wide spectrum of gram-positive and gram-negative bacteria, including most urinary tract pathogens. Subsequently, the antitubercular effects of isoniazid (**9**) [54-85-3], pyrazinamide (**10**) [98-96-4] and ethionamide (**11**) [536-33-4] were described, based on earlier research that had uncovered the weak tuberculostatic activity of the structurally related compound, nicotinamide (**12**) (12–15). Optimization of the antimycobacterial activity of a series of ethylenediamine derivatives by Wilkinson and colleagues at Lederle Laboratories ultimately led to a structurally distinct antituberculous compound, ethambutol (**13**) [74-55-5] (16).

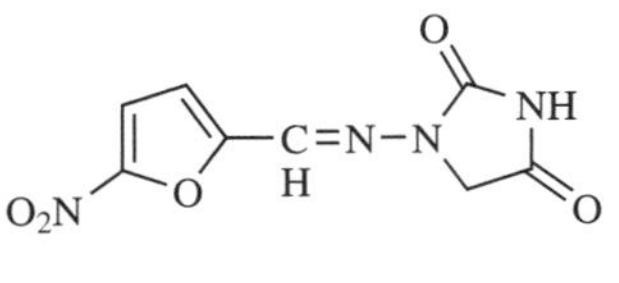

(7) 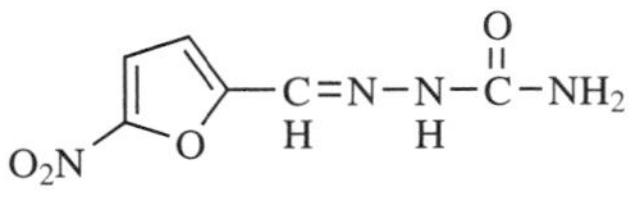   (8) 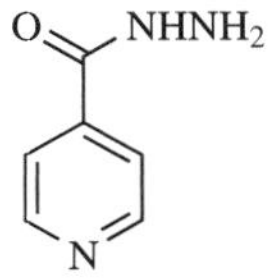   (9)

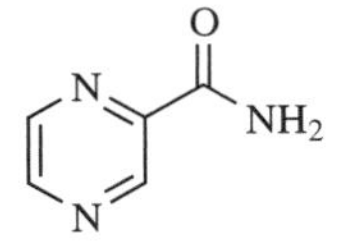

(10) 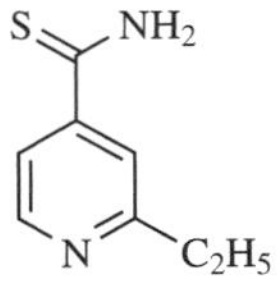   (11) 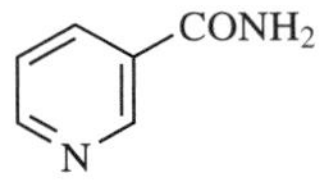   (12) 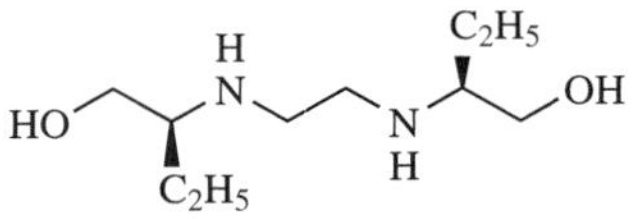   (13)

In the 1950s and 1960s, Hitchings and Elion of Wellcome Research Labs showed that selective inhibition of parasitic enzymes rather than the host enzymes could be exploited for chemotherapeutic ends. Through a series of iterative modifications of the non-selective dihydrofolate reductase inhibitor amethopterin (methotrexate) (**14**) [59-05-2], Hitchings discovered the simplified structure, pyrimethamine (**15**) [58-14-0], which is 2000-fold more selective for the dihydrofolate reductase from the malarial parasite, *Plasmodium berghei*, than the analogous mammalian enzymes (17). Despite increasing resistance to its action, pyrimethamine remains an important drug for the treatment of malaria in the tropical world. Further modification of the basic diaminopyrimidine scaffold led to trimethoprim (**16**) [738-70-5], which is particularly potent and selective for the dihydrofolate reductase from bacteria (18).

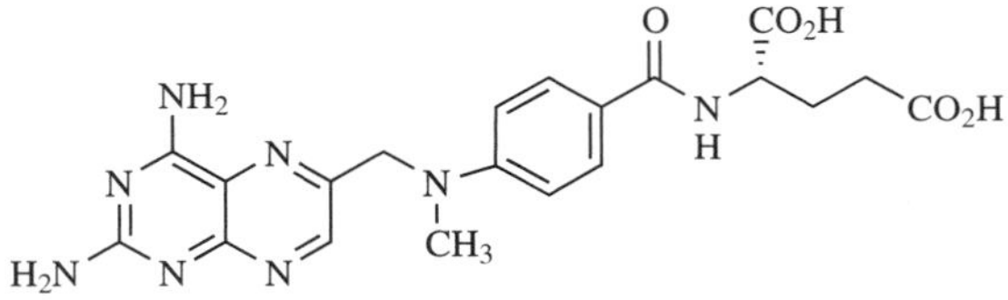

(14)

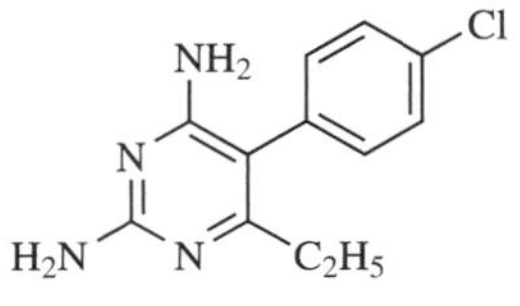

(15) 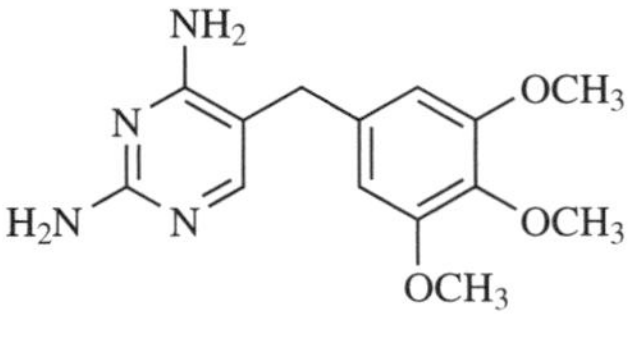   (16)

The next significant advance in the medicinal chemistry of antibacterial agents occurred in the late 1950s due to the astute observations of Lesher at Sterling-Winthrop Research Institute. As part of a study directed toward the characterization of by-products in the synthesis of the antimalarial drug chloroquine, Lesher and his colleagues isolated 7-chloro-1-ethyl-1,4-dihydro-4-oxo-3-quinolinecarboxylic acid (**17**) [16600-24-1], which showed weak *in vitro* antibacterial activity (19). Capitalizing on this discovery the Sterling chemists prepared a number of analogues, ultimately leading to the 1,8-naphthyridine derivative nalidixic acid (**18**) [389-08-2] (20). Although the antibacterial and pharmacokinetic properties of nalidixic acid limited its clinical use to gram-negative infections of the urinary tract, Lesher's discovery opened up a rich vein in antibacterial research, such that the quinolone class of antibacterial agents (including the 4-quinolones and 1,8-naphthyridine-4-ones) now accounts for 19.6% of the global antibacterial agents and antibiotics market (21). Structural modification of the quinolone nucleus afforded analogs, such as ciprofloxacin (**19**) [85721-33-1] and levofloxacin (**20**) [100986-85-4], with excellent oral bioavailability, good tissue distribution, prolonged serum half-lives and improved safety margins (22). Many of the newer quinolone antibacterial agents have increased activity against gram-positive bacteria (23).

(**17**)   (**18**)   (**19**)

(**20**)

The dramatic increase in the prevalence of multidrug-resistant bacterial pathogens during the 1970s and 1980s was the stimulus for further research into new antibacterial agents at several pharmaceutical companies (24). The oxazolidinones emerged as a viable new class of synthetic antibacterial agents following broad screening of the DuPont Corporation compound library. Analogues such as DuP 105 (**21**) [96800-41-8] and DuP 721 (**22**) [104421-21-8] were attractive as potential drug development candidates because they were active against the important hospital pathogen, methicillin-resistant *Staphylococcus aureus* (25). A concerted effort at Pharmacia Corporation to improve the safety and aqueous solubility of the class led to the market introduction in 2000 of linezolid (**23**) [165800-03-3], with activity against vancomycin-resistant enterococci (26). Linezolid represented the first new class of antibacterial agent to gain FDA approval in 35 years (27).

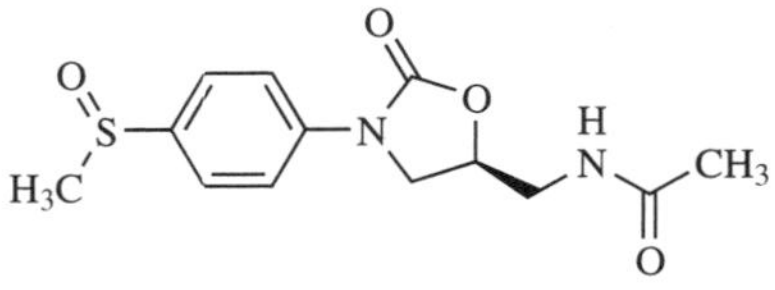

(21) 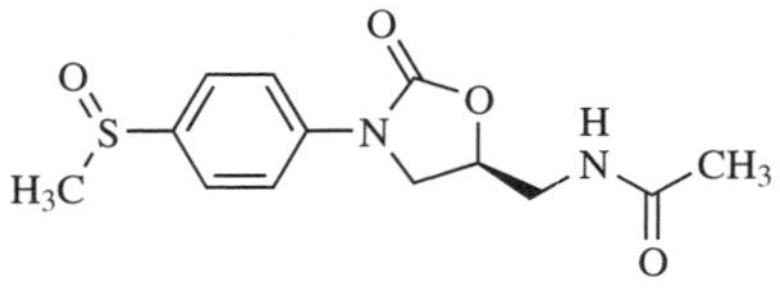   (22)

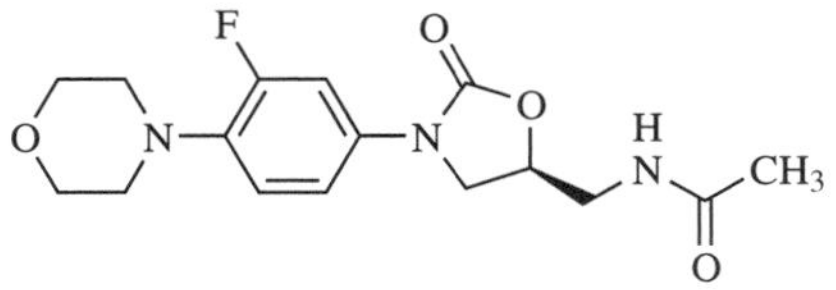

(23)

It is noteworthy that Domagk (1939) and Hitchings and Elion (1988) each received the Nobel Prize in Physiology or Medicine for their fundamental contributions to antibacterial chemotherapy. Table 1 lists years of historical significance to the development of selected synthetic antibacterial agents.

3. Nomenclature

Antibacterial agents are identified by three different types of names:

1. The chemical name is usually long and cumbersome and is based on conventional chemical nomenclature rules.
2. The generic name frequently has a common stem for a specific class of agents. For example, the generic names for the quinolone family end in

Table 1. **Year of Disclosure or Market Introduction of Selected Antibacterial Agents**

Antibacterial agent	CAS Registry Number	Year	
		Disclosure	Introduction
sulfamidochrysoidine	[103-12-8]	1932	1936
sulfapyridine	[144-83-2]	1938	
acetyl sulfisoxazole	[80-74-0]		1949
isoniazid	[54-85-3]		1952
pyrazinamide	[98-96-4]		1952
nitrofurantoin	[67-20-9]		1953
pyrimethamine	[58-14-0]	1950	
trimethoprim	[738-70-5]	1956	
nalidixic acid	[389-08-2]	1962	1965
trimethoprim–sulfamethoxazole	[8064-90-2]		1969
norfloxacin	[70458-96-7]	1978	1983
ciprofloxacin	[85721-33-1]	1983	1986
levofloxacin	[100986-85-4]	1987	1993
linezolid	[165800-03-3]	1995	2000

"-oxacin." Since this is a nonproprietary name, more than one brand name drug can have the same generic name.

3. The brand (trade) name is a proprietary name given by the manufacturer and is often based on commercial considerations.

The following example shows the difference between the three types of names for the same compound.

1. Chemical name—(S)-9-fluoro-2,3-dihydro-3-methyl-10-(4-methyl-1-piperazinyl)-7-oxo-7H-pyrido[1,2,3-de]-1,4-benzoxazine-6-carboxylic acid
2. Generic name—levofloxacin
3. Trade name—Levaquin

Generic names are usually preferred in scientific communications, and will be used when applicable in this report.

4. Classification of Antibacterial Agents

Antibacterial agents can be classified according to their molecular features. Agents within the same chemical family usually act by the same mechanism of action. However, since several chemical classes may exert the same, or closely related, mode of action, antibacterial agents may also be broadly classified according to the bacterial target affected. It is also possible to classify agents according to the therapeutic indication. For the purposes of this survey, antibacterial agents will be classified either according to the clinical indication for which the drug is used (eg, antitubercular agents), or according to the salient molecular features (eg, oxazolidinones).

4.1. Antitubercular Agents. The synthetic first-line antitubercular agents can be subdivided into two broad categories according to structure. The first group, consisting of isoniazid (**9**), pyrazinamide (**10**), and ethionamide (**11**) contain a heteroaryl hydrazide, amide, or thioamide. Isoniazid and ethionamide are metabolized by mycobacteria to electrophilic intermediates, which then inhibit the synthesis of mycolic acids essential to bacterial viability (28,29). Pyrazinamide has been shown to inhibit fatty acid synthesis by preventing the formation of precursors needed for the synthesis of mycolic acids (30). Conversion to pyrazinoic acid by a bacterial pyrazinamidase appears to be necessary, as mutations leading to the loss of enzymatic activity are a major mechanism of resistance. These agents all exhibit activity against *Mycobacterium tuberculosis*. Ethionamide also inhibits the growth of other slowly growing mycobacteria. Despite chemical similarities, these three agents do not always exhibit cross-resistance (31).

The second structural type of antitubercular agents is represented by ethambutol (**13**), which contains a symmetrical diamino-dihydroxy aliphatic chain. It has been proposed that ethambutol prevents mycobacterial cell wall synthesis by inhibiting the production of arabinan (32,33).

4.2. Nitrofurans. The nitrofuran class of antibacterial agents contain a 5-nitro-2-furanyl moiety. Compounds in this family (**7,8**) are usually hydrazone derivatives of 5-nitro-2-furancarboxaldehyde [698-63-5], of general structure (**24**). However, several compounds are olefinic derivatives, such as (**25**). An example of this type of nitrofuran is 3-(5-nitro-2-furyl)acrylamide (**26**) [710-25-8]. It has been shown that nitrofurans are converted by bacterial reductases to reactive intermediates that can inhibit a number of bacterial enzymes, including those responsible for DNA and ribonucleic acid (RNA) synthesis and carbohydrate metabolism (34). Due to the multiple mechanisms of action, resistance to the nitrofurans has not been a major concern. This class is active against a wide spectrum of gram-positive and gram-negative organisms, including enterococci and *Escherichia coli*, respectively, but has dropped out of widespread use in the United States for severe infections due to the discovery and development of new classes of agents.

(24)   (25)   (26)

4.3. Oxazolidinones. The oxazolidinone class of antibacterial agents, exemplified by linezolid (**23**), contains a 3-aryl-5-acetamidomethyloxazolidin-2-one (**27**) pharmacophore essential for biological activity. These agents selectively bind to the P site of the 23S RNA component of the 50S ribosomal subunit, thus inhibiting protein synthesis at an early stage of translation, possibly by inhibiting translocation of fMet-tRNA (35). Oxazolidinones have microbiological activity against a variety of susceptible and multidrug-resistant gram-positive organisms.

(27)

4.4. Quinolones. The 4-quinolone class of antibacterial agents (**28**) contains a 3-carboxylic acid attached to the core quinolone (Q = CR$_2$) or naphthyridone (Q = N) nucleus. In addition, the N-1 nitrogen is arylated or alkylated. The moiety at C-7, appended to the core via a carbon or a nitrogen atom, generally contains a basic amine, which enhances the antibacterial spectrum as well as improving *in vivo* efficacy. Substitution of fluorine at C-6 (Y = F) usually leads to a compound with enhanced potency against gram-positive organisms when compared to the analogous des-fluoro analogue (Y = H). The quinolones target the essential bacterial type II topoisomerases, DNA gyrase and topoisomerase IV, the relative potency depending on the organism and the specific compound.

Quinolones are broad-spectrum bactericidal agents against a variety of gram-positive and gram-negative pathogens, including some anaerobic bacteria and intracellular pathogens.

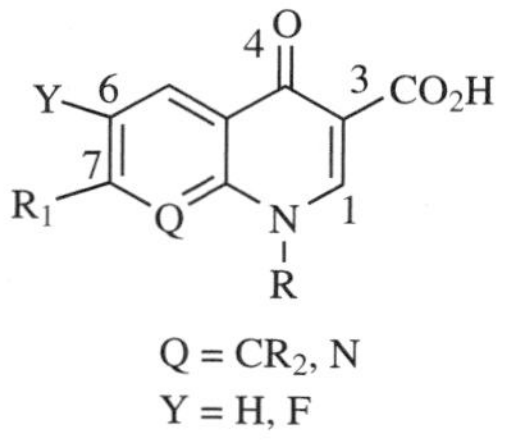

Q = CR$_2$, N
Y = H, F

(**28**)

4.5. Sulfonamides and 2,4-Diaminopyrimidines.

The sulfonamide class of antibacterial agents (**29**) includes N-1 derivatives of *para*-aminobenzene-sulfonamide [sulfanilamide (**5**); R = R$_1$ = H]. Sulfonamides compete with *para*-aminobenzoic acid (PABA) (**30**) [150-13-0] for incorporation into folic acid in a reaction catalyzed by dihydropteroate synthase (36). The 5-substituted-2,4-diaminopyrimidines inhibit the enzyme dihydrofolate reductase, the next step in the biosynthesis of tetrahydrofolic acid. Combination therapy of sulfonamides and 5-substituted-2,4-diaminopyrimidines (**31**), such as pyrimethamine (**15**) and trimethoprim (**16**), is frequently employed, based on George Hitchings' concept of using two metabolite analogues for "sequential blocking" of enzymes in a biochemical pathway (37). In particular, the drug combinations, sulfadoxine-pyrimethamine and trimethoprim-sulfamethoxazole [8064-90-2] have been used to treat uncomplicated malaria and bacterial urinary tract infections, bronchitis and otitis media, respectively. These agents are bacteriostatic. Emergence of resistance to sulfonamides as well as the introduction of new classes of more potent antibacterial agents has diminished the clinical usefulness of this class.

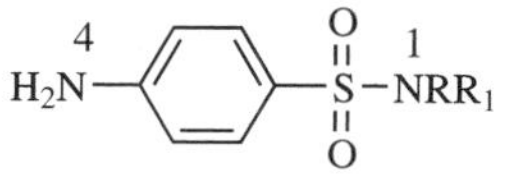

(**29**) 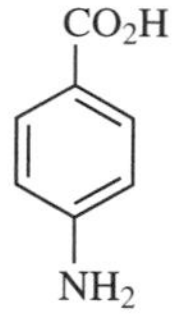   (**30**) 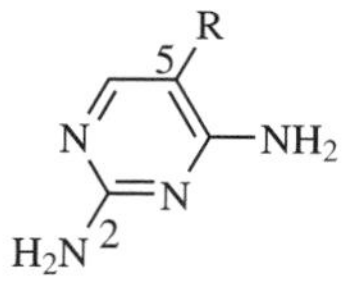   (**31**)

5. Preparation and Manufacture

5.1. Antitubercular Agents.

Isoniazid, pyrazinamide and ethionamide, related in structure, are manufactured by similar routes. Isoniazid (**9**) is prepared by condensation of ethyl isonicotinate (**32**) [1570-45-2] with hydrazine hydrate [7803-57-8] (eq. 1) (38), or alternatively by heating 4-cyanopyridine [100-48-1] with hydrazine hydrate in aqueous alkaline solution (39). A variation of this procedure involves the reaction of isonicotinic acid [55-22-1] with hydrazine hydrate in the presence of a catalyst, such as alumina [1344-28-1], titanium tetrabutoxide [5593-70-4], or a sulfonic acid cation exchanger (40,41).

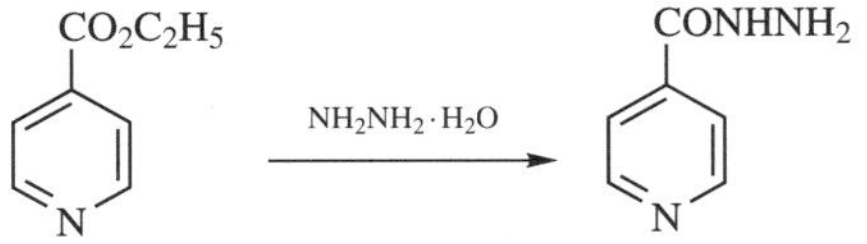

$$(1)$$

Pyrazinamide (**10**) is produced by ammonolysis of an alkyl ester of pyrazinoic acid (**35**) (eq. 2) (42,43). Alternatively, pyrazinamide can be obtained from the reaction of pyrazinecarbonitrile (**37**) [19847-12-2] with aqueous ammonia or by hydrolysis of pyrazinecarbonitrile under acidic or alkaline conditions (44–46). The alkyl pyrazinoates are produced by acid-catalyzed esterification of pyrazinoic acid (**34**) [98-97-5] in the presence of a lower alkanol (47). Pyrazinoic acid, in turn, is prepared by two general methods: 1) potassium permanganate [7722-64-7] oxidation of quinoxaline (**33**) [91-19-0], followed by decarboxylation of the intermediate pyrazine-2,3-dicarboxylic acid [89-01-0], and 2) oxidation of methylpyrazine (**36**) [109-08-0] with selenious acid [7783-00-8] in pyridine or, alternatively, reaction of ethylpyrazine [13925-00-3] with potassium permanganate (42,48). Pyrazinecarbonitrile is readily prepared by ammonoxidation of methylpyrazine (49).

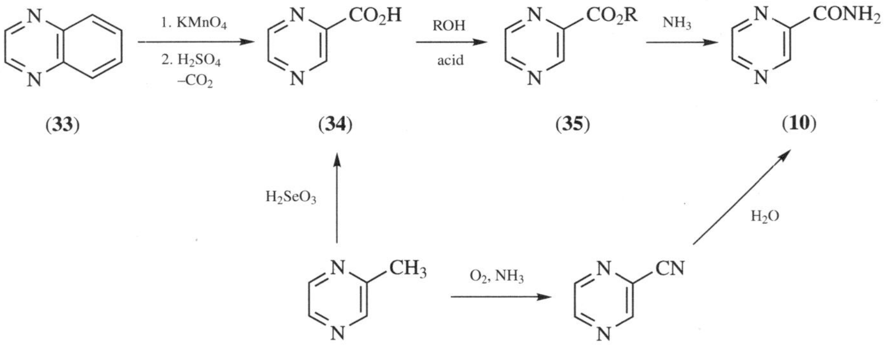

$$(2)$$

Two general methods are available for the preparation of ethionamide (**11**). Both routes converge at the key intermediate, 4-cyano-2-ethylpyridine (**42**) [1531-18-6], which is converted to ethionamide by treatment with hydrogen sulfide gas (50). In the first method, radical alkylation of methyl isonicotinate (**38**) [2459-09-8] with dipropionyl peroxide [3248-28-0] solution affords a mixture of 3-ethyl and 2-ethyl isomers (**39**) [13341-16-7] and (**40**) [1531-16-4], which can be converted to the corresponding 4-cyanopyridine derivatives (**41**) [13341-18-9] and (**42**) by condensation with ammonia, followed by dehydration of the resulting amide in the presence of alumina (eq. 3). Distillation of the mixture at atmospheric pressure affords 4-cyano-3-ethylpyridine (**41**) and the desired 4-cyano-2-ethylpyridine (**42**) in a ratio of 1:1.7 (50). Alternatively, 4-cyano-2-ethylpyridine can be prepared directly by radical alkylation of 4-cyanopyridine. The reaction provides a 4:1 mixture of 4-cyano-2-ethylpyridine and 4-cyano-2,6-diethylpyridine [37581-44-5], which can be separated by distillation. 4-Cyanopyridine, in turn, is obtained by ammonoxidation of 4-picoline [108-89-4] (51).

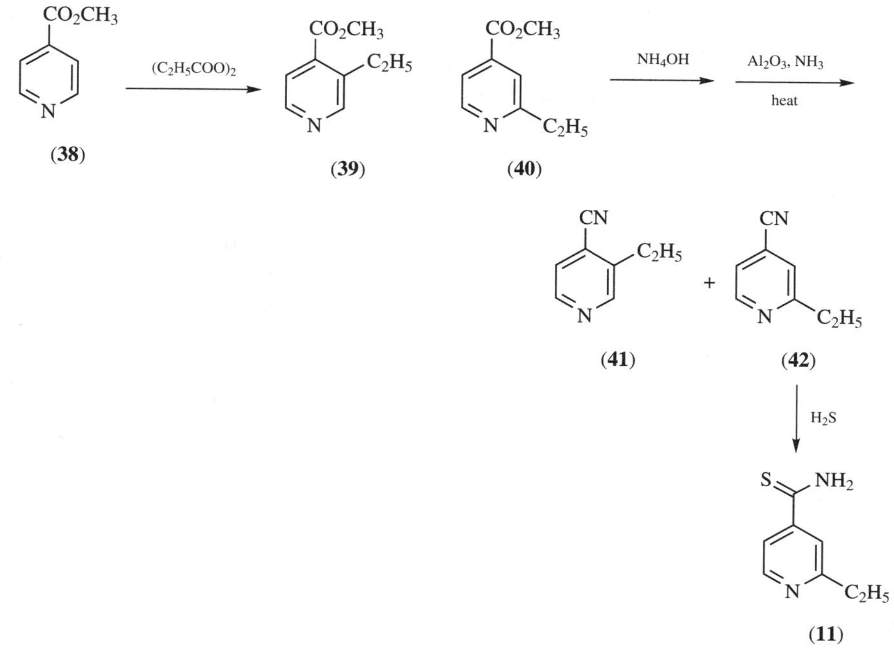

(3)

In the second general method, 2-ethylpyridine (**43**) [100-71-0] is converted to 4-cyano-2-ethylpyridine (**42**) through a series of steps, including oxidation to 2-ethylpyridine-N-oxide [4833-24-3], chlorination to give 4-chloro-2-ethylpyridine (**44**) [3678-65-7], treatment with an alkali metal bisulfite or pyrosulfite to give the intermediate 2-ethylpyridine-4-sulfonic acid [939-96-8], and finally reaction with an alkali metal cyanide (eq. 4) (52).

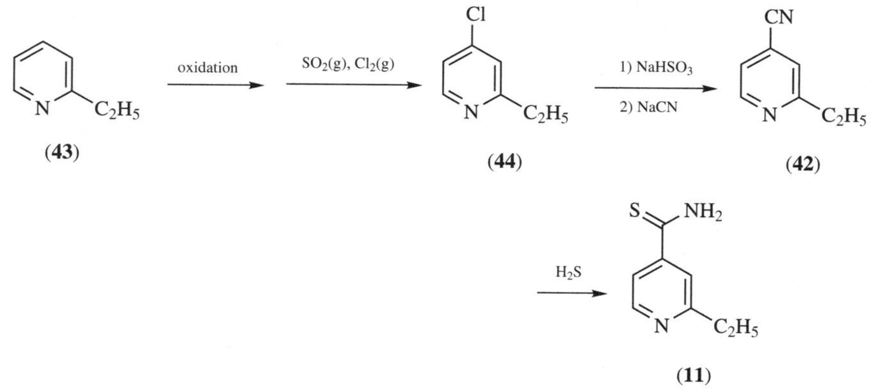

(4)

A number of methods for the manufacture of ethambutol (**13**), (+)-N,N'-bis[1-(hydroxymethyl)propyl]ethylenediamine, have been described. The vast majority begins with (+)-2-amino-1-butanol (**45**) [5856-62-2], obtained from resolution of racemic 2-amino-1-butanol [96-20-8] with L-glutamic acid [56-86-0] (53). Direct alkylation of (+)-2-amino-1-butanol with 1,2-dichloroethane [107-06-2] has been reported to give a 42% yield of ethambutol (54). Other methods appear to be more amenable to large-scale synthesis, however. Condensation of (+)-2-amino-1-butanol (**45**) with 3-pentanone [96-22-0] to give (+)-2,2,4-triethyloxazolidine (**46**) [28507-97-3] followed by alkylation with 1,2-dibromoethane [106-93-4] and hydrolysis gives ethambutol (eq. 5) (55). Alternatively, ethambutol can be prepared by reductive alkylation of (+)-2,2,4-triethyloxazolidine (**46**) with glyoxal hydrate [631-59-4] in the presence of hydrogen gas and palladium on carbon as catalyst (56). Hydrogenolysis of D-4,4'-diethyl-2,2'-bisoxazoline (**49**) [36697-75-3] over Pt-Rh catalysts or lithium aluminum hydride reduction of (+)-4,4'-diethyl-2,2'-bisoxazolidine (**47**) [4486-39-9] also affords ethambutol (eq. 6) (57,58). D-4,4'-Diethyl-2,2'-bisoxazoline is prepared from (+)-2-amino-1-butanol by condensation with diethyl oxalate [95-92-1] followed by bis-cyclodehydration of the resulting glyoxalamide (**48**) [61051-11-4] with sulfuryl chloride [7791-25-5] (57). (+)-4,4'-Diethyl-2,2'-bisoxazolidine can be obtained from (+)-2-amino-1-butanol by treatment with polyglyoxal [25266-42-6] in the presence of catalytic iodine [7553-56-2] (59).

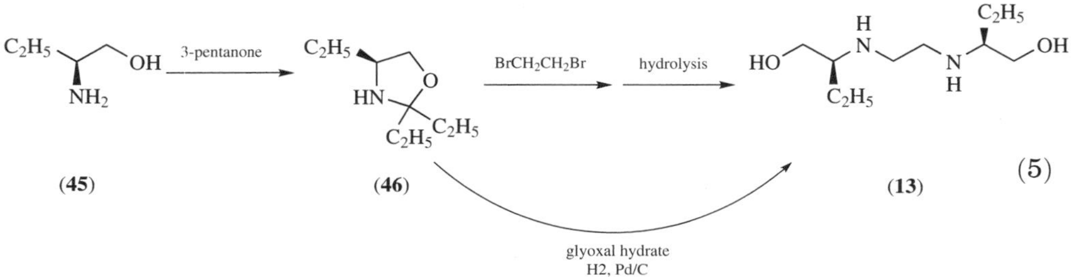

(5)

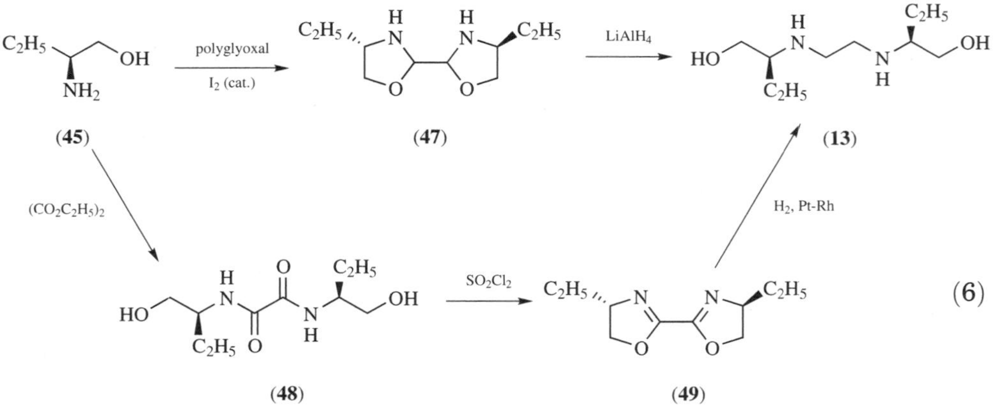

(6)

Another approach to the preparation of ethambutol (**13**) uses butadiene monoepoxide (**50**) [930-22-3] (60) as an inexpensive source of the carbon atoms of the molecule (eq. 7). Treatment of butadiene monoepoxide with phosgene [75-44-5] followed by reaction with benzylamine [100-46-9] produces 2-chloro-3-butenyl benzylcarbamate (**51**) [50297-20-6], which on reaction with potassium hydroxide or sodium hydroxide in ethanol affords 2-benzylamino-3-buten-1-ol

(**52**) [50838-63-6] through displacement of chloride by the carbamate nitrogen followed by cyclic carbamate hydrolysis. Resolution with (+)-dibenzoyltartaric acid [17026-42-5] to give (+)-2-benzylamino-3-buten-1-ol (**53**) [50297-23-9], followed by reaction with 1,2-dibromoethane, and hydrogenation over palladium on carbon gives ethambutol (**61**).

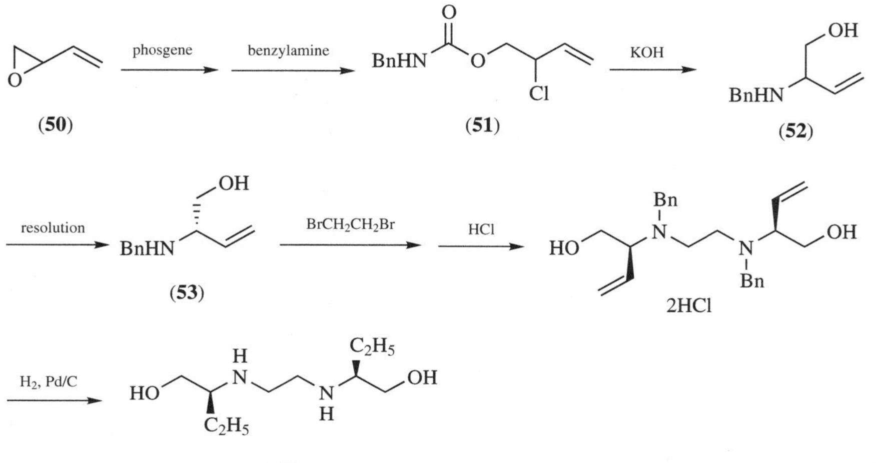

5.2. Nitrofurans. The majority of the nitrofurans (**24**) are commercially prepared by the condensation of either 5-nitro-2-furancarboxaldehyde (**54**) or 5-nitro-2-furancarboxaldehyde diacetate (**55**) [92-55-7] with the appropriate hydrazine derivative (eq. 8). Nitrofurans (**25**) are prepared in a similar manner utilizing a carbon nucleophile in place of the hydrazine derivative.

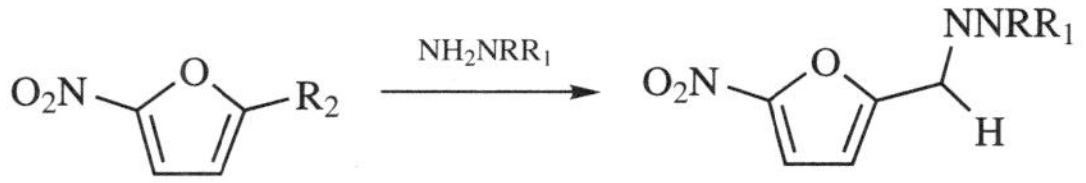

(**54**), R_2=CHO
(**55**), R_2=CH(OC(O)CH$_3$)$_2$

Nitrofurans (**54**) and (**55**) are commercially available but can also be prepared by nitration of 2-furancarboxaldehyde or the diacetate utilizing a variety of conditions.

5.3. Oxazolidinones. In the typical drug discovery route for the synthesis of the oxazolidinone class of antibacterial agents, the oxazolidinone ring is formed by reaction of the anion of the appropriately substituted Cbz-protected aniline (**56**) with (*R*)-glycidyl butyrate (**57**) [60456-26-0] (eq. 9) (26). This reaction proceeds via epoxide opening by the carbamate anion, followed by cyclization to the oxazolidinone. The liberated benzyl alcohol anion cleaves the butyrate ester to the alcohol. The resulting alcohol (**58**) is elaborated into amine (**59**), which is subsequently acetylated to afford (**27**).

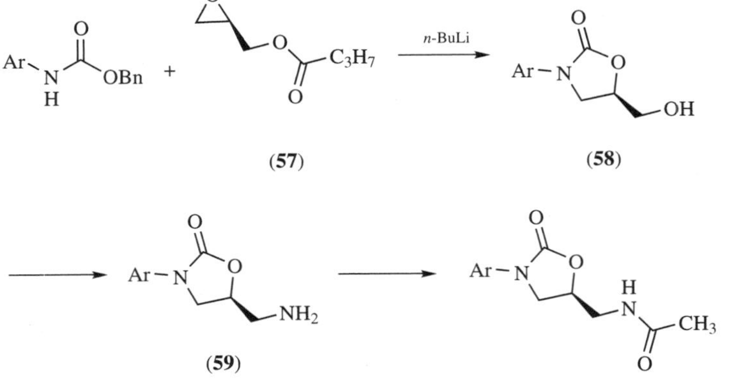

(9)

Several recent process patent applications have disclosed the condensation of carbamate (**56**) with nitrogen-containing three carbon reagents in place of (*R*)-glycidyl butyrate. Perrault and Gadwood describe the use of either (*S*)-Boc-protected glycidyl amine (**60**) [161513-47-9] or (*S*)-Boc-protected 1-amino-3-halo-2-propanol (**61**) to produce the Boc-protected aminomethyl oxazolidinone (**62**). Removal of the protecting group, followed by acetylation affords (**27**). A one-step process for the direct preparation of (**27**) from carbamate (**56**) utilizing (*S*)-*N*-{2-(acetyloxy)-3-chloropropyl}acetamide (**62**) [53460-78-9] or a similar derivative has also been disclosed (63).

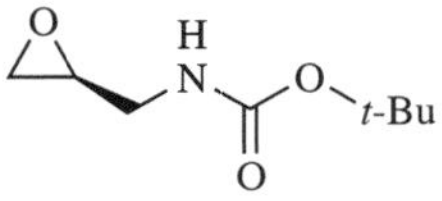

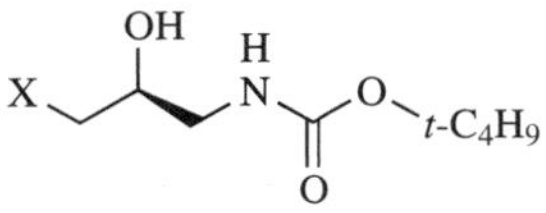

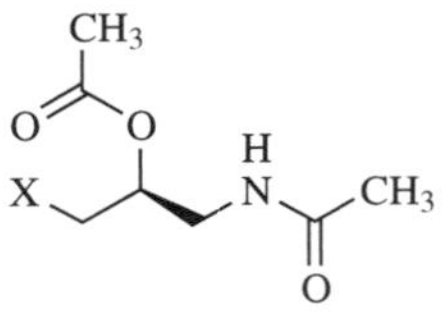

5.4. Quinolones. There are two common methods for the synthesis of the quinolone core structure. The earlier method relies on a Gould-Jacobs cyclization reaction between the appropriate aniline (**63**) and diethyl ethoxymethylenemalonate (**64**) [87-13-8] (eq. 10). Thermal cyclization of the resulting anilinomethylenemalonate (**65**) affords N-1 unsubstituted quinolone nucleus (**66**) (64). N-1 alkylation followed by hydrolysis and nucleophilic aromatic substitution with the appropriate amine produces the drug. Norfloxacin and other analogs of this type have been synthesized in this manner. However, this method is unsatisfactory for N-1 aryl substituents or for substituents that would be derived from an unreactive alkyl halide, such as cyclopropyl. In addition, there is the potential for the formation of regioisomers during the cyclization of unsymmetrical anilines.

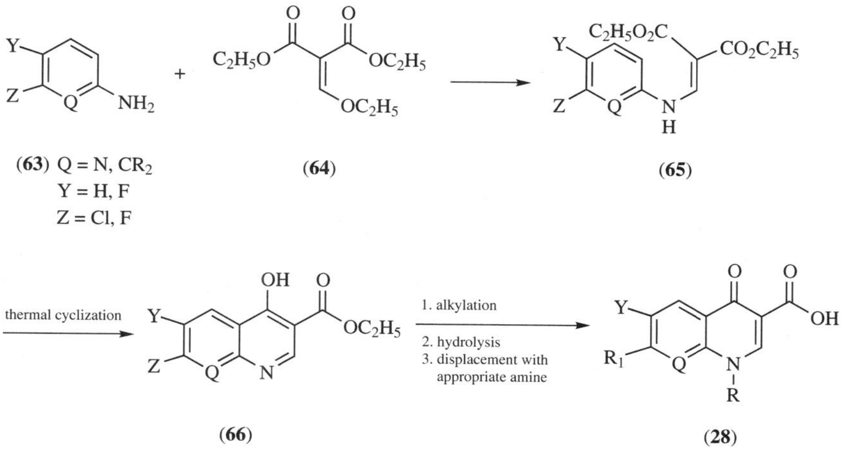

(63) Q = N, CR$_2$
Y = H, F
Z = Cl, F

(64)

(65)

(10)

(66)

(28)

The most widely utilized methodology allows for greater flexibility and produces a wide variety of quinolones including the tricyclic analogues (eqs. 11 and 12). The ring closure occurs by an intramolecular nucleophilic aromatic substitution reaction of enamine (**69**) (65). In turn, enamine (**69**) may be prepared by two slightly different routes. Acid chloride (**67**) may be transformed into enamine (**69**) in several steps via the enol ether (**68**) or, alternatively, it may be converted to enamine (**69**) in one step by reaction with the appropriately substituted 3-aminoacrylate (**70**) (66). Ciprofloxacin, levofloxacin and numerous other quinolones have been synthesized via this type of nucleophilic aromatic substitution reaction.

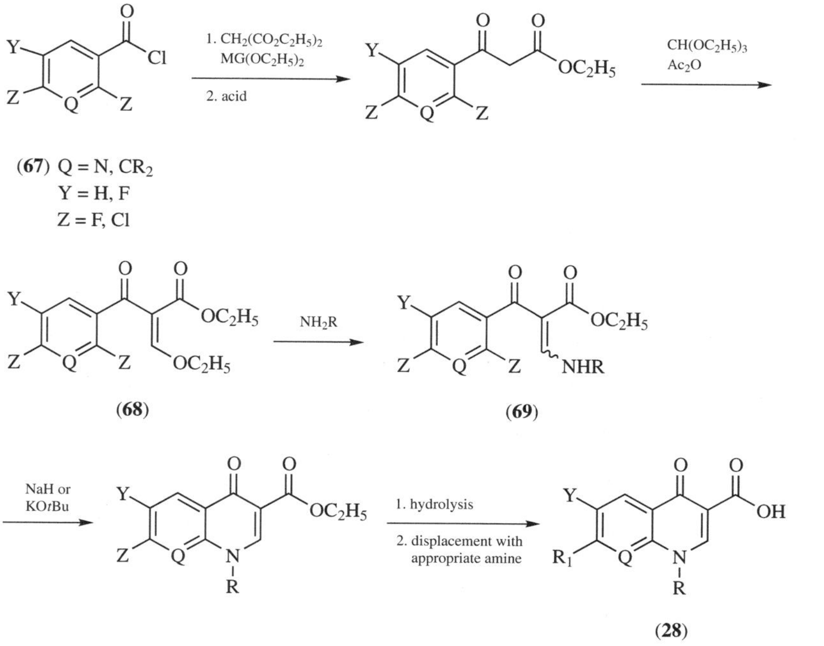

(67) Q = N, CR$_2$
Y = H, F
Z = F, Cl

(68)

(69)

(11)

(28)

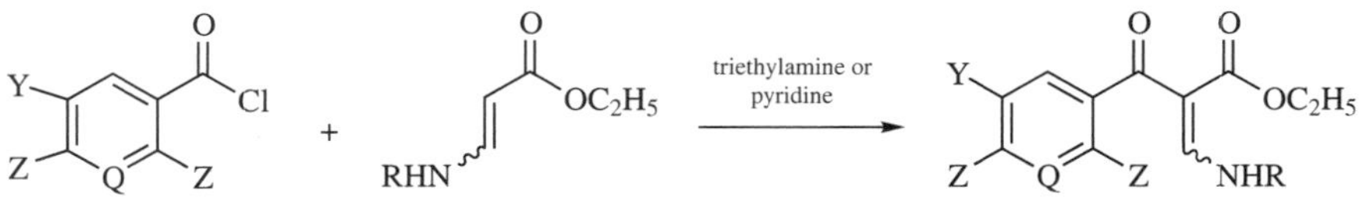

(**67**) Q = N, CR$_2$ (**70**)   (**69**)   (12)
      Y = H, F
      Z = F, Cl

5.5. Sulfonamides and 2,4-Diaminopyrimidines.

The sulfonamides (**29**) are usually prepared by the reaction of *N*-acetylbenzenesulfonyl chloride (**71**) [121-60-8] with the appropriate amine and an equivalent of base (or 2 equiv of the appropriate amine), followed by basic hydrolysis of the acetamide functionality (**67**) (eq. 13). The sulfonyl chloride is synthesized by chlorosulfonation of acetanilide [103-84-4].

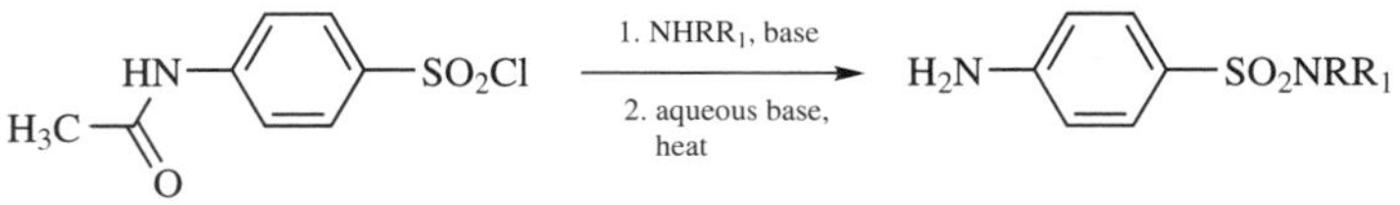

(**71**)   (**29**)   (13)

6. Economic Aspects

Worldwide sales of antiinfective agents for the treatment of bacterial diseases were estimated at 23.0 billion dollars for the 12-month period ending September 2002 (21). For the same period, sales of synthetic antibacterial agents reached $5.2 billion, constituting 22.6% of the global bacterial diseases antiinfective market (21) (Table 2). The increasing use of the quinolones for the treatment of community-acquired infections over the last decade has played a key role in expanding market share for the synthetic antibacterial agents. In particular, quinolone sales surpassed $4.5 billion for the year-ending September 2002,

Table 2. **World Market for Synthetic Antibacterial Agents**[a]

Structural classification	Sales (millions)[b]	% Change[c]
quinolones	4,593	7.4
trimethoprim combinations	189	−7.1
systemic nitrofurans[d]	144	10.2
antitubercular drugs[e]	141	2.4
oxazolidinones	108[f]	125.0
systemic sulfonamides	13	−22.0

[a] World market for the 12-month period ending September 2002 estimated at $5.2 billion (21).
[b] In U.S. dollars. See Ref. (21).
[c] Relative to the previous 12-month period
[d] Represents U.S. sales of Macrobid (nitrofurantoin hydrate/macrocrystals) for 2001 (68).
[e] May include sales of the antibiotic rifampin as part of a single ingredient or multidrug therapeutic regimen.
[f] See Ref. (69).

Table 3. **Relative Share of the Synthetic Antibacterial Agents Market for 2001**[a]

Company	Market share, %
Bayer	37
Ortho-McNeil	20
Daiichi	8
Bristol-Myers Squibb	6
Aventis	6
Procter & Gamble[b]	3
Pharmacia	2
Hoffmann-La Roche	2
Others	16

[a] See Ref. 69.
[b] Represents U.S. sales of Macrobid (nitrofurantoin hydrate/macrocrystals) for 2001 (68).

which represented 88.5% of sales for the synthetic antibacterial agent category and 19.6% of the overall bacterial diseases antiinfective market.

In contrast, annual sales of most of the other classes of synthetic antibacterial agents were comparatively low. The market for systemic sulfonamides and trimethoprim combinations continues to decrease due to resistance emergence in key bacterial pathogens. Sales of antitubercular drugs (including the antibiotic rifampicin) have been escalating as tuberculosis has reached epidemic proportions in portions of Eastern Europe, Asia, and Africa. Nevertheless, the market for antitubercular drugs remains modest due to their relatively low cost and the inadequate distribution networks in many developing nations. Annual U.S. sales of the systemic nitrofuran antibacterial agent, nitrofurantoin, continue to increase due to its effectiveness in the treatment of urinary tract infections. Sales of linezolid, the sole marketed oxazolidinone antibacterial agent, have increased since its introduction as the drug is approved for new clinical indications and in additional countries.

As would be expected from the above, companies engaged in the manufacture and/or sales of quinolone antibacterial agents dominate the synthetic antibacterial agents market (Table 3).

7. Therapeutic Utility

7.1. Antitubercular Agents. Treatment of infections caused by *M. tuberculosis* is most effective when multiple drugs are used for at least 4 months, due to the slow growth of mycobacteria and their propensity to develop resistance during monotherapy. Regimens recommended by The American Thoracic Society Medical Section of the American Lung Association (70) and The Tuberculosis Committee of the Infectious Disease Society of America in conjunction with the Division of Tuberculosis Elimination of the Centers for Disease Control and Prevention employ the use of three drug combinations unless that particular geographic region reports more than 4% of TB isolates resistant to isoniazid, in which case a four drug regimen is recommended (71). The most common combinations are isoniazid, rifampicin, and pyrazinamide, with the addition of either ethambutol or streptomycin depending upon geographical resistance profiles. A 6-month course of therapy is recommended unless multidrug-resistant bacteria are detected or treatment failure is observed, at which time at least two new agents are added to the regimen and treatment is continued for up to 18–24 months. Note that rifampicin and streptomycin are considered to be antibiotics, and, as such, are discussed in detail elsewhere. If a fully drug-susceptible strain is involved, a two drug regimen may be employed, such as rifampicin and isoniazid. Isoniazid alone for up to 12 months, or a rifampicin–pyrazinamide regimen for 2 months, has been recommended for prophylactic treatment of TB-infected asymptomatic HIV-infected patients (71,72).

7.2. Nitrofurans.   Urinary tract infections are the major area in which nitrofurans are used. Nitrofurantoin is specifically indicated for treatment of urinary tract infections caused by susceptible strains of *E. coli*, enterococci, *S. aureus*, *Klebsiella*, and *Enterobacter* species. Nitrofurantoin is not indicated for the treatment of more complicated renal infections, such as pyelonephritis or perinephric abscesses (73). Certain nitrofuran derivatives have been used in veterinary practice to treat or prevent protozoal and bacterial infections in both nonfood and food-producing animals. Use of these carcinogenic and teratogenic drugs, however, results in unacceptably elevated residues in edible tissues, leading to an FDA ban on the use of nitrofurans in food-producing animals in May 2002 (74).

7.3. Oxazolidinones.   Linezolid, the first and only oxazolidinone currently approved for therapeutic use by the regulatory agencies, represents the first new structural class of antibacterial agent in 35 years. Because it targets a stage of protein synthesis different from other agents, it does not demonstrate cross-resistance with any other antibiotic or antibacterial agent. Its *in vitro* antibacterial activity against susceptible and resistant gram-positive bacteria allows for its use in infections caused by organisms such as methicillin-resistant *S. aureus*, vancomycin-resistant enterococci, and multidrug-resistant *Streptococcus pneumoniae*, organisms whose increase in prevalence has become a major health issue over the past 10 years. Linezolid is specifically indicated for the treatment of adult patients with infections caused by linezolid-susceptible strains of vancomycin-resistant *Enterococcus faecium*, including bacteremia (73). Linezolid can be used in both nosocomial and community-acquired pneumonia caused by *S. aureus* or *S. pneumoniae*, with combination therapy indicated if gram-negative organisms are present. It has been approved for treatment of both complicated and uncomplicated skin and skin structure infections caused by *S. aureus* or *Streptococcus pyogenes*.

7.4. Quinolones.   Quinolones are broad-spectrum agents with antibacterial activity against both gram-positive and gram-negative bacteria, including anaerobic and intracellular pathogens. Their target of bacterial DNA topoisomerase means that they exhibit minimal cross-resistance with most other antibacterial agents, and generally retain activity against the penicillin-resistant streptococci that are becoming highly prevalent in community-acquired infections. Quinolones have been approved for multiple therapeutic indications, dependent upon the individual agent (73). Agents such as levofloxacin (**20**), gatifloxacin (**72**) [112811-59-3], moxifloxacin (**73**) [151096-09-2], and gemifloxacin (**74**) [175463-14-6] are often considered to be "respiratory quinolones" with approved use for community-acquired infections such as acute maxillary sinusitis, acute bacterial exacerbation of chronic bronchitis, and community-acquired pneumonia. In addition, some quinolones have been approved for treatment of more serious infections including both complicated and uncomplicated skin and skin structure infections and nosocomial pneumonia. They may also be used to treat genitourinary tract infections including complicated and uncomplicated urinary tract infections (mild to moderate), acute pyelonephritis, prostatitis, various urethral and cervical infections, pelvic inflammatory disease, and uncomplicated cystitis.

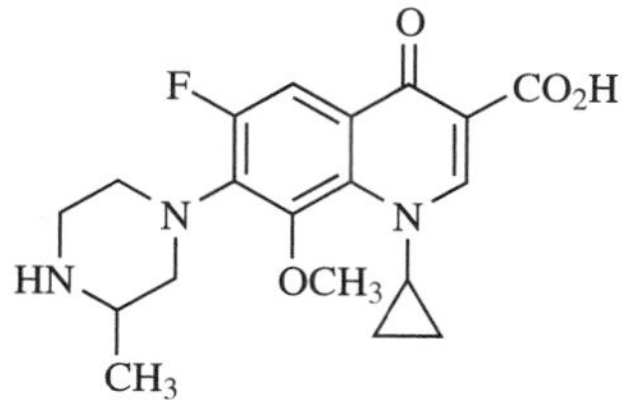

(72)

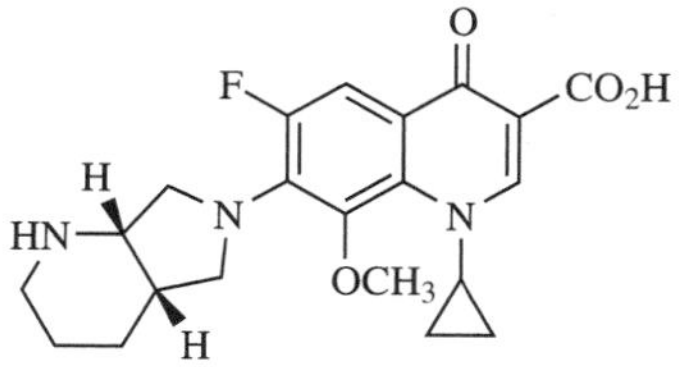

(73)

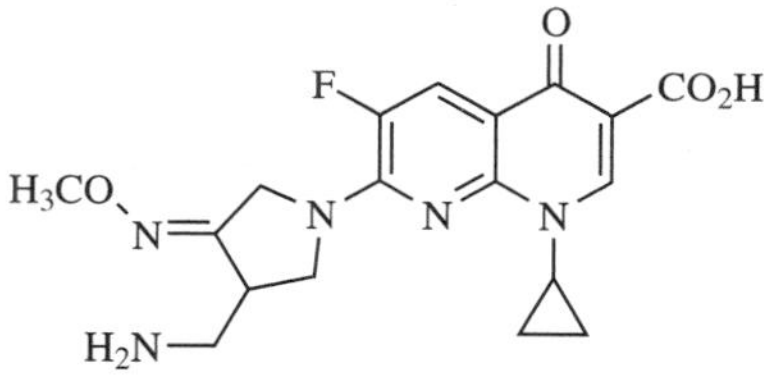

(74)

Additionally, six quinolones are marketed exclusively for use in veterinary medicine: danofloxacin (**75**) [112398-08-0], difloxacin (**76**) [98106-17-3], enrofloxacin (**77**) [93106-60-6], marbofloxacin (**78**) [115550-35-1], orbifloxacin (**79**) [113617-63-3], and sarafloxacin (**80**) [98105-99-8]. The human drugs, ciprofloxacin (**19**), ofloxacin (**81**) [82419-36-1], and trovafloxacin (**82**) [147059-72-1], are occasionally used in companion animal medicine (75).

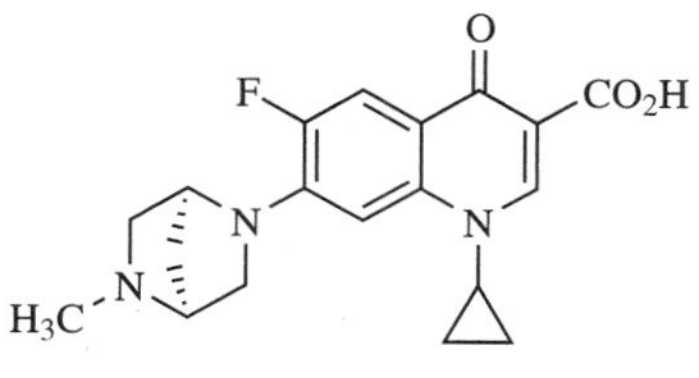

(75)

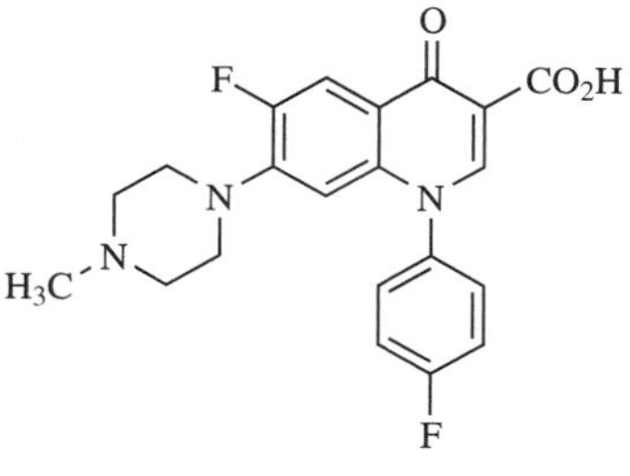

(76)

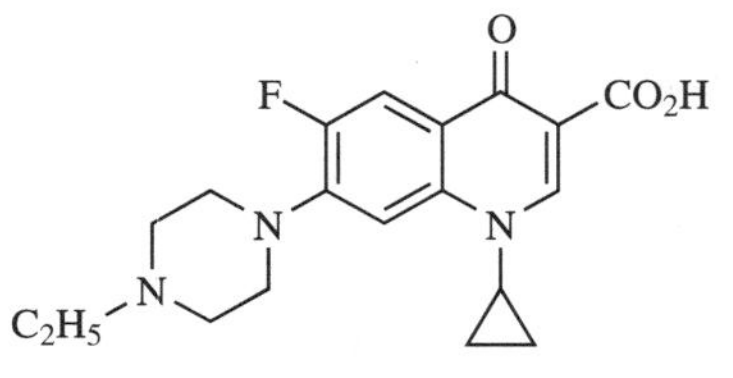

(77)

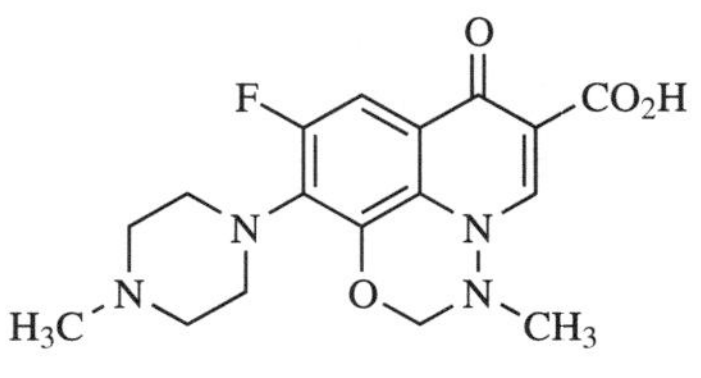

(78)

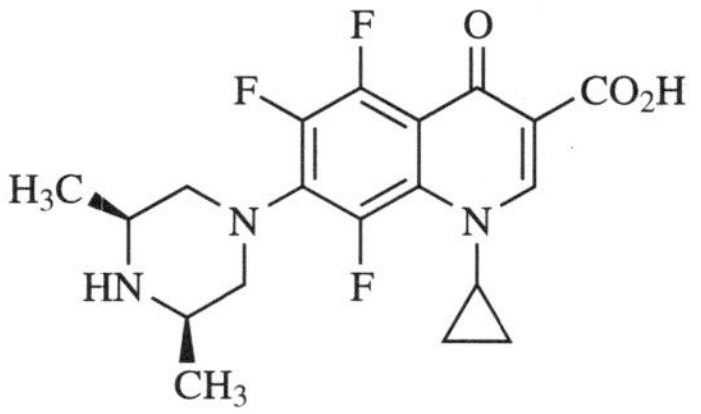

(79)

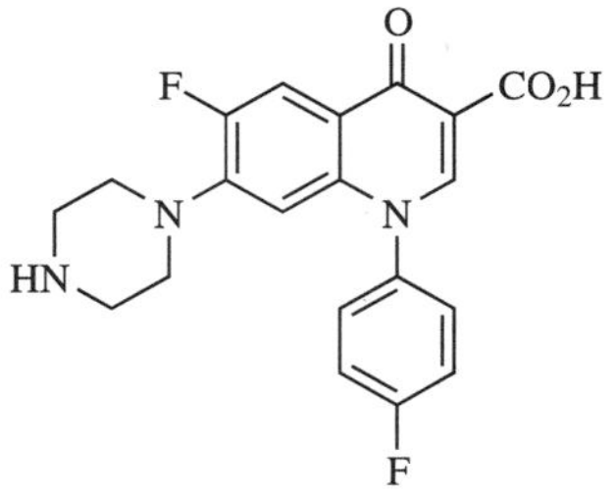

(80)

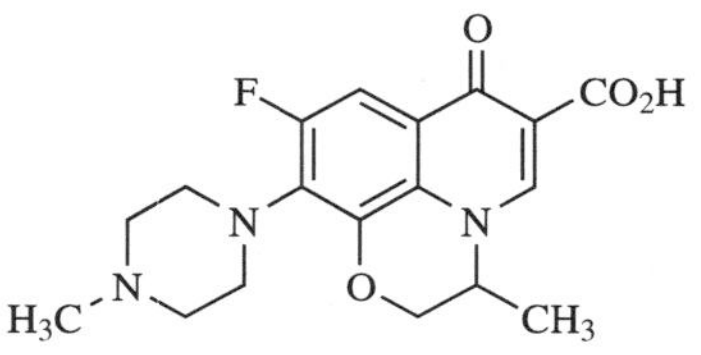

(81)

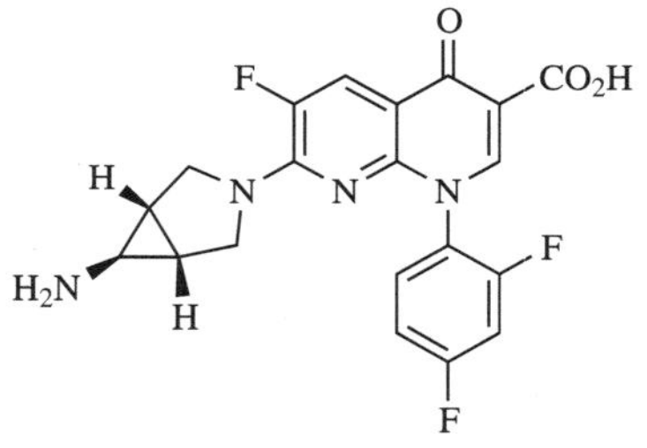

(82)

7.5. Sulfonamides and 2,4-Diaminopyrimidines. These agents have been surpassed as first line agents for most bacterial infections. However, sulfonamides alone are still utilized for the treatment of urinary tract infections due to susceptible enteric bacteria. The combination of a sulfonamide with trimethoprim, a dihydrofolate reductase inhibitor, can be used for treatment of a number of less serious microbial infections including urinary tract infections, acute otitis media due to susceptible strains of *S. pneumoniae* or *Haemophilus influenzae*, and acute exacerbations of chronic bronchitis in adults due to susceptible strains of *S. pneumoniae* or *H. influenzae*. This combination is also used for treatment of travelers' diarrhea in adults due to susceptible strains of enterotoxigenic *E. coli*, for shigellosis, and for enteritis caused by susceptible strains of *Shigella flexneri* and *Shigella sonnei*. The combination of trimethoprim and sulfamethoxazole is the treatment of choice for *Pneumocystis carinii* pneumonia, and is also used as prophylaxis against this common fungal infection in immunocompromised or acquired immune deficiency syndrome (AIDS) patients (73).

Sulfonamide–trimethoprim combinations have been utilized for the treatment of bacterial disease in a variety of animal species (76–78).

BIBLIOGRAPHY

1. A. Albert, *Selective Toxicity: The Physico-Chemical Basis of Therapy*, 5th Ed., Chapman and Hall, London, 1973, p. 134.
2. P. Ehrlich and S. Hata, *Die Experimentelle Chemotherapie der Spirillosen*, Springer, Berlin, 1910.
3. G. Domagk, *Deut. Med. Wochschr.* **61**, 250 (1935).
4. L. Colebrook and M. Kenny, *Lancet* **1**, 1279 (1936).
5. J. Trefouel, F. Nitti, and D. Bovet, *Compt. Rend. Soc. Biol.* **120**, 756 (1935).
6. H. Horstmann, *Therapie Gegenwart* **117**, 418 (1978).
7. M. C. Dodd and W. B. Stillman, *J. Pharmacol. Exp. Therapeutics* **82**, 11 (1944).
8. U.S. Pat. 2,319,481 (May 18, 1943), W. B. Stillman, A. B. Scott, and J. M. Clampit (to Norwich Pharmacal Co.).
9. O. Dann and E. F. Moller, *Chem. Ber.* **80**, 23 (1947).
10. D. R. McCalla and D. Voutsinos, *Mutation Res.* **26**, 3 (1974).
11. N. Gao, Y. C. Ni, J. R. Thornton-Manning, P. P. Fu, and R. H. Heflich, *Mutation Res.* **225**, 181 (1989).
12. J. Bernstein, W. A. Lott, B. A. Steinberg, and H. L. Yale, *Am. Rev. Tuberc.* **65**, 357 (1952).
13. L. Malone, A. Schurr, H. Lindh, D. McKenzie, J. S. Kiser, and J. H. Williams, *Am. Rev. Tuberc.* **65**, 511 (1952).
14. E. F. Rogers, W. J. Leanza, H. J. Becker, A. R. Matzuk, R. C. O'Neill, A. J. Basso, G. A. Stein, M. Solotorovsky, F. J. Gregory, and K. Pfister, 3rd, *Science* **116**, 253 (1952).
15. F. Grumbach, N. Rist, D. Libermann, M. Moyeux, S. Cals, and S. Clavel, *Compt. Rend.* **242**, 2187 (1956).
16. R. G. Shepard, C. Baughn, M. L. Cantrall, B. Goodstein, J. P. Thomas, and R. G. Wilkinson, *Ann. N. Y. Acad. Sci.* **135**, 686 (1966).
17. E. A. Falco, L. G. Goodwin, G. H. Hitchings, I. M. Rollo, and P. B. Russell, *Brit. J. Pharmacol.* **6**, 185 (1951).
18. B. Roth, E. A. Falco, G. H. Hitchings, and S. R. M. Bushby, *J. Med. Pharm. Chem.* **5**, 1103 (1962).
19. M. P. Wentland, in D. C. Hooper and J. S. Wolfson, eds., *Quinolone Antimicrobial Agents*, American Society for Microbiology, Washington, D.C., 1993, p. xiii.
20. G. Y. Lesher, E. J. Froelich, M. D. Gruett, J. H. Bailey, and R. P. Brundage, *J. Med. Pharm. Chem.* **5**, 1063 (1962).
21. *IMS Data Base*, IMS Health, London, NW1 6JB, U.K. (Jan. 2, 2003).
22. P. Ball, *J. Antimicrobial Chemother.* **46**, 17 (2000).
23. G. M. Eliopoulos, *Drugs* **58**, 23 (1999).
24. A. Tomasz, *New Engl. J. Med.* **330**, 1247 (1994).
25. W. A. Gregory, D. R. Brittelli, C. L. J. Wang, M. A. Wuonola, R. J. McRipley, D. C. Eustice, V. S. Eberly, A. M. Slee, M. Forbes, and P. T. Bartholomew, *J. Med. Chem.* **32**, 1673 (1989).
26. S. J. Brickner, D. K. Hutchinson, M. R. Barbachyn, P. R. Manninen, D. A. Ulanowicz, S. A. Garmon, K. C. Grega, S. K. Hendges, D. S. Toops, C. W. Ford, and G. E. Zurenko, *J. Med. Chem.* **39**, 673 (1996).
27. J. F. Plouffe, *Clin. Inf. Dis.* **31**, S144 (2000).
28. K. Johnsson, D. S. King, and P. G. Schultz, *J. Am. Chem. Soc.* **117**, 5009 (1995).
29. B. Phetsuksiri, A. R. Baulard, A. M. Cooper, D. E. Minnikin, J. D. Douglas, G. S. Besra, and P. J. Brennan, *Antimicrobial Agents Chemother.* **43**, 1042 (1999).
30. E. K. Schroeder, O. N. De Souza, D. S. Santos, J. S. Blanchard, and L. A. Basso, *Curr. Pharmaceutical Biotechnol.* **3**, 197 (2002).

31. J. Grange, in F. O'Grady, H. Lambert, R. G. Finch, and D. Greenwood, eds., *Antibiotic and Chemotherapy: Anti-infective Agents and Their Use in Therapy*, Churchill Livingstone, New York, 1997, pp 499–512.

32. M. Kishimoto, J. Kaneko, T. Miura, S. Adachi, S. Kojio, I. Taneike, and T. Yamamoto, *Niigata Igakkai Zasshi* **115**, 193 (2001).

33. P. Chakrabarti, *Proc. Nat. Acad. Sci., India, Section B: Biol. Sci.* **67**, 169 (1997).

34. D. R. Guay, *Drugs* **61**, 353 (2001).

35. H. Aoki, L. Ke, S. M. Poppe, T. J. Poel, E. A. Weaver, R. C. Gadwood, R. C. Thomas, D. L. Shinabarger, and C. M. Ganoza, *Antimicrobial Agents Chemother.* **46**, 1080 (2002).

36. J. W. A. Petri, in J. G. Hardman, L. E. Limbird, and A. Goodman Gilman, eds., *Goodman and Gilman's The Pharmacological Basis of Therapeutics*", McGraw-Hill, New York, 2001, pp. 1171–1187.

37. G. H. Hitchings, *Trans. R. Soc. Trop. Med. Hyg.* **46**, 467 (1952).

38. U. P. Basu and S. P. Dutta, *Ind. Chim. Belge* **32**, 1224 (1967).

39. Ger. (East) Pat. 63493 (Sept. 5, 1968), H. Seefluth, K. K. Moll, H. Baltz, L. Bruesehaber, and G. Schrattenholz.

40. U.S.S.R. Pat. 1,197,396 A1 (May 20, 1995), V. G. Voronin, V. N. Shumov, V. P. Sergovskaya, I. D. Muravskaya, N. F. Karaseva, I. I. Tyulyaev, V. S. Drizhov, and T. P. Arkhipova (to Filial Vsesoyuznogo Nauchno-Issledovatelskogo Khimiko-Farmatsevticheskogo Instituta im.Sergo Ordzhonikidze, USSR).

41. U.S. Pat. 3,951,996 (Apr. 20, 1976), R. H. Stanley and B. L. Shaw (to British Titan Ltd.).

42. E. Felder and D. Pitre, *Anal. Profiles Drug Substances* **12**, 433 (1983).

43. Jpn. Kokai Tokkyo Koho 3,205,975 (Sept. 4, 2001), M. Hatayama, Y. Fukuda, S. Ishihara, and M. Takebayashi (to Nippon Soda Co, Japan).

44. Jpn. Kokai Tokkyo Koho 57,011,971 A2 (Jan. 21, 1982) (to Mitsuwaka Pure Chemicals Co., Ltd., Japan).

45. Eur. Pat. Appl. 122,355 (Oct. 24, 1984), J. Zergenyi and B. Raz, (to Servipharm A.-G., Switz.).

46. Jpn. Kokai Tokkyo Koho 62,111,971 A2 (May 22, 1987), O. Oka (to Koei Chemical Industry Co., Ltd., Japan).

47. Jpn. Kokai Tokkyo Koho 3,205,972 (Sept. 4, 2001), M. Takebayashi, M. Hatayama, Y. Fukuda, and S. Ishihara (to Nippon Soda Co, Japan).

48. H. Gainer, *J. Org. Chem.* **24**, 691 (1959).

49. U.S. Pat. 3,555,021 (Jan. 12, 1971), R. H. Beutel, P. Davis, and E. F. Schoenewaldt (to Merck and Co., Inc.).

50. U.S. Pat. 3,364,222 (Jan. 16, 1968), H. D. Eilhauer and G. Reckling (to VEB Leuna-Werke "Walter Ulbricht").

51. C.-H. Wang, F.-Y. Hwang, J.-M. Horng, and C.-T. Chen, *Heterocycles* **12**, 1191 (1979).

52. Ger. (East) Pat. 45049 (Apr. 15, 1966) D. Schunke and H. G. Schwark.

53. F. H. Radke, R. B. Fearing, and S. W. Fox, *J. Am. Chem. Soc.* **76**, 2801 (1954).

54. R. G. Wilkinson, R. G. Shepherd, J. P. Thomas, and C. Baughn, *J. Am. Chem. Soc.* **83**, 2212 (1961).

55. Brit. Pat. 1,188,054 (Apr. 15, 1970) (to Societa Farmaceutici Italia).

56. Brit. Pat. 1,234,349 (June 3, 1971) (to Societa Farmaceutici Italia).

57. Brit. Pat. 1,327,315 (Aug. 22, 1973) (to PLIVA Tvornica Farmaceutskih i Kemijekih Proizvoda).

58. Austrian Pat. 323,122 (Jun. 25, 1975) (to PLIVA Tvornica Farmaceutskih i Kemijskih Proizvoda).

59. Austrian Pat. 366,379 (Apr. 13, 1982), L. H. Schlager (to Gerot-Pharmazeutika G.m.b.H., Austria).

60. U.S. Pat. 5,081,096 (Jan. 14, 1992) J. R. Monnier and P. J. Muehlbauer (to Eastman Kodak Co., USA).

61. U.S. Pat. 3,847,991 (Nov. 12, 1974) L. Bernardi, M. Foglio, and A. Temperilli.

62. U.S. Pat. Appl. 09/982,157 (Oct. 17, 2001) W. R. Perrault and R. C. Gadwood (to Pharmacia & Upjohn Company, USA).

63. U.S. Pat. Appl. 10/122,852 (Apr. 15, 2002) W. R. Perrault, B. A. Pearlman, and D. B. Godrej (to Pharmacia & Upjohn Company, USA).

64. R. G. Gould, Jr., and W. A. Jacobs, *J. Am. Chem. Soc.* **61**, 2890 (1939).

65. D. T. W. Chu, *J. Heterocyclic Chem.* **22**, 1033 (1985).

66. K. Grohe and H. Heitzer, *Liebigs Ann. Chem.*, 871 (1987).

67. E. H. Northey, *The Sulfonamides and Allied Compounds*, Reinhold Publishing Corp, New York, 1948, pp. 517–577.

68. M. Marketos, *Top 200 Drugs by Retail Sales in 2001*, Drugtopics.com, http://www.drugtopics.com/be_core/content/journals/d/data/2002/0218 (Feb. 3, 2003).

69. *Evaluate Pharma Database*, Evaluate Plc, London, E1 6PX, UK (Jul. 11, 2002).

70. American Thoracic Society Medical Section of the American Lung Association, *Am. J. Respir. Crit. Care Med.* 149 (1994).

71. C. R. J. Horsburgh, S. Feldman, and R. Ridzon, *Clin. Infect. Dis.* **31**, 633 (2000).

72. C. T. Wang, *Can. Family Phys.* **45**, 2397 (1999).

73. *Physicians' Desk Reference*, Medical Economics Company, 2001 Electronic Edition, Williams & Wilkins, Inc. (Jan. 5, 2003).

74. *Fed. Reg.* **67**, 5470 (2002).

75. R. D. Walker, *Austr. Veterinary J.* **78**, 84 (2000).

76. G. V. Ling and A. L. Ruby, *J. Am. Veterinary Medical Assoc.* **174**, 1003 (1979).

77. D. C. Hirsh, S. S. Jang, and E. L. Biberstein, *J. Am. Veterinary Med. Assoc.* **197**, 594 (1990).

78. E. Van Duijkeren, A. G. Vulto, and A. S. J. P. A. M. Van Miert, *J. Veterinary Pharmacol. Therapeut.* **17**, 64 (1994).

Mark J. Macielag
Karen Bush
Michele A. Weidner-Wells
Johnson & Johnson Pharmaceutical
Research & Development, L.L.C.

ANTIBIOTIC RESISTANCE

1. Introduction

It is widely accepted that bacteria as living organisms came to existence over 3.5 billion years ago. As these organisms became compelled to interact with other living entities, they became more complex and evolved the biochemical means for influencing the existence of each other. One of these evolutionary developments was the advent of biochemical pathways for production of antibiotics. In essence, if growth of a competitor were to be influence, more resources would be available for growth of the original organism. As such, multiple pathways for generation of "secondary metabolite", which include molecules that have antibacterial properties, have evolved (1,2). A number of these secondary metabolites with antibacterial properties have been discovered over the past few decades. The structures of many of them have been altered by chemists to expand their properties or to impart desirable chemical traits to them. Many of these molecules have found clinical use over the years. In this report, we provide an overview of the important classes of antibiotics of both natural and synthetic origins and we will describe what is known about the mechanisms by which nature gives rise to resistance to them.

2. Molecular Targets for Antibiotics

Many of the first antibiotics discovered in the past 60 years have been natural products from microbial systems. To date, antibiotics that trace their origins to natural products dominate the armamentarium of clinically useful antibiotics. These are molecules that interfere with the biochemical processes of bacteria with some specificity, hence they are useful in mammalian hosts.

A total of over 70 bacterial genomes have been sequenced to date. It would appear that ~1000–5000 genes are found in most of these organisms (3). It has been proposed that somewhere between 20 and 200 or so genes are critical for survival of a broad spectrum of bacteria (3–6). The proteins encoded by these genes are potential targets for antibiotics, if inhibitors for them could be delivered to the site. Furthermore, other type of antibiotics may interfere with assemblies of these gene products or with the structural components that result from their actions, such as the cell wall, bacterial envelope, or ribosome.

Known antibiotics interfere with a small number of biochemical processes coinciding with these critical genes. These processes include metabolic pathways, disruption of the integrity of the cytoplasmic membrane, inhibition of protein biosynthesis, inhibition of DNA biosynthesis, and disruption of the biosynthesis of the cell wall, of which the last three targets are especially important. Whereas it is beyond the scope of this article to discuss all of these processes, we summarize the important processes that are disrupted by the clinically important antibacterials. Figure 1 gives the structures of several important antibacterials.

The bacterial cell wall is an important target for antibacterials, in part because it is a uniquely bacterial structure with a biosynthetic pathway for its assembly that does not find any parallels in other organisms. The cell wall provides structural rigidity and morphology to bacteria. It is a polymeric structure made up of repeat units of N-acetyl muramic acid (MurNAc)-N-acetyl glucosamine (GlcNAc). Though there are some variations, most bacteria have a five amino acid chain (a pentapeptidyl consisting of L-alanine-D-glutanate-diamino pimelate-D-alanine-D-alanine) attached via the amino group of the L-alanine to the MurNAc segment (Fig. 2). The pentapeptide has uniquely bacterial features such as D-Glu with an amide bond to diaminopimelate (DAP) via its side-chain carboxylate. In turn, DAP is linked to the dipeptide D-Ala-D-Ala (Fig. 2). The substituted MurNAc-GlcNAc disaccharide segments is linked with a neighboring disaccharide segment in a reaction catalyzed by the transglycosylases (TGs). The disaccharide building block as a pyrophosphoryl-undecaprenol ester serves

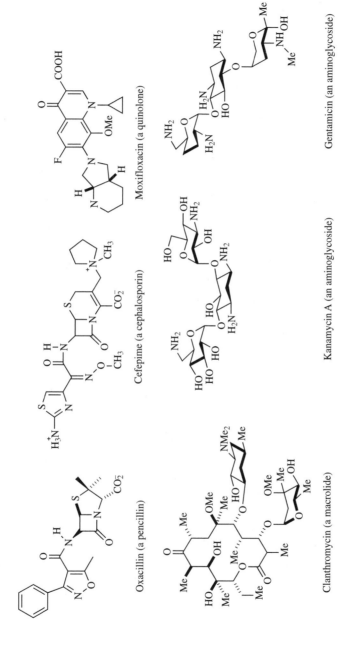

Fig. 1. Chemical structures of some of the antibiotics used for treatment of bacterial infections.

Oxacillin (a pencillin)

Cefepime (a cephalosporin)

Moxifloxacin (a quinolone)

Clanthromycin (a macrolide)

Kanamycin A (an aminoglycoside)

Gentamicin (an aminoglycoside)

25

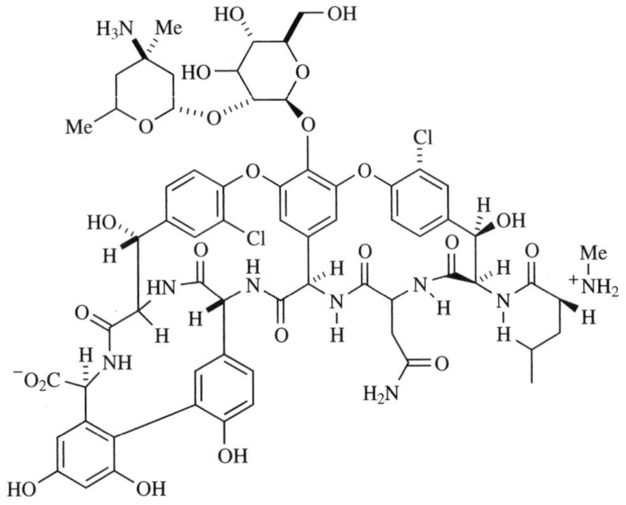

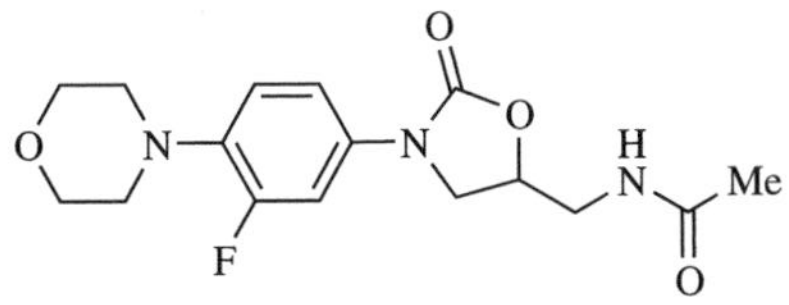

Linezolid (an oxazolidinone)

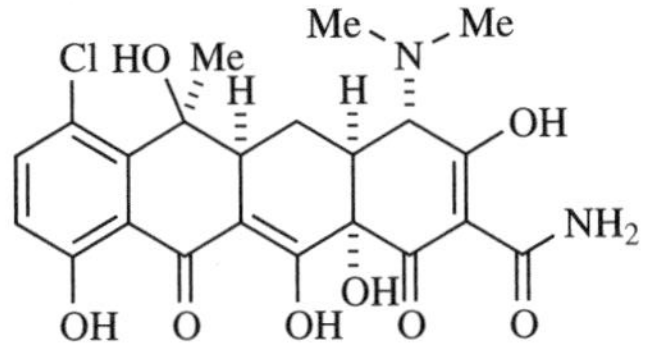

Chlortetracycline (a tetracycline)

Vancomycin (a glycopeptide)

Fig. 1 (*Continued*)

26

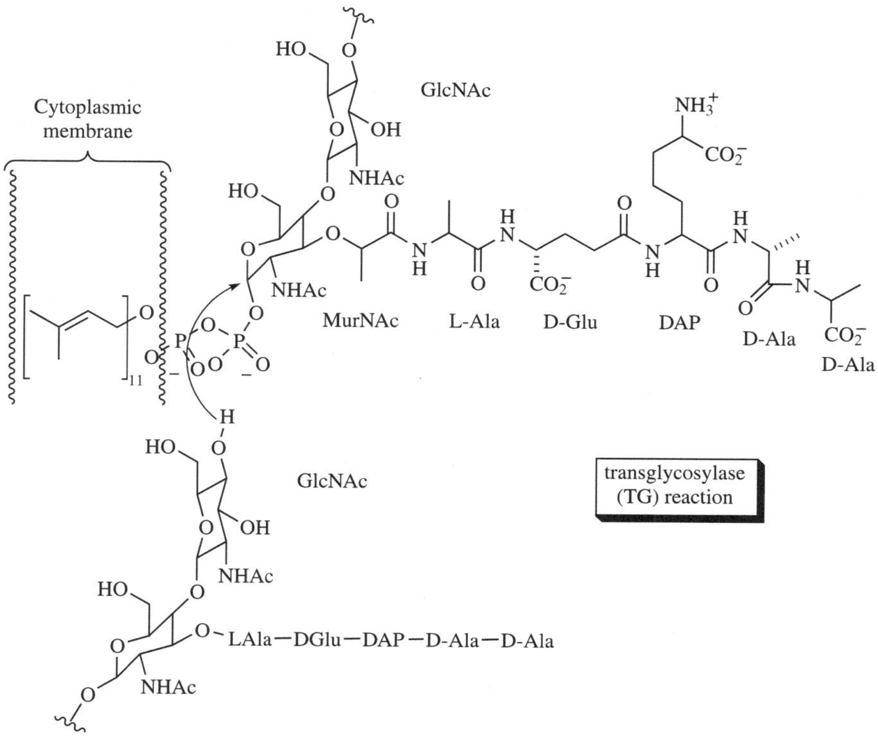

Fig. 2. The mechanism of tranglycosylase results in polymerization of the building blocks for bacterial peptidoglycan.

as a substrate for transglycosylases in this polymerization reaction, the product of which is referred to as the peptidoglycan. Subsequent to polymerization, cross-linking of the cell wall is required, ie, the peptidyl portions are cross-linked to each other via a peptide bond. Since there is no source of energy, such as adenosine triphosphate (ATP), for peptide bond formation outside the cytoplasm, where the cell wall is assembled, nature opted to exchange an amide bond in the peptides. This reaction is carried out by transpeptidases (TP) (Fig. 3), which proceeds through an acyl-enzyme species involving an active site serine residue. How the peptides from the two strands are sequestered in the active site to give the cross-linking reaction was elucidated recently (7). The tranglycosylase and transpeptidase activities are often found in bifunctional enzymes that are anchored to the surface of the cytoplasmic membrane. Both these activities are targets for commonly used antibiotics. For example, penicillin mimics the structure of the acyl-D-Ala-D-Ala portion of peptidoglycan (8). By so-doing, penicillin acylates the same active site serine in transpeptidase, resulting in a stable enzyme-modified species that accounts for the lethal action of these antibiotics (9–11).

If the pentapeptide was made inaccessible to the transpeptidases, then cross-linking would not take place and bacteria would die. This is the strategy

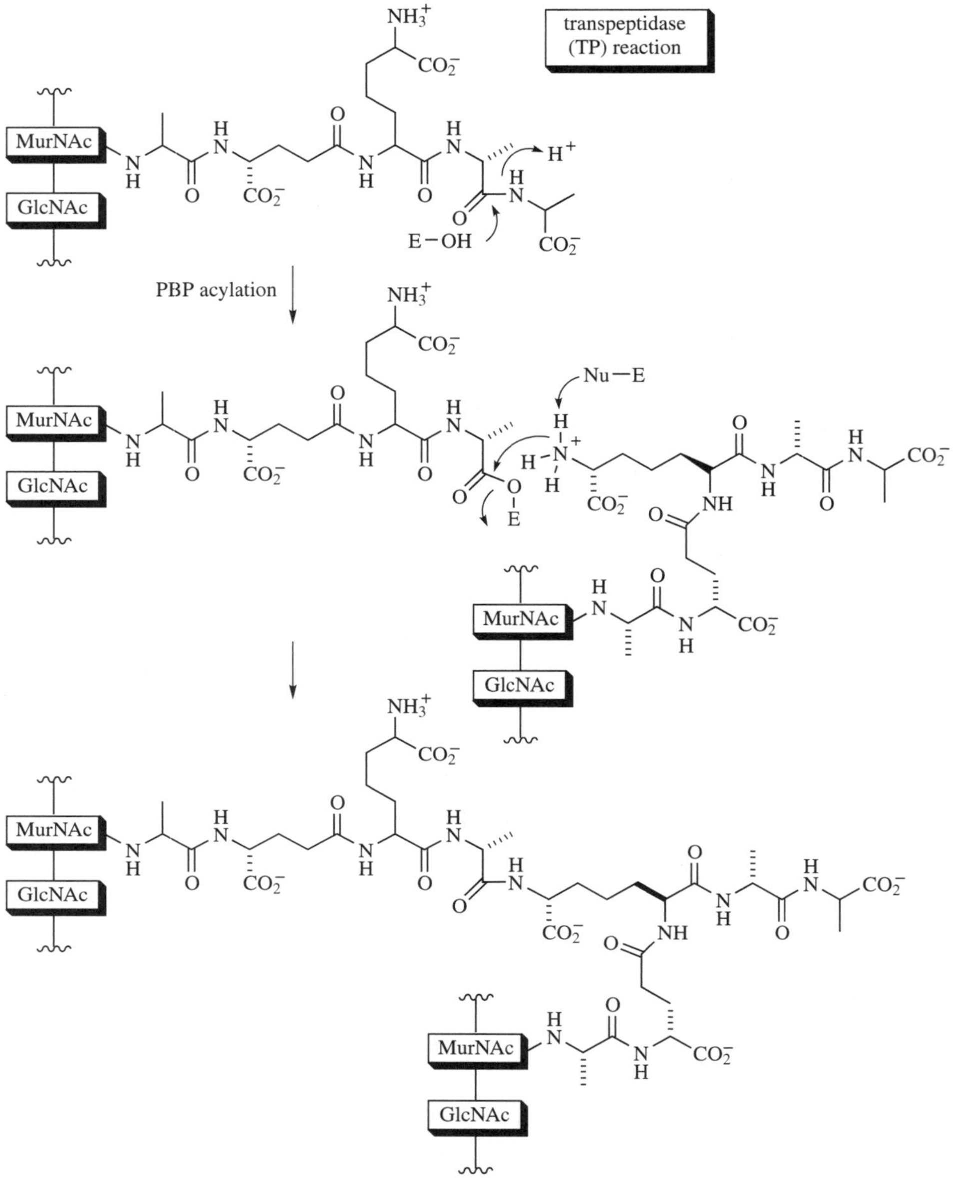

Fig. 3. The two-step reaction of DD-transpeptidases that gives rise to cross-linked cell wall in bacteria.

for glycopeptide antibiotics such as vancomycin, ristocetin, and teicoplanin, which coordinate to the D-Ala-D-Ala portion of the pentapeptide through five hydrogen bonds (Fig. 4) (12). Recent work from the Kahne lab has shown that by modifying the saccharide groups attached to vancomycin's peptide backbone, the target of these derivatives is altered. In contrast to the case of the parental

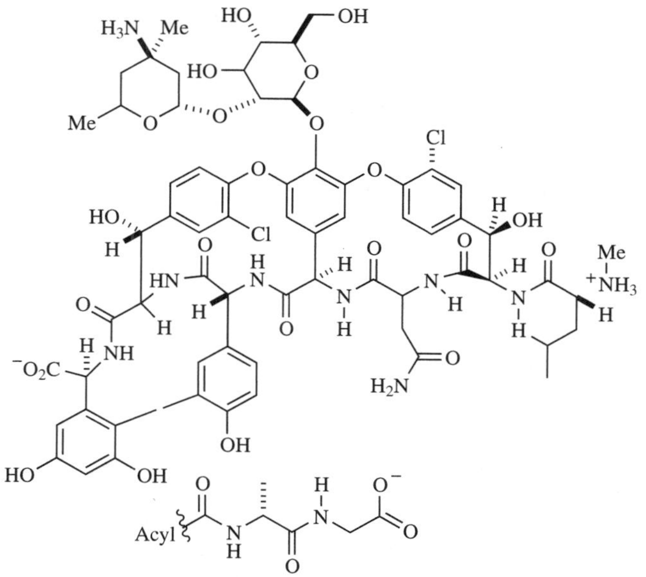

Fig. 4. The binding interactions between vancomycin (top) and the acyl-D-Ala-D-Ala portion of the pentapeptide of the bacterial peptidoglycan. The hydrogen bonds are indicated by dotted lines.

molecule, they seem to kill bacteria by inhibiting the transglycosylation step and furthermore, the modified saccharide groups themselves also posses substantial antibacterial activity (13,14).

Transglycosylases and transpeptidases are clearly important targets for antibiotics, and since they are present on the surface of the cytoplasmic membrane, they are more readily accessible by the respective inhibitors. However, assembly of MurNAc-GlcNac-pyrophosphorylundecaprenol ester, the substrate for transglycosylase, takes place in the cytoplasm. This process requires 10 enzymes, which themselves are targets for antibiotic development (15,16). However, there are not many examples of inhibitors for these enzymes that show antibacterial property, since the molecules should traverse the bacterial envelope to reach their target, which is often difficult.

Biosynthesis of DNA and its repair processes have been targeted by the quinolone class of antibiotics. The quinolones form a stable ternary complex with DNA and an enzyme, the DNA gyrase. When this stable complex is encountered by the replication fork, DNA replication cannot proceed further (17–20). It is worth noting that the parental member of the quinolone class, nalidixic acid, was not isolated as an antibiotic, but rather was a synthetic compound that was shown to possess antibacterial property. A number of derivatives of nalidixic acid, such as the fluoroquinolones ciprofloxacin, moxifloxacin (Fig. 1), and sparfloxacin have become important clinically used drugs.

Protein biosynthesis and the ribosomal assembly have been the target of many different classes of antibiotics, principally because their functions are very central to the life processes of bacteria. The protein biosynthetic process involves multiple steps that take place in the ribosome. Much of the surface of the ribosome is involved in these processes, which gives opportunities for binding of many different classes of small molecules to the same sites to interfere with the biochemical events. Nature has done so aptly, as there are multiple classes of antibiotics that are known to bind to ribosome. A full survey of these antibiotics is beyond the scope of this article, but the important classes of these antibiotics are macrolides, aminoglycosides, tetracyclines, and oxazolidinones. The first three are of natural product origin, whereas the oxazolidinones (eg, linezolid) were synthetic molecules discovered to have antibacterial activity during directed screening (Fig. 1).

The recent determination of the X-ray structure of the ribosome has been a major advance in understanding the protein synthesis machinery (21–25). A few publications have addressed the structural aspects of complexes of antibiotics with ribosome (26,27). These studies have revealed that antibiotics (eg, chloramphenicol, clindomycin) and macrolides (eg, erythromycin, clarithromycin, and roxithromycin) interact with the residues of 23S ribosomal RNA at the peptidyl transferase cavity. On the other hand, tetracycline interacts mainly with the small ribosomal subunit (30S) at the decoding center.

3. Biochemical Strategies for Resistance to Antibiotics

Development of antibiotic resistance is very complex. It is the result of a series of genotypic and phenotypic interactions of the biological systems of the host, pathogen, and antibiotic. Mutagenesis and gene acquisition are two important mechanisms in bacterial survival in the face of antibiotic or other life threatening challenges. There are many factors that effect the appearance and spread of acquired antibiotic resistance. Among these, the mutation frequency and the biological cost of resistance have become of increasing importance in understanding antibiotic resistance. The mutation frequency measures all the mutations present in a given population regardless of the status of the bacterial growth at which the mutation appears. Mutations happen randomly throughout the genome, and the rate by which the resistant mutants form will depend on the size of the genome and the bacterial population. When the mutation impairs a given gene product, the organism may die. However, should the mutation not be lethal, then it creates an incremental change in the organism. In a recent report, we have reported that as much as 10^6 or more mutants per milliliter of growth might exist in actively growing populations of bacteria (28). The resistance mutations that occur during antibiotic-induced stress generally are associated with loss of bacterial fitness (biological cost) (29). For these mutants to be selected in the face of the antibiotic challenge, other mutations, second-site mutations, are needed to counterbalance the effects of the resistance mutations (30). These second-site mutations usually compensate the biological cost on bacterial fitness without loss of resistance and are distinct under different growth conditions (31–33). There are several cases in the literature that demonstrate that antibiotic resistance is associated with a biological loss of the bacterial fitness. The emergence of *Staphylococcus aureus* gentamicin-resistant small colony variants was shown to be the result of selection for gentamicin resistance. When these variants were exposed to cycles of antibiotic-free medium, they returned to a sensitive parental phenotype (34). Also, it is believed that resistance, rather than virulence, selects for the clonal spread of methicillin-resistant *S. aureus* (35). In the same context, because only certain Salmonella strains express resistance to cephalosporins and β-lactamase inhibitors mediated by AmpC-type enzymes, Morosini and co-workers (36) argued that the maintenance and expression of the *ampC* gene may be too costly for Salmonella to support its normal growth and virulence.

Bateria have evolved many mechanisms for acquiring resistance genes. These mechanisms enable bacteria to move DNA sequences from cell to cell via conjugation and transformation, or from one genome to another via classical recombination, transposition, and site-specific recombination (37). The site-specific recombination mechanism is important in acquisition and spread of the bacterial resistance. It involves dissemination of resistance genes via gene cassettes and integrons, a very common process in gram-negative bacteria (38). Gene cassettes [500–1000 base pairs (bp) in size] generally consist of a target recombination sequence (*attC* site) normally associated with a single reading frame coding for an antibiotic resistance determinant. Resistance gene cassettes have been found for each class of known antibiotics. They quite often are acquired from more sophisticated genetic structures such as integrons (39). Integrons possess a recombination site (*aatI*) at which the gene cassettes are integrated. This site contains the gene for an integrase, a promoter, and a ribosome-binding site much like a cloning and expression vector (40). In many instances, several resistant gene cassettes are found in an integron. These genetic structures are known as multiresistant integrons. To date, there are 63 antibiotic-resistance gene cassettes identified in multiresistant integrons (41). The number of the gene cassettes organized in an integron can be as many as 200 (41). Such genetic structures are known as superintegrons and were first identified in the *Vibrio cholerae* genome (42). Super integrons also contain genes with other functions beyond resistance. Recent studies have revealed that super integrons are ancient structures, widespread in proteobacteria, serving as gene acquisition machines and most likely they are the source of modern gene cassettes and multiresistant integrons (41).

Since there are many different antibiotic agents, and each organism may experience a different selection event, there are multiple and disparate mechanisms for resistance. However, the most common mechanisms for selection fall into several categories, as listed in Table 1. For the purpose of this article we will describe the resistance to the most commonly used antibiotics.

The first β-lactam antibiotic to be used clinically was penicillin G (mid-1940s). This molecular class, including other β-lactam antibiotics, has enjoyed exceptional success clinically because it inhibits a step such as cross-linking of the cell wall, that is unique to bacteria. Barring the allergic response by a small fraction of the population to these antibiotics, these molecules generally are not toxic to the host. However, their clinical success resulted in extensive use of these antibiotics, which in turn contributed to the appearance and dissemination of mechanisms of resistance. It is important to note that at least four distinct mechanisms for resistance to β-lactam antibiotics have been documented.

Table 1. **Major Bacterial Resistance Mechanisms Identified to Date**

Resistance mechanisms	Class of antibiotics	Examples
reduced permeability	β-lactams, fluoroquinolones, folate inhibitors	penicillins, cephalosporins, norfloxacin, ofloxacin, ciprofloxacin, trimethoprim, sulfamethoxazole, fosfomycin
efflux mechanism	tetracyclines, fluoroquinolones, chloramphenicol, macrolides, aminoglycosides, β-lactams, quinolones, novobicin	tetracycline, minoclycline, doxycycline, ciprofloxacin, ofloxacin, chloramphenicol, erythromycin, lincosamide, penicillin, cephalosporin, imipenem
target modification	β-lactams, fluoroquinolones, aminoglycosides, tetracyclines, folate inhibitors, glycopeptides	penicillins, cephalosporins, norfloxacin, ofloxacin, ciprofloxacin, gentamycin, to bramycin, amikacin, streptomycin, rifamycin, tetracycline, doxycycline, trimethoprim, sulfamethoxazole, vancomycin, teicoplanin, mupirocin, fusidic acid
target bypass	sulfonamides, trimethoprim	sulfamethoxazole, sulfadiazine
target amplification	β-lactams,	penicillins, cephalosporins
resistance enzyme	β-lactams, aminoglycosides, macrolides, chloramphenicol	penicillins, cephalosporins, carbapenems, neomycin, tobramycin, amikacin, gentamycin, lincosamide, erythromycin, clindamycin, chloramphenicol
biofilm formation	most classes of antibiotics	

The most common mechanisms of resistance to β-lactams is through the expression of β-lactamases. These enzymes hydrolyze the β-lactam moiety of the drug, rendering it inactive. The success of this strategy is underscored by the fact that over 350 such enzymes have been identified from clinical strains (43). These enzymes fall into four structural classes, all of which appear to follow a distinct catalytic mechanism (44–46). It has been argued that four distinct progenitor proteins gave rise to the four classes of β-lactamases in disparate evolutionary steps (47). Golemi and co-workers (48) documented that the OXA-10 β-lactamase from *Pseudomonas aeruginosa* is sequestered in the periplasmic space of this organisms in a minimum concentration of 4 µM. This concentration is produced by ~1200 molecules of the enzyme per bacterial cell, each of which is able to turn over ~1500 molecules of cloxacillin—a penicillin—per second (a total of 1.8×10^6 cloxacillin molecules are turned over per second per resistant bacterium). It is self-evident that these enzymes are formidable barriers to the antibiotics effects of these pharmaceutical agents.

The second mechanism of resistance to β-lactam antibiotic is the evolutionary acquisition of DD-transpeptidases—the target enzymes—with reduced affinity for these drugs. The prime example of these is the case of methicillin-resistant *S. aureus* (MRSA), which is a scourge of hospitals. The aforementioned transglycosylases and transpeptidases are collectively referred to as penicillin-binding proteins (PBPs). In a single acquisition event of unknown origin, a DD-transpeptidase was introduced to *S. aureus* that has the ability to perform the functions of other PBPs in this organism (49). Hence, inhibition of the four known native staphylococcal PBPs in *S. aureus* by a β-lactam atibiotic is overcome by the availability of this new enzyme, PBP2′, which is not readily inhibited by these drugs (50,51). Other examples of low-affinity PBPs, are the chromosomal PBPs found in *enterococci*: PBP3r and PBP5 in *Enterococci hirae*, PBP in *Enterococci faecalis* and PBP4 *Enterococci faecium*.

β-Lactam antibiotics must reach the outer surface of the cytoplasmic membrane to inhibit the PBPs. Hence, in gram-negative bacteria, β-lactam antibiotic has to penetrate the outer membrane to reach its target. This penetration takes place through the channel-forming proteins, namely, porins. These proteins transverse the outer membrane and are the portals through which the nutrients enter the cell. Porins hav been known to undergo mutations such that penetration by the antibiotic is slowed down. This is a means for resistance to imipenem, a member of the carbapenem class of β-lactam antibiotics (52,53). In some other cases, the decrease in permeability of β-lactams has been related to the mutational loss of major porins (54). This mechanism for resistance is also seen in combination with hyperexpression of antibiotic-modifying enzymes (54) and is not common, as alteration in these protein portals into the bacterium would have implications for penetration of nutrients and the survival of the organism.

The fourth mechanism of resistance to β-lactam antibiotics was discovered only recently (55). It has been reported that there exists an LD-transpeptidase that is capable of carrying out the cross-linking reaction not with the penultimate D-Ala residue, but rather with the third amino acid (DAP, in Fig. 2). Such cross-linking occurs at low levels in sensitive strains of *Escherichia coli* and *Escherichia faceium* but Mainardi and co-workers (55) reported that *in vitro* selection for resistance to ampicillin in *E. faecium* has shifted the dependence on transpeptidases in favor of β-lactam-insensitive LD-transpeptidase (100%), hence bypassing the function of the ubiquitous β-lactam-sensitive DD-transpeptidases, the traditional targets of these antibiotics (9).

Vancomycin has been in use since 1958, but only in the early 1980s did clinicians start using it heavily against hospital-acquired (nosocomial) infections. Vancomycin has been considered to be the antibiotics of last resort against gram-positive infections, especially the ones caused by the methicillin-resistant S.aureus (56,57). Over the past 10 years, a number of Enterococcus strains with high-level inducible resistance to vancomycin and its analogues have been identified (58). Five plasmid-borne genes are found to be necessary to induce high level of vancomycin resistance, vanR, vanS, vanH, vanA, and vanX (59). These genes are responsible for the alteration of vancomycin target: the peptidoglycan precursors in resistant strain end in D-Ala-D-Lac (Lac for lactate) instead of D-Ala-D-Ala as in sensitive strains. The conversion of the amide bond to ester entails the loss of an important hydrogen bond to the glycopeptide in the complex, which has been estimated to result in a 1000-fold reduced affinity for complex formation (60). The VanR and VanS proteins comprise a two-component regulatory system that regulates the transcription of vanRS and vanHAX genes. VanA, gene product of vanA, is a ligase that synthesizes D-Ala-D-Lac, which is added to the UDP-MurNAc-tripeptide. VanH reduces pyruvate to D-Lac, the substrate for VanA. VanX hydrolyzes D-Ala-D-Ala produced by the chromosomally D-Ala-D-Ala ligase, thereby reducing the pool of D-Ala-D-Ala, which would otherwise compete with D-Ala-D-Lac for incorporation into the peptidoglycan precursor.

Vancomycin resistance is also seen in methicillin-resistant S. aureus clinical strains (VRSA). Such strains would appear to lack the enterococcal van genes, which suggest the possibility for other mechanisms in resistance to vancomycin (61). Studies on vancomycin resistance in MRSA have associated this resistance with overproduction of penicillin-binding protein 2 (PBP2) and/or cell-wall thickening (61,62). Overexpression of PBP2 would allow more peptidoglycan precursors to be incorporated into the cross-linked cell wall synthesis (ie, higher degree of cross-linking), thus less amount of this precursor will be available as a target of vancomycin binding. Cell-wall thickening, would also prevent penetration of vancomycin into the cell wall, thus the level of the antibiotic reaching the target would be less than is needed to kill the bacteria.

Aminoglycosides are another class of antibiotics used against the infections caused by gram-positive and gram-negative bacteria. They bind to the 30S subunit of the bacterial ribosome. As with any other class of antibiotics, their antimicrobial properties are compromised by bacterial resistance. Methylation of the ribosomal binding site is known to cause resistance to gentamicin, as an example of altered target (63). This mechanism is observed only in aminoglycoside-producing organisms. Altered uptake of aminoglycoside, a rare resistance mechanism, is exhibited by anaerobes and organisms such as P. aeruginosa (64). However, the most common mechanism for resistance to aminoglycosides is by their structure modification by three families of enzymes collectively referred to as aminoglycoside-modifying enzymes (66). As shown in Figure 5, three types of reactions have been documented for these activities, namely, N-acetyltransferase, O-phosphotransferase, and O-adenyltransferase reactions. In each case, the

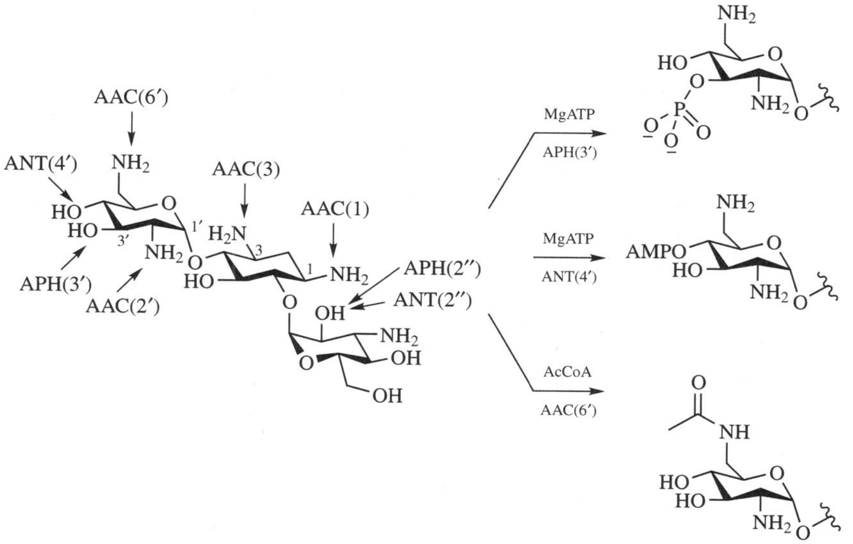

Fig. 5. Sites of modification in aminoglycosides antibiotics, as depicted for kanamycin B.

multifunctional aminoglycoside is chemically modified to derivatives that are devoid of antimicrobial activity, in light of the fact that the affinity of the ribosomal site for each is dramatically reduced (67). Note that these enzymes are cytoplasmic, as is the ribosome. However, the enzymes are produced in high concentrations (68), and they often carry out their respective reactions in rapid reactions (64,65).

Erythromycin, a macrolide antibiotic, is commonly used against grampositive bacteria. Macrolide antibiotics inhibit bacterial protein synthesis by binding to the 23S rRNA of the 50S ribosomal subunit (26). Resistance to macrolides developed soon after the introduction of erythromycin to the clinic in 1953. The first resistant clinical isolates to macrolides were *S. aureus*, but subsequently resistance transferred to other organisms. Bacteria have developed three mechanisms that protect them from the action of the macrolides: target site alteration (69), antibiotic modification (70), and altered antibiotic transport (71). Target site alteration is the most common mechanism of resistance in the organisms that produce this antibiotic (eg, *Streptomyces erythreus*). In these organisms the 23S rRNA is posttranscriptionally modified by an adenine-specific *N*-methyltransferase (methylase). These enzymes are encoded by a class of genes known as *erm* (erythromycin ribosome methylation), which mono- and dimethylate the exocyclic *N*-6 position of a highly conserved adenine nucleotide (A2058 according to *E. coli* numbering) within the peptidyl transferase loop (72,73), which is important for binding of the macrolide antibiotics. To date, >30 *erm*-related genes have been identified from different bacterial sources that range from clinical pathogens to actinomycetes (71). The methyltransferase enzymes have been classified into two classes. The first class includes the Erm enzymes that only monomethylate, such as Lrm from *Streptomyces lividans*, Clr from *Streptomyes caelestis*, and TlrD from *Streptomyces fradiae*. The second class includes those that predominately dimethylate adenine, such as ErmC from *S. aureus*, ErmE from *Saccharopolyspora erythrea*, and ErmSF from *S. fradiae*. Erm proteins can be expressed constitutively or induced by the presence of low levels of the antibiotics. It is interesting to note that resistance to erythromycin has always been seen associated with resistance to chemically distinct, but functionally overlapping antibiotic families such as lincosamide and streptogramin B. This type of resistance has been referred to as MLS$_B$ resistance, with the initials referring to the three types of antibiotics (71). Removal of macrolides from the cytoplasm of bacteria by efflux pumps is another mechanism of low-level resistance.

Since the discovery of the *erm* genes, another means of resistance involving alteration of rRNA structure has been identified. This mechanism involves single-base substitution at A2058 in the 23S rRNA. Generally, pathogenic bacteria that develop macrolide resistance through mutations at this position possess only one or two rRNA operons (*rrn,*) as in the case of *Helicobacter pylori* and *Mycobacterium* species. Species with more copies of *rrn* operons, such as *Enterococcus, Streptococcus*, and *Staphylococcus* confer resistance by expression of Erm enzymes or efflux pumps (69).

Linezoild (Zyvox-Pharmacia) is the first oxazolidinone antimicrobial approved by the U.S. FDA (April 2000) for clinical use against infections caused by multiresistant gram-positive bacteria, including MRSA, vancomycin-resistant enterococci (VRE) and penicillin-resistant *Streptococcus pneumonia* (74). This antibiotic inhibits initiation of protein synthesis by preventing the formation of a ternary complex among tRNA^{fMet}, mRNA, and the ribosome (75). Spontaneous resistance to linezolid in *S. aureus* and *S. epidermis* develops at a rate of $<10^{-9}$ (76). Studies of linezolid-resistant clinical isolates and laboratory-derived linezolid-resistant strains of MRSA and VRE have revealed single-point mutations clustered in the DNA region encoding the central loop of domain V of 23S rRNA (77–79). MRSA resistant to linezolid has been shown too possess either G2576T/U or G2447U mutations (78,80). The VRE strains resistant to linezolid have developed single-point mutations at positions G2528U (*E. faecalis*), G2576U (*E. faecalis* and *E. faecium*, and G2505A (*E. faecium*) (79). Interestingly, a laboratory-developed linezolid-resistant *E. faecalis* strain had acquired three other mutations in addition to C2512U, G2513U, and C2610G (79). In laboratory studies, *E. coli* has also been shown to develop resistance to linezolid attributed to G2032A/U/C mutations in the 23S rRNA.

Bacteria have an intrinsic mechanism for protection from any toxic compounds in their environment. The gram-negative bacteria and gram-positive mycobacteria combine two mechanisms of resistance. First, the outer membrane and the mycolate-containing cell wall, respectively, produce effective permeability barriers. Second, the antibiotics that make it through the first outer membrane barrier are pumped out by the multidrug resistance efflux pumps (MDR). In gram-negative bacteria, MDR pumps interact with outer membrane channels and accessory proteins, forming multisubunit complexes that extrude antibiotics directly into the medium, bypassing the outer-membrane barrier.

The outer membrane of gram-negative bacteria is a barrier to many antibiotics. It consists of an inner leaflet of glycerophospholipids, which has high fluidity owing to the presence of unsaturated fatty acids, and an outer leaflet of lipopolysaccharides. The lipopolysaccharides lack unsaturated fatty acids, hence they are more rigid and less permeable. The outer membrane is traversed by proteins known as porins, through which nutrients enter the cell. These porin channels have a limited opening with an exclusion limit of 600–1000 Da.

The implication of the reduced cell wall permeability alone in bacterial resistance could not justify the high level of resistance observed for many antibiotics. Rather, the synergistical action of outer-membrane barrier and the active efflux pumps, such as ArcB of *E. coli*, MexB of *P. aeruginosa*, and MtrD of *N. gonorrhoeae*, produces effective drug resistance. By actively pumping out antibiotic molecules, these systems prevent intracellular accumulation of the antibiotics in such levels that is necessary to exert the lethal activities. The efflux pumps have a broad range of substrates (Table 1), thus they have become a serious problem in the treatment of many infectious diseases (81). They are associated with both intrinsic and acquired resistance to antibiotics. MDR efflux pumps such as MexAB-OprM and MexXY-OprF in *P. aeruginosa* are reported to be constitutively expressed in wild-type strains, thus contributing to the intrinsic resistance of this organism to a number of antimicrobial agents, including tetracyclines, chloramphenicol, quinolones, novobiocin, macrolides, trimethoprim, β-lactams, and β-lactamase inhibitors (82). Recently, it has been reported that MDR pump AmrAB-OprA in *Burkholderia pseudomallei* and MexXY-OprF in *P. aeruginosa* are involved in extrusion of aminoglycosides directly into the external medium (83,84). Acquired MDR can arise via three mechanisms: (*1*) amplification and mutation of genes encoding the MDR proteins that alters expression and the activity of the transporters; (*2*) mutations in the regulatory proteins that lead to the increased expression of multidrug transporters, eg, mutations in a repressor gene *mex*R leads to the hyperexpression of *mexAB-oprM* in MDR clinical isolates; and (*3*) transfer of resistance genes on transposons or plasmids (85).

Microorganisms have the ability to irreversibly attach to and grow on a surface and produce extracellular polysaccarides that facilitate attachment and matrix formation (86). Such matrix association of cells is known as biofilms. Biofilms may form on any surface, but their formation on the surface of indwelling medical devices, tooth enamel, heart valves or the lung, and middle ear is of biomedical concern. Many different organisms develop biofilms, including pathogenic bacteria such as *K. pneumonia*, *P. aeruginosa*, *S. aureus*, *E. faecalis*, and fungi (87).

Biofilm-associated organisms have altered phenotypes with respect to growth rate and gene transcription (88) due to biofilm composition and structure (87). Biofilm consist of microcolonies held together by an extracellular matrix (polysaccharides). Its structure is hetrogeneous, with water channels that allow transport of essential nutrients and oxygen to the cells within the biofilm. Studies have shown that growth of biofilm causes a decrease in antimicrobial susceptibility, which might be intrinsic or acquired (89). Intrinsic resistance of biofilm can be related to the multicellular structure of the biofilm, which can slow down drug diffusion or possibly the matrix itself may react with the drug. Furthermore, biofilm-associated organisms have reduced growth rates that might as well minimize antimicrobial intake rate. Plasmid transfer through conjugation allows acquisition of antimicrobial resistance mechanisms, but several studies have shown that mechanisms of antibiotic resistance such as the efflux pumps, modifying enzymes, and target alteration do not seem to have a major impact on biofilm antibiotic resistance. Clearly, resistance in biofilms is more complicated: Multiple resistance mechanisms can act in concert. Adherence of bacteria to implanted medical devices or damaged tissues in the form of biofilm and inherent resistance contribute to duration of bacterial infections. As a result, biofilm formation has become a serious clinical problem (90).

BIBLIOGRAPHY

1. D. H. Williams, M. J. Stone, P. R. Hauck, and S. K. Rahman, *J. Nat. Prod.* **52**, 1189 (1989).
2. M. J. Stone and D. H. Williams, *Mol. Microbiol.* **1992**, 29 (1992).
3. L. P. Kotra, S. Vakulenko, and S. Mobashery, *Microbes Infect.* **2**, 651 (2000).
4. A. E. Allsop, *Curr. Opin. Microbiol.* **1**, 530 (1998).
5. D. T. Moir, K. J. Shaw, R. S. Hare, and G. F. Vovis, *Antimicrob. Agents Chemother.* **43**, 439 (1999).
6. D. J. C. Knowles and F. King, *Adv. Exp. Med. Biol.* **456**, 71 (1998).
7. W. L. Lee, M. A. McDonough, L. P. Kotra, Z. H. Li, N. R. Silvaggi, Y. Takeda, J. A. Kelly, and S. Mobashery, *Proc. Natl. Acad. Sci. USA.* **98**, 1427 (2001).
8. D. J. Tipper and J. L. Strominger, *Proc. Natl. Acad. Sci. USA* **54**, 1500 (1965).
9. J. A. Kelly, P. C. Moews, J. R. Knox, J. M. Frere, and J. M. Ghuysen, *Science* **218**, 479 (1982).
10. J. A. Kelly, A. P. Kuzin, P. Charlier, and F. Fonze, *Cell. Mol. Life Sci.* **54**, 353 (1998).
11. S. Pares, N. Mouz, Y. Pelliot, R. Hackenbeck, and O. Dideberg, *Nat. Struct. Biol.* **3**, 284 (1996).
12. D. H. Williams, *Acc. Chem. Res.* **17**, 364, (1984).
13. M. Ge, Zh. Chen, H. R. Onishi, J. Kohler, L. L. Silver, R. Kerns, F. Seketsu, C. Thompson, and D. Kakne, *Science* **284**, 507 (1999).
14. U. S. Eggert, R. Natividad, B. V. Falcone, A. A. Branstrom, R. C. Goldman, T. J. Silhavy, and D. Kahne, *Science* **294**, 361 (2001).
15. T. D. H. Bugg and C. T. Walsh, *Nat. Prod. Rep.* **9**, 199 (1992).
16. K. K. Wong and D. L. Pompliano, in B. P. Rosen, and S. Mobashery, (4), Vol. 456, 1998, p. 197.
17. X. S. Pan and L. M. Fisher, *Antimicrob. Agents Chemother.* **42**, 2810 (1998).
18. V. E. Anderson and N. Osheroff, *Curr. Pharm. Des.* **7**, 337 (2001).
19. L. Ferrero and co-workers, *Mol. Microbiol.* **13**, 641 (1994).
20. A. B. Khodursky, E. L. Zechiedrich, and N. R. Cozzarelli. *Proc. Natl. Acad. Sci. USA* **92**, 1180 (1995).
21. J. H. Cate, M. M. Yusupov, G. Zh. Yusupova, T. N. Earnest, and H. F. Noller, *Science* **285**, 2095 (1999).
22. G. C. Agalarov, G. Sridhar-Prasad, P. M. Funke, C. D. Stout, and J. R. Williamson, *Science* **288**, 107 (2000).
23. B. T. Wimberly, D. E. Brodersen, W. M. Clemons, Jr., R. J. Morgan-Warren, A. P. Carter, C. Vonrhein, T. Hartsch, and V. Ramakrishnan, *Nature (London)*, **407**, 327 (2000).
24. M. M. Yusupov, G. Zh. Yusupova, A. Baucom, K. Lieberman, T. N. Earnest, J. H. D. Cate, and H. F. Noller, *Science* **292**, 883 (2001).
25. J. M. Ogle, D. E. Brodersen, M. C. William, Jr., M. J. Tarry, A. P. Carter, and V. Ramakrishnan, *Science* **292**, 897 (2001).

26. F. Schlunzen, R. Zarivach, J. Harms, A. Bashan, A. Tocilj, R. Albrecht, A. Yonath, and F. Franceschi, *Nature(London)* **413**, 814 (2001).

27. M. Pioletti, F. Schlunzen, J. Harms, R. Zarivach, M. Gluhmann, H. Avila, A. Bashan, H. Bartels, T. Auerbach, C. Jacobi, T. Hartsch, A. Yonath, and F. Franceschi, *EMBO J.* **20**, 1829 (2001).

28. S. Mobashery and E. Azucena, *Encyclopedia Life Science London*, Nature Publishing Group, UK, 2000.

29. R. E. Lenski, *Int. Microbiol.* **4**, 265 (1998).

30. M. G. Reynolds, *Genetics* **156**, 1471 (2000).

31. J. Björkman, D. Hughes, and D. I. Anderson, *Proc. Natl. Acad. Sci. USA* **95**, 3949 (1998).

32. J. Björkman, P. Samuelsson, D. I. Anderson, and D. Hughes, *Mol. Mirobiol.* **31**, 53 (1999).

33. J. Björkman, I. Nagaev, O. G. J. Berg, D. Hughes, and D. I. Anderson, *Science* **287**, 1479 (1999).

34. R. C. Massey, A. Buckling, and S. J. Peacock, *Curr. Biol.* **11**, 1810 (2001).

35. B. Shopsin, B. Mathema, X. Zhao, J. Martínez, J. Kornblum, and B. N. Kreiswirth, *Microb. Drug Resist.* **3**, 239 (2000).

36. M. I. Morosini, J. A. Ayala, F. Baquero, J. L. Martínez, and J. Blèzquez, *Antimicrob. Agents Chemother.* **44**, 3137 (2000).

37. P. M. Bennet, *J. Antimicrob. Chemother.* **43**, 1 (1999).

38. R. M. Hall, in D. J. Chadwick, and J. Goode, eds., *Antibiotic Resistance: Origins, Evolution, Selection and Spread*, Ciba Foundation Symposium 270, John Wiley & Sons, Inc., Chichester, 1996, p. 192.

39. C. M. Collis and R. M. Hall, *Antimicrob. Agents Chemother* **39**, 155 (1995).

40. D. A. Rowe-Magnus and D. Mazel, *Curr. Opin. Microbiol.* **2**, 483 (1999).

41. D. A. Rowe-Magnus, A.-M. Guerout, P. Ploncard, B. Dychinco, J. Davies, and D. Mazel, *Proc. Natl. Acad. Sci. USA*, **98**, 652 (2001).

42. D. Mazel, B. Dychinco, V. A. Webb, and J. Davies, *Science*, **280**, 605 (1998).

43. *http://www.lahey.org*

44. L. P. Kotra, J.-P. Samama, and S. Mobashery, in K. Lewis, A. A. Salyers, H. W. Taber, and R. G. Wax, eds., *Bacterial Resistance to Antimicrobials*, Dekker, New York, 2001, P. 123.

45. D. Golemi, L. Meveyraud, S. Vakulenko, S. Trainer, A. Ishiwata, L. P. Kotra, J.-P. Samama, and S. Mobashery, *J. Am. Chem. Soc.* **122**, 6132 (2000).

46. A. Patera, L. C. Blaszczak, and B. K. Shoichet, *J. Am. Chem. Soc.* **122**, 10504 (2000).

47. I. Massova and S. Mobashery, *Antimicrob. Agents Chemother.* **42**, 1 (1998).

48. D. Golemi, L. Maveyraud, S. Vakulenko, S. Trainer, J.-P. Samama, and S. Mobashery, *Proc. Natl. Acad. Sci. USA*, **98**, 14280 (2001).

49. M. G. Pinho, S. R. Filipe, H. de. Lencastre, and A. Tomasz, *J. Bacteriol.* **183**, 6525 (2001).

50. S. Roychoudhury, J. E. Dotzlaf, S. Ghag, and W. K. Yeh, *J. Biol Chem.* **269**, 12067 (1994).

51. M. G. Pinho, H. de. Lencastre, and A. Tomasz, *Proc. Natl. Acad. Sci. USA*, **98**, 10886 (2001).

52. L. Martinez-Martinez, M. C. Conejo, A. Pascual, S. Hernandez-Alles, S. Ballesta, E. Ramirez De. Arellano-Ramos, V. J. Benedi, and E. J. Perea, *Antimicrob. Agents Chemother.* **9**, 2534 (2000).

53. M. M. Ochs, M. Bains, and R. E. Hancock, *Antimicrob. Agents Chemother.* **7**, 1983 (2000).

54. L. Martinez-Martinez, M. C. Conejo, A. Pascual, S. Hernandez-Alles, S. Ballesta, E. Ramirez De. Arellano-Ramos, V. J. Benedi, and E. J. Perea, *Antimicrob. Agents Chemother.* **44**, 2534 (2000).

55. J. L. Mainardi, R. Legrand, M. Arthur, B. Schoot, J. Van Heijenoort, and L. Gutman, *J. Biol. Chem.* **275**, 16490 (2000).

56. M. P. Wilheim. *Mayo Clin. Proc.* **66**, 1165 (1991).
57. D. H. Williams and B. Bardsley, *Angew. Chem. Int. Ed. Engl.* **38**, 1173 (1999).
58. M. N. Swartz, *Proc. Natl. Acad. Sci. USA.* **91**, 2420 (1994).
59. M. Arthur and P. Courvalin, *Antimicrob. Agents Chemother.* **37**, 1563 (1993).
60. C. T. Walsh, S. L. Fisher, I. S. Park, M. Prahalad, and Z. Wu, *Chem. Biol.* **3**, 21 (1996).
61. H. Hanaki, K. Kuwahara-Arai, S. Boyle-Vavra, R. S. Daum, H. Labischinski, and K. Hiramatsu, *J. Antimicrob. Chemother.* **42**, 199 (1998).
62. B. Moreira, S. Boyle-Vavra, B. L. M. de Jonge, and R. S. Daum, *Antimicrob. Agents Chemother.* **41**, 1788 (1997).
63. J. M. Musser, *Clin. Microbiol. Rev.* **8**, 496 (1995).
64. G. D. Wright, P. R. Thompson, *Front. Bioscience.* **4**, D9-D21 (1999).
65. G. D. Wright, A. M. Berghuis, and S. Mobashery, *Adv. Exp. Med. Biol.* **456**, 27 (1998).
66. J. Davies and G. D. Wright, *Trends Microbiol.* **6**, 234, (1997).
67. B. Llano-Sotelo, E. Azucena, L. P. Kotra, S. Mobashery, and C. S. Chow, *Chem. Biol.* **9**, 455 (2002).
68. J. Siregar, K. Miroshnikov, and S. Mobashery, *Biochemistry*, **39**, 12681 (1995).
69. B. Vester and S. Douthwaite, *Antimicrob. Agents Chemother.* **45**, 1 (2001).
70. B. Weisblum, *Drug Resistance Updates.* **1**, 29 (1998).
71. B. Weisblum, *Antimicrob. Agents Chemother.* **39**, 577 (1995).
72. C. D. Denoya and D. Dubnau, *J. Biol. Chem.* **264**, 2615 (1989).
73. R. Skinner, E. Cundliffe, and F. J. Schmidt, *J. Biol. Chem.* **258**, 12702 (1983).
74. H. B. Fung, H. L. Kirschenbaum, and B. O. Ojofeitimi, *Clin. Ther.* **23**, 356 (2001).
75. S. M. Swaney, H. Aoki, M. C. Ganoza, and D. L. Shinabarger, *Antimicrob. Agents Chemother.* **42**, 3251 (1998).
76. G. W. Kaatz and S. M. Seo, *Antimicrob. Agents Chemoter.* **40**, 799 (1996).
77. R. D. Gonzales, P. Schrekenberger, C. Graham, S. Kelkar, K. DenBesten, and J. P. Quinn, *Lancet*, **357**, 1179 (2001).
78. S. Tsiodras, H. S. Gold, G. Sakoulas, G. M. Eliopoulos, C. Wennersten, L. Vankataraman, R. C. Moellering, Jr., and M. J. Ferraro, *Lancet* **358**, 207 (2001).
79. J. Prystowsky, F. Siddiqui, J. Chosay, D. L. Shinabarger, J. Millichap, L. R. Peterson, and G. A. Noskin, *Antimicrob. Agents Chemother*, **45**, 2154 (2001).
80. S. M. Swney, D. L. Shinabarger, R. D. Schaadt, J. H. Bock, J. L. Slightom, and G. E. Zurenko, *Absrt, 38th Intersci. Conf. Antimicrob. Agents, Chemother.*, abstr. C-104, 1998.
81. H. Nikaido, *Cell Develop. Biol.* **12**, 215 (2001).
82. X.-Y. Li, H. Nikaido, and K. Poole, *Antimicrob. Agents Chemoter.* **39**, 1948 (1995).
83. R. A. Moore, D. DeShazer, S. Reckseidler, A. Weissman, and D.E. Woods, *Antimicrob. Agents Chemother.* **43**, 465 (1999).
84. J. R. Aires, T. Köhler, H. Nikaido, and A. Plésiat, *Antimicrob. Agents Chemother.* **43**, 2624 (1999).
85. M. Putman, H. W. vanVeen, W. N. Konnigs, *Microbiol. Mol. Biol. Rev.* **64**, 673 (2000).
86. H. M. Lappin-Scott and C. Bass, *Am. J. Infect. Control.* **4**, 250 (2001).
87. R. M. Donlan, *Clin. Infect. Dis.* **33**, 1387 (2001).
88. M. Whiteley, M. G. Bangera, R. E. Bumgarner, M. R. Parsek, G. M. Teitzel, S. Lory, and E. P. Greenberg, *Nature (London)* **413**, 860 (2001).
89. P. S. Stewart and J. W. Costerton, *Lancet* **358**, 135 (2001).
90. M. Wilson, *Sci. Prog.* **84**, 235 (2001).

DASANTILA GOLEMI-KOTRA
SHAHRIAR MOBASHERY
Wayne State University

ANTIMONY AND ANTIMONY ALLOYS

1. Introduction

Antimony [7440-36-0], Sb, belongs to Group 15 (VA) of the Periodic Table which also includes the elements arsenic and bismuth. It is in the second long period of the table between tin and tellurium. Antimony, which may exhibit a valence of +5, +3, 0, or −3 (see ANTIMONY COMPOUNDS), is classified as a nonmetal or metalloid, although it has metallic characteristics in the trivalent state. There are two stable antimony isotopes that are both abundant and have masses of 121 (57.25%) and 123 (42.75%).

2. History and Occurrence

Antimony and the natural sulfide of antimony were known at least as early as 4000 BC. The sulfide was used as an eyebrow paint in early biblical times, and a vase found at Tello, Chaldea, was reported to be cast antimony. Copper articles covered with a thin coating of metallic antimony, found in Egypt and dating from the period 2500–2200 BC, indicate that the early Egyptians were using antimony. Pliny (50 AD) gave it the name *stibium*; Geber used the term *antimonium*; and as late as the time of Lavoisier both terms continued to be used for the sulfide. Early alchemists referred to antimony sulfide as the "wolf of metals" because it devours all metals except gold. *Triumph Wagen des Antimonii*, written by Basil Valentine in the fifteenth century, is recognized as one of the first significant accounts of antimony and its chemistry. Both Agricola (1559) and Biringuccio (~1550) mention the liquation of antimony ores. The latter enumerates the following uses for antimony: as an alloy to increase the tone of bell metal; in the production of pewter and glass and metal mirrors; as medication for ulcers; and as yellow pigment for painting earthenware, and tinting enamels and glass. A scientific treatise on the element was written by Nicolas Lemery (1645–1715).

The crustal abundance of antimony (1) is ~0.2 g/t. Antimony ore bodies are small and scattered throughout the world. The word antimony (from the Greek *anti* plus *monos*) means "a metal not found alone" and, in fact, native antimony is seldom found in nature because of its high affinity for sulfur and metallic elements such as copper, lead, and silver. Over 100 naturally occurring minerals of antimony have been identified (2–4). Occasionally native metallic antimony is found; however, the most important source of the metal is the mineral stibnite [1317-86-8] (antimony trisulfide), Sb_2S_3. In areas where stibnite has been exposed to oxidation, it is converted to oxides of antimony. The important oxide minerals are stibiconite [12340-12-4] $Sb_2O_4 \cdot H_2O$; cervantite, Sb_2O_4, or $Sb_2O_3 \cdot Sb_2O_5$?; valentinite [1317-98-2], orthorhombic Sb_2O_3, and senarmontite [12412-52-1], cubic Sb_2O_3; kermesite [12196-98-0], $2\,Sb_2S_3 \cdot Sb_2O_3$, an oxysulfide ore, is also of commercial importance.

World reserves of antimony (5) are estimated to be 2.1 million metric tons. Approximately 95% of the world's primary antimony was mined in China (85%),

Table 1. **World Mine Production, Reserves, and Reserve Base**[a,b]

Country	Mine production		Reserves	Reserve Base
	2000	2001[c]		
United States	W	300	80,000	90,000
Bolivia	2,800	3,000	310,000	320,000
China	100,000	95,000	900,000	1,900,000
Kyrgyzstan	200	200	120,000	150,000
Russia	5,000	3,000	350,000	370,000
South Africa	5,000	5,000	240,000	250,000
Tajikistan	2,000	2,000	50,000	60,000
other countries	3,000	6,000	25,000	75,000
World total (may be rounded)	*118,000*	*115,000*	*2,100,000*	*3,200,000*

[a] Ref. 6.
[b] Data in metric tons.
[c] Estimated.

Bolivia 12%, Russia (4%), and the Republic of South Africa (4%), China has the world's largest reserves. See Table 1 for world mine production, reserves, and reserve base for 2000–2001. Most of the antimony produced in the United States is from the complex antimony deposits found in Idaho, Nevada, Alaska, and Montana. These deposits consist of stibnite and other sulfide minerals containing base metals and silver or gold. Ores of the complex deposits are mined primarily for lead, copper, zinc, or precious metals; antimony is a by-product of the treatment of these ores.

3. Properties

Physical properties of antimony are given in Table 2. Antimony, a silvery white, brittle, crystalline solid, is a poor conductor of electricity and heat. Two unstable allotropes, a black and a yellow modification, have been observed (7). The black modification is amorphous and forms on rapid quenching of antimony vapor. The yellow modification, covalent and similar to yellow arsenic (8,9), is formed by the low (−90°C) temperature oxidation of stibine using oxygen or chlorine. Explosive antimony, sometimes referred to as a third modification, can be produced from the electrolysis of antimony chloride, iodide, or bromide under special conditions, and is believed to be in a strained amorphous state. When scratched or bent, it explodes mildly, giving the crystalline form. Sudden heating of the electrodeposit to about 125°C produces an explosion; however, gradual heating leads safely to the crystalline form. Some electrodeposits drying at room temperature have been observed to ignite and burn, though incompletely, to form the oxide. On solidification, pure antimony contracts 0.79 ± 0.14 vol% (10).

Antimony is ordinarily quite stable and not readily attacked by air or moisture. It burns emitting a bluish light when heated to redness in air. Under controlled conditions antimony reacts with oxygen to form the oxides Sb_2O_3, Sb_2O_4,

Table 2. **Physical Properties of Antimony**

Property	Value
CAS Registry Number	
Sb	[7440-36-0]
^{121}Sb	[14265-72-6]
^{123}Sb	[14119-16-5]
at wt	121.75
mp, °C	630.8
bp, °C	1753
density at 25°C, kg/m^3	6684
crystal system	hexagonal (rhombohedral)
lattice constant, nm	
a	0.4307
c	1.1273
hardness, Mohs' scale	3.0–3.5
latent heat of fusion, J/mol^a	19,874
latent heat of vaporization, J/mol^a	195,250
coefficient of linear expansion at 20°C, μm/(m·°C)	8–11
electrical resistivity at 0°C, μΩ·cm	39
magnetic susceptibility at 20°C, cgs	-99.0×10^{-6}
specific heat at 25°C, J/(mol·K)[1]	25.2
thermal conductivity at 0°C, W/(m·K)	25.5
capture cross-section for thermal neutrons, at 2200 m/s, 10^{-28} m^2/atom	
121.75 (at wt)	5.40 ± 0.60
121 (isotope)	6.25 ± 0.20
123 (isotope)	4.28 ± 0.16

^{a} To convert J to cal, divide by 4.184

and Sb_2O_5, Antimony tetroxide [1332-81-6] may be considered a stoichiometric compound of composition $Sb_2O_3 \cdot Sb_2O_5$, or antimony(III) antimonate(V).

Certain conditions of pH, oxidation potential, and temperature promote the corrosion or dissolution of antimony in aqueous systems (11). In nonoxidizing, ie, unaerated solutions, antimony is stable and does not dissolve over a wide pH range. As more air, or oxidizing agent, is allowed into the solution, however, antimony begins to oxidize to the tri- and pentavalent states. In the trivalent state, antimony exists in the form SbO^+, oxoantimony [22877-95-8], or Sb_2O_3, antimony(III) oxide [1309-64-4], which dissolves in solution as antimonious acid, $HSbO_2$, and antimonite [27264-01-3], SbO_2^-. The pH of the oxidized solution determines which complex is predominant. Stronger oxidants such as nitric acid, mercury oxide, sodium peroxide, or hydrogen peroxide, oxidize antimony to its pentavalent state. Under these conditions, the species SbO_2^+, antimony(V) oxide [1314-60-9], Sb_2O_5, and SbO_3^-, antimonate [15600-59-6] exist.

Nitric acid oxidizes antimony forming a gelantinous precipitate of a hydrated antimony pentoxide (9). With sulfuric acid an indefinite compound of low solubility, probably an oxysulfate, is formed. Hydrofluoric acid forms fluorides or fluocomplexes with many insoluble antimony compounds. Hydrochloric acid in the absence of air does not readily react with antimony. Antimony also forms complex ions with organic acids.

Antimony reacts vigorously with chlorine to form tri- and pentachlorides. It combines with sulfur in all proportions and forms the compounds antimony red [1345-04-6], Sb_2S_3, and golden antimony [1315-04-4], Sb_2S_5. Antimony itself does not react directly with hydrogen. However, antimony hydride [7803-52-3] (stibine), SbH_3, which is extremely poisonous, may be formed by the reaction of metal antimonides, for example, zinc antimonide [12039-40-6], Zn_3Sb_2, and acid, the reduction of antimony compounds in hydrochloric acid with zinc, aluminum, or other reducing metals, and the electrolysis of acid or alkaline solutions using an antimony cathode (see ANTIMONY COMPOUNDS).

4. Process Metallurgy

The antimony content of commercial ores ranges from 5 to 60%, and determines the method of treatment, either pyrometallurgical or hydrometallurgical. In general, the lowest grades of sulfide ores, 5–25% antimony, are volatilized as oxides; 25–40% antimony ores are smelted in a blast furnace; and 45–60% antimony ores are liquated or treated by iron precipitation. The blast furnace is generally used for mixed sulfide and oxide ores, and for oxidized ores containing up to ~40% antimony; direct reduction is used for rich oxide ores. Some antimony ores are treated by leaching and electrowinning (4) to recover the antimony. The concentrates may be leached directly or converted into a complex matte first. The most successful processes use an alkali hydroxide or sulfide as the solvent for antimony (see METALLURY, SURVEY).

4.1. Oxide Volatilization.   Removal of antimony as the volatilized trioxide is the only pyrometallurgical method suitable for low grade ores. Combustion of the sulfide components of the ore supplies some of the heat; hence, fuel requirements are minor. There are many variations of the volatilization process, the principles employed being the same but the equipment differing. In all cases, the sulfur is burned away and removed from the waste gases, whereas the volatile antimony trioxide is recovered is flues, condensing pipes, a baghouse, Cottrell precipitator, or a combination of the above. Roasting and volatilization are effected almost simultaneously by heating the ore, mixed with coke or charcoal, under controlled conditions in equipment such as a shaft furnace, rotary kiln, converter, or roaster. If the volatilization conditions are too oxidizing, the nonvolatile antimony tetroxide may form and the recovery of antimony, as antimony trioxide, is diminished. Usually, the oxide produced in this manner is impure and can be reduced to metal. However, special attention to choice of charge, volatilization conditions, and selection of product results in a high grade oxide that is suitable for use in ceramics and other applications.

A process has been developed to recover antimony and arsenic from speiss and other materials (12). The speiss is roasted along with a source of solid sulfur and coal or coke at a temperature of 482–704°C for a sufficient time to volatilize arsenic and antimony oxides. The arsenic can then be separated from the antimony through careful control of the off-gas temperature and oxygen potential (13).

4.2. Liquation. Antimony sulfide is readily but inefficiently separated from the gangue of comparatively rich sulfide ore by heating the gangue to 550–600°C in perforated pots placed in a brick furnace. The molten sulfide is collected in lower containers. A more efficient method uses a reverberatory furnace and continuous liquation; however, a reducing atmosphere must be provided to prevent oxidation and loss by volatilization. The residue, containing 12–30% antimony, is usually treated by the volatilization process to recover additional antimony. The liquated product, called crude or needle antimony, is sold as such for applications requiring antimony sulfide, or is converted to metallic antimony by iron precipitation or careful roasting to the oxide followed by reduction in a reverberatory furnace.

4.3. Oxide Reduction. The oxides of antimony are reduced with charcoal in reverberatory furnaces. An alkaline flux consisting of soda, potash, and sodium sulfate, is commonly used to minimize volatilization and dissolve residual sulfides and gangue. Part of the slag is frequently reused. Loss of antimony from the charge by volatilization is high (12–20% or more), even with use of ample slag and careful control. This necessitates the use of effective Cottrell precipitators or baghouses, and considerable recycling of oxide.

4.4. Iron Precipitation. Rich sulfide ore or liquated antimony sulfide (crude antimony) is reduced to metal by iron precipitation. This process, consisting essentially of heating molten antimony sulfide in crucibles with slightly more than the theoretical amount of fine iron scrap, depends on the ability of iron to displace antimony from molten antimony sulfide. Sodium sulfate and carbon are added to produce sodium sulfide, or salt is added to form a light fusible matte with iron sulfide and to facilitate separation of the metal. Because the metal so formed contains considerable iron and some sulfur, a second fusion with some liquated antimony sulfide and salt follows for purification.

4.5. Blast Furnace Smelting. Intermediate (25–40%) grades of oxide or sulfide or mixed ores, liquation residues, mattes, rich slags, and briquetted fines or flue dusts are processed in water-jacketed blast furnaces. In general, the blast-furnace practice used for lead is followed, employing a high smelting column, comparatively low air pressure, and separation of slag and metal in a forehearth. Slag, usually running under 1% antimony, is desired because it tends to reduce volatilization losses.

4.6. Leaching Followed by Electrolysis. At the Sunshine Mining Co. (14, 15), a silver–copper–antimony concentrate containing 15–20% antimony is batch-leached in hot concentrated sodium sulfide, Na_2S, solution. Leaching is carried out in mild-steel tanks heated with steam coils and equipped with agitators. Four metric tons of sodium sulfide solution (~300 g/L) are charged with each metric ton of tetrahedrite [12054-35-2], $Cu_{12}S_{13}Sb_4$, concentrate. The material is leached for 14 at 100°C. At completion of the leach step, the slurry is pumped to a thickener. The clear solution containing sodium thioantimonate [13776-84-6], Na_3SbS_4, is decanted and sent to the electrowinning department. The silver and copper present in the tetrahedrite are not affected by the Na_2S leach and report to the thickener underflow. This residue is washed, filtered, and shipped to a smelter.

Electrolysis, also a batch operation, is conducted in cells each consisting of nine cathodes and eight anodes. The electrodes are fabricated from mild-steel.

The busbars carry a current of 1500 A resulting in a current density of 3 A/m^2 (28 A/ft^2). The voltage drop across each cell is typically 3 V. During electrolysis, the electrolyte can become fouled with sodium compounds including the polysulfides, thiosulfates, and sulfates that hinder the deposition of antimony on the cathode. To reduce the effect of these oxidation products, two separate electrolyte solutions, an anolyte and a catholyte, are used. The two solutions are kept separate by placing the anodes in steel baskets that have canvas sides. The canvas acts as a diaphragm that permits the flow of electric current but reduces the migration of harmful oxidation products to the catholyte. Fresh anolyte is made up of NaOH and barren electrolyte. Pregnant catholyte solution (~60 g/L Sb) is added to the electrolytic cells while withdrawing an equivalent volume of barren catholyte (~10 g/L Sb). The cathode metal obtained is 95% pure antimony.

The Bunker Hill Co. (16) and ASARCO, Inc. (17) have developed processes for the leaching and electrowinning of antimony from tetrahedrite ores. As of 1998, only Sunshine Mining Co. was electrowinning antimony metal.

The filtered sodium thioantimonate solution obtained from the leaching of stibnite with sodium sulfide may also be reduced directly to metal by elemental sodium (18). Yields in excess of 95% of 95.5 pure antimony are claimed (19).

4.7. By-Product and Secondary Antimony.   Antimony is often found associated with lead ores. The smelting and refining of these ores yield antimony-bearing flue, baghouse, and Cottrell dusts, drosses, and slags. These materials may be treated to recover elemental antimony or antimonial lead from which antimony oxide or sodium antimonate may be produced.

Recycling of antimony provides a large proportion of the domestic supply of antimony. Secondary antimony is obtained from the treatment of antimony-bearing lead and tin scrap such as battery plates, type metal, bearing metal, antimonial lead, etc. The scrap are charged into blast furnaces, reverberatory furnaces, or rotary furnaces, and an impure lead bullion or lead alloy is produced. Pure lead or antimony is then added to meet the specifications of the desired lead–antimony alloy.

4.8. Refining.   The metal produced by a simple pyrometallurgical reduction is normally not pure enough for a commercial product and must be refined. Impurities present are usually lead, arsenic, sulfur, iron, and copper. The iron and copper concentrations may be lowered by treating the metal with stibnite or a mixture of sodium sulfate and charcoal to form an iron-bearing matte that is skimmed from the surface of the molten metal. The metal is then treated with an oxidizing flux consisting of caustic soda or sodium carbonate and niter (sodium nitrate) to remove the arsenic and sulfur. Lead cannot be readily removed from antimony, but material high in lead may be used in the production of antimony-bearing, lead-based alloys. The yield of refined antimony from the matting and fluxing technique is 85–90%.

Impure metal may be refined by electrolysis (20), although this procedure is not as economical as the pyrometallurgical treatment. An electrolyte containing antimony fluoride and sulfuric acid gives the best result. Most of the anode impurities are lowered by electrolysis. However, if the concentrations of copper and arsenic are high in the anode they codeposit with the antimony and are present as significant impurities in the cathode metal. The arsenic and sulfur content of the cathode metal is further lowered by melting in an oxidizing flux. The purity of electrolytic antimony under favorable conditions exceeds 99.9%.

The purity of refined antimony, normally referred to as regulus, is usually judged by the appearance of a dendritic fern or starlike pattern on the surface of the metal. This product is produced by casting the antimony in molds into which a small amount of a starring slag has been poured; the antimony freezes while surrounded by the still molten slag. The starring flux has a melting point below that of antimony. An effective flux contains a 60–40 mixture of sodium sulfide and potassium carbonate, although other formulations can be used satisfactorily (2,21). Though this practice is generally accepted, it is quite unreliable in judging purity as highly contaminated antimony can also be starred by controlling the casting conditions and the flux composition.

4.9. Alternative Methods of Production.   In general, metal chlorides are more easily volatilized than metal oxides because of the relatively high vapor pressures. The advantages of chlorine-based pyrometallurgical processes over conventional oxidizing processes are the efficient removal of impurities and cost effectiveness. Antimony has been successfully removed from concentrates containing copper, silver, and gold using a chloridizing roast (22).

Microbiological leaching of copper and uranium has been commercially developed and research has indicated that microorganisms may be used to oxidize complex antimony sulfide minerals (23,24). If this technology is developed commercially, it may allow for the exploitation of many low grade antimony deposits.

5. Economic Aspects

The United States is not self-sufficient in its requirements for antimony and is heavily dependent on imports of both ore and metal. See Table 3 for United States statistics for the period 1997–2001.

5.1. Production.   In 2001, mine production in the United States accounted for only a small percentage of the annual domestic supply of antimony. An important component of the domestic supply is the recovery of antimony from old scrap; such as that recovered from the recycling of scrapped batteries (25). However, the percentage has decreased because of the increased use of low maintenance batteries, which use lead alloys containing less or no antimony; and the downsizing of automotive batteries that require less antimony per battery. Other factors that influence the supply of secondary antimony are the percentage of available batteries being recycled, the demand for batteries, and the prices of lead. The remainder of the U.S. supply of antimony is imported (6).

Producers of primary metal and oxide in the United States are Amspec Chemical Corp., Anzon ore., Laurel Industries Inc., Sunshine Mining Co., and U.S. Antimony Corp. (5).

5.2. Imports and Exports.   The availability of economical foreign sources of antimony, mainly from China, Mexico, South Africa, Belgium, and Bolivia has resulted in an increase in the quantity of antimony imported for consumption. The U.S. imports of antimony ore and concentrates and antimony

Table 3. **Salient Statistics for Antimony, United States**[a,b]

	Year				
	1997	1998	1999	2000	2001[c]
production					
mine (recoverable antimony)[d]	356	498	450	340	300
smelter:					
primary	26,400	24,000	23,800	20,900	18,000
secondary	7,550	7,710	8,220	7,920	7,500
imports for consumption	39,300	34,600	36,800	37,600	39,000
exports of metal, alloys, oxide, and waste and scrap[e]	3,880	4,170	3,190	1,080	1,500
shipments from Government stockpile	2,930	4,160	5,790	4,536	4,500
consumption, apparent[f]	46,600	42,700	36,500	49,376	49,800
price, metal, average, cents per pound[g]	98	72	63	66	65
stocks, year end	10,800	10,600	10,900	10,300	10,300
employment, plant, number[c]	100	80	75	70	70
net import reliance[h] as a percentage of apparent consumption	83	81	82	84	86

[a] Ref. 6.
[b] Data in metric tons.
[c] Estimated
[d] Data for 1997–2000 from the U.S. Securities and Exchange Commission 10-K report
[e] Gross weight.
[f] Domestic mine production + secondary production from old scrap + net import reliance.
[g] New York dealer price for 99.5–99.6% metal, c.i.f. U.S. ports.
[h] Defined as imports−exports + adjustments for government and industry stock changes.

oxide by country are listed in Table 4 (1999–2000). Much of the antimony imported by the United States comes from China. Export information of metal, alloys, oxide, and waste and scraps for the years 1997–2001 is listed in Table 3. The U.S. government stockpile sales continued. The Defense Logistics Agency (DLA) planned for the disposal of 5000 tons during fiscal 2002 (6).

5.3. Industrial Consumption. Reported consumption of primary antimony in 2000 was 24% above that in 1999 (15). Flame retardants accounted for much of the increase. The estimated distribution of antimony uses (2001) was flame retardants, 55%, transportation (including batteries), 18%, chemicals, 10%, glass, 7%, and others, 10%. Reported industrial consumption of primary antimony in the United States for 1999–2000 is listed in Table 5.

5.4. Prices. The price of antimony continued to decline during the first one-half of 2001. Prices started the year at $0.68–0.73/lb ($0.30–0.33/kg). By mid-summer, the price slipped to $0.58–62/lb ($0.26–0.28/kg). Prices fell because of continued world oversupply, especially from growing exports from China (6).

6. Specifications

Antimony is available as cast cakes, ingots, broken pieces, granules, shot, and single crystals. ASTM has published standards for two grades of antimony ingots

Table 4. **U.S. Imports for Consumption of Antimony, by Class and Country**[a,b]

Country	1999 Gross weight, metric tons	1999 Antimony content[c] metric tons	1999 Value, $ 10³	2000 Gross weight, metric tons	2000 Antimony content[c] metric tons	2000 Value, 10³
Antimony ore and concentrate						
Australia	1,660	1,070	710	1,750	1,150	751
Austria	95	66	307	140	98	392
Bolivia				220	144	101
China	436	398	508	1,000	1,000	1,550
France				6	3	18
Hong Kong	59	53	52	62	60	65
Mexico	1,340	1,290	1,770	937	903	1,170
Russia				499	315	193
Thailand				20	16	12
Total	*3,590*	*2,870*	*3,350*	*4,630*	*3,690*	*4,250*
Antimony oxide						
Belgium	3,290	2,730	5890	3,690	3,070	6,560
Bolivia[d]	1,770	1,470	2,110	1,150	957	1,220
Chile[d]	275	229	328			
China	9,470	7,860	11,800	13,100	10,900	17,300
France	233	193	329	66	54	230
Germany	16	14	277	47	39	802
Guatemala	249	207	428	77	64	132
Hong Kong	420	349	523	453	376	622
Japan	127	105	700	33	27	274
Kyrgyzstan				224	186	247
Mexico	3,560	2,950	4,710	5,530	4,590	7,660
Netherlands	178	148	193			
South Africa	3,220	2,680	938	3,830	3,180	999
Taiwan				29	24	53
Thailand				60	50	11
United Kingdom	224	202	699	176	146	298
Total	*23,100*	*19,100*	*28,900*	*28,500*	*23,700*	*36,500*

[a] Ref. 5, Source U.S. Census Bureau.

[b] Data are rounded to no more than three significant digits; may not add to totals shown.

[c] Antimony ore and concentrate content reported by the U.S. Census Bureau. Antimony oxide content is calculated by the U.S. Geological Survey.

[d] Antimony oxide from these countries believed to "crude" and would probably be shipped to refineries for upgrading.

(26). Grade A has a minimum antimony content of 99.8% and the following impurity maximums: arsenic 0.05%; sulfur 0.10%; lead 0.15%; and others 0.05% each. Grade B material is composed of 99.5% antimony as a minimum with maximum impurity levels of arsenic 0.1%, sulfur 0.1%, lead 0.2%; and others 0.1% each. ASTM standards are also available for pewter, babbits, and solders (27).

Table 5. **Reported Industrial Consumption of Primary Antimony in the United States, by Product**[a,b]

Product	1999[c]	2000[c]
Metal products		
antimonial lead	1,110[d]	864
bearing metal and bearings	29[d]	42
solder	136[d]	136
other[e]	1,170[d]	1,660
Total	*2,440*	*2,700*
Nonmetal products		
ammunition primers	23	26
ceramics and glass	1,120	862
pigments	1,020	620
plastics	1,580	1,960
other[f]	198	647
Total	*3,940*	*4,110*
Flame retardants		
adhesives	140	332
plastics	6,370[d]	8,920
rubber	391	382
textiles	229	221
other[g]	14	70
Total	*7,140*[d]	*9,930*
Grand total	*13,500*[d]	*16,700*

[a] Ref. 5.
[b] In metric tons of antimony context.
[c] Data are rounded to no more than three significant digits; may not add to totals shown.
[d] Revised.
[e] Includes ammunition, cable covering, casting, sheet and pipe, and type metal.
[f] Includes fireworks and rubber products.
[g] Includes paper and pigments.

7. Environmental Concerns

Antimony is a common air pollutant that occurs at an average concentration of 0.001 µg/m³ (28). Antimony is released into the environment from buring fossil fuels and from industry (29). In the air, antimony is rapidly attached to suspended particles and thought to stay in the air for 30–40 days (29). Antimony is found at low levels in some lakes, rivers, and streams, and may accumulate in sediments. Although antimony concentrations have been found in some freshwater and marine invertebrates, it does not biomagnify in the environment (29,30). The impact of antimony and antimony compouds on the environment has not been extensively studied to date.

Antimony may enter the human body through the consumption of meats, vegetables, and seafood which all contain ~0.2–1.1 ppb antimony.

8. Recycling

The bulk of secondary antimony has been recovered as antimonial lead. Most of this material was generated and then consumed by the battery industry. Changing trends have resulted in lesser amounts produced (6).

Antimony and its compounds have been designated as priority pollutants by the EPA (31). As a result users, transporters, generators, and processors of antimony-containing material must comply with regulations of the Federal Resource Conservative and Recovery Act (RCRA).

9. Health and Safety Factors

Although metallic antimony may be handled freely without danger, it is recommended that direct skin contact with antimony and its alloys be avoided. Properly designed exhaust ventilation systems and/or approved respirators are required for operations that create dusts or fumes. As with other heavy metals, orderly housekeeping practice and good personal hygiene are necessary to prevent ingestion of (or exposure to) antimony.

The Occupational Safety and Health Administration (OSHA) permissible exposure limit (PEL), the NIOSH recommended exposure limit (REL), and the ACGIH threshold limit value-weighted average (TLV–TWA) for antimony and its compounds 0.5 mg/m^3. The ACGIH has listed antimony trioxide as a suspected source of human carcinogenicity with an A2 designation (32).

Antimony is not known to cause cancer, birth defects, or affect reproduction in humans. However, antimony has been shown to cause lung cancer in laboratory animals that inhaled antimony-containing dusts and prolonged exposure to antimony can cause irritation of the eyes, skin, lungs, and stomach, in the form of vomiting and diarrhea. Heart problems can also result from overexposure to antimony (33).

Stibine (SbH$_3$), a highly toxic gas, can form when nascent hydrogen is present with antimony metal. OSHA and the ACGIH, have recommended a permissable exposure limit for employees exposed to stibine of 0.1 ppm as a time weighted average. Adequate safeguards against overexposure to this gas are advised when handling antimony and its alloys.

10. Uses

Antimony in the unalloyed state is extremely brittle and is not easily fabricated. For this reason, the use of the pure metal is restricted to ornamental applications.

10.1. Antimony Alloys. Approximately one-half of the total antimony demand is for metal used in antimony alloys. Antimonial lead is a term used to describe lead alloys containing antimony in proportions of up to 25%. Most commercial lead–antimony alloys have antimony contents <11%. The compositions of several important antimony alloys are given in Table 6.

Table 6. **Compositional Ranges of Antimony Alloys**

Material	Composition, wt %			
	Sb	Sn	Pb	Others
type metal	2–25	2–15	balance	Cu(0–2)
battery grids				
conventional	2.5–6	0.25–1	balance	
low maintenance	1.5–2.5	0.25–1	balance	
maintenance-free	0	0.25–1	balance	Ca(0.04–0.075)
babbitt metal				
tin base	4–8.5	83–90	balance	Cu(3–8.5); As(0.1)
lead base	9–17.5	0.5–11	balance	Cu(0.5); As(0.25–1.5)
cable covering	1–6	0.25–1	balance	
sheet and pipe	4–15	0.25–1	balance	
collapsible tubes	2–3	0.25–1	balance	
solder	0–6	1–100	balance	Ag(0–6); Cu(0.1–5)
pewter	1–8	balance	0–0.05	Cu(0–3)
britannia metal	2–10	balance	0–9	Cu(0.2–5); Zn(0–5)
bullets, shrapnel	0.5–12	0.25–1.0	balance	

The largest application for antimonial lead is its use as a grid metal alloy in the lead acid storage battery (see BATTERIES, SECONDARY, LEAD–ACID). In the manufacture of grid metal, antimony imparts fluidity, increased creep resistance, and fatigure strength, as well as electrochemical stability to the lead that is particularly advantageous for battery plates required in heavy-duty cycling. A disadvantage of using antimony in grid alloy is that high antimony levels increase the self-discharge characteristics, cause high gassing, and poisoning of the negative electrode resulting from ion migration. The release of gases means that the battery must be vented, which then promotes the loss of electrolyte through evaporation.

Demand for high performance SLI batteries has led to the development of smaller, lighter batteries that require less maintenance. The level of antimony is being decreased from the conventional 3–5% to 1.75–2.75% to minimize the detrimental effects. Lead alloys that contain no antimony have also been introduced. Hybrid batteries use a low antimony–lead alloy in the positive plate and a calcium–lead alloy in the negative plate.

Tin–antimony–copper and lead–antimony–tin white bearing alloys, commonly referred to as babbitt, are used to reduce friction and wear in machinery and help prevent failure by seizure or fatigue. These alloys exhibit good rubbing characteristics even under extreme operating conditions such as high loads, fatigue, or high temperatures. Tin babbits have greater corrosion resistance than lead babbitts. However, lead babbitts are generally cheaper than tin babbitts. Addition of antimony to babbitts increases strength and hardness. The choice of babbitt depends on the application and resultant desired properties. The use of antimony in babbits has been declining because technological advances have reduced the thickness of babbitt on backing material, and there is competition from aluminum–tin alloys and nonmetallic substitutes.

Soldering is a method of joining two metallic surfaces by flowing between them a low melting point alloy. Many different alloys are used for this purpose; however, tin–lead alloys are the most widely used. Other metals are added in small amounts, depending on the desired properties. antimony is added to increase the hardness of tin–lead alloys. Rising concern over the contamination of drinking water by lead has resulted in the use of lead-free alloys for soldering copper pipes. A similar trend is occurring in the canning industry. The demand for antimony in solders used in the electronic, semiconductor, and automobile industries is still strong (see SOLDERS AND BRAZING ALLOYS).

Type metal, another tin–antimony–lead alloy, is used primarily in relief or letterpress printing. Antimony is added to increase hardness, minimize shrinkage, permit sharp definition, and reduce the melting point of the alloy. There has been a substantial decrease in the use of type metals as a result of the emergence of less expensive typesetting techniques.

Antimony hardens the lead used in the manufacture of small arms ammunition. Antimony alloyed with lead is also used in cable covering, sheet and pipe, and collapsible tubes. In these applications, antimony is utilized to increase strength and inhibit corrosion.

Precision duplication, durability, and metallic beauty have made antimonial alloys, such as pewter and britannia metal, desirable for decorative castings. Several different tin- and lead-base antimony alloys are used in the jewelry industry. These alloys are typically cast in rubber or silicone molds.

Antimony may be added to copper-base alloys such as naval brass, Admiralty Metal, and leaded Muntz metal in amounts of 0.02–0.10% to prevent dezincification. Additions of antimony to ductile iron in an amount of 50 ppm, preferably with some cerium, can make the graphite fully nodular to the center of thick castings and when added to gray cast iorn in the amount of 0.05%, antimony acts as a powerful carbide stabilizer with an improvement in both the wear resistance and thermal cycling properties (34) (see CARBIDES).

Carbon (qv) impregnated with antimony gives a dense nonporous material with a low tendency to seizure or galling that may be useful in bearings and seals under high loads and velocities at temperatures up to 500°C (35).

10.2. Semiconductor and Solar Cells. High purity (up to 99.9%) antimony has a limited but important application in the manufacture of semiconductor devices. It may be obtained by reduction of a chemically purified antimony compound with a high purity gaseous or solid reductant, or by thermal decomposition of stibine. The reduced metal may be further purified by pyrometallurgical and zone melting techniques. When alloyed with Group 13 (IIIA) elements, the Group 15 (V) semiconductors, aluminum antimonide [25152-52-7], AlSb, gallium antimonide [12064-03-8], GaSb, and indium antimonide [1312-41-0], InSb, are formed. These intermetallic semiconductor materials exhibit optoelectronic behavior, ie, they emit or absorb electromagnetic radiation, and are utilized in such applications as infrared in devices, diodes, and Hall-effect components. High efficiency solar cells have been produced that are comprised of two layers: one of gallium arsenide and the other of gallium antimonide (see SOLAR ENERGY).

Antimony is also used as a dopant in *n*-type semiconductors. It is a common additive in dopants for silicon crystals with impurities, to alter the electrical conductivity. Interesting semiconductor properties have been reported for cadmium antimonide [12050-27-0], CdSb, and zinc antimonide [12039-35-9], ZnSb. The latter has good thermoelectric properties. Antimony with a purity as low as 99.9 + % is an important alloying ingredient in the bismuth telluride [1304-82-1], Bi_2Te_3, class of alloys, which are used for thermoelectric cooling.

In the computer industry, read/write optical discs, capable of storing over 250 megabytes, utilize a thin coating consisting of a germanium, tellurium, and antimony compound (see IMAGING TECHNOLOGY).

10.3. Antimony Compounds.   The greatest use of antimony compounds is in flame retardants (qv) for plastics, paints, textiles, and rubber. Antimony compounds used in flame retardants are antimony pentoxide, sodium antimonate [15593-75-6], $Na[Sb(OH)_6]$, and, most importantly, antimony trioxide. These compounds, when used alone, are poor flame retardants; however, when combined with halogen compounds, they produce mixtures that are effective.

Antimony trioxide and sodium antimonate are added to specialty glasses as decolorizing and fining agents, and are used as opacifiers in porcelain enamels. Antimony oxides are used as white pigments in paints, whereas antimony trisulfide and pentasulfide yield black, vermillion, yellow, and orange pigments. Camouflage paints contain antimony trisulfide, which reflects infrared radiation. In the production of red rubber, antimony pentasulfide is used as a vulcanizing agent. Antimony compounds are also used in catalysts, pesticides, ammunition, and medicines (see ANTIMONY COMPOUNDS).

BIBLIOGRAPHY

"Antimony and Antimony Alloys" in *ECT* 1st ed., Vol. 2, pp. 50–59, by B. W. Gosner and E. M. Smith, Battelle Memorial Institute; in *ECT* 2nd ed., Vol. 2, pp. 562–570, by S. C. Carapella, Jr., ASARCO Incorporated; in *ECT* 3rd ed., Vol. 3, pp. 96–105, by S. C. Carapella, Jr., ASARCO Incorporated; in *ECT* 4th ed., Vol. 3, pp. 367–381, by T. Li, G. F. Archer, and S. C. Carapella, Jr., ASARCO Incorporated; "Antimony and Antimony Alloys" in *ECT* (online) , posting date: December 4, 2000, by T. Li, G. F. Archer, and S. C. Carapella, Jr., A SARCO Incorporated.

CITED PUBLICATIONS

1. R. Fairbridge, ed., *Encyclopedia of Geochemistry and Environmental Sciences*, Vol. IV, Van Nostrand Reinhold Co., New York, 1972, p. 252.
2. C. Y. Wang, *Antimony*, Charles Griffin Co., London, 1952.
3. H. Quiring, *Die Metallischen Rohstoffe*, Vol. 7, Antimon Ferdinand Enke, Stuttgart, Germany, 1945.
4. *Antimony Materials Survey*, U.S. Department of the Interior, U.S. Bureau of Mines, Washington, D.C., Mar. 1960.
5. J. F. Carlin, Jr., "Antimony," in *U.S. Geological Survey Mineral Yearbook*, U.S. Geological Survey, Reston, Va., 2000.
6. J. F. Carlin, Jr., "Antimony," in *Mineral Commodity Summaries*, U.S. Geological Survey, Reston, Va., Jan. 2002.
7. J. W. Mellor, *Comprehensive Treatise on Inorganic and Theoretical Chemistry*, Vol. 9, Longmans, Green, & Co., Inc., New York, 1929, 339–586.

8. W. J. Maeck, *U.S. Atomic Energy Commission*, NAS-NSSOSS, Clearing House Scientific & Technical Information, NBS, U.S. Dept. Commerce, Springfield, Va., Feb. 1961, p. 10.

9. M. C. Sneed and R. C. Brasted, *Comprehensive Inorganic Chemistry*, Vol. 5, D. Van Nostrand Co. Inc., Princeton, N.J., 1956, pp. 111–152.

10. A. D. Kirschenbaum and J. A. Cahill, *Trans. Am. Soc. Met.* **55**, 849 (1962).

11. M. Pourbaix, *Atlas of Electrochemical Equilibria in Aqueous Solutions* (translated by J. A. Franklin), Pergamon Press, London, 1966.

12. U.S. Pat. 4,891,061 (Jan. 2, 1990), T. P. Clement, II, J. R. Wettlaufer, and J. A. Scott (to ASARCO Inc.).

13. U.S. Pat. 4,808,221 (Feb. 28, 1989), T. P. Clement, II, T. Li, and J. P. Hager (to ASARCO Inc.).

14. W. C. Holmes, *Eng. Min. J.* **145**, 54 (1944).

15. W. D. Gould, *Eng. Min. J.* **156**, 91 (1955).

16. *Min. World* **4**(6), 3 (1942).

17. U.S. Pat. 3,969,202 (July 13, 1976), A. E. Albrethsen, M. L. Hollander, and W. H. Wetherill (to ASARCO Inc.).

18. W. Wendt, *Met. Ind.* **77**, 276 (Dec. 15, 1950).

19. *Metal Industry Handbook and Directory 1961*, Iliffe Books Ltd., London, 1961, p. 3.

20. D. Schlain, J. D. Prater, and S. Revitz, *J. Electrochem. Soc.* **95**, 145 (1949).

21. C. Y. Wang and G. C. Riddell, *Trans. Am. Inst. Min. Metall. Pet. Eng.* **159**, 446 (1944).

22. Å. Holmström, *Scand. J. Metallurgy* **17**, 248–258 (1988).

23. G. I. Karavaiko and co-workers, in R. W. Lawrence, R. M. R. Branion, and H. G. Ebner, eds., *Fundamental and Applied Biohydrometallurgy*, Elsevier, New York, 1986, pp. 115–126.

24. I. D. Fridman and E. E. Savari, *Sov. J. Nonferrous Met.* **26**(1), 102–105 (1985).

25. T. O. Llewellyn and co-workers, *Minerals Yearbook, Antimony*, U.S. Department of the Interior, U.S. Bureau of Mines, Washington, D.C., 1974, 1976, 1981, 1989, and 1990.

26. *Annual Book of ASTM Standards, Part 8—Nonferrous Metals*, American Society for Testing and Materials, Philadelphia, Pa., 1981, pp. 164–165.

27. *Annual Book of ASTM Standards, Section 2—Nonferrous Metals*, American Society for Testing and Materials, Philadelphia, Pa., 1990, pp. 9, 15, and 419.

28. K. Frantzen in R. D. Haribson, ed., *Hamilton &* Hardy's Industrial Toxicology, 5th ed., Mosby Yearbook, St. Louis, Mo, 1998, pp. 25–27.

29. U.S. Environmental Protection Agency (USEPA), *Health and Environmental Effects Profile for Antimony Oxides*, Office of Health and Environmental Assessment, Environmental Criteria and Assessment Office, Office of Research and Development, USEPA, Washington, D.C., 1985.

30. U.S. Environmental Protection Agency (USEPA), *Ambient Aquatic Life Water Quality Criteria for Antimony (III)*, draft, Environmental Research Laboratories, Office of Research and Development, USEPA, Washington, D.C., 1988.

31. *Analysis of Clean Water Act Effluent Guidelines Pollutants. Summary of the Chemicals Regulated by Industrial Point Source Category*, U.S. EPA, Washington, D.C., 40 CFR Parts 400–475, 1991.

32. L. Gallicchio, B. A. Fowler, and E. F. Madden, "Arsenic, Antimony, and Bismuth," in E. Bingham, B. Cohrssen, and C. H. Powell, eds., *Patty's Toxicology*, 5th ed., Vol. 2, John Wiley & Sons, Inc., New York, 2001, Chapt. 36, pp. 770–779.

33. *Toxicological Profile for Antimony and Compounds*, Syracuse Research Corporation, Agency for Toxic Substances & Disease Registry, U.S. Public Health Services, under contract #205–88–0608, Oct. 1990.

34. R. H. Aborn, *Am. Foundry Soc.* **84**, 503 (1976).

35. D. Rai, and co-workers, *Electric Power Research Institute Publication*, Vol. 2, EPRI EA-3356, EPRI, Palo Alto, Calif., 1984.
36. V. Belogorskii and co-workers, *Russ. J. Nonferrous Met.* **16**, 59 (1975).

GENERAL REFERENCES

C. H. Mathewson, ed., *Modern Use of Non-Ferrous Metals*, 2nd ed., American Institute of Mining, Metallurgical, and Petroleum Engineers (AIME), New York, 1953.
C. R. Hayward, *An Outline of Metallurgical Practices*, 3rd ed., D. Van Nostrand Co., Inc., Princeton, N.J., 1952.
G. A. Roush, *Strategic Mineral Supplies*, 1st ed., McGraw-Hill Book Co., Inc., New York, 1939, pp. 238–273.
R. L. Kulpaca and J. C. Archibald, Jr., *J. Metals* **5**(6), 786 (1953).
Trends in Usage of Antimony, Publication NMAB274, National Academy of Sciences—National Academy of Engineering, Washington, D.C., Dec. 1970.
The Economics of Antimony, 7th ed., Roskill Information Services Ltd., London, 1990.
N. Lemery, *Cours Chem.*, W. Kettilby, London, 1977.

T. Li
ASARCO Inc.

ANTIMONY COMPOUNDS

1. Introduction

Antimony [7440-36-0] is the fourth member of the nitrogen family and has a valence shell configuration of $5s^25p^3$. The utilization of these orbitals and, in some cases, of one or two $5d$ orbitals permits the existence of compounds in which the antimony atom forms three, four, five, or six covalent bonds.

The valence bond theory in its most elementary form predicts that trivalent compounds of antimony should have pyramidal structures derived from the $5p$ orbitals and that the $5s$ electrons should act as an inert pair. Many trivalent derivatives of antimony, however, have intervalency angles significantly larger than the 90° angle predicted by this model. The size of these angles as well as the general chemical behavior of the trivalent compounds suggest that sp^3 hybridized antimony orbitals are being employed and that the lone pair occupies one of the tetrahedral positions. The fact that the bond angles are often considerably less than the regular tetrahedral value of 109.5° may be ascribed to repulsion by the lone pair.

Pentacoordinate compounds of antimony usually exhibit trigonal bipyramidal geometry corresponding to the sp^3d hybridized antimony orbitals of valence bond theory. The antimony atom in the octahedral sp^3d^2 valence state is present in numerous complex anions of the SbX_6^- type and in neutral complexes of pentavalent halides with electron-donating molecules such as alcohols, ethers, and nitriles. Ions of the type SbX_5^{2-} are also known and possess a square pyramidal configuration in which the antimony atom is located slightly below the basal plane of the pyramid. A lone pair of electrons presumably occupies one of the octahedral positions and repels the Sb–X bonding pairs.

2. Inorganic Compounds of Antimony

2.1. Stibine. Stibine [7803-52-3], SbH_3; mp, $-88°C$; bp, $-18°C$; density of the liquid at its bp, 2.204 g/mL; sp gr at $18°C$ with respect to air as 1.000, 4.344, is a colorless, poisonous gas having a disagreeable odor (1). It is the only well-characterized binary compound of antimony and hydrogen, although distibine [14939-42-5], Sb_2H_4, has been reported. The vapor pressure, heat of vaporization, heat capacity, density, viscosity, heat of formation, free energy of formation, surface tension, and thermal conductivity of stibine as a function of temperature have been reported (2). The formation is endothermic, 145.1 kJ/mol (34.7 kcal/mol); the compound decomposes slowly at room temperature and readily at $200°C$, to give metallic antimony and hydrogen. The decomposition is autocatalytic and under certain conditions can be explosive. The molecule is trigonal pyramidal (3). Stibine is readily soluble in organic solvents such as carbon disulfide or ethanol, and is slightly soluble in water. In aqueous solution there is no measurable tendency to form a stibonium ion analogous to NH^+_4 and PH^+_4.

Stibine may be prepared by the treatment of metal antimonides with acid, chemical reduction of antimony compounds, and the electrolysis of acid or alkaline solutions using a metallic antimony cathode:

$$Zn_3Sb_2 + 6\,H_3O^+ \longrightarrow 3\,Zn^{2+} + 2\,SbH_3 + 6\,H_2O$$

$$SbO_3^{3-} + 9\,H_3O^+ + 3\,Zn \longrightarrow SbH_3 + 3\,Zn^{2+} + 12\,H_2O$$

The classical synthesis involves the dissolution of a 33% Sb–67% Zn alloy by hydrochloric acid; the evolved gases contain up to 14% stibine. A detailed procedure using a Sb–Mg alloy has also been described (4). Aluminum hydride or alkali metal borohydrides have been used to reduce antimony(III) in acidic aqueous solution to produce stibine. A 23.6% yield of stibine, based on the borohydride used, has been reported (5). A 78% yield based on Sb has been obtained by gradually adding a solution that is 0.4 M in $SbCl_3$ and saturated in NaCl, to aqueous $NaBH_4$ at mol ratios of $NaBH_4$:$SbCl_3 > 10$ (6).

Stibine is readily oxidized and may be ignited in the presence of air or oxygen to form water and antimony trioxide; at lower temperatures metallic antimony and water are slowly formed. Sulfur and selenium react with stibine at $100°C$ in the presence of light to form antimony trisulfide [1345-04-6], Sb_2S_3, and antimony selenide [1315-05-5], Sb_2Se_3, respectively. At elevated temperatures stibine reacts with most metals to give antimonides. Heavy metal salts react with stibine to produce dark, metallic-appearing precipitates. In the case of silver nitrate, silver antimonide is first formed, and this reacts in turn with additional silver nitrate to produce metallic silver and antimony trioxide:

$$2\,SbH_3 + 12\,Ag^+ + 15\,H_2O \longrightarrow 12\,Ag + Sb_2O_3 + 12\,H_3O^+$$

High purity stibine is used as an *n*-type, gas-phase dopant for Si in semiconductors (2). Low temperature distillation of stibine at <53.3 kPa (400 torr) yields a product that on decomposition gives metallic antimony having less than $8 \times 10^{-4}\%$ impurity (6). A method for determining quantities of stibine in the neighborhood of 0.1 mg/m^3 in air has been reported (7).

Stibine may be inadvertently formed by acidified reducing agents reacting with antimony-containing materials. It is an extremely poisonous gas which causes blood destruction and damage to the liver and kidneys (8).

2.2. Metallic Antimonides. Numerous binary compounds of antimony with metallic elements are known. The most important of these are indium antimonide [1312-41-0], InSb, gallium antimonide [12064-03-8], GaSb, and aluminum antimonide [25152-52-7], AlSb, which find extensive use as semiconductors. The alkali metal antimonides, such as lithium antimonide [12057-30-6] and sodium antimonide [12058-86-5], do not consist of simple ions. Rather, there is appreciable covalent bonding between the alkali metal and the Sb as well as between pairs of Na atoms. These compounds are useful for the preparation of organoantimony compounds, such as trimethylstibine [594-10-5], $(CH_3)_3Sb$, by reaction with an organohalogen compound.

2.3. Antimony Trioxide. Antimony(III) oxide (antimony sesquioxide) [1309-64-4], Sb_2O_3, is dimorphic, existing in an orthorhombic modification; valentinite [1317-98-2] is colorless (sp gr 5.67) and exists in a cubic form; and senarmontite [12412-52-1], Sb_4O_6, is also colorless (sp gr 5.2). The cubic modification is stable at temperatures below 570°C and consists of discrete Sb_4O_6 molecules. The molecule is similar to that of P_4O_6 and As_4O_6 and consists of a bowed tetrahedron having antimony atoms at each corner united by oxygen atoms lying in front of the edges. This solid crystallizes in a diamond lattice with an Sb_4O_6 molecule at each carbon position.

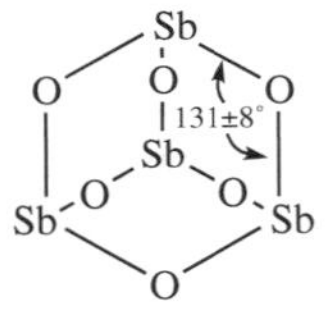

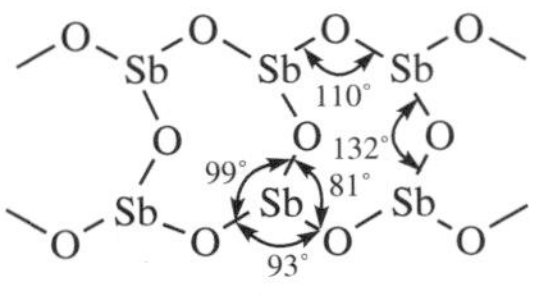

At higher temperatures the stable form is valentinite, which consists of infinite double chains. The orthorhombic modification is metastable below 570°C; however, it is sufficiently stable to exist as a mineral. Antimony trioxide melts in the absence of oxygen at 656°C and partially sublimes before reaching the boiling temperature, 1425°C. The vapor at 1500°C consists largely of Sb_4O_6 molecules, but these dissociate at higher temperatures to form Sb_2O_3 molecules.

Common methods of preparation include direct combination of metallic antimony with air or oxygen, roasting of antimony trisulfide, and alkaline hydrolysis of an antimony trihalide and subsequent dehydration of the resulting hydrous oxide; when heated too vigorously in air, some of the Sb(III) is converted to Sb(V).

Antimony trioxide is insoluble in organic solvents and only very slightly soluble in water. The compound does form a number of hydrates of indefinite composition which are related to the hypothetical antimonic(III) acid (antimonous acid). In acidic solution antimony trioxide dissolves to form a complex series of polyantimonic(III) acids; freshly precipitated antimony trioxide dissolves in strongly basic solutions with the formation of the antimonate ion [29872-00-2], $Sb(OH)^{3-}_6$, as well as more complex species. Addition of suitable metal ions to these solutions permits formation of salts. Other derivatives are made by heating antimony trioxide with appropriate metal oxides or carbonates.

Antimony trioxide has numerous practical applications (9). Its principal use is as a flame retardant in textiles and plastics (see FLAME RETARDANTS; FLAME RETARDANTS IN TEXTILES). It is also used as a stabilizer for plastics, as a catalyst, and as an opacifier in glass (qv), ceramics (qv), and vitreous enamels (qv).

2.4. Antimony Tetroxide. Antimony(III,V) oxide, antimony dioxide [1332-81-6], SbO_2 and Sb_2O_4, occurs in two modifications. Orthorhombic antimony tetroxide has long been known as the mineral cervantite [1332-81-6], α-Sb_2O_4, (colorless, sp gr 4.07). More recently a monoclinic modification, β-Sb_2O_4, has been recognized. In both dimorphs half of the antimony is in the +3 oxidation state, half in the +5 state (10,11). The antimony environments are quite similar in both modifications. The Sb(V) atoms are surrounded by a slightly distorted octahedron of oxygens, and the Sb(III) atoms are coordinated to four oxygens, all on the same side of the Sb(III). The α-modification may be formed by heating Sb_2O_3, valentinite, in air between 460 and 540°C; β-Sb_2O_4 is obtained either by heating α-Sb_2O_4 at 1130°C in dry air or in oxygen (12) or by heating antimonic(V) acid above 900°C (11). At higher temperatures the solid vaporizes without first undergoing any transformation; the recondensed vapors consist of a mixture of β-Sb_2O_4 and antimony trioxide (11).

Antimony tetroxide finds use as an oxidation catalyst, particularly for the dehydrogenation of olefins.

2.5. Antimony Pentoxide Hydrates. Antimonic acid (antimony(V) acid) [12712-36-6], and antimony(V) oxide [1314-60-9], $Sb_2O_5 \cdot nH_2O$, are both hydrates of Sb_2O_5. Commercial antimony pentoxide is either hydrated Sb_2O_5 or at times β-Sb_2O_4. Material having the approximate composition $Sb_2O_5 \cdot 3.5H_2O$ may be prepared by hydrolysis of antimony pentachloride or by acidification of potassium hexahydroxoantimonate(V) [12208-13-8], $KSb(OH)_6$, followed by filtration and drying to constant weight in air at room temperature. This substance is a white solid which loses water upon heating and becomes yellow in color. This loss of water fails to correspond to definitive ratios of $H_2O:Sb_2O_5$, nor is the composition Sb_2O_5 attained. At about 700°C the material is anhydrous and white in color; this is an antimony oxide [12165-47-8], Sb_6O_{13}, containing both Sb(III) and Sb(V) and having a cubic pyrochlore-type structure.

Hydrated antimony pentoxide (antimonic acid) is essentially insoluble in nitric acid solutions, only very slightly soluble in water, but dissolves in aqueous KOH. Numerous hydrated antimonate(V) salts have been reported in which the Sb(V) atom is octahedrally surrounded by six OH groups. Among these are derivatives of magnesium, cobalt, and nickel that have formulas $M(SbO_3)_2\cdot$ $12H_2O$, and a compound referred to as sodium pyroantimonate [10049-22-6], $Na_2H_2\ Sb_2O_7\cdot5H_2O$. X-ray studies show that these are actually $M(H_2O)_6$ $[Sb(OH)_6]_2$ and sodium hexahydroxantimonate(V) [12339-41-2], $Na[Sb(OH)_6]$, respectively. The latter compound is one of the least soluble sodium salts known and is useful in sodium analysis. Numerous polyantimonate(V) derivatives are prepared by heat treatment of mixtures of antimony trioxide and other metal oxides or carbonates. Of these, $K_3Sb_5O_{11}$ [12056-59-6] and $K_2Sb_4O_{11}$ [52015-49-3] have been characterized by x-ray. These consist of three-dimensional networks of SbO_6 in which corners and edges are shared with K^+ ions located in tunnels through the network (13). Simple species such as SbO_4^{3-} and $Sb_2O_7^{2-}$, analogous to orthophosphate and pyrophosphate, apparently do not exist.

Antimonic acid has been used as an ion-exchange material for a number of cations in acidic solution. Most interesting is the selective retention of Na^+ in 12 M HCl, the retention being 99.9% (14). At lower acidities other cations are retained, even K^+. Many oxidation and polymerization catalysts are listed as containing Sb_2O_5.

2.6. Antimony Trifluoride. Antimony(III) fluoride [7783-56-4], SbF_3, is a white, crystalline, orthorhombic solid; vapor pressure at the mp, 26.34 kPa (0.26 atm); Sb–F bond energy, 437.4 kJ (104.5 kcal) (15). The molecule shows a very distorted octahedral arrangement. Antimony trifluoride is extremely soluble in water, the solubility being increased by the presence of hydrofluoric acid. It is also very soluble in polar solvents such as methanol, 154 g/100 mL, and acetone. Table 1 lists physical constants for the antimony halides.

Antimony(III) fluoride may be prepared by treating antimony trioxide or trichloride with hydrofluoric acid. Pure SbF_3 is then obtained by carefully evaporating all of the water from the crude product, which is subsequently sublimed. SbF_3 does not hydrolyze as readily as do the other antimony trihalides. When heated in open air at 100°C, a crystalline solid quickly forms of composition $Sb_3O_2(OH)_2F_3$, which, upon further heating, is transformed into antimony oxide fluoride [11083-22-0], SbOF. This compound may also be prepared by heating 1:1 mixtures of Sb_2O_3 and SbF_3. There are three known crystalline modifications.

In the presence of excess fluoride, antimony trifluoride forms numerous types of complex ions, eg, SbF_4^-, SbF_5^{2-}, $Sb_4F_{13}^-$, and $Sb_2F_7^-$; but SbF_6^{3-} is unknown (16).

Antimony trifluoride is used as a fluorinating agent to replace nonmetal chloride with fluorine. Tri- and difluoromethyl groups are readily formed from the corresponding chlorine groups, but CH_2Cl and CHCl groups are usually unaffected. Antimony trifluoride can also be used to effect the replacement of chlorine bonded to other elements; thus trichloromethylphosphonous dichloride [3582-11-4], CCl_3PCl_2, can be converted to trichloromethylphosphonous difluoride [1112-03-4], CCl_3PF_2.

Uses of SbF_3 have been reported in the manufacture of fluoride glass and fluoride glass optical fiber preform (17) and fluoride optical fiber (18) in the preparation of transparent conductive films (19) (see FIBER OPTICS).

2.7. Antimony Trichloride. Antimony(III) chloride [10025-91-9], $SbCl_3$, is a colorless, crystalline solid, readily soluble in hydrochloric acid; water, ca 9%

Table 1. Physical Constants of the Antimony Halides

Parameter	Antimony trifluoride	Antimony trichloride	Antimony tribromide	Antimony triiodide[a]	Antimony pentafluoride	Antimony pentachloride
formula	SbF_3	$SbCl_3$	$SbBr_3$	SbI_3	SbF_5	$SbCl_5$
CAS Registry Number	[7783-56-4]	[10025-91-9]	[7789-61-9]	[7790-44-5]	[7783-70-2]	[7647-18-9]
mp, °C	291 ± 1	73.2	96.0 ± 0.5	170.5	6	3.2 ± 0.1
bp, °C	346 ± 10	222.6	287	401	150	68[b], 140[c]
Δ_f° at 298°C, kJ/mol[d]	−915.5	−382.2	−259.4	−100.4		−450.8 ± 6.2
S° at 298°C, J/(mol·K)[d]	127	184	207	216 ± 1		263 ± 12
ΔH_{fusion}, kJ/mol[d,e]	21.4			$22.7_{444} \pm 0.2$		
ΔS_{fusion}, J/(mol·K)[d,e]	38.2			$51.5_{444} \pm 0.4$		
ΔH_{vap}, kJ/mol[d,e]	$102.8_{298} \pm 1.3$	46.72_{496}	53.2_{560}			43.45_{449}
ΔS_{vap}, J/(mol·K)[d,e]	$175.8_{298} \pm 2.5$	93.3_{496}	94.9_{560}			95.44_{449}
C_p, J/mol·K[d]		108[f]		96[f], 144[g]		

[a] The ΔH_{subl} at 298°C is 101.6 ± 0.4 kJ/mol (24.3 ± 0.1 kcal/mol).

[b] At a pressure of 1.82 kPa.

[c] Decomposes at atmospheric pressure, 101.3 kPa.

[d] To convert from J to cal, divide by 4.184.

[e] At the temperature in °C indicated by the subscript.

[f] Value given is for solid.

[g] Value given is for liquid.

61

at 25°C, increasing with temperature; $CHCl_3$, 22%; CCl_4, 13%; benzene; CS_2; and dioxane.

Antimony trichloride may be prepared by chlorination of antimony metal, Sb_2O_3, or Sb_2S_3, or by reaction of Sb_2O_3 with concentrated HCl. $SbCl_3$ hydrolyzes readily, giving hydrous Sb_2O_3 with excess water, but when limited quantities of water are used, a large number of partially hydrolyzed products has been claimed, eg, $SbOCl$, Sb_2OCl_4, $Sb_4O_5Cl_2$, $Sb_4O_3(OH)_3Cl_2$, $Sb_8O_{11}Cl_2$, and Sb_8OCl_{22}, some of which are listed in Table 2. The hydrolysis product most frequently obtained and best characterized is tetraantimony dichloride pentoxide [12182-69-3], $Sb_4O_5Cl_2$, which is initially precipitated as a thick white solid, changing to well-defined colorless crystals. By carefully controlled hydrolysis $SbOCl$ is obtained, which, upon further dilution with water, changes to $Sb_4O_5Cl_2$.

In many situations $SbCl_3$ behaves as a Lewis acid. In the presence of excess Cl^-; and suitable cations, numerous chloroantimonate(III) ions are formed, eg, $SbCl_6^{3-}$, $SbCl_5^{2-}$, $SbCl_4^-$, $Sb_2Cl_7^{2-}$, $Sb_2Cl_7^{3-}$, $Sb_2Cl_9^{3-}$, and $Sb_2Cl_{11}^{5-}$. The first two of these are simple discrete ions. A large number of adducts have been formed by reaction of $SbCl_3$ with organic bases, eg, $SbCl_3 \cdot (C_2H_5)_2O$, $SbCl_3 \cdot H_2NC_6H_4NO_2$, $SbCl_3 \cdot (CH_3)_3N$, $2SbCl_3 \cdot (CH_3)_3N$, and $SbCl_3 \cdot 2CH_3COCH_3$. Isolable adducts are also formed with aromatic hydrocarbons, eg, $2SbCl_3 \cdot C_6H_6$ and $SbCl_3 \cdot C_6H_6$. In a few situations $SbCl_3$ apparently acts as an electron donor; thus the carbonyl complexes $Ni(CO)_3SbCl_3$ and $Fe(CO)_3(SbCl_3)_2$ have been isolated (20). Adducts and compounds are listed in Table 2.

Antimony trichloride is used as a catalyst or as a component of catalysts to effect polymerization of hydrocarbons and to chlorinate olefins. It is also used in hydrocracking of coal (qv) and heavy hydrocarbons (qv), as an analytic reagent for chloral, aromatic hydrocarbons, and vitamin A, and in the microscopic identification of drugs. Liquid $SbCl_3$ is used as a nonaqueous solvent.

Table 2. **Inorganic Antimony Compounds and Adducts**

Compound	CAS Registry Number	Formula
trifluorotetraoxotriantimonic(III) acid	[65229-25-6]	$Sb_3O_2(OH)_2F_3$
antimony chloride oxide	[7791-08-4]	$SbOCl$
diantimony tetrachloride oxide	[65229-26-7]	Sb_2OCl_4
tetraantimony dichloride pentoxide	[12182-69-3]	$Sb_4O_5Cl_2$
tetraantimony trihydroxydichlorotrioxide		$Sb_4O_3(OH)_3Cl_2$
antimony trichloride diethyl ether	[10025-91-9]	$SbCl_3 (C_2H_5)_2O$
antimony trichloride aniline	[21645-17-0]	$SbCl_3 H_2NC_6H_5$
antimony trichloride trimethylamine	[65186-11-0]	$SbCl_3 (CH_3)_3N$
bis(antimony trichloride) trimethylamine	[65186-12-1]	$2SbCl_3 (CH_3)_3N$
antimony trichloride bisacetone	[65186-13-2]	$SbCl_3 2CH_3COCH_3$
bis(antimony trichloride) benzene	[1123-15-5]	$2SbCl_3 C_6H_6$
(antimony trichloride)tricarbonylnickel	[65208-44-8]	$Ni(CO)_3SbCl_3$
bis(antimony trichloride)tricarbonyliron	[65208-43-7]	$Fe(CO)_3(SbCl_3)_2$
antimony pentachloride bis(iodine chloride)	[65186-14-3]	$SbCl_5 2ICl$
antimony pentachloride tris(iodine chloride)	[65186-15-4]	$SbCl_5 3ICl$
antimony pentachloride sulfur tetrachloride	[15597-82-7]	$SbCl_5 SCl_4$
bis(hexachloroantimonic(III) acid) nonahydrate	[65208-45-9]	$HSbCl_6 4.5H_2O$
antimony pentabromide diethyl ether	[29702-86-1]	$SbBr_5 O(C_2H_5)_2$

2.8. Antimony Tribromide and Triiodide. Antimony(III) bromide [7789-61-9], $SbBr_3$, is a colorless, crystalline solid having a pyramidal dimorphic molecular structure and an acicular (α-$SbBr_3$) and a bipyramidal (β-$SbBr_3$) habit.

Antimony(III) iodide [7790-44-5], SbI_3, forms red rhombohedral crystals, intermediate in structure between a molecular and an ionic crystal. In SbI_3 vapor there is no indication of association.

Both antimony tribromide and antimony triiodide are prepared by reaction of the elements. Their chemistry is similar to that of $SbCl_3$ in that they readily hydrolyze, form complex halide ions, and form a wide variety of adducts with ethers, aldehydes, mercaptans, etc. They are soluble in carbon disulfide, acetone, and chloroform. There has been considerable interest in the compounds antimony bromide sulfide [14794-85-5], antimony iodide sulfide [13868-38-1], ISSb, and antimony iodide selenide [15513-79-8] with respect to their solid-state properties, ferroelectricity, pyroelectricity, photoconduction, and dielectric polarization.

2.9. Antimony Pentafluoride. Antimony(V) fluoride [7783-70-2], SbF_5, is a colorless, hygroscopic, viscous liquid that has SbF_6 units with cis-fluorines bridging to form polymeric units. ^{19}F nmr shows that at low temperatures there are three different types of F atoms (21). Contamination with a small amount of HF markedly decreases the extent of polymerization. The vapor density at 150°C corresponds to the trimer. The solid is a cis-fluorine-bridged tetramer (22).

Antimony pentafluoride may be prepared by fluorination of SbF_3 or by treatment of $SbCl_5$ with HF. In the latter method the fifth chlorine is removed with difficulty; failure to remove the chlorine completely results in contamination of the distilled SbF_5 with Sb(III) (20).

Antimony pentafluoride is a strong Lewis acid and a good oxidizing and fluorinating agent. Its behavior as a Lewis acid leads to the formation of numerous simple and complex adducts. It reacts vigorously with water to form a clear solution from which antimony pentafluoride dihydrate [65277-49-8], $SbF_5 \cdot 2H_2O$, may be isolated. This is probably not a true hydrate, but may well be better formulated as $[H_3O][SbF_5OH]$.

Antimony pentafluoride reacts with iodine to form bis(antimony pentafluoride) iodide [12324-61-7], $Sb_2F_{10}I$, and antimony pentafluoride iodide [12324-57-1], SbF_5I; with nitrosyl fluoride to give a very stable compound nitrosyl hexafluoroantimonate(V) [16941-06-3], $NOSbF_6$; with sulfur to give a dark blue solution from which antimony pentafluoride sulfur can be isolated; and with NO_2 to form nitrosyl pentafluoronitratoantimonate(V) [26117-73-7], $NO[SbF_5(NO_3)]$. Combinations of Sb(V) and Sb(III) fluorides give fluorides of the general formula $SbF_5(SbF_3)_n$ where n may be 2, 3, 4, or 5. In combination with hydrofluoric acid or other fluorides the hexafluoroantimonate(V) ion [17111-95-4], SbF_6^-, is formed. This is frequently used as a negative counterion for compounds containing rather unstable, highly oxidizing, or highly fluorinated cations. The SbF_6^- anion has been shown to have an octahedral structure and may be hydrolyzed to $Sb(OH)_6^-$. Combinations of SbF_5, with either HSO_3F alone or with HSO_3F and SO_3, have extremely high acidities (23,24). A 1:1 mixture of HSO_3F and SbF_5 is frequently used for stabilization of carbocations and has been referred to as magic acid.

Treatment of graphite with SbF_5 liquid or vapor results in intercalation of SbF_5 between the graphite layers, and at 70°C a blue-black solid, antimony graphite fluoride [56126-99-9], $C_{13}Sb_2F_{10}$, is formed (25). This modifies the fluorinating properties of SbF_5, and such materials are used as specific fluorinating agents. Several mixed pentahalides are known. Thus fluorination of $SbCl_3$ yields antimony trichloride difluoride [31244-70-9], $SbCl_3F_2$, and chlorination of SbF_3 gives antimony dichloride trifluoride [7791-16-4], $SbCl_2F_3$. The latter compound has been shown to consist of $SbCl_4^+$ and $Sb_2Cl_2F_9^-$ (26).

Antimony pentafluoride is used as a catalyst in conjunction with IF_5 for the production of telomers, used in stain resistant products. Antimony pentafluoride is a powerful oxidizer and also a moderate fluorinating reagent, capable of fluorinating PCl_3, $SiCl_4$ and WCl_6 to PF_3, SiF_4, and WF_6. It has been used extensively in conjunction with HSO_3F or solutions of SO_3 in HSO_3F to produce "super acid" systems. Alkenes react to form stable carbocations by SbF_5, either neat or in Freon 113. Neat SbF_5 is capable of converting aldehydes to oxocarbonium ions, RCO^+. Antimony pentafluoride yields intercalation compounds when combined with graphite. The resulting material is a black powder that is relatively stable to moisture. It is useful for the exchange of fluorine with organic chloride (27).

2.10. Antimony Pentachloride. Antimony(V) chloride [7647-18-9], $SbCl_5$, is a colorless, hygroscopic, oily liquid that is frequently yellow because of the presence of dissolved chlorine; it cannot be distilled at atmospheric pressure without decomposition, but the extrapolated normal boiling point is 176°C. In the solid, liquid, and gaseous states it consists of trigonal bipyramidal molecules with the apical chlorines being somewhat further away than the equatorial chlorines (20).

Antimony pentachloride is usually prepared by chlorination of molten $SbCl_3$. It undergoes partial dissociation to Cl_2 and $SbCl_3$; ΔH_{496} for this equilibrium is -76.69 kJ/mol (-18.33 kcal/mol) and ΔS_{496} is -145 J/mol (-34.7 cal/mol) (28).

Antimony pentachloride is a strong Lewis acid and a useful chlorine carrier. Chlorine is lost in a number of chemical reactions resulting in the formation of adducts. Thus iodine is chlorinated to form ICl, which in turn combines with additional $SbCl_5$ to give $SbCl_5 \cdot 2ICl$ and $SbCl_5 \cdot 3ICl$. A similar reaction occurs with sulfur, giving $SbCl_5 \cdot SCl_4$. These adducts are listed in Table 2. In the presence of excess hydrochloric acid or metal chlorides the hexachloroantimonate(V) ion [17949-89-2], $SbCl_6^-$, is formed. The strong acid, $HSbCl_6 \cdot 4.5H_2O$, as well as many hexachloroantimonate salts can be isolated. With a 1:1 mol ratio of water to $SbCl_5$, antimony pentachloride monohydrate [14215-03-3], $SbCl_5 \cdot H_2O$, which is insoluble in $CHCl_3$, is formed; with a 4:1 mol ratio, chloroform soluble antimony pentachloride tetrahydrate [52940-44-0], $SbCl_5 \cdot 4H_2O$, can be isolated. Numerous examples can be cited where the Lewis acid $SbCl_5$ forms 1:1 adducts with oxygen- and nitrogen-containing bases, eg, sulfur nitride, ethers, nitriles, alcohols, or esters. Some of these are given in Table 2.

2.11. Antimony Trisulfide. Antimony(III) sulfide (antimony sesquisulfide) [1345-04-6], SbS_3, exists as a black crystalline solid, stibnite [1317-86-8], and as an amorphous red to yellow-orange powder. Stibnite melts at 550°C and has $\Delta H^{\circ}_{f,298}$, -175 kJ/mol (-41.8 kcal/mol)); S°_{298}, 182 J/(182 mol·K) [43.5 cal/(43.5 mol·K)]; for the amorphous solid $\Delta H^{\circ}_{f,298}$ is -147 kJ/mol (-35.1 kcal/mol) (29). The crystal structure of stibnite contains two distinctly different antimony sites and consists of two parallel Sb_4S_6 chains that are linked together to form crumpled sheets (two per unit cell).

Amorphous Sb_2S_3 can be prepared by treating an $SbCl_3$ solution with H_2S or with sodium thiosulfate, or by heating metallic antimony or antimony trioxide with sulfur. Antimony trisulfide is almost insoluble in water but dissolves in concentrated hydrochloric acid or in excess caustic. In the absence of air, Sb_2S_3 dissolves in alkaline sulfide solutions to form the thioantimonate(III) ion [43049-98-5]; SbS_2^-, in the presence of air the tetrathioantimonate(V) ion [17638-29-8], SbS_4^{3-}, is formed. The lemon-yellow crystalline salt, $Na_3SbS_4 \cdot 9H_2O$, known as Schlippe's salt [1317-86-8], contains the tetrahedral tetrathioantimonate(V) ion.

Antimony trisulfide is used in fireworks, in certain types of matches, as a pigment, and in the manufacture of ruby glass.

2.12. Antimony Pentasulfide. Antimony pentasulfide [1315-04-4], Sb_2S_5, is a yellow to orange to red amorphous solid of indefinite composition. It is frequently given the formula Sb_2S_5, but actually consists of Sb(III) with a variable quantity of sulfur (30). The product is prepared commercially by the conversion of Sb_2S_3 to tetrathioantimonate(V) by boiling with sulfur in alkaline solution. The antimony pentasulfide is liberated as a yellowish orange precipitate when the resulting mixture is acidified with hydrochloric acid.

The product is commercially known as golden sulfide of antimony, and is used in vulcanization to produce a red variety of rubber. The material is also used as a pigment and in fireworks.

2.13. Antimony(III) Salts. Concentrated acids dissolve trivalent antimony compounds. From the resulting solutions it is possible to crystallize normal and basic salts, eg, antimony(III) sulfate [7446-32-4], $Sb_2(SO_4)_3$; antimonyl sulfate [14459-74-6], $(SbO)_2SO_4$; antimony(III) phosphate [12036-46-3], $SbPO_4$; antimony(III) acetate [6923-52-0], $Sb(C_2H_3O_2)_3$; antimony(III) nitrate [20328-96-5], $Sb(NO_3)_3$; and antimony(III) perchlorate trihydrate [65277-48-7], $Sb(ClO_4) \cdot 3H_2O$. The normal salts all hydrolyze readily.

2.14. Hexafluoroantimonates. Hexafluoroantimonic acid [72121-43-8], $HSbF_6 \cdot 6H_2O$, is prepared by dissolving freshly prepared hydrous antimony pentoxide in hydrofluoric acid or adding the stoichiometric amount of 70% HF to SbF_5. Both of these reactions are exothermic and must be carried out carefully.

The superacid systems $HSO_3F \cdot SbF_5$ [33843-68-4] and $HF \cdot SbF_5$ [16950-06-4] (fluoroantimonic acid) are used in radical polymerization (31) and in carbocation chemistry (32). Addition of SbF_5 drastically increases the acidities of HSO_3F and HF (33,34).

Anhydrous salts, $MSbF_6$, where M = H, NH_4, and alkali metal, and $M(SbF_6)_2$, where M is an alkaline-earth metal, can be prepared by the action of F_2 on MF or MF_2 and SbF_3 (35) by the oxidation of Sb(III) with H_2O_2 or alkali metal peroxide in HF (36), by the action of HF on a mixture of $SbCl_5$ and MF where $M = NH_4$, Li, Na, K, Ru, Cs, Ag, and Tl (37). These compounds can be used as photoinitiators for the production of polymers (38).

2.15. Compounds Containing Sb–O–C or Sb–S–C Linkages. A large number of compounds have been prepared in which the antimony atom is linked to carbon through an oxygen or sulfur atom. The simplest of these compounds are esters of the hypothetical antimonic acid [13453-11-7], H_3SbO_3, or thioantimonic acid [65277-44-3], H_3SbS_3. The esters of H_3SbO_3 can be prepared by refluxing antimony trioxide with the appropriate alcohol and removing the water formed by means of anhydrous copper sulfate. They may also be obtained by ester exchange. For example, tributyl antimonate [2155-74-0], $C_{12}H_{27}O_3Sb$, is formed by the interaction of the triethyl ester, triethyl antimonate [10433-06-4], $C_6H_{15}O_3Sb$, and butyl alcohol. Esters of thioantimonic(III) acid are easily synthesized by the reaction of antimony trichloride and a mercaptan. A series of these thioantimonic esters have been prepared in the search for compounds of medicinal value (39).

By far the largest group of compounds containing the Sb—O—C linkage are those obtained by reaction of an antimony oxide with an α-hydroxy acid, o-dihydric phenol, sugar alcohol, or some other polyhydroxy compound containing at least two adjacent hydroxyl groups. The best known compound of this type is antimony potassium tartrate (tartar emetic) [28300-74-5] prepared by refluxing potassium hydrogen tartrate with freshly precipitated antimony trioxide.

Tartar emetic has been used as an antiparasitic agent in medicine, as an insecticide, and as a mordant in the textile and leather industries.

Tartar emetic was the subject of controversy for many years, and a variety of incorrect structures were proposed. In 1966, x-ray crystallography showed that tartar emetic contains two antimony(III) atoms bridged by two tetranegative D-tartrate residues acting as double bidentate ligands to form dipotassium bis[D-μ-(2,3-dihydroxybutanedioato)]diantimonate [28300-74-5] (40).

The four 5-membered chelate rings are nearly planar, and the oxygen atoms about the antimony atoms occupy four corners of a distorted square pyramid, the apex of which is presumably occupied by an unshared electron pair. Later work (41) has also shown that the three water molecules in the structure are hydrogen bonded to each other and to the carboxyl oxygen atoms and they connect the anions in infinite sheets. Essentially the same tartrato-bridged binuclear anion has been found in the racemic salts $(NH_4)_2Sb_2(D,L-C_4H_2O_6)_2 \cdot 4H_2O$ (42) and $K_2Sb_2(D,L-C_4H_2O^6)_2 \cdot 3H_2O$ (43), and in the ammonium (44) and the tris(o-phenanthroline)iron(II) (45) analogues of tartar emetic. It has been suggested that the formation of a bridged $meso$-tartrato dimer of antimony(III) requires an unfavorable eclipsed conformation for the bridging ligands (46), as $meso$-tartrate complexes similar to tartar emetic have never been prepared.

Two five-membered chelate rings per antimony atom are present in antimony hydrogen bis(thioglycolate) [65277-45-4], $C_4H_7O_4S_2Sb$, a compound prepared by the interaction of antimony trioxide and thioglycolic acid in aqueous solution (47). The coordination around the antimony has been described as a distorted trigonal bipyramid in which the two axial apices are occupied by oxygen atoms, and two of the equatorial apices are occupied by sulfur atoms. The third equatorial position is presumably occupied by an unshared electron pair which is responsible for the deformation of the bipyramid. Each unit of antimony hydrogen bis(thioglycolate) is joined to two other units by hydrogen bonds, forming endless zig-zag chains.

Lactic, malic, mandelic, and oxalic acids also give antimony(III) derivatives in which two molecules of acid are associated with one atom of antimony. Studies of the reaction between antimony trioxide and lactic or oxalic acid as a function of pH have suggested the following structures for the oxalate and lactate complexes (48,49).

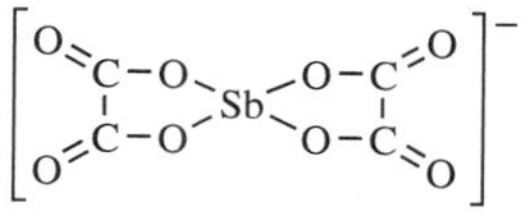

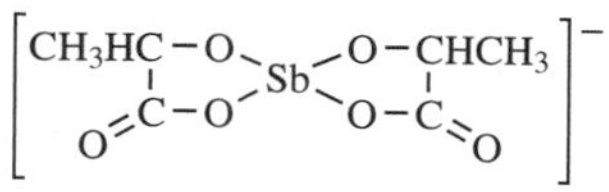

$$\left[\begin{array}{c} \text{O}{=}\text{C}{-}\text{O}\diagdown{\diagup}\text{O}{-}\text{C}{=}\text{O} \\ \text{Sb} \\ \text{O}{=}\text{C}{-}\text{O}\diagup{\diagdown}\text{O}{-}\text{C}{=}\text{O} \end{array} \right]^{-}$$

$$\left[\begin{array}{c} \text{CH}_3\text{HC}{-}\text{O}\diagdown{\diagup}\text{O}{-}\text{CHCH}_3 \\ \text{Sb} \\ \text{O}{=}\text{C}{-}\text{O}\diagup{\diagdown}\text{O}{-}\text{C}{=}\text{O} \end{array} \right]^{-}$$

The reaction of toluene-3,4-dithiol(3,4-dimercaptotoluene) and antimony trichloride in acetone yields a yellow solid $Sb_2(tdt)_3$, where tdt is the toluene-3,4-dithiolate anionic ligand (50). With the disodium salt of maleonitriledithiol ((Z)-dimercapto-2-butenedinitrile), antimony trichloride gives the complex ion $[Sb(mnt)_2]^-$, where mnt is the maleonitriledithiolate anionic ligand. This complex has been isolated as a yellow, crystalline, tetraethylammonium salt. The structures of these antimony dithiolate complexes have apparently not been unambiguously determined.

Antimony pentoxide also reacts with a variety of dihydroxy compounds. Thus pyrocatechol yields a crystalline substance in which three molecules of the diol are associated with one atom of antimony (51). The configuration of this substance has not been established, but the following structure seems reasonable:

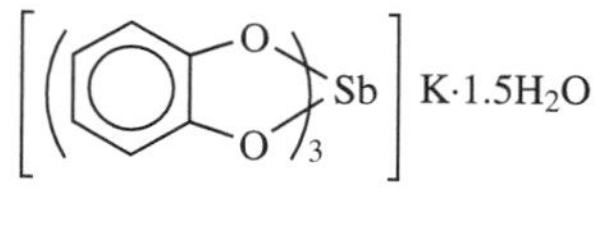

$$\left[\left(\begin{array}{c} \text{O} \\ \text{O} \end{array} \right)_{\!3} \!\!\text{Sb} \right] \text{K} \cdot 1.5\text{H}_2\text{O}$$

A number of complex derivatives of antimony pentoxide with polyhydroxy compounds have been investigated as drugs. The most important of these substances is known as antimony sodium gluconate [16037-91-5], $C_{12}H_{20}O_{17}Sb_2 \cdot 9H_2O \cdot 3Na$, which is prepared by the reaction of antimony pentoxide, gluconic acid, and sodium hydroxide (52).

3. Organoantimony Compounds

A wide variety of compounds containing the Sb—C bond is known. Organoantimony compounds can be broadly divided into Sb(III) and Sb(V) compounds. The former may contain from one to four organic groups, and the Sb(V) compounds from one to six organic groups. With a few exceptions, the nomenclature used here is that proposed by the International Union of Pure and Applied Chemistry (53). There are a number of heterocyclic compounds in which one or more antimony atoms are members of the heterocycle. Such compounds have been thoroughly discussed (54). A more recent but less comprehensive report on heterocyclic antimony compounds has also been published (55). The synthesis of organoantimony compounds has been described (56). Organoantimony(III) compounds (57) and organoantimony(V) compounds containing four, five, or six C–Sb bonds (57) and three C–Sb bonds (57), respectively, have been summarized, and a summary of organoantimony(V) compounds containing one, two, or three C–Sb bonds has also been published (58). Two lists of all organoantimony compounds prepared or studied between 1937 and 1968 have been published (59). Another monograph, published in 1970, critically discusses organoarsenic, -antimony, and -bismuth chemistry (60). Work on organoantimony compounds is reviewed annually, and the use of organoantimony and organobismuth compounds in organic synthesis has been reviewed (61).

3.1. Primary and Secondary Stibines. Relatively few primary ($RSbH_2$) and secondary (R_2SbH) stibines are known. Methylstibine [23362-09-6], CH_5Sb, ethylstibine [68781-03-3], C_2H_7Sb, isopropylstibine, C_3H_9Sb, and butylstibine [68781-04-4], $C_4H_{11}Sb$, have been prepared by the reduction of the corresponding alkyldichlorostibines using lithium aluminum hydride or sodium borohydride (62). All of the alkylstibines are thermally unstable, easily oxidizable, colorless liquids with strong alliaceous odors. Decomposition products include hydrogen and nonvolatile black solids analyzing for $(RSb)_x$. Reaction of the alkylstibines with hydrogen chloride produces lustrous, pale green polymeric solids also analyzing for $(RSb)_x$ (63). Diethylstibine, $C_4H_{11}Sb$, (64), di-*tert*-butylstibine, $C_8H_{19}Sb$, (65), and dicyclohexylstibine [1011-94-5], $C_{12}H_{23}Sb$, (66) have been obtained by reduction of the corresponding dialkylhalostibines with lithium aluminum hydride. Dimethylstibine [23362-10-9], C_2H_7Sb, was first prepared by the interaction of dimethylbromostibine [53234-94-9], C_2H_6BrSb, and LiHB $(OCH_3)_2$ at temperatures below $-40°C$ (67). Treatment of dimethylstibine with hydrogen chloride yields hydrogen:

$$(CH_3)_2SbH + HCl \longrightarrow H_2 + (CH_3)_2SbCl$$

Phenylstibine [58266-50-5], C_6H_7Sb, has been obtained by the reduction of phenyldiiodostibine [68972-61-2], $C_6H_5I_2Sb$, (68) or phenyldichlorostibine [5035-52-9], $C_6H_5Cl_2Sb$, (69) with lithium borohydride. It has also been prepared by the hydrolysis or methanolysis of phenylbis(trimethylsilyl)stibine [82363-95-9], $C_{12}H_{23}Si_2Sb$ (70). Diphenylstibine [5865-81-6], $C_{12}H_{11}Sb$, can be prepared by the interaction of diphenylchlorostibine [2629-47-2], $C_{12}H_{10}ClSb$, with either lithium borohydride (71) or lithium aluminum hydride (72). It is also formed by hydrolysis or methanolysis of diphenyl(trimethylsilyl) stibine [69561-88-2], $C_{15}H_{19}SbSi$ (70). Dimesitylstibine [121810-02-4] has been obtained by the protonation of lithium dimesitylstibide with trimethylammonium chloride (73). The x-ray crystal structure of this secondary stibine has also been reported.

The aromatic primary and secondary stibines are readily oxidized by air, but they are considerably more stable than their aliphatic counterparts. Diphenylstibine is a powerful reducing agent, reacting with many acids to liberate hydrogen (74). It has also been used for the selective reduction of aldehydes and ketones to the corresponding alcohols (75). At low temperatures, diphenylstibine undergoes an addition reaction with ketene (76):

$$(C_6H_5)_2SbH + CH_2\!=\!C\!=\!O \longrightarrow (C_6H_5)_2Sb\underset{\underset{O}{\|}}{C}CH_3$$

3.2. Tertiary Stibines.

A large number of trialkyl- and triarylstibines are known (57). They are usually prepared by the interaction of a reactive organometallic compound and an antimony trihalide, a halostibine, or a dihalostibine. The type of organometallic compound most widely employed in these syntheses is the Grignard reagent (77,78). Organolithium (79,80), organocadmium (81,82), organoaluminum (83), and organomercury (84) compounds have also been used. Triarylstibines can be readily prepared from an aryl halide, an antimony trihalide, and sodium (85).

Another excellent method for preparing tertiary stibines involves the interaction of an organostibide and an alkyl or aryl halide (86,87). This method is of particular value in preparing unsymmetrical tertiary stibines. For example, an interesting hybrid ligand has been obtained by the following reaction carried out in liquid ammonia (88):

$$CH_3SeC_6H_4Br + NaSb(CH_3)_2 \longrightarrow CH_3SeC_6H_4Sb(CH_3)_2 + NaBr$$

Trialkylstibines are sensitive to oxygen, and in some cases they ignite spontaneously in air. Trimethylstibine [594-10-5], C_3H_9Sb, may explode on contact with atmospheric oxygen (89). Triarylstibines usually do not react with air, and they are quite stable thermally (90). Trialkylstibines are powerful reducing agents and can convert halides of phosphorus or antimony to the corresponding elements (91). Triarylstibines are much less reactive as reducing agents, but they are readily oxidized by halogens, interhalogens, pseudohalogens, sulfur, and fuming nitric acid (92,93).

Tertiary stibines have been widely employed as ligands in a variety of transition metal complexes (94), and they appear to have numerous uses in synthetic organic chemistry (61), eg, for the olefination of carbonyl compounds (95). They have also been used for the formation of semiconductors by the metal–organic chemical vapor deposition process (96), as catalysts or cocatalysts for a number of polymerization reactions (97), as ingredients of light-sensitive substances (98), and for many other industrial purposes.

3.3. Halostibines, Dihalostibines, and Related Compounds.

Alkyldichloro- and alkyldibromostibines are readily prepared by the alkylation of the corresponding antimony trihalide with an organolead reagent (62,99):

$$R_4Pb + 3\ SbX_3 \longrightarrow 3\ RSbX_2 + PbX_2 + RX$$

The alkylation can also be accomplished using tetraalkyltin compounds. Alkyldiiodostibines are formed in about 20% yield via the interaction of alkylmagnesium iodides and antimony trichloride (62). Dialkylchlorostibines are obtained in good yields by the cleavage of tetraalkyldistibines using sulfuryl chloride (86):

$$R_2SbSbR_2 + SO_2Cl_2 \longrightarrow 2\ R_2SbCl + SO_2$$

Dialkylbromo- and dialkyliodostibines can similarly be prepared by the cleavage of tetraalkyldistibines using equimolar amounts of bromine or iodine (X_2) (100):

$$R_2SbSbR_2 + X_2 \longrightarrow 2\ R_2SbX$$

The thermal decomposition of trialkylantimony dihalides has been used for the preparation of chloro-, bromo-, and iodostibines (101):

$$R_3SbX_2 \xrightarrow{\Delta} R_2SbX + RX$$

The interaction of triarylstibines and antimony trichloride or tribromide is a convenient and efficient method for preparing aryldihalo- and diarylhalostibines (99,102,103):

$$Ar_3Sb + 2\ SbX_3 \longrightarrow 3\ ArSbX_2$$

$$2\ Ar_3Sb + SbX_3 \longrightarrow 3\ Ar_2SbX$$

Compounds $ArSbX_2$ and Ar_2SbX, in which X is Cl or Br and Ar is an aryl group, have also been obtained by the reduction of the corresponding stibonic or stibinic acids in hydrochloric or hydrobromic acid. The usual reducing agent is sulfur dioxide catalyzed by iodide ion, although stannous chloride has also been employed (104). The reduction of unsymmetrical diarylantimony trihalides is probably the best method for the synthesis of unsymmetrical chloro- and bromostibines (105). Aryldiiodostibines can be prepared by the reaction of the corresponding oxides with hydriodic acid (106):

$$\frac{1}{x}\ (ArSbO)_x + 2\ HI \longrightarrow ArSbI_2 + H_2O$$

Diaryliodostibines are usually obtained by the metathetical reaction of the chlorostibines with sodium iodide (106,107). Diphenylfluorostibine [6651-55-4], $C_{12}H_{10}FSb$, can be prepared from an organosilicon species (108):

$$2\ (NH_4)_2(C_6H_5SiF_5) + SbF_3 \longrightarrow (C_6H_5)_2SbF + 2\ (NH_4)_2SiF_6$$

Alkyldihalo- and dialkylhalostibines are highly reactive substances which are rapidly oxidized in air. Some are spontaneously inflammable (109). The aromatic counterparts are less susceptible to air oxidation but are readily oxidized by halogens. Alkaline hydrolysis (104,110,111) of the dihalo- and halostibines yields compounds of the types $(RSbO)_x$ and $R_2SbOSbR_2$, respectively, whereas reaction of the stibine with sodium sulfide (110) gives the analogous sulfur compounds. An interesting method for obtaining the bis(diarylantimony) oxides is by thermal disproportionation of the corresponding polymeric oxides (104,112):

$$\frac{4}{x}\,(ArSbO)_x \longrightarrow (Ar_2Sb)_2O + Sb_2O_3$$

A number of compounds of the types $RSbY_2$ and R_2SbY, where Y is an anionic group other than halogen, have been prepared by the reaction of dihalo- or halostibines with lithium, sodium, or ammonium alkoxides (113,114), amides (115), azides (116), carboxylates (117), dithiocarbamates (118), mercaptides (119,120), or phenoxides (113). Dihalo- and halostibines can also be converted to compounds in which an antimony is linked to a main group (121) or transition metal (122).

3.4. Distibines and Distibenes. A considerable number of tetraalkyl- and tetraaryldistibines have been investigated. These are usually obtained by the reduction of a dialkyl- or diarylhalostibine with sodium hypophosphite (106,107) or magnesium (103,111). Distibines can also be prepared by the treatment of a metal dialkyl- or diarylstibide with a 1,2-dihaloethane (65,66,72,86, 123)

$$2\,R_2SbM + XCH_2CH_2X \longrightarrow R_2SbSbR_2 + CH_2{=}CH_2 + 2\,MX$$

where M = Li, Na, or K and X = Cl or Br. Distibines undergo a variety of interesting reactions (64–67,80,100,103,106) and have also attracted attention because a number of these substances are thermochromic (80,123,124).

Although distibenes, the antimony analogues of azo compounds, have never been isolated as free, monomeric molecules (125), a tungsten complex, tritungsten pentadecacarbonyl[μ_3-η^2-diphenyldistibene] [82579-41-7], $C_{27}H_{10}O_{15}\,Sb_2W_3$, has been prepared by the reductive dehalogenation of phenyldichlorostibine (126):

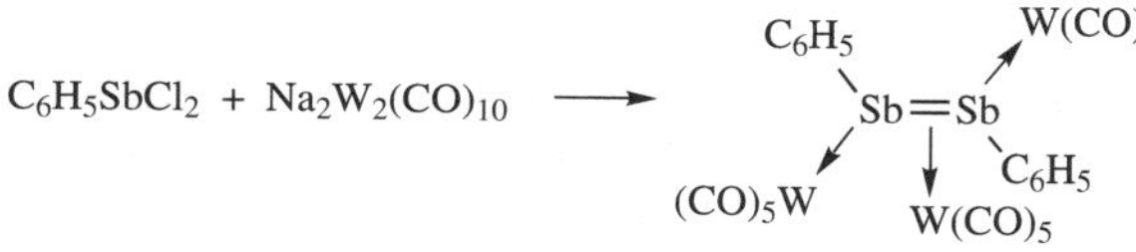

As expected, the Sb–Sb bond distance in this complex is significantly shorter than the Sb–Sb single bond distance. Chromium and tungsten complexes of dialkyldistibines have also been isolated (127).

3.5. Cyclic and Polymeric Substances Containing Antimony–Antimony Bonds. A number of organoantimony compounds containing rings of four, five, or six antimony atoms have been prepared. The first such compound to be adequately characterized, tetrakis-1,2,3,4-*tert*-butyltetrastibetane [47191-73-5], $C_{16}H_{36}Sb_4$, was obtained by the interaction of a dialkylstibide and iodine (65):

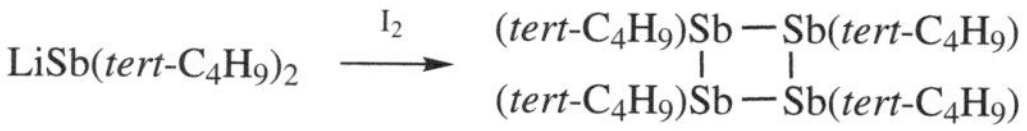

It has been prepared by the dehalogenation of *tert*-butyldichlorostibine [67877-43-4], $C_4H_9Cl_2Sb$, with magnesium (128). The corresponding five-membered ring compound, $(tert\text{-}C_4H_9Sb)_5$, is also formed in this reaction, but it has not yet been isolated in pure form (129). A tristibirane can also be formed by the dehalogenation of a dichlorostibine, but this substance has likewise been obtained only in admixture with a tetrastibetane (130). Compounds containing rings of six antimony atoms have been prepared by the slow air oxidation of bis(trimethylsilyl)phenylstibine [82363-95-9], $C_{12}H_{23}SbSi_2$, dissolved in 1,4-dioxane, benzene, or toluene (131):

$$6\ C_6H_5Sb[Si(CH_3)_3]_2 + 3\ O_2 + \text{solvent} \longrightarrow (C_6H_5Sb)_6 \cdot \text{solvent} + 6\ [(CH_3)_3Si]_2O$$

Polymeric substances, $(RSb)_x$ or $(ArSb)_x$, have been obtained by the decomposition of primary stibines (62,69), the reaction of primary stibines with hydrogen chloride (63), the treatment of primary stibines with dibenzylmercury (132), or the reduction of dihalostibines (133–135). Most of these polymers have not been well characterized.

3.6. Antimonin and its Derivatives. Antimonin(stibabenzene) [289-75-8], C_5H_5Sb, the antimony analogue of pyridine, can be prepared by the dehydrohalogenation of a cyclic chlorostibine using 1,5-diazabicyclo[4.3.0]non-5-ene (136):

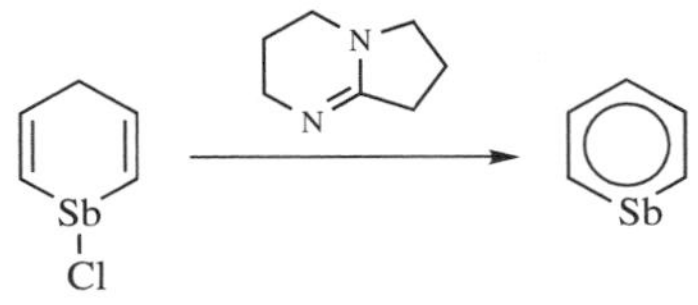

A number of derivatives of antimonin are also known (136,137). The potential aromaticity of this ring system has aroused considerable interest and has been investigated with the aid of spectroscopy as well as *ab initio* molecular orbital calculations (138). There seems to be no doubt that antimonin does possess considerable aromatic character.

3.7. Stibonic and Stibinic Acids. The stibonic acids, $RSbO(OH)_2$, and stibinic acids, $R_2SbO(OH)$, are quite different in structure from their phosphorus and arsenic analogues. The stibonic and stibinic acids are polymeric compounds of unknown structure and are very weak acids. IUPAC classifies them as oxide hydroxides rather than as acids. Thus $C_6H_5SbO(OH)_2$ is named phenylantimony dihydroxide oxide [535-46-6]; the *Chemical Abstracts* name is dihydroxyphenylstibine oxide [535-46-6], $C_6H_7O_3Sb$.

Methylstibonic acid, the only alkylstibonic acid known with certainty, was not reported until 1990 (139). Previous attempts to obtain alkylstibonic acids were unsuccessful (139). The methyl compound was prepared by two methods, the hydrolysis of a tetraalkoxymethylantimony compound or the oxidation of dimethoxymethylstibine [54553-25-2], $C_3H_9O_2Sb$, with hydrogen peroxide:

$$CH_3Sb(OR)_4 + 3\ H_2O \xrightarrow[0^\circ C]{CH_2Cl_2} CH_3SbO_3H_2 + 4\ ROH$$

$$CH_3Sb(OCH_3)_2 + H_2O_2 + H_2O \longrightarrow CH_3SbO_3H_2 + 2\ CH_3OH$$

Dialkylstibinic acids can be readily prepared. The preferred method is the aqueous hydrolysis of trialkoxydialkylantimony compounds (140):

$$R_2Sb(OR')_3 + 2\,H_2O \longrightarrow R_2SbO(OH) + 3\,R'OH$$

Dimethylstibinic acid [35952-95-5], $C_2H_7O_2Sb$, diethylstibinic acid [35952-96-6], $C_4H_{11}O_2Sb$, dipropylstibinic acid [35952-97-7], $C_6H_{15}O_2Sb$, and dibutylstibinic acid [35952-98-8], $C_8H_{19}O_2Sb$, have been prepared in this manner. Except for the dimethyl compound, they can be readily recrystallized from organic solvents. Aromatic stibonic acids are readily prepared by the diazo reaction:

$$ArN_2Cl + SbCl_3 \longrightarrow ArSbCl_4 + N_2$$

$$ArSbCl_4 + 3\,H_2O \longrightarrow ArSbO(OH)_2 + 4\,HCl$$

where Ar represents aryl. The arylstibonic acids prepared by this procedure are invariably contaminated with inorganic antimony compounds. Purification is usually effected by dissolving the crude product in hydrochloric acid and adding ammonium chloride or an amine hydrochloride, whereupon crystalline salts of the type $[R_4N][ArSbCl_5]$ separate from solution. These may be recrystallized and then hydrolyzed to the pure stibonic acids. The diarylstibinic acids are also usually prepared by the diazo reaction by substituting an aryldihalostibine for antimony trichloride:

$$ArN_2Cl + Ar'SbCl_2 \longrightarrow ArAr'SbCl_3 + N_2$$

$$ArAr'SbCl_3 + 2\,H_2O \longrightarrow ArAr'SbO(OH) + 3\,HCl$$

Diphenylstibinic acid [22811-63-8], $C_{12}H_{11}O_2Sb$, can be readily prepared in good yield from triphenylstibine by an oxidative cleavage reaction in which the stibine is heated with alkali and hydrogen peroxide (141). Stibonic and stibinic acids have had few industrial uses. A patent covering the use of stibonic or stibinic acids as catalysts for the condensation-polymerization of ethylene glycol and terephthalic acid has been issued (142). Another patent describes the use of diphenylstibinic acid as a cocatalyst with triisobutylaluminum for the polymerization of epichlorohydrin (143). The use of stibonic acids as catalysts for the epoxidation of alkenes by hydrogen peroxide is the subject of another patent (144). Anhydrides of arylstibonic acids, $(ArSbO_2)_n$, obtained by heating stibonic acids in vacuo at 100°C, have proved to be effective catalysts for the polymerization of oxiranes (145).

3.8. Stibine Oxides and Related Compounds. Both aliphatic and aromatic stibine oxides, R_3SbO, or their hydrates, $R_3Sb(OH)_2$, are known. Thus both dihydroxotrimethylantimony [19727-41-4], $C_3H_{11}O_2Sb$, and trimethylstibine oxide [19727-40-3], C_3H_9OSb, have been prepared. The former may be readily obtained by passing an aqueous solution of dichlorotrimethylantimony [13059-67-1], $C_3H_9Cl_2Sb$, through an anionic-exchange resin (146). When heated to 110°C, the dihydroxy compound loses one mole of water to form the oxide. Aliphatic stibine oxides containing larger alkyl groups have been obtained by oxidation of tri-alkylstibines using mercury(II) oxide (147). The trialkylstibine oxides are hygroscopic crystalline solids. Molecular weights in solution are usually two to three times the values calculated for monomeric structures. The Mössbauer spectrum of trimethylstibine oxide was consistent with a trigonal-bipyramidal structure having three methyl groups in equatorial positions (148). The use of dihydroxotrimethylantimony as a catalyst for the epoxidation of alkenes by aqueous hydrogen peroxide has been the subject of a patent (149).

The structure of triphenylstibine oxide [4756-75-6], $C_{18}H_{15}OSb$, has been the subject of considerable controversy. Apparently it can exist in two different forms; as prismatic crystals, mp 221–222°C, and as an amorphous powder. The structure of the crystalline form was shown by x-ray diffraction to be a dimer containing a planar four-membered ring with Sb–O–Sb bonds (150,151). It can be prepared by a number of synthetic methods including the thermal decomposition of hydroxotetraphenylantimony [19638-16-5], $C_{24}H_{21}OSb$, (152) or methoxotetraphenylantimony [14090-94-9], $C_{25}H_{23}OSb$ (153). The amorphous form of triphenylstibine oxide, termed poly(triphenylstibine oxide) [36562-85-3], $(C_{18}H_{15}OSb)_x$, is insoluble in water and in organic solvents; its structure is unknown. A number of papers have been published on the use of triphenylstibine oxide as a catalyst for various reactions which may be of considerable commercial value. It has been used as a catalyst for the condensation of diamines and carbon dioxide to form cyclic urea compounds (153), for the ring-opening polymerization of ethylene oxide or propylene oxide (154), and for the formation of 2-oxazolidinones from carbon dioxide and 2-aminoalcohols (155). It has been claimed that triphenylstibine oxide and triphenylstibine sulfide are of value as catalyst modifiers in the polymerization of propene by Ziegler catalysts (156,157), and a number of patents have been issued on the use of the oxide as a catalyst in various industrial processes. In addition to triphenylstibine oxide, other triarylstibine oxides have been synthesized by the oxidation of the corresponding triarylstibines with hydrogen peroxide (158). Dihydroxotriphenylantimony [896-29-7], $C_{18}H_{17}O_2Sb$, has been reported a number of times in the chemical literature. There has been, however, no investigation utilizing analytical instrumental techniques on this compound. Dihydroxotris(2,4,6-trimethylphenyl)antinomy [113002-54-3], $C_{27}H_{35}O_2Sb$, is the only compound of this class, the structure of which has been confirmed by x-ray diffraction (158).

Both trialkyl- and triarylstibine sulfides and selenides are known. Trimethylstibine sulfide [15082-93-6], C_3H_9SSb, has been prepared from trimethylstibine oxide and hydrogen sulfide (159). It is monomeric in benzene and chloroform. Trialkylstibine sulfides and selenides have been prepared from trialkylstibines and sulfur or selenium, respectively (160). Unlike triphenylstibine oxide, the structure of triphenylstibine sulfide is tetrahedral, as shown by both Mössbauer and x-ray diffraction studies (148). A patent covering the synergistic use of triphenylstibine sulfide with aromatic amines as antioxidants in lubricating oils has been issued (161).

3.9. Pentacovalent Antimony Halides and Related Compounds.

Antimony halides of the types $RSbX_4$, R_2SbX_3, R_3SbX_2, and R_4SbX, where X is a halogen, are known, but compounds of the first type have only been isolated and characterized where R is aryl. Tetrachloromethylantimony has been prepared at

$-70°C$, but not isolated (162):

$$CH_3SbCl_2 + SO_2Cl_2 \xrightarrow{-70°C} CH_3SbCl_4 + SO_2$$

$$CH_3SbCl_4 + (CH_3)_4NN_3 \xrightarrow{-70°C} [CH_3SbCl_4N_3][(CH_3)_4N]$$

The aryl compounds are unstable substances which decompose on standing and are hydrolyzed in moist air. The chlorides are readily prepared by the action of hydrochloric acid on the corresponding arylstibonic acids. Tetraacetatophenylantimony [116122-86-27], $C_{14}H_{17}O_8Sb$, has been prepared:

$$C_6H_5SbCl_2 + SO_2Cl_2 + 4\,Ag(O_2CCH_3) \xrightarrow{CH_2Cl_2} C_6H_5Sb(O_2CCH_3)_4 + 4\,AgCl + SO_2$$

The Sb atom in this tetraacetato compound is hexacoordinate with three monodentate acetate groups and one symmetrically bonded bidentate acetate group.

Both dialkyl- and diaryltrihaloantimony compounds are known, although only a few dialkyl compounds have been described. The trichlorides have been obtained by the chlorination of either dialkylchlorostibines (163) or tetraalkyldistibines (164) with sulfuryl chloride. Dimethyltrichloroantimony [7289-79-4], $C_2H_6Cl_3Sb$, is dimeric in the solid state but is monomeric in solution (165). The dimer exists in two different forms, covalent and ionic (166). The covalent form contains bridging chlorine atoms; the ionic form possesses the structure $[(CH_3)_4Sb][SbCl_6]$. The diaryl compounds, both symmetrical and unsymmetrical, are best prepared from a diazonium halide and an aryldihalostibine:

$$ArN_2Cl + Ar'SbCl_2 \longrightarrow ArAr'SbCl_3 + N_2$$

When a diazonium salt is allowed to react with antimony pentachloride or an aryltetrachloroantimony compound, the onium salts $[ArN_2][SbCl_6]$ or $[ArN_2][Ar'SbCl_5]$, respectively, are formed. These can be decomposed in an organic solvent by the addition of a powdered metal such as iron or zinc, with the formation of a diaryltrichloroantimony compound:

$$2\,[ArN_2][SbCl_6] + 3\,Fe \longrightarrow Ar_2SbCl_3 + 2\,N_2 + SbCl_3 + 3\,FeCl_2$$

This is known as the Nesmeyanov reaction. The trichloro compounds are somewhat more stable than the tetrachloro compounds and can usually be readily recrystallized. These are also formed from diarylstibinic acids and hydrochloric acid, and, because they usually possess sharp melting points, they can be used for the characterization of the corresponding stibinic acids. Trichlorodiphenylantimony [21907-22-2], $C_{12}H_{10}Cl_3Sb$, crystallizes from hydrochloric acid as trichlorophenylantimony monohydrate [18762-79-9], $(C_6H_5)_2SbCl_3\cdot H_2O$, but this readily loses water on heating to form the anhydrous trichloro compound which exists in the solid state as a dimer (167). The hydrate can also be prepared from $SbCl_5$ and $(C_6H_5)_4Sn$ (168). In addition to the trichlorides and tribromides, the mixed compounds, for example, dibromochlorodiphenylantimony [71191-19-0], $(C_6H_5)_2SbBr_2Cl$, and bromodichlorodiphenylantimony [71191-18-9],

$(C_6H_5)_2\text{-SbBrCl}_2$, have been prepared (169):

$$(C_6H_5)_2\text{SbCl} + \text{Br}_2 \xrightarrow[\text{CH}_2\text{Cl}_2]{-196^\circ\text{C}} (C_6H_5)_2\text{SbBr}_2\text{Cl}$$

$$(C_6H_5)_2\text{SbBr} + \text{Cl}_2 \xrightarrow[\text{CH}_2\text{Cl}_2]{-90^\circ\text{C}} (C_6H_5)_2\text{SbBrCl}_2$$

In contrast to the trichloro compound, these mixed halo compounds, as well as tribromodiphenylantimony [62170-61-0], $(C_6H_5)_2\text{SbBr}_3$, are monomeric in the solid state.

In addition to the trihalo compounds, triacetatodiphenylantimony [93833-20-6], $C_{18}H_{19}O_2\text{Sb}$, has been prepared (170):

$$(C_6H_5)_2\text{SbCl}_3 + 3\,\text{Ag(O}_2\text{CCH}_3) \longrightarrow (C_6H_5)_2\text{Sb(O}_2\text{CCH}_3)_3 + 3\,\text{AgCl}$$

The best known of the halides are the trialkyldihalo- and triaryldihaloantimony compounds. The dichloro, dibromo, and diiodo compounds are generally prepared by direct halogenation of the corresponding tertiary stibines. The difluoro compounds are obtained by metathasis from the dichloro or dibromo compounds and silver fluoride. The diiodo compounds are the least stable and are difficult to obtain in a pure state. The trialkyl- and triaryldichloro- and dibromoantimony compounds are all crystalline solids which are stable at room temperature that but decompose on heating:

$$\text{R}_3\text{SbX}_2 \xrightarrow{\Delta} \text{R}_2\text{SbX} + \text{RX}$$

The difluoro compounds, however, do not undergo this thermal decomposition.

Dichlorotriphenylantimony has been suggested as a flame retardant (171,172) and as a catalyst for the polymerization of ethylene carbonate (173). Dibromotriphenylantimony has been used as a catalyst for the reaction between carbon dioxide and epoxides to form cyclic carbonates (174) and for the oxidation of α-keto alcohols to diketones (175).

In addition to the trialkyldihalo- and triaryldihaloantimony compounds, mixed dihalo compounds such as chloroiodotriphenylantimony [7289-82-9], $(C_6H_5)_3\text{SbClI}$, have been reported (176). It has been shown, however, that such compounds disproportionate in solution to give a mixture of starting material plus products (177):

$$2\,\text{R}_3\text{SbXY} \rightleftharpoons \text{R}_3\text{SbX}_2 + \text{R}_3\text{SbY}_2$$

Other trialkyl and triaryl compounds of the type R_3SbY_2, where Y is a pseudohalide or a group such as NO_3, ClO_4, or $\frac{1}{2}\text{SO}_4$, have been prepared. They are usually obtained from the dihalides by metathesis with a silver salt, eg:

$$\text{R}_3\text{SbCl}_2 + 2\,\text{AgNO}_3 \longrightarrow \text{R}_3\text{Sb(NO}_3)_2 + 2\,\text{AgCl}$$

Compounds of the type $R_3Sb(OSO_2R')_2$, and $R_3Sb(O_2CR')_2$, where R is an alkyl or an aryl group, have been prepared from a dihydroxide or oxide and the appropriate acid:

$$R_3Sb(OH)_2 + 2\ R'SO_3H \longrightarrow R_3Sb(OSO_2R')_2 + 2\ H_2O$$

$$(Ar_3SbO)_2 + 4\ R'CO_2H \longrightarrow 2\ Ar_3Sb(O_2CR')_2 + 2\ H_2O$$

Hydrolysis of trialkyl- and triaryldihaloantimony compounds generally leads to the isolation of compounds of the type $(R_3SbX)_2O$ rather than compounds of the type $R_3Sb(OH)X$. However, hydroxoiodobis(2,6-dimethylphenyl)antimony [112515-20-5], $(2,6(CH_3)_2C_6H_3)_2Sb(OH)I$, (178) and four cyclohexyl compounds have been prepared (179): chlorohydroxotricyclohexylantimony [85362-32-9], $C_{18}H_{34}ClOSb$, bromohydroxotricyclohexylantimony [85362-33-0], $C_{18}H_{34}BrOSb$, acetatohydroxotricyclohexylantimony [85362-34-1], $C_{20}H_{37}O_3Sb$, and hydroxonitratotricyclohexylantimony [85362-35-2], $C_{18}H_{34}NO_4Sb$.

Tetraalkyl and tetraaryl compounds, R_4SbX, are well-known and are often referred to as stibonium salts. There is evidence, however, that most of the tetraaryl compounds contain pentacovalent antimony. The perchlorate $[(C_6H_5)_4Sb]$ (ClO_4), however, is ionic (180). The tetraalkyl halides are readily prepared by quaternization of the corresponding tertiary stibines:

$$R_3Sb + RI \longrightarrow R_4SbI$$

The tetraaryl compounds can be prepared by employing anhydrous aluminum chloride (181):

$$(C_6H_5)_3Sb + C_6H_5Cl \xrightarrow[230^\circ C]{AlCl_3} (C_6H_5)_4SbCl$$

Both the tetraalkyl and tetraaryl compounds can be prepared by cleavage of the corresponding pentaalkyl- or pentaarylantimony by a halogen or a hydrogen halide:

$$R_5Sb + HX \longrightarrow R_4SbX + RH$$

The cleavage of pentamethylantimony [15120-50-0], $C_5H_{15}Sb$, with HN_3, HCN, or HSCN gives the corresponding tetramethylantimony azides, cyanides, or thiocyanates (182); cleavage with one molecular equivalent of a carboxylic acid gives compounds of the type $(CH_3)_4SbO_2CR$ (183). Tetraphenylantimony iodide [13903-91-8], $C_{24}H_{20}ISb$, has been found to catalyze the condensation of oxetanes with carbon dioxide to give dioxanones (184) and the cycloaddition of oxiranes or oxetanes with diphenylketene to give γ- or δ-lactones (185).

3.10. Stibonium Ylids and Related Compounds.   In contrast to phosphorus and arsenic, only a few antimony ylids have been prepared. Until quite recently triphenylstibonium tetraphenylcyclopentadienylide [15081-36-4], $C_{47}H_{35}Sb$, was the only antimony ylid that had been isolated and adequately characterized (186). A new method, utilizing an organic copper compound as a catalyst, has resulted in the synthesis of a number of new antimony ylids (187):

$$\underset{Y}{\overset{X}{\diagdown}}C=N_2 + (C_6H_5)_3Sb \xrightarrow[C_6H_6,\ 80°C]{[(CF_3CO)_2CH]_2Cu} \underset{Y}{\overset{X}{\diagdown}}C=Sb(C_6H_5)_3 + N_2$$

Among the ylids prepared by this method are those in which X and Y are $C_6H_5SO_2$, $4–CH_3C_6H_4SO_2$, or CH_3CO or where X is CH_3CO and Y is C_6H_5CO. These ylids are solids, stable in a dry atmosphere, but readily hydrolyzed in protic solids by traces of moisture. Attempts to carry out the Wittig reaction with the stibonium ylids containing sulfonyl or carbonyl substituents, using highly reactive 2,4-dinitrobenzaldehyde as the substrate, were unsuccessful (188). Closely related to the ylids are imines of the type $R_3Sb=NR'$, where R is either an alkyl or an aryl group. The alkyl compounds, where R is ethyl or propyl and Ar is phenyl or 4-tolyl, have been prepared from trialkylstibines and arenesulfonyl azides (189):

$$R_3Sb + ArSO_2N_3 \longrightarrow R_3Sb{=}NSO_2Ar + N_2$$

The reaction of triphenylstibine with chloramine-T leads to the formation of a tosylimine (190):

$$(C_6H_5)_3Sb + CH_3C_6H_4SO_2N(Na)Cl \longrightarrow (C_6H_5)_3Sb{=}NSO_2C_6H_4CH_3 + NaCl$$

3.11. Organoantimony Compounds with Five Sb–C Bonds. A number of pentaalkyl- and pentaalkenylantimony compounds have been prepared from tetraalkyl- or tetraalkenylstibonium halides and alkyl or alkenyllithium or Grignard reagents, for example:

$$(CH_3)_4SbBr + CH_3Li \longrightarrow (CH_3)_5Sb + LiBr$$

Rather than using the stibonium halide, a trialkyl- or trialkenyldihaloantimony compound can be used, as in the preparation of pentavinylantimony [65277-46-5], $C_{10}H_{15}Sb$:

$$(CH{=}CH)_3SbBr_2 + 2\ CH_2{=}CHMgBr \longrightarrow (CH_2{=}CH)_5Sb + 2\ MgBr_2$$

Pentaarylantimony compounds can be readily prepared in a similar fashion:

$$(C_6H_5)_3SbF_2 + 2\ C_6H_5MgBr \longrightarrow (C_6H_5)_5Sb + MgBr_2 + MgF_2$$

Pentaphenylantimony [2170-05-0], $C_{30}H_{25}Sb$, has attracted considerable attention because it possesses square-pyramidal rather than the expected trigonal-bipyramidal geometry, both in the solid state and in solution. The cyclohexane solvate $(C_6H^5)^5Sb·1/2\,C_6H_{12}$ and penta-4-tolylantimony [51017-91-5], $C_{35}H_{35}Sb$, however, both possess trigonal-bipyramidal geometry. In addition to compounds of the type R_5Sb, mixed compounds of the type $R_4R'Sb$ or $R_3R'_2Sb$, where R and R' may be alkyl, alkenyl, alkynyl, or aryl groups, are known. Thus triethyldimethylantimony [67576-92-5], $C_8H_{21}Sb$, has been prepared (191):

$$(C_2H_5)_3SbCl_2 + 2\ CH_3Li \xrightarrow[-10^\circ C]{(C_2H_5)_2O} (C_2H_5)_3Sb(CH_3)_2 + 2\ LiCl$$

Methyltetraphenylantimony [33756-93-3], $C_{25}H_{23}Sb$, is readily prepared (192):

$$(C_6H_5)_4SbF + CH_3MgBr \longrightarrow (C_6H_5)_4SbCH_3 + MgBrF$$

Compounds containing aryl and alkynyl groups have also been prepared (193):

$$Ar_3SbBr_2 + 2\ C_6H_5C\equiv CLi \longrightarrow Ar_3Sb(C\equiv CC_6H_5)_2 + 2\ LiBr$$

Antimony trioxide is the most important of the antimony compounds. It is used primarily in flame retardant formulations. These formulations uses include as retardants, in children's clothing, toys, aircraft, and automobile seat covers.

In 1999, production of antimony trioxide was 30.8×10^6 kg (68×10^6 lb). In 2003, production of flame retardants from antimony trioxide is expected to reach 34×10^6 kg (75×10^6 lb) (194).

Table 3 and 4 give United States import and export information on antimony oxide by country (29).

Table 3. **U.S. Exports of Antimony Oxide, by Country**[a]

Country	2000			2001		
	Gross weight, t	Antimony content, t[b]	Value × 10^3\$	Gross weight, t	Antimony content, t	Value, × 10^3\$
Argentina	115	95	302	83	69	272
Australia	128	106	254	72	60	145
Belgium	13	11	40	19	16	26
Brazil	98	81	386	277	230	727
Canada	1,730	1,440	3,930	1,380	1,140	3,240
China	134	111	264	11	9	112
Colombia	118	98	214	67	56	133
France	50	42	130	28	23	76
Germany	102	85	438	68	56	178
Indonesia	0	0	0	6	5	13
Italy	0	0	0	5	4	20
Japan	130	108	509	41	34	214
Korea, Republic of Mexico	55	46	135	15	12	38
	3,820	3,170	5,680	4,360	3,620	6,930
Singapore	77	64	158	74	61	225
Spain	48	40	181	56	46	237
Taiwan	29	24	78	20	17	53
Turkey	62	51	189	83	69	239
United Kingdom	402	334	834	194	161	700
other	157[c]	130[c]	479[c]	242	199	715
Total	*7,280*	*6,040*	*14,200*	*7,090*	*5,880*	*14,300*

[a] Ref. 8, and the U.S. Census Bureau, data are rounded to no more than three significant digits; may not add to totals shown.

[b] Antimony content is calculated by the U.S. Geological Survey.

[c] Revised

Table 4. **U.S. Imports for Consumption of Antimony Oxide, by Class and Country**[a]

Country	2000			2001		
	Gross weight, t	Antimony content, t[b]	Value × 10³$	Gross weight, t	Antimony content, t[b]	Value, × 10³$
Belgium	3,690	3,070	6,560	3,770	3,130	6,450
Bolivia[c]	1,150	957	1,220	40	33	49
China	13,100	10,900	17,300	11,000	9,150	14,600
France	66	54	230	14	11	61
Germany	47	39	802	24	20	362
Guatemala	77	64	132	0	0	0
Hong Kong	453	376	622	790	656	966
Japan	33	27	274	69	57	429
Kyrgyzstan	224	186	247	0	0	0
Mexico	5,530	4,590	7,660	8,080	6,710	15,600
South Africa	3,830	3,180	999	3,750	3,110	900
Taiwan	29	24	53	41	34	63
Thailand	60	50	11	0	0	0
United Kingdom	176	146	298	60	50	65
Total	*28,500*	*23,700*	*36,500*	*27,700*	*23,000*	*39,500*

[a] Ref. 8 and U.S. Census Bureau, data are rounded to no more than three significant digits; may not add to totals shown.
[b] Antimony ore and concentrate content reported by the U.S. Census Bureau. Antimony oxide content is calculated by the U.S. Geological Survey.
[c] Antimony oxide from this country believed to be "crude" and would probably be shipped to refineries for upgrading.

4. Analytical Methods

A wide variety of titrimetric methods for the determination of antimony in the macro and semimicro range are available. Potassium bromate in strongly acid solution is probably the most widely used oxidimetric titrant. The end point can be determined either potentiometrically or by the use of an indicator. Organic dyes such as methyl red are used as indicators, but amaranth [915-67-3], $C_{20}H_{11}N_2Na_3O_{10}S_3$, has been reported as the indicator of choice (195). Potassium dichromate in hydrochloric acid–acetic acid solution with ferroin as the indicator has also been used (196). Other oxidimetric reagents used for the titration of antimony(III) include potassium iodate in acid solution, iodine in the presence of sodium tartrate and bicarbonate, and potassium permanganate in acid or alkaline solution. A number of organic compounds have been recommended as oxidimetric titrants. Such compounds are usually N-haloimides, and N-chlorophthalimide (197) and N-bromosuccinimide [128-08-5], C_4H_4Br NO_2, (198) have been used for this purpose. Sodium diethyldithiocarbamate has also been used for titrating antimony(III); the end point is determined potentiometrically (199). Because arsenic interferes with most methods used for determining antimony, the separation of the two elements is of great importance. It is possible to remove the arsenic as the trichloride by boiling a hydrochloric acid solution containing arsenic and antimony in the trivalent state. However, a method for determining both arsenic and antimony employs cerium(IV) sulfate as the titrant and ferroin as the indicator (200). A spectrofluorimetric method for determining antimony, based on the reduction of cerium(IV) to fluorescent cerium(III), has also been described (201). This method can also be used for determining antimony(III) in the presence of antimony(V).

A widely used colorimetric method for the estimation of microgram quantities of antimony is based on the reaction of antimony(V) with rhodamine B [81-88-9], $C_{28}H_{31}ClN_2O_3$, in hydrochloric acid solution to form a colored complex that is extracted with organic solvents and measured spectrophotometrically (202). For the determination of antimony in trace amounts, methods employing neutron activation or atomic absorption have been widely used. A comparison of the two methods has been reported (203). Both methods gave satisfactory results when used to determine specified values of antimony in several different biological materials. An excellent description of the determination of antimony involving the generation of stibine by sodium borohydride, followed by atomic absorption analysis, has also been reported (204).

5. Health and Safety Factors

OSHA has a TWA standard on a weight of Sb basis of 0.5 mg/m^3 for antimony in addition to a standard TWA of 2.5 mg/m^3 for fluoride. Most antimony compounds are poisonous by ingestion, inhalation, and intraperitoneal routes locally antimony compounds irritate the skin and mucous membranes (205). NIOSH has issued a criteria document on occupational exposure to inorganic fluorides. Antimony pentafluoride is considered by the EPA to be an extremely hazardous substance and releases of 0.45 kg or more reportable quantity (RQ) must be reported. Antimony trifluoride is on the CERCLA list and releasing of 450 kg or more RQ must be reported.

6. Environmental Impact

Antimony is a common air pollutant that occurs at an average concentration of 0.001 μg/m^3. Antimony is released into the environment from burning fossil fuels and from industry. In the air, antimony is rapidly attached to suspended particles and thought to stay in the air for 30 to 40 days. Antimony is found at low levels in some lakes, rivers, and streams, and may accumulate in sediments. Although antimony concentrations have been found in some freshwater and marine invertebrates, it does not biomagnify in the environment. The impact of antimony and antimony compounds on the environment has not been extensively studied to date (206).

BIBLIOGRAPHY

"Antimony Compounds," in *ECT* 1st ed., Vol. 2, pp. 59–64, by I. E. Campbell; in *ECT* 2nd ed., Vol. 2, pp. 570–588, by G. O. Doak, Leon Freedman, and G. Gilbert Long, North Carolina State of the University of North Carolina at Raleigh; in *ECT* 3rd ed., Vol. 3, pp. 105–128, and *ECT* 4th ed., Vol. 2, pp. 382–412 by Leon Freedman, G. O. Doak, and G. Gilbert Long, North Carolina State University; "Antimony Compounds" under "Fluorine compounds, Inorganic" in *ECT* 1st ed., Vol. 6, pp. 676–677, by F. D. Loomis and C. E. Inman, Pennsylvania Salt Manufacturing Co.; "Antimony" under "Fluorine Compounds, Inorganic" in *ECT* 2nd ed., Vol. 9, pp. 549–551, by W. E. White, Ozark-Mahoning Co., a subsidiary of the Pennwalt Corp.; in *ECT* 4th ed., Vol. II, pp. 290–294, by Tariq Mahmood and Charles B. Lindhal, Elf Atochem North America, Inc. "Antimony Compounds" in *ECT* (online), posting date: December 4, 2000, Leon D. Freedman, G. O. Doak, G, Gilbert Long, North Carolina State University.

CITED PUBLICATIONS

1. D. T. Hurd, *An Introduction to the Chemistry of the Hydrides*, John Wiley & Sons, Inc., New York, 1952, pp. 132–134.
2. C. L. Yaws and co-workers, *Solid State Technol.* **17**, 47 (1974).
3. A. W. Jache, G. S. Blevins, and W. Gordy, *Phys. Rev.* **97**, 680 (1955).
4. P. W. Schenk, *G. Brauer's Handbook of Preparative Inorganic Chemistry*, Vol. I, 2nd ed., Academic Press, New York, 1963, p. 591.
5. L. H. Berka, *The Chemistry of Stibine*, M.S. dissertation UCRL-8781, University of California, Berkeley, 1959, pp. 3–16.
6. A. D. Zorin and co-workers, *J. Appl. Chem. USSR (Engl. Transl.)* **47**, 1233 (1974).
7. G. Kh. Sorokin and S. A. Lomonosov, *Ind. Lab. USSR (Engl. Transl.)* **40**, 28 (1974).
8. L. Bretherick, *Hazards in the Chemical Laboratory*, 4th ed., The Royal Society of Chemistry, London, UK, 1986, p. 181.
9. J. F. Carlin, Jr., *Minerals Yearbook*, U.S. Geological Survey, Reston, Va., 2001.
10. G. G. Long, J. G. Stevens, and L. H. Bowen, *Inorg. Nucl. Chem. Lett.* **5**, 799 (1969).
11. D. J. Stewart and co-workers, *Can. J. Chem.* **50**, 690 (1972).
12. D. Rogers and A. C. Skapski, *Proc. Chem. Soc. (London)*, 400 (1964).
13. H. Y.-P. Hong, *Acta Crystallogr.* **B30**, 945 (1974).
14. S. S. Krishnan and D. R. Crapper, *Radiochem. Radioanal. Lett.* **20**, 279 (1975).
15. D. Cubicciotti, *High Temp. Sci.* **1**, 268 (1969).
16. C. J. Adams and A. J. Downs, *J. Chem. Soc., A*, 1534 (1971).
17. Eur. Pat. 331,483 A2 (Sept. 6, 1989), K. Fujiura, Y. Ohishi, M. Fujiki, T. Kanamori, and S. Takahashi.
18. Jpn. Pat. 6011239 A2 (Jan. 21, 1983), (to Nippon Telegraph & Telephone Public Co.).
19. Jpn. Pat. 63314713 A2 (Dec. 22, 1988), N. Sonoda and N. Sato.
20. L. Kolditz, *Halogen Chem.* **2**, 115 (1967).
21. T. K. Davies and K. C. Moss, *J. Chem. Soc., A*, 1054 (1970).
22. A. J. Edwards and P. Taylor, *Chem. Commun.*, 1376 (1971).
23. R. C. Thompson and co-workers, *Inorg. Chem.* **4**, 1641 (1965).
24. R. J. Gillespie and T. E. Peel, *J. Am. Chem. Soc.* **95**, 5173 (1973).
25. J. Melin and A. Herold, *C. R. Acad. Sci., Ser. C* **280**, 641 (1975).
26. H. Preiss and L. Kolditz, *Proceedings of the 2nd Seminar on Crystallochemistry of Coordination and Metallorganic Compounds*, 1973, p. 90.
27. *Antimony Pentafluoride*, Preliminary Technical Data Sheet, Ozark Fluorine Specialities, Folcroft, Pa., 2003.
28. H. Oppermann, *Z. Anorg. Allg. Chem.* **356**, 1 (1967).
29. D. D. Wagman and co-workers, *N.B.S. Technical Note 270-3*, U.S. Dept. of Commerce, Government Printing Office, Washington, D.C., 1968, pp. 99–102.
30. G. G. Long, J. G. Stevens, and L. H. Bowen, *Inorg. Nucl. Chem. Lett.* **5**, 21 (1969).

31. K. K. Laali, E. Geleginter, and R. Filler, *J. Fluorine Chem.* **53**(1), 107–126 (1991).
32. G. A. Olah, A. Germain, and H. C. Lin, *J. Am. Chem. Soc.* **97**(19), 5481–5488 (1975).
33. R. J. Gillespie and T. E. Peel, *J. Am. Chem. Soc.* **95**, 5173 (1973).
34. R. J. Gillespie, in V. Gold, ed., *Proton Transfer Reactions*, Chapman and Hall, London, 1975, p. 27.
35. Jpn. Pat. 62027306 A2 (Feb. 5, 1987), Y. Mochida and co-workers.
36. Jpn. Pat. 62108730 A2 (May. 20, 1987), Y. Mochida and co-workers.
37. Ger. Offen. DE 3432221 A1 (Mar. 13, 1986), A. Guenther.
38. U.S. Pat. 4,136,102 (Jan. 23, 1979), J. V. Crivello (to General Electric Co.).
39. L. W. Clemence and M. T. Leffler, *J. Am. Chem. Soc.* **70**, 2439 (1948).
40. H.-C. Mu, *K'o Hsueh T'ung* **17**, 502 (1966).
41. M. E. Gress and R. A. Jacobson, *Inorg. Chim. Acta* **8**, 209 (1974).
42. G. A. Kiosse, N. I. Golovastikov, and N. V. Belov, *Dokl. Akad. Nauk SSSR* **155**, 545 (1964).
43. B. Kamenar, D. Grdenić, and C. K. Prout, *Acta Crystallogr.* **B26**, 181 (1970).
44. G. A. Kiosse and co-workers, *Dokl. Akad. Nauk SSSR* **177**, 329 (1967).
45. A. Zalkin, D. H. Templeton, and T. Ueki, *Inorg. Chem.* **12**, 1641 (1973).
46. R. E. Tapscott, R. L. Belford, and I. C. Paul, *Coord. Chem. Rev.* **4**, 323 (1969).
47. I. Hansson, *Acta Chem. Scand.* **22**, 509 (1968).
48. A. C. Nanda and S. Pani, *J. Indian Chem. Soc.* **31**, 588 (1954).
49. B. C. Mohanty and S. Pani, *J. Indian Chem. Soc.* **31**, 593 (1954).
50. G. Hunter, *J. Chem. Soc., Dalton Trans.*, 1496 (1972).
51. R. Weinland and R. Scholder, *Z. Anorg. Allg. Chem.* **127**, 343 (1923).
52. S. Datta and T. N. Ghosh, *Sci. Cult.* **11**, 699 (1946).
53. J. Rigaudy and S. P. Klesney, *IUPAC Nomenclature of Organic Chemistry, Sections A–F, and H*, Pergamon Press, Oxford, UK, 1979, pp. 382–408.
54. F. G. Mann, *The Heterocyclic Derivatives of Phosphorus, Arsenic, Antimony and Bismuth*, 2nd ed., Wiley-Interscience, New York, 1970.
55. R. E. Atkinson in A. R. Katritzky and C. W. Rees, eds., *Comprehensive Heterocyclic Chemistry*, Vol. 1, Pergamon Press, Oxford, UK, 1984, pp. 539–561.
56. S. Samaan, *Houben–Weyl Methoden der Organischen Chemie: Metallorganischen Verbindungen, As, Sb, Bi*, Georg Thieme Verlag, Stuttgart, Germany, 1978, Band XIII, Teil 8.
57. M. Wieber, *Gmelin Handbook of Inorganic Chemistry, Sb Organoantimony Compounds, Part 1*, 8th ed., Springer-Verlag, Berlin, Germany, 1981, 1982, 1986.
58. M. Mirbach and M. Wieber in Ref. 57, Part 5, 1990.
59. M. Dub, *Organometallic Compounds*, Vol. III, 2nd ed., Springer-Verlag, New York, 1968, 1st Supplement, 1972.
60. G. O. Doak and L. D. Freedman, *Organometallic Compounds of Arsenic, Antimony, and Bismuth*, John Wiley & Sons, Inc., New York, 1970.
61. L. D. Freedman and G. O. Doak, *Chem. Met.-Carbon Bond*, **5**, 397 (1989).
62. A. L. Rheingold, P. Choudhury, and M. F. El-Shazly, *Synth. React. Inorg. Met.-Org. Chem.* **8**, 453 (1978).
63. P. Choudhury, M. F. El-Shazly, C. Spring, and A. L. Rheingold, *Inorg. Chem.* **18**, 543 (1979).
64. K. Issleib and B. Hamann, *Z. Anorg. Allg. Chem.* **339**, 289 (1965).
65. K. Issleib, B. Hamann, and L. Schmidt, *Z. Anorg. Allg. Chem.* **339**, 298 (1965).
66. K. Issleib and B. Hamann, *Z. Anorg. Allg. Chem.* **332**, 179 (1964).
67. A. B. Burg and L. R. Grant, *J. Am. Chem. Soc.* **81**, 1 (1959).
68. E. Wiberg and K. Mödritzer, *Z. Naturforsch Teil B* **B12**, 128 (1957).
69. K. Issleib and A. Balszuweit, *Z. Anorg. Allg. Chem.* **418**, 158 (1975).

70. M. Ates, H. J. Breunig, and S. Gülec, *Phosphorus, Sulfur Silicon Relat. Elem.* **44**, 129 (1989).
71. E. Wiberg and K. Mödritzer, *Z. Naturforsch. Teil B* **B12**, 131 (1957).
72. K. Issleib and B. Hamann, *Z. Anorg. Allg. Chem.* **343**, 196 (1966).
73. A. H. Cowley, R. A. Jones, C. M. Nunn, and D. L. Westmoreland, *Angew Chem., Int. Ed. Engl.* **28**, 1018 (1989).
74. A. N. Nesmeyanov, A. E. Borisov, and N. V. Novikova, *Izv. Akad. Nauk SSSR, Ser. Khim.*, 815 (1967).
75. Y.-Z. Huang, Y. Shen, and C. Chen, *Tetrahedron Lett.* **26**, 5171 (1985).
76. B. N. Laskorin and V. V. Yakshin, *Dokl. Akad. Nauk SSSR* **206**, 653 (1972).
77. W. J. C. Dyke, W. C. Davies, and W. J. Jones, *J. Chem. Soc.*, 463 (1930).
78. G. S. Hiers, *Organic Synthesis, Collective Volumes*, Vol. 1, 2nd ed., John Wiley & Sons, Inc., New York, 1941, p. 550.
79. T. V. Talalaeva and K. A. Kocheshkov, *Zh. Obshch. Khim.* **16**, 777 (1946).
80. A. J. Ashe III, C. M. Kausch, and O. Eisenstein, *Organometallics* **6**, 1185 (1987).
81. A. N. Nesmeyanov and L. G. Makarova, *Zh. Obshch. Khim.* **7**, 2649 (1937).
82. D. Naumann, W. Tyrra, and F. Leifeld, *J. Organomet. Chem.* **333**, 193 (1987).
83. L. I. Zakharkin and O. Yu. Okhlobystin, *Dokl. Akad. Nauk SSSR* **116**, 236 (1957).
84. E. A. Ganja, C. D. Ontiveros, and J. A. Morrison, *Inorg. Chem.* **27**, 4535 (1988).
85. S. P. Olifirenko, *Visn. L'viv. Derzh. Univ. Ser. Khim.* **6**, 100 (1963).
86. H. A. Meinema, H. F. Martens, and J. G. Noltes, *J. Organomet. Chem.* **51**, 223 (1973).
87. G. O. Doak and L. D. Freedman, *Synthesis*, 328 (1974).
88. E. G. Hope, T. Kemmitt, and W. Levason, *J. Chem. Soc., Perkin Trans. 2*, 487 (1987).
89. J. Seifter, *J. Pharmacol. Exp. Ther.* **66**, 366 (1939).
90. J. I. Harris, S. T. Bowden, and W. J. Jones, *J. Chem. Soc.*, 1568 (1947).
91. R. R. Holmes and E. F. Bertaut, *J. Am. Chem. Soc.* **80**, 2983 (1958).
92. G. O. Doak, G. G. Long, and L. D. Freedman, *J. Organomet. Chem.* **4**, 82 (1965).
93. P. Raj, R. Rastogi, and Firojee, *Indian J. Chem.* **26A**, 682 (1987).
94. C. A. McAuliffe and W. Levason, *Studies in Inorganic Chemistry*, Vol. 1, *Phosphine, Arsine and Stibine Complexes of the Transition Elements*, Elsevier, Amsterdam, The Netherlands, 1979.
95. C. Chen, Y.-Z. Huang, Y. Shen, and Y. Liao, *Heteroatom Chem.* **1**, 49 (1990).
96. R. D. Dupuis, *Science* **226**, 623 (1984).
97. T. Yoshimura, T. Masuda, and T. Higashimura, *Macromolecules* **21**, 1899 (1988).
98. A. S. Kholmanskii, E. A. Kuz'mina, and V. F. Tarasov, *Zh. Fiz. Khim.* **58**, 2095 (1984).
99. M. Wieber, D. Wirth, and I. Fetzer, *Z. Anorg. Allg. Chem.* **505**, 134 (1983).
100. H. J. Breunig and H. Jawad, *J. Organomet. Chem.* **243**, 417 (1983).
101. H. J. Breunig and W. Kanig, *Phosphorus Sulfur* **12**, 149 (1982).
102. M. Nunn, D. B. Sowerby, and D. M. Wesolek, *J. Organomet. Chem.* **251**, C45 (1983).
103. M. Ates, H. J. Breunig, A. Soltani-Neshan, and M. Tegeler, *Z. Naturforsch., B: Anorg. Chem., Org. Chem.* **41B**, 321 (1986).
104. G. O. Doak and H. H. Jaffé, *J. Am. Chem. Soc.* **72**, 3025 (1950).
105. I. G. M. Campbell and A. W. White, *J. Chem. Soc.*, 1184 (1958).
106. F. F. Blicke and U. O. Oakdale, *J. Am. Chem. Soc.* **55**, 1198 (1933).
107. F. F. Blicke, U. O. Oakdale, and F. D. Smith, *J. Am. Chem. Soc.* **53**, 1025 (1931).
108. S. P. Bone and D. B. Sowerby, *J. Chem. Soc., Dalton Trans.*, 1430 (1979).
109. G. T. Morgan and G. R. Davies, *Proc. Roy. Soc., Ser. A* **110**, 523 (1926).
110. H. J. Breunig and H. Kischkel, *Z. Naturforsch, B: Anorg. Chem., Org. Chem.* **36B**, 1105 (1981).
111. H. J. Breunig and H. Jawad, *Z. Naturforsch, B: Anorg. Chem., Org. Chem.* **37B**, 1104 (1982).

112. H. H. Jaffé and G. O. Doak, *J. Am. Chem. Soc.* **71**, 602 (1949).
113. N. Baumann and M. Wieber, *Z. Anorg. Allg. Chem.* **408**, 261 (1974).
114. V. L. Foss, N. M. Semenenko, N. M. Sorokin, and I. F. Lutsenko, *J. Organomet. Chem.* **78**, 107 (1974).
115. H. A. Meinema and J. G. Noltes, *Inorg. Nucl. Chem. Lett.* **6**, 241 (1970).
116. J. Müller and co-workers, *Z. Naturforsch. Teil B* **40B**, 1320 (1985).
117. W. T. Reichle, *J. Organomet. Chem.* **18**, 105 (1969).
118. E. J. Kupchik and C. T. Theisen, *J. Organomet. Chem.* **11**, 627 (1968).
119. H. Preut, F. Huber, and K.-H. Hengstmann, *Acta Crystallogr., Sect. C: Cryst. Struct. Commun.* **C44**, 468 (1988).
120. D. N. Kravtsov, B. A. Kvasov, S. I. Pombrik, and É. I. Fedin, *Izv. Akad. Nauk SSSR, Ser. Khim.*, 927 (1974).
121. H. Schumann, T. Östermann, and M. Schmidt, *J. Organomet. Chem.* **8**, 105 (1967).
122. A. M. Arif, A. H. Cowley, N. C. Norman, and M. Pakulski, *Inorg. Chem.* **25**, 4836 (1986).
123. A. J. Ashe III, E. G. Ludwig, Jr., and H. Pommerening, *Organometallics* **2**, 1573 (1983).
124. A. J. Ashe III, *Adv. Organomet. Chem.* **30**, 77 (1990).
125. F. Klages and W. Rapp, *Chem. Ber.* **88**, 384 (1955).
126. G. Huttner, U. Weber, B. Sigwart, and O. Scheidsteger, *Angew. Chem. Suppl.*, 411 (1982).
127. U. Weber, G. Huttner, O. Scheidsteger, and L. Zsolnai, *J. Organomet. Chem.* **289**, 357 (1985).
128. H. J. Breunig and W. Kanig, *Chem.-Ztg.* **102**, 263 (1978).
129. H. J. Breunig and H. Kischkel, *Z. Anorg. Allg. Chem.* **502**, 175 (1983).
130. H. J. Breunig and A. Soltani-Neshan, *J. Organomet. Chem.* **262**, C27 (1984).
131. H. J. Breunig, A. Soltani-Neshan, K. Häberle, and M. Dräger, *Z. Naturforsch. Teil B* **41B**, 327 (1986).
132. A. L. Rheingold and P. Choudhury, *J. Organomet. Chem.* **128**, 155 (1977).
133. G. Chobert and M. Devaud, *Electrochim. Acta* **25**, 637 (1980).
134. H. J. Breunig and W. Kanig, *J. Organomet. Chem.* **186**, C5 (1980).
135. M. Ates and co-workers, *Chem. Ber.* **122**, 473 (1989).
136. A. J. Ashe III, *Top. Curr. Chem.* **105**, 125 (1982).
137. A. J. Ashe III, T. R. Diephouse, and M. Y. El-Sheikh, *J. Am. Chem. Soc.* **104**, 5693 (1982).
138. K. K. Baldridge and M. S. Gordon, *J. Am. Chem. Soc.* **110**, 4204 (1988).
139. M. Wieber and J. Walz, *Z. Naturforsch., B: Chem. Sci.* **45B**, 1615 (1990); Ref. 65, p. 284.
140. H. A. Meinema and J. G. Noltes, *J. Organomet. Chem.* **36**, 313 (1972).
141. G. O. Doak and J. M. Summy, *J. Organomet. Chem.* **55**, 143 (1973).
142. U.S. Pat. 3,642,702 (Feb. 15, 1972), J. J. Ventura and J. G. Natoli (to M & T Chemicals, Inc.).
143. Ger. Offen. 2,102,102 (Sept. 7, 1972), H. G. J. Overmars and A. Van Elven (to Schering A.-G).
144. Eur. Pat. Appl. EP 74,259 (Mar. 16, 1983), Y. Kuriyama, M. Kakuda, and S. Nitoh (to Mitsubishi Gas Chemical Co., Inc.).
145. R. Nomura, Y. Wada, and H. Matsuda, *J. Polym. Sci., Part A: Polym. Chem.* **26**, 627 (1988).
146. G. G. Long, G. O. Doak, and L. D. Freedman, *J. Am. Chem. Soc.* **86**, 209 (1964).
147. G. N. Chremos and R. A. Zingaro, *J. Organomet. Chem.* **22**, 637 (1970).
148. J. Pebler, F. Weller, and K. Dehnicke, *Z. Anorg. Allg. Chem.* **492**, 139 (1982).
149. Ger. Offen. 2,605,041 (Aug. 26, 1976), M. Pralus, J. P. Schirmann, and S. Y. Delavarenne (to Ugine Kuhlmann).

150. J. Bordner, G. O. Doak, and T. S. Everett, *J. Am. Chem. Soc.* **108**, 4206 (1986).
151. G. Ferguson and co-workers, *Acta Crystallogr., Sect. C: Cryst. Struct. Commun.* **C43**, 824 (1987).
152. W. E. McEwen, G. H. Briles, and D. N. Schulz, *Phosphorus* **2**, 147 (1972).
153. R. Nomura, M. Yamamato, and H. Matsuda, *Ind. Eng. Chem. Res.* **26**, 1056 (1987).
154. R. Nomura, Y. Shiomura, A. Ninagawa, and H. Matsuda, *Makromol. Chem.* **184**, 1163 (1983).
155. H. Matsuda and co-workers, *Ind. Eng. Chem. Prod. Res. Dev.* **24**, 239 (1985).
156. N. M. Karayannis and S. S. Lee, *Makromol. Chem.* **186**, 1871 (1985).
157. N. M. Karayannis, H. M. Khelghatian, and S. S. Lee, *Makromol. Chem.* **187**, 863 (1986).
158. F. Huber, T. Westhoff, and H. Preut, *J. Organomet. Chem.* **323**, 173 (1987).
159. M. Shindo, Y. Matsumura, and R. Okawara, *J. Organomet. Chem.* **11**, 299 (1968).
160. R. A. Zingaro and A. Merijanian, *J. Organomet. Chem.* **1**, 369 (1964).
161. U.S. Pat. 4,032,462 (June 28, 1977), B. W. Hutten and J. M. King (to Chevron Research Co.).
162. K. Dehnicke and H.-G. Nadler, *Z. Anorg. Allg. Chem.* **426**, 253 (1976).
163. H.-G. Nadler and K. Dehnicke, *J. Organomet. Chem.* **90**, 291 (1975).
164. H. A. Meinema, H. F. Martens, and J. G. Noltes, *J. Organomet. Chem.* **51**, 223 (1973).
165. K. Dehnicke and H.-G. Nadler, *Chem. Ber.* **109**, 3034 (1976).
166. W. Schwarz and H.-J. Guder, *Z. Naturforsch., B: Anorg. Chem. Org. Chem.* **33B**, 485 (1978).
167. J. Bordner, G. O. Doak, and J. R. Peters, Jr., *J. Am. Chem. Soc.* **96**, 6763 (1974).
168. I. Haiduc and C. Silvestru, *Inorg. Synth.* **23**, 194 (1985).
169. S. P. Bone and D. B. Sowerby, *J. Chem. Soc., Dalton Trans.*, 715 (1979).
170. D. B. Sowerby, M. J. Begley, and P. L. Millington, *J. Chem. Soc., Chem. Commun.*, 896 (1984).
171. O. Horak, J. Havranek, and J. Vladyka, *Kunststoffe* **72**, 493 (1982).
172. O. Horak, *Plasty Kauc.* **24**, 271 (1987).
173. T. Otsu, K. Endo, K. Hozawa, and M. Komatsu, *Mem. Fac. Eng., Osaka City Univ.* **26**, 101 (1985).
174. A. Ninagawa, H. Matsuda, and R. Nomura, *Kenkyu Hokoku-Asahi Garasu Kogyo Gijutsu Shoreikai* **39**, 117 (1981).
175. K. Ohkata, H. Ohnari, and K. Akiba, *Nippon Kagaku Kaishi*, 1267 (1987).
176. A. D. Beveridge, G. S. Harris, and F. Inglis, *J. Chem. Soc. A*, 520 (1966).
177. C. G. Moreland, M. H. O'Brien, C. E. Douthit, and G. G. Long, *Inorg. Chem.* **7**, 834 (1968).
178. G. Ferguson, G. S. Harris, and A. Khan, *Acta Crystallogr., Sect. C: Cryst. Struct. Commun.* **C43**, 2078 (1987).
179. Y. Kawasaki, Y. Yamamoto, and M. Wada, *Bull. Chem. Soc. Jpn.* **56**, 145 (1983).
180. G. Ferguson, C. Glidewell, D. Lloyd, and S. Metcalfe, *J. Chem. Soc., Perkin Trans. 2*, 731 (1988).
181. G. O. Doak, G. G. Long, and L. D. Freedman, *J. Organomet. Chem.* **12**, 443 (1968).
182. H. Schmidbaur, K.-H. Mitschke, J. Weidlein, and St. Cradock, *Z. Anorg. Allg. Chem.* **386**, 139 (1971).
183. H. Schmidbaur, K.-H. Mitschke, and J. Weidlein, *Z. Anorg. Allg. Chem.* **386**, 147 (1971).
184. H. Matsuda and A. Baba, *Kenkyu Hokoku-Asahi Garasu Kogyo Gijutsu Shoreikai* **50**, 195 (1987).
185. M. Fujiwara, M. Imada, A. Baba, and H. Matsuda, *J. Org. Chem.* **53**, 5974 (1988).
186. D. Lloyd and M. I. C. Singer, *Chem. Ind. (London)*, 787 (1967).
187. C. Glidewell, D. Lloyd, and S. Metcalfe, *Tetrahedron* **42**, 3887 (1986).

188. G. Ferguson and co-workers, *J. Chem. Soc., Perkin Trans. 2*, 1829 (1988).

189. Z. I. Kuplennik, Zh. N. Belaya, and A. M. Pinchuk, *Zh. Obshch. Khim.* **51**, 2711 (1981).

190. G. Wittig and D. Hellwinkel, *Chem. Ber.* **97**, 789 (1964).

191. N. Tempel, W. Schwarz, and J. Weidlein, *J. Organomet. Chem.* **154**, 21 (1978).

192. G. Doleshall, N. A. Nesmeyanov, and O. A. Reutov, *J. Organomet. Chem.* **30**, 369 (1971).

193. K. Akiba, T. Okinaka, M. Nakatani, and Y. Yamamoto, *Tetrahedron Lett.* **28**, 3367 (1987).

194. *Buyers' News*, The freedonea Group, Purchasing.com, Nov. 1998, accessed Feb. 2003.

195. H. S. Gowda and S. Gurumurthy, *Indian J. Chem.* **21A**, 550 (1982).

196. S. G. Viswanath and M. K. N. Yenkie, *Chem. Anal. (Warsaw)* **28**, 43 (1983).

197. N. Jayasree and P. Indrasenan, *Talanta* **32**, 1067 (1985).

198. H. S. Gowda, R. Shakunthala, and U. Subrahmanya, *Indian J. Chem.* **20A**, 823 (1981).

199. W. S. Selig and G. L. Roberts, *Microchem J.* **34**, 140 (1986).

200. K. Sriramam, B. S. R. Sarma, N. R. Sastry, and A. R. K. V. Prasad, *Talanta* **28**, 963 (1981).

201. M. A. Al-Hajjaji and co-workers, *Anal. Lett.* **19**, 283 (1986).

202. S. Williams, ed., *Official Methods of Analysis of the Association of Official Analytical Chemists*, 14th ed., Association of Official Analytical Chemists, Arlington, Va., 1984, p. 449.

203. P. P. Coetzee and H. Pieterse, *S. Afr. J. Chem.* **39**, 85 (1986).

204. N. E. Parisis and A. Heyndrickx, *Analyst (London)* **111**, 281 (1986).

205. R. J. Lewis, Sr., "*Sax's* Dangerous Properties of Industrial Materials, Vol. 2, John Wiley & Sons, Inc., New York, 2000, p. 280.

206. L. Galliclhio, B. A. Fowler, E. F. Madden in E. Bingham, B. Cohrssen, and C. H. Powell, eds., *Patty's Toxicology*, Vol. 2, 5th ed., Wiley, New York, 2001, p. 779.

Leon D. Freedman
G. O. Doak
G. Gilbert Long
North Carolina State University
Tariq Mahmood
Charles B. Lindhal
Elf Atochem North America, Inc.

ANTIOBESITY DRUGS

1. Medical and Economic Aspects of Obesity

Obesity is an increasingly prevalent, complex disease with multiple etiologies and profound medical consequences. In addition to being a cosmetic problem, overweight and obesity are associated with an enhanced likelihood of developing chronic conditions including hypertension, hyperlipidemia, coronary heart disease, diabetes, cancer, gall bladder disease, and arthritis (1). In the 1990s, an estimated 55% of the U.S. adult population was either overweight [body mass index] (BMI) 25–29.9] or obese (BMI ≥30) (2,3) and ~280,000–325,000 deaths were attributed to obesity annually (4). The U.S. public spends heavily on weight control products and services. For example, in 1990, the total market including diet drinks, low calorie foods, meal replacements, health clubs, weight loss clinics, medically supervised weight loss programs, and various over-the-counter remedies amounted to $45.8 billion. In contrast, by 1998 prescription drugs accounted for only $184 million, or ~6% (5).

To define the obese state in a clinical setting, it is necessary to have a means of estimating the amount of adipose (fat) tissue relative to lean body mass. Large clinical studies typically employ measures of skin-fold thickness, waist/hip ratio, waist circumference, or more commonly, BMI=(weight in kg)/(height in m)^2] as a quantitative measure of obesity (6). Commonly accepted classifications for stages of obesity based on BMI are summarized in Table (1) (7).

In addition to the obvious role of the environment, there is a significant predisposing genetic component to obesity as indicated in a number of twin and adoption studies (8,9). Although there are rare cases of extreme obesity that can be ascribed to a defect in a single gene, a considerable body of evidence supports a polygenic contribution to most forms of obesity. Intense efforts are under way to identify candidate genes, since it is estimated that genetic factors account for 70% of the variation in weight between individuals (10,11). Obesity is rare in many third world populations until the people become westernized, adopting energy-rich diets and sedentary lifestyles. These individuals may be adapted to endure episodes of scarcity and are able to utilize food very efficiently. Those harboring these "thrifty genes" may be more prone to developing obesity when given access to abundant nutrients.

Individuals with a BMI in the range of 25–30 should begin a diet and exercise program with behavior modification. Those with at least one risk factor, such as a family history of heart disease, smoking, or high blood pressure, should be started on pharmacological intervention. People with a higher BMI, and particularly those with a BMI of >40, are candidates for increasingly aggressive treatment (12). An important consideration is the location of the excess fat. Although obesity is a serious health risk factor for both sexes, the abdominal or android pattern of fat distribution typical of overweight males carries a higher risk of life-threatening complications than the gluteal or gynoid fat distribution typical of overweight females (13,14).

Table 1. **Weight Classification by BMI**[a]

NHLBI[b] Terminology	BMI, kg/m^2	WHO[c] classification
underweight	<18.5	underweight
normal	18.5–24.9	normal range
overweight	25.0–29.9	preobese
obesity class 1	30.0–34.9	obese class 1
obesity class 2	35.0–39.9	obese class 2
obesity class 3	>40.0	obese class 3

[a] Reproduced with permission from Ref. 2.
[b] National Heart, Lung and Blood Insitute.
[c] World Health Organization.

The balance between energy intake and energy expenditure is finely regulated in most people, and they maintain relatively consistent body weights for long periods despite variations in their day-to-day nutrient consumption. Although the mechanism is not understood at present, individuals may have a metabolic set point that provides a homeostatic drive toward weight maintenance. For many obese patients, this set point is inappropriately high and favors an increase in metabolic efficiency and restoration of body mass (15,16) after weight loss. One attractive approach to treating obesity would be to find a way to reset this metabolic set point to a lower BMI.

2. Treatment of Obesity

2.1. Introduction. Obesity is a difficult condition to treat. Dietary restriction of caloric intake is the first line therapy and is optimally combined with an exercise program to promote loss of fat relative to lean body mass (12). Drug treatments that help to suppress appetite, increase energy expenditure, or decrease fat absorption are available and may be beneficially used in conjunction with a comprehensive weight loss program. For the extremely obese (BMI >40), gastric by pass surgery has shown promise (17). The majority of formerly obese patients eventually regain their excess weight lost through diet, and thus a truly successful program must include long-term behavior modification.

2.2. Anorectics (Appetite Suppressants). Appetite suppressants are widely used as an adjunct to dietary restriction, and sympathomimetic amines have traditionally been used for this purpose. The sympathetic or adrenergic nervous system operates in juxtaposition to the parasympathetic nervous system to maintain homeostasis in response to physical activity and physical or psychological stress. Sympathomimetic neurotransmission is generally mediated by norepinephrine (1) released from presynaptic storage granules upon stimulation. A second endogenous sympathomimetic agent, epinephrine (2) is released systemically from the adrenal glands during emergencies as part of the "fight or flight" response. A large variety of peripheral organ functions are affected by the sympathetic nervous systems including heart rate, cardiac contractile force, blood pressure, bronchopulmonary tone, and metabolism. In the central nervous system (CNS), sympathetic nervous stimulation results in increases in wakefulness and psychomotor activity, and reduction of appetite (18).

(1)   (2)

Compounds structurally related to the endogenous sympathomimetic amines have classically been employed as appetite suppressants. These agents, of which amphetamine is the prototypical example, generally retain the phenethylamine but lack the catechol moiety present in 1 and 2. As a consequence, they are well absorbed after oral administration and readily distribute into the central nervous system, where they exert their anorectic effects at hypothalamic appetite control centers. A component of their efficacy in promoting weight loss may be an increase in metabolic rate through stimulation of thermogenesis or physical activity.

These compounds act primarily by indirect mechanisms involving displacement of norepinephrine from presynaptic nerve storage vesicles or by prevention of its reuptake rather than by a direct effect at the receptor level (18). To a lesser extent, certain agents of this class affect dopaminergic or serotoninergic neurons. The overall pharmacological profile of members of the non-catecholaminergic sympathomimetic amine class depends on their individual tissue distribution and their precise mechanism of action. While they vary in degree, all members of this group share similar liabilities of cardiovascular side effects, the potential for CNS stimulation, the development of tolerance, and abuse potential. Introduction of an oxygen atom on the β-carbon of the side chain tends to reduce their CNS stimulant properties without decreasing their anorectic activity. Following the Federal Controlled Drug Act of 1970, many of these drugs were classified into one of five schedules in decreasing order of their abuse potential. The Controlled Substances Act of 1970 classified drugs into schedules depending on their abuse potential: Schedule II, high potential for abuse; Schedule III, some potential for abuse, Schedule IV, low potential for abuse.

Phenylpropanolamine. As indicated in Table 2, phenylpropanolamine is the one member of this class, that is available over the counter (OTC). It is present in a number of common diet aids and nasal decongestants with such well known names as Dexatrim, Accutrim, Contac, and Dimetapp. Since it has two asymmetric carbon atoms, phenylpropanolamine has four theoretically possible stereoisomers. The compound sold as phenylpropanolamine in the United States is (±)-norephedrine and consists of equal amounts of D-(−)- and L-(+)-norephedrine 3 and 4, respectively, which are shown in the Fisher projection to illustrate the erythro relationship between their amino and hydroxyl groups. The diastereomers with the opposite (threo) relationship between these groups are referred to as D-(−)- and L-(+)-norpseudoephedrine, 5 and 6, respectively. In Europe and Australia the material sold as cathine and often referred to as phenylpropanolamine in the literature, is 6. Although also used as an anorectic agent, 6 has greater CNS stimulant effects and a different toxicological profile from (±)-norephedrine (19). This situation has led to considerable confusion in the literature

Table 2. **Phenylpropanolamine**

	phenylpropanolamine hydrochloride	CASRN: [37577-28-9] (base) CASRN: [154-41-6] (hydrochloride)

[R*,S*]-(±)-α-(1-aminoethyl)benzenemethanol hydrochloride
formula: $C_9H_{13}NO$ mol.wt.: 151.21
brand name: Accutrim manufacturer: Novartis Consumer Health
brand name: Dexatrim manufacturer: Thompson Medical
dose: 75 mg 1×daily

over side effects and efficacy of individual isomers.

$$
\begin{array}{cccc}
\text{H—OH} & \text{HO—H} & \text{H—O} & \text{HO—H} \\
\text{H—NH}_2 & \text{H}_2\text{N—H} & \text{H}_2\text{N—H} & \text{H—NH}_2 \\
\text{CH}_3 & \text{CH}_3 & \text{CH}_3 & \text{CH}_3 \\
(3) & (4) & (5) & (6)
\end{array}
$$

(±)-Norephedrine is pharmacologically similar to other indirect acting sympathomimetic amines, but it has little CNS stimulant activity. After high doses, it does affect peripheral α- and β-adrenergic sympathetic receptors, resulting in cardiovascular symptoms, but these are generally not troublesome at normal clinical doses. Its efficacy as a nasal decongestant is related to vasoconstrictor effects on blood vessels in the nasopharynx. (±)-Norephedrine is well absorbed after oral administration and is excreted largely unchanged in the urine over the course of 24 hrs. A number of clinical trials have focused on its efficacy in appetite suppression, and these have been summarized (20–23). Because most of these trials were of short duration and rarely were run in comparison with other anorectic drugs, it is difficult to assess the relative utility of (±)-norephedrine with respect to the pharmacologically related prescription drugs in the longer term treatment of obesity. Informal surveys indicate the product is largely used to assist in short-term weight loss programs following a period of unusual weight gain, as might occur, for example, following a vacation. In October 2000, in light of a Yale University study indicating that use of phenylpropanolamine slightly increases the risk of hemorrhagic stroke in women, the Food and Drug Administration (FDA) advisory committee voted 13–0 that phenylpropanolamine cannot be classified as safe. Further action on the part of the FDA banning this drugs from OTC products is likely (24).

Sympathomimetic Amines. The sympathomimetic amines listed in Table 3 are related to amphetamine and are still available on the U.S. market for use in weight loss programs, although their sales are minimal (<$1 million/yr). These include methamphetamine (Desoxyn Gradumet, Schedule II (20), phendimetrazine (Bontril, Schedule III), diethylproprion hydrochloride (Tenuate, Schedule IV), and mazindol (Sanorex, Schedule IV). These compounds are indicated for short-term use when other therapies have been unsuccessful. They are relatively effective over the course of at least several weeks, and although tolerance is considered to be a problem, it does not always occur.

Phentermine. Phentermine (Fastin, Adipex-P, Ionamin), a sympathomimetic amine with low abuse potential, is still prescribed widely for the short-term treatment of obesity. It was launched in 1959 and came into widespread use in the mid-1990s in a combination with fenfluramine commonly known as fen-phen. By itself, phentermine acts at dopaminergic and adrenergic neurons to exert its anorexic effect (23,25,26). Since it is also an inhibitor of the enzyme monoamine oxidase -A (MAO-A), which plays an important role in the clearance of serotonin following its release from presynaptic neurons (27), an enhancement of

Table 3. Sympathomimetic Amines

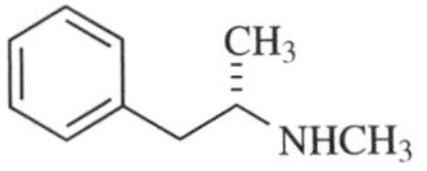

methamphetamine hydrochloride

(S)-N,α-dimethyl benzeneethanamine hydrochloride
formula: $C_{10}H_{15}N$

brand name: Desoxyn Gradumet
dose: 5, 10 or 15 mg 1×daily
Schedule II controlled substance

CASRN: [537-46-2] (base)
CASRN: [51-57-0] (hydrochloride)

mol. wt: 149.23 (base)
mol. wt. 185.69 (hydro-chloride)
manufacturer: Abbott Pharmaceuticals

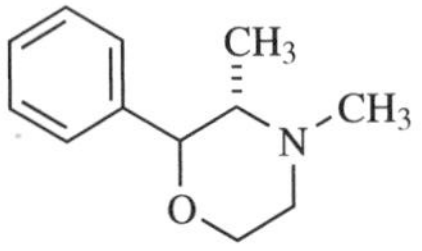

---

phendimetrazine hydrochloride

(2S,3S)-3,4-dimethyl-2- phenylmorpholine Tartrate
formula: $C_{12}H_{17}NO$

brand name: Bontril and Bontril Slow Release
dose: 105 mg, 1×daily (slow release formulation equivalent to 35 mg, 3×daily
Schedule III controlled substance

CASRN: [634-03-7] (base)
CASRN: [50-58-8] (tartrate)

mol. wt. 191.27 (base)
mol. wt. 341.36 (tartrate)
manufacturer: Carnick Laboratories

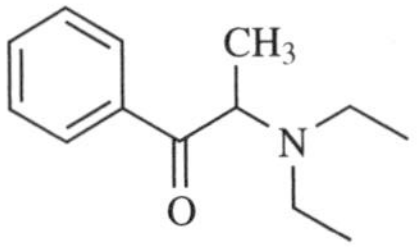

---

diethylpropion hydrochloride

2-(diethylamino)-1-phenyl-1-propanone hydrochloride
formula: $C_{13}H_{19}NO$

brand name: Tenuate
dose: 25 mg 3×daily or 75 mg of Tenuate Dospan 1×daily

CASRN: [90-84-6] (base)
CASRN: [139-80-5] (hydrochloride)

mol wt. 205.30 (base)
mol. wt. 241.78 (hydro-chloride)
manufacturer: Aventis[a]

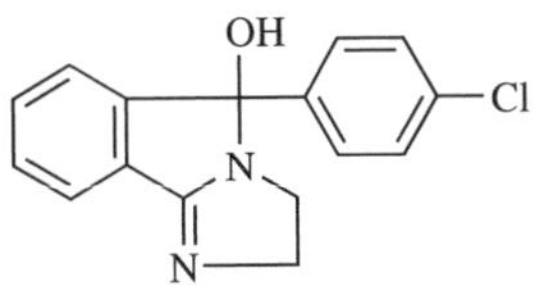

---

mazindol

5-(4-chlorophenyl)-2,5- dihydro-3H-imidazo[2,1-a] isoindol-5-ol
formula: $C_{16}H_{13}ClN_2O$

brand name: Sanorex
dose: 1–3 mg 1×daily

CASRN: [22232-71-9] (base)
CASRN: [58535-70-9] (hydrochloride)

mol. wt. 284.74 (base)
mol. wt. 321.20 (hyrdro-chloride)
manufacturer: Novartis

[a] Formerly Hoechst Marion Roussel

92

Table 4. **Phentermine**

generic name: phentermine hydrochloride	CASRN: [122-09-8] (base)
	CASRN: [1197-21-3] (hydrochloride)
α,α-dimethylbenzeneethan amine hydrochloride	
formula: $C_{10}H_{15}N$	mol. wt. 149.23 (base)
	mol. wt. 185.69 (hydrochloride)
brand name: Adipex-Pr[a]	manufacturer: Gate Pharmaceuticals
brand name: Fastin[a]	manufacturer: Smith Kline Beecham Pharmaceuticals
brand name: Ionamin[b]	manufacturer: Medeva Pharmaceuticals
dose: 15–30 mg 1×daily	

[a] Contains 30 mg of phentermine hydrochloride equiv to 24 mg of the free base.
[b] Contains either 15 or 30 mg of phentermine complexed with a cation-exchange resin.

serotonergic neurotransmission may also be a component of its activity. In rats, the compound is extensively metabolized by parahydroxylation (28), whereas in humans the 4-hydroxy derivative is only a minor metabolite (29). The results of several small clinical trials of up to 6-months duration have been summarized and indicate that the drug promotes a statistically significant weight loss with mild side effects related to its CNS stimulatory activity (25). The relatively low rate of metabolism of the drug in humans accounts for its long duration of action. Although sales of phentermine in the United States have fallen since the withdrawal of fenfluramine, it is still widely used, and in 1999 there were ~4 million prescriptions written with sales amounting to $54 million (30). Table 4 gives some basic data on phentermine.

Fenfluramine and Dexfenfluramine. Fenfluramine (Pondimin) is a racemate; the (+)-enantiomer, dexfenfluramine (Redux), is twice as effective as the racemate as an appetite suppressant and has been shown in clinical trials to be relatively free of side effects (31–32). Rat studies indicate that in contrast to amphetamine, which tends to delay the onset of meals and decrease protein intake, fenfluramine decreases the rate of eating and promotes early meal termination with no effect on meal pattern or nutrient selection (33–34). Fenfluramine was approved in the United States for the treatment of obesity in 1973, and dexfenfluramine was approved in 1996. A key paper by Weintraub (35) summarizing the results of a 4-year trial of a combination with fenfluramine with phentermine in patients sparked a dramatic popularization of the "fen-phen" combination; an estimated 18 million patients were treated with fen-phen between 1992 and 1997 (36). Reports of primary pulmonary hypertension (37) and valvular heart disease (38–40) associated with drug therapy prompted the withdrawal of fenfluramine and dexfenfluramine (Table 5) from the world markets in 1997 and resulted in one of the largest product liability lawsuits ever. In 1999 and 2000, American Home Products took a $12.25 billion charge against earnings to cover the costs of the settlement (41). Possible explanations for these rare side effects involving local serotonergic effects on the heart and lungs have been proposed (42), but the cause is still under investigation (43).

Sibutramine. Sibutramine (Meridia) is an inhibitor of both norepine-phrine and serotonin uptake with a much weaker effect on dopamine

Table 5. Mixed Sympathomimetic and Serotonergic Agents

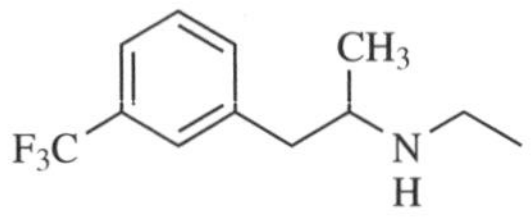

generic name: fenfluramine hydrochloride

N-ethyl-α-methyl-(3-trifluoromethyl)benzeneethanamine hydrochloride

formula: $C_{12}H_{16}F_3N$

brand name: Pondimin

dose: 75 mg 1×daily

CASRN: [458-24-2] (base)
CASRN: [404-82-0] (hydrochloride)

mol. wt.: 231.26 (base)
mol. wt.: 267.72 (hydrochloride)
manufacturer: American Home Products[a]

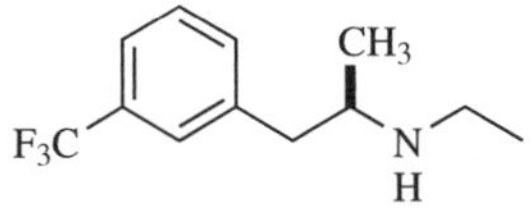

generic name: dexfenfluramine

formula: $C_{12}H_{16}F_3N$

brand name: Redux

(R)-N-ethyl-α-methyl-(3-trifluoromethyl)benzeneethanamine hydrochloride

CASRN: [37577-24-5] (base)
CASRN: [3616-78-2] (hydrochloride)

mol. wt.: 231.26 (base)
mol. wt.: 267.72 (hydrochloride)
manufacturer: American Home Products[a]

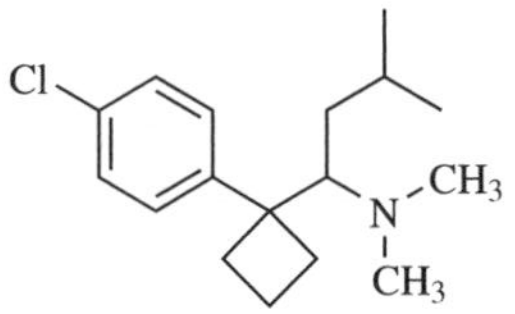

generic name: sibutramine hydrochloride monohydrate

1-(4-chlorophenyl)-N,N-dimethyl-α-(2-methylpropyl)cyclobutanemethanamine hydrochloride hydrate

formula: $C_{17}H_{28}ClN$

brand name: Meridia
dose: 10–15 mg, 1×daily

CASRN: [106650-56-0] (base)
CASRN: [12494-59-9] (hydrochloride hydrate)

mol. wt.: 279.86 (base)
mol. wt. 334.33 (hydrochloride hydrate)
manufacturer: Knoll Phamaceuticals

[a] Fenfluramine and Dexfenfluramine were withdrawn from the world wide markets in 1997.

reuptake (44,45). It affects appetite more consistently than selective serotonin reuptake inhibitors such as fluxetine, which are relatively ineffective in promoting weight loss (46). Results of feeding studies in rats treated simultaneously with sibutramine and various selective adrenergic and serontonin receptor antagonists are consistent with the reuptake inhibition mechanism and indicate that a combination of effects at adrenergic α_1 and β_1 and serotonergic 5-HT$_{2a/2c}$ receptors are responsible for its hypophagic activity (47). Sibutramine is also capable of blunting the homeostatic decrease in energy expenditure that normally accompanies weight loss (48). It was introduced in the U.S. market in 1998 after extensive clinical trials that demonstrated a modest effect on weight loss after treatments ranging from 2 months to 1 yr. Side effects were mainly increased blood pressure, dry mouth, and constipation. According to the summary provided in the *Physician*'s Desk Reference, ~60% of patients who initially responded to sibutramine treatment with a weight loss of 4 lb in the first month were able to achieve a 5% overall weight loss after 6 months (49).

The bioavailability of sibutramine is 77% in humans and it undergoes extensive first-pass metabolism to two primary demethylated metabolites, which are largely responsible for its biological activity. Thus, the parent drug has a clearance rate of 1760 L/h and a $t_{1/2}$ of 1.1 h, whereas the N-desmethyl metabolite, (7) and the N,N-didesmethyl metabolite, (8) have terminal elimination half-lives of 14 and 16 h, respectively. The metabolites reach a steady state after 4 days of dosing (48). Sibutramine is racemic. The major metabolites are derived from the (R)-isomer, as demonstrated recently by total synthesis and X-ray crystallographic analysis, and are more potent anorectic agents than the corresponding (S)-isomers (50,51).

There were 1.3 million prescriptions for sibutramine in the United States during 1999 with total sales amounting to $102 million, a decrease over the corresponding numbers for 1998, the year of launch (30).

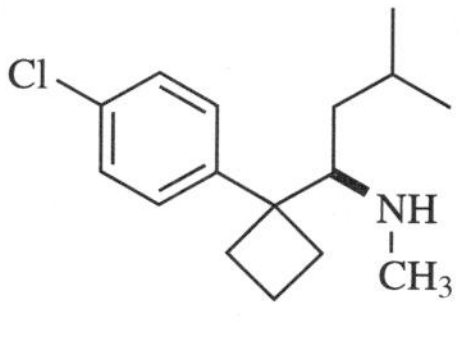

(7) 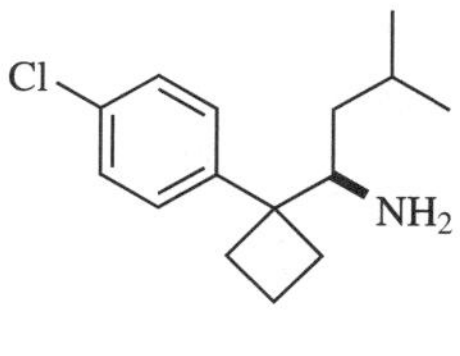   (8)

2.3. Lipid Adsorption Inhibitors. *Orlistat.* Dietary fat occurs largely in the form of triglycerides that must be hydrolyzed through the action of lipases, primarily pancreatic lipase, in the digestive tract prior to absorption. When maintenance or induction of weight loss is desired with a minimal impact on meal composition, an attractive approach to limiting caloric intake is to minimize the absorption of fat by inhibiting pancreatic lipase. Lipstatin is a pancreatic lipase inhibitor isolated from *Streptomyces toxytricini* (52,53). The corresponding tetrahydro derivative, orlistat (Xenical), has been shown to decrease fat absorption in mice, rats, and monkeys and to lower the rate of body weight gain in rats

Table 6. **Orlitstat**

	generic name: Orlistat CASRN: [96829-58-2]
	N-formyl-*L*-leucine (1*S*)-1-[[(2*S*,3*S*)-3-hexyl-4-oxo-2-oxetanyl)methyl]dodecyl ester
	formula: mol. wt.: 495.73
	$C_{29}H_{53}NO_5$
	brand name: manufacturer: Hoffmann-La Roche
	Xenical Inc
	dose: 120 mg
	3×daily at
	meal times

on a high calorie diet (54–56). Its complex structure, combining a δ-lactone with four stereocenters has inspired a number of total syntheses (57–63).

In humans, a 120-mg oral dose of orlistat prior to a meal inhibits 30% of fat absorption. When administered to patients for up to 1 yr in conjunction with a reduced-calorie diet, significantly more of drug-treated patients than placebo-treated controls achieved a 5–10% weight loss. Patients who achieved weight loss during 1 yr of therapy and were maintained on orlistat for a second year were able to maintain their weight loss in the absence of strict dietary control (62). In a study of adult-onset diabetic subjects, 1 yr of orlistat treatment led to the expected weight loss accompanied by improvements in fasting glucose, low density lipoprotein (LDL) levels, and HbA1c (64, 65).

The drug acts locally in the gastrointestinal (GI) tract to reversibly inhibit lipases through attack of a lipase serine hydroxyl group on the γ-lactone carbonyl to give an inactivated acyl enzyme. Less than 2% of an oral dose is absorbed, and systemic side effects are virtually unknown. The side effects, which do occur, are a consequence of its mechanism of action and relate to the presence of excess fat in the GI tract. These symptoms include occasional oily stools, flatus with discharge, and fecal urgency. These symptoms are generally mild and decrease in frequency with continued dosing.

Orlistat (Table 6) was introduced to the U.S. market in April 1999 and is the first marketed product to act at the level of fat absorption. Approximately 1.5 million prescriptions were written during its first 8 months on the market, and sales were $146 million (30).

3. Future Developments

As noted, there is a significant unmet medical need for better modalities for the management of overweight and obesity together with their consequent morbidity. The size of the patient population in developed countries, and the staggering direct and indirect costs, have prompted an intense effort on the part of academic as well as pharmaceutical company laboratories to understand the driving forces governing nutrient absorption, satiety, and energy utilization. A key stimulus for this work was the 1994 discovery of the protein leptin, a 167 amino acid hormone produced in mammalian fat tissue (66) and released to the general circulation as a function of the rate of glucose utilization (67). Leptin signaling is an important determinant of feeding behavior; mutations in the genes coding for leptin or its receptor in both mouse and hummans (68–70) lead to profound hyperphagia and obesity. Circulating leptin levels are proportional to overall adipose tissue mass but decrease markedly during fasting as a component of energy homeostasis. Research into the central pathways mediating leptin's role has substantially advanced our understanding of the mechanisms involved in appetite regulation.

3.1. Centrally Acting Drugs.   Concentrated in the arcuate nucleus in the hypothalamus, leptin receptors mediate both inhibitory and stimulatory signals, projecting into the lateral hypothalamic/perifornical areas and paraventricular nucleus, respectively, to inhibit food intake (71). Clinical trials of leptin itself as an anorectic agent were disappointing (72); however, investigation of neurons expressing leptin receptors has led to the identification of several potential drug targets. So far, these include neuropeptide Y (NPY), Agouti-related peptide (AGRP), α-melanocyte-stimulating hormone (α-MSH), melanin-concentrating hormone (MCH), and orexins A and B. All these are peptidic neurotransmitters that interact with specific G-protein-coupled receptors (GPCR) to either stimulate (NPY, AGRP, orexins, MCH) or inhibit (α-MSH) food intake (73). Since GPCRs are generally excellent targets for small molecular weight drugs, it is likely that ligands for each will be discovered and tested in the near future.

Regeneron Pharmceuticals has recently reported the results of a phase II clinical trial of Axokine, a modified form of ciliary neurotrophic factor in 170 severely obese patients. Patients on 1 μg/kg lost an average of 8.9% of their body weight over the 12 weeks of dosing. The mechanism of action of this drug is not known, although there is speculation that it may be acting in the central nervous system. Phase III studies are planned for 2001 (74).

The effectiveness of serotonergic agents such as fenfluramine and sibutramine has prompted further investigation into novel serotonergic drugs with an emphasis on the $5\text{-}HT_{2c}$ subclass of receptors (75) with the anticipation that such drugs will be more selective, hence safer.

3.2. Peripherally Acting Drugs.   Drugs that stimulate or block peripheral adrenergic receptors have been in widespread use for some time, e.g., as antihypertensives, antianginal agents, and bronchodilators. Detailed investigations with some of these led to the identification of the β_3-adrenergic receptor, which mediates catecholamine-induced lipolysis in brown adipose tissue leading to increases in thermogenesis. Compounds tested clinically to date were insufficiently selective for the human β_3-adrenergic receptor, but newer, more promising compounds are being evaluated (76,77).

The product of oxidative phosphorylation is adenosine triphosphate (ATP), which in turn is stoichiometrically coupled to various enzymatic pathways. The driving force for ATP synthesis is the oxidatively mediated transfer of protons across the inner mitochondrial membrane. In the late 1990s, uncoupling proteins (UCP-1, UCP-2, and UCP-3), were discovered that promote the leakage of protons across mitochondrial membranes, with consequent heat production (78, 79). One component of the thermogenic activity of β_3-agonists is their ability to promote increased expression of UCP-1 in brown adipose tissue. Other compounds that are capable of modulating the function of members of the uncoupling protein family could be of interest for their ability to increase energy expenditure.

Protein tyrosine phosphatase-1b (PTP-1b) has become an exciting target for drug discovery based on a paper describing a mouse model in which the gene coding for PTP-1b was knocked out. These animals were healthy and fertile, had increased insulin sensitivity, and were surprisingly resistant to diet-induced obesity (80,81). These findings have prompted an intense search within the pharmaceutical industry for a small-molecule PTP-1b inhibitor that might serve to effectively treat both type 2 diabetes and obesity.

Locally released gut neuropeptides such as cholecystokinin (CCK) and glucagon-like peptide 1 (GLP-1) regulate food intake, probably through stimulation of vagal afferent fibers. Although tolerance and local side effects have limited development of CCK agonists, Glaxo Smith Kline has one in phase II clinical trial and reports preliminary indications of efficacy. On the other hand, both analogues of GLP-1 and compounds that inhibit its hydrolysis by endopeptidases are being developed clinically, although the efficacy of such agents has not yet been established (82,83).

Finally, compounds that inhibit fat absorption or utilization may be of interest. In addition to gut lipases, targets of such drugs could include the intestinal fatty acid transport protein FATP4 (84); the enzyme acyl coenzyme (Co-A): diacylglycerol transferase (DGAT), which catalyzes a key step in triglyceride synthesis (85); and the enzyme complex fatty acid synthetase, which is responsible for the conversion of malonyl Co-A to palmitoyl coenzyme-A. Inhibitors of the latter complex have been shown to cause profound weight loss in mice, possibly mediated in part by inhibition of the central release of neuropeptide Y (86).

4. Summary

The available appetite suppressants based on stimulation of the sympathetic nervous system have fallen out of favor for the treatment of morbid obesity with the exception of sibutramine. The introduction of the fat absorption inhibitor orlistat has offered a novel approach, free of the side effects and tolerance development associated with present centrally acting agents. Although drug treatment is not likely to replace diet and behavior modification for the control of obesity, the profound medical need combined with intensive research on both peripheral and central pathways involved in regulation of nutrient absorption, meal size, and energy utilization is certain to lead to new opportunities for the control and maintenance of desirable body weight. Ultimately, the measure of success in the overall treatment of obesity will be a reduction in the associated morbidity.

BIBLIOGRAPHY

"Appetite-Suppressing Agents" in *ECT* 3rd ed., Vol. 3, pp. 184–193, by S. T. Ross (Smith, Kline & French Laboratories); "Antiobesity Drugs" in *ECT* 4th ed., Vol. 3, pp. 412–423, by Jefferson W. Tilley, Hoffmann-La Roche Inc., "Antiobesity Drugs" in *ECT* (online), posting date: December 4, 2000, by Jefferson W. Tilley, Hoffmann-La Roche Inc.

CITED PUBLICATIONS

1. P. G. Kopelman, *Nature (London)* **404**, 635–643 (2000).
2. R. J. Kuczmarski, M. D. Carroll, K. M. Flegal, and R. P. Troiano, *Obesity Res.* **5**, 542–548 (1997).
3. K. M. Flegal, M. D. Carroll, R. J. Kuczmarski, and C. L. Johnson, *Int. J. Obesity* **22**, 39–47 (1998).
4. D. B. Allison, K. R. Fontaine, J. E. Manson, J. Stevens, and T. B. Van Itallie, *J. Am. Med. Assoc.* **282**, 1530–1538 (1999).
5. The U.S. Weight Loss & Diet Control Market; 1999, Marketdata Enterprises, Inc., 2807 Busch Blvd., Suite 110, Tampa, Fla. 33618; A. M. Wolf, G. A. Colditz, *Pharm Econ* **5**, 34–37 (1994).
6. W. C. Willett, W. H. Dietz, and G. A. Colditz, *N. Engl. J. Med.* **341**, 427–434 (1999).
7. A. Must, J. Spadano, E. H. Coakley, A. E. Field, G. Colditz, and W. H. Dietz, *J. Am. Med. Assoc.* **282**, 1523–1529 (1999).
8. C. Bouchard and co-workers, *N. Engl. J. Med.* **322**, 1477–1482 (1990).
9. A. J. Stunkard, J. R. Harris, N. L. Pedersen, and G. E. McClearn *N. Engl. J. Med.*, **322**, 1483–1487 (1990).
10. A. G. Comuzzie and D. B. Allison, *Science* **280**, 1374–1377 (1998).
11. G. S. Barsh, S. Farooqi, and S. O'Rahilly, *Nature (London)* **404**, 644–651 (2000).
12. Expert Panel on the Identification, Evaluation and Treatment of Overweight in Adults, *Am. J. Clin. Nutr.* **68**, 899–917 (1998).
13. C. T. Montague and S. O'Rahilly, *Diabetes* **49**, 883–887 (2000).
14. M. E. J. Lean, T. S. Han, and J. C. Seidell, *Lancet* **351**, 853–856 (1998).
15. R. L. Leibel, M. Rosenbaum, and J. Hirsch, *N. Engl. J. Med.* **332**, 621–628 (1995).
16. M. W. Schwartz, D. G. Baskin, K. J. Kaiyala, and S. C. Woods, *Am. J. Clin. Nutr.* **69**, 584–596 (1999).
17. J. F. Munro, I. C. Stewart, P. H. Seidelin, H. S. MacKenzie, and N. G. Dewhurst, *Ann. N.Y. Acad. Sci.* **499**, 305–312 (1987).
18. B. B. Hoffman and R. J. Lefkowitz in J. G. Hardman, L. E. Limbird, P. B. Molinoff, R. W. Ruddon, and A. G. Gilman, ed, *Goodman & Gilman The Pharmacological Basis of Theraputics*, 9th ed., McGraw-Hill, New York, 1996, pp. 199–248.
19. M. Weiner in J. P. Morgan, D. V. Kagan, and J. S. Brody, ed., *Phenylpropanolamine: Risks, Benefits, and Controversies*, Greenwood Press, Westport, Conn., 1985, pp. 25–36.
20. J. P. Morgan in Ref. 19, pp. 180–194.
21. M. Weintraub in Ref. 19, pp 53–79.
22. H. Blumberg and J. P. Morgan, in Ref. 19, pp. 80–93.
23. T. Silverstone, *Drugs* **43**, 820–836 (1992).
24. S. G. Stolberg, *New York Times*, 1 (10/20/00,).
25. *Physician's* Desk Reference, 54th ed., Medical Economics Company, Inc., Montvale, N.J., 2000 pp. 1107–1108.
26. A. Balcioglu and R. J. Wurtman, *Int. J. Obes* **22**, 325–328 (1998).
27. I. H. Ulus, T. J. Maher, and R. J. Wurtman, *Biochem. Pharmacol.* **59**, 1611–1621 (2000).
28. M.-A. Mori, H. Uemura, M. Kobayashi, T. Miyahara, and H. Kozuka, *Xenobiotica* **23**, 709–716 (1993).
29. A. K. Cho, *Res. Comm. Chem. Path. Pharmacol.* **7**, 67–78 (1974).
30. Data courtesy of IMS Health, 660 West Germantown Pike, Plymouth Meeting, Pa. 19642.

31. N. Finer, D. Craddock, R. Lavielle, and H. Keen *Curr. Ther. Res.* **38**, 847–854 (1985).

32. G. Enzi, G. Crepaldi, E. M. Inelmen, R. Bruni, and B. Baggio *Clin. Neuropharmacol.* **11** Suppl. 1, S173–S178 (1988).

33. J. J. Wurtman and R. J. Wurtman *Curr. Med. Res. Opinion* **6** (Suppl. 1), 28–33 (1979).

34. J. E. Blundell, C. J. Latham, E. Moniz, R. A. McArthur, and P. J. Rogers *Curr. Med. Res. Opinion* **6** Suppl. 1, 34–52 (1979).

35. M. Weintraub, *Clin. Pharmacol. Therap.* **51**, 642–646 (1992).

36. J. P. Kassirer and M. Angell, *N. Engl. J. Med.* **338**, 52–54 (1998).

37. L. Abenhaim, Y. Moride, F. Brenot, S. Rich, J. Benichou, X. Kurz, T. Higenbottam, C. Oakley, E. Wouters, M. Aubier, G. Simonneau, and B. Begaud, *N. Engl. J. Med.* **335**, 609–616 (1996).

38. H. M. Connolly, J. L. Crary, M. D. McGoon, D. D. Hensrud, B. S. Edwards, W. D. Edwards, and H. V. Schaff, *N. Engl. J. Med.* **337**, 581–588 (1997).

39. J. M. Gardin, D. Schumacher, G. Constantine, K. Davis, C. Leung, and C. L. Reid, *J. Am. Med. Assoc.* **283**, 1703–1709 (2000).

40. M. K. Kancherla, H. I. Salti, T. A. Mulderink, M. Parker, R. O. Bonow, and D. J. Mehlamn, *Am. J. Cardiol.* **84**, 1335–1338 (1999).

41. *American Home Products Corporation*, Annual Report, 1999 p. 43.

42. P. J. Wellman and T. J. Maher, *Int. J. Obes.* **23**, 723–732 (1999).

43. R. B. Rothman, J. R. Redman, S. K. Raatz, C. A. Kwong, J. E. Swanson, and J. P. Bantle, *Am. J. Cardiol.* **85**, 913–915 (2000).

44. G. P. Luscombe, N. A. Slater, M. B. Lyons, R. D. Wynne, M. L. Scheinbaum, and W. R. Buckett, *Physchopharmacology* **100**, 345–349 (1990).

45. D. J. Heal, A. T. J. Frankland, J. Gosden, L. J. Hutchins, M. R. Prow, G. P. Luscombe, and W. R. Buckett, *Physchopharmacology* **107**, 303–309 (1992).

46. A. S. Ward, S. D. Comer, M. Haney, M. W. Fischman, and R. W. Foltin, *Physiol. Behav.* **66**, 815–821 (1999).

47. H. C. Jackson, M. C. Bearham, L. J. Hutchins, S. E. Mazurkiewicz, A. M. Needham, and D. J. Heal, *Br. J. Pharmacol.* **121**, 1613–1618 (1997).

48. D. L. Hansen, S. Toubro, M. J. Stock, I. A. Macdonald, A. Astrup, *Int. J. Obes.* **23**, 1016–1024 (1999).

49. *Physician*'s Desk Reference, 54th ed., Medical Economics Company, Inc., Montvale, N.J., 2000, pp. 1509–1513.

50. S. D. Glick, R. E. Haskew, I. M. Maisonneuve, J. N. Carlson, and T. P. Jerussi, *Eur. J. Pharmacol.* **397**, 93–102 (2000).

51. Q. K. Fang, C. H. Senanayake, Z. Han, C. Morency, P. Grover, R. E. Malone, H. Butler, S. A. Wald, and T. S. Cameron, *Tetrahedron Asymmetry* **10**, 4477–4480 (1999).

52. E. K. Weibel, P. Hadvary, E. Hochuli, E. Kupfer, and H. Lengsfeld *J. Antibiot.* **40**, 1081–1085 (1987).

53. E. Hochuli, E. Kupfer, R. Maurer, W. Meister, Y. Mercadal, and K. Schmidt *J. Antibiot.* **40**, 1085–1089 (1987).

54. S. Hogan, A. Fleury, P. Hadvary, H. Lengsfeld, M. K. Meier, J. Triscari, and A. C. Sullivan *Int. J. Obesity* **11** Suppl. 3, 35–42 (1987).

55. E. Fernandez and B. Borgström *Biochim. Biophys. Acta* **1001**, 249–255 (1989).

56. P. Hadvary, H. Lengsfeld, and H. Wolfer *Biochem. J.* **256**, 357–361 (1988).

57. P. Barbier and F. Schneider, *J. Org. Chem.* **53**, 1218–1221 (1988).

58. N. K. Chada, A. D. Batcho, P. C. Tang, L. F. Courtney, C. M. Cook, P. M. Wovkulich, and M. R. Uskokovic, *J. Org. Chem.* **56**, 4714–4718 (1991).

59. S. C. Case-Green, S. G. Davies, and C. J. R. Hedgecock, *Synlett* 781–782 (1991).

60. S. Hanessian and A. Tehim, *J. Org. Chem.* **58**, 7768–7781 (1993).
61. A. Pommier and J.-M. Pons, *Synthesis* 1294–1300 (1994).
62. P. J. Parsons and J. K. Cowell, *Synlett.* 107–109 (2000).
63. A. K. Ghosh and C. Liu, *Chem. Comm.* 1743–1744 (1999).
64. *Physician's* Desk Reference, 54th edn, Medical Economics Company, Inc., Montvale, N.J., 2000, pp. 2693–2696.
65. K. M. Hvizdos and A. Markham, *Drugs* **58**, 743–760 (1999).
66. Y. Zhang, R. Proenca, M. Maffel, M. Barone, L. Leopold, and J. M. Friedman, *Nature (London)* **372**, 425–432 (1994).
67. W. M. Mueller, and co-workers, *Endocrinology* **139**, 551–558 (1998).
68. A. Strobel, T. Issad, L. Camoin, M. Ozata, and A. D. Strosberg, *Nature Genet.* **18**, 213–215 (1998).
69. C. T. Montague and co-workers, *Nature (London)* **387**, 903–908 (1997).
70. K. Clement, C. Vaisse, N. Lahlou, S. Cabrol, V. Pelloux, D. Cassuto, M. Gourmelen, C. Dina, J. Chambaz, J.-M. Lacorte, A. Basdevant, P. Bougneres, Y. Lebouc, P. Froguel, and B. Guy-Grand, *Nature (London)* **392**, 398–401 (1998).
71. M. W. Schwartz, S. C. Woods, D. Porte, R. J. Steeley, and D. G. Baskin, *Nature (London)* **404**, 661–671 (2000).
72. S. B. Heymsfield, A. S. Greenberg, K. Fujioka, R. M. Dixon, R. Kushner, T. Hunt, J. A. Lubina, J. Patane, B. Self, P. Hunt, and M. McCamish, *J. Am. Med. Assoc.* **282**, 1568–1575 (1999).
73. J. Proietto, B. C. Fam, D. A. Ainslie, and A. W. Thorburn, *Exp. Opin. Invest. Drugs* **9**, 1317–1326 (2000).
74. Regeneron Pharmaceuticals company communication. See http://regeneron.com.
75. K. Nonogaki, A. M. Strack, M. Dallman, and L. H. Tecott, *Nature Med.* **4**, 1152–1156 (1998).
76. A. D. Strosberg and F. Pietri-Rouxel, *Trends Pharm. Sci.* **17**, 373–381 (1996).
77. C. P. Kordik and A. B. Reitz, *J. Med. Chem.* **42**, 181–201 (1999).
78. T. Gura, *Science* **280**, 1369–1370 (1998).
79. B. B. Lowell and B. M. Spiegelman, *Nature (London)* **404**, 652–660 (2000).
80. M. Elchebly, P. Payette, E. Michaliszyn, W. Cromlish, S. Collins, A. L. Loy, D. Normandin, A. Cheng, J. Himms-Hagen, C.-C. Chan, C. Ramachandran, M. J. Gresser, M. L. Tremblay, and B. P. Kennedy, *Science* **283**, 1544–1548 (1999).
81. B. P. Kennedy, C. Ramachandran, and C.-D. Pointe, *Biochem. Pharmacol.* **60**, 977–883 (2000).
82. M. A. Nauck, *Acta Diabetol* **35**, 117–129 (1998).
83. A. Flint, A. Raben, A. Astrup, and J. J. Holst, *J. Clin. Invest.* **101**, 515–520 (1998).
84. A. Stahl, D. J. Hirsch, R. E. Gimeno, S. Punreddy, P. Ge, N. Watson, S. Patel, M. Kotler, A. Raimondi, L. A. Tartaglia, and H. F. Lodish, *Mol. Cell.* **4**, 299–308 (1999).
85. S. J. Smith, S. Cases, D. R. Jensen, H. C. Chen, E. Sande, B. Tow, S. A. Sanan, J. Raber, R. H. Eckel, and R. V. Farese, *Nature. Genet.* **25**, 87–90 (2000).
86. T. M. Loftus, D. E. Jaworsky, G. L. Frehywot, C. A. Townsend, G. V. Ronnett, M. D. Lane, and F. P. Kuhajda, *Science* **288**, 2379–2381.

JEFFERSON W. TILLEY
Hoffmann-La Roche Inc.

ANTIOXIDANTS, POLYMERS

1. Introduction

Antioxidants are used to retard the reaction of organic materials, such as synthetic polymers, with atmospheric oxygen. Such reaction can cause degradation of the mechanical, aesthetic, and electrical properties of polymers; loss of flavor and development of rancidity in foods; and an increase in the viscosity, acidity, and formation of insolubles in lubricants. The need for antioxidants depends upon the chemical composition of the substrate and the conditions of exposure. Relatively high concentrations of antioxidants are used to stabilize polymers such as natural rubber and polyunsaturated oils. Saturated polymers have greater oxidative stability and require relatively low concentrations of stabilizers. Specialized antioxidants that have been commercialized meet the needs of the industry by extending the useful lives of the many substrates produced under anticipated conditions of exposure. In 2000, ~500 million pounds of antioxidants were sold in polymer applications with a value of $1.3 billion (1). On average, the growth rate of antioxidants is ~4%, roughly tracking the growth of the global polymers markets (2).

2. Mechanism of Uninhibited Autoxidation

The mechanism by which an organic material (RH) undergoes autoxidation involves a free-radical chain reaction is shown below (3–5):

Initiation

$$RH \longrightarrow \text{free radicals, eg, } R\cdot,\ ROO\cdot,\ RO\cdot,\ HO\cdot \tag{1}$$

$$ROOH \longrightarrow RO\cdot + OH\cdot \tag{2}$$

$$2\,ROOH \longrightarrow RO\cdot + ROO\cdot + H_2O \tag{3}$$

$$ROOR \longrightarrow 2\,RO\cdot \tag{4}$$

Propagation

$$R\cdot + O_2 \longrightarrow ROO\cdot \tag{5}$$

$$ROO\cdot + RH \longrightarrow ROOH + R\cdot \tag{6}$$

Termination

$$2\,R\cdot \longrightarrow R{-}R \tag{7}$$

$$ROO\cdot + R\cdot \longrightarrow ROOR \tag{8}$$

$$2\,ROO\cdot \longrightarrow \text{nonradical products} \tag{9}$$

2.1. Initiation. Free-radical initiators are produced by several processes. The high temperatures and shearing stresses required for compounding, extrusion, and molding of polymeric materials can produce alkyl radicals by homolytic chain cleavage. Oxidatively sensitive substrates can react directly with oxygen, particularly at elevated temperatures, to yield radicals.

It is virtually impossible to manufacture commercial polymers that do not contain traces of hydroperoxides. The peroxide bond is relatively weak and cleaves homolytically to yield radicals (eqs. 2 and 3). Once oxidation has started, the concentration of hydroperoxides becomes appreciable. The decomposition of hydroperoxides becomes the main source of radical initiators.

The absorption of ultraviolet (uv) light produces radicals by cleavage of hydroperoxides and carbonyl compounds (eqs. 10–12).

$$ROOH \xrightarrow{h\nu} RO\cdot + OH\cdot \tag{10}$$

$$\underset{\text{R--C--R}}{\overset{\text{O}}{\|}} \xrightarrow{h\nu} \underset{\text{R--C}\cdot + R\cdot}{\overset{\text{O}}{\|}} \tag{11}$$

$$\underset{/}{\overset{\backslash}{}}NOR + R'OO\cdot \longrightarrow \underset{/}{\overset{\backslash}{}}NO\cdot + R{=}O + R'OH \tag{12}$$

Most polymer degradation caused by the absorption of uv light results from radical-initiated autoxidation.

Direct reaction of oxygen with most organic materials to produce radicals (eq. 13) is very slow at moderate temperatures. Hydrogen-donating antioxidants (AH), particularly those with low oxidation–reduction potentials, can react with oxygen (eq. 14), especially at elevated temperatures (6).

$$RH + O_2 \longrightarrow R\cdot + HO_2\cdot \tag{13}$$

$$AH + O_2 \longrightarrow A\cdot + HO_2\cdot \tag{14}$$

2.2. Propagation. Propagation reactions (eqs. 5 and 6) can be repeated many times before termination by conversion of an alkyl or peroxy radical to a nonradical species (7). Homolytic decomposition of hydroperoxides produced by propagation reactions increases the rate of initiation by the production of radicals.

The reaction rate of molecular oxygen with alkyl radicals to form peroxy radicals (eq. 5) is much higher than the reaction rate of peroxy radicals with a hydrogen atom of the substrate (eq. 6). The rate of the latter depends on the dissociation energies (Table 1) and the steric accessibility of the various carbon–hydrogen bonds; it is an important factor in determining oxidative stability (8).

Polybutadiene and polyunsaturated fats, that contain allylic hydrogen atoms oxidize more readily than polypropylene, which contains tertiary hydrogen atoms. A linear hydrocarbon such as polyethylene, that has secondary hydrogens is the most stable of these substrates.

Autocatalysis. The oxidation rate at the start of aging is usually low and increases with time. Radicals, produced by the homolytic decomposition of

Table 1. **Dissociation Energies of Carbon–Hydrogen Bonds**[a]

R–H	D_{R-H} kJ/mol[b]	Bond type
$CH_2{=}CHCH_2{-}H$	356	allylic
$(CH_3)_3C{-}H$	381	tertiary
$(CH_3)_2CH{-}H$	395	secondary

[a] Ref. (8).
[b] To convert kJ to kcal, divide by 4.184.

hydroperoxides and peroxides (eqs. 2–4) accumulated during the propagation and termination steps, initiate new oxidative chain reactions, thereby increasing the oxidation rate.

Metal-Catalyzed Oxidation. Trace quantities of transition metal ions catalyze the decomposition of hydroperoxides to radical species and greatly accelerate the rate of oxidation. Most effective are those metal ions that undergo one-electron transfer reactions, eg, copper, iron, cobalt, and manganese ions (9). The metal catalyst is an active hydroperoxide decomposer in both its higher and its lower oxidation states. In the overall reaction, two molecules of hydroperoxide decompose to peroxy and alkoxy radicals (eq. 5).

$$ROOH + M^n \longrightarrow ROO\cdot + M^{(n-1)+} + H^+ \tag{15}$$

$$ROOH + M^{(n-1)+} \longrightarrow RO\cdot + M^{n+} + OH^- \tag{16}$$

$$2\,ROOH \longrightarrow ROO\cdot + RO\cdot + H_2O \tag{3}$$

Termination. The conversion of peroxy and alkyl radicals to nonradical species terminates the propagation reactions, thus decreasing the kinetic chain length. Termination reactions (eqs. 7 and 8) are significant when the oxygen concentration is very low, as in polymers with thick cross-sections where the oxidation rate is controlled by the diffusion of oxygen, or in a closed extruder. The combination of alkyl radicals (eq. 7) leads to cross-linking, which causes an undesirable increase in melt viscosity and molecular weight.

3. Radical Scavengers

Hydrogen-donating antioxidants (AH), such as hindered phenols and secondary aromatic amines, inhibit oxidation by competing with the organic substrate (RH) for peroxy radicals. This shortens the kinetic chain length of the propagation reactions.

$$ROO\cdot + AH \xrightarrow{k_{17}} ROOH + A\cdot \tag{17}$$

$$ROO\cdot + RH \xrightarrow{k_6} ROOH + R\cdot \tag{6}$$

Because k_{17} is $\gg k_6$, hydrogen-donating antioxidants generally can be used at low concentrations. The usual concentrations in saturated thermoplastic polymers range from 0.01 to 0.05%, based on the weight of the polymer. Higher concentrations, ie, ~0.5–2%, are required in substrates that are highly sensitive to oxidation, such as unsaturated elastomers and acrylonitrile–butadiene–styrene (ABS).

3.1. Hindered Phenols.   Even a simple monophenolic antioxidant, such as 2,6-di-*tert*-butyl-*p*-cresol [128-37-0] **(1)**, has a complex chemistry in an autooxidizing substrate as seen in Figure 1 (10).

Stilbenequinones such as **(5)** absorb visible light and cause some discoloration. However, upon oxidation phenolic antioxidants impart much less color than aromatic amine antioxidants and are considered to be nondiscoloring and nonstaining.

The effect substitution on the phenolic ring has on activity has been the subject of several studies (11–13). Hindering the phenolic hydroxyl group with at least one bulky alkyl group in the ortho position appears necessary for high

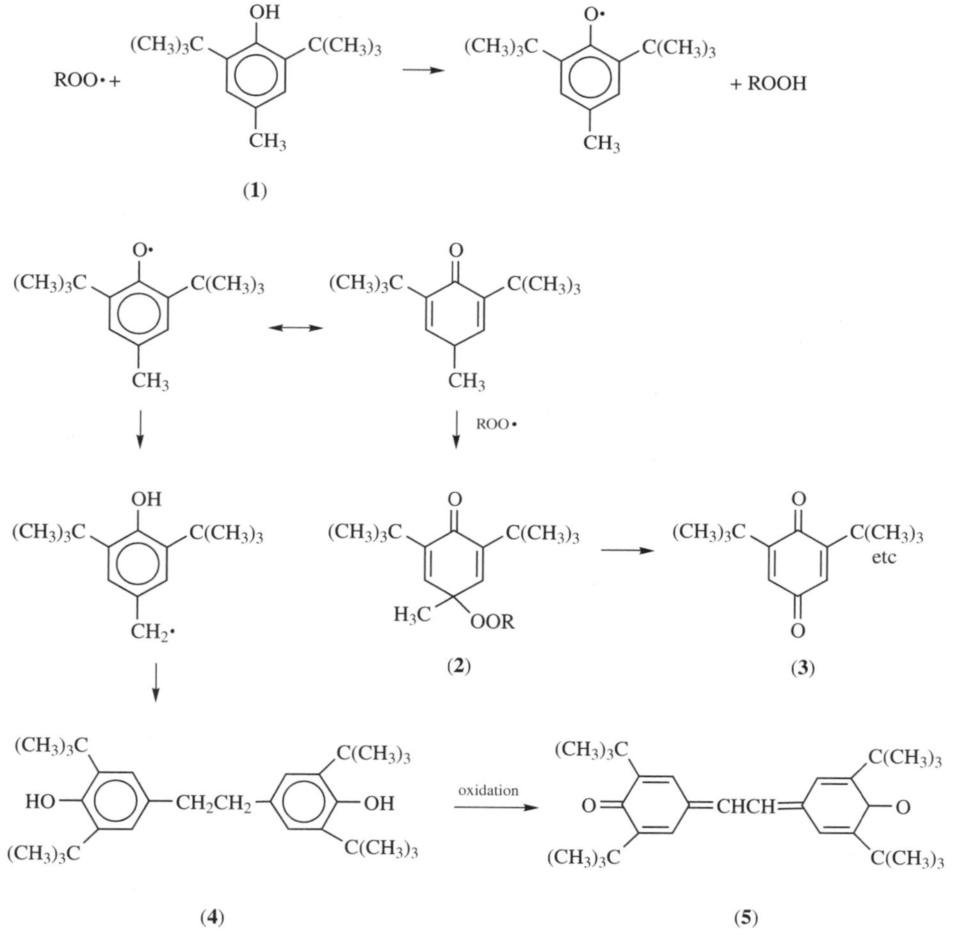

Fig. 1.   Chemical transformations of 2,6-di-*tert*-butyl-*p*-cresol in an oxidizing medium (10).

antioxidant activity. Nearly all commercial antioxidants are hindered in this manner. Steric hindrance decreases the ability of a phenoxyl radical to abstract a hydrogen atom from the substrate and thus produces an alkyl radical (14) capable of initiating oxidation (eq. 18).

$$A\cdot + RH \longrightarrow AH + R\cdot \tag{18}$$

Replacing a methyl with a tertiary alkyl group in the para position usually decreases antioxidant effectiveness. The formation of antioxidants such as (4) by dimerization is precluded because all benzylic hydrogen atoms are replaced by methyl groups. A strong electron-withdrawing group on the aromatic ring, like cyano or carboxy, decreases the ability of the phenol to donate its hydrogen atom to a peroxy radical of the substrate and reduces antioxidant effectiveness.

The usefulness of a hindered phenol for a specific application depends on its radical-trapping ability, its solubility in the substrate, and its volatility under test conditions. Table 2 shows the importance of volatility to stabilizer performance. Equimolar quantities of alkyl esters (6) of 3,5,di-*tert*-butyl-4-hydroxy-hydrocinnamic acid were evaluated in polypropylene (pp) at 140°C using two different procedures (15). When tested in an air stream, only the octadecyl ester (6), where $n = 18$, was effective in stabilizing the polymer. Under these conditions, the lower homologues were lost by volatilization. The oxygen-uptake test, carried out in a closed system that minimizes evaporative loss, showed that homologues were effective to varying degrees. The differences in effectiveness can probably be attributed to differences in the solubility of various homologues in the amorphous phase of the pp. When dodecane, a liquid in which all the compounds are soluble, was used as a substrate instead of pp, the antioxidant activities were relatively close.

Table 2. Influence of Antioxidant (6) Volatility on Effectiveness at 140°C

		Time to failure in PP, h		Time to failure in dodecane, h
n^a	$t_{1/2}$, h^b	O$_2$ Uptake	Air stream	
1	0.28	95	2	25
6	3.60	312	2	23
12	83.0	420	2	20
18	660.0	200	165	20

^{a} n The letter the number of carbon atom in the ester chain of (6).

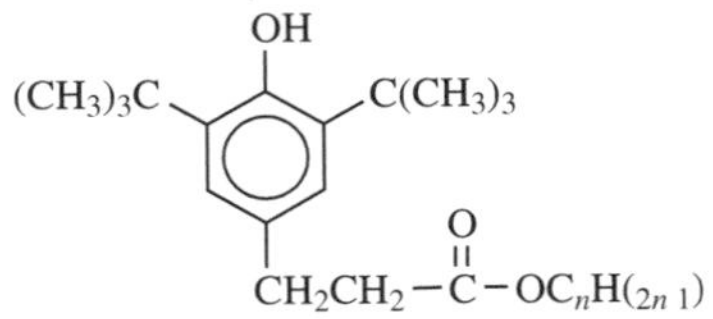

(6)

^{b} Antioxidant half-life in pp exposed to a nitrogen stream at 140°C.

Introducing long aliphatic chains into a stabilizer molecule decreases volatility and increases solubility in hydrocarbon polymers, which improves performance. However, it also increases the equivalent weight of the active moiety. Di-, tri-, and polyphenolic antioxidants combine relatively low equivalent weights with low volatility. Commercially important di-, tri-, and polyphenolic stabilizers include 2,2′-methylenebis(6-*tert*-butyl-*p*-cresol) [85-60-9] (**7**), 1,3,5-trimethyl-2,4,6-tris(3′5′-di-*tert*-butyl-4-hydroxybenzyl)benzene [1709-70-2] (**8**), and tetrakis[methylene(3,5-di-*tert*-butyl-4-hydroxyhydrocinnamate)]methane [6683-19-8] (**9**).

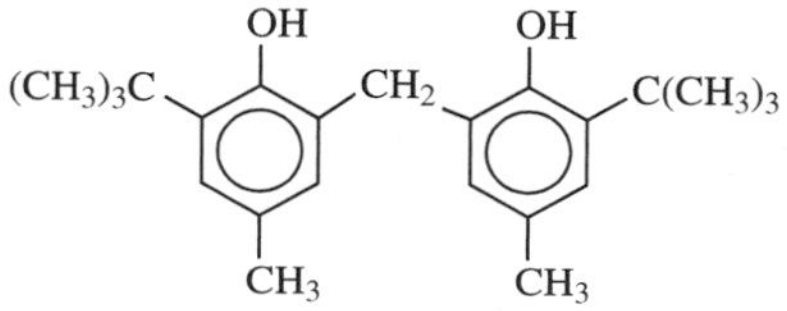

(7)

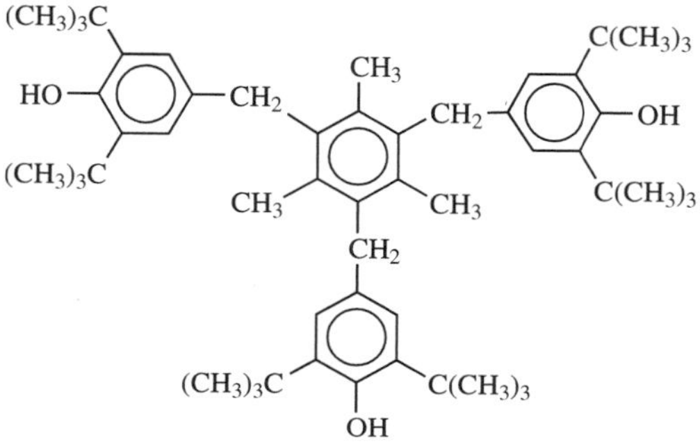

(8)

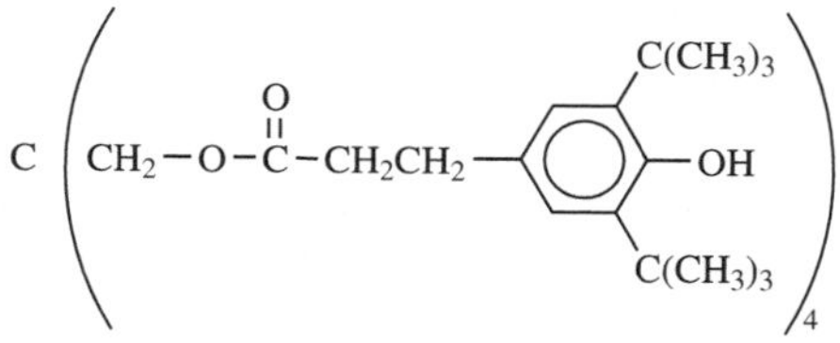

(9)

3.2. Aromatic Amines. Antioxidants derived from *p*-phenylenediamine and diphenylamine are highly effective peroxy radical scavengers. They are more effective than phenolic antioxidants for the stabilization of easily oxidized organic materials, such as unsaturated elastomers. Because of their intense staining effect, derivatives of *p*-phenylenediamine are used primarily for elastomers containing carbon black (qv).

N,N′-Disubstituted-*p*-phenylenediamines, such as *N*-phenyl-*N′*-(1,3-dimethylbutyl)-*p*-phenylenediamine [793-24-8] (**10**), are used in greater quantities than other classes of antioxidants. These products protect unsaturated elastomers against oxidation as well as ozone degradation (see Rubber Chemicals).

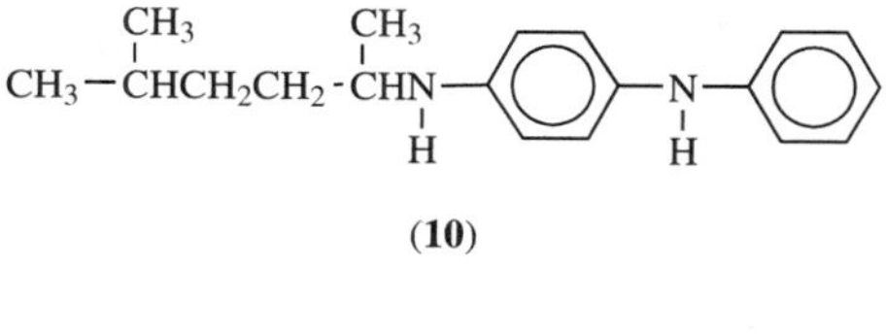

(10)

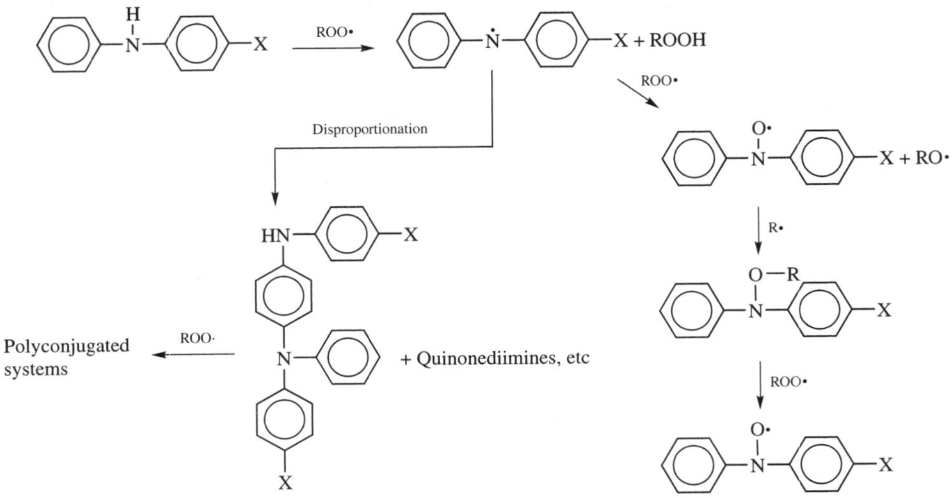

Fig. 2. Oxidation of aromatic amine antioxidants (**10**).

Low concentrations of alkylated paraphenylenediamines such as N,N'-di-sec-butyl-p-phenylenediamine [69796-47-0] N,N'-Di-sec-butyl-p-phenylenediamine [69796-47-0] are added to gasoline to inhibit oxidation.

Figure 2 shows some of the reactions of aromatic amines that contribute to their activity as antioxidants and to their tendency to form highly colored polyconjugated systems.

Alkylated diphenylamines (**11**) and derivatives of both dihydroquinoline (**12**) and polymerized 2,2,4-trimethyl-1,2-dihydroquinoline [26780-96-1] (**13**) develop less color than the p-phenylenediamines and are classified as semistaining antioxidants. Derivatives of dihydroquinoline are used for the stabilization of animal feed and spices.

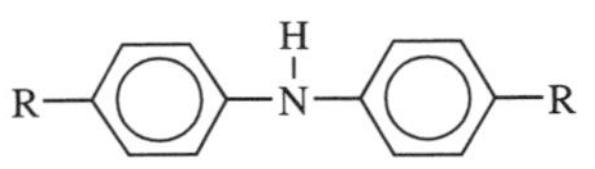

(**11**) 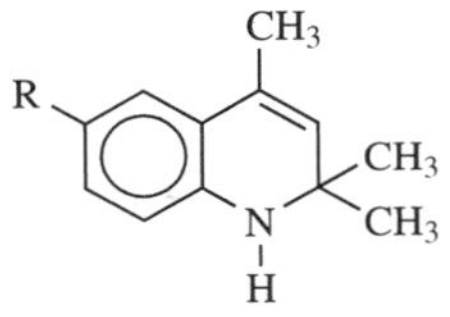   (**12**)

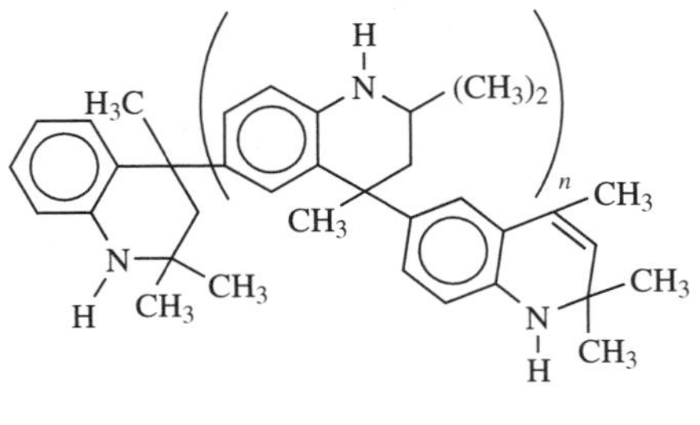

(**13**)

where $n = 0–6$

4,4′-Bis(α,α-dimethylbenzyl)diphenylamine [1008-67-1] (**14**) has only a slight tendency to stain and has been recommended for use in plastics as well as elastomers.

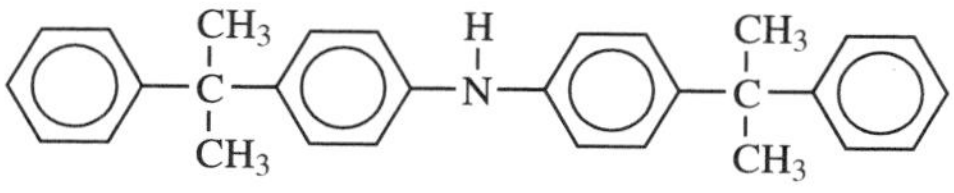

(**14**)

3.3. Hindered Amines.   Hindered amines are extremely effective in protecting polyolefins and other polymeric materials against photodegradation. They usually are classified as light stabilizers rather than antioxidants.

Most of the commercial hindered-amine light stabilizers (HALS) are derivatives of 2,2,6,6-tetramethylpiperidine (**15**) [768-66-1] (**16**).

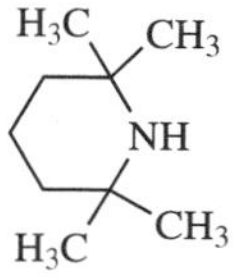

(**15**)

These stabilizers function as light-stable antioxidants to protect polymers. Their antioxidant activity is explained by the following sequence (**17**):

$$\text{\textbackslash NH} \xrightarrow{\text{radicals } O_2} \text{\textbackslash NO·} \qquad (19)$$

$$\text{\textbackslash NO· + R·} \longrightarrow \text{\textbackslash NOR} \qquad (20)$$

$$\text{\textbackslash NOR + R′OO·} \longrightarrow \text{\textbackslash NO· + R=O + R′OH} \qquad (21)$$

According to this mechanism, hindered-amine derivatives terminate propagating reactions (eqs. 5 and 6) by trapping both the alkyl and peroxy radicals. In effect, NO competes with O_2, and NOR competes with RH. Since the nitroxyl radicals are not consumed in the overall reactions, they are effective at low concentrations.

3.4. Hydroxylamines.   A relatively new stabilizer chemistry, commercially introduced in 1996 (**18**), based on the hydroxylamine functionality, can serve as a very powerful hydrogen atom donor and free-radical scavenger (**19**), as illustrated in Figure 3.

This hydroxylamine chemistry is extremely powerful on an equivalent weight basis in comparison to conventional phenolic antioxidants and phosphite melt-processing stabilizers. However, in terms of it's free-radical scavenger capability, however, it is more effective during melt processing of the polymer, and not during long term thermal stability (ie, below the melting point of the polymer). This finding is quite interesting based on the similarity between the

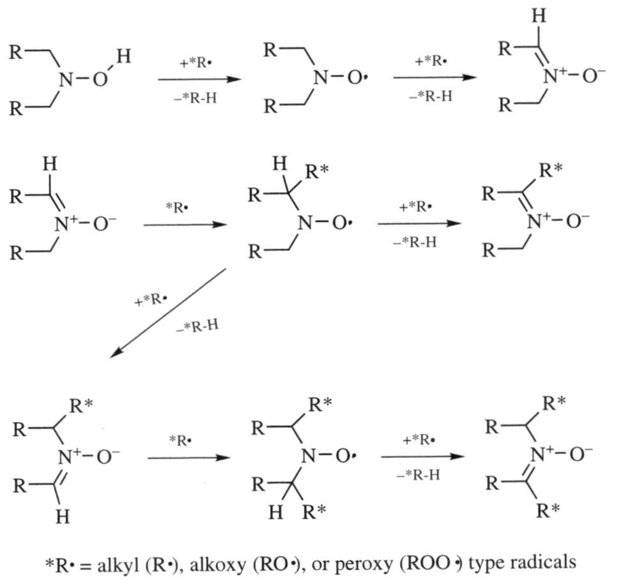

*R• = alkyl (R•), alkoxy (RO•), or peroxy (ROO•) type radicals

Fig. 3. Free-radical decomposition mechanism for hydroxylamines.

type of free-radical scavenging chemistry that hydroxylamines and phenols are both capable of providing. However, the temperature range is different, which is discussed further below.

The only commercial hydroxylamine product used in polyolefins and other selected polymers is a product by process based on the oxidation of bis tallow amine [14325-92-2].

There is a similar commercially available product based on similar chemistry. It is a product by process based on the oxidation of methyl-bis-tallow amine [204933-93-7]. The oxidation product of the methyl-bis-tallow amine is a tri-alkylamineoxide, a precursor to hydroxylamine stabilization chemistry. The trialkylamineoxide is converted to a hydroxylamine during the initial melt compounding of the polymer via a Cope elimination reaction, as shown below in equation 22.

$$R_2(CH_3)N^+ - O^- + heat \longrightarrow R(CH_3)N{-}OH + R{-}CH{=}CH_2 \qquad (22)$$

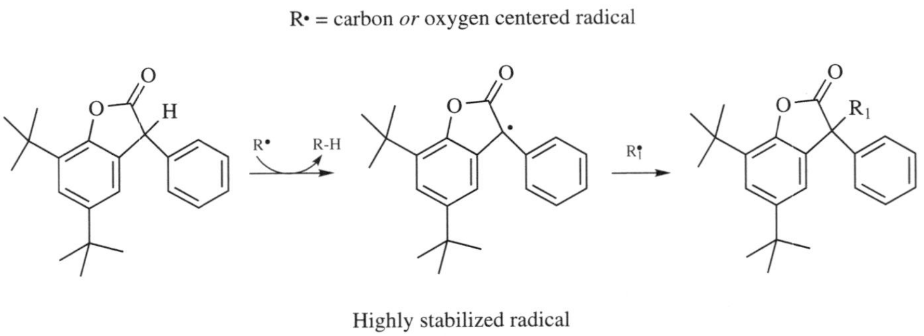

R• = carbon *or* oxygen centered radical

Highly stabilized radical

Fig. 4. Proposed stabilization mechanism of arylbenzofuranones.

3.5. Benzofuranones. In 1997, a fundamentally new chemistry was introduced, that not only inhibits the autoxidation cycle, but attempts to shut it down as soon as it starts (20). The exceptional stabilizer activity of the class of benzofuranones is due to the ready formation of a stable benzofuranyl radical by donation of the weakly bonded benzylic hydrogen atom; see Figure 4.

The resonance stabilized benzofuranyl (lactone) radicals can either reversibly dimerize or react with other free radicals. Model experiments have demonstrated that this class of chemistry behaves as a powerful hydrogen atom donor and are also effective scavengers of carbon centered and oxygen centered free radicals (21); (see Figure 5).

While the sterically hindered phenols react preferentially with oxygen-centered radicals like peroxy and alkoxy rather than with carbon-centered radicals, benzofuranones can scavenge both types of radicals. Accordingly, a benzofuranone, can be repeatedly positioned at key locations around autoxidation cycle to inhibit the proliferation of free radicals. In addition, the scavenging of carbon-centered radicals is representative of a mode of stabilization that is more like "preventive maintenance", in comparison to more traditional stabilizers such as phenols and phosphites, which operate in something more like a "damage control" mode.

Benzofuranones are similar to hydroxylamines in that on an equivalent weight basis, they are more powerful than conventional phenolic antioxidants or phosphite-based melt-processing stabilizers. Once again, note that even though benzofuranones are capable of providing free-radical scavenging chemistry similar to phenolic antioxidants, the effective temperature domain is typically above the melting point of the polymer; eg, during melt processing (similar to hydroxylamines). This will be discussed further below.

4. Peroxide Decomposers

Thermally induced homolytic decomposition of peroxides and hydroperoxides to free radicals (eqs. 2–4) increases the rate of oxidation. Decomposition to nonradical species removes hydroperoxides as potential sources of oxidation initiators. Most peroxide decomposers are derived from divalent sulfur and trivalent phosphorus.

4.1. Divalent Sulfur Derivatives. A dialkyl ester of thiodipropionic acid (**16**) is capable of decomposing at least 20 mol of hydroperoxide (22). Some of the reactions contributing to the antioxidant activity of these compounds are shown in Figure 6 (23).

According to Figure 6, hydroperoxides are reduced to alcohols, and the sulfide group is oxidized to protonic and Lewis acids by a series of stoichiometric reactions. The sulfinic acid (**21**), sulfonic acid (**23**), sulfur trioxide, and sulfuric acid are capable of catalyzing the decomposition of hydroperoxides to nonradical species.

When used alone at low temperatures, dialkyl thiodipropionates are rather weak antioxidants. However, synergistic mixtures with hindered phenols are highly effective at elevated temperatures and are used extensively to stabilize polyolefins, ABS, impact polystyrene (IPS), and other plastics.

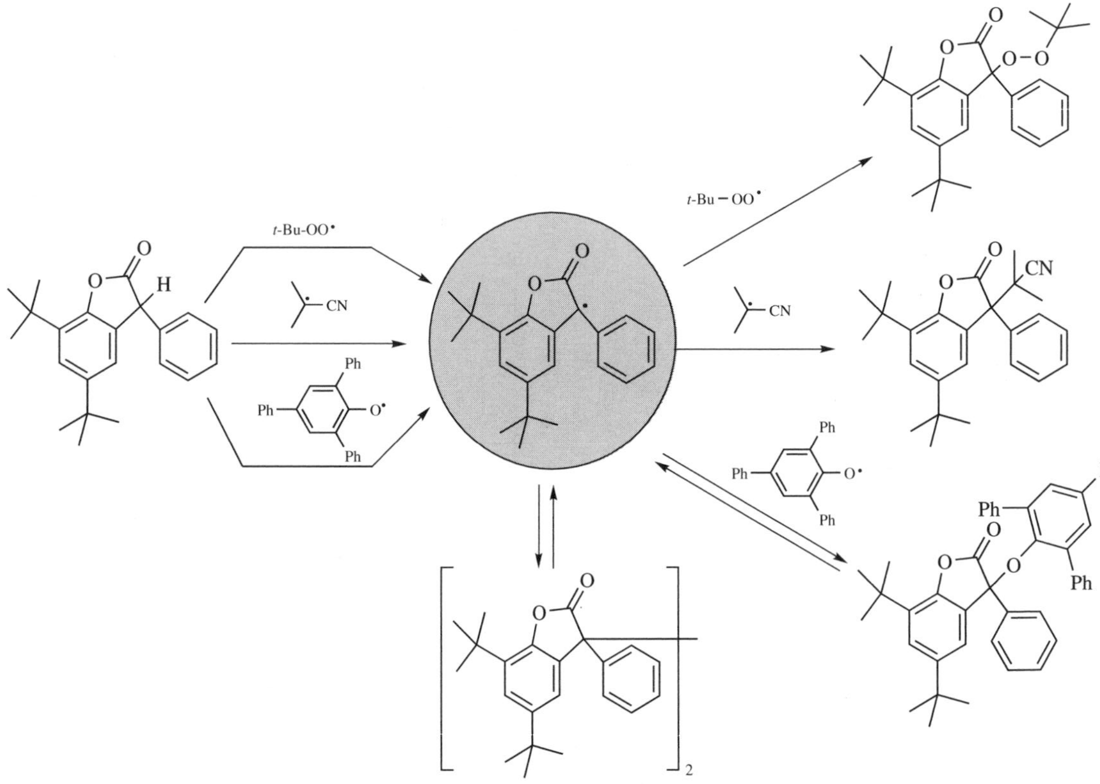

Fig. 5. Carbon-centered free-radical trapping reactions with benzofuranones.

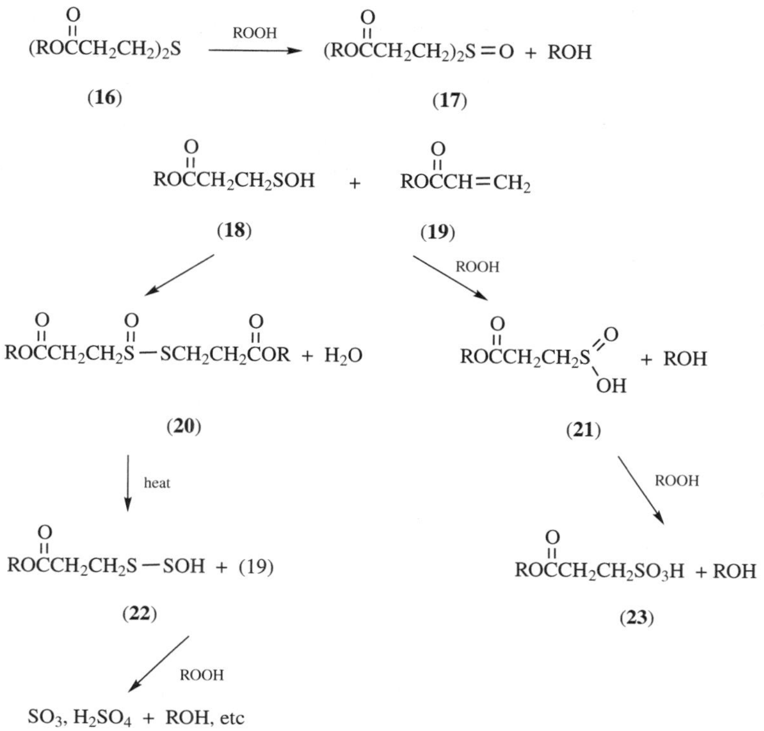

$$\text{(ROCCH}_2\text{CH}_2)_2\text{S} \quad \xrightarrow{\text{ROOH}} \quad \text{(ROCCH}_2\text{CH}_2)_2\text{S}=\text{O} + \text{ROH}$$

$$\text{(16)} \qquad\qquad\qquad\qquad \text{(17)}$$

$$\text{ROCCH}_2\text{CH}_2\text{SOH} \quad + \quad \text{ROCCH}=\text{CH}_2$$

$$\text{(18)} \qquad\qquad\qquad \text{(19)}$$

ROOH

$$\text{ROCCH}_2\text{CH}_2\text{S}-\text{SCH}_2\text{CH}_2\text{COR} + \text{H}_2\text{O} \qquad \text{ROCCH}_2\text{CH}_2\text{S}\!\!\stackrel{=\!\text{O}}{\diagdown}_{\text{OH}} + \text{ROH}$$

$$\text{(20)} \qquad\qquad\qquad\qquad\qquad \text{(21)}$$

heat

$$\text{ROCCH}_2\text{CH}_2\text{S}-\text{SOH} + \text{(19)} \qquad\qquad \text{ROCCH}_2\text{CH}_2\text{SO}_3\text{H} + \text{ROH}$$

$$\text{(22)} \qquad\qquad\qquad\qquad\qquad \text{(23)}$$

ROOH

$$\text{SO}_3, \text{H}_2\text{SO}_4 + \text{ROH, etc}$$

Fig. 6. Decomposition of hydroperoxides by esters of thiodipropionic acid.

Esters of thiopropionic acid tend to decompose at high processing temperatures, and their odor makes them unsuitable for some food-packaging applications.

4.2. Trivalent Phosphorus Compounds. Trivalent phosphorus compounds reduce hydroperoxide to alcohols:

$$\text{ROOH} + \text{PR}_3' \longrightarrow \text{ROH} + \text{O}=\text{PR}_3' \qquad\qquad (23)$$

These compounds are used most frequently in combination with hindered phenols for a broad range of applications in rubber and plastics. They are also able to suppress color development caused by oxidation of the substrate and the phenolic antioxidant. Unlike phenols and secondary aromatic amines, phosphorus-based stabilizers generally do not develop colored oxidation products.

Esters of phosphorous acid derived from aliphatic alcohols and unhindered phenols, eg, tris(nonylphenyl)phosphate (**24**), hydrolyze readily and special care must be taken to minimize decomposition by exposure to water or high humidity. The phosphorous acid formed by hydrolysis is corrosive to processing equipment, particularly at high temperatures.

The hydrolysis of phosphites is retarded by the addition of a small amount of a base such as triethanolamine. A more effective approach is the use of hindered phenols for esterification. Relatively good resistance to hydrolysis is shown by two esters derived from hindered phenols: tris(2,4-di-*tert*-butylphenyl)-phosphite [31570-04-4] (**25**) and tetrakis(2,4-di-*tert*-butylphenyl) 4,4'-biphenyle-nediphosphonite [38613-77-3] (26). A substantial research effort over the last decade to develop hydrolytically stable phosphites while retaining the excellent hydroperoxide decomposing activity has resulted in the introduction of a number of new commercial products such as the dicumyl phosphite [154862-43-8] (**27**).

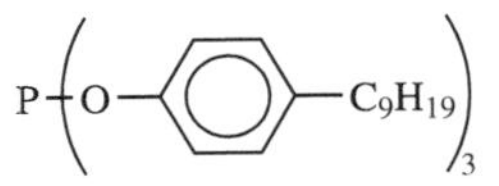

(**24**)

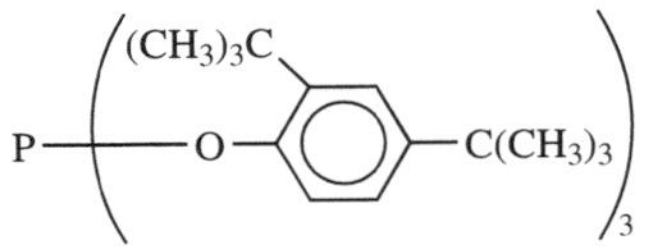

(**25**)

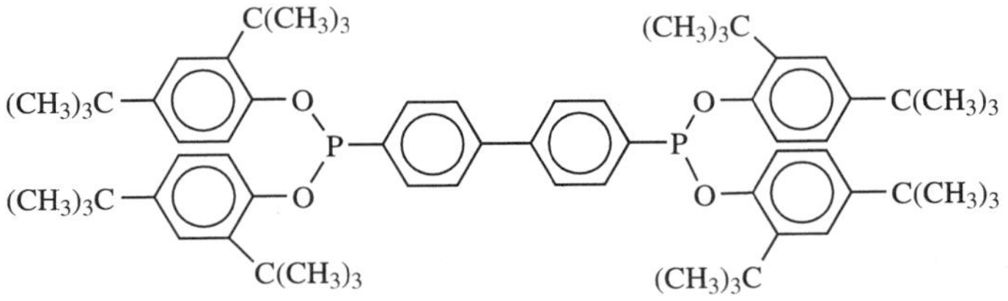

(**26**)

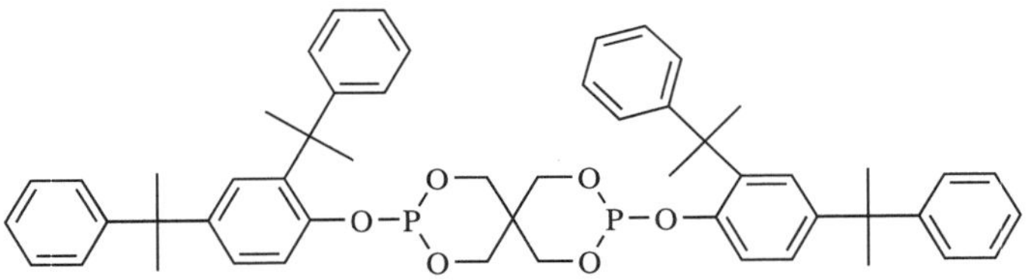

(**27**)

4.3. Hydroxylamines. As mentioned above, hydroxylamines are very effective as free radical scavengers. They are also noted for their ability to decompose hydroperoxides (24); shown in equation 24 and illustrated in Figure 7.

$$R_2N - OH + R^*OOH \rightarrow RR{-}CH - N^+{-}O^- + R^*OH + H_2O \qquad (24)$$

Hydroxylamines serve as a sequential source of hydrogen atoms, reducing hydroperoxide to their corresponding alcohol. In the course of this reaction, the hydroxylamine is converted to a nitrone.

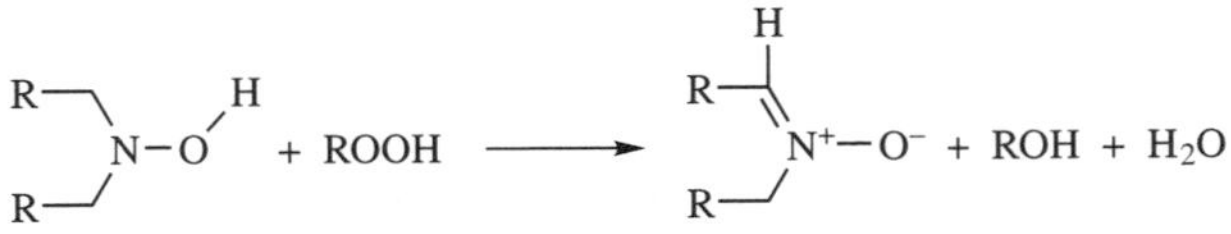

Fig. 7. Hydroperoxide decomposition mechanism for hydroxylamines.

5. Metal Deactivators

The ability of metal ions to catalyze oxidation can be inhibited by metal de-activators (25). These additives chelate metal ions and increase the potential difference between their oxidized and reduced states, which decreases the ability of the metal to produce radicals from hydroperoxides by oxidation and reduction (eqs. 15 and 16). Complexation of the metal by the metal deactivator also blocks its ability to associate with a hydroperoxide, a requirement for catalysis (26).

Examples of commercial metal deactivators used in polymers areoxalyl bis (benzylidene)hydrazide [6629-10-3] (**28**), *N,N'*-bis(3,5-di-*tert*-butyl-4-hydroxy hydrocinnamoylhydrazine [32687-78-8] (**29**), 2,2'-oxamidobis-ethyl(3,5-di-*tert*-butyl-4-hydroxyhydrocinnamate) [70331-94-1] (**30**), *N,N'*-(disalicylidene)-1,2-propanediamine [94-91-7] (**31**), and ethylenediaminetetra-acetic acid [60-00-4] (**32**) and its salts and critic acid (**33**) (Fig. 8).

6. Effect of Temperature

As mentioned above, certain types of antioxidants provide free-radical scaveng-ing capability; albeit over different temperature ranges. Figure 9, illustrates this in a general fashion for representative classes of stabilizers, over the tempera-ture range of 0–300°C.

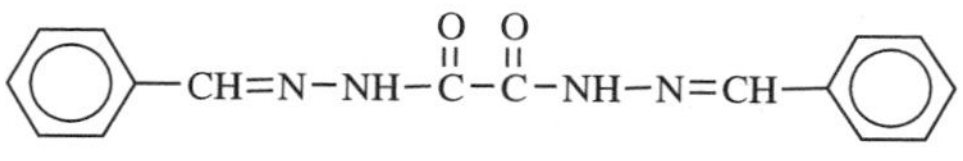

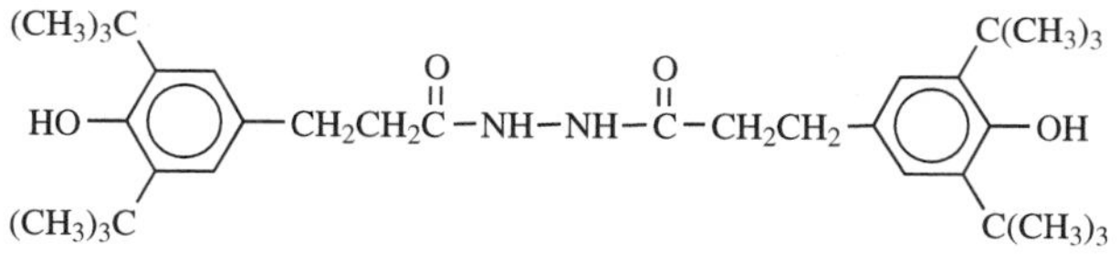

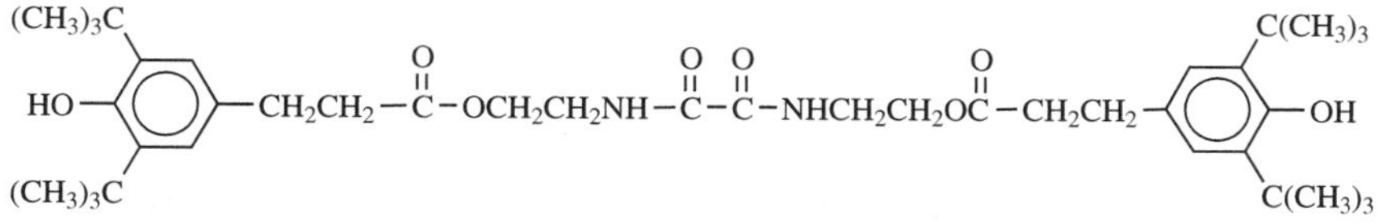

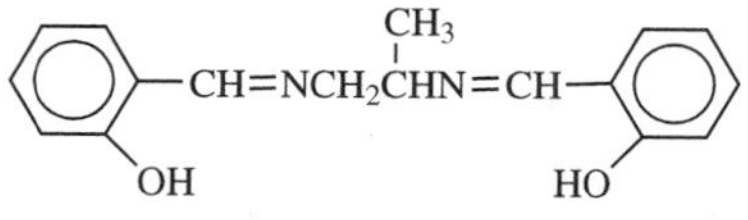

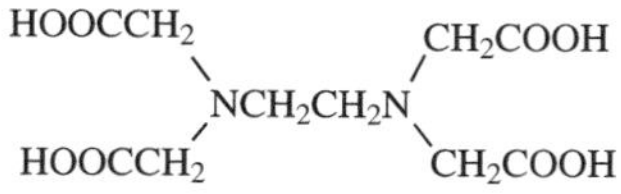

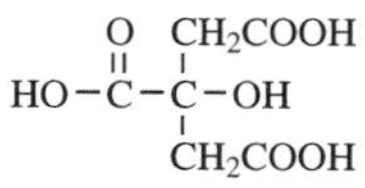

Fig. 8. Commercial metal deactivators.

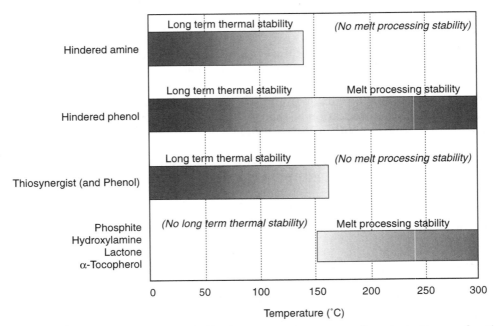

Fig. 9. General representation of effective temperature ranges for selected types of anti-oxidants.

As a representative example, hindered phenols are capable of providing long term thermal stability below the melting point of the polymer, as well as melt-processing stability above the melting point of the polymer. As such most (if not all) hindered phenols are useful across the entire temperature range.

Thiosynergists, in combination with a hindered phenol, contribute to long term thermal stability, primarily below the melting point of the polymer. In extreme cases, where peroxides have built up in the polymer, thiosynergist can be shown to have a positive impact during melt processing. This finding, however, is not the norm, and this type of melt processing efficacy has been left out of the figure.

Hindered amines, commonly thought of as being useful for uv stabilization, are also useful for long term thermal stability below the melting point of the polymer. This effectiveness is due the fact that hindered amines work by a free-radical scavenging mechanism, but they are virtually ineffective at temperatures >150°C. Therefore, hindered amines, when used as a reagent for providing long term thermal stability, should always be used in combination with an effective melt-processing stabilizer.

Phosphites, hydroxylamines, and lactones, are most effective during melt processing; either through free-radical scavenging or hydroperoxide decomposition. They are not effective as long term thermal stabilizers. These type of stabilizers also help with long term thermal stability. By sacrificing themselves during melt processing, they lessen the workload on the phenolic antioxidant, allowing more to remain intact to help with long term thermal stability.

One anomaly that should be pointed out are the hindered phenols based on tocopherols. Even though tocopherols, such as Vitamin E [10191-41-01], fall into the general class of hindered phenols, they behave more as melt processing stabilizers, and less as reagents for providing long term thermal stability, at least with regard to polymer stabilization.

7. Antioxidant Blends

In practical application, it is reasonable to use more than one type of antioxidant in order to meet the requirements of the application, such as melt-processing stability as well as long term thermal stability. The most common combination of stabilizers used, particularly in polyolefins, are blends of a phenolic antioxidant and a phosphite melt-processing stabilizer. Another common combination is a blend of a phenolic antioxidant and a thioester; especially for applications that require long term thermal stability. These common phenol-based blends have been used successfully in many different types of end-use applications. The combination of phenolic, phosphite, and lactone moieties represents an extremely efficient stabilization system since all three components provide a specific function.

For color critical applications requiring "phenol free" stabilization, synergistic mixtures of hindered amines (for both uv stability as well as long term thermal stability) with a hydroxylamine or benzofuranone (for melt processing), with or without a phosphite, can be used to avoid discoloration typically associated with the overoxidation of the phenolic antioxidant(27). The use of "phenol free" stabilization systems is very effective in color critical products such as polyolefin films and fibers as well as selcted thermoplastic polyolefin (TPO) applications.

7.1. Synergism Mixtures of Antioxidants. A mixture of antioxidants that function by different mechanisms might be synergistic and provide a higher degree of protection than the sum of the stabilizing activities of each component. The most frequently used synergistic mixtures are combinations of radical scavengers and hydroperoxide decomposers.

Typically, blend titration experiments are performed at a set loading of additives, starting with 100% of component A and 0% of component B. A series of formulations are designed to shift to the other extreme with 0% component A and 100% component B. By measuring a series of performance parameters, the optimum ratio of A to B can be determined. This type of work is time consuming, but in the end, the optimum ratio can be identified with real data. If three or more components are being assessed at the same time, statistically designed experiments are often useful in terms of sorting out the data.

7.2. Antagonistic Mixtures of Antioxidants. Mixtures of antioxidants can also work against each other. Chemistries that interfere with each other may not necessarily be obvious until the evidence is presented. For example, to ensure long term thermal stability and good light stability, one might use a combination of a phenolic antioxidant and a divalent sulfur compound for thermal stability and a hindered amine for light stability.

Unfortunately, the oxidation products of the sulfur compound can be quite acidic and can complex the hindered amine as a salt, preventing the hindered amine from entering into its free-radical scavenging cycle. This antagonism has been generally known for quite a while (28) and has recently been discussed (29).

Other types of antagonistic chemistry often involve relatively strong acids or bases (either Brønstead or Lewis) that can interact with the antioxidants in such a way as to divert them into transformation chemistries that have nothing to do with polymer stabilization.

8. Application of Antioxidants in Polymers

Nearly all polymeric materials require the addition of antioxidants to retain physical properties and to ensure an adequate service life. The selection of an antioxidant system is dependent upon the polymer and the anticipated end use.

8.1. Polyolefins. Low concentrations of stabilizers ($< 0.01\%$) are often added to polyethylene and polypropylene after synthesis and prior to isolation to retard oxidation of the polymer before they are exposed to sources of oxygen or air. Higher concentrations are added downstream during the conversion of the reactor product to a pelletized form. The antioxidant components and concentrations are selected by the manufacturer to yield general purpose grades, or can be optimized to meet a specific end-use application.

In downstream applications, these polymers can be subjected to temperatures as high as 300°C, during cast film extrusion and thin wall injection molding. In these type of demanding applications, processing stabilizers are used to decrease both the change in viscosity (molecular weight) of the polymer melt and the development of color. A phosphite, such as tris(2,4-di-*tert*-butylphenyl)phosphite (**25**) or bis(2,4-di-*tert*-butylphenyl)pentaerythritol diphosphite [26741-53-7], in combination with a phenolic antioxidant such as octadecyl 3,5-di-*tert*-butyl-4-hydroxyhydrocinnamate (**6**), may be used. Concentrations usually range from 0.01 to 0.5% depending on the polymer and the severity of the processing conditions. For long term exposure, a persistent antioxidant like tetrakis[methylene(3,5-di-*tert*-butyl-4-hydroxyhydrocinnamate)]methane (**9**), at a concentration of 0.1–0.5%, may be added to the base stabilization package.

A sulfur-containing synergistic mixture can be used to obtain an extended service life at a decreased cost.

The synergistic effect of a hydroperoxide decomposer, eg, dilauryl thiodipropionate [123-28-4] (**34**), and a radical scavenger, eg, tetrakis[methylene(3,5-di-*tert*-butyl-4-hydroxyhydrocinnamate)]methane (**9**), in protecting polypropylene during an oxygen-uptake test at 140°C is shown in Table 3.

$$S(CH_2CH_2-\overset{\overset{\textstyle O}{\|}}{C}-OC_{12}H_{25})_2$$

(**34**)

The sum of the individual activities of these antioxidants was 20 days, whereas a mixture of the two stabilizers protected the polymer for 45 days (30).

Table 3. **Synergism between a Hindered Phenol and a Thiosynergist**[a]

Additive, %		Induction period, days
Radical scavenger[b]	Hydroperoxide decomposer[c]	
0.0	0.3	4
0.1	0.0	16
0.1[d]	0.3[d]	45

[a] Ref. 21.
[b] Tetrakis[methylene(3,5-di-*tert*-butyl-4-hydroxyhydro-cinnamate)]methane.
[c] Dilauryl thiodipropionate.
[d] Mixture.

Oligomeric hindered amine light stabilizers are effective thermal antioxidants for polypropylene. Thus 0.1% of *N,N'*-bis(2,2,6,6-tetramethyl-4-piperadinyl)-1,6-hexanediamine polymer, with 2,4,6-trichloro-1,3,5-triazine and 2,4,4-trimethyl-2-pentaneamine [70624-18-9] (**35**) (Fig. 10), protects polypropylene multifilaments against oxidation when exposed at 120°C in a forced-air oven for 47 days (31). The simple hindered phenol 3,5-di-*tert*-butyl-4-hydroxytoluene [128-37-0] (0.1%) affords protection for only 14 days. Other examples of hindered amine light stabilizers are [82451-48-7] (**36**) and [106990-43-6] (**37**).

The stabilization of polyolefins used to insulate copper conductors requires the use of a long term antioxidant plus a copper deactivator. Both *N,N'*-bis(3,5-di-

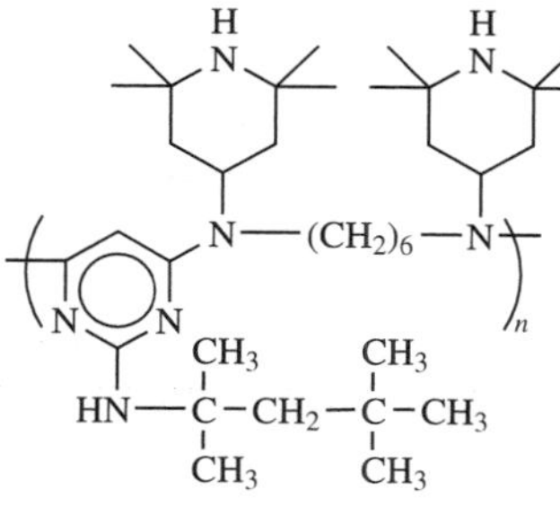

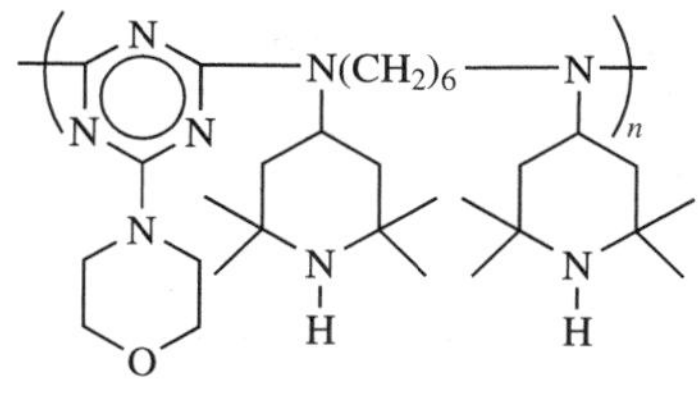

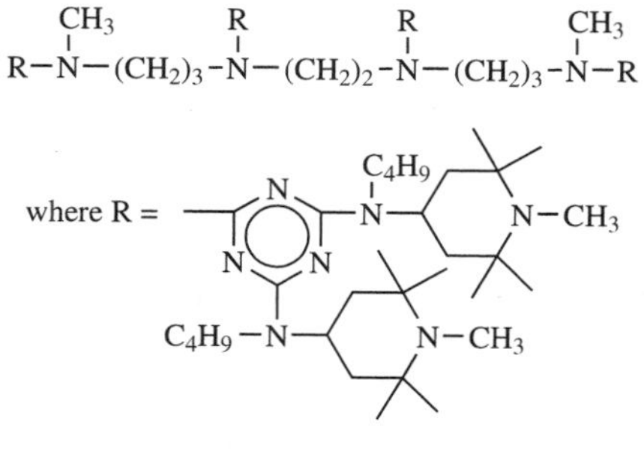

Fig. 10.   Hindered 1,6-hexanediamine antioxidants.

tert-butyl-4-hydroxycinnamoyl)hydrazine (**29**) and 2,2′-oxamidobisethyl(3,5-di-*tert*-butyl-4-hydroxycinnamate) (**30**) are bifunctional antioxidants that have built-in metal deactivators. Oxalyl bis(benzylidenehydrazide) (**28**) is an effective copper deactivator as part of an additive package that includes an antioxidant.

8.2. Polyamides. Due to their excellent mechanical properties at high temperatures, polyamides, particularly mineral and glass-filled grades, are finding increased usage in demanding applications such as automotive under-the-hood application. Only a few publications dealing with stabilization of polyamides are found in the literature (32). The aliphatic polyamides differ in their structure, PA 6,6, PA 6, PA 4,6, PA 11, and PA 12 being the most common types. The oxidative stability of the various types is dependent on the density of the amorphous phase and the degree of crystallinity because these two factors control oxygen migration into the polymer matrix (33). Aromatic polyamides are rather insensitive to oxidative degradation.

The traditional stabilization system for aliphatic polyamides are copper salts. Typical systems are based on low levels of copper (< 50 ppm) and iodide or bromide salts (34). The mechanism of stabilization is not well understood but may be due to hydroperoxide decomposition initiated by metal ions (35). These systems are effective in polyamides whereas in other polymers, such as polylefins, small amounts of oxidized copper can often as prodegradants. Good dispersion of the copper is critical to good performance. Since copper salts are water soluble, they can be leached from the polymer in certain applications (eg, aqueous dye baths) leading to reduced efficacy and environmental issues. Aromatic amines are effective long term stabilizers but due to their discoloring nature, their use is limited to carbon black filled systems. Phenolic antioxidants, when added during the polycondensation reaction, contribute to improved initial color and long term thermal stability, particularly at lower end use temperatures. The relative merits of the various stabilization systems are shown in Table 4.

8.3. Styrenics. Unmodified styrenics such as crystal polystyrene are relatively stable and under most end-use conditions it is not necessary to add antioxidants. Low levels (0.1%) of a hindered phenolic antioxidant are added to protect the polymer during repeated processing of scrap. Styrene–acrylonitrile copolymers (SAN) discolor during processing >220°C. While this is primarily a

Table 4. **Stabilization Systems**

Antioxidant system	Advantages	Disadvantages
Cu/halogen salts	best performance at elevated temperatures (>150°C) very low levels required	discoloration leaching in aqueous environment must be well dispersed
aromatic amines	good long term thermal stabilizer performance	discoloration needs high concentration
phenolics	best performance at lower temperatures when copper salts cannot be used good initial and long term color performance	less effective at high temperature conditions such as under-the-hood

nonoxidative process related to the acrylonitrile comonomer, the color can be suppressed to some extent through the addition of a combination of a phenolic antioxidant and a phosphite (36). High impact polystyrene (HIPS) is more susceptible to oxidative degradation due to the presence of an unsaturated butadiene rubber phase. Antioxidants can be added either during the manufacturing process to protect the rubber phase during polymerization and monomer stripping or post-polymerization. Color and impact properties are typically better if the antioxidant is added during polymerization but care must be taken to avoid adverse effects on the kinetics through interaction with the peroxide catalysts. This is particularly problematic if phosphites are added. The polymer ABS is a graft copolymer produced primarily in an emulsion process followed by a compounding step in which the high rubber content polymer is blended with SAN in various ratios to achieve the desired end-use properties. Antioxidants are required during the coagulation and drying steps to protect the high rubber content particles. The best performance is achieved through the use of a combination of hindered phenol (0.25%) and thioester synergist (0.5%) (37). The stability of the finished ABS is directly related to the butadiene content (38).

8.4. Polyesters. Poly(ethylene terephthalate) (PET) requires little to no antioxidant during thermal processing. In some cases, phosphites are added to improve the color of regrind. During polycondensation, pentavalent phosphorus compounds such as triphenylphosphate or trimethylphosphate may be added. These additives form complexes with the transesterification catalyst residues (manganese, tin, zinc), yielding a polymer with reduced hydroxyl end groups. This affords better hydrolytic stability during end use and reduced discoloration prior to the condensation reaction (39). Poly(butylene terephthalate) (PBT), because of its higher hydrocarbon content, is more susceptible to oxidative degradation than PET. The combination of a phenolic antioxidant (0.05–0.10%) and a phosphite (0.1%) is typically used to stabilize PBT.

8.5. Polycarbonate. Polycarbonate (PC) is susceptible to photooxidation, and antioxidants are necessary to maintain the low color and high transparency critical to its end-use applications. Phosphites (0.1%) are used to minimiz color development during processing. It has been shown that the inherent stability of PC is related to the level of phenolic end groups (40). These levels can increase as a result of humidity induced hydrolysis catalyzed by acid. The phosphite chosen must be very stable to avoid the generation of catalytic amounts of phosphorus acids.

8.6. Polyacetal. Polyacetals thermal decompose by an acid-catalyzed depolymerization process starting at the chain ends. The polymer structure is stabilized by end capping of the polymer and introducing comonomers to interrupt the unzipping. The process is autocatalytic since the liberated formaldehyde is easily oxidized to formic acid, which is a prodegradant. Formaldehyde scavengers and phenolic antioxidants are typically used in polyacetal formulations (41).

8.7. Polyurethanes. The oxidative stability of polyurethanes (PURs) is highly dependent on the chemical nature of both the polyol component and the isocyanate. Thus, PUR derived from a polyester polyol is typically more stable than one derived from a polyether polyol. The methylene group adjacent to the ether linkage in polyether polyols is easily oxidized to hydroperoxides during storage. If not inhibited, decomposition of these built-up hydroperoxides can occur catastrophically during the highly exothermic reaction of the polyol and the isocyanate. Blends of a hindered phenol (0.2%) and an aromatic amine (0.1%) are typically used as storage stabilizers in polyether systems providing PUR foams with excellent color. Similar systems are used in polyester polyols, but at lower use levels (42).

8.8. Elastomers.   Polyunsaturated elastomers are sensitive to oxidation. Stabilizers are added to the elastomers prior to vulcanization to protect the rubber during drying and storage. Nonstaining antioxidants such as butylated hydroxytoluene (**1**), 2,4-bis(octylthiomethyl)-6-methylphenol [110553-27-0], 4,4′-bis(α,α-dimethylbenzyl)diphenylamine (**14**), or a phosphite such as tris(non-ylphenyl)phosphite (**24**) may be used in concentrations ranging from 0.01 to 0.5%.

Staining antioxidants such as N-isopropyl-N'-phenyl-p-phenylenediamine [101-72-4] (**38**) are preferred for the manufacture of tires. These potent antioxidants also have antiozonant activity and retard stress cracking of the vulcanized rubber. Carbon black (qv), used in tires for reinforcement, hides the color developed by the antioxidant. According to use requirements, up to 3% of an amine antioxidant having antiozonant activity is added prior to vulcanization.

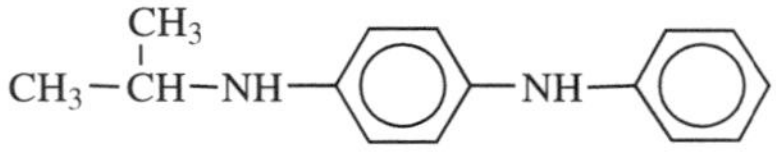

(**38**)

When staining tendencies of the substituted paraphenylene diamines cannot be tolerated, semistaining amine antioxidants are used to provide some protection to the elastomers. The semistaining antioxidants include polymerized 2,2,4-trimethyl-1,2-dihydroquinolines (**13**) and substituted diphenylamines such as 4,4′(-bisα,α-dimethylbenzyl)diphenylamine (**14**). These compounds, however, provide no protection against ozone.

Antioxidants resistant to extraction by lubricants and gasoline are preferred for the stabilization of elastomers used in automotive applications such as gaskets and tubing. Aromatic amine antioxidants, such as N-phenyl-N'-(p-toluenesulfonyl)-p-phenylenediamine [100-93-6] (**39**), with low solubility in hydrocarbons, are extracted slowly from elastomers and are used for these applications.

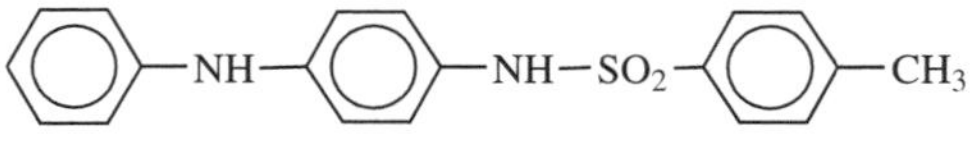

(**39**)

Binding the antioxidant chemically to the elastomer chain by copolymerization or grafting is a better solution to this problem. The addition of N-(4-anilino-phenyl)methacrylamide [22325-96-8] (**40**) to a polymerization recipe for NBR rubber produces a polymer with a built-in antioxidant resistant to extraction (43).

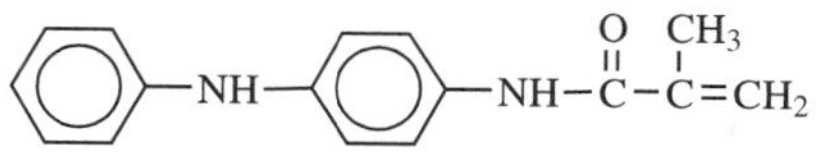

$$\langle\!\!\bigcirc\!\!\rangle\!-\!NH\!-\!\langle\!\!\bigcirc\!\!\rangle\!-\!NH\!-\!\overset{\overset{\displaystyle O}{\|}}{C}\!-\!\overset{\overset{\displaystyle CH_3}{|}}{C}\!=\!CH_2$$

(40)

Raw NBR containing 1.5% of the built-in antioxidant retained 92% of its original resistance to oxidation after exhaustive extraction with methanol. NBR containing a conventional aromatic amine antioxidant (octylated diphenylamine) retained only 4% of its original oxidative stability after similar extraction.

It is also possible to graft an aromatic amine antioxidant bearing a sulfhydryl group on to the backbone of an elastomer.

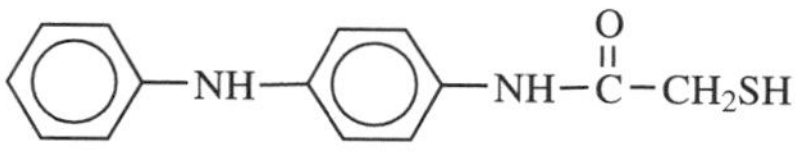

$$\langle\!\!\bigcirc\!\!\rangle\!-\!NH\!-\!\langle\!\!\bigcirc\!\!\rangle\!-\!NH\!-\!\overset{\overset{\displaystyle O}{\|}}{C}\!-\!CH_2SH$$

(41)

When 4-(mercaptoacetamido)diphenylamine [60766-26-9] (**41**) is added to EPDM rubber and mixed in a torque rheometer for 15 min at 150°C, 87% of it chemically binds to the elastomer (44). The mechanical and thermal stress placed on the polymer during mixing ruptures the polymer chain, producing free radicals that initiate the grafting process.

8.9. Poly(vinyl chloride).   While reasonably stable with respect to oxidative degradation, poly(vinyl chloride) (PVC) is very susceptible to thermal degradation. Protection of PVC from thermal degradation leading primarily to dehydrohalogenation reactions is out of the scope of this article. Some additives, used primarily as HCl scavengers, also have antioxidant properties. Aralkylphosphites in particular are used as components of thermal stabilizing mixtures that also exhibit antioxidant properties, leading to improved resin color.

8.10. Fuels and Lubricants.   Gasoline and jet engine fuels contain unsaturated compounds that oxidize on storage, darken, and form gums and deposits. Radical scavengers such as 2,4-dimethyl-6-*tert*-butylphenol [1879-09-0], 2,6-di-*tert*-butyl-*p*-cresol (1), 2,6-di-*tert*-butylphenol [128-39-2], and alkylated paraphenylene diamines are used in concentrations of ~5–10 ppm as stabilizers.

The catalytic activity of copper as an oxidant in fuels and lubricants can be inhibited by the use of a metal deactivator such as *N*,*N*'-disalicylidene-1,2-diaminopropane (**31**) at a concentration of 5–10 ppm.

Lubricants for gasoline engines are required to withstand harsh conditions. The thin films of lubricants coating piston walls are exposed to heat, oxygen, oxides of nitrogen, and shearing stress. Relatively high concentrations of primary antioxidants and synergists are used to stabilize lubricating oils. Up to 1% of a mixture of hindered phenols, of the type used for gasoline, and secondary aromatic amines, such as alkylated diphenylamine and alkylated phenyl-α-napththylamine, are used as the primary antioxidants. About 1% of a synergist, zinc dialkyldithiophosphonate, is added as a peroxide decomposer. Zinc dialkyldithiophosphates (**42**) are cost effective multifunctional additives. They interrupt oxidative chains by trapping radicals by electron donation, decompose peroxides, and serve as a corrosion and wear inhibitors.

$$RO-\underset{\underset{OR}{|}}{\overset{\overset{S}{\|}}{P}}-S-Zn-S-\underset{\underset{OR}{|}}{\overset{\overset{S}{\|}}{P}}-OR$$

(**42**)

R = C$_3$H$_7$, C$_8$H$_{17}$

Both zinc and phosphorus deactivate the catalysts used to control emissions and governmental regulations limit their concentrations to < 0.1%.

9. Test Methods

9.1. Polymers.   There are a variety of test methods available for monitoring the oxidative stability in the polymer as it is exposed to different types of stresses. These changes can be physical or aesthetic. Physical transformations of the polymer might involve changes in molecular weight, molecular weight distribution, or crystallinity. Aesthetic transformations of the polymer might involve changes such as discoloration or changes in the surface appearance due to microscopic cracks that affect gloss.

These changes can be measured using simple testing equipment and procedures described in ASTM methods (eg, tensile strength, impact resistance, color development; oxidative induction time, oven aging, etc).

Estimation of oxidative stability under conditions of use is more difficult to measure. To decrease the time required for oxidative failure, specimens are exposed at temperatures higher than those anticipated in use. Oven aging the polymer at temperatures below the melting point of the polymer is a method for assessing the long term thermal stability of a given substrate. A range of test temperatures can be used to create an Arrhenius plot as test temperature (°C or kelvin) versus time to failure (eg, embrittlement, or 50% retention of tensile strength or elongation). If the apparent activation energy remains constant, a plot of the logarithm of failure time against the reciprocal of the absolute temperature results in a straight line. Extrapolation to temperature of use provides an estimate of failure time.

Due to different reactions and the raw materials used, the apparent activation energy of the overall process can deviate considerably from linearity and an extrapolation can lead to serious errors in estimating failure time (45).

Representative measures of aesthetic properties would include color development, loss of gloss, increase in haze, chalking, loss of surface smoothness, exudation and/or blooming of additives or low molecular weight polymer, staining, and the like.

To assess the retention of physical or aesthetic properties, it is important to understand that increasing surface/volume ratio increases susceptibility to oxidation as does the rate of loss of the antioxidant by volatilization (46).

In the past, oxygen-uptake measurements at elevated temperatures were used to determine oxidation resistance. Today, other types of testing, such as oxidative induction time, OIT (by differential scanning calorimetry, DSC) or chemiluminescence are used.

For oxygen uptake experiments, the time required to absorb a specified volume of oxygen is an indication of failure when the change in the rate of oxygen absorption is not sufficient to permit an accurate measurement of the induction period. However, the amount of oxygen absorbed and the loss of desired properties must be correlated. This type of test is not reliable for estimating service life under conditions in which the polymer is exposed to air movement.

Oxidative induction time or oxidative induction temperature experiments are used to assess the relative oxidative stability of the polymer. The experiments are typically conducted under severe testing conditions, typically at elevated temperatures (significantly above the melting point of the polymer) using oxygen as the oxidant. Note that this method is not useful for predicting long term thermal stability; however, these methods can be useful for quickly assessing the relative potency of different types of stabilizer systems (47).

For chemiluminescence experiments, the procedure is similar to the aforementioned oxidative induction time experiments, in that the time to catastrophic oxidative of the polymer is measured; in this case, below the melting point of the polymer. During this catastrophic oxidation, there are certain chemistries taking place that result in chemiluminescence. The amount of light that is given off during this process is measured by a charge coupled device. This method has been developing over the last decade, can provide useful information about the oxidative stability of the polymer in a much shorter time period than conventional oven aging, but under more realistic oxidative environment in comparison to oxidative induction time (48).

9.2. Lubricants. A sequence of tests has been devised to evaluate antioxidants for use in automotive crankcase lubricants. The Indiana Stirring Oxidation Test (ISOT) JISK2514 is an example of a laboratory screening test. The oil is stirred at 165.5°C in the presence of air. Copper and iron strips are used as metal catalysts. The development of sludge, viscosity, and acidity are determined periodically. Failure time is determined when the development of acidity requires 0.4-mg KOH/g for neutralization. Formulated lubricants are then evaluated for performance in engines (see Oldsmobile Sequence III D) (49). Candidate lubricants containing the antioxidants are tested in fleets of automobiles for thousands of miles. As described above for polymers, modified oxygen uptake tests can also be used with lubricants to measure oxidative stability (50).

The effectiveness of antioxidants as preservatives for fats and oils is evaluated by determining the rate of peroxide development using the active oxygen method (AOM) (51). The development of a rancid odor is used to evaluate the stability of food items (Schaal oven stability test) (52).

10. Ancillary Properties

In reality, there is more to antioxidants than providing stability to the polymer by quenching free radicals and decomposing hydroperoxides. Other key issues besides rates of reactivity and efficiency include performance parameters such as volatility, compatibility, color stability, physical form, taste or odor, regulatory issues associated with food contact applications, and polymer performance versus cost (53).

10.1. Volatility. Most additives are melt compounded into the polymer after the polymerzation stage. The melt compounding and downstream conversion (into shaped articles) processing steps represent significant heat histories. Storage of the product can also be quite warm in certain climates. It is important that the antioxidant, as well as its transformation products that may also provide stability, not volatilize from the polymer. Many commercial antioxidants have been designed with higher molecular weights to address this issue.

10.2. Compatibility. Antioxidants should be soluble in the polymeric matrix. If they are not soluble, they should at least migrate or diffuse slowly. If the solubility limit of the antioxidant in the polymer is exceeded, exudation will occur. Exudation or blooming involves the migration of the additive out of the polymer matrix onto the surface as a very thin film. Blooming of the antioxidant can also diminish surface gloss, create stickiness, eliminate blocking (cling) of film surfaces to one another, negatively affect printability.

10.3. Color Stability. It is important that antioxidant do not provide unwanted color due to the transformation chemistries associated with preventing oxidation. Some antioxidants are prone to forming color by their very nature, while other antioxidants discolor only when they have been overoxidized. Pigments can be used to mask the subtle changes in the base color of the polymer.

10.4. Physical Form. Because of hazards associated with dusting, many antioxidants are now being offered commercially in dust free forms, such as granules or pellets. Liquid or molten antioxidants are another interesting alternative, as long as they are compatible in the polymer matrix. Still, some polymer manufactures need the additives as fine powders in order to achieve good premixing with their reactor product before melt compounding into the traditional pellet form.

10.5. Taste and Odor. For applications that involve food contact, home or personal use, taste, and odor are key issues. The human nose is very sensitive, more so than some analytical instruments, able to detect some compounds at the part per billion (ppb) level.

11. Health and Safety Factors

Safety is assessed by subjecting the antioxidant to a series of animal toxicity tests, eg, oral, inhalation, eye, and skin tests. Mutagenicity tests are also carried out to determine possible or potential carcinogenicity. Granulated and liquid forms of antioxidants are receiving greater acceptance to minimize the inhalation of dust and to improve flow characteristics.

A number of antioxidants have been regulated by the U.S. Food and Drug Administration as indirect additives for polymers used in food contact applications (primarily food packaging) under Title 21 of the U.S. Code of Federal Regulations (21CFR), Part 175 (Adhesives and Coatings) and/or Part 177 (Polymers). Acceptance is determined by subchronic or chronic toxicity in more than one

Table 5. **Commonly Used Antioxidants by Class**

Chemical Name	CAS No.	FDA Reg.[a]	Suggested substrates[b]	Suppliers
Monophenols				
alpha tocopherol	[10191-41-01]	x	PO, PUR	Ciba SC, BASF
2,6-di-*tert*-butyl-4-methylphenol	[128-37-0]	x	PA, PES, PO, POM, PUR, PVC, RU, PS	Great Lakes, Merisol, PMC, Crompton
octadecyl, 3,5-di-*tert*-butyl-4-hydroxycinnamate	[2082-79-3]	x	CE, PA, PO, PUR, PVC, PS	Ciba SC, Crompton, Great Lakes, GE Specialty
isooctyl, 3,5-di-*tert*-butyl-4-hydroxycinnamate	[126-43-61-0]	x	PUR	Ciba SC, Crompton
Bisphenols				
2,2'-methylenebis(4-methyl-6-*tert*-butyl-phenol)	[119-47-1]	x	CE, PA, POM, PUR, PVC, RU, PS	Cytec, RT Vanderbilt, Great Lakes, Ferro
2,2'-methylenebis(4-ethyl-6-*tert*-butyl-phenol)	[88-24-4]	x	PA, PO, PVC	Cytec
4,4'-butylidenebis(6-*tert*-butyl-3-methylphenol)	[85-60-9]	x	CE, EVA, PA, PES, PO, PVC, RU, PS	Flexsys
N,N'-hexamethylene bis(3,5-di-*tert*-butyl-4-hydroxyhydrocinnamide)	[23128-74-7]	x	PA, PES, PA, RU	Ciba SC, Great Lakes
1,6-hexamethylenebis(3,5-di-*tert*-butyl-4-hydroxyhydrocinnamate)	[35074-77-2]	x	CE, PES, PO, POM, PUR, PVC, PS	Ciba SC
benzenepropanoic acid, 3-(1,1-dimethyl-ethyl)-4-hydroxy-5-methyl-, 2,4,8,10-tetraoxaspiro [5.5]undecane-3,9-diyl-bis(2,2-dimethyl-2,1- ethanediyl) ester	[90498-90-1]			Sumitomo
triethyleneglycol bis(3,5-di-*tert*-butyl-4-hydroxyhydrocinnamate)	[36443-68-2]	x	PA, POM, PVC, PS	Ciba SC, Great Lakes
calcium bis[O-ethyl(3,5-di-*tert*-butyl-4-hydroxybenzyl)-phosphonate	[65140-91-2]	x	PO, RU	Ciba SC
Polyphenols				
1,3,5-trimethyl-2,4,6-tris(3',5'-di-*tert*-butyl-4'-hydroxybenzyl)-benzene	[1709-7-2]	x	CE, PA, PES, PO, POM, PVC, RU, PS	Albemarle, Ciba SC, Great Lakes, Sigma 3V

127

Table 5 (Continued)

Chemical Name	CAS No.	FDA Reg.[a]	Suggested substrates[b]	Suppliers
3:1 condensate of 3-methyl-6-tert-butyl-phenol with crotonaldehyde	[1843-03-4]	x	CE, PA, PES, PO, POM, PUR, PVC, RU, PS	ICI Americas
tetrakis[methylene(3,5-di-tert-butyl-4-hydroxyhydrocinnamate)methane	[6683-19-8]	x	CE, PA, PES, PO, POM, PVC, PS	Ciba SC, Great Lakes, Crompton, GE Specialty Chem
1,3,5-tris(3,5-di-tert-butyl-4-hydroxyben-zyl)isocyanurate	[27676-62-6]	x	CE, PA, PES, PO, POM, PVC, PS	Ciba SC, RT Vanderbilt, Great Lakes Chem, Sigma 3V
3,5-di-tert-butyl-4-hydroxy-hydrocinnamic triester with 1,3,5-tris (2-hydroxyethyl)-s-triazine-2,4,6,(1H,3H,5H)-trione	[34137-09-2]	x	CE, PA, PO, PUR, PVC, RU, PS	Ciba SC, RT Vanderbilt
1,3,5-tris(4-tert-butyl-3-hydroxy-2,6-dimethylbenzyl)-s-triazine-2,4,6-(1H,3H,5H)trione	[40601-76-1]	x	CE, PA, PES, PO, POM, PUR, PVC, RU, PS	Cytec, Ciba SC, Great Lakes Chem
bis[3,3-bis(4-hydroxy-tert-butylphenyl)bu-tanoic acid] glycol ester	[32509-66-3]	x	CE, PA, PES, PO, POM, PVC, PS	Clariant
butylated reaction product of p-cresol and dicyclopentadiene	[31851-03-3]	x	CE, PA, PES, PO, POM, PVC, RU, PS	Goodyear
Phenolics with Dual Functionality				
4,4'-thiobis(6-tert-butyl-3-methylphenol)	[96-69-5]	x	PA, PES, PO, PVC, RU, PS	Flexsys, Sumitomo
4,4'-thiobis(2-methyl-6-butylphenol)	[96-66-2]		PO, RU	Albemarle
thiodiethylene bis(3,5-di-tert-butyl-4-hydroxycinnamate)	[41484-35-9]	x	CE, PA, PO, PUR, PVC, RU, PS	Ciba SC, Great Lakes
4,6-bis(octylthiomethyl)-o-cresol	[110553-27-0]	x	RU, PS	Ciba SC
Reaction product of nonylphenol, dodeca-nethiol and formaldehyde	[188793-84-2]	x	RU, PS	Goodyear
2,4-bis(n-octylthio-6-(4-hydroxy-3,5-di-tert-butylanilino)-1,3,5-triazine	[991-84-4]	x	RU, PS	Ciba SC
2-(1,1-dimethylethyl)-6-[3-(1,1-dimethy-lethyl)-2-hydroxy-5-methylphenyl]-methyl-4-methylphenylacrylate	[61167-58-6]	x	PS	Sumitomo, Ciba SC

128

Metal Deactivators

	CAS		Applications	Suppliers
N,N'-bis(3,5-di-*tert*-butyl-4-hydroxyhydro-cinnamoyl)hydrazine	[32687-78-8]	x	PA, PO, RU	Ciba SC, Great Lakes
2,2'-oxamidobisethyl(3,5-di-*tert*-butyl-4-hydrocinnamate)	[70331-94-1]	x	PA, PO, RU	Crompton
Arylamines				
oxalic acid, bis(benzylidenehydrazide)	[6629-10-3]	x	PA, PO, RU	Eastman
4,4'-bis(dimethylbenzyl)-diphenyl amine	[10081-67-1]	x	RU, PO, PS, PUR	Crompton
N-phenyl-alpha-naphthylamine	[90-30-2]	x	PA, PES, PO, RU	Bayer, Flexsys, RT Vanderbilt
N-phenyl-N'-isopropyl-*p*-phenylenedia-mine	[101-72-4]	x	RU	Bayer, Flexsys, RT Vanderbilt, Crompton
polymerized 2,2,4-trimethyl-1,2-dihydro-quinoline	[26780-96-1]		RU	Flexsys, RT Vanderbilt, Crompton
octylated diphenylamine	[68411-46-1]	x	RU, PUR	Ciba SC, Crompton, Great Lakes, RT Vanderbilot
Thioethers				
didodecyl-3,3'-thiodipropionate	[123-28-4]	x	PA, PO, RU	Cytec, Crompton, Evans Chemetic
distearyl-3,3'-thiodipropionate	[693-36-7]	x	PA, PO, RU	Cytec, Crompton, Evans Chemetic
pentaerythritol tetrakis(3-dodecylthio)-propionate	[29598-76-3]	x	PA, PO	Crompton
S,S'distearyldisulfide	[2500-88-1]	x	PA, PO, RU	Clariant
Phosphites/Phosphonites				
tris nonylphenylphosphite	[26523-78-4]	x	CE, PES, PO, PUR, PVC, RU, PS	Crompton, GE Specialty, Dover
tris (2,4-di-*tert*-butylphenyl) phosphite	[31570-04-4]	x	CE, PES, PO, PUR, PS	Ciba SC, GE Specialty, Great Lakes, Crompton
distearyl pentaerythritoldis phosphite	[3806-34-6]	x	CE, PES, PO, PUR, PS, RU	GE Specialty
bis(2,4-di-*tert*-butylphenyl)-pentaerythri-tol diphosphite	[26741-53-7]	x	CE, PES, PO, PUR, PS, RU	GE Specialty, Ciba SC, Great Lakes
2,4,6-tri-*tert*-butylphenyl-2-butyl-2-ethyl-1,3-propanediol phosphite	[161717-32-4]	x	PO, PS	GE Specialty
bis(2,4-dicumylphenyl) pentaerythritol diphosphite	[154862-43-8]	x	CE, PES, PO, PUR, PS, RU	Dover
tetrakis(2,4-di-*tert*-butylphenyl)4,4'-biphenylenediphosphonite	[119345-01-6]	x	CE, PES, PO, PUR, PS, RU	Clariant, Ciba SC, Great Lakes

Table 5 *(Continued)*

Chemical Name	CAS No.	FDA Reg.[a]	Suggested substrates[b]	Suppliers
Hindered Amines				
poly[6-[(1,1,3,3-tetramethylbutyl) amino]-1,3,5-triazine-2,4-diyl][[2,2,6,6-tetramethyl-4-piperidinyl)imino]-1,6-hexanediyl[(2,2,6,6-tetramethyl-4-piperidinyl)imino]	[71878-19-8]	x	PO, EVA	Ciba SC
poly[6-[1-morpholinol-1,3,5-triazine-2,4-diyl][[2,2,6,6-tetramethyl-4-piperidinyl)imino]-1,6-hexanediyl[(2,2,6,6-tetramethyl-4-piperidinyl)imino]	[82451-48-7]	x	PO, EVA	Cytec
butanedioic acid, dimethylester, polymer with 4-hydroxy-2,2,6,6-tetramethyl-1-piperidine ethanol	[65447-77-0]	x	PO, EVA	Ciba SC
7-oxa-3,20-diazadispiro[5.1.11.2] heneicosan-21-one, 2,2,4,4-tetramethyl-, hydrochloride, reaction products with epichlorohydrin, hydrolyzed, polymd	[202483-55-4]	x	PO, EVA	Clariant
1,3-propanediamine, *N,N″*-1,2-ethanediyl-bis-, polymer with 2,4,6-trichloro-1,3,5-triazine, reaction products with *N*-butyl-2,2,6,6-tetramethyl-4-piperidinamine	[136504-96-6]	x	PO, EVA	Sigma 3V
1,3,5-triazine-2,4,6-triamine,*N,N*[1,2-ethane-diyl-bis[[(4,6-bis[butyl(1,2,2,6,6-pentamethyl-4-piperidinyl)amino]-3,1-pentamethyl-4-piperidinyl)]bis(*N′,N′*-dibutyl-*N′,N′*-propanediyl)]bis(1,2,2,6,6-pentamethyl-4-piperidinyl)-	[106990-43-6]	x	PO, EVA	Ciba SC

siloxanes and silicones, methyl-hydrogen, reaction products with 2,2,6,6-tetramethyl-4-(2-propenyloxy)piperidine	[182635990]	x	PO, EVA	Great Lakes

Processing Stabilizers

2(3*H*)-benzofuranone, 5,7-bis-(1,1-dimethylethyl)-3-hydroxy, reaction products with *o*-xylene	[181314-48-7]	x	PO, PUR	Ciba SC
1-octadecanamine, *N*-hydroxy-*N*-octadecyl-	[123250-74-8]	x	PO, RU	Ciba SC
di (rape oil)alkyl-*N*-methylamine oxide	[204933-93-7]	x	PO, RU	GE Specialty

[a] Regulated by the US Food and Drug Administration as an indirect food additive under Title 21 of the *U.S. Code of Federal Regulations* (21 CFR), Part 175 (Adhesives and Coatings) and/or Part 177 (Polymers).

[b] CE = cellulosics, EVA = ethylene vinylacetate copolymers, PA = polyamides, PES = polyesters, PO = polyolefins, POM = polyoxymethylenes, PUR = polyurethanes, polyols, RU = rubber, PS = polystyrenes, PVC = poly(vinyl chloride).

animal species and by the concentration expected in the diet, based on the amount of the additive extracted from the polymer by solvents that simulate food in their extractive effects. Materials are regulated by the FDA for use in plastics contacted by food stuffs to ensure a minimum risk to the consumer. Broad FDA regulation is increasingly a requirement for the successful introduction of a new antioxidant.

12. Cost Effectiveness

The point in using an antioxidant is to chose the apropriate type and level to adequately stabilize the polymer for a particular end-use application.

For example, if the material is a nondurable good, such as bags and food wrap, the type and concentration of antioxidant is chosen to minimize unnecessary costs associated with stabilizing the polymer. The antioxidant should be able to provide stabilization for the initial melt compounding and processing into the finished article. It is important to consider that scrap from the melt compounding needs to be recycled and unexpected shut downs and start-ups may occur.

On the other hand, if the material is a durable good, such as geomembranes, insultation for wire and cable, or gas and water transmission pipes, the type and concentration of the phenolic antioxidant is chosen to meet the long service life criteria. The costs associated with this type or level of antioxidant is worth the additional value of the final product.

12.1. Commercial Antioxidants.   Table 5 includes the main classes of antioxidants sold in the United States and the supplier's suggested applications. Some of these are mixtures rather than single components, which is especially true of alkylated amines and alkylated phenols. The extent of alkylation and the olefins used for alkylation can vary among manufacturers. Table 5 is not a complete listing of available antioxidants in the United States.

BIBLIOGRAPHY

"Antioxidants" in *ECT* 1st ed., Vol. 2, pp. 69–75, by A. E. Bailey; in *ECT* 2nd ed., Vol. 2, pp. 588–604, by L. R. Dugan, Jr., Michigan State University; "Antioxidants and Antiozonants" in *ECT* 3rd ed., Vol. 2, pp. 128–148, by P. P. Nicholas and A. M. Luxeder, The B. F. Goodrich Co., L. A. Brooks, R. T. Vanderbilt Co. and P. A. Hammes, Merck and Co.; "Antioxidants" in *ECT* 4th ed., Vol. 3, pp. 424–447, by M. Dexter, Consultant; "Antioxidants" in *ECT* (online), posting date: December 4, 2000, by M. Dexter, Consultant.

CITED PUBLICATIONS

1. Plastics Additives, *Chem. Eng. News*, 21 (December 4, 2000); search date March 14, 2001.
2. Plastics Additives, *Chemical Marketing Reporter*, 7 (June 19, 2000) search date March 14, 2001.

3. L. Bolland, *Quart. Rev. London* **3**, 1 (1949).
4. Bateman, *Quart. Rev. London* **8**, 147 (1954).
5. U. Ingold, *Chem. Rev.* **61**, 563 (1961).
6. R. Shelton and W. L. Cox, *Rubber Chem. Technol.* **27**, 671 (1954).
7. Barnard, L. Bateman, J. I. Cunneen, and J. F. Smith in L. Bateman, ed., *Chemistry and Physics of Rubber-like Substances*, Maclaren, London, 1963, p. 593.
8. A. Kerr, *Chem. Rev.* **66**, 465 (1966).
9. J. Chalk and J. F. Smith, *Trans. Faraday Soc.* **53**, 1214 (1957).
10. P. Pospisil in E. Scott, ed., *Developments in Polymer Stabilization*, Vol. 1, Applied Science Publishers, Ltd., London, 1979, pp. 6–11.
11. C. Nixon, H. B. Minor, and G. M. Calhoun, *Ind. Eng. Chem.* **48**, 1874 (1956).
12. H. Rosenwald, J. R. Hoatson, and J. A. Chenicek, *Ind. Eng. Chem.* **41**, 162 (1950).
13. I. Wasson and W. M. Smith, *Ind. Eng. Chem.* **45**, 197 (1953).
14. Scott, *Atmospheric Oxidation and Antioxidants*, Elsevier, Amsterdam, The Netherlands, 1965, p. 109.
15. Scott, *J. Appl. Polym. Sci.* **35**, 131 (1979).
16. J. Galbo, Light Stabilizers in *Encyclopedia of Polymeric Materials*, CRC Press, 1996, pp. 3616–3623.
17. P. Klemchuk and M. E. Gande, *Makromol. Chem. Macromols. Symp.* **28**, 117 (1989).
18. D. Horsey, "Hydroxylamines, a New Class of Low Color Stabilizers for Polyolfins", Additives '96, Houston, Feb 1996.
19. P. A. Smith, and S.E. Gloyer, *J. Org. Chem.*, **40** 2508 (1975).
20. C. Kröhnke, "A Major Breakthrough in Polymer Stabilization", Polyolefins X Conference, Houston, Feb. 1997.
21. P. Nesvadba, C. Kröhnke, "A New Class of Highly Active Phosphorus Free Processing Stabilizers for Polymers", Additives 97, 6th International. Conference., New Orleans, Feb. 1997.
22. R. Shelton in W. L. Hawkins, ed., *Polymer Stabilization*, Wiley-Interscience, New York, 1972, p. 83.
23. Scott, ed., *Development in Polymer Stabilization*, Vol. 4, Applied Science Publishers, Ltd., London, 1981, pp. 16–17; C. M. J. Armstrong, C. Husbands, and G. Scott, *Eur. Polym. J.* **15**, 241 (1979).
24. H. E. DeLaMare, and G.M. Coppinger, *J. Org. Chem.* **28** 1068 (1963).
25. Ref. (9), p. 1235.
26. Ref. (17), p. 83.
27. R. E. King, III, "Recent Advances in the Phenol Free Stabilization of Polypropylene Fiber", SPE Polyolefins RETEC, Houston, Feb. 2001.
28. K. B. Chakraborty and G. Scott, *Chem. Ind.* 237 (1978).
29. K. Kikkawa and Y. Nakahara, *Polym. Degrad. Stab.* **18**, 237 (1987).
30. D. Rysavy, *Kunststoffe* **60**, 118 (1970).
31. G. Tozzi, Cantatore, and F. Masina, *Text. Res. J.* **48**, 434 (1978).
32. B. Brassat and H. J. Buuysch, *Kunststoffe* **70**, 833 (1980).
33. P. Gijsman, D. Tummers, and K. Janssen, *Polymer Degrad. Stab.* **49**, 121 (1995).
34. M. I. Kohan, *Nylon Plastics Handbook*, Carl Hanser Verlag, Munich, 1995, pp. 58–59, 441.
35. K. Janssen, P. Gijsman, and D. Tummers, *Polymer Degrad. Stab.* **49**, 127 (1995).
36. F. Gugumus, in H. Zweifel, ed., *Plastics Additives Handbook*, 5th ed., Carl Hanser Verlag, Munich, 2001, p. 73.
37. H. Zweifel, *Stabilization of Polymeric Materials*, Springer-Verlag, Berlin 1998, pp. 88–90.
38. T. Hirai, *Jpn. Plastics*, October (1970).
39. Ref. (36), p. 85.

40. C. A. Pryde, and M. Y. Hellmann, *J. Appl. Polym. Sci.*, **25**, 2573 (1980).
41. F. R. Stohler, and K. Berger, *Angew. Makromol. Chem.* **176/177**, 327 (1990).
42. S. M. Andrews, Paper presented at the "Sixty Years of Polyurethanes", *International Symposium.*, University of Detroit-Mercy (1998).
43. E. Meyer, R. W. Kavchik, and F. J. Naples, *Rubber Chem. Technol.* **46**, 106 (1973).
44. G. Scott and H. Setoudeh, *Polym. Degrad. Stability* **5**, 81 (1983).
45. H. Hansen, D. R. Falcone, W. M. Martin, and W. I. Vroom, *Org. Coat. Plast. Prepr.* **34**, 97 (1974).
46. H. Gysling, *Adv. Chem. Ser.* **85**, 239 (1968).
47. J. R. Pauquet, R. V. Todesco, and W. O. Drake, "Limitations and Applications of Oxidative Induction Time (OIT) to Quality Control of Polyolefins", 42nd International Wire & Cable symposium, November 1993.
48. S. W. Bigger, P. K. Fearon, D. J. Whiteman, T. L. Phease, and N. C. Billingham *Poly. Prep.*, **42**(1) 375, 2001.
49. *ASTM Bull.*, 19103 (1990).
50. D. Chasan in J. Pospisil and P. P. Klemchuk, eds., *Oxidation Inhibition in Organic Materials*, Vol. I, CRC Press, Inc., Boca Raton, Fla., 33 and 31, pp. 291–326, 1990.
51. Active Oxygen Method, *Technical Data Bulletin ZG-159c*, Eastman Chemical Products, Inc., Kingsport, Tenn., Mar. 1985.
52. Storage Stability Tests for Evaluating Antioxidants in Fats and Oils, *Technical Data Bulletin ZG-194c*, Eastman Chemical Products, Inc., Kingsport, Tenn., Oct. 1987.
53. R. E. King, III, Antioxidants, in *Encyclopedia of Polymeric Materials*, CRC Press, 1996, pp. 306–313.

GENERAL REFERENCES

1. Refs. 14, 22, 36, and 49 are good general references.
2. M. Dexter in J. I. Kroschwitz, ed., *Encyclopedia of Chemical Technology*, 4th ed. Vol. 3, Wiley-Interscience, New York, 1992, pp. 424–447.
3. Voigt, *Die Stabilisierung der Kunstoffe Gegen Licht and Warme*, Springer-Verlag, Heidelberg, 1966.
4. K. Schwarzenbach, *Plastics Additives Handbook*, H. Zweifel, ed.; Hanser, Munich, 2000, pp. 1–137.
5. Pospisil and P. P. Klemchuk, eds., *Oxidation Inhibition in Organic Materials*, Vols. I and II, CRC Press Inc., Boca Raton, Fla., 1990.

RICHARD THOMAS
Ciba Specialty Chemicals
MARTIN DEXTER
Consultant
R. E. KING III
Ciba Specialty Chemicals

ANTIVIRAL AGENTS

1. Introduction

In the 10 years since the last publication of an article on antiviral agents (1), research on this topic has taken on an explosive course, largely because of the growing threat of the epidemic of AIDS (acquired immunodeficiency syndrome) in the western hemisphere, which not only intensified research on HIV (human immunodeficiency virus), but also on other opportunistic viral infections associated with HIV, such as HCV (hepatitis C virus), HBV (hepatitis B virus), and CMV (cytomegalovirus). Thanks to many rapid advances made from chemical, biochemical, as well as molecular biological fronts that led to effective anti-HIV therapies including the most successful combination drug regimen, the threat from HIV infection has been somewhat downplayed in recent years from the edict of a "death sentence" to that of a "manageable illness". On the other hand, a relatively less known virus such as the West Nile virus (WNV), or the less heeded viruses such as HCV and HBV, have suddenly taken up the center stage. In this context, this article will largely focus on viruses of current notoriety and public health concerns, while only brushing up on others that do not evoke alarm at the moment, but are still highly virulent and dangerous when given proper conditions for replication and proliferation. In view of the enormous and ever-expanding literature on antiviral agents in recent years, it will be a mammoth task to provide a comprehensive treatise on this subject, covering all of the known viral maladies and remedies. Therefore, four major viruses have been chosen, for which a relatively more detailed discussion will be given here concerning viral structure and replication as well as recent advances made in antiviral therapies. These four viruses include HIV, HBV, HCV, and WNV. Also, since nucleoside analogues have played a major role as therapeutics in combating these viruses, a special emphasis has been placed on this class of drugs. Before elaborately discussing the mentioned viruses as well as the available antiviral therapies for them individually, this article will briefly delve on the classification of viruses, the general process of viral replication, the potential targets for selective antiviral action, and a few selected natural and synthetic nucleosides as antiviral agents.

2. Classification of Viruses

In the early twentieth century, viruses were classified based on the hosts they infected. Thus, they were grouped into (a) plant viruses, (b) animal viruses, and (c) bacteriophages. The present-day broad classification of viruses is based on the genetic material they contain: DNA or RNA viruses (1). They may contain single-stranded DNA (parvoviruses), double-stranded DNA (herpesviruses), single-stranded RNA (poliovirus), or double-stranded RNA (reoviruses). The RNA viruses are unique in that they are the only living organisms that use RNA to store their genetic information. All other reproducing forms of life employ DNA. The more subtle classification of viruses, however, would include their hosts, chemical composition (including nucleic acid, protein, presence or absence of lipid envelope), shape, size, and symmetry. The major *animal viruses* can thus be subdivided into 18 categories: (*a*) herpesviruses, (*b*) papovaviruses, (*c*) adenoviruses, (*d*) poxviruses, (*e*) hepadnaviruses, (*f*) retroviruses, (*g*) orthomyxoviruses, (*h*) picornaviruses, (*i*) togaviruses, (*j*) rhabdoviruses, (*k*) paramyxoviruses, (*l*) reoviruses, (*m*) parvoviruses, (*n*) arenaviruses, (*o*) bunyaviruses, (*p*) filoviruses, and (*q*) coronaviruses.

Herpesviruses possess double-stranded, linear DNA that is 120,000–200,000 nucleotides long, icosahedron symmetry, protein coat, and lipid envelope. They include Herpes-simplex virus types 1 and 2 (HSV-1 and HSV-2), which cause recurrent cold sores and lesions (oral: type 1; genital: type 2). They also include Varicella-Zoster virus (VZV) that causes chicken pox and shingles, Epstein-Barr virus (EBV), which causes infectious mononucleosis, and is associated with selected cancers in China and Africa, and cytomegalovirus (CMV), which causes birth defects, and under special circumstances, pneumonia or hepatitis. Human CMV is one of the major opportunistic infections in HIV-infected patients as well as patients of the solid organ and bone marrow transplants.

Papovaviruses contain double-stranded, circular DNA that is 5000–8000 nucleotides long, icosahedral symmetry, and protein coat. They include human papillomaviruses, some of which cause oral or genital carcinomas, while others are responsible for benign genital tumors, polyomavirus that initiates tumors of wide variety in mouse, and simian virus 40 (SV 40), which is monkey virus that initiates tumors in rodents.

Adenoviruses possess double-stranded, linear DNA, 36,000–38,000 nucleotides long, with icosahedral geometry and protein coat. Human adenoviruses cause respiratory or enteric disease and infectious pinkeye. Some types of these viruses are capable of initiating tumors in rodents.

Poxviruses have double-stranded DNA, 130,000–280,000 nucleotides long. They are brick-shaped, and have lipids in the coat. The virion includes an RNA polymerase. Poxviruses that cause infections in man include smallpox, monkeypox, cowpox, tanapox, and *Molluscum contagiosum*.

Hepadnaviruses are part single-stranded and part double-stranded with a circular DNA, 3300–3400 nucleotides long, and possess nucleocapsid, protein coat, and lipid envelope. The virion includes DNA polymerase and reverse transcriptase. An important member of this family is the hepatitis B virus, which is responsible for causing serum hepatitis and liver cancer.

Retroviruses contain two linear, (+) single-strand, RNA molecules per virion, 3500–9000 nucleotides long, and a reverse transcriptase (RNA to DNA). They possess icosahedral shape, protein coat, and a lipid envelope. The retroviral family includes human T-cell leukemia virus-I (HTLV-I) that causes adult T-cell leukemia, human T-cell leukemia virus-II (HTLV-II), which is a possibly linked to hairy-cell leukemia, and human immunodeficiency virus types 1 and 2 (HIV-1 and HIV-2), which cause acquired immunodeficiency syndrome (AIDS). In addition, a variety of animal viruses, including Rous sarcoma virus and avian leukosis virus, are classified under retroviruses, and are known to be linked to cancers or immunodeficiencies in animals.

Orthomyxoviruses possess eight linear, (−) single-strand RNA molecules per virion, 13,600 nucleotides long, and a transcriptase (−RNA to +RNA).

They are helical in shape, and have a lipid envelope. *Influenza A Virus*, which causes the respiratory illness, belongs to this family.

Picornaviruses contain a (+) single-strand RNA genome, 7000 nucleotides long. They have icosahedral shape and a protein coat. Members of this family include poliovirus that causes infantile paralysis, rhinovirus that is responsible for common colds, and hepatitis A virus that causes infectious hepatitis.

Togaviruses contain a (+) single-strand RNA genome, 10,000–12,000 nucleotides, and have icosahedral shape, protein coat, and a lipid envelope. The flaviviruses used to be regarded as a genus within the *Togaviridae* family, but in 1984, they were provisionally reclassified as a distinct family (see below).

Rhabdoviruses contain a (−) single-strand RNA genome, 12,000 nucleotides long, and a transcriptase (−RNA to +RNA). They are bullet shaped, with a protein coat and a lipid envelope. Rabies virus, which causes rabies, is a member of this viral family.

Paramyxoviruses possess a (−) single-strand RNA genome, 15,900 nucleotides long. The virion includes a transcriptase (−RNA to +RNA). They have a helical shape, protein coat, and a lipid envelope. Mumps virus and measles virus (MV), which cause mumps and measles, respectively, along with respiratory synctial virus (RSV), which causes common cold-like upper respiratory tract infection in young children, represent this family.

Reoviruses have double-stranded RNA, and 10 chromosomes, 1000–4000 base-pairs. They have icosahedral shape and a protein coat. Rotaviruses, which cause infant enteritis, represent this family.

Parvoviruses are among the smallest, simplest eukaryotic viruses and were only discovered in the 1960s. They are widespread in nature; human parvovirus infections were only recognized in the 1980s. Essentially, they fall into two groups, defective viruses that are dependent on helper virus for replication and autonomous, replication-competent viruses. In all, >50 parvoviruses have been identified. They contain linear, nonsegmented, single-stranded DNA, ~5000 nucleotides long. Most of the strands packaged seem to be (−)sense, but adeno-assiciated viruses (AAVs) package equal amounts of (+) and (−) strands, and all seem to package at least a proportion of (+)sense strands. The virus particles are icosahedral, 18–26 nm diameter, and consist of protein (50%) and DNA (50%). Parvoviruses cause infections in a wide variety of birds and mammals, but 70–90% of most human populations are seropositive. The only known human parvovirus is referred to as B19. The most obvious symptom of B19 infection is a rubella-like rash.

Flaviviruses have single-stranded, positive sense RNA genomes that are 40–50 nm in diameter, 11 kilobase pairs in length. The virions are icosahedral and enveloped. Both hepatitis C virus (HCV) and West Nile virus (WNV), the two dreaded viruses of current notoriety in the western hemisphere, belong to the family of *Flaviviridae*. This family consists of arboviruses (ie, viruses borne by arthropods) that are classified into three genera: the flaviviruses, the pestiviruses, and the hepaciviruses. The flavivirus genus causes many human diseases like dengue fever, yellow fever, encephalitis, and hemorrhagic fevers. Although the natural reservoir of arboviruses is in avians and other animals, they are transmitted by arthropods like mosquitoes. Humans are infected with such viruses after being bitten by an infected arthropod. Pestiviruses only affect cattle. Hepaciviruses are the hepatitis C viruses, affecting ~3% of the global human population.

Arenaviruses were named after the Latin word arena, meaning sand, because of their granular interior. They are large RNA viruses that contain dense, ribosome-sized particles that give the appearance of sand particles when viewed by an electron microscope. The virus particles are spherical and have an average diameter of 110–130 nm. All are enveloped in a lipid (fat) membrane. The natural hosts of these viruses are generally rats, bats, and mice. The viruses are then shed in the feces and urine to contaminate food and water. When humans consume the infected foods, they contract the infection. Some of them cause meningitis and various hemmorrhagic fevers. Other infections include Lassa fever, lymphocytic choriomeningitis, and Argentinean and Bolivian hemorrhagic fevers.

Bunyaviruses are single stranded, (−)sense, RNA viruses in three circular segments of 7, 4, and 2 kilobases, with an envelope and helical symmetry, 90–100 nm in diameter. They have a lipid envelope through which glycoprotein spikes protrude. Within the family *Bunyaviridae* there are two types of viruses that cause disease in humans, the arthropod-borne viruses and the hantaviruses. Arthropod-borne viruses include the bunyaviruses, phleboviruses, and nairoviruses. Hantaviruses, named after the Hantaan river in Korea where it was first discovered in 1978, include six different species. Bunyavirus infection attacks the central nervous system leading to viral encephalitis. Phlebovirus infection leads to two different disease entities that are found at two different geographical locations. One of them, the Sandfly fever, clinically gives rise to fever, rash and arthralgia. The other one, the Rift Valley fever, may lead to complications of retinopathy, meningoencephalitis, haemorrhagic manifestations and hepatic necrosis. Infection with the nairovirus results in an influenza-like illness with fever and haemorrhagic symptoms. The clinical features of hantaviruses include haemorrhagic fever with renal syndrome (HFRS), or its milder form called the nephropathia epidemica, and hantavirus pulmonary syndrome (HPS), characterized by sudden onset of coughing, dyspnoea, pulmonary oedema, pleural effusion and shock.

Filoviruses contain single, unsegmented, (−) sense RNA, ~19 kilobases long. They appear in several shapes, a biological feature called pleomorphism. These shapes include long, sometimes branched filaments, as well as shorter filaments shaped like a "6", a "U", or a circle. Viral filaments may measure up to 14,000 nm in length, have a uniform diameter of 80 nm, and are enveloped in a lipid (fatty) membrane. They have strong structural and genetic similarities to both the rhabdoviruses and paramyxoviruses. They cause severe hemorrhagic fever in humans and nonhuman primates. So far, only two members of this virus family have been identified, including the Marburg virus and the Ebola virus.

Coronaviruses are irregularly shaped particles, ~60–220 nm in diameter, with an outer envelope bearing distinctive, 'club-shaped' peplomers that give its 'crown-like' appearance, and hence, its family name. Their genomes contain nonsegmented, single-stranded, (+) sense RNA, 27–31 kilobases long, the longest of any RNA virus. They infect a variety of mammals and birds. The exact number of human isolates are not known as many cannot be grown in culture. They cause common respiratory and occasional enteric infections in infants older than 12 months.

3. The General Process of Viral Infection and the Available Remedies

In order to discover site- or process-specific antiviral agents, it is important to understand the specific biochemical processes that occur during viral infection. In all, there are seven stages (2) in a typical viral infection process: (a) *adsorption*: The attachment of the virus to specific receptors on the cell surface; (b) *penetration*: The viral entrance into the cell by penetration through plasma membrane; (c) *uncoating*: The release of viral nucleic acid from the covering proteins; (d) *transcription*: The production of viral mRNA from the viral genome; (e) *translation*: The synthesis of viral proteins, including coat proteins and enzymes necessary for viral replication, as well as replication of viral nucleic acid (ie, the parental genome or complimentary strand); (f) *virion assembly*: The assembly of individual components of the viron (nucleic acid and structural proteins synthesized in stage e), and transportation to the site of nucleocapsid assembly, followed by autocatalytic assembly; (g) *release*: For viruses with icosahedral symmetry that do not have an envelope, this stage comes after disintegration of host cell as a result of the killing action of the infecting virus; for enveloped viruses, the assembled nucleocapsids move toward the modified membrane areas where the synthesized viral matrix protein replaces the cellular membrane proteins, and then nucleocapsids bud through the modified membrane, wrapping themselves into a portion of membrane in the process.

The preferred approach to combat viral diseases is the prevention of infection by active immunization. There are a number of successful vaccines for prophylaxis of some viruses such as polio, mumps, measles, influenza, encephalitis, hepatitis, and smallpox. On the other hand, there has been less success in the prevention of viruses such as the HIV, HSV, and RSV. Therefore the need for effective medicines to treat these viruses is urgent. Furthermore, millions of people around the globe are still suffering from a variety of viral diseases for which the vaccines already exist. For example, with >1 million child deaths per year, the measles virus (MV) ranks eight as the cause of death worldwide, especially in the developing countries (3). Despite large vaccination campaigns, MV is still resisting eradication, and there is no available therapeutic treatment (4). The MV infection causes a respiratory disease which is, more often than not, controlled solely by the immune response. The uncontrolled MV infection can lead to a severe immunosuppression that is responsible for additional opportunistic infections (5,6). Furthermore, in certain cases, MV establishes persistent infection of the brain leading to neurological complications (7). Also, some viruses are known to have very long latency period (8). Papova viruses may remain latent for years following childhood activation. These viruses reactivate and lead to viral diseases once T-lymphocyte hyporesponsiveness develops, either as a result of exogenous therapy, as in transplant recipients, or because of endogenous factors such as cancer or AIDS.

Two major virus-specific processes are normally targeted in order to develop selective antiviral agents: (a) early events, including adsorption, penetration and uncoating, and (b) later synthetic events that concern intracellular replication of the virus. In the fist stage of virus activity, heparin, an anionic polyelectrolyte of relatively high molecular weight (>5000 D) has shown to favor the formation of a complex with HSV that prevents virus from establishing an effective interaction with cell membrane (9,10). This phenomenon is ascribed to the unique anionic structure, an acid mucopolysaccharide (MW = 13,000 D), built by sulfated D-glucosamine and D-glucuronic acid units. Although there is no evidence heparin is toxic for the host cells, its antiherpes virus action is essentially nonspecific because of the ionic nature of the chemical-virus interaction. The electrostatic interaction can also be established between heparin and the positively charged groups projected out of cell membrane. Once virus successfully makes contacts with the host membrane, heparin is no longer effective. On the other hand, oligopeptides have comparably more potential as candidates in inhibiting the early activities of certain types of viruses. Sequence-specific oligopeptides that mimic the N-terminal region of the paramyxovirus F1 polypeptide (11) (16–19) are specific inhibitors of paramyxoviruses (12). Oligopeptides that mimic the N-terminal region of the orthomyxovirus polypeptide specifically inhibit influenza viruses (11). Recently, two synthetic proteins (DP-107 and DP-178) (13), which mimic the separate domains within the HIV-1 transmembrane (TM) protein-gp41, have been found to be stable and potent inhibitors of HIV-1 infection and fusion. This inhibitory effect can be intensified by increasing the length of the oligopeptide or by the presence of a carbobenzyloxy group on the N-terminal amino acid, whereas the esterification of the C-terminal amino acid decreases the activity. It was proposed that the antiviral effects of these oligopeptides are due to interference with binding of the N-termini of the viral envelope glycoproteins to specific receptors on the mammalian host cell membrane. Adamantane derivatives, amantadine hydrochloride (Symmetrel or Symadine) and rimantadine hydrochloride (α-methyl-1-adamantane methylamine hydrochloride or Flumadine) salts, are commercially available drugs for prevention and treatment of type A influenza viruses. Although Symmetrel is the first antiviral drug licensed in United States nearly 50 years ago, its mechanism of antiviral action remained unclear until recently. Early research using electron microscopy and pulse-labeling techniques revealed that amantadine acted at some point after late uncoating, but before initiation of viral RNA transcription (14). Rimantadine and amantadine have no inhibitory effect on the activity of viral polymerase; instead, the synthesis of the latter enzyme is prevented. Recently, the structure and function of the small protein-M2 in influenza A were elucidated (15). This protein has a single transmembrane helix that associates to form a tetramer *in vivo*, which forms proton-selective ion channels. This association is a pH dependent monomer–tetramer equilibrium. Upon binding of amantadine, the equilibrium shifts to tetrameric species. At higher pH, close to the pK_a of a histidine side chain where the protonation occurs within the transmembrane helices, the binding of amantadine is favored, which pushes the equilibrium toward the tetramer. It is suggested that amantadine competes with protons for binding to the deprotonated tetramer, thereby stabilizing the tetramer in a slightly altered conformation. It leads to the blockage of proton flux.

Other examples of antiviral agents targeted at the early events of viral activity include DIQA (3,4-dihydro-1-isoquinolineacetamide HCl), which has a broad-spectrum antiviral activity against lethal influenza A virus, echovirus, Columbia SK virus, herpes simplex virus, and rhino virus (16,17). The mechanism of its action involves the inhibition of virus penetration into the host cell membrane (16). Arildone (4-[6-(2-chloro-4-methoxyphenoxy)hexyl]-3,5-heptanedione), also a broad-spectrum antiviral agent against a number of RNA-containing and DNA-containing viruses, has been reported to interact directly with the polio virus capsid proteins in a way as to inhibit the uncoating of the virus and to prevent the subsequent virus-induced inhibition of cellular protein synthesis (18). It also showed inhibition of herpes virus DNA and virus-specific protein synthesis by acting on an early event in the virus replication cycle (19).

There exists a much larger pool of synthetic drugs that target the later events in a virus life cycle as compared to only a few synthetic or natural products that target the early events. These events are known to be virus specific. The viruses synthesize and utilize specific enzymes and proteins, and more importantly, the replication of viral genetic codes is also virus specific. The specificity in the synthesis of viral DNA or RNA is conferred by the virus-specific enzymes such as kinases, helicases, polymerases, transcriptases, reductases, etc. Viruses are more prone to mutations as compared with other microorganisms. The mutation rate of a virus is much higher than that of its host cell. Its mutants possess an excellent chance to be accommodated in the new host cell and escape from the host immune responses. On the other hand, high mutation rate means less selectivity toward substrates for the enzymes involved in DNA/RNA replication process. Once the potential drug candidate (an unnatural nucleotide analogue, for example) enters the catalytic site, it may disrupt or terminate the activity of enzymes. The unnatural nucleotides can be incorporated into DNA double helix, distort the DNA structure, and utimately stop the virus replication. For example, the nucleoside analogues idoxuridine (5-iodo-2′-deoxyuridine) (20,21), trifluridine (5-trifluoromethyl-2′-deoxyuridine) (22,23), and vidarabine

Idoxuridine   Trifluridine   Ara-A

(1-β-D-arabinofuranosyladenine or Ara-A) (20,21,24) appear to block replication in herpesviruses by three general mechanisms: first, as the monophosphates, they inhibit the formation of precursor nucleotides required for DNA synthesis; second, as triphosphates, they inhibit DNA polymerase; and third, the triphosphates are incorporated into DNA, which then does not function normally. For example, the DNA containing idoxuridine is more susceptible to strand breakage as well as to miscoded errors in RNA and protein synthesis.

There are also many compounds that fall outside the nucleoside family. A distinct example comes from the ever-growing fight against AIDS. In the late stage of a virus life cycle, the virus-specific processing of certain viral proteins by viral or cellular proteases is crucial. It was revealed that HIV expresses three genes as precursor polypepteins. Two of these gene products (designated as P55gag and p160 gag-pol proteins) undergo cleavage at several sites by a virally encoded protease to form structural proteins and enzymes required for replication (25). This fact has stimulated the research efforts to find safe and effective inhibitors for the viral protease. It is believed that the inhibitors should resemble a small portion of the substrate polyprotein structure but contain an isosteric replacement for the scissible (hydrolyzable) peptide bond that mimics the transition state for the hydrolysis of that bond, which is stable against cleavage. This has led to the discovery of several successful clinical candidates for HIV infection.

4. Nucleoside Analogues as Antiviral Agents

Initially, the term "nucleoside" was referred to the purine and pyrimidine N-glycosides derived from nucleic acid. However, after the discovery of pseudouridine (5-β-D-ribofuranosyluracil), a natural constituent of tRNA, it became a common practice to consider even those molecules whose heterocyclic rings are connected to the sugar moieties at the anomeric junctions through carbon–carbon single bonds. Such compounds are classified as C-nucleosides (26,27). Another interesting class of nucleosides, called L-nucleosides (28,29), are lately emerging as powerful antiviral compounds. The sugar parts of these nucleoside analogues possess the L- instead of the natural D-configuration. Furthermore, considering the possibility of existence of the carbon linking the base to the heterocycle into α- or β- anomeric form (β being the natural form), there exists an additional category of α-nucleosides. These different classes of nucleosides are contrasted below, using uridine as an example.

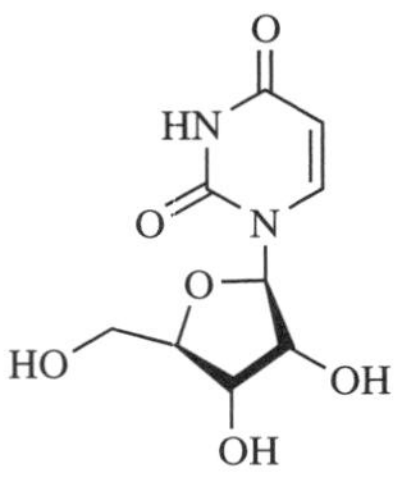

D-Uridine 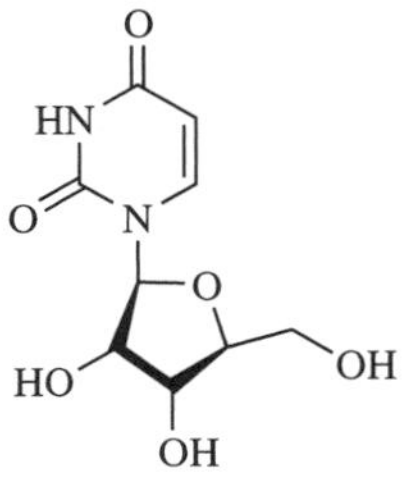   L-Uridine 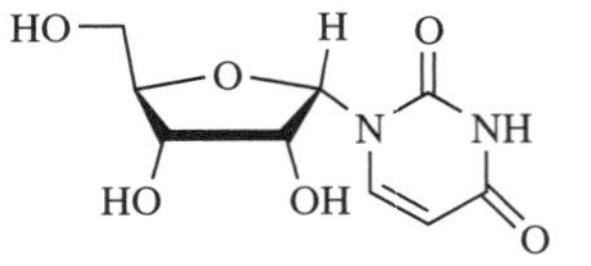   α-Uridine

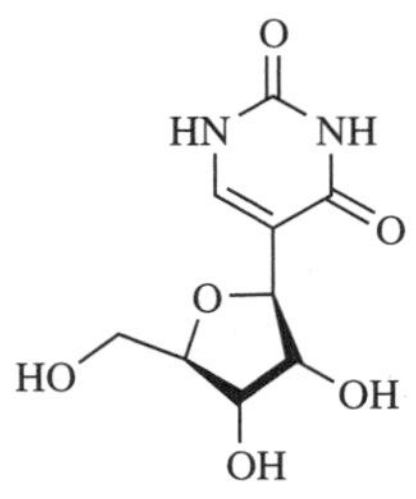

C-Uridine
(pseudouridine)

Among the naturally occurring nucleosides, sinefungin, an antifungal antibiotic isolated from *Streptomyces griseolus* (30), and its related metabolite A9145C (31), were found to be potent inhibitors of Newcastle disease virion, vaccinia virion mRNA (guanine-7-)-methyltransferase and vaccinia virion mRNA (nucleoside-2′)-methyltransferase. The structure of sinefungin is close to that of *S*-adenosylmethionine (AdoMet) with the methylthio group replaced by an aminomethylene group. They were found to be competitive inhibitors of the *S*-adenosyl-L-methionine-dependent enzymes. Neplanocin A, a carbocyclic analogue of adenosine with a unique cyclopentene structure, was isolated from the culture filtrate of *Ampullariella regularis* A11079 in 1980 (32). It has potent antitumor as well as antiviral activity. It acts primarily as an *S*-adenosylhomocysteine (AdoHcy) hydrolase inhibitor (33), which accounts for its broad-spectrum antiviral activities. AdoHcy hydrolase plays a key role in methylation reactions that depend on *S*-adenosylmethionine (AdoMet) as a methyl donor.

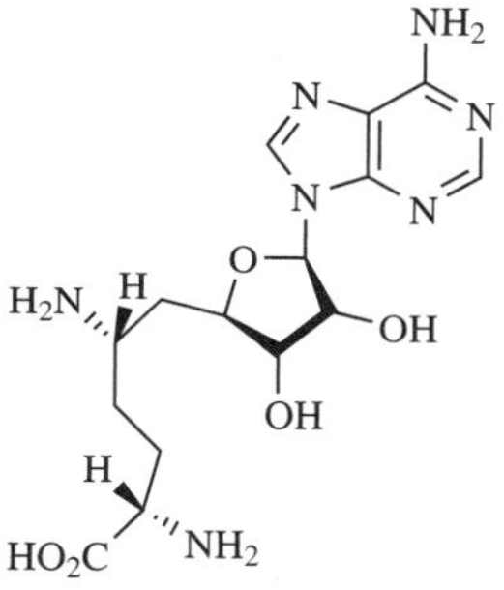

Sinefungin 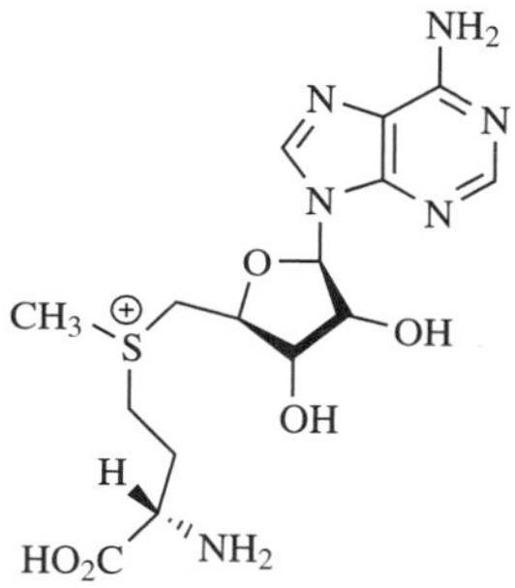   AdoMet 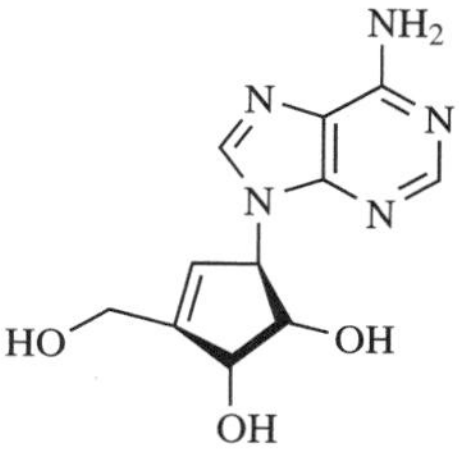   Neplanocin

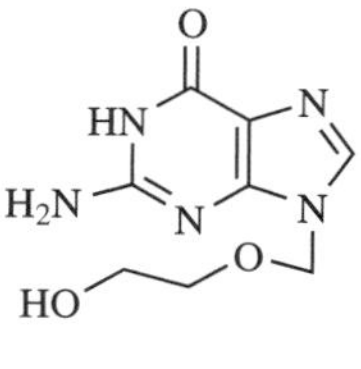

Acyclovir 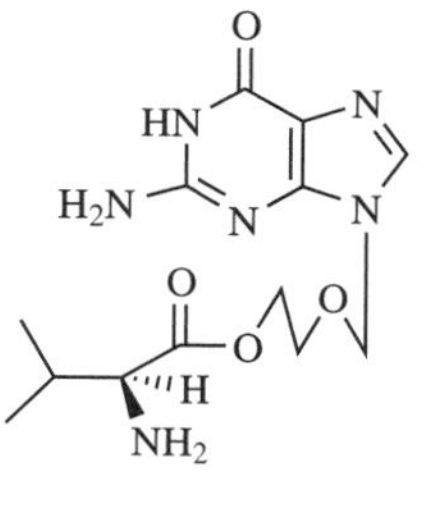   Valaciclovir

In contrast to the limited number of natural nucleosides, numerous synthetic nucleosides are now available for treating viral infections. This is because of the unlimited possibilities for modifications both at the carbohydrate and the base sites. Acyclovir or ACV [9-(2-hydroxyethoxymethyl)guanine] (34–43) and its oral prodrug–Valaciclovir (Val–ACV) (35,38,44–53) are the two most commonly prescribed drugs for the treatment of HSV infections. These compounds contain a unique structure as compared with the naturally occurring guanosine in that an acyclic side chain is designed to replace the cyclic ribose moiety. When tested in cell culture, the majority of isolates of HSV are sensitive to ACV. The ACV-resistant strains are rarely found in clinical practice among immunocompetent patients (<1% isolates). Resistant HSV occurs much more frequently, however, among immunocompromised patients during treatment (~5% isolates). Acyclovir and valaciclovir also show activity against human cytomegalovirus (HCMV) and varicella-zoster virus (VZV) (44,50,51,54,55). To date, acyclovir is the standard for the treatment of mucosal, cutaneous, and systemic HSV-1 and HSV-2 infections (including herpes encephalitis and genital herpes) and VZV infections. Valaciclovir was discovered to achieve substantially higher plasma levels of acyclovir than oral acyclovir itself. It has proven to be particularly useful in the treatment of herpes zoster and in the prevention of HCMV disease after renal transplantation (56). Other acyclic nucleoside analogues, besides ACV and Val–ACV, that are available for treatment of diseases caused by viruses belonging to the herpes family (comprising HSV-1, HSV-2, HCMV, and VZV), include penciclovir, famciclovir, ganciclovir, and cidofovir (35,36,45–47,49,53,57–74).

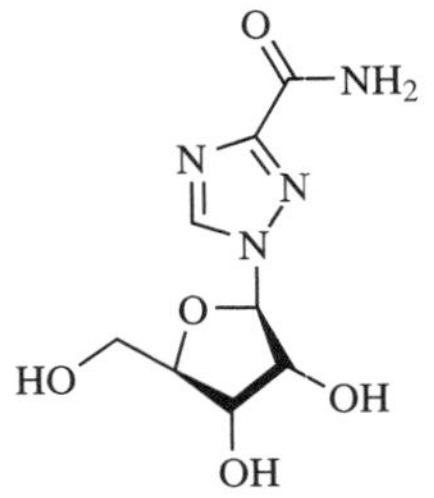

Ribavirin

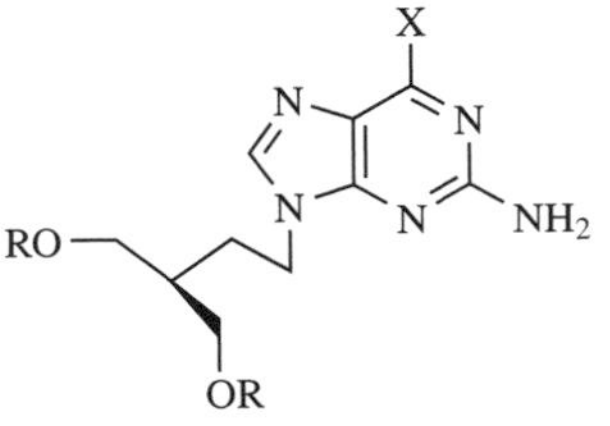

R = H, X = OH, Penciclovir
R = Ac, X = H, Famciclovir

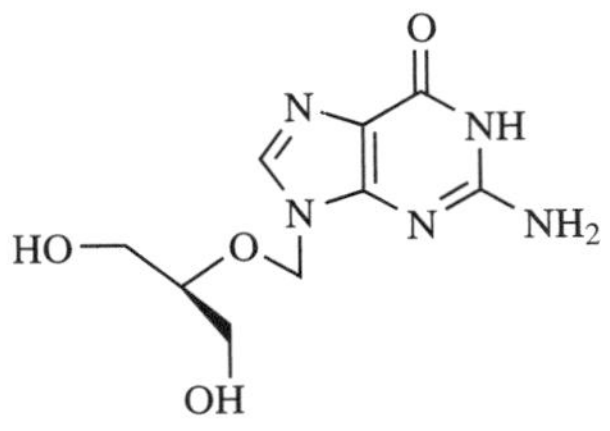

Ganciclovir

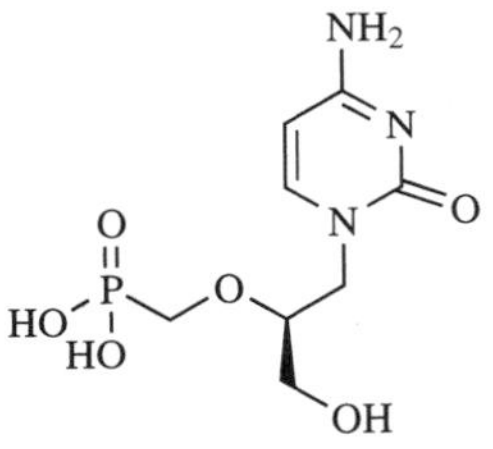

Cidofovir

Ribavirin, which was synthesized nearly 30 years ago (75) by the research group at ICN Pharmaceuticals, is a broad-spectrum antiviral agent (76–79). Instead of a usual purine or pyrimidine base, it has a five-membered triazole ring with a carboxamide substitution at position-3 of the heterocycle. A large number of RNA and DNA viruses are sensitive to ribavirin, including the respiratory syncytial virus, influenza, parainfluenza, herpes viruses, and RNA tumor viruses. It was found that 5′-monophosphate of ribavirin accounts for the antiviral action in mammalian cell culture (80). The 5′-monophosphate derivative of ribavirin was a potent inhibitor of the enzyme IMP dehydrogenase, thereby preventing the conversion of IMP to xanthine monophosphate (XMP) (81,82). XMP is required for guanosine triphosphate (GTP) synthesis. Thus, the antiviral activity of ribavirin might be due to the inhibition of GTP biosynthesis in virus-infected cells, which in turn results in the inhibition of viral nucleic acid synthesis. Several other possible mechanisms have been proposed (78,79), which include the recently described activity as an RNA mutagen (77,83–85,87,88,90). Ribavirin triphosphate (RTP) inhibits viral RNA polymerases. It also prevents the capping of viral mRNA by inhibiting guanyl N^7-methyltransferase. It was suggested that the phosphorylation of ribavirin was most likely accomplished by deoxyadenosine kinase (86). The kinetic studies from rat liver preparations showed that ribavirin and deoxyadenosine competitively inhibited the phosphorylation of each other. To date, success has been achieved with the aerosol use of ribavirin in treating respiratory syncytial virus infection in infants and young children (87–89). It also showed clinical benefit in treating severe and life-threatening infections caused by the Lassa fever virus (90). It is the first antiviral drug that is able to reduce mortality in a highly lethal systemic disease by more than 90%. Furthermore, ribavirin is the only approved nucleoside analogue for the treatment of hepatitis C virus (HCV) infections, but the approval is limited to combination therapy with interferon, another drug used against HCV. HCV currently threatens the global public health with more than 200 million people having been infected worldwide (85,91,92). However, there have been a few documented side effects associated with the use of ribavirin. The treatment results in a fall in transaminase levels and some decrease in hepatic inflammation.

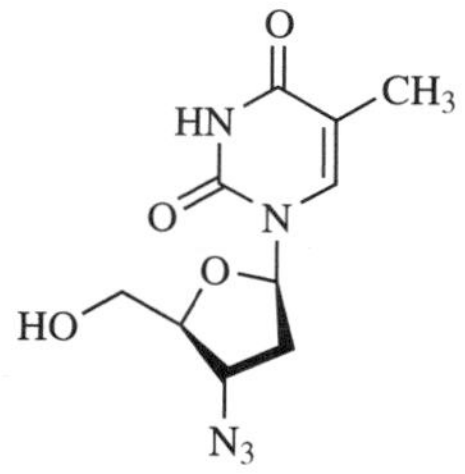

AZT

AZT (azidothymidine) and other 2′,3′-dideoxynucleoside analogues that are currently employed for treating HIV infections, are inhibitors of the HIV reverse transcriptase (HIV RT) (93,94). Their mechanism of action is believed to be the chain termination of nucleic acid synthesis during the RT-catalyzed reverse transcription of HIV RNA genome into its complementary DNA strand (95). The nucleoside analogue is first converted into its 5′-triphosphate derivative by the host kinases, which then is incorporated into the developing DNA strand opposite to an adenosine residue in the viral RNA template. The lack of the 3′-hydroxy group in AZT, which is crucial for the chain extension, prevents further incorporation of nucleotide building blocks beyond the point of insertion, thus leading to chain termination.

5. The Viruses of Current Health Concern and the Related Antiviral Therapy

As mentioned under Introduction, the following four viruses are currently of prime health concern worldwide: HIV, HBV, HCV and WNV, and so, this article will focus on these viruses and the progress being made on antiviral therapy to treat each of these viral infections.

5.1. Human Immunodeficiency Virus (HIV).

Perhaps no other virus in recent history has stirred more global panic and paranoia than HIV, an etiological agent causing the acquired immunodeficiency syndrome (AIDS) (96–98). The AIDS epidemic has made more impact on public health than even the black plague of the late Middle Ages. With >25 million people vanishing worldwide due to its infection since the early 1980s, and >40 million individuals currently infected with, HIV is one of the deadliest viruses ever to hit the mankind (99). Despite intense efforts from several research fronts including chemistry, biochemistry, biology, and biotechnology, and not to mention epidemiology and prevention measures, the fight to conquer HIV altogether still remains largely elusive. As proven techniques of viral attack seem inadequate against HIV, and chances for a suitable vaccine continue to be disappointing, the current research trend is to focus on the complete viral life cycle and the replication process for new targets. The ultimate success may lie in the power of modern molecular biology to explore every aspect of the HIV life cycle and every response of the human body toward viral invasion (97). The tools of biotechnology have greatly aided in sequencing the viral genome as well as the proteins that are associated with it. So, it is important to review the current status of knowledge on the viral structure and its life cycle (100–106) before delving into what is being targeted for antiviral therapy (107,108).

As classified earlier, HIV is a retrovirus consisting of two copies of a single-stranded RNA genome and a few replicative and accessory proteins within the boundaries of a lipoprotein shell (see Figure 1), known as the viral envelope. Embedded in the viral envelope is a complex protein known as *env*, which consists of an outer protruding cap glycoprotein (gp) 120, and a stem gp41. Within the viral envelope is an HIV protein called p17 (matrix), and within this is the viral core or capsid, which is made of another viral protein p24 (core antigen). The major elements contained within the viral core are two single strands of HIV–RNA, a protein p7 (nucleocapsid), and three enzyme proteins, p66 (reverse transcriptase), p11 (protease), and p31 (integrase) (100–106).

Infection begins when an HIV particle encounters a target T-Helper cell of the host containing a surface receptor molecule called CD4 (109). The virus particle uses gp120 to attach itself to the host cell membrane and then enters. Within the cell, the virus particle releases its RNA as well as the crucial enzyme *reverse transcriptase* (HIV RT), which converts the viral RNA into a cDNA copy. This new HIV–DNA then moves into the cell's nucleus where, with the help of the enzyme *integrase*, it is then inserted into the host cell's DNA. Once into the host cell's genes, HIV DNA is called a *provirus*. The HIV provirus is then replicated by the host cell, which then cranks out multiple copies of its own genome, and produces the mRNA necessary for creation of viral proteins. The fully replicated and packaged viral particles then bud out of the cell to infect fresh new cells.

In addition to HIV *reverse transcriptase* and *integrase*, the two key enzymes involved in the viral replication process as described above, a third enzyme called HIV *protease* is also a viable target for antiviral therapy. After replication within a host cell, when new viral particles are ready to break off to infect other cells,

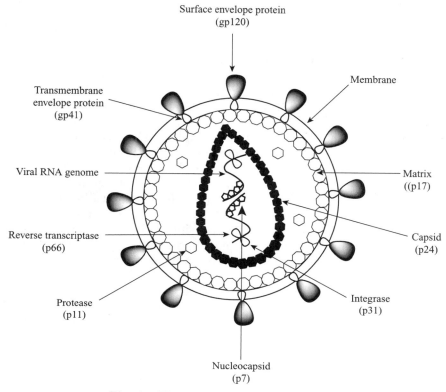

Fig. 1. The molecular structure of HIV.

protease plays a vital role in cutting longer protein strands into smaller parts needed to assemble a mature virus. These shorter polypeptides include three structural proteins, capsid (p24), matrix (p17), and nucleocapsid (p7), encoded by the viral *gag* gene, as well as three enzymes that are crucial for replication, including a reverse transcriptase (p51), an integrase (p31), and an RNAse H (p15), in addition to a new protease (p11), encoded by the viral *pol* gene. Thus, when protease is blocked, the new viral particles fail to mature.

Infection of the host T-cells by HIV involves a process of receptor interaction and membrane fusion. The viral *env* gene codes for the envelope protein (ENV) gp160. The latter is cleaved by HIV protease into two protein fragments called gp120 and gp41. While gp120 binds to the CD4 receptor on the surface of the host immune cells, gp41 mediates the fusion between viral and cellular membranes.

HIV protease thus plays a critical function in the HIV life cycle. Consequently, the detailed structural analysis of HIV protease has led to the discovery of protease inhibitors (110–115), one of the important components in the *highly active antiretroviral therapy*, commonly referred to as HAART Therapy (116–118) that consists of multiple drug regimen aimed at different targets in the viral life cycle. The "cocktail" regimen normally includes a protease inhibitor along with two HIV RT inhibitors, one a nucleoside and the other a nonnucleoside.

The HAART therapy has been highly successful in preventing AIDS-related deaths in the industrialized world, although it had relatively low impact in the developing world because of the high cost of medication.

5.2. Antiviral Therapy for HIV Infections. *Vaccines.* To date, over 60 phase I/II trials of 30 candidate vaccines against HIV have been conducted worldwide (98,119–133). Most initial approaches focused on the HIV envelope protein, produced in insect, bacteria, yeast, or mammalian cells, which was logical given that envelope is the primary target for neutralizing antibodies in HIV-infected persons. At least 13 different gp120 and gp160 envelope candidates have been evaluated in phase I/II trials, predominantly through NIAID-supported AIDS vaccine evaluation group. Most research focused on gp120 rather than gp140/gp160, as the latter are generally more difficult to produce and did not initially offer any clear advantage over gp120 forms. Overall, they have been safe and immunogenic in diverse populations, and have induced neutralizing antibody in nearly 100% recipients, but rarely induced CD8+ cytotoxic T lymphocytes (CTL) even when formulated in novel adjuvants that effectively induced CTL in mice, although mammalian-derived envelope preparations have been better inducers of neutralizing antibody than candidates produced in yeast and bacteria. CTLs recognize surface markers on other cells that have been labeled for destruction. In this way, CTLs help to keep virus-infected (or malignant) cells in check. The antibodies induced by these early envelope preparations rarely neutralized primary isolates of HIV.

In an effort to induce both CTL and antibody responses, the attention was turned to evaluating a combination vaccine approach in which two types of vaccines are used (124,125,129,131). Most commonly referred to as "prime-boost", this has involved an immunization (priming) with a recombinant viral vector followed by or combined with boosting doses of recombinant protein. Three recombinant attenuated vaccinia vectors and five recombinant canarypox vectors have been evaluated in phase I trials alone and in combination with a recombinant protein envelope boost. In general, vaccinia-immune individuals have not responded as well as vaccinia-naïve individuals to vaccinia vectors, although there has been no difference in the response of these groups to recombinant canarypox vectors. All recombinant viral vectors have been safe and immunogenic to date, and have been shown to prime the immune response to an envelope boost, thereby necessitating fewer doses of recombinant protein to reach maximum antibodies titers. However, the antibodies elicited in prime-boost protocols so far have a limited breadth of reactivity.

The availability of several recombinant canarypox vectors has provided interesting results that may prove to be generalizable to other viral vectors. Canarypox is the first candidate HIV vaccine that has induced cross-clade functional CTL responses. Increasing the complexity of the canarypox vectors by inclusion of more genes/epitopes has increased the percent of volunteers that have detectable CTL to a greater extent than did increasing the dose of the viral vector. Importantly, CTLs from volunteers were able to kill peripheral blood mononuclear cells infected with primary isolates of HIV, suggesting that induced CTLs could have biological significance.

Other strategies that have progressed to phase I trials in uninfected persons include peptides, lipopeptides, DNA, an attenuated Salmonella vector, lipopeptides, p24, etc. To date, none has proven as effective in eliciting human CTL and/or antibody as the recombinant canarypox-envelope combination. Merck has advanced a candidate DNA vaccine (125,129,134–136) containing a codon-optimized gag gene to phase I trials. In 2001, NIAID began phase I trials of a vaccine that contained DNA for the *gag* and *pol* genes. Gag and Pol are considered good candidates for developing an AIDS vaccine as they are relatively constant across different virus strains and account for a large percentage of total virus protein. Other approaches to improve the immunogenicity of DNA vaccines are being pursued and may enter phase I trials over the next few years.

In summary, clinical trials of candidate HIV vaccines have so far been only informative. In the absence of validated correlates of immune protection, larger trials of the most promising candidates will be needed. Furthermore, as promising candidates advance to efficacy trials, there does appear to be room for improvement. There is at least as much if not more known about the HIV genome than other pathogens for which vaccines have successfully been made. Advances in genomics and micro-array technologies will likely have multiple applications in the field of HIV vaccine development. For one, new DNA approaches in which combinations of DNA containing genes of different clades are currently in preclinical research. Methods that help identify optimal DNA sequences for inducing CTL in proposed trial populations with defined HLA backgrounds could help increase the immunogenicity of these and other approaches. In addition, there will be a need to apply new, highly sensitive techniques for HIV detection to determine true infection and to detect infection in small volume samples in high through put assays. Finally, the advent of micro-array technologies could prove to be useful in exploring and cataloguing immune response genes that are up or down regulated and that correlate with protection.

Chemotherapy. Currently, there are a total of 16 drugs that have been approved by the U.S. Food and Drug Administration (FDA) for the treatment of AIDS (93,94). Seven out of the 16 are nucleoside-based reverse transcriptase inhibitors (NRTI), three are nonnucleoside reverse transcriptase inhibitors (NNRTI), and 6 are protease inhibitors. NRTIs (Figure 2) include AZT (zidovudine) (93,137,138), ddC (93,139) (zalcitabine), ddI (didanosine) (93,139), d4T (stavudine) (93,140,141), 3TC (lamivudine) (93,137,142,143), abacavir (93,94,144–148), and Bis-POC-PMPA (tenofovir disoproxil) (93,94,147,149–152), while the approved NNRTIs (Figure 3) are nevirapine (viramune) (153), delavirdine (93,154,155), and efavirenz (sustiva; stocrin) (93,144,155–157). The six protease inhibitors that have been FDA-approved include saquinavir (93,94,112,113, 115,144,147,158,159), indinavir (93,94,112,115,147,160,161), ritonavir (93,94, 111–115,144,147,162,163), nelfinavir (93,94,112,113,115,147,161), amprenavir (93,94,112,113,147,162,164–170), and lopinavir (93,94,111–114,144,171,172) (Figure 4). The HIV reverse transcriptase (RT) (173–178), coded by the *pol* gene, is both RNA- and DNA-dependent polymerase. While HIV–RT makes a DNA copy of the viral RNA template, RNAse H of the virus chews away the RNA strand from the initially formed RNA–DNA hybrid. This will allow further synthesis of viral DNA duplex, which can then integrate into the host genome assisted by viral integrase. HIV–RT has been the key target of anti-AIDS drugs for a number of years, and still continues to be the major focus in the HAART therapy (179) described earlier. All of the nucleoside RT inhibitors

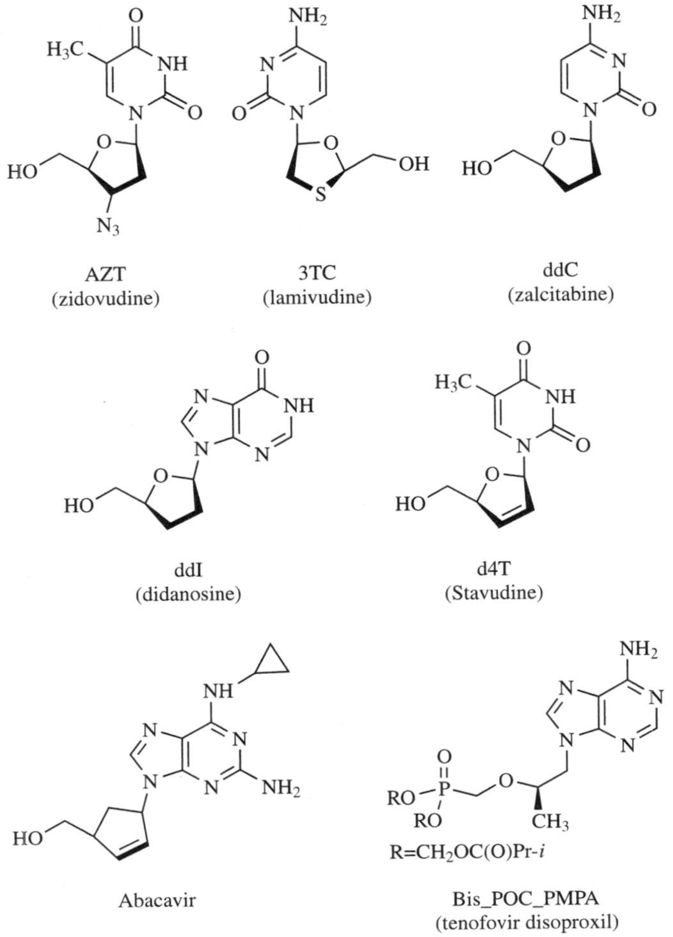

Fig. 2. FDA-approved nucleoside reverse transcriptase inhibitors (NRTIs) against HIV.

(NRTIs) (93,94,179,180) share a common mechanistic principle in that they are phosphorylated *in vivo* in the host cells, and subsequently are incorporated into the developing viral nucleic acids. This in turn results in nucleic acid chain termination since NRTIs lack the crucial 3′-hydroxy group that is necessary for chain extension.

Like NRTIs, NNRTIs (93,147,155,161,179–184) also target HIV–RT. However, unlike NRTIs, they bind RT at a secondary or allosteric site instead of at the active sites utilized by the NRTI. This causes conformational change in RT, which leads to alteration of the active site pocket. The change results in reduced binding of naturally occurring nucleosides and thus reduced viral cDNA elongation. NNRTIs are direct inhibitors of HIV reverse transcriptase that, unlike RTIs, are not incorporated into the viral DNA molecule. A major advantage of NNRTIs, therefore, is that these compounds require no phosphorylation by cellular enzymes in order to be active. NNRTIs work synergistically with

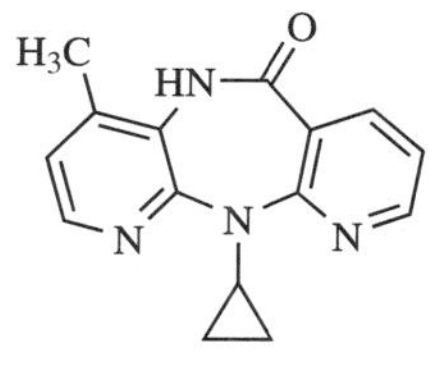

Nevirapine
(viramune)

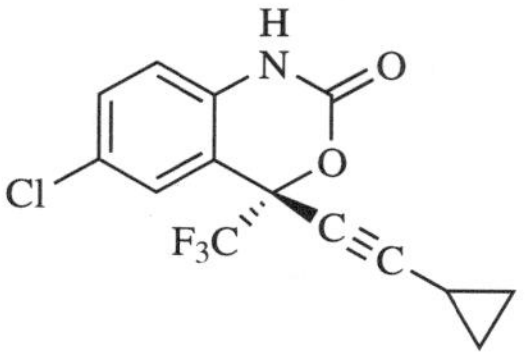

Efavirenz
(sustiva or stocrin)

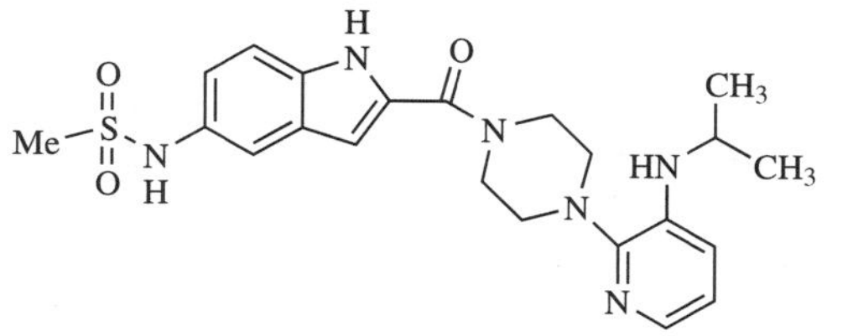

Delavirdine

Fig. 3. FDA-approved nonnucleoside reverse transcriptase inhibitors (NNRTIs) against HIV.

most NRTIs, and are demonstrating impressive efficacy in increasing immunologic markers and decreasing viral load markers in HIV-infected patients. While they are very potent antiretrovirals, they also suffer from a major drawback in that the resistance against them can develop quickly if the drugs are not taken exactly as prescribed, and once the resistance develops to one drug in the class, there will probably be a resistance to all the drugs in that class. Thus, NNRTIs appear to be highly cross-resistant. A mutation at position 103 on the HIV reverse transcriptase gene is known to confer resistance to all of the agents in the class. However, this mutation does not confer resistance to drugs in other classes.

Two other classes of drugs against HIV that are currently under clinical trials are *Integrase* (93,94,185–191) and *Fusion* Inhibitors (192–195). HIV integrase is the third key enzyme in HIV replication besides protease and reverse transcriptase. As described earlier, the integration of provirus into the host genome is catalyzed by virally encoded integrase which has multiple functions. First, it acts as an exonuclease to cut the complementary viral DNA produced by HIV–RT to the appropriate size. Second, it serves as an endonuclease to cut the host DNA so as to facilitate insertion of the provirus. Finally, it acts as a ligase to fuse the host and viral DNAs into a seamless whole. Currently, a few drugs are being developed for inhibition of this integration step. At the most recent fourteenth International AIDS conference held in Barcelona, Spain in July 2002, Merck and Co. presented the results of clinical and animal trials of its two HIV integrase inhibitors, called L870810 and L870812. Initial data indicate that they are safe and well-tolerated in healthy volunteers.

HIV Fusion Inhibitors are a new class of drugs that bind to the viral protein gp120 and prevent HIV from infecting host cells. The virus is basically frozen

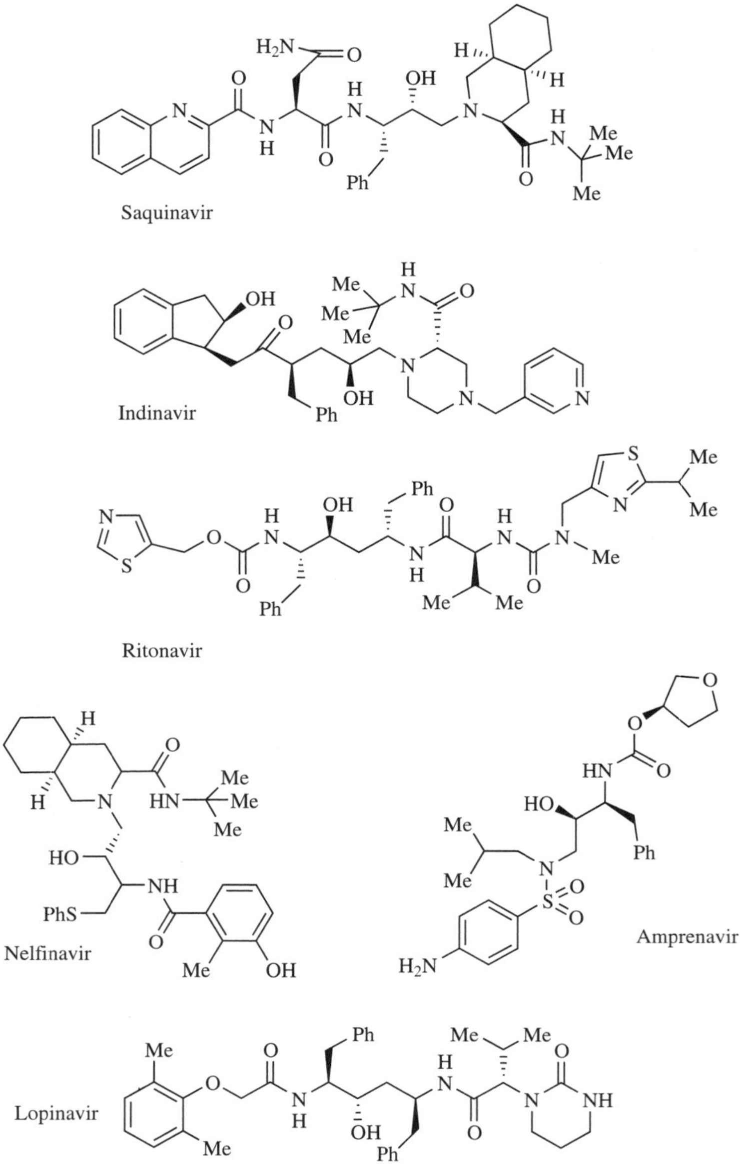

Fig. 4. FDA-approved protease inhibitors against HIV.

on the surface of the cell preventing it from entering the cell, and therefore cannot propagate in an HIV-infected person. The drug called T-20, which is being developed by Trimeris Research, belongs to this category. T-20 is a polypeptide (L-phenylalaninamide, *N*-acetyl-L-tyrosyl-L-threonyl-L-seryl-L-leucyl-L-isoleucyl-L-histidyl-L-seryl-L-leucyl-L-isoleucyl-L-.alpha.-glutamyl-L-.alpha.-glutamyl-L-seryl-L-glutaminyl-L-asparaginyl-L-glutaminyl-L-glutaminyl-L-.alpha.-glutamyl-L-lysyl-L-asparaginyl-L-.alpha.-glutamyl-L-glutaminyl-L-.alpha.-glutamyl-L-leucyl-L-leucyl-L-.alpha.-glutamyl-L-leucyl-L-.alpha.-aspartyl-L-lysyl-L-tryptophyl-L-alanyl-L-seryl-L-leucyl-L-tryptophyl-L-asparaginyl-L-tryptophyl-). Data from initial clinical trials show that it works effectively in patients for whom other types of drugs no longer work well. All patients in the clinical studies had already developed resistance to all three types of anti-HIV drugs now on the market, and many were developing full-blown AIDS. At the fourteenth International AIDS conference held in Barcelona, Spain in July 2002, T-20 was hailed as one of the most exciting things to happen since the discovery of protease inhibitors.

Another interesting antiviral therapy that is currently being actively pursued by a number of researchers is based on the enzyme ribonuclease H (RNAse H) of HIV (196). The latter is responsible for digesting the RNA strand of the initially formed RNA–DNA hybrid from the viral RNA template. This important property of RNAse H is exploited in promoting the enzyme-catalyzed destruction of the viral mRNA target via formation of RNA–DNA duplexes employing appropriately designed complementary antisense oligonucleotides (AON). While the success has so far been limited, many important criteria are emerging to enable to draw correlation between the structure of the hybrid and its property as a suitable substrate for RNAase H, as well as to reveal the crucial structural requirements for AONs to preserve their RNAse H potency (197).

In summary, despite enormous progress made in understanding its life cycle as well as its potentially viable targets for the development of antiviral therapies, HIV remains an elusive virus even in the face of the successful HAART therapy. The major obstacle in conquering HIV through therapy concerns the high level of viral mutagenicity and the consequent drug resistance. Its eerie ability to integrate into the host to kill the very cells that are normally mobilized to confront the invader makes HIV a formidable virus. Neither the efficacious long-term therapies nor the uniformly effective vaccines against HIV are yet close in sight, but the international research to fight the virus continues unabated.

5.3. Hepatitis B Virus (HBV). According to the estimates of World Health Organization (WHO), HBV has infected over two billion people worldwide, making it one of the most ubiquitous human pathogens on earth, and ranking third in the global illnesses behind venereal disease and chickenpox (198–206). Approximately 500 million of the infected people are chronic carriers, and about 1 million die each year from HBV-related chronic liver disease. Most people are from Asia, Africa, and the western Pacific, although >1.5 million are infected in the United States, and ~15,000 new cases are detected each year. Mother to infant transmission accounts for most cases in the undeveloped countries, whereas unsafe sex and body fluid contacts are the major forms of transmission in the developed countries. The viral transmission occurs primarily through blood and/or sexual contact, though other methods of transmission have also been suggested. Transmission is most efficient via percutaneous mode, whereas sexual transmission is somewhat inefficient. The virus is primarily found in the blood of infected individuals, and virus titres as high as 10 billion virions per milliliter of blood have been reported. HBV has also been detected in other body fluids including urine, saliva, nasopharyngeal fluids, semen, and menstrual fluids. However, HBV has so far not been detected in feces, perhaps due to inactivation and degradation within the intestinal mucosa or by the bacterial flora.

HBV is responsible for both acute and chronic hepatitis (198,207–210). Individuals infected with acute HBV show no apparent clinical signs of the disease, but at the end of the incubation period, a flu-like symptoms, such as fever, fatigue, and general discomfort, and in some cases jaundice, will occur. About 2–10% of the adult acute HBV carriers will become chronic carriers of the disease, but in infants this percentage is >90% via neonatal exposure. An average of 25–40% of the chronic carriers will develop liver cirrhosis and primary hepatocellular carcinoma (HCC) (211–220), which are the major causes of morbidity and mortality. In the last few decades, the correlation between HBV and the development of HCC has been well established, although the mechanism by which HBV transforms hepatocytes still remains elusive. Before HBV can transform a cell, the virus must first infect it. However, the mechanism through which HBV enters hepatocytes has not been resolved despite further understanding of the viral proteins involved. Vaccines are available against HBV, but they may not be 100% effective against all variants of HBV. Furthermore, there is no cure for individuals already infected. Much more research is needed before we fully understand and control the spread of this infectious agent. The HBV life cycle is depicted in Figure 5 (210,221–230). As mentioned above, the virus attachment and entry into the host cell, as well as the cellular receptor for the virus are as yet poorly understood. After the initial entry, the viral core particle is translocated into the host nucleus. The viral DNA then becomes matured, forming a covalently closed circular DNA (called *ccc*DNA or supercoiled DNA). The *ccc*DNA remains episomal and serves as a template for cellular RNA polymerase II, giving rise to many viral RNA transcripts. The largest of these RNA transcripts serves as both mRNA for the viral polymerase (HBV DNA polymerase) and pregenomic RNA, which is slightly larger than genomic size, and is translated and packaged into viral particles. Concurrently, the smaller RNA transcripts-PreS- and PreC- are translated into surface and core proteins of the virus. The viral DNA is synthesized using reverse transcription of the pregenomic RNA by HBV DNA polymerase. The initial synthesis of (−) strand DNA is followed by synthesis of a short (+) strand DNA in a remarkable process, unique to HBV, called *priming*. Priming involves a specific tyrosine residue located at the N-terminus of HBV polymerase, which forms a covalent bond with the initiating deoxynucleotide residue, normally a dGTP. Priming is templated by a bulge sequence in the stem-loop structure (epsilon or,) on the pregenomic RNA, and results in a short DNA oligomer, covalently linked to the polymerase. This covalent enzyme–DNA adduct is then translocated to the appropriate complementary sequence at the 3′-end of pregenomic RNA. The (−) strand is then elongated via reverse transcriptase activity of the polymerase. The newly synthesized partially double-stranded viral DNA is either recycled as a resource for *ccc*DNA or functions as the viral nucleic acids in the matured virions budding out of the host cells.

5.4. Antiviral Therapy for HBV Infections. *Vaccine.* HBV vaccine is the first successful *recombinant vaccine* against a human infectious disease, in particular, against a mucosal virus (127,213,216, 231–249). The original vaccine, prepared in 1978 and licensed in the United States in 1981, was based on the

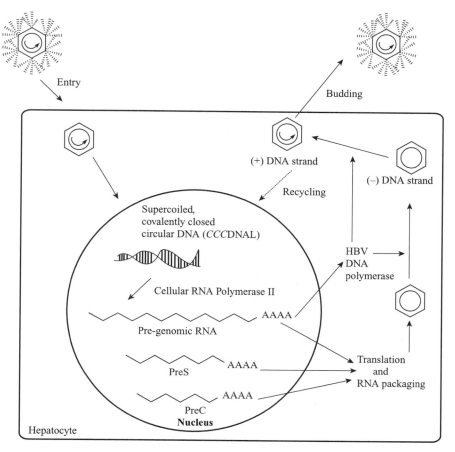

Fig. 5. The life cycle of the hepatitis B virus.

viral envelope protein (HBsAg). The latter was isolated and purified from the plasma of individuals infected with chronic HBV. The plasma vaccine has been replaced with the recombinant vaccine, prepared from yeast, mainly because the principal source of HbsAg-positive plasma was the same population that was also at the highest risk of contracting AIDS. Although the vaccine can help prevent the spread of HBV, it is not useful for those 350 million people who have already been chronically infected with the virus.

Anti-HBV Therapy. (*a*) *Interferons:* Interferons are a family of proteins— α, β, ω, and γ—that are induced in response to viral infections or double-stranded RNA. Interferon α is the most effective against HBV, and has been approved by FDA in 1991 for treatment of chronic HBV infections (212,213,215,250–255). Relapse of the disease after discontinuation of treatment, side effects, high expense of the drug, and the necessity to administer the drug only through injection are some of the limitations of interferon therapy. (*b*) *Nucleoside Analogues:* Nucleoside analogues are currently the most intensely studied anti-HBV agents (28,213,224,227,236,250–252,256–272). Some of the leading candidates include

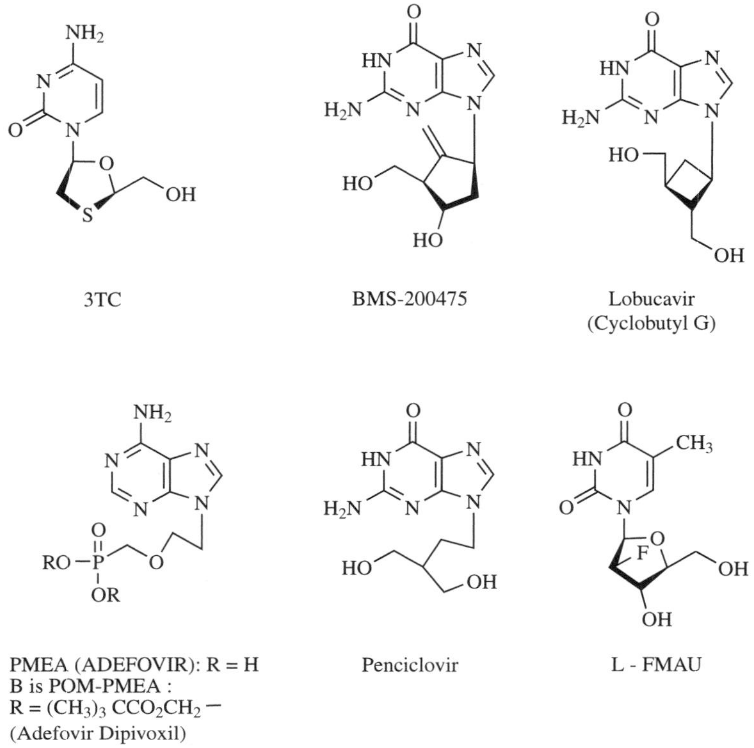

Fig. 6. Nucleoside inhibitors of replication of the hepatitis B virus (HBV).

3TC (lamivudine) (28,250,252,256,257,260–262,273–275), BMS-200475 (276–281), lobucavir (230,282–284), PMEA (adefovir) (285–289), adefovir dipivoxil (259,265,282,290–299), penciclovir (famciclovir) (257,300–306), and L-FMAU (227,262,282,307–317) as outlined in Figure 6. The clinical trials with a number of nucleoside analogues, however, were either unsuccessful or accompanied by severe toxicity. Lamivudine (3TC) is the first, effective, and reasonably well tolerated, oral treatment for chronic HBV infection, approved by FDA. The other approved drug for chronic HBV infection is adefovir dipivoxil. Lamivudine is an inhibitor of RT, and is in clinical use in HIV-infected individuals. The use of lamivudine on patients with HBV infection clearly shows the development of resistance arising from base-pair substitutions at a specific locus called YMDD of the viral DNA polymerase, resulting in significant clinical problem (275). So, the future of HBV chemotherapy may reside in combination drug therapy with newer, less toxic nucleoside analogues, along with other classes of agents including immunomodulators. With the discovery of 3TC, a nucleoside with a sugar moiety in the unnatural L-configuration, and its potent dual activity against both HIV and HBV, the interest in L-nucleoside analogues has taken on an explosive course. A number of L-nucleosides are currently undergoing preclinical and clinical trials against both HIV and HBV as well as other viral

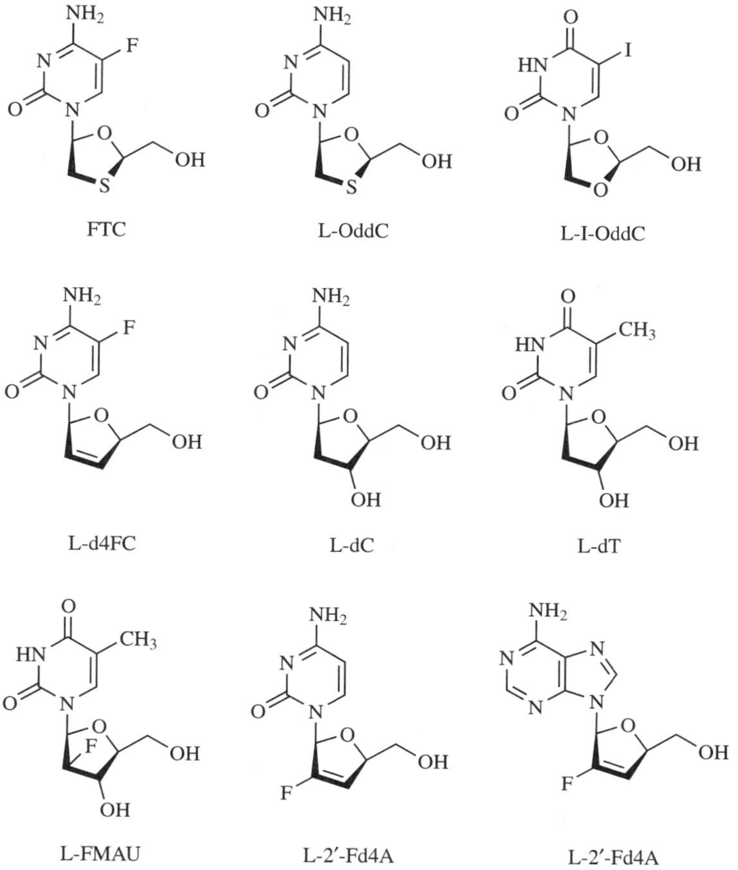

Fig. 7. L-Nucleoside analogues that are currently undergoing preclinical and clinical trials against HIV, HBV, and other viral infections.

infections, and are listed in Figure 7. Beneficial features of L-nucleosides include an antiviral activity comparable or sometimes greater than their natural D-counterparts, a more favorable toxicological profile, and more importantly, a greater metabolic stability due to their lower susceptibility to catabolic and hydrolytic enzymes. The synthesis and biology of L-nucleosides have been the object of many recent reviews (28,29,318).

Recently, we reported that the ring-expanded nucleosides **REN-1** (319) and **REN-2** (320) containing the imidazo[4,5-*e*][1,3]diazepine ring system, along with nucleoside **REN-3** (321), containing the imidazo[4,5-*e*][1,2,4]triazepine ring system (Figure 8), exhibit potent and selective *in vitro* anti-hepatitis B virus (anti-HBV) activity in cultured human hepatoblastoma 2.2.15 cells (322). The 50% effective concentration (EC_{50}) values for inhibition of extracellular virion release are 0.39, 0.13, and 4.2 μM, respectively. The compounds were also able to inhibit intracellular HBV DNA replication intermediates (RI) in 2.2.15 cells following 9 days of treatment. In addition, they exhibited very low cellular toxicities

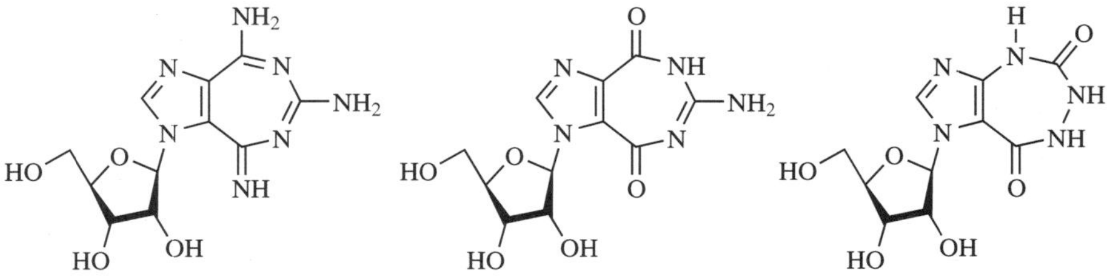

Fig. 8. Ring-expanded nucleosides with potent *in vitro* anti-hepatitis B virus (anti-HBV) activity with little, if any, toxicity.

with the respective selective indices (SI) of 1262, 18,526, and 439. The EC_{50} value for inhibition of virion DNA synthesis by **REN-2** suggested that it was two- to threefold less potent than 3TC. Comparison of the antiviral activity of **REN-1** and **REN-2** reveals that the replacement of the amino groups on the seven-membered heterocycle with oxygen increased the *in vitro* anti-HBV activity by threefolds. More importantly, this change in the structure resulted in a decrease in the *in vitro* cellular toxicity of **REN-2** ($CC_{50} = 2427$ μM) by fivefolds compared with toxicity of **REN-1** ($CC_{50} = 501$ μM) in confluent 2.2.15 cells. It was interesting to note that all three compounds showing antiviral activity were riboside analogues since 2′-deoxy analogues of both **REN-1** and **REN-2** were found to be inactive against HBV.

We then evaluated nucleosides **REN-1**, **REN-2**, and **REN-3** for their ability to inhibit viral RNA synthesis in 2.2.15 cells (322). Since HBV uses host cellular RNA polymerase II for the transcription of viral RNA from the covalently closed circular HBV–DNA during its replication, any effect on the synthesis of viral RNA by these compounds would mean interference with the cellular RNA polymerase which could lead to unacceptable cellular toxicity. Like 3TC, all three compounds showed no inhibition of the synthesis of viral 3.6 and 2.1 kb RNA by HBV in 2.2.15 cells. In spite of no suppression of viral RNA synthesis in the presence of these compounds, it was interesting to see that treatment of 2.2.15 cells with these compounds, unlike 3TC, did result in the reduction of viral protein synthesis especially that of the core antigen. The significance of this observation is not clear at the present time.

Another interesting and useful observation was that the antiviral activity exhibited by **REN-1**, **REN-2**, and **REN-3** was specific against HBV (322). These compounds were also tested against HIV, herpes simplex virus (HSV-1 and HSV-2), cytomegalovirus (CMV), VZV and EBV. They showed no antiviral activity against any of these viruses.

In vitro cellular toxicity of **REN-1**, **REN-2**, and **REN-3** was evaluated in several stationary and rapidly growing cell systems. Toxicity of **REN-2** was also studied in bone marrow precursor cells (by erythroid burst forming units and granulocyte macrophage CFU). In bone marrow precursor cells, **REN-2** had CC_{50} values that are comparable to those exhibited by 3TC. In rapidly growing human HFF cells and Daudi cells, all three compounds were found to be nontoxic up to 100 and 50 μM concentrations, respectively. In summary, ring-expanded nucleosides represented by compounds **REN-1**, **REN-2**, and **REN-3** carry excellent promise as therapeutic agents against chronic HBV infections, and to that end, the efforts are currently underway in our laboratory.

5.5. Hepatitis C Virus (HCV).

The hepatitis C virus (HCV) is one of the most dreadful infectious diseases of modern times, which has currently infected >175 million people worldwide and >5 million in the United States (323–332). What makes it so dreadful is that most people do not even know that they have been infected with the virus as it can remain dormant for scores of years in the infected individual without revealing any signs or symptoms of the disease. Some estimates say the number of HCV-infected individuals may be four times the number of those infected with the AIDS virus, the main differences being that hepatitis C does not kill as quickly as AIDS. Until 1989, HCV was known by the name non-A, non-B hepatitis, when scientists at Chiron, Inc. succeeded in isolating portions of the HCV genome and conclusively demonstrated that the virus was indeed responsible for the noted pathogenicity that did not fit the category of either the A or B type hepatitis, and so classified it as type C. Subsequently, the complete genomes of various HCV isolates were cloned and sequenced by several research groups.

HCV is one of the major causes of chronic liver disease in the United States (323,324). It accounts for ~15% of acute viral hepatitis, 60–70% of chronic hepatitis, and up to 50% of cirrhosis, end-stage liver disease, and liver cancer. Hepatitis C causes an estimated 8,000–10,000 deaths annually in the United States. Hepatitis C is the major reason for liver transplants in the United States, accounting for 1000 of the procedures annually. A conspicuous characteristic of hepatitis C is its tendency to cause chronic liver disease. At least 75% of patients with acute hepatitis C ultimately develop chronic infection, and most of these patients have accompanying chronic liver disease. But chronic hepatitis C varies greatly in its course and outcome. At one end of the spectrum are patients who have no signs or symptoms of liver disease and completely normal levels of serum liver enzymes. Liver biopsy usually shows some degree of chronic hepatitis, but the degree of injury is usually mild, and the overall prognosis may be good. At the other end of the spectrum are patients with severe hepatitis C who have symptoms, HCV–RNA in serum, and elevated serum liver enzymes, and who ultimately develop cirrhosis and end-stage liver disease. In the middle of the spectrum are many patients who have few or no symptoms, mild-to-moderate elevations in liver enzymes, and an uncertain prognosis. Some patients learn they have hepatitis C only after a routine physical or when they donate blood and a blood test shows elevated liver enzymes. Further testing for HCV antibodies using the enzyme immunoassay (EIA) test and a supplemental test such as the "Western blot" or HCV–RNA detection can positively identify the infection. A liver biopsy shows disease manifested by damage already done to the liver. It is estimated that at least 20% of patients with chronic hepatitis C develop cirrhosis, a process that takes 10–20 years. After 20–40 years, a smaller percentage of patients with chronic disease develop liver cancer.

The virus is transmitted primarily by blood and blood products (323,324). The majority of infected individuals have either received blood transfusions prior to 1990 (when screening of the blood supply for HCV was implemented) or have used intravenous drugs. Sexual transmission between monogamous couples is rare but HCV infection is more common in sexually promiscuous individuals. Perinatal transmission from mother to fetus or infant is also relatively low but possible. Many individuals infected with HCV have no obvious risk factors. Most of these persons have probably been inadvertently exposed to contaminated blood or blood products.

HCV is also considered an opportunistic infection in HIV-infected individuals, and about one quarter of them are also infected with HCV (333–347). Since HCV is transmitted primarily by large or repeated direct percutaneous (ie, passage through the skin by puncture) exposures to contaminated blood, coinfection with HCV is common (50–90%) especially among HIV-infected injection drug users. Also, HCV infection progresses more rapidly to liver damage in HIV-infected persons. HCV infection may also impact the course and management of HIV infection. Prevention of HCV infection for those not already infected and reducing chronic liver disease in those who are infected are important concerns for HIV-infected individuals and their health care providers.

HCV is a member of the family of RNA viruses called *Flaviviridae* (348–351) to which also belongs the West Nile virus, another frightful virus of current notoriety in the United States and the western hemisphere as stated in the introduction. The viruses of the *flaviviridae* family are small, enveloped, spherical particles of 40–50 nm in diameter with single-stranded, positive sense RNA genomes (352–354). They are known to be the cause of severe encephalitic, hemorrhagic, hepatic, and febrile illnesses in humans. The viral genome encodes a polyprotein of 3000–4000 amino acids that is processed by host-cell and viral proteases into three structural (C, prM, and E) and seven non-structural (NS) proteins (see Figure 9) (352,354,355). Among these proteins the NS3 appears to be the most promising target for antiviral agents because of the multiple enzymatic activities associated with this protein. NS3 exhibits serine protease-, RNA-stimulated nucleoside triphosphatase (NTPase)-, and RNA helicase activities (356–358). The catalytic domain of the chymotrypsin-like NS3 protease has been mapped to the NH2-terminus region of the NS3, whereas the NTPase and the helicase activities are associated with the COOH-terminus of NS3

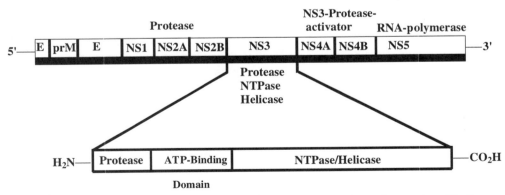

Fig. 9. Schematic representation of the structure of flaviviridae polyprotein with the expanded NS3 region. The enzymatic activities associated with the nonstructural proteins are shown.

(355,356). Helicases are capable of unwinding duplex RNA and DNA structures by disrupting the hydrogen bonds that keep the two strands together (359,360). This unwinding activity is essential for the virus replication. Recently reported "knock out" experiments demonstrated unambiguously that the switch-off of the helicase activity abolishes the virus propagation of bovine diarrhea virus (BVDV) and of dengue fever virus (DENV). According to the data, the inhibition of the helicase activity associated with NS3 protein may be an effective tool for reduction of virus replication. The NS5 region, on the other hand, is associated with the viral RNA-dependent RNA polymerase activity. The NTPase/helicase activities of NS3, along with the RNA-dependent RNA polymerase activity of NS5, are thought to be the essential components of the viral replicase complex (361), and therefore, are the potential targets for development of antiviral therapy.

5.6. Antiviral Therapy for HCV Infections. *Vaccines.* There is no vaccine for HCV and vaccines for hepatitis A and B do not provide immunity against hepatitis C (248,362–373). There are various strains of HCV and the virus undergoes mutations. Consequently, it will be difficult to develop a vaccine. Also, there is no effective immune globulin preparation. Furthermore, despite the discovery of HCV by molecular biological methods and the successful sequencing of the entire genome, a permissive cell culture system for propagating HCV has yet to be established. Although breakthroughs have been made recently in the development of model systems for studying viral RNA replication, no cell lines are yet available for producing the infectious virus. A non-primate animal model also does not exist. As a result, the production of specific drugs against HCV has been impeded although excellent diagnostic methods for it have been developed. An encouraging news, nevertheless, is the recent discovery that mutations in a protein of certain strains of HCV will allow these strains to replicate more vigorously in human cell culture. The *in vitro* assay based on this finding, called the HCV-replicon system (373), is a big step forward in improving an essential tool for studying the virus and suggests a starting point for the design of effective vaccines.

Currently Available Treatments. (a) *Interferons:* All current treatment protocols for hepatitis C are based on the use of various preparations of interferon alpha, which are administered by intramuscular or subcutaneous injection (323,374–384). Interferon alpha is a naturally occurring glycoprotein that is secreted by cells in response to viral infections. It exerts its effects by binding to a membrane receptor. Receptor binding initiates a series of intracellular signaling events that ultimately leads to enhanced expression of certain genes. This leads to the enhancement and induction of certain cellular activities including augmentation of target cell killing by lymphocytes and inhibition of virus replication in infected cells. Interferon alfa-2a (Roferon-A; Hoffmann-La Roche), inteferon alpha-2b (Intron-A; Schering-Plough) and interferon alfacon-1 (Infergen; Intermune) are all approved in the United States for the treatment of adults with chronic hepatitis C as single agents. More recently peginterferon alpha, sometimes called pegylated interferon, has been available for the treatment of chronic hepatitis C. There are two preparations of peginterferon alpha that have been studied in patients with hepatitis C: peginterferon alpha-2b (Peg-Intron; Schering-Plough) and peginterferon alpha-2a (Pegasys; Hoffmann-La Roche). The differences between these two preparations are subtle and most data suggest that they are equivalent with regards to efficacy and side effect profile. Peginterferon alphas differ from the older, unmodified interferon alphas in that a polyethylene glycol molecule is attached to the interferon molecule. As a result its elimination from the body is slowed and higher, more constant blood levels of interferon alpha are achieved with less frequent dosing. In contrast to unmodified interferon alpha, which must be injected three times a week to treat chronic hepatitis C, peginterferon alpha needs to be injected only once a week. With peginterferon alpha-2a alone, ~30–40% of patients achieve a sustained response to treatment for 24–48 weeks (332,385). *(b) Nucleoside Analogue Ribavirin and Combination Therapy:* As mentioned earlier under general description of nucleoside analogues as antiviral agents, ribavirin is a synthetic nucleoside containing a five-membered triazole ring, which has shown activity against a broad spectrum of viruses (77,79,90,376,382,383,386–389). In several studies, oral ribavirin was examined as a single agent for the treatment of adults with chronic hepatitis C. Although decreases in serum alanine transaminase (ALT) activities were seen with treatment (390–392), the overall results of these studies were discouraging as sustained-responses were rarely achieved. Therefore, FDA did not approve ribavirin alone for hepatitis C. In the United States, it was first approved in aerosol form for the treatment of a certain type of respiratory virus infection in children. Because of its partial effectiveness, ribavirin was studied in subsequent trials in combination with interferon alpha (376,383,384, 393–395). It was discovered that the addition of ribavirin to interferon alpha-2b is superior to interferon alpha-2b alone in the treatment of chronic hepatitis C, especially in achieving a sustained response in patients not previously treated with interferon. This led to FDA approval of this combination therapy of interferon alpha-ribavirin for this indication in December 1998. Most recently, the FDA has approved the combination of peginterferon alpha plus ribavirin for the treatment of chronic hepatitis C. For eligible patients with chronic hepatitis C, a peginterferon alpha plus ribavirin is likely to be the best treatment option for the near future. Clinical trials have show that the sustained response rate is ~50% of patients given this combination for 24–48 weeks.

The treatment using interferon alpha with or without ribavirin is, nevertheless, associated with may side effects. During treatment, patients must be monitored carefully for side effects including flu-like symptoms, depression, rashes, other unusual reactions and abnormal blood counts. Furthermore, ribavirin is associated with a significant risk of abnormal fetal development, and women of childbearing potential should not begin therapy until a report of a negative pregnancy test has been obtained and not become pregnant during treatment. In general, the patient probably needs to have blood tests approximately once a month, and somewhat more frequently at the beginning of treatment. In addition, patients considered for treatment with interferon alpha-2b plus ribavirin should not have the complications of serious liver dysfunction and such subjects should only be considered for treatment of hepatitis C in specialized studies. Thus, it appears that more research is needed to develop safer, more effective and cheaper drugs against HCV.

Current Research Trends in Mechanism-Based HCV Inhibitors. The current intensive effort to discover novel therapies to treat HCV infection is aimed primarily at specific processes that are essential to HCV replication.

These include viral RNA replication, which uses the NS3 helicase/NTPase and the NS5B RNA-dependent RNA polymerase (RdRp); virus translation, controlled by regulatory elements such as the 5′-nontranslated region (5′NTR) that contains the internal ribosome entry site (IRES); and processing of the viral protein by the NS2–NS3 and NS3–NS4A proteases.

Based on the solved crystal structure of HCV RNA NTPase/helicase and of DNA NTPase/helicases from *Escherichia coli* and *Bacillus stearothermophilus*, two alternative mechanisms of the duplex unwinding reaction (see above) have been postulated (359,396–398). Both models predict that the enzymes bind and hydrolyse NTP by a well characterized NTP binding pocket. The energy released is used for the "march" of the enzyme along the DNA or RNA structures and the unwinding reaction results from capturing single strand (ss) regions that arise due to thermal fluctuations at the fork (359,396). Alternatively, the energy could be transferred to the fork and used for disruption of the hydrogen bonds that keep the strains together (359,396). Consistent with the proposed models, the following mechanisms of inhibition of the helicase activity could be considered: (*a*) inhibition of the NTPase activity by interference with NTP binding (399,400), (*b*) inhibition of NTPase activity by an allosteric mechanism (399), and (*c*) inhibition of the coupling of NTP hydrolysis to unwinding reaction (400). Additional mechanistic possibilities include interference in the interaction of helicase with its RNA or DNA substrate via (*d*), competitive blockade of substrate binding site (401), or by (*e*) inhibition of the unwinding by steric inhibition of translocation of the enzyme along the polynucleotide chain (402). There are even more mechanistic possibilities by which the helicase activity could be inhibited. Binding studies of Porter and coworkers (403,404) revealed a putative nucleoside-binding site within the HCV NTPase/helicase molecule. The function and location of the second binding site remains unknown. Nevertheless, there is accumulating evidence that the NTPase and helicase activities of the viral super family II (SFII) enzymes might be modulated by occupation of these putative nucleoside-binding sites. For example, ribavirin-5′-triphosphate (RTP), which is a potent, classical, competitive inhibitor of the NTPase activity of the WNV and HCV NTPase/helicases at lower ATP concentrations ($<K_M$), failed to inhibit the ATPase activity at higher ATP concentrations ($\gg K_M$), and instead, even stimulated the enzyme activity (400,405). By contrast, the RTP inhibits moderately the helicase activity of both enzymes by a mechanism that is independent of the ATP concentrations (405). The phenomenon results most probably from occupation of a second nucleoside binding site by RTP (400). Thus, intense research efforts are currently being directed at designing inhibitors of HCV helicase and NTPase.

HCV RNA-dependent RNA polymerase (RdRp) is also a good target for anti-HCV therapy in that its activity is essential for viral replication and infectivity. The biochemical properties of NS5B have been characterized extensively (406). A detailed view of HCV NS5B was revealed by the crystal structures of the RdRp (407,408). Although canonical polymerase features exist, HCV NS5B adopts a unique molecular structure that resembles a "thumb–palm–finger" that is different from other known DNA and RNA polymerases, highlighting the attractiveness of the HCV polymerase as a drug target. Recently, a benzo-1,2,4-thiadiazine derivative has been reported to be a potent inhibitor of HCV RdRp (409). As nucleosides and nucleotides are anticipated to modulate the activities of HCV helicase, NTPase, as well as poymerase, the future drugs against HCV are likely to be based on analogues of nucleosides/-tides.

The highly conserved regions in the internal ribosome entry site (IRES) of the HCV RNA genome, its distinctive translational–initiation mechanism and its essential role in mediating the unusual translational–initiation and replication processes of HCV make these elements an attractive target for compounds that inhibit transcription and translation of the HCV RNA. The specific sites and subdomains for interfering with IRES function have been identified. As the tertiary structures of these important subdomains are now available, it is possible to apply structure-based methods for the discovery of inhibitors of HCV protein synthesis and replication. Recently, it has been reported that Vitamin B12 stalls the 80 S ribosomal complex on HCV IRES (410).

The metal-dependent cysteine protease NS2–NS3 catalyzes cleavage between NS2 and NS3 in an autoproteolytic manner (370). The amino-terminal portion of NS2 is responsible for membrane association, whereas its carboxy terminus, which overlaps with NS3, is believed to catalyze the cleavage of the NS2–NS3 site. The activity of the chymotrypsin-like serine protease that is encoded within the amino-terminal 180 amino acids of NS3 is indispensable for HCV infectivity in the chimpanzee model (411). The structure of the protease domain and the full-length NS3 protein were solved by X-ray crystallography (412). Efficient processing requires the NS3 protease in combination with the NS4A cofactor and a structural zinc molecule (370). The NS3 protease is prone to inhibition by specific penta- or hexapeptides derived from the amino-terminal NS3 cleavage products, which have provided the basis for lead optimization of peptidomimetic inhibitors (413–415). This class of optimized compounds has shown submicromolar potencies in in vitro enzymatic assays, as well as in the HCV-replicon system.

Another class of compounds being developed as HCV inhibitors are ribozymes, which inhibit viral replication by cleavage of the target HCV genomic RNA (416,417). Ribozymes are naturally occurring, short RNA molecules with endoribonuclease activity that can catalyze sequence-specific cleavage of RNA. Antisense oligonucleotides have been employed as an alternative to selectively target the HCV RNA genome. The target RNA is cleaved RNA by an RNaseH at the site of oligonucleotide hybridization, and results in inhibition of gene expression. A number of antisense oligonucleotides have been designed to bind to the stem–loop structures in the HCV IRES, and have been shown to be effective in inhibiting HCV replication in cell-culture assays (418,419). ISIS 14803 is a 20-mer, 5′-methylcytidine phosphorothioate antisense oligonucleotide that is in a Phase II clinical trial at present in patients with chronic HCV infections (420).

5.7. West Nile Virus (WNV). With an alarming increase in the number of cases of infection in wild birds, horses, pets, and humans, the WNV is currently gaining a wide attention in the United States and the western hemisphere (421–425). A number of Science Focus and News Focus articles have appeared in recent issues of Science magazine (426), in addition to countless news stories in popular magazines and newspapers. Three years after the 1999 outbreak of the WNV in New York City, which sickened 62 people, most of them elderly, and killed 7, the virus has been detected in >60 bird species and about a dozen mammals, and has spread to 44 states and the District of Columbia. As of September

2002, the Centers for Disease Control and Prevention have verified 1295 human cases of WNV, resulting in 54 deaths. WNV is mainly a bird virus that is spread by mosquitoes. Humans, horses, as well as a dozen other mammals are its dead-end hosts. Crows are the virus's most conspicuous hosts as they have been dying en masse with WNV infection. Most humans infected with WNV do not even know it, or they experience only mild, flu-like symptoms. Those over 65, and individuals with weakened immune systems, are especially vulnerable to WNV although recent cases have brought down the age barrier to as low as 50. Three months after the initial outbreak, 70% of the survivors still reported muscle weakness, 75% suffered from memory loss, 60% from confusion, and more than one-half could no longer live at home, although most were healthy, active, and lived normal lives before the WNV attack. Many of the patients end up with lingering neurological damage as often occurs with encephalitic infections.

Since there are currently no approved drugs or vaccines against WNV infection, the focus has been mainly on prevention. Given that mosquitoes are associated with WNV transmission, the key to preventing or controlling future outbreaks of WNV among horses and other animals is to control mosquito populations. Because horses and pets could be infected the same way people are, the key to prevention is to prevent mosquito bites. Products to prevent fleas and ticks have no effect on mosquitoes. There are over-the-counter products, however, available to repel mosquitoes. Similar recommendations would apply for other pets, livestock, or poultry should illness due to WNV in those types of animals come to be commonly recognized. In 2001, a license was issued by the USDA–APHIS Center for Veterinary Biologics, Inc. for an equine WNV vaccine, and so, vaccination is now available as an option for horses (427).

As noted above, both WNV and HCV belong to the same family of viruses called *Flaviviridae*. However, unlike HCV, WNV can be isolated from clinical specimens by tissue culture methods, and therefore, is used as a close mimic of HCV in experimental models. Also, since the structures of the viral genome, the encoded polyprotein, and the protease-processed structural and NS protein fragments are also very similar to those of HCV, the same viral targets as described above for HCV are currently being investigated by several research groups including ours (428,429). These targets include the WNV NTPase/helicase of the NS3 region as well as the viral RNA-dependent RNA polymerase of the NS5 region.

5.8. Antiviral Therapy for WNV Infections. There are no currently approved drugs or vaccines for treating or preventing the disease in humans, although a vaccine has recently been approved for horses as described above. Although ribavirin was initially reported to halt the viral replication, the need to use very high doses of the drug proved too toxic to be clinically useful (430). Furthermore, since WNV is still a rare virus affecting humans, there is not enough incentive for drug companies to develop anti-WNV drugs, but this scenario is likely to change as more and more cases of infection emerge. In a recent study, we have demonstrated that some imidazo[4,5-*d*]pyridazine nucleoside analogues act as inhibitors of WNV NTPase/helicase, and moderately reduce the unwinding activity of the enzyme (428). A comparable inhibitory potency was also observed in tissue culture systems, suggesting that this enzyme is indeed a viable target for inhibition of WNV replication. We have also recently

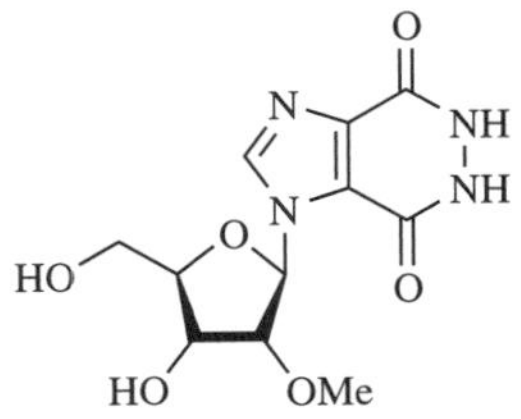

I

Imidazo[4,5-*d*]pyridazine

($IC_{50} = 30 \mu M$)

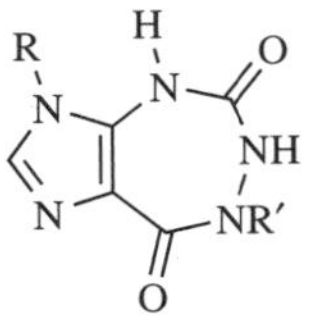

II

Imidazo[4,5-*e*][1,2,4]triazepine

($IC_{50} = 1.3$–$3.5 \mu M$)

(a) R = H, R′ = CH$_2$Ph
(b) R = R′ = CH$_2$Ph

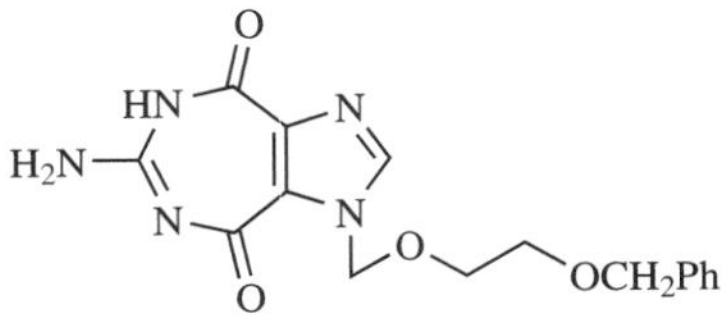

III

Imidazo[4,5-*e*][1,3]diazepine

($IC_{50} = 5.0$–$11.0 \mu M$)

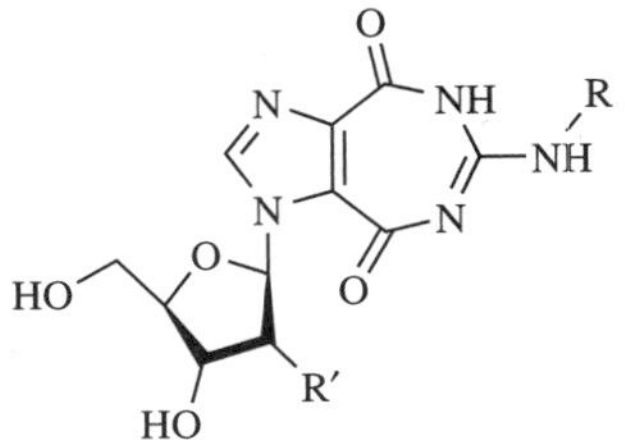

IV

Imidazo[4,5-*e*][1,3]diazepine

(a) R = (CH$_2$)$_{11}$CH$_3$, R′ = OH ($IC_{50} = 1.0$–$3.0 \mu M$)
(b) R = (CH$_2$)$_{13}$CH$_3$, R′ = OH ($IC_{50} = 3.0$–$10.0 \mu M$)
(c) R = (CH$_2$)$_{17}$CH$_3$, R′ = OH ($IC_{50} = 5.0 \mu M$)
(d) R = (CH$_2$)$_{11}$CH$_3$, R′ = H ($IC_{50} = 3.0$–$10 \mu M$)

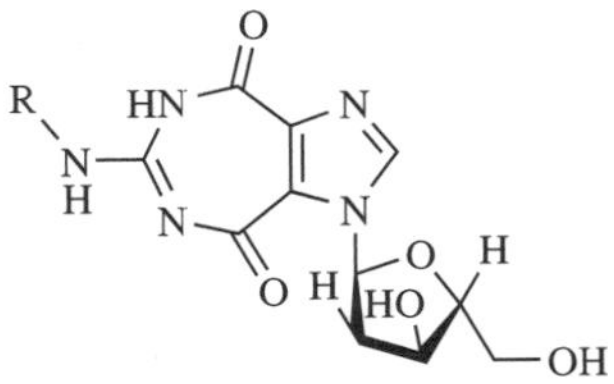

V

Imidazo[4,5-*e*][1,3]diazepine

(a) R = (CH$_2$)$_{11}$CH$_3$
($IC_{50} = 3$–$10 \mu M$)
(b) R = H ($IC_{50} = 20$–$50 \mu M$)

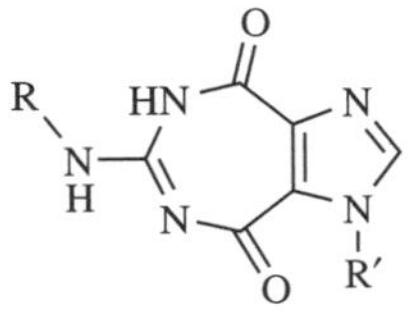

VI

Imidazo[4,5-*e*][1,3]diazepine

(a) R = (CH$_2$)$_{11}$CH$_3$, R′ = H
($IC_{50} = 3$–$10 \mu M$)
(b) R = (CH$_2$)$_{11}$CH$_3$,
R′ = CH$_2$Ph-*p*-OMe
($IC_{50} = 5.0 \mu M$)

Fig. 10. *In vitro* inhibitors of the helicase activity of WNV NTPase/helicase, containing the imidazopyridazine, imidazodiazepine, and imidazotriazepine ring systems. The term IC_{50} represents the concentration of the inhibitor required to reduce the unwinding activity of the enzyme by 50% of that observed in the absence of the inhibitor.

discovered (429) that a variety of 5:7 fused heterocyclic bases, nucleosides, and nucleotides resembling ring-expanded (fat) purine structure are excellent *in vitro* inhibitors of the helicase activity of WNV NTPase/helicase and/or WNV RNA-dependent RNA polymerase (RdRp). Listed in Figure 10 are a few such inhibitors of the helicase activity of WNV NTPase/helicase, along with their respective IC_{50} values. The helicase activity was assessed, using a DNA substrate and ATP, as a function of increasing concentration of inhibitors. The term IC_{50} represents the concentration of the inhibitor required to lower the original unwinding activity observed in the absence of the inhibitor by 50%. Likewise, listed in Figure 11 are the structures and the corresponding IC_{50} values of inhibitors of the WNV RdRp activity *in vitro*. The term IC_{50} here reflects the concentration of the inhibitor required to reduce the WNV polymerase activity by 50%.

As is evident from IC_{50} values in Figure 10, both ring-expanded heterocyclic bases as well as nucleosides possess potent anti-WNV activity. Since compounds of general formula **IV** containing shorter than the C-12 side chain at position-6 failed to exhibit any significant activity, the presence of an adequately hydrophobic group at this junction appears to be necessary for activity. Most surprising was the fact that the sugar moiety is not absolutely necessary for activity as heterocyclic bases with the appropriately hydrophobic functionalities at either the seven- or the five-membered ring were just as or even more active than their nucleoside counterparts, as revealed by the activities of compounds of general formula **II**, **III**, and **VI**. The relatively somewhat less critical role of the sugar moiety in the observed anti-WNV activity was further revealed by the compounds of general formula **V**, which possess the unnatural α-anomeric configuration at the base–sugar junction. Finally, in view of the observed tight complex between nucleoside **IVd** and a DNA substrate that was completely stable in the presence of 0.5% sodium dodecyl sulfate (SDS), the mechanism of action of ring-expanded heterocyclic bases and nucleosides is currently believed to involve their interaction with the nucleic acid substrate of WNV helicase through binding to the major or minor groove of the double helix, and the consequent modulation of the enzyme activity. The substrate binding can result in either the inhibition or the enhancement of helicase activity, which was indeed found to be the case as a few other ring-expanded nucleosides tested were found to be the activators, rather than inhibitors, of the helicase activity of WNV NTPase/helicase.

The observed inhibition of WNV RdRp activity by ring-expanded heterocycles and nucleosides listed in Figure 11 also exhibited a parallel trend in that the presence, type, or configuration of the sugar moiety was relatively less critical as opposed to the type and location of the substituent on the heterocyclic ring. Thus, compounds **IVe** and **IVf**, which contained shorter than the C-12 side chain at position-6, and which were inactive against the helicase activity of WNV NTPase/helicase, are now found to be active against the polymerase activity of WNV RdRp. Once again, both the α-anomeric nucleoside **V** and the heterocycles **VI** with the appropriate hydrophobic substituents at position 1 and/or 6 exhibited remarkable inhibition of WNV RdRp activity.

Recently, Chu and co-workers reported (431) the anti-WNV activity of L-neplanocin analogues in tissue culture systems. While the parent L-neplanocin

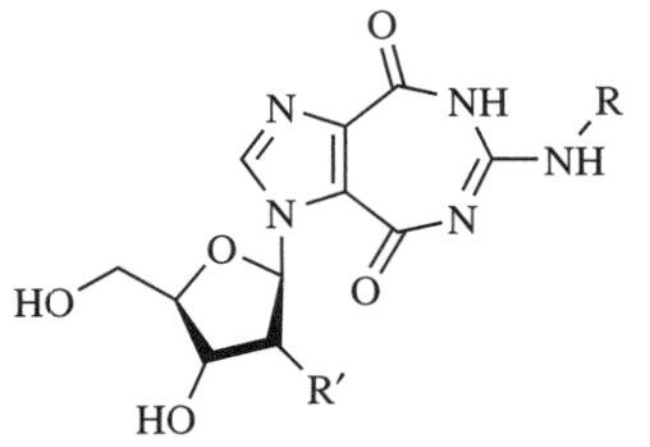

IV

Imidazo[4,5-e][1,3]diazepine

(e) R = $(CH_2)_3CH_3$, R′ = OH   (IC_{50} = 10–30 μM)
(f) R = $(CH_2)_9CH_3$, R′ = OH   (IC_{50} = <<3.0 μM)

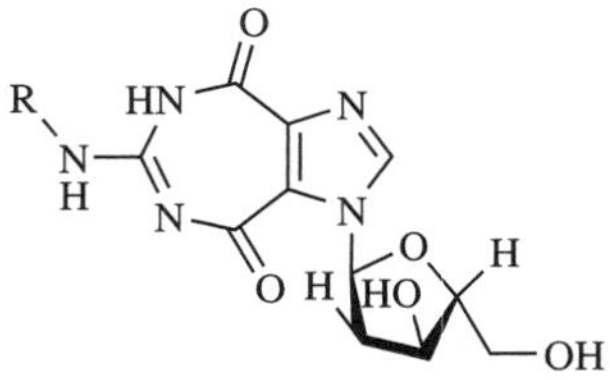

V

Imidazo[4,5-e][1,3]diazepine

R = $(CH_2)_{11}CH_3$ (IC_{50} <<3.0 μM)

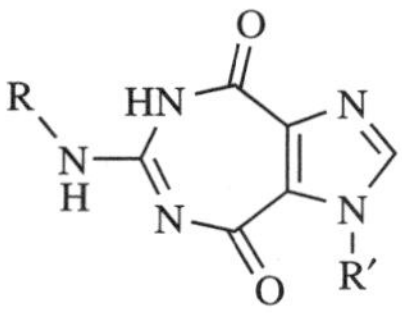

VI

Imidazo[4,5-e][1,3]diazepine

(a) R = $(CH_2)_{11}CH_3$, R′ = H
    (IC_{50} << 3.0 μM)
(b) R = $(CH_2)_{11}CH_3$,
    R′ = CH_2Ph-p-OMe
    (IC_{50} << 3.0 μM)

Fig. 11. *In vitro* inhibitors of the polymerase activity of WNV RNA-dependent RNA polymerase (RdRp) activity. The term IC_{50} represents the inhibitor concentration required to reduce the polymerase activity by 50%.

itself was inactive, both its cytosine (EC_{50} = 0.06 μM; IC_{50} = 0.08 μM in CEM cells) and 5-fluorocytosine (EC_{50} = 5.34 μM; IC_{50} = 51.4 μM in CEM cells) analogues exhibited potent anti-WNV activity, but unfortunately, the compounds also suffered from significant cellular toxicity.

6. Conclusion

The molecular structure, life cycle, mode of infection, and replication process of four major viruses of current health scare, including HIV, HBV, HCV, and WNV, have been discussed at length with cursory references to other human viruses. Also elaborated on are the prophylactic as well as postinfection remedies that are both currently approved and under clinical development, along with viable, mechanism-based targets for future development of antiviral therapies. While no vaccine nor total cure is yet available against HIV infection, great strides have been made in antiviral therapy to enable classification of AIDS as a manageable illness from that of an "absolute death sentence" only a few years ago. With regard to HBV infection, a vaccine is now available, but the initial clinical trials with a number of nucleoside analogues as anti-HBV agents were disappointing in light of severe toxicities associated with them. Nevertheless, a number of other nucleoside analogues belonging to the family of unnatural L-nucleosides and RENs that are currently under development appear to be promising. The ultimate success in treating HBV may lie in the combination drug therapy similar to the successful HAART therapy applied against HIV infection. Unfortunately, neither vaccines nor good drugs are currently available for treating HCV or WNV, the two viruses belonging to the family of *flaviviridae*, but a vast array of information is being rapidly accumulated on the structural biology and molecular virology of the two viruses to afford development of suitable antiviral therapies against them in the near future.

7. Acknowledgments

This paper is dedicated to Professor Nelson J. Leonard on the occasion of his 87th birthday. The work was supported by a grant (No. 9 R01 AI 55452) from the National Institute of Allergy and Infectious Diseases (NIAID) of the National Institutes of Health (NIH), Bethesda, Maryland. The author gratefully acknowledges Dr. Peter Borowski of Bernhard-Nocht Institute für Tropenmedizin, Hamburg, Germany for the preliminary *in vitro* anti-helicase/NTPase and anti-polymerase (RdRp) screening data of ring-expanded nucleosides (RENs) against *flaviviridae*, and to Dr. Brent Korba of the Division of Molecular Virology & Immunology, Georgetown University Medical Center, Rockville, Maryland, for the *in vitro* screening data of some the RENs against HBV. The continuous encouragement and support of Dr. Christopher Tseng, the Program Officer of Antiviral Research and Antimicrobial Chemistry of the Virology Branch of NIAID is also gratefully acknowledged.

BIBLIOGRAPHY

"Antiviral Agents", in *ECT* 4th ed., Vol. 3, pp. 576–607, by G. R. Revankar, Triplex Pharmaceutical Corporation and R. K. Robins, ICN Nucleic Acid Research Institute; "Antiviral Agents" in *ECT* (online), posting date: December 4, 2000, by Ganapathi R. Revankar, Triplex Pharmaceutical Corporation and Roland K. Robins, ICN Nucleic Acid Research Institute.

CITED PUBLICATIONS

1. A. J. Levine, *Viruses*, Scientific American Library, New York, 1992.
2. J. N. Delgado and W. A. Remers, *Textbook of Organic Medicinal and Pharmaceutical Chemistry*; 10th ed., Lippincott-Raven Publishers, Philadelphia, 1998, p. 329.
3. D. Naniche, A. Yeh, D. Eto, M. Manchester, R. M. Friedman, and M. B. Oldstone, Evasion of host defenses by measles virus: Wild-type measles virus infection

interferes with induction of alpha/beta interferon production. *J. Virol.* **74**, 7478–7484 (2000).

4. N. Zhang, H.-M. Chen, R. K. Sood, K. Kalicharran, A. I. Fattom, R. B. Naso, D. L. Barnard, R. W. Sidwell, and R. S. Hosmane, *Bioorg. Med. Chem. Lett.* **12**, 3391–3394 (2002).

5. J. W. Chien and J. L. Johnson, *Postgrad. Med.* **107**, 67–70, 73–74 (2000).

6. Ref. 5, pp. 77–80.

7. J. Schneider-Schaulies, S. Niewiesk, S. Schneider-Schaulies, and V. ter Meulen, *J. Neurovirol.* **5**, 613–622 (1999).

8. T. Takasu, *Nippon Rinsho—Jpn. J. Clin. Med.* **55**, 783–786 (1997).

9. A. H. Rux, H. Lou, J. D. Lambris, H. M. Friedman, R. J. Eisenberg, and G. H. Cohen, *Virology.* **294**, 324–332 (2002).

10. A. Vaheri, E. Ikkala, E. Saxen, and K. Penttinen, *Acta Pathol. Microbiol. Scand.* **62**, 340–348 (1964).

11. D. M. Lambert, S. Barney, A. L. Lambert, K. Guthrie, R. Medinas, D. E. Davis, T. Bucy, J. Erckson, G. Merutka, and S. R. Petteway, Jr., *Proc. Natl. Acad. Sci. USA* **93**, 2186–2191 (1996).

12. P. W. Choppin, C. D. Richardson, D. C. Merz, W. W. Hall, and A. Scheid, *J. Infect. Dis.* **143**, 352–363 (1981).

13. C. T. Wild, D. C. Shugars, T. K. Greenwell, C. B. McDanal, and T. J. Matthews, *Proc. Natl. Acad. Sci. USA* **91**, 9770–9774 (1994).

14. J. S. Oxford and S. Patterson, *Developments in Antiviral Therapy*, Academic Press, London, 1980, pp. 119–131.

15. A. Okada, T. Miura, and H. Takeuchi, *Biochemistry* **40**, 6053–6060 (2001).

16. Q. Z. Yao and R. W. Compans, *Virology* **223**, 103–112 (1996).

17. T. F. Wild and R. Buckland, *J. Gen. Virol.* **78**, 107–111 (1997).

18. C. D. Richardson, A. Scheid, and P. W. Choppin, *Virology* **105**, 205–222 (1980).

19. R. E. Dutch, R. N. Hagglund, M. A. Nagel, R. G. Paterson, and R. A. Lamb, *Virology* **281**, 138–150 (2001).

20. B. Bean, *Clin. Microbiol. Rev.* **5**, 146–182 (1992).

21. B. W. Fox, *J. Antimicrob. Chemoth.* **7**, (3) 23–32.

22. J. R. Wingard, R. K. Stuart, R. Saral, and W. H. Burns, *Antimicrob. Agents Ch.* **20**, 286–290 (1981).

23. P. Collins and D. J. Bauer, *J. Antimicrob. Chemother.* **3**, 73–81 (1977).

24. T. W. North and S. S. Cohen, *Pharmacol. Therapeut.* **4**, 81–108 (1979).

25. A. Wlodawer and J. W. Erickson, *Annu. Rev. Biochem.* **62**, 543–585 (1993).

26. U. Hacksell and G. D. Daves, Jr., *Prog. Med. Chem.* **22**, 1–65 (1985).

27. K. Gerzon, D. C. DeLong, and J. C. Cline, *Pure Appl. Chem.* **28**, 489–497 (1971).

28. G. Gumina, G. Y. Song, and C. K. Chu, *FEMS Microbiol. Lett.* **202**, 9–15 (2001).

29. P. Wang, J. H. Hong, J. S. Cooperwood, and C. K. Chu, *Antivir. Res.* **40**, 19–44 (1998).

30. L. D. Bobeck, G. M. Clem, M. M. Wilson, and J. E. Westhead, *Antimicrob. Agents Ch.* **3**, 49–56 (1973).

31. C. S. Pugh, R. T. Borchardt, and H. O. Stone, *J. Biol. Chem.* **253**, 4075–4077 (1978).

32. M. Hayashi, S. Yaginuma, N. Muto, and M. Tsujino, *Nucleic Acids Symp. Ser.* **8**, s65–67 (1980).

33. E. De Clercq and M. Cools, *Biochem. Biophys. Res. Commun.* **129**, 306–311 (1985).

34. H. J. Schaeffer, B. L., P. de Miranda, G. B. Elion, J. B. D. and P. Collins, *Nature (London)* **272**, 583–585 (1978).

35. E. De Clercq, G. Andrei, R. Snoeck, L. De Bolle, L. Naesens, B. Degreve, J. Balzarini, Y. Zhang, D. Schols, P. Leyssen, C. Ying, and J. Neyts, *Nucleos. Nucleot. Nucl.* **20**, 271–285 (2001).

36. L. Naesens and E. De Clercq, *Herpes.* **8**, 12–16 (2001).

37. K. Chu, D. W. Kang, J. J. Lee, and B. W. Yoon, *Arch. Neurol.* **59**, 460–463 (2002).

38. E. Schmutzhard, *J. Neurol.* **248**, 469–477 (2001).

39. S. Drake, S. Taylor, D. Brown, and D. Pillay, *Brit. Med. J.* **321**, 619–623 (2000).

40. S. Leflore, P. L. Anderson, and C. V. Fletcher, *Drug Safety.* **23**, 131–142 (2000).

41. D. H. Emmert, *Am. Fam. Physician.* **61**, 1697–1706, 1708 (2000).

42. C. P. Kaplan and K. P. Bain, *Brain Injury.* **13**, 935–941 (1999).

43. S. Efstathiou, H. J. Field, P. D. Griffiths, E. R. Kern, S. L. Sacks, N. M. Sawtell, and L. R. Stanberry, *Antivir. Res.* **41**, 85–100 (1999).

44. P. Wutzler, *Intervirology.* **40**, 343–356 (1997).

45. A. M. Fillet, *Drug. Aging.* **19**, 343–354 (2002).

46. R. Snoeck and E. De Clercq, *Curr. Opin. Infect. Dis.* **15**, 49–55 (2002).

47. E. De Clercq, L. Naesens, L. De Bolle, D. Schols, Y. Zhang, and J. Neyts, *Rev. Med. Virol.* **11**, 381–395 (2001).

48. D. T. Leung and S. L. Sacks, *Drugs* **60**, 1329–1352 (2000).

49. R. Snoeck, *Int. J. Antimicrob. Ag.* **16**, 157–159 (2000).

50. D. Ormrod and K. Goa, *Drugs* **59**, 1317–1340 (2000).

51. D. Ormrod, L. J. Scott, and C. M. Perry, *Drugs* **59**, 839–863 (2000).

52. A. R. Bell, *Adv. Exp. Med. Biol.* **458**, 149–157 (1999).

53. R. Snoeck, G. Andrei, and E. De Clercq, *Drugs* **57**, 187–206 (1999).

54. J. Otero, E. Ribera, J. Gavalda, A. Rovira, I. Ocana, and A. Pahissa, *Eur. J. Clin. Microbiol. Infect. Dis.* **17**, 286–289 (1998).

55. K. S. Erlich, *Western J. Med.* **166**, 211–215 (1997).

56. D. Lowance, H. H. Neumayer, C. M. Legendre, J. P. Squifflet, J. Kovarik, P. J. Brennan, D. Norman, R. Mendez, M. R. Keating, G. L. Coggon, A. Crisp, and I. C. Lee, *New Engl. J. Med.* **340**, 1462–1470 (1999).

57. C. L. Dekker and C. G. Prober, *Pediatr. Infect. Dis. J.* **20**, 1079–1081 (2001).

58. B. Randolph, *J. Dent. Child.* **68**, 189–190 (2001).

59. E. C. Villarreal, Current and potential therapies for the treatment of herpesvirus infections. *Fortschritte der Arzneimittelforschung-Progress in Drug Research-Progres des Recherches Pharmaceutiques.* (2001) *Spec,* 185–228.

60. E. C. Villarreal, Current and potential therapies for the treatment of herpesvirus infections. *Fortschritte der Arzneimittelforschung-Prog Drug Res-Progres des Rech Pharmaceutiques.* (2001) **56**, 77–120.

61. M. R. Holdiness, *Contact Dermatitis.* **44**, 265–269 (2001).

62. R. Snoeck, G. Andrei, and E. D. Clercq, *Expert Opin. Invest. Drug.* **9**, 1743–1751 (2000).

63. R. J. Whitley, *Contrib. Microbiol.* **3**, 158–172 (1999).

64. S. L. Sacks and B. Wilson, *Adv. Exp. Med. Biol.* **458**, 135–147 (1999).

65. H. E. Kaufman, *Prog. Retin. Eye Res.* **19**, 69–85 (2000).

66. D. Salmon-Ceron, *HIV Med.* **2**, 255–259 (2001).

67. M. D. Khare and M. Sharland, *Expert Opin. Pharmacother.* **2**, 1247–1257 (2001).

68. J. K. McGavin and K. L. Goa, *Drugs* **61**, 1153–1183 (2001).

69. W. D. Rawlinson, *Med. J. Australia.* **175**, 112–116 (2001).

70. V. C. Emery, *J. Clin. Virol.* **21**, 223–228 (2001).

71. D. T. Tendero, *Clin. Lab.* **47**, 169–183 (2001).

72. M. Maschke, O. Kastrup, and H. C. Diener, *CNS Drugs* **16**, 303–315 (2002).

73. E. Bogner, *Rev. Med. Virol.* **12**, 115–127 (2002).

74. P. Reusser, *Support Care Cancer* **10**, 197–203 (2002).

75. J. T. Witkowski, R. K. Robins, R. W. Sidwell, and L. N. Simon, *J. Med. Chem.* **15**, 1150–1154 (1972).

76. R. W. Sidwell, J. H. Huffman, G. P. Khare, L. B. Allen, J. T. Witkowski, and R. K. Robins, *Science* **177**, 705–706 (1972).

77. S. Crotty, C. Cameron, and R. Andino, *J. Mol. Med.* **80**, 86–95 (2002).

78. R. C. Tam, J. Y. Lau, and Z. Hong, *Antivir. Chem. Chemoth.* **12**, 261–272 (2001).

79. J. Y. Lau, R. C. Tam, T. J. Liang, and Z. Hong, *Hepatology* **35**, 1002–1009 (2002).

80. D. G. Streeter, J. T. Witkowski, G. P. Khare, R. W. Sidwell, R. J. Bauer, R. K. Robins, and L. N. Simon, *Proc. Natl. Acad. Sci. USA* **70**, 1174–1178 (1973).

81. P. Franchetti and M. Grifantini, *Curr. Med. Chem.* **6**, 599–614 (1999).

82. D. F. Smeet, M. Bray, and J. W. Huggins, *Antivir. Chem. Chemoth.* **12**, 327–335 (2001).

83. A. M. Contreras, Y. Hiasa, W. He, A. Terella, E. V. Schmidt, and R. T. Chung, *J. Virol.* **76**, 8505–8517 (2002).

84. S. Crotty, D. Maag, J. J. Arnold, W. Zhong, J. Y. Lau, Z. Hong, R. Andino, and C. E. Cameron, *Nat. Med.* **6**, 1375–1379 (2000).

85. J. D. Graci and C. E. Cameron, *Virology* **298**, 175–180 (2002).

86. D. G. Streeter, L. N. Simon, R. K. Robins, and J. P. Miller, *Biochemistry.* **13**, 4543–4549 (1974).

87. A. Greenough, *Curr. Opin. Pulm. Med.* **8**, 214–217 (2002).

88. M. A. Staat, *Semin. Respir. Infect.* **17**, 15–20 (2002).

89. R. B. Wright, W. J. Pomerantz, and J. W. Luria, *Emerg. Med. Clin. N. Am.* **20**, 93–114 (2002).

90. H. Schmitz, B. Kohler, T. Laue, C. Drosten, P. J. Veldkamp, S. Gunther, P. Emmerich, H. P. Geisen, K. Fleischer, M. F. Beersma, and A. Hoerauf, *Microbes Infect.* **4**, 43–50 (2002).

91. C. E. Cameron and C. Castro, *Curr. Opin. Infect. Dis.* **14**, 757–764 (2001).

92. D. Maag, C. Castro, Z. Hong, and C. E. Cameron, *J. Biol. Chem.* **276**, 46094–46098 (2001).

93. E. De Clercq, *Biochim. Biophys. Acta* **1587**, 258–275 (2002).

94. E. De Clercq, *Curr. Med. Chem.* **8**, 1543–1572 (2001).

95. E. Papadopulos-Eleopulos, V. F. Turner, J. M. Papadimitriou, D. Causer, H. Alphonso, and T. Miller, *Curr. Med. Res. Opin.* **15**, S1–45 (1999).

96. P. Piot, M. Bartos, P. D. Ghys, N. Walker, and B. Schwartlander, *Nature (London)* **410**, 968–973 (2001).

97. M. S. Lesney, *Mod. Drug Discov.* **5**, 31–37 (2002).

98. M. M. Thomson, L. Perez-Alvarez, and R. Najera, *Lancet Infect. Dis.* **2**, 461–471 (2002).

99. Anonymous, *AIDS Care* **14**, 144 (2002).

100. T. Wilk and S. D. Fuller, *Curr. Opin. Struct. Biol.* **9**, 231–243 (1999).

101. S. G. Sarafianos, K. Das, J. Ding, P. L. Boyer, S. H. Hughes, and E. Arnold, *Chem. Biol.* **6**, R137–146 (1999).

102. S. Doublie, M. R. Sawaya, and T. Ellenberger, *Structure.* **7**, R31–35 (1999).

103. B. G. Turner and M. F. Summers, *J. Mol. Biol.* **285**, 1–32 (1999).

104. Q. J. Sattentau, *Structure* **6**, 945–949 (1998).

105. R. Wyatt and J. Sodroski, *Science* **280**, 1884–1888 (1998).

106. X. Li, Y. Quan, and M. A. Wainberg, *Cell. Mol. Biol.* **43**, 443–454 (1997).

107. H. Jonckheere, J. Anne, and E. De Clercq, *Med. Res. Rev.* **20**, 129–154 (2000).

108. M. Gotte, X. Li, and M. A. Wainberg, *Arch. Biochem. Biophys.* **365**, 199–210 (1999).

109. C. Pinter, S. Beltrami, H. Stoiber, D. R. Negri, F. Titti, and A. Clivio, *Expert Opin. Inv. Drugs.* **9**, 199–205 (2000).

110. R. Mitsuyasu, *AIDS* **13**, S19–27 (1999).

111. N. A. Qazi, J. F. Morlese, and A. L. Pozniak, Lopinavir/ritonavir (ABT-378/R). *Expert Opin. Pharmacother.* **3**, 315–327 (2002).

112. R. P. van Heeswijk, A. Veldkamp, J. W. Mulder, P. L. Meenhorst, J. M. Lange, J. H. Beijnen, and R. M. Hoetelmans, *Antivir. Ther.* **6**, 201–229 (2001).
113. G. J. Moyle and D. Back, *HIV Med.* **2**, 105–113 (2001).
114. E. M. Mangum and K. K. Graham, *Pharmacotherapy* **21**, 1352–1363 (2001).
115. S. Ren and E. J. Lien, Development of HIV protease inhibitors: A survey. *Fortschritte der Arzneimittelforschung-Progress in Drug Research-Progres des Recherches Pharmaceutiques.* 2001, *Spec*, pp. 1–34.
116. M. A. French, *AIDS Reader.* **9**, 548–549, 554–545, 559–562 (1999).
117. D. A. Carrasco and S. K. Tyring, *Dermatol. Clin.* **19**, 757–772 (2001).
118. M. S. Saag, **15**, S4–10 (2001).
119. G. Borkow and Z. Bentwich, *Clin. Diagn. Lab. Immun.* **9**, 505–507 (2002).
120. B. Ensoli and A. Cafaro, *Virus Res.* **82**, 91–101 (2002).
121. R. C. Gallo and A. Garzino-Demo, *Cell. Mol. Biol.* **47**, 1101–1104 (2001).
122. J. Chinen and W. T. Shearer, *J. Allergy Clin. Immun.* **110**, 189–198 (2002).
123. S. Kinloch-de Loes and B. Autran, *J. Infection.* **44**, 152–159 (2002).
124. N. Imami and F. Gotch, *Clin. Exp. Immun.* **127**, 402–411 (2002).
125. R. L. Edgeworth, J. H. San, J. A. Rosenzweig, N. L. Nguyen, J. D. Boyer, and K. E. Ugen, *Immun. Res.* **25**, 53–74 (2002).
126. D. D. Ho and Y. Huang, *Cell* **110**, 135–138 (2002).
127. P. Vandepapeliere, *Lancet Infect. Dis.* **2**, 353–367 (2002).
128. A. McMichael and T. Hanke, The quest for an AIDS vaccine: Is the CD8$^+$ T-cell approach feasible? *Nature Rev. Immun.* **2**, 283–291 (2002).
129. H. L. Robinson, *Nature Rev. Immun.* **2**, 239–250 (2002).
130. N. L. Letvin, *J. Clin. Inv.* **110**, 15–20 (2002).
131. B. Gaschen, J. Taylor, K. Yusim, B. Foley, F. Gao, D. Lang, V. Novitsky, B. Haynes, B. H. Hahn, T. Bhattacharya, and B. Korber, *Science* **296**, 2354–2360 (2002).
132. P. J. Weidle, T. D. Mastro, A. D. Grant, J. Nkengasong, and D. Macharia, *Lancet* **359**, 2261–2267 (2002).
133. M. J. Newman, B. Livingston, D. M. McKinney, R. W. Chesnut, and A. Sette, *Front. Biosci.* **7**, d1503–1515 (2002).
134. P. Lundholm, A. C. Leandersson, B. Christensson, G. Bratt, E. Sandstrom, and B. Wahren, *Virus Res.* **82**, 141–145 (2002).
135. T. Hanke, *Curr. Mol. Med.* **1**, 123–135 (2001).
136. A. M. Schultz and J. A. Bradac, *AIDS* **15**, S147–158 (2001).
137. M. Nolan, M. G. Fowler, and L. M. Mofenson, *J. Acq. Immun. Def. Synd.* **30**, 216–229 (2002).
138. H. Mitsuya, K. J. Weinhold, P. A. Furman, M. H. St Clair, S. N. Lehrman, R. C. Gallo, D. Bolognesi, D. W. Barry, and S. Broder, *Proc. Natl. Acad. Sci. USA* **82**, 7096–7100 (1985).
139. H. Mitsuya and S. Broder, *Proc. Natl. Acad. Sci. USA* **83**, 1911–1915 (1986).
140. J. Zemlicka, *Biochim. Biophys. Acta* **1587**, 276–286 (2002).
141. T. S. Lin, R. F. Schinazi, and W. H. Prusoff, *Biochem. Pharmacol.* **36**, 2713–2718 (1987).
142. K. S. Anderson, *Biochim. Biophys. Acta* **1587**, 296–299 (2002).
143. R. F. Schinazi, C. K. Chu, A. Peck, A. McMillan, R. Mathis, D. Cannon, L. S. Jeong, J. W. Beach, W. B. Choi, and S. Yeola, *Antimicrob. Agents Ch.* **36**, 672–676 (1992).
144. A. M. van Rossum, P. L. Fraaij, and R. de Groot, *Lancet Infect. Dis.* **2**, 93–102 (2002).
145. M. A. Garcia-Viejo, M. Ruiz, and E. Martinez, *Expert Opin. Inv. Drugs* **10**, 1443–1456 (2001).
146. G. Moyle, *HIV Med.* **2**, 154–162 (2001).
147. E. De Clercq, *Farmaco* **56**, 3–12 (2001).

148. S. M. Daluge, S. S. Good, M. B. Faletto, W. H. Miller, M. H. St Clair, L. R. Boone, M. Tisdale, N. R. Parry, J. E. Reardon, R. E. Dornsife, D. R. Averett, and T. A. Krenitsky, *Antimicrob. Agents Ch.* **41**, 1082–1093 (1997).
149. K. E. Squires, *Antivir. Ther.* **6**, 1–14 (2001).
150. B. G. Gazzard, *Int. J. Clin. Pract.* **55**, 704–709 (2001).
151. L. K. Naeger and M. D. Miller, *Curr. Opin. Inv. Drugs* **2**, 335–339 (2001).
152. J. Balzarini, S. Aquaro, C. F. Perno, M. Witvrouw, A. Holy, and E. De Clercq, *Biochem. Biophys. Res. Commun.* **219**, 337–341 (1996).
153. D. Podzamczer and E. Fumero, *Expert Opin. Pharmacother.* **2**, 2065–2078 (2001).
154. J. Q. Tran, J. G. Gerber and B. M. Kerr, *Clin. Pharmacokinet.* **40**, 207–226 (2001).
155. G. Campiani, A. Ramunno, G. Maga, V. Nacci, C. Fattorusso, B. Catalanotti, E. Morelli, and E. Novellino, *Curr. Pharm. Design.* **8**, 615–657 (2002).
156. G. L. Plosker, C. M. Perry, and K. L. Goa, *Pharmacoeconomics* **19**, 421–436 (2001).
157. P. Keiser, *J. Acq. Immun. Def. Synd.* **29**, S19–27 (2002).
158. S. Grub, P. Delora, E. Ludin, F. Duff, C. V. Fletcher, R. C. Brundage, M. W. Kline, N. R. Calles, H. Schwarzwald, and K. Jorga, *Clin. Pharmacol. Ther.* **71**, 122–130 (2002).
159. J. Gill and J. Feinberg, *Drug Safety.* **24**, 223–232 (2001).
160. E. Florence, W. Schrooten, K. Verdonck, C. Dreezen, and R. Colebunders, *Ann. Rheum. Dis.* **61**, 82–84 (2002).
161. G. Moyle, *Drugs* **61**, 19–26 (2001).
162. V. Soriano and C. de Mendoza, *HIV Clin. Trials.* **3**, 249–257 (2002).
163. G. Moyle, *AIDS Reader.* **11**, 87–98; quiz 107–108 (2001).
164. E. De Clercq, *Nature Rev. Drug Discov.* **1**, 13–25 (2002).
165. B. M. Sadler and D. S. Stein, *Ann. Pharmacother.* **36**, 102–118 (2002).
166. M. Pirmohamed and D. J. Back, *Pharmacogenomics J.* **1**, 243–253 (2001).
167. J. M. Gatell, *J. HIV Ther.* **6**, 95–99 (2001).
168. G. A. Balint, *Pharmacol. Therapeut.* **89**, 17–27 (2001).
169. A. Velazquez-Campoy and E. Freire, *J. Cell. Biochem.-Suppl. Suppl*, 82–88 (2001).
170. S. Kodoth, S. Bakshi, P. Scimeca, K. Black, and S. Pahwa, *AIDS Patient Care Stds.* **15**, 347–352 (2001).
171. M. B. Abbott and R. H. Levin, *Pediatr. Rev.* **22**, 357–359 (2001).
172. D. J. Porche, *J. Assoc. Nurses AIDS Care* **12**, 101–104 (2001).
173. M. H. el Kouni, *Curr. Pharmaceut. Design.* **8**, 581–593 (2002).
174. C. Mao, E. A. Sudbeck, T. K. Venkatachalam, and F. M. Uckun, *Biochem. Pharmacol.* **60**, 1251–1265 (2000).
175. K. Parang, L. I. Wiebe, and E. E. Knaus, *Curr. Med. Chem.* **7**, 995–1039 (2000).
176. M. Jung, S. Lee, and H. Kim, *Curr. Med. Chem.* **7**, 649–661 (2000).
177. G. Matthee, A. D. Wright, and G. M. Konig, *Planta Med.* **65**, 493–506 (1999).
178. T. B. Ng, B. Huang, W. P. Fong, and H. W. Yeung, *Life Sci.* **61**, 933–949 (1997).
179. M. Flepp, V. Schiffer, R. Weber, and B. Hirschel, *Swiss Med. Wkly.* **131**, 207–213 (2001).
180. V. Soriano and C. de Mendoza, *HIV Clin. Trials.* **3**, 237–248 (2002).
181. S. Maddocks and D. Dwyer, *Pediatr. Drugs* **3**, 681–702 (2001).
182. A. L. Pozniak, *HIV Med.* **1**, 7–10 (2000).
183. B. G. Gazzard, *HIV Med.* **1**, 11–14 (2000).
184. M. Nelson, *Int. J. STD AIDS.* **12**, 1–2 (2001).
185. L. Tarrago-Litvak, M. L. Andreola, M. Fournier, G. A. Nevinsky, V. Parissi, V. R. de Soultrait, and S. Litvak, *Curr. Pharmaceut. Design.* **8**, 595–614 (2002).
186. V. Nair, *Rev. Med. Virol.* **12**, 179–193 (2002).
187. J. Raulin, *Prog. Lipid Res.* **41**, 27–65 (2002).
188. A. K. Field, *Curr. Opin. Mol. Therapeut.* **1**, 323–331 (1999).

189. R. Craigie, *J. Biol. Chem.* **276**, 23213–23216 (2001).
190. Z. Debyser, P. Cherepanov, W. Pluymers, and E. De Clercq, *Method. Mol. Biol.* **160**, 139–155 (2001).
191. N. Neamati, C. Marchand, and Y. Pommier, *Adv. Pharmacol.* **49**, 147–165 (2000).
192. S. Jiang, Q. Zhao, and A. K. Debnath, *Curr. Pharmaceut. Design.* **8**, 563–580 (2002).
193. D. M. Eckert and P. S. Kim, *Ann. Rev. Biochem.* **70**, 777–810 (2001).
194. A. Pozniak, *J. HIV Ther.* **6**, 91–94 (2001).
195. E. De Clercq, *Drugs R D.* **2**, 321–331 (1999).
196. M. M. Mangos and M. J. Damha, in R. S. Hosmane, ed., *Current Topics in Medicinal Chemistry: Recent Developments in Antiviral Nucleosides, Nucleotides and Oligonucleotides*; Bentham Science Publishers, Ltd., Karachi, 2002, pp. 1147–1171.
197. E. Zamaratski, P. I. Pradeepkumar, and J. Chattopadhyaya, *J. Biochem. Biophys. Methods* **48**, 189–208 (2001).
198. Y. Poovorawan, P. Chatchatee, and V. Chongsrisawat, *J. Gastroen. Hepatol.* **17**, S155–166 (2002).
199. K. Kidd-Ljunggren, Y. Miyakawa, and A. H. Kidd, *J. Gen. Virol.* **83**, 1267–1280 (2002).
200. M. L. Funk, D. M. Rosenberg, and A. S. Lok, *J. Viral Hepatitis.* **9**, 52–61 (2002).
201. P. Simmonds, *J. Gen. Virol.* **82**, 693–712 (2001).
202. M. C. Yu, J. M. Yuan, S. Govindarajan, and R. K. Ross, *Can. J. Gastroenterol.* **14**, 703–709 (2000).
203. A. S. Lok, *J. Hepatol.* **32**, 89–97 (2000).
204. F. X. Bosch, J. Ribes, and J. Borras, *Sem. Liver Dis.* **19**, 271–285 (1999).
205. P. Bonanni, *Vaccine* **16**, S17–22 (1998).
206. G. L. Davis, *Southern Med. J.* **90**, 866–870; quiz 871 (1997).
207. J. A. O'Connor, *Adolescent Med. State Art Rev.* **11**, 279–292 (2000).
208. R. P. Perrillo, *Gastroenterology* **120**, 1009–1022 (2001).
209. C. Seeger and W. S. Mason, *Microbiol. Mol. Biol. Rev.* **64**, 51–68 (2000).
210. F. J. Mahoney, *Clin. Microbiol. Rev.* **12**, 351–366 (1999).
211. H. E. Blum, *Digest. Dis.* **20**, 81–90 (2002).
212. G. L. Davis, *Rev. Gastroenterol. Disord.* **2**, 106–115 (2002).
213. C. Pramoolsinsup, *J. Gastroen. Hepatol.* **17**, S125–145 (2002).
214. M. Torbenson and D. L. Thomas, *Lancet Infect. Dis.* **2**, 479–486 (2002).
215. Y. Shiratori, H. Yoshida, and M. Omata, *Expert Rev. Anticancer Ther.* **1**, 277–290 (2001).
216. J. H. Kao and D. S. Chen, *Lancet Infect. Dis.* **2**, 395–403 (2002).
217. J. H. Kao and D. S. Chen, *J. Formosan Med. Assoc.* **101**, 239–248 (2002).
218. Z. Y. Tang, *World J. Gastroenterol.* **7**, 445–454 (2001).
219. H. Dominguez-Malagon and S. Gaytan-Graham, *Ultrastruct. Pathol.* **25**, 497–516 (2001).
220. C. Rabe, B. Cheng, and W. H. Caselmann, *Digest. Dis.* **19**, 279–287 (2001).
221. W. C. Maddrey, *J. Med. Virol.* **61**, 362–366 (2000).
222. K. Deres and H. Rubsamen-Waigmann, *Infection* **27**, S45–51 (1999).
223. J. Torresi and S. Locarnini, *Gastroenterology* **118**, S83–103 (2000).
224. O. Hantz, J. L. Kraus, and F. Zoulim, *Curr. Pharmaceut. Design* **6**, 503–523 (2000).
225. A. H. Malik and W. M. Lee, *Ann. Intern. Med.* **132**, 723–731 (2000).
226. F. Zoulim, *Antivir. Res.* **44**, 1–30 (1999).
227. E. De Clercq, *Int. J. Antimicrob. Agents.* **12**, 81–95 (1999).
228. F. Zoulim and C. Trepo, *Intervirology* **42**, 125–144 (1999).
229. J. H. Hong, Y. Choi, B. K. Chun, K. Lee, and C. K. Chu, *Arch. Pharm. Res.* **21**, 89–105 (1998).

230. J. M. Colacino and K. A. Staschke, The identification and development of antiviral agents for the treatment of chronic hepatitis B virus infection. *Fortschritte der Arzneimittelforschung-Progress in Drug Research-Progres des Recherches Pharmaceutiques.* **1998**, 50, 259–322.

231. M. Rizzetto and A. R. Zanetti, *J. Med. Virol.* **67**, 463–466 (2002).

232. R. Montesano, *J. Med. Virol.* **67**, 444–446 (2002).

233. W. M. Cassidy, *Minerva Pediatr.* **53**, 559–566 (2001).

234. M. L. Michel and D. Loirat, *Intervirology* **44**, 78–87 (2001).

235. M. Lu and M. Roggendorf, *Intervirology* **44**, 124–131 (2001).

236. M. F. Yuen and C. L. Lai, *Lancet Infect. Dis.* **1**, 232–241 (2001).

237. R. E. Vryheid, E. S. Yu, K. M. Mehta, and J. McGhee, *Asian Am. Pac. Island. J. Health.* **9**, 162–178 (2001).

238. G. Webster and A. Bertoletti, *Mol. Immunol.* **38**, 467–473 (2001).

239. O. B. Engler, W. J. Dai, A. Sette, I. P. Hunziker, J. Reichen, W. J. Pichler, and A. Cerny, *Mol. Immunol.* **38**, 457–465 (2001).

240. P. Beutels, *Health Econ.* **10**, 751–774 (2001).

241. S. E. Robertson, M. V. Mayans, A. El-Husseiny, J. D. Clemens, and B. Ivanoff, *Vaccine* **20**, 31–41 (2001).

242. S. Feldman, *Pediatr. Infect. Dis. J.* **20**, S23–29 (2001).

243. G. Leroux-Roels, T. Cao, A. De Knibber, P. Meuleman, A. Roobrouck, A. Farhoudi, P. Vanlandschoot, and I. Desombere, *Acta Clin. Belg.* **56**, 209–219 (2001).

244. V. Raj, *Clin. Cornerstone* **3**, 24–36 (2001).

245. S. A. Gall, *Infect. Dis. Obstet. Gynecol.* **9**, 63–64 (2001).

246. R. Schirmbeck and J. Reimann, *Biol. Chem.* **382**, 543–552 (2001).

247. M. P. Cooreman, G. Leroux-Roels, and W. P. Paulij, *J. Biomed. Sci.* **8**, 237–247 (2001).

248. R. S. Koff, *Infect. Dis. Clin. N. Am.* **15**, 83–95 (2001).

249. D. Shouval, *Indian J. Gastroenterol.* **20**, C55–58 (2001).

250. M. Lagget and M. Rizzetto, *Curr. Pharmaceut. Design.* **8**, 953–958 (2002).

251. Y. F. Liaw, *J. Gastroen. Hepatol.* **17**, 406–408 (2002).

252. M. Rizzetto and M. Lagget, *Forum* **11**, 137–150 (2001).

253. A. C. Lyra and A. M. Di Bisceglie, *Minerva Med.* **92**, 431–434 (2001).

254. F. Bonino, F. Oliveri, P. Colombatto, B. Coco, D. Mura, G. Realdi, and M. R. Brunetto, *J. Hepatol.* **31**, 197–200 (1999).

255. V. Baffis, I. Shrier, A. H. Sherker, and A. Szilagyi, *Ann. Intern. Med.* **131**, 696–701 (1999).

256. S. H. Chen, *Curr. Med. Chem.* **9**, 899–912 (2002).

257. G. V. Papatheodoridis, E. Dimou, and V. Papadimitropoulos, *Am. J. Gastroenterol.* **97**, 1618–1628 (2002).

258. N. Leung, *J. Gastroen. Hepatol.* **17**, 409–414 (2002).

259. R. P. Perrillo, *Curr. Gastroenterol. Rep.* **4**, 63–71 (2002).

260. M. Rizzetto, *J. Med. Virol.* **66**, 435–451 (2002).

261. L. M. Wolters, H. G. Niesters, and R. A. de Man, *Eur. J. Gastroen. Hepatol.* **13**, 1499–1506 (2001).

262. Y. C. Cheng, *Antivir. Chem. Chemother.* **12**, 5–11 (2001).

263. M. V. Galan, D. Boyce, and S. C. Gordon, *Expert Opin. Pharmacother.* **2**, 1289–1298 (2001).

264. K. P. Fischer, K. S. Gutfreund, and D. L. Tyrrell, *Drug Resist. Updates* **4**, 118–128 (2001).

265. K. A. Staschke and J. M. Colacino, Drug discovery and development of antiviral agents for the treatment of chronic hepatitis B virus infection. *Fortschritte der*

Arzneimittelforschung-Progress in Drug Research-Progres des Recherches Pharmaceutiques. **2001**, *Spec*, 111–183.

266. F. Zoulim, *J. Clin. Virol.* **21**, 243–253 (2001).
267. D. Mutimer, *J. Clin. Virol.* **21**, 239–242 (2001).
268. A. Regev and E. R. Schiff, *Adv. Intern. Med.* **46**, 107–135 (2001).
269. J. N. Zuckerman and A. J. Zuckerman, *J. Infection.* **41**, 130–136 (2000).
270. A. S. Befeler and A. M. Di Bisceglie, *Infect. Dis. Clin. N. Am.* **14**, 617–632 (2000).
271. J. S. Freiman and G. W. McCaughan, *J. Gastroen. Hepatol.* **15**, 227–229 (2000).
272. T. Shaw and S. A. Locarnini, *J. Viral Hepatitis* **6**, 89–106 (1999).
273. E. Sokal, *Expert Opin. Pharmacother.* **3**, 329–339 (2002).
274. F. Nakhoul, R. Gelman, J. Green, E. Khankin, and Y. Baruch, *Transplant. Proc.* **33**, 2948–2949 (2001).
275. Y-F. Liaw, *Antivir. Chem. Chemother.* **12**, 67–71 (2001).
276. S. J. Yoo, H. O. Kim, Y. Lim, J. Kim, and L. S. Jeong, *Bioorg. Med. Chem.* **10**, 215–226 (2002).
277. P. L. Marion, F. H. Salazar, M. A. Winters, and R. J. Colonno, *Antimicrob. Agents Chemother.* **46**, 82–88 (2002).
278. G. Yamanaka, T. Wilson, S. Innaimo, G. S. Bisacchi, P. Egli, J. K. Rinehart, R. Zahler, and R. J. Colonno, *Antimicrob. Agents Chemother.* **43**, 190–193 (1999).
279. E. V. Genovesi, L. Lamb, I. Medina, D. Taylor, M. Seifer, S. Innaimo, R. J. Colonno, D. N. Standring, and J. M. Clark, *Antimicrob. Agents Chemother.* **42**, 3209–3217 (1998).
280. M. Seifer, R. K. Hamatake, R. J. Colonno, and D. N. Standring, *Antimicrob. Agents Chemother.* **42**, 3200–3208 (1998).
281. S. F. Innaimo, M. Seifer, G. S. Bisacchi, D. N. Standring, R. Zahler, and R. J. Colonno, *Antimicrob. Agents Chemother.* **41**, 1444–1448 (1997).
282. G. C. Farrell, *Drugs* **60**, 701–710 (2000).
283. F. Yao and R. G. Gish, *Curr. Gastroenterol. Rep.* **1**, 20–26 (1999).
284. M. Berenguer and T. L. Wright, *Proc. Assoc. Am. Physician.* **110**, 98–112 (1998).
285. J. Delmas, O. Schorr, C. Jamard, C. Gibbs, C. Trepo, O. Hantz, and F. Zoulim, *Antimicrob. Agents Chemother.* **46**, 425–433 (2002).
286. D. G. Brust, Low-dose adefovir for the treatment of chronic hepatitis B in HIV infected people. *GMHC Treatment Issues: the Gay Men's Health Crisis Newsletter of Experimental AIDS Therapies* **2001**, 15, 8–11.
287. D. Mutimer, B. H. Feraz-Neto, R. Harrison, K. O'Donnell, J. Shaw, P. Cane, and D. Pillay, *Gut* **49**, 860–863 (2001).
288. M. K. Bijsterbosch, C. Ying, R. L. de Vrueh, E. de Clercq, E. A. Biessen, J. Neyts, and T. J. van Berkel, *Mol. Pharmacol.* **60**, 521–527 (2001).
289. S. Hatse, Mechanistic study on the cytostatic and tumor cell differentiation-inducing properties of 9-(2-phosphonylmethoxyethyl)adenine (PMEA, adefovir)-collected publications. *Verhandelingen-Koninklijke Academie voor Geneeskunde van Belgie.* **2000**, Vol. 62, pp. 373–384.
290. H. Yang, C. E. Westland, W. E. t. Delaney, E. J. Heathcote, V. Ho, J. Fry, C. Brosgart, C. S. Gibbs, M. D. Miller, and S. Xiong, *Hepatology* **36**, 464–473 (2002).
291. C. Delaugerre, A. G. Marcelin, V. Thibault, G. Peytavin, T. Bombled, M. V. Bochet, C. Katlama, Y. Benhamou, and V. Calvez, *Antimicrob. Agents Chemother.* **46**, 1586–1588 (2002).
292. Y. Benhamou, M. Bochet, V. Thibault, V. Calvez, M. H. Fievet, P. Vig, C. S. Gibbs, C. Brosgart, J. Fry, H. Namini, C. Katlama, and T. Poynard, *Lancet* **358**, 718–723 (2001).
293. K. M. Walsh, T. Woodall, P. Lamy, D. G. Wight, S. Bloor, and G. J. Alexander, *Gut* **49**, 436–440 (2001).

294. E. De Clercq, *J. Clin. Virol.* **22**, 73–89 (2001).
295. R. J. Gilson, K. B. Chopra, A. M. Newell, I. M. Murray-Lyon, M. R. Nelson, S. J. Rice, R. S. Tedder, J. Toole, H. S. Jaffe, and I. V. Weller, *J. Viral Hepatitis* **6**, 387–395 (1999).
296. M. G. Peters, G. Singer, T. Howard, S. Jacobsmeyer, X. Xiong, C. S. Gibbs, P. Lamy, and A. Murray, *Transplantation.* **68**, 1912–1914 (1999).
297. M. G. Pessoa and T. L. Wright, *J. Gastroen. Hepatol.* **14**, S6–11 (1999).
298. M. Tsiang, J. F. Rooney, J. J. Toole, and C. S. Gibbs, *Hepatology* **29**, 1863–1869 (1999).
299. E. De Clercq, *Intervirology* **40**, 295–303 (1997).
300. C. L. Lai, M. F. Yuen, C. K. Hui, S. Garrido-Lestache, C. T. Cheng, and Y. P. Lai, *J. Med. Virol.* **67**, 334–338 (2002).
301. R. Shapira, N. Daudi, A. Klein, D. Shouval, E. Mor, R. Tur-Kaspa, G. Dinari, and Z. Ben-Ari, *Transplantation* **73**, 820–822 (2002).
302. S. Tang, S. K. Ho, K. Moniri, K. N. Lai, and T. M. Chan, *Transplantation* **73**, 148–151 (2002).
303. D. Mutimer, D. Pillay, P. Shields, P. Cane, D. Ratcliffe, B. Martin, S. Buchan, L. Boxall, K. O'Donnell, J. Shaw, S. Hubscher, and E. Elias, *Gut* **46**, 107–113 (2000).
304. M. P. Manns, P. Neuhaus, G. F. Atkinson, K. E. Griffin, S. Barnass, J. Vollmar, Y. Yeang, and C. L. Young, *Transplant Infect. Dis.* **3**, 16–23 (2001).
305. M. Berenguer, M. Prieto, M. Rayon, M. Bustamante, D. Carrasco, A. Moya, M. A. Pastor, M. Gobernado, J. Mir, and J. Berenguer, *Am. J. Gastroenterol.* **96**, 526–533 (2001).
306. N. Rayes, D. Seehofer, U. Hopf, R. Neuhaus, U. Naumann, W. O. Bechstein, and P. Neuhaus, *Transplantation* **71**, 96–101 (2001).
307. S. Menne, C. A. Roneker, B. E. Korba, J. L. Gerin, B. C. Tennant, and P. J. Cote, *J. Virol.* **76**, 5305–5314 (2002).
308. P. Krishnan, Q. Fu, W. Lam, J. Y. Liou, G. Dutschman, and Y. C. Cheng, *J. Biol. Chem.* **277**, 5453–5459 (2002).
309. T. Yamamoto, S. Litwin, T. Zhou, Y. Zhu, L. Condreay, P. Furman, and W. S. Mason, *J. Virol.* **76**, 1213–1223 (2002).
310. W. A. Tao, L. Wu, R. G. Cooks, F. Wang, and J. A. Begley, *J. Med. Chem.* **44**, 3541–3544 (2001).
311. R. Chin, T. Shaw, J. Torresi, V. Sozzi, C. Trautwein, T. Bock, M. Manns, H. Isom, P. Furman, and S. Locarnini, *Antimicrob. Agents Chemother.* **45**, 2495–2501 (2001).
312. Y. C. Cheng, *Cancer Lett.* **162**, S33-S37 (2001).
313. K. Lee and C. K. Chu, *Antimicrob. Agents Chemother.* **45**, 138–144 (2001).
314. S. F. Peek, P. J. Cote, J. R. Jacob, I. A. Toshkov, W. E. Hornbuckle, B. H. Baldwin, F. V. Wells, C. K. Chu, J. L. Gerin, B. C. Tennant, and B. E. Korba, *Hepatology* **33**, 254–266 (2001).
315. I. Kocic, *Curr. Opin. Inv. Drugs* **1**, 308–313 (2000).
316. C. Ying, E. De Clercq, W. Nicholson, P. Furman, and J. Neyts, *J. Viral Hepatitis* **7**, 161–165 (2000).
317. J. Du, Y. Choi, K. Lee, B. K. Chun, J. H. Hong, and C. K. Chu, *Nucleosides Nucleotides.* **18**, 187–195 (1999).
318. G. Gumina, Y. Chong, H. Choo, G.-Y. Song, and C. K. Chu, L-nucleosides: Antiviral activity and molecular mechanism, in R. S. Hosmane, ed., *Current Topics in Medicinal Chemistry: Recent Developments in Antiviral Nucleosides, Nucleotides and Oligonucleotides*, Bentham Science Publishers Ltd.: Karachi, 2002, pp. 1065–1086.
319. L. Wang, A. Bhan, and R. S. Hosmane, *Nucleosides Nucleotides* **13**, 2307–2320 (1994).

320. H. M. Chen, R. Sood, and R. S. Hosmane, *Nucleosides Nucleotides* **18**, 331–335 (1999).

321. R. S. Hosmane, V. S. Bhadti, and B. B. Lim, *Synthesis*, 1095–1100 (1990).

322. R. K. Sood, V. S. Bhadti, A. I. Fattom, R. B. Naso, B. E. Korba, E. R. Kern, H. M. Chen, and R. S. Hosmane, *Antivir. Res.* **53**, 159–164 (2002).

323. K. Iosue, *Nurse Practitioner* **27**, 32–33, 37–38, 40 passim; quiz 50–31 (2002).

324. M. J. Alter, D. Kruszon-Moran, O. V. Nainan, G. M. McQuillan, F. Gao, L. A. Moyer, R. A. Kaslow, and H. S. Margolis, *New Engl. J. Med.* **341**, 556–562 (1999).

325. M. I. Memon and M. A. Memon, *J. Viral Hepatitis* **9**, 84–100 (2002).

326. H. J. Alter and L. B. Seeff, *Sem. Liver Dis.* **20**, 17–35 (2000).

327. S. M. Lemon and D. L. Thomas, *New Engl. J. Med.* **336**, 196–204 (1997).

328. T. J. Liang, B. Rehermann, L. B. Seeff, and J. H. Hoofnagle, *Ann. Intern. Med.* **132**, 296–305 (2000).

329. M. P. Manns, J. G. McHutchison, S. C. Gordon, V. K. Rustgi, M. Shiffman, R. Reindollar, Z. D. Goodman, K. Koury, M. Ling, and J. K. Albrecht, *Lancet* **358**, 958–965 (2001).

330. J. G. McHutchison, S. C. Gordon, E. R. Schiff, M. L. Shiffman, W. M. Lee, V. K. Rustgi, Z. D. Goodman, M. H. Ling, S. Cort, and J. K. Albrecht, *New Engl. J. Med.* **339**, 1485–1492 (1998).

331. G. Ramadori and V. Meier, *Eur. J. Gastroen. Hepatol.* **13**, 465–471 (2001).

332. S. Zeuzem, S. V. Feinman, J. Rasenack, E. J. Heathcote, M. Y. Lai, E. Gane, J. O'Grady, J. Reichen, M. Diago, A. Lin, J. Hoffman, and M. J. Brunda, *New Engl. J. Med.* **343**, 1666–1672 (2000).

333. C. L. Cooper and D. W. Cameron, *Clin. Infect. Dis.* **35**, 873–879 (2002).

334. R. Bruno, P. Sacchi, M. Puoti, V. Soriano, and G. Filice, *Am. J. Gastroenterol.* **97**, 1598–1606 (2002).

335. O. Prakash, A. Mason, R. B. Luftig, and A. P. Bautista, *Front. Biosci.* **7**, e286–300 (2002).

336. A. H. Talal, P. W. Canchis, and I. Jacobson, *Curr. Gastroenterol. Rep.* **4**, 15–22 (2002).

337. S. Pol, A. Vallet-Pichard, and H. Fontaine, *J. Viral Hepatitis* **9**, 1–8 (2002).

338. S. J. Cotler and D. M. Jensen, *Clin. Liver Dis.* **5**, 1045–1061 (2001).

339. R. Rodriguez-Rosado, M. Perez-Olmeda, J. Garcia-Samaniego, and V. Soriano, *Antivir. Res.* **52**, 189–198 (2001).

340. E. Tedaldi and P. Bean, *Am. Clin. Lab.* **20**, 26–32 (2001).

341. J. Sasadeusz, *Intern. Med. J.* **31**, 418–421 (2001).

342. T. W. Waldrep, K. K. Summers, and P. A. Chiliade, *Pharmacotherapy* **20**, 1499–1507 (2000).

343. M. Bonacini and M. Puoti, *Arch. Intern. Med.* **160**, 3365–3373 (2000).

344. M. A. Poles and D. T. Dieterich, *Clin. Infect. Dis.* **31**, 154–161 (2000).

345. M. S. Sulkowski, E. E. Mast, L. B. Seeff, and D. L. Thomas, *Clin. Infect. Dis.* **30**, S77–84 (2000).

346. J. K. Rockstroh, R. P. Woitas, and U. Spengler, *Eur. J. Med. Res.* **3**, 269–277 (1998).

347. C. A. Sabin, *AIDS Patient Care Stds.* **12**, 199–207 (1998).

348. P. Leyssen, E. De Clercq, and J. Neyts, *Clin. Microbiol. Rev.* **13**, 67–82, (2000) table of contents.

349. S. Sherlock, *J. Viral Hepatitis* **6**, 1–5 (1999).

350. G. V. Ludwig, and L. C. Iacono-Connors, *In Vitro Cell. Dev. Biol. Animal* **29A**, 296–309 (1993).

351. E. G. Westaway, M. A. Brinton, S. Gaidamovich, M. C. Horzinek, A. Igarashi, L. Kaariainen, D. K. Lvov, J. S. Porterfield, P. K. Russell, and D. W. Trent, *Intervirology* **24**, 183–192 (1985).

352. N. Kato, *Acta Med. Okayama* **55**, 133–159 (2001).
353. S. Steffens, H. J. Thiel, and S. E. Behrens, *J. Gen. Virol.* **80**, 2583–2590 (1999).
354. H. Miller, R. P., *Proc. Natl. Acad. Sci. USA* **87**, 2057–2061 (1990).
355. J. F. Bazan, R. F., *Virology* **171**, 637–639 (1989).
356. P. Galinari, D. Brennan, C. Nardi, M. Brunetti, L. Tomei, and C. Steinkühler, R. D. F., *J. Virol.* **72**, 6758–6769 (1998).
357. A. E. Gorbalenya, E. V. Koonin, A. P. Donchenko, M. B. V., *Nucleic Acids Res.* **17**, 4713–4730 (1989).
358. A. E. Gorbalenya and E. V. Koonin, *Curr. Opin. Struct. Biol.* **3**, 419–429 (1993).
359. J. Kim, K. Morgenstern, J. Griffith, J. Dwyer, M. Thomson, M. Murcko, C. Lin, and P. Caron, *Structure* **6**, 89–100 (1998).
360. T. C. Hodgman, *Nature (London)* **333**, 22–23 (1988).
361. M. Kapoor, L. Zhang, M. Ramachandra, J. Kusukawa, K. E. Ebner, and R. Padmanabhan, *J. Biol. Chem.* **270**, 19100–19106 (1995).
362. C. Brinster and G. Inchauspe, *Intervirology* **44**, 143–153 (2001).
363. J. Bukh, X. Forns, S. U. Emerson, and R. H. Purcell, *Intervirology* **44**, 132–142 (2001).
364. A. M. Prince and M. T. Shata, *Clin. Liver Dis.* **5**, 1091–1103 (2001).
365. N. N. Zein, *Expert Opin. Inv. Drugs* **10**, 1457–1469 (2001).
366. M. M. Jonas, *Clin. Liver Dis.* **4**, 849–877 (2000).
367. Q. M. Wang and B. A. Heinz, Recent advances in prevention and treatment of hepatitis C virus infections. *Fortschritte der Arzneimittelforschung-Progress in Drug Research-Progres des Recherches Pharmaceutiques.* **2000**, 55, 1–32.
368. J. I. Cohen, *Med. Hypoth.* **55**, 353–355 (2000).
369. M. Lechmann and T. J. Liang, *Sem. Liver Dis.* **20**, 211–226 (2000).
370. R. De Francesco, P. Neddermann, L. Tomei, C. Steinkuhler, P. Gallinari, and A. Folgori, *Sem. Liver Dis.* **20**, 69–83 (2000).
371. S. Locarnini, *J. Viral Hepatitis* **7**, 5–6 (2000).
372. M. Houghton, *Curr. Top. Microbiol. Immunol.* **242**, 327–339 (2000).
373. R. Bartenschlager and V. Lohmann, *Antivir. Res.* **52**, 1–17 (2001).
374. Y. He and M. G. Katze, *Viral Immunol.* **15**, 95–119 (2002).
375. M. Willems, H. J. Metselaar, H. W. Tilanus, S. W. Schalm, and R. A. de Man, *Transplant Int.* **15**, 61–72 (2002).
376. L. Amati, L. Caradonna, T. Magrone, M. L. Mastronardi, R. Cuppone, R. Cozzolongo, O. G. Manghisi, D. Caccavo, A. Amoroso, and E. Jirillo, *Curr. Pharmaceut. Design.* **8**, 981–993 (2002).
377. G. Ideo and A. Bellobuono, *Curr. Pharmaceut. Design.* **8**, 959–966 (2002).
378. N. W. Leung, *J. Gastroen. Hepatol.* **17**, S146–154 (2002).
379. P. J. Pockros, *Expert Opin. Inv. Drugs* **11**, 515–528 (2002).
380. Y. S. Wang, S. Youngster, M. Grace, J. Bausch, R. Bordens, and D. F. Wyss, *Adv. Drug Deliv. Rev.* **54**, 547–570 (2002).
381. R. P. Myers, C. Regimbeau, T. Thevenot, V. Leroy, P. Mathurin, P. Opolon, J. P. Zarski, and T. Poynard, *Cochrane Database Systemat. Rev.* CD000370 (2002).
382. J. G. McHutchison, *J. Gastroen. Hepatol.* **17**, 431–441 (2002).
383. L. J. Scott and C. M. Perry, *Drugs* **62**, 507–556 (2002).
384. M. Cornberg, H. Wedemeyer, and M. P. Manns, *Curr. Gastroenterol. Rep.* **4**, 23–30 (2002).
385. E. J. Heathcote, M. L. Shiffman, W. G. Cooksley, G. M. Dusheiko, S. S. Lee, L. Balart, R. Reindollar, R. K. Reddy, T. L. Wright, A. Lin, J. Hoffman, and J. De Pamphilis, *New Engl. J. Med.* **343**, 1673–1680 (2000).
386. P. J. Gavin and B. Z. Katz, *Pediatrics* **110**, e9 (2002).
387. R. A. Willson, *J. Clin. Gastroenterol.* **35**, 89–92 (2002).

388. L. L. Kjaergard, K. Krogsgaard, and C. Gluud, *Cochrane Database Systemat. Rev.* CD002234 (2002).
389. L. L. Kjaergard, K. Krogsgaard, and C. Gluud, *Cochrane Database Systemat. Rev.* CD002234 (2002).
390. P. Marcellin, M. Martinot, N. Boyer, and S. Levy, *Clin. Liver Dis.* **3**, 843–853 (1999).
391. G. L. Davis, *Curr. Gastroenterol. Rep.* **1**, 9–14 (1999).
392. M. P. Civeira and J. Prieto, *J. Hepatol.* **31**, 237–243 (1999).
393. E. Gane and H. Pilmore, *Transplantation* **74**, 427–437 (2002).
394. L. Benson, A. Birkel, L. Caldwell, V. Stafford-Fox, and B. Casarico, *J. Am. Acad. Nurs. Pract.* **12**, 364–373 (2000).
395. J. Collier and R. Chapman, *Biodrugs* **15**, 225–238 (2001).
396. N. Yao, T. Hesson, M. Cable, Z. Hong, A. Kwong, H. Le, and P. Weber, *Nature Struct. Biol.* **4**, 463–467 (1997).
397. H. S. Subramanya, L. E. Bird, J. A. Brannigan, and D. B. Wigley, *Nature (London)* **384**, 379–383 (1996).
398. K. Theis, P. Chen, M. Skorvaga, B. van Houten, and C. Kisker, *EBMBO J.* **24**, 6899–6907 (1999).
399. P. Borowski, R. Kuehl, O. Mueller, L.-H. Hwang, J. Schulze zur Wiesch, H. S. *Eur. J. Biochem.* **266**, 715–723 (1999).
400. P. Borowski, O. Mueller, A. Niebuhr, M. Kalitzky, L.-H. Hwang, H. Schmitz, A. M. Siwecka, and T. Kulikowski, *Acta Biochim. Polon.* **47**, 173–180 (2000).
401. C.-L. Tai, W.-K. Chi, D.-S. Chen, L-H. H., *J. Virol.* **70**, 8477–8484 (1996).
402. L. Lun, P.-M. Sun, C. Trubey, N. B., *Cancer Chemother. Pharmacol.* **42**, 447–453 (1998).
403. D. Porter, *J. Biol. Chem.* **273**, 7390–7396 (1998).
404. D. Porter, *J. Biol. Chem.* **273**, 14247–14253 (1998).
405. P. Borowski, A. Niebuhr, O. Mueller, M. Bretner, K. Felczak, T. Kulikowski, and H. Schmitz, *J. Virol.* **75**, 3220–3229 (2001).
406. V. J.-P. Leveque and Q. M. Wang, *Cell. Mol. Life Sci.* **59**, 909–919 (2002).
407. S. Bressanelli, L. Tomei, A. Roussel, I. Incitti, R. L. Vitale, M. Mathieu, R. De Francesco, and F. A. Rey, *Proc. Natl. Acad. Sci. USA* **96**, 13034–13039 (1999).
408. C. A. Lesburg, M. B. Cable, E. Ferrari, Z. Hong, A. F. Mannarino, and P. C. Weber, *Nature Struct. Biol.* **6**, 937–943 (1999).
409. D. Dhanak, K. J. Duffy, V. K. Johnston, J. Lin-Goerke, M. Darcy, A. N. Shaw, B. Gu, C. Silverman, A. T. Gates, M. R. Nonnemacher, D. L. Earnshaw, D. J. Casper, A. Kaura, A. Baker, C. Greenwood, L. L. Gutshall, D. Maley, A. DelVecchio, R. Macarron, G. A. Hofmann, Z. Alnoah, H. Y. Cheng, G. Chan, S. Khandekar, R. M. Keenan, and R. T. Sarisky, *J. Biol. Chem.* **277**, 38322–38327 (2002).
410. S. S. Takyar, E. J. Gowans, and W. B. Lott, *J. Mol. Biol.* **319**, 1–8 (2002).
411. A. A. Kolykhalov, K. Mihalik, S. M. Feinstone, and C. M. Rice, *J. Virol.* **74**, 2046–2051 (2000).
412. N. H. Yao, P. Reichert, S. S. Taremi, W. W. Prosise, and P. C. Weber, *Structure* **7**, 1353–1363 (1999).
413. B. W. Dymock, P. S. Jones, and F. X. Wilson, *Antivir. Chem. Chemother.* **11**, 79–96 (2000).
414. P. Ingallinella, D. Fattori, S. Altamura, C. Steinkuhler, U. Koch, D. Cicero, R. Bazzo, R. Cortese, E. Bianchi, and A. Pessi, *Biochemistry* **41**, 5483–5492 (2002).
415. R. M. Zhang, J. P. Durkin, and W. T. Windsor, *Bioorg. Med. Chem. Lett.* **12**, 1005–1008 (2002).
416. N. Usman and L. R. M. Blatt, *J. Clin. Invest.* **106**, 1197–1202 (2000).
417. S-L. Tan, A. Pause, Y. Shi, and N. Sonenberg, *Nature Rev. Drug Discov.* **1**, 867–881 (2002).

418. H. Zhang, R. Hanecak, V. Brown-Driver, R. Azad, B. Conklin, M. C. Fox, and K. P. Anderson, *Antimicrob. Agents Chemother.* **43**, 347–353 (1999).
419. V. Brown-Driver, T. Eto, E. Lesnik, K. P. Anderson, and R. C. Hanecak, *Antisense Nucl. Acid Drug Dev.* **9**, 145–154 (1999).
420. G. Witherell, *Curr. Opin. Invest. Drugs* **2**, 1523–1529 (2001).
421. J. T. Roehrig, M. Layton, P. Smith, G. L. Campbell, R. Nasci, and R. S. Lanciotti, *Curr. Top. Microbiol. Immunol.* **267**, 223–240 (2002).
422. B. Murgue, H. Zeller, and V. Deubel, *Curr. Top. Microbiol. Immunol.* **267**, 195–221 (2002).
423. N. Komar, *Rev. Sci. Techn.* **19**, 166–176 (2000).
424. Z. Hubalek, *Viral Immunol.* **13**, 415–426 (2000).
425. J. O. Lundstrom, *J. Vector Ecol.* **24**, 1–39 (1999).
426. West Nile watch. *Science* **293**, 1413 (2001).
427. B. S. Davis, G. J. Chang, B. Cropp, J. T. Roehrig, D. A. Martin, C. J. Mitchell, R. Bowen, and M. L. Bunning, *J. Virol.* **75**, 4040–4047 (2001).
428. P. Borowski, M. Lang, A. Haag, H. Schmitz, J. Choe, H. M. Chen, and R. S. Hosmane, *Antimicrob. Agents Chemother.* **46**, 1231–1239 (2002).
429. N. Zhang, H.-M. Chen, V. Koch, H. Schmitz, C.-L. Liao, M. Bretner, V. S. Bhadti, A. I. Fattom, R. B. Naso, R. S. Hosmane, and P. Borowski, *J. Med. Chem.* in press.
430. I. Jordan, T. Briese, N. Fischer, J. Y. Lau, and W. I. Lipkin, *J. Infect. Dis.* **182**, 1214–1217 (2000).
431. G. Y. Song, V. Paul, H. Choo, J. Morrey, R. W. Sidwell, R. F. Schinazi, and C. K. Chu, *J. Med. Chem.* **44**, 3985–3993 (2001).

RAMACHANDRA S. HOSMANE
University of Maryland

AQUACULTURE

1. Introduction

One definition of aquaculture is the rearing of aquatic organisms under controlled or semicontrolled conditions (1). Another, used by the Food and Agriculture Organization (FAO) of the United Nations, is that aquaculture is, "the farming of aquatic organisms, including fish, molluscs, crustaceans, and aquatic plants" (2). Included within those broad definitions are activities in fresh, brackish, marine, and even hypersalinewaters. The term *mariculture* is often used in conjunction with aquaculture in the marine environment.

Public sector aquaculture involves production of aquatic animals to augment or establish recreational and commercial fisheries. Public sector aquaculture is widely practiced in North America and to a lesser extent in other parts of the world. The FAO definition of aquaculture also indicates that farming implies ownership of the organisms being cultured, which would seem to exclude public sector aquaculture.

In recent years, aquaculture has been increasingly used as a means of aiding in the recovery of threatened and endangered species. Those efforts are currently public sector activities, although there is interest in the private sector to become involved. As global awareness of endangered species issues grows, recovery programs for aquatic threatened and endangered species may arise in many more countries. Going hand in hand with attempts to recover endangered species are enhancement stocking programs aimed at releasing juvenile animals to rebuild stocks of aquatic animals that have been reduced due to overfishing. Examples of enhancement programs currently in existence include the stocking of cod in Norway, flounders in Japan, and red drum in the United States.

The bulk of global production from aquaculture is utilized directly as human food, with public aquaculture playing a minor role in many nations or being absent. Private aquaculture is not only about human food production, however. In some regions, well-developed private sector aquaculture is involved in the production of bait and ornamental fishes and invertebrates.

Aquatic plants are cultured in many regions of the world. In fact, aquatic plants, primarily seaweeds, account for nearly 23% of the world's aquaculture production (3). Most of the information available in the literature relates to the production of such aquatic animals as molluscs, crustaceans, and finfish.

The origins of aquaculture are rooted in ancient China and may date back some 4000 years. Today, Asia dominates the world in aquaculture production, with China producing over 10 million metric tons in 1992 and Japan and India each producing well over 1 million metric tons (4). By 1996, China's aquaculture production accounted for over 67% of the global total and had reached over 23 million metric tons (3). India and Japan continued to rank second and third globally in 1996 with 6.7 and 3.1% of global production, respectively (3). In North America, the culture of fish began in the mid-nineteenth century and grew rapidly in the public sector after the establishment of the U.S. Fish and Fisheries Commission in 1871 (5). Private aquaculture existed as a minor industry for many decades, coming into prominence in the 1960s. Since then the United States has become one of the leaders in aquaculture research and development, although production, while significant at over 400,000 metric tons by 1992 (4), amounted to only about 2% of the world's total of nearly 19 million metric tons. The contribution of U.S. aquaculture to the global total had dropped to 1.5% by 1996 (3). The United States commercial aquaculture industry is dominated by channel catfish, (*Ictalurus punctatus*), trout, salmon, minnows, oysters, mussels, clams, and crawfish. A number of other fishes and invertebrates are also being reared. Included are tilapia, (*Oreochromis* spp.), striped bass (*Morone saxatilis*), and hybrid striped bass (*M. saxatilis x M. chrysops*), red drum (*Sciaenops ocellatus*), goldfish (*Carassius auratus*), tropical fishes, and shrimp. In the public sector, hatcheries produce large numbers of such species as salmon, trout, largemouth (*Micropterus salmoides*) and smallmouth bass (*M. dolomieui*), sunfish (*Lepomis* spp.), crappie (*Pomoxis* spp.), northern pike (*Esox lucius*) muskellunge (*E. masquinongy*), walleye (*Stizostedion vitreum vitreum*), and catfish (*Ictalurus* spp.) for stocking or growout.

Aquaculture production continues to grow annually, but increasing competition for suitable land and water, problems associated with wastewater from aquaculture facilities, disease outbreaks, and potential shortages of animal protein for aquatic animal feeds are having, or may have, negative effects on future growth. New technology, including the application of genetic engineering approaches to improving performance and disease resistance in aquatic species, along with the development of water reuse (recirculating) systems and the establishment of offshore facilities, may provide the impetus for a resurgence of growth in the industry.

2. Economics

The production of aquatic animals for recreation, in nations where that type of aquaculture exists, is typically funded through user fees such as fishing licenses that support hatcheries and the personnel to run them. In order for most private aquaculture companies to get started, outside funding is required. Funding may come through banks and other commercial lending sources or from venture capitalists. The high risks associated with aquaculture have made it difficult for many startup firms to obtain bank loans, although that situation is changing as bankers become more knowledgeable and comfortable with underwriting aquaculture ventures.

A key factor in obtaining funding support for aquaculture is development of a sound business plan. The plan needs to demonstrate that the prospective culturist has identified all costs associated with establishment of the facility and its day-to-day operation. One or more suitable sites should have been identified and the species to be cultured selected before the business plan is submitted. Cost estimates should be verifiable. Having actual bids for a specific task at a specific location; eg, pond construction, well drilling, building construction, and vehicle costs helps strengthen the business plan.

Land costs vary enormously both between and within countries. Compare the cost of coastal land in south Florida where it might be possible to consider rearing shrimp with that of Mississippi farmland suitable for catfish farming. The former might be thousands of dollars for every meter of ocean front, while the latter may be obtained for one or two thousand dollars per hectare.

The amount of land required varies as well, not only as a function of the amount of production that is anticipated, but also on the type of culture system that is used. It may take several hectares of static culture ponds to produce the same biomass of animals as one modest size raceway through which large volumes of water are constantly flowed. Construction costs vary from one location to another. Local labor and fuel costs must be factored into the equation. The experience of contractors in building aquaculture facilities is another factor to be considered.

The need for redundancy in the culture system needs to be assessed. Failure of a well pump that brings up water to supply a static pond system may not be a serious problem in countries where new pumps can be purchased in a nearby town. However, it can be disastrous in developing countries where new pumps and pump parts are often not available, but must be ordered from another country. Several weeks or months may pass before the situation can be remedied unless the culturist maintains a selection of spares.

The business plan needs to provide projections of annual production. Based on those estimates and assumed food conversion rates (food conversion is calculated by determining the amount of feed consumed by the animals for each kilogram of weight gain), an estimate of feed costs can be made. For many aquaculture ventures, between 40 and 50% of the variable costs involved in aquaculture can be attributed to feed.

Aquaculturists may elect to purchase animals for stocking or maintain their own broodstock and hatchery. The decision may rest on such factors as the availability and cost of fry fish, post-larval fish, oyster spat, or other early life history stages in the location selected for the aquaculture venture.

Land purchases and many of the costs associated with facility development can be accomplished with long-term loans of 15 to 30 years. Equipment such as pumps and trucks are usually depreciated over a few years and are funded with shorter-term loans. Operating expenses for such items as feed, chemicals, fuel, utilities, salaries, taxes, and insurance may require periodic short-term loans to keep the business solvent. The projected income should be based on a realistic estimate of farmgate value of the product and an accurate assessment of anticipated production. Each business plan should project income and expenses projected over the term of all loans in order to demonstrate to the lending agency or venture capitalist that there is a high probability the investment will be repaid.

3. Regulation

The extent to which governments regulate aquaculture varies greatly from one nation to another. In some parts of the world, particularly in developing nations, there has historically been little or no regulation. Inexpensive land and labor, low taxes, excellent climates conducive to rapid growth of aquatic species, and a lack of government interference have drawn many aquaculturists to underdeveloped countries, most of which are in the tropics. Unregulated expansion of aquaculture in some countries has led to pollution problems, destruction of valuable habitats such as mangrove swamps, and has enhanced the spread of disease from one farm to another. The need for imposing regulations is now becoming evident around the world. Response to that need varies considerably from one nation to another.

In developed countries there may or may not be a standardized set of national regulations. The United States is an example of a mixture of local, state, and federal regulations. Permits from a county, state, or federal agency may be required for drilling wells, pumping water, releasing water, use of exotic species, constructing facilities, etc. In the United States, most permits can be obtained at the local or state level. In some instances the federal government has delegated permitting authority to the states when state regulations are as rigorous or more so than national regulations. Federal agencies become involved when aquaculture projects are conducted in navigable waters (U.S. Army Corps of Engineers) in the Exclusive Economic Zone, or might impact threatened or endangered species (U.S. Fish and Wildlife Service).

In general, it is easier to establish an aquaculture facility on private land than in public waters such as a lake or coastal embayment. Prospective aquaculturists who want to establish facilities inpublic waters may be confronted at public hearings by outraged citizens who do not want to see an aquaculture facility in what they consider to be their water. The issue is highly contentious in some nations (eg, the United States). In other countries, aquaculture in public waters is seen as not only a good use of natural resources, but can be considered an amenity (eg, Japan).

Obtaining permits is often not simple. Few states have one office that can accommodate the prospective aquaculturist. In most cases it is necessary to contact a number of state, and often federal agencies to apply for permits. Public hearings may be required before permits are approved. The process can take months or even years to complete. The costs involved in going through the process may be prodigious. After the expenditure of considerable amounts of time and money, there is no guarantee that the permits will ultimately be granted.

Most states now have an aquaculture coordinator, usually housed in the state department of agriculture, who can assist prospective aquaculturists in finding a path through the permitting process. Anyone considering development of an aquaculture facility should become educated on the permitting process of the state or nation in which the facility will be developed. In cases where the process is involved, it should be initiated well in advance of the anticipated time of actual facility construction.

4. Species under Cultivation

This article emphasizes aquatic animal production, but many hundreds of thousands of people are involved, worldwide, in aquatic plant production. The quantity of brown seaweeds, red seaweeds, green seaweeds, and other algae produced in 1996 was estimated at over 7.7 million metric tons (Table 1). Miscellaneous aquatic plants such as watercress and water chestnuts contributed an additional 600,000 metric tons. Microscopic algae and cyanobacteria are sometimes marketed as food or as a nutritional supplement (eg, *Spirulina* sp.). In addition,

Table 1. **World Aquaculture Production in 1996 for Selected Aquaculture Species or Species Groups**[a]

Species group	Production (10^3 tons)
Cyprinids	11,504
Tilapia and other cichlids	801
Atlantic salmon	556
Rainbow trout	380
Milkfish	365
Eels	216
Channel catfish	214
Pacific salmon	88
Japanese seabream	78
Oysters	3,067
Clams, cockles, and arkshells	1,777
Scallops and pectens	1,275
Mussels	1,179
Marine shrimp	915
Crabs	119

[a]Ref. 3.

an undocumented quantity of algae (mostly of the single-celled variety) is produced for use as food for filter-feeding aquatic animals (primarily molluscs and zooplankton). Planktonic organisms such as rotifers are reared on algae and then used to feed the young of crustaceans and fishes that do not accept prepared feeds.

Animal aquaculture is concentrated on finfish, molluscs, and crustaceans. Sponges, echinoderms, tunicates, turtles, frogs, and alligators are also being cultured, but production is insignificant in comparison with the three principal groups. Common and scientific names of many of the species of the finfish, molluscs, and crustaceans currently under culture are presented in Table 2. Included are examples of bait, recreational, and food animals.

Table 2. **Common and Scientific Names of Selected Aquaculture Species**

Type of organism	Common name	Scientific name
finfish	African catfish	*Clarias gariepinus*
	Atlantic halibut	*Hippoglossus hippoglossus*
	Atlantic salmon	*Salmo salar*
	Bighead carp	*Aristichthys nobilis*
	Bigmouth buffalo	*Ictiobus bubalus*
	Black crappie	*Pomoxis nigromaculatus*
	Blue catfish	*Ictalurus furcatus*
	Blue tilapia	*Oreochromis aureus*
	Bluegill	*Lepomis macrochirus*
	Brook trout	*Salvelinus fontinalis*
	Brown trout	*Salmo trutta*
	Catla	*Catla catla*
	Channel catfish	*Ictalurus punctatus*
	Chinook salmon	*Oncorhynchus tshawytscha*
	Chum salmon	*Oncorhynchus keta*
	Coho salmon	*Oncorhynchus kisutch*
	Common carp	*Cyprinus carpio*
	Fathead minnow	*Pimephales promelus*
	Gilthead sea bream	*Sparus aurata*
	Goldfish	*Carassius auratus*
	Grass carp	*Ctenopharyngodon idella*
	Largemouth bass	*Micropterus salmoides*
	Milkfish	*Chanos chanos*
	Mossambique tilapia	*Oreochromis mossambicus*
	Mrigal	*Cirrhinus mrigala*
	Mud carp	*Cirrhina molitorella*
	Muskellunge	*Esox masquinongy*
	Nile tilapia	*Oreochromis niloticus*
	Northern pike	*Esox lucius*
	Pacu	*Colossoma metrei*
	Pink salmon	*Oncorhynchus gorbuscha*
	Plaice	*Pleuronectes platessa*
	Rabbitfish	*Siganus* spp.
	Rainbow trout	*Oncorhynchus mykiss*
	Red drum	*Sciaenops ocellatus*
	Rohu	*Labeo rohita*
	Sea bass	*Dicentrarchus labrax*
	Shiners	*Notropis* spp.

Table 2   (*Continued*)

Type of organism	Common name	Scientific name
	Silver carp	*Hypophthalmichthys molitrix*
	Smallmouth bass	*Micropterus dolomieui*
	Sole	*Solea solea*
	Steelhead	*Oncorhynchus mykiss*
	Striped bass	*Morone saxatilis*
	Walking catfish	*Clarias batrachus*
	Walleye	*Stizostedion vitreum vitreum*
	White crappie	*Pomoxis annularis*
	Yellow perch	*Perca flavescens*
	Yellowtail	*Seriola quinqueradiata*
molluscs	American oyster	*Crassostrea virginica*
	Bay scallop	*Aequipecten irradians*
	Blue mussel	*Mytilus edulis*
	Northern quahog	*Mercenaria mercenaria*
	Pacific oyster	*Crassostrea gigas*
	Southern quahog	*Mercenaria campechiensis*
crustaceans	Freshwater shrimp	*Macrobrachium rosenbergii*
	Blue shrimp	*Litopenaeus stylirostris*
	Kuruma shrimp	*Marsupenaeus japonicus*
	Pacific white shrimp	*Litopenaeus vannamei*
	Red swamp crawfish	*Procambarus clarkii*
	Tiger shrimp	*Penaeus monodon*
	White river crawfish	*Procambarus acutus acutus*
	White shrimp	*Litopenaeus setiferus*
algae (seaweeds)	California giant kelp	*Macrocystis pyrifera*
	Eucheuma	*Eucheuma cottoni*
	False Irish moss	*Gigartina stellata*
	Gracilaria	*Gracilaria* sp.
	Irish moss	*Chondrus crispus*
	Laminaria	*Laminaria* spp.
	Nori or laver	*Porphyra* spp.
	Wakame	*Undaria* spp.

Various species of carp and other members of the family Cyprinidae lead the world in terms of quantity of animals produced. In 1996 the total was over 11.5 million metric tons (see Table 1). China is the leading carp producing nation, and is the world's leading aquaculture nation overall (Table 3). Significant amounts of carp are also produced in India and parts of Europe.

Fishes in the family Salmonidae (trout and salmon) are in high demand, with the interest in salmon being greatest in developed nations. Salmon, mostly Atlantic salmon (*Salmo salar*), are produced in Canada, Chile, Norway, New Zealand, Scotland, and the United States. Fishes in the family Cichlidae, which includes several cultured species of tilapia, are reared primarily in the tropics, but have been widely introduced throughout both the developed and developing world.

Catfish are not a major contributor to aquaculture production globally, but the channel catfish (*Ictalurus punctatus* industry dominates United States aquaculture. United States catfish production, primarily channel catfish, was 214,154 metric tons in 1996 (4). (3)

Table 3. **Top Aquaculture Producing Nations in 1996 and Percentage of Total Global Production**[a]

Nation	Country Rank	Percentage Share
China	1	67.1
India	2	6.7
Japan	3	3.1
Indonesia	4	2.6
Thailand	5	1.9
United States	6	1.5
Bangladesh	7	1.5
Korean Republic	8	1.4
Philippines	9	1.3
Norway	10	1.2
France	11	1.1
Taiwan	12	1.0
Spain	13	0.9
Chile	14	0.8
all others		7.9

[a]Ref. 3.

Among the invertebrates, most of the world's production is associated with mussels, oysters, shrimp, scallops, and clams. Red swamp crawfish culture is of considerable importance in the United States, but amounted to only 23,581 metric tons in 1996 (4) (3); insignificant compared to some other invertebrate species.

Small amounts of crabs, lobsters, and abalone are being cultured in various nations, and production has been on the increase in recent years. All three bring good prices in the marketplace but have drawbacks associated with their culture. Crabs and lobsters are highly cannibalistic. Rearing them separately to keep them from consuming one another during molting has precluded their economic culture in nearly every instance. Abalone eat seaweeds and can only be reared in conjunction with a concurrent seaweed culture facility or in regions where large supplies of suitable seaweeds are available from nature. In some instances the value of the seaweed for direct human consumption may be the highest and best use of the plants.

In all, there are perhaps 100 species of aquatic animals under culture. Many researchers have turned their attention to species for which there is demand by consumers, but for which the technology required for commercial production is not available, under development, or in the early stages of being tested commercially. Examples are dolphin fish, also known as mahimahi (*Coryphaena hippurus*), Pacific halibut (*Hippoglossus stenolepis*), summer flounder (*Paralichthys dentatus*), winter flounder (*Pseudopleuronectes americanus*), American lobster (*Homarus americanus*), and blue crab (*Callinectes sapidus*). Each of the species mentioned is marine and has small eggs and larvae. Providing the first feeding stages with acceptable food has been a common problem, as has the fragility of the early life stages of many species, and the problem of cannibalism.

Fishes with large eggs, such as trout, salmon, catfish, and tilapia, were among the first to be economically successful in modern times. However, small eggs do not necessarily mean that sophisticated research is required to develop the technology required for successful culture. Carp, which have been cultured in China for millennia, have extremely small eggs. At the time the methodology for carp culture was developed, there were no research scientists, although there must have been dedicated farmers who used their common sense and trial-and-error methods to establish carp aquaculture.

5. Culture Systems

At one extreme aquaculture can be conducted with a small amount of intervention from humans and the employment of little technology. At the other is total environmental control and the use of computers, molecular genetics, and complex modern technology. Many aquaculturists operate between the extremes. The range of culture approaches can be described as running from extensive to intensive, or even hyper-intensive, with extensive systems being relatively simple and intensive systems being complex to very complex. In general, as the level of culture intensity increases, stocking density, and as a consequence, production per unit area of culture system or volume of water, increases.

The most extensive types of aquaculture involve minimal human intervention to promote increases in natural productivity. Good examples can be found relative to oyster and pond fish culture. With respect to oysters, one of the most extensive forms of culture involves placing oyster, clam, or other types of shell (cultch) on the bottom in intertidal areas that are known to have good oyster reproduction, but limited natural cultch material. The additional substrate may subsequently be colonized by oyster larvae (spat), thereby potentially increasing productivity. The next level of intensity might involve placing bags of cultch out in nature to collect spat in a productive area that already has sufficient quantities of natural cultch. After spat settlement, the bags of shells would be moved to an area where limited natural substrate availability has led to low productivity. The growout area may be held as a common resource, or it may be made commercially available on a leasehold basis.

The next step in increasing oyster culture intensity might involve hatchery production and settling of spat on cultch. Once again, the cultch would be later distributed over a bed leased or owned by the oyster culturist (Fig. 1). Control of predators such as starfish and oyster drills could easily be a part of culture at all levels.

The highest level of intensity with respect to oysters involves hatchery production of spat and the rearing of them suspended from rafts, long-lines, or as cultchless oysters in trays. In the raft and longline techniques cultch material to which oyster spat or larval mussels are attached is strung on ropes (longlines) suspended from floating rafts or from other ropes held parallel to the water surface with buoys. The lines of growing oysters suspended from the ropes (called strings) are of such a length that the young shellfish are kept within the photic zone and not allowed to touch the bottom where starfish, oyster drills, and other predators can attack. Scallops, which do not attach to cultch, are sometimes grown in bags suspended from long lines or rafts. Similarly, young mussels can be held in proximity to strings with fine mesh materials that retain the

Fig. 1. Bags of oyster shell used as cultch for the settling of oyster spat. The spat-laden shell is ultimately distributed on leased oyster beds.

animals in place until they attach (by means of what is called a byssus thread) to the string.

The stocking of ponds, lakes, and reservoirs to increase the production of desirable fishes that depend on natural productivity for their food supply and are ultimately captured by recreational fishermen or for subsistence is another example of extensive aquaculture. Some would consider such practices as lying outside of the realm of aquaculture, but since the practice involves human intervention and often employs fishes produced in hatcheries, recreational or subsistence level stocking is associated with, if not a part of aquaculture. Similarly, stocking new ponds or water bodies which have been drained or poisoned to eliminate undesirable species prior to restocking, can lead to increased production of desirable species.

Most of the aquaculture practiced around the world is conducted in static ponds (Fig. 2). Ponds range in size but production units are generally 0.1 to 10 ha in area. The intensity of aquaculture in ponds can range from a few kg/ha to thousands of kg/ha of annual production. Ponds may be of the watershed type where they are constructed in a manner that takes advantage of rain runoff for pond filling, or they may be constructed with levees above the surrounding terrain elevation. The latter requires that the vast majority of the water used to fill the ponds comes from wells or other sources, not runoff. Excavated ponds with the top of their levees at the original ground level are also an option, but if those occur on flat landscapes the amount of runoff available for pond filling is generally insufficient and other sources of water will be required.

Aquaculture ponds, unlike farm ponds, should be fitted with drains. While many aquaculturists do not drain their ponds on an annual or more frequent basis during harvesting, periodic draining is required. In some cases, harvesting

Fig. 2.   Aquaculture ponds are often rectangular in shape. They should be equipped with plumbing for both inflow and drainage of water.

is conducted by capturing animals that are flushed out through the pond drain. Drain structures may be elaborate or quite simple.

Pond levees typically have side slopes of 2:1 or 3:1 (2 or 3 units of measure laterally for each unit of measure in height). Steeper slopes (e.g. 1:1) make construction more difficult and are too steep for easy entry and egress by personnel who are required to enter the pond. If the side slope is less steep than recommended, aquatic plants tend to become a significant problem by invading the shallow and extensive shoreline areas of the pond.

Fertilization of ponds to increase productivity is the next level of intensity with respect to fish culture, followed by provision of supplemental feeds. Fertilization promotes the production of natural foods within a pond. Included are phytoplankton, zooplankton, and benthic organisms. In some cases, aquatic vegetation is also encouraged and its production will benefit from fertilization. In most aquaculture operations, the production of rooted or floating aquatic plants is discouraged. Supplemental feeds are those that provide some additional nutrition but cannot be depended upon to supply all the required nutrients. Provision of complete feeds, those that do provide all of the nutrients required by the fish, translates to another increase in intensity. Associated with one or more of the stages described might be the application of techniques that lead to the maintenance of good water quality. Examples are continuous water exchange, mechanical aeration, and the use of various chemicals used to adjust such factors as pH, alkalinity, and hardness.

With the application of increased technology and control over the culture system, intensity continues to increase. Utilization of specific pathogen-free animals, provision of nutritionally complete feeds, careful monitoring and control of water quality, and the use of animals bred for good performance, can lead to

Fig. 3. A commercial trout facility in Idaho, U.S. Linear raceways are commonly used for the production of trout from fry to either release or market and for salmon from fry to smolt size.

impressive production levels. The United States channel catfish industry is a good example. During the early 1960s pond production levels were typically about 1500 kg/ha/yr. As better feeds and management practices were developed, production increased to an average of about 3000 kg/ha/yr by the next decade. In the 1980s, nutritionally complete feeds were perfected, better methods for prediction and amelioration of water quality problems had been developed, and diseases were being better avoided or controlled. Catfish farmers typically produced 4000 kg/ha/yr and some, who used aeration and exchanged water during part of the growing season, were able to produce 10,000 kg/ha/yr or more.

Where water is plentiful and inexpensive, raceway culture is an attractive option and one which allows for production levels well in excess of what is possible in ponds. Trout are frequently reared in linear raceways from hatching to market size. Linear raceways are essentially channels that are longer than they are wide, and are usually no deeper than 1–2 m (Fig. 3). Water flows in one end and out the other. The total volume of the raceway may be exchanged as often as every few minutes. High density raceways used in production facilities are commonly constructed of poured concrete. Small raceways of the type used in hatcheries and research facilities may be constructed of fiberglass or other resilient materials. Water is introduced at one end and flows by gravity through the raceway to exit the other end. Circular raceways, called tanks (Fig. 4), are also used by aquaculturists. Tanks are usually no more than 2 m deep and may be from less than 1 m to as much as 10 m in diameter. Concrete tanks can be found, but most are constructed of fiberglass, metal, or wood that is sealed and covered with epoxy or some other waterproof material. Plastic liners are commonly used in metal or wood tanks to prevent leakage, and in the case of metal, to avoid exposing the aquaculture animals to trace element toxicity.

Fig. 4. Circular tanks are a raceway option. They can be placed outdoors or used in conjunction with indoor water systems. Circular tanks are commonly used in recirculating systems.

Water is introduced into tanks at the surface in most cases. It may be sprayed in (to enhance aeration) or introduced in a manner that does not cause turbulance. Venturi drains, located internally or externally, will collect and remove solids that settle to the bottom of a tank with the drain water.

Linear raceways are commonly used by trout and salmon culturists both for commercial production and for hatchery programs conducted by government agencies. Large numbers of state and federal salmon hatcheries in British Columbia, Canada and in the states of Washington, Idaho and Oregon, along with governmental and private hatcheries in Alaska, collect and fertilize eggs, hatch them, and rear the young fish to the smolt stage at which time they become physiologically adapted to enter seawater.

Commercial salmon culturists can rear their fish to market size in freshwater raceways although most salmon are grown from smolt to market size or adulthood in the marine environment, either as free roaming fish or in confinement. Since salmon instinctively return to spawn in the waters where they were hatched, it is possible to establish hatcheries and smolt-rearing facilities that take advantage of the homing instinct. The technique is known as ocean ranching. When the fish that had been released as smolts return to spawn, sufficient numbers of adults are collected for use as broodfish to continue the cycle. The remainder may be harvested by the aquaculturist or by commercial fishermen after which the fish are processed and marketed.

Salmon, steelhead trout, and a variety of marine fishes are currently being reared in net-pens (Fig. 5). The typical salmon net-pen is several meters on each side and may be as much as 10 m deep or deeper (1). Smaller units, called cages, are sometimes used by freshwater aquaculturists, primarily in freshwater, but also in the marine environment in some instances. Cages tend to have volumes

Fig. 5. Marine net-pens such as the ones shown here in Puget Sound, Washington (U.S.), are used for the rearing of salmon by commercial fish farmers.

of no more than a few cubic meters. Unlike net-pens which have rigid frames at the surface from which netting is hung, cages have rigid frameworks on all sides covered with netting, welded wire mesh, or plastic webbing.

Net-pen technology was developed in the 1960s, but has only been widely employed commercially for salmon production since the 1980s when the Norwegian salmon farming industry was developed. The Japanese began producing large numbers of sea bream and yellowtail in net-pens during the 1960s. Other nations have employed the technology as well. Most net-pens are located in protected waters since they are easily damaged or destroyed by storms.

Competition by various user groups for space in protected coastal waters in much of the world has led to strict controls and in some cases prohibitions against the establishment of inshore net-pen facilities. As a result, there is growing interest in developing the technology to move offshore. Various designs for offshore net-pens have been developed and a few have been tested (Fig. 6). A number of different designs, including systems that are semi- or totally submersible, have been able to withstand storm waves of at least 6 m, but the costs of those systems are very high compared with inshore net-pens, so commercial viability has yet to be demonstrated.

The highest levels of intensity that can be found in aquaculture systems are associated with totally closed systems, often called recirculating systems. In these systems, all water passing through the chambers in which the finfish or shellfish are held is continuously treated and reused. Once filled initially, closed systems can theoretically be operated for long periods of time without water replacement. In practice, it is necessary to add some water to such systems to make up for that lost to evaporation, splashout, and in conjunction with solids removal. Most of the recirculating systems in use today are operated in a mode

Fig. 6.   A salmon net-pen in Scotland designed for use offshore, and in this case, exposed coastal waters.

between entirely closed and completely open. In many a significant percentage of replacement water is added either continuously or intermittently on a daily basis. Such partial recirculating systems may exchange from a few percent to several hundred percent of system volume each day.

The heart of a recirculating water system is the biofilter, a device that contains solid media on which bacteria that help purify the water become established (Fig. 7). Fish and aquatic invertebrates produce ammonia as their primary metabolic waste product. If not removed or converted to a less toxic chemical, ammonia can quickly reach lethal levels. Two genera of bacteria are responsible for ammonia removal in biofilters. The first, *Nitrosomonas*, converts ammonia (NH_3) to nitrite (NO_2^-). The second, *Nitrobacter*, converts nitrite to nitrate (NO_3^-). Nitrite is highly toxic to aquatic animals, although nitrate can be allowed to accumulate to relatively high levels. If both genera of bacteria are active, the conversion from ammonia through nitrite to nitrate is so rapid that nitrite levels remain within the safe range.

Other than the biofilter and culture chambers, recirculating systems typically also employ one or more settling chambers or mechanical filters to remove solids such as unconsumed feed, feces, and mats of bacteria that slough from the biofilter into the water. Each recirculating system requires a mechanical means of moving water from component to component. That usually means mechanical pumping, though air-lifts can also be used.

Control of circulating bacteria and oxidation of organic matter can be obtained through ozonation of the water. Ozone (O_3) is highly toxic to aquatic organisms. Ozone must be allowed to dissipate prior to exposing the water to the aquaculture animals. With time, and with the assistance ofaeration, ozone can be driven off or converted to molecular oxygen. Various commercial firms market ozone generators and can assist aquaculturists in selecting the proper equipment to meet system needs.

Fig. 7. A bead filter, one of many types of biological filters, shown in association with a laboratory-scale recirculating water system. Small plastic beads inside the fiberglass chamber provide surface area for colonization by bacteria that convert ammonia to nitrate.

Ultraviolet (UV) light has also been used to sterilize the water in aquaculture systems. The effectiveness of UV decreases with the thickness of the water column being treated, so the water is usually flowed past UV lights as a thin film (alternatively, the water may flow through a tube a few cm in diameter that is surrounded by UV lights). UV systems require more routine maintenance than ozone systems. UV bulbs lose their power with time and need to be changed periodically. In addition, organic materials exposed to UV light foul the surface of the transparent quartz (sometimes plastic) tubes past or through which the water flows, thereby causing a film between the water and the UV source that reduces the penetration of the UV light.

Recirculating systems often feature other types of apparatus, such as foam strippers and supplemental aeration. The technology for denitrifying nitrate to nitrogen gas has developed to the point that it may find a place in commercial culture systems in the near future. Computerized water-quality monitoring systems that will sound alarms and call emergency telephone numbers to report system failures to the culturists are also finding increased use.

The technology involved makes recirculating systems expensive to construct and operate. Redundancy in the system, ie, providing backups for all critical components, and automation are important considerations. When a pump fails, for example, the failure must be instantly communicated to the culturist and the culturist must have the ability to keep the system operating while the problem is being addressed. Loss of a critical component for even a few minutes can result in the loss of all animals within the system.

Recirculating systems can make aquaculture feasible in locations where conditions would not otherwise be conducive to successful operations. Such systems can also be used to reduce transportation costs by making it possible to grow animals near markets. In areas where there are concerns about pollution or the use of exotic species, closed systems provide an alternative approach to more extensive types of operations.

Another approach to aquaculture is enhancement, which involves spawning and rearing aquatic organisms to a size large enough that the organisms will have a good chance of survival in nature. Ocean ranching (previously mentioned) is a form of enhancement that is normally conducted with anadromous species. Enhancement in a broader context could be conducted with virtually any aquatic species of economic importance. Once the organisms reach marketable size (that is, when they recurit to a fishery), they could be harvested by commercial or recreational fishermen who would pay a license fee for the opportunity to fish on the enhanced species. This is already being done in conjunction with the recreational red drum fishery in Texas. Before enhancement programs are put into place, research should be conducted to determine the size at which the target organisms should be released and to determine that the addition of large numbers of organisms to a natural ecosytem will not be detrimental to naturally occuring species or overwhelm the food base that exists in the area slated for stocking.

6. Water Sources and Quality

Sources of water for aquaculture include municipal supplies, wells, springs, streams, lakes, reservoirs, estuaries, and the ocean. The water may be used directly from the source or it may be treated in some fashion prior to use (see WATER).

Many municipal water sources are chlorinated with sufficiently high levels of chlorine to be toxic to aquatic life. Chlorine can be removed by passing the water through activated charcoal filters or through the use of sodium thiosulfate metered into the incoming water. Municipal water is usually not used in aquaculture operations that utilize large quantities of water, either continuously or periodically, because of the initial high cost of the water and the cost of pretreatment to remove chlorine.

If polled, most aquaculturists would probably indicate a preference for well water. Both freshwater and saline wells are common sources of water for aquaculture. The most common pretreatments include temperature alteration (either heating or cooling); aeration to add oxygen or to remove or oxidize substances such as carbon dioxide, hydrogen sulfide, and iron; and increasing the salinity (in mariculture systems). Pretreatment may also include adjusting pH, hardness, and alkalinity through the application of appropriate chemicals.

To heat or cool water requires large amounts of energy. A major consideration in locating an aquaculture facility is to have not only a sufficient supply of water, but to have water at or near the optimum temperature for growing the species that has been selected. The vast supply of spring water of almost perfect temperature in the Hagerman Valley of Idaho supports the majority of the trout production in the United States. Where geothermal water is available, tropical species can be grown in locations where ambient winter temperatures would otherwise not allow them to survive.

Another large cost associated with incoming water is associated with its movement. Many aquaculture facilities that utilize surface waters and those that obtain their water from wells other than artesian wells are required to pump the water into their facilities. Pumping costs can be a major expense, particularly when the facility requires continuous inflow.

Surface water can sometimes be obtained through gravity flow by locating aquaculture facilities at elevations below those of adjacent springs, streams, lakes, or reservoirs. Coastal facilities may be able to obtain water through tidal flow.

The most common treatment of incoming surface water is removal of particulate matter. This can be effected through the use of settling basins orfiltration. Particle removal may involve the reduction or elimination of suspended inorganic material such asclay, silt, and sand. It may also involve removal of organic material, including living organisms. Organisms that enter aquaculture facilities if not filtered from the incoming water include phytoplankton and zooplankton, plants and plant parts, macroinvertebrates, and fishes. Some of the organisms, if not removed, can survive and grow to become predators on, or competitors with, the target aquaculture species. Very small organisms, such as bacteria, can be removed mechanically. However, other forms of sterilization, such as ozonation and the use of UV radiation, are more efficient and effective.

For many freshwater species that can be characterized as warmwater (such as channel catfish and tilapia) or coldwater (such as trout), the conditions outlined in Table 4 should provide an acceptable environment. So-called midrange species are those with an optimum temperature for growth of about 25C (examples are walleye, northern pike, muskellunge, and yellow perch). Typically they do well under the conditions, other than temperature, specified in Table 4 for coldwater species. Some species have higher or lower tolerances than others. For example, tilapia can tolerate temperatures in excess of 34C, but have poor tolerance for low temperature. Most tilapia species die when the temperature falls below about 12C. Tilapia have a remarkably high tolerance for ammonia compared with such species as trout and salmon, which have a high tolerance for cold water, but cannot tolerate water temperatures much above 20C. Marine fish may be able to tolerate a wide range of salinity (such euryhaline species include flounders, red drum, salmon, and some species of shrimp), or they may have a narrow tolerance range (they are called stenohaline species, examples of which are dolphin, halibut, and lobsters). Recommended water quality conditions for marine fish production systems are presented in Table 3.

Table 4. **General Water Quality Requirements for Cold and Warmwater Aquatic Animals in Fresh Water**[a]

Variable	Acceptable level or range	
	Coldwater	Warmwater
temperature, °C	<20	26–30
alkalinity, mg/L	10–400	50–400
dissolved oxygen, mg/L	>5	≥5
hardness, mg/L	10–400	50–400
pH	6.5–8.5	6.5–8.5
total ammonia, mg/L	<0.1	<1.0
ferrous iron, mg/L	0	0
ferric iron, mg/L	0.5	0–0.5
carbon dioxide, mg/L	0–10	0–15
hydrogen sulfide, mg/L	0	0
cadmium, µg/L	<10	<10
chromium, µg/L	<100	<100
copper, µg/L	<25	<25
lead, µg/L	<100	<100
mercury, µg/L	<0.1	<0.1
zinc, µg/L	<100	<100

[a]Refs. 1,6,7

The water quality criteria for each species should be determined from the literature or through experimentation when literature information is unavailable. Synergistic effects that occur among water quality variables can have an influence on the tolerance a species has under any given set of circumstances. Ammonia is a good example. Ionized ammonia (NH_4^+) is not particularly lethal to aquatic animals, but unionized ammonia (NH_3) can be toxic even when present at a fraction of a part per million (depending on species). The percentage of unionized ammonia in the water at any given total ammonia concentration changes in relation to such factors as temperature and pH. As either temperature or pH increase, so does the percentage of unionized ammonia relative to the level of total ammonia.

Another example is dissolved oxygen (DO). The amount of DO water can hold at saturation is affected by both temperature and salinity. The warmer and/or more saline the water, the lower the saturation DO level. Oxygen saturation is also affected by atmospheric pressure, decreasing markedly as elevation increases.

Biocides should not be present in water used for aquaculture. Typical sources of herbicides and pesticides are runoff from agricultural land, contamination of the water table, and spray drift from crop-dusting activity. Excessive levels ofphosphorus andnitrogen may occur where runoff from fertilized land enters an aquaculture facility either from surface runoff orgroundwater contamination. Trace metal levels should be low as indicated in Tables 4 and 5.

Most aquaculture facilities release water constantly or periodically into the environment without passing it through a municipal sewage treatment plant. The effects of those effluents on natural systems have become a subject of intense scrutiny in recent years and have, in some instances, resulted in opposition to

Table 5. **Suggested Water Quality Conditions for Marine Fish Production Facilities**[a]

Variable	Acceptable level or range
temperature, °C	1–40 (depends on species)
salinity, g/kg	1–40 (depends on species)
dissolved oxygen, mg/L	>6
pH	<7.9 – 8.2
total ammonia, µg/L as NH_3	<10
iron, µg/L	100
carbon dioxide, mg/L	<10
hydrogen sulfide, µg/L	<1
cadmium, µg/L	<3
chromium, µg/L	<25
copper, µg/L	<3
mercury, µg/L	<0.1
nickel, µg/L	<5
lead, µg/L	<4
zinc, µg/L	<25

[a]Ref. 8

further development of aquaculture facilities in those locales. There have even been demands that some existing operations should be shut down.

Regulation of aquaculture varies greatly both between and within nations. Some governmental agencies with jurisdiction over aquaculture have placed severe restrictions on the levels of nutrients such as phosphorus and nitrogen that can be released into receiving waters. Regulations on levels of suspended solids in effluent water are also common. The installation of settling ponds and constructed wetlands, exposure of the water to filter feeding animals that will remove solids, and mechanical filtration have all been used to treat effluents. Reduction or removal of dissolved nutrients through tertiary treatment is possible, but is generally not economically feasible at present. Research is currently underway to develop feeds containing reduced levels of nutrients or to provide nutrients in forms that can better be utilized by the culture animals. The goal in both approaches is to reduce discharges of nutrients to the environment through excretion.

7. Nutrition and Feeding

There are cases in which intentional fertilization is commonly used by aquaculturists in order to produce desirable types of natural food for the species under culture. This can also cause problems associated with excessive levels of nutrients and unwanted nuisance species. Examples of this situation include inorganic fertilizer applications in ponds to promote phytoplankton and zooplankton blooms that provide food for young fish such as channel catfish, the development of algal mats through fertilization of milkfish ponds, and the use of organicfertilizers (from livestock and human excrement) in Chinese carp ponds to encourage the growth of phytoplankton, macrophytes, and benthic invertebrates. In the latter instance, various species of carp with different food habits are stocked to ensure that all of the types of natural foods produced as a result of fertilization are consumed.

Provision of live foods is currently necessary for the early stages of many aquaculture species because acceptable prepared feeds have yet to be developed. Algae are routinely cultured for the early stages of molluscs produced in hatcheries. Once the molluscs are placed in growout areas, natural productivity is depended upon to provide the algae and other microorganisms upon which the shellfish feed.

In cases where zooplankton are reared as a food for predatory larvae or fry, it may be necessary to maintain three cultures. Though wild zooplankton have been used successfully in some instances (eg, in Norway wild zooplankton have been collected and fed to larval Atlantic halibut, *Hippoglossus*), the normal process involves culturing algae to feed to zooplankton that are fed to young shrimp hippoglossus or fish.

The most popular live foods for first feeding animals such as shrimp and marine fishes that have small eggs and larvae are rotifers and brine shrimp nauplii. After periods ranging from several days to several weeks, depending on the species being reared, the aquaculture animals will become sufficiently large to accept pelleted feeds and can be weaned onto prepared diets. Problems associated with utilizing prepared feeds from first feeding include difficulty in providing very small particles that contain all the required nutrients, loss of soluble nutrients into the water from small particles before the animals consume the feed, and in some cases, the fact that prepared feeds do not behave the same as live foods when placed in the water. For species that are sight feeders, behavior of the food is an important factor.

Some of the most popular aquaculture species accept prepared feeds from first feeding. Included are catfish, tilapia, salmon, and trout. All of the fishes listed have relatively large eggs (several a few mm diameter) that develop into fry that have large yolk sacs. The nutrients in the yolk sac lead to production of first-feeding fry with well-developed digestive tracts that produce the enzymes required to efficiently digest diets that contain the same types of ingredients used for larger animals.

Fish nutritionists have, over the last few decades, successfully determined the nutritional requirements of many aquaculture species and have developed practical feed formulations based on those requirements. For species such as Atlantic salmon, various species of Pacific salmon and trout, common carp, channel catfish and tilapia sufficient information exists to design diets precisely suited to each species. There is always interest among aquaculturists to develop new species. In each instance, the nutritional requirements of the new species must be investigated. Although there are many similarities among aquatic animals, diets that produce the best growth at the least cost vary significantly and can only be formulated when the nutritional requirements are known. Determination of those requirements may require several years of research, although suitable diets based on existing formulations can be employed while the research is being conducted.

Requirements for energy, protein, carbohydrates, lipids, vitamins and minerals have been determined for the species commonly cultured (9). As a rule of thumb, trout and salmon diets will, if consumed, support growth and survival in virtually any aquaculture species. Such diets often serve as the control against which experimental diets are compared.

Since feeds contain other substances than those required by the animals of interest, knowledge is also required of antinutritional factors in feedstuffs and on the use of additives. Certain feed ingredients contain chemicals that retard growth or may actually be toxic. Examples are gossypol in cottonseed meal and trypsin inhibitor in soybean meal. Restriction on the amount of the feedstuffs used is one way to avoid problems. In some cases, as is true of trypsin inhibitor, proper processing can destroy the antinutritional factor in the feed ingredient. In this case, heating of soybean meal is the method of choice.

Animals that do not readily accept pelleted feeds may be enticed to do so if the feed carries an odor that induces ingestion. Color development is an important consideration in aquarium species and some animals produced for human food. External coloration is desired in aquarium species. Pink flesh in cultured salmonids is desired by the consuming public. Coloration, whether external or of the flesh, can be achieved by incorporating ingredients that contain pigments or by adding extracts or synthetic compounds. One class of additives used to impart color is the carotenoids.

Prepared feeds are marketed in various sizes from very fine small particles (fines) through crumbles, flakes, and pellets. Pelleted rations may be hard, semimoist, or moist. Hard pellets typically contain less than 10% water and can be stored under cool, dry conditions for at least 90 days without deterioration of quality. Semimoist pellets are chemically stabilized to protect them from degradation if they are properly stored, while moist pellets must be frozen if they are not used immediately after manufacture. Moist feeds are produced in machines similar to sausage grinders.

Hard pellets are the type preferred if the species under culture will accept them. Semimoist feeds are most commonly used with young fishes and species that find hard pellets unpalatable. Moist feeds, which contain high percentages of fresh fish, are usually available only in the vicinity of fish-processing plants.

The most widely used types of prepared feeds are produced by pressure pelleting or extrusion. Pressure pelleting involves pushing the ground and mixed feed ingredients through holes in a die that is a few centimeters thick to produce spaghettilike strands of the desired diameter. The strands are cut to length as they exit the die. Steam is often injected into the pellet mill in a location that exposes the feed mixture to moist heat just before the mix enters the die. This improves binding and extends pellet water stability.

Extruded pellets are produced by exposing the ground and mixed ingredients to much higher heat and pressure and for a longer time than is the case with pressure pellets. In the extrusion process the ingredients undergo some cooking that can be beneficial in reducing the levels of certain antinutritional factors, such as trypsin inhibitor. There may be concomitant losses of heat labile nutrients such as vitamin C, so overfortification or the addition of heat stable forms of certain ingredients to obtain the desired level in the final product may be required.

Crumbles are formed by grinding pellets to the desired sizes. Specialty feeds such as flakes can be made by running newly manufactured pellets through a press or through use of a double drum dryer. The latter type of flakes begin as a slurry of feed ingredients and water. When the slurry is pressed between the hot rollers of the double drum dryer, wafer thin sheets of dry feed are produced that are then broken into small pieces. The different colors observed in some tropical fish foods represent a mixture of flakes, each of which contains one or more different additives that impart color.

Pressure pellets sink when placed in water, whereas under the proper conditions, floating pellets can be produced through the extrusion process. That is accomplished when the feed mixture contains high levels of starch that expands and traps air as the cooked pellets leave the barrel of the extruder. This gives the pellets a density of less than 1.0. Floating pellets are desirable for species that come to the surface to feed since the aquaculturist can visually determine that the fish are actively feeding and can control daily feeding rates based on observed consumption.

Sinking extruded pellets are used for shrimp and other species that will not surface to obtain food. Shrimp consume very small particles, so they will nibble pieces from a pellet over an extended period of time. For that reason, both pressure and extruded pellets need to have high water stability. Extruded feeds, whether sinking or floating, may remain intact for up to 24 hours after being placed in the water. Pressure pellets begin to disintegrate after a few minutes, unless supplemental binders are incorporated into the feed mixture. As previously indicated, the use of steam in conjunction with pressure pelleting also enhances pellet stability.

Nearly all aquaculture feeds contain at least some animal protein since theamino acid levels in plant proteins typically cannot meet the requirements of most aquatic animals. Fish meal is the most commonly used source of animal protein in aquaculture feeds, though blood meal, poultry by-product meal, and meat and bone meal have also been successfully used. Commonly used plant proteins include corn meal, cottonseed meal, peanut meal, rice, soybean meal, and wheat. A number of other ingredients have also been used, many of which are only locally available. Most formulations contain a few percent of added fat from such sources as fish oil, tallow, or more commonly, oilseed oils such as corn oil and soybean oil. Complete rations contain added vitamins, and minerals. Purified amino acids, binders, carotenoids, and antioxidants are other components found in many feeds. Growth hormone andantibiotics are sometimes used. Regulations on the incorporation of hormones along with other chemicals and drugs into aquatic animal feeds are in place in the United States and some other countries (Table 6). Few such regulations have been promulgated in developing nations.

Feeding practices vary from species to species. It is important not to overfeed since waste feed not only means wasted money, it can also lead to degradation of water quality. Most species require only three to four percent of body weight in dry feed daily for optimum growth. Very young animals are an exception. They are fed at a higher rate because they are growing rapidly and consume a greater daily percentage of body weight than older animals. It is important to have food readily available to them. Food should be spread evenly over the culture chamber area so the young animals do not have to expend a great deal of energy searching for a meal. Feeding rates as high as 50% of body weight daily are not uncommon for young animals. Since total biomass is small, even

Table 6. **Therapeutants and Disinfecting Agents Approved for Use in United States Aquaculture**[a]

Name of compound	Use of compound
Therapeutants	
copper	antibacterial for shrimp
formalin	parasiticide for various species
furanace (Nifurpyrinol)	antibiotic for aquarium fishes
oxytetracycline (Terramycin)	antibiotic for fishes and lobsters
sodium chloride	osmoregulatory enhancer for fishes
sulfadimethoxine (Romet)	antibacterial for salmonids and catfish
trichlorofon (Masoten)	parasiticide for baitfish and goldfish
Disinfectants	
calcium hypochlorite (HTH)	used in raceways and on equipment
didecyl dimethyl ammonium chloride (Sanaqua)	used in aquaria and fish-holding equipment
povidone–iodine compounds (Argentyne, Betadine, Wescodyne)	disinfection of fish eggs

[a] Ref. 11.

in intensively stocked units such as raceways, the economic cost is not high. Water quality in raceways can be maintained by siphoning out waste feed periodically. In ponds, any unconsumed feed acts as fertilizer and the quantities used are not high enough to affect water quality adversely.

Young animals may be fed several times daily. Examples include the standard practices of feeding fry channel catfish every three hours and young northern pike as frequently as every few minutes. Keeping carnivorous species such as northern pike satiated helps reduce the incidence of cannibalism.

8. Reproduction and Genetics

Species such as carp, salmon, trout, channel catfish, and tilapia have been bred for many generations in captivity though they usually differ little in appearance or genetically from their wild counterparts. A few exceptions exist, such as the leather carp, a common carp strain selectively bred to produce only one row of scales, and the Donaldson trout, a strain of rainbow trout developed over numerous generations to grow more rapidly to larger size and with a stouter body than its wild cousins.

Selective breeding has long been practiced as a mean of improving aquaculture stocks. In some instances it has not been possible or it is quite difficult and expensive to produce broodstock and spawn them in captivity, so culturists continue to rear animals obtained from nature. Most of the species that are being reared in significant quantities around the world are produced in hatcheries using either captured or cultured broodstock. Milkfish is a notable exception. The species has been spawned in captivity, but most of the fish reared in confinement are collected as juveniles in seines and sold to fish culturists. Wild shrimp post-larvae continue to be used to stock ponds in some parts of the world though hatcheries may also be available in the event sufficient numbers of wild post-larvae are unavailable in a given year.

Spawning techniques vary widely from one species to another. Tilapia and catfish are typically allowed to spawn in ponds. Fertilized eggs can be collected from the mouths of female tilapia, but it is common practice to collect schools of fry after they are released from the mother's mouth to forage on their own. Catfish lay eggs in adhesive masses. Spawning chambers such as milk cans and grease cans are placed in ponds and may be examined every few days for the presence of egg masses. Some catfish farmers allow the eggs to hatch in the pond, though most collect eggs and incubate them in a hatchery. Adult Pacific salmon die after spawning. Females are usually sacrificed by cutting open the abdomen to release the eggs. Milt is obtained by squeezing the belly of males. Trout and Atlantic salmon can be reconditioned to spawn annually. Eggs are usually obtained from those species in the same fashion as from male Pacific salmon (1).

Unlike catfish, tilapia, trout and salmon, that produce several hundred to several thousand eggs per female, marine species typically produce large numbers of very small eggs. Hundreds of thousands to millions of eggs are produced by such species as halibut, flounders, red drum, striped bass, and shrimp. Catfish, salmon, and trout spawn once a year, while tilapia and some marine species spawn repeatedly if the proper environmental conditions are maintained (1). Red drum, for example, spawn every few days for periods of several months when light and temperature and are properly controlled (10).

Fish breeders have worked with varying degrees of success to improve growth and disease resistance in a number of species. Asgenetic engineering techniques are adapted to aquatic animals, dramatic and rapid changes in the genetic makeup of aquaculture species may be expected. However, since it is virtually impossible to prevent escapement of aquacultured animals into the natural environment, potential negative impacts of such organisms on wild populations cannot be ignored.

In some species, one sex may grow more rapidly than the other. A prime example is tilapia, which mature at an early age (often within six months of hatching). At maturity, submarketable females divert large amounts of food energy to egg production. Also, since they are mouth brooders (holding the eggs and fry within their mouths for about two weeks) and repeat spawners (spawning about once a month if the water temperature is suitable), the females grow very slowly once they mature. All-male, or predominantly male, populations of tilapia can be produced by feeding androgens to fry, which are still undifferentiated sexually. Various forms of testosterone have been used effectively in sex reversing tilapia and other fishes.

In species such as flatfishes, females may grow more rapidly than males and ultimately reach much larger sizes. For them, producing all-female populations for growout might be beneficial.

9. Diseases and Their Control

Aquatic animals are susceptible to a variety of diseases including those caused by viruses, bacteria, fungi, and parasites. A range of chemicals and vaccines has been developed for treating the known diseases, although some conditions have resisted all control attempts to date. In some nations, severe restrictions on the use of therapeutants has impaired that ability of aquaculturists to control disease outbreaks. The United States is a good example of a nation in which the variety of treatment chemicals is limited by government regulators (Table 6).

Managing conditions in the culture environment to keep stress to a minimum is one of the best methods of avoiding diseases. Vaccines have been developed against several diseases and more are under development. Selective breeding of animals with disease resistance has met with only limited success. Good sanitation and disinfection of contaminated facilities are important avoidance and control measures. Some disinfectants are listed in Table 6. Pond soils can be sterilized with burnt lime (CaO), hydrated lime [Ca(OH)$_2$], or chlorine compounds (12).

When treatment chemicals have to be employed, they may be incorporated in the food, used in dips, flushes and baths, or allowed to remain in the water for extended periods. Since one of the first responses of aquatic animals to disease is reduction or cessation of feeding, treatments with medicated feeds must be initiated as soon as development of an outbreak is suspected. Antibiotics, such as terramycin, can be dissolved in the water, but may be less effective than when given orally.

Vaccines can be administered through injections, orally, or by immersion. Injection is the most effective means of vaccinating aquatic animals but it is stressful, time-consuming, and expensive. The time and expense may be acceptable for use in conjunction with broodfish and other valuable animals. Oral administration of vaccines may be ineffective as many vaccines are deactivated in the digestive tract of the animals the vaccines are intended to protect. Dip treatment by which the vaccines enter the animals through diffusion from the water are not generally as effective as injection but can be used to vaccinate large numbers of animals in short periods of time.

10. Harvesting, Processing, and Marketing

Harvesting techniques vary depending on the type of culture system involved. Seines are often used to capture fish from ponds, or the majority of the animals can be collected by draining the pond through netting. Fish pumps are available that can physically transfer aquatic animals directly onto hauling trucks from ponds, raceways, cages and net-pens without causing skin abrasions, broken fins, or other damage.

Aquaculturists may harvest, and even process their own crops, although custom harvesting and hauling companies are often available in areas where the aquaculture industry is sufficiently developed to support them. Some processing plants also provide harvesting and live-hauling services.

Some species, with channel catfish being a good example, can develop off-flavors. A characteristic off-flavor in catfish is often described as an earthy, musty, or muddy flavor. The problem is associated with the chemical geosmin and related compounds that are produced by certain types of algae (1). Processors often require that a sample fish from each pond scheduled for harvest be brought to the plant about two weeks prior to harvest for a taste test. Subsequent samples are taken to the processor three days before and during the day of processing. If off-flavor is detected, the fish will be rejected. Once the source of geosmin is no longer present, the fish will metabolize the compound. The process may involve moving the fish into clean well water or by merely waiting until the algae bloom dissipates, after which the geosmin will be rapidly metabolized. Within a few days the fish can be retested and if no off-flavor is detected, they can be harvested and processed.

Centralized processing plants specifically designed to handle regional aquaculture crops are established in areas where production is sufficiently high. In coastal regions, aquacultured animals are often processed in plants that also service capture fisheries.

Marketing can be done by aquaculturists who operate their own processing facilities. Most aquaculture operations depend on a regional processing plant to market the final product. In all cases aquaculturists should note well that their job is not complete until the product reaches the consumer in prime condition.

BIBLIOGRAPHY

"Aquaculture" in *ECT* 3rd ed., Vol. 3, pp. 194–213, by Howard P. Clemens and Michael Conway, University of Oklahoma; "Aquaculture" in *ECT* 4th ed., Supplement, pp. 22–47, by Robert Stickney, Texas A & M University; "Aquaculture" in *ECT* (online), posting date: December 4, 2000, by Robert Stickney, Texas A & M University.

CITED PUBLICATIONS

1. R. R. Stickney, *Principles of Aquaculture*, John Wiley & Sons, Inc., New York, 1994.
2. FAO, *Agriculture Production Statistics*, 1974–1993, Food and Agriculture Organization of the United Nations, 1995.
3. FAO, *FAO Fisheries Circular 815, revision 2*, Food and Agriculture Organization of the United Nations, 1991.
4. *Aquaculture Buyer's Guide '95 and Industry Directory*, Vol. **8**, 1995 Aquaculture Magazine, Asheville, NC.
5. R. R. Stickney, *Aquaculture in the United States: A Historical Review*, John Wiley & Sons, Inc., New York, 1996.
6. R. G. Piper, I. B. McElwain, L. E. Orme, J. P. McCraren, L. G. Fowler, and J. R. Leonard, *Fish Hatchery Management*, U.S. Fish and Wildlife Service, Washington, D.C., 1982.
7. C. E. Boyd, *Water Quality in Ponds for Aquaculture*, Alabama Agricultural Experiment State, Auburn University, 1990.
8. J. Huegenin and J. Colt, *Design and Operating Guide for Aquaculture Seawater Systems*, Elsevier, New York, 1989.
9. National Research Council, *Nutrient Requirements of Fish*, National Academy Press, Washington, D.C., 1993.
10. A. Henderson-Arzapalo, *Rev. Aquat. Sci.* **6**, 479 (1992).
11. F. P. Meyer and R. A. Schnick, *Rev. Aquatic Sci.* **1**, 693 (1989); a review of chemicals used for the control of fish diseases.
12. C. E. Boyd, *Bottom Soils, Sediment, and Pond Aquaculture*, Chapman and Hall, New York, 1995.

RORBERT STICKNEY
Texas A&M University

AQUACULTURE CHEMICALS

1. Introduction

Intensive or extensive culture of aquatic animals requires chemicals that control disease, enhance the growth of cultured species, reduce handling trauma to organisms, improve water quality, disinfect water, and control aquatic vegetation, predaceous insects, or other nuisance organisms. The Aquaculture chemical needs for various species have been described for rainbow trout, *Oncorhynchus mykiss* (1); Atlantic and Pacific salmon, *Salmo salar* and *Oncorhynchus* sp. (2); channel catfish, *Ictalurus punctatus* (3); striped bass, *Morone saxatilis* (4); milk-fish, *Chanos chanos* (5); mollusks (6); penaeid *Penaeus* shrimp (7); and a variety of other freshwater and marine species (8).

Laws and regulations on the use of chemicals in aquaculture vary by country and serve to ensure safe and effective use and protection of humans and the environment. Regulations and therapeutants or other chemicals that are approved or allowed for use in the United States, Canada, Europe, Japan, Chile, and Australia are presented below.

2. Regulation of Aquaculture Chemicals in the United States

In the United States, the U.S. Food and Drug Administration (FDA) and the U.S. Environmental Protection Agency (EPA) regulate the application of chemicals to organisms or to their environments. FDA controls the use of drugs and anesthetics and EPA controls the application of chemicals and pesticides to the environment. In cases that involve treatments to control pathogens that are present in the water, the jurisdiction becomes unclear and has been changed over time. Each agency develops appropriate guidelines and policies to implement the laws for its field of responsibility (9,10).

Therapeutant approval requires information on human safety, efficacy against target organisms, toxicity to nontarget organisms, residues in food-producing animals, and effects on the environment. Drugs for food-producing animals require information on metabolites and residue dynamics to establish tolerance levels and withdrawal times. Withdrawal time is the period of time that must pass after the last treatment or exposure to a certain drug, chemical, or pesticide before an animal can be consumed. Residue studies may not be necessary if the actual holding time of an aquatic animal is significantly longer than the required withdrawal period for a major species; examples include the treatment of eggs, fry, or small fingerlings long before they are harvested or slaughtered for food as adults. Studies to generate the required data must be conducted according to good laboratory practices (GLP) requirements and each drug must be produced and formulated with good manufacturing practices (GMP). Requirements of GLP and GMP significantly increase the cost of developing new drugs (9,10).

Most pharmaceutical and chemical companies do not try to gain approvals of aquaculture drugs because of increased costs for approval requirements, limited sales and profit potential, diversity of cultured aquatic animals and diseases, and lack of uniform testing guidelines and data requirements among countries (9). To be profitable, an aquaculture drug must be marketed worldwide, but few nations accept data on aquaculture drugs that were tested in a foreign country and the required tests are too expensive to repeat in every country. Data on mammalian safety, environmental fate, residues, and metabolism are the most expensive to obtain and are not available for most of the drugs known to be effective on aquatic organisms. In fact about $3 million of the estimated $3.5 million spent to register a new drug in the United States is for these types of studies. There is a great need for international harmonization and cooperation to get drugs and chemicals registered for aquatic species (9,11). To that end, two international harmonization workshops were held in 1997 and 1998 and several special sessions on cooperative efforts at international aquaculture conferences have been organized (9).

3. Registered Aquaculture Chemicals in the United States

3.1. Antibacterials. Few therapeutants are registered in the United States for use on any cultured aquatic species (Table 1). In the most critical area of antibacterials, only two (Terramycin for Fish and Romet-30) are approved and available. Only the aquatic species and diseases listed on the label may be treated unless a veterinarian prescribes extra-label use for other aquatic species and their diseases. Terramycin (oxytetracycline) is legal to use against *Aeromonas*, *Hemophilus*, and *Pseudomonas* in salmonids and catfishes and gaffkemia in lobster (10,12). Romet-30 (sulfadimethoxine, ormetoprim) is labeled for the control of furunculosis in salmon and enteric septicemia in catfishes (10,12). Data are available for the control of enteric redmouth disease with Romet but have not been submitted to FDA for label inclusion. A range of bacteria has developed resistance to both compounds, and broad-spectrum antibacterial agents are needed as replacements.

3.2. Fungicides. Formalin is the only fungicide approved by FDA for use on eggs of all fish at 1,000–2,000 mg/L for 15 min (12). Delivery apparatus has been developed to reduce human exposure toformalin. Fish culturists in the Pacific Northwest have successfully used formalin to control fungal infections on salmon broodstock, but controlled research that indicates formalin is a good fungicide on adult fish of many species is lacking.

When it approved the New Animal Drug Application (NADA) of formalin, FDA ruled that use of formalin for fisheries was safe for humans and the environment. They ruled that effluents from fish treatments at 250 mg/L should be diluted 10 times and from egg treatments 100 times if 1,000–2,000 mg/L were used (12). Before approving the drug, FDA also addressed carcinogenicity by stating it was not concerned about human exposure from either water or fish treated with formalin. The U.S. Fish and Wildlife Service (USFWS) has procedural guidelines that should protect workers from harmful levels of formalin. Calculations based on treatment levels demonstrated that a fishery worker is exposed to not more than 0.117 mg/L formalin in the air, well below the levels

Table 1. Chemicals Registered[a] or Allowed[b] for Use in Aquaculture in the United States by the FDA or the EPA[c]

Product (trade or alternative name)	CAS Registry Number	Molecular formula	Use pattern	Tolerance[d]	Withdrawal time[e]	Comments
			Therapeutants			
acetic acid, glacial (vinegar)	[64-19-7]	$C_2H_4O_2$	parasiticide for fish.—1000–2000 mg/L for 1–10 min	none established	none established	low regulatory priority
calcium chloride	[10043-52-4]	$CaCl_2$	osmoregulatory enhances for fish.—up to 150 mg/L	none established	none established	low regulatory priority
calcium oxide	[1305-78-8]	CaO	protozoacide—2000 mg/L for 5 s	none established	none established	low regulatory priority
formalin (Parasite-F, Parasite-S, Formalin-F)	[50-00-0]	CH_2O	parasiticide for all fish—25 mg/L in ponds; up to 250 mg/L for 1 h in tanks and raceways; fungicide for all fish eggs—1000–2000 mg/L for 15 min in egg-treatment tanks; protozoacide for penaeid shrimp—50–100 mg/L for up to 4 h in tanks and raceways; 25 mg/L in ponds	none required	none required	
hydrogen peroxide	[7722-84-1]	H_2O_2	fungicide for all fish and their eggs—250–500 mg/L	none required	none required	low regulatory priority
magnesium sulfate (Epsom salts)	[7487-88-9]	MgO_4S	monogenetic trematodes and external crustacean parasites—30,000 mg/L for 5–10 min	none established	none established	low regulatory priority
oxytetracycline (Terramycin for fish)	[79-57-2]	$C_{22}H_{24}N_2O_9$	antibacterial for salmonids and catfish and lobster—2.75–4.125 g/50 kg[f] per day for 10 d in feed	2 ppm in salmonids, catfish, and lobsters	21 d	
papain	[9001-73-4]		remove gelatinous matrix of fish eggs.— 0.2% solution	none established	none established	low regulatory priority
potassium chloride	[7447-40-7]	KCl	osmoregulatory enhancer for fish—10–2,000 mg/L	none established	none established	low regulatory priority

Table 1 (Continued)

Product (trade or alternative name)	CAS Registry Number	Molecular formula	Use pattern	Tolerance^d	Withdrawal time^e	Comments
sodium chloride (salt)	[7647-14-5]	NaCl	osmoregulatory enhancer for fish—0.5–1% for indefinite period; 3% for parasiticide.—10–30 min	none established		low regulatory priority
sulfadimethoxine and ormetoprim (Romet-30, Romet-B)	[122-11-2]	$C_{12}H_{14}N_4O_4S$	antibacterial for salmonids and catfish—50 mg/kg per day for 5 d	0.1 ppm in salmonids and catfish	6 wks for salmonids, 3 d for catfish	has potential for use in other aquatic species
sulfamerazine (Sulfamerazine in Fish Grade)	[6981-18-6] [127-79-7]	$C_{14}H_{18}N_4O_2$ $C_{11}H_{12}N_4O_2S$	antibacterial for salmonids—11 g/50 kg^g per day for 14 d in feed; discontinued after 14 d	zero in edible tissues of trout	3 wks	no longer available
thiamine hydrochloride	[67-03-8]	$C_{12}H_{18}Cl_2N_4OS$	prevent thiamine deficiency in salmonids—up to 100 mg/L; for up to 1000 mg/L for up to 1 h	none established	none established	low regulatory priority
Disinfecting agents calcium hypochlorite (Olin HTH Dry Chlorinator Granular)	[7778-54-3]	$Ca(ClO)_2$	disinfectant and sanitizer in fish tanks, raceways, and on utensils—200 mg/L for 1 h; control of algae and bacteria in ponds—5–10 mg/L residual chlorine for 12–24 h	none established		
didecyldimethylammonium chloride (Sanaqua)	[7173-51-5]	$C_{22}H_{48}ClN$	disinfection of aquarium and fish-holding equipment—59 mL in 15 L^h water for 10 min; disinfection in fish disease control institutions—104 mL in 15 L for 10 min	none established	none established	do not use directly on fish or other aquatic life

Drug	CAS number	Chemical formula	Application and recommended treatment	Tolerance	Regulatory status
povidone-iodine compounds	[25655-41-8]		disinfection of fish eggs— 50 mg/L for 30 min during water hardening 100 mg/L for 10 min after water hardening	none established	low regulatory priority
quaternary ammonium compounds			disinfection of water, gear, and tanks—2 mg/L for 1 h	none established	not for use directly on fish or other aquatic life
(benzalkonium chloride)	[68424-85-1]	$C_{27}H_{42}ClNO_2$			
(benzethonium chloride)	[121-54-0]				
Water treatment compounds and dyes					
calcium chloride	[10043-52-4]	$CaCl_2$	increase water calcium concentration to ensure proper egg hardening— to increase 10–20 mg/L calcium carbonate	none established	low regulatory priority
fluorescein sodium	[518-47-8]	$C_{20}H_{12}O_5$	dye to check water flows or dilution—0.1 mg/L	none established	exempted from registration by EPA
lime, slaked lime (calcium hydroxide)	[1305-62-0]	$Ca(OH)_2$	pond sterilant— 0.33 L/m2 (1.338 L/acre) of quick lime; 0.2 kg/m2 (1.784 lb/acre) of slaked lime	none established	generally recognized as safe
(calcium oxide)	[1305-78-8]	CaO		none established	
(calcium carbonate)	[471-34-1]	$CaCO_3$			
oxytetracycline (Terramycin)	[79-57-2]	$C_{22}H_{24}N_2O_9$	dye to mark fish	2 ppm in salmonids and catfish	7 d for oral, 15 d for injectable compounds
potassium permanganate	[7722-64-7]	$KMnO_4$	oxidizer and detoxifier— 2 mg/L	none established	FDA ruled that there is no public health concern when used as directed
rhodamine B and WT	[81-88-9]	$C_{28}H_{31}ClN_2O_3$	dye to check water flows or dilution—20 µg/L	none established	exempted from registration by EPA / exempted from registration by EPA

213

Table 1 (Continued)

Product (trade or alternative name)	CAS Registry Number	Molecular formula	Use pattern	Tolerance[d]	Withdrawal time[e]	Comments
sodium methanesulfonate (Amquel)	[2386-57-4]	$CH_4O_3S \cdot Na$	detoxifier of chlorine, ammonia, and chloramines	none established		FDA ruled that the compound is not a drug and therefore is not under FDA jurisdiction
Spawning aids						
human chorionic gonadotropin (Chorulon)			spawning aid by intramuscular injection—50–510 IU/lb body wt for males 67–1816 IU/lb body wt for females	none required		veterinanian prescription only
Anesthetics						
carbon dioxide gas	[124-38-9]	CO_2	anesthetic on fish— 200–400 mg/L for 4 min	none established		low regulatory priority
sodium bicarbonate (baking soda)	[144-55-8]	$NaHCO_3$	anesthetic for fish— 142–642 mg/L for 5 min	none established		low regulatory priority
tricaine methane sulfonate (Finquel; MS-222)	[886-86-2]	$C_{10}H_{15}NO_5S$	anesthetic for fish and amphibians—15–66 mg/L for 6–48 h for sedations; 50–330 mg/L for 1–40 min for anesthesia	none established	21 d	food fish use
Herbicides and algicides						
acid blue and acid yellow (Aquashade)			algicide and herbicide	none established		also used to control off-flavors in fish
copper, elemental (Aquatrine, etc)	[7440-50-8]	Cu	algicide and antibacterial	none established	fish and shrimp may be harvested immediately	do not use in water containing trout; used as antibacterial on penaeid shrimp

214

Common name (trade names)	CAS number	Formula	Use	Tolerance	Waiting period	Comments
copper sulfate (Calco Copper Sulfate, etc)	[7758-98-7]	$CuSO_4$	algicide and herbicide	none established		low alkalinity could cause the chemical to be hazardous to fish
2,4-D (Aquacide, Aqua-Kleen, etc)	[94-75-7]	$C_8H_6Cl_2O_3$	herbicide	1 ppm in fish and shellfish		
dichlobenil (Acme Norosac G-10, Casoron-10G)	[1194-65-6]	$C_7H_3Cl_2N$	herbicide	none established		nonfood fish use
diquat dibromide (Aqua-Clear, etc)	[2764-72-9]	$C_{12}H_{12}Br_2N_2$	algicide and herbicide	0.1 ppm in fish and shellfish		EPA proposed a maximum contaminant level (MCL) of 0.02 ppm for drinking water
endothall (Aquathol Granular, Aquathol K, etc)	[45-73-2]	$C_8H_{10}O_5$	herbicide	0.2 ppm in potable water	3 d	EPA proposed a MCL of 0.1 ppm for drinking water
fluridone (Sonar)	[59156-60-4]	$C_{19}H_{14}F_3NO$	herbicide	0.5 ppm in fish and crayfish		not for use in tidewater or brackish water
glyphosate (Rodeo)	[1071-83-6]	$C_3H_8NO_5P$	herbicide	0.25 ppm in fish		EPA proposed a MCL of 0.1 ppm for drinking water
Piscicides						
antimycin (Fintrol Concentrate)	[27220-56-0]	$C_{26}H_{36}N_2O_9$	piscicide	none established		general fish toxicant; can also be used to selectively remove scaled fish from catfish ponds
rotenone (Chem-Fish Synergized, Prentox, etc)	[83-79-4]	$C_{23}H_{22}O_6$	piscicide	none established		general fish toxicant

Table 1 (*Continued*)

Product (trade or alternative name)	CAS Registry Number	Molecular formula	Use pattern	Tolerance[d]	Withdrawal time[e]	Comments
niclosamide (Bayluscide)	[1420-04-8]	$C_{13}H_8Cl_2N_2O_4$	piscicide; molluscicide	none established		piscicide only for use on sea lamprey in the Great Lakes area; molluscicide to control snails (vectors of swimmers itch, etc)

[a] A registered compound is an available commercial product bearing an EPA or FDA label specifying its allowed uses. Refs. 10 and 12.

[b] An allowed product does not necessarily have an EPA or FDA label because some other classification or designation may allow its use in aquatic situations (eg, Low Regulatory Priority drugs).

[c] Ref. 10, 12.

[d] Tolerance refers to residue levels of a drug or chemical that are permitted by regulatory agencies in food eaten by humans.

[e] Withdrawal time is the period of time that must pass after the last treatment or exposure to a certain drug, chemical, or pesticide before an animal can be consumed. None has been established if there is no notation.

[f] 2.75–4.125 g/50kg= 2.5–3.75 g/100lb.

[g] 11g/50 kg=10 g/100lb.

[h] 59mL in15 L≈2 fl oz in 4 gal.

established by the U.S. Occupational Safety and Health Administration to protect workers.

Salt applied as equal parts of unionized sodium chloride and calcium chloride at 20 g total per L for 1 h, three times a week, has also been used to control fungal infections on eggs. The salt combination is first applied one day after fertilization to the first pick of eggs. The Center for Veterinary Medicine (CVM) categorized these compounds as unapproved drugs of Low Regulatory Priority (10). CVM is unlikely to object to the use of these drugs if the following conditions are met: (1) the drugs are used for the prescribed indications, including species and life stage where specified, (2) the drugs are used at the prescribed dosages, (3) the drugs are used according to good management practices, (4) the product is of an appropriate grade for use in food animals, and (5) an adverse effect on the environment is unlikely.

3.3. Parasiticides. Formalin is the only parasiticide currently approved for use on all fish and penaeid shrimp (12). It is registered for use on all fish at concentrations up to 250 mg/L for 1 h in tanks and raceways and 15 to 25 mg/L for an indefinite period in ponds and for penaeid shrimp at 50–100 mg/L for up to 4 h in tanks and raceways and 25 mg/L in ponds (12). A second chemical, trichlorfon (Masoten) was registered for use on nonfood fishes by EPA but is not currently available. Vinegar (glacial acetic acid) and salt (sodium chloride) are also used to control external parasites on fishes and CVM classifies these compounds as unapproved drugs of Low Regulatory Priority (10).

3.4. Disinfectants. Several disinfecting agents can be used in hatcheries and two are of particular interest. Because they are considered as unapproved drugs of Low Regulatory Priority by FDA, povidone-iodine compounds can be used to disinfect the surface of eggs (10). Benzalkonium chloride [68424-85-1] and benzethonium chloride (quaternary ammonium compounds) are allowed at 2 mg/L by FDA to disinfect water.

3.5. Water Treatment Compounds. Like the disinfecting agents, several water treatment compounds are used in aquaculture. Of particular interest ispotassium permanganate which is exempted from registration by EPA when used as an oxidizer or detoxifier and can control certain parasites, external bacteria, and possibly fungi (9).

3.6. Spawning Aids. One spawning aid is approved in the United States, human chorionic gonadotropin (Chorulon) (12). The drug is administered by intramuscular injection at 50-510 IU/lb body weight for males and 67-1816 IU/lb body weight for females.

3.7. Anesthetics. Tricaine methanesulfonate (MS-222) is the only currently approved anesthetic and requires a 21-day withdrawal time (10,12). The withdrawal time for MS-222 is of special concern to FDA when the broodfishes of salmon or other species are taken immediately after spawning for pet or human food. Both carbon dioxide and sodium bicarbonate [144-55-8] have also been used as anesthetics and are classified as unapproved drugs of Low Regulatory Priority by FDA; however, both chemicals are difficult to use with consistent results and involve long induction and recovery periods (9).

3.8. Herbicides. An array of herbicides is registered for use in aquatic sites, but copper sulfate and diquat dibromide are of particular interest because they also have therapeutic properties (9,10).Copper sulfate has been used to control bacteria, fungi, and certain parasites, including *Ichthyophthirius* (ich). Diquat dibromide can control columnaris disease, but it also exhibits fungicidal properties (9)

3.9. Piscicides.   The two piscicides, antimycin androtenone, are both used in ponds to control nuisance fish. Antimycin is used selectively to remove scaled fishes from catfish ponds, and rotenone is used as a general fish toxicant (9,10). Observations by catfish farmers indicate that antimycin at low concentrations also acts as a therapeutant against external parasites.

4. Regulation and Registration of Aquacultural Chemicals Outside the United States

The control of aquaculture drugs varies among countries from no regulation to restrictive regulations. Generally, few requirements are needed for a therapeutant to be licensed or registered in South America, Africa, and most of Asia. Seafood-exporting countries are increasingly concerned because importing countries may no longer accept products without a guarantee that the products contain no chemical residues of concern.

4.1. Canada.   Except for environmental studies, requirements for registration data in Canada are similar to requirements in the United States. However, Canada has significantly different regulations and approval processes. Canadian aquaculturalists use drugs (Table 2) that are either licensed for other food animals and prescribed by veterinarians or used in an emergency under the direction of the Canadian Bureau of Veterinary Drugs (BVD) (13).

Table 2. **Chemicals Licensed for Use in Canadian Aquaculture**

Product (trade or alternative name)	CAS Registry Number	Molecular formula
Antibacterials		
florfenicol (Aquaflor)	[73231-34-2]	$C_{12}H_{14}Cl_2FNO_4S$
oxytetracycline[a] (Terramycin AQUA)	[79-57-2]	$C_{22}H_{24}N_2O_9$
sulfadimethoxine and ormetoprim[b]	[122-11-2]	$C_{12}H_{14}N_4O_4S$
(Romet-30 and Romet-B)	[6981-18-6]	$C_{14}H_{18}N_4O_2$
Sulfadiazine and trimethoprim	[68-35-9]	$C_{10}H_{10}N_4O_2S$
(Tribrissen)	[738-70-5]	$C_{14}H_{18}N_4O_3$
Disinfectants		
hydrogen peroxide	[7722-84-1]	H_2O_2
Antifungals / Antiparasiticides		
formalin (AquaLife Parasite – s)	[50-00-0]	CH_2O
Antiparasiticides		
azamethiphos (salmosan)		
Anesthetics		
metomidate (AquaLife marinil)		
tricaine methane sulfonate	[886-86-2]	$C_{10}H_{15}NO_5S$
(AquaLife TMS powder)		

[a] Licensed only for use in the treatment of fish and lobster. Withdrawal time is at least 40 d at $>10°C$ and at least 80 d at $<10°C$.
[b] Withdrawal time is 42 d at $>10°C$ and 81 d at $<10°C$.

The BVD is concerned about the lack of data on the pharmacokinetics of fishes, especially the difference in uptake of drugs at a range of temperatures (14). Chloramphenicol and tributyltin compounds are two classes of compounds that cannot be used in Canadian aquaculture. Canadian regulations also differ from the United States in that they have no minor-use policy or classifications such as Low Regulatory Priority drugs.

4.2. Europe. The European Agency for the Evaluation of Medicinal Products (EMEA) regulates the approvals of all Veterinary Medicinal Products in Europe and establishes the Maximum Residue Limit (MRL) for each animal drug (16). Those chemicals that should have established MRLs were banned from use if they were not established by January 1, 2000 (16). Requirements for MRLs for drugs are divided into four groups or annexes: Annex I = fixed MRL, Annex II = No MRL needed, Annex III = temporary MRL, and Annex IV = No MRL can be established (11). Banned from use are drugs such as the nitrofurans and chloramphenicol that have been placed in Annex IV. A MRL is required before a member country can evaluate a drug for approval. Although MRLs are European-wide, a license will depend upon the efforts of member countries and the drug sponsors (16). Under current regulations, veterinary medicines allowed for use in various European countries (Table (3)) are either fully licensed for aquacultural use (oxytetracycline, oxolinic acid) or can be prescribed by veterinarians if (1) the drugs are licensed for use on other food animals or in humans, (2) only a limited number of animals are treated, and (3) a 500 degree day withdrawal time is observed (16–19). Fish specific MRL approvals are available only for amoxicillin, potentiated sulfonamides, oxolinic acid, flumequine, sarafloxacin, oxytetracycline, and thiamphenicol (16).

4.3. Japan. In Japan, registration of drugs for aquatic species requires the same data as those required for drugs used on terrestrial animals. The Ministry of Agriculture, Forests, and Fisheries and the Ministry of Welfare control the use of chemicals in aquaculture in Japan (20). The preclinical data requirements include product chemistry, toxicity (acute, sub acute, special) using rats and mice, safety to target animals, and metabolism. The requirements for clinical data include availability and residues. As of April 2001, more chemicals were registered for aquacultural use in Japan than in any other country (Table 4).

4.4. Chile. The Servico Agricola y Ganadero has recently increased its scrutiny of drugs used in aquaculture. New approvals have and will become even more difficult. The agency will accept foreign data but some data are required to be generated in Chile. Table 5 lists the five drugs (i.e., antibacterials) currently approved in Chile (22). Other drugs being used or under consideration for approval include amoxicillin, benzocaine, chloramine-T, and MS-222.

4.5. Australia. In the past ten years, Australia has increased its aquaculture production and as a result has begun to register drugs and chemicals for that use. The National Registration Authority for Agricultural and Veterinary Chemicals has registered the following: benzocaine as a sedative and anesthetic for finfish and abalone, formalin to control protozoan and metazoan ectoparasites on fish and epicommensal ciliates on shrimp, flubendazole to control gill flukes on ornamental fish, leutinizing hormone releasing hormone analogue to induce spawning in finfish broodstock, methyltestosterone to produce female salmonid fish stocks, and trifluralin as a selective herbicide for prawn

Table 3. **Chemicals Authorized or Allowed for Use in Aquaculture in Certain European Countries**

Product (trade or alternative name)	CAS Registry Number	Molecular formula	Withdrawal time
Antibacterials			
amoxicillin	[26787-78-0]	$C_{16}H_{19}N_3O_5S$	30–150 d[a]
florfenicol	[73231-34-2]	$C_{12}H_{14}Cl_2FNO_4S$	2 d in France
flumequine	[42835-25-6]	C14H12FNO3	2 d in France
oxolinic acid (Aqualinic Powder, etc)	[14698-29-4]	$C_{13}H_{11}NO_5$	6 d in France 500 ° d in UK[b]
oxytetracycline (Terramycin, etc)	[79-57-2]	$C_{22}H_{24}N_2O_9$	30–40 d at >9 or >10°C; 60 d at <9 or <10°C 400 degree d in UK[b]
sarafloxacin	[98105-99-8]	$C_{20}H_{17}F_2N_3O_3$	150° d
sulfadiazine	[68-35-9]	$C_{10}H_{10}N_4O_2S$	40 d at >10 – 12°C; 60–90 d at
and trimethoprim (Tribrissen, etc)	[738-70-5]	$C_{14}H_{18}N_4O_3$	<8 – 10°C
			350–400 ° d in the UK[b]
sulfamerazine	[127-79-7]	$C_{11}H_{12}N_4O_2S$	30 d at>9°C; 60 d at 200 degree d in the UK[b]
sulfadimethoxine	[122-11-2]	$C_{12}H_{14}N_4O_4S$	49 d only in Germany
and trimethoprim	[738-70-5]	$C_{14}H_{18}N_4O_3$	
thiamphenicol	[15318-45-3]	$C_{12}H_{15}Cl_2NO_5S$	
Topicals			
chloramine-T	[127-65-1]	$C_7H_7ClNNaO_2S$	
formalin	[50-00-0]	CH_2O	
azjamethiphos (Salmosan)			24 h
burserelin			
cypermethrin			100° d
diflubenzuron			Harest 2 years
emamectin (Slice)			
fenbendazole			
fearragillin	[23110-15-8]	$C_{26}H_{34}O_7$	500 ° d
hydrogen peroxide	[7722-84-1]	H_2O_2	24 h or none
praziquantel			
teflubenzuron			100° d
povidone–iodine compounds	[25655-41-8]		
sodium chloride	[7647-14-5]	NaCl	
trichlorfon	[52-68-6]	$C_4H_8C_{13}O_9P$	14–30 d
Anesthetics			
chlorobutanol	[57-15-8]	$C_4H_7Cl_3O$	
carbon dioxide	[124-38-9]	CO_2	
tricaine methanesulfonate	[886-86-2]	$C_{10}H_{15}NO_5S$	
Spawning Hormones			
Somatosalm			1 year

[a] Compounds used only in conjunction with other drugs.

[b] Degree days are number of days past treatment multiplied by the water temperature (°C).

Table 4. **Drugs Approved for Aquaculture Use in Japan, April 2001**[a]

Product	CAS Registry Number	Molecular formula	Use pattern, mg/ kg per day[b]	Withdrawal time, d
		Antibacterials		
amoxicillin	[26787-78-0]	$C_{16}H_{19}N_3O_5S$	20–40	
ampicillin	[69-53-4]	$C_{16}H_{19}N_3O_4S$	5–20	5
bicozamycin benzoate				
doxycycline	[17086-28-1]	$C_{22}H_{24}N_2O_8H_2O$	20–50	20
erythromycin	[114-07-8]	$C_{37}H_{67}NO_{13}$	25–50	30
florfenicol	[73231-34-2]	$C_{12}H_{14}Cl_2FNO_4S$	10	
flumequine	[42835-25-6]	$C_{14}H_{12}FNO_3$	20	
josamycin	[16846-24-5]	$C_{42}H_{69}NO_{15}$	30–50	
leucomycin		$C_{42}H_{69}NO_{15}$		
lincomycin	[1392-21-8]	$C_{18}H_{34}N_2O_6S$	20–40	20
myroxacin				
nifurstylenic acid			50	2
novobiocin	[303-81-1]	$C_{31}H_{36}N_2O_{11}$	50	
oleandomycin	[3922-90-5]	$C_{35}H_{61}NO_{12}$	25	30
oxolinic acid	[14698-29-4]	$C_{13}H_{11}NO_5$	25–30d	14–30
oxytetracycline	[79-57-2]	$C_{22}H_{24}N_2O_9$	50	25–30
phosphomycin calcium				
spiramycin	[8025-81-8]	$C_{43}H_{74}N_2O_{14}$	25–40	30
sulfadimethoxine	[122-11-2]	$C_{12}H_{14}N_4O_4S$	50–100	30
sulfamonomethoxine	[1220-83-3]	$C_{11}H_{12}N_4O_3S$	100–200	15–30
sulfamonomethoxine	[1220-83-3]	$C_{11}H_{12}N_4O_3S$	10–20	15
and ormetoprim	[6981-18-6]	$C_{14}H_{18}N_4O_2$		
sulfisozole	[73247-57-1]		100–200	10–15
tetracycline	[60-54-8]	$C_{22}H_{24}N_2O_8$	55–110	10
thiamphenicol	[15318-45-3]	$C_{12}H_{15}Cl_2NO_5S$	20–50	15
tobicillin				
		Anesthetics		
eugenol	[97-53-0]	$C_{10}H_{12}O_2$	20–50c	
		Paracite repellants		
hydrogen peroxide	[7722-84-1]	H_2O_2		
lysozyme	[9001-63-2]			
praziquantel	[55268-74-1]	$C_{19}H_{24}N_2O_2$		
		Disinfectants		
povidone–iodine compounds	[25655-41-8]			
trichlorfon	[52-68-6]	$C_4H_8Cl_3O_4P$	0.2–0.3c	5

[a] Ref. 21.
[b] Unless otherwise noted.
[c] mg/L solution.
[d] 25–30 mg/kg per day or 5–10 mg/L.

larvae mycosis (see Table 6 for details). Certain chemicals have been exempted from the need for registration: calcium carbonate [471-34-1], $CaCO_3$; calcium hydroxide [1305-62-0], $Ca(OH)_2$; calcium oxide [1305-78-8], CaO; magnesium carbonate; calcium sulfate [7778-18-9], CaO_4S; zeolite [1318-02-1], $Na_2O \cdot Al_2O_3 \cdot (SiO_2)X \cdot (H_2O)Y$; aluminum sulfate [10043-01-3], $Al_2O_{12}S_3$; ferric chloride [7705-08-0], Cl_3Fe; and inorganic and organic fertilizers (11,23).

Table 5. **Antibacterials Approved for Aquaculture Use in Chile**[a]

Product	CAS Registry Number	Molecular formula
florfenicol	[73231-34-2]	$C_{12}H_{14}Cl_2FNO_4S$
flumequine	[42835-25-6]	$C_{14}H_{12}FNO_3$
oxolinic acid	[14698-29-4]	$C_{13}H_{11}NO_5$
oxytetracycline	[79-57-2]	$C_{22}H_{24}N_2O_9$
sarafloxacin	[98105-99-8]	$C_{20}H_{17}F_9N_3O_3$

[a] Ref. 22.

Table 6. **Chemicals Registered for Aquaculture Use in Australia**[a]

Product	CAS Registry Number	Molecular formula
benzocaine	[94-09-7]	$C_9H_{11}NO_2$
formalin	[50-00-0]	CH_2O
flubendazole		
leutinizing hormone releasing hormone analogue		
methyltestosterone	[58-18-4]	$C_{25}H_{38}O_2$
trifluralin	[1582-09-8]	$C_{13}H_{16}F_3N_3O_4$

[a] Ref. 23.

5. Promising Chemicals for Registration for Aquaculture

More therapeutants and vaccines may soon be added to the medicine chest of fish farmers. A variety of chemicals have potential for registration and use in aquaculture (9,11,24,25).

5.1. Antibacterials. Research has been conducted on three important external and systemic antibacterial compounds in the United States. Various registration data are or have been generated by the U.S. Geological Survey's Upper Midwest Environmental Sciences Center at La Crosse, Wisconsin (UMESC), and USFWS on chloramine-T with funds from the Federal-State Aquaculture Drug Approval Partnership (a project of the International Association of Fish and Wildlife Agencies = IAFWA Project) (26–28). Chloramine-T is used mainly to control bacterial gill disease on fry and fingerlings of salmonids but is effective for other external bacterial diseases of a variety of fishes (9,25,28). (see CHLORAMINES AND BROMAMINES). Hydrogen peroxide was identified by UMESC as a good candidate to control external bacterial infections on all fish. Research is underway at UMESC with funds from the IAFWA Project to complete the technical sections needed for approval (9,28). Efforts by UMESC and USFWS with funds from the IAFWA Project should allow the expansion and extension of oxytetracycline to other fish species and other diseases (9,28).

Sarafloxacin [98105-99-8], $C_{20}H_{17}F_2N_3O_3$, a broad-spectrum antibacterial in a class of compounds called fluoroquinolones, was under development by Abbott Laboratories (Chicago, Ill.) for worldwide use by the aquaculture industry until the U.S. Centers for Disease Control and Prevention (CDC) presented concerns about the use of all fluoroquinolones in animal health because of the perceived potential for developing pathogen resistance to drugs used in humans. It is doubtful that CVM will allow any approvals for sarafloxacin or any fluoroquinolone (e.g., enrofloxacin) for aquaculture uses (9,28). (see ANTI-BACTERIAL AGENTS, SYNTHETIC, QUINOLONES).

The University of Idaho and USFWS, with funds from the Bonneville Power Administration, was also gathering data for registration of erythromycin when CDC also became concerned about the use of any human drug in animals because of antimicrobial resistance concerns. Erythromycin is intended for control of bacterial kidney disease in salmonid fingerlings that can also be transmitted by broodstock to the eggs (9). It is not known if this approval will ever be completed (see ANTIBIOTICS, MACROLIDES).

Florfenicol, an oral antibacterial is being considered for development in the United States by the sponsor for use on catfish and salmonids. Efforts are underway through a grant from the Multi-State Conservation Grant Program to generate efficacy data on other species and other diseases on florfenicol by USFWS and UMESC (28). Amoxicillin is another antibacterial that has potential for development for control of streptococcal infections in tilapia and hybrid striped bass (9).

5.2. Fungicides and Parasiticides.   UMESC is working on several fronts to improve the availability of fungicides and parasiticides. UMESC, with funding from Bonneville Power Administration screened and tested promising candidates for replacement of malachite green as both fungicides and parasiticides (9). Although several compounds show promise for controlling fungi, the best fungicide candidate was identified as hydrogen peroxide. UMESC has developed data with funds from the IAFWA Project to gain an approval for use of hydrogen peroxide to control fungal infections on all fish and their eggs. Hydrogen peroxide was also identified by UMESC as a good candidate to control external parasitic infestations on all fish. Research is underway at UMESC with funds from the IAFWA Project to complete the technical sections needed for approval. Pyceze is another candidate fungicide that was identified in the United Kingdom and is now being developed worldwide by the sponsor (9).

The approval of formalin as a fungicide now extends to all fish eggs and may soon extend to all fish. The approval of formalin as a parasiticide was also extended through efforts by UMESC with funds from the IAFWA Project and efforts by Auburn University with funds from the National Research and Support Program Number Seven (28).

The Harry K. Dupree Stuttgart National Aquaculture Research Center, Stuttgart, Arkansas (SNARC) has developed data on copper sulfate with funds from the IAFWA Project for the control of *Ichthophthirius*, an external protozoan that causes significant losses in the catfish industry. Potassium permanganate is another chemical being researched at SNARC with funding from the IAFWA Project for use as a control for *Ichthyophthirius* (28). Other promising parasiticides include praziquantel, fumagillin, and sea lice control agents (e.g., azamethiphos, cypermethrin, emamectin, and hydrogen peroxide) (9). (See ANTIPARASITIC AGENTS, ANTHELMINTICS).

5.3. Disinfectants.   Promising disinfectants include ultraviolet (uv) light andozone [10028-15-6], O_3. High doses of uv light have prevented the transmission of the epizootic epitheliotropic virus (EEV) disease among lake trout (*Salvelinus namaycush*) (25). Uv sterilization can only be successful when water volumes and suspended solids are low (29). Ozone is also considered to be a good candidate to both control EEV and other diseases, and remove chemicals and wastes from aquacultural effluents (25). However, ozone is considered to be a food additive by FDA and, therefore, must meet registration requirements that are almost as stringent as the registration requirements for drugs.

5.4. Anesthetics.   Ethyl aminobenzoate [94-09-7] (benzocaine), $C_9H_{11}NO_2$, was a candidate anesthetic but it would have required additional mammalian safety studies, did not have a sponsor, and probably would not have allowed the use of spawned-out broodstock carcasses to be used for pet or human food. AQUI-S, an anesthetic developed in New Zealand has great potential as a zero withdrawal drug and is being developed by its sponsor for worldwide approval. In the United States, UMESC with funds from the IAFWA Project is developing the anesthetic for use on all fish (9,28). Electronarcosis is an alternative to chemical anesthesia that uses varying electrical frequencies to rapidly anesthetize fishes and allow gentle recovery. Electronarcosis has been used effectively on tilapia (*Oreochromis* sp.) and common carp (*Cyprinus carpio*) and the technique is being tested with other fishes (30,31).

5.5. Herbicides and Piscicides.   Registrants reregistered aquatic herbicides under the 1988 Pesticide Law in the United States. Reregistration of piscicides was funded mainly by public agencies. Data collection for the reregistration of rotenone was completed by UMESC. Antimycin must be reregistered soon but funds have not been available to complete the data requirements and it may be suspended or canceled if the data are not generated soon. Control of Bayluscide (niclosamide) was transferred from the Mobay Corp. (now Bayer Corporation) to USFWS after Mobay determined that the estimated $2.5–3.5 million for reregistration was not economically feasible. Bayluscide is used by USFWS to control the sea lamprey (*Petromyzon marinus*) in the Laurentian Great Lakes and to manage and control nuisance mollusks in both aquaculture and natural waters. Bayluscide also has the potential to control digenetic trematodes of finfishes and snail vectors of several parasites, including *Bolbophorus* sp. (25,32).

BIBLIOGRAPHY

"Aquaculture," in *ECT* 3rd ed., Vol. 3, pp. 194–213, H. P. Clemens and M. Conway, University of Oklahoma; in *ECT* 4th ed., Vol. 3, pp. 608–623, by Rosalie A. Schnick, U. S. Fish and Wildlife Science; "Aquaculture Chemicals" in *ECT* (online), posting date: December 4, 2000, by Robert Stickey, Texas A & M University.

CITED PUBLICATIONS

1. R. A. Busch, *Vet. Hum. Toxicol.* **29**(Suppl. 1), 45 (1987).
2. L. W. Harrell, *Vet. Hum. Toxicol.* **29**(Suppl. 1), 49 (1987).

3. M. H. Beleau and J. A. Plumb, *Vet. Hum. Toxicol.* **29**(Suppl. 1), 52 (1987).

4. J. P. McCraren, *The Aquaculture of Striped Bass*, Sea Grant Publication No. UM-SG-MAP-84-01, University of Maryland, College Park, Md., 1984.

5. G. Lio-Po in J. V. Juario, R. P. Ferraris, and L. V. Benitz, eds., *Proceedings of the Second International Milkfish Aquaculture Conference, October 4–6, 1983*, Iloilo City, Philippines, 1984.

6. C. Brown, *Vet. Hum. Toxicol.* **29**(Suppl. 1), 35 (1987).

7. R. R. Williams and D. V. Lightner, *J. World Aquacult. Soc.* **19**, 188 (1988).

8. C. J. Sindermann and D. V. Lightner, *Disease Diagnosis and Control in North American Marine Aquaculture*, Elsevier, New York, 1988.

9. R. A. Schnick, *Use of Chemicals in Fish Culture: Past and Future*, Chapt. 1, pp. 1–14 in D. J. Smith, W. H. Gingerich, and M. G. Beconi-Barker, eds., Xenobiotics in Fish, Kluwer Academic/Plenum Publishing Corp., New York, 1999.

10. Joint Subcommittee on Aquaculture, *Guide to Drug, Vaccine, and Pesticide Use in Aquaculture*, Texas Agricultural Extension Service, College Station, Tex., 1994.

11. R. A. Schnick, and co-workers *Bull. Euro. Assoc. Fish Path.* **17**, 251 (1997).

12. U. S. Food and Drug Administration (FDA), *Food and Drugs, Code Fed. Reg.* **21**, Parts 500–599 (2000).

13. G. E. Brooks, *Aquacult. Assoc. Canada Bull.* **89–4**, 39 (1989).

14. G. R. Johnson and D. J. Rainnie, *Aquacult. Assoc. Canada Bull.* **89–4**, 43 (1987).

15. R. Linehan, personal communication, April 2001.

16. D. J. Alderman, personal communication, April 2001.

17. J. Shepherd, paper presented to the *Aquaculture Feed and Veterinary Products: Worldwide Business Opportunities for Feed, Pharmaceutical, and Chemical Companies Conference*, Stamford, Conn., June 6–8, 1990.

18. L. A. Brown, *Vet. Hum. Toxicol.* **29**(Suppl. 1), 54 (1987).

19. E.-M. Bernoth, *J. Vet. Med. Ser. B.* **37**, 401 (1990).

20. T. Sano and H. Fukuda, *Aquaculture* **67**, 59 (1987).

21. S. Ishihara, personal communication, April 2001.

22. D. Farcas, personal communication, April 2001.

23. K. Allan, personal communication, April 2001.

24. D. J. Alderman in J. F. Muir and R. J. Roberts, eds., *Recent Advances in Aquaculture*, Vol. 3, Timber Press, Portland, Oreg., 1988, p. 1.

25. F. P. Meyer and R. A. Schnick, *Rev. Aquat. Sci.* **1**, 693 (1989).

26. R. A. Schnick, *Trans. 61st No. Am. Wildlife Nat. Res. Conf.* **61**, 6 (1996).

27. R. A. Schnick, W. H. Gingerich, and K. H. Koltes, *Fisheries* **21**(5), 4 (1996).

28. B. R. Griffin, R. A. Schnick, and W. H. Gingerich, *Aquacult. Mag.* **26** (3), 56 (2000).

29. G. Barnabe', *Aquaculture*, Vol. **1**, Ellis Horwood, London, UK, 1990.

30. W. T. Barham, A. J. Schoonbee, and J. G. J. Visser, *Onderstepoort J. Vet. Res.* **54**, 617 (1987).

31. *Ibid.*, **56**, 215 (1989).

32. B. R. Griffin, personal communication, May 2001.

Rosalie A. Schnick
Aquaculture Chemicals

AROMA CHEMICALS

1. Introduction

Aroma chemicals are an important group of organic molecules used as ingredients in flavor and fragrance compositions (see FLAVORS; PERFUMES). Aroma chemicals consist of natural, nature-identical, and artificial molecules. Natural products are obtained directly from the plant or animal sources by physical procedures. Nature-identical compounds are produced synthetically, but are chemically identical to their natural counterparts. Artificial flavor substances are compounds that have not yet been identified in plant or animal products for human consumption.

There are ca. 3000 different molecules that find use in the production of flavor and fragrance compositions. Synthetic ingredients play a major part as components due to their convenient availability and the relatively lower costs compared to natural molecules from isolation of relatively limited natural sources.

The world flavors and fragrance market data show that the share of aroma chemicals is $\approx 19\%$. The share of essential oils (see OILS, ESSENTIAL)and other natural products is roughly 17%. The most important part of the flavor and fragrance market (called F&F market) are flavor compounds (31%) and fragrance compounds (33%). The total of the F&F market is nearly divided into 50% flavors and 50% fragrances (Table 1) (1).

The estimated value of the entire F&F market is US$13–14 Bn (2000). The market value of aroma chemicals is considered as US$ 2.0 Bn (2000), which corresponds to the share of 15% of the total market.

The regional distribution of the F&F market in % per region is shown in Figure 1. The largest market for flavors and fragrances is North America/NAFTA, followed by Western Europe and Asia-Pacific (estimation 2000). The leading global suppliers of flavors and fragrances in 2000 are IFF (incl. BBA), Givaudan, Quest Intl., Firmenich, Takasago, Haarman & Reimer, T. Hasegawa, Sensient Technologies, Dragoco, and Mane.

2. Odors Descriptors

The odors of single chemical compounds (aroma chemicals) are very difficult to describe unequivocally. The odors of complex mixtures called compounds are often impossible to describe unless one of the components is so characteristic

Table 1. **F&F Market**

Product	Estimated share (%)	Market value (US$Bn)
fragrances	33	4.5
flavors	31	4.1
aroma chemicals	19	2.5
essential oils and other natural products	17	2.3

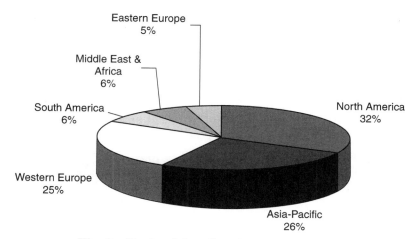

Fig. 1.   Regional destribution of F&F market.

that it determines the odor or flavor of the composition. Although an objective classification is not possible, an odor can be described by adjectives such as flowery, fruity, woody, or hay-like, which will relate to natural occurring or other well-known products with such odors characteristics.

A few terms (2) used to describe odors are listed in Table 2, with a few examples.

Table 2. **Terms to Describe Odors**

Odor	Description	Examples
aldehydic	note of the long-chain fatty aldehydes, eg, fatty–sweaty, ironed laundry, sea-water	*n*-Decanal *n*-Octanal
animalic	typical notes from the animal kingdom, eg, musk, castoreum, skatol, civet, ambergis	Indole Ambrox
balsamic	heavy, sweet odors, eg, cocoa, vanilla, cinnamon	Vanillin isobutyrate Cinnamaldehyde
camphoraceous	reminiscent of camphor	2-Adamantanone (+)-Isoborneol

Table 2 *(Continued)*

Odor	Description	Examples
citrus	fresh, stimulating odor of citrus fruits such as lemon or orange	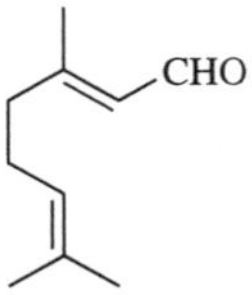 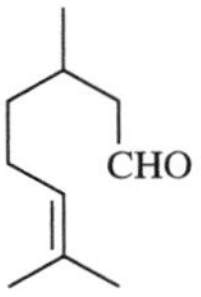 Citral Citronellal
earthy	humus-like, reminiscent of humid earth	 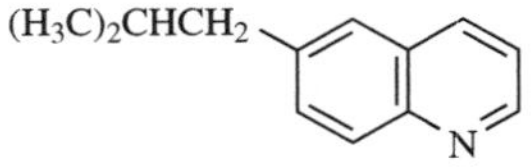 2- Ethylfenchol $(H_3C)_2CHCH_2$— 6-Isobutylquinoline
fatty	reminiscent of animal fat and tallow	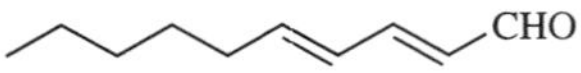 $CH_3(CH_2)_8CO_2(CH_2)_4CH_3$ Amyl decanoate (*E,E*)-2,4-Decadienal
floral, flowery	generic terms for odors of various flowers	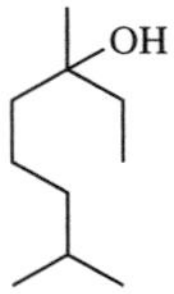 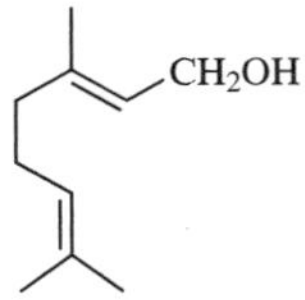 Tetrahydrolinalool Geraniol
fruity	generic terms for odors of various fruits	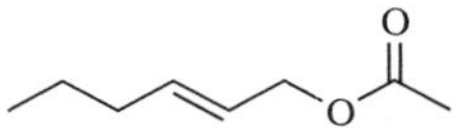 $CH_3CO_2(CH_2)_7CH_3$ *n*-Octyl acetate *trans*-2-Hexenylacetate
green	typical odor for freshly cut grass and leaves	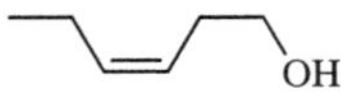 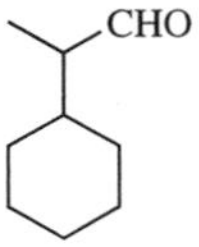 cis-3-Hexenol 2-(Cyclohexyl)-propanal
herbaceous	noncharacteristic, complex odor of green herbs with, eg, sage, minty, eucalyptus-like, or earhty nuances	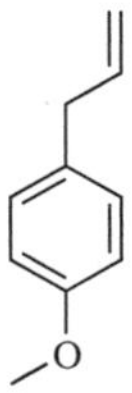 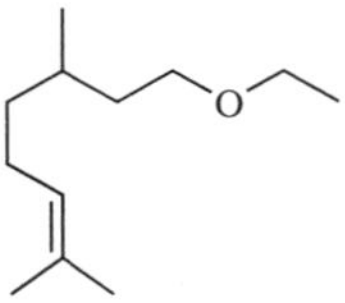 Estragole Citronellylethyl ether
medicinal	odor reminiscent of disinfectants, eg, phenol, lysol, methyl salycilate	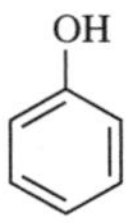 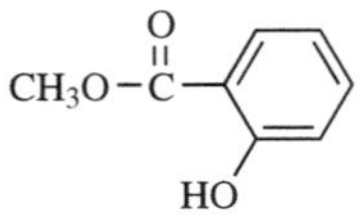 Phenol Methyl salicylate

Table 2   (*Continued*)

Odor	Description	Examples
metallic	typical odor observed near metal surfaces, eg, brass or steel	2,5-Dimethyl-2-vinyl-4-hexenenitrile Benzyl methyl disulfide
minty	peppermint-like odor	(–)-Menthol Menthone
mossy	typical note reminiscent of forests and seaweed	Ethyl 2-hydroxy-4-methoxy-6-methyl-benzoate 3-Methoxy-5-methyl phenol
powdery	odor identified with toilet powders, sweet-diffusive	2-(1-Cyclohexenyl)-cyclohexanone Methyl-β-naphthyl ketone
resinous	aromatic odor of tree exudates	2-Isopropyl-5-methyl-2-hexenal 3-Phenylpropionic acid
spicy	generic term for odors of various spices	Carvacrol 2,4-Dimethyl-acetophenone
waxy	odor resembling that of candle wax	*n*-Decanal Citronellyl-isobutyrate

Table 2 (*Continued*)

Odor	Description	Examples	
woody	generic term for the odor of wood, eg, cedarwood, sandalwood	α-Cedrene	cis-p-tert-Butyl-cyclohexylacetate

3. General Production Routes

Aroma chemicals are specific molecules of particular aroma, which can be obtained (1) by isolation from natural sources, with or without chemical modifications, using natural molecules as precursors for many aroma chemicals (partial synthesis); (2) from petrochemical raw materials; or (3) by synthesis from cyclic and aromatic precursors.

For example, cedarwood oils obtained from plants like *Cedrus atlantica*, *Chamaecyparis funebris*, or *Juniperus mexicana*, contain aromatic molecules, eg, (*E*)−(+)−α−atlantone, α−thujone, or (+)−cedrol:

E-(+)-α-Atlantone   α-Thujone   (+)-Cedrol

Acetylation of (+)-cedrol gives cedryl acetate, a woody-earthy odorous molecule, applied in woody compounds for all purposes.

(+)-Cedrol   Cedryl acetate

3.1. The Use of Natural Molecules as Precursors. One of the most useful sources for natural molecules as chemical precursors is turpentine oil, originated from *Pinus* sp. The oil contains 60−70% of α-pinene and β-pinene, along with other natural molecules, i.e, α-phellandrene, γ-terpinene, anethole, caryophyllene, 3-carene, and camphene (see Figs. 2 and 3).

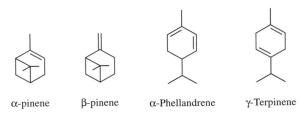

α-pinene   β-pinene   α-Phellandrene   γ-Terpinene

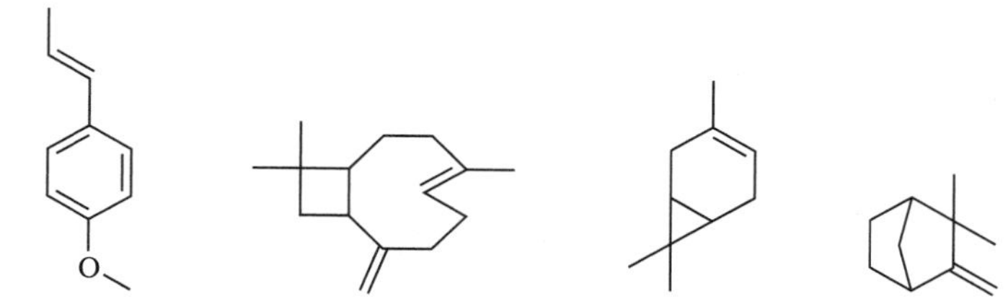

Anethole Caryophyllene 3-Carene Camphene

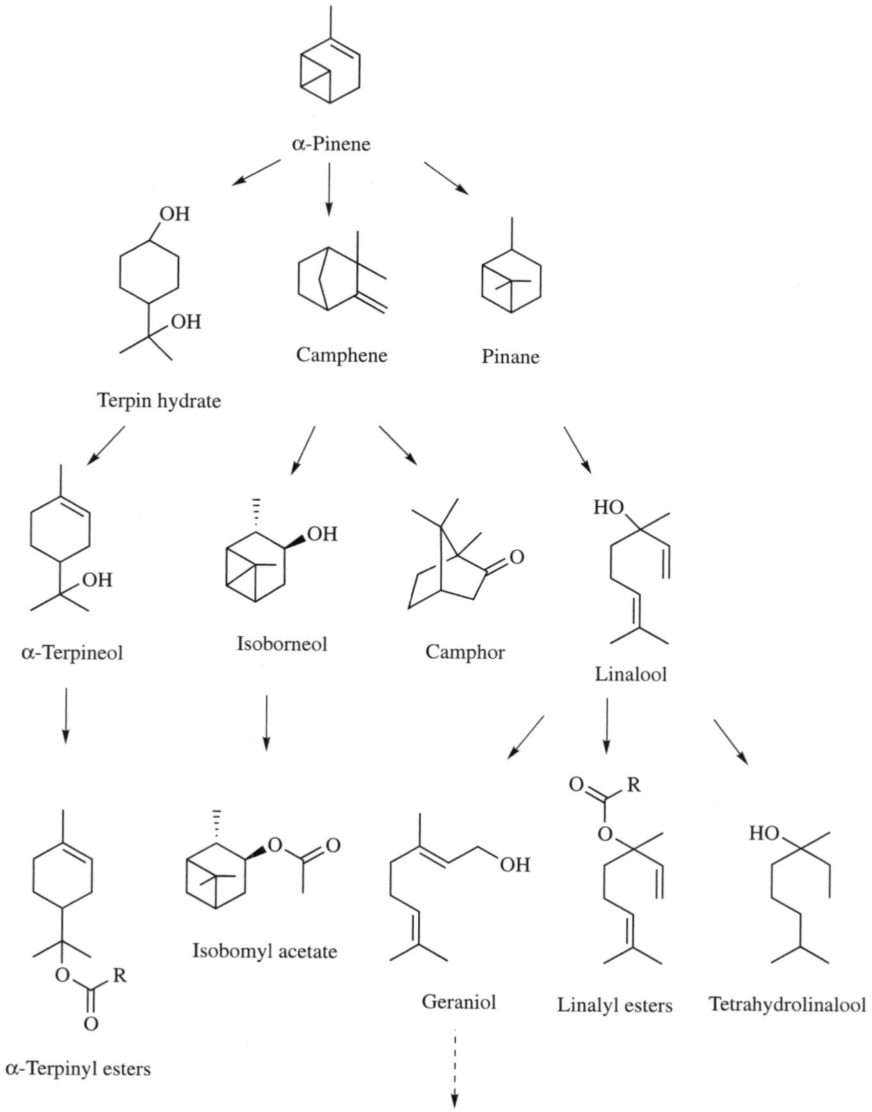

Fig. 2. α-Pinene as a natural precursor for aroma chemicals.

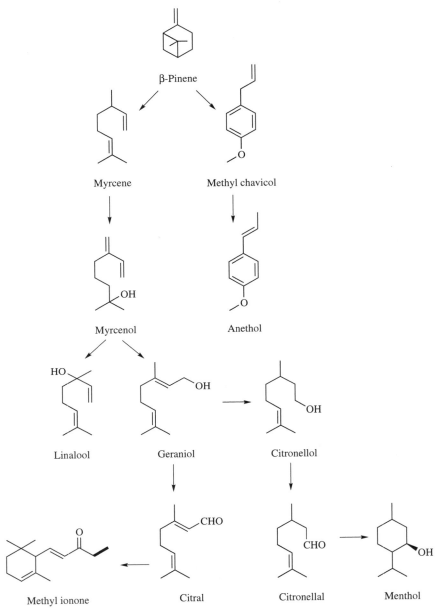

Fig. 3.   β-Pinene as a natural precursor for aroma chemicals.

3.2. The Use of Petrochemicals as Precursors.   Synthesis from petrochemical precursors of one-to-five carbon atoms, ie, carbon monoxide/ formaldehyde, acetylene, isobutylene, and isoprene, represents one of the most important routes to produce aroma chemicals.

Aromatic molecules, e.g. benzene, toluene, xylenes, phenol, cresols, and naphthalene, are also important precursors for aroma chemicals (see Figs 4–6).

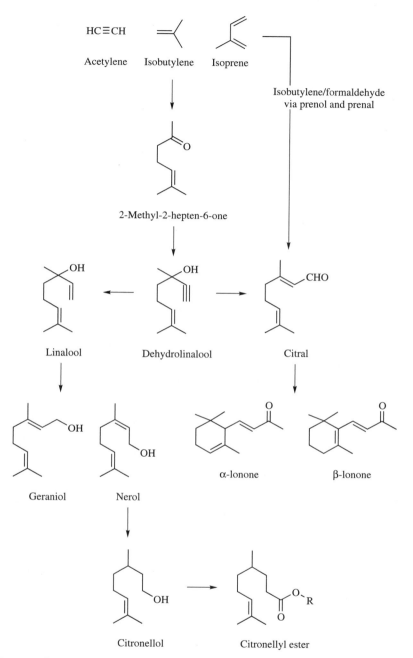

Fig. 4. General overview: Petrochemicals as a source for aroma chemicals.

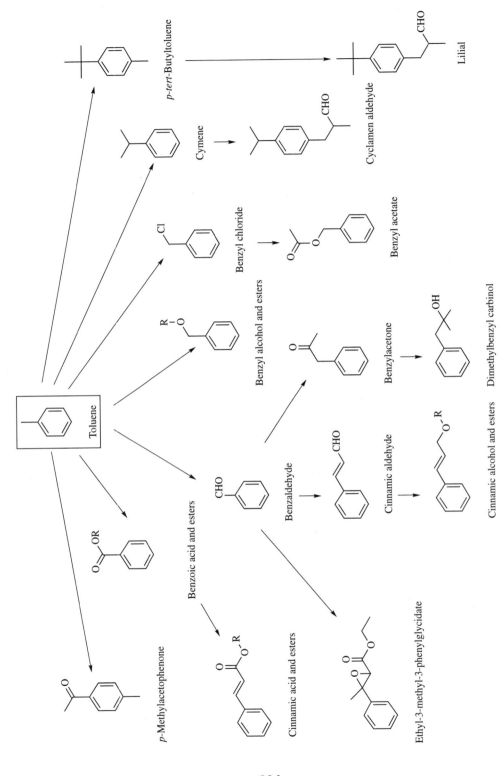

Fig. 5. Aroma chemicals derived from toluene.

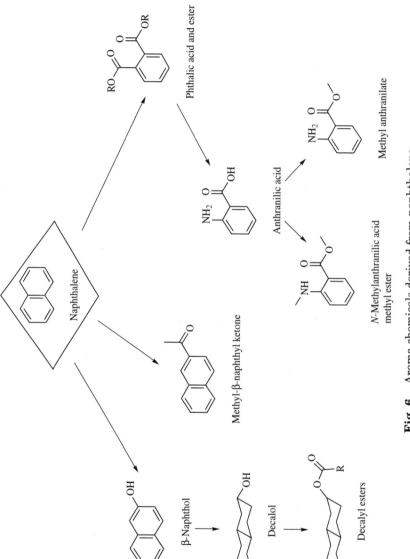

Fig. 6. Aroma chemicals derived from naphthalene.

4. Functional Groups of Aroma Chemicals

As mentioned before, over 3000 specific chemical molecules are used in the F&F industry, but only a few hundreds are produced in a scale between 20 and 50 mt year. These molecules include most of the functional groups, from aliphatic molecules to heterocyclic ones, according to the following list:

- Hydrocarbons (aliphatic, acyclic terpenes, cyclic terpenes, benzenoids)
- Alcohols (aliphatic, alicyclic, cyclic)
- Ethers
- Aldehydes and ketones (including acetals and ketals)
- Carboxylic acids
- Esters and lactones
- Nitriles
- Amines
- Nitroaromatic compounds
- Thio compounds
- Heterocyclic molecules

The following sections contain selected examples of each functional group, the chemical structure, and organoleptic characteristcs.

4.1. Hydrocarbons. Hydrocarbons include simple aliphatic molecules, terpenes—both acyclic and cyclic, and benzene rings.

Unsaturated Aliphatic Non-Terpenes

1-Heptene	green, diary-like, creamy, apple, vegetable and strawberry flavor. Application in diary flavorings; cream, tomato, apple, vegetable, strawberry; general fresh green fruity notes (3) green, occurs in fresh apples
1-Decyne	
(*E,E,Z*)-1,3,5-Undecatriene	oily, waxy, slightly fruity, peppery aroma, galbanum-like, green, musty, with an earthy rooty, fatty meat-like nuance flavor

Terpenes. Terpenes (see TERPENOIDES; OILS, ESSENTIAL; FLAVORS; PERFUMES) are a group of plant originated natural products, which are usually composed of usually two, three, four, five, six or eight units of C_5 atoms. These units are formally derived from 2-methyl-1,3-butadiene (isoprene).

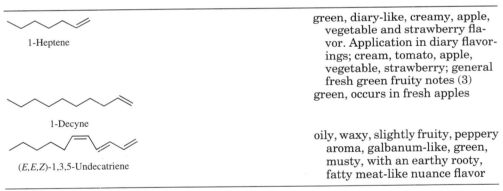

2-Methyl-1,3-butadiene (Isoprene)

These molecules are named as follows:

Name	Number of isoprene units	Number of carbon atoms
monoterpenes	2	10
sesquiterpenes	3	15
diterpenes	4	20
sesterterpenes	5	25
triterpenes	6	30
tetraterpenes	8	40

Acyclic Monoterpenes

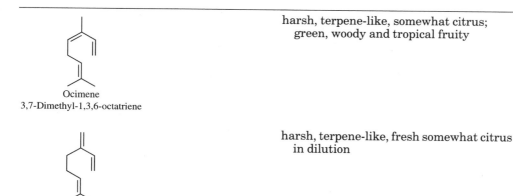

harsh, terpene-like, somewhat citrus; green, woody and tropical fruity

Ocimene
3,7-Dimethyl-1,3,6-octatriene

harsh, terpene-like, fresh somewhat citrus in dilution

Myrcene
7-Methyl-3-methylene-1,6-octadiene

Cyclic Monoterpenes

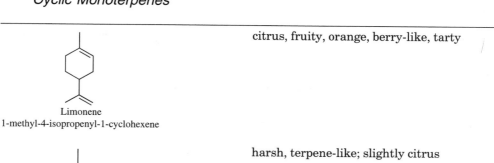

citrus, fruity, orange, berry-like, tarty

Limonene
1-methyl-4-isopropenyl-1-cyclohexene

harsh, terpene-like; slightly citrus

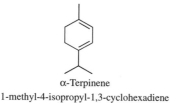

α-Terpinene
1-methyl-4-isopropyl-1,3-cyclohexadiene

Terpinolene
1-methyl-4-isopropylidene-1-cyclohexene

harsh, terpene-like, slightly citrus

α-Phellandrene
1-methyl-4-isopropyl-1,5-cyclohexadiene

terpenic, citrus lime with a fresh green note

4.2. Bicyclic Monoterpenes

α-Pinene
2,6,6-trimethylbicyclo[3.1.1]hept-2-ene

harsh, terpene-like, coniferous

β-Pinene
6,6-dimethyl-2-methylenebicyclo[3.1.1]-heptane

harsh, terpene-like, coniferous

Camphene
2,2-dimethyl-3-methylenebicyclo[2.2.1]-heptane

harsh, fresh camphoraceous, terpene-like

Δ-3-Carene

harsh, terpene-like, coniferous

Acyclic Sesquiterpenes

citrus, herbaceous

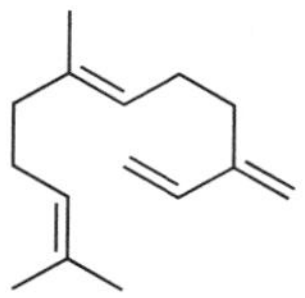

Farnesene
(3,7,11-trimethyl-1,3,6,10-dodeca-tetraene)

Bicyclic Sesquiterpenes

Spicy, woody, dusty, oily; pepper-like, camphoraceous, with a citrus background

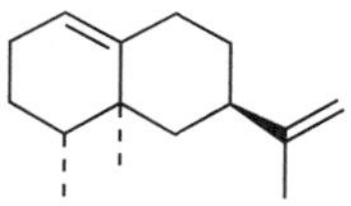

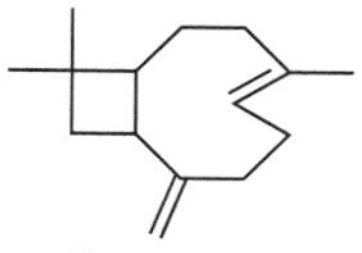

Caryophyllene
4,11,11-trimethyl-8-methylene-bicyclo-
[7.2.0]undec-4-ene

paraffin, oily, somewhat citrus, grapefruit-like

Valencene
5,6-dimethyl-8-iso-propenylbicyclo- [4.4.0]-
dec-1-ene

Terpenes are formed in nature via the "two carbons metabolism", a process enabled by acetyl coenzyme A (CoA), which is produced from pyruvic acid. Acetyl—CoA forms mevalonic acid, which loses one carbon atom by decarboxylation to yield a C_5 unit—isopentenyl pyrophosphate:

Acetyl-S-Coenzyme A

Acetoacetyl-S-Coenzyme A

Mevalonic Acid

Mevalonic acid

1. ATP
Mevalonic kinase
2. $-CO_2$

Isopentenyl pyrophosphate

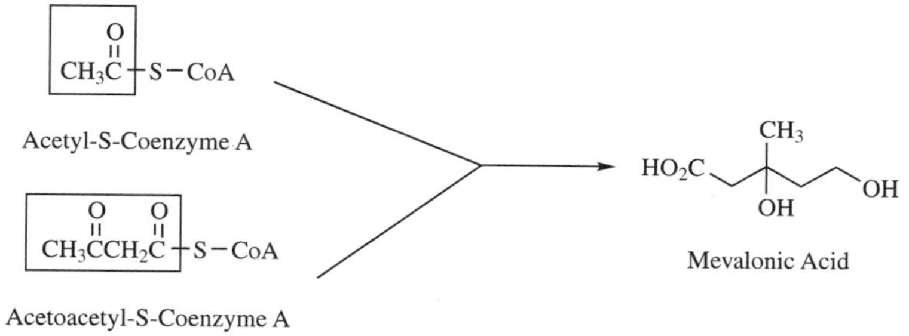

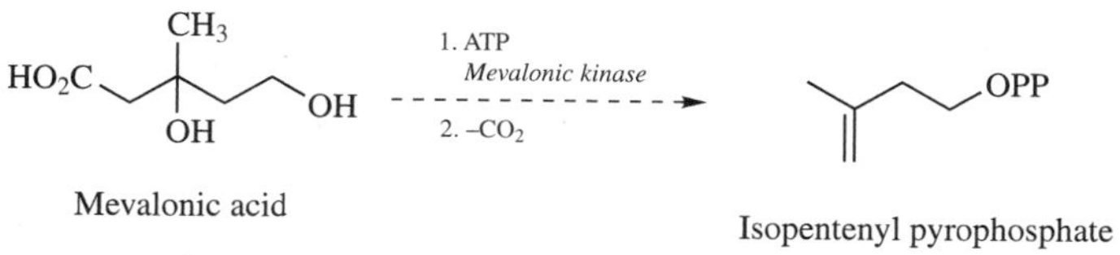

Two units of isopentenyl pyrophosphate are combined with one C_{10} atom-unit—geranyl pyrophosphate, which loses its pyrophosphate group to form a unstabile intermediate—geranyl carbocation:

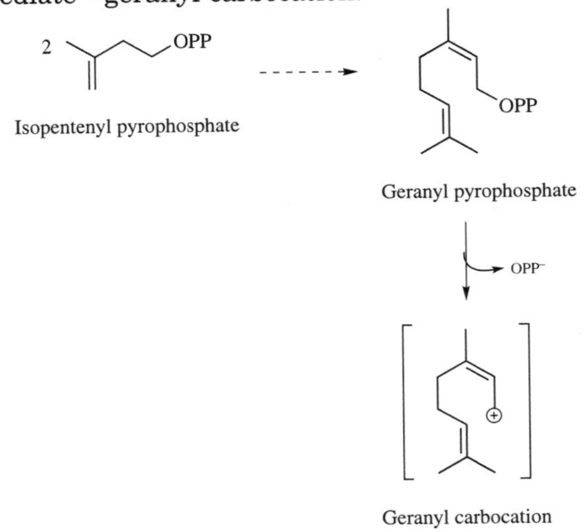

Isopentenyl pyrophosphate

Geranyl pyrophosphate

OPP^-

Geranyl carbocation

The geranyl carbocation can be stabilized by the following possibilities:

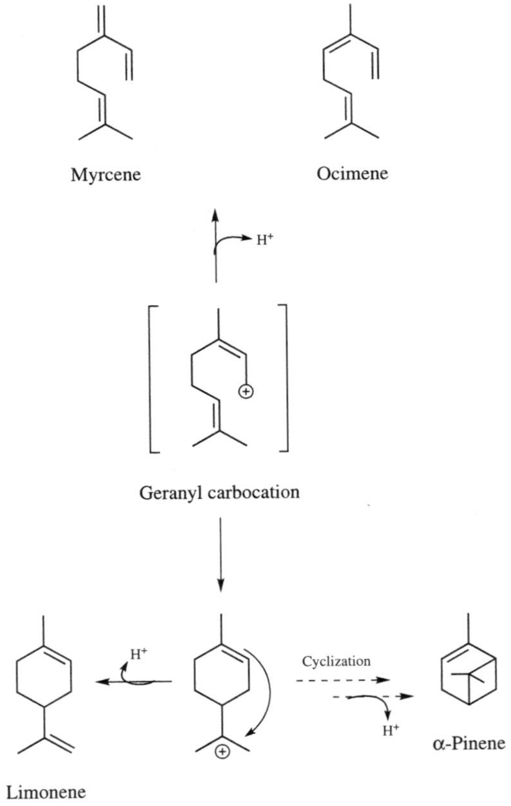

Myrcene

Ocimene

H^+

Geranyl carbocation

Cyclization

H^+

H^+

Limonene

α-Pinene

Benzenoids

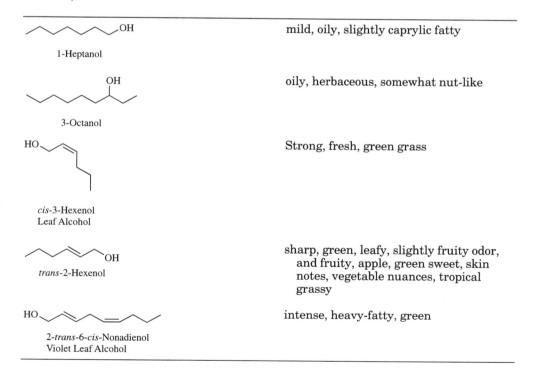

p-Cymene	harsh, gasoline, terpene-like
Diphenylmethane	aromatic oily, spicy on dilution

4.3. Alcohols. The alcohol function is found in simple aliphatic molecules, in acyclic and cyclic terpenes, and in molecules containing benzene rings. Phenols are also contained in this group of aroma chemicals.

Aliphatic Alcohols

1-Heptanol	mild, oily, slightly caprylic fatty
3-Octanol	oily, herbaceous, somewhat nut-like
cis-3-Hexenol Leaf Alcohol	Strong, fresh, green grass
trans-2-Hexenol	sharp, green, leafy, slightly fruity odor, and fruity, apple, green sweet, skin notes, vegetable nuances, tropical grassy
2-*trans*-6-*cis*-Nonadienol Violet Leaf Alcohol	intense, heavy-fatty, green

Alcohols—Acyclic Terpenes

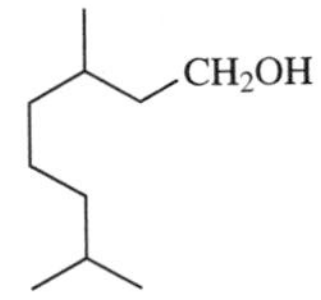

fresh floral, rosy, fatty

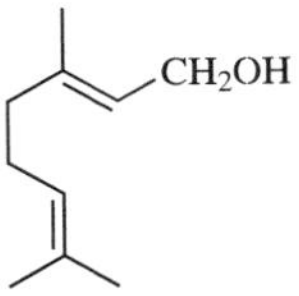

Tetrahydrogeraniol
3,7-dimethyloctanol

floral rose, citrus-like, fruity, slightly fatty

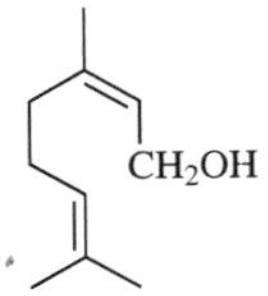

Geraniol [(*E*)-isomer]
3,7-dimethyl-(*E*)-2,6-octadienol

floral rose, geranium; fruity, pear; citrus-
lemon

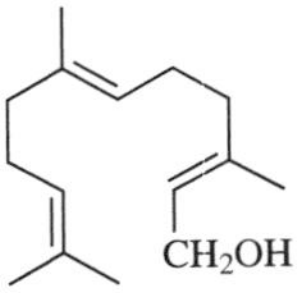

Nerol [(Z)-isomer]
3,7-dimethyl-(Z)-2,6-octadienol

delicate, fresh green; floral (muguet)

(*E,E*)-Farnesol
3,7,11-trimethyl-2,6,10-dodecatrieno

Alcohols—Cyclic Terpenes

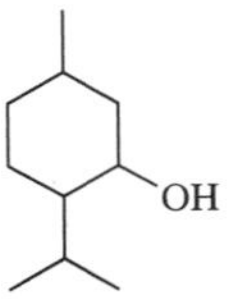

fresh, minty, with a dusty and earthy note

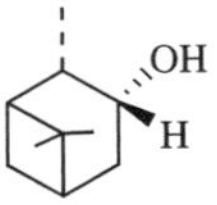

Menthol (8-p-menthen-3-ol)

natural camphoraceous, pine-needlelike

l-(−)-Borneol

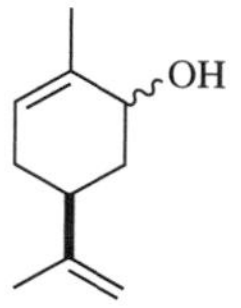

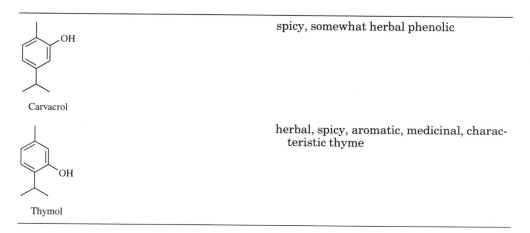

spearmint, caraway

L-Carveol

Alcohols Containing Benzene Rings

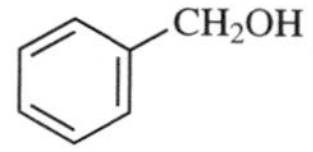

CH$_2$OH

chemical, fruity with balsamic nuances

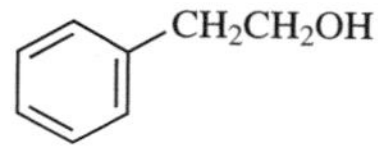

Benzyl alcohol

CH$_2$CH$_2$OH

mild, warm honey, fruity, sweet floral-rose

Phenethyl alcohol

Phenols

spicy, somewhat herbal phenolic

OH

Carvacrol

herbal, spicy, aromatic, medicinal, characteristic thyme

OH

Thymol

Preparation Methods of Alcohols
1. From Natural Sources
Conversion of α- and β-pinene to alcohols

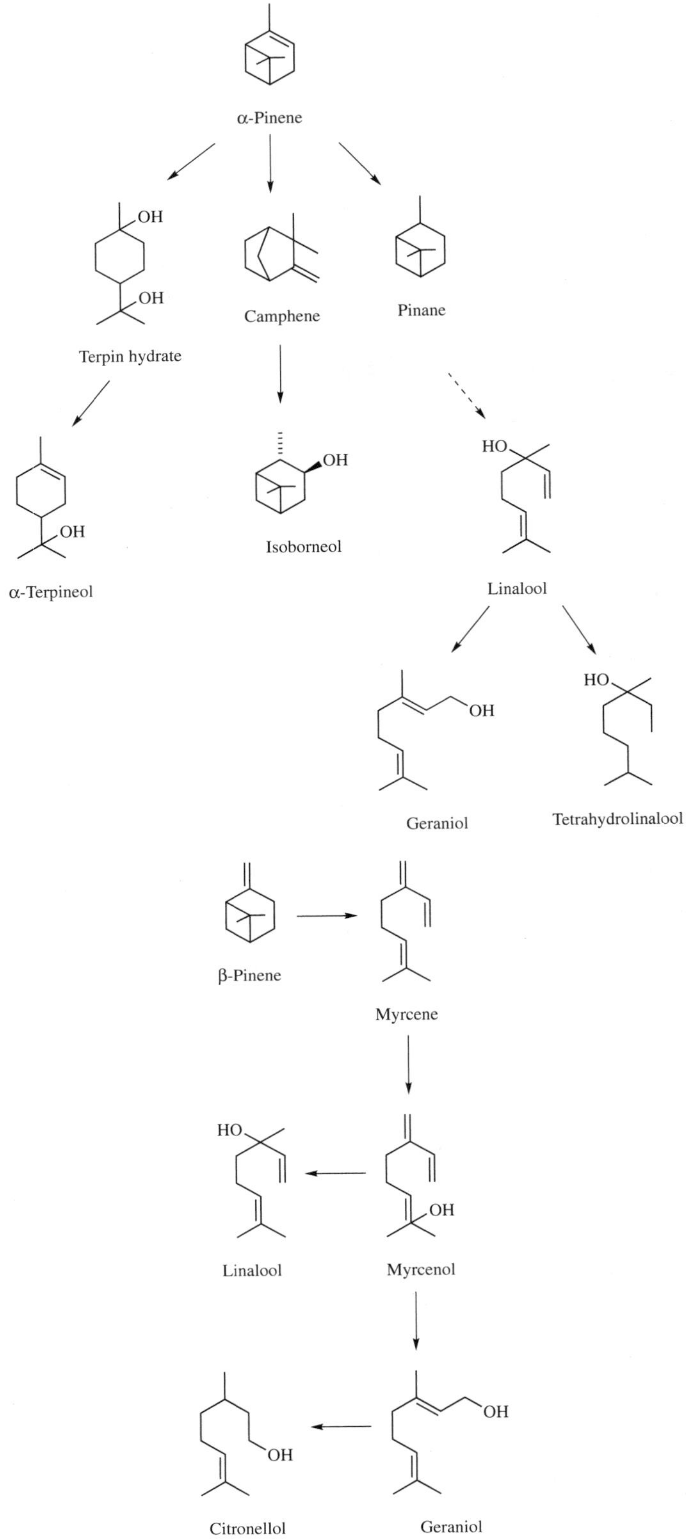

α-Pinene

Terpin hydrate

Camphene

Pinane

α-Terpineol

Isoborneol

Linalool

Geraniol

Tetrahydrolinalool

β-Pinene

Myrcene

Linalool

Myrcenol

Citronellol

Geraniol

244

2. **From chemical precursors.** The starting materials are isoprene, acetylene, formaldehyde, and acetone, which are used for the production of one of the possible the key intermediates for linalool and geraniol - 6-methyl-5-hepten-2-one.

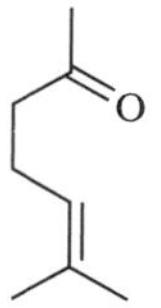

6-Methyl-5-hepten-2-one

6-Methyl-5-hepten-2-one is synthesized in several routes (4):

1. **From acetylene and acetone.** Addition of acetylene to acetone, yielding 3-methyl-1-butyn-3-ol, which undergoes hydrogen addition to obtain 3-methyl-1-buten-3-ol, in presence of palladium catalyst:

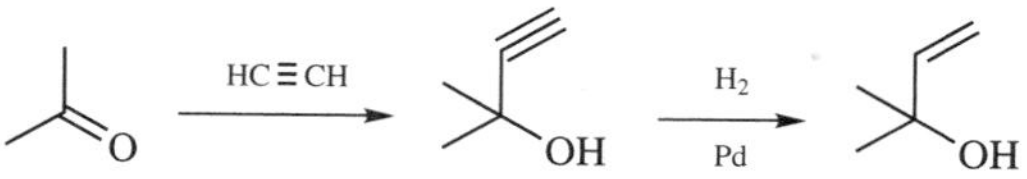

3-Methyl-1-butyn-3-ol reacts with diketene or with ethylacetoacetate as following:

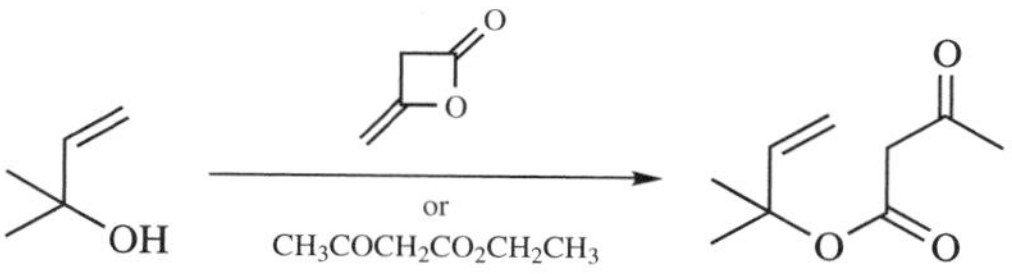

2. This acetaoacetate derivative, undergoes Carroll rearrangement, accompanied by decarboxylation to give the desired product:

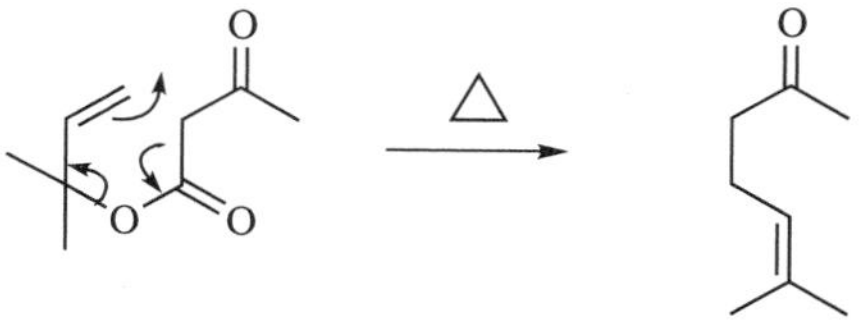

6-Methyl-5-hepten-2-one

3. **By Claisen Rearangement.** In this route, 6-methyl-5-hepten-2-one is prepared by reaction of 3-methyl-1-buten-3-ol with isopropenylmethyl ether, followed by Claisen rearangement:

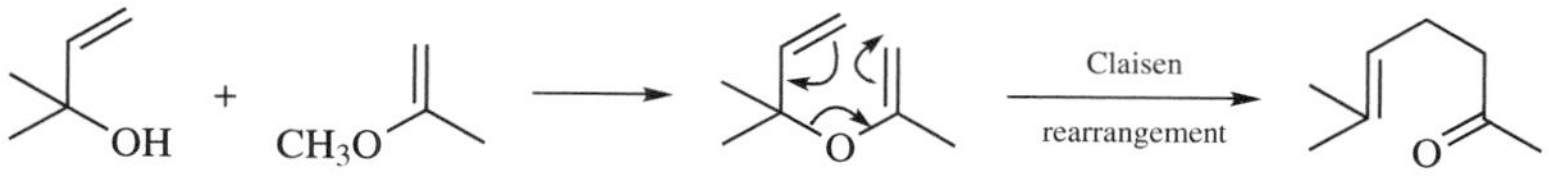

4. **From Acetone and Isoprene.** In this route, hydrochloric acid is added to isoprene to obtain 3-methyl-2- butenylchloride. Reaction of the hydrogen chloride with acetone, in the presence of catalytic amount of organic

base, yields the desired product:

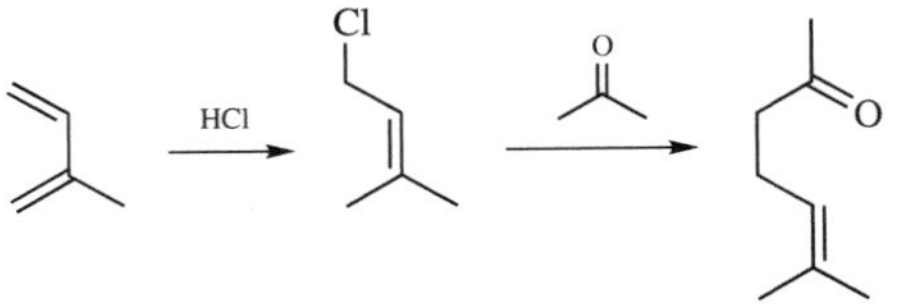

5. From isobutylene and formaldehyde. In this process 6-methyl-5-hepten-2-one is prepared via isoprenol by isomerization of 2-methyl-1-hepten-6-one. The starting material can be prepared in two steps from isobutylene and formaldehyde. The formed 3-methyl-3-buten-1-ol reacts with acetone to yield the desired product:

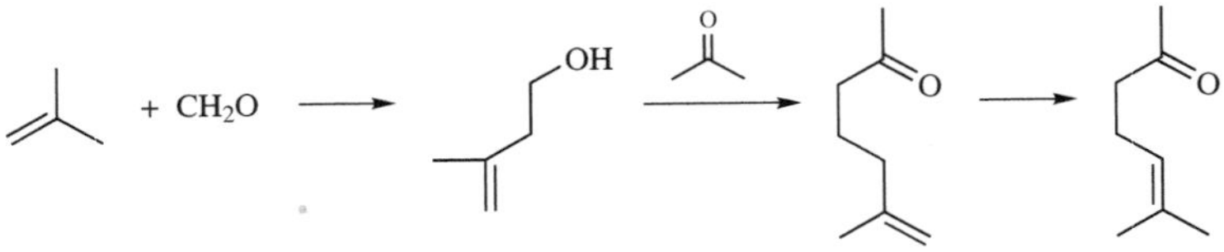

6-methyl-5-hepten-2-one, the main intermediate to linalool, can be further converted to important aroma chemicals such as geraniol, tetrahydrolinalool, methyl ionones, and others.

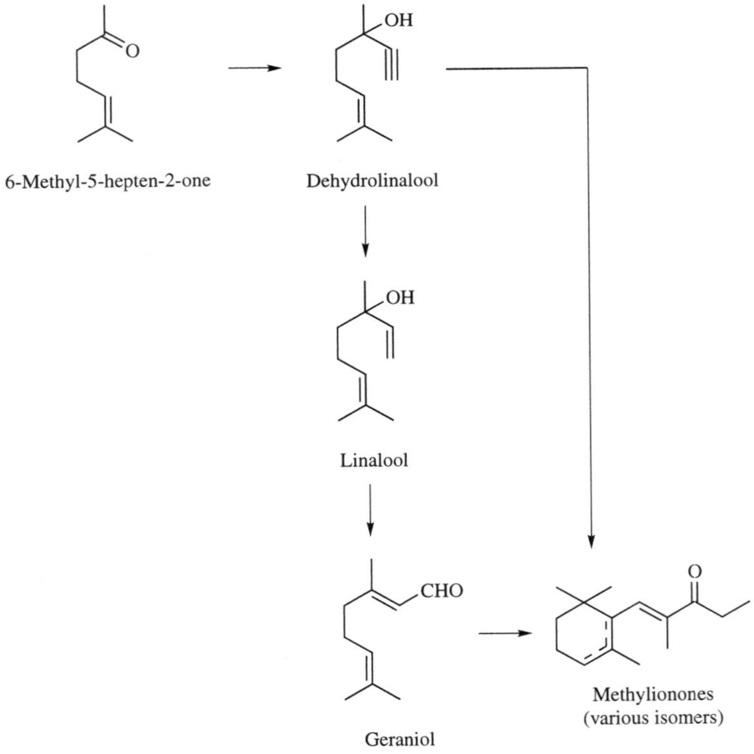

6-Methyl-5-hepten-2-one   Dehydrolinalool

Linalool

Geraniol

Methylionones
(various isomers)

4.4. Ethers. The ether function is found both in aliphatic and aromatic molecules.

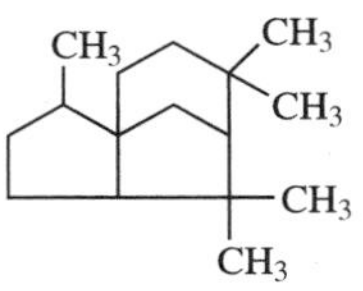

 Cedrylmethyl ether	A colorless liquid with a fine cedarwood odor and a distinct amber nuance
 Diphenyl ether	Aromatic, floral on dilution, rose-like
 trans-Anethole	sweet, warm, herbaceous, strong anise-, licorice-, root beer-like
 Carvacryl ethyl ether	spicy, herbaceous, leafy
 Estragole	sweet, herbaceous, anise like

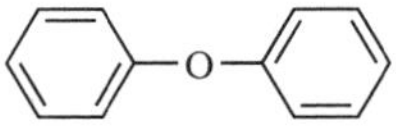

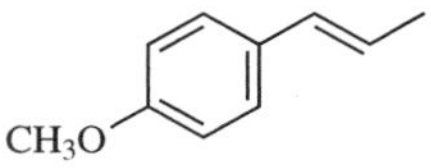

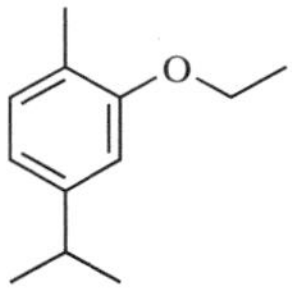

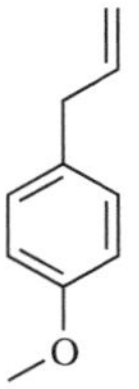

4.5. Aldehydes and Ketones
Saturated Aldehydes

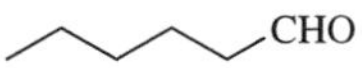

 Hexanal	aldehydic green, slightly fruity; somewhat green apple-like
 Nonanal	aldehydic, peely, floral (somewhat rosy), orange
 Decanal	soft fatty; slightly green-fruity; cream, milk, cheese-like and green melon

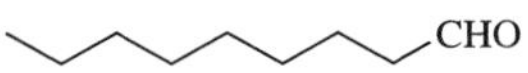

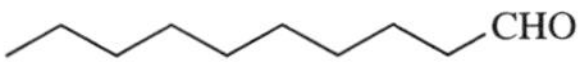

Monounsaturated Aldehydes

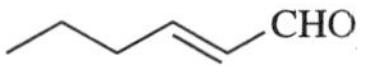 *trans*-2-Hexenal	strongly leafy green, slightly spicy, bitter almond-like
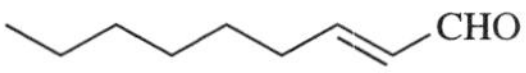 *trans*-2-Nonenal	green, soapy, cucumber/melon-like with an aldehydic fatty nuance
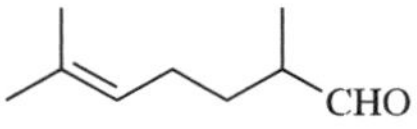 2,6-Dimethyl-5-heptenal	fresh, watery fruity (melon-like), with herbal notes.

Diunsaturated Aldehydes

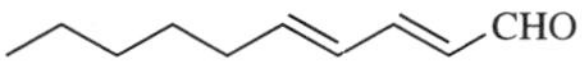 *trans,trans*-2,4-Decadienal	powerful fatty, aldehydic, somewhat citrus
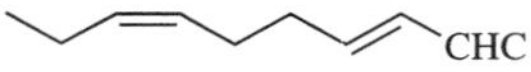 2-*trans*-6-*cis*-Nonadienal	powerful green cucumber, melon, violet leaf; aldehydic with a fresh vegetable note

Terpene Aldehydes

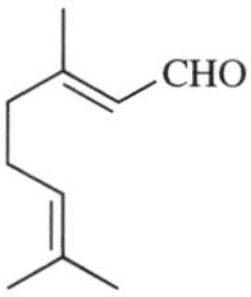 Geranial	fresh lemon-like, citrus and fruity
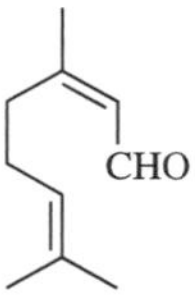 Neral	Fresh, natural, citrus, slightly fruity-herbal.
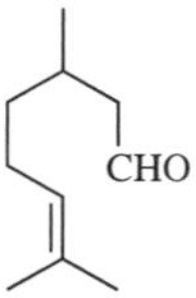 Citronellal	citrus, green, fruity, perfumistic, aldehydic, soapy.

Aldehydes Containing Benzene Ring

Heliotropin — sweet aromatic, somewhat vanilla, characteristic heliotropic

Cuminaldehyde — green, herbal, spicy; characteristic cumin

Cyclamen aldehyde — fresh, watery, floral, cyclamen-like

p-*tert*-Butyl-α-methyl dihydrocinnamic aldehyde — fresh, light, green, floral, reminiscent of lily-of-the valley; notes of muguet

Vanillin — intensive sweet, tenacious creamy, characteristic vanilla aroma

Ketones

L-Carvone — fresh, herbal; characteristic spearmint note

Nootkatone — full grapefruit character; slightly woody

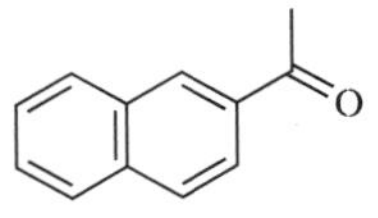

β-Methylnaphthylketone

powdery, sweet aromatic, floral; on dilution resembling neroli

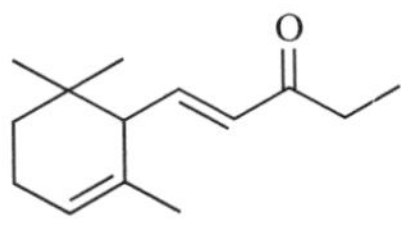

α-*n*-Methylionone

floral, woody; violet-like

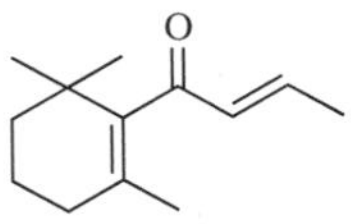

β-Damascone

fruity-floral, slightly woody, herbal; somewhat raspberry connotation

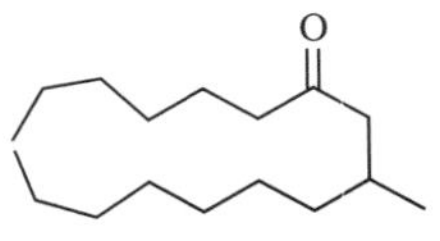

3-Methylcyclopentadecanone

natural, erogenic, animal-like musk

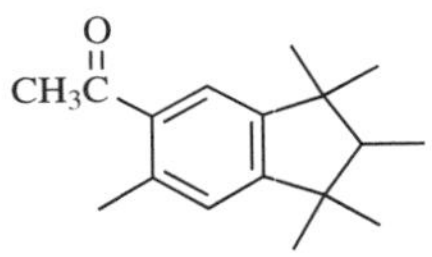

5-Acetyl-1,1,2,3,3,6-hexamethylindan

nitro-free musk compounds, herbal, and floral aspects

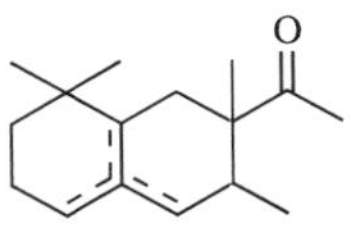

3-Acetyl-3,4,10,10-tetramethylbicyclo[4.4.0]decane

woody, amber

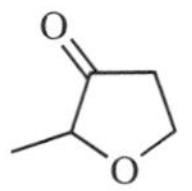

2-Methyltetrahydrofuran-3-one

breadlike, buttery top-note; nutty and astringent with a slight creamy almond nuance flavor; sweet, somewhat fruity, caramellic

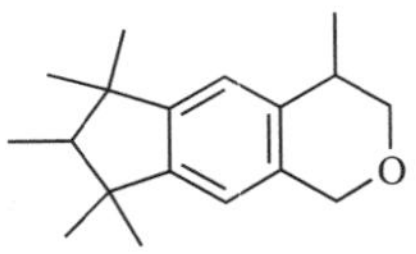

1,3,4,6,7,8-Hexahydro-4,6,6,7,8,8-hexamethylcyclopenta-(*g*)-2-benzopyran

powerful and clean musk, approaching the aspects of macrocylic musks

Last isochormanic system drawn, 1,3,4,6,7,8-hexahydro-4,6,6,7,8,8-hexa-methylcyclopenta-(*g*)-2-benzopyran, which was developed in the middle 1960 (5) by Beets and Heeringa from IFF is know also commercially as eg, Galaxolide, Abbalide. This molecule is syntehsized as following; There is a condensation–cyclization stage of *tert*-amyl alcohol and α-methyl styrene in acidic conditions to obtain the indane system, followed by a Friedel–Crafts reaction with propylene oxide to get the side chain. The side chain is finally closed to the isochromanic system using formaldehyde:

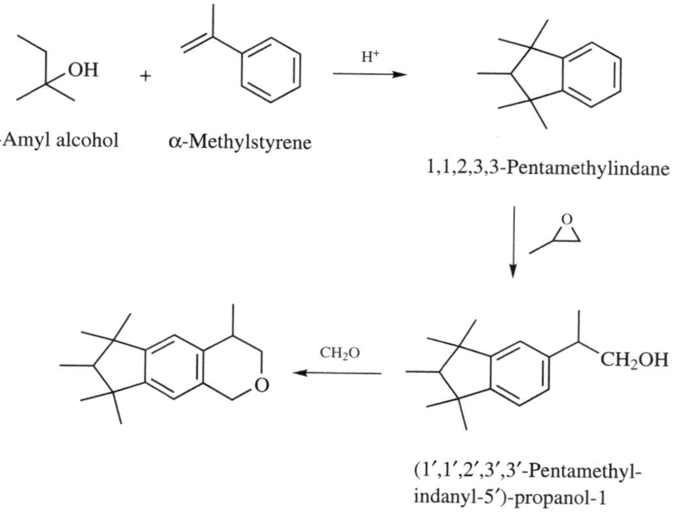

t-Amyl alcohol   α-Methylstyrene   1,1,2,3,3-Pentamethylindane

(1′,1′,2′,3′,3′-Pentamethyl-indanyl-5′)-propanol-1

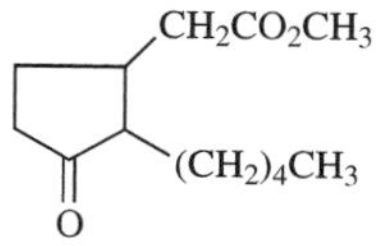

Methyl dihydrojasmonate

extremely persistent and powerful floral, fruity; characteristic of natural jasmin flower

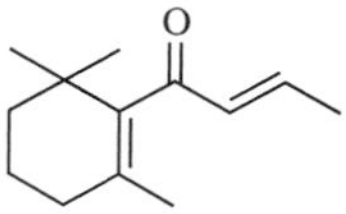

β-Damascone

fruity-floral, slightly woody, herbal; some-what raspberry connotation

Methyl dihydrojasmonate is known also commercially by the names Hedione, Claigeon.

Methyl dihydrojasmonate is synthesized by the following route (6), namely, Michael addition of diethyl malonate to the pentyl cyclopentenone to obtain the second side chain, followed by hydrolysis and decarboxylation, and finaly esterification:

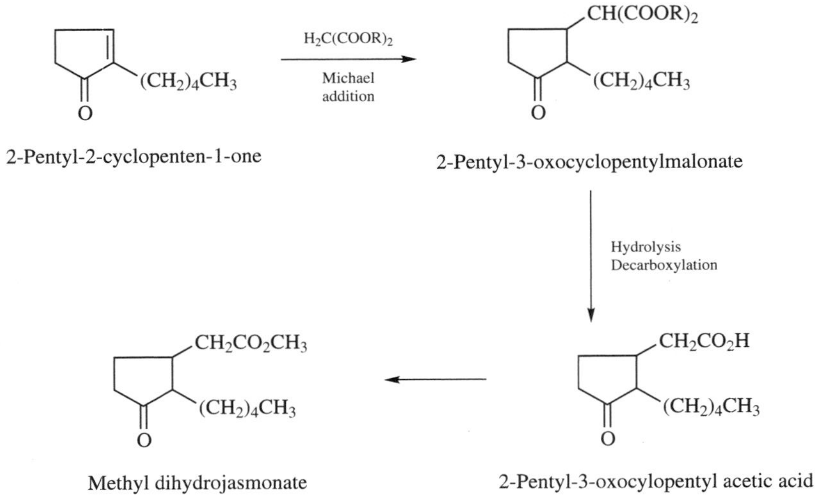

$H_2C(COOR)_2$

Michael addition

$CH(COOR)_2$

$(CH_2)_4CH_3$

2-Pentyl-2-cyclopenten-1-one

2-Pentyl-3-oxocyclopentylmalonate

Hydrolysis
Decarboxylation

$CH_2CO_2CH_3$

$(CH_2)_4CH_3$

Methyl dihydrojasmonate

CH_2CO_2H

$(CH_2)_4CH_3$

2-Pentyl-3-oxocylopentyl acetic acid

Diketones. The diketones used as aroma chemicals are mostly α-diketones.

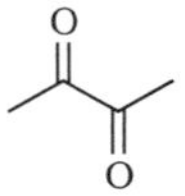 2,3-Butanedione (diacetyl)	sweet, strongly buttery, creamy, milky
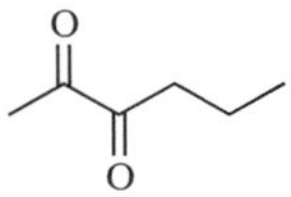 2,3-Hexanedione	creamy, sweet odor, buttery and cheesy
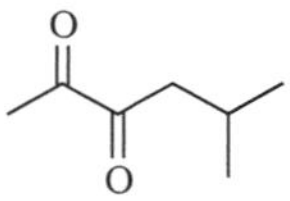 5-Methyl-2,3-hexanedione	buttery, cheesy, "oily", somewhat fruity.
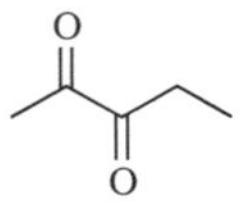 2,3-Heptanedione	"Oily", buttery, cheesy, pungent.
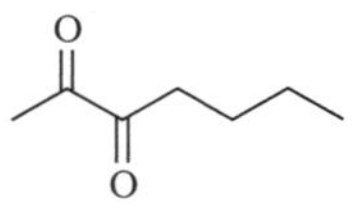 2,3-Heptanedione	strong penetrating buttery, cheesy to slightly animal. In dilution: sweet "oily" berry note
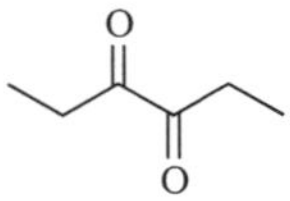 3,4-Hexanedione	burnt caramellic flavor, and aromatic, burnt, caramelic

Acetals and Ketals

dry, green-floral, fruity, citrus peel

Phenylacetaldehyde dimethylacetal

sweet aromatic, honey, brown, somewhat floral (hyacinth-like)

Phenylacetaldehyde diisobutylacetal

strongly fruity, slightly floral; apple-, pear-, and berry-like

Ethyl acetoacetate ethylene glycol ketal

4.6. Carboxylic Acids

Saturated Carboxylic Acids (see Table 3)

Table 3. **Saturated Carboxylic Acids**

Name	Organoleptic characteristics	Structure
formic acid	pungent, acidic, sour, astringent with a fruity depth	HCO_2H
acetic acid	sour, vinegar-like	CH_3CO_2H
propionic acid	sour, fruity on dilution	$CH_3CH_2CO_2H$
butyric acid	penetrating, reminiscent of rancid butter	$CH_3(CH_2)_2CO_2H$
valeric acid	strongly acidic, caprylic, cheese-like	$CH_3(CH_2)_3CO_2H$
caproic acid	acidic, caprylic, fatty	$CH_3(CH_2)_4CO_2H$
oenanthic acid	caprylic, fatty, green	$CH_3(CH_2)_5CO_2H$
caprylic acid	caprylic, fatty, oily	$CH_3(CH_2)_6CO_2H$
pelargonic acid	oily, fatty, caprylic; cheesy with a mild creamy background	$CH_3(CH_2)_7CO_2H$
capric acid	sour, fatty aroma	$CH_3(CH_2)_8CO_2H$
undecylic acid	fatty, fruity aspects	$CH_3(CH_2)_9CO_2H$
lauric acid	mild fatty	$CH_3(CH_2)_{10}CO_2H$
myristic acid	faint oily, fatty	$CH_3(CH_2)_{12}CO_2H$
palmitic acid	faint oily aroma	$CH_3(CH_2)_{14}CO_2H$
stearic acid	fatty, stearinic	$CH_3(CH_2)_{16}CO_2H$

Unsaturated Carboxylic Acids

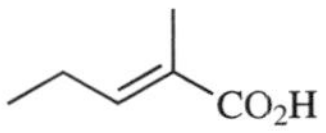

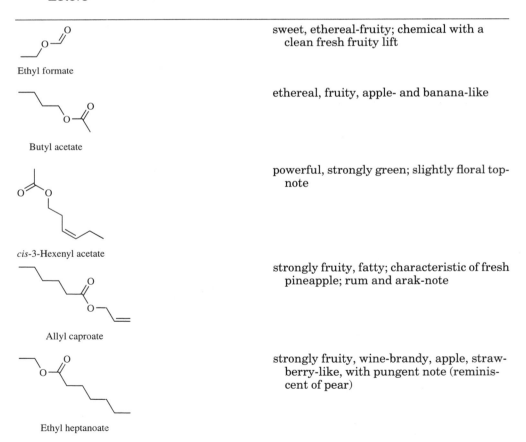

(*E*)-2-Methyl-2-pentenoic acid

acidic, fruity, somewhat cooked strawberry connotation

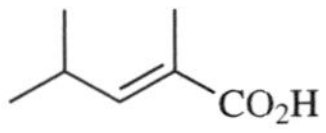

2,4-Dimethyl-2-pentenoic acid

acidic, caprylic, somewhat boiled strawberry connotations

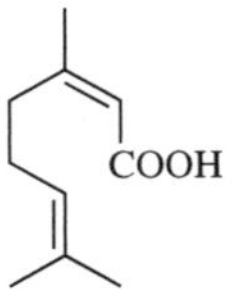

cis-Geranic acid

Green, floral, weedy, woody aroma

4.7. Carboxylic Acids Derivatives
Esters

Ethyl formate

sweet, ethereal-fruity; chemical with a clean fresh fruity lift

Butyl acetate

ethereal, fruity, apple- and banana-like

cis-3-Hexenyl acetate

powerful, strongly green; slightly floral top-note

Allyl caproate

strongly fruity, fatty; characteristic of fresh pineapple; rum and arak-note

Ethyl heptanoate

strongly fruity, wine-brandy, apple, strawberry-like, with pungent note (reminiscent of pear)

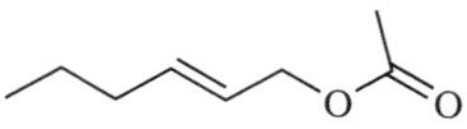

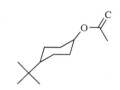

trans-2-Hexenylacetate

sweet, green, fresh, waxy and fruity; banana- and apple-like

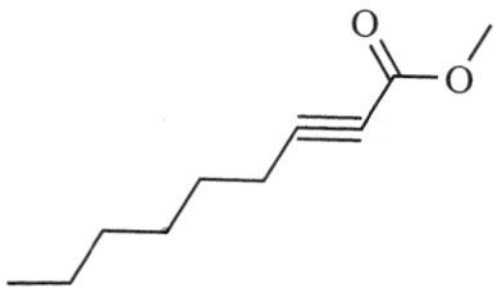

Methyl-2-nonynoate

green, violet-leaf

p-*tert*-Butylcyclohexyl acetate

sweet and rich woody, pleasant floral, with fruity note

Terpenic Esters

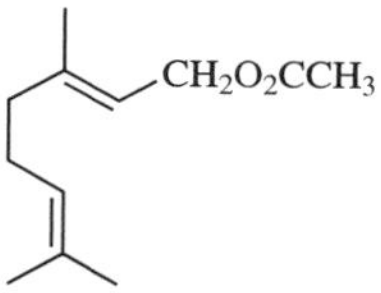

Geranyl acetate

$CH_2O_2CCH_3$

sweet fruity-floral, rose-, and lavender-like

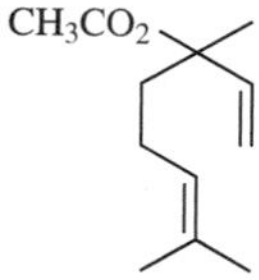

Linalyl acetate

CH_3CO_2

freshly floral; bergamot-, petitgrain-, lavender-, and cologne-like

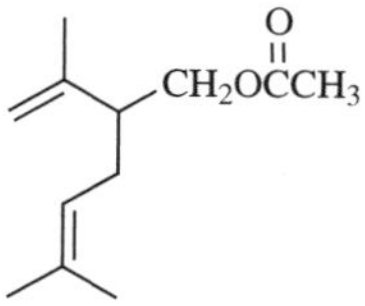

Lavandulyl acetate

CH_2OCCH_3

fresh, floral-herbal, slightly fruity; lavender-like

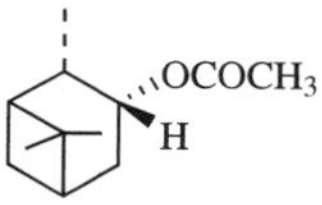

Bornyl acetate

$OCOCH_3$

natural pineneedle-like, coniferous, camphoraceous, slightly minty

Lactones

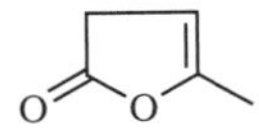

α-Angelica lactone

nutty, maple, caramel, sweet, herbaceous

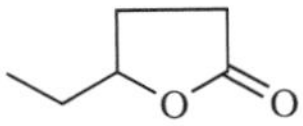

γ-Hexalactone

sweet, creamy, vanilla-like with green lactonic powdery nuances

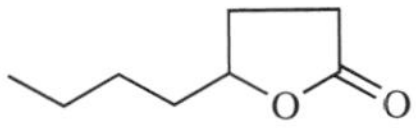

γ-Octalactone

sweet creamy with coconut character

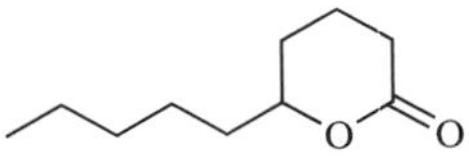

δ-Decalactone

sweet, dairy, creamy, fatty with a fruity nuance; coconut- and peach-like

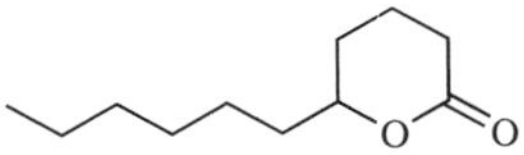

δ-Undecalactone

creamy, fatty, somewhat fruit-like, peach, coconut

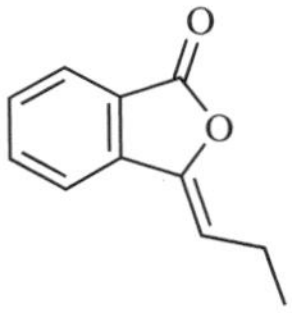

3-Propylidene phthalide

powerful, warm, spicy, strongly celery-like

Ethylene brassylate

musk-like and oil-like scent, classical macrocyclic musk with herbal connotations

Benzylic and Homobenzylic Esters

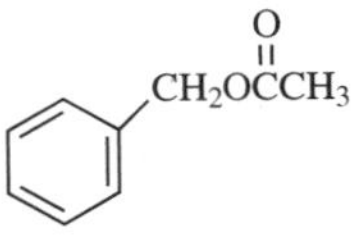

Benzyl acetate

green, dry-powdery, fruity, somewhat milky and estery

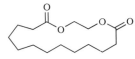

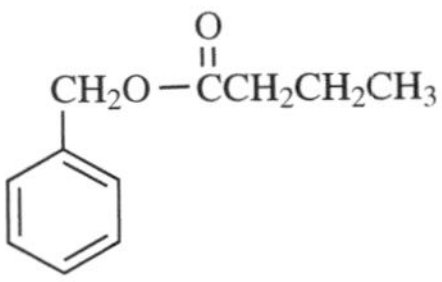

Benzyl butyrate

sweet aromatic, floral fruity, plum-like

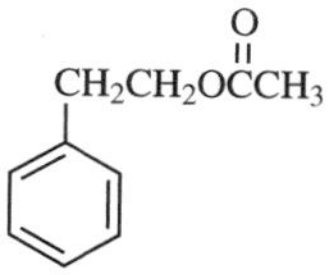

$$CH_2CH_2OCCH_3$$

honey, sweet, floral

Phenethyl acetate

Benzoate and Homobenzoate Esters

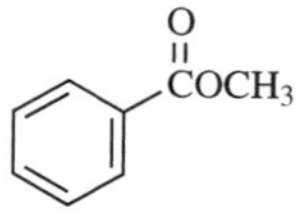

$$COCH_3$$

heavy sweet, slightly floral-fruity; berry-like

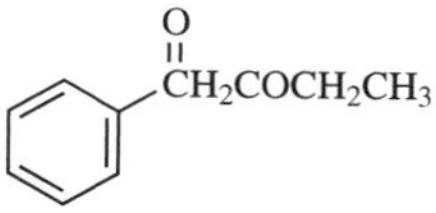

Methyl benzoate

$$CH_2COCH_2CH_3$$

sweet aromatic, honey, waxy, fruity

Ethyl phenylacetate

Cinnamate Esters

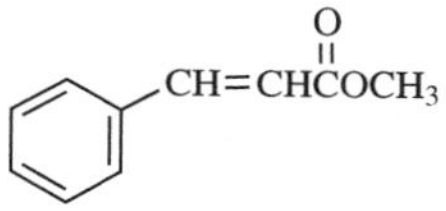

$$CH=CHCOCH_3$$

fruity, balsamic, somewhat strawberry-like

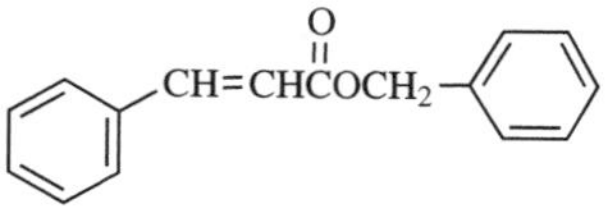

Methyl cinnamate

$$CH=CHCOCH_2-$$

sweet, floral, fruit, spicy; coumarin, balsamic, honey

Benzyl cinnamate

Salicylate Esters

OH

long lasting, green floral, leathery note

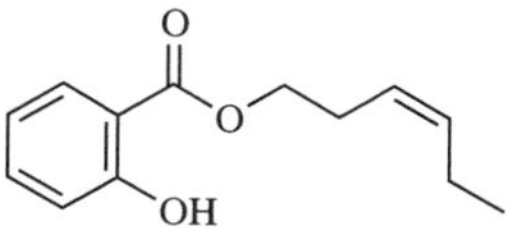

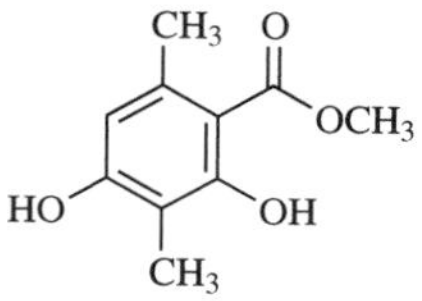

cis-3-Hexenyl-salicylate

$$CH_3 \quad O$$
$$OCH_3$$
HO   OH
$$CH_3$$

character-impact compound of oak- and treemoss; true moss-character

Methyl-3-methylorselinate

Nitriles

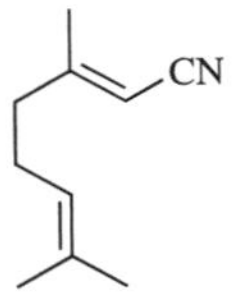

Geranyl nitrile

fresh, citrus, floral; lemon note of citral

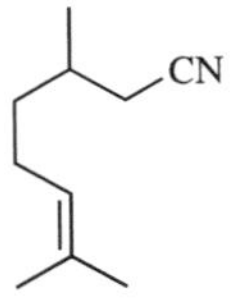

Citronellyl nitrile

fresh, lemon odor with greenish accent, citrus and herbal notes

---

Amines

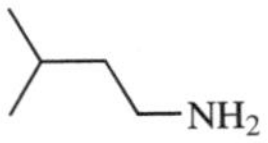

Isopentylamine

fishy, ammonia-like; in low concentration somewhat fermented

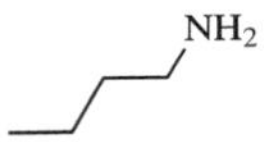

n-Butylamine

strong amine-like, fishy, on dilution slightly cheese like

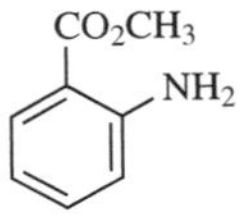

Methyl anthranilate

orangeflower-like, sweet fruity, tangerine and grape-note

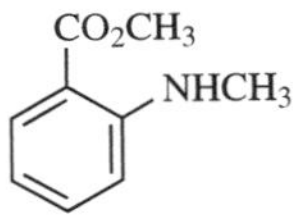

Dimethyl anthranilate

mandarin- and grape-like, tangerine note; somewhat orange-blossom

---

Nitroaromatic Compounds

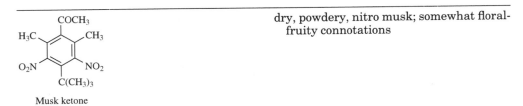

Musk ketone

dry, powdery, nitro musk; somewhat floral-fruity connotations

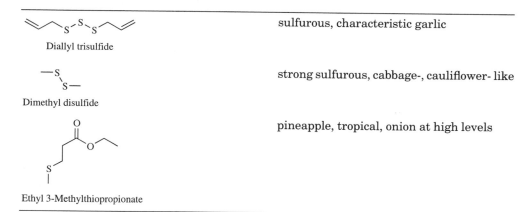

strong nitromusk, with fruity (pear-like) note

Musk ambrette

4.8. Thio Compounds

sulfurous, characteristic garlic

Diallyl trisulfide

strong sulfurous, cabbage-, cauliflower- like

Dimethyl disulfide

pineapple, tropical, onion at high levels

Ethyl 3-Methylthiopropionate

4.9. Heterocyclic Compounds

Nonaromatic Compounds Containing Oxygen, Nitrogen, or Sulfur

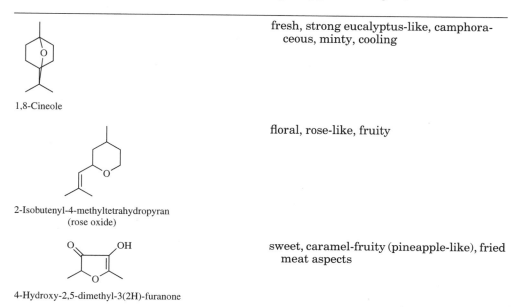

fresh, strong eucalyptus-like, camphoraceous, minty, cooling

1,8-Cineole

floral, rose-like, fruity

2-Isobutenyl-4-methyltetrahydropyran
(rose oxide)

sweet, caramel-fruity (pineapple-like), fried meat aspects

4-Hydroxy-2,5-dimethyl-3(2H)-furanone

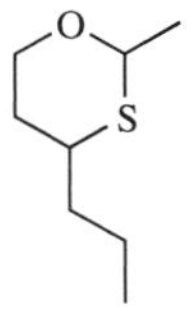

sufurous, on dilution herbal, fruity

2-Methyl-4-propyl-1,3-oxathiane

Furans

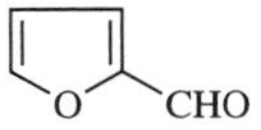

Furfural

sweet caramel-like, nutty, baked bread, almond

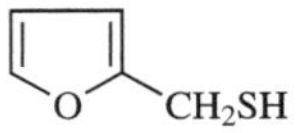

Furfuryl mercaptan

on dilution strong coffee-like

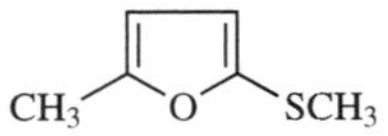

2-Methyl-5-(methylthio)furan

sulphurous, burnt, roasted (coffee-like) on dilution

Pyrrols and Indoles

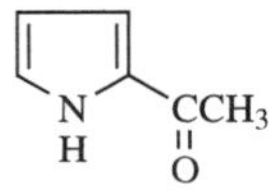

2-Acetyl pyrrole

sweet musty, nutty and tea-like

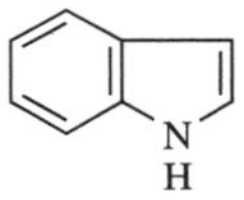

Indole

animalic, musk, cheese, slightly fecal on dilution

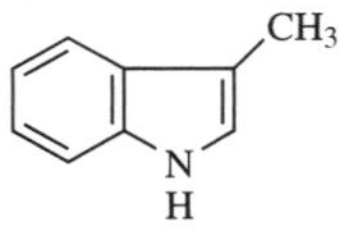

Skatole

putrid, sickening, mothballs, decayed, fecal

Pyridines and Quinolines

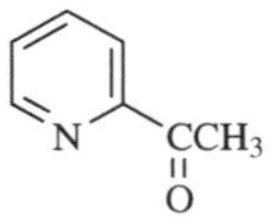

2-Acetylpyridine

heavy oily, fatty, dusty, nutty, reminiscent of hazelnut and popcorn

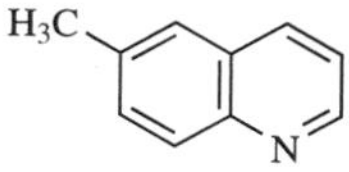

H₃C

6-Methylquinoline

narcotic, earthy, green

Pyrazines and Quinoxalines

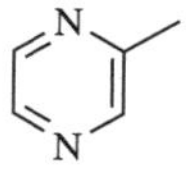

2-Methylpyrazine

musty, nutty, roasted, cocoa, peanut

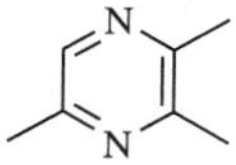

2,3,5-Trimethylpyrazine

burnt roasted, earthy, tobacco-like

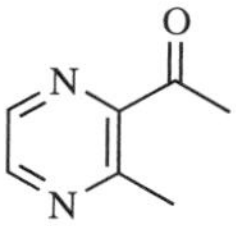

2-Acetyl-3-methylpyrazine

roasted potatoes, nutty, vegetable, and cereal

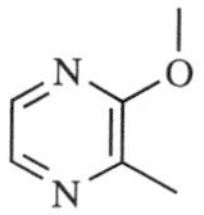

2-Methoxy-3-methylpyrazine

roasted peanuts

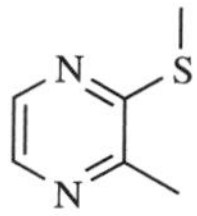

2-Methlthio-3-methyl-pyrazine

nutty, sweet, weakly green

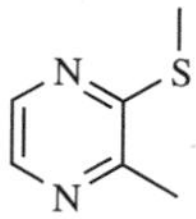

2-Methlthio-3-methyl-pyrazine

dusty, roasted

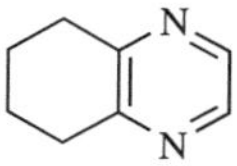

5,6,7,8-Tetrahydroquinixaline

narcotic, fishy; on dilution fried and roasted aspects

Thiazoles

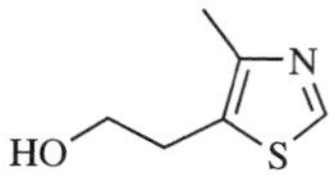

HO

5-(2-Hydroxyethyl)-4-methylthiazole

meaty, nutty

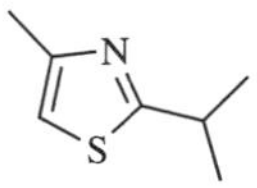

2-Isopropyl-4-methylthiazole

green, vegetable character; nut-like, fruity

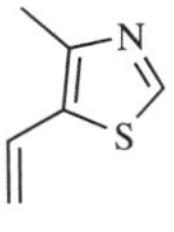

4-Methyl-5-vinylthiazole

nutty, musty, earthy, cocoa powder like

BIBLIOGRAPHY

1. http://www.leffingwell.com/index.htm
2. Flavors and Fragrances Report, SRI International, 1992
3. Most of the flavors data and descriptions mentioned in this article, are taken from "FRM - Flavour Raw Materials by BACIS" and "PMP - Fragrance Raw Materials by BACIS".
4. K. Bauer, D. Garbe, and H. Surburg, *Common Fragrance and Flavor Materials*, 3rd Revised Edn., VCH, Germany, 1997.
5. K. Bauer, D. Garbe, and H. Surbur, *Common Fragrance and Flavor Materials*, 3rd Revised Edn., VCH, Germany, 1997.
6. K. Bauer, D. Garbe, and H. Surburg, *Common Fragrance and Flavor Materials*, 3rd Revised Edn., VCH, Germany, 1997.

MICHAEL ZVIELY
Frutarom, Ltd.

ARSENIC AND ARSENIC ALLOYS

1. Introduction

Arsenic [7440-38-2], although often referred to as a metal, is classified chemically as a nonmetal or metalloid and belongs to Group 15 (VA) of the Periodic Table (as does antimony). The principal valences of arsenic are $+3$, $+5$, and -3. Only one stable isotope of arsenic having mass 75 (100% natural abundance) has been observed.

Elemental arsenic normally exists in the α-crystalline metallic form which is steel-gray in appearance and brittle in nature, and in the β-form, a dark-gray amorphous solid. Other allotropic forms, ie, yellow, pale reddish-brown to dark brown, have been reported (1), but the evidence supporting some of these allotropes is meager. Metallic arsenic, heated under ordinary conditions, does not exhibit a discrete melting point but sublimes. Molten arsenic can be obtained by heating under pressure.

2. Occurrence

Arsenic is widely distributed about the earth and has a terrestrial abundance of ~5 g/t (2). Over 150 arsenic-bearing minerals are known (1). Table 1 lists the most common minerals. The most important commercial source of arsenic, however, is as a by-product from the treatment of copper, lead, cobalt, and gold ores. The quantity of arsenic usually associated with lead and copper ores may range from a trace to 2–3%, whereas the gold ores found in Sweden contain 7–11% arsenic. Small quantities of elemental arsenic have been found in a number of localities.

World resources of copper and lead contain ~11 million tons of arsenic. Substantial resources of arsenic occur in copper ores in northern Peru and the Philippines and in copper–gold ores in Chile. World gold resources, particularly in Canada, contain substantial resources of arsenic. World reserves and reserve base are thought to be ~20 and 30 times, respectively, annual world production (see Table 2). U.S. reserve base in estimated to be ~80,000 tons (3).

Table 1. **Naturally Occurring Arsenic-Bearing Minerals**

Name	CAS Registry Number	Formula	Name	CAS Registry Number	Formula
loellingite	[12255-65-1]	$FeAs_2$	sperrylite	[12255-87-7]	$PtAs_2$
safforlite	[12044-43-8]	$CoAs_2$	arsenopyrite (mispickel)	[1303-18-0]	FeAsS
niccolite	[1303-13-5]	NiAs	cobaltite	[1303-15-7]	CoAsS
rammels bergite	[1303-22-6]	$NiAs_2$	enargite	[14933-50-7]	Cu_3AsS_4
realgar	[12044-30-3]	AsS	gersdorffite	[12255-11-7]	NiAsS
orpiment	[12255-89-9]	As_2S_3	glaucodot	[12198-14-0]	(Co,Fe)AsS

Table 2. **World Production, Reserves, and Reserve Base, 10^3 t[a]**

Country	Production[b]	
	2000	2001[c]
Belgium	1,500	1,500
Chile	8,200	8,000
China	16,000	16,000
France	1,000	1,000
Kazakhstan	1,500	2,000
Mexico	2,400	2,600
Russia	1,500	1,500
other countries	1,800	2,000
World total (may be rounded)	33,900	35,000

[a] Ref. 3.
[b] As arsenic trioxide.
[c] Estimated.

3. Properties

Physical properties of α-crystalline metallic arsenic are given in Table 3. The properties of β-arsenic are not completely defined. The density of β-arsenic is 4700 kg/m^3; it transforms from the amorphous to the crystalline form at 280°C; and the electrical resistivity is reported to be 107 Ω cm.

Metallic arsenic is stable in dry air, but when exposed to humid air the surface oxidizes, giving a superficial golden bronze tarnish that turns black upon further exposure. The amorphous form is more stable to atmospheric oxidation. Upon heating in air, both forms sublime and the vapor oxidizes to arsenic trioxide [1327-53-3], As_2O_3. Although As_4O_6 represents its crystalline makeup, the oxide is more commonly referred to as arsenic trioxide. A persistent garliclike odor is noted during oxidation.

Elemental arsenic combines with many metals to form arsenides. When heated in the presence of halogens it forms trihalides; however, pentahalides with the exception of AsF_5 (4) and the unstable $AsCl_5$ are not readily formed. It reacts with sulfur to form the compounds As_2S_3, AsS, As_2S_5, and complex mixtures in various proportions (see ARSENIC COMPOUNDS).

Arsenic vapor [12187-88-5], As_4, does not combine directly with hydrogen to form hydrides. However, arsine (arsenic hydride) [7784-42-1], AsH_3, a highly poisonous gas, forms if an intermetallic compound such as AlAs is hydrolyzed or treated with HCl. Arsine may also be formed when arsenic compounds are reduced using zinc in hydrochloric acid. Heating to 250°C decomposes arsine into its elements.

Metallic arsenic is not readily attacked by water, alkaline solutions, or nonoxidizing acids. It reacts with concentrated nitric acid to form orthoarsenic acid

Table 3. **Physical Properties of Arsenic**

Property	Value
atomic weight	74.9216
mp at 39.1 MPa^a, °C	816
bp, °C	615^b
density at 26°C, kg/m^3	5,778
latent heat of fusion, J/(mol · K)^c	27,740
latent heat of sublimation, J/(mol · K)^c	31,974
specific heat at 25°C, J/(mol · K)^c	24.6
linear coefficient of thermal expansion at 20°C, μm/(m °C)	5.6
electrical resistivity at 0°C, μ Ω cm	26
magnetic susceptibility at 20°C	$-5.5 \times 10^{-6\,d}$
nuclear absorption cross section^e	4.3 ± 0.2
crystal system	hexagonal (rhombohedral)
lattice constants at 26°C, nm	$a = 0.376$, $c = 1.0548$
hardness, Mohs' scale	3.5

^{a} To convert MPa to psi, multiply by 145.

^{b} Sublimes.

^{c} To convert to cal/(mol · K), divide by 4.184.

^{d} From cgs system units.

^{e} Thermal neutrons 2200 m/s of arsenic mass 75.

[7778-39-4], H_3AsO_4. Hydrochloric acid attacks arsenic only in the presence of an oxidant.

Arsenic may be detected qualitatively as a yellow sulfide, As_2S_3, by precipitation from a strongly acidic HCl solution. Other members of this group that are normally precipitated with hydrogen sulfide do not interfere if the solution contains 25% or more hydrochloric acid. Trace quantities of arsenic may be detected by first converting to arsine (5). The arsine is decomposed by heating the gas in a small tube and an arsenic mirror is formed (Marsh test). Alternatively, the arsine may be allowed to react with test paper impregnated with mercuric chloride (Gutzeit test).

4. Metallurgy

Metallic arsenic can be obtained by the direct smelting of the minerals arseno-pyrite or loellingite. The arsenic vapor is sublimed when these minerals are heated to about 650–700°C in the absence of air. The metal can also be prepared commercially by the reduction of arsenic trioxide with charcoal. The oxide and charcoal are mixed and placed into a horizontal steel retort jacketed with fire-brick that is then gas-fired. The reduced arsenic vapor is collected in a water-cooled condenser (6). In a process used by Boliden Aktiebolag (7), the steel retort, heated to 700–800°C in an electric furnace, is equipped with a demountable air-cooled condenser. The off-gases are cleaned in a scrubber system. The yield of metallic arsenic from the reduction of arsenic trioxide with carbon and carbon monoxide has been studied (8) and a process has been patented describing the gaseous reduction of arsenic trioxide to metal (9).

The demand for metallic arsenic is limited and thus arsenic is usually marketed in the form of the trioxide, referred to as white arsenic, arsenious oxide, arsenious acid anhydride, and also by the generally accepted misnomer arsenic.

Arsenic trioxide was recovered from smelting or roasting of nonferrous metal or as concentrates in at least 16 countries in 2000.

Arsenic trioxide is readily volatilized during the smelting of copper and lead concentrates, and is therefore concentrated with the flue dust. Crude flue dust may contain up to 30% arsenic trioxide, the balance being oxides of copper or lead, and perhaps of other metals such as antimony and zinc. This crude flue dust is further upgraded by mixing with a small amount of pyrite or galena concentrate and roasting (10). The pyrite or galena is added to prevent arsenites from forming during roasting and to obtain a clinkered residue which can be returned for additional processing. The gases and vapors are passed through a cooling flue which consists of a series of brick chambers or rooms called kitchens. The temperature of the gas and vapor is controlled so that they enter the first kitchen at 220°C and by the time the gas and vapor reach the last kitchen they are cooled to 100°C or less. The arsenic trioxide vapor that condenses in these chambers is of varying purity analyzing from 90–95%. A higher purity product is obtained by resubliming the crude trioxide, an operation normally carried out in a reverberatory furnace. The vapors pass first through a settling chamber and then through ~39 kitchens that cover a length of ~68.6 m. The temperature of the settling chamber is kept at ~295°C, which is above the condensation temperature of the trioxide. A black, amorphous mass containing about 95% As_2O_3 condenses in the kitchens nearest the furnace and is reprocessed. The bulk of the trioxide is condensed in the kitchens having temperature ranges of 180–120°C giving a product of 99–99.9% purity. The dust which exits from the kitchens at a temperature of 90–100°C is caught in the baghouse. It assays ~90% As_2O_3 and may be sold as a crude grade or reprocessed. (A typical flow sheet for the production of arsenic trioxide at a copper smelter is depicted in Fig. 1.)

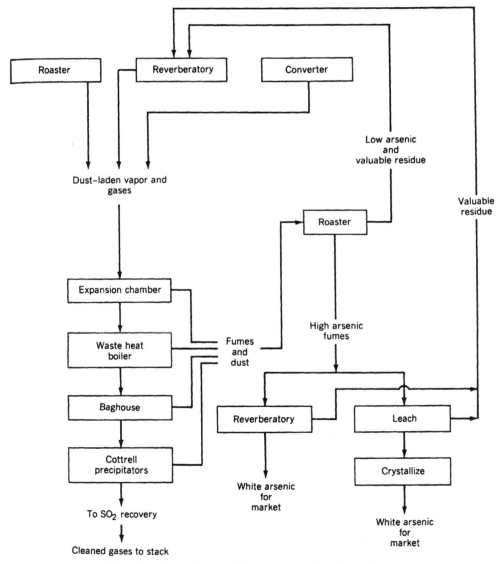

Fig. 1. Arsenic trioxide copper smelter-flow sheet.

In addition to pyro-refining, crude oxide analyzing 80–95% can be refined by a wet process (7). In this process, advantage is taken of the fact that the solubility of arsenic trioxide in water increases with temperature as follows:

Temperature, °C	Solubility, g/L
0	12.1
20	18.1
98.5	81.8

Most of the common impurities present in crude arsenic have a low solubility.

The crude oxide is pressure-leached in a steam-heated autoclave using water or circulating mother liquor. The arsenic trioxide dissolves, leaving behind a residue containing a high concentration of heavy metal impurities and silica. The solution is vacuum-cooled and the crystallization is controlled so that a coarse oxide is obtained that is removed by centrifuging. The mother liquor is recycled. The oxide (at least 99% purity) is dried and packaged in a closed system.

The refined arsenic trioxide is analyzed for purity and also tested for solubility, a term referring to its rate of reactivity with nitric acid; this test is important if the arsenic is used in the manufacture of insecticides and herbicides. The product is graded for marketing as white soluble having 99% min As_2O_3, white insoluble, or crude having 95% min As_2O_3.

Minor quantities of arsenic trioxide have been obtained from the roasting of arsenopyrite, but the presence of copious amounts of SO_2 in the gas and vapor stream requires the use of lead-lined kitchens (10).

5. Economic Aspects

The United States had no domestic production of arsenic in 2000 and needed to rely on imports for its needs. The United States did import some arsenic metal, but most of its imports were in the form of arsenic trioxide. China remains the principal supplier to the United States. See Table 4 for U.S. import data (11).

The estimated value of arsenic consumed by the United States during 2001 was ~\$20 million (3).

Economic statistics for the United States are given in Table 5.

5.1. Prices. During the 1970s demand for arsenic was growing. During the late 1970s, regulations related to exposure and emissions were adopted. Arsenic prices peaked in 1980 at \$6612/t (\$3.00/lb). In 2001, it was \$110.2/t (\$0.50/lb) (3,12).

Table 4. **U.S. Imports for Consumption of Arsenicals**[a,b]

Class and country	1999 Quantity metric tons	1999 Value $\times 10^3$ $	2000 Quantity metric tons	2000 Value $\times 10^3$ $
arsenic trioxide				
Belgium	724	429	576	356
Bolivia	280	159	212	118
Canada			1	2
Chile	8,870	3,340	9,110	3,620
China	15,500	8,380	15,400	7,800
France	1,410	862	1,340	871
Germany	3	34	4	15
Hong Kong	70	42	406	232
Mexico	1,680	1,090	1,900	1,330
Morocco	148	90	2,180	1,150
Spain				
Switzerland				
Vietnam	437	251		
Total	*29,100*	*14,700*	*31,100*	*15,500*
arsenic acid	4	24		
arsenic acid, arsenic metal				
France				
China	1,240	1,600	612	694
Germany	16	3,210	21	3,410
Hong Kong			41	36
Japan	45	3,580	157	5,660
Taiwan				
United Kingdom	c	7	c	2
Total	*1,300*	*8,390*	*830*	*9,800*

[a,b] Ref. 11. Data are rounded to no more than three significant digits; may not add to totals shown.
[c] Less than 1/2 unit.

Table 5. **United States Economic Statistics for Arsenic**[a]

Statistic	t^b 1993	t^b 1998	Year 1999	Year 2000	Year 2001^b
production					
imports for consumption:					
metal, t	909	997	1,300	830	1,200
compounds, t	22,800	29,300	22,100	23,600	24,000
exports, metal, t	61	177	1,350	41	60
estimated consumption, t^b	23,700	30,100	22,000	24,000	25,000
value, cents per pound, average[c]					
metal (China)	32	57	59	51	50
trioxide (Mexico)	31	32	29	32	31
net import reliance[d] as a percentage of apparent consumption	100	100	100	100	100

[a] Ref. 3.
[b] Estimated to be the same as net imports.
[c] From U.S. Census Bureau import data. To convert to $/t, multiply by 2.204.62.
[d] Defined as imports—exports + adjustments for government and industry stock changes.

6. Recycling

Arsenic was not recovered from consumer end-product scrap. However, process water and contaminated runoff collected at wood treatment plants were reused in pressure treatment, and gallium arsenide scrap from the manufacture of semiconductor devices was reprocessed for gallium and arsenic recovery. In the United States, no arsenic was recovered from arsenical residues and dusts at nonferrous smelters, although some of these materials are processed for recovery of other metals (3).

7. Health and Safety Factors

The toxicity of arsenic ranges from very low to extremely high depending on the chemical state. Metallic arsenic and arsenious sulfide [1303-33-9], As_2S_3, have low toxicity. Arsine is extremely toxic. The toxicity of other organic and inorganic arsenic compounds varies (13).

Arsenic is classified as a carcinogen by the International Agency for Research on Cancer (IARC) (14). An association between high and lengthy exposures to inorganic arsenic compounds and cancer has been reported (15), but evidence supporting this relationship is equivocal (16). Ulceration and perforation of the nasal septum is caused by airborne As_2O_3 if proper precautions are not observed. However, these injuries have not been associated with malignancy (17).

The handling of arsenic in the workplace should be in compliance with the Occupational Safety and Health Administration (OSHA) regulations: the maximum permissible exposure limit for arsenic in the workplace is 10 $\mu g/m^3$ of air as determined as an average over an 8-h period (18).

The National Institute for Occupational Safety and Health (NIOSH) has established an exposure level of 0.002 mg/m^3 as a recommended ceiling (15 min) for inorganic arsenic; no level for organic arsenic and an REL of 0.002 mg/m^3 as a 15-min celling concentration (19).

The American Conference of Governmental Industrial Hygienists (ACGIH) has established air TLV for arsenic, elemental and inorganic compounds (except arsine), at 0.01 mg/m^3).

Precaution should be taken to avoid accidental generation of arsine gas; the maximum permitted exposure is 0.05 ppm in air per 8-h period five days per week (20). Disposal of arsenical products should be in compliance with federal and local government environmental regulations.

8. Environmental Concerns

The location and extent of arsenic in ground water was the subject of a U.S. Geological Survey study in 2000 (21). The presence of arsenic in ground water is due largely to minerals dissolving. Data on 19,000 samples of potable water showed that the arsenic concentration was lower that the 50 μ/L, which was the EPA standard at that time. Ten percent of the samples exceeded to 10 μ/L, which is the World Health Organization standard.

The United States EPA has adopted a new standard for drinking water of 10 μ/L based on a study by the National Academy of Sciences (http://www.epa.gov/safewater/arsenic/html).

For a detailed discussion of this topic see ARSENIC, ENVIRONMENTAL IMPACT, HEALTH EFFECTS, AND TREATMENT METHODS.

9. Uses

Table 6 gives the United States demand patterns for specific end uses.

The use of many arsenical chemicals are subject to registration and must comply with federal and local government environmental regulations.

9.1. Wood Preservative. The largest use for arsenic (as arsenic trioxide) is in the production of wood preservatives. The demand will probably continue to correlate closely with housing construction, renovations, and replacements of existing structures using pressure-treated lumber (3,22,23). In 2000, the three principal U.S. producers of arsenical wood preservatives wer Hudson Corp. (Smyrna, Ga.), Chemical Specialties, Inc. (Harrisburg, NC) Osmose Wood Preserving, Inc. (Buffalo, N.Y.) (11).

9.2. Semiconductor Applications. A limited but important demand for metallic arsenic of 99.99% and greater (exceeding 99.999 + %) purities exists in semiconductor applications (see SEMICONDUCTORS). In 2001, are estimated 30 t/process used in the United States. This high purity arsenic may be prepared by the reduction of a highly purified arsenic compound using a high purity gaseous or solid reductant.

High purity (HP) arsenic, when alloyed with aluminum, gallium, and indium, form the III–V semiconductor compounds, aluminum arsenide [22831-42-1], AlAs, gallium arsenide [1303-00-0], GaAs, and indium arsenide [1303-11-3], InAs, respectively. These compounds or variations such as GaAlAs, GaAsP, and InGaAs, are used in device manufacture. Gallium aluminum arsenide, GaAlAs, is used in the manufacture of solar cells having efficiencies exceeding 20% (see SOLAR ENERGY). $GaAs_xP_y$ is used in the manufacture of light emitting diodes (LEDs) of red light; yellow LEDs are produced by increasing the phosphorus content (see LIGHT GENERATION, LIGHT EMITTING DIODES). GaAs infrared (ir) emitters and ir detectors have use in fiber optic applications (see FIBER

Table 6. **U.S. Arsenic Demand Pattern, metric tons, arsenic content**[a,b]

Use	1996	1997	1998	1999	2000
agricultural chemicals	950	1,400	1,500	1,100	1,000
glass	700	700	700	700	700
wood preservatives	19,200	20,000	27,000	19,000	21,000
nonferrous alloys and electronics	250	900	1,000	1,300	800
other	300	300	300	300	300
total	*21,400*	*23,700*	*30,100*	*22,000*	*24,000*

[a,b] Ref. 11. Data are rounded to no more than three significant digits; may not add to totals shown.

OPTICS). GaAs is also used in the manufacture of microwave devices (see MICRO-WAVE TECHNOLOGY), integrated circuits (qv), lasers (qv), laser windows, and optoelectronic devices. Indium arsenide has been used to produce Hall effect and infrared devices (see INFRARED AND RAMAN SPECTROSCOPY; MAGNETOHYDRODYNAMICS). InGaAs is used as lasers and photodetectors (qv). GaAs is also used in the manufacture of microwave devices (see MICROWAVE TECHNOLOGY), integrated circuits (qv), lasers (qv), laser windows, and optoelectronic devices. Indium arsenide has been used to produce Hall effect and infrared devices (see INFRARED AND RAMAN SPECTROSCOPY; MAGNETOHYDRODYNAMICS). InGaAs is used as lasers and photodetectors (qv).

HP arsenic is used in the manufacture of photoreceptor arsenic-selenium alloys for xerographic plain paper copiers (see ELECTROPHOTOGRAPHY). The level of arsenic may be 0.5, 5.0, or 35% present as arsenic triselenide [1303-36-2], As_2Se_3.

Arsenic from the decomposition of high purity arsine gas may be used to produce epitaxial layers of III–V compounds, such as InAs, GaAs, AlAs, etc, and as an n-type dopant in the production of germanium and silicon semiconductor devices. A group of low melting glasses based on the use of high purity arsenic (27–30) were developed for semiconductor and ir applications.

9.3. Other. Other uses for arsenic metal are as an additive to improve corrosion resistance and tensile strength in copper alloys and as a minor additive (0.01–0.5% 0 to increase strength of posts and grids in lead storage batteries (11). Arsenic acid is used by the glass industry as a fining agent to disperse air bubbles (11,24,25). Arsenic is also used in some herbicides (11,26).

10. Alloys

Arsenic metal is used primarily in alloys in combination with lead and, to a lesser extent, copper.

Trace quantities of arsenic are added to lead-antimony grid alloys used in lead–acid batteries (31) (see BATTERIES, LEAD ACID). The addition of arsenic permits the use of a lower antimony content, thus minimizing the self-discharging characteristics of the batteries that result from higher antimony concentrations.

No significant loss in hardness and casting characteristics of the grid alloy is observed (32,33).

Arsenic added in amounts of 0.1–3% improves the properties of lead-base babbitt alloys used for bearings (see BEARING MATERIALS). Arsenic (up to 0.75%), has been added to type metal to increase hardness and castability (34). Addition of arsenic (0.1%) produces a desirable fine-grain effect in electrotype metal without appreciably affecting the hardness or ductility. Arsenic (0.5–2%) improves the sphericity of lead ammunition. Automotive body solder of the composition 92% Pb, 5.0% Sb, and 2.5% Sn, contains 0.50% arsenic (see SOLDERS AND BRAZING ALLOYS).

Minor additions of arsenic (0.02–0.5%) to copper (qv) and copper alloys (qv) raise the recrystallization temperature and improve corrosion resistance. In some brass alloys, small amounts of arsenic inhibit dezincification (35), and minimize season cracking.

Table 7. **Arsenical Copper Alloys**

Alloy number	Composition, wt %						
	Cu	As	Pb	Fe	Sn	P	Al
142	99.4	0.015–0.50				0.015–0.040	
366[a]	58–61	0.02–0.10	0.40–0.9	0.15	0.25		
443[a]	70–73	0.02–0.10	0.07	0.06	0.9–1.2		
465[a]	59–62	0.02–0.10	0.20	0.10	0.5–1.0		
687[a]	76–79	0.02–0.10	0.07	0.06			1.8–2.5

[a] Zinc constitutes the remainder of composition.

Phosphorized deoxidized arsenical copper (alloy 142 (36)) is used for heat exchangers and condenser tubes. Copper-arsenical leaded Muntz metal (alloy 366), Admiralty brass (alloy 443), naval brass (alloy 465), and aluminum brass (alloy 687), all find use in condensers, evaporators, ferrules, and heat exchanger and distillation tubes. The composition of these alloys is listed in Table 7.

BIBLIOGRAPHY

"Arsenic" in *ECT* 1st ed., Vol. 2, pp. 113–118, by G. A. Roush, "Mineral Industry, Arsenic"; in *ECT* 2nd ed., Vol. 2, pp. 711–717, by S. C. Carapella, Jr., American Smelting and Refining Company; in *ECT* 3rd ed., Vol. 3, pp. 243–250, by S. C. Carapella, Jr., American Smelting and Refining Company; in *ECT* 4th ed., Vol. 3, pp. 624–633, by S. C. Carapella, Jr., Consultant; "Arsenic and Arsenic Alloys" in *ECT* (online), posting date: December 4, 2000, by S. C. Carapella, Consultant.

CITED PUBLICATIONS

1. J. W. Mellor, *Comprehensive Treatise on Inorganic and Theoretical Chemistry*, Vol. 9, Longmans, Green & Co., Inc., New York, 1930, 3–9, 16–19.
2. *American Institute of Physics Handbook*, McGraw-Hill Book Co., Inc., New York, 1957, Sect. 7, Chapt. 9.
3. R. G. Reese, Jr., Arsenic *Mineral Commodity Summaries*, U.S. Geological Survey, Reston, Va., Jan. 2002.
4. M. C. Sneed and R. C. Brasted, *Comprehensive Inorganic Chemistry*, Vol. 5, D. Van Nostrand Co., Inc., Princeton, N.J., 1956, p. 135.
5. N. H. Furman, ed., *Scott's* Standard Methods of Chemical Analysis, Vol. 1, 6th ed., D. Van Nostrand Co., Inc., Princeton, N.J., 1962, 106–137.
6. C. H. Jones, *Chem. Met. Eng.* **23**, 957 (1920).
7. S. Wallden and H. Hilmer, *Ulmanns Encyclopadie der Technischen Chemie*, Verlag Chemie, GmbH, Weinheim, Germany, 1974, 53–55.
8. R. C. Vickery and R. W. Edwards, *Metallurgia* **36**, 3 (1947).
9. U.S. Pat. 3,567,370 (March 2, 1971), S. T. Henriksson (to Boliden Akliebolag).
10. W. C. Smith in D. M. Liddell, ed., *Handbook of Non Ferrous Metallurgy*, Vol. 2, McGraw-Hill Book Co., Inc., New York, 1945, pp. 94–103.
11. R. G. Reese, Jr., Arsenic *Minerals Yearbook 2000*, U.S. Geological Survey, Reston, Va., Jan. 2002.

12. R. G. Reese, Jr., Arsenic *Year End Metal Prices*, U.S. Geological Survey, Reston, Va., Jan. 1998.

13. *Occupational Exposure to Inorganic Arsenic*, U.S. Dept. of HEW National Institute of Safety and Health, Washington, D.C., 1973.

14. IARC Monographs on the Evaluation of the Carcinogenic Risks of Chemicals to Humans 23, *Some Metals and Metallic Compounds*, IARC, Lyon, France, 1980.

15. *Fed. Regist.* **40**(14), OSHA, Department of Labor, Jan. 21, 1975.

16. R. J. Bauer, in W. H. Lederer and R. J. Fensterheim, eds., *Arsenic: Industrial, Biomedical, Environmental Perspectives*, Van Nostrand Reinhold Co., New York, 1983, 45–154, 166–169, 203–209, 245–254.

17. P. Drinker and T. Hatch, *Industrial Dust*, 2nd ed., McGraw-Hill Book Co., Inc., New York, 1954.

18. *Code of Federal Regulations*, Title 29, Part 1910.1018, U.S. Food and Drug Administration, Washington, D.C., revised July 11, 1988.

19. L. Gallicchio, B. A. Fowler, and E. F. Madden, in E. Bingham, B. Cohrssen, and C. H. Powell, eds., *Patty's Toxicology 5th ed.*, Vol. 2, John Wiley & Sons, Inc., New York, 2001, pp. 747–770.

20. *Code of Federal Regulations*, Title 29, Part 1910, OSHA, Washington, D.C., May 21, 1971.

21. M. Focazio and co-workers, A Retrospective Analysis of the Occurrence of Arsenic in Ground-water Resources of the United States and Limitations in the Driniking Water Supply Characterizations, *U.S. Geological Survey Water-resources Investigations Rept. 99-4279*, U.S. Geological Survey, Reston, Va., 2000.

22. D. D. Nicholas, ed., *Preservative and Preservative Systems*, Vol. II, Syracuse University Press, Syracuse, N.Y., 1973, pp. 66–84.

23. W. J. Baldwin in Ref. 16, pp. 99–110.

24. S. deLajarte, *Arsenic in Glass*, Arsenic Development Committee, Rue LaFayette, Paris, France, Mar. 1969.

25. Ref. 16, pp. 45–55.

26. U.S. Pat. 3,130,035 (Apr. 21, 1964), W. H. Culver (to Pennsalt Chemicals Corp.).

27. S. Flaschen, D. Pearson, and W. Northover, *J. Am. Ceram. Soc.* **43**, 274 (1960); *J. Appl. Phys.* **31**, 219 (1960).

28. U.S. Pat. 2,883,292; 2,883,295 (Apr. 25, 1959) and U.S. Pat. 3,241,986 (Mar. 22, 1966), (to Servo Corp.).

29. U.S. Pat. 3,154,424 (Oct. 27, 1964), L. Bailey and co-workers (to Texas Instruments).

30. R. Hilton, *Appl. Optics* **5**, 1877 (1966).

31. W. Hofmann, *Lead and Lead Alloys*, English translation by Lead Development Association, Springer-Verlag, Berlin, Germany, 1970, pp. 349–356.

32. U.S. Pat. 3,801,310 (Apr. 2, 1974), S. Nijhawan (to Varta Aktiegellschaft).

33. Ger. Offens. 2,312,322; 2,412,320; 2,412,321 (Sept. 19, 1974), K. Peters (to Electric Power Storage Ltd.).

34. *Metals Handbook*, 8th ed., American Society for Metals, Ohio, 1961, 1061–1062.

35. E. E. Langenegger and F. Robinson, *Corrosion* **25**, 137 (1969).

Copper and Copper Alloy Data Permanent File No. 1, Copper Development Association, London, England, 1968.

GENERAL REFERENCES

W. H. Lederer and R. J. Fensterheim, eds., *Arsenic: Industrial, Biomedical, Environmental Perspectives*, Van Nostrand Reinhold Co., Inc., N.Y., 1983.

R. Reddy, ed., *Arsenic Metallurgy-Fundamental and Applications*, The Metallurgical Society (AIME), Warrendale, Pa., 1988.

H. Quiring, *Die Metallischen Rohstoffe-Arsen*, Vol. 8, Ferdinand Enke, Stuttgart, Germany, 1946.

H. C. Beard, *The Radiochemistry of Arsenic*, NAS-NS 3002, U.S. Atomic Energy Commission, Washington D.C., Jan. 1960.

O. Herneryd, O. Sundstrom, and A. Norro, *J. Metals* **6**(3), 330 (Mar. 1954).

S. C. CARAPELLA, JR.
Consultant

ARSENIC— ENVIRONMENTAL IMPACT, HEALTH EFFECTS, AND TREATMENT METHODS

1. Introduction

Arsenic, a cancer causing substance, is present in a variety of forms in soil, water, air, and food. As a naturally occurring element in the earth's crust, arsenic enters into aquifers and wells through natural activities, and to the water cycle as a result of anthropogenic activities. The four arsenic species commonly reported are arsenite [As(III)], arsenate [As(V)], monomethyl arsenic acid (MMA), and dimethyl arsenic acid (DMA). It is generally known that As(III) is more toxic than As(V) and inorganic arsenicals are more toxic than organic derivatives. In oxygen-rich environments, where aerobic conditions persist, arsenate [As(V)] is prevalent and exists as a monovalent ($H_2AsO_4^-$) or divalent ($HAsO_4^-$) anion, whereas, arsenite [As(III)] exists as an uncharged molecule (H_3AsO_3) and anionic ($H_2AsO_4^-$) species in moderately reducing environment where anoxic conditions persist (1). Despite the fact that inorganic forms are predominant in natural waters, presence of MMA and DMA has also been reported (2), and their existence is due to microbial metabolism of inorganic arsenic.

Extensive arsenic contamination of surface and subsurface waters has been reported in many parts of the world (3–10), thereby threatening the health of a number of people in the affected areas. Due to human health concerns, arsenic standard for drinking water has been lowered in many countries. Such an action might impose a considerable burden on water utilities in respect of compliance and cost, in their effort to adopt an effective technology to remove arsenic from drinking water. A better understanding of the occurrence of arsenic species in subsurface waters and their behavior in water treatment processes can assist the water utility managers to select an appropriate technology that could help to solve the problems for arsenic removal from drinking water. Furthermore, it will provide a basis for evaluating the treatment costs, and aid the researchers and epidemiologists to estimate the risk of arsenic intake by humans. This article provides the details about the occurrence of arsenic, health effects due to arsenic exposure, and available treatment technologies for arsenic removal, so that a better understanding of problems and possible solutions can be obtained.

2. Occurrence of Arsenic in the Environment

Arsenic is mainly transported to the environment by water. Arsenic contamination of subsurface waters is believed to be geological, and high arsenic concentrations in groundwater may result from dissolution of, or desorption from iron oxide, and oxidation of arsenic pyrites (8). In addition, the occurrence of arsenic in groundwater depends on factors such as redox conditions, ion exchange, precipitation, grain size, organic content, biological activity, and characteristics of the aquifer (11). The severity of arsenic pollution of groundwater is reported in Bangladesh, where most of the people rely on wells as a source of drinking water. Until recently, occurrence of arsenic in Bangladesh was believed to be due to pyrite oxidation; however, recent studies showed that the causative mechanism of arsenic release to groundwater was reductive dissolution of arsenic-rich Fe oxyhydroxide and the reduction was driven by microbial degradation of organic matter, which was present in concentrations as high as 6% C (12).

2.1. Natural Sources. The natural weathering processes contribute ~40,000 tons of arsenic to the global environment annually, while twice this amount is being released by human activities (13). The primary natural sources are weathering of rocks, geothermal, and volcanic activity; rocks are the major reservoirs for arsenic, and soils and oceans are the remaining natural sources of arsenic. Arsenic ranks twentieth in crystal abundance, and is the major constituent of at least 245 minerals (14). These minerals are mostly ores containing sulfide, along with copper, nickel, lead, cobalt, or other minerals. The most common arsenic containing minerals are arsenic pyrites (FeAsS), realgar (AsS), lollingite ($FeAs_2$, $FeAs_3$, and $FeAs_5$), and orpiment (As_2S_3). Depending on the type of rocks, arsenic concentration varies, with sedimentary rocks having a higher level of arsenic than igneous and metamorphic rocks. The average concentration of As in igneous and sedimentary rocks is 2 mg/kg, and in most rocks it ranges from 0.5 to 2.5 mg/kg (15). The mining operations of coal containing arsenic increase the potential for soil contamination with arsenic. Soils in areas close to or derived from sulfur ore deposits may contain concentrations as high as 8000-mg As/kg soil (16); however, the mean levels of As in soils are usually 5-mg As/kg soil (17). The elemental arsenic has several allotropic forms, in which only gray arsenic is stable; it is a brittle, crystalline semimetallic solid that sublimes at 615°C (1 atm) without melting. Elemental arsenic tarnishes in air and burns with a bluish flame while heating; it gives off an odor of garlic and dense white fumes of AS_2O_3 (18).

2.2. Anthropogenic Sources. Anthropogenic activities such as mining and smelting activities, and the use of pesticides and fossil fuels have resulted in a dramatic effect on natural environmental arsenic levels. In addition, arsenic and arsenic compounds are used in pigments and dyes, preservatives of animal hides and wood, pulp and paper production, electroplating, battery plates, dye and soaps, ceramics and in the manufacture of semiconductors, glass, and various pharmaceutical substances (19). Chromated copper arsenate (CCA), an inorganic arsenic compound that is used to treat lumber, accounts for ~90% of the arsenic used annually by industry in the United States. Arsenic is a contaminant of concern at 916 of the 1467 National Priorities List (superfund) of hazardous waste sites in the United States (20).

Up to the mid-1900s, inorganic compounds, usually as Pb, Ca, Mg, and zinc arsenate, were used extensively as pesticides in orchards (21). In coal, arsenic is mainly present at concentrations from <1 to >90 mg/kg. During combustion, arsenic compounds in coal are volatilized and may condense on the surface of the fly ash particles, thereby increasing the arsenic content in the fly ash. In the leaching experiments with fly ash, it was found that both As(III) and As(V) species leached from the fly ash (22). Out of the total arsenic added to the soils from anthropogenic activities, ~23% is contributed by coal fly ash and bottom ash, 14% by atmospheric fallout, 10% by mine tailings, 7% by smelters, 3% by agriculture, and 2% by manufacturing, urban, and forestry wastes (23).

3. Arsenic Exposure and Health Effects

3.1. Arsenic Exposure.

All humans are exposed to low levels of arsenic through drinking water, air, food, and beverages. Consumption of food and water are the major sources of arsenic exposure for the majority of the affected people. At present a large population worldwide has been exposed significantly to high arsenic levels in drinking water. In addition, workers involved in the operations of mining and smelting of metals, pesticide production and application, production of pharmaceutical substances, and glass manufacturing have a high level of occupational exposure to arsenic (24). The use of arsenic-containing compounds such as potassium arsenite (Fowler's solution) in medical treatment for treating various illnesses caused skin and internal cancers. Though Fowler's solution is not in use at present, some arsenicals are still prescribed as medicine in Asian countries (25).

The arsenic compounds that are known to be present in food and water, and that affect the health of human individuals upon ingestion are shown in Table 1. The food products that come from the marine environment have a high level of arsenic concentration than other food products; AsB, a nontoxic arsenic species

Table 1. **Arsenicals Present in Water and Food**

arsenic compound	Chemical formula
arsenious acid [As(III)]	H_3AsO_3
arsenic acid [As(V)]	H_3AsO_4
monomethylarsonic acid (MMA)	$H_2(CH_3)AsO_3$
dimethylarsinic acid (DMA)	$H(CH_3)_2AsO_2$
trimethylarsine oxide (TMAO)	$(CH_3)_3\,AsO$
tetramethylarsonium ion	$(CH_3)_4As^+$
arsenobetaine (AsB)	$(CH_3)_3As^+$ CH_2COOH

Table 2. **Arsenic Levels in Food and Marine Species**

Food and marine species	Total As	Inorganic As	Reference
meat: beef and pork	0.15 mg/kg		28
lobsters	4.7–26 mg/kg		18
prawns	5.5–20.8 mg/kg		18
crabs	3.5–8.6 mg/kg		18
canned tuna	1100 ng/g		29
marine fish	0.19–65 mg/kg		30
shellfish	0.2–125.9 mg/kg		30
freshwater fish	0.007–1.46 mg/kg		30
fats and oils	19 ng/g		32
potatoes	2.3 µg[a]		28
rice	NR[b]	74 ng/g	32
flour	NR[b]	11 ng/g	32
spinach	NR[b]	6.1 ng/g	32
peas	NR[b]	4.5 ng/g	32
carrots	NR[b]	3.9 ng/g	32
onions	NR[b]	3.3 ng/g	32

[a] As daily dietary intake.
[b] NR = not reported.

mainly present in seafoods such as fish and shrimp, is readily excreted in the urine (26). The studies by Schoof and co-workers (27) showed that unpolished rice had a higher inorganic arsenic concentration (74 µg/kg) than corn and flour. Among the four kinds of fruits and vegetables tested, spinach and grapes had the highest inorganic As concentration. The arsenic concentration present in the food products and marine species is shown in Table 2. The Food and Drug Administration's (FDA) total diet study showed that the average adult's total arsenic intake in the United States was ~53 µg/day (31), whereas the average daily dietary ingestion of total arsenic by Canadians was estimated to be 38.1 µg (32). The characterization studies by the United States Environmental Protection Agency (USEPA) showed that ~20 percent of the daily intake of dietary arsenic was in the inorganic form (31). Upon ingestion of inorganic arsenicals, it is methylated in the human body and the metabolites of ingested arsenic are eliminated by the kidney and excreted in urine within 1–3 days. DMA is the predominant metabolite excreted in urine and faeces of animals and humans exposed to inorganic arsenic (33).

3.2. Arsenic Toxicity and Health Effects. Arsenic is considered as a notorious poison because of its toxicity. The carcinogenic effect of arsenic compounds was first noted in the eighteenth century, when patients treated with arsenicals were found to have an unusual number of skin tumors (34). The toxicity of arsenic depends on its speciation. The toxicity of arsenite is 25–60 times higher than that of arsenate (35), and the toxicity decreases in the order of arsine > inorganic As(III) > organic As(III) > inorganic As(V) > organic As(V) > arsonium compounds and elemental arsenic. Recent studies (35, 36) showed that arsenite was more prevalent in groundwater than arsenate. Ingestion of inorganic arsenic can result in both cancer (skin, lung, and urinary bladder) and non-cancer effects (26). Acute and chronic toxicity due to the drinking of arsenic contaminated water has been well documented through population-based studies that showed the capacity of arsenite [As(III)] and arsenate [As(V)] to adversely affect numerous organs in the human body (3,10,37).

3.3. Cancer Effects. The people living in Asian, South American, and Mexican countries with exposure to high level of arsenic concentration in drinking water are reported to have increased risks of skin, bladder, and lung cancer. The association of arsenic in drinking water and skin cancer was first reported in Taiwanese people (3). The Taiwanese study population was large, numbering 40,421 inhabitants in 37 villages. The results of the studies showed that prevalence of skin cancer was noted among the people; high incidence of skin cancer was observed among the elderly people (age >60 years). Based on the Taiwanese data, the USEPA estimated the lifetime risk of developing skin cancer as 1 or 2 per 1000 people for each microgram of inorganic arsenic per liter of drinking water (28). Recently, National Research Council (NRC) has stated that the total cancer risk due to the consumption of drinking water with 50-µg/L arsenic will be 1 in 100 (26).

An additional strong evidence that drinking arsenic-contaminated water causes cancer is from Chile, where the population studied was nearly 10 times larger than that of the Taiwanese study population. In Northern Chile, nearly 7.3% of all deaths among those aged 30 years and over were due to internal cancers (bladder and lung cancer) caused by drinking arsenic contaminated water (37). In both the Taiwanese and Chilean studies, the people were exposed to a high level of arsenic (>500 µg/L) in drinking water. Increased risks of bladder and lung cancer were noted among men and women in Argentina, even though the average arsenic concentration was 170 µg/L (38).

3.4. Non-Cancerous Effects. Non-cancerous effects have been reported in humans after exposure to drinking inorganic arsenic contaminated water. Inorganic arsenic in drinking water may affect many organs including central and peripheral nervous systems, dermal, cardiovascular, gastrointestinal, and respiratory systems. The most common ailments such as keratoses and hyperpigmentation may occur after 5–15 years of arsenic exposure equivalent to 700 µg/day for a 70 kg adult (26). Hyperpigmentation was the most common ailment (183.5/1000) among the affected people in Taiwan (3). Further, long-term exposure to high inorganic arsenic in drinking water caused black foot disease in Taiwan.

Dermatitis, a skin lesion of arseniasis is prevalent in Bangladesh; an increased prevalence of bronchitis has also been observed among the exposed populations in Bangladesh (7). A recent survey conducted by the Dhaka Community Hospital (DCH) in 80% of the total area of Bangladesh showed that people were affected with melanosis (93.5%), keratosis (68.3%), hyperkeratosis (37.6%), dipigmentation (39.1%), and cancer (0.8%) (10). Data from the population based and clinical case studies showed that there was a dose-response relationship for ingested arsenic water and several non-cancerous effects (26).

4. Arsenic Determination

In the past, measurement of total elemental concentrations was considered to be sufficient for environmental considerations. Since the element occurs in different species and the species have different properties, a determination of total concentration of an element alone may not provide adequate information about the physical/chemical forms of the element and its toxicological properties. Therefore, it is essential to determine the individual species of an element, enabling one to obtain realistic information about the toxicity and transformation of the species. The term speciation refers to the determination of different oxidation states of an element that prevail in a certain specimen or to the identification and quantification of the biologically active compounds to which the element is bound (39).

The presence of As(III) and As(V) in different proportions in water supplies may produce different toxic effects. Often it is documented in the literature that the measurement of total arsenic concentration is insufficient to assess the risks of As exposure in human populations. Several instrumental methods have been used to determine the concentration of arsenic and its species. Such methods include hydride generation atomic absorption spectrometry (GF–AAS), graphite furnace atomic absorption spectrometry (HG–AAS), inductively coupled plasma (ICP–AES), ICP–mass spectrometry (ICP–MS), and high-performance liquid chromatography (HPLC). A number of techniques are available for the speciation of arsenic; recently Edwards and co-workers (40) established an arsenic speciation protocol that can be applied to water treatment plant in situ.

Ficklin (41) used the anion exchange resin of 100–200 mesh size and a glass column in speciating arsenite and arsenate present in samples. One percent HCl was used to acidify the samples before resin treatment. Edwards and co-workers (40) used an anion exchange resin of 50–100 mesh size, and 0.05% H_2SO_4 (v/v) to acidify the samples before resin treatment; the column (polypropylene) used was twice in diameter compared to that of Ficklin (41). Thirunavukkarasu and co-workers (42) used a speciation protocol (Fig 1) similar to that of Edwards and co-workers (40) except that samples were acidified with nitric acid (trace metal grade) instead of sulfuric acid. Speciation recovery studies of samples preserved with nitric acid showed that recoveries of As(III) were in the range of 81.2–105.2%. Further, speciation studies with a natural water sample showed that the particulate and soluble arsenic contributed 11.4 and 88.8% of the total arsenic present in the natural water, respectively. The fractions of As(III) and As(V) present in the soluble arsenic were 47.3 and 52.7%, respectively (42).

5. Treatment Technologies for Arsenic Removal

Various treatment methods have been reported in the literature to remove arsenic effectively from the drinking water. Such treatment methods include coagulation/filtration (43–50), adsorption on activated alumina (51–57), adsorption on activated carbon (52), adsorption on ion-exchange resin (41,57,58), adsorption on hydrous ferric oxides (59–62), adsorption on various iron oxides (42,63–67), and adsorption/filtration by manganese greensand (68, 69). After careful review, the USEPA suggested ion-exchange, activated alumina, reverse osmosis, modified coagulation/filtration, and modified lime softening as the best available technologies (BAT) based on arsenate removal; however, it put forth the importance on iron-based coagulation assisted microfiltration, iron oxide-coated sand (IOCS), manganese greensand filtration, and granular ferric hydroxide (GFH)

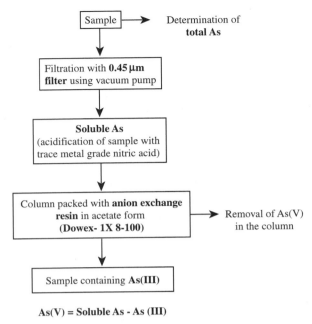

Fig. 1. Arsenic speciation protocol (42).

for arsenic removal, for which rigorous testing is necessary to validate the technologies (31,70). The removal efficiencies of the BATs are shown in Table 3.

The selection of an appropriate treatment process for a specific water supply system will depend on many factors such as concentration of arsenic, source water composition of other constituents, pH, and cost effectiveness. Though coagulation/filtration is a simple method, the disadvantages are the production of large amounts of sludge-containing arsenic (hazardous in nature) that will pose serious problems for safe disposal. In recent years, there has been an overwhelming research effort to develop an innovative technology to achieve a low level of arsenic in drinking water supplies. Iron salts are generally added as coagulants in conventional treatment processes. At a particular pH, the iron forms precipitates, referred to as Fe oxides, which are good adsorbents for As removal.

Table 3. **The Removal Efficiencies of the Best Available Technologies (USEPA 2000)**

Treatment technology	Maximum percent removal[a]
ion exchange	95
activated alumina	90
reverse osmosis	>95
modified coagulation/filtration	95
modified lime softening	80
electrodialysis reversal	85

[a] Percent removals based on As(V) removal.

Since adsorption processes are most effective, and iron compounds are widely available, it is obvious that iron-based material should be investigated in detail for arsenic removal from drinking water. Fixed-bed treatment systems, such as adsorption and ion-exchange, are getting increasingly popular for arsenic removal in small-scale treatment systems because of their simplicity, ease of operation and handling, regeneration capacity and sludge-free operation.

5.1. Arsenic Removal by the Coagulation/Filtration Process. Coagulation/filtration processes are mainly used in large-scale water utilities. It is a simple treatment process, in which chemicals are added to form precipitate or flocs that are removed by a subsequent sedimentation or filtration process. Based on the type and initial concentration of the contaminant, either precipitation or coprecipitation or both play an important role in the removal during coagulation (71). Alum and iron(III) salts are mainly used as coagulants in drinking water treatment for arsenic removal, and numerous studies have been conducted to evaluate the performance of these coagulants, especially for arsenic removal. In a coagulation process, arsenic removal is dependent on adsorption and coprecipitation of arsenic onto metal hydroxides (48).

In both the laboratory and field coagulation experiments, ferric chloride produced the best arsenic removal compared to ferrous and aluminum sulfate (43). In laboratory experiments, arsenic removal (82%) obtained with ferric chloride was nearly 2.5 times higher than the removal achieved with aluminum sulfate. Based on experimental results, Gulledge and O'Connor (44) reported that As(V) adsorption on ferric hydroxide exceeded the adsorption on aluminium hydroxide, and an increased coagulant dosage resulted in an increase in arsenic removal. In the pH range of 5–7, >90% removal of As(V) was achieved with a 30-mg/L dose of ferric sulfate. The results of these studies showed that ferric chloride coagulation achieved better removal in the pH range studied than alum coagulation. Similarly, the results of the studies by Sorg and Logston (45) showed that ferric sulfate achieved a better arsenic removal than alum in the pH range of 5–8.

Arsenate removal from groundwater by iron and alum coagulation/filtration treatment showed that iron coagulation was more effective than alum, and the removals achieved using iron coagulants were not pH dependent between 5.5 and 8.5 (46). In these studies, nearly 100% removal was achieved with iron coagulants (30 mg/L) in the entire pH range, when the raw-water initial arsenate concentration was 300 µg/L. Cheng and co-workers (47) concluded that for the source waters tested, enhanced coagulation was effective for arsenic removal and less ferric chloride than alum, on a mass basis, was needed to achieve the same removal. Also, in the studies with ferric chloride, pH between 5.5 and 7.0 had no significant effect on arsenate removal.

Edwards (48) reported that As(III) removal by coagulation was primarily controlled by the coagulant dose and was relatively unaffected by the solution pH. It was also found that ferric coagulants were effective in the removal of As(V) at pH < 7.5, and iron was more effective than alum in removing both As(V) and As(III) at pH > 7.5. The data compiled by Edwards (48) showed that >90 percent arsenate removal was achieved in the coagulation studies, when ferric chloride was used at >20 mg/L or alum at >40 mg/L. Scott and co-workers (49) reported that arsenic removal of 81–96% was achieved when source water was treated with 3–10 mg/L of ferric chloride, and concluded that ferric chloride was more effective than alum in removing arsenic.

In the coagulation experiments with ferric chloride over the pH range of 4.0–9.0, Hering and co-workers (50) observed that pH had no significant effect on arsenate removal, and nearly 100% arsenate removal was achieved in the entire pH range studied. However, they observed that pH did have an effect on arsenite removal in the studies. They also demonstrated that using ferric chloride as a coagulant at pH 7.0, both arsenite and arsenate removals were independent of the initial concentration.

5.2. Arsenic Removal by Activated Alumina.

Activated alumina (AA) treatment is a physical/chemical process by which ions in the drinking water are removed by the oxidized AA surface. Activated alumina treatment is considered to be an adsorption process, even though the reactions involved in the process involve actually an exchange of ions. Several studies have demonstrated that AA is an effective treatment for the removal of arsenic from drinking water. However, factors such as pH, competing ions, arsenic oxidation state, and empty bed contact time (EBCT) have significant effects on the removal of arsenic using AA (57).

The highest arsenic removal was achieved at a pH closer to 6 (53), and arsenic removal decreased as the pH increased beyond 6 (56). In contrast, the results of the studies by Vagliasindi and co-workers (58) showed that arsenate adsorption onto AA was relatively insensitive to pH in the range of 5.5–8.5. The studies conducted by Frank and Clifford (54) observed that the bed volumes achieved for up to an effluent arsenate level of 50 µg/L were high compared to the bed volumes achieved for arsenite removal, which indicated arsenate adsorption was faster than arsenite adsorption.

Similarly, in both the laboratory and pilot plant studies (57), the bed volumes achieved for arsenate removals were higher than for arsenite removal, and this was mainly due to the oxidization of As(III) to As(V). In the pilot plant studies on arsenic removal using AA (56), the results showed that arsenic run length was directly proportional to EBCT, and high bed volumes were achieved at high EBCTs. Further, it was observed that at a pH of 7.5 the arsenic removal capacity of AA to 10-µg/L arsenic varied between 0.19- and 0.35-g As/kg AA at different EBCTs. The presence of sulfate and chloride had a significant effect on arsenic removal using AA (53), whereas the results of the studies by Vagliasindi and co-workers (58) showed that presence of sulfate had little effect on arsenate adsorption and the presence of chloride had no effect on arsenate adsorption.

5.3. Arsenic Removal by Ion-Exchange.

Although ion exchange is an efficient treatment system for arsenic removal from drinking water, its application is limited to small and medium scale point-of-entry (POE) systems because of its high treatment cost as compared to other treatment technologies (72). Ion-exchange is an ion selective process, which removes As(V) significantly but does not remove As(III). It is an effective process for arsenic removal, if the source water contains <500-mg/L total dissolved solids and <150-mg/L sulfate; preoxidation of As(III) to As(V) is necessary (57).

The factors that affect the efficiency of the ion-exchange process include competing ions such as sulfate and total dissolved solids, EBCT and regenerant strength (57). In the ion exchange studies for arsenic removal, Vagliasindi and Benjamin (73) found that high bed volumes were achieved at high EBCTs for different source waters tested. The results of the column studies showed that the column continued to run for longer time until arsenate breakthrough occurred. However, they reported that the treatment efficiency was affected by the source water composition, and the presence of sulfate drastically reduced the adsorption capacity of the resin. In the batch studies using an ion exchange resin, Vagliasindi and co-workers (58) reported that arsenate adsorbed strongly onto the resin, when the source water had the lowest total organic carbon and sulfate among the different source waters tested.

5.4. The Role of Iron Oxides in Arsenic Removal.

Iron oxides, oxyhydroxides, and hydroxides (all are called iron oxides) consist of Fe in association with O and/or OH. They differ in composition, in the nature of Fe, and in crystal structure. The basic structural unit of all Fe oxides is an octahedran, in which each Fe atom is surrounded either by six O or by both O and OH ions. The O and OH ions form layers that are either approximately hexagonally close-packed (hcp), or cubic close-packed (ccp). The hcp forms are termed as α-phases, and ccp forms are termed as γ-phases. The α-phases are more stable than γ-phases (74).

There are 16 iron oxides, and these iron oxides play an important role in a variety of industrial applications, including pigments for the paint industry, catalyst for industrial synthesis, and raw material for iron and steel industry (74). The application of iron oxide has been extended to remove metals from water and wastewater (63,75). Arsenic removal with iron oxides had been investigated (59–69,42). It is generally assumed that arsenate [As(V)] has a stronger affinity than arsenite [As(III)] on iron oxide surfaces. However, recent studies by Raven and co-workers (65) showed that at high initial As concentration, arsenite adsorption on ferrihydrite was higher than arsenate adsorption throughout the pH range of 3.0–11.0. They used a suspension containing a known amount of ferrihydrite to study the removal of As(III) and As(V) in synthetic water. Pierce and Moore (59) reported that adsorption of As(III) onto amorphous ferric hydroxide increased with pH up to a maximum pH of 7.0. Pierce and Moore (60) found that the rate of adsorption of As(V) onto amorphous ferric hydroxide was much faster than that of As(III) in the initial phase (1h) of contact with the adsorbent. They also recommended that for maximum arsenic removal, pH 7.0 was optimum for As(III) and pH 4.0 was optimum for As(V).

In the studies using hydrous ferric hydroxide (HFO) for arsenic removal, Hsia and co-workers (61) observed that the amount of As(V) adsorbed onto the iron oxide surface increased as the equilibrium concentration of arsenate in the solution increased. They also reported that at high initial As(V) concentration, nearly 100% of arsenate was adsorbed over the pH range of 4.0–9.0. In similar batch studies with HFO, Wilkie and Hering (62) found that As(V) was adsorbed stronger than As(III), and the adsorption of As(III) over the pH range of 4.0–9.0 was not strongly dependent on pH. They reported that adsorption of As(III) onto HFO increased as the pH increased up to a maximum at pH 7.0, whereas complete As(V) removal was observed over the pH range examined in the studies with the solution, which had an initial arsenic [As(III) or As(V)] concentration of 1.33 μM and a background electrolyte of 0.01 M $NaNO_3$. In a separate study, the initial arsenic concentration was varied from 0.033 to

1.33 µM and they reported that As(III) adsorption was high (maximum removal of 96%) at a low initial As(III) concentration, and one of the reasons was attributed to the partial oxidation of As(III) on HFO surface.

In the column studies using IOCS, Joshi and Chaudhury (64) reported that the bed volumes achieved at the value of 10 µg/L for arsenic in drinking water were in the range of 163–184 and 149–165 per cycle for As(III) and As(V), respectively. They added that 94–99% of arsenic was recovered at the end of each cycle during regeneration, and virtually no iron was detected in the effluent. The rate of influent feed maintained in the tests performed by Joshi and Chaudhury (64) was low, and insufficient while scaling up to pilot or small water facilities. Thus, the performance of IOCS will be different and have an impact on the arsenic removal efficiency at normal filtration rates that are being maintained in water facilities.

Benjamin and co-workers (63) studied the removal of arsenite in the column studies by using IOCS as an adsorbent. The influent had an initial As(III) concentration of 75 µg/L and an EBCT of 5 min was maintained in the column studies. They reported that the column removed 100% arsenite up to 650 bed volumes treated, which indicated extreme binding capacity of arsenite onto IOCS. In kinetic studies using IOCS prepared in a high temperature coating process, the results showed that 85–90% of arsenic was removed in the initial phase (1 h) of contact, and >95% removal was obtained for both As(III) and As(V) in the pH range of 5.0–7.6 after 6 h of contact (67). It was also reported that the bed volumes achieved up to 5 µg/L of As(III) in the effluent were 1380 in the column studies (Fig 2), and the column continued to remove As(III) to a value <5 µg/L for a period of 50 h, where the influent had an arsenite [As(III)] concentration of 300 µg/L. Driehaus and co-workers (66) reported that the results obtained

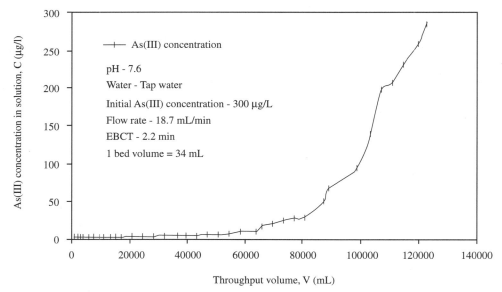

Fig. 2. Arsenite concentration remaining in solution in the column studies (1 bed volume = 34 mL) (67).

from the fixed adsorber tests with GFH for arsenic removal were encouraging, and nearly 30000 bed volumes were treated with the effluent As concentration at 10 µg/L.

Recent studies (42,63–67) showed that iron-based materials are effective in reducing arsenic to a low level in drinking water, which indicated that filtration systems containing iron-based materials offer an excellent choice among the treatment systems available for arsenic removal in small water facilities. Though the use of GFH to remove arsenic from drinking water was proven to be successful at pilot-scale facilities in Germany (66), the detailed cost economics has to be worked out while extending the applicability of GFH to small water facilities. On the other hand, IOCS filtration systems that are tested at the laboratory-scale level (42,63,64,67) may offer the best choice in respect of arsenic removal efficiency and cost economics, for which testing is necessary at the pilot-scale level. In addition, conducting experiments at various initial arsenic concentrations [both As(III) and As(V)] using IOCS is necessary to study the effect of initial arsenic concentration on the adsorptive capacity of IOCS, and arsenic adsorption behavior. Further research and evaluation is required to develop cost-effective treatment technologies for arsenic removal, especially for rural communities in developing countries. Research is also needed to develop and integrate environmentally acceptable disposal of treatment system residuals into the overall treatment philosophy.

6. Legislation and Economic Aspects

The World Health Organization (WHO) standard for arsenic has been lowered from 50 to 10 µg/L. The German drinking water standard had been lowered to 10 µg/L, and the Commission of European Community is aiming at a standard of 2–20 µg/L (Driehaus and co-workers 1998). The existing arsenic standard for drinking water in Canada and Australia is 25 and 7 µg/L, respectively. Recently, USEPA adopted a new arsenic standard for drinking water at 10 µg/L after an additional review by the National Academy of Sciences (76). Since the majority of the affected people worldwide live in small communities, it is essential to develop an appropriate treatment technology that would solve the problems for small communities. Simplicity and cost are the two major factors that should influence the selection of a treatment system for arsenic removal in small communities dependent on groundwater for their drinking water supply. Though several technologies were proven to be successful in the removal of arsenic from drinking water at laboratory and pilot-scale studies, the practical applicability of such systems to small communities has not been fully tested and exploited.

Although coagulation-assisted microfiltration, ion exchange systems may be suitable for large communities, systems based on adsorption/filtration process are appropriate and advantageous to small communities, especially in developing countries such as Bangladesh, because of simplicity, ease of construction, cost-effectiveness, and operation and maintenance. Currently, simple household purification systems such as bucket filtration systems containing sand and iron fillings are used in most of the arsenic affected areas in Bangladesh as short-term measures; however, permanent measures are necessary.

BIBLIOGRAPHY

1. J. F. Ferguson and J. Gavis, *Water Res.*, **6**, 1259 (1972).
2. R. S. Braman and C. C. Foreback, *Science*, **182**, 1247 (1973).
3. W. P. Tseng, H. M. Chu, S. W. How, J. M. Fong, C. S. Lin, and S. Yeh, *J. Nat. Cancer Inst.*, **40**, 453 (1968).
4. M. E. Cebrian, A. Albores, M. Aguilar, and E. Blakely, *Human Toxicol.*, **2**, 121 (1983).
5. A. Chatterjee, D. Das, B. K. Mandal, T. Roy Chowdhry, G. Samanta, and D. Chakraborti, *Analyst*, **120**, 643 (1995).
6. S. Niu, S. Cao, and E. Shen, in C. O. Abernathy, R. L. Calderon, and W. R. Chappell, eds., *Arsenic exposure and health effects*, Chapmann and Hall, London, pp. 78–83, 1997.
7. R. K. Dhar, B. K. Biswas, G. Samanta, B. K. Mandal, D. Chakraborti, S. Roy, A. Jaffer, A. Islam, G. Ara, S. Kabir, A. W. Khan, S. A. Ahmed, and S. A. Hadi, *Curr. Sci.*, **73**, 48 (1997).
8. A. H. Welch, D. R. Helsel, M. J. Focazio, and S. A. Watkins, in W. R. Chappell, C. O. Abernathy, and R. L. Calderon, eds., *Arsenic exposure and health effects*, Elsevier Science, Elmsford, N. Y., pp. 9–17 1999.
9. I. Koch, J. Feldman, L. Wang, P. Andrewes, K. J. Reimer, and W. R. Cullen, *The Sci. Total Environ.*, **236**, 101 (1999).
10. M. M. Karim, *Wat. Res.*, **34**, 304 (2000).
11. F. N. Robertson, *Environ. Geochem. Health*, **11**, 171 (1989).
12. R. T. Nickson, J. M. McArthur, P. Ravenscroft, W. G. Burgess, and K. M. Ahmed, *Appl. Geochem.*, **15** 403 (2000).
13. C. R. Paige, W. J. Snodgrass, R. V. Nicholson, and J. M. Scharer, *Water Environ. Res.*, **68**, 981 (1996).
14. E. A. Woolson, in B. A. Fowler, ed., *Biological and environmental effects of arsenic*, Elsevier Science, Elmsford, N. Y., pp. 51 1983.
15. A. Kabata-Pendias, and H. Pendias, *Trace elements in soils and plants*, CRC press, Boca Raton, Fl., pp. 315, 1984.
16. I. Thornton, and M. Farago, in C. O. Abernathy, R. L. Calderon, and W. R. Chappell, eds., *Arsenic exposure and health effects*, Chapmann and Hall, London, pp. 1–16, 1997.
17. ATSDR, Agency for Toxic Substances and Disease Registry, Toxicological profile for arsenic, U.S. Public Health Service, Atlanta, Ga, 1997.
18. K. A. Francesconi, and J. S. Edmonds, in J. O. Nriagu, ed., *Arsenic in the Environment, Part 1: Cycling and Charecterization*, John Wiley & Sons, Inc., New York, pp. 189–219, 1994.
19. J. M. Azcue and J. O. Nriagu, in J. O. Nriagu, ed., *Arsenic in the Environment, Part 1: Cycling and Charecterization*, John Wiley & Sons, Inc., New York, pp. 1–15, 1994.
20. ATSDR, Agency for Toxic Substances and Disease Registry, Toxicological profile for arsenic, U.S. Public Health Service, Atlanta, Ga., 1998.
21. D. Chisholm, *Can. J. Plant. Sci.*, **52**, 584 (1972).
22. V. T. Breslin and I. W. Duedall, *Mar. Chem.*, **13**, 341 (1983).
23. D. K. Bhumbla and R. F. Keefer, in J. O. Nriagu, ed., *Arsenic in the Environment, Part 1: Cycling and Charecterization*, John Wiley & Sons, Inc., New York, pp. 51–82 1994.

24. U.S. Public Health Service, Toxicological profile for arsenic, U.S. Public Health Service, Washington D.C., 1989.
25. R. E. Grissom, C. O. Abernathy, A. S. Susten, and J. M. Donohue, in W. R. Chappell, C. O. Abernathy, and R. L. Calderon, eds., *Arsenic exposure and health effects*, Elsevier Science, Elmsford, N. Y., pp. 51–60, 1999.
26. National Research Council, Arsenic in drinking water, National Academy Press, Washington, D.C., 1999.
27. R. A. Schoof and co-workers, in W. R. Chappell, C. O. Abernathy, and R. L. Calderon, eds., *Arsenic exposure and health effects*, Elsevier, New York, pp. 81–88, 1999.
28. USEPA, Special report on ingested inorganic arsenic, EPA/625/3-87/013, Washington, D.C. 1988.
29. L. J. Yost, R. A. Schoof, and R. Aucoin, *Human Ecol. Risk Assess.*, **4**, 137 (1998).
30. J. M. Donohue and C. O. Abernathy, in W. R. Chappell, C. O. Abernathy, and R. L. Calderon, eds., *Arsenic exposure and health effects*, Elsevier Science, Elmsford, New York, pp. 89–98 1999.
31. USEPA, http://www.epa.gov/safewater/arsenic.html (June 22, 2000).
32. R. W. Dabeka and co-workers, *J. AOAC Int.*, **76**, 14 (1993).
33. M. Vahter, in W. R. Chappell, C. O. Abernathy, and C. R. Cothern, eds., *Arsenic exposure and health*, Science and Technology Letters, Northwood, pp. 171–180, 1994.
34. J. Hutchinson, *Br. Med. J.*, **2**, 1280 (1887).
35. N. E. Korte and Q. Fernando, *Crit. Rev. Environ. Control*, **21**, 1 (1991).
36. D. K. Chada, in P. Bhattacharya and A. H. Welch, eds., *Proceedings of the 31st International Geological Congress, Arsenic in groundwater of sedimentary aquifers*, Rio de Janeiro, Brazil, pp. 33–35, 2000.
37. A. H. Smith, M. Goycolea, R. Haque, and M. L. Biggs, *Am. J. Epidemiol.*, **147**, 660 (1998).
38. A. Hopenhayn-Rich, M. L. Biggs, A. H. Smith, *Int. J. Epidemiol.*, **27**, 561 (1998).
39. M. Burguera and J. L. Burguera, *Talanta*, **44**, 1581 (1997).
40. M. Edwards, S. Patel, L. Mcneill, H. W. Chen, M. Frey, A. D. Eaton, R. C. Antweiler, and H. E. M. Taylor, *J. AWWA*, **90**, 103 (1998).
41. W. H. Ficklin, *Talanta*, **30**, 371 (1983).
42. O. S. Thirunavukkarasu, T. Viraraghavan, and K. S. Subramanian, *Water Qual. Res. J. Can*, **36**, 55 (2001).
43. Y. S. Shen, *J. AWWA*, **65**, 543 (1973).
44. H. J. Gulledge and T. J. O'Connor, *J. AWWA*, **65**, 548 (1973).
45. T. J. Sorg and G. S. Logsdon, *J. AWWA*, **70**, 379 (1978).
46. T. J. Sorg, *Proc. of the AWWA WQTC*, Denver, Colorado, 1993.
47. R. C. Cheng, S. Liang, H. C. Wang, and M. D. Beuhler, *J. AWWA*, **86**, 79 (1994).
48. M. Edwards, *J. AWWA*, **86**, 64 (1994).
49. N. K. Scott, F. J. Green, D. H. Do, and J. S. McLean, *J. AWWA*, **87**, 114 (1995).
50. J. G. Hering, P. Y. Chen, J. A. Wilkie, M. Elimelech, and S. Liang, *J. AWWA*, **88**, 155 (1996).
51. E. Bellack, *J. AWWA*, **63**, 454 (1971).
52. S. K. Gupta and K. Y. Chen, *J. Water Poll. Control Fedn.*, **50**, 493 (1978).
53. E. Rosenblum and D. Clifford, EPA-600/52-83-107, Cincinnati, Ohio, (1984).
54. P. Frank and D. Cifford, EPA-600-52-86/021, Cincinnati, Ohio (1986).
55. S. W. Hathway and J. F. Rubel, *J. AWWA*, **79**, 61 (1987).
56. J. Simms and F. Azizian, *Proceedings of the AWWA WQTC*, Denver, Colo, 1997.
57. D. A. Clifford, in R. D. Letterman, ed., *Water quality and treatment*, AWWA, McGraw-Hill Book Co. Inc., 5th ed., New York, 1999.
58. F. G. A. Vagliasindi, M. Henley, N. Schulz, and M. M. Benjamin, *Proceedings of the AWWA WQTC*, Denver, Colorado, pp. 1829–1853, 1996.

59. L. M. Pierce and B. C. Moore, *Environ. Sci. Technol.*, **14**, 214 (1980).

60. L. M. Pierce and B. C. Moore, *Wat Res.*, **16**, 1247 (1982).

61. T. H. Hsia, S. L. Lo, C. F. Lin, and D. Y. Lee, *Colloids Surfaces A: Physicochem. Eng. Aspects.* **85**, 1 (1994).

62. J. A. Wilkie and J. G. Hering, *Colloids Surfaces A: Physicochem. Eng. Aspects.* **107**, 97 (1996).

63. M. M. Benjamin, R. S. Sletten, R. P. Bailey, and T. Bennet, *Wat Res.*, **30**, 2609 (1996).

64. A. Joshi and M. Chaudhuri, *J. Environ. Eng.*, **122**, 769 (1996).

65. K. P. Raven, A. Jain, and R. H. Loeppert, *Environ. Sci. Technol.*, **32**, 344 (1998).

66. W. Driehaus, M. Jekel, and U. Hildebrandt, *J. Water SRT - Aqua*, **47**, 30 (1998).

67. T. Viraraghavan, O. S. Thirunavukkarasu, and K. S. Subramanian, in M. M. Sozanski, ed., *Proceedings of the Fourth International Conference on Water Supply and Water Quality*, Krakow, Poland, pp.1013 2000.

68. K. S. Subramanian, T. Viraraghavan, T. Phommavong, and S. Tanjore, *Water Qual. Res. J. Can.*, **32**, 551 (1997).

69. T. Viraraghavan, K. S. Subramanian, and J. A. Aruldoss, *Wat Sci. Tech.*, **40**, 69, (1999).

70. USEPA, Technologies and costs for removal of arsenic from drinking water, Draft report, EPA-815-R-00-012, Washington D.C., 1999.

71. L. D. Benefield and J. S Morgan, R. D. Letterman, ed., *Water quality and treatment AWWA*, McGraw-Hill Book Co., Inc., 4th ed., New York, 1990.

72. USEPA, Arsenic removal from drinking water by ion exchange and activated alumina plants, EPA/600/R-00/088, Cincinnati, Ohio, 2000.

73. F. G. A Vagliasindi and M. M. Benjamin, *Proceedings of the AWWA annual conference*, Denver, Colorado, 1997.

74. R. M. Cornell and U. Schwertmann, *The iron oxides-structure, properties, reactions, occurrence and uses*, VCH Publishers, New York, 1996.

75. M. Edwards and M. M. Benjamin, *J. Water Poll. Control Fedn.*, **61**, 1523 (1989).

76. USEPA, http://www.epa.gov/safewater/arsenic.html (October 31, 2001).

O. S. THIRUNAVUKKARASU
T. VIRARAGHAVAN
University of Regina

ASBESTOS

1. Introduction

Asbestos is a generic term referring to six types of naturally occurring mineral fibers that are or have been commercially exploited. These fibers belong to two mineral groups: serpentines and amphiboles. The serpentine group contains a single asbestiform variety: chrysotile. There are five asbestiform varieties of amphiboles: anthophyllite asbestos, grunerite asbestos (amosite), riebeckite asbestos (crocidolite), tremolite asbestos, and actinolite asbestos. Usually, the term asbestos is applied only to those varieties that have been commercially exploited (1,2). That does not preclude the occurrence of other asbestos-like minerals, however. Magnesioriebeckite with an asbestiform habit was mined in Bolivia and potassian winchite with an asbestiform habit was found in western Texas. Additionally, richterite asbestos has been synthesized in the laboratory (3).

The asbestos varieties share several properties: (*1*) they occur as bundles of fibers that can be easily separated from the host matrix or cleaved into thinner fibers (1,4); (*2*) the fibers exhibit high tensile strengths (1); (*3*) they show high length: diameter (aspect) ratios, with a minimum of 20 and up to 1000 (1,4); (*4*) they are sufficiently flexible to be spun; and (*5*) macroscopically, they resemble organic fibers such as cellulose (2,4). Since asbestos fibers are all silicates, they exhibit several other common properties, such as incombustibility, thermal stability, resistance to biodegradation, chemical inertia toward most chemicals, and low electrical conductivity. The usual definition of asbestos fiber excludes numerous other fibrous minerals that may possess an asbestiform habit but do not exhibit all of the properties of asbestos. A few examples of these fibrous minerals are sepiolite, erionite (rod-like and fibrous habits), and nemolite. Other minerals also may occasionally crystallize with a fibrous habit (3,4).

The mineralogical designations of the various asbestos fibers, their most common alternative designations, and the main sources of these fibers are reported in Table 1. The fractional breakdown of the recent world production of the various fiber types shows that the industrial applications of asbestos fibers have now shifted almost exclusively to chrysotile. Amosite and crocidolite are no longer being mined although some probably is still being sold from stock. Current use of amosite and crocidolite is estimated to be less than a few hundred tons annually. Actinolite asbestos, anthophyllite asbestos, and tremolite asbestos may be still mined in small amounts for local use; production probably is <100 tons annually.

Table 1. Asbestos Fiber Production

Mineral species[a]	Other designations	Major sources[b]	World production, 2000, %
chrysotile	white asbestos	Russia, Canada, China, Brazil, Kazakhstan, and Zimbabwe	>99%
cummingtonite-grunerite	amosite (brown asbestos)	deposits in South Africa	none
riebeckite	crocidolite (blue asbestos)	deposites in South Africa	none
anthophyllite		deposits in Bulgaria, Finland, India, South Africa, United States	insignificant
actinolite		deposits in South Africa and Taiwan	insignificant
tremolite		deposits in India, Italy, Korea, Pakistan, South Africa	insignificant

[a]Chrysotile is in the serpentine mineral group; all others are amphiboles.
[b]In order of production only for chrysotile.

2. History

Early uses of asbestos exploited the reinforcement and thermal properties of asbestos fibers. The first recorded application can be traced to Finland (~2500 BC), where anthophyllite from a local deposit was used to reinforce clay utensils and pottery (5). Numerous early references also can be found describing the use of asbestos fibers for the fabrication of lamp wicks and crematory clothing. Other applications of asbestos fibers in heat- or flame-resistant materials have been sporadically reported. At the end of the seventeenth century, Peter the Great of Russia initiated the fabrication of asbestos paper, using chrysotile fibers extracted from deposits in the Ural mountains. The use of asbestos fibers on a true industrial scale began in Italy early in the nineteenth century with the development of asbestos textiles (4,6). By the end of the nineteenth century, significant asbestos deposits had been identified throughout the world and their exploitation had begun in Canada (1878), South Africa (1893, 1908–1916), and the USSR (1885) (7).

From the beginning of the twentieth century, the demand for asbestos fibers grew in a spectacular fashion for numerous applications, in particular for thermal insulation (8). The development of the Hatschek machine in 1900 for the continuous fabrication of sheets from an asbestos–cement composite also opened an important field of industrial application for asbestos fibers as did the development of the automobile industry for asbestos brakes, clutches, and gaskets.

World War II supported the growth of asbestos fiber production for military applications, typically in thermal insulation and fire protection. Such applications were later extended into residential or industrial constructions for several decades following the war.

During the late 1960s and 1970s, the finding of health problems associated with long-term heavy exposure to airborne asbestos fibers led to a large reduction in the use of asbestos fibers. In most of the current applications, asbestos fibers are contained within a matrix, typically cement or organic resins.

The world production of asbestos fibers reached a maximum in 1977 of 4.8×10^6 tons, decreasing to 1.9×10^6 tons in 2000. The major producing countries of chrysotile asbestos are Russia (39%), Canada (18%), China (14%), Brazil (9%), Kazakhstan (7%), and Zimbabwe (6%). In 2000, active mining operations of asbestos fibers are found in 21 countries (9).

3. Geology and Fiber Morphology

The genesis of asbestos fibers as mineral deposits required certain conditions with regard to chemical composition, nucleation, and fiber growth; such conditions must have prevailed over a period sufficiently long and perturbation-free to allow a continuous growth of the silicate chains into fibrous structures (10). Some of the important geological or mineralogical features of the industrially significant asbestos fibers are summarized in Table 2. More emphasis is given to chrysotile in the following section owing to its total dominance in the industry over the years.

Table 2. **Geological Occurence of Asbestos Fibers**

	Chrysotile [12001-29-5]	Amosite [19172-73-5]	Crocidolite [12001-28-4]	Tremolite [14567-73-8]
mineral species	chrysotile	cummingto-nite-gruner-ite	riebeckite	tremolite
structure	as veins in serpentine and mass fiber deposits	lamellar, coarse to fine, fibrous and asbestiform	fibrous in ironstones	long, prismatic, and fibrous aggregates
origin	alteration and metamorphism of basic igneous rocks rich in magnesium silicates	metamorphic	regional metamorphism	metamorphic
essential composition	hydrous silicates of magnesia	hydroxy silicate of Fe and Mg	hydroxy silicate of Na, Mg, and Fe	hydroxy silicate of Ca and Mg

Only three varieties of amphibole fibers will be discussed because (1) crocidolite and amosite were the only amphiboles with significant industrial uses in recent years; and (2) tremolite, although having essentially no industrial application, may be found as a contaminant in other fibers or in other industrial minerals (e.g., chrysotile and talc).

3.1. Chrysotile. Chrysotile belongs to the serpentine group of minerals, varieties of which are found in ultra basic rock formations located in many places in the world (11). Chrysotile accounts for only a small percentage of the minerals found in these rock types. Chrysotile fibers are found as veins in serpentines, in serpentinized ultramafic rocks, and in serpentinized dolomitic marbles. It has been suggested that the ultrabasic rocks, containing olivine, Mg-rich pyroxenes, and amphiboles are first altered by hydrothermal processes to form the serpentine minerals; in a later metamorphic event, the serpentines are partially redissolved and crystallized as chrysotile fibers. Clearly, the genesis of each chrysotile deposit must have involved specific features related to the composition of the precursor minerals, the stress and deformations in the host matrix, the water content, the temperature cycles, etc. Nonetheless, it is generally observed that the chemical composition of the fibrous phase is closely related to that of the surrounding rock matrix (12).

Growth of chrysotile fibers at right angles to the walls of cracks (cross-vein) in massive serpentine formations led to the most common type of chrysotile deposit. Most of the industrial chrysotile fibers are extracted from deposits where fiber lengths can reach several centimeters, but most often do not exceed 1 cm (12). Figure 1 illustrates the typical aspect of cross-vein chrysotile fibers (also called cross-fiber) separated from the host serpentine rock; a nonfibrous variety of serpentine, antigorite [12135-86-3], having a chemical composition nearly identical to that of chrysotile, is also pictured in Figure 1. In some occurrences, the relative motions (slipping) of blocks in the host rock, during or after fiber growth, lead to veins in which the fibers are inclined or parallel

Fig. 1. Asbestos fibers (chrysotile, crocidolite, and amosite) as separated from host rock and their massive varieties (antigorite, riebeckite, cummingtonite- grunerite) (11). Courtesy of R. T. Vanderbilt Company, Inc.

to the vein axes (slip fibers). In still other local conditions, dispersed aggregates of short fibers are found with no preferential orientation; in such cases the fiber content of the rock can be very high, up to 50%. These are called mass-fiber deposits.

Chrysotile is a hydrated magnesium silicate and its stoichiometric chemical composition may be given as $Mg_3Si_2O_5(OH)_4$ [12001-29-5]. However, the geothermal processes that yield the chrysotile fiber formations usually involve the codeposition of various other minerals. These mineral contaminants comprise: brucite [$Mg(OH)_2$]; [1317-43-7], magnetite (Fe_3O_4); [1309-38-2], calcite ($CaCO_3$); [13397-26-7], dolomite [$(Mg,Ca)(CO_3)_2$]; [16389-88-1], chlorites [$(Mg,Al,Fe)_{12}Si_8O_{20}(OH)_{16}$], and talc [$Mg_6Si_8O_{20}(OH)_4$]; [14807-96-6]. Other iron-containing minerals may also be found in chrysotile, eg, pyroaurite, brugnatellite, and pyroxenes. In commercial fibers, dust particles from the host rock generated during the mining and milling processes are most inevitably present (3). Elemental analysis data for several chrysotiles and amphiboles are presented in Table 3.

Chrysotile fibers can be extremely thin, the unit fiber having an average diameter of $\sim$25 nm (0.025 μm). Industrial chrysotile fibers are aggregates of these unit fibers that usually exhibit diameters from 0.1 to 100 μm; their lengths range from a fraction of a millimeter to several centimeters, though most of the chrysotile fibers used are <1 cm.

Table 3. **Elemental Analysis of Asbestos Fibers**[a]

Variety and source	SiO$_2$ (silica)	FeO (ferrous oxide)	Fe$_2$O$_3$ (ferric oxide)	Al$_2$O$_3$ (alumina)	MgO (magnesia)	CaO (lime)	MnO (manganese oxide)	Na$_2$O (sodium oxide)	K$_2$O (potassium oxide)	H$_2$O, adsorbed	H$_2$O$^+$, combined
Chrysotile											
Quebec	40.2	1.0	0.5	2.9	39.9	1.1	0.1	0.1	0.1	0.8	13.4
Zimbabwe	39.7	0.7	0.3	3.2	40.3	1.1	0.3	0.1	0.1	0.6	12.2
Ural Mts.	38.1	1.3	1.4	5.0	37.7	2.2	0.1	0.1	0.1	0.8	11.1
Crocidolite											
Cape Province	50.9	20.5	16.9		1.1	1.5	0.1	6.2	0.2	0.2	2.2
Australia	52.8	14.9	18.6	0.2	4.6	1.1	trace	6.0	0.1	0.2	2.8
Bolivia	55.7	3.8	13.0	4.0	13.1	1.5	trace	6.9	0.4	trace	1.8
Amosite											
Transvaal	49.4	40.6	0.1		6.7	0.7	0.7	0.1	0.2	0.1	1.9
Tremolite											
Pakistan	55.1	2.0	0.3	1.1	25.7	11.5	0.1	0.3	0.2	3.5	0.2

[a]Ref. 13.

3.2. Amphiboles. The amphibole group of minerals is widely found throughout the earth's crust. Their chemical composition can vary widely. Of the amphiboles, only a few varieties have an asbestiform habit and the latter occur in relatively low quantities. The geological origin of amphibole asbestos fibers appears to be quite varied (3). In the case of the crocidolite deposit of South Africa (Transvaal), the amphibole fibers formed during secondary chemical reactions that took place as the banded ironstone host rock consolidated from a gel of iron hydroxide and colloidal silica. Crocidolite fiber veins formed, presumably with magnetite particles acting as nucleating agents. The presence of mechanical stress appears to be necessary for the formation of crocidolite. It appears that fibers form in places where there is shearing (slip fibers) or rock dialatium (cross fiber). One would not expect fibers to form in rock that is not mechanically disturbed. The amosite deposit found in similar rock formations is the result of a high temperature metamorphic process (3).

The chemical composition of amphiboles readily reflects the complexity of the environment in which they formed. The average chemical composition of amphibole minerals may be represented as

$$A_{0-1}, B_2 C_5 T_8 O_{22} (OH, O, F, Cl)_2 \quad \text{where} \quad A = Na, K$$

$$B = Na, Ca, Mg, Fe^{2+}, Mn, Li$$

$$C = Al, Fe^{2+}, Fe^{3+}, Ti, Mg, Mn, Cr$$

$$T = Si, Al$$

A, B, C each represent cationic sites within the crystal structure (2,4,10). From this generic representation, the chemical composition of the amphibole asbestos can be given as follows:.

Name	Composition	CAS Registry Number
grunerite (amosite)	$Na_2(Fe^{2+}, Mg)_3 Fe^{3+}{}_2 Si_8 O_{22}(OH)_2$	[12172-73-5]
riebeckite (crocidolite)	$Na_2(Fe^{2+}, Mg)_3 Fe^{3+}{}_2 Si_8 O_{22}(OH)_2$	[12001-28-4]
anthophyllite	$Mg_7 Si_8 O_{22}(OH)_2$	[17068-78-9]
tremolite	$Ca_2 Mg_5 Si_8 O_{22}(OH)_2$	[14567-73-8]
actinolite	$Ca_2(Mg, Fe^{2+})_5 Si_8 O_{22}(OH)_2$	[12172-67-7]

From their respective compositions, the amphibole fibers can be viewed as a series of minerals in which one cation is progressively replaced by another at a given site. For example, tremolite and actinolite may be seen related; the magnesium in tremolite is partly replaced by divalent iron in the C position to yield actinolite. Similarly, in the cummingtonite–grunerite amphiboles, the substitution involves the replacement of magnesium in positions B and C by divalent iron (2,4). The change of designation from cummingtonite to grunerite is, by convention, at a Mg/Fe ratio of 0.5 (10).

The two most important amphibole asbestos minerals are amosite and crocidolite, and both are hydrated silicates of iron, magnesium, and sodium (crocidolite only). The appearance of these fibers and of the corresponding nonfibrous

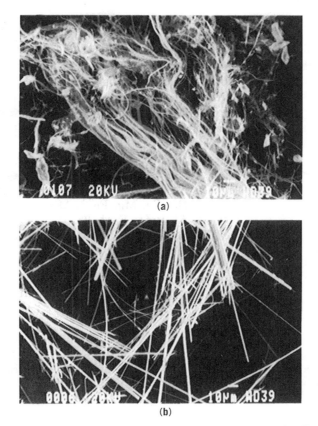

Fig. 2. Electron micrographs of asbestos fibers: (**a**) chrysotile; (**b**) crocidolite.

amphiboles is shown in Figure 1. Although the macroscopic visual aspect of clusters of various types of asbestos fibers is similar, significant differences between chrysotile and amphiboles appear at the microscopic level. Under the electron microscope, chrysotile fibers are seen as clusters of fibrils, often entangled, suggesting loosely bonded, flexible fibrils (Fig. 2a). Amphibole fibers, on the other hand, often appear individually, rather than in fiber bundles (Fig. 2b).

4. Crystal Structure of Asbestos Fibers

The microscopic and macroscopic properties of asbestos fibers stem from their intrinsic, and sometimes unique, crystalline features. As with all silicate minerals, the basic building blocks of asbestos fibers are the silicate tetrahedra that may occur as double chains $(Si_4O_{11})^{6-}$, as in the amphiboles, or in sheets $(Si_4O_{10})^{4-}$, as in chrysotile (4) (Fig. 3).

4.1. Chrysotile. In the case of chrysotile, an octahedral brucite layer having the formula $[Mg_6O_4(OH)_8]^{4-}$ is intercalated between each silicate tetrahedra sheet, as illustrated in Figure 4. The silicate and brucite layers share

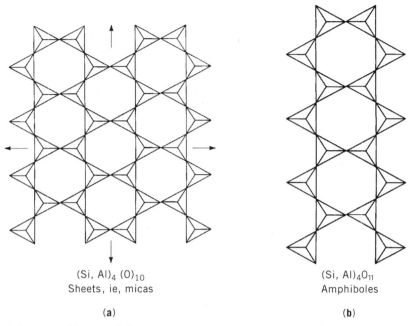

$(Si, Al)_4 (O)_{10}$
Sheets, ie, micas

$(Si, Al)_4 O_{11}$
Amphiboles

(a)   (b)

Fig. 3.   Silicate backbones of chrysotile and amphiboles: (**a**) the sheet silicate structure of chrysotile, analogous to that of micas; (**b**) the double-chain silicate structure found in amphiboles (4). Courtesy of Oxford University Press.

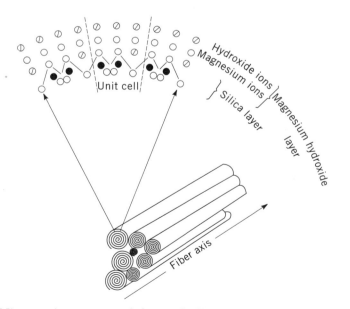

Fig. 4.   Microscopic structure of chrysotile fibers (3). Reprinted with permission.

oxygen atoms that would normally be separated by distances of 0.305 nm in the silicate layer and 0.342 nm in the brucite layer (4). This mismatch of $O-O$ distances induces a curvature of the sheets, the ideal radius of which has been calculated as 8.8 nm (14). The curvature of the sheets propagates along a preferred axis leading to the formation of the tubular structure found in chrysotile. The concentric sheets forming the fibers have a curvature radius from 2.5 to 3.0 nm for the internal layers up to ~25 nm for the external layers, yielding unit fibers (fibrils) with external diameters ranging between 20 and 50 nm (15). Electron microscopy studies (15) have also shown that, in the unit chrysotile fiber cross section, the layers may appear in a concentric or spiral arrangement.

The stacking of the tetrahedral and octahedral sheets in the chrysotile structure has been shown to yield three types of chrysotile fibers (3,4):

clino-chrysotile: monoclinic stacking of the layers, x parallel to fiber axis, most abundant form.

ortho-chrysotile: orthorhombic stacking of the layers, x parallel to fiber axis.

para-chrysotile: two layer structure, 180° rotation of two-layer structures, y parallel to fiber axis.

The extent of substitution of magnesium and silicon by other cations in the chrysotile structure is limited by the structural strain that would result from replacement with ions having inappropriate radii. In the octahedral layer (brucite), magnesium can be substituted by several divalent ions, Fe^{2+}, Mn^{2+}, or Ni^{2+}. In the tetrahedral layer, silicon may be replaced by or Al^{3+} or rarely Fe^{3+}. Most of the other elements that are rarely found in vein fiber samples, or in industrial asbestos fibers, are associated with interstitial mineral phases. Typical compositions of bulk chrysotile fibers from different locations are given in Table 3.

4.2. Amphiboles. The crystalline structure common to amphibole minerals consists of two ribbons of silicate tetrahedra placed back to back as shown in Figure 5. The plane of anionic valency sites created by this double ribbon arrangement is neutralized by the metal cations. The crystal structure has 16 cationic sites of four different types; these sites can host a large variety of metal cations without substantial disruption of the lattice (see Section 3.2 and Fig. 5).

In contrast to chrysotile fibers, the atomic crystal structure of amphiboles does not inherently lead to fiber formation. The formation of asbestiform amphiboles must result from multiple nucleation and specific growth conditions. The difference between asbestiform and massive amphibole minerals is obvious on the macroscopic scale, although the crystalline structures of the two varieties do not exhibit substantial differences. The asbestiform amphiboles tend to have a larger number of crystal defects (Wadsley defects and twinning and chain width disorder) than the nonasbestiform varieties. The frequency and width of these defects vary with amphibole type (3,10). Amphibole minerals, in general, are characterized by prismatic cleavage planes parallel to the c axis that intersect at an angle of ~56°. Hence, in the crushing of massive, nonfibrous amphiboles, microscopic fragments are found having the appearance of asbestos fibers. However, the statistical average of their aspect ratio is considerably lower than that of the asbestiform amphiboles.

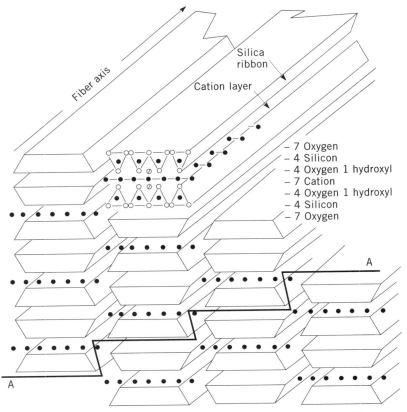

Fig. 5. Microscopic structure of amphibole fibers (22). Reprinted with permission.

5. Properties of Asbestos Fibers

Asbestos fibers used in most industrial applications consist of aggregates of smaller units (fibrils), which is most evident in chrysotile that exhibits an inherent, well-defined unit fiber. Diameters of fiber bundles in bulk industrial samples may be in the millimeter range in some cases; fiber bundle lengths may be several millimeters to 10 cm or more.

The mechanical processes employed to extract the fibers from the host matrix, or to further separate (defiberize, open) the aggregates, can impart significant morphological alterations to the resulting fibers. Typically, microscopic observations on mechanically opened fibers reveal fiber bends and kinks, partial separation of aggregates, fiber end-splitting, etc. The resulting product thus exhibits a wide variety of morphological features, some of which can be seen in Figure 2.

Morphological variances occur more frequently with chrysotile than amphiboles. The crystal structure of chrysotile, its higher flexibility, and interfibril adhesion (3) allow for a variety of intermediate shapes when fiber aggregates are subjected to mechanical shear. Amphibole fibers are generally more brittle and accommodate less morphological deformation during mechanical treatment.

5.1. Fiber Length Distribution. For industrial applications, the fiber length and length distribution are of primary importance because they are closely related to the performance of the fibers in matrix reinforcement. Various fiber classification methods have thus been devised. Representative distributions of fiber lengths and diameters can be obtained through measurement and statistical analysis of microphotographs (15); fiber length distributions have also been obtained from automated optical analyzers (16). Typical fiber length distributions obtained from these approaches are illustrated in Figure 6 for chrysotile fibers. As in the cases shown there, industrial asbestos fiber samples usually contain a rather broad distribution of fiber lengths.

5.2. Physicochemical Properties. The industrial applications of chrysotile fibers take advantage of a combination of properties: fibrous morphology, high tensile strength, resistance to heat and corrosion, low electrical

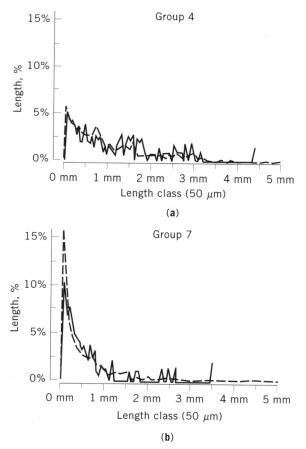

Fig. 6. Fiber length distribution for (**a**) a long sample (group 4) and (**b**) a short sample (group 7) of chrysotile; successive length classes separated by 50 μm. —, Automated measurement; —, microphotographic measurement.

Table 4. **Physical and Chemical Properties of Asbestos Fibers**

Property	Chrysotile	Amosite	Crocidolite	Tremolite
color	usually white to grayish green; may have tan coloration	yellowish gray to dark brown	cobalt blue to lavender blue	gray-white, green, yellow, blue
luster	silky	vitreous to pearly	silky to dull	silky
hardness, Mohs	2.5–4.0	5.5–6.0	4.0	5.5
specific gravity	2.4–2.6	3.1–3.25	3.2–3.3	2.9–3.2
optical properties	biaxial positive parallel extinction	biaxial positive parallel extinction	biaxial negative oblique extinction	biaxial negative oblique extinction
refractive index	1.53–1.56	1.63–1.73	1.65–1.72	1.60–1.64
flexibility	high	fair	fair to good	poor, generally brittle
texture	silky, soft to harsh	coarse but somewhat pliable	soft to harsh	generally harsh
spinnability	very good	fair	fair	poor
tensile strength, MPa[a]	1100–4400	1500–2600	1400–4600	<500
resistance to:				
acids	weak, undergoes fairly rapid attack	fair, slowly attacked	good	good
alkalies	very good	good	good	good
surface charge, mV (zeta potential)	+13.6 to +54[b]	−20 to −40	−32	
decomposition temperature, °C	600–850	600–900	400–900	950–1040
residual products	forsterite, silica, eventually enstatite	Fe and Mg pyroxenes, magnetite, hematite, silica	Na and Fe pyroxenes, hematite, silica	Ca, Mg, and Fe pyroxenes, silica

[a]To convert MPa to psi, multiply by 145.
[b]Chrysotile fibers tend to become negative after weathering and/or leaching.

conductivity, and high friction coefficient. In many applications, the surface properties of the fibers also play an important role; in such cases, a distinction between chrysotile and amphiboles can be observed because of differences in their chemical composition and surface microstructure. Technologically relevant physical and chemical properties of asbestos fibers are given in Table 4.

Thermal Behavior. Asbestos fiber minerals are hydrated silicates so their behavior as a function of temperature is related first to dehydration (or dehydroxylation) reactions. In the case of chrysotile, the crystalline structure is stable up to ~550°C [depending on the heating period (17)], where the dehydroxylation of the brucite layer begins. This process is completed near 750°C and is characterized by a total weight loss of 13%. The resulting magnesium silicate recrystallizes to form forsterite and silica in the temperature range 800–850°C, as an exothermic process. The weight loss as a function of heating temperature [thermogravimetry (tga)] and the reaction heats [differential thermal analysis (dta)] are illustrated for chrysotile in Figure 7a. The strongly endothermic dehydration process enhances the high temperature thermal insulation properties of chrysotile asbestos.

The behavior of amphibole fibers under continuous heating is similar to that observed with chrysotile, although the temperatures of dehydroxylation and recrystallization processes are different. The amphiboles have a lower water (hydroxyl) content and their dehydroxylation reaction begins between 400 and 600°C, depending on the amphibole type; this reaction leads to a weight loss of ~2%. The latter is illustrated in Figure 7b, together with a dta recording of the associated thermal events. The products resulting from the thermal decomposition of the amphiboles are pyroxenes, magnetite, hematite, and silica.

In the presence of oxygen, the thermal decomposition of amphiboles is associated with an oxidation of divalent iron to trivalent iron, which may lead to an increase in the sample weight. The oxidation process also induces an obvious color alteration: The fibers acquire the characteristic ferric oxide color.

Asbestos fibers, in particular chrysotile fibers, can undergo substantial thermal degradation during mechanical grinding. In high energy attrition equipment, such as ball mills, the high localized impact energies can cause the crystalline structure to become amorphous, which is particularly true for dry milling or milling in organic solvents. However, milling in aqueous slurries is less detrimental, water offering some protection against high impact degradation (18).

Tensile Strength. The inherent tensile strength of a single asbestos fiber, based on the strength of Si—O—Si bonds in the silicate chain, should be near 10 GPa (1.45×10^6 psi) (19). However, industrial fibers exhibit substantially lower values, because of the presence of various types of structural or chemical defects.

The measured tensile strength of chrysotile fibers has been reported in the range 1.1–4.4 GPa (160,000–640,000 psi). The accurate determination of this parameter is difficult since the measurement performed on a fiber aggregate is influenced by interfibril adhesion, discontinuities in some of the unit fibers, mineral inclusions, etc. Consequently, higher tensile strength results are obtained for measurements done on short and thin fibers (20,21). The tensile strengths of amosite and crocidolite are comparable to that of chrysotile. With amphiboles, the tensile strength is highly influenced by the iron content since iron—oxygen bonds located in the fiber axes, particularly those involving trivalent Fe, are particularly strong. The trend observed of increasing tensile strength of amphiboles from tremolite, to amosite, to crocidolite is directly related to the iron content of these fibers (22).

The variation in tensile strength of asbestos fibers as a function of temperature also sharply distinguishes chrysotile and amphiboles. Chrysotile retains (and even slightly increases) its tensile strength up to 500°C, until the dehydroxylation reaction begins; it drops sharply at higher temperatures. Amphiboles, on the other hand, exhibit a decreasing tensile strength beginning at ~200°C. For example, at 350°C, crocidolite has lost 50% of its initial tensile strength (22).

Asbestos Fibers in Aqueous Media. Although asbestos fibers cannot be viewed as water-soluble silicates, prolonged exposure of chrysotile or amphiboles to water (especially at high temperature) leads to slow progressive leaching of

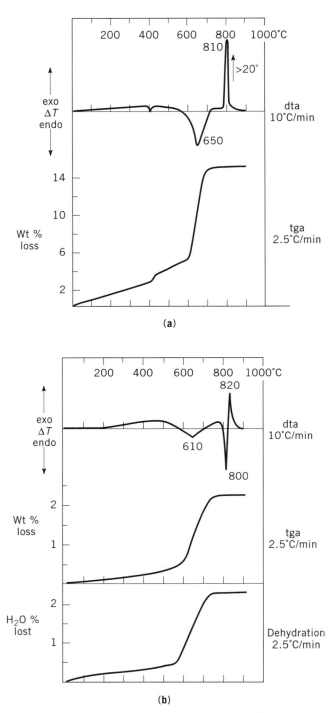

Fig. 7. Thermal analysis curves for asbestos: (**a**) chrysotile; (**b**) crocidolite (in inert atmosphere) (3). Reprinted with permission.

both their metal and silicate components (23). In the case of chrysotile fibers (in a given amount of water), the brucite layer will, fairly rapidly, dissolve in part, with concomitant increase in the pH of the solution. The equilibrium pH value for an aqueous chrysotile slurry reaches 10.0–10.5. Free brucite, present as contaminant in the fibers, also contributes to the pH increase.

In contact with solutions of mineral acids, organic acids, or magnesium complexing agents, the rate of dissolution of the brucite layers is increased. When carried out to the extreme, the leaching of magnesium leads to a weight loss of 58%. The residual silica is largely amorphous and, although the fibers retain an apparent fibrous morphology, their mechanical resistance is drastically reduced.

The dispersion of amphiboles in concentrated HCl solutions also leads to partial leaching, the rate of which depends on the metal cations present. With crocidolite, only small amounts of magnesium and sodium are extracted in these conditions, whereas amosite liberates substantial quantities of iron and magnesium. Overall, tremolite appears to exhibit the highest resistance to acid leaching (3).

On the other hand, both chrysotile and the amphiboles exhibit a high degree of chemical inertia toward strong alkalies over extended periods. At high temperatures, reactions with alkalies [NaOH, KOH, $Ca(OH)_2$] become significant over relatively short periods; eg, crocidolite was reported to be attacked by potassium hydroxide >100°C (23).

Other Bulk Physical Properties. The hardness of asbestos fibers is comparable to that of other crystalline or glassy silicates. Compared to glass fibers, amphiboles have similar hardness values, while chrysotile shows lower hardness values (22).

The friction coefficients of asbestos fibers are also different for chrysotile and amphiboles (when measured against the same material). Compared to glass fibers, the friction coefficients decrease in the order: chrysotile, amphiboles, glass fibers.

The high electrical resistivity of asbestos fibers is well known and has been widely exploited in electrical insulation applications. In general, the resistivity of chrysotile is lower than that of the amphiboles, particularly in high humidity environments because of the availability of soluble ions. For example, the electrical resistivity of chrysotile decreases from 1 to 2100 MΩ/cm in a dry environment to values of 0.01–0.4 MΩ/cm at 91% relative humidity. Amphiboles, on the other hand, exhibit resistivity between 8000 and 900,000 Ω/cm.

With respect to magnetic properties, the intrinsic magnetic susceptibility of pure chrysotile is very weak. However, the presence of associated minerals such as magnetite, as well as substitution ions (Fe, Mn), increases the magnetic susceptibility to values ~1.9–3.5×10^{-6}/g Oe. With amphiboles, the magnetic susceptibility is much higher, mainly because of the high iron content; typically, amosite and crocidolite exhibit susceptibility values of 69–71×10^{-6}/g Oe (24).

5.3. Surface Properties. *Surface Area.* The specific surface area of industrial asbestos fibers obviously depends on the extent of their defiberization (opening), and is usually between 1 and 30 m^2/g. As measured by BET (Brunauer, Emmett, and Teller) nitrogen adsorption, chrysotile fibers exhibit surface areas between 15 and 30 m^2/g. With regard to amphibole fibers, surface areas of

1.8–9 m^2/g have been reported for crocidolite and 1.3–5.5 m^2/g for amosite (25). Such a difference originates, in part, in the relative sizes of the unit fibers of chrysotile and the amphiboles; however, the results also reflect the ability of nitrogen molecules to diffuse between the unit fibers of a larger aggregate. In relation to magnesium acid leaching of chrysotile discussed earlier, the surface area of an amorphous silica resulting from extensive acid leaching may reach 450 m^2/g (26).

Also, the adsorption of anionic or neutral surfactants on chrysotile fibers in aqueous dispersions enhances fiber separation, with a concomitant increase of surface area (27). Such effects have not been reported for amphibole fibers.

Surface Charge in Aqueous Media. Because of dissolution–ionization effects, the surface of asbestos fibers in aqueous dispersions adopts an electrostatic charge. In the case of chrysotile, partial dissolution of the brucite layer leads to a positive surface charge (or potential), which is strongly influenced by the solution pH. The isoelectric point of chrysotile (pH of zero surface charge) has been reported as 11.8. In the case of amphibole fibers, the surface charge seems dominated by the silica component, and is generally observed to be negative, increasing toward 0 as the pH is decreased. Since the progressive leaching of magnesium from the external brucite layer of chrysotile gradually exposes silica, the surface potential rapidly decreases early in the leaching reaction (3).

Adsorption and Surface Chemical Grafting. As with silica and many other silicate minerals, the surface of asbestos fibers exhibits a significant chemical reactivity. In particular, the highly polar surface of chrysotile fibers promotes adsorption (physi- or chemisorption) of various types of organic or inorganic substances (23). Moreover, specific chemical reactions can be performed with the surface functional groups (OH groups from brucite or exposed silica).

The chemical reaction of coupling agents, such as organosilane compounds, on chrysotile yields fibers with a highly hydrophobic surface. Through an adequate choice of coupling agents (organosilanes or others), the fiber matrix interactions in composite materials can thus be optimized (28). Surface chemical modifications can be carried out in fiber slurries or through gas–solid reactions; eg, the gas-phase reaction of POCl$_3$ with chrysotile, leads to a surface coverage with insoluble magnesium phosphate (29). The surface properties and reactivity of chrysotile fibers can also be modified by cationic substitution, eg, replacing magnesium by aluminum (30) or other metal cations (31).

6. Analytical Methods and Identification of Asbestos Fibers

The identification of asbestos fibers can be performed through morphological examination, together with specific analytical methods to obtain the mineral composition and structure. Morphological characterization in itself usually does not constitute a reliable identification criterion (1). Hence, microscopic examination methods and other analytical approaches are usually combined.

6.1. Microscopic Methods. The use of microscopic methods is preferred in cases where limited quantities of sample are available, typically in analyzing fibers recovered from sampling of airborne dust. With fibers having lengths in excess of 5 μm, optical microscopy with light polarization (polarizing light microscopy) has proven a very powerful technique. The optical properties of the different types of asbestos fibers, combined with information on fiber shapes, enable positive identification of all varieties of asbestos fibers. This identification can be carried out even when the fibers are mixed with their nonfibrous analogue or with various other materials (32). Often, refractive index, color, pleochroism, birefringence, orientation, etc cannot be measured accurately, thus other methods of analysis must be employed.

Asbestos fiber identification also can be achieved through transmission or scanning electron microscopy (TEM, SEM) techniques that are especially useful with very short fibers, or with extremely small samples (see MICROSCOPY). With appropriate peripheral instrumentation, these techniques can yield the elemental composition of the fibers using energy dispersive X-ray fluorescence, and the crystal structure from selected area electron diffraction (SAED). Both chemical composition and crystal structures are required for positive identification.

6.2. Instrumental Methods for Bulk Samples. With bulk fiber samples, or samples of materials containing significant amounts of asbestos fibers, a number of other instrumental analytical methods can be used to help in the identification of asbestos fibers. The elemental characterization of minerals can be accomplished using methods, such as X-ray fluorescence (XRF) and X-ray photoelectron spectroscopy (XPS). The X-ray diffraction technique (XRD) enables the analyst to identify the crystal structure of the various types of asbestos fibers, as well as the nature of other minerals associated with the fibers (12,33,34).

Thermoanalytical methods (DTA, TGA) often aid in the identification of the type of asbestos fibers (Fig. 7). For example, the strong exotherm observed with chrysotile at 830°C can be used as a routine indicator for determining the chrysotile content of talc (3,8). Thermal methods also are useful for determining certain mineral contaminants of asbestos fibers, eg, brucite and calcite in chrysotile.

Infrared (IR) spectroscopy also is used to analyze samples containing asbestos fibers. Absorption bands in the IR spectrum associated with asbestos fibers are in the $3600-3700$-cm^{-1} range (specific hydroxyl bands) and the $600-800$ and $900-1200$-cm^{-1} ranges (specific absorption bands for various silicate minerals) (3). Because other minerals, including nonasbestiform amphiboles, also absorb within these wavelength bands, IR spectroscopy is not always definitive for identifying asbestos.

Each of these microscopic and bulk methods of analysis provide clues to resolve the identity of the asbestos type. As noted earlier, however, both crystallographic and elemental data are required for a positive identification.

7. Production

A breakout, by major producing countries, of the world asbestos production is shown from 1920 to 2000 in Table 5 and total world production from 1920 to 2000 shown in Figure 8. During the early 1930s, there was a brief period of stagnation in world asbestos production, much of which can be attributed to reduced consumption associated with the economic depression in the United States. At that time, Canada was the major producing country and was highly dependent on U.S. markets for sales. By 1935, production began to increase slowly until the

Table 5. **World Production of Asbestos by Principal Producing Countries, 1920–2000**[a,b], tons

Country	1920	1930	1940	1950	1960	1970	1980	1990	2000
Brazil	—	—	454	844	3,538	16,329	169,173	205,000	170,000
Canada	146,999	199,255	284,406	794,100	1,014,647	1,507,420	1,323,053	725,000	340,000
China	5	286	18,157	—	81,647	172,365	131,700	221,000	260,000
South Africa	5,853	15,868	22,543	79,300	159,544	290,318	277,734	146,000	18,909
U.S.S.R.[c]	—	49,063	—	217,725	598,743	1,065,943	2,070,000	2,400,000	875,000
United States (sold or used by producers)	1,356	3,491	16,509	38,496	41,026	113,683	80,079	W	5,260
Zimbabwe	15,491	31,080	46,093	64,888	121,529	79,832	250,949	161,000	110,000
Other	5,296	7,957	39,838	94,648	189,326	244,109	397,312	152,000	120,831
Total	175,000	307,000	428,000	1,290,000	2,210,000	3,490,000	4,7000,000	4,010,000	1,900,000

[a]World totals, U.S. data, and estimated data are rounded to three significant digits; may not add to totals shown.
[b]W. Withheld. — Zero.
[c]Soviet Republics combined, 1930 to 1990; Russia, 1920; Russia (750,000 tons) and Kazakhstan (125,000 tons), 2000.

Source: U.S. Geological Survey and U.S. Bureau of Mines Minerals Yearbook chapters on asbestos

World Production of Asbestos, 1920-2000

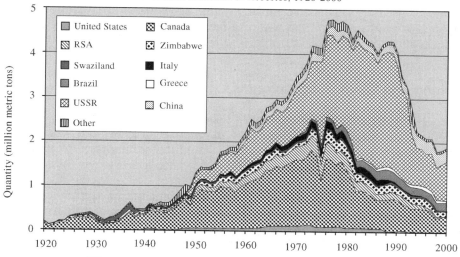

Fig. 8.   World production of asbestos, 1920–2000.

onset of World War II. Canada accounted for the bulk of the increased production tonnage. During World War II, production declined in most regions of the world except Canada, and southern Africa. United States demands absorbed much of the increased production from these regions during the war. Following World War II, world production increased again, fueled by expanding economies, first in the United States and elsewhere later. However, growing opposition to the use of asbestos, which began in earnest in the early 1970s, soon brought this growth to a halt. After 1977, there was a downturn in world production and consumption. In the United States, asbestos regulations became increasingly strict with sizable reductions in permissible exposure limits being enacted. Liability also became a major issue. Companies that mined asbestos and those that manufactured asbestos products faced an increasing number of class-action lawsuits and had difficulties obtaining liability insurance. Compounding the problem was the slow shift to asbestos substitutes or alternative products [ie, ductile iron or poly(vinyl chloride) (PVC) pipe instead of asbestos-cement pipe] in response to public demand. As a result, U.S. consumption peaked in 1973 at ~801,000 tons. World production and consumption peaked at ~4.8×10^6 tons in 1977. In 2000, U.S. consumption was 14,600 tons and world production was 1.9×10^6 tons.

Shifts in production by the main producing countries occurred during this time. Canada was the dominant producer during the first half of the century. By 1980, the former Soviet Union had become, and still remains, the largest producing region. Brazil, China, South Africa, and Zimbabwe also rose from relatively obscurity to become major asbestos producers. Current production has declined in all major producing countries except China due to the opposition to the use of asbestos. Brazil, Canada, China, the former Soviet Union republics of Russia and Kazakhstan, and Zimbabwe now account for >90% of the world production (9). Most of China's production, as well as the limited production of many other countries, is used in local industrial applications. Essentially all production is now chrysotile. Production of amosite and crocidolite ceased in the mid-1990s. Small amounts of actinolite asbestos, anthophyllite asbestos, and tremolite asbestos probably are produced for local use in a few countries such as India, Pakistan, and Turkey.

8. Mining and Milling Technologies

The finding and mapping of chrysotile asbestos ore deposits usually relies on magnetometric surveys largely because magnetite is associated with asbestos deposits, except in the case of ore bodies located in sedimentary formations. As in other mining operations, core drilling is used for a precise evaluation of the grade and volume of the ore body (35).

The choice of a particular mining method depends on a number of parameters, typically the physical properties of the host matrix, the fiber content of the ore, the amount of sterile materials, the presence of contaminants, and the extent of potential fiber degradation during the various mining operations (36). Most of the asbestos mining operations are of the open pit type, using bench drilling techniques.

The fiber extraction (milling) process must be chosen so as to optimize recovery of the fibers in the ore, while minimizing reduction of fiber length. Since the asbestos fibers have a chemical composition similar to that of the host rock, the separation processes must rely on differences in the physical properties between the fibers and the host rock rather than on differences in their chemical properties (36).

In dry milling operations, which are currently the most widely used, the ore is first crushed to a nominal size and then dried (Fig. 9). Fiber extraction then begins through a series of crushing operations, each followed by a vacuum aspiration of the ore running on a vibrating screen. On the latter, the fibers released from the ore have a tendency to move to the surface and, because of their aerodynamic properties, they can be readily collected into a vacuum system. The fibers recovered from consecutive vibrating screens are brought to cyclone separators, and the air is filtered to remove the finer suspended fibers.

Generally, the consecutive crushing–aspiration steps liberate fibers that increasingly shorter. Longer fibers in wide veins are easier to release and thus are recovered in the early phases of the milling operation. In this way, a primary classification, or grading of the fibers can be achieved. Various secondary processing systems may be used to further separate the fiber aggregates and to remove nonfibrous mineral dust. All fiber extraction and classification operations are usually carried out under negative pressure to minimize airborne dust in the working environment.

Wet milling operations, where the asbestos is dispersed in water and not dried until after the final separation process is completed, offer advantages in dust control and the separation of mineral contaminants from the fiber product. However, wet process technology currently is used in only a few small-scale milling operations. With the decline of the asbestos market since the 1970s, the economics of constructing new wet processing facilities, or even dry milling facilities, was not favorable. For the most part, any new capacity that came on

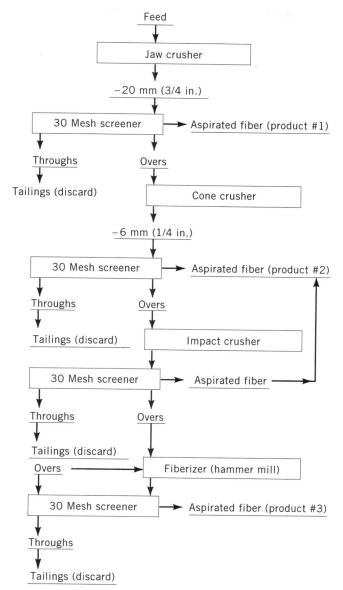

Fig. 9. Schematic of a typical asbestos milling flowline (30 mesh = 590 μm) (35).

line worldwide, with the possible exception of China, in the past 20 years has involved the reopening or expansion of old milling operations.

9. Fiber Classification and Standard Testing Methods

In the beneficiation of asbestos fibers, several parameters are considered critically important and are used as standard evaluation criteria (37): length (or length distribution), degree of opening and surface area, performance in cement reinforcement, and dust and granule content. The measurement of fiber length is important since the length determines the product category in which the fibers will be used and, to a large extent, their commercial value.

9.1. Dry Classification Method. The most widely accepted method for chrysotile fiber length characterization in the industry is the Quebec Standard (QS), test, which is a dry sieving method (on vibrating screens) that enables the fractionation of an asbestos fiber sample into four fractions of decreasing sizes (>2 mesh, <2 and ≥ 4 mesh, <4 and ≥ 10 mesh, and <10 mesh, with mesh defined as the number of openings per linear inch). This standard test has been used as a basis for developing a grade system comprising nine main fiber grades. The grading system identifies the longest fibers as grade 1 and the shortest as grade 9. Grades 1 and 2 designate crude asbestos fibers that are hand picked and essentially contain large clusters or aggregates of fibers. Grade 1 fibers are those with lengths $\geq \frac{3}{4}$ in. (1.9 cm). Grade 2 fibers are those with lengths $\frac{3}{4}$ in. and $\geq \frac{3}{8}$ in. (0.95 cm).

Fiber grades between 3 and 9 (decreasing lengths) are known as milled asbestos fibers and represent most of the production. The fiber grades 3–7 are determined using the QS test, whereas grades 8 and 9, containing very short fibers, are usually determined according to their bulk densities.

Other fiber classification schemes have been devised for chrysotile fibers, but historically the QS grade system has been used as a reference; other classification schemes usually have correspondence scales for conversion to the QS values. Amosite can be classified according to the QS grade system, but crocidolite requires a different scheme mainly due to the harshness of these fibers.

9.2. Wet Classification Method. A second industrially important fiber length evaluation technique is the Bauer–McNett (BMN) classification. In this method, a fiber slurry is circulated through a series of four grids with decreasing opening size (positioned vertically), thus yielding five fractions ($+4$, $+14$, $+35$, $+200$, and -200 mesh). A similar method, using smaller samples and horizontal grids has also been developed and referred to as the Turner–Newall classification.

Currently, the BMN classification and the QS test are the most widely used fiber classification techniques. Whereas there are qualitative relationships between QS and BMN, there is no quantitative correspondence. It is readily understood that these standard tests do not provide accurate definition of the fiber lengths; the classification also reflects the hydrodynamic behavior (volume) of the fibers, which, because of their complex shapes, is not readily predictable. Other classification techniques that provide some insight on fiber lengths are the Ro–Tap test, the Suter–Webb Comb, and the Wash test.

9.3. Other Fiber Evaluation Methods. The extent of fiber separation (fiber openness) is an important evaluation criterion that is commonly measured by several techniques, viz, air permeability, adsorbent gas volume, bulk density, and resilience (compression and recovery). The adsorption and retention of kerosene is also used as a measure of fiber openness and fiber adsorption capacity (37).

The filtration behavior of asbestos fibers is evaluated through a drainage test using, generally, a saturated calcium sulfate solution. This test is designated as "freeness" and reflects the water drainage rate related to asbestos–cement fabrication processes (37).

The reinforcing capacity of asbestos fibers in a cement matrix constitutes another key criterion for the evaluation of asbestos fibers. Preparing samples of asbestos–cement composites that, after a standard curing period, are tested for flexural resistance assesses this property. The measured rupture moduli are converted into a parameter referred to as the fiber strength unit (37).

Finally, other properties of asbestos fibers may be evaluated depending on the potential application. Typically, the grits and spicule content, the magnetic susceptibility (magnetic rating), the content in soluble chlorides, and the humidity level may be of particular interest in specific applications.

10. Industrial Applications

Asbestos fibers have been used in a broad variety of industrial applications. In the peak period of asbestos consumption in industrialized countries, some 3000 applications, or types of products, have been listed. Because of recent restrictions and changes in end use-markets, most of these applications have been abandoned and the remainder is pursued under strictly regulated conditions.

The main properties of asbestos fibers that can be exploited in industrial applications (11) are their thermal, electrical, and sound insulation; nonflammability; matrix reinforcement (cement, plastic, and resins); adsorption capacity (filtration, liquid sterilization); wear and friction properties (friction materials); and chemical inertia (except in acids). These properties have led to several main classes of industrial products or applications (8).

Loose asbestos fibers, or formulations containing asbestos fibers for spray coatings, were widely used in the building industry for fire protection and heat or sound insulation during and following World War II. Such applications used mainly chrysotile or amosite, although crocidolite was commonly used in Europe, this practice was discontinued in the 1970s because of health concerns.

Asbestos fibers also have been widely used for the fabrication of papers and felts for flooring and roofing products, pipeline wrapping, electrical insulation, etc. Asbestos textiles, comprising yarn, thread, cloth, tape, or rope, also found wide application in thermal and electrical insulation and friction products in brake or clutch pads. In recent years, use of asbestos in these applications has decreased significantly. Production of asbestos roofing and flooring felts has been discontinued.

The reinforcing properties of asbestos fibers have been widely exploited in asbestos–cement products mostly by the construction and water industries. Asbestos–cement products such as board, pipe, and sheet represent by far the largest worldwide industrial consumption of asbestos fibers, an estimated 80% of the market in 1988 (38). With market changes since 1988, asbestos–cement products now probably account for >98% of fiber sales.

Asbestos fibers also have been used to reinforce plastic products made from PVC, phenolics, polypropylene, nylon, etc. Reinforcement of thermoset and thermoplastic resins by asbestos fibers was used to develop products for the automotive, electronic, and printing industries. Except for some specialty products, the use of asbestos in plastics has essentially ceased.

The combination of asbestos fibers with various types of natural or synthetic resins has led to the development of a variety of products and applications. Among those, the incorporation of asbestos fibers (mainly chrysotile) into rubber matrices yields materials that were widely used for fabrication of packings and gaskets. Complex formulations, comprising short asbestos fibers (usually chrysotile), resins, and other fillers and modifiers, have been developed as friction materials for brake linings and pads. Asbestos fibers also have found broad application as reinforcing agents in coatings, sealings, and adhesive formulations. Of these applications, brake linings and pads, gaskets, and roof coatings comprise the bulk of the consumption.

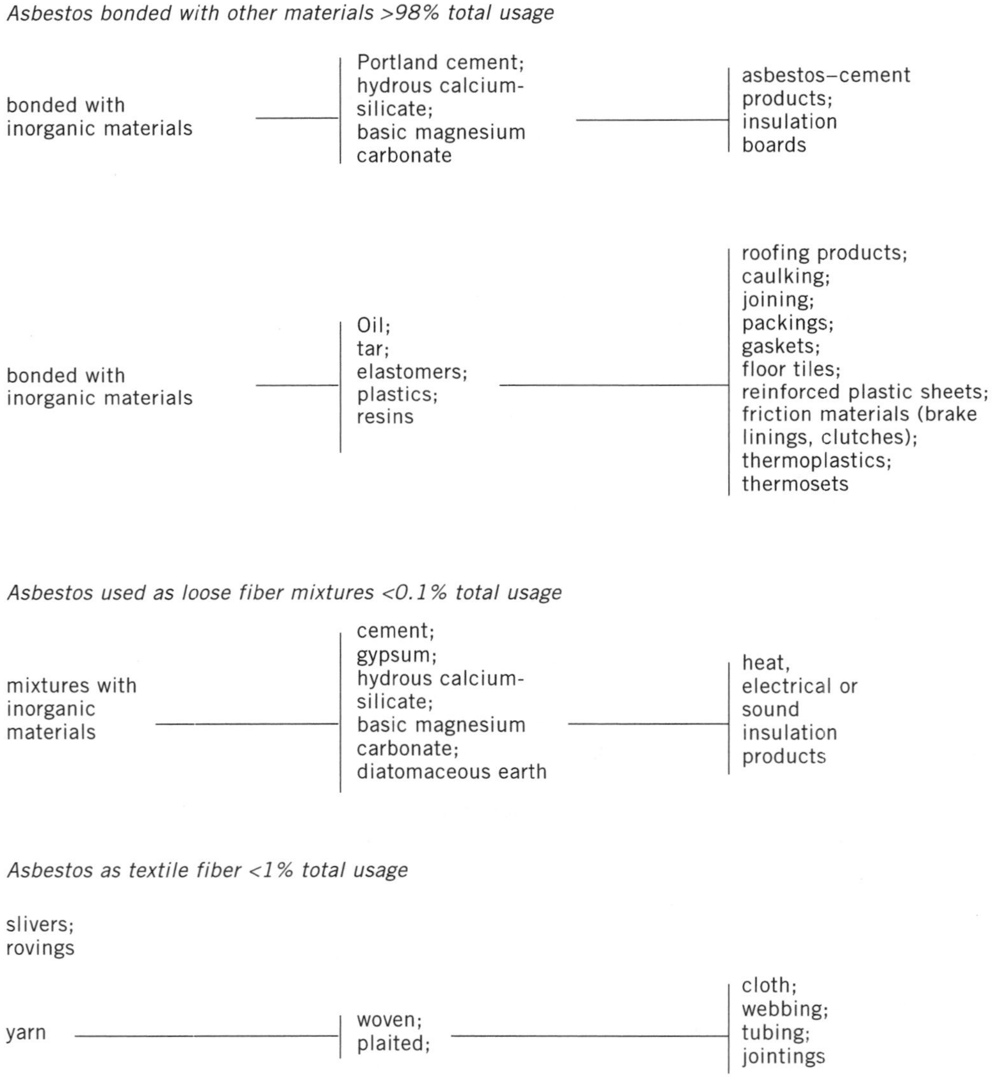

Asbestos bonded with other materials >98% total usage

bonded with inorganic materials	Portland cement; hydrous calcium-silicate; basic magnesium carbonate	asbestos–cement products; insulation boards

bonded with inorganic materials	Oil; tar; elastomers; plastics; resins	roofing products; caulking; joining; packings; gaskets; floor tiles; reinforced plastic sheets; friction materials (brake linings, clutches); thermoplastics; thermosets

Asbestos used as loose fiber mixtures <0.1% total usage

mixtures with inorganic materials	cement; gypsum; hydrous calcium-silicate; basic magnesium carbonate; diatomaceous earth	heat, electrical or sound insulation products

Asbestos as textile fiber <1% total usage

slivers;
rovings

yarn	woven; plaited;	cloth; webbing; tubing; jointings

Fig. 10.   Utilization of asbestos fibers by process. Percent, by category, in 2000. Product groups in bold represent major end uses within a process category (modified from 39).

Table 6. **Uses of Asbestos**[a]

Grade[b]	Length specifications, mm	Uses[c]
long fiber		
No. 1 crude	19	textiles
No. 2 crude	9.5–19	textiles, insulation
No. 3	6–9.5	textiles, packings, brake linings, clutch facings, electrical, high pressure and marine insulation
medium fiber		
No. 4	3–6	asbestos–cement pipe, brake linings, plastics
No. 5	3–6	asbestos–cement pipe and sheets, molded products, paper products, brake linings and gaskets
No. 6	3	asbestos–cement products, brake linings and gaskets, plaster, backing for vinyl sheets
short fiber		
No. 7	3	molded brake linings and clutch facings, plastics, filler in vinyl and asphalt floor tiles, asphalt compounds, caulking compounds, thermal insulation, gaskets, paints and drilling mud additive
No. 8	d	similar to No. 7

[a]Ref. 36.
[b]Quebec Standard.
[c]Many uses shown are discontinued; major use in 2000 in bold.
[d]Loose density specifications.

Finally, the combined reinforcing effect and high absorption capacity of asbestos fibers were exploited in a variety of applications to increase dimensional stability, typically in vinyl or asphalt tiles and asphalt road surfacing. Asbestos is no longer used in these applications.

Figure 10 summarizes the various classes of application for asbestos fibers in combination with other materials. The diagram shows that in recent years, most industrial applications involve asbestos fibers bonded within an organic or inorganic matrix. Asbestos–cement products account for the bulk of the world's asbestos usage. In the United States, the major use is in roofing compounds (62%), followed by gaskets (22%), and friction products (11%). Small amounts of asbestos also are used to manufacture some insulation products and woven and plastic products (9).

Each type or group of products usually requires a selected asbestos fiber grade (or range of grades). Table 6 lists various types of chrysotile applications in relation to QS fiber grades.

11. Alternative Industrial Fibers and Materials

Considerable effort has been devoted to finding alternative fibers or minerals to replace asbestos fibers in their applications. Such efforts have been motivated by various reasons, typically, availability and cost, and more recently, health and liability concerns. During World War II, some countries lost access to asbestos

Table 7. **Asbestos Substitutes and Relative Costs**[a]

Minerals	Synthetic mineral fibers	Synthetic organic fibers
	< 2 $/kg[b]	
attapulgite	mineral wool	
diatomite	glass wool	
mica		
perlite		
sepiolite		
talc		
vermiculite		
wollastonite		
asbestos, grades 3–7		
	2–10 $/kg	
	steel fibers	polypropylene (PP)
	continuous filament glass	poly(vinyl alcohol) (PVA)
	alkali-resistant glass	polyacrylonitrile (PAN)
	aluminosilicates	
	10–20 $/kg	
	continuous filament glass	polytetrafluoroethylene (PTFE)
	>20 $/kg	
	alumina fibers	polybenzimidazole (PBI)
	silica fibers	aramid fibers
	graphite fibers	pitch and PAN carbon fibers

[a]In U.S. $, 1989. Ref. 40.
[b]The natural organic fiber, cellulose (pulp), also falls in the < $2/kg range.

fiber supplies and had to develop substitute materials. Also, in the production of fiber-reinforced cement products, many developing countries focused on readily available, low-cost cellulose fibers as an alternative to asbestos fibers. Since the 1980s, however, systematic research has been pursued in several industrialized countries to replace asbestos fibers in many current applications because of perceived health risks.

The substitution of asbestos fibers by other types of fibers or minerals must, in principle, comply with three types of criteria (40): the technical feasibility of the substitution; the gain in the safety of the asbestos-free product relative to the asbestos-containing product; and the availability of the substitute and its comparative cost.

In some applications, particularly those that rely on several characteristic features of asbestos fibers, the substitution has presented a significant challenge. For example, in fiber–cement composites, the fibers must exhibit high tensile strength, good dispersibility in Portland cement pastes, and high resistance to alkaline environments. Likewise, in friction materials, several properties of the asbestos fibers are important: high affinity toward resins, good heat resistance, high friction coefficient, and low abrasion of the opposing surface. In such applications, the replacement of asbestos fibers has required a combination of several materials. Table 7 lists some of the materials and fibers that have been suggested or used in the development of asbestos-free products along with an estimate of the cost ranges of asbestos fibers and several types of substitution materials.

Various asbestos substitution strategies have been followed and a wide range of asbestos-free products have been developed. For example, in bulk thermal insulation and sprayed insulation coating applications, synthetic mineral fibers (glass or slag fibers) and cellulose fiber products have replaced natural asbestos fibers. Clothing made from aramid fibers or aluminized glass fibers has replaced asbestos in most textiles (see HIGH PERFORMANCE FIBERS). Asbestos in floor tile has been replaced by a combination of synthetic fibers and nonfibrous fillers. Asbestos packings have been replaced by various materials including aramid and glass fiber, graphite mixtures, and cellulose fibers.

In fiber–cement construction materials, several alternatives are being practiced, either using cellulosic fibrous products or synthetic organic fibers such as polypropylene or polyacrylonitrile or alternative products such as cast iron, PVC, or PP pipe. The use of asbestos–cement products still is considerable owing to low cost, availability, low technology requirements, and performance.

For friction material applications, composite materials comprising glass or metallic fibers with other minerals have been developed. In these applications, aramid and graphite fibers are effective, although the cost of these materials restricts their use to heavy duty or high technology applications (see CARBON FIBERS). Substitution has been more successful in disk brake pads as compared with brake shoes (40).

Efforts to substitute for asbestos have been fairly successful as evidenced by the fact that peak world asbestos production was 4.8×10^6 tons in 1977, declining to 1.9×10^6 tons in 2000. This value represents a decline of 60% in the use of asbestos despite growth in every market in which asbestos was used during the 25-year period. The availability of adequate replacement materials, the cost-to-performance ratio of such materials, and the uncertainty of long-term health risks of these replacement materials have limited the extent of substitution of asbestos fibers by other fibers in some product applications.

12. Health and Safety Factors

The relationship between workplace exposure to airborne asbestos fibers and respiratory diseases is one of the most widely studied subjects of modern epidemiology (41–43). Asbestos-related health concerns were first raised at the beginning of the century in the United Kingdom, which appears to have been the first country to regulate the asbestos-user industry (44). It wasn't until the early 1960s, however, that, researchers firmly established a correlation between worker excess exposure to asbestos fibers and respiratory cancer diseases (45). This finding triggered a significant research effort to unravel important issues such as the influence of fiber size, shape, crystal structure, and chemical composition; the relationship between exposure levels and diseases; the consequence of exposure to asbestos fibers in different types of industries, or from different types of products; and the development of technologies to reduce worker exposure.

The research efforts resulted in a consensus in some areas, although controversy still remains in other areas. It is widely recognized that the inhalation of long (considered usually as >5 μm), thin, and durable fibers in high concentrations over a long period of time can induce or promote lung cancer. It is also widely accepted that asbestos fibers can be associated with three types of diseases (46): asbestosis: A lung fibrosis resulting from long-term, high level exposures to airborne fibers; lung cancer: Usually resulting from long-term high level exposures and often correlated with asbestosis; mesothelioma: A rare form of cancer of the lining (mesothelium) of the thoracic and abdominal cavities.

A further consensus developed within the scientific community regarding the relative carcinogenicity of the different types of asbestos fibers. There is strong evidence that the genotoxic and carcinogenic potentials of asbestos fibers are not identical; in particular mesothelial cancer is most strongly associated with amphibole fibers (47–51).

The replacement of asbestos fibers by other fibrous materials has raised similar health issues about substitute materials. However, lung cancer has a latency period of ~25 years and fiber exposure levels to substitutes are far lower than those that prevailed half a century ago with asbestos. Consequently, the epidemiological data on most substitutes is insufficient to establish statistically significant correlations between exposure and pulmonary disease (52). A possible exception is slag fibers for which several studies on worker populations are available over extended periods (53); some results show a substantial increase in lung cancer occurrence. Consequently, the toxicity of asbestos substitute fibers remains a subject of active investigation.

12.1. Regulation. The identification of health risks associated with long-term, high level exposure to asbestos fibers, together with the fact that large quantities of these minerals were used (several million tons annually) in a variety of applications, prompted the enactment of regulations to limit the maximum exposure of airborne fibers in workplace environments. The exposure limits may be defined as average or peak values, measured either as a weight or as a number of fibers-per-unit-volume. The International Labor Organization adopted the following definition (Convention 162, article 2d and R172): "the term respirable asbestos fibers means asbestos fibers having a diameter of less than 3 µm and a length-to-diameter ratio greater than 3:1. Only fibers of a length greater than 5 µm shall be taken into account for purposes of measurement." In general, exposure limits vary considerably among countries with the range being from 0.1 fibers per cubic centimeter (f/cc) for 4- or 8-h time-weighted-average exposure to 2 f/cc.

In accordance with demonstrated differences between the various asbestos fiber types, workplace regulations in many countries specify different exposure limits for chrysotile and the amphiboles (54). Typically, exposure limits for amphibole asbestos are one-half to one-tenth that for chrysotile. Also, the European Union voted to ban the use of amphibole asbestos and its use has been discontinued. Moreover, to alleviate established, or apprehended, risk from substitute fibers, regulations often specify maximum exposure limits for synthetic fibers (55).

Some countries have opted for a broader approach and have adopted regulations that also minimize exposure of the general public to environmental asbestos fibers, ie, by banning or restricting asbestos imports and types of applications. Countries that have banned (either a complete bans or a ban with exemptions) or are phasing out the use of asbestos and in some cases, asbestos products, in the next few years include Argentina, Austria, Belgium, Chile, Denmark, Finland,

France, German, Italy, Netherlands, Norway, Poland, Saudi Arabia, Sweden, Switzerland, and the United Kingdom. The European Union voted to ban asbestos use in most applications by its members by 2005.

The obvious trend with asbestos is toward stricter regulations or the banning of its use. Banning has occurred for the most part in developed countries where substitute materials or alternative products are readily available and it is economically feasible to use asbestos substitutes. In lesser developed countries where economics and the level of industrial development is a factor, asbestos substitutes are not yet considered to be a suitable option in many cases.

Despite the current bans and continued opposition to the use of asbestos, markets for asbestos probably will exist long into the future. Consumption can be expected to decline as substitutes and alternative products gain favor in the remaining world markets, which, however, is a process that probably will occur over a period of decades. Even then, there probably will remain specialized applications for asbestos, particularly for matrix-based products.

BIBLIOGRAPHY

"Asbestos" in *ECT* 1st ed., Vol. 2, pp. 134–142, by M. S. Badollet, Johns-Manville Research Center; in *ECT* 2nd ed., Vol. 2, pp. 734–747 by M. S. Badollet, Consultant; in *ECT* 3rd ed., Vol. 3, pp. 267–283, by W. C. Streib, Johns-Manville Corporation, in *ECT* 4th ed., Vol. 3, pp. 659–688, by C. R. Jolicoeur, J. Alary, and A. Sokov, Universite de Sherbrooke, Canada; "Asbestos" in *ECT* (online), posting date: December 4, 2000, by C.R. Jolicoeur, J. Alary, and A. Sokov, Universite de Sherbrooke, Canada.

CITED PUBLICATIONS

1. M. Ross, R. A. Kuntze, and R. A. Clifton, in B. Levadie, ed., *Definition for Asbestos and Other Health Related Silicates*, ASTM STP 834, American Society for Testing and Materials, Philadelphia, Pa., 1984, pp. 139–147.
2. W. J. Campbell and co-workers, *Selected Silicates Minerals and Their Asbestiform Varieties*, IC 8751, U.S. Bureau of Mines, Washington, D.C., 1977, pp. 5–17, 33.
3. A. A. Hodgson, *Scientific Advances in Asbestos, 1967 to 1985*, Anjalena Publication, Crowthorne, UK, 1986, pp. 10, 14, 17, 23, 42, 53, 76, 78, 85, 95–99, 107–117.
4. H. C. W. Skinner, M. Ross, and C. Frondel, *Asbestos and Other Fibrous Materials*, Oxford Press, New York, 1988, pp. 21–23, 25, 31, 34, 35, 42–66.
5. A. Europaeus-Äyräpää, *Acta Archaeol.* **1**, 169 (1930).
6. J. Alleman and B. Mossman, *Sci. Am.*, July 1997, pp. 70–75.
7. O. Bowles, *Asbestos—The Silk of the Mineral Kingdom*, The Ruberoid Co., New York, 1946, pp. 13–22.
8. M. A. Bernarde, ed., *Asbestos: The Hazardous Fiber*, CRC Press, Boca Raton, Fla., 1990, pp. 4, 30–38, 41, 80.
9. R. L. Virta, *Asbestos*, U.S. Geological Survey Minerals Yearbook 2000 preprint, 2001, 7 p.
10. E. J. W. Whittaker, in R. L. Ledoux, ed., *Short Course in Mineralogical Techniques of Asbestos Determination*, Vol. 4 (Mineralogical Association of Canada, Quebec, Canada, May 1979), pp. 16, 24–28, 30.

11. *Report of the Royal Commission on Matters of Health and Safety Arising from the Use of Asbestos in Ontario*, J. S. Dupré, chairman, Vol. 1, Ontario Government Book-Store, Toronto, Ontario, Canada, 1984, pp. 78, 89–93.

12. F. J. Wicks, in Ref. 10, pp. 58, 65–72.

13. *Asbestos Factbook*, Asbestos, Willow Grove, Pa., 1970.

14. E. J. W. Whittaker, *Acta Crystallogr.* **10**, 149 (1957).

15. K. Yada, *Acta Crystallogr.* **27**, 659 (1971).

16. J. F. Alary, Y. Côté, and C. Jolicoeur, unpublished data, 1991.

17. C. Jolicoeur and D. Duchesne, *Can. J. Chem.* **59**(10), 1521 (1981).

18. E. Papirer and P. Roland, *Clays and Clay Miner.* **29**, 161 (1981).

19. E. Orowan, *Rep. Prog. Phys.* **12**, 186 (1948–1949).

20. R. G. Bryans and B. Lincoln, *Paper 6.24, 3rd International Conference on the Physics and Chemistry of Asbestos Minerals*, Québec, Canada, 1975.

21. R. G. Bryans and B. Lincoln, *Paper 3.6, 2nd International Conference on the Physics and Chemistry of Asbestos Minerals*, Université de Louvain, Louvain, Belgium, 1971.

22. A. A. Hodgson, in L. Michaels and S. S. Chissick, eds., *Asbestos: Properties, Applications and Hazards*, Vol. 1, John Wiley & Sons, Inc., New York, 1979, pp. 89–90, 93.

23. S. Speil and J. P. Leinerveber, *Environ. Res.* **2**(3), 166 (1969).

24. G. Stroink, R. A. Dunlop, and D. Hutt, *Can. Mineral.* **19**, 519 (1981).

25. A. A. Hodgson and C. A. White, *Papers 2-10, The Physics and Chemistry of Asbestos Minerals, Oxford Conference on Asbestos Minerals*, Oxford, UK, 1967.

26. Z. Johan and co-workers, *Adv. Org. Geochem.*, 883–903 (1973).

27. T. Otouma and S. Take, *Paper 5.21*, in Ref. 19.

28. L. Zapata and co-workers, *Paper 3.4*, in Ref. 20.

29. U.S. Pat. 4,356,057 (Oct. 26, 1982), J. M. Lalancette and J. Dunnigan (to Société Nationale de L'amiante).

30. E. Papirer, *Clays and Clay Miner.* **29**, 69 (1981).

31. U.S. Pat. 3,689,430 (Sept. 5, 1972), P. C. Yates (to E. I. du Pont de Nemours & Co., Inc.).

32. W. C. McCrone, *The Asbestos Particle Atlas*, Ann Arbor Science Publishers, Ann Arbor, Mich., 1980, p. 33.

33. F. A. Settle, ed., *Handbook of Instrumental Techniques for Analytical Chemistry*, Prentice-Hall, New Jersey, 1997, 995 p.

34. M. J. Keane and co-workers, *J. Toxicol. Environ. Health*, **57**(8), 529 (1994).

35. M. Cossette and P. Delvaux, in Ref. 7, pp. 79–80.

36. *An Appraisal of Minerals Availability for 34 Commodities*, U.S. Bureau of Mines Bulletin 692, Washington, D.C., 1987, p. 35.

37. *Chrysotile Asbestos Test Manual*, 3rd ed., Asbestos Textile Institute and the Quebec Asbestos Mining Association, Quebec, Canada, 1974 (revised 1978).

38. *The Economics of Asbestos*, 6th ed., Roskill Information Services Ltd., London, 1990, p. 11.

39. J. Harrod and V. Thorpe, *Asbestos, Politics and Economics of a Lethal Product*, International Federation of Chemical, Energy and General Workers' Unions, Geneva, Switzerland, 1984.

40. A. A. Hodgson, ed., *Alternatives to Asbestos, The Pros and Cons*, John Wiley & Sons, Inc., New York, 1989, p. xi.

41. *Asbestos and Other Natural Mineral Fibers*, WHO Environmental Health Criteria 53, World Health Organization, Geneva, Switzerland, 1986.

42. J. Bignon, J. Peto, and R. Saracci, eds., *Nonoccupational Exposure to Mineral Fibres*, IARC Scientific Publication No. 90, International Agency for Research on Cancer, Lyon, France, 1989.

43. *Occupational Exposure Limit for Asbestos*, Report of a WHO Meeting, Oxford, UK, Apr. 10–11, 1989.
44. R. Murray, *Br. J. Ind. Med.* **47**, 361 (1990).
45. I. J. Selikoff, J. Churg, and E. C. Hammond, *J. Am. Med. Assoc.* **188**, 142 (1964).
46. J. Peto and co-workers, *Environ. Health Perspect.* **70**, 51 (1986).
47. Robert Nolan et al., eds., *The Health Effects of Chrysotile Asbestos, Special Publication 5*, The Canadian Mineralogist, Ottawa, Ontario, 2001.
48. M. J. Gardner and C. A. Powell, *J. Soc. Occup. Med.* **36**(4), 124 (1986).
49. R. Wilson and B. Price, *paper 5.3*, in Ref. 47.
50. Graham Gibbs, *paper 4.1*, in Ref. 47.
51. John Dement, Carcinogenicity of Asbestos-Differences by Fiber Type?, presented at 2001 Asbestos Health Effects Conference, May 24–25, 2001, Oakland, CA, Environmental Protection Agency.
52. M. Camus, *paper 4.8*, in Ref. 45.
53. "International Symposium on Man-made Mineral Fibers in the Working Environment", Copenhagen, Denmark, Oct. 28–29, 1986, *Ann. Occup. Hyg.* **31**, 4B (1987).
54. *Summary of Main Features of Asbestos/Health Regulations at the Workplace*, AIA Information Memorandum (AIM) No. 3/80, Asbestos International Association, Epson, Surrey, UK, Nov. 1990.
55. *Safety in the Use of Mineral and Synthetic Fibers*, Working Documents (Part 1) of ILO Meeting of Experts, Geneva, Switzerland, Apr. 17–25, 1989, International Labor Organization, 1989, p. 47.

ROBERT L. VIRTA
U.S. Geological Survey

ATOMIC FORCE MICROSCOPY—AFM

1. Principles of Atomic Force Microscopy

The atomic force microscope (AFM) was invented in 1986 at Stanford University by Binnig, Quate, and Gerber (1) as a means of measuring interatomic forces. Their invention was based on a commonly used tool called a profilometer. In this device, a stylus attached to a spring is placed in contact with any relatively flat sample to be examined. The stylus is dragged along the surface of the sample and the deflection of the spring, measured with a variety of techniques, is translated into an image of the sample surface. To create the AFM, Binnig and co-workers (1) used an extremely small cantilever as the spring, a diamond as the stylus or "tip", and a second tip to measure and control the position of the cantilever via a tunneling current. The tunneling mechanism, being exponential in tip–sample distance, made it possible to measure displacements as small as 10^{-3} nm. Since a change of 0.1 nm in the tunneling gap can change the tunneling current by one order of magnitude, a slight lateral drift of the second tip relative to the cantilever will give rise to a false reading of the position of the second tip, rendering the AFM system unstable. Because of this and other reasons, it was

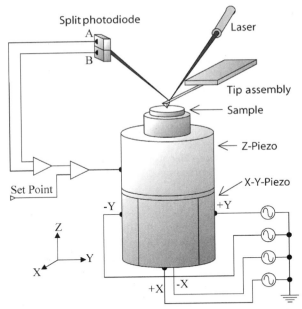

Fig. 1. A schematic diagram of a typical AFM system showing the piezoelectric tube scanner with its electric contacts, the sample mounted on top of the scanner, the tip assembly, and the optical elements.

found necessary to develop alternative methods for monitoring the deflection of the cantilever.

1.1. Contact Mode of Operation. Among the methods for monitoring the displacement of the cantilever in modern AFM systems, one finds optical deflection, optical interference, and cantilever-mounted strain gauges (2). Most current systems, however, are based on the optical deflection technique as shown in Figure 1. Here, a laser beam is reflected off the back surface of the cantilever and is incident on a split (usually four quadrants) photodiode. While the cantilever is in its relaxed position with its tip far from the sample, the photodiode is aligned such that the light falls equally on each quadrant. During operation of the AFM, the difference in the intensity of the signal produced by each quadrant measures the deflection of the cantilever. The most important advantage of this method relative to that using a tunneling current is that the deflection of the cantilever is now measured across a large area, adding stability and reproducibility. Also, one can attach an optical microscope above the cantilever and observe the position of the tip as the sample is raster scanned beneath it. This technique is extremely sensitive, and the resolution of the system is limited only by the shape of the tip, contamination of the sample, and cantilever thermal fluctuations. To produce an AFM image in what is called the contact mode, the sample and tip are brought into contact using a stepper motor until the cantilever deflects slightly. The sample is then raster scanned beneath the tip by applying appropriate triangle waves to the x- and y-contacts of the piezoelectric tube on which the sample is mounted. These plus the z-contact provide a fine control over all three directions of motion allowing accuracies of a fraction of an angstrom in the z direction while providing up to 100-μm travel in both lateral directions.

A typical method for obtaining an image of the sample topography is to map the cantilever deflection as the tip is scanned across the surface, assuming the photodiode voltage is calibrated in terms of the tip deflection for the particular cantilever in use. Though simple and fast, this "constant height" contact mode allows the tip-sample force to vary with the deflection of the cantilever according to Hooke's law, $F_{\text{cant}} = -k\,z$, where k is the spring constant of the cantilever and z is its deflection. Imaging of a soft sample may then show an incorrect representation of the topography as the force, and therefore penetration depth into the sample, changes.

A more accurate technique for imaging is called the "constant force" contact mode. In this mode, the initial deflection is set at a value chosen by the user, thus determining the force of contact. As the tip comes into contact with a feature on the surface of the sample, the deflection of the tip is measured and compared to the set-point deflection. The bias to the z contact of the piezoelectric tube is then adjusted to raise or lower the sample until the cantilever returns to the set-point deflection. This bias is displayed on the screen, calibrated in units of distance, to produce the topography image of the sample. With such feedback electronics, a constant deflection of the tip, and therefore force, is maintained during the whole scan leading to an accurate image of the topography even on soft samples. This technique also has the advantage that the system calibration is dependent on the z contact of the piezoelectric tube, which is rather constant, as opposed to properties of the cantilever, which varies from cantilever to cantilever. Note, however, that the imaging rate is limited by the speed of the feedback electronics and the spring constant of the cantilever.

1.2. Cantilever Assemblies. Modern AFM cantilever assemblies, made from a variety of materials, can be purchased from several vendors. Most common today are microfabricated silicon or silicon-nitride cantilevers. Figure 2 is an FIB image of an *Olympus* brand AFM cantilever. On the left-hand side of the image is a portion of the spring-like cantilever that is typically 100–300 μm long with a spring constant of 0.01–100 N/m, depending on the application. On the right-hand side of Figure 2, pointing down, is the tip whose radius of curvature is typically 10–50 nm with a cone-angle of ~15°. Such tips can often be processed to improve their sharpness, commonly down to a radius of 5–10 nm, providing a resolution of ~1–2 nm. The tips can be coated with various materials to harden them or to make them conducting or magnetic. Much work is also being invested in the development of single-wall carbon nanotube-based tips. These may provide a durable structure with a radius of curvature of 2–3 nm (3). Not shown in this figure is the tip substrate, a large area used to hold and manipulate the cantilever.

1.3. Tip-Sample Interactions. To fully understand the operation of the AFM, it is important to consider the forces acting between the tip and the sample. However, the flexibility of the cantilever complicates a direct measurement of these forces. To simplify the analysis, assume a tip attached to a cantilever with infinite stiffness such that the tip–sample separation is controlled directly by the substrate–sample separation. Later, this restriction will be relaxed.

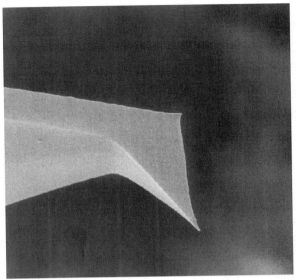

Fig. 2. An image of an *Olympus*-brand AFM cantilever with its pointed tip.

The three dominant tip–sample interactions involve the van der Waals (F_{vdW}), capillary (F_c), and ionic (F_I) forces, counteracted by the force due to the deflection of the cantilever (F_{cant}),

$$F_{cant} = F_{vdW} + F_c + F_I \tag{1}$$

Figure 3 shows a typical plot of the three forces and their sum as a function of the tip–sample separation. At large separations, these forces are negligible. At

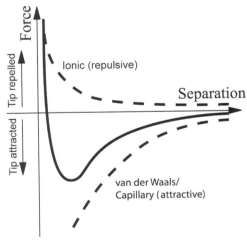

Fig. 3. Typical attractive and repulsive forces acting on the tip of a cantilever as a function of its separation from a sample under ambient conditions.

smaller separations, in vacuum or other inert environments, the van der Waals force attracts the tip toward the sample. This force is generally given in terms of the Hamaker constant, A, defined as

$$A = \frac{3}{4}kT\left(\frac{\epsilon_1 - \epsilon_2}{\epsilon_1 + \epsilon_2}\right)^2 + \frac{3I}{16\sqrt{2}}\frac{(n_1^2 - n_2^2)^2}{(n_1^2 + n_2^2)^{3/2}} \tag{2}$$

Here ϵ_1, ϵ_2, n_1, and n_2 are the dielectric constants and refractive indexes of the tip and sample and I is their ionization potential. For a sphere of radius R, separated by D from a flat surface, the force is given by

$$F_{vdW} = AR/6D^2 \tag{3}$$

For example, using $A = 10^{-19}$ J, $R = 10$ nm, and $D = 0.2$ nm (tip–sample contact) yields a force of attraction of 4.1 nN.

In air, where the tip and sample are generally covered with a thin layer of physisorbed water vapor, a capillary bridge forms between the two surfaces that creates an attractive force between the tip and the sample. This force is given by (4)

$$F_c = 4\pi R\gamma\cos\theta \tag{4}$$

Here, γ is the surface tension of the liquid and θ the angle subtended by the capillary bridge. This force is typically $\sim$10 nN and it is commonly the case that a small concentration of a vapor forming a capillary bridge will dominate the tip–sample force.

As the tip approaches to within $\sim$0.2 nm of the sample, or about the distance of a chemical bond, the ionic antibonding force begins to repel the tip. The ionic repulsion increases until this repulsion balances the attraction of the van der Waals and capillary forces, and the tip is said to be in contact with the sample. Further decreasing of the substrate sample separation results in an upward force on the tip. This contact force is usually kept in the range of 1–100 nN.

Although extremely useful conceptually, the type of curve shown in Figure 3 is difficult to obtain at the nanometer scale because the finite deflection of the flexible cantilever, required to measure the force, is comparable to the tip–sample separation. However, some information about the tip–sample interactions can still be obtained with a flexible cantilever by measuring the deflection of the cantilever as a function of the substrate–sample separation. Figure 4 shows such a typical curve called a force curve.

When the substrate–sample separation is large, the cantilever is not deflected. As the substrate is brought closer to the sample, attractive forces between the tip and sample increase until they are sufficient to deflect the cantilever slightly in the direction of the sample. By bending the cantilever and bringing the tip closer to the sample, the forces increase further, causing the cantilever to bend farther toward the sample. A run-away process thus snaps the tip toward the sample until the repulsive ionic force and the Hooke's law restoring

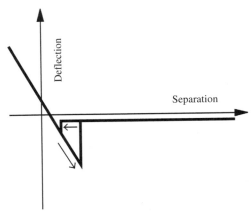

Fig. 4.   The deflection of a cantilever as a function of the separation between its base and the surface of a sample. Note the hysteresis loop associated with the various forces acting on the tip.

force of the cantilever balance the attraction. At this point, the tip is in contact with the sample and further reduction of the tip–substrate separation causes a cantilever relaxation and eventually an upward deflection. On a hard sample, the cantilever will now deflect approximately as a linear function of the separation distance, and this deflection can be directly translated into a tip–sample force using Hooke's law. On a soft sample, however, the tip penetrates the sample that translates into a smaller cantilever deflection for the same change in separation distance.

As the separation is now increased, the process is reversed. However, in this case, the attractive forces act to keep the tip in contact with the sample until the cantilever's bending force overcomes the attractive forces upon which the tip snaps away from the sample. This process occurs at a greater separation than the snap-in process, thus creating a hysteresis loop.

In this manner, the measured deflection shown on the left-hand side of Figure 4 translates directly into a real measurement of tip–sample forces. Note, however, that this curve is not a measure of the tip–sample force as a function of tip–sample separation. Because the cantilever bends, the tip–sample separation is not the same as the substrate–sample separation, which the user can control. Although called a force curve, a more accurate term for Figure 4 might be a deflection curve as the interaction forces can only be measured from this curve while the tip is in contact with the sample.

1.4. Resolution.   A scanning tunneling microscope (STM), a predecessor of the AFM that requires a conducting sample, operates via a tunneling current that is exponentially dependant on the tip–sample separation distance. Therefore, the predominant contributor of current to the signal is from the very bottom of the lowest atom on the tip, with no contributions coming from atoms farther away from the sample. This tool is therefore capable of providing true atomic resolution. The tip–sample interaction of the AFM, however, is not as sensitive to distance (5,6), so atoms other then the end-most atom can contribute to the image, the number of which is dependent on the sharpness of the tip. Thus,

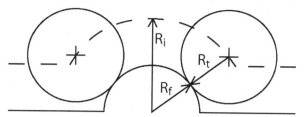

Fig. 5. An example of a tip–sample topographic convolution effect. Here, two positions of the scanning tip straddle a topographic bulge.

while many AFM images of atoms are available, most of these are due to the summation of the interactions of many atoms. The AFM, in general, cannot detect the presence or absence of individual atoms. Only under highly controlled conditions is the AFM capable of true atomic resolution (7). For large structures, in general, the resolution is mostly limited by tip–sample topographic convolution effects.

This convolution is demonstrated below in Figure 5, showing a tip of radius R_t imaging a feature of radius R_f on a sample. The resulting image of the feature has a radius of curvature of $R_i = R_t + R_f$. Thus, since we tend to think of the tip as an infinitesimal point, it appears as though this point has imaged a much larger feature.

When imaging surfaces on larger scales, say a 10–100-μm scan, with a common resolution of 512 pixels per scan line, visible features can be no smaller than ~20–200 nm. In this case, the convolution can generally be negligible. For smaller scans, a sharp tip is necessary to produce reliable images.

1.5. Noncontact Mode of Operation.   The contact mode AFM operates solely within the repulsive region of the tip–sample force curve. This implies that the tip continuously applies a normal force as high as 10 μN as well as applying a shear force as a result of the tip being dragged across the surface of the sample. This mode of operation can be highly destructive to the sample, particularly in the case of soft samples such as organic or biological material.

In the noncontact mode of operation, the tip is within the attractive region of the tip–sample force curve and the interaction forces are much smaller and therefore more difficult to measure (8,9). Such an operation requires a phase-sensitive method. To that end, the tip is oscillated with an amplitude of several nanometers near its resonance frequency. This is commonly accomplished using an additional piezocrystal or a magnetic field. The force derivatives acting on the tip modify the cantilever's effective spring constant, and hence its resonance frequency and amplitude of vibration. In this manner, variations in the oscillation amplitude indicate the tip–sample separation. As with the "constant-force" mode of operation, feedback electronics is employed to adjust the sample z position until the oscillation amplitude returns to a set value. Because the tip is no longer dragged across the sample, shear forces are completely eliminated. Additionally, the normal force is limited to the much weaker attractive forces, which are on the order of 10^{-12} N. In this mode, stiffer cantilevers, such as 10 N/m, are generally desired to allow for high oscillation frequencies, typically 100–400 kHz, and to prevent the tip from snapping into the sample.

Along with variations in amplitude, it is also possible to produce images based on variations in the phase lag between the cantilever driving force and the response of the cantilever.

1.6. Intermittent-Contact Mode Atomic Force Microscopy (FM). Although the noncontact mode of operation significantly reduces the sample damage associated with the contact mode of operation, the repulsive interaction of the latter provides a higher force and therefore higher speed. In the case of samples with large topographic features, it may be advantageous to operate in the intermittent-contact mode of operation (10–13). In this mode of operation, the tip is still vibrated. However, here the tip is repeatedly brought from the attractive region into the repulsive region of the tip–sample force curve. Therefore, while increasing the interaction force to that near the contact mode of operation, the shear forces of the contact mode of operation are still avoided. This mode is commonly called the Tapping Mode by Digital Instruments or Intermittent Mode by other companies.

1.7. Manufacturers. Atomic force microscopy systems and components are manufactured by several companies. The largest supplier is currently the Veeco Metrology Group, which was formed by aquiring Digital Instruments and WyKo, Topometrix, Thermomicroscopes, Advanced Surface Microscopy, JPK Instruments, K-TEK International, MikroMasch, Molecular Imaging Corporation, NanoDevices, Nanosurf, NANOSENSORS, Novascan Technologies, NT-MDT, Omicron Vacuumphysik, Quesant Instruments, RHK Technology, Surface Imaging Systems, and Triple-O Microscopy.

2. Variants of Atomic Force Microscopy

The three basic modes of operation of the AFM provide different techniques for mapping the topography of a surface. However, the AFM has proven to be a useful base for a wide range of related techniques for mapping other qualities of a sample including, eg, electrical properties, thermal conductivity, elasticity, and friction.

2.1. Scanning Capacitance Microscopy. One technique used for electrical characterization of semiconductor materials is scanning capacitance microscopy (SCM) (14). This technique maps the tip–sample capacitance at a given voltage at every point in the scan using the contact mode of operation. Such a measurement can yield the local doping level of semiconductor substrate.

In general, the capacitance in the vicinity of a conducting tip AFM is determined by applying an ac bias across the tip–sample and measuring the current flow with a lock-in amplifier. Theoretically, the total capacitance of a metal–oxide–semiconductor (MOS) system, C_{tot}, is given as the series capacitance of the oxide, C_{ox}, and the depletion region, C_D. A standard C–V curve is obtained by applying a slowly sweeping dc bias to adjust the depletion depth and a small ac bias to measure the capacitance. In the high frequency ac limit, where charge carriers do not have time to collect under inversion, C_D can be approximated in terms of the depletion depth, $C_D = \varepsilon_{Si}/x_D$, where the depletion depth, x_d, is given by (15)

$$X_D = \sqrt{\frac{2\varepsilon_{Si}\psi_s}{qn_p/p_p}} \tag{5}$$

Here, ψ_s is the potential drop between the silicon bulk and the silicon/oxide surface and n_p/p_p the ratio of minority to majority carriers in the substrate. In the low frequency ac limit, where minority carriers collect in the inversion region, C_D is given by (15)

$$x_D = \frac{\varepsilon_{\text{Si}}}{\sqrt{2}L_D} \frac{1 - e^{-\beta\psi_s} + n_p/p_p(e^{\beta\psi_s} - 1)}{\sqrt{(e^{-\beta\psi_s} + \beta\psi_s - 1) + n_p/p_p(e^{\beta\psi_s} - \beta\psi_s - 1)}} \quad (6)$$

where L_D is the Debye length and $\beta = q/KT$. In both cases, the surface potential, ψ_s, is given in terms of the applied dc bias. Early experiments in SCM imaging were performed at a fixed dc bias in the high frequency limit, providing a measure of C_{tot}. In modern systems, a second medium-frequency bias is added to measure dC_{tot}/dV to provide a greater signal-to-noise ratio by measuring the signal at a frequency above the most prevailing noise contributions.

2.2. Tunneling AFM. A standard technique for characterizing MOS devices on large-scale areas consists of obtaining $I-V$ curves across mmeter-size metal disks fabricated on the surface of the sample. Here, the bias swings are larger than the work function of the media involved (~4 V), namely, under the Fowler-Nordheim current emission regime. These $I-V$ curves provide information on charge trapping, dielectric strength, and oxide degradation. In its standard form, however, this technique lacks the required resolution. Therefore, a new form of AFM has been developed that uses a conducting tip and a constant tip–sample bias that gives rise to a varying tunneling current through a thin dielectric film. Such an AFM is called a tunneling atomic force microscope (TAFM) or TUNA by Digital Instruments. Here the tunneling current depends on the local thickness of the probed dielectric, making it possible to obtain dual maps of the topography of the top and bottom surfaces of the dielectric, as well as a map of the thickness or electrical quality of the dielectric, all at a resolution equaling approximately the thickness of the dielectric (16). The utility of the TAFM has been demonstrated by several groups who use it routinely for the characterization of the local thickness and dielectric strength of thin silicon oxide and nitride films, and carbon contamination of oxides on silicon (16–25).

Electron Tunneling. MOS devices are typically fabricated on doped silicon that is thermally oxidized to produce a few nanometers-thick oxide. To control the generation of a conducting channel in the silicon beneath the oxide, a metal or polysilicon gate is fabricated above the oxide. If a large enough bias is placed across the thin barrier layer, such as the oxide in the MOS structure, quantum mechanical tunneling of electrons occurs through the layer. This process has been studied extensively (26–33). The total tunneling current through the dielectric, I, is determined by integrating over all electrons with the probability of any electron tunneling through the barrier determined by the WKB approximation (34,35),

$$I = A_{\text{eff}} \frac{e^2}{2\pi h \Delta d^2} \left[\bar{\phi} \exp\left(\sqrt{\frac{8m_{\text{eff}}}{\hbar^2}} \Delta d \sqrt{e\bar{\phi}}\right) - (V_{\text{ox}} + \bar{\phi}) \exp\left(-\sqrt{\frac{8m_{\text{eff}}}{\hbar^2}} \Delta d \sqrt{e(V_{\text{ox}} + \bar{\phi})}\right) \right] \quad (7)$$

Here, A_{eff} is the effective tunneling area, e the electron charge, ϕ is the average barrier height, V_{ox} is the bias across the oxide that is dependent on the flat-band

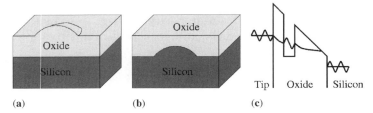

Fig. 6. A depiction of possible dielectric structures and their expression as revealed by AFM and TAFM images of the same area: Case (**a**) shows a silicon−oxide interface that is flat yet the oxide has a bulge. In case (**b**), the silicon−oxide interface has a bulge yet the oxide is flat. In case (**c**), there is a localized lowering of the dielectric conduction band, possibly due to an impurity.

voltage and thus the substrate doping, Δd is the distance through which the electron must tunnel, m_{eff} is the effective mass of the electron in the oxide, and $\hbar$ is Planck's constant. Although this equation is derived for parallel-plate metal electrodes and is being applied to a case where one electrode is a semiconductor and the other is a curved-tip surface, it is still accurate enough for a simple treatment.

Image Interpretation. The TAFM uses a conducting probe scanned over a dielectric of interest to produce images containing features that are interpreted as variations in one or more of the parameters mentioned earlier. In most cases, the dominant variation is assumed to be the change in the oxide thickness, d.

Figure 6 illustrates how features observed in TAFM images can be interpreted in terms of dielectric features. Figure 6**a** shows a case where the silicon−oxide interface is flat yet the oxide has a bulge. The thickened oxide reduces the current flow at a constant applied bias, which can be detected by both tunneling and topography maps. A defect on the surface that is buried, such as that shown in Figure 6**b**, increases the current flow at a constant applied bias and so will be observed by the tunneling system but not by the AFM. Finally, the feature in Figure 6**c** depicts a localized lowering of the conduction band, possibly due to an impurity, effectively decreasing the tunneling length and therefore the tunneling resistivity. This local effect may again not be observed as a topographic feature, as observed by the AFM. To distinguish between these three cases, one needs therefore to analyze both AFM and TAFM maps.

Current Feedback. Although a constant bias applied between tip and sample allows some imaging to be performed, it is easy to conceive of cases where this mode is not desirable. The reason is that since the tunneling equations are depend exponentially on the ratio of the applied bias and the oxide thickness, a sample containing both thin and thick regions of the oxide can cause the current to rapidly swing from below the noise level to above the range of the current detector, producing only a small range of sensitivity. On the other hand, maintaining a constant tunneling current across the oxide yields a direct, linear map of the thickness of the oxide. Note that such an interpretation breaks down at extremely thin oxides. Note, on the other hand, that forcing a current through an oxide can create charge-trapping centers that affect the tunneling current. Constant current imaging, however, makes this effect uniform over the surface

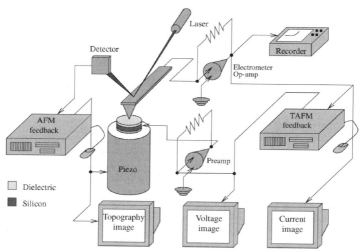

Fig. 7. A schematic diagram of the TAFM system showing the dual feedback electronics, one for the AFM and one for the TAFM.

of the sample, avoiding the large currents associated with a constant-voltage imaging of thin oxides. Care must be taken in controlling the current density during imaging while maintaining a reasonable scan speed. The TAFM system discussed here operates in a constant (fA-range) current mode where a second feedback electronics controls the applied tip–sample bias.

Experimental System.   Figure 7 is a schematic diagram of the TAFM system. The system uses a conventional AFM (36) to map the surface of a dielectric with areas that can be as large as 100×100 μm². For obtaining TAFM maps, a standard silicon or silicon–nitride cantilever is replaced with a conducting cantilever and the area is scanned while applying a tip–sample bias. A second, independent, computer-controlled feedback system (37) simultaneously monitors the tip–sample tunneling current through the dielectric and adjusts the applied positive sample bias to maintain a constant 100-fA set-point current. This current is detected by a preamplifier that is capable of sensing currents as small as 10 fA with a bandwidth of 1 kHz. This amplifier can be based, eg, on the Burr Brown OPA128LM electrometer op-amp. As the tip or sample is raster scanned, the bias required to maintain a constant current is recorded and displayed as a local resistivity map of the oxide with a resolution approximating the thickness of the film. Imaging is usually performed by injecting electrons from the silicon–dielectric interface rather than the tip, providing a reasonably uniform barrier height. The effective electron mass and contact area are also assumed to be constant. Thus, features observed in TAFM images can be attributed to changes in the dielectric thickness and the doping of the substrate.

This system is also capable of producing $I–V$ curves at a chosen location by using the AFM to position the tip at a particular location, ramping the applied bias and recording the current. The bias is ramped back to 0 V once a preset current is reached. The AFM feedback electronics maintains a constant contact tip–sample area by maintaining a constant force on the tip thus producing a

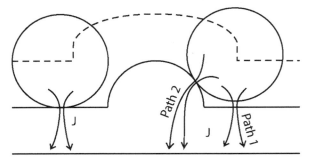

Fig. 8. An example of a TAFM tip–sample topographic and electronic convolution. Here, tunneling between the tip and sample can occur also from the sides of the tip.

conventional AFM image. The entire system is enclosed within a bell jar for environmental control and a Faraday cage for electrical noise reduction.

TAFM Tip Convolution. In the same manner as AFM imaging, TAFM imaging can suffer from tip–sample topographic convolution. However, where AFM convolution is a relatively simple addition of the radii of the probed surface feature and the tip in a given direction, the TAFM convolution is more complex, as shown in Figure 8. When the tip is in direct contact with a flat surface surrounding a feature, the current flows in a continuum of paths directly down through the dielectric as shown on the left-hand side of Figure 8. Upon contacting the feature, the tip lifts off the flat surface, as shown on the right-hand side of Figure 8. At this point, one might imagine that if the tip lifts 1.5 nm above a 5-nm thick feature, than the resulting current would indicate a 6.5-nm thick feature. There are two complications that prevent this from being the case. One is the addition of a vacuum barrier in the original path and the other is the addition of a second continuum of paths contributing to the total current. These two groups of paths are labeled in Figure 8 as Path 1 and Path 2.

The deviation of the current from values expected from a simple increase in the dielectric feature to 6.5 nm is a complex balance between the added vacuum barrier shown as Path 1 and the added path length shown as Path 2. A complete solution to this problem would require a self-consistent accounting of these two sets of paths including consideration of oxide charging. Nevertheless, it is reasonable to assume that as the tip lifts from the surface, there is a shift from Path 1 to Path 2 until Path 2 is the viable one at the top of the feature. Further analysis of this problem, however, is beyond the scope of this work.

Example. Consider an example of a TEM image of a gate oxide with a thinning defect, as shown in Figure 9 (Produced by Siemens). Conventionally, obtaining a transmission electron microscopy (TEM) image requires sectioning the sample and thereby destroying it. The tunneling map of the oxide, however, Figure 10, obtained by TAFM, depicts this defect as a peak, indicating lowered tunneling resistivity were the oxide narrows. Note that unlike the TEM, the TAFM image was produced without damaging the probed device.

Conclusion. The TAFM has been developed as a method for locating and characterizing defects in thin dielectrics. Its ability to do so has been demonstrated on both a manufacturing defect, in this case a MOS capacitor, and

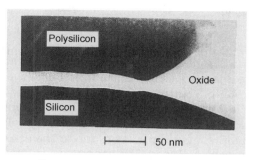

Fig. 9. A TEM image revealing an oxide defect. Note that here the sample had to be cut exactly at the defect site.

on contaminated samples. Although currently useful as a tool for qualitative analysis, continued development would help to improve quantitative analysis. In particular, the issue of sample charging needs to be addressed further as this is a significant effect on the imaging process. For example, it might be possible to determine under what conditions charging could be reduced or eliminated. Alternately, it may be possible to account for this effect rather than eliminating it. Solving the problem of charging should dramatically improve the TAFM calibration and make the tool even more useful.

2.3. Scanning Spreading Resistance Microscopy.   Scanning spreading resistance microscopy (38–42) is another technique for profiling doped semiconductor structures using a conductive-tip AFM in the contact mode of operation. Here, one applies a bias across the tip–sample and the resulting current through the sample is measured with a logarithmic amplifier as the tip is scanned. The current, for a given bias, is proportional to the product of free carrier concentrations and mobility near the location of the tip. These are determined by the doping concentration, and thus a high resolution map of the surface of a semiconductor structure can be generated by moving the tip in controlled fashion across the surface, assuming all dopants are electrically active.

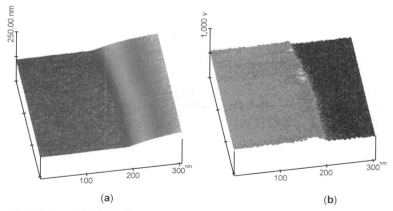

Fig. 10. AFM (**a**) and TAFM (**b**) images of a defective oxide. The TAFM image reveals the defect as a change in tunneling, which is transparent to the AFM.

Typically, this method is applied to the cross-section of semiconductor device structures.

2.4. Lateral Force Microscopy.   One of the problems associated with the operation of a contact-mode AFM is a twisting of the cantilever as it is pulled across the sample. The degree of twisting is a function of the tip–sample friction that in turn is a function of the tip and sample materials. Lateral force microscopy (43) scans the sample such that this twisting is along the axis of the cantilever. Using an additional set of photodetectors, namely, a quadrant structure, yields a measure of changes in friction due to the sample material.

2.5. Electric/Magnetic Force Microscopy.   Electric force microscopy is based on the use of a conducting-tip AFM to map the tip–sample electric field, while the magnetic counterpart uses a tip whose apex is magnetized (44–47). With the AFM operating in the noncontact mode, these forces, which are stronger than the van der Waals force, act through their derivatives to modify the resonant frequency of the vibrating cantilever. The modification of the resonance frequency for a constant driving frequency gives rise to changes in the cantilever amplitude of vibration that is proportional to the local electric or magnetic field.

2.6. Kelvin Probe Microscopy.   Kelvin probe microscopy (48–50) is a technique for measuring the relative work function of a sample and tip. To perform this measurement, an ac bias is applied between a conducting tip and sample to vibrate the cantilever. In this way, the AFM is in a noncontact or intermittent-contact mode. The force between the cantilever and the sample is given by

$$F = 1/2dC/dz(V_{dc} + V_{ac}\sin\omega t)^2$$
$$= 1/2dC/dz(V_{dc}^2 + 2V_{dc}V_{ac}\sin\omega t + 1/2V_{ac}^2 - 1/2V_{ac}^2\cos 2\omega t) \tag{8}$$

Here, C is the tip–sample capacitance, V_{ac} is the applied tip–sample ac bias, and V_{dc} is the applied dc bias plus the work functions of the tip and sample. From this equation, one observes that the strength of the harmonic at the frequency of the applied ac bias is dependent on the strength of the dc component. Thus, by applying an additional dc bias that negates these work functions, the ac term can be canceled so that the tip oscillates at twice the frequency of the applied ac bias. This is performed by using a lock-in amplifier to measure the signal at the frequency of the applied ac bias that in turn generates the negating signal that is a measure of the relative work functions.

2.7. Force Modulation Microscopy.   This AFM variant (51) is operated in a constant-force contact mode with a small oscillation that is applied to the tip at a frequency higher than the frequency response of the feedback electronics of the AFM. However, the amplitude of oscillation of the tip, which is in direct contact with the sample, is a function of the elasticity of the sample. Therefore, a topographic image is obtained from the slow feedback loop while an elasticity map of the sample is obtained from the faster oscillation amplitude using a lock-in amplifier.

2.8. Scanning Thermal Microscopy.   Scanning thermal microscopy (52,53) is a method for measuring the thermal conductance of a sample that exhibits usually a lower resolution than obtained by other AFM techniques. In this AFM variant, a resistive heating element is bent into a V-shaped structure that is used as the AFM tip. A current applied to the element heats it to a temperature above that of the sample where some of this heat is dissipated by the thermal conductance of the sample. This effect can be expressed as $Q = -k (T_{tip} - T_{samp})$, where Q is the heat generated by the tip, which is proportional to the bias applied to the element. Therefore, the bias required to maintain a given temperature is indicative of the sample thermal conductivity.

3. Examples of Atomic Force Microscopy Applied to Semiconductors

Dielectric breakdown of thin gate oxides fabricated on silicon wafers is, in general, a result of a large variety of defects including, eg, particulate contamination, pinholes, and surface roughness. One particular limiting factor affecting the quality of these oxides results from the presence of a variety of metallic islands introduced during wafer processing. Low concentrations of copper in particular have been observed to cause defects when the copper was deposited from hydrofluoric acid (HF) solutions in ultrathin oxides (54–57). For example, for 3-nm oxides, $<1 \times 10^{10}$ atoms/cm^2 copper deposited on the silicon surface from contaminated HF are required prior to oxidation to adversely affect time-dependent dielectric breakdown (TDDB) measurements (58).

Copper deposition from HF occurs via an electrochemical redox reaction (59–66). The Cu^{2+} ions in HF solution, upon contact with a silicon surface, are reduced and deposited at nucleation sites on the silicon. The oxidation reaction takes the form of etching of the silicon surface and as the evolution of hydrogen, which is catalyzed by the presence of copper on the silicon surface. It is the second reaction that is the dominant cathodic reaction (66,67),

$$\text{Cathodic reactions}: \quad \begin{aligned} Cu^{2+} + 2e^- &\rightarrow Cu \\ Cu + 2H^+ + 2e^- &\rightarrow Cu + H_2 \end{aligned} \tag{9}$$
$$\text{Anodic reaction}: \quad Si + 6HF \rightarrow H_2SiF_6 + 2H^+ + H_2 + 2e^-$$

The reduction in oxide quality due to copper-contaminated HF solutions has been found to be due to the roughening of the silicon surface prior to oxidation (58,67). Removal of the copper itself with a hydrochloric peroxide mixture (HPM or RCA-2) clean, used to remove noble metals, fails to improve the oxide quality significantly implying that the roughening of the surface due to the dissolution of silicon is the cause of oxide failure, not the presence of copper formations.

3.1. AFM Imaging of Silicon Etching. An important question, however, is Where does this etching occur? Researchers have observed that silicon dissolution occurs primarily next to copper precipitation (60,61). Figure 11 is an AFM image of a sample contaminated with an HF solution containing high levels of copper. The image shows pits resulting from significant amounts of silicon being removed. This figure is also one of the first demonstrations of a "moat", namely, a ring of etched silicon surrounds a region of material that is higher than

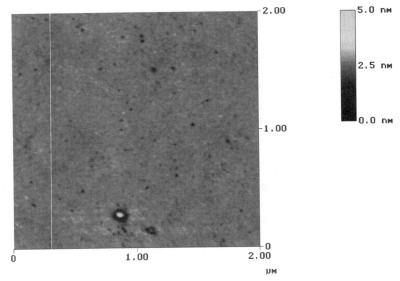

Fig. 11. An AFM image of copper-etched pits surrounded by moats.

the surrounding region. Common sense would suggest that the feature rising above the background consists of a copper deposit. However, AFM examinations of contaminated samples that have been cleaned in either HF or HCl to remove copper show that the surface features remain intact. This result indicates that silicon dissolution occurs not directly beneath the copper deposit but in the area surrounding it, while the silicon beneath the copper is protected from etching. Such a process is shown schematically in Figure 12, where the dashed line

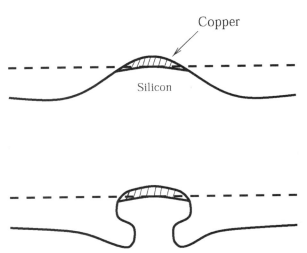

Fig. 12. Two cross-sections of features associated with the cleaning process where the dashed line represents the original silicon height. The top figure refers to a low contamination level that produces little undercutting. The bottom figure refers to a high contamination level that produces a noticeable undercutting.

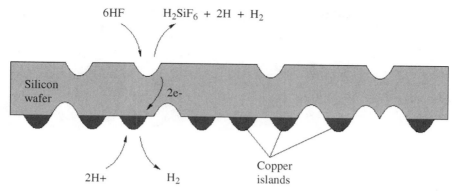

Fig. 13. A diagram of copper-catalyzed silicon etching depicting the silicon wafer, the copper islands and the electrochemsitry involved in the process.

represents the original silicon height. At lower contamination levels, little undercutting occurs and the small copper deposit causes the formation of a large silicon feature beneath it. At higher contamination levels, undercutting occurs and eventually the copper deposit is completely removed, leaving a pit in its place.

Although most silicon dissolution resulting from copper contamination is local, some dissolution also occurs at large distances from the precipitation sites. In particular, the electrochemical reactions suggest that the presence of metallic copper on the unpolished side of a wafer will catalyze hydrogen evolution on the polished side. As a result, the accompanying silicon dissolution will occur on both surfaces of the wafer, as illustrated in Figure 13. Using AFM, verified experimentally that the presence of copper on the unpolished side gave rise to roughening of the polished side, and that roughening caused oxide fails. It was also shown that the presence of hydrochloric acid (HCl) in an HF solution dissolved metallic copper on the wafer before the surface could become significantly rough (68).

Experiment. Photoresist was spun on the front sides of several silicon wafers to protect them from copper deposition. The wafers were then dipped in Cu contaminated HF depositing a small amount of copper on the unpolished sides. After rinsing, some wafers were dipped in uncontaminated HF and some in uncontaminated HF and HCl. This caused the redox reaction to occur. In particular, this includes the etching of silicon on the front side of the wafer. After another rinse, some wafers were thermally oxidized to a thickness of 3 nm for oxide defect analysis.

Results. The resulting wafers were characterized using a Digital Instruments tapping-mode AFM. Figure 14 depicts typical AFM images of (**a**) a raw wafer, w0, (**b**) contaminated wafer without HCl, w2, and (**c**) contaminated wafer with HCl, w3. A z height of 2 nm is represented by a black-to-white contrast in these images. These results show slight roughness of the surface of the reference wafer, however, there is obvious increases in roughening on the test wafer placed in uncontaminated HF. Note that although the additional roughness in Figure 14**b** appears to be due to "particles", the vertical and lateral scales must be considered. The apparent lateral scale of these features is 30–50 nm

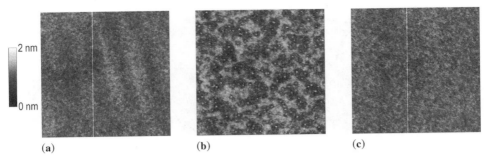

Fig. 14. Three randomly chosen AFM images covering an area of 3×3 µm. (**a**) A raw wafer, (**b**) a contaminated wafer without HCl, and (**c**) a contaminated wafer with HCl.

while the apparent vertical scale is <2 nm. Since the shape of each feature is different, tip convolution can be ruled out. It is concluded that what appears as particles is simply surface roughness. However, the HCl appears to significantly reduce this effect. It appears from Figure 14**b** that the etching process commonly produces a central island surrounded by a moat, further supporting the process suggested in Figure 12, since it is known that no copper is present on the front surface of the wafer.

Figure 15 shows a two-dimensional (2D) isotropic power spectral density (PSD) of all of the data. A technique based on a Fourier transform of the image, the PSD plots the contribution to the image as a function of wavelength and is an excellent method for quantifying surface roughness. Values of PSD on the right-hand side of the plot indicate the presence of low frequency, long wavelength features. In AFM measurements, high values here are typically an indication of drift or nonlinearities in the scanning piezoelectric tube and should not be compared between scans. Peaks on the left-hand side of the PSD plot indicate the presence of high frequency, short wavelength features. These are likely due to incompletely damped vibrations present during scanning.

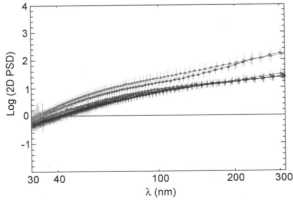

Fig. 15. The averaged power spectral densities of a group of samples. For details, see the text.

The PSD of AFM images are the most accurate between these extremes. For the 1- and 3-μm size images obtained in this experiment, this region is between ~30 and 300 nm and is shown in Figure 15. The lower grouping of curves represents data taken from the raw wafers and the wafers placed in HF where HCl was also present. The middle and top curves represent data taken from the test wafers placed in HF with no HCl and shows extensive roughening of the surface, particularly at wavelengths near 300 nm. This figure shows a 10-fold increase in PSD over a wide frequency range for the test wafers placed in clean HF over the raw wafers. However, the presence of HCl in the HF clearly counters the effect of the copper.

Time-dependent dielectric breakdown measurements of the oxidized wafers without HCl present during the clean confirm a marked decrease in time-to-breakdown on contaminated wafers, similar to earlier results where wafers were contaminated on their polished side (58). In this experiment, however, this degradation is observed as a result of surface roughening of the unpolished side. The use of HCl, again, was found to nearly eliminate this effect.

3.2. TAFM Characterization of Copper Contaminated Gate Oxides.

The TAFM system was used to electrically examine several samples of silicon contaminated with large quantities of copper. Of particular interest, was the direct effect of the large copper deposits on the electrical uniformity of the oxide as compared to the global weakening of the oxide due to silicon dissolution that was examined in the previous experiment. In this experiment, copper-contaminated wafers were thermally oxidized to 4–20 nm in thickness. Figure 16 depicts AFM and TAFM images of the oxide that were obtained simultaneously. Note that the AFM image quality is poor at this stage in the development of the TAFM tool. Due to the high density of defects, it is difficult to interpret the TAFM image; however, some points are worth noting. First, the highly prevalent small black dots indicate regions of high conductivity. It is likely that these result from the increased surface roughness. Second, it also appears that there are regions of lower conductivity indicated by the small white dots. These are larger copper contaminants similar to those observed in the previous sample that are slightly visible in AFM image as well.

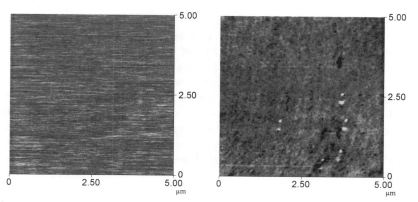

Fig. 16. AFM and TAFM images of a copper-contaminated oxide obtained simultaneously from the same area.

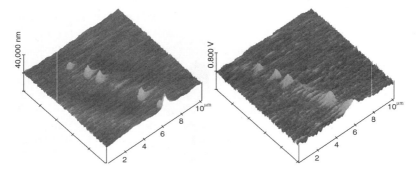

Fig. 17. AFM and TAFM images of a contaminated oxide obtained simultaneously from the same area.

Figure 17 shows another pair of simultaneous AFM and TAFM images. Here again it was found that the islands required a higher bias before the onset of tunneling and therefore would not contribute to oxide breakdown. The AFM images of the oxidized samples showed that the islands protruded ~4 nm above the surface of the already 4-nm thick oxide. The bias required to tunnel through these islands suggests that the oxide thickness was ~6 nm on top of the islands.

4. Summary

This article covered three topics: The principles of AFM, variants of AFM, and examples of AFM applied to semiconductors. The first topic discussed the three main modes of operation of AFM, namely, the contact, noncontact, and intermittent-contact (tapping) modes. Also discussed were cantilever assemblies, tip–sample interactions, and resolution and manufacturers of AFMs. Here, only the most basic concepts were addressed, concepts that will familiarize the reader with the basic phenomena involved in the very rich field of AFM. New insights and interpretations accompanied by a large body of experimental evidence have been published on the operation of the AFM since its inception in 1984, as evidenced by the publication of a large number of articles and books.

The second topic described scanning spreading resistance microscopy, lateral force microscopy, electric and magnetic force microscopy, kelvin probe microscopy, force modulation microscopy, and scanning thermal microscopy. Again, these are the most prevalent variants of AFM that can be found in every well-equipped laboratory. All of these variant are by now commercial products that can be purchased as systems operating either in ambient or under ultrahigh vacuum conditions. Also, some of these products are tailored to applied research applications and others to industrial applications that usually require more automation and the ability to handle larger samples.

The third topic deals with examples of AFM applied to semiconductors. Here, this particular choice was made because of two reasons. The first reason is that the list of examples of results obtained with AFM, even if one were to collect only the most obvious ones, is remarkably rich and could not be fit into this article. The second reason for presenting this particular choice is that the utilization of variants of AFM in new fields of technology appears at an ever growing pace. It was felt, therefore, that it will serve an important purpose to highlight one of the novel applications of AFM, the market-driving industry of semiconductors. Indeed, the shrinking of devices fabricated on a silicon chip to nanometer-scale size requires quantum mechanic considerations for the description of their operation. To characterize and check such minute structures, one has to resort to either electron microscopy or AFM. We therefore decided that it will be important to give examples of applications of AFM to the semiconductors industry.

It is hoped that this article will be found useful in presenting a tutorial of the basic principles of AFM and at the same time will wet the appetite of the reader by describing a state-of-the-art, fast-progressing application that undoubtedly will be superceded by newer applications in the near future.

BIBLIOGRAPHY

1. G. Binnig, C. F. Quate, and C. Gerber, *Phys. Rev. Lett.* **56**, 930 (1986).
2. D. Sarid, *Exploring Scanning Probe Microscopy using Mathematica*, John Wiley & Sons, Inc., Interscience, 1997.
3. H. Dai, J. H. Hafner, A. G. Rinzler, D. T. Colbert, and R. E. Smalley, *Nature (London)* **384**, 147 (1996).
4. U. Hartman, "Theory of Non-contact Force Microscopy", *Scanning Tunneling Microscopy III*, Springer-Verlag, p. 293, 1993.
5. D. J. Kellerr and F. S. Franke, *Surf. Sci.* **294**, 409 (1993).
6. J. Vesenka, R. Miller, and E. Henderson, *Rev. Sci. Instrum.*, **65**, 1 (1994).
7. F. J. Giessibl, *Science* **267**, 6871 (1995).
8. H. Ueyama, M. Ohta, Y. Sugawara, and S. Morita, *Jpn. J. Appl. Phys., Part 2* **34**, L1086 (1995).
9. S. Kitamura and M. Iwatsuki, *Jpn. J. Appl. Phys., Part 2* **34**, L145 (1995).
10. H. G. Hansma and J. H. Hoh, *Ann. Rev. Biophy. Biomol. Structure* (1994).
11. H. G. Hansma and co-workers, **15**, 296 (1993).
12. P. K. Hansma and co-workers,, *Appl. Phys. Lett.*, **64**, 1738 (1994).
13. C. A. J. Putman and co-workers, *Appl. Phy. Lett.*, **64**, 2454, (1994).
14. C. C. Williams, W. P. Hough, and S. A. Rishton, *Appl. Phys. Lett.* **55**, 203 (1989).
15. S. M. Sze, *Physics of Semiconductor Devices*, John Wiley & Sons, Inc., New York, 1981.
16. T. G. Ruskell, R. K. Workman, D. Chen, D. Sarid, S. Dahl, and S. Gilbert, *Appl. Phys. Lett.* **68**, 93 (1996).
17. M. P. Murrell, M. E. Welland, S. J. O'Shea, T. M. H. Wong, J. R. Barnes, and A. W. McKinnon, *Appl. Phys. Lett.* **62**, 786 (1993).
18. S. J. O'Shea, R. M. Atta, and M. E. Welland, *Rev. Sci. Instrum.* **66**, 2503 (1995).
19. B. Ebersberger, C. Boit, H. Benzinger, and E. Gunther, "Thickness Mapping of Thin Dieletrics with Emission Microscopy and Conductive Atomic Force Microscopy for Assessment of Dielectrics Reliability," 1996 International Reliability Symposium April 29–May 2, Dallas, Tex.
20. B. Ebersberger, C. Boit, H. Benzinger, and E. Guenther, *34th Annual IEEE International Reliability Physics Proceedings*, 1996 p. 126.
21. S. J. O'Shea, R. M. Atta, M. P. Murrell, and M. E. Welland, *J. Vac. Sci. Technol. B* **13**, 1945 (1995).

22. M. E. Welland and M. P. Murrell, *Scanning* **15**, 251 (1993).

23. F. J. Feigl, D. R. Young, D. J. DiMaria, S. Lai, and J. Calise, *J. App. Phys.* **52**, 9 (1981).

24. E. H. Nicollian, C. N. Berglund, P. F. Schmidt, and J. M. Andrews, *J. App. Phys.* **42**, 13 (1971).

25. A. Olbrich, B. Ebersberger, and C. Boit, "Nanoscale Electrical Characterization of Thin Oxides with Conducting Atomic Force Microscopy", *36th Annual IEEE International Reliability Physics Symposium*, 1998.

26. K. L. Jensen, *J. Vac. Sci. Technol. B* **13**, 516 (1995).

27. R. Waters and B. Van Zeghbroeck, *App. Phys. Lett.* **75**, 2410 (1999).

28. H. M. Gupta and M. B. Morais, *J. Appl. Phys.* **68**, 176 (1990).

29. Q. Huang, *J. Appl. Phys.* **78**, 6770 (1995).

30. Z. A. Weinberg, *Sol. Stat. Electron.* **20**, 11 (1977).

31. Z. A. Weinberg, *J. Appl. Phys.* **53**, 5052 (1982).

32. G. Krieger and R. M. Swanson, *J. Appl. Phys.* **52**, 5710 (1981).

33. M. Lenzlinger and E. H. Snow, *J. Appl. Phys.* **40**, 278 (1969).

34. D. Bohm, *Quantum Theory*, Prentice Hall Inc., New York, 1951.

35. J. G. Simmons, *J. Appl. Phys.* **34**, 1793 (1963).

36. Nanoscope III AFM system, Digital Instruments, Inc., 520 E. Montecito St., Santa Barbara, Calif. 93103.

37. Controlled with software and hardware from Intelligent Instrumentation, 6550 S. Bay Colony Drive, MS130, Tucson, Ahiz. 85706.

38. Eur. Pat. 90,201,853, (July 9, 1990), U.S. Pat. 5,585,734, (Dec. 17, 1996), M. Meuris, W. Vandervorst, and P. de Wolf.

39. P. De Wolf, T. Clarysse, and W. Vandervorst, *J. Vac. Sci. Technol. B.* **16**, 320 (1998).

40. T. Clarysse and W. Vandervorst, *J. Vac. Sci. Technol B.* **16**, 260 (1998).

41. C. Shafai, D. Thomson, M. Simard-Normandin, G. Mattiussi, and P. J. Scanlon, *Appl. Phys. Lett.* **64**, 342 (1994).

42. P. De Wolf, M. Geva, T. Hantschel, W. Vandervorst, and R. B. Bylsma, *Appl. Phys. Lett.* **73**, 2155 (1998).

43. C. D. Frisbie, L. F. Rozsnyai, A. Noy, M. S. Wrighton, and C. M. Lieber, *Science* **265**, 2071 (1998).

44. P. Grutter, *MSA Bull.* **24**, 416 (1994).

45. C. Schonenberger, S. F. Alvarado, S. E. Lambert, and I. L. Sanders, *J. Appl. Phys.* **67**, 12 (1990).

46. T. Ohkubo, J. Kishigami, K. Yanagisawa, and R. Kaneko, *IEEE Trans. J. Mag. Jpn.* **8**, 245 (1993).

47. R. Proksch, *Magnetic Force Microscopy*, Ph.D. Thesis, University of Minnesota Department of Physics, 1993.

48. H. Jacobs, H. Knapp, and A. Stemmer, *Rev. Sci. Instrum.* **70**, 1756 (1999).

49. H. Jacobs, P. Leuchtmann, O. Homan, and A. Stemmer, *J. Appl. Phys.* **84**, 1168 (1998).

50. P. Schmutz and G. Frankel, *J. Electrochem. Soc.* **145**, 2285 (1998).

51. P. Maivald, H. J. Butt, S. A. C. Gould, C. B. Prater, B. Drake, J. A. Gurley, V. B. Elings, and P. K. Hansma, *Nanotech.* **2**, 103 (1991).

52. M. Maywald, R. J. Pylkki, and L. J. Balk, *Scanning Microsc.* **8**, 181 (1994).

53. A. Hammiche, M. Reading, H. M. Pollock, M. Song, and D. J. Hourston, "Localised thermal analysis using a miniaturized resistive probe," *Rev. Sci. Instrum.* **67**, 4268 (1996).

54. S. Verhaverbeke, M. Meuris, P. W. Mertens, M. M. Heyns, A. Philopossian, D. Gräf, and A. Schnegg, *IEEE IEDM Tech. Dig.*, **71** (1991).

55. D. Ballutaud, P. De Mierry, and M. Aucouturier, *Appl. Surf. Sci.* **47**, 1 (1991).

56. W. R. Aderhold, N. Shah, S. Bogen, A. Bauer, and E. P. Burte, *Mater. Res. Soc. Symp. Proc.* **429**, 275 (1996).
57. Bert Vermeire, Lichyn Lee, and Harold G. Parks, *IEEE Trans. Semiconduct. Manufact.* **11**, 232 (1998).
58. B. Vermeire, C. A. Peterson, H. G. Parks, and D. Sarid, *Proc. Electrochem. Soc.* **99**, 69 (2000).
59. Harold G. Parks, Ronald D. Schrimpf, Bob Craigin, Ronald Jones, and Paul Resnick, *IEEE Trans. Semi. Manufacturing* **7**, 249 (1994).
60. Hitashi Morinaga, Makoto Suyama, and Tadahiro Ohmi, *J. Electrochem. Soc.* **141**, 2834 (1994).
61. L. Torcheux, A. Mayeux, and M. Chemla, *J. Electrochem. Soc.* **142**, 2037 (1995).
62. I. Teerlinck, P. W. Mertens, H. F. Schmidt, M. Meuris, and M. M. Heyns, *J. Electrochem. Soc.* **143**, 3323 (1996).
63. Oliver M.R. Chyan, Jin-Jian Chen, Hsu Y. Chien, Jennifer Sees, and Lindsey Hall, *J. Electrochem. Soc.* **143**, 1 92 (1996).
64. V. Bertagna, F. Rouelle, G. Revel, and M. Chemla, *J. Electrochem. Soc.* **144**, 4175 (1997).
65. F. W. Kern, Jr., M. Itano, I. Kawanabe, M. Miyashita, R. Rosenburg, and T. Ohmi, "Metallic contamination of semiconductor devices from processing chemicals the recognized potential," presented at the 37th Annual IES Meeting, San Diego, Calif. May 6–10, 1991.
66. G. Li, E. A. Kneer, B. Vermeire, H. G. Parks, S. Raghavan, and J. S. Jeon, *J. Electrochem. Soc.* **145**, 241 (1998).
67. S. Kunz, S. Marthon, and F. Tardif, *Proc. Electrochem. Soc.* **97**, 120 (1997).
68. I. Teerlinck, P. W. Mertens, R. Vos, M. Meuris, and M. M. Heyns, *Proc. Electrochem. Soc.* **96**, 250 (1996).

CHARLES ANTHONY PETERSON
Intel Corporation
DROR SARID
Optical Sciences Center, University of Arizona

B

BARIUM

1. Introduction

Barium [7440-39-3], Ba belongs to Group 2. (IIA) of the Periodic Table. Calcium, strontium, and barium belong to the alkaline earth metals and form a closely allied series in which the chemical and physical properties of the elements and their compounds vary systematically with increasing size, ionic and electropositive nature, and specific density. The properties are greatest for barium.

In 1774, Scheele determined that barium oxide was a distinct oxide or "earth," and named it terra ponderosa because of its high density (1). Later, this name was changed to barote from the Greek word meaning heavy. Later still, the name of the oxide was modified to baryta to conform to the nomenclature recommended by Lavoisier, and from this the name barium was derived.

After many unsuccessful attempts, Davy produced barium as a mercury amalgam in 1808 by electrolyzing barium chloride in the presence of a mercury cathode. Attempts were made to isolate the pure metal by distilling the mercury, but it is doubtful that metal of high purity was ever obtained. Early in the twentieth century, high purity barium was prepared by heating barium amalgam in a stream of hydrogen, thereby converting the barium to a hydride and simultaneously volatilizing the last traces of mercury. The pure metal was then obtained by thermal decomposition of the hydride followed by condensation of the volatile barium vapor (1).

Barium is prepared commercially by the thermal reduction of barium oxide with aluminum. Barium metal is highly reactive, a property that accounts for its principal uses as a getter for removing residual gases from vacuum systems and as a deoxidizer for steel and other metals.

In metallic form, barium reacts readily with water to release hydrogen:

$$Ba + 2\,H_2O \longrightarrow Ba(OH)_2 + H_2$$

In aqueous solution it is present as an ion with +2 charge.

2. Ocurrence

In its natural form, barium never occurs as the metal because of reactivity, but is almost always found as the ore barite [13462-86-71], $BaSO_4$, which is also known as heavy spar. A smaller deposit is found as barium carbonate, $BaCO_3$ (witherite) [14941-39-0] barium carbonate can easily be decomposed by heating (calcination) to BaO. Barium oxide is used commercially for the production of barium metal.

3. Physical Properties

Pure barium is a silvery white metal, although contamination with nitrogen lead to a yellowish color. The metal is relatively soft and ductile and may be worked readily. It is fairly volatile (though less than magnesium) and this property is used to advantage in commercial production. Barium has a body-centered entered cubic (bcc) crystal structure at atmospheric pressure, but undergoes structural phase transitions at high pressure (2,3). Barium also exhibits hig ressure induced superconductivity at low temperatures (4,5) and is an essential component of several high temperature superconductors eg, $YBa_2Cu_3O_7$ (6).

It is not easy to obtain samples of ultrahigh purity, and therefore accurate measurements of some physical properties of barium metal are difficult to carry out. In fact, the values for some physical properties are still the subject of controversy. Physical properties of barium are listed in Table 1.

4. Chemical Properties

Barium has a valence electron configuration of $6s_2$ and characteristically forms divalent compounds. It is an extremely reactive metal and its compounds possess large free energies of formation. At room temperature, it combines readily and exothermically with oxygen and the halogens. It reacts vigorously with water, liberating hydrogen and forming barium hydroxide [17194-00-2], $Ba(OH)_2$. At elevated temperatures, barium combines with hydrogen to form barium hydride [13477-09-3], BaH_2, and with nitrogen to form barium nitride [12047-79-9], Ba_3N_2. With nitrogen and carbon, barium forms a cyanide that is thermally stable.

Finely divided barium is susceptible to rapid, violent combination with atmospheric oxygen. Therefore, in powdered form it must be considered pyrophoric and very dangerous to handle in the presence of air or other oxidizing gases. Barium powder must be stored under dry argon or helium to avoid the possibility of violent explosions. Large pieces of barium, however, oxidize relatively slowly and present no explosion hazard if kept dry.

Most barium compounds are not as thermodynamically stable as the corresponding compounds of magnesium and calcium, and therefore, can be reduced

Table 1. **Physical Properties of Barium**[a]

Physical properties	Value
atomic number	56
relative atomic mass A_r	137.34
mass number (natural abundance, %) of stable isotopes:	130 (0.101),
	132 (0.097),
	134 (2.42),
	135 (6.59),
	136 (7.81),
	137 (11.3),
	138 (71.7)
density at 20°C, g/cm^3	3.74
melting point °C	726.2
boiling point, at 101.3 kPa, °C[b]	1637
hardness (mohs scale)	1.25
crystal structures	bcc
lattice constant α_0 at 20°C nm	0.5025
coefficient of thermal expansion, α_1, (mean, 0–100°C)	$1.8 \times 10^{-5} \mathrm{K}^{-1}$
modulus of elasticity N/m^2	1.265×10^{10}
heat of fusion, ΔH_m, kJ/mol[c]	7.98
heat of vaporization, ΔH_v, kJ/molc	140.3 kJ/mol
specific heat capacity c at 20°C J/hg·K	192 J
at 900°C, J/hg·K	230 J kg

Vapor pressure at						
temperature, °C	630	730	1050	1300	1520	1637
pressure, kPa	0.00133	0.0133	0.133	13.3	53.3	101.3

Electrical resistivity, Ω cm	40×10^{-6}
commercial purity	
extra high purity	30×10^{-6}
liquid barium at mp	314×10^{-6}
Thermal coefficient of electrical resistivity do/dT (mean, 0–100°C)	$6.5 \times 10^{-3} \mathrm{K}^{-1}$

[a] Ref. 7
[b] To convert kPa to mm Hg, multiply by 7.5.
[c] To convert J to cal, divide by 4.18.

by these metals. However, rather than producing pure barium, barium alloys are formed. Barium combines with most metals, forming a wide range of alloys and intermetallic compounds. Among the phase systems that have been b tter characterized are those with Ag, Al, Bi Hg, Pb, Sn, Zn, and the other Group 2 (II A) metals (8).

Barium reduces the oxides, halides, and sulfides of most of the less reactive metals, thereby producing the corresponding metal. However, calcium metal can, in most cases, be used for similar purposes and is usually preferred over barium because of lower cost per equivalent weight and nontoxicity (see ACTINIDES AND TRANSACTINIDES).

5. Manufacture and Processing

Barium metal is produced commercially by the reduction of barium oxide with a less reactive, nonvolatile element, usually aluminum (9–15):

$$4BaO + 2\,Al \longrightarrow 3\,Ba + BaAl_2O_4$$

The barium oxide is mixed with aluminum granules, and the mixture briquetted and charged into long tubular retorts of heat-resistant steel. These are evacuated and heated to $\approx 1100°C$ in the segment containing the charge, while the other end is kept cool. Molten aluminum and aluminum vapor react with the solid barium oxide, releasing barium vapor, which condenses on the cooler part of the apparatus (16). It is collected and cast into chill molds under argon.

Production of ultrapure barium metal has been investigated on a laboratory scale. Redistillation (17,18), zone recrystallization (19,20), and combinations of these techniques (21) have been studied. Impurity levels of <100 ppm have been attained.

Barium production requires large amounts of energy for two reasons: the high temperatures required in the process itself and the energy-intensive raw materials employed, the calcined BaO and the electrolytically produced aluminum.

Because of its high reactivity, production of barium by such processes as electrolysis of barium compounds solution or high temperature carbon reduction is impossible. Electrolysis of an aqueous barium solution yields $Ba(OH)_2$, whereas carbon reduction of an ore such as BaO produces barium carbide [50813-65-5], BaC_2, which is analogous to calcium carbide (see CARBIDES). Attempts to produce barium by electrolysis of molten barium salts, usually $BaCl_2$, met with only limited success (22), perhaps because of the solubility of Barium in $BaCl_2$ (23).

6. Shipment, Storage, and Handling

6.1. Shipment. The barium crowns are usually broken into smaller pieces and can be sold in this form or cast or extruded into bars or wire. Usually, the metal is packaged filled plastic bags inside argon-filled steel containers.

Barium is commercially available in bars up to 20 kg or in rods 22 mm in diameter and 40 mm in length (24). The rods can be cut into small pieces or extruded into wires.

Barium is packaged in airtight steel drums containing up to 100 kg of the metal under paraffin oil. Smaller amounts (1–10 kg) are packaged in tin cans, and even smaller samples are packaged in hermetically sealed glass bottles (25,26).

Barium is classed as a flammable solid and cannot be mailed. If it comes into contact with water, there is always the danger of explosion because of the liberated hydrogen. Therefore barium should always be stored in a dry, well-ventilated place and contact with moisture and air avoided. Protective glasses and safety gloves should be worn while handling barium. Burning barium can be extinguished with sand, aluminum oxide, etc.

Transport classification (24):

GGVE, GGVS, RID, ADR: class 4.3, Fig. 11b
IMDG-Code: class 4.3 UN-No. 1400 PG.II
ICAO: class 4.3 UN-No. 1400 PG.II/Drill-Code 4W

6.2. Storage. Store barium metal in a sealed container away from water, acids, or organic compounds. Protect containers against physical damage. Avoid damaging container.

6.3. Handling. Provide adequate exhaust ventilation to meet exposure limit requirements. An exhaust filter system may be required to avoid environmental contamination. Wear a positive pressure air-supplied respirator in situations where there may be a potential for airborne exposure. Wear impervious clothing including gloves to prevent contact with skin, also used goggles and on face shold.

7. Economic Aspects

Chemetall GmbH, Germany, is the leading producer of barium metal. Chemetall GmbH covers the global demand of berium metal and BaAl₄. Only little is known about its production in Russia and China.

Production or consumption figures are not available.

Price levels were not published in 1999. Price average value for barite are in 2001 was ~$25/t, mine (27).

8. Grades, Specifications, and Quality Control

Barium metal is marketed in purities from 95% for technical applications to 99.5% for high purity applications.

Assays and purities of commercial products are derived by subtracting the sum of analyzed impurity levels from unity. Alkali metal impurities are analyzed by emmission spectroscopy, whereas alkalin earth metals are determined by atomic absorption. Other metals and anions can be determined by photometric methods. Chloride is established argentometrically.

9. Analytical Methods

Volatile barium compounds impart a pale green color to flames, and this is an effective, simple qualitative test for barium (455.4,493.4,553.6, and 611.1 nm). Barium is separated from magnesium, strontium, and calcium by precipitation from a dilute solution in nitric or hydrochloric acid with a solution of potassium dichromate in acetic acid. Barium is determined gravimetrically by precipitation of the small quantities are determined spectrometrically (29).

The metallic impurities in commercial barium (see Table 2) are determined by atomic absorption and flame emission spectroscopy. Trace impurities are best determined by inductively coupled emission spectroscopy (ICP) (30). The carbon content in barium is determined by combustion; nitrogen, by the Kjeldahl method; and hydrogen, by vacuum hot extraction. Vacuum hot extraction is not useful for analysis of oxygen. Neutron activation analysis based on the reaction $^{16}O(n, p) \rightarrow {}^{16}N$ is the recommended method for determination of oxygen in alkaline earth metals (31).

10. Environmental Concerns

The terrestrial abundance of barium is ~250 g/t (32). The estimated average barium concentration in the soil is 500 g/t (33). Measured concentrations range betweeen 100 and 3000 g/t (34).

Table 2. **Chemical Analysis of Commerical Barium**[a]

Element		%
barium	(incl. Sr)	99.2 ± 0.30
strontium	max.	0.8
calcium	max.	0.25
aluminum	max.	0.06
carbon	max.	0.06
magnesium	max.	0.02
nitrogen	max.	0.02
iron	max.	0.02
chlorine	max.	0.01
lithium, sodium, potassium	max.	0.01

[a] Ref. 24.

Barium occurs in seawater in a concentration of 6 µg/L (35). This level is due to reaction between barium and sulfate ions also present in the ocean. The precipitated barium sulfate forms a permanent part of the sediment on the ocean floor (36). In fresh water, the barium content depends on the occurence of barium and the concentration of anions that form barium salts of low solubility such as sulfate and carbonate ions. Values between 7 and 15,000 µg/L (average: 50 µg/L) have been reported (37).

Studies of drinking water give a wide range of values between only traces and 10,000 µg/L in the United States (38–44) in Canada (45) and 1–20 µg/L in municipal drinking water in Sweden (37).

Barium levels in the air are not well documented. An estimate of a mean value for the United States 0.05 µg/L (46). There is no correlation between the degree of industrilization and the barium concentration in the air. Higher levels are found in areas with high natural dust levels. Anthropogenic emissions are primarily industrial. Other atmospheric emissions result from the handling of barium Co or materials containing barium compounds, such as welding wires (47).

11. Recycling and Disposal

Consult with environmental regulatory agencies for guidance on acceptable disposal practices.

12. Health and Safety Factors

Barium metal reacts with water and acids to form hydrogen gas, barium oxide, and barium hydroxide; the reaction is exothermic. If barium metal contacts moisture in the eyes, on the skin, or in the respiratory tract, severe corrosive irritation may result. Inhalation of dust or fume may cause severe respiratory irritation, cough, difficulty in breathing, and chemical pnemonitis. Contact with skin causes irritation and possible corrosive damage.

The substance is severely irritating to eyes and may injure eye tissue if not promptly removed.

Ingestion may cause acute irritation or burns to the mouth, throat, and stomach; barium may cause vomiting. Preexisting chronic respiratory, skin, or eye diseases may be aggravated.

The symptoms of inhalation include severe irritation of respiratory tract. Skin and eye contact symptoms are severe irritation.

There are no known adverse health effects resulting from long-term exposure to barium metal.

Barium metal poisoning is virtually unknown in industry, although the potential exists when the soluble barium compound forms are used. When ingested or given orally, the soluble, ionized barium compounds exert a profound effect on all muscles and especially smooth muscles, markedly increasing their contractility. The heart rate is slowed and may stop in systole. Other effects are increased intertinal peristalsis, vascular constriction, bladder contraction, and increased voluntary muscle tension.

For soluble barium compounds an exposure limit of 0.5 mg/m^3 (as Ba) has been established by both TLV (1989) and MAK (1996) commissions.

12.1. First Aid

Eye contact. Flush eyes with a steady stream of water for at least 15 min. Lift upper and lower eye lids frequently. Get prompt medical attention.

Skin contact. Immediately remove and isolate contaminated clothing. Carefully brush off material from skin and wash affected area thoroughly with water. Call a physician if irritation develops. Inhalation. Remove to fresh air. If symptoms develop, seek immediate medical attention. If not breathing, give artificial respiration.

Ingestion. Call a poison control center, emergency room, or physician if barium is ingested. Unless advised otherwise, induce vomiting by giving either syrup of ipecac followed by two glasses of water. If a soluble barium compound has been swallowed, get medical attention. If the person is drowsy or unconscious, DO NOT GIVE ANYTHING BY MOUTH or leave alone. Never give anything to drink to a person who is convulsing or has no gag reflex. Loosen tight fitting clothing, clear the airway, and keep the person warm.

Note to physician: Treatment should be directed at preventing absorption, administering to the symptoms as they occur, and providing supportive therapy.

12.2. Fire Fighting and Explosion Hazards.

Do not use water, foam, or halogenated hydrocarbons such as Halon or carbon tetrachloride to extinguish fire. Use only dry chemical/dolomite (powdered limestone), or an appropriate metal-fire-extinguishing dry powder, such as Met-L-X or Totalit M. For large fires, withdraw from the area and let the fire burn.

Firefighters should wear self-contained breathing apparatus with full face piece operated in the pressure-demand or positive-pressure mode. Firefighters should move containers from the fire area if this can be done without risk. Do not use water or foam. Use dry powder only.

Water reacts dangerously with barium metal and is not recommended as an extinguishing agent for fires. If water must be used, prevent it from coming into direct contact with barium metal. If contact is unavoidable, apply the water in flooding amounts to safely absorb the heat that will be generated.

Barium metal is extremely dangerous when wet. Barium metal forms barium hydroxide and hydrogen gas resulting in an explosion hazard when wet.

Barium metal forms BaO when it burns. It reacts with wet extinguishing agents such as water, halogens, and possibly carbon dioxide.

12.3. Accidental Release Measures. Do not touch spilled barium metal. Wear protective apparel. Do not smoke or place flame or ignition sources near spill area. Do not allow water to touch spilled barium metal or to get inside containers. Use a cover to prevent water or rain from dissolving spilled barium metal or to prevent its spreading. Isolate hazard area and keep nonessential personnel away from spill or leak site. Shovel small dry spills into a dry container and cover it tightly. Move containers away from spill to a safe area. Take up small spills with sand or an absorbent and contain as described above. Dike the flow of large barium metal and water spills with soil, sandbags, or concrete. Keep the waste form entering drains and open sewers. Wear full protective gear.

13. Uses

The major use of barium is the production of barium–aluminum alloy—evaporation getters (gas absorbers) in CRTs (cathode ray tubes) for television sets and computer monitors to generate and to maintain high vacuum by reaction with detrimental gases. Barium is used as getter material in X-ray and emitter tubes and in sodium vapor lamps.

Because of its low vapor pressure and its reactivity toward gases, such as oxygen, nitrogen, hydrogen, carbon dioxide, and water vapor barium is an ideal getter material. The market demand of barium depends strongly on the demand for CRTs. The demand for CRTs has been increasing in the past few years. This is a result of increased demand for personal computers requiring color monitors.

Barium is also used to improve performance of lead alloy grids of acid batteries.

The deoxidizing and reducing properties of barium find numerous minor applications in the metal refining and alloying industry.

Many other uses of barium have been described in the literature, but they are of minor importance. Barium increases the creep resistance in lead–tin soldering alloys therefore it has been used in bearing alloys. Instead of strontium or sodium, barium has been used as a modifying agent for "silumin" where barium refines the structure of the eutectic aluminum–silicon alloy.

BIBLIOGRAPHY

"Barium" treated in *ECT* 1st ed. under "Alkaline Earth Metals and Alkaline Earth Metal Alloys," Vol. 1, pp. 458–463, by C. L. Mantell, Consulting Chemical Engineer; in *ECT* 2nd ed., Vol. 3, pp. 77–80, by L. M. Pidgeon, University of Toronto; "Barium" in *ECT* 3rd ed., Vol. 3, pp. 457–463, by C. J. Kunesh, Pfizer, Inc.; in *ECT* 4th ed., Vol. 3, pp. 902–908, by Claudio Boffito, Seas Getters SpA; "Barium" in *ECT* (online), posting date: December 4, 2000, by Claudio Boffito, Seas Getters SpA.

CITED PUBLICATIONS

1. J. W. Mellor, *Comprehensive Treatise on Inorganic and Theoretical Chemistry*, Vol. 3, Longmans, Green & Co., Inc., New York, 1923, 619–631.
2. J. P. Bastide, C. Susse, and R. Epain, *C. R. Acad. Sci.* **267**, 857 (1968).

3. S. Akimoto and co-workers, *High Temp. High Pressures* **7**, 287 (1975).

4. M. A. Il'ina and E. S. Itskevich, *JETP Lett.* **11**, 32 (1970).

5. M. A. Il'ina, E. S. Itskevich, and E. M. Dizhur, *Zh. Eksp. Teor. Fiz.* **61**, 2357 (1971).

6. M. K. Wu co-workers, *Phys. Rev. Lett.*, **58** 908 (1987).

7. J. Evers and A. Weiss, *J. Less Common Met.* **30**, 83 (1973).

8. F. Emley, in D. M. Considine, ed., *Chemical and Process Technology Encyclopedia*, McGraw-Hill, Book Co., Inc., New York, 1974, p. 151.

9. W. J. Kroll, *U. S. Bur. Mines, Inf. Circ.*, 7327 (1945).

10. E. Fukuda and H. Yokomizo, *J. Electrochem. Soc. Jpn.* **20**, 430 (1952).

11. M. Orman and E. Zembala, *Prac. Inst. Met.* **4**, 437 (1952).

12. G. G. Gvelisiani and V. A. Pazukhin *Ref. Zh. Metall.* 1018 (1956).

13. H. Sawamoto, T. Oki, and T. Umemura, *Mem. Fac. Eng. Nagoya Univ.* **12**, 123 (1960).

14. G. N. Kozhevnikov, *Tsvetn. Metall.* **36**, 53 (1963).

15. G. G. Gvelisiani and N. P. Mgaloblishvili, *Tr. Gruz. Inst. Met.* **14**, 205 (1965).

16. H. Seliger, *Freiberg. Forschungsh. B* **34** (1959) 80.

17. T. E. Brown and K. A. McEwen, *J. Phys. D* **3**, 980 (1970).

18. J. Evers and co-workers, *J. Less Common Met.* **30**, 83 (1974).

19. V. N. Vigdorovich and co-workers, *Izv. Vyssh. Uchebn. Zaved. Tsvetn. Metall.* **1**, 86 (1973).

20. A. V. Vokhobov, V. G. Khudaiberdiev, and M. K. Nasyrova, *Russ. Metall.* **4**, 116 (1974).

21. A. V. Vokhobov, V. N. Vigdorovich, and V. G. Khudaiberdiev, *Izv. Vyssh. Uchebn. Zaved. Tsvetn. Metall.* **4**, 115 (1973).

22. C. L. Mantell, in C. A. Hampel, ed., *Rare Metals Handbook*, 2nd ed., Reinhold Publishing Corp., London, 1961, p. 26.

23. K. Grjotheim and H. G. Nebell, *Acta Chem. Scand.* **22**. 1159 (1968).

24. Chemical GmbH, *Lieferprogramm Sondermetalle*, Frankfurt, 2000.

25. IMDG, Class 4.3, p. 4147; UN 1400; 4.3/1A RID, ADR, GGVS, GGVE.

26. R. G. Lewes, ed., *Sax's Dangerous Properties of Industrial Materials*, Vol. 2, 10th ed., John Wiley & Sons, New York, 2000, p. 343.

27. J. P. Searls, "Barite" *Mineral Commodity Summaries*, U. S. Geological Survey, Jan. 2002.

28. Ges. Deutscher Metallhütten- und Bergleute: *Analyse der Metalle*, 2nd ed., vol. II/1, Springer-Verlag, Berlin 1961, pp. 139–145.

29. L Edelbeck and P. W. West, *Anal. Chim. Acta* **52**, 447 (1970).

30. G. Tölg, *Fresenius' Z. Anal. Chem. 1*, (1979) Ref. 28, pp. 22–23.

31. L. Melnick, L. Lewis, and B. Holt, eds., "Determination of Gaseous Elements in Metals," *Monographs oh Analytical Chemistry*, Vol. 40, John Wiley & Sons, Inc., New York 1974.

32. B. Mason *Principles of Geochemistry* John Wiley & Sons, Inc., New York, 1952, p. 41.

33. R. R. Brooks, "Pollution Through Trace Elements", in J. O. M. Bockris, ed., *Environmental Chemistry*, Plenum Press, New York 1978, pp. 429–476

34. W. O. Robinson, R. R. Whetstone, and G. Edington, "The Occurrence Of Barium In Soils And Plants", *U.S. Dept. Agric. Tech.* **1013**, 429–476 (1950). H. A. Schroeder, "Barium" *American Petroleum Institute, Air quality Monograph* **70**, 12 (1970).

35. H. A. Schroeder, I. H. Tipton, A. P. Nason, and H. A. Schroeder, "Trace Metals in Man: Strontium and Barium", *J. Chron. Dis.* **25**, 491 (1972).

36. K. Wolgemuth, and W. S. Broecker, *Earth Planet. Sci. Lett.* **8**, 372 (1970).

37. A. L. Reeves, "Barium", in L. Friberg, G. F. Nordberg, and B. Velimir, eds., *Handbook on the Toxicology of Metals—Volume II: Specific Metals*, Elsevier Science Publishers, Amsterdam, 1986, pp 84–93.

38. C. M. Durfor and E. Becker, "Public Water Supplies of the 100 largest Cities in the United States," Water Supply Paper No. 1812, U.S. Department of the Interior, Government Printing Office, Washington D.C. 1962.

39. P. R. Barnett, M. W. Skougstadt and K. J. Miller, *J. Am. Water Works Assoc.* **2** 60 (1969).

40. L. McCabe, in A. Saponzik, ed., "Problems of Trace Metals in Water Supplies—an Overview," *Proceedings of the 16th Water Quality Conference on Trace Metals in Water Supplies: Occurence, Significance and Control*, University of Illinois, Urbana, IL, 12–13 Feb. 1974, pp. 1–9.

41. L. J. McCabe, and co-workers *J. Am. Water Works Assoc.* **62** 670 (1970)

42. E. J. Calabrese, *J. Environ. Health* **39**(5) 366 (1977).

43. *J. Am. Water Aorks Assoc.* **38**(117), 67 (1985).

44. K. S. Subramanian and J. C. Meranger *At. Spectrosc.* **5**, 34 (1984).

45. K. S. Subramanian and J. C. Meranger *At. Spectrosc.* **5**, 34 (1984).

46. ICPS Environmental Health Criteria 107, Barium, World Health Organization, Geneva 1990, p. 14.

47. *Health Hazard Evaluation Determination Report: Mark Steel Corporation, Salt Lake City, Utah*, NIOSH Report No. 78–93–536, National Institute for Occupational Health and Safety, Cincinnati, Ohio 1978.

GENERAL REFERENCES

Barium Bibliography, Mineral Products Division, Food Machinery and Chemical Corporation, New York, 1961 (esp. pp. 95–115 and 119–238).

W. J. Kroll, *U.S. Bur. Mines Inf. Circ.*, 7327 (1945).

C. L. Mantell, in C. A. Hampel, ed., *Rare Metals Handbook*, 2nd ed., Reinhold Publishing Corp., London, 1961, pp. 15–31.

J. W. Mellor, *Comprehensive Treatise on Inorganic and Theoretical Chemistry*, Vol. 3, Longmans, Green & Co., Inc., New York, 1923, pp. 619–652.

CLAUDIO BOFFITO
Saes Getters SpA

BARIUM COMPOUNDS

1. Introduction

The first report concerning barium compounds occurred in the early part of the seventeenth century when it was noted that the ignition of heavy spar gave a peculiar green light. A century later, Scheele reported that a precipitate formed when sulfuric acid was added to a solution of barium salts. The presence of natural barium carbonate, witherite [14941-39-0], $BaCO_3$, was noted in Scotland by Withering.

In its natural form, barium [7440-39-3], Ba, never occurs as the metal but is almost always found as the ore barite [13462-86-7], $BaSO_4$. More than 90% of all barium is actually used as the ore, albeit after preliminary beneficiation. In the U.S., nearly 95% of the barite sold in 2001 was used for oil- and gas-well drilling fluids (muds). The other 5% was used as filler and/or for extender uses and the manufacture of all other barium chemicals (1). Witherite, the only other significant natural barium ore, is not mined commercially.

Barium is a member of the alkaline-earth group of elements in Group 2 (IIA) of the period table. Calcium [7440-70-2], Ca, strontium [7440-24-6], Sr, and barium form a closely allied series in which the chemical and physical properties of the elements and their compounds vary systematically with increasing size, the ionic and electropositive nature being greatest for barium (see CALCIUM AND CALCIUM ALLOYS; CALCIUM COMPOUNDS; STRONTIUM AND STRONTIUM COMPOUNDS). As size increases, hydration tendencies of the crystalline salts increase; solubilities of sulfates, nitrates, chlorides, etc, decrease (except fluorides); solubilities of halides in ethanol decrease; thermal stabilities of carbonates, nitrates, and peroxides increase; and the rates of reaction of the metals with hydrogen increase.

In metallic form, barium is very reactive, reacting readily with water to release hydrogen. In aqueous solution it is present as an ion with a +2 charge. Barium acetate, chloride, hydroxide, and nitrate are water-soluble, whereas barium arsenate, chromate, fluoride, oxalate, and sulfate are not. Most water-insoluble barium salts dissolve in dilute acids; barium sulfate, however, requires strong sulfuric acid.

Compared to the hydroxides of calcium and strontiuim, barium hydroxide is the most water-soluble and also the strongest base. Additionally, barium hydroxide is more difficult to convert to the oxide by heating than are the corresponding hydroxides of calcium and strontium. Barium oxide is more readily converted to the peroxide than are the oxides of the other alkaline earths.

The large size, ionic radius = 0.143 nm, and electronic configuration [Xe] $6s^2$, of the barium(II) ion [22541-12-4], Ba^{2+}, makes isomorphous substitution possible only with strontium, Sr^{2+}, 0.127 nm, and generally not with other members of Group IIA such as Ca^{2+}, 0.106 nm, and Mg^{2+}, 0.078 nm. Among the other elements that occur with barium in nature, substitution is common only for potassium, K^+, 0.144 nm, but not for the smaller ions of Na, Fe, Mn, Al, and Si (2). For a discussion of barium bromate [13967-90-3], $Ba(BrO_3)_2$, see BROMINE COMPOUNDS; for barium chlorate [13477-000-4], $Ba(ClO_3)_2 \cdot H_2O$, see CHLORINE OXYGEN ACIDS AND SALTS, CHLORIC ACID AND CHLORATES; for barium chromate [10294-40-3], $BaCrO_4$, see CHROMIUM COMPOUNDS; and for barium cyanide [542-62-1], $Ba(CN)_2$, see CYANIDES. For a discussion of barium ferrite [11138-11-7] and [12409-27-7] ($BaFe_{12}O_{13}$), see FERRITES; and for barium hydride [13477-09-3], BaH_2, see HYDRIDES.

2. Barite

Barite [13462-86-7], natural barium sulfate, $BaSO_4$, commonly known as barytes, and sometimes as heavy spar, till, or cawk, occurs in many geological environments in sedimentary, igneous, and metamorphic rocks. Commercial deposits are of three types: vein and cavity filling deposits; residual deposits; and bedded deposits. Most commercial sources are replacement deposits in limestone, dolomitic sandstone, and shales, or residual deposits caused by differential weathering that result in lumps of barite enclosed in clay. Barite is widely distributed and has minable deposits in many countries.

Mineralogically, barite crystallizes in the dipyramidal class of the orthorhombic system. Although the barite of many deposits fractures unevenly or has an apparent cleavage along planes because of separation between successively deposited layers, well-formed crystals are mostly tabular and cleave along three different planes. The barite in most deposits occurs in irregular masses, nodules, rosettelike aggregates, and in laminated to massive beds of fine crystallinity. Barite is most commonly associated with quartz [14808-60-7], chert, jasperoid, calcite [13397-26-7], dolomite [17069-72-6], siderite [14476-16-5], rhodochrosite [14476-12-1], celestite [14291-02-2], gluorite, various sulfide minerals, and their oxidation products. In most mines, barite is the primary material being mined, yet barite is also a common gangue mineral in many types of ore deposits including those for lead, zinc, gold, silver, fluorite, and rare-earth minerals (3).

Barite is a moderately soft crystalline mineral, Mohs' hardness 3–3.5; sp gr 4.3–4.6; n_D 1.64. The ore is white opaque to transparent, but impurities can produce pale shades of yellow, green, blue, brown, red, or gray-black. The most important impurities are Fe_2O_3, Al_2O_3, SiO_2, and $SrSO_4$, all of which are undesirable in chemical-grade barite. When the barite is used for drilling mud, the iron content can be permitted to be much higher than for other uses.

Residual barite is usually mined by open-pit methods. Bedded and vein deposits may be mined by either open-pit or underground methods, depending on the characteristics of each deposit. Some barite is sufficiently high, up to 96% $BaSO_4$, grade that it can be shipped without beneficiation. Many deposits require concentration and beneficiation can involve any one or a combination of gravity separations. "Primary barite," the first marketable product, includes crude run-of-mine barite, flotation concentrates, and material concentrated by other beneficiated processing such as washing, jigging, or magnetic separation. Chemical and glass manufacturers prefer coarser material; for chemicals, ca 4760–840 µm (4–20 mesh) is preferred, and for glass, ca 590–105 µm (30–140 mesh). Barite to be used in well drilling is ground dry to 44 µm (−325 mesh).

2.1. Production and Consumption. About 80% of the world's barite production is used as a weighting agent for the muds circulated in rotary drilling of oil and gas wells (see PETROLEUM, DRILLING FLUIDS AND OTHER OIL RECOVERY CHEMICALS). Table 1 shows the U.S. production–consumption balance. The 2001 demand for barite increased nearly 17% over that recorded in 2000 (4).

Technological developments such as 3D-seismic surveying and new drilling techniques resulted in more production from fewer wells. The use of oil-based and synthetic drilling fluids based on water also resulted in a reduction of demand for barites. However, improved market conditions could see a significant increase in the number of active rigs, which could have a positive effect on barite demand (5). At present the natural gas industry appears to be the main market for barite consumption (4).

World mine production, reserves, and reserve base are listed in Table 2.

2.2. Uses. Drilling muds are aqueous suspensions of clay and barite used in the petroleum industry. During drilling operations, the muds are pumped into a well through the hollow drill stem, passing through the tip of the bit, and back up the space between the stem and the walls of the hole. The mud is effective in lubricating and cooling the drill bit, in sealing the walls to prevent the caving of the hole, in suspending the drill cuttings and carrying them to the surface, and, by establishing a hydrostatic column head of weighted fluid, helps to restrain high gas and oil pressures, reducing the tendency for blowouts. For this last reason, muds having a specific gravity as great as 2.5 are used and barite is the material of choice because of its high density, chemical inertness, relative nonabrasiveness, and widespread availability at reasonable cost.

Table 1. **U.S. Barite Production–Consumption Balance**[a]

Statistics	1997	1998	1999	2000	2001[b]
sold or used, mine	692	476	434	392	400
imports for consumption:					
crude barite	2210	1850	836	2070	2670
ground barite	31	20	17	16	20
other	12	13	18	15	10
exports	22	15	22	36	40
consumption, apparent[c] (crude barite)	2920	2340	1280	2460	2960
consumption[d] (ground and crushed)	2180	1890	1370	2100	2600
price, average value, dollars per ton, mine	22.45	22.70	25.60	25.10	25.00
employment, mine and mill, number[b]	380	410	300	330	340
net import reliance[e] as a percentage of apparent consumption	76	80	66	84	87

[a] From Ref. 1.
[b] Estimated.
[c] Sold or used by domestic mines - exports + imports.
[d] Domestic and imported crude barite sold or used by domestic grinding establishments.
[e] Defined as imports - exports + adjustments for government and industry stock changes.

Finely ground barite which may be bleached, usually by sulfuric acid, or unbleached, is used as a filler or extender in paints (qv), especially in automotive undercoats, where its low oil absorption, easy wettability in oils, and good sanding properties are advantageous (see FILLERS). It is also used as a filler in plastics

Table 2. **World Mine Production, Reserves, and Reserve Base of Barite**[a]

Country	Mine production		Reserves[c]	Reserve base[c]
	2000	2001[b]		
United States	392	400	6,000	60,000
Algeria	50	50	9,000	15,000
Bulgaria	120	120	10,000	20,000
China	3,500	3,800	30,000	150,000
France	75	75	2,000	2,500
Germany	120	120	1,000	1,500
India	550	650	53,000	80,000
Iran	185	190	NA	NA
Korea, North	70	70	NA	NA
Mexico	127	120	7,000	8,500
Morocco	350	320	10,000	11,000
Russia	60	60	2,000	3,000
Thailand	50	50	9,000	15,000
Turkey	130	120	4,000	20,000
United Kingdom	70	70	100	600
Other countries	350	250	12,000	160,000
World total (rounded)	*6,200*	*6,600*	*160,000*	*550,000*

[a] From Ref. 1
[b] Estimated
[c] NA = not available

and rubber products. In nonasbestos brake linings, the barite filler acts as a heat sink (see BRAKE LININGS AND CLUTCH FACINGS). In floor mats and carpet backings made of polyurethane foam, barite imparts sound-deadening characteristics and improves processing qualities (see URETHANE POLYMERS). In furniture manufacture where polyurethane foam is used for recoil and density properties, the unique chemical inertness of barite in combination with its high density plays a primary role.

In the glass (qv) and ceramic industry (see CERAMICS), barite can be used both as a flux, to promote melting at a lower temperature or to increase the production rate, and as an additive to increase the refractive index of glass. The viscosity of barite-containing glass often needs to be raised. Alumina in the form of feldspar is sometimes used. To offset any color produced by iron from the barite addition, more decolorizer may be needed. When properly used, barytes help reduce seed, increase toughness and brilliancy, and reduce annealing time. Barite is also a raw material for the manufacture of other barium chemicals.

3. Barium Acetate

Barium acetate [543-80-6], $Ba(C_2H_3O_2)_2$, crystallizes from an aqueous solution of acetic acid and barium carbonate or barium hydroxide. The level of hydration depends on crystallization temperature. At $<24.7°C$ the trihydrate, density 2.02 g/mL is formed; from 24.7 to 41°C barium acetate monohydrate [5908-64-5], density 2.19 g/mL precipitates; and above 41°C the anhydrous salt, density 2.47 g/mL results. The monohydrate becomes anhydrous at 110°C. At 20°C, 76 g of the monohydrate dissolves in 100 g of water. Barium acetate is used in printing fabrics, lubricating grease, and as a catalyst for organic reactions.

4. Barium Bromide

Barium bromide [10553-31-8], $BaBr_2$, mp 854°C, density 4.781 g/mL, also exists as barium bromide dihydrate [7791-28-8], $BaBr_2 \cdot 2H_2O$, dehydration temperature 120°C, density 3.58 g/mL. The solubility, wt %, of $BaBr_2$ in water is

Temperature, °C	−20	0	20	40	60	80
Solubility, wt % $BaBr_2$	45.6	49.5	51.0	53.2	55.1	57.4

Barium bromide is very soluble in methanol, yet almost insoluble in ethanol. Reported uses of barium bromide include: fabrication of phosphors, for example from BaF_2, $BaBr_2 \cdot 2H_2O$ and $EuBr_3$ (6); as a crystallization nucleating agent to control supercooling of $CaBr_2$ solutions (7); and in the production of halide glasses having ir transmission properties (8). Glass-transition temperature is sometimes influenced by $BaBr_2$ content.

5. Barium Carbonate

Most barium compounds are prepared from reactions of barium carbonate [513-77-9], $BaCO_3$, which is commercially manufactured by the "black ash" process from barite and coke in a process identical to that for strontium carbonate production. Depending on the co-product, soda ash and/or carbon dioxide are also consumed.

Precipitated or synthetic barium carbonate is the most commercially important of all the barium chemicals except for barite. Barium carbonate is an unusually dense material, that is almost insoluble in water and only slightly soluble in carbonated water. It does dissolve in dilute hydrochloric, nitric, and acetic acids and is also soluble in ammonium nitrate and ammonium chloride solutions.

5.1. Manufacture. An outline of the black ash process for $BaCO_3$ manufacture is shown in Figure 1. It is from the appearance of the product exiting the thermal reduction step that the process derives its name.

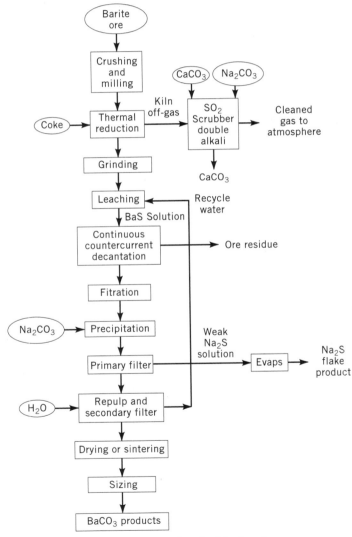

Fig. 1. Flow diagram for the black ash process.

Barite ore is reduced in size to between 20 and 200 mesh (840 to 74 μm). Sizing represents a compromise between increases in reaction rate and dust loss. Coke is fed to the kiln at a coke to ore ratio of about 0.20:0.25.

Thermal Reduction.   Thermal reduction is usually accomplished in a high temperature countercurrent rotary kiln. "Hot zone," a region near the kiln spill, temperature is usually controlled at 1100–1200°C. The reaction rate has been shown to be only slightly lower at 1050°C than at 1130°C (9). About 6% of the feed $BaSO_4$ remains unreacted after 30 min at 1050°C. Reaction completion is approached in less than 10 min at 1100°C (10).

The chief reduction reaction in the kiln is

$$BaSO_4 + 4\,CO \rightleftharpoons BaS + 4\,CO_2 \tag{1a}$$

BaS is leached from the black ash by hot water. The resulting solution is filtered and then treated with soda ash or carbon dioxide or a combination of the two to precipitate fine $BaCO_3$ crystals, which in turn are filtered and dried. Sulfide values can be recovered as H_2S, NaHS, Na_2S, or elemental sulfur.

Carbon dioxide, detrimental to high BaS yields, is repressed according to the Boudouard equilibrium

$$C + CO_2 \rightleftharpoons 2CO \tag{1b}$$

Side reactions of barium with silicate impurities in the ore have been noted (1,11–13). These reactions can cause appreciable loss of barium values by forming water-insoluble barium compounds (14). For example, barium sulfate can form the orthosilicate

$$2\,BaSO_4 + SiO_2 \longrightarrow Ba_2SiO_4 + 2\,SO_3 \tag{2}$$

whereas if carbon is not present, this reaction does not occur below 1100°C. In the presence of carbon the reaction initiates at 1005°C.

Any $BaCO_3$ formed in the kiln can undergo the following reactions at temperatures below 1100°C to form the metasilicate (13)

$$BaCO_3 + SiO_2 \longrightarrow BaSiO_3 + CO_2 \tag{3}$$

and subsequently the orthosilicate

$$BaSiO_3 + BaCO_3 \longrightarrow Ba_2SiO_4 + CO_2 \tag{4}$$

In the hot leaching step, barium metasilicate is insoluble, and the orthosilicate yields only one-half of its barium value to give barium hydroxide

$$Ba_2SiO_4 + H_2O \longrightarrow BaSiO_3 + Ba(OH)_2 \tag{5}$$

If coke is not present in sufficient quantity, more silicates are formed, presumably because of increased CO_2 concentration which leads to greater amounts of $BaCO_3$ being formed (1).

Barium is reported in the kiln spill in three different forms: water-soluble, eg, BaS, BaO; acid-soluble, eg, $BaCO_3$, aluminate, silicate, ferrate; and insoluble, eg, unreduced barite.

Reduction of $BaSO_4$ appears to begin about 900°C (15). The presence of iron or iron oxide can catalyze the barium (9) and also strontium reduction reaction rates. However, iron impurity can also increase the acid-soluble content of the black ash (9).

Calcareous minerals such as gypsum [13397-24-5], when added in stoichiometric amounts relative to the barite impurities, reduce acid-soluble barium losses (16).

Off-gases from the kiln are scrubbed for SO_2, SO_3, dust, and organic volatiles. In addition to equation 2, reactions which can generate SO_x are

$$BaS + 3\,CO_2 \longrightarrow BaO + 3\,CO + SO_2 \tag{6}$$

and

$$BaSO_4 + CO \longrightarrow BaO + CO_2 + SO_2 \tag{7}$$

Black ash wt % composition is typically BaS, 65–72, SiO_2, 10, $BaSO_4$, 3, metal oxides, 0.5, $BaSiO_3$, 16, and carbon, 5.5. The amount of black ash produced is about 65% by weight of the total ore and coke fed.

Significant waste heat may be recovered from the high (about 600°C) kiln off-gas. Pre-heating combustion air or feed ore improves the energy efficiency of the process. Reduction of barite in a fluid bed with CO and/or hydrogen has been performed on an experimental scale.

Leaching.   Details of a typical countercurrent decantation system used in the leaching of barium black ash are shown in Figure 2. Black ash from the ball milling step is fed to a series of staged make-up tanks in which the dissolution of BaS begins. The dissolving media is an aqueous solution of 6–7% BaS from the overflow of the second of three clarifiers. Barium sulfide on contact with water

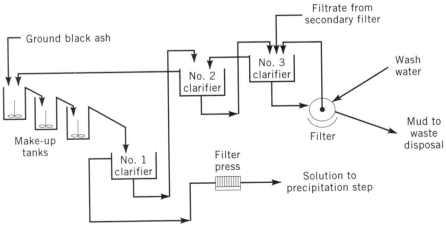

Fig. 2.   Schematic of leaching circuit.

hydrolyzes to the hydroxide and hydrosulfide according to

$$2\,BaS + 2\,H_2O \longrightarrow Ba(OH)_2 + Ba(SH)_2 \tag{8}$$

Leaching is 90–95% complete as the solids exit in the underflow of the first clarifier. Final washing of the solids (mud) using fresh water takes place in a rotary filter from whence the mud, consisting of 65% H_2O, 10–15% $BaSO_4$ and silicates, and 10–15% coke, is delivered to landfill at an approved site.

Overflow from the first clarifier, typically a 20% BaS solution, is filtered and sent on to the precipitation department. Settling of solids in the clarifiers can be enhanced by various flocculating agents (qv), preferably weakly anionic polyacrylamides (17).

The net percentage of solids in the clarifier underflows increases in going from the first to the second to the third clarifier. Typical values are 11, 15, and 21% solids, respectively. Clarifier operations have been investigated by computer simulation studies (17).

Precipitation.   Filtered overflow from the first clarifier, 20% BaS solution, is fed to an agitated tank where, on tight control, carbonate values are added in slight excess of stoichiometric requirements. The excess carbonate suppresses soluble barium which would otherwise later precipitate in equipment.

Carbonate values are usually supplied from soda ash or carbon dioxide or an equimolar mix. Possible reactions are

$$BaS + Na_2CO_3 \longrightarrow BaCO_3 + Na_2S \tag{9}$$

or

$$BaS + 0.5\,Na_2CO_3 + 0.5\,CO_2 + 0.5\,H_2O \longrightarrow BaCO_3 + NaHS \tag{10}$$

or

$$BaS + CO_2 + H_2O \longrightarrow BaCO_3 + H_2S \tag{11}$$

An additional possibility is

$$BaS + NaHCO_3 \longrightarrow BaCO_3 + NaHS \tag{12}$$

The H_2S in equation 11 can, if desired, be converted into elemental sulfur in a Claus furnace.

A photomicrograph of barium carbonate formed by precipitation using pure soda ash (eq. 9), is shown in Figure 3. The average particle size is 1.2 μm. The exclusive use of soda ash results in a barium carbonate having included sodium that cannot be reduced below a certain level by repeated washings. The sodium can be detrimental if the $BaCO_3$ is to be used for barium titanate production.

Filtration.   The slurry resulting from the precipitation step is filtered in a primary stage on rotary filters separating the product, $BaCO_3$, from the co-product, eg, Na_2S or NaHS in solution. The $BaCO_3$ wet cake is repulped in fresh water and filtered on a second-stage rotary filter. The aqueous stream

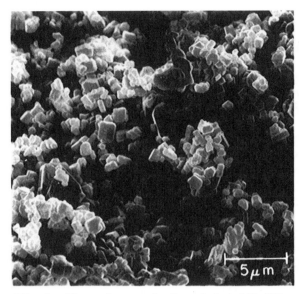

Fig. 3.   Photomicrograph of precipitated $BaCO_3$.

from this separation, carrying some sulfide values, is then used in the third clarifier of the leaching step (see Figs. 1 and 2).

Drying and Calcining.   A rotary dryer operating at a 600–800°C spill temperature may be used to form a hard, sintered, granular barium carbonate product. This form finds favor in glass manufacture. More powdery products can be produced at rotary dryer spill temperatures of less than 400°C (14). Spray-dried barium carbonate, including added dispersants (qv), provides a finer pigment-grade material which is used in ferrite production (see FERRITES).

Sulfide Solution Concentration.   The by-product of the black ash process has considerable bearing on the ultimate economic viability of barium production. It is desirable to be able to vary by-products to maximize earnings. Sodium sulfide [1313-82-2], Na_2S, sold as 60% flake, and sodium bisulfide [16721-80-5], NaHS, sold as 45% solution or as 70% flake, are typically co-produced.

When solid soda ash is used to supply all the carbonate values in the precipitation step (eq. 9), a ca 10% Na_2S solution results from the primary filtration step which can be concentrated to 40% Na_2S in a three-effect evaporator train. Final concentration to 60% Na_2S occurs in a high vacuum single-effect evaporator. This concentrated solution can then be fed to a flaker to produce a 60% sodium sulfide flake which is sold as co-product.

Both sodium sulfide and the bisulfide are used in the flotation process for copper minerals and as a depilatory for animal hides (see COPPER; COPPER ALLOYS; LEATHER). Also, sodium polysulfide can be produced from Na_2S, and elemental sulfur can be produced if H_2S is generated as an intermediate.

5.2. Supply and Demand.   The U.S. imported 18,600 t of barium carbonate in 2001, down from 26,200 t in 2000. The value in 2001 was \$ 9.2×10^6 (4).

5.3. Uses.   There are several different grades of barium carbonate manufactured to fit the specific needs of a wide variety of applications: very fine, highly reactive grades are made for the chemical industry; coarser and more readily handleable grades are mainly supplied to the glass industry.

The main use for barium carbonate is in the manufacture of glass. This is the most important use for a barium compound after well drilling fluids (5).

In 2000, 50% of the barium carbonate produced was used in glass manufacturing. Other uses include manufacturing of brick and clay products, barium chemicals, barium ferrites, and in the production of photographic papers.

Barium carbonate acts as a flux for silica. It is added to textile fiber glass, crown and flint optical glass, laboratory glassware, decorative glass, and frits for ceramic material. Barium carbonate has the lowest melting point of all the alkaline earths, and, in the presence of sodium carbonate, glass melting reactions may begin at less than 400°C. Barium oxide from the carbonate becomes part of the silicate structure and imparts many useful properties. It increases the glass durability, adds weight and density, increases the refractive index imparting a brilliance to the glass, and perhaps most significantly, absorbs x rays. Hence, barium carbonate is used in television glass manufacturing as an x-ray screening agent. Barium carbonate is used especially in black and white television face plates, and in small quantities in color face plates, where strontium carbonate is the principal screening agent.

Glass-grade barium carbonate has a high bulk density so that it does not become airborne when charged to the furnace. It also has a particle size distribution which has less tendency to segregate from other materials in the glass mixture.

Barium carbonate prevents formation of scum and efflorescence in brick, tile, masonry cement, terra cotta, and sewer pipe by insolubilizing the soluble sulfates contained in many of the otherwise unsuitable clays. At the same time, it aids other deflocculants by precipitating calcium and magnesium as the carbonates. This reaction is relatively slow and normally requires several days to mature even when very fine powder is used. Consequently, often a barium carbonate emulsion in water is prepared with carbonic acid to further increase the solubility and speed the reaction.

In the oil-well drilling industry, the barite suspension used as drilling mud can be destabilized by the presence of soluble materials such as gypsum. Addition of barium carbonate precipitates the gypsum, inhibits coagulation, and thus permits the mud to retain the desired consistency and dispersion.

Barium carbonate of finely controlled particle size reacts in the solid state when heated with iron oxide to form barium ferrites. Magnetically aligned barium ferrite [11138-11-7] powder can be pressed and sintered into a hard-core permanent magnet which is used in many types of small motors. Alternatively, ground up magnetic powder can be compounded into plastic strips which are used in a variety of appliances as part of the closure mechanism.

Barium carbonate also reacts with titania to form barium titanate [12047-27-7], $BaTiO_3$, a ferroelectric material with a very high dielectric constant (see FERROELECTRICS). Barium titanate is best manufactured as a single-phase composition by a solid-state sintering technique. The asymmetrical perovskite structure of the titanate develops a potential difference when compressed in specific crystallographic directions, and vice versa. This material is most widely used for its strong piezoelectric characteristics in transducers for ultrasonic technical applications such as the emulsification of liquids, mixing of powders and paints, and homogenization of milk, or in sonar devices.

Barium carbonate is also used in the manufacture of photographic paper to generate barium sulfate which imparts a flat white appearance (see PHOTOGRAPHY).

6. Barium Chloride

Both anhydrous barium chloride [10361-37-2], $BaCl_2$, mol wt 208.25, density 3.856 g/mL, and barium chloride dihydrate [10326-27-9], $BaCl_2 \cdot 2H_2O$, mol wt 244.28, density 3.097 g/mL, are produced from a filtered aqueous solution formed by the reaction of hydrochloric acid and $BaCO_3$ or BaS. If BaS is used, the H_2S generated must be appropriately handled.

The solubility of $BaCl_2$ in water is

Temperature, °C	0	20	40	60	80	100
Solubility, wt % $BaCl_2$	24.0	26.3	28.9	31.7	34.4	26.2

Anhydrous $BaCl_2$ exists as monoclinic or cubic crystals. The transition to cubic occurs at 925°C. Barium chloride melts at 962°C; the dihydrate, which has monoclinic crystals, loses water at 113°C. Barium chloride, which is very hygroscopic, is sold in moisture-proof bags and steel or fiber drums.

Barium chloride finds use in the production of barium colors, such as the diazo dyes barium lithol red [50867-36-2] and barium salt of Red Lake C [5160-02-1], a mordant for acid dyes and dying of textiles. Other uses include aluminum refining and boiler water treatment.

$BaCl_2$ is used in heat treating baths because of the eutectic mixtures it readily forms with other chlorides. The melting points of some eutectic mixtures are: $BaCl_2 \cdot 2KCl$, 672–680°C; $BaCl_2$–NaCl, 39 mol % $BaCl_2$, 654°C; $BaCl_2$–$CaCl_2$, 631°C. $BaCl_2$ is also used to set up porcelain enamels for sheet steel (see ENAMELS, PORCELAIN OR VITREOUS; STEEL), and it is used to produce blanc fixe [7727-43-7].

The U.S. imported 341 t of barium chloride in 2001 and was valued at $ 291,000. This figure is down from the 2000 demand of 1,240 t valued at $ 752,000 (4).

7. Barium 2-Ethylhexanoate

Barium 2-ethylhexanoate [4696-54-2], $Ba(C_8H_{16}O_2)_2$, also known as barium octanoate or barium octoate is usually used in synergistic combination with cadmium or zinc organic salts as a thermal stabilizer for PVC (18,19). It is often available preformulated with other additives such as the cadmium or zinc salts. Owing to cadmium's toxicity, the development of noncadmium stabilizers is growing, and use of barium octanoate is declining. Barium ethylhexanoate is a liquid; barium stearate [6865-35-6], $Ba(C_{18}H_{36}O_2)_2$, is a powder that also serves as a thermostabilizer for PVC. Organophosphites are often added to barium–zinc stabilizers to improve stabilizing effectiveness.

8. Barium Fluoride

Barium fluoride [7782-32-8], BaF_2, is a white crystal or powder. Under the microscope crystals may be clear and colorless. Reported melting points vary from 1290 (20) to 1355°C (21), including values of 1301 (22) and 1353°C (23). Differences may result from impurities, reaction with containers, or inaccurate temperature measurements. The heat of fusion is 28 kJ/mol (6.8 kcal/mol) (24), the boiling point 2260°C (25), and the density 4.9 g/cm^3. The solubility in water is about 1.6 g/L at 25°C and 5.6 g/100 g (26) in anhydrous hydrogen fluoride. Several preparations for barium fluoride have been reported (27–29).

High purity BaF_2 can be prepared from the reaction of barium acetate and aqueous HF (30), by dissolving the impure material in 2–12N HCl and recrystallizing at −40°C (31), by vaccum distillation of the metal fluoride impurities from BaF_2 melt (32), by purification of the aqueous acetate solution by ion exchange followed by fluorination (33), by solvent extraction using dithiocarbamate and CCl_4 (34–36), and by solvent extraction using acetonitrile (37).

A typical analysis of the commercial product is 99% with a loss on ignition of 0.9%; sulfates as SO_4, 0.2%; hexafluorosilicate as SiF_4, 0.02%; heavy metals as lead, 0.02%; and iron, 0.005%.

Barium fluoride is used commercially in combination with other fluorides for arc welding (qv) electrode fluxes. However, this usage is limited because of the availability of the much less expensive naturally occurring calcium fluoride.

Other reported uses of barium fluoride include the manufacture of fluorophosphate glass (38); stable fluoride glass (39); fluoroaluminate glass (40); fluorozirconate glass (41); infared transmitting glass (42); in oxidation-resistant ceramic coatings (43); in the manufacture of electric resistors (44); as a superconductor with copper oxide (45); as a fluoride optical fiber (46) (see FIBER OPTICS; GLASS); and in a high repetition rate uv excimer laser (47).

9. Barium Hydrosulfide

Barium hydrosulfide [25417-81-6], $Ba(HS)_2$, is formed by absorption of hydrogen sulfide into barium sulfide solution. On addition of alcohol, barium hydrosulfide tetrahydrate [12230-74-9], $Ba(HS)_2 \cdot 4H_2O$, crystallizes as yellow rhombic crystals that decompose at 50°C. Solid barium hydrosulfide is very unstable. Its solubility in water is

Temperature, °C	0	20	40	60	80	100
Solubility, wt % $Ba(HS)_2$	32.6	32.8	34.5	36.2	39.0	43.7

10. Barium Hydroxide

Barium hydroxide is the strongest base and has the greatest water-solubility of the alkaline-earth elements. Barium hydroxide (barium hydrate, caustic baryta)

exists as the octahydrate [12230-71-6], $Ba(OH)_2 \cdot 8H_2O$, the monohydrate [22326-55-2], $Ba(OH)_2 \cdot H_2O$, or as the anhydrous[17194-00-2] material, $Ba(OH)_2$. The octahydrate and monohydrate have sp gr 2.18 and 3.74, respectively. The mp of the octahydrate and anhydrous are 77.9°C and 407°C, respectively.

Solubility of $Ba(OH)_2$ in water is strongly temperature dependent in the range >40°C.

Temperature, °C	0	20	40	60	80
Solubility, wt % $Ba(OH)_2$	1.65	3.74	7.60	17.32	50.35

These solutions are highly alkaline and can effectively remove CO_2 or other acidic gases from ambient atmosphere. The octahydrate is also soluble in methanol, but only slightly soluble in ethanol.

The octahydrate is prepared by dissolving BaO in hot water for several hours, filtering off undissolved impurities, then cooling the solution to effect crystallization. Vapor pressure data for the octahydrate (48) is

Temperature, °C	17.6	32.1	40.7	50.7	57.8	64.8
Pressure, kPa	0.48	1.32	2.36	5.08	8.01	12.56

The monohydrate can be produced by vacuum drying of the octahydrate (49).

Barium hydroxide is used in the manufacture of barium greases and plastic stabilizers such as barium 2-ethylhexanoate, in papermaking, in sealing compositions (see SEALANTS), vulcanization accelerators, water purification, pigment dispersion, in a formula for self-extinguishing polyurethane foams, and in the protection of objects made of limestone from deterioration (see FINE ART EXAMINATION AND CONSERVATION; LIME AND LIMESTONE). Uses of the octahydrate include: use as a low temperature latent heat storage material in combination with Na or KNO_3 or NaOAc (50); use as a nucleating agent to reduce supercooling of $CaBr_2$ solution and removal of CO_2 and $^{14}CO_2$ by passing through a bed of solid mono- or octahydrate.

The U.S. imported 3,780 t of barium oxide, hydroxide, and peroxide valued at \$ 3.3×10^6 in 2001. In 2000, the U.S. imported 5,290 t valued at \$ 4.8×10^6 (4).

11. Barium Iodide

Barium iodide dihydrate [7787-33-9], $BaI_2 \cdot 2H_2O$, crystallizes from hot aqueous solution. Below 30°C, barium iodide hexahydrate [13477-15-1], $BaI_2 \cdot 6H_2O$, crystallizes. The dihydrate, sp gr 5.15, loses water to form anhydrous barium iodide [13718-50-8], BaI_2, at 150°C, mp 740°C. Barium iodide discolors on exposure to air containing carbon dioxide, and decomposes to barium carbonate and iodine. BaI_2 is soluble in water.

Temperature, °C	−20	0	25	60	120
Solubility, wt % BaI_2	58.60	62.5	68.8	70.70	74.30

BaI_2 is also soluble in alcohol and acetone. BaI_2 is useful in making other iodides. BaI_2 has been cited for producing ir transparent glasses that are useful in power transmission from CO and CO_2 lasers (qv) from the ZnI_2–CsI–BaI_2 system (51); as a catalyst promoter, for such catalysts as rhodium(III) chloride, in carbonylation reactions (52,53); as being useful in chemical vapor deposition as a precursor in forming the superconducting composition $YBa_2Cu_3O_{7-x}$; as a sintering aid for aluminum nitride (54), and in phosphor formulations for cathode-ray tubes.

12. Barium Metaborate

Barium metaborate monohydrate [23436-05-7], $Ba(BO_2)_2 \cdot H_2O$, has a sp gr of 3.25–3.35, a fusion pt of 900–1050°C, n_D of 1.55–1.60, solubility of 0.3% in water at 21°C, and pH 9.8–10.3 for the saturated solution. It can be prepared from the reaction of a solution of BaS and sodium tetraborate.

$Ba(BO_2)_2 \cdot H_2O$, used in flame retardant plastic formulations as a synergist for phosphorus or halogen compounds and as a partial or complete replacement for antimony oxide (see FLAME RETARDANTS), is excellent as an afterglow suppressant. The low refractive index of $Ba(BO_2)_2$ results in greater transparency and brighter colors in formulated plastics (55). Barium metaborate has been reported in paint formulations to convey insecticidal properties (56) (see INSECT CONTROL TECHNOLOGY). Barium metaborate crystals (the low temperature beta form) have been grown by Czocharski techniques (57,58) and are used in nonlinear optics for high power uv lasers and can be applied to uv photolithography (see LITHOGRAPHY; NONLINEAR OPTICAL MATERIALS). $Ba(BO_2)_2$ has been reported to be used in antibacterial coatings for aluminum heat exchanger surfaces of air conditioners (59). Use of the metaborate as a sintering aid for $BaTiO_3$ has been studied (60).

The barium oxide borate [12007-55-5], $BaO \cdot 2B_2O_3$, and dibarium oxide triborate [13840-10-3], $2BaO \cdot 3B_2O_3$, have also been reported.

13. Barium Nitrate

Barium nitrate [10022-31-8], $Ba(NO_3)_2$, occurs as colorless crystals; mp 592°C; sp gr 3.24. Its solubility in water is

Temperature, °C	0	25	40	60	80	100	135
Solubility, wt % $Ba(NO_3)_2$	4.72	9.27	12.35	16.9	21.4	25.6	32.0

Barium nitrate is prepared by reaction of $BaCO_3$ and nitric acid, filtration and evaporative crystallization, or by dissolving sodium nitrate in a saturated solution of barium chloride, with subsequent precipitation of barium nitrate. The precipitate is centrifuged, washed, and dried. Barium nitrate is used in pyrotechnic green flares, tracer bullets, primers, and in detonators. These make use of its property of easy decomposition as well as its characteristic green flame. A small amount is used as a source of barium oxide in enamels.

The U.S. imported 5,010 t of barium nitrate in 2001. Value of this quantity was $ 6.10×10^6. The U.S. imported 4,930 t in 2000, valued at $ 4.5×10^6 (14).

14. Barium Nitrite

Barium nitrite [13465-94-6], $Ba(NO_2)_2$, crystallizes from aqueous solution as barium nitrite monohydrate [7787-38-4], $Ba(NO_2)_2 \cdot H_2O$, which has yellowish hexagonal crystals, sp gr 3.173, solubility 54.8 g $Ba(NO_2)_2$/100 g H_2O at 0°C, 319 g at 100°C. The monohydrate loses its water of crystallization at 116°C. Anhydrous barium nitrite, sp gr 3.234, melts at 267°C and decomposes at 270°C into BaO, NO, and N_2. Barium nitrite may be prepared by crystallization from a solution of equivalent quantities of barium chloride and sodium nitrite, by thermal decomposition of barium nitrate in an atmosphere of NO, or by treating barium hydroxide or barium carbonate with the gaseous oxidiation products of ammonia. It has been used in diazotization reactions.

15. Barium Oxide and Peroxide

Barium oxide [1304-28-5], BaO, occurs as colorless cubic or hexagonal crystals; mp 1923°C; sublimation ca 2000°C; bp ca 3088°C; sp gr (cubic) 5.72, (hexagonal) 5.32. Barium oxide is highly reactive toward carbon dioxide in the presence of water vapor, and is converted to the hydroxide and carbonate when exposed to air. Upon additions of small amounts of water, or in a carbon dioxide atmosphere, absorption is so rapid that large amounts of heat are liberated, raising the temperature to a red heat. The heat of formation of barium hydroxide from barium oxide and water is 102 kJ/mol (24.4 kcal/mol) and that of barium carbonate from barium oxide and carbon dioxide is 264 kJ/mol (63.1 kcal/mol). Consequently, accumulation of barium oxide and the peroxide, dust, and dirt presents a fire hazard. When a fire occurs, large concentrations of carbon dioxide from burning organic matter cause further incandescence because the carbon dioxide is absorbed by the barium oxide. Such fires spread rapidly and are difficult to extinguish.

Of the alkaline-earth carbonates, $BaCO_3$ requires the greatest amount of heat to undergo decomposition to the oxide. Thus carbon in the form of coke, tar, or carbon black, is added to the carbonate to lower reaction temperature from about 1300°C in the absence of carbon to about 1050°C. The potential for the reverse reaction is decreased by removing the CO_2 as shown in equation 1b.

Barium oxide, which can react directly with oxygen to give the peroxide (61), is soluble in methanol and ethanol forming the alkoxides (see ALKOXIDES, METAL).

BaO is used to impart improved strength to porcelain (62), as a solid base catalyst, in specialty cements, and for drying gases.

The U.S. imported 3,780 t of barium oxide, hydroxide, and peroxide in 2001, valued at $ 3.3×10^6. In 2000, the U.S. imported 5,290 t valued at $ 4.8×10^6 (4).

When heated in air or oxygen to 500°C, barium oxide is converted readily to barium peroxide [1304-29-6], BaO_2. Upon further heating to 700°C, the peroxide decomposes to BaO and oxygen. This reaction was used for many years to make pure oxygen via the Brin process. Other preparations for the peroxide such as the reaction between $BaCl_2$ and H_2O_2 in alkaline solution, or between BaO and H_2O_2 have been reported.

Reported uses of BaO_2 include in the cathodes of fluorescent lamps, formation of $YBa_2Cu_3O_{7-x}$ superconducting phase from CuN_3, BaO_2, and Y_2O_3 (52), and as a drying agent forlithographic inks.

16. Barium Sodium Niobium Oxide

Barium sodium niobium oxide [12323-03-4], $Ba_2NaNb_5O_{15}$, finds application for its dielectric, piezoelectric, nonlinear crystal and electro-optic properties (63, 64). It has been used in conjunction with lasers for second harmonic generation and frequency doubling. The crystalline material can be grown at high temperature, mp ca 1450°C (65).

17. Barium Sulfate

Barium sulfate [7727-43-7], $BaSO_4$, occurs as colorless rhombic crystals, mp 1580°C (dec); sp gr 4.50; solubility 0.000285 g/100 g H_2O at 30°C and 0.00118 at 100°C. It is soluble in concentrated sulfuric acid, forming an acid sulfate; dilution with water reprecipitates barium sulfate. Precipitated $BaSO_4$ is known as blanc fixe, prepared from the reaction of aqueous solutions of barium sulfide and sodium sulfate.

Because of its extreme insolubility, barium sulfate is not toxic; the usual antidote for poisonous barium compounds is to convert them to barium sulfate by administering sodium or magnesium sulfate. In medicine, barium sulfate is widely used as an x-ray contrast medium (see IMAGING TECHNOLOGY; X-RAY TECHNOLOGY). It is also used in photographic papers, filler for plastics, and in concrete as a radiation shield. Commercially, barium sulfate is sold both as natural barite ore and as a precipitated product. Blanc fixe is also used in making white sidewall rubber tires or in other rubber applications.

18. Barium Sulfide

Impure barium sulfide with 20–35% contaminants is produced in large volume by the black ash kiln. Pure barium sulfide [21109-95-5], BaS, occurs as colorless cubic crystals, sp gr 4.25 and as hexagonal plates of barium sulfide hexahydrate [66104-39-0], $BaS \cdot 6H_2O$. BaS melts at 2227°C. Solubility in water is reported (49) as

Temperature, °C	0	20	40	60	80	90	100
Solubility, wt % BaS	2.8	7.3	13.0	21.7	33.3	40.2	37.6

BaS hydrolyzes to $Ba(OH)_2$ and barium hydrosulfide. Cooling of an aqueous BaS solution can precipitate the double salt barium hydroxide sulfide hydrate [42821-46-5], $Ba(OH)_2 \cdot Ba(SH)_2 \cdot xH_2O$.

Barium sulfide solutions undergo slow oxidation in air, forming elemental sulfur and a family of oxidized sulfur species including the sulfite, thiosulfate, polythionates, and sulfate. The elemental sulfur is retained in the dissolved liquor in the form of polysulfide ions, which are responsible for the yellow color of most BaS solutions. Some of the more highly oxidized sulfur species also enter the solution. Sulfur compound formation should be minimized to prevent the compounds made from BaS, such as barium carbonate, from becoming contaminated with sulfur.

BaS is used in the manufacture of lithophone[8006-32-4], useful as a white pigment in paints, according to

$$BaS + ZnSO_4 \longrightarrow BaSO_4 \text{ (ppt)} + ZnS \text{ (ppt)}$$

BaS has been used in the production of thin-film electroluminescent phosphors, often activated with Eu^{2+} (66, 67). Similarly, infrared-triggered phosphors may be fabricated from Ba or Sr sulfide, a dopant such as EuO and a fusible salt such as LiF (68). BaS has also been mentioned as a nucleating agent in combination with $BaCl_2 \cdot 2H_2O$ for a $CaCl_2 \cdot 6H_2O$ heat storage system (69), and in vulcanization of carbon black-filled neoprene rubbers (70).

19. Barium Titanate

The basic crystal structure of barium titanate [12047-27-7], $BaTiO_3$, the so-called perovskite structure, after the mineral, $CaTiO_3$, leads to unique, outstanding dielectric properties. The barium and oxygen ions together form a face-centered cubic lattice, the titanium ions fitting into octahedral interstices. The minimum energy positions for the titanium ion are off-center and give rise to an ordered electrical dipole, which can be made permanent during production by the proper heating and cooling regime (see CERAMICS; FERROELECTRICS; TITANIUM COMPOUNDS).

Barium titanate has widespread use in the electronics industry. Its high dielectric constant and the ease with which its electrical properties can be modified by combination with other materials make it exceptionally suitable for a variety of items, ie, miniature capacitors (see CERAMICS AS ELECTRICAL MATERIALS). Several mentions of barium titanate for use as a semiconductive ceramic have appeared in the patent literature (71–73).

Barium titanate is usually produced by the solid-state reaction of barium carbonate and titanium dioxide. Dielectric and piezoelectric properties of $BaTiO_3$ can be affected by stoichiometry, microstructure, and additive ions that can enter into solid solution. In the perovskite lattice, substitutions of Pb^{2+}, Sr^{2+}, Ca^{2+}, and Cd^{2+} can be made for part of the barium ions, maintaining the ferroelectric characteristics. Similarly, the Ti^{4+} ion can partially be replaced with Sn^{4+}, Hf^{4+}, Zr^{4+}, Ce^{4+}, and Th^{4+}. The possibilities for forming solution alloys in all these structures offer a range of compositions, which present a wide range of dielectric, temperature dependence, and other characteristics. At the same time, impurities such as SiO_2, Al_2O_3, Na_2O, etc, or anything affecting the crystal lattice, can also alter the dielectric properties and must be carefully controlled, both in raw materials and within the production process itself. Non-stoichiometry originating from the $BaO:TiO_2$ ratio or additional impurities may result in semiconductive ceramics upon firing.

Barium titanate ceramic products are used for application in numerous sonic and ultrasonic devices, such as underwater sonar, guided missiles, acoustic mines, ultrasoniccleaning, measuring instruments, eg, for flaw detection, liquid-level sensing, and thickness gauges, sound reproduction, filters, and ultrasonic machining (see LIQUID-LEVEL MEASUREMENTS; NONDESTRUCTIVE TESTING; SURFACE AND INTERFACE ANALYSIS). Its most important application is for ceramic capacitors of disk, tube, and multilayer designs (74).

20. Yttrium–Barium–Copper Oxide

Yttrium–barium–copper oxide, $YBa_2Cu_3O_{7-x}$, is a high T_c material which has been found to be fully superconductive at temperatures above 90 K, a temperature that can be maintained during practical operation. The foremost challenge is to be able to fabricate these materials into a flexible form to prepare wires, fibers, and bulk shapes. Ultrapure powders of yttrium–barium–copper oxide that are sinterable into single-phase superconducting material at low temperatures are required, creating a worldwide interest in high purity barium chemicals.

A number of promising new routes to chemically synthesize ultrapure, ultrahomogeneous particles of controlled particle size and particle size distribution are currently under development. A modified sol–gel method has been developed to produce thermoplastic gels that are compatible with fiber spinning technology (see SOL–GEL TECHNOLOGY), as has a process which avoids the formation of barium carbonate, a troublesome impurity at the grain boundaries which seriously detracts from the critical current. In this latter process, hyponitrites and hydrated oxides derived from the hydrolysis of organometallic solutions are subjected to a series of thermal treatments involving decomposition, oxidation, and finally annealing. Materials which can be decomposed and converted to the orthorhombic superconducting form are produced (75,76).

21. Analytical Methods

The classical analytical method of determination of barium ion is gravimetric, by precipitating and weighing insoluble barium sulfate. Barium chromate, which is more insoluble than strontium chromate in a slightly acidic solution, gives a fairly good separation of the two elements.

Alkaline-earth metals are often determined volumetrically by complexometric titration at pH 10, using Eriochrome Black T as indicator. The most suitable complexing titrant for barium ion is a solution of diethylenetriamine-pentaacetic acid (DTPA). Other alkaline earths, if present, are simultaneously titrated, and in the favored analytical procedure calcium and strontium are determined separately by atomic absorption spectrophotometry, and their values subtracted from the total to obtain the barium value.

Barium can also be determined by x-ray fluorescence (XRF) spectroscopy, atomic absorption spectroscopy, and flame emission spectroscopy. Prior separation is not necessary. XRF can be applied directly to samples of ore or products to yield analysis for barium and contaminants. All crystalline barium compounds can be analyzed by x-ray diffraction.

22. Health and Safety Aspects

The average adult human body contains 22 mg Ba, of which 93% is present in bone (77). The remainder is widely distributed throughout the soft tissues of the body in very low concentrations. Accumulation of barium also takes place in the pigmented parts of the eyes.

22.1. Environmental Levels and Exposures. Barium constitutes about 0.04% of the earth's crust (77). Agricultural soils contain Ba^{2+} in the range of several micrograms per gram. The Environmental Protection Agency, under the Safe Drinking Water Act, has set a limit for barium of 1 mg/L for municipal waters in the United States.

Generally, barium content of food parallels calcium content in a ratio from $1/10^2$ to $1/10^5$ (77). Milk contains about 45–136 micrograms Ba per gram Ca; wheat and oatmeal contain 1300 and 2320–8290 micrograms Ba per gram Ca, respectively. The average daily intake of barium may be as high as 1.33 mg in the diet of the general population. Dietary barium intake has been estimated to originate 25% from milk, 25% from flour, 25% from potatoes, and 25% from miscellaneous high barium foods, such as nuts, consumed in small quantities.

22.2. Toxicity. The toxicity of barium compounds depends on solubility (77–79). The free ion is readily absorbed from the lung and gastrointestinal tract. The mammalian intestinal mucosa is highly permeable to Ba^{2+} ions and is involved in the rapid flow of soluble barium salts into the blood. Barium is also deposited in the muscles where it remains for the first 30 h and then is slowly removed from the site (80). Very little is retained by the liver, kidneys, or spleen and practically none by the brain, heart, and hair.

22.3. Soluble Compounds. The mechanism of barium toxicity is related to its ability to substitute for calcium in muscle contraction. Toxicity results from stimulation of smooth muscles of the gastrointestinal tract, the cardiac muscle, and the voluntary muscles, resulting in paralysis (77). Skeletal, arterial, intestinal, and bronchial muscle all seem to be affected by barium.

22.4. Oral Exposure. Following oral exposure to high doses of barium compounds, symptoms of nausea, vomiting, colic, and diarrhea result. Other symptoms include excessive salivation, hypertension, tremors, tingling of the extremities, twitching of facial muscles, and paralysis of the tongue and pharynx with loss or impairment of speech (79,81). Barium stimulation of arterial muscles causes an elevation in blood pressure resulting from vasoconstriction. Severe cases of poisoning cause loss of tendon reflexes, heart fibrillation, and general muscular paralysis, including the respiratory muscles, leading to death (77). Acute barium toxicity results in low serum potassium levels and leukocytosis in both experimental animals and humans (79,82). Symptoms of chronic barium poisoning are similar, but of lesser severity.

The more soluble forms of barium such as the carbonate, chloride, acetate, sulfide, oxide, and nitrate, tend to be more acutely toxic (80). Mean lethal doses for ingested barium chloride were 300–500 mg/kg in rats and 7–29 mg/kg in mice (77).

Administration of 5 ppm barium, the acetate, to mice in the drinking water in a life-time study had no observable effects on longevity, mortality, and body weights, or on the incidence of tumors (83). Longterm studies in rats exposed to Ba^{2+} in drinking water containing 5 mg/L, as acetate, or 10–250 mg/L, as chloride, resulted in no measurable toxic effects (77).

Water-insoluble barium salts are poorly absorbed. In fact, barium sulfate is used as a contrast material for x-ray examination of the gastrointestinal tract based on its limited solubility and low toxicity (82). Barium sulfate fed to mice at various levels up to 8 ppm dietary Ba (~ 1.14 mg/kg·d as Ba^{2+}) for three generations had no significant effects on growth, mortality, morbidity, or reproductive performance (83).

22.5. Inhalation Exposure.   Workers exposed to barium carbonate dust for 7–27 years did not reveal any specific chronic poisoning (49). The American Conference of Governmental Industrial Hygienists (ACGIH) and the Occupational Safety and Health Administration (OSHA) have recommended a threshold limit value (TLV) of 0.5 mg/m^3 as Ba, as a time weighted average, for soluble compounds of barium. This time weighted average is an average airborne exposure in any 8-h work shift of a 40-h work week which should not be exceeded.

The OSHA limit for barium sulfate dust in air is 10 mg/m^3. NIOSH currently recommends that a level of 50 mg/m^3 be considered immediately dangerous to health (84).

In humans, inhaled insoluble barium salts are retained in the lung (77,79). Inhalation of high concentrations of the fine dusts of barium sulfate can result in the formation of harmless nodular granules in the lungs, a condition called baritosis (79). Baritosis produces no specific symptoms and no changes in pulmonary function. The nodulates disappear upon cessation of exposure to the barium salt. However, it is possible that barium sulfate may produce benign pneumoconiosis because, unlike barium carbonate, barium sulfate is poorly absorbed (49).

The threshold of a toxic dose in adult humans is about 0.2–0.5 g Ba; the lethal dose in untreated cases is 3–4 g Ba, LD_{50} about 66 mg/kg (77). The fatal-dose of barium chloride for humans is reported to be between 0.8 and 0.9 g (0.55–0.60 g of Ba) (80). However, for most of the acid-soluble salts of barium, doses greater than 1 g have been tolerated (81). Lethal doses are summarized in Table 3. Dusts of barium oxide are considered potential dermal and nasal irritants (82).

22.6. Treatment.   Treatment of poisoning from soluble barium salts may be preventive or curative (77,81). Preventive treatment involves inhibition of intestinal absorption by administering such soluble sulfates as magnesium or sodium, causing precipitation of barium sulfate in the alimentary tract. Curative treatment involves counteracting the paralytic effect of the Ba^{2+} ion on the muscle by intravenous infusion of a potassium salt.

Table 3. **Acute Lethal Doses of Soluble Barium Compounds**[a]

Compound	Route of exposure	Species	Toxicity, LD_{50}, mg/kg
$BaCO_3$	oral	rat	418
	oral	mouse	200
	oral	human	11[b]
			29[b]
	oral	human	57[c]
$BaNO_3$	oral	rat	355
	intravenous	mouse	8.5
$BaCl_2$	oral	rat	118
	oral	human	11.4[c]
BaO	subcutaneous	mouse	50
BaF_2	oral	rat	250

[a] Ref. 85.

[b] Value is toxic dose low. TD_{LO}, the lowest dose of a substance introduced by any route other than inhalation, over any given period of time to which humans or animals have been exposed and reported to produce any nonsignificant toxic effect in humans or to produce nonsignificant tumorigenic or reproductive effects in animals or humans.

[c] Value is lethal dose low, LD_{LO}, the lowest dose of a substance introduced by any other route other than inhalation, over any given period of time and reported to have caused death in humans or animals.

BIBLIOGRAPHY

"Barium Compounds" in *ECT* 1st ed., Vol. 2, pp. 307–323, by L. Preisman, Pittsburgh Plate Glass Co., and "Barium Hydride," by L. W. Davis, Metal Hydrides, Inc.; in *ECT* 2nd ed., Vol. 3, pp. 80–98, by L. Preisman, Pittsburgh Plate Glass Co.; in *ECT*, 3rd ed., Vol. 3, pp. 463–479, by T. Kirkpatrick, Sherwin-Williams Co.; in *ECT* 4th ed., Vol. 3, pp. 904–931, by Patrick M. Dibello, James L. Manganaro, and Elizabeth R. Aguinaldo, FMC Corporation; "Barium Fluoride" under "Fluorine Compounds, Inorganic," in *ECT* 1st ed., Vol. 6, p. 677, by F. D. Loomis, Pennsylvania Salt Manufacturing Co.; in *ECT* 2nd ed., Vol. 9, p. 551, by W. E. White, Ozark-Mahoning Co.; "Barium" under "Fluorine Compounds, Inorganic," in *ECT*, 3rd ed., Vol. 10, p. 684, by C. B. Lindahl, Ozark-Mahoning Co.; in *ECT* 4th ed., Vol. 11, pp. 298–299, by Tariq Mahmood and Charles B. Lindahl, Elf Atochem North America; "Barium Compounds" in *ECT* (online), posting date: December 4, 2000, by Patrick M. Dibello, James L. Manganaro, Elizabeth R. Aguinaldo, FMC Corporation.

CITED PUBLICATIONS

1. J. P. Searls, *"Barite"*, Mineral Commodity Summaries, U.S. Geological Survey, Reston, Va., Jan. 2002.
2. F. A. Cotton and G. Wilkinson, *Advanced Inorganic Chemistry*, John Wiley & Sons, Inc., New York, 1980, 271–273.
3. *Industrial Minerals and Rocks*, 4th ed., American Institute of Mining, Metallurgical, and Petroleum Engineers, Inc., New York, 1975, pp. 427–442.
4. "Barite", *Minerals Yearbook*, U.S. Geological Survey, Reston, Va., 2001.
5. "Barytes", Roskill Reports on Metals and Minerals, Jan. 2000.

6. U.S. Pat. 4,534,884 (Aug. 13, 1985), A. Satoshi, N. Takashi, and T. Kenji (to Fuji Photo Film Co.).
7. U.S. Pat. 4,690,769 (Sept. 1, 1987), G. A. Lane and H. E. Rossow (to The Dow Chemical Company).
8. Jpn. Pat. 60,246,242 (Dec. 5, 1985) (to Nippon Sheet Glass KK).
9. W. Lobunez, *FMC Report No. 5729-R*, Princeton, N.J., July 13, 1972.
10. W. Lobunez, *FMC Report No. 5686-R*, Princeton, N.J., Mar. 13, 1973.
11. W. E. Brownell, *J. Am. Ceram. Soc.* **46**, 1225–1228 (1963).
12. T. G. Akhmetov, *Chem. Ind.* **48**, 288 (1973).
13. T. Yamaguchi and co-workers, *J. Inorg. Nucl. Chem.* **34**, 2739 (1972).
14. D. J. Muyskens, D. W. Tunison, and E. Rau, *Engineering Encyclopedia.*
15. D. W. Tunison, *FMC Report No. 5527-R*, Princeton, N.J., 1971.
16. K. Kadic and M. Klofec, *Chem. Prum.* **31**(10), 519–521 (1981).
17. J. L. Manganaro, R. C. Schmidt, and W. K. Lau, *FMC Report ICG/T-79-075*, Princeton, N.J., June 12, 1979.
18. H. Andreas, in R. Gachter and H. Muller, eds., *Plastics Additives Handbook*, Hanser Publications, 1985, Chapt. 4.
19. O. S. Kauder, in J. T. Lutz, ed., *Thermoplastics Polymer Additives*, Marcel Dekker Inc., New York, 1989, Chapt. 12.
20. I. Barin and O. Knache, *Thermochemical Properties of Inorganic Substances*, Springer Verlag, Berlin, 1973.
21. H. Kojima, S. G. Whiteway, and C. R. Masson, *Can. J. Chem.* **46**, 2698 (1968).
22. I. Jackson, *Phys. Earth Planet. Inter.* **14**, 143 (1977).
23. B. Porter and E. A. Brown, *J. Am. Ceram. Soc.* **45**, 49 (1962).
24. G. Petit and A. Cremieo, *C. R. Acad. Sci.* **243**, 360 (1956).
25. O. Ruff and L. LeBoucher, *Z. Anorg. Chem.* **219**, 376 (1934).
26. A. W. Jache and G. H. Cady, *J. Phys. Chem.* **56**, 1106 (1952).
27. SU 1325018 Al (July 23, 1985), V. A. Bogomolov and co-workers.
28. SU 998352 Al (Feb. 23, 1983), A. A. Luginina and co-workers.
29. A. A. Lugina and co-workers *Zh. Neorg. Khim* 1981, **26**(2), 332–336.
30. Jpn. Pat. 90-144378 (June 4, 1990), K. Kobayashi, K. Fujiura, and S. Takahashi.
31. EP 90-312689 (Nov. 21, 1990), J. A. Sommers, R. Ginther, and K. Ewing.
32. A. M. Garbar, A. N. Gulyaikin, G. L. Murskii, I. V. Filimonov, and M. F. Churbanov, *Vysokochist, Veshchestva* (6), 84–85 (1990).
33. A. M. Garbar, A. V. Loginov, G. L. Murskii, V. I. Rodchenkov, and V. G. Pimenov, *Vysokochist, Veshchestva* (3), 212–213 (1989).
34. Jpn. Pat. 01028203 A2 (Jan. 30, 1989), K. Kobayashi (to Heisei).
35. K. Kobayashi, *Mater. Sci. Forum*, **32–33**(5), 75–80 (1988).
36. DE 3813454 Al (Nov. 3, 1988); Jpn. Pat. 87-100025 (Apr. 24, 1987), H. Yamashita and H. Kawamoto.
37. J. Guery and C. Jacoboni, in Ref. 35, pp. 31–35.
38. V. D. Khalilev, V. G. Cheichovskii, M. A. Amanikov, and Kh. V. Sabirov, *Fiz. Khim. Stekla* **17**(5), 740–743 (1991).
39. Y. Wang, *J. Non. Cryst. Solids* **142**(1–2), 185–188 (1992).
40. H. Hu, F. Lin, and J. Feng, *Guisudnyan Xuebao* **18**(6), 501–505 (1990).
41. M. N. Brekhovskikh, V. A. Fedorov, V. S. Shiryaev, and M. F. Churbanov, *Vysokochist Veshchestva* (**1**), 219–223 (1991).
42. A. Jha and J. M. Parker, *Phys. Chem. Glasses* **32**(1), 1–12 (1991).
43. EP 392822 A2 17 (Oct. 17, 1990), L. M. Niebylski.
44. Jpn. Pat. 63215556 A2 (Sept. 8, 1988); 6321553A (Sept. 28, 1988) T. Honda, T. Yamada, K. Onigata, and S. Tosaka (to Showa).

45. S. R. Ovshinsky, R. T. Young, B. S. Chao, G. Fournier, and D. A. Pawlik *Rev. Solid State Sci.* **1**(2), 207–219 (1987).
46. J. Chen and co-workers, *J. Non-Cryst. Solids* **140**(1–3), 293–296 (1992).
47. U. S. Pat. Appl. 20020122451 (Sept. 5, 2002), R. W. Sparrow (To Corning Inc.).
48. E. W. Washburn, ed., *International Critical Tables of Numerical Data, Physics, Chemistry, and Technology (ICT)*, Vol. 7, McGraw-Hill, New York, 1930.
49. W. Gerhatz, ed., *Ullmann's* Encyclopedia of Industrial Chemistry, Vol. A3, VCH, New York, 1985.
50. R. Kniep, S. Mann, and A. Zachos, *Sol. Energy* **36**, 291 (1986).
51. K. Kadonao and co-workers, *J. Non-Cryst. Solids* **116**(1), 33–38.
52. K. Shinoda, K. Yasuda, and D. Miyatani, *Nippon Kayaku (Kaishi)* **5**, 724–729 (1988).
53. U.S. Pat. 4,705,890 (Nov. 10, 1987), G. Steinmetz and M. Rule (to Eastman Kodak Co.).
54. Jpn. Pat. 61,021,977 (Jan. 30, 1986), H. Taniguchi and N. Kuramoto (to Tokuyama Soda KK).
55. J. Green and co-workers, in H. S. Katz and J. V. Milewski, eds., *Handbook of Fillers for Plastics*, Van Nostrand Reinhold Co., New York, 1987, Chapt. 18.
56. A. Z. Gomaz, *Pigm. Resin Technol.* **18**(4), 4–8 (1989).
57. Jpn. Pat. 63,215,598 (Sept. 8, 1988), N. Oonishi (to Agency of Ind. Sci. Tech.).
58. U.S. Pat. 4,793,894 (Dec. 27, 1988), J. C. Jacco and G. M. Loiacono (to North American Philips Corp.).
59. Jpn. Pat. 61,168,675 (July 30, 1986), T. Watanabe (to Mitsubishi Heavy Ind KK, Nakabishi Eng KK).
60. D. Kolar, M. Trontely, and L. Marsel, *J. Phylics. Colloq.* **C11, C1**(447), **Ci**(450) (1986).
61. T. Moeller, *Inorganic Chemistry*, Wiley-Interscience, New York, 1952.
62. H. Schubert, *Sprechsael* **120**(6), 529–531 (1987).
63. Jpn. Pat. 61,106,463 (May 24, 1986), K. Nagata and K. Hikita (to Mitsubishi Mining Cement).
64. S. V. Kruzhalov and co-workers, *Pis'ma Zh. Tech. Fiz.* **8**(12), 756 (1982).
65. L. G. Van Uitert, *Platinum Metals Rev.* **14,647**, 118–121 (1970).
66. W. O. 8,908,921 A1 (Sept. 21, 1989), H. Hirano, N. Iwase, and N. Koshino (to Fujitsu Ltd.).
67. Jpn. Pat. 63,000,995 A2 (Jan. 5, 1988), K. Takahashi, Y. Onuki, and A. Kondo (to Toyo Soda M. Lq. KK).
68. U.S. Pat. 4,705,952 (Nov. 10, 1987), J. Lindmayer (to Quantex Corp.).
69. EP 240,583 (Oct. 14, 1987), J. Lindmayer (to Kubota Tekko KK).
70. A. A. Nosuikov, C. A. Blokh, and V. V. Bondovenko, *Izv. Vyssh. Uchebn. Zaved., Khim. Khim. Tekhnol.* **29**(12), 121–125 (1986).
71. U.S. Pat. Appl. 20030033784 (Jan. 30, 2003), M. Kawamoto (to Murata Manufacturing Co., Ltd.).
72. U.S. Pat. Appl. 20010026865 (Oct. 4, 2001), T. Miyoshi (to Murata Manufacturing Co., Ltd.).
73. U.S. Pat. Appl. 20010008866 (July 19, 2001), M. Kawamoto.
74. W. D. Kingery, H. K. Bowen, and D. R. Uhlmann, *Introduction to Ceramics*, 2nd ed., John Wiley & Sons, Inc., New York, 1960.
75. J. R. Gaines, Jr., *Am. Ceram. Soc. Bull.* **68**(4), 857–869.
76. D. R. Ulrich, "Chemical Processing of Ceramics," *C & EN* (Jan. 1, 1990).
77. A. L. Reeves, in L. Friberg, G. F. Nordberg, and V. B. Vouk, eds., *Handbook on the Toxicology of Metals*, Vol. 2, 2nd ed., Elsevier, New York, 1986, 84–94.
78. C. D. Klaaseen, M. O. Amdur, and J. Doull, eds., *Casarett and Doull's* Toxicology: The Basic Science of Poisons, 3rd ed., Macmillan Publishing Co., New York, 1986, 623–624.

79. B. Venugopal and T. D. Luckey, *Metal Toxicity in Mammals*, Vol. 2, Plenum Press, New York, 1978, 63–67.
80. G. D. Clayton and F. E. Clayton, eds., *Patty's* Industrial Hygiene and Toxicology, Vol. 2A, 3rd ed., John Wiley & Sons, Inc., New York, 1981, pp. 1531–1537.
81. R. E. Gosselin, R. P. Smith, and H. C. Hodge, *Clinical Toxicology of Commercial Products*, Williams and Wilkins, Baltimore, Md., 1984, Section III-61-3.
82. *Documentation of the Threshold Limit Values and Biological Exposure Indices*, American Conference of Governmental Industrial Hygienists (ACGIH), 5th ed., 1986, 47–48.
83. E. J. Underwood, *Trace Elements in Human and Animal Nutrition*, 4th ed., Academic Press, New York, 1977, 434–436.
84. M. B. Genter, in E. Bingham, B. Cohrssen, and C. H. Powell, eds., *Patty's* Toxicology, 5th ed., Vol. 2, John Wiley & Sons, Inc., New York, 2001, p. 242.
85. *Registry of Toxic Effects of Chemical Substances*, 1985–1986.

PATRICK M. DIBELLO
JAMES, L. MANGANARO
ELIZABETH R. AGUINALDO
FMC Corporation

TARIQ MAHMOOD
CHARLES B. LINDAHL
Elf Atochem North America Inc.

BARRIER POLYMERS

1. Introduction

Barrier polymers are used for many packaging and protective applications. As barriers they separate a system, such as an article of food or an electronic component, from an environment. That is, they limit the introduction of matter from the environment into the system or limit the loss of matter from the system or both. In many cases, the environment is simply room air, but the environment can be very different, such as in the case of protecting a submerged system from water.

All polymers are barriers to some degree; however, no polymer is a perfect barrier. Polymers only limit the movement of substances. When a polymer limits the movement enough to satisfy the requirements of a particular application, it is a barrier polymer for that application. Hence, barrier polymer finds definition in the application. A given polymer may be a good barrier for protecting one system from a component of an environment, but a poor barrier for protecting another system from a different component of the environment. This description of a barrier polymer is much wider than the traditional operational definitions that focus narrowly on oxygen permeabilities and water–vapor transmission rates (WVTR). This article discusses barrier polymers and the permeation process within the context of barrier applications. It also covers units of measure, physical factors affecting permeation, typical barrier structures, prediction and measurement of barrier properties, uses, and health and safety factors.

2. The Permeation Process

Barrier polymers limit movement of substances, hereafter called permeants. The movement can be through the polymer or, in some cases, merely into the polymer. The overall movement of permeants through a polymer is called permeation, which is a multistep process. First, the permeant molecule collides with the polymer. Then, it must adsorb to the polymer surface and dissolve into the polymer bulk. In the polymer, the permeant "hops" or diffuses randomly as its own thermal kinetic energy keeps it moving from vacancy to vacancy while the polymer chains move. The random diffusion yields a net movement from the side of the barrier polymer that is in contact with a high concentration or partial pressure of the permeant to the side that is in contact with a low concentration of permeant. After crossing the barrier polymer, the permeant moves to the polymer surface, desorbs, and moves away.

Permeant movement is a physical process that has both a thermodynamic and a kinetic component. For polymers without special surface treatments, the thermodynamic contribution is in the solution step. The permeant partitions between the environment and the polymer according to thermodynamic rules of solution. The kinetic contribution is in the diffusion. The net rate of movement is dependent on the speed of permeant movement and the availability of new vacancies in the polymer.

A few simple equations describe most applications of barrier polymers. Equation 1 is an adaptation for films of Fick's first law.

$$\frac{\Delta M_x}{\Delta t} = \frac{PA\Delta p_x}{L} \tag{1}$$

where $\Delta M_x/\Delta t$ is the steady-state rate of permeation of permeant x through a polymer film, P is the permeability coefficient (commonly called the permeability), A is the area of the film, Δp_x is the difference in pressure of the permeant on the two sides of the film, and L is the thickness of the film.

The permeability is the product of the diffusion coefficient D and the solubility coefficient S.

$$P = DS \tag{2}$$

The diffusion coefficient, sometimes called the diffusivity, is the kinetic term that describes the speed of movement. The solubility coefficient, which should not be called the solubility, is the thermodynamic term that describes the amount of permeant that will dissolve in the polymer. The solubility coefficient is a reciprocal Henry's law coefficient, as shown in equation 3.

$$C_{eq} = Sp_x \tag{3}$$

where C_{eq} is the equilibrium concentration of permeant that will dissolve in a polymer when the permeant has a partial pressure of p_x.

A polymer can have a low permeability because it has a low value of D or a low value of S or both. A low value of D can result from either static or dynamic effects. Static effects include molecular packing in the amorphous phase, orientation, and the amount of crystallinity. Molecular packing affects the way permeant molecules move through the free volume or vacanies in the polymer host. When there is a small amount of free volume, movement is limited. Symmetric monomers lead to good packing, and hence lower diffusion rates. For example, vinylidene chloride, which is symmetric 1,1-dichloroethene [75-35-4], leads to a good barrier polymer because the adjacent molecular chains can pack together well. The analogue vinyl chloride [75-01-4], which is asymmetric monochloroethene, leads to a polymer that does not pack as well. The simple chain of high density polyethylene (HDPE) packs much better than the branched chain of low density polyethylene (LDPE). Hence, HDPE has better barrier properties than LDPE.

Orientation sometimes leads to lower permeability values (better barrier properties). Orientation can increase packing density, which lowers the diffusion coefficient D; it can also increase the difficulty of hopping or diffusing in a direction perpendicular to the film. In the latter case, movement in general may be fast, but movement through the film is limited. However, mere stretching does not always lead to orientation of the molecular chains. In fact, stretching can lead to void formation, which increases permeability.

Increased crystallinity can reduce permeability values because the crystal regions of a polymer are impenetrable in most semicrystalline polymers. Hence, the average value of the solubility coefficient S is reduced. It also means that movement must occur around the crystallites, which means that a longer distance must be traveled, thus lowering the effective value of D.

The dynamic effects on the diffusion coefficient are related to polymer chain mobility. Permeant movement is enhanced when there are fluctuations in the free volume of the polymer. As chains move with thermal motion, opportunities for diffusion or hopping increase. Spaces between polymer molecules that are too small for passage at one point in time can be adequate later as the molecules move. Ultimately, this effect is related to temperature in general and often to the glass transition temperature T_g in particular. As the temperature increases, thermal motion increases, and the probability of larger openings occurring increases. Below T_g, thermal motion is limited to short range; small openings occur at low rates. Above T_g, thermal motion is enhanced, longer range motion is common, and larger openings occur at usable rates. All the polymer–polymer interactions that affect T_g affect the permeability, including dipole–dipole interactions and efficient packing.

The common model of a polymer as a collection of noodles in a bowl is not adequate for diffusion. A better model is a collection of long, active worms. Movement through the collection is enhanced if the worms are widely spaced (static effect) and if they have agitated motion (dynamic effect).

A low solubility coefficient can result when the permeant does not condense readily or does not interact strongly with the polymer. Generally, those permanent gases that have low boiling temperatures have low solubility coefficients in polymers. They are inherently adverse to existing in a condensed state either as a liquid or in a polymer solution. Hence, molecules like oxygen have solubility coefficients that are many orders of magnitude lower than the solubility coefficients of heavy flavor molecules. When no specific interactions, such as dipole–dipole interactions or hydrogen bonding, occur between the polymer and the permeant, solution is minimal. All those effects that control solutions in general apply to the polymer solution. These effects are not fully understood, but progress is being made.

3. Units of Measure

To understand permeability or barrier property values, it is necessary to define the units of measure. These units are complicated and many different sets of units are in common use. Furthermore, from time to time the units of permeability are presented in confused or incorrect fashion in the literature.

The dimensions of permeability become clear after rearranging equation 1 to solve for P. The permeability must have dimensions of quantity of permeant (either mass or molar) times thickness in the numerator with area times a time interval times pressure in the denominator. Table 1 contains conversion factors for several common unit sets with the permeant quantity in molar units. The unit $nmol/(m \cdot s \cdot GPa)$ is used herein for the permeability of small molecules because this unit is SI, which is preferred in current technical encyclopedias, and it is only a factor of 2, different from the commercial permeability unit, $[cc(STP) \cdot mil]/(100 \ in.^2 \cdot d \cdot atm)$ where cc(STP) is a molar unit for absorbed permeant (nominally cubic centimeters of gas at standard temperature and pressure). The molar character is useful for oxygen permeation, which could ultimately involve a chemical reaction, or carbon dioxide permeation, which is often related to the pressure in a beverage bottle.

For permeation of flavor, aroma, and solvent molecules another metric combination of units is more useful, namely, $(kg \cdot m)/(m^2 \cdot s \cdot Pa)$. In this unit, the permeant quantity has mass units. This is consistent with the common practice of describing these materials. Permeability values in these units often carry a cumbersome exponent; hence, a modified unit, an MZU $(10^{-20} \ kg \cdot m)/(m^2 \cdot s \cdot Pa)$, is used herein. The conversion from this permeability unit to the preferred unit for small molecules depends on the molecular weight of the permeant. Equation 4 expresses the relationship where MW is the molecular weight of the permeant in daltons (g/mol).

$$\text{permeability in MZU} \times (10/MW) = \text{permeability in } nmol/(m^3 GPa) \quad (4)$$

The solubility coefficient must have units that are consistent with equation 3. In the literature, S has units $cc(STP)/(cm^3 \cdot atm)$, where cm^3 is a volume of polymer. When these units are multiplied by an equilibrium pressure of permeant, concentration units result. In preferred SI units, S has units of $nmol/(m^3 \cdot GPa)$.

$$\begin{aligned} &\text{solubility coefficient in } cc(STP)/(cm^3 atm) \times (4.04 \times 10^{14}) \\ &= \text{solubility coefficient in } nmol/(m^3 GPa) \end{aligned} \quad (5)$$

Table 1. Permeability Units[a] with Conversion Factors

Multiply $\longrightarrow$
to obtain $\downarrow$

	$\dfrac{\text{nmol}}{\text{m·s·GPa}}$	$\dfrac{\text{cc·mil}}{100\ \text{in.}^2\text{·d}}$	$\dfrac{\text{cc·mil}}{\text{m}^2\text{·d·atm}}$	$\dfrac{\text{cc·cm}}{\text{cm}^2\text{·s·atm}}$	$\dfrac{\text{cc·cm}}{\text{cm}^2\text{·s·cm Hg}}$	$\dfrac{\text{cc·20 } \mu\text{m}}{\text{m}^2\text{·d·atm}}$
$\dfrac{\text{nmol}}{\text{m·s·GPa}}$	1	2	0.129	4.390×10^{10}	3.336×10^{12}	0.1016
$\dfrac{\text{cc·mil}}{100\ \text{in.}^2\text{·d}}$	0.50	1	6.452×10^{-2}	2.195×10^{10}	1.668×10^{12}	5.08×10^{-2}
$\dfrac{\text{cc·mil}}{\text{m}^2\text{·d·atm}}$	7.75	15.50	1	3.402×10^{11}	2.585×10^{13}	0.787
$\dfrac{\text{cc·cm}}{\text{cm}^2\text{·s·atm}}$	2.278×10^{-11}	4.557×10^{-11}	2.939×10^{-12}	1	76.00	2.315×10^{-12}
$\dfrac{\text{cc·cm}}{\text{cm}^2\text{·s·cm Hg}}$	2.998×10^{-13}	5.996×10^{-13}	3.860×10^{-14}	1.316×10^{-2}	1	3.046×10^{-14}
$\dfrac{\text{cc·20 } \mu\text{m}}{\text{m}^2\text{·d·atm}}$	9.84	19.68	1.27	4.32×10^{11}	3.283×10^{13}	1

[a] Throughout the *Encyclopedia* cm^3 (or mL) is used in preference to cc. However, the advantage of using cc here is an obvious visual aid in the complex units and there are further comments regarding cc vs cm^3 in the text.

Table 2. **Water Vapor Transmission Rate Units with Conversion Factors**

Multiply $\longrightarrow$ to obtain $\downarrow$	$\frac{nmol}{m \cdot s}$	$\frac{g \cdot mil}{100\ in.^2 \cdot d}$	$\frac{g \cdot cm}{m^2 \cdot d}$
$\frac{nmol}{m \cdot s}$	1	0.253	6.43
$\frac{g \cdot mil}{100\ in.^2 \cdot d}$	3.95	1	25.40
$\frac{g \cdot cm}{m^2 \cdot d}$	0.155	3.94×10^{-2}	1

In the mass units for flavor, aroma, and solvent molecules, the solubility coefficient has units $kg/(m^3 \cdot Pa)$. Equation 6 shows how to convert from the mass units to the molar units.

$$\text{solubility coefficient in } kg/(m^3 Pa) \times (10^{21}/MW)$$
$$= \text{solubility coefficient in } nmol/(m^3 GPA) \tag{6}$$

The diffusion coefficient has been commonly reported in cm^2/s in many applications. Hereafter, m^2/s will be used since it uses the basic SI units.

The water–vapor transmission rate is another descriptor of barrier polymers. Strictly, it is not a permeability coefficient. The dimensions are quantity times thickness in the numerator and area times a time interval in the denominator. These dimensions do not have a pressure dimension in the denominator as does the permeability. Common commercial units for WVTR are $(g \cdot mil)/(100\ in.^2 \cdot d)$. Table 2 contains conversion factors for several common units for WVTR. This text uses the preferred $nmol/(m \cdot s)$. The WVTR describes the rate that water molecules move through a film when one side has a humid environment and the other side is dry. The WVTR is a strong function of temperature because both the water content of the air and the permeability are directly related to temperature. For the WVTR to be useful, the water–vapor pressure difference for the value must be reported. Both these facts are recognized by specifying the relative humidity and temperature for the WVTR value. This enables the user to calculate the water–vapor pressure difference. For example, the common conditions are 90% relative humidity (rh) at $37.8°C$, which means the pressure difference is 5.89 kPa (44 mm Hg). The WVTR may be converted to a permeability by dividing by the specified pressure difference of the water vapor.

4. Small Molecule Permeation

4.1. Permanent Gases. Table 3 lists the permeabilities of oxygen [7782-44-7], nitrogen [7727-37-9], and carbon dioxide [124-38-9] for selected barrier and nonbarrier polymers at 20°C and 75% rh. The effect of temperature and humidity are discussed later. For many polymers, the permeabilities of nitrogen, oxygen, and carbon dioxide are in the ratio 1:4:14.

The traditional definition of a barrier polymer required an oxygen permeability $<2\ nmol/(m \cdot s \cdot GPa)$ [originally, $<(cc \cdot mil)/(100\ in.^2 \cdot d \cdot atm)$] at room temperature. This definition was based partly on function and partly on conforming

Table 3. **Permeabilities of Selected Polymers**[a]

Polymer	Gas permeability nmol/(m · s · GPa)		
	Oxygen	Nitrogen	Carbon dioxide
vinylidene chloride copolymers	0.02–0.30	0.005–0.07	0.1–1.5
ethylene–vinyl alcohol (EVOH) copolymers, dry at 100% rh	0.014–0.095, 2.2–1.1		
nylon-MXD6[b]	0.30		
nitrile barrier polymers	1.8–2.0		6–8
nylon-6	4–6		20–24
amorphous nylon (Selar[c] PA 3426)	5–6		
poly(ethylene terephthalate) (PET)	6–8	1.4–1.9	30–50
poly(vinyl chloride) (PVC)	10–40		40–100
high density polyethylene	200–400	80–120	1200–1400
polypropylene	300–500	60–100	1000–1600
low density polyethylene	500–700	200–400	2000–4000
polystyrene	500–800	80–120	1400–3000

[a] Ref. 1; see Table 1 for unit conversion.
[b] Trademark of Mitsubishi Gas Chemical Co.
[c] Trademark of Du Pont.

to the old commercial unit of permeability. The old commercial unit of permeability was created so that the oxygen permeability of Saran Wrap brand plastic film, a trademark of The Dow Chemical Company, would have a numerical value of 1.

However, the traditional definition of a barrier polymer is a good starting point for food packaging. Table 4 contains the approximate oxygen tolerances of several food groups. This table and equation 1 can be used to analyze a hypothetical packaging situation. For a package with an area of 0.045 m^2 (70 in.^{2}) and a wall thickness of 50.8 μm (2 mil), the oxygen permeability

Table 4. **Estimated Maximum Oxygen Tolerance of Various Foods**[a]

Food or beverage	Maximum oxygen tolerance, ppm
beer (pasteurized)	1–2
typical autoclaved low acid foods	1–3
canned milk	
canned meats and vegetables	
canned soups	
baby foods	
fine wine	2–5
coffee (fresh ground)	2–5
tomato-based products	3–8
high acid fruit juices	8–20
carbonated soft drinks	10–40
oils and shortenings	20–50
salad dressings, peanut butter	30–100
liquor, jams, jellies	50 – 100+

Table 5. Diffusion and Solubility Coefficients for Oxygen and Carbon Dioxide in Selected Polymers at 23°C, Dry[a]

Polymer	Oxygen		Carbon dioxide	
	D, m^2/s	S, nmol/(m$^3 \cdot$GPa)[b]	D, m^2/s	S, nmol/(m$^3 \cdot$GPa)[b]
vinylidene chloride copolymer	1.2×10^{-14}	1.01×10^{13}	1.3×10^{-14}	3.2×10^{13}
EVOH copolymer[c]	7.2×10^{-14}	2.4×10^{12}		
acrylonitrile barrier polymer	1.0×10^{-13}	1.0×10^{13}	9.0×10^{-14}	4.4×10^{13}
PET	2.7×10^{-13}	2.8×10^{13}	6.2×10^{-14}	8.1×10^{14}
PVC	1.2×10^{-12}	1.2×10^{13}	8.0×10^{-13}	9.7×10^{13}
polypropylene	2.9×10^{-12}	1.1×10^{14}	3.2×10^{-12}	3.4×10^{14}
high density polyethylene	1.6×10^{-11}	7.2×10^{12}	1.1×10^{-11}	4.3×10^{13}
low density polyethylene	4.5×10^{-11}	2.0×10^{13}	3.2×10^{-11}	1.2×10^{14}

[a] Refs. (4–9).
[b] For unit conversion, see equation 5.
[c] 42 mol % ethylene.

needs to be <1.9 nmol/(m·s·GPa) to give 100 days of protection in air (2.1×10^{-5} GPa O_2) to 500 g of contents that could withstand up to 20 ppm (wt/wt) oxygen. A lower oxygen permeability is needed if the application requires greater oxygen protection, a longer storage time, or a thinner wall.

Poly(ethylene terephthalate) [25038-59-9], with an oxygen permeability of 8 nmol/(m·s·GPa), is not considered a barrier polymer by the old definition; however, it is an adequate barrier polymer for holding carbon dioxide in a 2-L bottle for carbonated soft drinks. The solubility coefficients for carbon dioxide are much larger than for oxygen. For the case of the PET soft drink bottle, the principal mechanism for loss of carbon dioxide is by sorption in the bottle walls as 500 kPa (5 atm) of carbon dioxide equilibrates with the polymer (3). For an average wall thickness of 370 µm (14.5 mil) and a permeability of 40 nmol/ (m·s·GPa), many months are required to lose enough carbon dioxide (15% of initial) to be objectionable.

The diffusion and solubility coefficients for oxygen and carbon dioxide in selected polymers have been collected in Table 5. Determination of these coefficients is neither common, nor difficult. Methods are discussed later. The values of S for a permeant gas do not vary much from polymer to polymer. The large differences that are found for permeability are due almost entirely to differences in D.

4.2. Polymers with Good Barrier-to-Permanent Gases.
Those polymers that are good barriers to permanent gases, especially oxygen, have important commercial significance. They represent a diverse set of chemical structures shown in Figure 1.

Vinylidene chloride copolymers are available as resins for extrusion, latices for coating, and resins for solvent coating. Comonomer levels range from 5 to 20 wt %. Common comonomers are vinyl chloride, acrylonitrile, and alkyl acrylates. The permeability of the polymer is a function of type and amount of comonomer. As the comonomer fraction of these semicrystalline copolymers is increased, the melting temperature decreases and the permeability increases. The permeability of vinylidene chloride homopolymer has not been measured.

Vinylidene chloride copolymers are marketed under a variety of trade names. Saran is a trademark of The Dow Chemical Company for vinylidene chloride copolymers. Other trade names include Daran (W. R. Grace), Amsco Res (Union Oil), and Serfene (Morton Chemical) in the United States; Haloflex (Imperial Chemical Industries, Ltd.), Diofan (BASF), Ixan (Solvay and Cie SA), and Polyidene (Scott-Bader) in Europe.

Hydrolyzed ethylene–vinyl acetate copolymers [24937-78-8], commonly known as EVOH copolymers [25067-34-9], are usually used as extrusion resins, although some may be used in solvent-coating applications. As the ethylene fraction of these semicrystalline copolymers increases, the melting temperature decreases, the permeabilities increase, and the sensitivity to humidity decreases. The permeabilities as a function of polymer composition and humidity are shown in Figure 2. Vinyl alcohol homopolymer [9002-89-5] has a very low oxygen permeability in dry conditions; however, the polymer is water-soluble. Trade names for these barrier polymers include Eval, Soarnol, Selar OH, and Clarene. Table 6 lists the compositions of some commonly used EVOH copolymers.

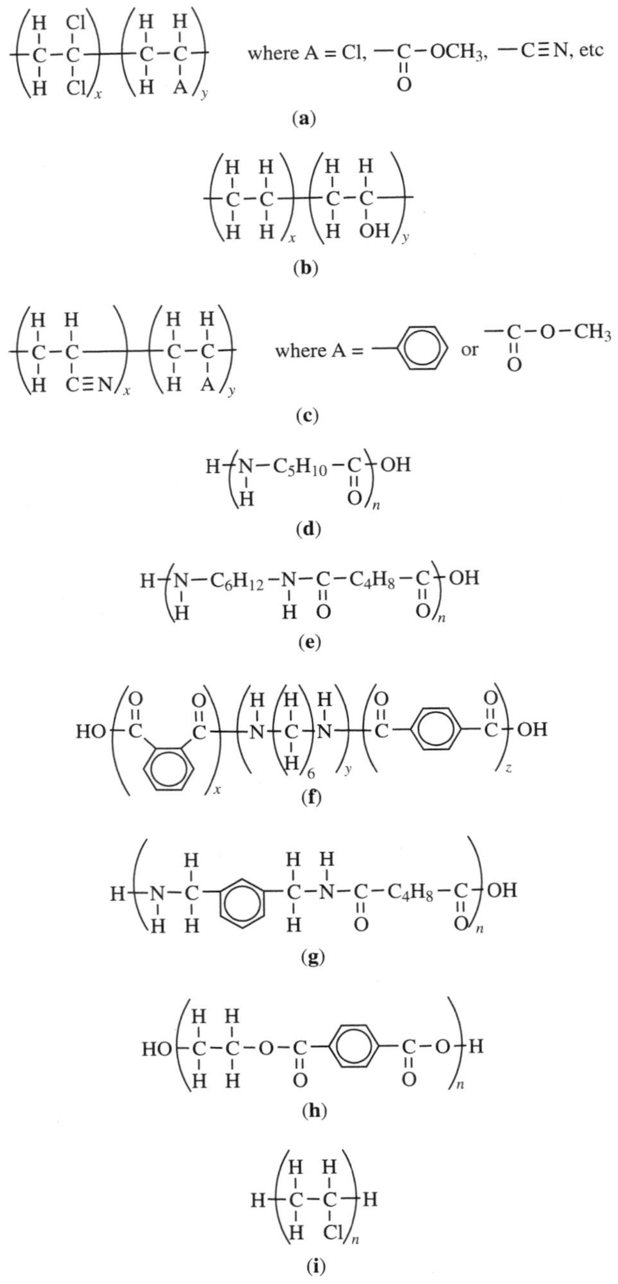

$$\left(\begin{array}{cc} H & Cl \\ | & | \\ -C-C- \\ | & | \\ H & Cl \end{array}\right)_x \left(\begin{array}{cc} H & H \\ | & | \\ -C-C- \\ | & | \\ H & A \end{array}\right)_y \qquad \text{where A} = \text{Cl}, \; -\overset{O}{\underset{\|}{C}}-OCH_3, \; -C\equiv N, \text{ etc}$$

(**a**)

$$\left(\begin{array}{cc} H & H \\ | & | \\ -C-C- \\ | & | \\ H & H \end{array}\right)_x \left(\begin{array}{cc} H & H \\ | & | \\ -C-C- \\ | & | \\ H & OH \end{array}\right)_y$$

(**b**)

$$\left(\begin{array}{cc} H & H \\ | & | \\ -C-C- \\ | & | \\ H & C\equiv N \end{array}\right)_x \left(\begin{array}{cc} H & H \\ | & | \\ -C-C- \\ | & | \\ H & A \end{array}\right)_y \qquad \text{where A} = -\bigcirc \; \text{ or } \; -\overset{O}{\underset{\|}{C}}-O-CH_3$$

(**c**)

$$H \left(\begin{array}{c} N-C_5H_{10}-\overset{O}{\underset{\|}{C}} \\ | \\ H \end{array}\right)_n OH$$

(**d**)

$$H \left(\begin{array}{c} N-C_6H_{12}-N-\overset{O}{\underset{\|}{C}}-C_4H_8-\overset{O}{\underset{\|}{C}} \\ | \qquad\qquad | \\ H \qquad\qquad H \end{array}\right)_n OH$$

(**e**)

$$HO \left(\begin{array}{cc} \overset{O}{\underset{\|}{C}} & \overset{O}{\underset{\|}{C}} \end{array}\right)_x \left(\begin{array}{c} H \\ | \\ N- \end{array}\begin{array}{c} H \\ | \\ C \\ | \\ H \end{array}_6 \begin{array}{c} H \\ | \\ -N \end{array}\right)_y \left(\begin{array}{cc} \overset{O}{\underset{\|}{C}} & \overset{O}{\underset{\|}{C}} \end{array}\right)_z OH$$

(**f**)

$$H \left(\begin{array}{c} H \\ | \\ N-C \\ | \\ H \end{array} \bigcirc \begin{array}{c} H \; H \\ | \; | \\ C-N-\overset{O}{\underset{\|}{C}}-C_4H_8-\overset{O}{\underset{\|}{C}} \\ | \\ H \end{array}\right)_n OH$$

(**g**)

$$HO \left(\begin{array}{cc} H & H \\ | & | \\ C-C-O-\overset{O}{\underset{\|}{C}} \\ | & | \\ H & H \end{array} \bigcirc \overset{O}{\underset{\|}{C}}-O \right)_n H$$

(**h**)

$$H \left(\begin{array}{cc} H & H \\ | & | \\ C-C \\ | & | \\ H & Cl \end{array}\right)_n H$$

(**i**)

Fig. 1. Chemical structures of barrier polymers. (**a**) Vinylidene chloride copolymers; (**b**) hydrolyzed EVOH; (**c**) acrylonitrile barrier polymers; (**d**) nylon-6; (**e**) nylon-6,6; (**f**) amorphous nylon (Selar PA 3426), $y = x + z$; (**g**) nylon-MXD6; (**h**) PET; and (**i**) PVC.

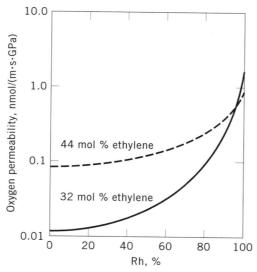

Fig. 2. Oxygen permeabilities of EVOH copolymers at 20°C (10). See Table 1 for unit conversions.

Copolymers of acrylonitrile [107-13-1] are used in extrusion and molding applications. Commercially important comonomers for barrier applications include styrene and methyl acrylate. As the comonomer content is increased, the permeabilities increase as shown in Figure 3. These copolymers are not moisture-sensitive. Table 7 contains descriptions of three high nitrile barrier polymers. Barex and Cycopac resins are rubber-modified to improve the mechanical properties.

Polyamide polymers can provide good-to-moderate barrier-to-permeation by permanent gases. Nylon-6 [25038-54-4] and nylon-6,6 [32131-17-2] have been available for many years. Nylon-6 and nylon-6,6 are moderate oxygen barriers

Table 6. **Composition of EVOH Copolymers**

Manufacturer	Name[a]	Ethylene, mol %
Du Pont	Selar OH 3003, 3007	30
	Selar OH 3803	38
	Selar OH 4416	44
Nippon Goshei	Soarnol D, DT	29
	Soarnol E, ET	38
Eval Company of America	Eval L	27
	Eval F	32
	Eval H	38
	Eval E	44
	Eval K[b]	38
Solvay	Clarene L	30
	Clarene P	36

[a] Names are trademarks of the individual manufacturer.
[b] Blend of 32 mol % and 44 mol % copolymers.

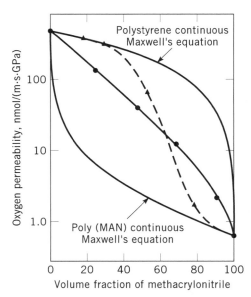

Fig. 3. Oxygen permeabilities of block (▲) and random (●) styrene–methacrylonitrile (MAN) copolymers [33961-16-9] (11). See Table 1 for unit conversions.

that are slightly humidity sensitive. They are, however, excellent barriers against migration of flavor and aroma compounds. New polyamides are being introduced. One of these is an amorphous nylon, Selar PA (Du Pont), which is a terpolymer of hexamethylenediamine [124-09-4] and a mixture of phthalic acids or acid chlorides (12). Nylon-MXD6 (Mitsubishi Gas Chemical Company) is a copolymer of m-xylenediamine [1477-55-0] and adipic acid [124-04-9]. The amorphous nylons are unusual because the oxygen permeability actually decreases slightly as the humidity increases. Nylon-MXD6 [25805-74-7] is more typical; its oxygen permeability increases as the humidity increases. Research is active in the area of polyamide chemistry for new barrier polymers.

Two often, used polymers have adequate properties for some applications. Poly(ethylene terephthalate) is used to make films and bottles. This polymer is commonly made from ethylene glycol and dimethyl terephthalate [120-61-6] or

Table 7. **Compositions of High Nitrile Barrier Polymers**

Polymer[a]	Manufacturer	Chemical composition[b]
Lopac	Monsanto Co.	100% copolymer of 70% acrylonitrileand 30% styrene
Barex	Sohio	90% copolymer of 74% acrylonitrile and 26% methyl acrylate + 10% butadiene rubber graft
Cycopac	Borg-Warner Chemicals	90% copolymer of 74% acrylonitrile and 26% styrene + 10% butadiene rubber graft

[a] Names are trademarks of the individual manufacturer.
[b] Data from U.S. Food and Drug Administration (FDA) regulations for corresponding materials.

Table 8. **Permeability of Household Films at 25°C to Selected Permeants**

	Film type		
Permeant[a]	Plasticized vinylidene chloride copolymer[b]	Plasticized PVC[c]	Polyethylene[d]
oxygen, nmol/(m·s·GPa)	1.9	220	640
water vapor[e], nmol/(m·s)	0.055	0.30	0.19
d-limonene, MZU[f]	130	1.1×10^5	3.3×10^5
dipropyl disulfide, MZU[f]	1.1×10^4	3.3×10^6	6.8×10^6

[a] See Tables 1 and 2 for unit conversions.
[b] SaranWrap brand plastic film (trademark of The Dow Chemical Company).
[c] Reynolds plastic wrap (trademark of Reynolds Metals Co.).
[d] Handi-Wrap II brand plastic film (trademark of DowBrands, Inc.).
[e] Measured at 37.8°C, 90% rh.
[f] MZU $= (10^{-20}$ kg·m)/(m^2·s·Pa).

from ethylene glycol [107-21-1] and terephthalic acid [100-21-0]. PET is a moderate barrier-to-permanent gases; however, it is an excellent barrier to flavors and aromas. The oxygen permeability decreases slightly with increasing humidity. Poly(vinyl chloride) is a moderate barrier-to-permanent gases. Plasticized PVC is used as a household wrapping film. The plasticizers greatly increase the permeabilities. Table 8 lists the permeabilities of selected permeants in the three main household wrapping films.

Water–Vapor Transmission. Table 9 lists WVTR values for selected polymers. Comparison of Tables 3 and 9 shows that often there is a reversal of roles. Those polymers that are good oxygen barriers are often poor water–vapor barriers and vice versa, which can be rationalized as follows. Barrier polymers often rely on dipole–dipole interactions to reduce chain mobility and, hence, diffusional movement of permeants. These dipoles can be good sites for hydrogen

Table 9. **WVTR of Selected Polymers**[a]

Polymer	WVTR, nmol/(m·s)[b]
vinylidene chloride copolymers	0.005–0.05
HDPE	0.095
polypropylene	0.16
LDPE	0.35
EVA, 44 mol% ethylene[c]	0.35
PET	0.45
PVC	0.55
EVA, 32 mol % ethylene[c]	0.95
nylon-6,6, nylon-11	0.95
nitrile barrier resins	1.5
polystyrene	1.8
nylon-6	2.7
polycarbonate	2.8
nylon-12	15.9

[a] At 38°C and 90% rh unless otherwise noted (13).
[b] See Table 2 for unit conversions.
[c] Measured at 40°C.

bonding. Water molecules are attracted to these sites, leading to high values of S. Furthermore, the water molecules enhance D by interrupting the attractions and chain packing. Polymer molecules without dipole–dipole interactions, such as polyolefins, dissolve very little water and have low WVTR and permeability values. The low values of S more than compensate for the naturally higher values of D.

5. Large Molecule Permeation

The permeation of flavor, aroma, and solvent molecules in polymers follows the same physics as the permeation of small molecules. However, there are two significant differences. For these larger molecules, the diffusion coefficients are much lower and the solubility coefficients are much higher. This means that steady-state permeation may not be reached during the storage time of some packaging situations. Hence, large molecules from the environment might not enter the contents, or loss of flavor molecules would be limited to sorption into the polymer. However, since the solubility coefficient is large, the loss of flavor could be important solely from sorption in the polymer. Furthermore, the large solubility coefficient can lead to enough sorption of the large molecule that plasticization occurs in the polymer, which can increase the diffusion coefficient.

Table 10 contains some selected permeability data including diffusion and solubility coefficients for flavors in polymers used in food packaging. Generally, vinylidene chloride copolymers and glassy polymers such as polyamides and EVOH are good barriers to flavor and aroma permeation whereas the polyolefins are poor barriers. Comparison to Table 5 shows that the large molecule diffusion coefficients are 1000 or more times lower than the small molecule coefficients. The solubility coefficients are as much as 1 million times higher. Equation 7 shows how to estimate the time to reach steady-state permeation t_{ss} if the diffusion coefficient and thickness of a film are known.

$$t_{ss} = \frac{L^2}{4D} \tag{7}$$

For d-limonene diffusion in a 50-μm thick vinylidene chloride copolymer film, steady-state permeation is expected after 2000 days. For a 50-μm thick LDPE film, steady-state permeation is expected in <1 h. If steady-state permeation is not achieved, the effective penetration depth L^* for simple diffusion, after time t has elapsed, can be estimated with equation 8.

$$L^* = 2(Dt)^{1/2} \tag{8}$$

For a food container, the amount of sorption could be estimated in the following way. For simple diffusion, the concentration in the polymer at the food surface could be estimated with equation 3, which would require a knowledge of the partial pressure of the flavor in the food. This knowledge is not always available, but methods exist for estimating this when the food matrix

Table 10. Examples of Permeation of Flavor and Aroma Compounds in Selected Polymers at 25°C,^{a} Dry^b

Flavor/aroma (compound)	Cas Registry Number	Permeant formula	P, MZU^c	D, m^2/s	S, kg/ (m$^3 \cdot$ Pa)^d
Vinylidene chloride copolymer					
ethyl hexanoate	[123-66-0]	$C_8H_{16}O_2$	570	8.0×10^{-18}	0.71
ethyl 2-methylbu-tyrate	[7452-79-1]	$C_7H_{14}O_2$	3.2	1.9×10^{-17}	1.7×10^{-3}
hexanol	[111-27-3]	$C_6H_{14}O$	40	5.2×10^{-17}	7.7×10^{-3}
trans-2-hexenal	[6728-26-3]	$C_6H_{10}O$	240	1.8×10^{-17}	0.14
d-limonene	[5989-27-5]	$C_{16}H_{16}$	32	3.3×10^{-17}	9.7×10^{-3}
3-octanone	[106-68-3]	$C_8H_{16}O$	52	1.3×10^{-18}	0.40
propyl butyrate	[105-66-8]	$C_7H_{14}O_2$	42	4.4×10^{-18}	9.4×10^{-2}
dipropyl disulfide	[629-19-6]	$C_6H_{14}S_2$	270	2.6×10^{-18}	1.0
EVA copolymer					
ethyl hexanoate			0.41	3.2×10^{-18}	1.3×10^{-3}
ethyl 2-methylbu-tyrate			0.30	6.7×10^{-18}	4.7×10^{-4}
hexanol			1.2	2.6×10^{-17}	4.6×10^{-4}
trans-2-hexenal			110	6.4×10^{-17}	1.8×10^{-2}
d-limonene			0.5	1.1×10^{-17}	4.5×10^{-4}
3-octanone			0.2	1.0×10^{-18}	2.0×10^{-3}
propyl butyrate			1.2	2.7×10^{-17}	4.5×10^{-4}
LDPE					
ethyl hexanoate			4.1×10^6	5.2×10^{-13}	7.8×10^{-2}
ethyl 2-methylbu-tyrate			4.9×10^5	2.4×10^{-13}	2.3×10^{-2}
hexanol			9.7×10^5	4.6×10^{-13}	2.3×10^{-2}
trans-2-hexenal			8.1×10^5		
d-limonene			4.3×10^6		
3-octanone			6.8×10^6	5.6×10^{-13}	1.2×10^{-1}
propyl butyrate			1.5×10^6	5.0×10^{-13}	3.0×10^{-2}
dipropyl disulfide			6.8×10^6	7.3×10^{-14}	9.3×10^{-1}
HDPE					
d-limonene			3.5×10^6	1.7×10^{-13}	2.5×10^{-1}
menthone	[1074-95-9]	$C_{10}H_{18}O$	5.2×10^6	9.1×10^{-13}	4.7×10^{-1}
methyl salicylate	[119-36-8]	$C_8H_8O_3$	1.1×10^7	8.7×10^{-14}	1.6
Polypropylene					
2-butanone	[78-93-3]	C_4H_8O	8.5×10^3	2.1×10^{-15}	4.0×10^{-2}
ethyl butyrate	[105-54-4]	$C_6H_{12}O_2$	9.5×10^3	1.8×10^{-15}	5.3×10^{-2}
ethyl hexanoate			8.7×10^4	3.1×10^{-15}	2.8×10^{-1}
d-limonene			1.6×10^4	7.4×10^{-16}	2.1×10^{-1}

^{a} Values for vinylidene chloride copolymer and EVA are extrapolated from higher temperatures.
^{b} Permeation in the vinylidene chloride copolymer and the polyolefins is not affected by humidity; the permeability and diffusion coefficient in the EVA copolymer can be as much as 1000 times greater with high humidity (14–17).
^{c} MZU = $(10^{-20}\,\text{kg} \cdot \text{m})/(\text{m}^2 \cdot \text{s} \cdot \text{Pa})$; see equation 4 for unit conversions.
^{d} See equation 6 for unit conversions.

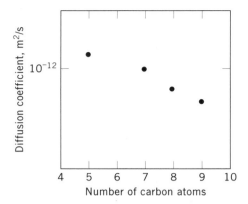

Fig. 4. Diffusion coefficient at 30°C of esters in a low density polyethylene film (18).

is water-dominated. The concentration in the polymer at the depth of penetration is zero. Hence, the average concentration C_{avg} is as from equation 9.

$$C_{avg} = 0.5Sp_x \tag{9}$$

The quantity sorbed ΔM_{sorb}, is simply the volume of polymer affected times the average concentration. The volume affected is the package area times L^*. The result is equation 10.

$$\Delta M_{sorb} = A(Dt)^{1/2}Sp_x \tag{10}$$

In this equation, A and t are observable. The pressure of the flavor must be obtained from an external reference, and D and S must be measured or estimated. If steady state is achieved, the quantity sorbed is given by equation 11. In either case the quantity sorbed can be important.

$$\Delta M_{sorb} = 0.5ALSp_x \tag{11}$$

The scalping of flavor and aroma by a package can be minimized by placing a barrier material as near as possible to the food. The ingress of undesirable permeants from the environment can be minimized by placing a barrier polymer between the food and the environment, not necessarily near the food.

Figures 4 and 5 show how the diffusion coefficient and solubility coefficient vary for a series of linear esters in LDPE film. The trends are generally true for other permeants in other films. As the size of the permeant increases, the diffusion coefficient decreases and the solubility coefficient increases. Since the increase in solubility coefficient is larger than the decrease in the diffusion coefficient, the permeability actually increases as the permeant size increases.

6. Physical Factors Affecting Permeability

Several physical factors can affect the barrier properties of a polymer. These include temperature, humidity, orientation, and cross-linking.

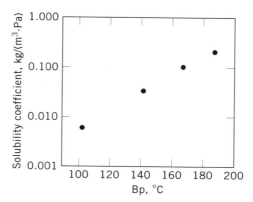

Fig. 5. Solubility coefficient at 30°C vs boiling point of ester in a low density polyethylene film (18). For unit conversion see equation 6.

6.1. Temperature. The permeability varies with temperature according to equation 12, where P_o is a constant, E_p is the activation energy for permeation, R is the gas constant, and T is the absolute temperature.

$$P = P_o\exp\left(-E_p/RT\right) \tag{12}$$

The temperature dependence of the permeability arises from the temperature dependencies of the diffusion coefficient and the solubility coefficient. Equations 13 and 14 express these dependencies where D_o and S_o are constants, E_d is the activation energy for diffusion, and ΔH_{sol} is the heat of solution for the permeant in the polymer.

$$D = D_o\exp(-E_d/RT) \tag{13}$$

$$S = S_o\exp(-\Delta H_{sol}/RT) \tag{14}$$

The activation energy for diffusion is always positive and increases with increasing permeant size. Figure 6 shows this for diffusion in PVC. The heat of solution is near zero for small permanent gases such as oxygen; however, it becomes increasingly negative with increasing permeant size because the main contribution to the heat of solution is the heat of condensation. Equation 15 shows the relationship among the three exponential factors.

$$E_p = E_d + \Delta H_{sol} \tag{15}$$

Although, in principle, this sum could be negative and give a negative activation energy for the permeability, no examples are known. The permeability increases with increasing temperature.

Figure 7 shows generally how the permeability varies with temperature. Equation 12 is useful above and below the T_g, but not at the T_g. The main contribution to this effect is the diffusion coefficient. Above T_g, the diffusion process

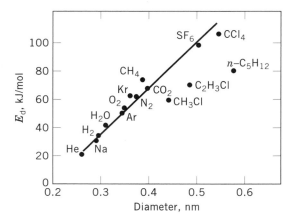

Fig. 6. Activation energy for diffusion in PVC as a function of penetrant mean diameter (19). To convert J to cal, divide by 4.184.

is controlled by motions of the polymer chains; hence, the activation energy is large. Below T_g, the diffusion process is controlled by the movement of the permeant; therefore, the activation energy is smaller. There is growing evidence with theoretical support that, above T_g, E_d decreases slightly with increasing temperature. Hence, the curve in Figure 7 would be concave downward above T_g.

For oxygen, the permeabilities increase ~10% per degree in polymers that are above their T_g such as vinylidene chloride copolymers and polyolefins. The

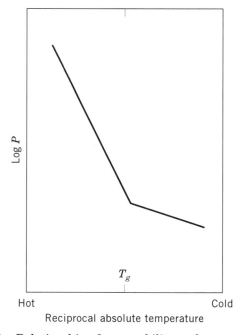

Fig. 7. Relationship of permeability and temperature.

Table 11. **Effect of Orientation on Oxygen Permeability for Certain Polymers**

Polymer	Degree of orientation, %	Oxygen permeability,[a] nmol/(m·s·GPa)[b]
polypropylene	0	300
	300	160
polystyrene	0	840
	300	600
polyester	0	20
	500	10
copolymer of 70%	0	2.0
acrylonitrile and	300	1.8
30% styrene		

[a] At 23°C.
[b] See Table 1 for unit conversions.

permeabilities increase ∼5% per degree in polymers that are below their T_g such as acrylonitrile copolymers, EVOH, and PET.

6.2. Humidity. When a polymer equilibrates with a humid environment, it absorbs water. The water concentration in the polymer might be very low as in polyolefins or it might be several weight percent as in EVA copolymers. Absorbed water does not affect the permeabilities of some polymers including vinylidene chloride copolymers, acrylonitrile copolymers, and polyolefins. Absorbed water increases the permeabilities in some polymers including EVA copolymers and most polyamides. Figure 2 shows how the oxygen permeability in an EVA copolymer increases with increasing humidity. A few polymers show a slight decrease in the oxygen permeability with increasing humidity. These include PET and the new amorphous nylon.

6.3. Orientation. The effect of orientation on the permeability of polymers is difficult to assess because the words orientation and elongation or strain have been used interchangeably in the literature. Diffusion in some polymers is unaffected by orientation; in others, increases or decreases are observed. The circumstances of the orientation (elongation) are important. Table 11 shows the effect of orientation on the oxygen permeability of four polymers. In each case a slight reduction in permeability occurs with orientation. In another study, however, no difference in oxygen or nitrogen permeability was observed between biaxially oriented polystyrene films and films that were cast from toluene solutions (20). Vinylidene chloride copolymer films show no difference in oxygen permeabilities between biaxially oriented (700%) and unoriented films. When increases in permeability have been observed, the creation of microcracks has been suspected.

6.4. Cross-Linking. Cross-linking has been shown to decrease the diffusion coefficient. The effect is greater for large permeants. To be effective, the cross-links need to be closer than ∼30 − 40 nm. Such a high extent of cross-linking is a severe handicap for commercial applications; hence, little research is being done in this area for barrier materials.

7. Barrier Structures

Barrier polymers are often used in combination with other polymers or substances. The combinations may result in a layered structure either by coextrusion, lamination, or coating. The combinations may be blends that are either miscible or immiscible. In each case, the blend seeks to combine the best properties of two or more materials to enhance the value of a final structure.

7.1. Layered Structures.   Whenever a barrier polymer lacks the necessary mechanical properties for an application or the barrier would be adequate with only a small amount of the more expensive barrier polymer, a multilayer structure via coextrusion or lamination is appropriate. Whenever the barrier polymer is difficult to melt process or a particular traditional substrate such as paper or cellophane [9005-81-6] is necessary, a coating either from latex or a solvent is appropriate. A layered structure uses the barrier polymer most efficiently since permeation must occur through the barrier polymer and not around the barrier polymer. No short cuts are allowed for a permeant. The barrier properties of these structures are described by the permeance P^*, which is described in equation 16 where P_i and L_i are the permeabilities and thicknesses of the layers.

$$\frac{1}{P^*} = \frac{L_1}{P_1} + \frac{L_2}{P_2} + \cdots \frac{L_i}{P_i} \tag{16}$$

The permeance can be used in equation 17 (which is a modification of eq. 1) to estimate package performance.

$$\frac{\Delta M_x}{\Delta t} = P^* A \Delta p_x \tag{17}$$

Figure 8 shows how the permeance varies for a hypothetical two-layer sheet as the relative thicknesses shift for a barrier layer with a permeability of 0.1

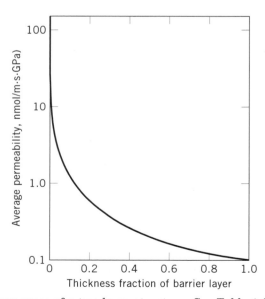

Fig. 8.   Average permeance of a two-layer structure. See Table 1 for unit conversions.

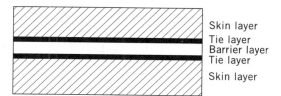

Fig. 9.   Symmetric five-layer barrier coextruded sheet.

units and a nonbarrier layer with a permeability of 100 units. Figures 9 and 10 show profiles of coextruded sheet. Figure 9 shows a typical symmetric five-layer sheet, which is the simplest case and three extruders are necessary. One extruder supplies the outer layers that give structural integrity and toughness, eg, polypropylene. Another extruder supplies the barrier layer, and a third extruder supplies the adhesive tie layers that hold the structure together. A fourth extruder would be necessary if the outermost layers are different. Figure 10 shows an asymmetric sheet with a recycle layer. An additional extruder is needed to supply this layer to the coextrusion die. Both Figures 9 and 10 are drawn to scale. Figure 10 represents a recycle layer that is 30–50% of the total structure. Recycle at 30–50% has been demonstrated and reported for containers with a barrier layer of a vinylidene chloride copolymer (21–23). Development with other barrier polymers is continuing.

A typical coextruded sheet has a total thickness of 1270 μm (50 mil). After the container is formed by one of several methods, the sheet might then yield a wall thickness of 510 μm. The barrier layer in the container is typically in the range of 25–75 μm.

A thin layer of a barrier polymer can be coated onto a substrate. For example, a water-borne latex can be used to coat paper or a polymer such as polypropylene or PET. Commonly, two coats are applied to give a total barrier thickness of 5 μm. Two coats are useful so that minor holes in the first layer can be covered with the second layer. Adhesion of the coating is important for all substrates and hold out (minimizing wasteful soaking-in) is important for paper. Hence, sometimes the first layer is different from the second layer or the substrate receives a surface treatment before coating. A typical application is a vinylidene chloride copolymer latex on oriented polypropylene film for packaging snack foods. The most significant commercial application of solvent coating with barrier polymers is a special vinylidene chloride resin dissolved in a polar solvent coated onto cellophane or PET. Ethylene–vinyl alcohol polymers are potentially applicable for

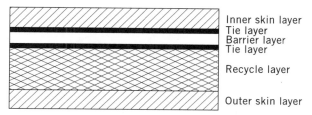

Fig. 10.   Barrier sheet with recycle layer.

solvent coating. Inorganic coatings on polymers have the potential to enhance the barrier. However, commercial development is not significant.

7.2. Immiscible Blends. When two polymers are blended, the most common result is a two-phase composite. The most interesting blends have good adhesion between the phases, either naturally or with the help of an additive. The barrier properties of an immiscible blend depend on the permeabilities of the polymers, the volume fraction of each, phase continuity, and the aspect ratio of the discontinuous phase. Phase continuity refers to which phase is continuous in the composite. Continuous for barrier applications means that a phase connects the two surfaces of the composite. Typically, only one of the two polymer phases is continuous, with the other polymer phase existing as islands. It is possible to have both polymers be continuous. The aspect ratio L/W refers to the shape of the particles in the discontinuous phase. It is the average dimension of this phase parallel to the plane of the film L divided by the average dimension perpendicular to the film W. Plates in the plane of the film would have a high aspect ratio. Spheres or cubes would have an aspect ratio equal $=1$.

Figure 11 shows the theoretical permeabilities that are expected for a two-phase blend of polymers. The two solid curves represent calculations based upon Maxwell's equation (24) for an aspect ratio of 1 for the discontinuous phase. The dotted line is a prediction of the permeability using Nielsen's model (25) when a barrier polymer with an aspect ratio of 8 is discontinuous in a nonbarrier matrix. Figure 12 shows the expected result of a phase inversion for a two-polymer blend. The discontinuous phase is assumed to have an aspect ratio of 1. At some critical

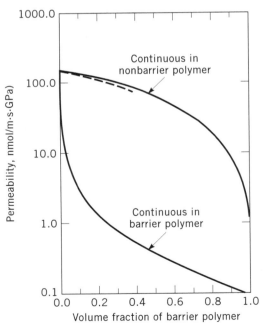

Fig. 11. Calculated permeabilities for a two-phase blend using Maxwell's result. Discontinuous phase has aspect ratio of 1.0. See Table 1 for unit conversion.

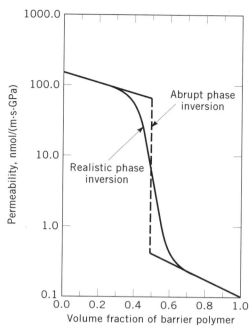

Fig. 12. Permeabilities for a two-phase blend with a phase inversion. Discontinuous phase has aspect ratio of 1.0. See Table 1 for unit conversion.

composition, the composite switches from being continuous in one polymer to being continuous in the other. Figure 12 is really a special case of Figure 11. Selar RB is a blend of polyethylene and nylon-6. Polyethylene is the majority constituent and forms the continuous phase. The product has its best barrier when it can be used in processes that impart orientation to the product. This gives a high aspect ratio to the nylon-6 and enhanced barrier to the article. Blends of polyethylene and EVOH are being developed.

Block copolymers of styrene and methacrylonitrile appear to behave like immiscible blends (11). Another case of immiscible blends is an inert filler in a polymer matrix. Although the inorganic fillers are assumed to have permeabilities nearly equal to zero, the filler must be wetted by the polymer to be effective at lowering the permeability of the host polymer. Table 12 shows the effect of a surface treatment on calcium carbonate to enhance wetting by polyethylene. With good wetting the permeability is reduced; with poor wetting the permeability is increased. Selar OH Plus resins are an example of filled polymers. By using 23.1 wt% mica flakes in an EVA alcohol copolymer, the permeability can be reduced with optimum processing to as low as 20% of the permeability of the unfilled polymer. Crystallinity in polymers acts like inert filler and lowers the permeability. However, since the amount of crystallinity in a given polymer is rarely under the control of the user, it is not considered here. Some orientation efforts can lower the permeability if the aspect ratio of the crystallites is increased.

Table 12. **Effect of Calcium Carbonate Fillers on Oxygen Permeability of LDPE**

CaCO$_3$, volume %	Type CaCO$_3$	Oxygen permeability,[a] nmol/(m·s·GPa)[b]
0		960
15	untreated	~ 2000
25	untreated	~ 4000
15	treated	500
25	treated	300

[a] At 23°C.
[b] See Table 1 for unit conversion.

7.3. Miscible Blends. Sometimes a miscible blend results when two polymers are combined. A miscible blend has only one amorphous phase because the polymers are soluble in each other. There may also be one or more crystal phases. Simple theory (26) has supported the empirical relation for the permeability of a miscible blend. Equation 18 expresses this relation, where P_{mb} is the permeability of the miscible blend and ϕ_1 and ϕ_2 are the volume fractions of polymer 1 and 2.

$$\ln P_{mb} = \phi_1 \ln P_1 + \phi_2 \ln P_2 \qquad (18)$$

This relationship is shown in Figure 13 where polymer 1 has a permeability 1000 times higher than that of polymer 2. Published data have small negative

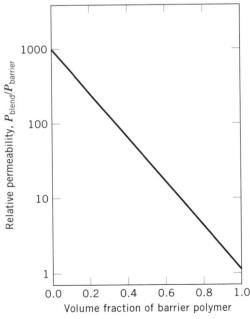

Fig. 13.   Theoretical permeabilities for miscible blends of two polymers.

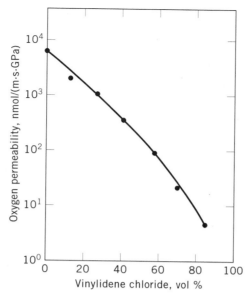

Fig. 14. Oxygen permeabilities of vinylidene chloride–*n*-butyl acrylate copolymers (27). See Table 1 for unit conversion.

deviations from this theoretical relationship. Part of the deviation can be explained by densification of the blend relative to the starting components. Random copolymers, which are forced (by covalent bonds) to imitate combinations of two materials, have permeabilities that are similar to miscible blends. However, the deviations from equation 18 tend to be positive. A series of styrene–MAN copolymers were studied (11) and slight positive deviations were found. Figure 14 shows the oxygen permeabilities of a series of vinylidene chloride–*n*-butyl acrylate copolymers [9011-09-0].

Plasticized polymers have been observed to behave like miscible blends. The permeabilities of oxygen, carbon dioxide, and water vapor in a vinylidene chloride copolymer increase exponentially with increasing plasticizer (4,5,28). About 1.6 parts plasticizer per hundred parts polymer is enough to double the permeability.

8. Predicting Permeabilities

Reasonable prediction can be made of the permeabilities of low molecular weight gases such as oxygen, nitrogen, and carbon dioxide in many polymers. The diffusion coefficients are not complicated by the shape of the permeant, and the solubility coefficients of each of these molecules do not vary much from polymer to polymer. Hence, all that is required is some correlation of the permeant size and the size of holes in the polymer matrix. Reasonable predictions of the permeabilities of larger molecules such as flavors, aromas, and solvents are not easily made. The diffusion coefficients are complicated by the shape of the permeant,

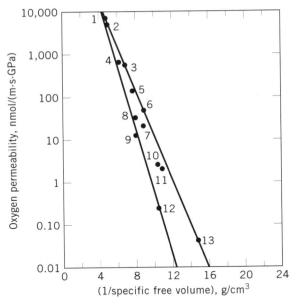

Fig. 15.   Oxygen permeability vs 1/specific free volume at 25°C (30). 1. Polybutadiene; 2. polyethylene (density 0.922); 3. polycarbonate; 4. polystyrene; 5. styrene-*co*-acrylonitrile; 6. PET; 7. acrylonitrile barrier polymer; 8. poly(methyl methacrylate); 9. PVC; 10. acrylonitrile barrier polymer; 11. vinylidene chloride copolymer; 12. polymethacrylonitrile; and 13. polyacrylonitrile. See Table 1 for unit conversions.

and the solubility coefficients for a specific permeant can vary widely from polymer to polymer.

The permachor method is an empirical method for predicting the permeabilities of oxygen, nitrogen, and carbon dioxide in polymers (29). In this method, a numerical value is assigned to each constituent part of the polymer. An average number is derived for the polymer, and a simple equation converts the value into a permeability. This method has been shown to be related to the cohesive energy density and the free volume of the polymer (2). The model has been modified to liquid permeation with some success.

A less empirical model based solely on the specific free volume of the polymer has been proposed (30). Once the density and the intrinsic volumes of the polymer components are known, the specific free volume can be calculated. A correlation between the specific free volume and the oxygen permeabilities of several polymers is shown in Figure 15. The model seems to give better predictions for polymers with higher permeabilities. This model is for low molecular weight gases such as oxygen, nitrogen, and carbon dioxide.

For larger molecules, independent predictions of the diffusion coefficients and the solubility coefficients are required. Figure 16 shows how the diffusion coefficient varies as a function of permeant size in PVC. The two sets of data represent glassy PVC, which is below its T_g and plasticized PVC, which is above its T_g. Other glassy polymers show the steep slope, and other rubbery polymers show the shallow slope. The points near the lines represent spherical

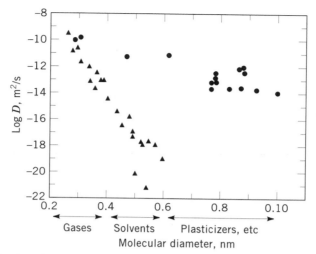

Fig. 16. Diffusivities of penetrants in rigid (▲) and plasticized (●) PVC vs molecular diameter at 30°C (31).

permeants whereas the points above the lines represent linear permeants. Predicting the diffusion coefficient for a permeant in a polymer requires knowing one other diffusion coefficient in the polymer. Figure 4 shows how the diffusion coefficient varies for permeant molecules slightly larger than shown in Figure 16.

The solubility coefficients are more difficult to predict. Although advances are being made, the best method is probably to use a few known solubility coefficients in the polymer to predict others with a simple plot of S vs $(\delta_{\mathrm{poly}} - \delta_{\mathrm{perm}})^2$, where δ_{poly} and δ_{perm} are the solubility parameters of the polymer and permeant, respectively. When insufficient data are available, S at 25°C can be estimated with equation 19, where $\kappa = 1$ and the resulting units of cal/cm^3 are converted to kJ/mol by dividing by the polymer density and multiplying by the molecular mass of the permeant and by 4.184 (16).

$$S = \kappa\left(\delta_{\mathrm{poly}} - \delta_{\mathrm{perm}}\right)^2 \qquad (19)$$

The boiling temperature of a permeant can be used to predict the solubility coefficient only when the solubility coefficients of other permeants of the same chemical family are known.

9. Measuring Barrier Properties

Measuring the barrier properties of polymers is important for several reasons. The effects of formulation or process changes need to be known, new polymers need to be evaluated, data are needed for a new application before a large investment has been made, and fabricated products need to have performance verified. For some applications a full range of data is necessary, including P, D, and S plus the effects of temperature and humidity.

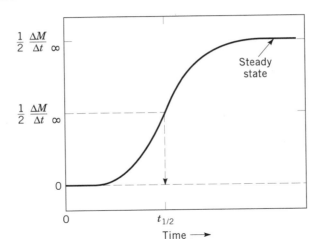

Fig. 17. Determining the diffusion coefficient from instantaneous transmission rates.

9.1. Oxygen Transport. The most widely used methods for measuring oxygen transport are based upon the Ox-Tran instrument (Modern Controls, Inc.). Several models exist, but they all work on the same principle. The most common application is to measure the permeability of a film sample. Typically, oxygen is introduced on one side of the film, and nitrogen gas sweeps the other side of the film to a coulometric detector. The detector measures the rate that oxygen comes through the film. The detector response at steady state can easily be converted to $\Delta M_{\text{oxygen}}/\Delta t$ (eq. 1). Simple algorithms come with the instruction manual for calculating the permeability. The detector response can be monitored as a function of time to determine D, which is demonstrated in Figure 17. During the transient time before steady state is achieved, the transmission rate increases. First $t_{1/2}$, the time for the transmission rate to rise to 50% of the steady-state rate, must be found. Then equation 20 (32) is used where L is the film thickness. After P and D are known, S can be calculated.

$$D = \frac{L^2}{7.2t_{1/2}} \tag{20}$$

The instrument has a heating element that allows for measurements at selected temperatures. The humidity can be controlled by passing the test gases through bubblers, monitoring the humidity, and adjusting the flow rates. This method requires some sophistication and is preferred when such testing is routine. The sandwich method is preferred for infrequent testing (33). With this method, pads that have been treated with saturated salt solutions are placed beside the film and encapsulated with thin polyethylene layers. The polyethylene offers almost no barrier to the oxygen and does a good job of holding the humidity near the test film.

Containers can be tested with the Ox-Tran instrument. However, care must be taken to obtain a good seal around the base of the container. A generous amount of epoxy glue is recommended.

9.2. Water Transport.   Two methods of measuring WVTR are commonly used. The newer method uses a Permatran-W (Modern Controls, Inc.). In this method a film sample is clamped over a saturated salt solution, which generates the desired humidity. Dry air sweeps past the other side of the film and past an infrared detector, which measures the water concentration in the gas. For a calibrated flow rate of air, the rate of water addition can be calculated from the observed concentration in the sweep gas. From the steady-state rate, the WVTR can be calculated. In principle, the diffusion coefficient could be determined by the method outlined in the previous section. However, only the steady-state region of the response is serviceable. Many different salt solutions can be used to make measurements at selected humidity differences; however, in practice, $CaSO_4 \cdot 7H_2O$ is used nearly always because it gives a humidity of 90% as required by the traditional commercial applications. Typical experiments take between several hours and a day to complete.

The other method is the ASTM cup method (34). In this method, a desiccant is placed in a waterproof dish. The dish is covered with the experimental film and placed in an environmental chamber. The temperature and humidity are set for the conditions of interest, typically 37.8°C and 90% rh. At regular intervals, the dish is removed and weighed. After a few days enough data have been gathered to describe a steady-state rate of weight gain, and the WVTR can be calculated. Typical experiments take about a week to complete.

9.3. Carbon Dioxide Transport.   Measuring the permeation of carbon dioxide occurs far less often than measuring the permeation of oxygen or water. A variety of methods are used; however, the simplest method uses the Permatran-C instrument (Modern Controls, Inc.). In this method, air is circulated past a test film in a loop that includes an infrared detector. Carbon dioxide is applied to the other side of the film. All the carbon dioxide that permeates through the film is captured in the loop. As the experiment progresses, the carbon dioxide concentration increases. First, there is a transient period before the steady-state rate is achieved. The steady-state rate is achieved when the concentration of carbon dioxide increases at a constant rate. This rate is used to calculate the permeability. Figure 18 shows how the diffusion coefficient can be determined in this type of experiment. The time lag t_L is substituted into equation 21. The solubility coefficient can be calculated with equation 2.

$$D = \frac{L_2}{6t_L} \tag{21}$$

9.4. Flavor and Aroma Transport.   Many methods are used to characterize the transport of flavor, aroma, and solvent molecules in polymers. Each has some value, and no one method is suitable for all situations. Any experiment should obtain the permeability, the diffusion coefficient, and the solubility coefficient. Furthermore, experimental variables might include the temperature, the humidity, the flavor concentration, and the effect of competing flavors.

Any of the mathematical methods discussed previously to calculate P, D, and S in a single experiment are applicable to flavor permeation. Two cautions

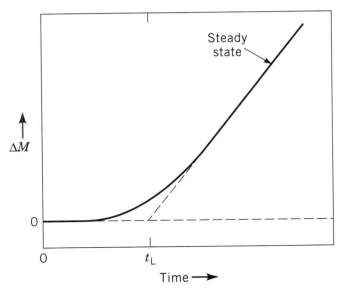

Fig. 18.   Determining the diffusion coefficient from cumulative transmission.

are necessary. First, since the diffusion coefficients are typically low, the time to reach steady state can be long. The instrument must give stable responses for long times or something must be done to speed the experiment. The latter can be accomplished by using a thinner film or by raising the temperature. Second, the solubility coefficients are usually high. This means that the supply of permeant molecules could become depleted or the film could become altered. Recognizing the potential for depleting the flavor reservoir or for altering the experimental film is important. Using a thinner film or a constant supply of fresh permeant can help. Also working at high flavor concentrations should be done only when modeling performance that involves high concentrations.

Experimental data that show the effect of environmental variables have been given earlier. The most commonly misconstrued facet of the data is the effect of concentration on the permeability. The permeabilities of flavors in polymers have been correctly reported to be both independent and strongly dependent on the flavor concentration. However, to assume that there is disagreement would be incorrect. At very low concentrations, as commonly observed in foods, the permeabilities are independent of concentration. However, at high concentrations, as observed in some foods and in neat flavor oils, the permeabilities are quite concentration-dependent. The entire trend has been observed (35,36). Figure 19 shows that the permeability of ethyl propionate [105-37-3] in dry PVA is not affected by permeant concentration until the partial pressure of the permeant exceeds ~30% of the equilibrium vapor pressure. Above that concentration, the permeability is strongly dependent on the permeant concentration. This topic is important to include here because many techniques lack the sensitivity to work at low permeant concentrations. Then, at high concentrations, great variation is observed from experiment to experiment and from experimenter to experimenter. The variation does not necessarily diminish

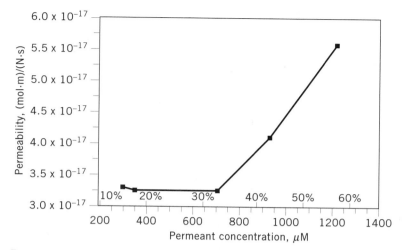

Fig. 19. Permeation of ethyl propionate at various concentrations through PVA at 25°C, 0% rh. Percentages on *x*-axis represent partial pressure of the permeant as a fraction of the equilibrium vapor pressure.

the quality of the measurement; it merely reflects the strong concentration dependence.

10. Safety and Health Factors

The use of safe materials is vital for barrier applications, particularly for food, medical, and cosmetics packaging. Suppliers of specific barrier polymers can provide the necessary details, such as material safety data sheets, to ensure safe processing and use of barrier polymers.

11. Applications

The primary application for barrier polymers is food and beverage packaging. Barrier polymers protect food from environmental factors that could compromise both taste and shelf life. They also help retain desirable flavors and aroma. Barrier polymers are also used for packaging medical products, agricultural products, cosmetics, and electronic components and in moldings, pipe, and tubing.

BIBLIOGRAPHY

"Barrier Polymers" in *ECT* 3rd ed., Vol. 3, pp. 480–502, by S. Steingiser, S. P. Nemphos, and M. Salame, Monsanto Company; in *ECT* 4th ed., Vol. 3, pp. 931–962, by Phillip De-Lassus, The Dow Chemical Company; "Barrier Polymers" in *ECT*, (online), posting date: December 4, 2000, by Phillip DeLassus, The Dow Chemical Company.

CITED PUBLICATIONS

1. *Permeability of Polymers to Gases and Vapors*, Brochure P302-335-79, The Dow Chemical Company, Midland, Mich., 1979.
2. M. Salame, *J. Plast. Film Sheeting* **2**, 321 (1986).
3. P. T. DeLassus, D. L. Clarke, and T. Cosse, *Mod. Plast.*, 86 (Jan. 1983).
4. P. T. DeLassus, *J. Vinyl Technol.* **1**, 14 (1979).
5. P. T. DeLassus and D. J. Grieser, *J. Vinyl Technol.* **2**, 195 (1980).
6. R. Sezi and J. Springer, *Colloid Polym. Sci.* **259**, 1170 (1981).
7. M. Salame, *J. Polym. Sci., Symposium No. 41*, 1 (1973).
8. D. W. Van Krevelen, *Properties of Polymers*, 3rd ed., Elsevier, Amsterdam, The Netherlands, 1990, Chapt. 18, pp. 535–583.
9. H. Iwasaki and K. Hoashi, *Kobunshi Ronbunshu*, **34**(11), 785 (1977).
10. A. L. Blackwell, *J. Plast. Film Sheeting* **1**, 205 (1985).
11. A. E. Barnabeo, W. S. Creasy, and L. M. Robeson, *J. Polym. Sci. Polym. Chem. Ed.* **13**, 1979 (1975).
12. T. D. Krizan, J. C. Coburn, and P. S. Blatz, in W. J. Koros, ed., *Barrier Polymers and Structures, ACS Symposium Series No. 423*, American Chemical Society, Washington, D.C., 1990, pp. 111–125.
13. L. E. Gerlowski, in W. J. Koros, ed., *Barrier Polymers and Structures, ACS Symposium Series No. 423*, American Chemical Society, Washington, D.C., 1990, Chapt. 8, pp. 177–191.
14. P. T. DeLassus and co-workers, in J. H. Hotchkiss, ed., *Food and Packaging Interactions, ACS Symposium Series No. 365*, American Chemical Society, Washington, D.C., 1988, Chapt. 2, pp. 11–27.
15. P. T. DeLassus, G. Strandburg, and B. A. Howell, *TAPPI J.* **71**(11), 177 (1988).
16. G. Strandburg, P. T. DeLassus, and B. A. Howell, in W. J. Karos, ed., *Barrier Polymers and Structures, ACS Symposium Series No. 423*, American Chemical Society, Washington, D.C., 1990, pp. 333–350.
17. G. Strandburg, P. T. DeLassus, and B. A. Howell, in S. A. Risch and J. H. Hotchkiss, eds., *Food and Packaging Interactions II, ACS Symposium Series, No. 473*, American Chemical Society, Washington, D.C., 1991, pp. 133–148.
18. Ref. 16, Chapt. 18.
19. A. R. Behrens and H. B. Hopfenberg, *J. Memb. Sci.* **10**, 283 (1982).
20. A. F. Burmester, T. A. Manial, J. S. McHattie, and R. A. Wessling, *Polym. Prepr. Am. Chem. Soc. Div. Polym. Chem.* **27**(2), 414 (1986).
21. K. Krumm, *Proc. TAPPI Polym. Laminations Coatings Conf.*, 313 (1987).
22. M. Yamada, *Proc. Europak '87*, 121 (1987).
23. A. J. Westlie and G. D. Oliver, *Proc. Future Pak '87*, 213 (1987).
24. J. C. Maxwell, *Electricity and Magnetism*, Vol. 1, 3rd ed., Dover, New York, 1981, p. 440.
25. L. E. Nielsen, *J. Macromol. Sci. Chem.* **A1**(5), 929 (1967).
26. D. R. Paul, *J. Membrane Soc.* **18**, 75 (1984).
27. P. T. DeLassus, K. L. Wallace, and H. J. Townsend, *Polym. Prepr. Am. Chem. Soc. Div. Polym. Chem.* **26**(1), 116 (1985).
28. P. T. DeLassus, *J. Vinyl Technol.* **3**, 240 (1981).
29. M. Salame, *Polym. Prepr. Am. Chem. Soc. Div. Polym. Chem.* **8**, 137 (1967).
30. W. M. Lee, *ACS Organic Coatings and Plastics Chemistry* **39**, 341 (1978).
31. A. R. Behrens, *Polym. Prepr. Am. Chem. Soc. Div. Polym. Chem.* **30**(1), 5 (1989).
32. K. D. Ziegel, H. K. Frensdorff and D. E. Blair, *J. Polym. Sci. Part A-2* **7**, 809 (1969).
33. R. A. Wood, *J. Test. Eval.* **12**(3), 149 (1984).

34. *1990 Annual Book of ASTM Standards*, ASTM Designation: E96-80, Vol. 15.09, ASTM, Philadelphia, Pa., pp. 811–818.

35. M. G. R. Zobel, *Polymer Testing* **5**, 153 (1985).

36. J. Landois-Garza and J. H. Hotchkiss, in J. H. Hotchkiss, ed., *Food and Packaging Interactions*, ACS Symposium Series No. 365, American Chemical Society, Washington, D.C., 1988, Chapt. 4, pp. 42–58.

GENERAL REFERENCES

W. A. Combellick, in J. I. Kroschwitz, ed., *Encyclopedia of Polymer Science and Engineering*, 2nd ed., Vol. 2, Wiley-Interscience, New York, 1985, pp. 176–192.

J. Comyn, ed., *Polymer Permeability*, Elsevier Applied Science Publishers, Ltd., Barking, UK, 1985.

J. Crank and G. S. Park, eds., *Diffusion in Polymers*, Academic Press, London, 1968.

R. M. Felder and G. S. Huvard, in R. A. Fava, ed., *Methods of Experimental Physics*, Vol. 16, Part C, Academic Press, New York, 1980, Chapt. 17, pp. 315–377.

PHILLIP DeLASSUS
The University of Texas—Pan American

BATTERIES, INTRODUCTION

1. Introduction

Batteries are storehouses for electrical energy "on demand". They range in size from large house-sized batteries for utility storage, cubic foot-sized batteries for automotive starting, lighting, and ignition, down to tablet-sized batteries for hearing aids and paper-thin batteries for memory protection in electronic devices. The historical development of batteries starting with Galvani and Volta has been summarized (1–5) and useful texts are available for a more detailed discussion of the topics covered herein (5–12).

In bulk chemical reactions, an oxidizer (electron acceptor) and fuel (electron donor) react to form products resulting in direct electron transfer and the release or absorption of energy as heat. By special arrangements of reactants in devices called batteries, it is possible to control the rate of reaction and to accomplish the direct release of chemical energy in the form of electricity on demand without intermediate processes.

Figure 1 schematically depicts an electrochemical reactor in which the chemical energy stored in the electrodes is manifested directly as a voltage and current flow. The electrons involved in the chemical reactions are transferred from the active materials undergoing oxidation to the oxidizing agent by means of an external circuit. The passage of electrons through this external circuit generates

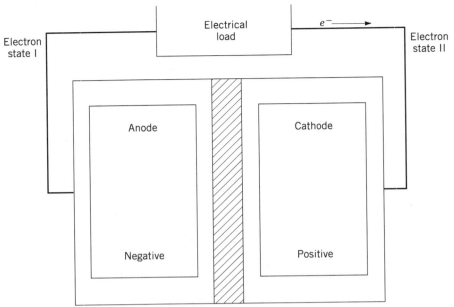

Fig. 1. Schematic representation of a battery system also known as an electrochemical transducer where the anode, also known as electron state I, may be comprised of lithium, magnesium, zinc, cadmium, lead, or hydrogen, and the cathode, or electron state II, depending on the composition of the anode, may be lead dioxide, manganese dioxide, nickel oxide, iron disulfide, oxygen, silver oxide, or iodine.

an electric current, thus providing a direct means for energy utilization without going through heat as an intermediate step. As a result, electrochemical reactors can be significantly more efficient than Carnot cycle heat engines.

The relationship between current flow and chemical reactions was established by Faraday who demonstrated that the amount of chemical change was directly proportional to the quantity of charge passed (It) and to the equivalent weight of the reacting material.

$$g = \frac{MIt}{nF} \tag{1}$$

where g is the mass of material (g) that reacts during electrolysis, M is the molecular weight (g/mol), n is the number of charges transferred in the electrode reaction (equiv/mol), t is the time of electrolysis, F is Faraday's constant (26.8 A·h/g − equiv), and I is the current (amperes) passed. Equation 1 can be expressed in current per unit area i,

$$i = I/A = \frac{nFg}{MtA} \tag{2}$$

The names for the electrodes in a battery are taken from the Greek *hodes* (pathway) for the path of electron flow. The anode, from the Greek *anodos* (up and out), or negative electrode of the cell is defined as the electrode from which an electron leaves the cell and oxidation occurs. The cathode, from the Greek *kathodos* (down and in or descent), or positive electrode of the cell is the electrode from which electrons enter the cell and reduction occurs. These terms, anode and cathode, work well for primary batteries but are not adequate for secondary or rechargeable ones. Since electron flow changes direction on charge and discharge, whereas, the electrodes retain their polarity, the terms negative and positive electrode are best used to describe the terminals of rechargeable systems.

The three main types of batteries are primary, secondary, and reserve. A primary battery is used or discharged once and discarded. Secondary or rechargeable batteries can be discharged, recharged, and used again. Reserve batteries are normally special constructions of primary battery systems that store the electrolyte apart from the electrodes, until put into use. They are designed for long-term storage before use. Fuel cells (qv) are not discussed herein.

Useful definitions include

Battery is one or more electrically connected electrochemical cell having terminals/contacts to produce electrical energy.

Primary battery is an electrolytic cell or group of cells for the generation of electric energy intended to be used until exhausted and then discarded.

Secondary battery is an electrolytic cell or group of cells for the generation of electric energy in which the cell, after being discharged, may be restored to its original charged condition by an electric current flowing in the direction opposite to the flow of current when the cell was discharged.

Reserve battery is a primary battery that is stored dry to improve the shelf life of the cell or battery and activated just before use by the addition of an electrolyte, or a battery in which the electorlyte is solid or "frozen" at room temperature and becomes conductive on heating. The electrolyte is usually stored in the battery.

Anode is the negative electrode of a primary cell associated with chemical reactions that release electrons into the external circuit.

Cathode is the positive electrode of a primary cell associated with chemical reactions that gain electrons from the external circuit

Electrolyte is a material that provides ionic conductivity between the positive and negative electrodes of a cell.

Separator is a physical barrier between the positive and negative electrodes incorporated into most cell designs to prevent electrical shorting. The separator can be a gelled electrolyte or a microporous plastic film or other porous inert material filled with electrolyte. Separators must be permeable to ions and inert in the battery environment.

Active mass is the material that generates electrical current by means of a chemical reaction within the battery.

Open circuit voltage is the voltage across the terminals of a cell or battery when no external current flows. It is usually close to the thermodynamic voltage for the system.

Closed circuit voltage is the voltage of a cell or battery when the battery is producing current into the external circuit.

Discharge is an operation in which a battery delivers electric energy to an external load.

Charge is an operation in which the battery is restored to its original charged condition by reversal of the current flow.

Internal impedance is the impedance that a battery or cell offers to alternating current flow at a given frequency.

Internal resistance is the resistance that a battery or cell offers to current flow.

Faraday constant (F) is the amount of charge (C) that transfers when an equivalent weight of active mass reacts, $96,485.3$ C/g-equiv $= 26.8015$ Ah /g-equiv).

2. Economic Aspects

The U.S. primary battery market is divided according to the chemical system used in the batteries, whereas the secondary battery market is usually divided according to usage. The 2000 estimate of the total battery market is given in Table 1. The lead–acid battery accounts for >64% of the secondary battery market.

3. Thermodynamics

Batteries can be thought of as miniature chemical reactors that convert chemical energy into electrical energy on demand. The thermodynamics of battery systems follow directly from that for bulk chemical reactions (13). For the general reaction

$$aA + bB \rightleftharpoons cC + dD \tag{3}$$

Table 1. **Estimated 2002 World Battery Market, $ × 10^6**

primary	
alkaline manganese	7,400
carbon–zinc	5,800
lithium, alkaline button,	
medical and military	3,300
total primary	16,500
rechargeable	
lead acid	16,300
Ni–Cd (sealed)	1,200
Ni–MH (Sealed)	1,400
lithium-ion	3,200
other (vented nickel, etc.)	3,200
total rechargeable	25,300
total market	41,800

the basic thermodynamic equations for a reversible electrochemical transformation are given as

$$\Delta G = \Delta H - T\Delta S \tag{4}$$

$$\Delta G^\circ = \Delta H^\circ - T\Delta S^\circ \tag{5}$$

where ΔG is the Gibbs free energy, or the energy of a reaction available for useful work, ΔH is the enthalpy, or the energy released by the reaction, ΔS is the entropy, or the heat associated with the organization of material, and T is the absolute temperature. The superscript $^\circ$ is used to indicate that the value of the function is for the material in the standard state at 25°C and unit activity. Although the Helmholtz free energy ΔA is used to describe constant volume situations found in battery systems, the use of the Gibbs free energy ΔG is adequate to describe practical battery systems.

The terms ΔG, ΔH, and ΔS are state functions and depend only on the identity of the materials and the initial and final states of the reaction. Tables of thermodynamic quantities are available for most known materials (see also THERMODYNAMIC PROPERTIES) (14).

Because ΔG is the net useful energy available from a given reaction, in electrical terms, the net available electrical energy from a reaction is given by

$$-\Delta G = nFE \tag{6}$$

and

$$-\Delta G^\circ = nFE^\circ \tag{7}$$

where n is the number of electrons transferred in the reaction, F is Faraday's constant, E is the voltage or electromotive force (emf) of the cell, and E° is the voltage at 25°C and at unit activity. The voltage is unique for each group of reactants comprising the battery system. The amount of electricity produced is determined by the total amount of materials involved in the reaction. The voltage may be thought of as an intensity factor, and the term nF may be considered a capacity factor.

The more negative the value of ΔG, the more energy or useful work can be obtained from the reaction. Reversible processes yield the maximum output. In irreversible processes, a portion of the useful work or energy is used to help carry out the reaction. The cell voltage or emf also has a sign and direction. Spontaneous processes have a negative free energy and a positive emf; the reaction, written in a reversible fashion, goes in the forward direction.

From equations 4 and 6, it can be shown that

$$\Delta G = \Delta H - nFT(\partial E/\partial T)_P \tag{8}$$

$$\Delta S = nF(\partial E/\partial T)_P \tag{9}$$

$$-\Delta H = nF\left[E - T(\partial E/\partial T)_P\right] \tag{10}$$

where $(\partial E/\partial T)_P$ is the temperature coefficient of the emf of a reversible cell at constant pressure. Equations 8–10 permit the calculation of heats of reaction from simple measurements of the voltage of an electrochemical cell as a function of temperature.

Once the values of thermodynamic functions, ΔH_{T1}, ΔS_{T1}, are known at a given temperature T_1 the value for the function can be calculated at any other temperature T_2 by

$$\Delta H_{T2} = \Delta H_{T1} \int_{T_1}^{T_2} \Delta C_P \, dT \quad \text{and} \quad \Delta S_{T2} = \Delta S_{T1} \int_{T_1}^{T_2} \Delta C_P \, dT \qquad (11)$$

where ΔC_P is the heat capacity at constant pressure.

The relationship between the chemical equilibrium constant K and the Gibbs free energy is

$$-RT \ln K = \Delta G^\circ \qquad (12)$$

The Van't Hoff isotherm identifies the free energy relationship for bulk chemical reactions.

$$\Delta G = \Delta G^\circ + RT \ln \frac{[\Pi A^s(\text{products})]}{[\Pi A^s(\text{reactants})]} \qquad (13)$$

By combining equation 6 and 7 with the Van't Hoff isotherm, the Nernst equation for electrochemical reactions is obtained

$$E = E^\circ - \frac{RT}{nF} \ln \frac{[\Pi A^s(\text{products})]}{[\Pi A^s(\text{reactants})]} \qquad (14)$$

where ΠA^s (products) is the product of the activities of the products, each raised to its stoichiometric power, and ΠA^s (reactants) is the corresponding term for the reactants, R is the gas constant, T is the absolute temperature, and E° is the reversible cell voltage under standard conditions. It is important to realize that only the surface composition of the electrode in contact with the electrolyte should be considered in calculating E°. Material buried deep within an electrode, completely covered with active material, and without direct electronic contact to the current collector, does not affect E° and the measured values of cell voltage.

The activity or effective concentration of the reactants and products often differs from the actual concentration of the material. The activity is related to the chemical potential μ_i of the species i by

$$\mu_i = (\partial G/\delta n_i)_{T,P,nj \neq 1} \qquad (15)$$

where n_i is the mole fraction of species i. Then for species A

$$\mu_A = \mu_A^\circ + RT \ln A_A \qquad (16)$$

where $A_A = f_A x_A$, the activity of species A; f_A is the activity coefficient; and x_A is the concentration. In dilute solutions, $f_A \longrightarrow 1$ as $x_A \longrightarrow 0$. In concentrated solutions the activity coefficient often decreases through a minimum as the concentration increases.

The modern definition of electricity involves the flow of current from a higher potential to a lower one. This places materials having a high electron energy level as the negative material (anodes) in an electrochemical cell and those having a lower electron energy level as the positive material (cathodes). Whereas the potentials of electrodes can be calculated using a Born–Haber cycle, the absolute electrode potentials of single electrodes cannot be measured experimentally. Only differences in the potential between two electrodes can be measured. For convenience, the potential of the hydrogen electrode at 25°C, 101.3 kPa (1 atm) H_2 pressure, and unit activity of hydrogen ions is chosen as the reference point for potential measurements. The standard hydrogen electrode (SHE) is defined as the zero point on the potential scale.

$$H_2 - 2e^- = 2H^+ \qquad E° = 0.000 \text{ V} \tag{17}$$

Table 2 illustrates the potential scale for selected battery reactions. The sign for the electrode potential indicates whether the electrode is negative or positive to the potential of the hydrogen electrode. Reactions, in both acid and basic media, are included to indicate the influence of pH on the electrode potential. In a battery, the electrode with the more negative potential becomes

Table 2. Selected Standard Potentials for Battery in Acid and Basic Media[a,b]

Electrode reaction	$E°$
Standard potentials in acid media	
$Li^+ + e^- = Li$	−3.045
$Na^+ + e^- = Na$	−2.714
$Mg^{2+} + 2\,e^- = Mg$	−2.356
$Al^{+3} + 3e^- = Al$	−1.676
$Zn^{+2} + 2\,e^- = Zn$	−0.7626
$S + 2\,e^- = S^{-2}$	−0.447
$Cd^{+2} + 2\,e^- = Cd$	−0.4025
$PbSO_4 + 2\,e^- = Pb + SO_4{}^{2-}$	−0.3505
$2\,H^+ + 2\,e^- = H_2$	0.0000
$AgCl + e^- = Ag + Cl^-$	0.2223
$O_2 + 2\,H^+ + 2\,e^- = H_2O_2$	0.695
$\gamma\text{-}MnO_2 + H^+ + e^- = MnOOH$	1.0
$Br_2 + 2\,e^- = 2\,Br^-$	1.06
$O_2 + 4\,H^+ + 4\,e^- = 2\,H_2O$	1.229
$PbO_2 + 4\,H^+ + 4\,e^- = Pb + 2\,H_2O$	1.698
Standard potentials in basic media	
$Zn(OH)_4{}^{2-} + 2\,e^- = Zn + 4OH^-$	−1.285
$CdO + H_2O + 2\,e^- = Cd + 2\,OH^-$	−0.783
$\gamma\text{-}MnO_2 + H_2O + e^- = MnOOH + OH^-$	0.36
$Ag_2O + H_2O + e^- = 2Ag + 2\,OH^-$	0.342
$O_2 + 2\,H_2O + 4\,e^- = 4\,OH^-$	0.401
$NiOOH + H_2O + e^- = Ni(OH)_2 + OH^-$	0.41

[a] Ref. 15.
[b] Potentials are given in volts (V).

the battery's negative electrode. The voltage of a battery is the algebraic difference between the individual electrode potentials on the scale. For example, the voltage of a zinc [7440-66-6]–silver(I) oxide [20667-12-3] cell in alkaline media is 1.627 V = 0.342 V − (− 1.255) V. Several excellent compilations of electrode potentials are available (15,16). Similar scales of electrode potentials can be constructed for each nonaqueous electrolyte.

When the reactants and products of an electrode reaction share a single phase or crystal habitat, or form an intercalate, the voltage of that electrode depends on the ratio of concentration of product to reacting species. This results in a sloping discharge curve that is typical of cells having manganese dioxide [1313-13-9], MnO_2, titanium disulfide [12039-13-3], TiS_2, molybdenum disulfide [1317-33-5], MoS_2, etc, electrodes. When the reactants and products of an electrode reaction form separate phases, the voltage of the electrode is constant during discharge. This is illustrated in electrodes of mercury [7439-97-6]–mercuric oxide [21908-53-2], ie, Hg/HgO, cadmium [7440-43-9]–cadmium hydroxide [21041-95-2], ie, Cd/Cd(OH)$_2$, and lead dioxide [1309-60-0]–lead sulfate [7446-14-2], PbO_2/$PbSO_4$. These electrodes are often used as reference electrodes in experimental studies because of their invariant voltages.

Because batteries directly convert chemical energy to electrical energy in an isothermal process, they are not limited by the Carnot efficiency. The thermodynamic efficiency ϵ for electrochemical processes is given by:

$$\epsilon = \frac{\Delta G^\circ}{\Delta H^\circ} = 1 - \frac{T \Delta S^\circ}{\Delta H^\circ} \tag{18}$$

In electrochemical units, equation 18 becomes

$$\epsilon = 1 - \frac{nFT(\partial E/\partial T)_P}{\Delta H^\circ} \tag{19}$$

Whenever energy is transformed from one form to another, an inefficiency of conversion occurs. Electrochemical reactions having efficiencies of 90% or greater are common. In contrast, Carnot heat engine conversions operate at $\sim$ 40% efficiency. The operation of practical cells always results in less than theoretical thermodynamic prediction for release of useful energy because of irreversible (polarization) losses of the electrode reactions. The overall electrochemical efficiency is, therefore, defined by

$$\text{Eff}_{\text{electrochem}} = \int \frac{(E'I\,dt)}{\Delta G} \tag{20}$$

where E' is the closed circuit terminal voltage when the net current I is flowing at time t.

The performance of a battery is often designed to be limited by one electrode in order to achieve special performance characteristics, such as overcharge protection and safety. The Coulombic efficiency of the active mass is of particular interest in battery design and performance.

$$\text{Eff}_{\text{Coulombic}} = It/Q \tag{21}$$

where Q is the quantity of coulombs expected from a given amount of active mass based on the equivalent weight and the amount of material. Utilization of the active mass may vary during the course of battery operation and depends on the cutoff voltage, rate of discharge–charge, and the nature of the electrode structure.

In addition to reversible heat absorbed, or released, as a result of the entropy of reaction, heat is released in batteries during operation because of the irreversibility of the reaction processes involved in converting chemical into electrical energy. The amount of heat q generated during battery operation is given by

$$q = \frac{T\Delta S}{nF} + I(E - E')\,dt \tag{22}$$

The total heat released is the sum of the entropy contribution plus the irreversible contribution. This heat is released inside the battery at the reaction site. Heat release is not a problem for low rate applications; however, high rate batteries must make provisions for heat dissipation. Failure to accommodate heat can lead to thermal runaway and other catastrophic situations.

4. Electrolytes

Electrolytes are a key component of electrochemical cells and batteries. Electrolytes are formed by dissolving an ionogen into a solvent. When salts are dissolved in a solvent such as water, the salt dissociates into ions through the action of the dielectric solvent. Strong electrolytes, ie, salts of strong acids and bases, are completely dissociated in solution into positive and negative ions. The ions are solvated but positive ions tend to interact more strongly with the solvent than do the anions. The ions of the electrolyte provide the path for the conduction of electricity by movement of charged particles through the solution. The electrolyte also provides the physical separation of the positive and negative electrodes needed for electrochemical cell operation.

Electrical conduction in electrolytic solutions follows Ohm's law:

$$I = \kappa V \tag{23}$$

where I is the current in amperes, V is the voltage drop in volts across the electrolytic resistor of conductivity κ in $(\Omega \cdot cm)^{-1}$. The conductivity κ is the reciprocal of the resistivity ρ in $\Omega \cdot cm$. The resistance of the electrolyte depends on the length l and cross-sectional area A of the conductor.

$$R = \rho l/A \tag{24}$$

The Debye-Hückel theory of electrolytes, based on the electric field surrounding each ion, forms the basis for modern concepts of electrolyte behavior (17,18). The two components of the theory are the relaxation and the electrophoretic effect. Each ion has an ion atmosphere of equal opposite charge surrounding it. During movement, the ion may not be exactly in the center of its ion atmosphere, thereby, producing a retarding electrical force on the ion. A finite time is required to reestablish the ion atmosphere at any new location. Thus the ion atmosphere produces a drag on the ions in motion and restricts their freedom of movement, which is termed a relaxation effect. When a negative ion moves under the influence of an electric field, it travels against the flow of positive ions and solvent moving in the opposite direction, which is termed an electrophoretic effect. The Debye–Hückel theory combines both effects to calculate the behavior of electrolytes. The theory predicts the behavior of dilute (≤ 0.05 m) solutions but does not portray accurately the behavior of concentrated solutions found in practical batteries.

Ions of an electrolyte are free to move about in solution by Brownian motion and, depending on the charge, have specific direction of motion under the influence of an external electric field. The movement of the ions under the influence of an electric field is responsible for the current flow through the electrolyte. The velocity of migration of an ion v_i is given by

$$v_i = -z_i u_i F d\phi/dx \tag{25}$$

where z_i is the charge number of the ion, $d\phi/dx$ is the electric field gradient, and u_i is the mobility of the ion. Solutions are electrically neutral so

$$\Sigma z_i c_i = 0 \tag{26}$$

When current I is passed through an electrolyte, the total current is given by

$$I = \Sigma z_i c_i \tag{27}$$

Each ion has its own characteristic mobility. The total conductivity of the electrolyte is the sum of the conductivities of the positive and negative ions, which is known as Kohlrausch's law of independent migration of ions.

Because both positive and negative ions move under the influence of an electric field, albeit in opposite directions, the fraction of the current carried by an ion is given by the ratio of the mobility of the ion and the total mobility of all the ions in solution. This ratio is called the transference number t of the ion. Thus

$$t_+ = u_+/(u_+ + u_-) t_- = u'_-/(u_+ + u_-) \tag{28}$$

where

$$t_+ + t_- = 1 \tag{29}$$

The term equivalent conductance Λ is often used to describe the conductivity of electrolytes. It is defined as the conductivity of a cube of solution having a cross-section of $1/\text{cm}^2$ and containing 1 equiv of dissolved electrolyte.

$$\Lambda = k/C^* \quad \text{or} \quad \Lambda = 1000 \, k/C \tag{30}$$

where C^* and C are the concentration in equiv/mL and equiv/L, respectively. It can be easily shown that

$$\Lambda = F(u_+ + u_-) \tag{31}$$

Transport properties of the electrolyte, as well as electrode reactions, have a significant impact on battery operation. The electrode reactions and ionic transference that occur during discharge result in considerable modifications to the solution composition at each electrode compartment. The negative and positive electrode compartments can lose or gain electrolyte and solvent depending on the transference numbers of the ions and the electrode reactions. The composition of the electrolyte in the separator between the two compartments generally remains unchanged.

Battery electrolytes are concentrated solutions of strong electrolytes and the Debye–Hückel theory of dilute solutions is only an approximation. Typical values for the resistivity of battery electrolytes range from about 1 ohm·cm for sulfuric acid [7664-93-9], H_2SO_4, in lead–acid batteries, and for potassium hydroxide [1310-58-3], KOH, in alkaline cells to $\sim$100 ohm·cm for organic electrolytes in lithium [7439-93-2], Li, batteries.

The physical picture in concentrated electrolytes is more aptly described by the theory of ionic association (19,20). As the solutions become more concentrated, the opportunity to form ion pairs held by electrostatic attraction increases. This tendency increases for ions with smaller ionic radii and in the lower dielectric constant solvents used for lithium batteries. A significant amount of ion-pairing and triple-ion formation exists in the high concentration electrolytes used in batteries. The ions are solvated, causing solvent molecules to be highly oriented and polarized. In concentrated solutions the ions are close together and the attraction between them increases ion-pairing of the electrolyte. Solvation can tie up a considerable amount of solvent and increase the viscosity of concentrated solutions.

Lithium batteries must use nonaqueous electrolytes, usually combinations of solvents, for stability because lithium reacts readily with water. Many of these electrolytes dissolve a high concentration of solute but are relatively poor conductors. The cause appears to be related to the increase of ion-pair and triple-ion formation in the lower dielectric constant, relative to water, solvents and to the increased viscosity of concentrated solutions. For example, a 1 M, lithium perchlorate [7791-03-9], $LiClO_4$, solution in propylene carbonate (PC) [108-32-7], $C_4H_6O_3$, has only $\sim$ 20% of the conductivity expected because $\sim$80% of the ions are involved in ion-pairing and triple-ion formation. Tables of electrolyte properties are available for most aqueous electrolytes(21,22) and selected nonaqueous electrolytes (23,24).

Each electrolyte is stable only within certain voltage ranges. Exceeding these limits results in decomposition. The stable range depends on the solvent, electrolyte composition, and purity level. In aqueous systems, hydrogen and oxygen form when the voltage limit is exceeded. In the nonaqueous organic solvent-based systems used for lithium batteries, exceeding the voltage limit can result in polymerization or decomposition of the solvent system. It is especially important to remove traces of water from the nonaqueous electrolytes as water can catalyze the electrolytic decomposition of the organic solvent.

In addition to the liquid conductors described above, two types of solid-state ionic conductors have been developed; one involves inorganic compounds and the other is based on polymeric materials. Several inorganic solids have been found to have excellent conductivity resulting wholly from ionic motion in the crystal lattice. Conductivity is related to specific crystal structures in which one ion, usually the cation, can move freely through the lattice. In the case of the solid materials, silver rubidium iodide, $AgRb_4I_5$, the silver ion moves freely through the lattice and $t_{Ag^+} = 1$. The voltage of batteries constructed with these materials is limited only by the decomposition voltage of the electrolyte. One solid electrolyte, lithium iodide [10377-51-2], LiI, has found application in heart-pacer batteries even though it has a fairly low conductivity.

A second type of solid ionic conductor, which is based around polyether compounds such as poly(ethylene oxide) [25322-68-3] (PEO), has been discovered and characterized (25,26). These materials follow equations 23–31 as opposed to the electronically conducting polyacetylene [26571-64-2] and polyaniline type materials. The polyethers can complex and stabilize lithium ions in organic media. They also dissolve salts such as $LiClO_4$ to produce conducting solid solutions. The use of these materials in rechargeable lithium batteries is under development for stationary energy storage and electric vehicle applications at Avistor Corp.(27,28).

A third type of "solid" ionic conductor is based on the ability of some polymers to absorb organic electrolytes, while maintaining a solid physical dimension. These are called "plasticized" or "gel" polymer electrolytes. These electrolytes have good conductivity at room temperature in contrast to the pure polymer electrolytes that have good conductivity only >60°C.

5. Electrical Double Layer

When two conducting phases come into contact with each other, a redistribution of charge occurs as a result of any electron energy level difference between the phases. If the two phases are metals, electrons flow from one metal to the other until the electron levels equilibrate. When an electrode, ie, electronic conductor, is immersed in an electrolyte, ie, ionic conductor, an electrical double layer forms at the electrode–solution interface resulting from the unequal tendency for distribution of electrical charges in the two phases. Because overall electrical neutrality must be maintained, this separation of charge between the electrode and solution gives rise to a potential difference between the two phases, equal to that needed to ensure equilibrium.

On the electrode side of the double layer, the excess charges are concentrated in the plane of the surface of the electronic conductor. On the electrolyte side of the double layer, the charge distribution is quite complex. The potential drop occurs over several atomic dimensions and depends on the specific reactivity and atomic structure of the electrode surface and the electrolyte composition. The electrical double layer strongly influences the rate and pathway of electrode reactions. The reader is referred to several excellent discussions of the electrical double layer at the electrode–solution interface (29–31).

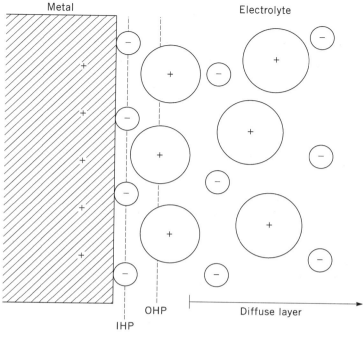

Fig. 2.   Representation of the electrical double layer at a metal electrode–solution interface for the case where anions occupy the IHP and cations occupy the OHP.

Figure 2 schematically depicts the structure of the electrode–solution interface. The inner-Helmholtz plane (IHP) refers to the distance of closest approach of specifically adsorbed ions, generally anions to the electrode surface. In aqueous systems, water molecules adsorb onto the electrode surface. The outer-Helmholtz plane (OHP) refers to the distance of closest approach of nonspecifically adsorbed ions, generally cations. The interactions of the ions of the OHP with the surface are not specific and have the character of longer range Coulombic interactions. Cations that populate the OHP are usually solvated and are generally larger in size than the anions.

To a first approximation, the ions in both Helmholtz layers can be considered point charges. They induce an equal and opposite image charge inside the conductive electrode. When the electrode is negative to the point of zero charge, cations populate the IHL.

When the electrode is positive to the point of zero charge, anions occupy the IHP. Anions are smaller in size and are solvated to a lesser degree than are the cations. Anions are more likely to adsorb specifically or bond to the electrode surface. If this is the case, the electrode can reach a considerable negative potential before strongly bonded anions are repelled from the inner layer. When an ion bonds to the surface it partially transfers charge to the surface. This charge transfer reduces the effective charge of the ion and causes the potential difference produced by the ion at the surface to be less than would be expected by a point charge an atomic distance from the surface.

The region of the gradual potential drop from the Helmholtz layer into the bulk of the solution is called the Gouy or diffuse layer (32,33). The Gouy layer has similar characteristics to the ion atmosphere from electrolyte theory. This layer has an almost exponential decay of potential with increasing distance. The thickness of the diffuse layer may be approximated by the Debye length of the electrolyte.

Electrical double layers are not confined to the interface between conducting phases. Solid particles of active mass, or of conductive additives of colloidal size, can acquire an electric charge by specific adsorption of cations or anions from the electrolyte or by reaction of surface moieties with components of the solution. The resulting excess charge on the particle is neutralized by a diffuse or Gouy layer in the solution. The electrokinetic properties of the interface and zeta potential concepts are based on the characteristics of the Gouy layer. Migration of the colloidal-sized solid particles can occur under the influence of an applied electric field during battery operations.

Electrically, the electrical double layer may be viewed as a capacitor with the charges separated by a distance of the order of molecular dimensions. The measured capacitance ranges from about two to several hundred microfarads per square centimeter depending on the structure of the double layer, the potential, and the composition of the electrode materials. Figure 3 illustrates the behavior of the capacitance and potential for a mercury electrode where the double layer capacitance is $\sim$ 16 μF/cm^2 when cations occupy the OHP and $\sim$ 38 μF/cm^2 when anions occupy the IHP. The behavior of other electrode materials is judged to be similar.

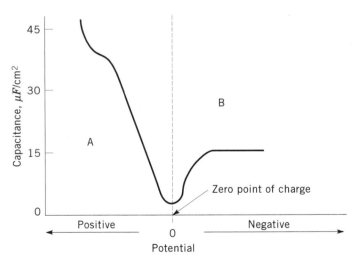

Fig. 3. Capacitance–potential relationship at a mercury electrode for a nonspecific absorbing electrolyte where regions A and B represent inner-layer anions and cations, respectively.

6. Kinetics and Transport

6.1. Activation Processes. Reactions must occur at a reasonable rate to be useful in battery applications. The rate or ability of battery electrodes to produce current is determined by the kinetic processes of electrode operations, not by thermodynamics. Thermodynamics describes the characteristics of reactions at equilibrium when the forward and reverse reaction rates are equal. Electrochemical reaction kinetics (34–37) follow the same general considerations as those of bulk chemical reactions. Two differences are a potential drop that exists between the electrode and the solution because of the electrical double layer at the electrode interface, and the reaction that occurs at interfaces that are two-dimensional (2D) rather than in the three-dimensional (3D) bulk.

Electrode kinetics lend themselves to treatment using the absolute reaction rate theory or the transition state theory (38–40). In these treatments, the path followed by the reaction proceeds by a route involving an activated complex where the element determining the reaction rate, ie, the rate–limiting step, is the dissociation of the activated complex. The general electrode reaction may be described as

$$A = C^{n+} + ne^- \tag{32}$$

where n is the number of electrons in the reaction. By using reaction rate theory, it can be shown that the current flow results from a change in the activation energy barrier introduced by a departure from the equilibrium potential drop across the electrode–solution interface. The free energies of activation for the forward ΔG_f and reverse ΔG_r reactions are given by

$$\Delta G_f = \Delta G^* - \alpha n\, FE' \tag{33}$$

$$\Delta G_r = \Delta G^{**} + (1 - \alpha)n\, FE' \tag{34}$$

where E' is the potential of the electrode in its operating environment, α is the transfer coefficient or the fraction of the potential drop through the electrical double layer that operates on the activated complex, and $\Delta G^{**} = \Delta G^* + \Delta G°$. The rate of electrochemical reactions are expressed as a current flow I in coulombs per second (amperes), rather than as a change in concentration per unit time. The direction of the potential change from the equilibrium potential E determines the direction of current flow. In terms of net current flow where $i = I/\mathrm{cm}^2$, it can be shown that

$$i = i_f - i_r = nFk_1 A_A e^{(-\Delta G^* - \alpha nFE^*/RT)} - nF k_{-1} A_C e^{(-\Delta G^{**} + (1-\alpha)nFE^*RT)} \tag{35}$$

where A_A and A_C are the activities of reactants and products, k_1 and k_{-1} are the specific rate constants, and ΔG^* and ΔG^{**} are the free energy of activation for the forward and reverse reactions, respectively. The free energy of activation is strongly influenced by the fraction of the potential drop across the electrode–solution interface that acts on the activated complex. Thus the structure of the electrical double layer plays a key role in the kinetics and rate of electrode reactions.

At equilibrium, there is no net current flow, the rate of the forward and reverse reactions are equal and $E' = E$. The rate of the forward and reverse reactions at equilibrium

$$i_f = i_r = i_0 \tag{36}$$

expressed as a current flow i_o are given the special term exchange current. Because electrode reactions are in dynamic equilibrium, the term i_o is the rate of the reaction at the equilibrium potential E.

The exchange current is directly related to the reaction rate constant, to the activities of reactants and products, and to the potential drop across the double layer. The larger i_0, the more reversible the reaction and, hence, the lower the polarization for a given net current flow. Electrode reactions having high exchange currents are favored for use in battery applications.

By taking the polarization η as the departure from equilibrium potential ($\eta = E' - E$), it follows from equations 35 and 36 that

$$i = i_0\{e^{-(\alpha nF\eta/RT)} - e^{((1-\alpha)nF\eta/RT)}\} \tag{37}$$

The expected behavior of a current–potential diagram is given in Figure 4. These plots constitute the classic method for investigating electrode kinetics and for characterizing battery reactions. The value of α, and the relationship of i_o with concentration, provide valuable insight into the kinetics of the electrode reactions. At low current drains, the polarization in battery reactions is almost always activation energy (charge-transfer) controlled. When the electrode

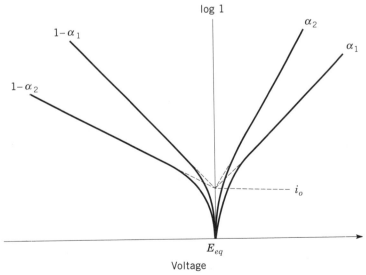

Fig. 4. Currrent–potential behavior of an electrode reaction based on equation 37. Tafel behavior is noted at high currents for two different values of α.

reaction occurs and the polarization is larger than ~ 0.05 V, the term for the reverse reaction in equation 37 may be neglected and equation 37 can be simplified.

$$i = i_0 \{e^{-(\alpha n F \eta / RT)}\} \tag{38}$$

or

$$\eta = a - b \log i \tag{39}$$

where $a = (2.303\, RT/\alpha nF) \log i_0$ and $b = 2.303\, RT/\alpha nF$. Equation 39 is the well-known Tafel equation that describes the voltage–current behavior for processes under activation control, and is illustrated in Figure 4 by the linear portion of the plot. The kinetic parameters for many electrode reactions have been summarized (41).

6.2. Transport Processes.   The velocity of electrode reactions is controlled by the charge-transfer rate of the electrode process or by the velocity of the approach of the reactants to the reaction site. The movement or transport of reactants to and from the reaction site at the electrode interface is a common feature of all electrode reactions. Transport of reactants and products occurs by diffusion, by migration under a potential field, and by convection. The complete description of transport requires a solution to the transport equations. A full account is given in texts and discussions on hydrodynamic flow (42,43) (see FLUID MECHANICS). Molecular diffusion in electrolytes is relatively slow. Although the process can be accelerated by stirring, enhanced mass transfer (qv) by stirring or convection is not possible in most battery designs. Natural convection from density changes does occur but does not greatly enhance transport in battery operation. Lead acid batteries, used for motive power and stationary applications, are given a gassing overcharge on a regular basis. The gas evolution stirs up the electrolyte and equalizes the sulfuric acid concentration in the electrolyte.

Whenever the local concentration of a reacting component in a battery departs significantly from its equilibrium value, the rate of reaction becomes controlled by the transport of that component to the reaction site. The polarization resulting from these concentration changes η_c is given by

$$\eta_c = \frac{RT}{nF} \ln (C_e / C) \tag{40}$$

where C_e is the concentration at the electrode surface and C is the concentration in the bulk of the electrolyte.

The limiting current density i_l for the transport of species i to the reacting site is given from Fick's law by

$$i_l = \frac{DnFC_B}{\delta} \tag{41}$$

where δ is the thickness of the stationary diffusion layer, D is the diffusion coefficient, C_B is the concentration on the bulk of the solution, and n and F are as previously defined. Allowing for electromigration, equation 41 becomes

$$i_l = \frac{DnFC_B}{\delta(1 - t_i)} \tag{42}$$

where t_i is the transference number of the reacting species. The limiting current calculation corresponds to the case where the concentration at the electrode surface is zero and the thickness of the diffusion layer is very thin. For many aqueous electrolytes the limiting current can be estimated by

$$i_l = 0.025 \; C \tag{43}$$

where C is in g-equiv/L. The relationship between concentration polarization and limiting current density is given by

$$\eta_c = \frac{RT}{nF} \ln \left(1 - i/i_e\right) \tag{44}$$

Whereas, the above discussion on concentration polarization was developed for electrolyte-side supply of reactants, concentration polarization can also arise from surface diffusion, diffusion into the solid structure (intercalation) of the active mass, and diffusion of products away from the reaction site. The deposition and dissolution reactions can involve movement of surface atoms to and from the deposition site to the equilibrium position in the lattice.

The detailed mechanism of battery electrode reactions often involves a series of chemical and electrochemical or charge-transfer steps. Electrode reaction sequences can also include diffusion steps on the electrode surface. Because of the high activation energy required to transfer two electrons at one time, the charge-transfer reactions are believed to occur by a series of one electron-transfer steps illustrated by the reactions of the zinc electrode in strongly alkaline medium (44).

$$Zn + OH^- = Zn(OH) + e^- \tag{45}$$

$$Zn(OH) + OH^- = Zn(OH)_2^- \tag{46}$$

$$Zn(OH)_2^- + OH^- = (Zn)OH_3^- + e^- \tag{47}$$

$$Zn(OH)_3^- + OH^- = Zn(OH)_4^{2-} \tag{48}$$

In this reaction sequence, equation 47, the formation of the complex ion, $Zn(OH)_3^-$, is the rate determining step. Once the solubility of zincate, $Zn(OH)_4^{2-}$, is exceeded, zinc hydroxide [20427-58-1], $Zn(OH)_2$, precipitates. The crystal form that falls out of solution depends on the concentration of the alkali.

7. Experimental Techniques

In the thermodynamic treatment of electrode potentials, the assumption was made that the reactions were reversible, which implies that the reactions occur infinitely slowly. This is never the case in practice. When a battery delivers current, the electrode reactions depart from reversible behavior and the battery voltage decreases from its open circuit or equilibrium voltage E. Thus the voltage during battery use or discharge E' is lower than the voltage measured under open circuit or reversible conditions E by a quantity called the polarization η.

$$\eta = E' - E \tag{49}$$

Likewise, the battery voltage is increased from its open circuit voltage on charge.

Three sources for the departure of battery operation from equilibrium voltage are activation, concentration, and ohmic polarizations. The regions controlled by each of these change as the battery is discharged. *Activation polarization* arises from a kinetic hinderance in one or more of the detailed reaction processes. Slow or rate-limiting steps of the reaction mechanism may occur in the charge-transfer process or in the chemical steps preceding or following the charge-transfer process. An activation barrier and an activation energy characterize these processes. On initiation or cessation of current flow, activation polarization generally builds up or decays in an exponential manner. *Concentration polarization* is associated with the decrease in availability of reacting species at the reaction site in the electrode–solution interface. Changes in concentration occur when the transport rate of reactants and/or products is slower than the rate of reaction. On initiation or cessation of current flow, the concentration polarization builds up or decays in a complex manner and is slower to recover than either activation and ohmic polarization. *Ohmic polarization* arises from the electrical resistances of the electrolyte, current collectors, the active mass, the conductive additives, and the contact of the collector to the active mass. It may also be associated with a resistive film formation on the electrode surface. Ohmic polarization follows Ohm's law. On cessation of current flow, the ohmic polarization disappears instantaneously.

A variety of experimental techniques are used to study electrochemical and battery reactions (37,45–47). The direct measurement of the instantaneous current–voltage characteristics or power curve, and the discharge curve at various discharge rates, are the two most common techniques. Indeed, the discharge performance of a typical battery is characterized by these curves as depicted in Figure 5. The characteristics of the power curve varies, depending on the state of charge of the battery.

The impedance behavior of a battery can reveal a significant amount of information about battery operation characteristics (38,48). The impedance of an electrode or battery is given by

$$Z = R + jX \tag{50}$$

where $X = \omega L - 1/(\omega C), j = \sqrt{-1}$, ω is the angular frequency ($2\pi f$), L the inductance, and C is the capacitance. Each of the characteristic polarizations has a distinctive impedance behavior. A schematic of a battery circuit, and the corresponding Argand diagram, illustrating the behavior of the simple processes, are shown in Figure 6. In ideal behavior, activation processes exhibit a semicircular behavior with frequency that is characteristic of relaxation processes;

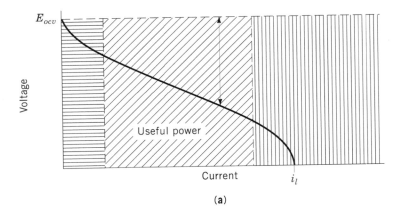

(a)

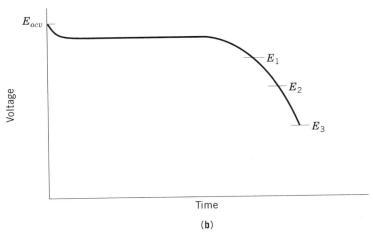

(b)

Fig. 5. Discharge behavior of a battery where E_{ocv} is the open circuit voltage; (**a**) current–potential or power curve showing activation, ohmic, and concentration polarization regions where the double headed arrow represents polarization loss and (**b**) voltage–time profile depicting different end point voltages for the discharge.

concentration processes exhibit a 45° behavior characteristic of diffusion processes; and ohmic polarizations are independent of frequency. Battery electrodes have large surface areas and, therefore, exhibit large capacitances. It is common for larger cells to have a capacitance of farads and a resistance of milliohms in measurements of battery impedances.

8. Practical Battery Systems

Most battery electrodes are porous structures in which an interconnected matrix of solid particles, consisting of both nonconductive and electronically conductive materials, is filled with electrolyte. When the active mass is nonconducting,

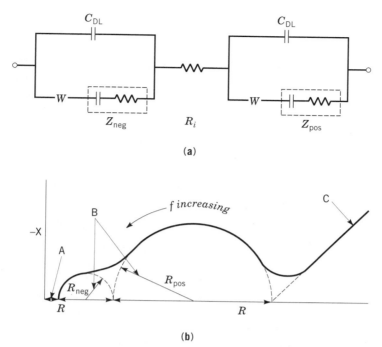

Fig. 6. (**a**) Simple battery circuit diagram where C_{DL} represents the capacitance of the electrical double layer at the electrode–solution interface, Z_{meg} and Z_{pos} are the inpedance associated with the electrode reaction processes at the anode and cathode respectively, W depicts the Warburg impedance for diffusion processes, and R_i is internal resistance. (**b**) The corresponding Argand diagram of the behavior of impedance with frequency, f, for an idealized battery system, where the characteristic behavior of A, ohmic; B, activation; and C, diffusion or concentration (Warburg behavior) processes are shown. The parameter X is as defined in equation 50.

conductive materials, usually carbon or metallic powders, are added to provide electronic contact to the active mass. The solids occupy 60–80% of the volume of a typical porous battery electrode. Most battery electrode structures do not have a well-defined planar surface but have a complex surface extending throughout the volume of the porous electrode. Macroscopically, the porous electrode behaves as a homogeneous unit

When a battery produces current, the sites of current production are not uniformly distributed on the electrodes (48). The nonuniform current distribution lowers the expected performance from a battery system and causes excessive heat evolution and low utilization of active materials. Two types of current distribution, primary and secondary, can be distinguished. The primary distribution is related to the current production based on the geometric surface area of the battery construction. Secondary current distribution is related to current production sites inside the porous electrode itself. Most practical battery constructions have nonuniform current distribution across the surface of the electrodes. This primary current distribution is governed by geometric factors such as height (or length) of the electrodes, the distance between the electrodes, the resistance of the anode and cathode structures, by the resistance of the electrolyte, and by the polarization resistance or hinderance of the electrode reaction processes.

Cell geometry, such as tab/terminal positioning and battery configuration, strongly influence primary current distribution. The monopolar construction is most common. Several electrodes of the same polarity may be connected in parallel to increase capacity. The current production concentrates near the tab connections unless special care is exercised in designing the current collector. Bipolar construction, wherein the terminal or collector of one cell serves as the anode and cathode of the next cell in pile formation, leads to greatly improved uniformity of current distribution. Several representations are available to calculate the current distribution across the geometric electrode surface (49–51).

Whereas, current producing reactions occur at the electrode surface, they also occur at considerable depth below the surface in porous electrodes. Porous electrodes offer enhanced performance through increased surface area for the electrode reaction and through increased mass-transfer rates from shorter diffusion path lengths. The key parameters in determining the reaction distribution include the ratio of the volume conductivity of the electrolyte to the volume conductivity of the electrode matrix, the exchange current, the diffusion characteristics of reactants and products, and the total current flow. The porosity, pore size, and tortuosity of the electrode all play a role. Figure 7 illustrates the reaction distribution in porous electrodes. The location of the reaction sites is seen to be strongly dependent on the characteristics of the electrode structure and reactions.

The effectiveness of a porous electrode over a plane surface electrode is given by the product of the active surface area S in cm^2/mL and the penetration depth L_P of the reaction process into the porous electrode.

$$\text{Effectiveness} = SL_P \tag{51}$$

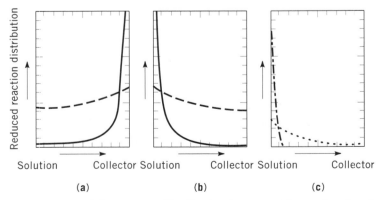

Fig. 7. Representation of the current distribution in porous electrodes showing the effect of conductivities of the electrolyte and electrodes where for (**a**) $K_{\text{electrolyte}} \ll K_{\text{matrix}}$ and (**b**) $K_{\text{matrix}} \gg K_{\text{electrolyte}}$ (52) (———) represents low current and (−) high current; and for (**c**) the effect of exchange current (53) where (···) represents low i_0 and (— – —) high i_0.

An effectiveness value > 1 indicates that the porous electrode is more effective than an electrode of the same geometric surface area, and that the reaction extends into the porous electrode structure.

Mathematical formulations, based on models of primary and secondary battery systems, permit rapid optimization in the design of new battery configurations. Models to describe and predict porous electrode performance in the lead–acid battery system have been developed (54–56) and used in industry to speed the development of new battery configurations giving superior performance. The high rate performance of the present starting, lighting, and ignition (SLI) automotive batteries have evolved directly from coupling collector designs with the porous electrode compositions identified from modeling studies. Modeling has proven useful in primary as well as rechargeable as well as in the development of new battery systems as well (57,58).

In practice, the Peukert equation, equation 52, finds wide use in reporting and comparing battery performance (59).

$$It^n = \text{constant} \tag{52}$$

where I is the current, t the time to a given discharge voltage, and n is a constant. If the system is reversible, $n = 1$. However, because battery systems have limitations in reaction rate, diffusion, and internal resistance, $n > 1$ for most practical systems. Figure 8 shows the Peukert curve constructed from the performance of a given battery to different cutoff voltages. Other equations for describing and predicting battery performance have also been developed (60,61).

The positive electrode in a battery system is most often a metal oxide, but it may also be a metal sulfide or halide. Generally, these materials are relatively poor electrical conductors and exhibit extremely high ohmic polarizations (impedances) if not combined with supporting electronic conductors such as graphite

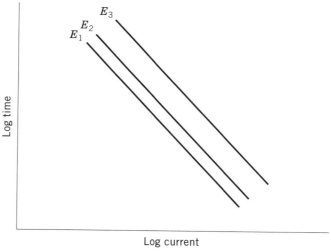

Fig. 8. Peukert diagram of the performance of a battery based on the discharge curves from Figure 5**b**, where $E_1 > E_2 > E_3$.

[7782-42-5], lead [7439-92-1], silver [7440-22-4], copper [7440-50-8], or nickel [7440-02-0] in the form of powder, rod, mesh, wire, grid, or other configurations. In almost all cases, the negative electrode is a metallic element of sufficient conductivity to require only minimal supporting conductive structures. Exceptions are the oxygen (air, positive) and hydrogen gas (negative) electrodes that require a substantial conductive, catalytically active, surface support that also serves as current collector.

Although there are a multitude of chemical reactions that can store and release energy, only a few have the characteristics requisite for use in commercial batteries. A set of criteria can be established to characterize reactions suitable for battery development(62,63). The principal features necessary for battery reactions include (1) Mechanical and chemical stability, ie, the reactants or active masses and cell components must be stable over time (5 years or more) in the operating environment and must reform in their original condition on recharge; (2) energy content, ie, the reactants must have sufficient energy content to provide a useful voltage and current level; (3) power density, ie, the reactants must be capable of reacting at rates sufficient to deliver useful rates of electricity; (4) temperature range, ie, the reactants must be able to maintain energy, power, and stability over a normal operating environment; (5) safety, ie, the battery must be safe in the normal operating environment as well as

Table 3. Specialty Batteries: Military, Medical

Common name	Cell reaction	Nominal voltage	Nominal Wh/l	Nominal Wh/l	Comments
thionyl chloride	$Li + 2\,SOCl_2 = 4\,LiCl + 2\,S + SO_2$	3.6	1000	550	oil well logging
lithium sulfur dioxide	$2\,Li + 2\,SO_2 = Li_2S_2O_4$	2.8	425	330	military radios
lithium iron sulfide	$2\,Li + FeS = Li_2S + Fe$	1.8	40	120	military fuse battery excellent high rate 450°C operation
lithium iodine	$2\,Li + I_2 = 2\,LiI$	2.8	1000	300	heart pacers high energy storage
lithium SVO	$3.5\,Li + AgV_2O_{5.5} = Li_{3.5}\,AgV_2O_{5.5}$	3.2	780	270	heart defibrilators excellent pulse capability
seawater	$Mg + 2\,AgCl = MgCl_2 + 2\,Ag$	1.5	360	150	sonobouys weather balloons
nickel hydrogen	$H_2 + 2\,NiOOH = 2\,Ni(OH)_2$	1.2	90	60	long cycle life spacecraft power
silver zinc	$Zn + AgO = ZnO + Ag$	1.8	180	90	rechargeable good high rate performance limited cycle life

under mild abusive conditions; and (*6*) cost, ie, the reactants and the materials of construction should be inexpensive and in good supply as well as manufacturing employing high speed automated production processes. Figures 9 and 10 contain characteristics of large-scale commercial primary and secondary battery systems. The alkaline $Zn-MnO_2$ battery sets the standard for primary battery systems. This battery was first introduced commercially in 1958 and replaced the carbon–zinc battery as the principal primary battery system in ~1985. Lithium batteries were first developed in the early 1970s for military applications and the commercial market began to develop in the late 1970s, largely in Japan. They now power most watches and cameras. The lead–acid battery dominates the rechargeable battery market for automotive and stationary energy storage applications: sales of lead–acid batteries account for over 64% of the rechargeable battery market. The lithium-ion and nickel hetal hydride systems have shown high growth rates based on portable electronic devices such as cellular telephone and notebook applications. Table 3 contains performance parameters for specialty batteries and Table 4 describes promising rechargeable battery systems in various stages of research and commercial development. Table 5 gives the theoretical energy content of selected high energy systems that are under consideration for use in electric vehicle propulsion and other applications where these high energy battery systems are required. Several of the systems are more energetic than gasoline. They also are significantly greater than for any commercial rechargeable battery system.

Table 4. Battery Systems in Various Stages of Research and Development and Early Stages of Commercialization

Common name	Cell reaction	Nominal voltage	Nominal Wh/l	Nominal Wh/l	Comments
zinc-bromine	$Zn + Br_2 = ZnBr_2$	1.6	95	60	low cost, high rate capability limited cycle life
zinc air	$2\,Zn + O_2 = 2\,ZnO$	1.4	150	180	electrically or mechaniclly rechargeable system
sodium-sulfur	$2\,Na + 3\,S = Na_2S_3$	2.1	120	160	solid-Al_2O_3 electrolyte operates at high temperatures
aluminum-air	$4\,Al + 3\,O_2 + 2\,H_2O = 2\,Al_2O_3(H_2O)$	1.6	360	250	mechanically rechargeable
"super iron"	$2\,BaFeO_4 + 3\,Zn = Fe_2O_3 + ZnO + 2\,BaZnO_2$	1.7	NA	319	primary or rechargeable possible
lithium polymer	$8\,Li + V_6O_{13} = Li_8V_6O_{13}$	2.4	333	420	lithium system solid polymer electrolyte
ultracapacitor	electrical double layer charge and discharge	2.5	12	12	nonaqueous electrolyte excellent pulse capability

Table 5. **Theoretical High Energy Systems**

System/reaction	Cell voltage, V	Wh/kg
$2\,Li + H_2O_2 = 2\,LiOH$	3.98	4400
$4\,Li + O_2 + 2\,H_2O = 4\,LiOH$	4.47	3240
$2\,Mg + O_2 + 2\,H_2O = 2\,Mg(OH)_2$	3.09	2850
$4\,Al + 3\,O_2 + 2\,H_2O = 2\,Al_2O_3 \cdot 2\,H_2O$	2.7	2791
$2\,Li + Cl_2 = 2\,LiCl$	3.95	2510
gasoline (octane) $+ 12.5\,O_2 = 8\,CO_2 + 9\,H_2O$		2300
$2\,H_2 + O_2 = 2\,H_2O$	1.23	1940

BIBLIOGRAPHY

"Batteries, Primeary (Introduction)" in *ECT* 2nd ed., Vol. 3, pp. 99–110, by E. B. Yeager, Western Reserve University, and J. F. Yeager, Union Carbide Corp., "Batteries and Electric Cells, Secondary (Introduction)" in *ECT* 2nd ed., Vol. 3, pp. 161–171, by W. W. Jakobi, Gould-National Batteries, Inc.; "Batteries and Electric Cells, Primary (Introduction)" in *ECT* 3rd ed., Vol. 3, pp. 569–591, by J. B. Doe, ESB Technology Co. "Batteries and Electric Cells, Secondary (Introduction)" in *ECT* 3rd ed., vol. 3, pp. 569–591, by J. B. Doe, ESB Technology Co; "Batteries (Introduction)" in *ECT* 4th ed., Vol. 3, pp. 963–991, by Ralph Brodd, Gould Inc; "Batteries, Introduction" in *ECT* (online), posting date: December 4, 2000, by Ralpha Brodd, Gould. Inc.

CITED PUBLICATIONS

1. C. A. Vincent, B. Scrosati, M. Lazzari, and F. Bonino, *Modern Batteries*, 2nd ed., Edward Arnold, Ltd., London, 1997.
2. D. Linden, ed., *Handbook of Batteries and Fuel Cells*, 2nd ed., McGraw-Hill Book Co., New York, 1995.
3. G. W. Heise and N. C. Cahoon, eds., *The Primary Battery*, John Wiley & Sons, Inc., New York, 1965, Vol. 1, 1; 1976, Vol. 2.
4. S. U. Falk and A. J. Salkind, *Alkaline Storage Batteries*, John Wiley & Sons, Inc., New York, 1969.
5. H. Bode, *Lead Acid Batteries*, John Wiley & Sons, Inc., New York, 1976.
6. J. P. Gabano, *Lithium Batteries*, Academic Press, New York, 1983.
7. G. Pistoia, ed., *Lithium Batteries*, Elsevier Science, New York, 1993.
8. M. Wakihara and O. Yamamoto, eds., *Lithium Ion Batteries*, Wiley-VCH, New York, 1998.
9. V. S. Bagotsky and K. Muller, *Fundamentals of Electrochemistry*, Plenum Press, New York, 1993.
10. K. B. Oldham and J. C. Myland, *Principles of Electrochemical Science*, Academic press, 1994.
11. J. Koryta and J. Dvorak, *Principles of Electrochemistry*, John Wiley & Sons, Inc., New York, 1987.
12. J. O'M. Bockris and A. K. N. Reddy, and M. Gamboa-Aldeco, *Modern Electrochemistry: Fundamentals of Electrodics*, Plenum Press, New York, 2000.
13. I. M. Klotz, *Chemical Thermodynamics*, Benjamin Press, New York, 1967.

14. D. R. Lide, *Handbook of Chemistry and Physics*, 72nd ed., CRC Press, Boca Raton, Fla., 1992.
15. A. J. Bard, R. Parsons, and J. Jordan, eds., *Standard Potentials in Aqueous Solution*, Marcel Dekker, Inc., New York, 1985.
16. M. Pourbaix, *Atlas of Electrochemical Equilibria in Aqueous Solutions*, Pergamon Press, Inc., Elmsford, N.Y., 1969.
17. P. Debye and E. Hückel, *Z. Phys. (Lepizig)* **24**, 185, 305 (1923).
18. R. A. Robinson and R. H. Stokes, *Electrolytic Solutions*, 2nd ed., Butterworths, London, 1959.
19. N. Bjerrum, *Kgl. Danske Vidensk. Selskab* **7**(9) (1926).
20. R. M. Fuoss and F. Accascina, *Electrolytic Conductance*, Interscience Publishers, New York, 1959.
21. H. S. Harned and B. B. Owen, *The Physical Chemistry of Electrolyte Solutions*, 3rd ed., Reinhold Publishing Corp., New York, 1958.
22. B. E. Conway, *Electrochemical Data*, Elsevier, Amsterdam, The Netherlands, 1952; H. V. Venkatasetty, ed., *Lithium Battery Technology*, Wiley-Interscience, New York, 1984.
23. G. J. Janz and R. P. T. Thomkins, *Nonaqueous Electrolyte Handbook*, Academic Press, New York, 1972, Vol. **1**; 1974, Vol. **2**.
24. D. Aurbach, ed., *NonAqueous Electrochemistry*, Marcel Dekker, New York, 1999.
25. D. E. Furton, J. M. Parker, and P. V. Wright, *Polymer* **14**, 589 (1973).
26. F. M. Grey, *Solid Polymer Electrolytes: Fundamentals and Technological Applications*, John Wiley & Sons, Inc., New york, 1991.
27. M. Armand, J. M. Chabagno, and M. Duclot, in P. Vashishta, ed., *Fast Ion Transport in Solids*, North Holland, New York, 1979, p. 131.
28. D. Kuller and B. Kapfer, Lecture notes in EnV 14, "Energy Storage Devices—Latest Realities—Lithium Polymer Batteries for Electric Vehicles".
29. J. A. V. Butler, *Electrical Phenomena at Interfaces*, Butterworths, London, 1951.
30. D. C. Grahame, *Chem. Rev.* **41**, 441 (1947).
31. D. M. Mohilner, in A. Bard, ed., *Electroanalytical Chemistry*, Marcel Dekker, New York, 1966, Vol. **1**, p. 241.
32. G. Gouy, *J. Phys.* **9**, 457 (1910).
33. D. L. Chapman, *Philos. Mag.* **25**, 475 (1913).
34. K. Vetter, *Electrochemical Kinetics*, Academic Press, New York, 1967.
35. R. A. Marcus, *Ann. Rev. Phys. Chem.* **15**, 155 (1964).
36. T. Erdey-Gruz, *Kinetics of Electrode Processes*, Wiley-Interscience, New York, 1972.
37. A. J. Bard and L. R. Faulkner, *Electrochemical Methods, Fundamentals and Applications*, Second Edition, John Wiley & Sons, Inc., New York, 1999.
38. W. J. Albery, *Electrode Kinetics*, Clarendon, Oxford, 1975.
39. S. Glasstone, K. J. Laidler, and H. Eyring, *The Theory of Rate Processes*, McGraw-Hill Book Co., Inc., New York, 1941.
40. W. C. Gardner, Jr., *Rates and Mechanisms of Chemical Reactions*, Benjamin, New York, 1969.
41. N. Tanaka and R. Tamamushi, *Electrochim. Acta* **9**, 963 (1964).
42. V. C. Levich, *Physicochemical Hydrodynamics*, Prentice-Hall, Englewood Cliffs, N.J., 1962.
43. J. Newman, *Electrochemical Systems*, Prentice-Hall, Englewood Cliffs, N.J., 1973.
44. J. O'M. Bockris, Z. Nagy, and A. Damjanovic, *J. Electrochem. Soc.* **119**, 285 (1972).
45. R. J. Brodd and A. Kozawa, in E. B. Yeager and A. J. Salkind, eds., *Techniques of Electrochemistry*, John Wiley & Sons, Inc., New York, 1978, Vol. **3**.
46. G. Halpert, in E. B. Yeager and A. J. Salkind, eds., *Techniques of Electrochemistry*, John Wiley & Sons, Inc., New York, 1978, Vol. **3**.

47. R. J. Brodd, *Proc. 5th Ann. Battery Conf.*, California State University, Long Beach, Calif., 1990.
48. R. J. Brodd, *Electrochem. Tech.* **6**, 289 (1968).
49. C. W. Shephard, *J. Electrochem. Soc.* **112**, 252 (1965).
50. C. Wagner, *J. Electrochem. Soc.* **98**, 116 (1951).
51. J. A. Klingert, S. Lynn, and C. W. Tobias, *Electrochim. Acta* **9**, 297 (1961).
52. J. Newman and C. W. Tobias, *J. Electrochem. Soc.* **109**, 1183 (1962).
53. E. A. Grens II and C. W. Tobias, *Ber. Bunsenges. Phys. Chem.* **68**, 236 (1964).
54. J. Newman and W. Tiedemann, *AIChE J.* **21**, 25 (1975).
55. W. H. Tiedemann and J. S. Newman, in S. Gross, ed., *Proc. Symp. Batt. Design Optim.* The Electrochemical Society, Pennington, N.J., Vol. 79-1, 1979, pp. 23–49.
56. W. H. Tiedemann and J. S. Newman, in K. Bullock and D. Pavlov, eds., *Proc. Symp. Adv. Lead–Acid Batt.* The Electrochemical Society, Pennington, N. J., Vol. 84–14, 1984, pp. 336–349.
57. Z. Mao and R. E. White, *J. Electrochem. Soc.* **139**, 1105 (1992).
58. M. Doyle and J. Newman, *J. Electrochem. Soc.* **143**, 1890 (1996).
59. W. Peukert, *Z. Electrotech.* **18**, 287 (1897).
60. R. Selim and P. Bro, *J. Electrochem. Soc.* **118**, 829 (1971).
61. C. M. Shephard, *J. Electrochem. Soc.* **112**, 657 (1965).
62. R. J. Brodd and R. M. Wilson, in G. W. Heise and N. C. Cahoon, eds., *The Primary Battery*, John Wiley & Sons, Inc., New York, 1976, Vol. **2**, pp. 369–428.
63. R. J. Brodd, in U. Landau, E. B. Yeager, and D. Kortan, eds., *Electrochemistry in Industry*, Plenum Press, New York, 1982, pp. 153–180.

RALPH BRODD
Broddarp of Nevada, Inc.

BATTERIES, PRIMARY CELLS

1. Introduction

Primary cells are galvanic cells designed to be discharged only once and attempts to recharge them can present possible safety hazards. The cells are designed to have the maximum possible energy in each cell size because of the single discharge. Thus, comparison between battery types is usually made on the basis of the energy density in $W \cdot h/cm^3$. The specific energy, $W \cdot h/kg$, is often used as a secondary criterion for primary cells especially when the application is weight sensitive as in space applications. The main categories of primary cells are carbon–zinc, known as heavy-duty and general purpose; alkaline, cylindrical and miniature; lithium; and reserve or specialty cells.

2. Carbon–Zinc Cells

Carbon–zinc batteries are the most commonly found primary cells worldwide and are produced in almost every country. Traditionally there is a carbon

[7440-44-0] rod, for cylindrical cells, or a carbon-coated plate, for flat cells, to collect the current at the cathode and a zinc [7440-66-6] anode. There are two basic versions of carbon–zinc cells: the Leclanché cell and the zinc chloride [7646-85-7], $ZnCl_2$, or heavy-duty cell. Both have zinc anodes, manganese dioxide [1313-13-9], MnO_2, cathodes, and include zinc chloride in the electrolyte. The Leclanché cell also has an electrolyte saturated with ammonium chloride [12125-02-9], NH_4Cl. Additional undissolved ammonium chloride is usually added to the cathode, whereas the zinc chloride cell has at most a small amount of ammonium chloride added to the electrolyte. Both types are dry cells in the sense that there is no excess liquid electrolyte in the system. The zinc chloride cell is often made using synthetic manganese dioxide and gives higher capacity than the Leclanché cell, which uses inexpensive natural manganese dioxide for the active cathode material. The MnO_2 is only a modest conductor. Thus the cathodes in both types of cell contain 10–30% carbon black in order to distribute the current. Because of the ease of manufacture and the long history of the cell, this battery system can be found in many sizes and shapes.

Leclanché cells are usually made using paste separators in which electrolyte solution and various types of cereal are cooked until thick. Some designs use a cold-set paste however. The paste is metered into the cell and the prepressed cathode body is inserted, forcing the paste into a separating layer. Figure **1a** shows a cutaway view of a typical pasted cylindrical cell. Paper-lined cells have a starch or modified starch-coated paper separator, which is much thinner and more conductive than the starch paste separator. This starch-coated paper separator is typically used in premium Leclanché cells and zinc chloride cells. In these cells, the separator is inserted into the zinc can and the cathode mix is extruded into the can, followed by insertion of the carbon rod into the cathode mix. Figure **1b** shows a cutaway of a typical paper-lined cell. Flat cells are used to stack up in multicell batteries, most commonly for 6- or 9-volt batteries. Figure 2 shows a cutaway of a typical flat cell, as used in a 9-volt battery. Note that in this cell the zinc plate is carbon-coated to serve as the collector for the cathode of the cell beneath as the cells are stacked.

The Leclanché cell was patented in 1866 (2). It consisted of an amalgamated zinc rod for the anode, a clay pot filled with a mix of manganese dioxide and carbon, and a carbon rod or plate driven into the mix as the current collector. The electrolyte was an aqueous solution of ammonium chloride that filled the glass container for the whole assembly. The cathode was open to the air allowing workers to replace the zinc rod whenever necessary yet continue to operate the cell. The MnO_2 was regarded as simply a depolarizer, ie, as acting to reduce the polarization of the cathode reaction that was thought to be the production of hydrogen. The term depolarizer is sometimes used in the battery field in this incorrect sense. In a later development (3) a zinc can was used as a container for the entire contents of the cell and thickeners were placed in the electrolyte, making the first version of the dry cell.

2.1. Cell Chemistry. Work on the mechanism of the carbon–zinc cell has been summarized (4), but the dynamics of this system are not entirely understood. The electrochemical behavior of electrolytic (EMD), chemical (CMD), and natural (NMD) manganese dioxide is slightly different. Battery-grade NMD is most commonly in the form of the mineral nsutite [12032-72-3], $H_2O \cdot xMnO_2$,

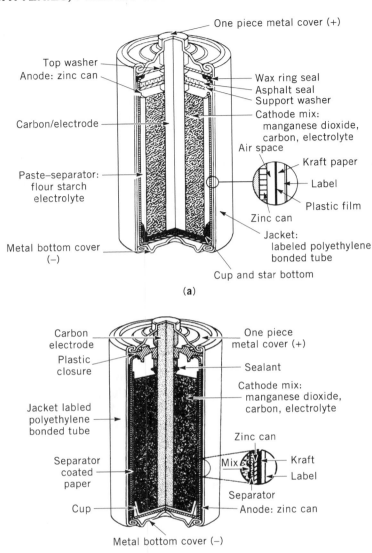

Fig. 1.   Cutaway view of (**a**) a "D"-size pasted Leclanché cell, and (**b**) a "D"-size paper-lined carbon-zinc cell (1). Courtesy of Eveready Battery Co.

which is a structural intergrowth of the minerals ramsdellite [12032-73-4], MnO_2, and pyrolusite [14854-26-3], MnO_2 (5). Occasionally, the pyrolusite or cryptomelane [12325-71-2], KMn_8O_{16}, form is used in batteries. The pure ramsdellite form is quite rare and expensive and is not used. The mineral as well as the synthetic types of MnO_2 have manganese oxidation states less than four and contain bound protons, generally considered to be present as hydroxyl ions, as well as internal and adsorbed water. NMD normally contains higher levels of impurities than the other types, so that the overall activity or theoretical capacity is

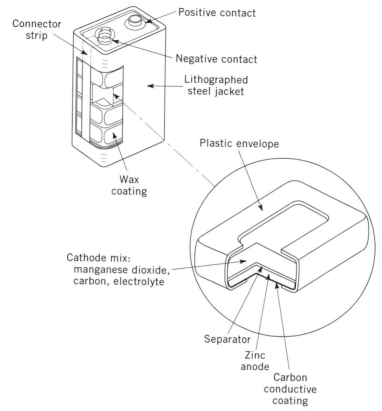

Fig. 2. Cutaway view of a Leclanché flat cell used in multicell batteries (1). Courtesy of Eveready Battery Co.

considerably lower. However, the cost of NMD is considerably less; the mineral is relatively abundant and no chemical or electrochemical processing is required. Thus NMD is the material of choice for low priced carbon–zinc cells. Premium cells are made with some proportion of the higher priced, higher activity CMD or EMD in the cathode. CMD, which is made by a rather complex chemical process, often has comparable activity to EMD on a weight basis and has a lower cost, but its high surface area and poor packing capability makes it less favorable in some applications. The most active form is EMD, which is also the most expensive. It is made by electrolytic deposition of MnO_2 from a bath of $MnSO_4$ in sulfuric acid (see MANGANESE COMPOUNDS).

The mechanism of the cathode reaction for all three types of MnO_2 can best be described by two approximately one-electron steps.

$$MnO_2 + H_3O^+ (\text{or } NH_4^+) + e^- \longrightarrow MnOOH + H_2O(\text{or } NH_3) \qquad (1)$$

and

$$MnOOH + H_3O^+ (\text{or } NH_4^+) + e^- \longrightarrow Mn(OH)_2 + H_2O(\text{or } NH_3) \qquad (2)$$

where NH^+_4 and NH_3 are appropriate to Leclanché cells and H_3O^+ and H_2O to zinc chloride cells. Equation 1 is believed to occur in a single overall phase with the gradual incorporation of protons and electrons. It appears that the end product of equation 1 is an intergrowth material in which the manganese ions occupy the same octahedral sites as in MnO_2 and the remaining octahedral sites are filled with protons, although the hexagonal close packed oxide ion sublattice is somewhat distorted. This kind of homogeneous reaction is similar to a reaction occurring in solution and the potential is related to the proton activity, expressed in mole fraction, within the MnO_2.

As the reduction of the cathode proceeds through equation 1 and the potential decreases, the reduction equation 2 becomes more favorable. First the divalent $Mn(OH)_2$ is solubilized to some extent in the mildly acidic electrolyte.

$$Mn(OH)_2 + 2\,H_3O^+ \longrightarrow Mn^{2+} + 4\,H_2O \tag{3}$$

The manganous ion [16397-91-4], Mn^{2+}, in solution then reacts with higher valent manganese oxide and zinc ions in solution to form a new phase called hetaerolite [12163-55-2], $ZnMn_2O_4$, or hydrohetaerolite, a poorly defined hydrated form.

$$MnO_2 + Mn^{2+} + Zn^{2+} + 2\,H_2O \longrightarrow ZnMn_2O_4 + 4\,H^+ \tag{4}$$

Hetaerolite is the most stable form of trivalent manganese at low cathode potentials and this sequence of reactions contributes to the familiar recovery of potential on open circuit that gives rise to higher cell capacity on intermittent discharge compared to continuous discharge. A second mode of potential recovery is the equilibration of the pH, resulting from the slow kinetics of zinc complex formation in solution and the dependence of the potential of the MnO_2 electrode on the pH.

In most cylindrical carbon–zinc cells, the zinc anode also serves as the container for the cell. The zinc can is made by drawing or extrusion. Mercury [7439-97-6] has traditionally been incorporated in the cell to improve the corrosion resistance of the anode, but the industry is in the process of removing this material because of environmental concerns. Corrosion prevention is especially important in cylindrical cells because of the tendency toward pitting of the zinc can which leads to perforation and electrolyte leakage. Other cell types, such as flat cells, do not suffer as much from this problem.

The anode reaction depends on the electrolyte used, but the charge-transfer step is

$$Zn \longrightarrow Zn^{2+} + 2e^- \tag{5}$$

In Leclanché cells, the high concentration of ammonium chloride leads to the formation of insoluble diammine zinc chloride through the reaction in the electrolyte.

$$Zn^{2+} + 2\,NH_3 + 2\,Cl^- \longrightarrow Zn(NH_3)_2Cl_2 \tag{6}$$

where the sources of the ammonia are the cathode reactions of equations 1 and 2. The initial form of the zinc ions is predominantly $ZnCl_4^{2-}$ because of the high concentration of ammonium chloride. As the reaction proceeds and NH_4Cl is consumed, however, lower chlorocomplexes, $ZnCl_3^-$, $ZnCl_2$, and $ZnCl^+$, are formed. The final stage in the anode reaction is the precipitation of basic zinc chloride.

$$5\,Zn^{2+} + 2\,Cl^- + 17\,H_2O \longrightarrow ZnCl_2 \cdot 4Zn(OH)_2 \cdot H_2O + 8\,H_3O^+ \qquad (7)$$

In the zinc chloride cell, precipitated basic zinc chloride is the primary anode product because of the low concentration of ammonium chloride in the cell. Water and zinc chloride are consumed in equations 1 and 7 and must be provided in adequate amounts for the cell to discharge efficiently. Usually more carbon is used in zinc chloride cells than in Leclanché cells in order to increase the electrolyte absorptivity of the cathode and thus allow the use of a larger volume of electrolyte. Also, the use of a thin paper separator, which decreases internal resistance, allows less space for water storage than the thick, pasted separator construction traditionally used in Leclanché cells.

Corrosion and other unwanted reactions cause difficulties in the design of carbon–zinc cells. Fortunately, the overpotential for hydrogen evolution on zinc is quite high in both mildly acidic and alkaline solutions. In some cases, basic ZnO is added to zinc chloride electrolyte to raise the pH and lower the corrosion rate. Good control of heavy metal impurities in the cathode is also important, to avoid these metals being solubilized by the electrolyte and subsequently deposited on the zinc to become low overpotential sites for hydrogen evolution. Another type of corrosion is the direct reaction of oxygen from air entering the cell. To protect against this loss, the cell is sealed so that access of air is restricted. The cell must not be hermetically sealed, however, as hydrogen and other gases, mainly carbon dioxide, must be able to escape from the cell as they are formed. The source of CO_2 is the cathode and results from the excellent oxidative properties of MnO_2. Thus, if a starch paste separator is used, some glucose becomes available from starch hydrolysis and readily reacts with MnO_2. The carbon used as conductor in the cathode also reacts with MnO_2, but more slowly. The higher the voltage the more easily the MnO_2 reacts with these materials. Thus premium cells have special separator materials such as methyl cellulose-coated paper to minimize this reaction.

The porous carbon rod is often the main pathway of escape for the gases formed in the cell. This pathway also allows ingress of oxygen to the cell limiting the shelf life of the system. The use of shrink tube outer wrapping and other devices have, however, improved the leakage property dramatically over prior generation cells.

2.2. Performance. Carbon–zinc cells perform best under intermittent use and many standardized tests have been devised that are appropriate to such applications as light and heavy flashlight usage, radios, cassettes, and motors (toys). The most frequently used tests are American National Standards Institute (ANSI) tests (6). The tests are carried out at constant resistance and the results reported in minutes or hours of service. Figure 3 shows typical results under a light load for different size cells, whereas Figure 4 shows results for different types of R20 "D"-size cells under a heavy intermittent load.

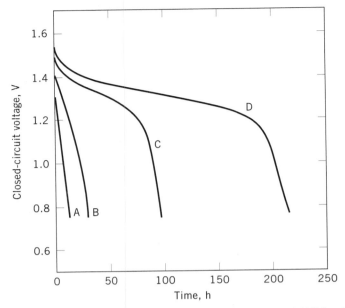

Fig. 3.   Hours of service on 40-Ω discharge for 4 h/d radio test at 21°C for A, RO3 "AAA"; B, R6 "AA"; C, R14 "C"; and D, R20 "D" paper-lined, heavy-duty zinc chloride cells.

To compare one battery with another, it is useful to compute the energy density from these data. Because the voltage declines with capacity, the average voltage during the discharge is used to compute an average current which is then multiplied by the service in hours to give the ampere-hours of capacity. Watt-hours of energy can be obtained by multiplying again by the average voltage.

The voltage at the end of discharge is also important in defining these quantities because of the sloping discharge curve. Typical computed values for R20

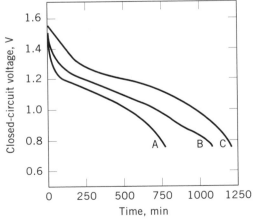

Fig. 4.   Hours of service on 4-Ω discharge, 15 min/h, 8 h/d, LIF test (see Table 1) at 21°C for A, general purpose (pasted Leclanché) "D"-size; B, premium Leclanché (paper-lined Leclanché); and C, heavy-duty (paper-lined ZnCl$_2$) cells.

Table 1. **Specific Energies and Energy Densities of Carbon–Zinc Cells**[a]

Battery designation and type	Test, Ω[b,c]	W·h/mL	W·h/kg
	R20 "D"-size cells		
general purpose Leclanché	40 Ω radio	0.12	73
	LIF	0.07	41
heavy-duty zinc chloride	40 Ω radio	0.16	100
	LIF	0.15	90
	R6 "AA"-size zinc chloride cells		
general purpose	75 Ω radio	0.13	72
	LIF	0.10	51
heavy-duty	75 Ω radio	0.17	81
	LIF	0.15	68

[a] Calculations are based on data from reference 7.
[b] The radio test is one 4-h continuous discharge daily.
[c] Each discharge performed at 2.25 Ω on the light industrial flashlight (LIF) test: one 4-min discharge at 1-h intervals for eight consecutive hours each day, with 16-h rest periods; ie, 32 min of discharge per day.

"D" and R6 "AA" cells of the Leclanché and zinc chloride battery types are given in Table 1. For heavy duty cells, the specific energies (W·h/kg) are higher for the larger R20 cells than for the R6 cells, whereas the energy densities (W·h/mL) are higher for the R6 cells. This is because the can weight makes a much greater contribution to the total weight of the system for small cells. Thus the smaller cathode of the R6 cell allows a more efficient design of the cell. These relationships are not exactly preserved in the data for the general purpose cells in Table 1, because the R6 cell used in this test is made from the more efficient zinc chloride cell design.

The effect of the discharge rate is especially pronounced for the general purpose cells. On intermittent tests, the heavy-duty cell operates at high efficiency even at high rate. On continuous test at high rate, heavy-duty cells provide 60–70% of the intermittent service, whereas general purpose cells give only 30–50% of the intermittent service values.

3. Cylindrical Alkaline Cells

Primary alkaline cells use sodium hydroxide [1310-73-2] or potassium hydroxide [1310-58-3] as the electrolyte. They can be made using a variety of chemistries and physical constructions. Early alkaline cells were of the wet cell type, but the alkaline cells of the 1990s are mostly of the limited electrolyte, dry cell type. Most primary alkaline cells are made using zinc as the anode material; a variety of cathode materials can be used. Primary alkaline cells are commonly divided into two classes, based on type of construction: the larger, cylindrically shaped batteries, and the miniature, button-type cells. Cylindrical alkaline batteries are mainly produced using zinc–manganese dioxide chemistry, although some cylindrical zinc–mercury oxide cells are made.

Cylindrical alkaline cells are zinc–manganese dioxide cells having an alkaline electrolyte, which are constructed in the standard cylindrical sizes, R20 "D",

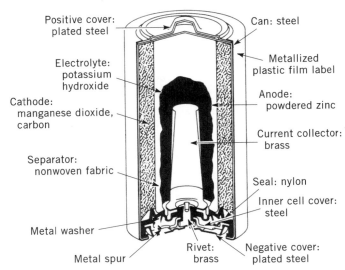

Fig. 5. Cutaway view of typical R20 or "D"-size alkaline manganese battery showing components and the corresponding materials of construction (8). Courtesy of Eveready Battery Co.

R14 "C", R6 "AA", RO3 "AAA", as well as a few other less common sizes. They can be used in the same types of devices as ordinary Leclanché and zinc chloride cells. Moreover, the high level of performance makes them ideally suited for applications such as toys, audio devices, and cameras.

Figure 5 shows a cross-sectional diagram of a typical R20 size cylindrical alkaline cell. The battery is contained in a steel can, which also serves as the current collector for the cathode. Inside the can is a dense compacted cathode mass containing manganese dioxide, carbon, and sometimes a binder. The cathode has a hollow center lined with a separator to isolate the cathode from the anode. Within the separator-lined cavity is placed the anode mix consisting of zinc powder and alkaline electrolyte, together with a small amount of gelling material to immobilize the electrolyte and suspend the zinc powder. Contact to the anode is made by a metal pin or leaf inserted into the anode mix. The cell has a plastic seal assembly, to keep electrolyte from leaking out of the cell and to keep air from getting in. Contact to the anode collector is made through this seal. The seal contains a fail-safe vent that is activated when the battery internal pressure exceeds a certain level.

Cells using alkaline electrolytes have been known since the early days of batteries. Early alkaline cells included the zinc–copper oxide cell in the 1880s (9), the rechargeable iron–nickel oxide cells (10,11), and an early version of the zinc–manganese dioxide alkaline cell (12). Commercial zinc–mercuric oxide batteries were developed in the 1940s (13). Work on the use of the zinc anode in alkaline cells was very valuable to the subsequent development of the cylindrical alkaline general purpose cell in the late 1950s as was research into the use of manganese dioxide as a cathode material in small alkaline cells

(14). Commercial "B" batteries for portable radios came out of those investigations. "B" batteries contained a large number of individual cells connected in series to produce a high voltage but only a relatively low current. In the late 1950s the first multipurpose, cylindrical, zinc–manganese dioxide batteries were commercialized (15). Such batteries having high output capacity and high current carrying ability are now made by many manufacturers throughout the world. There is ongoing competition among manufacturers to improve the performance of cylindrical alkaline batteries.

3.1. Chemistry.   The alkaline cell derives its power from the reduction of the manganese dioxide cathode and the oxidation of the zinc anode. The reactions

$$Zn + 2\,OH^- \longrightarrow ZnO + H_2O + 2e^-\,\text{anode} \tag{8}$$

$$2e^- + 2\,MnO_2 + 2\,H_2O \longrightarrow 2\,MnOOH + 2\,OH^-\,\text{cathode} \tag{9}$$

give, as the overall reaction,

$$Zn + 2\,MnO_2 + H_2O \longrightarrow ZnO + 2\,MnOOH \tag{10}$$

Both the anode and cathode equations are more complicated than is indicated by the equations shown.

3.2. Anode.   *Anode Corrosion Reaction.*   Zinc might at first appear to be an unusual choice for battery anode material, because the metal is thermodynamically unstable in contact with water

$$Zn + H_2O \longrightarrow ZnO + H_2 \quad \Delta G = -81.142 \text{ kJ/mol}(-19.39 \text{ kcal/mol}) \tag{11}$$

However, the reaction with water can be made to be extremely slow. Because the alkaline electrolyte is corrosive toward human tissue as well as toward the materials in devices, it is more important to have a good seal toward preventing electrolyte leakage in an alkaline battery than in a carbon–zinc cell. The formation of a good seal is, however, incompatible with the formation of a noncondensable gas like hydrogen.

The reaction of zinc and water is not a simple homogeneous one. Rather it is a heterogeneous electrochemical reaction, involving a mechanism similar to that of a battery. There are two steps to the reaction: zinc dissolves at some locations as shown in equation 8 while hydrogen gas is generated at other sites.

$$2\,e^- + 2\,H_2O \longrightarrow H_2 + 2\,OH^-\,\text{cathode} \tag{12}$$

These two reactions then add up to the overall zinc corrosion reaction (eq. 11).

In a battery, the anode and cathode reactions occur in different compartments, kept apart by a separator that allows only ionic, not electronic conduction. The only way for the cell reactions to occur is to run the electrons through an external circuit so that electrons travel from the anode to the cathode. But in the corrosion reaction the anode and cathode reactions, equations 8 and 12 respectively, occur at different locations within the anode. Because the anode is a single, electrically conductive mass, the electrons produced in the anode reaction travel easily to the site of the cathode reaction and the zinc acts like a battery where the positive and negative terminals are shorted together.

The rate at which the corrosion of the zinc proceeds depends on the rates of the two half reactions (eqs. 8 and 12). Equation 8, a necessary part of the desired battery reaction, fortunately represents a reaction that proceeds rather rapidly, whereas the reaction represented by equation 12 is slow. Ie, the generation of hydrogen on pure zinc is a sluggish reaction and thus limits the overall corrosion reaction rate.

Certain impurities on zinc can act as catalysts for the generation of hydrogen, thereby greatly increasing the corrosion rates. For this reason, zinc in alkaline cells must be of high purity, and careful control exercised over the level of the harmful impurities. Moreover, other components of the cell must not contain harmful levels of these impurities that might dissolve and migrate to the zinc anode.

Alternatively, there are also inhibitors that decrease the rate of hydrogen generation and thus decrease corrosion. Mercury, effective at inhibiting zinc corrosion, has long been used as an additive to zinc anodes. More recently, however, because of increased interest in environmental issues, the amount of mercury in alkaline cells has been reduced.

3.3. Zinc Oxidation Mechanism.   The oxidation reaction for the zinc anode (eq. 8) takes place in several steps (16, 17), ultimately resulting in the zincate ions [16408-25-6], $Zn(OH)_4^{2-}$, that dissolves in the electrolyte.

$$Zn + 4\,OH^- \longrightarrow Zn(OH)_4^{2-} + 2\,e^- \tag{13}$$

As a battery discharges the concentration of zincate in the electrolyte increases. Eventually, the concentration exceeds the solubility of zinc oxide [1314-13-2], ZnO, which can precipitate from the solution

$$Zn(OH)_4^{2-} \longrightarrow ZnO + H_2O + 2\,OH^- \tag{14}$$

This precipitation can be sluggish and in some cases the zincate concentration can increase to three or four times the equilibrium solubility value, after which precipitation of zinc hydroxide [20427-58-1] can occur.

$$Zn(OH)_4^{2-} \longrightarrow Zn(OH)_2 + 2\,OH^- \tag{15}$$

Dehydration of zinc hydroxide can then lead to formation of the thermodynamically more stable zinc oxide.

$$Zn(OH)_2 \longrightarrow ZnO + H_2O \tag{16}$$

In a cylindrical alkaline cell, the amount of electrolyte is limited and the electrolyte becomes saturated with zincate rather early in the discharge. Thus the cell produces the zinc oxide reaction product through most of its discharge.

Cathode Reaction.   There are many different types of manganese dioxide (18), having varying activity in batteries. The only type suitable for alkaline batteries is γ-MnO_2, the mineral form of which is nsutite. The chemical composition of γ-MnO$_2$ has been described (19) by the general formula $(MnO_2)_{2n}-3 \cdot (MnOOH)_{4-2n} \cdot mH_2O$, where the value of n ranges from 1.5 to 2. γ-MnO$_2$ is useful in batteries because of the unusual nature of its discharge reaction. When γ-MnO$_2$ is discharged the reaction product α-MnOOH (groutite structure) is formed and this material, when formed during discharge, has a structure so similar to that of the starting MnO$_2$ that the two substances can form a solid solution in which equilibrium is rapidly achieved. Thus when MnOOH is formed at the surface of MnO$_2$, there is a rapid diffusion of protons from the surface MnOOH into the interior, coupled with electron transfer; ie, the result is that the lower valent material initially formed on the surface effectively diffuses into the solid particle. In this way the entire particle is homogeneously reduced and the surface tends to have the same composition as the overall particle (20). The cathode material starts as nearly all manganese(IV) and ends as manganese(III).

The homogeneous phase reduction of γ-MnO$_2$ results in a discharge curve that differs from that of many other battery systems. In any battery system there can be a voltage drop from the open circuit value resulting from polarization effects. But for a typical heterogeneous (two-phase) reaction as exists in lead–acid batteries, Zn–HgO batteries, Zn–Ag$_2$O batteries, nickel cadmium batteries, etc, the open circuit recovery voltage is substantially constant throughout the discharge. In the homogeneous phase reduction of γ-MnO$_2$, however, the open circuit voltage itself changes as the cathode is reduced (Fig. 6). Thus the operating voltage of the alkaline zinc–manganese dioxide battery starts around 1.5 volts, and drops to below 1 volt by the time a 1-electron reduction of the MnO$_2$ is complete. At this point, when the cathode material approaches 100% manganese(III), the solubility of the manganese increases and this soluble manganese then combines with the zincate present in the electrolyte to form the insoluble

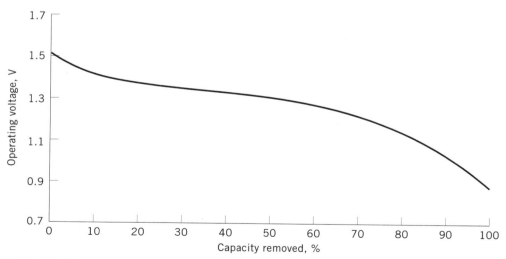

Fig. 6. Alkaline–manganese battery operating voltage as a function of remaining capacity (21). Courtesy of Eveready Battery Co.

Table 2. **Characteristics of Aqueous Primary Batteries**

Parameter	Carbon–zinc (Zn/MnO_2)	Alkaline manganese dioxide ($Zn–MnO_2$)	Mercuric oxide ($Zn–HgO$)	Silver oxide ($Zn–Ag_2O$)	Zinc–air ($Zn–O_2$)
nominal voltage, V	1.5	1.5	1.35	1.5	1.25
working voltage, V	1.2	1.2	1.3	1.55	1.25
specific energy, $W \cdot h/kg$	40–100	80–95	100	130	230–400
energy density, $W \cdot h/mL$	0.07–0.17	0.15–0.25	0.40–0.60	0.49–0.52	0.70–0.80
temperature range, °C					
storage	−40–50	−40–50	−40–60	−40–60	−40–50
operating	−5–55	−20–55	−10–55	−10–55	−10–55

hetaerolite ($ZnMn_2O_4$). In the process, water is released.

$$2\,MnOOH + Zn(OH)_4^{2-} \longrightarrow ZnMn_2O_4 + 2\,H_2O + 2\,OH^- \qquad (17)$$

For heavy drain discharges of alkaline cells, there is no useful capacity after this point because the rate of discharge of the $ZnMn_2O_4$ is quite slow. But for lighter drain discharges, further reduction of the cathode is possible. The reaction mechanisms are not entirely clear, but there is some evidence for the formation of a final reaction product resembling hausmannite [1309-55-3], Mn_3O_4.

3.4. Performance. Alkaline manganese-dioxide batteries have relatively high energy density, as can be seen from Table 2. This results in part from the use of highly pure materials, formed into electrodes of near optimum density. Moreover, the cells are able to function well with a rather small amount of electrolyte. The result is a cell having relatively high capacity at a fairly reasonable cost.

The performance of the cells is influenced not only by the relatively high theoretical capacity, but also by the fact that the cells can provide good efficiency at various currents over a wide variety of conditions. The high conductivity and low polarization of the alkaline electrolyte, combined with the good electronic and ionic conductivity in the electrodes, lead to high performance even under heavy drain conditions. Alkaline zinc–manganese dioxide batteries show less variation in output capacity with variation in discharge rate than do Leclanché or zinc chloride batteries. Thus alkaline cells have a moderate advantage over Leclanché cells in capacity on light drains (Fig. 7**a**) and they have a much greater advantage on heavy drains (Fig. 7**b**).

Whereas Leclanché and zinc chloride batteries perform much better on intermittent high rate discharge than on continuous high rate discharge, alkaline zinc–manganese dioxide cells do not show this effect to any substantial degree. The output of these alkaline cells at a given drain rate is similar regardless of whether the discharge is continuous or intermittent. Thus alkaline cells

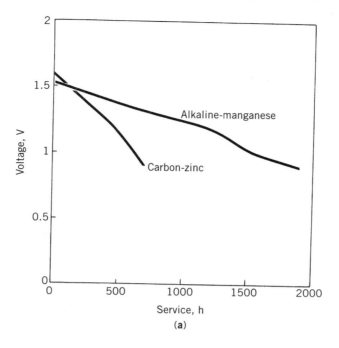

(a)

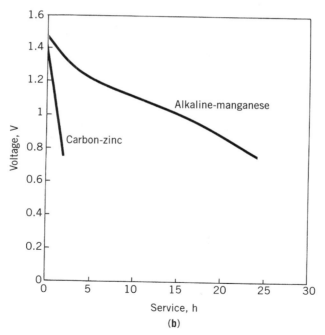

(b)

Fig. 7. Performance comparison of "D"-size alkaline–manganese vs carbon–zinc batteries at 21°C on (**a**) a light drain 150-Ω continuous test at 21°C, and (**b**) a heavy drain 2.2-Ω continuous test (8). Courtesy of Eveready Battery Co.

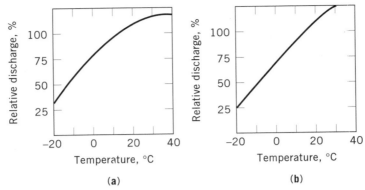

Fig. 8. Effect of temperature on relative discharge performance of a fresh "D"-size battery for service on simulated ratio use, 25-Ω 4-h/d test for (**a**) an alkaline–manganese battery undergoing 260 h of service, and (**b**) a carbon–zinc battery undergoing 70 h of service (22). Courtesy of Eveready Battery Co.

have an advantage over Leclanché cells on high rate intermittent tests, but have an even greater advantage on high rate continuous tests.

Batteries tend to perform more poorly as the operating temperature is decreased because of decreased conductivity of the electrolyte and slower electrode kinetics. Ultimately, freezing of the electrolyte occurs and the battery fails. Batteries tend to perform better at higher temperatures only up to the point that loss of performance occurs because of cell venting and drying out or parasitic reactions in the cell. Overall, alkaline cells have less performance loss at low and high temperatures than do Leclanché cells as can be seen in Figure 8.

The ability of a cell to retain capacity after long-term storage is another important aspect of performance. Cells can lose capacity during storage because of various parasitic reactions, such as zinc corrosion. Also, if the cell is not tightly

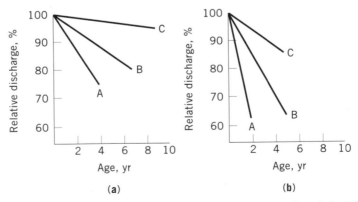

Fig. 9. Effect of time and storage temperatures, A, 40°C; B, 20°C; and C, 0°C, on relative discharge performance of fresh and aged "D"-size cells on simulated radio use, 25-Ω 4-h/d test for (**a**) alkaline–manganese, and (**b**) carbon–zinc batteries (22). Courtesy of Eveready Battery Co.

sealed, there can be transfer of water vapor into or out of the cell, transfer of carbon dioxide into the cell, etc. Alkaline zinc–manganese dioxide cells have good capacity retention on long-term storage. The use of high purity materials ensures very low rates of parasitic reactions in the cells. Moreover, the alkaline cells are well sealed so that there is minimal vapor transmission into or out of the cell. Figure 9 shows the capacity retention of alkaline cells and that of Leclanché cells after prolonged storage at different temperatures.

4. Miniature Alkaline Cells

Miniature alkaline cells are small, button-shaped cells, which use alkaline NaOH or KOH electrolyte, generally have zinc anodes but may have a variety of cathode materials. They are used in watches, calculators, cameras, hearing aids, and other miniature devices. Figure 10 shows the construction of a typical miniature alkaline cell. The cathode mix is placed into the can, followed by the insertion of a separator layer. The type of separator materials used and the number of layers of material vary according to the type of cathode used in the cell. The anode mix is then placed in contact with the anode cup. An insulating, sealing gasket electrically separates the anode cup from the can and prevents leakage of electrolyte from the cell. The miniature cells have a similar basic construction, but differ in cathode materials, separator materials, electrolyte, etc.

Cylindrical alkaline cells are made in only a few standard sizes and have only one important chemistry. In contrast, miniature alkaline cells are made in a large number of different sizes, using many different chemical systems. Whereas the cylindrical alkaline batteries are multipurpose batteries, used for a wide variety of devices under a variety of discharge conditions, miniature alkaline batteries are highly specialized, with the cathode material, separator type, and electrolyte all chosen to match the particular application.

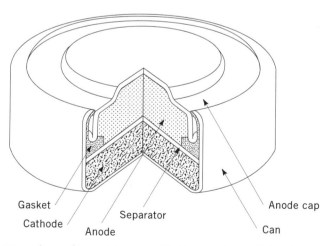

Fig. 10. Cutaway view of a miniature alkaline battery (21). Courtesy of Eveready Battery Co.

4.1. Zinc–Mercuric Oxide Batteries.   Miniature zinc—mercuric oxide batteries have a zinc anode and a cathode containing mercuric oxide [21908-53-2], HgO. The cathode reaction

$$HgO + H_2O + 2e^- \longrightarrow Hg + 2\,OH^- \tag{18}$$

gives, when combined with the anode reaction (eq. 8), the overall reaction

$$Zn + HgO \longrightarrow ZnO + Hg \tag{19}$$

Water is not used in the reaction. Therefore, these cells have a very high capacity, exceeding that of zinc–manganese dioxide batteries (Table 2).

The cathode reaction is rather simple: mercuric oxide is reduced to mercury metal which is a liquid and does not block off the surface of the cathode. Therefore, the reaction is a normal heterogeneous two-phase reaction, producing a substantially constant voltage during discharge, in contrast to the sloping voltage of a cell containing a manganese dioxide cathode. Thus the zinc–mercuric oxide cell has high capacity and a flat discharge curve as shown in Figure 11. Thick separators are used to keep the liquid mercury from bridging to the anode and short circuiting the cell.

Some cells are made using small amounts of manganese dioxide added to the cathode. These cells have somewhat more sloping discharge curves (Fig. 11) and are used in some devices which do not require as constant a voltage. The added manganese dioxide provides for improved high rate capability for the cells. In addition, such cells have a slightly more gradual voltage drop at the end of battery life, giving the consumer some advance warning that the battery should be replaced.

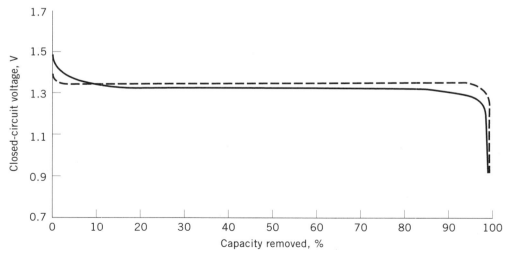

Fig. 11.   Typical discharge curve comparison for zinc–mercuric oxide batteries: (– – –) model 325, HgO, and (——) model E312E, HgO–MnO$_2$, cathodes (21).Courtesy of Eveready Battery Co.

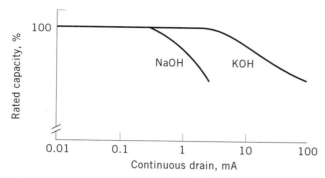

Fig. 12. Comparison of battery efficiency for miniature zinc–mercuric oxide cells containing KOH or NaOH electrolyte (21).Courtesy of Eveready Battery Co.

Miniature zinc–mercuric oxide batteries may be made with either KOH or NaOH as the electrolyte. Cells having KOH operate more efficiently than those having NaOH at high current drains (Fig. 12) because of the higher conductivity of KOH. On the other hand, batteries with KOH are more difficult to seal, cells with NaOH are more resistant to leakage.

Miniature zinc–mercuric oxide batteries function efficiently over a wide range of temperatures. The maximum useful temperature is about 55°C, and the minimum is −28°C for batteries having KOH electrolyte, or −10°C for batteries having NaOH electrolyte. Figure 13 shows the effect of temperature on efficiency of discharge for a typical zinc–mercuric oxide cell. Miniature zinc–mercuric oxide batteries also have good storage life. As shown in Figure 14, they maintain 90% of capacity even after storage at 20°C for five years.

Although the zinc–mercuric oxide battery has many excellent qualities, increasing environmental concerns has led to a de-emphasis in the use of this system. The main environmental difficulty is in the disposal of the cell. Both the mercuric oxide in the fresh cell and the mercury reduction product in the used cell have long-term toxic effects.

4.2. Zinc–Silver Oxide Batteries. Miniature zinc–silver oxide batteries have a zinc anode, and a cathode containing silver oxide [20667-12-3], Ag_2O. The cathode reaction

$$Ag_2O + H_2O + 2\,e^- \longrightarrow 2\,Ag + 2\,OH^- \tag{20}$$

gives, when combined with the anode reaction (eq. 8), the overall reaction

$$Zn + Ag_2O \longrightarrow ZnO + 2\,Ag \tag{21}$$

The overall reaction neither consumes nor produces water. As in the case of mercury cells these cells have very high capacity.

The cathode reaction involves reduction of silver oxide to metallic silver [7440-22-4]. The reaction is a two-phase, heterogeneous reaction producing a substantially constant voltage during discharge. Some manganese dioxide may be added to the cathode, as in the case of mercury oxide cells.

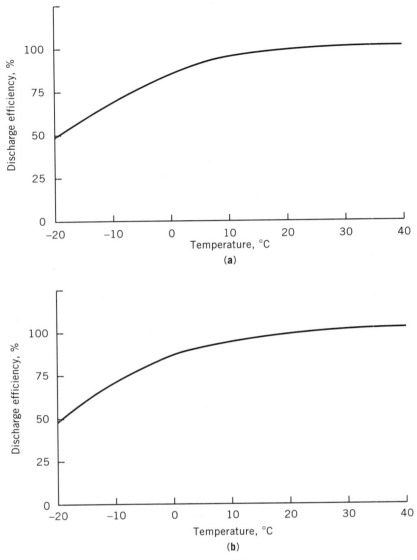

Fig. 13. Effect of temperature on discharge efficiency (**a**) at 270 mA·h of miniature zinc–mercuric oxide batteries type EP675E, and (**b**) at 175 mA·h of miniature zinc–silver oxide batteries type 357 (21). Courtesy of Eveready Battery Co.

Unlike some other cathode materials, such as manganese dioxide, which are quite insoluble, silver oxide has a fair degree of solubility in alkaline electrolyte. If the soluble silver species were allowed to be transported to the zinc anode it would react directly with the zinc, and as a result the cell would self-discharge. In order to prevent this from happening, zinc–silver oxide cells use special separator materials such as cellophane [9005-81-6], that are designed to inhibit migration of soluble silver to the anode.

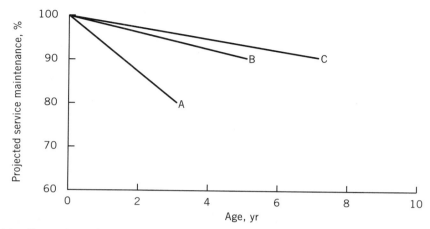

Fig. 14. Retention of discharge capacity of miniature zinc–mercuric oxide batteries after storage at temperatures of A, 40°C; B, 20°C; and C, 0°C (21). Courtesy of Eveready Battery Co.

Zinc–silver oxide cells have high energy density, nearly as high as for mercury cells. They operate at higher voltages than mercury cells, 1.5 vs 1.3 V, but for somewhat less time as shown in Figure 15. Miniature zinc–silver oxide batteries may be made using either KOH or NaOH electrolyte. Again, KOH-containing cells operate more efficiently than NaOH ones at high current drains (Fig. 16), whereas batteries having NaOH are easier to seal and are more resistant to leakage.

Miniature zinc–silver oxide batteries are commonly used in electronic watches and in other applications where high energy density, a flat discharge profile, and a higher operating voltage than that of a mercury cell are needed. These batteries function efficiently over a wide range of temperatures and are comparable to mercury batteries in this respect (Fig. 13**b**). Miniature zinc–silver

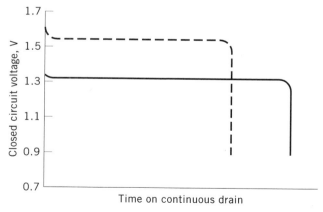

Fig. 15. Relative discharge curves for (---) zinc–silver oxide, and (—) zinc–mercuric oxide batteries. Cells are of equal volume (21). Courtesy of Eveready Battery Co.

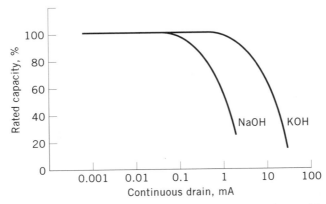

Fig. 16. Comparison of battery efficiency for miniature zinc–silver oxide cells containing KOH or NaOH electrolyte (21). Courtesy of Eveready Battery Co.

oxide batteries have good storage life as shown in Figure 17. The use of barrier separators prevents migration of soluble silver so that self-discharge is kept quite low, and the rate of service loss is comparable to that of mercury cells for storage at 20°C or at 0°C. Silver cells stored at elevated temperatures lose capacity at a slightly higher rate than do mercury cells.

4.3. Divalent Silver Oxide Batteries. It is possible to produce a silver oxide in which the silver has a higher oxidation state, approaching a composition of AgO. This material can provide both higher capacity and higher energy density than Ag_2O. Alternatively, a battery can be made with the same capacity as a monovalent silver cell, but with cost savings. However, some difficulties with regard to material stability and voltage regulation must be addressed.

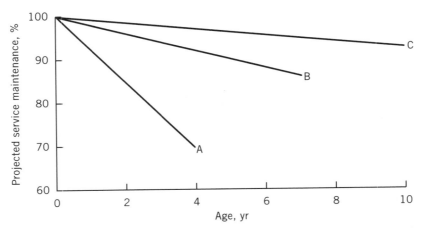

Fig. 17. Retention of discharge capacity of miniature zinc–silver oxide batteries after storage at temperatures of A, 40°C; B, 20°C; and C, 0°C (21). Courtesy of Eveready Battery Co.

The cathode reaction for divalent silver oxide [1301-96-8], AgO

$$AgO + H_2O + 2e^- \longrightarrow Ag + 2OH^- \tag{22}$$

gives the overall reaction when combined with equation 8

$$Zn + AgO \longrightarrow ZnO + Ag \tag{23}$$

The reaction of divalent silver oxide would normally tend to occur in two steps. In the first step divalent silver oxide would be reduced to monovalent silver oxide. Then, in the second step, the monovalent silver oxide would be further reduced to silver metal. These two reactions occur at different voltages so the battery would normally operate at a higher voltage during the first part of the discharge than during the latter part. Such discharge behavior, however, is unacceptable for many devices using the battery and it is desirable to make the cell discharge at a constant voltage. The material at the cathode reaction surface is caused to be all Ag_2O, and the interior AgO is protected from contact with the electrolyte. The voltage level of the monovalent silver oxide is thus obtained, but with the enhanced coulombic capacity of the divalent material. Several schemes have been developed to accomplish this behavior. Construction features of divalent silver oxide batteries are similar to those of monovalent silver oxide batteries.

4.4. Zinc–Manganese Dioxide Batteries. The combination of a zinc anode and manganese dioxide cathode, which is the dominant chemistry in large cylindrical alkaline cells, is used in some miniature alkaline cells as well. Overall, this type of cell does not account for a large share of the miniature cell market. It is used in cases where an economical power source is wanted and where the devices can tolerate the sloping discharge curve shown in Figure 18.

The chemistry is the same as for alkaline manganese–dioxide batteries. The construction features are typical of the other miniature alkaline batteries.

4.5. Zinc–Air Batteries. Zinc–air batteries offer the possibility of obtaining extremely high energy densities. Instead of having a cathode material placed in the battery when manufactured, oxygen from the atmosphere is used as cathode material, allowing for a much more efficient design. The construction of

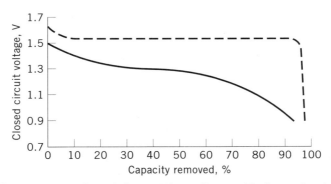

Fig. 18. Discharge curves for miniature zinc–silver oxide batteries (- - -), and zinc–manganese dioxide batteries (—) (21). Courtesy of Eveready Battery Co.

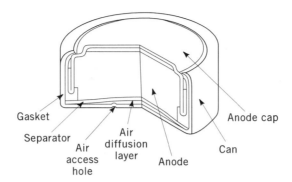

Fig. 19. Cutaway view of a miniature air cell battery (21). Courtesy of Eveready Battery Co.

a miniature air cell is shown in Figure 19. From the outside, the cell looks like any other miniature cell, except for the air access holes in the can. On the inside, however, the anode occupies much more of the internal volume of the cell. Rather than the thick cathode pellet, there is a thin layer containing the cathode catalyst and air distribution passages. Air enters the cell through the holes in the can and the oxygen reacts at the surface of the cathode catalyst. The air access holes are often covered with a protective tape, which is removed when the cell is placed in service.

The cathode reaction is

$$O_2 + 2\,H_2O + 4e^- \longrightarrow 4\,OH^- \tag{24}$$

and the overall cell reaction, after combining with equation 8, is

$$2\,Zn + O_2 \longrightarrow 2\,ZnO \tag{25}$$

The oxygen reaction is quite complex. Complete reduction from oxygen gas to hydroxide ion involves four electrons and requires several steps. Initially, oxygen is reduced to peroxyl ion [14691-59-9]

$$O_2 + H_2O + 2\,e^- \longrightarrow HO_2^- + OH^- \tag{26}$$

The peroxyl may be further reduced:

$$HO_2^- + H_2O + 2e^- \longrightarrow 3\,OH^- \tag{27}$$

or it may be catalytically decomposed, producing hydroxide and oxygen.

$$2\,HO_2^- \longrightarrow 2\,OH^- + O_2 \tag{28}$$

In a typical miniature air cell, the reaction proceeds according to equations 26 and 28. The performance of the air cathode cell depends largely on the ability to catalyze the reaction of equation 28. If that reaction is too slow, then during discharge large amounts of peroxide builds up in the cathode, resulting in large

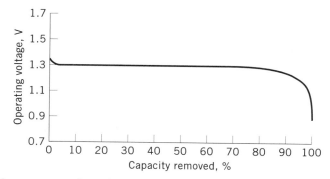

Fig. 20. Discharge curve for miniature zinc–air battery (21). Courtesy of Eveready Battery Co.

polarization, and low operating voltage. If the reaction is well catalyzed, the battery can operate at a higher, more useful voltage. The miniature air cell cathode typically contains carbon to provide a surface for the initial reduction of oxygen (eq. 26). Some types of carbon are also particularly effective at catalyzing the peroxyl decomposition and are used in air cells. Alternatively, small amounts of metal oxides can be used as catalysts.

Figure 20 shows a typical discharge curve for a miniature air cell which discharges at 1.2–1.3 V, with a flat discharge profile. Its discharge voltage is slightly lower than that of a mercury cell, and considerably lower than that of a silver cell, but its capacity is far greater than for any of the other miniature cells, including lithium cells (23).

The performance level of air cells is exceptional, but these are not general purpose cells. They must be used in applications where the usage is largely continuous, and where the discharge level is relatively constant and well-defined. The reason for these limitations lies in the fact that the cell must be open to the atmosphere and the holes that allow oxygen into the cell also allow other gases to enter or leave the cell. Carbon dioxide from the air can enter an unsealed cell and react with the electrolyte to form potassium carbonate which results in decreased cell performance. Water vapor can enter the cell if the outside humidity is high or leave the cell if the outside humidity is low. In the former case, the cell fills up with water and eventually bulges or leaks. In the latter case, the cell dries out and can no longer function.

The relationship between the rate of water-vapor diffusion into or out of the cell and the rate of oxygen diffusion into the cell has been studied (24). If the cell openings are large then larger amounts of oxygen diffuse into the cell, and the cell maintains relatively high discharge rates. But the same large openings also allow more rapid diffusion of water vapor and carbon dioxide, so that the useful life of the cell is shortened. Alternatively, if the cell openings are small, the diffusion of water into or out of the cell, and the diffusion of carbon dioxide into the cell, is slower, and the cell operates for a longer period of time. However, the smaller holes limit oxygen diffusion so that the cell is limited to lower currents. Therefore, an air cell must be carefully designed for a particular use such that the air access passages are just big enough to handle the required

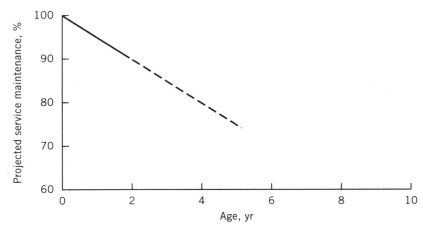

Fig. 21. Retention of discharge capacity of miniature zinc–air battery having an unopened sealed cell after storage at 20°C (- - -) projected data (21). Courtesy of Eveready Battery Co.

oxygen flux. The cell must be used essentially continuously after activation, and disposed of promptly after discharge. Miniature air cells are mainly used in hearing aids, where they are required to produce a relatively high current for a relatively short time period such as a few weeks. In this application they provide exceptional performance compared to other batteries.

Air cells are packaged with sealing tape over the air holes or in sealed containers. In the sealed state, they maintain their capacity well during storage as shown in Figure 21. Once the sealing tape is removed, or the cell is taken from its sealed package, it must be put into use.

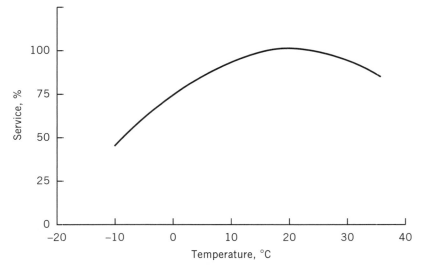

Fig. 22. Effect of temperature on discharge efficiency of miniature zinc–air batteries (21). Courtesy of Eveready Battery Co.

Miniature air cells can be used for continuous service over a temperature range of $-10°C$ to $+55°C$. However, performance at high temperatures suffers as a result of drying out of the cells. Figure 22 shows typical air cell service as a function of temperature.

5. Lithium Cells

Cells having lithium [7439-93-2] anodes are generally called lithium cells regardless of the cathode. They can be conveniently separated into two types: cells having solid cathodes and cells having liquid cathodes. Cells having liquid cathodes also have liquid electrolytes and in fact, at least one component of the electrolyte solvent and the active cathode material are one and the same. Cells having solid cathodes may have liquid or solid electrolytes but, except for the lithium–iodine system, those having solid electrolytes are not yet commercial.

All of the cells take advantage of the inherently high energy of lithium metal and its unusual film-forming property. This latter property, discussed in detail in reference 25, allows for lithium compatibility with solvents and other materials with which it is thermodynamically unstable, yet permits high electrochemical activity when the external circuit is closed. This is possible because the film formed with certain organic and inorganic materials is conductive to lithium ions, but not to electrons. Thus corrosion reactions occur only to the extent of forming thin, electronically insulating films on lithium which are both coherent and adherent to the base metal. The thinness of the film is important in order to allow reasonable rates of transport of lithium ions through the film when the circuit is closed. Many materials, such as water and alcohols, which are thermodynamically unstable to lithium, do not form this kind of passivating film (see THIN FILMS).

Much analytical study has been required to establish the materials for use as solvents and solutes in lithium batteries. References 26 and 27 may be consulted for discussions of electrolytes. Among the best organic solvents are cyclic esters, such as propylene carbonate [108-32-7] (PC), $C_4H_6O_3$, ethylene carbonate [96-49-1] (EC), $C_3H_4O_3$, and butyrolactone [96-48-0], $C_4H_6O_2$, and ethers, such as dimethoxyethane [110-71-4] (DME), $C_4H_{10}O_2$, the glymes, tetrahydrofuran (THF), and 1,3-dioxolane [646-06-0], $C_3H_6O_2$. Among the most useful electrolyte salts are lithium perchlorate [7791-03-9], $LiClO_4$, lithium trifluoromethanesulfonate [33454-82-9], $LiCF_3SO_3$, lithium tetrafluoroborate [14283-07-9], $LiBF_4$, and lithium hexafluoroarsenate [29935-35-1], $LiAsF_6$. A limitation of these organic electrolytes is the relatively low conductivity, compared to aqueous electrolytes. This limitation combined with the generally slow kinetics of the cathode reactions has forced the use of certain designs such as thin electrodes and very thin separators, in all lithium batteries. This usage led to the development of coin cells rather than button cells for miniature batteries and jelly or Swiss roll designs rather than bobbin designs for cylindrical cells. Figure 23 shows cutaway views of typical coin cells (Fig. 23**a**) and cylindrical jelly roll (Fig. 23**b**) lithium batteries.

Many of the cyclindrical cells have glass-to-metal hermetic seals, although this is becoming less common because of the high cost associated with this type of

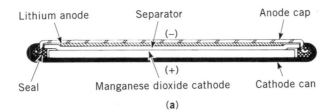

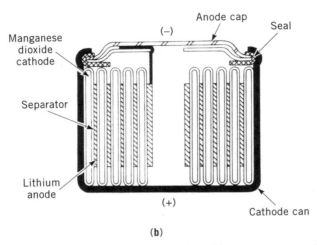

Fig. 23. Cutaway view of (**a**) 2016-size Eveready lithium manganese dioxide coin cell, and (**b**) jelly roll cylindrical lithium manganese dioxide cell (28). Courtesy of Eveready Battery Co.

seal. Alternatively, cylindrical cells have compression seals carefully designed to minimize the ingress of water and oxygen and the egress of volatile solvent. These construction designs are costly and the high price of the lithium cell has limited its use. However, the energy densities are superior.

5.1. Solid Cathode Cells. Solid cathode cells were investigated first and the very hydrophobic material, carbon monofluoride [3889-75-6], CF, was among the first cathode materials to be produced in commercial cells (29). The hydrophobicity minimized the need for drying the cathode material and the layerlike structure of the material facilitated the cathode reaction. Shortly thereafter, the solid electrolyte system, $Li-I_2$, which allowed *in situ* formation of a lithium iodide [10377-51-2], LiI, interface between the two materials, was developed (30) and eventually became the principal system to be used in heart pacers and other low drain medical applications. This system also was intrinsically hydrophobic. The $Li-MnO_2$ system, unsuccessfully investigated in the 1960s and early 1970s, was shown in 1975 to be viable (31) if certain types of MnO_2 were heat-treated to at least 300°C. The large amounts of water within and on the surface of the MnO_2 had caused cathode inefficiencies as well as problems with the lithium, because of migration through the electrolyte. The other solid cathode material in use in lithium cells is iron pyrite [1309-36-0], FeS_2. Whereas this material had been widely studied in high temperature systems, its use in room temperature systems relied heavily on the introduction of new electrolytes (32).

Lithium–Manganese Dioxide Cells. The $Li-MnO_2$ cell is becoming the most widely used 3-V lithium battery. The critical step in obtaining good performance is the heat treatment of the MnO_2 prior to incorporation in the cathode. There is still some controversy about the exact nature of the reduction of the cathode, although the reduction appears to occur in two steps. The first step, in close analogy with the aqueous reduction, is a homogeneous process that concludes with the formation of a partially lithiated material.

$$x Li^+ + x e^- + MnO_2 \longrightarrow Li_x MnO_2 \qquad (29)$$

where $x \approx 0.4$. This step is followed by a heterogeneous process to a new phase, the structure of which has not been determined.

$$(1-x) Li^+ + (1-x) e^- + Li_x MnO_2 \longrightarrow LiMnO_2 \qquad (30)$$

The anodic reaction is conventionally written as

$$Li \longrightarrow Li^+ + e^- \qquad (31)$$

where it is understood that the Li^+ ion is solvated predominantly by the strongest Lewis base among the solvents and that, prior to the dissolution/solvation step, the lithium ion must diffuse through the passivating film. The most common electrolytes used with the system are $LiClO_4$ or $LiCF_3SO_3$ in a mixture of PC and DME (usually 1:1).

The kinetics are not very sensitive to the electrolyte so the choice is largely dependent on safety, toxicity, and cost. The relatively slow kinetics of the system has necessitated the use of thin electrodes in order to obtain sufficient current carrying capability and these cells are designed as coin cells (Fig. 23**a**) or as jelly rolls (Fig. 23**b**) with alternating anode, separator, cathode, and another separator layer. These 3-V batteries are made in sizes not used for aqueous 1.5-V cells to help prevent their insertion in circuits designed for 1.5 V.

The energy density of the system depends on the type of cell as well as the current drain. Table 3 gives the specification for the various lithium systems. These coin cells have already been widely used in electronic devices such as calculators and watches, whereas the cylindrical cells have found applications in cameras.

Lithium-Carbon Monofluoride Cells. Although the Li–CF cell was the first lithium cell to be manufactured in quantity for consumer applications, it is becoming less popular because of the relatively high cost of the cathode material and the ability of the less expensive $Li-MnO_2$ cells to perform most of the same functions. The cathodes have traditionally used carbon as a conductor because carbon monofluoride is itself a poor conductor. Some interesting hybrid cathode cells using mixtures of MnO_2 and CF have also been studied (33). Carbon monofluoride has a laminar structure, which is similar to graphite, and when graphite reacts with fluorine to make CF, the flat planes of graphite become

Table 3. Characteristics of Lithium Primary Batteries

Cathode (cell system)	Voltage, V		Specific energy, $W \cdot h/kg$	Energy density, $W \cdot h/mL$	Operating temperature range, °C
	Nominal	Working			
Solid cathodes					
carbon monofluoride (Li–CF)	3	2.5–2.7	200	0.40	−20–60
manganese dioxide (Li–MnO$_2$)	3	2.7–2.9	200	0.40	−20–55
iron disulfide (Li–FeS$_2$)	1.5	1.3–1.7	125	0.40	−20–50
iodine (Li-I$_2$)	3	2.4–2.8	250	0.92	37
Liquid cathodes					
sulfur dioxide (Li–SO$_2$)	3	2.7–2.9	250	0.44	−55–70
thionyl chloride (Li–SOCl$_2$)	3.6	3.3–3.5	300	0.95	−50–70

puckered as layers of fluorine atoms become interspersed among the carbon layers. Another pure compound, C_2F, is also known and this has been shown to have battery activity as well (34).

The CF cathode reaction is believed to be a heterogeneous process, initiated by the insertion of lithium ions between the CF planes. It is completed by the extrusion of LiF and the collapse of the structure to carbon.

$$Li^+ + CF + e^- \longrightarrow LiF + C \tag{32}$$

Various lithium salts and butyrolactone or PC–DME mixtures are usually used as electrolytes. The close competitive performance of CF_x and MnO_2 cathodes is evidenced in Table 3. The construction of cells is also similar for the two systems. In addition to uses mentioned for the lithium manganese dioxide system, some unique applications such as lighted fishing bobbers have been developed for the Japanese market.

Lithium-Iron Disulfide Cells.   These cells were first manufactured in button cell sizes and because of the voltage similarity, they are used as direct replacements for Zn–Ag$_2$O cells (35). The lower conductivity of the organic electrolyte in Li–FeS$_2$ systems as compared to the KOH electrolyte in the Zn–Ag$_2$O cells, precludes use in the highest drain applications, but such uses are relatively few. Recently, "AA"-size Li–FeS$_2$ cylindrical batteries with a high surface area jelly roll construction and a plastic gasket compression seal have been introduced. The high area construction permits the use of these cells in almost all applications for which alkaline Zn–MnO$_2$ cells can be used and gives higher energy than that system. The compression seal helps to keep the cost at reasonable levels.

The chemistry of the Li–FeS$_2$ system is quite complex. There are at least two steps to the reaction at low discharge rates. The first reaction is an approximately two-electron reduction to a new phase which is a lithiated FeS$_2$ compound.

$$2\,Li^+ + FeS_2 + 2\,e^- \longrightarrow Li_2FeS_2 \tag{33}$$

The second step is also heterogeneous and involves the breakdown of the intermediate compound with further lithiation into lithium sulfide [12136-58-2] and finely divided iron [7439-89-6] particles.

$$2\,Li^+ + Li_2FeS_2 + 2\,e^- \longrightarrow 2\,Li_2S + Fe \tag{34}$$

The crystal structure of the intermediate is not well understood. The final iron phase is termed superparamagnetic because the particle size is too small to support ferromagnetic domains. At low rates, the discharge occurs in two steps separated by a small voltage difference. At high rates, however, the two steps become one, indicating that the first step is rate limiting, ie, the second step (eq. 34) occurs immediately after formation of the intermediate (eq. 33).

The performance of the cylindrical cell is superior to alkaline cells at all rates of discharge, but, because of the high area electrode construction, the performance advantage is magnified during high rate discharge. Table 3 shows the energy density and a comparison to Tables 1 and 2 shows the superiority of the cells to aqueous alkaline and Leclanché cells and the parity for $Zn-Ag_2O$ button cells.

Lithium–Iodine Cells. Iodine [7553-56-2] forms charge-transfer complexes with unsaturated compounds, many of which have reasonably high electronic conductivity. The iodine cathode takes advantage of this property by combining iodine and poly(2-vinylpyridine) (PVP) or other copolymers having unsaturated repeating units. In one process for making a $Li-I_2$ cell, the iodine and polymer, using excess iodine, are heated together to form the molten charge-transfer complex. Then the liquid is poured into the waiting cell which already contains a lithium anode and a cathode collector screen. After cooling and solidification of the cathode, the cell is hermetically sealed. In another process, the iodine–PVP mixture is reacted and then pelletized. The pellets are then pressed into the lithium to form a sandwich structure which is inserted into the waiting cell and hermetically sealed. The direct contact of the excess iodine and the lithium results in an immediate reaction to form an *in situ* LiI film which then protects the lithium from further chemical degradation.

The cathodic reaction is the reduction of iodine to form lithium iodide at the carbon collector sites as lithium ions diffuse to the reaction site. The anode reaction is lithium ion formation and diffusion through the thin lithium iodide electrolyte layer. If the anode is corrugated and coated with PVP prior to adding the cathode fluid, the impedance of the cell is lower and remains at a low level until late in the discharge. The cell eventually fails because of high resistance, even though the drain rate is low.

The $Li-I_2$ battery system has allowed simple heart pacers to operate for as long as 10 years compared to the 1.5–2 years of operation for alkaline batteries. This is mostly because of the very high stability of the lithium batteries allowing for trouble-free operation. Many other lithium batteries have been tried in this application, but the lithium iodine system is now almost universally used and the surgical gain is high. Table 3 gives the energy density of the system and Figure 24 shows a cutaway view of the letter "D" shaped cell which is used to fit the electronics pacer package. These cells are extremely expensive because of the great care taken in manufacturing and the careful testing, labeling, and monitoring of

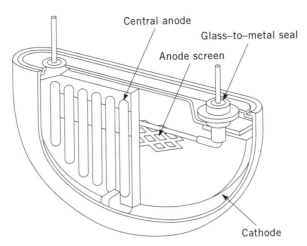

Fig. 24. Cutaway view of lithium–iodine pacemaker cell in case-grounded, central anode configuration (36). Courtesy of Plenum Press.

each cell. The system is very sensitive to moisture because corrosion of both the lithium anode and the stainless steel container occur in the presence of moist iodine vapor. However, a completely dry cell does not form any gas and has no fluids to leak and it is this stability which has made the Li–I$_2$ cell so useful for medical batteries.

Less expensive coin-type cells have also been developed for consumer electronics applications, but the severe current limitations have restricted the use of the cells.

5.2. Liquid Cathode Cells.   Liquid cathode cells were discovered at almost the same time as the successful solid cathode cells. A strongly oxidizing liquid such as SO$_2$, was determined to be suitable for direct contact with the strongly reducing lithium, because an excellent passivating film forms spontaneously on the lithium (37). Subsequently, a number of other liquid cathode cells were discovered, the most successful of which is the Li–SOCl$_2$ cell (38). Although liquid cathode cells have been used primarily in military applications, most recently a number of memory backup applications have been served by low rate lithium–thionyl chloride cells. These cells have high energy, a long shelf life, and a constant voltage profile.

Lithium–Sulfur Dioxide Cells.   Lithium–sulfur dioxide cells (39) generally use either acetonitrile [75-05-8] AN, C$_2$H$_3$N, or PC or a mixture of the two as cosolvents with the SO$_2$, usually in amounts as high as 50 vol %. This has the advantage of lowering the vapor pressure of the sulfur dioxide [7446-09-5] which at 25°C is about 300 kPa (3 atm). The liquids are usually chilled, however, to make management of the toxic SO$_2$ gas easier during cell filling and sealing, which are the last steps in the construction of cells. AN can be used as a cosolvent only because of the excellent film-forming properties of SO$_2$. Lithium is known to react extensively with AN in the absence of sulfur dioxide. The other properties of AN, relatively high dielectric constant and very high fluidity, enhance the electrolyte conductivity and permit construction of high rate cells. The cell reaction proceeds according to

$$2\,Li + 2\,SO_2 \longrightarrow Li_2S_2O_4 \tag{35}$$

where the cathode reaction, reduction of SO_2, takes place on a highly porous, >80%, carbon electrode. The lithium dithionite [59744-77-3] product is insoluble in the electrolyte and precipitates in place in the cathode as it is formed. The reaction involves only one electron per SO_2 molecule which, along with the high volume of cosolvent in the cell, limits the capacity and energy density. Lithium bromide [7550-35-8], LiBr, is the most commonly used salt, although sometimes $LiAsF_6$ is used, especially in reserve cell configurations. Because of the high rate capability, the cell has been widely used for military applications.

Safety has been a primary focus for the Li–SO_2 system, partly because of the high toxicities of the materials. Also, the potentially high rate of reaction during an accidental short circuit or other incident has led to some runaway reactions of high energy. However, tight control has allowed usage and the Li–SO_2 battery has become an important battery for military communications. In addition, the careful engineering of electrodes and vent mechanisms has controlled incidents. The cells have been subjected to extensive abuse testing.

All sulfur dioxide batteries have hermetic glass-to-metal seals to prevent loss of the highly volatile sulfur dioxide. They also have pressure operated vents to accomplish emergency evacuation of the cell under abuse to prevent explosions. Safety has also been improved by using a balanced cell design, ie, an equal capacity of lithium and sulfur dioxide, so that low rate discharge uses up all of the lithium. The performance of these cells is generally optimized for high current density operation and the battery is usually limited by blockage of the porous cathode structure with the insoluble lithium dithionite reaction product. Figure 25 gives a comparison of the energy output for lithium sulfur dioxide cells with that for lithium thionyl chloride liquid cathode cells. Both

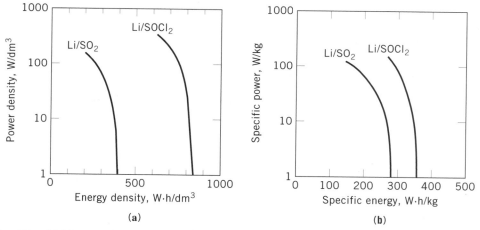

Fig. 25. Lithium–sulfur dioxide and lithium–thionyl chloride high rate batteries profile with (**a**) power density vs energy density, and (**b**) specific power vs specific energy.

types of batteries have high energy density and specific energy (Table 3), because of the efficient use of space characteristic of a liquid cathode cell. Many cell types and sizes have been manufactured and many of these cells have been combined in series-parallel networks in batteries. They have been mostly made under contract for government use or, in some cases, for original equipment use for electronic device manufacturers. It is difficult for a consumer to purchase these cells separately.

Lithium–Thionyl Chloride Cells.   Lithium–thionyl chloride cells have very high energy density. One of the main reasons is the nature of the cell reaction.

$$4\,Li + 2\,SOCl_2 \longrightarrow 4\,LiCl + S + SO_2 \tag{36}$$

The reaction involves two electrons per thionyl chloride [7719-09-7] molecule (40). Also, one of the products, SO_2, is a liquid under the internal pressure of the cell, facilitating a more complete use of the reactant. Finally, no cosolvent is required for the solution, because thionyl chloride is a liquid having only a modest vapor pressure at room temperature. The electrolyte salt most commonly used is lithium aluminum chloride [14024-11-4], $LiAlCl_4$. Initially, the sulfur product is also soluble in the electrolyte, but as the composition changes to a higher SO_2 concentration and sulfur [7704-34-9] builds up, a saturation point is reached and the sulfur precipitates.

The end of discharge occurs as the anode is used up because the cells are usually anode limited. However, if very high discharge rates are used, the collector, which is usually made of a porous carbon such as that in the SO_2 cell, becomes blocked with the insoluble lithium chloride [7447-41-8], LiCl. In either case, the discharge curve is flat over time, then declines abruptly at the end of life. Because of the high activity of the cathode and the good mass transfer of the liquid catholyte, the system is capable of very high rate discharges and operates at 3.6 V in open circuit. The anode is also very active except for a tendency to form thick films of LiCl during open circuit stand, much like the lithium iodine system. The film can cause delay effects during start-up of the cell.

Delay effects can be largely eliminated by treating the anode with various types of polymeric films, which decrease the rate of corrosion and LiCl formation markedly. One of the best films for this purpose is poly(vinyl chloride) [9002-86-2]. The wide liquid range of the catholyte solutions gives the system a good operating range, from very low (−50°C) to unusually high (>100°C) temperatures. The cells should not be stored at high temperatures, however, because corrosion reactions can occur which reduce the cell life or cause an enhanced delay effect.

Several applications are developing for the Li–$SOCl_2$ battery system. Because of the excellent voltage control, high energy density, and high voltage, the battery is finding increasing use on electronic circuit boards to supply a fixed voltage for memory protection and other standby functions. These cells are designed in low rate configuration which maximizes the energy density and cell stability. Military and space applications are also developing because of the wide range of temperature/performance capability as well as the high specific energy.

6. Reserve Batteries

Reserve batteries have been developed for applications that require a long inactive shelf period followed by intense discharge during which high energy and power, and sometimes operation at low ambient temperature, are required. These batteries are usually classified by the mechanism of activation which is employed. There are water-activated batteries that utilize fresh or seawater; electrolyte-activated batteries, some using the complete electrolyte, some only the solvent; gas-activated batteries where the gas is used as either an active cathode material or part of the electrolyte; and heat-activated or thermal batteries which use a solid salt electrolyte activated by melting on application of heat.

Activation of these batteries involves adding the missing component which can be done in a simple way, such as pouring water into an opening in the cell, for water-activated cells, or in a more complicated way by using pistons, valves, or heat pellets activated by gravitational or electric signals for the case of the electrolyte- or thermal-activation types. Such batteries may be stored for 10–20 years while awaiting use. Reserve batteries are usually manufactured under contract for various government agencies such as the Department of Defense, although occasional industrial or safety uses have been found. Many of the electrochemical systems involved in these batteries are beyond the scope of this article. Reference 41 contains further details on these systems.

The lithium-thionyl chloride, or the lithium–sulfur dioxide, system is often used in a reserve battery configuration in which the electrolyte is stored in a sealed compartment which upon activation may be forced by a piston or inertial forces into the interelectrode space. These high energy density systems have gained some of the applications of the older liquid ammonia [7664-41-7] reserve batteries, which usually had a magnesium [7439-95-4] anode and an m-dinitrobenzene [99-65-0], $C_6H_4N_2O_4$, cathode. Much of the engineering relating to flowing liquid ammonia into the battery was applied to the manufacture of lithium cells. An even closer analogy to the ammonia battery is the lithium–vanadium pentoxide battery which performs as a very high rate, high energy cell. Most applications for such batteries are in mine and fuse applications in military ordnance.

One variant of the liquid cathode reserve battery is the lithium–water cell in which water serves as both the liquid cathode and the electrolyte. A certain amount of corrosion occurs, but sufficient lithium is provided to compensate. The reaction product is soluble lithium hydroxide [1310-65-2], Li(OH). Sometimes a solid cathode material, such as silver oxide, or another liquid reactant, such as hydrogen peroxide [7722-84-1], H_2O_2, is used in combination with the lithium anode and aqueous electrolyte to improve the rate or decrease the amount of gas given off by the system. These cells are mostly used in the marine environment where water is available or compatible with the cell reaction product. Common applications are for torpedo propulsion and to power sonobuoys and submersibles. An older system still used for these applications is the magnesium–silver chloride–seawater-activated battery. This cell is much heavier than the corresponding lithium cell, but the buoyancy of the sea makes this less of a detriment than might first be thought. The magnesium–silver chloride cell is also useful for powering emergency communication devices for airplane crews whose planes have come down in the sea.

The last type of reserve cell is the thermally activated cell. The older types use calcium [7440-70-2] or magnesium anodes; newer types use lithium alloys as anodes. Lithium forms many high melting alloys such as those with aluminum, silicon, and boron. Furthermore, lithium can diffuse rapidly within the alloy phase permitting high currents to flow. The electrolyte for both the older and newer chemistries is usually the eutectic composition of lithium chloride and potassium chloride which melts at 352°C. This electrolyte has temperature dependent conductivities which are an order of magnitude higher than the best aqueous electrolytes. The high conductivity and the enhanced kinetics and mass transport allow the battery to be discharged at a very high rate of several A/cm^2 with complete discharge in 0.5 s. The cathodes for the older calcium anode cells are typically metal chromates such as calcium chromate [13765-19-0], $CaCrO_4$. The anode reaction product is calcium chloride [10043-52-4], whereas the cathode product is a mixed calcium chromium oxide of uncertain composition. One of the best cathodes for lithium alloy cells is FeS_2 which forms a system similar in reaction mechanism to that used in miniature $Li-FeS_2$ cells.

The heat pellet used for activation in these batteries is usually a mixture of a reactive metal such as iron or zirconium[7440-67-7], and an oxidant such as potassium perchlorate [7778-74-7]. An electrical or mechanical signal ignites a primer which then ignites the heat pellet which melts the electrolyte. Sufficient heat is given off by the high current to sustain the necessary temperature during the lifetime of the application. Many millions of these batteries have been manufactured for military ordnance as they have been employed in rockets, bombs, missiles, etc.

7. Economic Aspects

In 1989 the breakdown in sales of primary cells in the United States by category was alkaline, 65%; carbon–zinc, 30%, 21% of which was heavy-duty, 9% general purpose; specialty, 4%; and lithium, 1% (7). The total primary cell market in that year was $2813 million; with a unit volume of 2514 million cells. Over five years the average annual growth rate was 9.2% in terms of sales and 4.5% in terms of units. The growth of sales has been tied mostly to growth in the electronics industry, although the sales in battery-operated toys has also shown substantial growth.

The market outside of the United States reflects the historical dominance of carbon–zinc cells. For example, in Japan nearly half of all sales are carbon–zinc cells, about 33% are alkaline, and about 17% are lithium. The high proportion of lithium cells relative to U.S. sales reflects the important photographic market in Japan. Western European sales are similar to those in Japan, and Third World sales are almost totally dominated by carbon–zinc batteries.

There are hundreds of primary battery manufacturers, most of which are limited to a specialty product or a limited national market. However, a number of multinational battery suppliers have manufacturing facilities in many

Table 4. **Battery Companies Manufacturing Primary Cells**

Company	Cell type		
	Carbon–zinc	Alkaline	Lithium
North America			
Duracell International		x	x
Eastman Kodak		x	
Eveready Battery Co.	x	x	x
Ray-O-Vac International	x	x	x
European Economic Community			
Ever Ready (BEREC)	x	x	
SAFT			x
Varta Batterie AG	x	x	x
Fast East			
Matsushita	x	x	x
Sanyo	x	x	x
Toshiba	x	x	x
Yuasa	x	x	x

countries and a broad line of products. Table 4 lists the companies having sales of primary batteries of greater than $100 million per annum together with the regions of their headquarters and their product lines.

BIBLIOGRAPHY

"Batteries, Electric (Batteries, Primary)," in *ECT* 1st ed., Vol. 2, pp. 324–340, by J. J. Coleman and O. W. Storey, Burgess Battery Company; "Cells, Electric," Vol. 3, pp. 292–342, by W. J. Hamer, National Bureau of Standards; and "Cells, Electric," Suppl. 2, pp. 126–161, by C. K. Morehouse, R. Glicksman, and G. S. Lozier, Radio Corporation of America; "Batteries and Electric Cells, Primary (Primary Cells)," in *ECT* 2nd ed., Vol. 3, pp. 111–139, by J. F. Yeager, Union Carbide Corporation, E. B. Yeager, Western Reserve University, A. F. Daniel (Military Types), U.S. Army Signal Research and Development Laboratory; "Batteries, Primary (Primary Cells)," in *ECT* 3rd ed., Vol. 3, pp. 515–545, by H. Gu and D. N. Bennion, University of California, Los Angeles.

CITED PUBLICATIONS

1. *Eveready Battery Engineering Data*, Vol. 3, Eveready Battery Co., St. Louis, Mo., 1984.
2. Fr. Pat. 71,865 (1866), G. Leclanché.
3. U.S. Pat. 373,064 (1887), C. Gassner.
4. F. L. Tye, in T. Trans and M. Skyllas-Kazeos, eds., *Proceedings of the 7th Australian Electrochemistry Conference*, The Electrochemistry Division, The Royal Australian Chemical Institute, 1988, 37–48.
5. R. G. Burns and V. M. Burns, in A. Kozawa and R. J. Brodd, eds., *Proceedings of the Manganese Dioxide Symposium*, Vol. 1, Cleveland, I. C. Sample Office, Cleveland, Ohio, 1975, p. 306.

6. *American National Standard Specification for Dry Cells and Batteries*, ANSI C18. 1-1979, American National Standards Institute, Inc., New York, May 1979.

7. *The Dry Cell Battery Market, Packaged Facts*, The International/Research Co., New York, 1990.

8. *Eveready Battery Engineering Data*, Vol. 2A, Eveready Battery Co., St. Louis, Mo., 1990.

9. Fr. Pat. 143,644 (1881), F. de Lalande and G. Chaperon; U.S. Pat. 274,110 (1883).

10. Ger. Pat. 157,290 (1901), T. A. Edison; U.S. Pat. 678,722 (1901).

11. Sw. Pat. 15,567 (1901), W. Jungner; Ger. Pat. 163,170 (1901).

12. Ger. Pat. 24,552 (1882), G. Leuchs.

13. U.S. Pat. 2,422,045 (1947) and 2,576,266 (1951), S. Ruben.

14. W. S. Herbert, *J. Electrochem. Soc.* **99**, 190C (1952).

15. C. G. Saxe and R. J. Brodd, in A. J. Salkind, ed., *Proceedings of the Symposium on History of Battery Technology*, Vol. 87–14, The Electrochemical Society, Pennington, N.J., 1987, p. 47.

16. J. O'M. Bockris, Z. Nagy, and A. Damjanovic, *J. Electrochem. Soc.* **119**, 285 (1972).

17. J. Hendrikx, A. van der Putten, W. Visscher, and E. Barendrecht, *Electrochim. Acta* **29**, 81 (1984).

18. R. G. Burns and V. M. Burns, in B. Schumm, Jr., H. M. Joseph, and A. Kozawa, eds., *Proceedings of the Manganese Dioxide Symposium*, Vol. 2, Tokyo, I. C. Sample Office, Cleveland, Ohio, 1980, p. 97.

19. P. Ruetschi, R. Giovanoli, and P. Burki, in Ref. 5, p. 12.

20. A. Kozawa and J. F. Yeager, *J. Electrochem. Soc.* **112**, 959 (1965); A. Kozawa and R. A. Powers, *J. Electrochem. Soc.* **113**, 870 (1966).

21. *Eveready Battery Engineering Data*, Vol. 1A, Eveready Battery Co., St. Louis, Mo., 1988.

22. *Eveready Battery Engineering Data, Temperature Effects*, BE-282, 1988.

23. R. A. Putt and A. I. Attia, *Proceedings of the 31st Power Sources Symposium*, The Electrochemical Society, Pennington, N.J., 1984, p. 339.

24. J. W. Cretzmeyer, H. R. Espig, and R. S. Melrose, in D. H. Collins, ed., *Power Sources 6*, Academic Press, New York, 1966, p. 269.

25. E. Peled in J. P. Gabano, ed., *Lithium Batteries*, Academic Press, New York, 1983, p. 43.

26. G. E. Blomgren, Ref. 24, p. 13.

27. H. V. Venkatasetty, ed., *Lithium Battery Technology*, John Wiley & Sons, Inc., New York, 1984, pp. 1 and 13.

28. *Eveready Lithium Battery Engineering Data*, Eveready Battery Co., St. Louis, Mo., 1987.

29. U.S. Pat. 3,536,532 (1970) and 3,700,502 (1972), N. Watanabe and M. Fukuda (to Matsushita Electric Co.).

30. A. A. Schneider, J. R. Moser, T. H. E. Webb, and J. E. Desmond, *Proc. Ann. Power Sources Conf.* **24**, 27 (1970).

31. H. Ikeda, T. Saito, and H. Tamura, in Ref. 5, p. 384.

32. U.S. Pat. 3,996,069 (1976), M. L. Kronenberg (to Union Carbide Corp.).

33. U.S. Pat. 4,327,166 (1982), V. Z. Leger (to Union Carbide Corp.).

34. N. Watanabe, R. Hagiwara, and T. Nakajima, *J. Electrochem. Soc.* **131**, 1980 (1984).

35. M. B. Clark, in Ref. 25, p. 115.

36. C. F. Holmes, in B. B. Owens, ed., *Batteries for Implantable Biomedical Devices*, Plenum Press, New York, 1986, p. 156.

37. U.S. Pat. 3,567,515 (1971), D. L. Maricle and J. P. Mohns (to American Cyanamid).

38. Ger. Pat. 2,262,256 (1972), G. E. Blomgren and M. L. Kronenberg (to Union Carbide Corp.).

39. C. R. Walk, in Ref. 25, p. 281.
40. C. R. Schlaikjer, in Ref. 25, p. 303.
41. D. Linden, ed., *Handbook of Batteries and Fuel Cells*, McGraw-Hill, New York, 1984.

GENERAL REFERENCES

G. W. Heise and N. C. Cahoon, eds., *The Primary Battery*, Vol. 1 and 2, John Wiley & Sons, Inc., New York, 1971, 1976.

D. Linden, ed., *Handbook of Batteries and Fuel Cells*, McGraw-Hill, New York, 1984.

K. V. Kordesch, ed., *Batteries, Manganese Dioxide*, Vol. 1, Marcel Dekker, New York, 1974.

T. R. Crompton, *Battery Reference Book*, Butterworths, London, 1990.

J. P. Gabano, ed., *Lithium Batteries*, Academic Press, New York, 1983.

H. V. Venkatasetty, ed., *Lithium Battery Technology*, John Wiley & Sons, Inc., New York, 1984.

B. Boone Owens, ed., *Batteries for Implantable Biomedical Devices*, Plenum Press, New York, 1986.

R. J. Brodd, *Batteries for Cordless Appliances*, John Wiley & Sons, Inc., New York, 1987.

GEORGE BLOMGREN
JAMES HUNTER
Eveready Battery Company, Inc.

BATTERIES, SECONDARY CELLS

1. Alkaline Cells

Alkaline electrolyte storage battery systems are more suitable than others in applications where high currents are required, because of the high conductivity of the electrolyte. Additionally, in almost all of these battery systems, the electrolyte which is usually an aqueous solution containing 25–40% potassium hydroxide [1310-58-3], KOH, does not enter into the chemical reaction. Thus concentration and cell resistance are invariant with state of discharge and these battery systems give high performance and have long cycle life. The manufacturing value of alkaline storage batteries in 2002 was estimated to be in the range of $4.5–5 billion dollars (worldwide) representing approximately 17% of the value for all secondary batteries. Approximately one half of this production value was in small sealed cells. The remainder was in industrial designs. The latter includes approximately 40,000 nickel-metal hydride (Ni-MH) units for hybrid electric vehicles (HEVs). The annual battery supply for this application was forecasted [100] to increase to over a million units by the year 2008, representing a production value of $1 billion.

Positive electrode active materials have been made from the oxides or hydroxides of nickel, silver, manganese, copper, mercury, and from oxygen. Negative electrode active materials have been fabricated from various geometric forms of cadmium [7440-43-9], Cd, iron [7439-89-6], Fe, and zinc [7440-66-6], Zn, and from hydrogen [1333-74-0]. Two different types of hydrogen electrode designs are common: those used in space, which employ hydrogen as a gas, and those used in consumer batteries, where the hydrogen is used as a metallic hydride. As indicated in Table 1 and Figure 1, nine electrode combinations exist in some scale of commercial production. Five system combinations are in the research/development stage, and two have been abandoned before or after commercial production for reasons such as short life, high cost, low voltage, low energy density, and excessive maintenance.

The annual production value of small-sealed nickel cadmium cells is approximately 1.0×10^9. Environmental considerations relating to cadmium have necessitated changes in the fabrication techniques, as well as recovery of failed cells. Battery system designers have switched to nickel–metal hydride (MH) cells for high power applications. However, the highest discharge/recharge rates are still achieved with nickel–cadmium cells typically in "AA"-size cells, to increase capacity in the same volume and avoid the use of cadmium.

There are many methods of fabricating the electrodes for these cell systems. The earliest commercially successful developments used nickel hydroxide

Table 1. Rechargeable Alkaline Storage Battery Systems

System[a]	Historical name	Voltage, V	Production[b]
nickel–cadmium	Jungner	1.30	vl
nickel–iron	Edison	1.37	vs
nickel–zinc	Drumm	1.70	vs
nickel–hydrogen (H$_2$ or MH)		1.30	vl
silver–cadmium		1.38 and 1.16[c]	vs
silver–iron	Jirsa	1.45 and 1.23[c]	vs
silver–zinc	Andre	1.86 and 1.60[c]	s
silver–hydrogen		1.38 and 1.16[c]	vs
manganese–zinc[d]		1.52	vs
mercury–cadmium		0.92	r
air(oxygen)–zinc		1.60	r
air(oxygen)–iron		1.40	r
air(oxygen)–aluminum			vs
copper–lead		1.20	r
copper–cadmium	Darrieus Waddell-Entz,	0.45	n
copper–zinc	Edison-LeLande Lelande-Chaperon	0.85	n

[a] The substance named first represents the positive electrode; the substance named second is the negative electrode. In all cases except for air(oxygen) systems, the active electrode material is the oxide or the hydroxide of the named species.
[b] vl = $>100 \times 10^6$ A·h/yr product; l = $>25 \times 10^6$ A·h/yr; s = $>5 \times 10^6$ A·h/yr; vs = $<5 \times 10^6$ A·h/yr; r = research and development phase; and n = no longer in production.
[c] Silver oxide electrodes have two voltage plateaus.
[d] Secondary system designs.

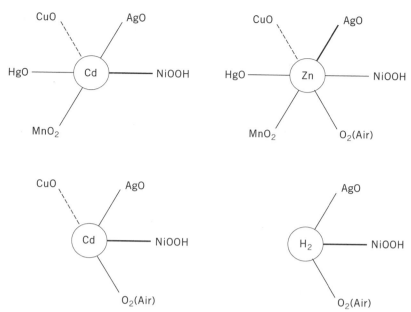

Fig. 1. Electrode combinations for alkaline storage batteries where the substance within the circle comprises the negative electrode and the combinations include (—) commercial products, (—) systems in research and development or limited production, and (- - -) historic system. The air (O_2)–H_2 system is commonly called a fuel cell.

[12054-48-7], $Ni(OH)_2$, positive electrodes. These electrodes are commonly called nickel electrodes, disregarding the actual chemical composition. Alkaline cells using the copper oxide–zinc couple preceeded nickel batteries but the CuO system never functioned well as a secondary battery. It was, however, commercially available for many years as a primary battery (see BATTERIES-PRIMARY CELLS).

The original alkaline battery, designed by Edison around 1900, contained nickel hydroxide mixed with graphite [7782-42-5] in perforated steel pockets as the positive active material and a high surface area iron powder as the negative. Both positive and negative electrodes were "pocket" plates, where the active material was contained in small, rectangular, boxlike pockets formed from finely perforated sheet steel. These plates very closely resembled the pocket plates in use in some current designs of nickel–cadmium batteries.

Because the nickel–iron cell system has a low cell voltage and high cost compared to those of the lead–acid battery, lead–acid became the dominant automotive and industrial battery system except for heavy-duty applications. Renewed interest in the nickel–iron and nickel–cadmium systems, for electric vehicles started in the mid-1980s using other cell geometries. This was supplanted in the 1990s by the use of NiMH batteries in hybrid electric vehicles.

In the early 1930s, production of nickel–cadmium batteries having thinner pocket-type plates and lower internal resistence became available. This grew to be an important design for truck, locomotive, and marine engine starting applications. Very high rate cell designs of nickel–cadmium cells became possible

Table 2. **Technology and Processing Contributions to Alkaline Battery Development**

Year	Procedure	Reference
1948	impregnation procedures for nickel and cadmium electrodes	6
1956	x-ray structure of nickel hydroxide electrode	7
1958	*in situ* x-rays of nickel and cadmium electrodes	8
	slurry-processed nickel-sintered electrode (SAFT)	9
1960	crystal structures of silver oxide	10,11
	structure of nickel hydroxide electrode	12
	in situ study of the nickel electrode	13
1961	microporous plastic-reinforced zinc electrode	14
1962	microporous plastic-bound cadmium electrode	15
1965	solid-state chemistry of nickel hydroxide	16
	crystal structures of nickel hydroxide	17
1966–1970	impregnation procedures for nickel electrodes	18–20
1972	electrochemical impregnation method for nickel electrodes	21
1974	nickel foam substrates	22
1975	plastic-bonded nickel electrodes	23
1978	controlled microgeometry electrodeposited electrodes	24
1980	nickel composite electrode	25
1981	nickel fiber mat substrates	26
1984	nickel felt substrates	27
	production of nickel fiber electrodes	28
1987	rechargeable MnO_2–zinc cell	29
1989	modified manganese dioxide for deep-cycling	30
1990	metal–hydride-type hydrogen electrodes used in commercial nickel–MH cells	31
1995	nickel–MH cells used in hybrid electric vehicles	32

with the development of the sintered nickel substrate in 1928 (2) and then in the late 1940s and early 1950s, the shortcomings of the sealed nickel–cadmium designs were overcome (3). The use of sealed nickel–cadmium cells grew rapidly and beginning in the early 1990s, designs of nickel–hydrogen, the hydrogen as hydride, cells captured a strong commercial and technical interest. Table 2 details the advances in alkaline battery technology from 1948 to 1990.

The nickel–zinc combination has a high cell voltage (about 1.75 V), which results in a very favorable energy density compared to that of nickel–cadmium or lead–acid. Additionally, zinc is relatively inexpensive and, in the absence of mercury additive, is environmentally benign. The nickel–zinc system was discussed as early as 1899 (4). There has been a resurgence of interest in the system for electric vehicles, but the problems of limited cycle life have not been completely overcome.

Silver [7440-22-4], Ag, as an active material in electrodes was first used by Volta, but the first intensive study using silver as a storage battery electrode was reported in 1889 (5) using silver oxide–iron and silver oxide–copper combinations. Work on silver oxide–cadmium followed. In the 1940s, the use of a semipermeable membrane combined with limited electrolyte was introduced by André in the silver oxide–zinc storage battery.

Many of the most recent applications for alkaline storage batteries require higher energy density and lower cost designs than previously available. Materials such as foam and/or fiber nickel [7440-02-0], Ni, mats as substrates, and new processing techniques including plastic bounded, pasted, or electroplated electrodes, have enabled the alkaline storage battery to meet these new requirements, while reducing environmental problems in the manufacturing plants. In addition, substantial technical efforts have been devoted to the recovery of used batteries. The most recent innovations in materials relate to the development of metal–hydride alloys for the storage and electrochemical utilization of hydrogen. Modifications to the chemical structure and/or the cell design of manganese dioxide [1313-13-9], MnO_2, electrodes have resulted in sufficient improvement to allow the reintroduction of the rechargeable MnO_2–zinc cell to the market as a lower cost, albeit lower performance, alternative to nickel–cadmium consumer size cells. Improvements in materials science and electrical circuits have lead to better separators, seals, welding techniques, feedthroughs, and charging equipment.

1.1. Nickel–Cadmium Cells. *Electrodes.* A number of different types of nickel oxide electrodes have been used. The term nickel oxide is common usage for the active materials that are actually hydrated hydroxides at nickel oxidation state 2+, in the discharged condition, and nickel oxide hydroxide [12026-04-9], $NiO·OH$, nickel oxidation state 3+, in the charged condition. Nickelous hydroxide [12054-48-7], $Ni(OH)_2$, can be precipitated from acidic solutions of bivalent nickel either by the addition of sodium hydroxide or by cathodic processes to cause an increase in the interfacial pH at the solution–electrode surface (see NICKEL AND NICKEL ALLOYS; NICKEL COMPOUNDS).

Several investigators have used combined approaches, particularly in the *in situ* precipitation of active material in the pores of sintered substrates, using cathodic polarization and caustic precipitation in simultaneous or nearly simultaneous steps. A considerable amount of the reported information on the chemistry, electrochemistry, and crystal structure of the nickel electrode has been obtained on thin films (qv) made by the anodic corrosion of nickel surfaces. However, such films do not necessarily duplicate the chemical and/or crystallographic condition of active material in practical electrodes. In particular, the high surface area, space charge region, and lattice defect structure are different. Some of the higher (3.5+) valence state electrochemical behavior seen in thin films has rarely been reproduced in practical electrodes.

The many varieties of practical nickel electrodes can be divided into two main categories. In the first, the active nickelous hydroxide is prepared in a separate chemical reactor and is subsequently blended, admixed, or layered with an electronically conductive material. This active material mixture is afterwards contained in a confining porous metallic structure or pasted onto a metallic mat or grid. Electrodes for pocket, tubular, pasted, and most button cells are made this way. The porous metallic structures such as the pocket or tube, ensure good particulate contact, prevent shedding, and confine expansion. Less expensive alternatives, those resulting from pasting on expanded metal, foam metal, punched corrugated sheet, or fiber metal nickel mat, do not confine expansion as well or provide the same amount of mechanical strength. Plastic reinforcement to minimize shedding and promote adherence to the substrate is often employed. The plastic mix can be achieved either by milling or extrusion of mixes at plastic flow temperatures or by using plastic–solvent combinations. The latter includes simple gels of water, KOH, and methylcellulose [9004-67-5].

Recent patent literature gives details on preparation of nickel positive electrodes using nickel hydroxide (33–35).

The other type of nickel electrode involves constructions in which the active material is deposited *in situ*. This includes the sintered-type electrode in which nickel hydroxide is chemically or electrochemically deposited in the pores of a 80–90% porous sintered nickel substrate that may also contain a reinforcing grid.

Almost all the methods described for the nickel electrode have been used to fabricate cadmium electrodes. However, because cadmium, cadmium oxide [1306-19-0], CdO, and cadmium hydroxide [21041-95-2], $Cd(OH)_2$, are more electrically conductive than the nickel hydroxides, it is possible to make simple pressed cadmium electrodes using less substrate (see CADMIUM AND CADMIUM ALLOYS; CADMIUM COMPOUNDS). These are commonly used in button cells.

Electrochemistry and Crystal Structure.   The solid-state chemistry of the nickel electrode is complex. Nickel hydroxide in the discharged state has a hexagonal layered lattice, where planes of Ni^{2+} ions are sandwiched between planes of OH^-; as shown in Figure 2. This structure, similar to that of cadmium iodide

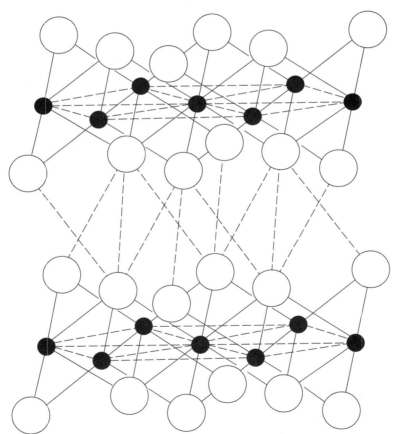

Fig. 2.   Nickel hydroxide structure where (•) represents nickel and (○) represents oxygen or hydroxyl.

[7790-80-9], CdI_2, is common to seven metal hydroxides including those of cadmium and cobalt. There are various hydrated and nonhydrated nickel hydroxides that have slightly different crystal habitats and electrochemical potentials. The most common form of charged material observed in batteries is NiOOH, density = 4.6 g/mL. In comparison, $Ni(OH)_2$ has a density of 4.15 g/mL. Thus the theoretical change in density on charge–discharge is less than 10%, and the kinetics involve only a proton transfer. This reaction can be written in simplified form as:

$$2\,\beta-\text{NiOOH}^* + \text{Cd} + 2\,H_2O \rightleftharpoons 2\,Ni(OH)_2^* + Cd(OH)_2$$

where the asterisks represent differing amounts of hydrated or absorbed water and/or electrolyte. It has also been established that species such as Li^+ and K^+ enter the nickel hydroxide structure to form a space–charge region and the actual reaction is believed to be more complicated than that shown. The interlayer separation of planes of nickel ions is increased by insertion of water or ions from the electrolyte. Different preparative and cycling conditions result in variations in defect crystal structures that affect electrochemical activity and the ability to retain charge (36). Additionally, electrode material oxidation states do not range precisely between +2 and +3, and because of hysteresis in the charge–discharge curves, direct measurement of an equilibrium potential is not possible.

The existence of $Ni(OH)_2$ and β-NiOOH as the usual discharged and charged materials was confirmed through x-ray diffraction powder patterns (7,8,37). The presence of other structural forms, in different electrolytes and under unusual cycling conditions, has also been observed. The crystalline-structural variability of $Ni(OH)_2$ has been described as being dependent on preparative procedure (16). Other phases form after prolonged overcharging of β-NiOOH in concentrated sodium hydroxide [1310-73-2], NaOH, or KOH. At elevated temperatures, in lithium hydroxide [1310-65-2], LiOH, electrolytes, $Ni(OH)_2$ is oxidized to a trivalent phase having the lithium nickelate [12031-65-1], $LiNiO_2$, crystalline structure and lithium additions to KOH electrolytes eliminate the formation of α-phase material. The sequence of these electrochemical processes has been summarized as: (1) an electrochemical exchange at the solid electrolyte interface involving a set of ionic-electronic lattice imperfections; (2) electron transport through the oxidized phase region to the metallic contact; (3) mass transport through the oxidized phase to the lower valence phase at the second interface; and (4) electrochemical exchange at the second solid-phase interface functioning as a source or sink for the ionic-electronic imperfections necessary to maintain the mass and charge balance.

A sequence of events taking place in the charge process has been described (16) demonstrating that during constant current charge, the surface double-layer region first produces a local interior space–charge field. As the semiconductor charge phase is extended into the interior, a distributed space–charge field is formed that is characterized by a combination of the semiconductor band structure and the mobility of dopant imperfections. When the state of charge increases, the magnitude of the space charge decreases. At the end of charge, there is only an ir (voltage) loss in the solid, and the applied field is mainly at the solid electrolyte interface. If charging is continued until a steady-state oxygen evolution is attained, the distributed space charge disappears. The semiconductor is characterized by a flat band potential and, if the nickel electrode is left on open circuit, the space charge-field gradually decays through atom movements in the solid phase, approximating a flat band potential condition.

The charge–discharge process cannot be satisfactorily represented by one equation (17). At least two reactions, based on different starting materials as well as different products, can be formulated. Both reactions are heterogeneous. In each of the reaction chains two distinct states exist in the oxidized phase. One of these can only be charged whereas the cathodic current are blocked. The other state, existing at a lower potential, can only be discharged but was blocked in the anodic direction. When the observed potential differences are extrapolated into the region of very low current densities, the difference between the two states is ca 60 mV in one of the reaction chains and ca 100 mV in the other. Thermodynamically this means that neither reaction series can be described by a single, reversible potential as in the manner of the $Cd–Cd(OH)_2$ electrodes. Rather, the semiconductor properties of oxidized nickel hydroxides play some part.

The description of the process is best illustrated as (17):

$$\beta\text{-Ni(OH)}_2 \xleftarrow[\text{in KOH}]{\text{transformation}} [\alpha\text{-Ni(OH)}_2] \cdot \tfrac{2}{3}H_2O$$

$$\text{reduction} \nearrow \quad \downarrow \text{oxidation} \qquad \text{oxidation} \nearrow \quad \uparrow \text{reduction}$$

$$\beta\text{-NiOOH} \quad \beta\text{-NiOOH} \xrightarrow[\text{in KOH}]{\text{overcharge}} \gamma\text{-NiOOH} \quad \gamma\text{-NiOOH}$$

$$\text{(more positive} \qquad\qquad \text{(more positive}$$
$$\text{by 100 mV)} \qquad\qquad\quad \text{by 60 mV)}$$

The nickel oxide modification obtained electrochemically in KOH electrolyte contained potassium ion and its nickel oxidation level are higher than that of $NiO_{1.5}$. Conclusions regarding the transitions between the reduced and oxidized products within the two series are that the redox process was not reversible and although the oxidized phases of the β- and the γ-nickel hydroxides differ in energy contents, differences in analyses and x-ray patterns are not significant.

Some γ-NiOOH has been shown to be formed in sintered nickel electrodes (38), and changes in water and KOH concentration during the cycling of nickel electrodes has been studied (12,39–41). Although there is some disagreement on the movement of water, KOH is adsorbed on the nickel electrode when the cell is charged and desorbed from the electrode when the cell is discharged.

The chemistry, electrochemistry, and crystal structure of the cadmium electrode is much simpler than that of the nickel electrode. The overall reaction is generally recognized as:

$$Cd + 2\,OH^- \underset{\text{charge}}{\overset{\text{discharge}}{\rightleftharpoons}} Cd(OH)_2 + 2\,e^-$$

However, there is a strong likelihood of a soluble intermediate in the formation of $Cd(OH)_2$. Cadmium has an appreciable solubility in alkaline solutions: $\sim 2 \times 10^{-4}$ mol/L in 8 M potassium hydroxide at room temperature. In general it is believed that the solution process consists of anodic dissolution of cadmium ions in the form of complex hydroxides (see CADMIUM COMPOUNDS).

In more recent studies involving cyclic line scan voltammetry of the nickel electrode, it was suggested that nickel can exist in the positive 2, 3, and 4 oxidation states (42). The structural parameters (43) using transmission extended x-ray absorption fine structure (exafs) and *in situ* electrodes confirmed earlier x-ray data (7,8), showing that the presence of cobalt [7440-48-4], Co, does not change crystal lattice parameters. The density and compressibility of nickel electrodes have been found to be highly variable (44). Cobalt additions appear to reduce the compressibility of nickel hydroxide, resulting in a firmer attachment of the active material to the substrate. However, a felt metal grid has been shown (45) to reduce shear failures of the electrode structure and minimize the need for stabilizing additives, such as cobalt.

In addition to the normal charge–discharge reaction, properly fabricated sealed nickel–cadmium cells have a mechanism for absorbing infinite amounts of overcharge. The cell must be fabricated with an excess amount of uncharged active material [$Cd(OH)_2$] in the cadmium electrode. When the nickel electrode nears full charge oxygen is evolved. This oxygen diffuses through open areas of the separator and reacts with the charged cadmium species forming cadmium oxide that hydrates to cadmium hydroxide (electrical charge reaction).

$$Cd(OH)_2 + 2\,e^- \longrightarrow Cd + 2\,OH^-$$

Therefore, the cadmium electrode is being electrically charged and chemically discharged at the same rate (chemical discharge reaction).

$$2\,Cd(OH)_2 \longleftarrow 2\,Cd + O_2 + 2\,H_2O$$

Sealed Cells. Most sealed cells are based on the principles appearing in patents of the early 1950s (3) where the virtues of limiting electrolyte, a separator that would absorb and retain electrolyte, and leaving free passage for the oxygen from the positive to the negative plate were described. First, the negative electrode has a surplus of uncharged active material so that the positive plate starts to produce oxygen before the negative plate is fully charged. The oxygen reacts with the negative active material, so that the negative electrode never becomes fully charged and consequently never evolves hydrogen. Second, the amount of electrolyte used is generally lower than can normally be absorbed in the electrodes and separators, facilitating the transfer of oxygen from the positive to the negative plate. Oxygen transport, at least to a certain extent is carried out in the gaseous stage. Third, the separators generally used can pass oxygen to the gaseous state for rapid transfer to the negative plate. Although both pocket and sintered electrodes of the nickel–cadmium type have been used in sealed-cell construction, the preponderant majority of cells in commercial production use sintered positive (nickel) electrodes, and either sintered or pasted negative (cadmium) electrodes.

Although the charged form of the anode is metallic cadmium, some traces of the discharged form always remain in the electrode even at prolonged overcharge. Indeed, as the anode is oxidized (discharged), x-ray diffraction peaks show lines from both Cd and $Cd(OH)_2$. These lines have sharper peaks than those obtained using the cathodic nickel hydroxides, confirming the more crystalline nature of cadmium.

Cadmium electrodes maintain a flat $E°$ potential throughout discharge, as is true of most electrodes in which the materials of the charged and discharged state form distinct independent crystalline forms. Because the conductivities of cadmium electrodes are high, and no ir drops or films are generated during discharge, they also display a flat working voltage curve.

The nickel–cadmium cell, unlike the lead–acid system, has a negative temperature voltage coefficient that is in the range of $-(0.2 - 0.4)$ mV/°C, depending on design factors and the nature and doping of the nickel active materials. Thus at higher temperatures the cell has a slightly lower open circuit voltage and this, combined with a reduction of internal resistance at higher temperatures, can result in a reduced back emf in charging. Therefore, special precautions must be taken in charging under constant voltage conditions to avoid the so-called runaway condition; ie, an increase in charge current that comes from a reduction in back emf resulting in an increase in temperature and then a new increase in charge current. Cells can be destroyed in this runaway condition unless current limiters are provided in the charge circuit. The runaway condition problem is not limited to the nickel–cadmium battery. However, it is more likely in those systems where the voltage change with temperature is negative and in those designs where there is very low internal resistance.

The capacity utilization of active material depends on cell design, discharge rate, temperature, and charging conditions. High rate ability depends on the degree of conducting support provided for the active material surface that is in contact with electrolyte, separator resistance, and state-of-charge. The maximum capacity that can normally be achieved from nickel active materials is 0.30 A·h/g calculated on the basis of $Ni(OH)_2$. The capacity of negative active material is ca 0.37 A·h/g based on $Cd(OH)_2$. In actual cell use, working capacities range from 60 to 90% of these values, depending on discharge rate and temperature.

Cell Fabrication Methods. Pocket Cells. A view of a pocket electrode nickel–cadmium cell is shown in Figure 3. The essential steps of positive (nickel) electrode construction are (1) cold-rolled steel ribbon is cut to proper width and is perforated using either needles or rolls; (2) the perforated steel ribbon is nickel-plated and usually annealed in hydrogen. The ribbon is formed into a trough shape, is filled with active material by either a briquetting or a powder-filling technique; (3) a second strip is formed into a lid that covers and locks with the filled trough; (4) the filled strips are cut to length and are arranged to form an electrode sheet by interleaving. This operation, carried out by means of rollers in a forming roll, is often combined with the pressing of a pattern into the electrode sheet in order to ensure good contact between ribbon and active material and to add mechanical strength to the construction; and (5) the electrode sheet is then cut to pieces of appropriate size and side bedding and lugs attached to form a metallic frame. The frame material is usually also cold-rolled steel ribbon.

The pockets are usually arranged horizontally in the electrodes as shown in Figure 3, but in a few cases vertical pockets are used. No significant difference has been observed between the two arrangements.

Fig. 3. View of pocket electrode nickel–cadmium cell.

Pocket-type cadmium electrodes are made by a procedure similar to that described for the positive electrode. Because cadmium active material is more dense than nickel active material, and because cadmium has a 2+ valence, cadmium electrodes, when fabricated to equal thicknesses, have almost twice the working capacity of the nickel electrode. A cell having considerably greater negative capacity provides for loss of negative capacity during life and avoids generation of hydrogen during charging. Thus in actual practice plates of equal thickness are used. Some manufacturers prefer to make the electrolyte transfer area larger for the negative electrode by increasing the area of perforation, usually by increasing the hole size. Pocket electrode plates usually have thicknesses of 0.7–6.3 mm.

After the individual pocket electrodes are fabricated, they are assembled into electrode groups. Electrodes of the same polarity are electrically and mechanically connected to each other and to a pole bolt as illustrated in Figure 3. Plates of opposite polarity are interleaved with separators. Ordinarily single-plates are used for each leaf in a plate group, but some manufacturers prefer to use two thin plates back to back to form one leaf in the positive group. Both bolting and welding methods are used in the assembly of electrodes into groups.

To complete the assembly of a cell, the interleaved electrode groups are bolted to a cover and the cover is sealed to a container. Originally, nickel-plated steel was the predominant material for cell containers but, more recently plastic containers have been used for a considerable proportion of pocket nickel-cadmium cells. Polyethylene, high impact polystyrene, and a copolymer of propylene and ethylene have been the most widely used plastics.

Steel containers are mechanically stronger than plastic and easier to fabricate in large sizes. They dissipate heat better and tend to keep the electrodes cooler during high temperature or high rate operations. However, cells assembled in steel containers must not have contact with each other during assembly to prevent intercell shorts. Plastic containers are the better option for most small and medium-sized cells because they require no protection against corrosion, they permit visual observation of the electrolyte level, they are lighter than steel containers, they can be closely packed into a battery, and small cells can be cemented or taped into batteries, eliminating a tray.

The cells are usually filled with an electrolyte solution of potassium hydroxide of density 1.18–1.23 g/mL, which may also contain lithium hydroxide. A potassium hydroxide solution of 1.20 g/mL freezes at $\sim-27°C$. Thus cells intended for climates colder than $-27°C$ have an electrolyte density of 1.25 g/mL or greater. In some designs there is a large volume of excess electrolyte above the electrodes in order to reduce the need for rewatering to once every few years. In these cells, the initial electrolyte density is lower than normal because the solution concentrates during operation. Concentrations might be as high as 1.26 g/mL at the time of rewatering (topping up). Lithium hydroxide has been shown to increase the life of positive pocket electrodes in cycling operation. However, the addition also increases the electrolyte resistivity, and is not ordinarily used in high rate starter batteries.

Individual cells are usually precycled before assembling into batteries. These early charge–discharge cycles, often called formation cycles, improve the capacity of the cell by increasing the surface area of the active material and effecting crystal structure changes.

The individual cells are assembled into batteries after a leakage test to check for faulty welding joints in steel containers or cracks and improper seals in the plastic encased cells. Cells that are to be delivered without electrolyte are emptied after the formation cycles. The steel-cased cells have to be separated by mechanical means to prevent intercell shorts and are often assembled into wooden crates. The cells in plastic cases can be cemented together or strapped with tape.

Tubular Cells. Although the tubular nickel electrode invented by Edison is almost always combined with an iron negative electrode, a small quantity of cells is produced in which nickel in the tubular form is used with a pocket cadmium electrode. This type of cell construction is used for low operating temperature environments, where iron electrodes do not perform well or where charging current must be limited.

Sintered Cells. The fabrication of sintered electrode batteries can be divided into five principal operations: preparation of sintering-grade nickel powder; preparation of the sintered nickel plaque; impregnation of the plaque with active material; assembly of the impregnated plaques (often called plates) into electrode groups and into cells; and assembly of cells into batteries.

A good powder for sintering purpose should be very pure (Fe, Cu, and S should be especially avoided) and should have a very low apparent density, in the range of 0.5–0.89 g/mL. An excellent powder for this purpose is made by the decomposition of nickel carbonyl [13463-39-3], $Ni(CO)_4$, (see CARBONYLS). Nickel carbonyl is a poisonous vapor obtained by passing carbon monoxide [630-08-0] over finely divided metallic nickel at about 200°C in rotary kilns using special gas-tight seals. The gas is condensed and the liquid carbonyl distilled to eliminate impurities. The purified liquid nickel carbonyl is injected

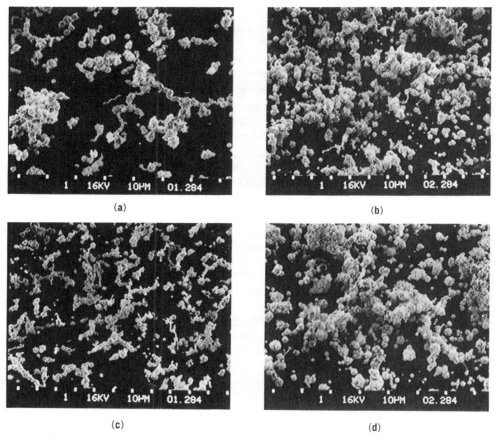

Fig. 4. Microstructure of carbonyl nickel powder at various magnifications (**a**) and (**c**) 1000x, (**b**) and (**d**) 1500x. Courtesy of Inco.

into large decomposers that have heated walls. The nickel carbonyl decomposes into carbon monoxide, which can then be recirculated, and a finely divided nickel powder. The structure of such powders is shown in Figure 4.

Sintering is a thermal process through which a loose mass of particles is transformed to a coherent body. It usually takes place at a temperature equal to two-thirds the melting point, or ca 800–1000°C for nickel. The sintered nickel structure without active material is called a plaque and it can be prepared by either dry or wet processes (see METALLURGY, POWDER).

In the dry process, nickel powders are sifted into ceramic or carbon molds with a reinforcing layer of wire mesh cloth or electroformed nickel. In most production processes part of the powder is sifted into the mold, a grid is inserted, and a second layer of sifted powder is added. The reinforcing grid is provided to reduce shrinkage during the sintering, reduce electrical resistivity, and provide extra mechanical strength. The molds are transferred to a sintering furnace that contains a reducing or inert atmosphere. Usually a reducing atmosphere is maintained by cracking ammonia ($2\,NH_3 \longrightarrow N_2 + 3\,H_2$) or partially dissociating natural gas. After 5–10 min in a hot zone, the molds are transferred to a cold zone in the furnace. Most furnaces for the dry processes are horizontal and of the moving-belt type.

To reduce labor and other expenses, most sintered nickel plaques are produced by a wet-slurry method. A nickel slurry is prepared by mixing a low density nickel powder with a viscous aqueous solution such as carboxymethylcellulose [9004-42-6] (CMC). Pure nickel gauze, a nickel-plated gauze, or a nickel-plated perforated steel strip is continuously carried through a container filled with the nickel paste and sintering is done in a horizontal furnace. The time of sintering in the furnace is ca 10–20 min.

Usually the plaques produced by either method are coined (compressed) in those areas where subsequent welded tabs are connected or where no active material is desired, eg, at the edges. The uncoined areas usually have a Brunauer-Emmet-Teller (BET) area in the range of 0.25–0.5 m^2/g and a pore volume >80%. The pores of the sintered plaque must be of suitable size and interconnected. The mean pore diameter for good electrochemical efficiency is 6–12 µm, determined by the mercury-intrusion method.

The process by which porous sintered plaques are filled with active material is called impregnation. The plaques are submerged in an aqueous solution, which is sometimes a hot melt in a compound's own water of hydration, consisting of a suitable nickel or cadmium salt and subjected to a chemical, electrochemical, or thermal process to precipitate nickel hydroxide or cadmium hydroxide. The electrochemical (46) and general (47) methods of impregnating nickel plaques have been reviewed.

In the original process for the positive electrode, the plaques were placed in a metal vessel, which was evacuated to <5.3 kPa (40 mm Hg), and a nearly saturated solution of nickel nitrate (density 1.6 g/mL) admitted. After a 5–15 min soaking period, the plaques were transferred at 101 kPa (1 atm) to a polarizing unit where they were cathodically polarized in hot caustic solution. After polarization the plates were washed and dried. These four steps were repeated four or five times until the desired weight gain of active material was achieved.

For the negative electrolyte, cadmium nitrate solution (density 1.8 g/mL) is used in the procedure described above. Because a small (3–4 g/L) amount of free nitric acid is desirable in the impregnation solution, the addition of a corrosion inhibitor prevents excessive contamination of the solution with nickel from the sintered mass (see CORROSION AND CORROSION CONTROL). In most applications for sintered nickel electrodes the optimum positive electrode performance is achieved when 40–60% of the pore volume is filled with active material. The negative electrode optimum has one-half of its pore volume filled with active material.

Other processes have been developed in which the impregnation is accomplished in one or two steps; the most promising is electrodeposition directly from nitrate solutions having pH controlled at 4–5. After electrodeposition, the plaques are either cathodically polarized in sodium hydroxide solution or electrochemically formed in sodium hydroxide to eliminate all traces of nitrate. The latter steps must proceed at low current densities to avoid blistering and shedding of the loaded plaques.

Some manufacturers add a small (10–20% of the positive loading) amount of cadmium to positive plates as an antipolar mass to prevent some of the problems of reversal in sealed cells. This practice may, however, create as many problems as it solves in that positive capacity is reduced proportionally to the quantity of antipolar mass added.

In most cases, the impregnation process is followed by an electrochemical formation where the plaques are assembled into large temporary cells filled with 20–30% sodium hydroxide solution, subjected to 1–3 charge–discharge cycles, and subsequently washed and dried. This eliminates nitrates and poorly adherent particles. It also increases the effective surface area of the active materials.

Formation also offers a convenient means of regulating the state of charge of the plates prior to cell assembly. This is important in sealed cell manufacture, where cell performance is optimized when 10–15% of the negative active mass is in the charged condition prior to initial cell charging. Some manufacturers have found that elimination of formation is feasible by extensive conversion of the nitrates in the impregnation process, followed by meticulous washing. However, charge retention can suffer to some degree because of traces of nitrate in the finished cells.

Cell Assembly. The methods for cell assembly, starting with the processed plaques depend on whether the cells are to be vented or sealed. For vented cells, processed plaques are usually compressed to 85–90% of their processed thickness allowing sufficient porosity for electrolyte retention and strengthening the plate structure. For sealed cells, sizing of the negative plaques is usually avoided because maximum surface area is important to oxygen recombination.

The next operation is the cutting of plaques into individual plates, usually through the coined areas (Fig. 5). Plate edges are sometimes coated with an

Fig. 5. Coined master plaque.

adherent plastic film to smooth any rough edges that could otherwise penetrate the thin separators used in sintered-plate cells. A tab of nickel or nickel-plated steel is welded to each plate. In some cases, this is accomplished by spot or projection welding at the coined area. For plates provided with perforated sheet grids the tabs are spot welded directly to the sheet at the proper point at the edge of the sheet. For both types of cells most of the requirements of the separators are high electronic resistance and low electrolytic resistance in the usual electrolyte (30% KOH solution). Separators should be as thin as possible and have small and evenly distributed pores. Additionally, they should be resistant to heat, KOH, gases formed on charge, and other components of the system.

The most important separator material property difference for vented and sealed cells is permeability to oxygen. In the charging of the vented cells, it is important to minimize oxygen transport from the positive to the negative plates. In float-charge applications, oxygen recombination at the negative plates results in generation of heat and ultimately to thermal runaway if excessive charge current is drawn with rising temperature. The opposite is required in sealed cells, where oxygen recombination is necessary to prevent an increase of internal pressure. Glycerol-free cellophane is used as part of the separator in the vented cells. Other components in typical vented cell separators are woven nylon, felted nylon, felted nylon–cellulose, felted (vinyl chloride–acrylonitrile) copolymer, microporous polypropylene, and irradiated polyethylene. Most sealed-cell separators are single-layer felted materials, usually nylon or polypropylene, chosen for the ability to retain electrolyte and prepared in such a manner as to allow oxygen permeability, eg, calendering of such materials should be avoided.

Another difference between vented and sealed cell manufacture is found in the ratio of negative to positive active material. Although both types of cells require an excess of negative material, in the vented type a ratio of 1.5 is sufficient for the high rates of discharge and low temperature conditions encountered. In the sealed type a typical ratio is 2, to provide for sufficient sites for the oxygen recombination, to provide for the residual negative capacity mentioned earlier, and to preclude the possibility of the negative plate reaching the fully charged state exemplified by H_2 evolution.

Both vented and sealed cells use the same basic electrolyte (30% KOH), but different amounts are required. The vented cell contains a considerable amount of free electrolyte to allow for decomposition and loss of water on charge and to allow for maximum performance on discharge. The sealed cell contains only enough electrolyte to fill the plate pores and to completely saturate the separator; excess electrolyte inhibits oxygen recombination by reducing the number of sites, ie, oxygen–cadmium–electrolyte interfaces, for oxygen recombination. In certain applications, 15–30 g/L lithium hydroxide is added to the electrolyte. Where elevated temperature operation is encountered, this addition improves charge acceptance, especially in sealed cells. Larger (50 g/L) amounts of lithium hydroxide are used in repeated cycling with constant voltage charging and in float-charge applications (vented cells). This larger amount of lithium hydroxide maintains capacity better than KOH alone but increases electrolyte resistance and adversely affects low temperature performance.

The presence of certain forms of cellulose (qv) in the electrolyte is beneficial for negative electrodes, probably by preventing the formation of large grain sizes of cadmium on charge. Cellophane in vented cell separators usually provides sufficient cellulose for this effect. Early sealed cells using a single layer of cellulosic filter paper as separator gave excellent performance but had reduced cycle life as a result of the degradation of the cellulose. In later sealed cells, where felted nylon or felted polypropylene was used as separator, the addition of small amounts of cellulose to the electrolyte, improved negative plate performance, but led to early separator degradation.

Assembly of vented cells begins by interleaving the electrodes and separators. A most widely used separator is a sandwich of woven nylon (0.08–0.10 mm), cellophane (0.05 mm), and woven nylon (0.08–0.10 mm) of the proper width, usually 0.6 cm wider than the plate height, and of a length to form a continuous barrier around the edges of all the plates. The electrode tabs of each polarity are brought together and connected to the cell terminals by spot welding or, in a few cases, by bolting. The stack is attached to the cell cover by inserting terminals through holes that are marked for polarity. When threaded terminal posts are used, sealing is accomplished by compressing a neoprene gasket or O-ring between a flat or groove on the horizontal surface of the terminal and the inside surface of the cover, with the proper torque being applied by means of a nut on the outside of the cover. In the case of a smooth terminal post, compression is effected by means of a snap ring and belleville washer arrangement.

The most common cell case and cover materials are nylon, polyolefin, polysulfone, and styrenic (ABS). After inserting the cell element, with cover attached, into the cell case, cover and case are cemented. For nylon, a most satisfactory adhesive is phenol. For styrene copolymers, either a solvent seal or an ultrasonic seal is effective (see ADHESIVES).

The cell filler caps in the covers are designed to prevent spillage and usually contain a venting mechanism to allow for escape of gas during charge and to prevent the carbon dioxide of the air from contacting the electrolyte. Most filler caps are a type of rubber-ring screw vent assembly provided with holes or slits that are sealed by the rubber ring. These are usually made of nylon or polystyrene, but in certain applications they are made of nickel-plated steel.

In filling cells with electrolyte, vacuum techniques are employed for uniformity and for hastening the complete wetting of plates and separator. A properly manufactured cell is ready for use immediately after the first charge following electrolyte fill. The first charge is ideally at the 10-h rate for 20 h. Higher rates can be used but should not exceed the 5-h rate.

Plastic-cased cells are assembled into batteries by placing them a small distance apart or in contact with heat-conducting fins in a specially treated steel container tray. The cell spacing depends on the battery application. For example, in aircraft starting batteries provision is required for circulating cooling air between cells. The steel boxes (battery cases) are previously prepared by coating all surfaces with a tough, insulating plastic (usually epoxy). Excellent results are obtained using a fluidized bed technique of application, which results in a uniformly thick coating. Intercell connectors are normally of nickel-plated copper, but in small sizes connectors are often of nickel or nickel-plated steel.

The prismatic sealed cells are made in a manner similar to the above. Most prismatic sealed cells use a metal case and cover. The most desirable case material is stainless steel, although nickel-plated steel can be used. Terminal feed-through is effected by ceramic-to-metal or glass-to-metal sealing techniques; and case-to-cover seal, by inert gas welding. Most prismatic sealed cells incorporate a high release, resealable vent in the cover that usually consists of a metallic spring working on an elastomer sealing disk or ring.

The prismatic sealed cells are not self-supporting and are normally used in battery operation where the battery case is used to constrain cell cases because internal cell pressures in the range of 690 kPa (100 psi) are common.

By far the majority of sealed cells are of the small cylindrical self-supporting type in the familiar "AA," "C," and "D" commercial sizes such as that shown in Figure 6. The element for these cells contains only one plate of each polarity; thus the electrodes are relatively long. The plates and the single-layer separator are rolled into a tight spiral, the plates arranged so that the positive tabs protrude from the element in one direction and the negative tabs in the opposite direction. The top and bottom of the element are protected and insulated by circular plastic disks having slits for the protruding tabs.

Elements are then inserted in nickel-plated steel cans, normally with the negative tab bent 90° to make contact with the can bottom. The tab is spot welded to the can bottom employing a long slender electrode that fits through the center hole of the element. The cell cover is welded to the positive tab. The cover

Fig. 6.   Partial cutaway of a coiled, sintered-plate, nickel–cadmium cell ("D"-size).

is nickel-plated steel with a nylon gasket around the edges. After adding electrolyte, the cover is pressed into the can so that the nylon gasket rests on a shelf formed into the can well by scoring (after insertion of element). The rim of the can and sometimes the upper part of the gasket is folded over either in a press or by a rolling, rotating tool.

The cover assembly usually contains a safety vent. The most common type consists of a steel or nickel diaphragm built into the cover and a bent point cut from the cover. At a certain internal cell pressure, the diaphragm moves to the bent point and is pierced. Some cylindrical sealed cells use one or more terminal feed-throughs employing glass-to-metal or ceramic-to-metal techniques.

Battery assembly using cylindrical cells varies, and cell-to-cell connections are spot welded after using either flat tabs or cup tabs. Cell-to-cell insulation is effected either by using plastic cell jackets (shrink-on) or by inserting cells in plastic modules with each cell occupying its own cavity.

Other Cells. Other methods to fabricate nickel–cadmium cell electrodes include those for the button cell, used for calculators and other electronic devices. This cell, the construction of which is illustrated in Figure 7, is commonly made using a pressed powder nickel electrode mixed with graphite that is similar to a pocket electrode. The cadmium electrode is made in a similar manner. The active material, graphite blends for the nickel electrode, are almost the same as that used for pocket electrodes, ie, 18% graphite.

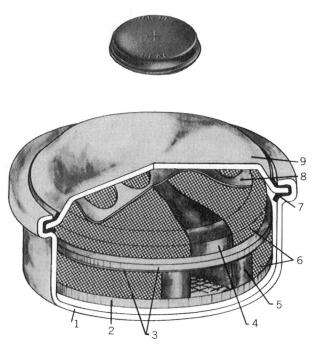

Fig. 7. Section of disk-type cell where: 1, is the cell cup; 2, is the bottom insert; 3, is the separator; 4, is the negative electrode; 5, is the positive electrode; 6, is the nickel wire gauze; 7, is the sealing washer; 8, is the contact spring; and 9, is the cell cover.

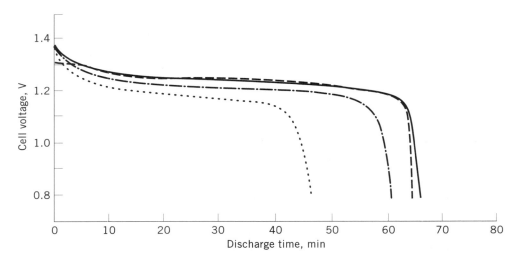

Fig. 8. Discharge capacity of small sealed nickel–cadmium cells where the initial charge is 0.1 C × 16 h at 20°C and the discharge is 1 C at temperatures of (· · ·) −20°C, (− · −) 0°C, (—) 20°C, and (– – –) 60°C. C is the current required to discharge the cell in one hour.

Lower cost and lower weight cylindrical cells have been made using plastic bound or pasted active material pressed into a metal screen. These cells suffer slightly in utilization at high rates compared to a sintered-plate cylindrical cell, but they may be adequate for most applications. The effect of temperature and discharge rate on the capacity of sealed nickel-cadmium cells is illustrated in Figure 8 and Table 3.

Applications.   In the U.S., rechargeable NiCd batteries provide power for about three-fourths of most common portable products (48). Uses are divided into three categories: pocket cells are used in emergency lighting, diesel starting, and stationary and traction applications where the reliability, long life, medium-high rate capability, and low temperature performance characteristics warrant the extra cost over lead–acid storage batteries; sintered, vented cells are used in extremely high rate applications, such as jet engine and large diesel engine starting; and sealed cells, both the sintered and button types, are used in computors, phones, cameras, portable tools, electronic devices, calculators, cordless razors, toothbrushes, carving knives, flashguns, and in space applications, where

Table 3. **Nominal Capacities of Consumer Nickel–Cadmium Cells**

Size	Capacity, A·h	
	Normal cells	High capacity cells
"AA"	0.6–0.7	0.9–1.0
"Sub-C"	1.2	1.7
"C"	2.0	2.3
"D"	4.0	5.0
"F"	7.0	

nickel–cadmium is optimum because it can be recharged a great number of cycles and given prolonged trickle overcharge. Cells of this category are generally made in sizes comparable to conventional dry cells, such as "D", "C", "AA", etc. In sizes larger than the "D" cell, sealed lead–acid cells offer a useful economic alternative for applications where the extra weight and space of the lead–acid is not critical.

In order to reduce costs, achieve higher energy density, and minimize environmental problems, a growing percentage of prismatic and cylindrical cells are now made by pasting on substrate. The newer substrates include nickel fiber mat, nickel foam, and nickel-plated graphite fiber. One manufacturer uses a nickel-plated plastic fiber substrate for the fabrication of large nickel–cadmium cells in prismatic configurations (28). These cells are reported to require less maintenance than the older design pocket cells (48). In small industrial sizes, pasted cells are offered as sealed batteries.

Charger Technology. Alkaline storage batteries are commonly charged from rectified d-c equipment, solar panels, or other d-c sources and have fairly good tolerance to ripple and transient pulses. Because the voltage of the nickel electrode is variable, the cutoff voltage is not a good control parameter for nickel–cadmium cells. It is, however, often used in vented cell chargers. For sealed nickel–cadmium cells, and other systems where a combination mechanism exists, a negative voltage slope detector is often incorporated into the charger control circuit. As the charge of a cell or battery progresses there is a slow rise in the unit voltage. When the nickel electrode approaches full charge, the oxygen evolved combines with the cadmium electrode reducing the overpotential on the cadmium electrode. This slight change in cell voltage can be detected by electronic voltage slope detectors and used to reduce the charge current or shut off the charge. A method for charging sterilizable rechargeable batteries has been reported (49).

Nickel–Iron Cells. The original tubular design nickel–iron battery developed by Edison has little commercial application. In the 1980s –1990s there was a renewed interest in a high rate sintered electrode design for electric vehicle applications (50–58). However, when interest shifted to hybrid electric vehicles, in the early 2000s, with lower battery capacity requirements, interest shifted to Ni-MH because of superior performance (at greatly increased cost).

Electrochemistry and Kinetics. The electrochemistry of the nickel–iron battery and the crystal structures of the active materials depends on the method of preparation of the material, degree of discharge, the age (life cycle), concentration of electrolyte, and type and degree of additives, particularly the presence of lithium and cobalt. A simplified equation representing the charge–discharge cycle can be given as:

$$2\,NiOOH^* + \alpha\text{–}Fe + 2\,H_2O \rightleftharpoons 2\,Ni(OH)_2^* + Fe(OH)_2$$

where the asterisks indicate adsorbed water and KOH. However, the discharge can be carried to a lower plateau for the iron electrode, although this is usually undesirable for life cycle, represented by:

$$8\,NiOOH^* + 3\,Fe + 4\,H_2O \rightleftharpoons 8\,Ni(OH)_2^* + Fe_3O_4$$

When discharges are carried out beyond the voltage range of the ferrous hydroxide [18624-44-7], $Fe(OH)_2$, plateau which is ca -0.90 to -0.85 V vs HgO, a second reaction in the voltage range of -0.65 to -0.5 V takes place (59,60).

$$3\ Fe(OH)_2 + 2\ OH^- \rightleftharpoons Fe_3O_4 + 4\ H_2O$$

Discharging to this lower cell voltage usually results in shorter cycle life. Enough excess iron should be provided in the cell design to avoid this problem.

Active iron in the metallic state is slowly attacked by the alkaline electrolyte according to

$$Fe + 2\ H_2O \longrightarrow Fe(OH)_2 + H_2$$

This reaction is accelerated by increased temperature, increased electrolyte concentration, and by the use of sodium hydroxide rather than potassium hydroxide in the electrolyte. It is believed that the presence of lithium and sulfur in the electrode suppress this problem. Generally, if the cell temperature is held below 50°C, the oxidation and/or solubility of iron is not a problem under normal cell operating conditions.

Electrode Structures. The classical iron active material for pocket and pasted iron electrodes was formed by roasting recrystallized ferrous sulfate [7720-78-7], $FeSO_4$, in an oxidizing atmosphere to ferric oxide [1309-37-1], Fe_2O_3, and then reducing the latter in hydrogen. The α-iron formed was then heated to a mixture of Fe_3O_4 and Fe. As such it was pure enough to be used for pharmaceutical purposes. For battery use, a small amount of sulfur, as FeS, was added as were other additives which were believed to increase the cycle life by acting as depassivating agents, ie, helping to reduce the tendency of iron to evolve hydrogen upon standing in alkaline electrolyte. Extensive studies on the stability of iron active material have been reported (61–63). Addition of cadmium and antimony salts have been claimed to decrease hydrogen evolution on stand by 50% (62). However, some blends also reduce electrode capacity. A blend of materials such as mucic acid [526-99-8], $C_6H_{10}O_8$, with indium [7440-74-6], In, increases the iron stability on stand without changing electrode capacity (63). The effects of electrolyte makeup and concentration on iron corrosion have also been studied (64).

A study of sintered iron electrodes claimed advantages of high rate capability, long life, and low hydrogen evolution (52). Fine carbonyl generated iron powder was coated on a nickel screen and sintered in a hydrogen–nitrogen atmosphere. Sintered raw plaques were believed suitable if the porosity was between 50 and 80%. The sinter had to be strong enough to leave a conductive skeleton after part of the iron was corroded to active material. This process is unlike sintered-type nickel electrodes in which active material is deposited directly into the pores of a very porous plaque. Sintered iron electrodes (49) were in pilot production in Sweden and the United States.

Sintered nickel electrodes used in nickel iron cells are usually thicker than those used in Ni/Cd cells. These result in high energy density cells, because very high discharge rates are usually not required.

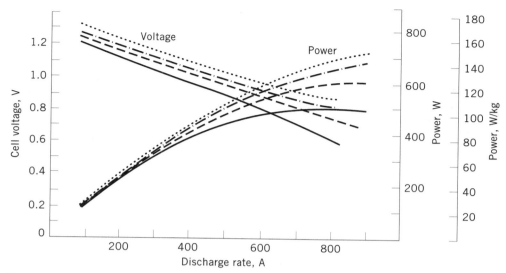

Fig. 9. Power and voltage characteristics of the nickel–iron cell where the internal resistance of the cell, R_i, is 0.70 mΩ, at various states of discharge: ($\cdots$) 8%; ($-\cdot-$) 32%; ($---$) 52%; and ($—$) 72%. Courtesy of Westinghouse Electric Corp.

Performance Characteristics.   The sintered nickel-sintered iron design battery has outstanding power characteristics at all states of discharge making them attractive to the design of electric vehicles (EV) which must accelerate with traffic even when almost completely discharged. Although the evolution of hydrogen is a problem preventing sealed cell design, introduction of automatic watering systems have ameliorated the maintenance time requirements. Typical performance curves are given in Figure 9.

1.2. Silver–Zinc Cells.   The silver–zinc battery has the highest attainable energy density of any rechargeable system in use except for lithium-ion cells as of this writing. In addition, it has an extremely high rate capability coupled with a very flat voltage discharge characteristic. Its use, in the early 1990s, was limited almost exclusively to the military for various aerospace applications such as satellites and missiles, submarine and torpedo propulsion applications, and some limited portable communications applications. The main drawback of these cells is the rather limited lifetime of the silver–zinc system. Life is normally limited to less than 200 cycles with a total wet-life of no more than about two years. The silver–zinc system also carries a very high cost and applications are justified only where cost is a minor factor. The high cost of silver battery systems is attributable to the cost of the active silver material used in the positive electrodes.

Cellophane or its derivatives have been used as the basic separator for the silver–zinc cell since the 1940s (65,66). Cellophane is hydrated by the caustic electrolyte and expands to approximately three times its dry thickness inside the cell exerting a small internal pressure in the cell. This pressure restrains the zinc anode active material within the plate itself and renders the zinc less available for dissolution during discharge. The cellophane, however, is also the principal limitation to cell life. Oxidation of the cellophane in the cell environment degrades the separator and within a relatively short time short circuits may occur in the cell. In addition, chemical combination of dissolved silver species in the electrolyte may form a conductive path through the cellophane.

A second lifetime limitation is the zinc anode. In spite of the separator and cell designs, some zinc material is solubilized during the charge–discharge reaction. Over a period of cycling there is a shift of active material, originally distributed evenly over the face of the electrode, to the center and bottom areas of the electrode (50). This shape change limits the life of the cell as exemplified by a fading of the capacity and a build-up of internal pressure that may eventually lead to a short circuit.

Reaction Mechanisms.   There is considerable difference of opinion concerning the specific cell reactions that occur in the silver–zinc battery. Equations that are readily acceptable are

$$2\,AgO + Zn + H_2O \rightleftharpoons Ag_2O + \begin{array}{c} Zn(OH)_2 \\ (ZnO + H_2O) \end{array} \qquad E^{\circ} = 1.85\ V$$

and

$$Ag_2O + Zn + H_2O \rightleftharpoons 2\,Ag + \begin{array}{c} Zn(OH)_2 \\ (ZnO + H_2O) \end{array} \qquad E^{\circ} = 1.59\ V$$

The charge and discharge of silver–zinc cells occurs at two voltage levels, representing the energy levels of AgO (an overall valence 2 silver oxide compound sometimes mislabeled as silver peroxide) and a lower energy level material Ag_2O. The AgO is often represented as Ag_2O_2, since crystallographic data indicates there are two different bond lengths in the structure.

Electrochemistry.   Silver–zinc cells have some unusual thermodynamic properties. The equations indicate that the higher valence silver oxide is AgO, silver(II) oxide [1301-96-8]. However, in the crystallographic unit cell, which is monolithic, there are four silver atoms and four oxygen atoms, and none of the Ag–O bonds conforms to a silver(II) bond length. Instead there are two Ag–O bonds of 0.218 nm corresponding to silver(I) and two Ag–O bonds of 0.203 nm corresponding to silver(III) (67). This structure has also been proposed on the basis of magnetic and semiconductor properties (67) and confirmed using neutron diffraction (68,69).

For the Ag–O material a reversible voltage of 1.856 V is obtained and the $(\partial E/\partial T)_P$ is positive, $+5.7 \times 10^{-5}$ V/°C. For the Ag_2O material a reversible voltage of 1.602 V is obtained in 11.6 N KOH but the $\partial E/\partial T$ is negative, -16.9×10^{-5} V/°C. This is about one-third the value of nickel–cadmium cells. However, because the negative coefficient compound is not present in the fully charged state, the high risk of thermal runaway described for nickel–cadmium cells does not exist. The high conductivity of the silver active material and the low internal resistance of the remaining components do, however, make thermal runaway a possibility to be considered in charger and system design.

Electrodes.   All of the finished silver electrodes have certain common characteristics: the grids or substrates used in the electrodes are usually made of silver, although in some particular cases silver-plated copper is used. Material can be in the form of expanded silver sheet, silver wire mesh, or perforated silver sheet. In any case, the intent is to provide electronic contact of the external circuit of the battery or cell and the active material of the positive plate. Silver is necessary to avoid any possible oxidation at this junction and the increased resistance that would result.

Finished electrodes need fairly good physical strength so that they can be handled easily during separator wrapping and cell assembly. This is usually accomplished by a sintering process. Finished electrodes should also have a relatively high surface area per unit weight of active material coupled with an apparent porosity of about 50–60% based on the active material. The high surface area of the active material is attained by high surface area starting materials. These can be finely divided powders of metallic silver or either the monovalent or divalent oxides of silver. Silver electrodes can attain coulombic efficiencies of up to 85% of theoretical when manufactured from high surface area active material.

There are three methods of silver electrode fabrication: (1) the slurry pasting of monovalent or divalent silver oxide to the grid, drying, reducing by exposure to heat, and then sintering to agglomerate the fine particles into an integral, strong structure; (2) the dry processing of fine silver powders by pressing in a mold or by a continuous rolling operation onto a silver grid followed by sintering; and, (3) the use of plastic-bonded active material formed by imbedding the active material (fine silver powder) in a plastic vehicle such as polyethylene, which can then be milled into flexible sheets. These sheets are cut to size, pressed in a mold on both sides of a conductive grid, and the pressed electrode subjected to sintering where the plastic material is fired off, leaving the metallic silver. Silver electrodes produced by these processes range from 0.18 to 1.52 mm in thickness. Electrodes prepared by method (1) usually cannot be below 0.76 mm in thickness whereas the thinner electrodes can be prepared by methods (2) and (3).

Silver electrodes prepared by any of the three methods are almost always subjected to a sintering operation prior to cell or battery assembly. Sintering is basically a heating operation at temperatures well below the melting point of pure silver, 960.5°C, during which an agglomeration of the particles occurs, greatly strengthening the electrodes produced. Sintering is normally carried out in electric muffle furnaces, either on an individual batch process or a continuous conveyor belt-type operation. Reducing atmospheres are not necessary for silver electrode sintering because the process is carried out well above the decomposition temperatures of the various silver oxides. Sintering process parameters vary, however, examples are: 537°C for 30 min and 732°C for 3 min. The longer exposure at a somewhat lower temperature is said to produce electrodes that are less susceptible to shedding. Such a procedure, however, requires a furnace with a long heating zone, is more expensive, and is not practical to most manufacturers.

Zinc electrodes for secondary silver–zinc batteries are made by one of three general methods: the dry-powder process, the slurry-pasted process, or the electroformed process. Current-carrying grids for zinc electrodes can be the same regardless of the process of plate manufacture chosen. Expanded metal, screen, or perforated metal is the generally accepted form for these grids. Silver is the material of preference. However, cost considerations often dictate that copper be used. In these cases silver-plated copper forms are usually employed to avoid the possibility of copper dissolving in the caustic solution during over-discharge of a cell.

The active material used in any of the processes for the manufacture of electrodes is a finely divided zinc oxide powder, USP grade 12. The zinc oxide active material is usually blended with from 1–4% mercuric oxide in the dry state for any of the dry processing procedures. Mercuric oxide is converted to mercury during charge which then amalgamates the zinc formed at the same time. This tends to suppress the evolution of hydrogen on the zinc electrode during charged stand, and is required in most military applications to avoid a hydrogen hazard.

In the dry powder process the active material mix is spread evenly in a mold, the grid and lead assembly inserted, and the entire electrode then pressed in a hydraulic press. Often binder additives such as poly(vinyl alcohol) (PVA) or fine Teflon powder are added to the active material mix to increase the cohesiveness of the finished electrodes. Cellulosic paper liners are also used, and they are inserted into the mold prior to introduction of the active mix and grid. After the active mix and grid are introduced, the paper liner is then folded over the top of the electrode and the electrode pressed as before. Porosities obtained by pressing electrodes in this manner are usually in the range of 50%, although in actual production these figures can range from about 35–60% depending on the rate of discharge required. In the slurry or paste method a paste is prepared by mixing water with the active material mix (zinc oxide plus mercuric oxide). Sometimes a binder such as CMC, or a flock such as short rayon or Dynel fibers is added to the paste to increase the cohesiveness of the pasted electrode. Pasting is usually done on large strips of grid material.

After pasting, the strips are air dried at relatively low heats, individual electrodes are cut from the strip, and the electrodes are pressed to desired thicknesses. The porosities and densities of the active material made by the pasting processes are approximately the same as those indicated for the dry pressed powder processes. One danger in pasting electrodes is the occurrence of sharp grid edges on the electrodes after cutting to size. Often a secondary operation of smoothing or trimming is required to avoid this problem.

The electroforming or deposition of zinc from solution uses a slurry of zinc oxide in strong caustic solution or actual metallic zinc anodes in caustic solution. In the slurry method, the zinc oxide is deposited on silver or copper grids that comprise the grids of the finished electrodes. In the metallic zinc anode method, the solutions are not depleted but rather the anodes are depleted. Both methods utilize large sheets onto which the active zinc material is deposited. Often individual plate leads are welded in position prior to deposition. In other cases active zinc must be scraped from the grid in a secondary operation and the plate leads then welded into position. Zinc plates prepared by these deposition methods usually have the active material in a voluminous, mossy state. After plating, the sheets of deposited material must be washed to completely free them of any traces of caustic solution to avoid fire hazards during production and the subsequent drying operation.

After the plates have been washed and dried thoroughly, they are pressed in a preliminary operation to the desired thickness. Individual electrodes are then cut from the sheets and a secondary pressing operation to final thickness is done. Often a secondary operation is required to remove sharp edges of electro-deposited zinc electrodes.

Silver–Zinc Separators. The basic separator material is a regenerated cellulose (unplasticized cellophane) which acts as a semipermeable membrane allowing ionic conduction through the separator and preventing the migration of active materials from one electrode to the other. Usually, multiple layers are used in cell fabrication.

A stronger separator is one made of sausage casing material (FSC), a regenerated cellulose similar to cellophane but including some fibrous material. FSC is usually extruded in tubes and electrodes are inserted into each end of the tube. The tube is folded to form the so-called U wrap.

Another method of extending the life of the cellulosic separators has been to incorporate a silver organic compound, eg, silver xanthate [6333-67-1], into the cellulosic separator. This is done by passing the separator material through a hot caustic solution containing a silver salt. The result is a deposition of a silver organic compound within the structure of the cellophane or FSC separator. This type of material resists the degradative effects of oxidation more than the untreated material. In general the positive electrodes are wrapped in the layers of separator. Normally, an absorber is used around each positive electrode to maintain an adequate supply of electrolyte. Absorbers are usually of nonwoven materials, such as polyamide or polypropylex felts. In some cases, nonwoven cellulosic felts are also used, but these tend to degrade more rapidly. The absorber wrapped positive electrodes are usually wrapped in several layers of cellophane or FSC and then folded to form Us. Negative electrodes are alternately stacked forming the cell assembly. Normally ca three or four layers of PUDO (battery-grade) cellophane 0.025 mm thick are used in so-called high rate designs. For lower rate cells where longer life is required the number of layers of cellophane may be as many as ten. Occasionally cells are constructed with the negative electrodes wrapped in the separator material. This is said to reduce the shape change effect on the negative electrodes and increase life of the cell. These so-called reverse-wrap cells are usually reserved for low rate applications only.

Electrolyte. The electrolyte in silver–zinc cells is 30–45% KOH. The lower concentrations in this range have higher conductivities and are preferred for high rate cells. Higher concentrations have a less deleterious effect on cellulosic separators and are preferable for extended life characteristics. The higher concentrations also have a greater capacity for dissolving zinc oxide and accelerate change in shape of the zinc electrode. In most cases, a concentration of about 40% KOH is considered as the optimum. Occasionally, some manufacturers use as electrolyte a saturated zincate solution at the particular concentration desired. This supposedly slows the effect of zinc dissolution. However, in actual use such additives have not been shown to have much beneficial effect. Other additives, such as LiOH, which are used in other alkaline systems, have no beneficial effect on silver–zinc batteries. In practice the electrolyte is added to the cell in the discharged condition, and a period of soaking is allowed for the separators to absorb electrolyte and the electrodes to become thoroughly impregnated with electrolyte.

Cell Hardware. Cell jars are constructed almost exclusively of injection-molded plastics, which are resistant to the strong alkali electrolyte. The most generally used materials are modified styrenes or copolymers of styrene and acrylonitrile (SAN). Another material that has been found to increase shock resistance of cells is ABS plastic (acrylonitrile–butadiene–styrene). All of these plastics can be injection-molded, are solvent-sealable and, in general, meet operating temperature ranges up to about 70°C. For applications that require greater resistance to temperature, some of the more recent plastics such as polysulfone and poly(phenylene oxide) (PPO) injection-moldable materials able to withstand operating temperatures up to 150°C are used.

Cell terminal connections are usually brought out by two-threaded terminals that protrude through the cell jar cover. They are usually steel, brass, or copper with a hollow construction. The plate leads are soldered in place in the center hollow portion of the terminal to effect an electrical contact and cell seal. The terminal itself is potted into the jar cover using epoxy-type potting compounds. Normally, terminal hardware is silver-plated. However, for corrosion resistance nickel-plating has been used.

Performance. Charging.   Charging of silver–zinc cells can be done by one of several methods. The constant-current method which is most common consists of a single rate of current usually equivalent to a full input within the 12–16-h period. A typical two-level constant-current charge curve results as shown in Figure 10. The first 20–30% of the A·h input is attained at slightly above the monovalent silver level of ca 1.65 V. After this input a sharp rise to the divalent level takes place, and the remainder of the charge is completed at a voltage somewhat above the divalent silver level of 1.90–1.95 V. When the silver electrodes have reached full charge, there is a sharp rise to about 2.0–2.05 V; at this point oxygen is liberated at the positive electrode. Normally the charge is terminated to avoid excessive overcharge of the zinc electrodes and further oxygen evolution. If the charge is continued beyond this point, charging voltage remains at approximately a 2.05–2.1 V level until all of the zinc material is fully charged. Then another rise in voltage occurs to ca 2.2–2.3 V, and both hydrogen and oxygen are liberated. Further charging results in electrolysis of the water in the electrolyte.

Other acceptable charging methods that have been used are the two-level charging, modified constant potential, and constant potential methods. The two-level method is one in which a relatively high rate of charge is used until

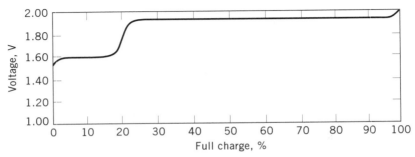

Fig. 10.   Constant-current charge curve for a high rate Ag–Zn cell at room temperature. Charging carried out at the 10-h rate.

a specified input is obtained, at which time the charging rate is lowered to a second level until a voltage cut-off is reached. In the modified constant potential method an initial charging rate is set, somewhat higher than would be used for a single-level constant, current charge; and then the charging current is allowed to drift downward as the battery voltage rises during charge. The constant potential systems are usually current-limited to avoid excessive inrush currents, and are essentially similar to modified constant potential methods in that the initial current is a very high value which rapidly decreases and approaches zero during final stages of charge.

Charge acceptance of the silver–zinc system is normally on the order of 95–100% efficient based on coulombic (ampere-hour output over input) values. This is true of any of the charging methods when carried out in the proper manner. Thus overcharge is rarely necessary in charging silver–zinc cells and batteries.

Discharge. Silver–zinc cells have one of the flattest voltage curves of any practical battery system known. However, there are two voltage plateaus. Even at rates as high as 10 minutes a fairly flat characteristic is obtained. The actual level of voltage is, of course, rate dependent. However, because of the high conductivity of the silver electrode, derating of voltage with higher rates is less than other systems. Figure 11 gives typical discharge curves for a high rate silver–zinc cell. At the low rates the initial part of the voltage discharge curve exhibits the higher level or peroxide voltage. After about 20–25% of the A·h capacity has been discharged, this drops to the monovalent level. As the rate increases, the proportion of the discharge capacity at the higher voltage level becomes less, and at very high rates a dip often occurs at the beginning of discharge as in the 10-min curve. The capacity of the silver–zinc cells is also rate dependent, but less derating occurs here than in most other batteries. For example, from Figure 11 it can be seen that even at the 10-min rate a high rate silver–zinc cell provides over 60% of its rated capacity.

Performance of silver–zinc cells is normally considered to be adequate in the temperature range of 10–38°C. If a wider temperature range is desired silver–zinc cells and batteries may be used in the range 0–71°C without any appreciable derating. Lower temperatures result in some reduction of cell voltage capacities at medium to high rates, and higher temperatures curtail life

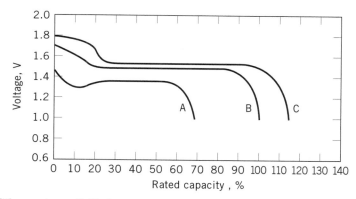

Fig. 11. Silver–zinc cell discharge curves at rates of A, 10 min; B, 1 h; and C, 10 h.

because of deterioration of the separator materials. Operation at temperatures below 0°C results in more serious derating and normally external heat is provided for operation in this range. Long use at temperatures above 71°C seriously curtails life.

Cell Life.   Silver–zinc cells are usually manufactured as either low or high rate cells. Low rate cells contain fewer and thicker electrodes and have many layers of separator (up to the equivalent of 10 layers of cellophane). High rate cells, on the other hand, contain many thinner electrodes and have separator systems of the equivalent of three to four layers of cellophane. Approximately 10–30 cycles can be expected for high rate cells depending on the temperature of use, the rate of discharge, and methods of charging. Low rate cells have been satisfactorily used for 100–300 cycles under the proper conditions. In general, the overall life of the silver–zinc cell with the separator systems normally in use is approximately 1–2 yr.

1.3. Other Silver Positive Electrode Systems.   *Silver–Cadmium Cells.*   The first silver–cadmium batteries were manufactured in 1900 for motor cars. Use of this electrochemical system was quite limited. Then in the late 1950s, there was interest for applications such as appliances, power tools, and scientific satellites when it was hoped that silver–cadmium batteries could offer an energy density close to silver–zinc batteries and a life characteristic approaching that of the nickel–cadmium system. In satellite applications the nonmagnetic property of the silver–cadmium battery was of utmost importance because magnetometers were used on satellites to measure radiation and the effects of magnetic fields of energetic particles. Satellites had to be constructed of nonmagnetic components in sealed batteries.

The overall reactions are

$$2\,AgO + Cd + H_2O \rightleftharpoons Ag_2O + Cd(OH)_2 \qquad E° = 1.38 \text{ V}$$

$$Ag_2O + Cd + H_2O \rightleftharpoons 2\,Ag + Cd(OH)_2 \qquad E° = 1.16 \text{ V}$$

The silver–cadmium cell exhibits a two-plateau voltage characteristic on charge and discharge. At moderate discharge rates at 25°C, approximately 20% of the A·h capacity is delivered at a nominal 1.2 V; and the remaining capacity is delivered at 1.08–0.9 V. The discharge voltage characteristics are very sensitive to current rates, amount of cycling, charged stand time, float-charging, and temperature. For example, cells that are continuously float-charged exhibit a flat discharge voltage and no loss of capacity. At moderate charge rates, the voltage characteristic consists of two levels, 1.3 V and 1.5 V, nominal. Either constant current or constant potential charging is used. During cycling, A·h efficiencies are >95% and W·h efficiencies are <75%.

The positive plates are sintered silver on a silver grid and the negative plates are fabricated from a mixture of cadmium oxide powder, silver powder, and a binder pressed onto a silver grid. The main separator is four or five layers of cellophane with one or two layers of woven nylon on the positive plate. The electrolyte is aqeous KOH, 50 wt%. In the aerospace applications, the plastic cases were encapsulated in epoxy resins. Most useful cell sizes have ranged from 3 to 15 A·h, but small (0.1 A·h) and large (300 A·h) sizes have been evaluated. Energy densities of sealed batteries are 26–31 W·h/kg.

Silver–cadmium satellite batteries have been used in cyclic periods of five hours or more with discharge times of 30–60 min. Based on nominal capacities, depths of discharge have been 8–30%. The electrical performance of silver–cadmium batteries degrades below 0°C and above 40°C. At low temperatures the main problems are capacity maintenance during cycling and a dip in voltage on initiation of discharge. Operational and test programs have shown cycle life periods of 3 yr at low temperatures. At temperatures of 40°C and 50°C, the cycle life is 1 yr and 0.2 yr, respectively. The cycle life at intermediate temperatures is 1.4–2.0 yr.

Another application for silver–cadmium batteries is propulsion power for submarine simulator-target drones. High current drains are required (average C, pulses up to 6C), and greater recyclability than the silver–zinc counterparts used in torpedo propulsion. Batteries designed for this use utilize vented cells, high temperature plastic cell jars, and cell designs having a large number of thin electrodes to maximize electrode surface area.

Silver–Iron Cells. The silver–iron battery system combines the advantages of the high rate capability of the silver electrode and the cycling characteristics of the iron electrode. Development has been undertaken (70) to solve problems associated with deep cycling of high power batteries for ocean systems operations.

Cells consisted of porous sintered silver electrodes and high rate iron electrodes. The latter were enclosed with a seven-layered, controlled-porosity polypropylene bag which serves as the separator. The electrolyte contains 30% KOH and 1.5% LiOH.

Initial experiments were conducted with 350–A·h cells that maintained capacity over 200 cycles. Conventional silver-zinc cells lose capacity in similar deep-cycle operations. Cell tests conducted under a variety of temperature and pressure conditions revealed no discernable effects up to pressures of 69 MPa (10,000 psi) using a flooded electrolyte system. Tests at 0–25°C showed less than 5% capacity loss resulting from temperature when tested at the 8-h discharge rate. The cells as produced have a discharge voltage of ca 1.1 V. Voltage capacity characteristics of a nominal 140–A·h silver–iron cell are shown in Figure 12. The discharge capacity of a typical cell is 145 – 155 A·h at discharge rates of 80–10 A. A battery of 24 series-connected cells (containing excess electrolyte) weighs 40 kg and provides an energy density of 3.5 kW·h. At the 3-h rate the energy density is 100 W·h/kg, or about three and a half times the energy density of comparable lead–acid batteries.

Applications have been found for these batteries in emergency power applications for telecommunications systems in tethered balloons. Unfortunately, the system is expensive because of the high cost of the silver electrode. Applications are, therefore, generally sought where recovery and reclamation of the raw materials can be made.

Small silver–iron sealed button cells have been produced in Sweden (70) for long-life operation of hearing aids, calculators, and electric razors. These units have energy densities similar to silver–zinc button cells but are reported to be capable of being fully charged and discharged up to five hundred times without leakage or change in performance. Comparable performance silver–zinc cells have limited cycle life of the order of 75–100 cycles.

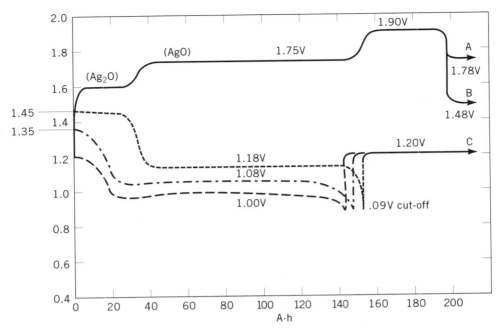

Fig. 12. Charge–discharge characteristics of a nominal $140-A \cdot h$ silver–iron cell where the charge (—) is at 25 A for 8 h, A represents a 0.25 A float charge; B an open circuit; and the discharge at open circuit after 1 h is shown for (—) 10 A, (– – –) 45 A, and (– – –) 80 A; and C represents the open circuit discharge.

1.4. Nickel–Zinc Cells.

Nickel–zinc cells offer some advantages over other rechargeable alkaline systems. The single-level discharge voltage, 1.60–1.65 V/cell is approximately 0.35–0.45 V/cell higher than nickel–cadmium or nickel–iron and approximately equal to that of silver–zinc. In addition, the use of zinc as the negative electrode should result in a higher energy density battery than either nickel–cadmium or nickel–iron and a lower cost than silver–zinc. In fact, nickel–zinc cells having energy densities in the range of $50-60$ W $\cdot$ h/kg have been successfully demonstrated.

Work in the 1930s resulted in a rechargeable nickel–zinc railroad battery utilizing standard Edison tubular or Jungner pocket nickel positives and electrodeposited zinc negatives. Cells were constructed utilizing physical separations between the plates and containing a large excess of KOH electrolyte. This Drumm battery exhibited a limited life caused by negative shape change and premature short circuits. Shape change is a phenomenon caused by zinc replating in a nonuniform manner during charge and resulting in a loss of negative capacity. Short circuits resulted from zinc dendrites formed during discharge that grew perpendicular to the face of the negative electrode until electrical contact was made with the positive. In the late 1950s and 1960s interest in the nickel–zinc system was again aroused as a possible substitute for silver–zinc batteries (71–75). By then the technology of nickel and zinc electrodes was well developed and led to a higher energy density nickel–zinc battery having tightly packed electrodes, barrier separators, and limited electrolyte. Most of this work related to vented batteries for low to medium rate applications, such as portable military communication applications.

Some efforts toward sealed battery development (76) were made. However, a third electrode, an oxygen recombination electrode was required to reduce the cost of the system. High rate applications such as torpedo propulsion were investigated (77) and moderate success achieved using experimental nickel–zinc cells yielding energy densities of 35 W·h/kg at discharge rates of 8 C. A commercial nickel–zinc battery is considered to be a likely candidate for electric bicycle development in China. Activity continues in electric vehicle design in several parts of the world (78,79). If the problems of limited life and high installation cost are solved, a nickel–zinc EV battery could provide twice the driving range for an equal weight lead–acid battery. Work is developmental; there is only limited production of nickel–zinc batteries.

Reaction Mechanism.   The overall reactions in the nickel–zinc cell can be represented by

$$2\,NiOOH + Zn + 2\,H_2O \rightleftharpoons 1\,2\,Ni(OH)_2 + Zn(OH)_2 \qquad E^\circ = +1.73\ V$$

Alternatively the discharged state of the zinc electrode is represented as ZnO.

Cell Construction.   Nickel–zinc batteries are housed in molded plastic cell jars of styrene, SAN, or ABS material for maximum weight savings. Nickel electrodes can be of the sintered or pocket type, however, these types are not cost effective and several different types of plastic-bonded nickel electrodes (78–80) have been developed.

Nickel hydrate, usually 5–10% cobalt added, serves as the active material and is mixed with a conductive carbon, eg, graphite. The active mass is mixed with an inert organic binder such as polyethylene or poly(tetrafluoroethylene) (TFE). The resultant mass is rolled into sheets on a compounding mill or pressed into electrodes as a dry powder on a nickel grid. The resultant electrodes offer a high energy density and low cost of fabrication. In performance, the plastic-bonded electrodes can support rates up to C/2, discharge capacity in 2 h, at voltages equivalent to those of sintered electrodes, but at higher rates some derating occurs. Life of plastic-bonded electrodes has not been fully evaluated, but a life of 500 cycles appears attainable.

Negative electrodes are fabricated of zinc oxide by any of the methods (pasting, pressing, etc) described. Binders, usually TFE, are used to reduce the solubility of the electrode in KOH. In addition, other techniques such as extended edges, inert extenders, contouring, and variable density have been tried in an effort to reduce shape change of the negative electrode upon cycling. The electrolyte is KOH, usually a 30–35% aqueous solution with the addition of LiOH at a level of 10–25 g/L to enhance nickel capacity throughout life.

Separators are both of the organic and inorganic type. During the 1960s most cells were built using cellophane or FSC (fibrous sausage casing) separators. Developments by NASA (81) and others in the area of inorganic films, consisting of a layered or film structure of a heavy metallic oxide such as zirconia, ZrO_2, or magnesia, MgO, bonded by an inert organic film and often layered with an organic resin-impregnated asbestos mat led to use of these materials as

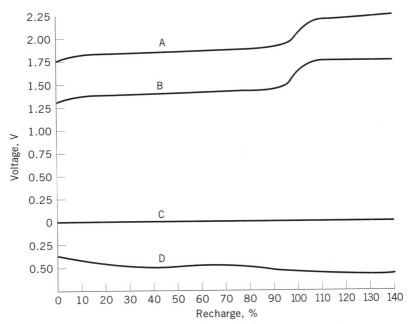

Fig. 13.   Nickel–zinc electrode potential on charge A, the Ni–Zn cell; B, the Zn vs Hg ref.; C, the Hg–HgO ref.; and D, the Ni vs Hg ref.

separators. In Ni–Zn electric vehicle batteries (82), over 800 cycles at 50% depth are claimed for these inorganic separators. Drawbacks are the relatively high cost of manufacture and the increased resistance per layer, which in effect limits discharge rate.

Performance.   The limited life of nickel–zinc batteries is the principal drawback to widespread use. Normally the nickel cathode is not a factor. Even plastic-bonded nickel electrodes perform for a greater number of cycles than either the zinc counter electrodes or the separator used and to a great extent it is the zinc electrode that limits the life of a nickel–zinc cell. In order to charge nickel electrodes sufficiently, a discreet amount of overcharge is necessary. Unfortunately, overcharge is not desirable for the zinc electrodes, as it promotes dendritic growth which will eventually penetrate the separator and cause short circuits. This can be alleviated somewhat by an overdesign in zinc capacity, but that reduces energy density. Typical cell and electrode voltage curves are shown in Figure 13 and cycle life data as a function of depth of discharge in Figure 14.

Nickel–zinc batteries containing a vibrating zinc anode has been reported (83). In this system zinc oxide active material is added to the electrolyte as a slurry. During charge the anode substrates are vibrated and the zinc is electroplated onto the surface in a uniform manner. The stationary positive electrodes (nickel) are encased in a thin, open plastic netting which constitutes the entire separator system.

The vibration serves a dual purpose. First, a macroturbulence is created which keeps the ZnO uniformly dispersed in the electrolyte. Second, the microturbulence created at the surface of the negative electrode minimizes the zincate

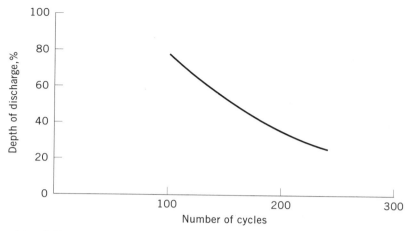

Fig. 14. Cell performance as a function of the number of cell recharge cycles.

concentration gradients, discouraging the formation of zinc dendrites and causing a uniform deposit of zinc. Any zinc dendrites that may form are brushed off against the plastic netting by the vibratory motion. Thus the problems of shape change, dendritic shorting, and separator failure are in theory solved by this system; the zinc electrode dissolves during discharge, replates during charge, and no separator is required. A disadvantage is reduced energy density, especially on a volume basis, as a result of the increased electrode spacing and the quantity of electrolyte required. The vibration hardware imposes an additional 5% weight penalty as well as increased cost. Alternatively, similar benefits have been reported from experiments in which the electrolyte is pulsed.

1.5. Nickel–Hydrogen Cells. There are two types of nickel–hydrogen cells; those that employ a gaseous H_2 electrode and those that utilize a metal hydride, MH.

Gaseous Hydrogen Systems. The nickel–hydrogen cell incorporating a gaseous hydrogen electrode is a hybrid consisting of one gaseous and one solid electrode. The nickel electrode is of the type used in a nickel–cadmium battery and the hydrogen electrode is a gas diffusion electrode of the type used in alkaline fuel cells (qv). These two electrodes are capable of extremely long, stable life. This system was developed to serve as a long life, lightweight battery for satellite applications that would be superior to the aerospace nickel–cadmium batteries (84–86).

The couple has a theoretical energy density of 172 W·h/kg and complete cells are capable of delivering 55 – 66 W·h/kg. The cell reaction is

$$NiOOH + \frac{1}{2} H_2 \rightleftharpoons Ni(OH)_2$$

However, the generation and migration of water in the half-cell reactions must be considered in the cell design. At the nickel electrode:

$$NiOOH + H_2O + e^- \rightleftharpoons Ni(OH)_2 + OH^-$$

and at the hydrogen electrode:

$$\frac{1}{2} H_2 + OH^- \rightleftharpoons H_2O + e^-$$

During charge the nickel hydroxide is converted to NiOOH, the charged state of nickel, and on the surface of the hydrogen electrode, hydrogen gas is evolved. By placing the electrode stack in a sealed container, the hydrogen is captured for subsequent reuse. During discharge the same hydrogen is reconsumed on the same electrode surface and the nickel electrode is reduced to provide electric energy.

Another desirable feature of this battery system is its capability of high rate of overcharge. During overcharge oxygen gas is generated on the surface of the nickel electrode. Simultaneously hydrogen continues to be evolved on the surface of the hydrogen electrode. Because there is a large area of catalyzed hydrogen electrode and ready access of the oxygen to diffuse to that surface, the oxygen readily recombines within the cell to form water and the cell pressure remains constant.

The primary packaging arrangements for this chemistry has been single cylindrical cells as shown in Figure 15, where a stack of disk electrodes are placed within a cylindrical outer housing having domed end caps. The housing

Fig. 15.   Cutaway view of a typical construction of a nickel–hydrogen cell.

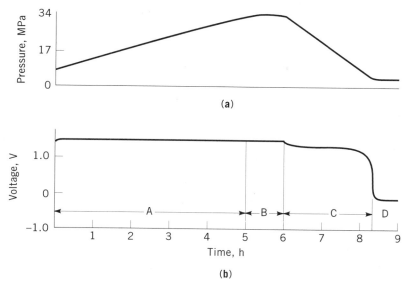

Fig. 16. 50 A·h nickel–hydrogen performance showing (**a**) pressure and (**b**) voltage curves where region A represents charging at 10 A, region B represents overcharge at 10 A, region C represents discharge at 25 A, and region D represents reversal at 25 A. To convert MPa to psi, multiply by 145.

also serves as a lightweight pressure vessel for hydrogen containment with all the free volume inside the housing used for hydrogen storage. Two insulated feed throughs are provided in the cell housing for electric contact to the positive and negative electrodes. The electrode stack is a repeating sequence of sintered nickel, absorber separator, typically asbestos, or other inorganic stable materials, teflon bonded platinum black fuel cell-type electrodes, and gas spacers. The electrolyte is typically 30–35% KOH with a LiOH additive for the nickel electrode.

Figure 16 shows a typical charge–discharge voltage and pressure profile for a 50 A·h cell. In this design the cell is precharged at 517 kPa (75 psi) hydrogen and the operating pressure range is from 517 to 4100 kPa (75–600 psi). Monitoring cell pressure, which is typically done with pressure transducers, enables the user to follow the cells state of charge. This is an additional desirable feature of this system. Figure 17 shows cell discharge characteristics as a function of rate, demonstrating the high rate capability of this battery. Figure 18 gives open circuit stand characteristics. This cell exhibits a greater self-discharge than the nickel–cadmium chemistry because of the reaction of pressurized hydrogen with the active nickel material.

Because this battery is only produced for special satellite applications, production quantities are limited. Rigorous quality inspections, expensive light-weight housing and seals, and the use of high loading platinum hydrogen electrodes all make this type of battery very expensive. For typical satellite applications multiple single-cells are connected in series and are packaged in an egg crate mounting. In this arrangement the waste heat generated during discharge

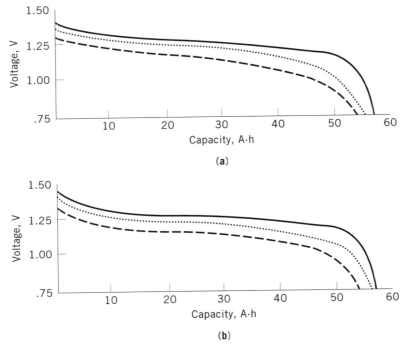

(a)

(b)

Fig. 17.   Discharge characteristics for 50 A·h nickel–hydrogen batteries at (—) 10 A, (···) 25 A, and (———) 50 A at (**a**) 20°C and (**b**) 0°C.

and overcharge is conducted from the stack out the cylinder wall through conduction rings to a radiator plate of the satellite.

Limited development efforts have been undertaken to develop a lower cost battery for terrestrial use utilizing reduced catalyst quantities and multicell

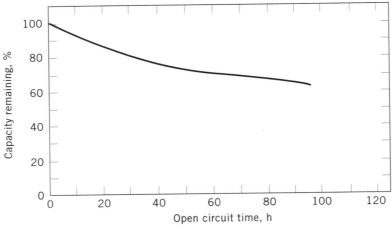

Fig. 18.   Self-discharge at ambient temperatures for a 35 A·h cell, NTS-2 prototype Sanyo Electric Co. cell.

packaging. Multicell common pressure vessel arrangements have also been considered for aerospace applications, but these configurations have not found market acceptance.

Metal Hydride Systems.   The success of the gaseous nickel–hydrogen system led to the investigation of replacing the gaseous hydrogen with metal hydrides in order to reduce the cell pressure and the volume required for hydrogen storage. A number of metal hydrides were developed for reversible hydrogen storage. Of particular interest were LaNi$_5$ and MmNi$_5$ (Misch metal [8049-20-5]) the isotherms of which are shown in Figures 19 and 20, and FeTi. In the initial efforts these materials were packed into the free volume of standard nickel hydrogen cells using standard catalyzed gas diffusion hydrogen electrodes. The metal hydrides absorb up to one hydrogen atom per metal atom and cells incorporating hydrides could be fabricated that were more compact than gas pressure cells. However these hydrides deactivated upon repeated deep-cycling and

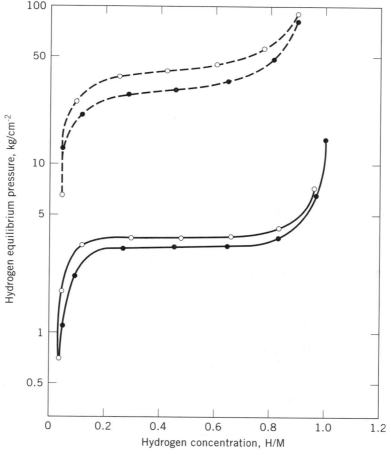

Fig. 19.   Pressure concentration curves of MmNi$_5$ (- - -) and LaNi$_5$ (—) at 45°C where open circles denote absorption and closed circles desorption of hydrogen. H/M represents the ratio in the hydride of the mole fraction of hydrogen to the mole fraction of the metal.

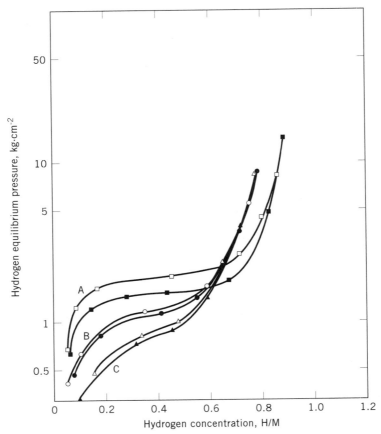

Fig. 20. Pressure constant temperature (PCT) curves of $MmNi_5$ alloy system where open symbols represent absorption and closed symbols represent desorption for A, $MmNi_{4.3}$ $Mn_{0.4}Al_{0.3}$; B, $MmNi_{3.8}Mn_{0.4}Al_{0.3}Co_{0.5}$; and C, $MmNi_{3.5}Mn_{0.4}Al_{0.3}Co_{0.75}$. H/M represents the ratio in the hydride of the mole fraction of hydrogen to the mole fraction of the metal.

gradually lost their hydrogen absorption ability presumably because of attack on the hydrides by water vapor or oxygen gas within the cell environment. Newer, more complex alloys have resolved most of these problems.

An alternative approach utilizes the hydride material as the hydrogen electrode. In this case, as the hydrogen is generated on the hydrogen electrode, it enters the hydride lattice for storage. On discharge the hydrogen leaves the hydride for reaction. The cell reactions are essentially the same as gaseous nickel hydrogen cells.

$$Ni(OH)_2 + OH^- \rightleftharpoons NiOOH + H_2O + e^-$$

$$M + H_2O + e^- \rightleftharpoons MH + OH^-$$

and overall

$$Ni(OH)_2 + M \rightleftharpoons NiOOH + MH$$

The overcharge reactions for the cell are the same as for nickel–cadmium and nickel–hydrogen cells. The oxygen generated on the nickel electrode at the end of charge and overcharge finds its way to the anode and reacts to form water in the $Ni–H_2$ case and $Cd(OH)_2$ in the Ni–Cd case.

A critical issue is the stability of the hydride electrode in the cell environment. A number of hydride formulations have been developed. Most of these are Misch metal hydrides containing additions of cobalt, aluminum, or manganese. The hydrides are prepared by making melts of the formulations and then grinding to fine powers. The electrodes are prepared by pasting and or pressing the powders into metal screens or felt. The additives are reported to retard the formation of passive oxide films on the hydrides.

A number of manufacturers started commercial production of nickel–MH cells in 1991 (31–35). The initial products are "AA"-size, "Sub-C", and "C"-size cells constructed in a fashion similar to small sealed nickel–cadmium cells. Ovonics also delivered experimental electric vehicle cells, 22 A·h size, for testing. The charge–discharge of "AA" cells produced in Japan (Matsushita) are compared in Figure 21.

From these data, the hydride cells contain approximately 30–50% more capacity than the Ni–Cd cells. The hydride cells exhibit somewhat lower high rate capability and higher rates of self-discharge than nickel–cadmium cells. Life is reported to be 200–500 cycles. Though not yet in full production it has been estimated that these cells should be at a cost parity to nickel–cadmium cells on an energy basis.

Several manufacturers were using nickel-metal hydride batteries to power their gasoline-electric hybrid and pure electric vehicles for 2004 and 2005 models. In the first quarter of 2002, more than 41,300 hybrid automobiles were operating on U.S. highways. An additional 5,200 battery electric automobiles, vans, and

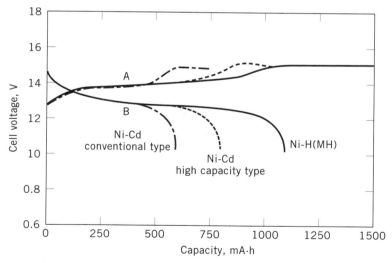

Fig. 21. Charge A and discharge B curves of (—) a Ni–H(MH) cell employing $MmNi_{3.55}Mn_{0.4}Al_{0.3}Co_{0.75}$ alloy compared with those of (———) conventional and (- - -) high capacity types of Ni–Cd cells (88).

light trucks were leased or sold. One Japanese-based automobile manufacturer was ramping up operations to produce 300,000 hybrid vehicles by 2007 (89).

1.6. Other Cell Systems. *Silver–Hydrogen Cells.* With the development of the nickel–hydrogen system limited attention was directed to the development of a silver–hydrogen cell (89,90). The main characteristics of interest were the potential for a higher gravametric energy density based on the ligher weight of the silver electrode vs that of the nickel. The cell reactions for this couple are

Hydrogen electrode

$$2\,H_2O + 2\,e^- \rightleftharpoons H_2 + 2\,OH^-$$

Silver electrode

$$Ag + OH^- \rightleftharpoons \frac{1}{2}\,Ag_2O + \frac{1}{2}\,H_2O + e^-$$

$$\frac{1}{2}\,Ag_2O + OH^- \rightleftharpoons AgO + \frac{1}{2}\,H_2O + e^-$$

Overall reaction

$$H_2 + AgO \rightleftharpoons Ag + H_2O$$

The packaging approach utilized for this battery is similar to that for nickel–hydrogen single cylindrical cells as shown in Figure 22. The silver electrode is typically the sintered type used in rechargeable silver–zinc cells. The hydrogen electrode is a Teflon-bonded platinum black gas diffusion electrode.

Because the silver oxide electrode is slightly soluble in the potassium hydroxide electrolyte the separator is of a barrier type to minimize silver diffusion to the opposite electrode. Figure 23 shows a charge–discharge profile of a 25 A·h cell that exhibits the two voltage plateaus seen for silver electrodes in the silver–zinc battery system. The silver–hydrogen cell exhibits a lower self-discharge than nickel–hydrogen cells as a result of the slower rate of reaction of hydrogen with silver oxide. The actual cells do not exhibit substantial differences in energy density when compared to those of nickel–hydrogen. Therefore this system is not being actively pursued.

Zinc–Oxygen Cells. On the basis of reactants the zinc–oxygen or air system is the highest energy density system of all the alkaline rechargeable systems with the exception of the $H_2 \cdot O_2$ one. The reactants are cheap and abundant and therefore a number of attempts have been made to develop a practical rechargeable system. The reactions of this system are as follows:

Zinc electrode

$$Zn + 2\,OH^- \rightleftharpoons Zn(OH)_2 + 2\,e^-$$

Oxygen electrode

$$\frac{1}{2}\,O_2 + H_2O + 2\,e^- \rightleftharpoons 2\,OH^-$$

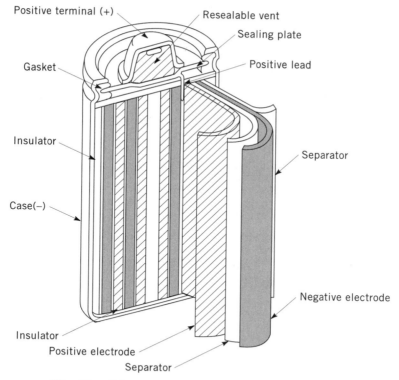

Fig. 22. Schematic diagram of Ni–H(MH) cell (88).

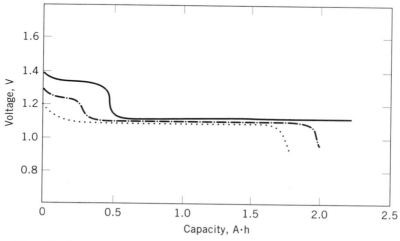

Fig. 23. Silver–hydrogen cell discharge characteristics where (···) represents a 0.5 h rate at 4 A or 80 mA/cm^2; (— · —) represents a 1 h rate at 2 A; and (—) represents a 4 h rate at 0.5 A or 10 mA/cm^2.

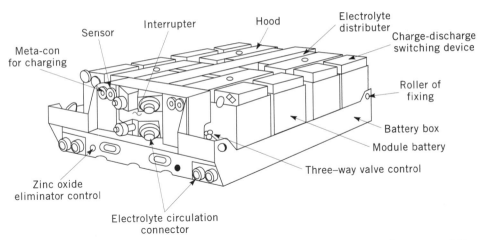

Fig. 24. Zinc–air 124-V battery system. Meta-con is a connector.Courtesy of Sanyo Electric Co.

Overall reaction

$$Zn + H_2O + \frac{1}{2} O_2 \rightleftharpoons Zn(OH)_2$$

In open cycle systems the oxygen reactant is released into the surrounding air during charge, and during the subsequent discharge oxygen from the surrounding air is consumed on a fuel cell-type electrode. The oxygen is delivered by forced or free confection depending on the system design. Whereas the zinc–oxygen cell has significant potential, it also has a number of inherent problems. First is the poor rechargeability of the normal zinc electrode; then there is the poor stability of the oxygen electrode when used in a bifunctional, ie, charge and discharge mode, contamination of the electrolyte by carbon dioxide from the air when used as an open cycle system, and poor retention of energy efficiency because of the irreversibility of the oxygen electrode reaction.

The system shown in Figure 24, in which the electrolyte was circulated through a multicell stack to improve the rechargeability of the zinc electrode has been studied by Sanyo Electric. However, such circulating systems are prone to electrolyte leakage and have common manifolds that cause internal self-discharge. Systems in which zinc particles or zinc-coated beads are circulated through the electrode stack and are utilized as a fluidized electrode have also been investigated by many organizations. In these systems the zinc particles contact the anode current collector to undergo reaction. In order to dissolve all the reaction product during discharge, it is necessary to use relatively large quantities of electrolyte so that a zincate solubility of approximately 200 g/L is not exceeded. The CGE (91) (Fig. 25) system utilized a tubular cell and a separate regeneration stack to deposit new zinc particles from the discharged electrolyte, which is saturated with dissolved zincate. By utilizing a separate regeneration stack the stability problem of the bifunctional oxygen electrode is avoided and regeneration can be carried out at any location so that the battery can be refueled

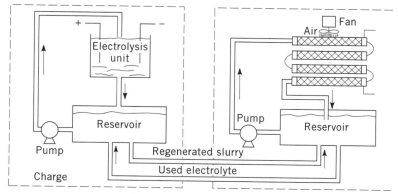

Fig. 25. Schematic diagram of the separate charge and discharge modules of the Génerale d'Electricité circulating zinc–air battery (91).

rapidly. A sealed system in a configuration similar to single cylindrical gaseous nickel–hydrogen cells has also been studied (92).

Iron–Air Cells. The iron–air system is a potentially low cost, high energy system being considered mainly for mobile applications. The iron electrode, similar to that employed in the nickel–iron cell, exhibits long life and therefore this system could be more cost effective than the zinc–air cell. Reactions include:

Iron electrode

$$Fe + 2\,OH^- \rightleftharpoons Fe(OH)_2 + 2\,e^-$$

Oxygen electrode

$$\frac{1}{2}\,O_2 + H_2O + e^- \rightleftharpoons 2\,OH^-$$

Overall reaction

$$Fe + \frac{1}{2}\,O_2 + H_2O \rightleftharpoons Fe(OH)_2$$

In the experimental systems studied the iron electrode has been of the sintered type and the oxygen–air electrodes have been of the bifunctional type.

A system as shown in Figure 26, which incorporates circulating electrolyte for thermal management and removal of gases generated during charge, has been developed (93). The iron electrode has a low hydrogen overvoltage and therefore the electrode evolves hydrogen during charge and to some extent during open circuit stand. A similar system, where the primary emphasis is on the development of a stable bifunctional air electrode is under investigation (94). A Teflon-bonded formulation consisting of a carbon-base, catalyzed with silver and other additives, is reported to be stable for up to 500 cycles. Because of the inefficiency of the iron electrode, and the irreversibility of the oxygen electrode, this system exhibits recharge energy efficiencies of less than 50%.

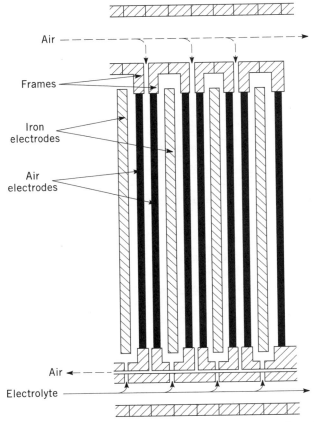

Fig. 26. Cross-section of SNDC iron–air battery pile (93).

Hydrogen–Oxygen Cells. The hydrogen–oxygen cell can be adapted to function as a rechargeable battery, although this system is best known as a primary one (see Fuel cells). The electrochemical reactions involve:

Electrodes

$$H_2 + 2\,OH^- \rightleftharpoons 2\,H_2O + 2\,e^-$$

$$\frac{1}{2}\,O_2 + H_2O + 2\,e^- \rightleftharpoons 2\,OH^-$$

Overall

$$H_2 + \frac{1}{2}\,O_2 \rightleftharpoons H_2O$$

During charge, water is electrolyzed to produce hydrogen and oxygen which are stored as pressurized gas. During discharge those gases electrochemically react to produce water. The reactants have a theoretical energy content of 770 W·h/kg and the interest that has been directed at this system has been

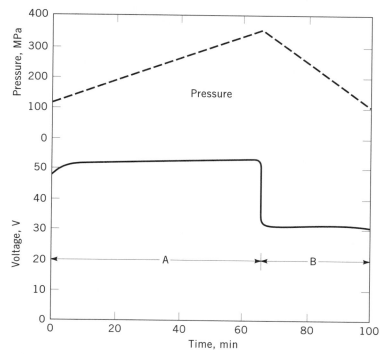

Fig. 27. Cycle of 34-cell regenerative hydrogen–oxygen fuel cell where A represents the charging region at 10 A, B represents discharging at 18.2 A. Both (—) voltage and (- - -) pressure changes are shown. To convert MPa to psig, multiply by 145.

primarily for light batteries for military and/or aerospace applications. In the 1960s a multicell bipolar stack contained inside a cylindrical pressure vessel was studied (95). The stack was internally manifolded to feed hydrogen and oxygen to separate compartments within the pressure vessel at operating pressures that ranged between 0.7 and 2.4 MPa (100–350 psi). Figure 27 shows a charge–discharge profile of the system.

Because of cross-gas leakage and other complexities, a single-cell was developed. In this configuration the electrodes were constructed in the form of a cylinder; the hydrogen gas was stored in the central compartment, and the oxygen gas in the space between the electrode core and the outer pressure vessel. These two compartments were sized in a two to one volume ratio to maintain the two gases at the same pressure as they are generated during charge. A flexible bellows was placed between the two compartments to compensate for slight differences in the compartments' volume and temperature effects. This system was not pursued however; it was dropped in favor of the simpler single-gas nickel–hydrogen system.

As primary alkaline fuel cells were developed for space applications, consideration was given to separate stack rechargeable designs. In these approaches, the product water formed during the discharge of a primary fuel cell is stored, then fed to a separate electrolizer stack during charge. The hydrogen and oxygen gas generated during charge is stored in separate pressure vessels. This

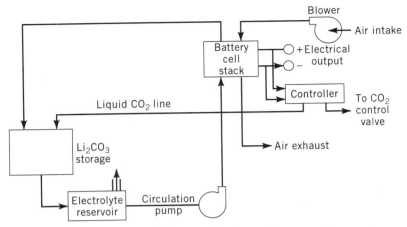

Fig. 28. Schematic of lithium–air automotive propulsion system.

approach overcomes the stability problem of the bifunctional oxygen electrode and the respective stacks can be optimized for their function. This system is still rather complex and bulky and has not yet been applied.

Mechanically Rechargeable Batteries. To avoid the time required for electric recharge, the problems of *in situ* electric recharge, or to utilize anodes that are not electrically rechargeable in aqueous electrolytes, mechanically rechargeable batteries have been studied. These systems are metal–air couples. The anodes that have received attention are zinc, lithium, and aluminum (97–99). Mechanically rechargeable zinc–air batteries were developed for military portable electronic equipment in the 1970s. These batteries were never deployed because of difficulties of leakage, problems of repeatedly replacing the anodes, and the development of lithium primary batteries having superior performance.

Lithium as an anode in alkaline electrolyte has been considered in the battery system shown in Figure 28. Even though lithium reacts directly with water, it was possible to operate the battery because of a protective lithium hydroxide film that forms on the anode. However, the film was not totally protective and units exhibited poor efficiency and were very complex.

The most significant results with these battery types focus on aluminum as the anode. Figure 29 shows an aluminum–air cell being developed for electric vehicle applications. The aluminum hydroxide reaction product would be returned to the factory to be reprocessed into fresh aluminum anodes. One set of anodes could yield up to 500 miles range before replacement. However, the corrosion reaction of the aluminum with the electrolyte is still a problem. Additionally, the system is complex and it is anticipated that replacing the anodes repeatedly will be as problematic as the zinc systems. The system has poor energy efficiency when consideration is given to the full cycle of electric generation, aluminum production, battery efficiency, and reprocessing of the battery reaction product back to aluminum.

1.7. Electrolyte. Potassium hydroxide is the principal electrolyte of choice for the above batteries because of its compatibility with the various electrodes, good conductivity, and low freezing point temperature. Potassium

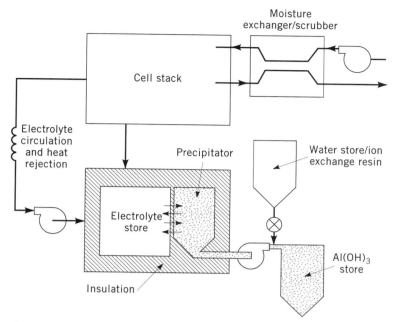

Fig. 29. Aluminum–air power cell system. The design provides for forced convection of air and electrolyte, heat rejection, electrolyte concentration control via $Al(OH)_3$ precipitation, and storage for reactants and products.

hydroxide is a white crystalline substance having a mol wt = 56.10; density = 2.044 g/mL, and mp = 360°C (see POTASSIUM COMPOUNDS). It is hygroscopic and very soluble in water. The most conductive aqueous solution at 25°C is at 27% KOH, but the conductivity characteristics are relatively flat over a broad range of concentrations.

The characteristics for aqueous KOH (97–99) solutions vary somewhat for battery electrolytes when additives are used. Furthermore, potassium hydroxide reacts with many organics and with the carbon dioxide in air to form carbonates. The build-up of carbonates in the electrolyte is to be avoided because carbonates reduce electrolyte conductivity and electrode activity in some cases.

1.8. Health and Safety Factors. The potassium hydroxide electrolyte used in alkaline batteries is a corrosive hazardous chemical. It is a poison and if ingested attacks the throat and stomach linings. Immediate medical attention is required. It slowly attacks skin if not rapidly washed away. Extreme care should be taken to avoid eye contact that can result in severe burns and blindness. Protective clothing and face shields or goggles should be worn when filling cells with water or electrolyte and performing other maintenance on vented batteries.

Alkaline batteries generate hydrogen and oxygen gases under various operating conditions. This can occur during charge, overcharge, open circuit stand, and reversal. In vented batteries free ventilation should be provided to avoid hydrogen accumulations surrounding the battery. A vented battery must never be placed in a sealed container for which it was not designed. As a result of operation, hydrogen–oxygen mixtures that are flammable or explosive can exist in the cells' head space. Ignition of this mixture, which is rare, can result in blowing off cell lids. This can occur from internal short circuits or external sparks that propagate back into the cell housing.

Alkaline batteries are capable of high current discharges and accidental short circuits should be avoided. Short circuiting can result in significant heat generation, electrolyte boiling, and cell rupture. High voltage cell strings also present an electric shock hazard; therefore tools should be insulated and operators should not wear rings.

Spontaneous low resistance internal short circuits can develop in silver–zinc and nickel–cadmium batteries. In high capacity cells heat generated by such short circuits can result in electrolyte boiling, cell case melting, and cell fires. Therefore cells that exhibit high resistance internal short circuits should not continue to be used. Excessive overcharge that can lead to dry out and short circuits should be avoided.

The European Union is evaluating a proposal to ban all NiCd batteries containing more then 0.002% cadmium by Jan. 1, 2008 and to increase collection rate for all spent industrial and automotive batteries to 95% by weight by Dec. 31, 2003. Some cadmium experts said that the proposal fails to differentiate between the different forms of cadmium with disparate toxicity and does not address the environmental effects of replacements (48).

1.9. Recycling. The most difficult aspect of NiCd battery recycling is collection of spent batteries. Although large industrial batteries, containing 20% cadmium are easy to collect and are recycled at the rate of 80%, the smaller NiCd batteries are usually discarded by the public. Thus, voluntary industry-sponsored collection programs and government agency programs are being devised to improve collection of these smaller batteries.

The most successful recycling program is operated by the Rechargeable Battery Recycling Corp. (RBRC) of Atlanta, Ga. RBRC was supported by more than 285 manufacturers and has a network of 26,000 collection centered across the United States and Canada. The RBRC recycling program contains several key elements that are specified in EPA regulations, Federal law, and in various state laws. These elements include uniform battery labeling, removability from appliances, a national network of collection systems, regulatory relief to facilitate battery collection, and widespread publicity to encourage public participation. Another successful program is run by INMETCO (a subsidiary of International Nickel). INMETCO's has a prepaid container program where companies purchase containers for collection and shipment of spent batteries. Because most of the industrial NiCd batteries are not allowed to be discarded in municipal dumps, they are recycled through collection programs (48,89).

2. Lead Acid Cells

The lead–acid battery is one of the most successful electrochemical systems and the most successful storage battery developed. About 87% of the lead [7439-92-1] (qv), Pb, consumption in the United States was for batteries in 2001.

The lead–acid battery consists of a number of cells in a container. These cells contain positive (PbO_2) and negative (Pb) electrodes or plates, separators to keep the plates apart, and sulfuric acid [7664-93-9], H_2SO_4, electrolyte. The battery reactions are highly reversible, so that the battery can be discharged and charged repeatedly. The number of charge–discharge cycles that can be obtained depends strongly on the use mode and can vary from several hundred to thousands of cycles.

Each cell has a nominal voltage of 2 V and capacities typically vary from 1 to 2000 ampere-hours. Lead–acid cells can be operated with coulombic efficiencies as high as 95% and with energy efficiencies greater than 80%. The many cell designs available for a wide variety of uses can be divided into three main categories: automotive, industrial, and consumer. Shipments of lighting and ignition (SLI) for cranking of internal combustion engines in North America totaled $10^6 \times 10^6$ units in 2001 (100). Industrial batteries are used for heavy-duty application such as motive and standby power. More recently, the use of batteries for utility peak shaving has been increasing. Consumer batteries for emergency lighting, security alarm systems, cordless convenience devices and power tools, and small engine starting is one of the fastest growing markets for the lead–acid battery (100).

In Figure 30, the cutaway view of the automotive battery shows the components used in its construction. An industrial motive power battery, shown in Figure 31 (101), is the type used for lift trucks, trains, and mine haulage. Both types of batteries have the standard free electrolyte systems and operate only in the vertical position. Although a tubular positive lead–acid battery is shown for

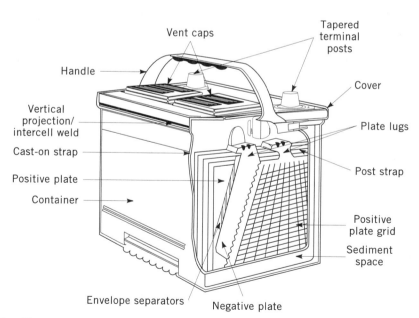

Fig. 30. Cutaway view of an automotive SLI lead–acid battery container and cell element. Courtesy of Johnson Controls, Inc.

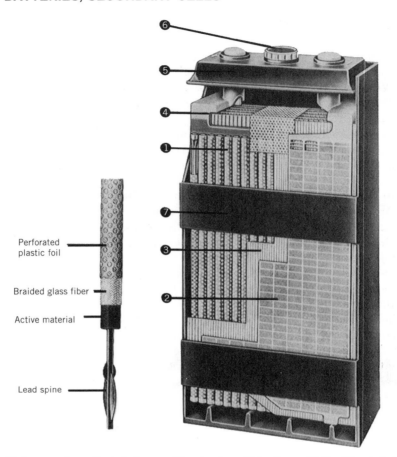

Perforated
plastic foil

Braided glass fiber

Active material

Lead spine

Fig. 31. Cutaway view of a tubular positive lead–acid battery. (1) Positive tubular plate; (2) negative plate; (3) separator; (4) connecting strap; (5) cell cover; (6) cell plug; and (7) cell container. Courtesy of A. B. Tudor, Sweden.

industrial applications, the flat plate battery construction (Fig. 30) is also used in a comparable size.

Two types of batteries having immobilized electrolyte systems are also made. They are most common in consumer applications, but their use in industrial and SLI applications is increasing. Both types have low maintenance requirements and usually can be operated in any position. They are sometimes called valve regulated or recombinant batteries because they are equipped with a one-way pressure relief vent and normally operate in a sealed condition with an oxygen recombination cycle to reduce water loss.

In the gelled electrolyte battery, the sulfuric acid electrolyte has been immobilized by a thixotropic gel. This is made by mixing an inorganic powder such as silicon dioxide [7631-86-9], SiO_2, with the acid (102). Other cells, such as the one shown in Figure 32, use a highly absorbent separator to immobilize the electrolyte (103).

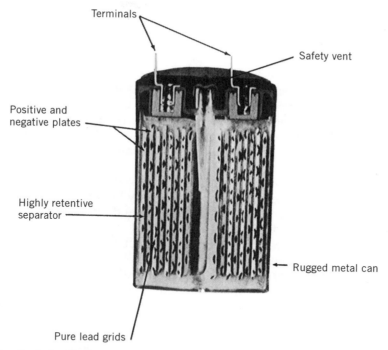

Fig. 32. Sealed cylindrical lead–acid cell.Courtesy of Gates Energy Products, Inc.

2.1. History. Gaston Planté developed the first working model of the lead–acid battery (104) in 1859, but electrochemical energy storage was not practical then because there was no efficient way of recharging a battery. The lead–acid battery generated interest because it could provide higher currents than primary batteries. The use of the system as a capacitor was also of great interest. Between 1860 and 1880, the principal commercial application of the Planté battery was in the telegraphic industry (105).

The first commercial application for a rechargeable energy storage device was electric lighting. Edison had developed a long-lasting incandescent bulb and improved the dynamo in 1879 (106). The invention of pasted plates (107) and the perforated lead current collector (108) in 1881 improved the energy storage capacity of the lead–acid battery and in the 1880s European and American companies were organized to commercialize electric lighting. Lighting provided a strong incentive for improvements in battery technology. Better grids, flat plates, and lead–antimony alloys soon followed (105). Electric vehicles, including boats, submarines, cars, and trams were developed and batteries were an important factor in the development of the power industry. Batteries were first used as transformers in electric power stations and later for load leveling. They provided the flexibility that was necessary for expansion of the industry without excessive capital investment. As power production improved and expanded, however, batteries were phased out because the maintenance costs were too high.

The invention of the self-starter in 1911 impacted heavily on the battery market (109). The battery-started, gasoline-driven vehicle was so successful that electric vehicles could not compete. Except for submarines and a few delivery trucks, electric vehicles disappeared and automotive starting became the largest market for the lead–acid battery. Indeed, the SLI application has not only grown steadily but it has also financed the majority of lead–acid battery research and development. Over the years, substantial advances have been made by using new materials such as rubber and plastics for battery containers, separators, and other components, to increase the power and energy density of batteries. Maintenance requirements have been reduced substantially, and machine automation has accelerated production rates and lowered costs. Better electronics for charge and discharge control have improved battery performance and life.

In the 1990s, the use of batteries in electric vehicles and for load leveling revived partly for environmental reasons and partly because of scarce energy resources. Improvements in battery performance and life, fewer maintenance requirements, and automatic control systems are making these applications feasible. Research and development is ongoing all over the world to develop improved lead–acid batteries as well as other systems to meet these needs.

2.2. Cell Thermodynamics. The chemical reaction of the lead–acid battery was explained as early as 1882 (110). The double sulfate theory has been confirmed by a number of methods (111–113) as the only reaction consistent with the thermodynamics of the system. The thermodynamics of the lead–acid battery has been reviewed in great detail (114).

At the cathode, or positive electrode, lead dioxide [1309-60-0], PbO_2, reacts with sulfuric acid to form lead sulfate [7446-14-2], $PbSO_4$, and water in the discharging reaction

$$PbO_2 + 3\,H^+ + HSO_4^- + 2e^- \rightleftharpoons PbSO_4 + 2\,H_2O \qquad (1)$$

and the reverse occurs as the battery charges. At the anode, or negative electrode, metallic lead reacts with bisulfate ion to form lead sulfate in the discharging reaction

$$PbSO_4 + H^+ + 2e^- \rightleftharpoons Pb + HSO_4^- \qquad (2)$$

and the reverse occurs as the battery charges. The sum of these two half-cell reactions is called the double sulfate reaction.

$$Pb + PbO_2 + 2\,H_2SO_4 \rightleftharpoons 2\,PbSO_4 + 2\,H_2O \qquad (3)$$

Lead sulfate is formed as the battery discharges, sulfuric acid is regenerated as the battery is charged. The open circuit voltage of the lead–acid battery is a function of the acid concentration and temperature. A review of this subject is available (115). The Nernst equation may be used to calculate the open circuit cell voltage. The battery voltage is then obtained by multiplying the cell voltage by the number of cells.

The Nernst equation for the cell voltage V is

$$V = V^\circ + 2.303(RT/F)\left(\log a_{H_2SO_4} - \log a_{H_2O}\right) \qquad (4)$$

Because Pb, PbO_2, and $PbSO_4$ are all solids having low solubilities, the activities of these substances are unity. At 25°C, the absolute temperature T is 298.15 K. The value of R, the gas constant, used is 8.3144 J/(moK). F, the Faraday constant, is 96,485 C/mol. The standard cell voltage $V°$ for the double sulfate reaction must be known as well as the activities of sulfuric acid and water at any given concentration or temperature.

The dependence of cell voltage on pressure is small and can be neglected in normal applications. At 25°C and an acid concentration of 3.74 m, $dV/dP = -3.32 \times 10^{-5}\ \mu\text{V/Pa}(-4.81 \times 10_{\mu}^{-3}\text{V/psi})$. Tables and equations for the pressure dependence are available (114).

Sulfuric acid dissociates only partially according to the following equilibria.

$$H_2SO_4 \rightleftharpoons H^+ + HSO_4^-\ K_1 = a_{H^+}a_{HSO_4^-}/a_{H_2SO_4} \tag{5}$$

$$HSO_4^- \rightleftharpoons H^+ + SO_4^{2-}\ K_2 = a_{H^+}a_{SO_4^{2-}}/a_{HSO_4^-} \tag{6}$$

Here the values of a are the activities of the designated ions in solution, and K_1 and K_2 are the equilibrium constants for the dissociation reactions. K_1 is infinity because dissociation to hydrogen and bisulfate ions is essentially complete. The best value for K_2 is probably 0.0102 (116). Thus sulfuric acid contains a mixture of hydrogen, bisulfate, and sulfate ions where the ratios of these ions vary with concentration and temperature.

The activity of any ion, $a = \gamma m$, where γ is the activity coefficient and m is the molality (mol solute/kg solvent). Because it is not possible to measure individual ionic activities, a mean ionic activity coefficient, $\gamma_{\pm}$, is used to define the activities of all ions in a solution. The convention used in most of the literature to report the mean ionic activity coefficients for sulfuric acid is based on the assumption that the acid dissociates completely into hydrogen and sulfate ions. This assumption leads to the following formula for the activity of sulfuric acid.

$$a_{H_2SO_4} = (a_{H^+})^2 a_{SO_4^{2-}} = (\gamma_{\pm}2m)^2 (\gamma_{\pm}m) = 4\gamma_{\pm}^3 m^3 \tag{7}$$

where m is the molality of sulfuric acid (mol acid/kg water).

To adjust the activities of sulfuric acid to the convention which assumes that the acid dissociates only partially into hydrogen and bisulfate ions, the following formula is used:

$$a'_{H_2SO_4} = a_{H_2SO_4}/K_2 \tag{8}$$

The activity coefficient of water is related to the osmotic coefficient by the formula:

$$2.303\log a_{H_2O} = -3m\phi/55.5 \tag{9}$$

where ϕ is the osmotic coefficient ofsulfuric acid of molality m.

The activity coefficients ofsulfuric acid have been determined independently by measuring three types of physical phenomena: cell potentials, vapor pressure, and freezing point. A consistent set of activity coefficients has been reported from 0.1 to 8 m at 25°C (113), from 0.1 to 4 m and 5 to 55°C (117), and from 0.001 to 0.02 m at 25°C (118). These values are all based on cell potential measurements. The activity coefficients based on vapor pressure measurements (119) agree with those from potential measurements when they are corrected to the same reference activity coefficient.

To calculate the open circuit voltage of the lead–acid battery, an accurate value for the standard cell potential, which is consistent with the activity coefficients of sulfuric acid, must also be known. The standard cell potential for the double sulfate reaction is 2.048 V at 25°C. This value is calculated from the standard electrode potentials: for the $(Pt)H_2|H_2SO_4(m)|PbSO_4|PbO_2(Pt)$ electrode 1.690 V (113), for the $Pb(Hg)|PbSO_4|H_2SO_4(m)|H_2(Pt)$ electrode 0.3526 V(19), and for the $Pb|Pb^{2+}|Pb(Hg)$ 0.0057 V (120).

Table 4 gives the calculated open circuit voltages of the lead–acid cell at 25°C at the sulfuric acid molalities shown. The corrected activities of sulfuric acid from vapor pressure data (119) are also given.

The temperature dependence of the open circuit voltage has been accurately determined (121) from heat capacity measurements (122). The temperature coefficients E_T are given in Table 5. The accuracy of these temperature coefficients does not depend on the accuracy of the open circuit voltages at 25°C shown in Table 4. Using the data in Tables 4 and 5, the open circuit voltage can be calculated from 0 to 60°C at concentrations of sulfuric acid from 0.1 to 13.877 m.

The mercurous sulfate [7783-36-0], Hg_2SO_4, mercury reference electrode, $(Pt)H_2|H_2SO_4(m)|Hg_2SO_4(Hg)$, is used to accurately measure the half-cell potentials of the lead–acid battery. The standard potential of the mercury reference electrode is 0.6125 V (113). The potentials of the lead dioxide, lead sulfate, and mercurous sulfate, mercury electrodes versus a hydrogen electrode have been measured (123,124). These data may be used to calculate accurate half-cell potentials for the lead dioxide, lead sulfate positive electrode from temperatures of 0 to 55°C and acid concentrations of from 0.1 to 8 m.

2.3. Lead Grid Corrosion. The corrosion of the lead grid at thelead dioxide electrode is one of the primary causes of lead–acid battery failure. Grid corrosion is complex for several reasons. First, the acid concentration and voltage at the lead dioxide electrode vary widely during battery operation. Second, the voltage, pH, and other ionic concentrations vary across the corrosion film. Different conditions across the film favor different reactions. Changing conditions mean multiple corrosion products form includinglead sulfate, basic lead sulfates, and lead oxides (125). Third, the lead oxides form a nonstoichiometric series which can be represented by the general formula, PbO_n, where $1 < n < 2$. Two polymorphs have been identified in lead corrosion films for both lead(II) oxide [1317-36-8], PbO, orthorhombic and tetragonal, and PbO_2, rhombic and tetragonal. PbO_2 may also be amorphous. A variety of intermediates, such as the intermediate lead oxide [1314-41-6], Pb_3O_4, and lead(III) oxide [1314-27-8], Pb_2O_3, have also been found.

The mechanisms of lead corrosion in sulfuric acid have been studied and good reviews of the literature are available (126–129). The main techniques used in lead corrosion studies have been electrochemical measurements, x-ray diffraction, and electron microscopy. Laser Raman spectroscopy and photoelec-

Table 4. **Thermodynamic Values for the Lead-Acid Cell at 25°C**

Molality, m	$\gamma_{\pm}$	αH_2O^a	αH_2SO_4a	V
0.1	0.245	9.963×10^{-1}	5.882×10^{-5}	1.798
0.2	0.193	9.928×10^{-1}	2.296×10^{-4}	1.833
0.3	0.169	9.892×10^{-1}	5.167×10^{-4}	1.854
0.5	0.144	9.819×10^{-1}	1.483×10^{-3}	1.881
0.7	0.131	9.743×10^{-1}	3.067×10^{-3}	1.900
1.0	0.121	9.618×10^{-1}	7.164×10^{-3}	1.922
1.5	0.117	9.387×10^{-1}	2.137×10^{-2}	1.951
2.0	0.118	9.126×10^{-1}	5.224×10^{-2}	1.975
2.5	0.123	8.836×10^{-1}	1.158×10^{-1}	1.996
3.0	0.131	8.516×10^{-1}	2.440×10^{-1}	2.016
3.5	0.143	8.166×10^{-1}	4.989×10^{-1}	2.035
4.0	0.157	7.799×10^{-1}	9.883×10^{-1}	2.054
4.5	0.173	7.422×10^{-1}	1.888×10	2.072
5.0	0.192	7.032×10^{-1}	3.541×10	2.090
5.5	0.213	6.643×10^{-1}	6.463×10	2.106
6.0	0.237	6.259×10^{-1}	1.148×10^{1}	2.123
6.5	0.263	5.879×10^{-1}	2.002×10^{1}	2.139
7.0	0.292	5.509×10^{-1}	3.421×10^{1}	2.154
7.5	0.323	5.152×10^{-1}	5.685×10^{1}	2.169
8.0	0.356	4.814×10^{-1}	9.255×10^{1}	2.183
8.5	0.393	4.488×10^{-1}	1.492×10^{2}	2.197
9.0	0.431	4.180×10^{-1}	2.334×10^{2}	2.211
9.5	0.472	3.886×10^{-1}	3.617×10^{2}	2.224
10.0	0.516	3.612×10^{-1}	5.490×10^{2}	2.236
11.0	0.610	3.111×10^{-1}	1.208×10^{3}	2.260
12.0	0.711	2.681×10^{-1}	2.480×10^{3}	2.283
13.0	0.819	2.306×10^{-1}	4.835×10^{3}	2.304
14.0	0.938	1.980×10^{-1}	9.072×10^{3}	2.324
15.0	1.065	1.698×10^{-1}	1.630×10^{4}	2.343
16.0	1.200	1.456×10^{-1}	2.828×10^{4}	2.361
17.0	1.338	1.252×10^{-1}	4.708×10^{4}	2.378
18.0	1.484	1.076×10^{-1}	7.621×10^{4}	2.394
19.0	1.634	9.250×10^{-2}	1.198×10^{5}	2.410
20.0	1.790	7.960×10^{-2}	1.836×10^{5}	2.424
21.0	1.951	6.860×10^{-2}	2.750×10^{5}	2.439
22.0	2.122	5.890×10^{-2}	4.072×10^{5}	2.453
23.0	2.302	5.060×10^{-2}	5.940×10^{5}	2.466
24.0	2.460	4.410×10^{-2}	8.233×10^{5}	2.478
26.0	2.805	3.310×10^{-2}	1.552×10^{6}	2.502
28.0	3.159	2.500×10^{-2}	2.767×10^{6}	2.524
30.0	3.499	1.910×10^{-2}	4.627×10^{6}	2.544

^{a} From Ref. 119.

trochemistry have been used to gain new insight into the corrosion process (129,130).

   2.4. Charge–Discharge Processes. An excellent review covers the charge and discharge processes in detail (129) and ongoing research on lead–acid batteries may be found in two symposia proceedings (131,132). Detailed studies of the kinetics and mechanisms of lead–acid battery reactions are published continually (133). Although many questions concerning the exact nature of the

Table 5. **Temperature Coefficients of the Lead−Acid Cell, E_T^a, mV**

Concentration, m	Temperature, °C									
	0	5	10	15	20	30	35	40	45	50
0.1	−6.102	−4.713	−3.410	−2.191	−1.055	0.975	1.871	2.689	3.431	4.098
0.2	−3.872	−2.943	−2.092	−1.319	−0.622	0.548	1.024	1.428	1.762	2.026
0.3	−2.414	−1.785	−1.231	−0.749	−0.339	0.270	0.472	0.607	0.676	0.681
0.4	−1.317	−0.914	−0.581	−0.319	−0.126	0.059	0.053	−0.017	−0.151	−0.347
0.5	−0.400	−0.193	−0.050	0.029	0.045	−0.106	−0.271	−0.495	−0.776	−1.115
0.6	0.351	0.399	0.387	0.316	0.187	−0.243	−0.542	−0.894	−1.301	−1.760
0.7	1.005	0.913	0.766	0.564	0.308	−0.360	−0.772	−1.233	−1.744	−2.304
0.8	1.606	1.385	1.113	0.791	0.420	−0.468	−0.982	−1.543	−2.150	−2.801
0.9	2.135	1.801	1.421	0.993	0.519	−0.563	−1.171	−1.821	−2.513	−3.248
1.0	2.620	2.180	1.698	1.173	0.607	−0.647	−1.334	−2.059	−2.823	−3.624
1.110	3.063	2.528	1.953	1.339	0.688	−0.725	−1.485	−2.281	−3.112	−3.977
1.388	4.020	3.280	2.508	1.704	0.867	−0.898	−1.826	−2.784	−3.771	−4.787
1.850	5.175	4.194	3.185	2.150	1.088	−1.113	−2.251	−3.414	−4.601	−5.812
2.220	5.702	4.612	3.496	2.355	1.189	−1.213	−2.449	−3.709	−4.990	−6.294
2.775	6.081	4.912	3.720	2.503	1.263	−1.286	−2.593	−3.923	−5.274	−6.647
3.172	6.168	4.982	3.771	2.537	1.280	−1.302	−2.627	−3.973	−5.341	−6.729
3.700	6.002	4.854	3.679	2.478	1.252	−1.276	−2.577	−3.901	−5.249	−6.620
4.626	5.256	4.270	3.250	2.199	1.115	−1.146	−2.322	−3.527	−4.762	−6.026
5.551	4.134	3.437	2.674	1.846	0.954	−1.016	−2.092	−3.228	−4.423	−5.675
6.167	3.651	3.056	2.393	1.661	0.863	−0.928	−1.919	−2.972	−4.087	−5.262
6.938	3.171	2.674	2.107	1.472	0.769	−0.834	−1.732	−2.693	−3.717	−4.801
7.929	2.760	2.340	1.854	1.301	0.682	−0.745	−1.553	−2.422	−3.350	−4.338
8.539	2.641	2.238	1.772	1.242	0.651	−0.711	−1.481	−2.308	−3.192	−4.131
9.291	2.602	2.192	1.725	1.201	0.628	−0.681	−1.413	−2.195	−3.028	−3.909
10.092	2.647	2.207	1.721	1.191	0.617	−0.659	−1.360	−2.102	−2.884	−3.706
11.101	2.725	2.249	1.737	1.192	0.612	−0.645	−1.322	−2.030	−2.769	−3.539
12.335	2.797	2.290	1.756	1.197	0.611	−0.636	−1.296	−1.981	−2.689	−3.421
13.877	2.843	2.313	1.764	1.195	0.607	−0.626	−1.269	−1.931	−2.611	−3.308

$^a E = E_{25°C} - E_T$ (mV) of the lead storage cell from the third law of thermodynamics.

reactions remain unanswered, the experimental data on the lead–acid cell are more complete than for most other electrochemical systems.

Significant progress is also being made in simulating battery charge and discharge processes using electrochemical models. Models of the lead–acid cell based on porous electrode theory have been developed (134–138). Discharge behavior can be predicted within 10% accuracy over a wide range of discharge rates. Recharge behavior has also been simulated (139); a detailed model of the changes in acid concentration resulting from convection in free electrolyte cells has been presented (140); a model for distribution of acid immobilized in a porous separator has been developed (141); and a model of the effect of lead sulfate supersaturation on battery behavior has been published (142). Such models can be used to predict the performance of large batteries from data on small cells, develop improved battery designs, and simulate the performance of any design in a new application. Battery development is therefore accelerated by reducing the time required to build and test battery prototypes.

At high discharge rates, such as those required for starting an engine, the voltage drops sharply primarily because of the resistance of the lead current collectors. This voltage drop increases with the cell height and becomes significant even at moderate discharge rates in large industrial cells. Researchers have measured this effect in industrial cells and have developed a model which has been used to improve grid designs for automotive batteries (143,144).

Self-Discharge Processes.   The shelf life of the lead–acid battery is limited by self-discharge reactions, first reported in 1882 (145), which proceed slowly at room temperature. High temperatures reduce shelf life significantly. The reactions which can occur are well defined (146) and self-discharge rates in lead–acid batteries having immobilized electrolyte (147) and limited acid volumes (148) have been measured.

The lead current collector in the positive lead–dioxide plate corrodes and the compounds which form are a function of the acid concentration and positive electrode voltage. Other reactions which take place at the positive electrode are shown.

Oxygen evolution

$$PbO_2 + H_2SO_4 \rightleftharpoons PbSO_4 + H_2O \; \frac{1}{2} O_2 \tag{10}$$

Oxidation of organics, R,

$$2\,PbO_2 + R + 2\,H_2SO_4 \rightleftharpoons 2\,PbSO_4 + 2\,H_2O + CO_2 + R' \tag{11}$$

Sulfation of PbO (in new cells)

$$PbO + H_2SO_4 \rightleftharpoons PbSO_4 + H_2O \tag{12}$$

Oxidation of additives such as antimony, in the grid alloy:

$$5\,PbO_2 + 2\,Sb + 6\,H_2SO_4 \rightleftharpoons (SbO_2)_2SO_4 + 5\,PbSO_4 + 6\,H_2O \tag{13}$$

Similar reactions can be written for other metallic additives. At the negative electrode two more reactions can occur.

Hydrogen evolution

$$Pb + H_2SO_4 \rightleftharpoons PbSO_4 + H_2 \tag{14}$$

Oxygen recombination

$$Pb + H_2SO_4 + \frac{1}{2} O_2 \rightleftharpoons PbSO_4 + H_2SO \tag{15}$$

The rate of the oxygen recombination reactions is very fast depending only on the rate at which the oxygen is transported to the lead in the negative electrode. If the cell electrode stack is fully saturated with acid, the oxygen recombination reaction proceeds very slowly. The rate depends on the solubility of oxygen in sulfuric acid. However, in batteries which are stored with limited acid volume, oxygen can recombine very rapidly. A small leak in the cell container results in rapid self-discharge of the negative electrode by atmospheric oxygen. Although hydrogen recombination at the positive electrode has been proposed, its rate appears to be insignificant (149).

The rates of the other self-discharge reactions are primarily determined by the acid concentration (147,148). High acid concentrations accelerate the gas evolution reactions at both electrodes. In contrast, grid corrosion appears to be faster at low acid concentrations and voltages. Sulfation of unconverted lead oxide is so rapid that it generally occurs before the battery is shipped from the factory. Batteries are sometimes recharged just before shipment to complete the conversion of lead oxide to lead dioxide in the positive plate. Organics are necessary additives to the separator and negative plate but are kept to a minimum to avoid excessive self-discharge and grid corrosion.

Overcharge Reactions. Water electrolysis during overcharge is an irreversible process. Oxygen forms at the positive electrode:

$$H_2O \rightleftharpoons 2\,H^+ + \frac{1}{2} O_2(g) + 2e^- \tag{16}$$

At the negative electrode, hydrogen ions react to form hydrogen gas:

$$2\,H^+ + 2e^- \rightleftharpoons H_2(g) \tag{17}$$

The net reaction being

$$H_2O \rightleftharpoons H_2(g) + \frac{1}{2} O_2(g) \tag{18}$$

Theoretically, water should decompose at a voltage below the voltage required to recharge a lead–acid battery. However, the rate of water electrolysis is much slower than the rate of the recharge reaction. Thus the lead–acid battery can operate with as little as 5% excess charge to compensate for water electrolysis. Use of lead–antimony alloys for the current collectors in lead–acid batteries increases water loss. Some of these batteries need regular maintenance by addition of water to replace the water lost on overcharge. Many newer designs, however, use either lower concentrations of antimony in the alloy or lead–calcium

alloys to reduce water loss (see LEAD ALLOYS). This is the basis for the maintenance-free batteries.

Some battery designs have a one-way valve for pressure relief and operate
on an oxygen cycle. In these systems the oxygen gas formed at the positive electrode is transported to the negative electrode where it reacts to reform water.
Hydrogen evolution at the negative electrode is normally suppressed by this
reaction. The extent to which this process occurs in these valve regulated
lead–acid batteries is called the recombination-efficiency. These processes are
reviewed in the literature (149–151).

The oxygen cycle is often described by equation 16 at the positive electrode,
and the reverse of equation 16 at the negative electrode,

$$\frac{1}{2} O_2 + 2\, H^+ + 2e^- \rightleftharpoons H_2O \tag{19}$$

When these half-reactions are summed, there is no net reaction. Thus the material balance of the cell is not altered by overcharge. At open circuit, equation 19
at the negative electrode is the sum of a two-step process, represented by equation 15 and

$$PbSO_4 + 2\, H^+ + 2e^- \rightleftharpoons Pb + H_2SO_4 \tag{20}$$

These equations are based on the thermodynamically stable species. Further
research is needed to clarify the actual intermediate formed during overcharge.
In reality, the oxygen cycle can not be fully balanced because of other side reactions, that include grid corrosion, formation of residual lead oxides in the positive
electrode, and oxidation of organic materials in the cell. As a result, some gases,
primarily hydrogen and carbon dioxide (152), are vented.

2.5. Material Fabrication and Manufacturing Processes. The lead–
acid battery is comprised of three primary components: the element, the container, and the electrolyte. The element consists of positive and negative plates
connected in parallel and electrically insulating separators between them. The
container is the package which holds the electrochemically active ingredients
and houses the external connections or terminals of the battery. The electrolyte,
which is the liquid active material and ionic conductor, is an aqueous solution of
sulfuric acid.

Element. The process of fabricating lead–acid battery elements as depicted in Figure 33 involves numerous chemical and electrochemical reactions
and several mechanical assembly operations. All of the processes involved
must be carefully controlled to ensure the quality and reliability of the product.

Plates. Plates are the part of the cell that ultimately become the battery
electrodes. The plates consist of an electrically conductive grid pasted with a lead
oxide–lead sulfate paste which is the precursor to the electrode active materials
which participate in the electrochemical charge–discharge reactions.

Lead is the basic raw material for the two components, both the active
material and grid, of both the positive and negative plates or electrodes. A finely
divided mixture of lead monoxide powder and metallic lead particles is the key
ingredient in preparing positive and negative active material. The physical and

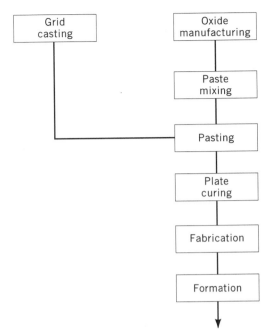

Fig. 33.   Flow sheet of plate processing and battery manufacturing.

chemical characteristics of this leady litharge, battery oxide, or grey oxide as it is called, have a profound impact on the performance and life expectancy of the battery. Battery oxide is manufactured from primary or secondary (recycled) lead of extremely high purity using either the Barton or ball mill process.

In the Barton (153) process, molten lead is metered into a reactor or a Barton pot where it impinges on a rotating paddle. The resulting droplets of lead are oxidized and conveyed out in an airstream. The ball mill process converts solid lead pieces directly into oxide through attrition by causing them to rub against each other within a rotating reactor or ball mill. The lead in the ball mill is the grinding media as well as a reactant. The particles or agglomerates produced by the molten lead Barton process are droplike or spherical whereas those produced by the solid state ball mill grinding process are flat or flakelike. The oxidation reaction in either process is controlled such that approximately 75% of the product is oxidized to lead monoxide, leaving about 25% as finely divided metallic lead to be oxidized in subsequent processes. In addition to chemical composition, particle size or reactivity of the material is measured and controlled. Process control variables include reactor temperature, air flow, and lead feed rate.

The oxide exiting either the Barton or ball mill reactor is conveyed by an air stream to separating equipment, ie, settling tank, cyclone, and baghouse, after which it is stored in large hoppers or drummed for use in paste mixing. Purity of the lead feed stock is extremely critical because minute quantities of some impurities can either accelerate or slow the oxidation reaction markedly. Detailed discussions of the oxide-making process and product are contained in references (154–156). A lightweight low resistance electrode plate for lead–acid batteries has been described (157).

Paste Mixing. The active materials for both positive and negative plates are made from the identical base materials. Lead oxide, fibers, water, and a dilute solution of sulfuric acid are combined in an agitated batch mixer or reactor to form a pastelike mixture of lead sulfates, the normal, tribasic, and tetrabasic sulfates, plus PbO, water, and free lead. The positive and negative pastes differ only in additives to the base mixture. Organic expanders, barium sulfate [7727-43-7], $BaSO_4$, carbon, and occasionally mineral oil are added to the negative paste. Red lead [1314-41-6] or minium, Pb_3O_4, is sometimes added to the positive mix. The paste for both electrodes is characterized by cube weight or density, penetration, and raw plate density.

Additives, ie, carbon or Pb_3O_4, are used to improve pasted plate conductivity and formed or charged plate characteristics. The expander used in the negative paste keeps it porous or spongelike throughout its life, rather than becoming a dense, tightly packed sheet (158,159). The barium sulfate in the negative paste is believed to act to stimulate the formation of fine seed crystals of $PbSO_4$ during discharge of the electrode (160–162). The physical characteristics of the paste are controlled by the mixing process (163,164). The amounts and order of constituent addition also play a role in the makeup of the finished paste.

During the paste mixing process the lead oxides are partially converted to basic lead sulfate (165). Some of these compounds are monobasic lead sulfate [12036-76-9], $PbO \cdot PbSO_4$, dibasic lead sulfate [65589-92-6], $2PbO \cdot PbSO_4$, predominantly tribasic lead sulfate [12202-17-4], $3PbO \cdot PbSO_4$, and, if high temperatures are encountered during the mixing step, tetrabasic lead sulfate [12065-90-6], $4PbO \cdot PbSO_4$. These compounds, because of their crystal morphology, bind tightly together on the grid, and facilitate plate processing and battery cycling. It is the degradation of this binding property resulting from an increased crystal growth (166) that causes failure of the positive plate by shedding of the active material. This active material drops through the separator ribs and collects in the sediment space (Fig. 30). This material is lost during electrochemical conversion and, subsequently, performance of the positive plate deteriorates over repeated cycling. The use of antimony [7440-36-0], Sb, in the positive grid increases the paste adhesion and cohesion. Some designs use tubular positive plates or separator-glass mat combinations to retard shedding.

The water added during the mixing operation functions as a lubricant and bulking agent to produce a paste of the proper consistency. During the plate drying operation, the evaporation of water gives the desired plate porosity. In addition to forming the binding material, the sulfuric acid expands the paste and thus gives it greater porosity.

During the acid addition step of the paste mixing process, considerable heat is generated because of the exothermic reaction of lead oxide and sulfuric acid. Care must be taken to prevent excessive loss of water before the paste can be applied to the grids. The mixer is usually cooled by water or air. After the paste has reached the desired uniformity and consistency for applying to the grid, the mixing is terminated. The paste density range is generally 3.7–4.0 g/mL for high initial capacity battery designs and for long cycle life product it may be as high as 4.3–4.6 g/mL. Muller or plough type mixers are used and a typical batch size is approximately 1100 kg. An excellent discussion of paste preparation and curing processes is available (167) and comprehensive review (168) describes the morphological features of the compounds used or evolved in the initial paste material.

Grids. The grid in a battery plate performs two vital functions. It acts as the mechanical support framework of the plate during manufacturing, and provides uniform, efficient current flow to and from the plate during formation and use. Grids are designed to have a lug or current collection point, current carrying arms, structural frame, and usually feet (Fig. 34).

The grid must possess sufficient stiffness to prevent damage or distortion during the casting, plate pasting, and battery assembly operations. During the life of the battery the grid must bear significant loads such as active material weight, its own weight, and stresses resulting from corrosion and volumetric changes caused by the sulfation of lead and lead dioxide. The most severe environment for the grid is in the positive plate. The positive active material operates in a potential range where lead is thermodynamically unstable. Thus the lead

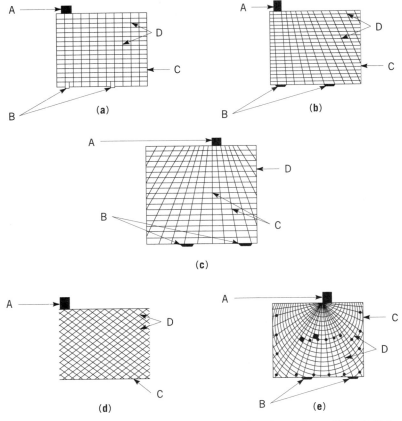

Fig. 34. Lead–acid battery grid design variations showing A lugs, B feet, C frames, and D current carrying wire for (**a**) rectilinear design, (**b**) corner lug radial, (**c**) center lug radial, (**d**) corner lug expanded metal, and (**e**) plastic/lead composite.

grid is continuously oxidizing to PbO and PbO_2 which affects the grid by reducing its strength and conductivity, and sometimes causing elongation of its members, commonly called grid growth.

Most grids are cast or mechanically expanded from cast or wrought strip. Historically, negative and positive grids have been the same; however, with the advent of specialized batteries and extreme performance demands, this is less likely. However if these plates are different, it is generally in configuration, alloy, and/or thickness.

The conductivity of the grid plays a substantial role in a battery's ability to meet high current demands. The importance of grid conductivity for lead–acid batteries has been discussed (100,169). Composition and configuration are important design factors impacting grid conductivity.

Potential distribution in grids (169–172) has been evaluated and the results put into practice (173). Innovative grid structures which incorporate element members positioned to optimize current flow patterns have been proposed (169,174,175) as well as economical lightweight grids for the negative plate made from plastics coated or woven with lead (176,177), or with plastic strengthening members and metallic conductors (178). All incorporate lead terminal connectors in their design for easily soldering to one another, or to current collectors.

Grid Material. Pure lead has low tensile and creep strength, and thus is easily deformed at room temperature. Because the grid supplies plate strength during manufacturing and life, it must be durable. This is accomplished by alloying lead with other elements. The primary alloying ingredients are antimony, calcium [7440-70-2], and tin [7440-31-5]. Numerous other elements, eg, arsenic [7440-38-2], selenium [7782-49-2], etc can and are being used in lesser amounts.

High antimony (2–12%) content lead alloys have been used widely in the manufacture of grids where antimony increases the structural integrity of the resulting grid. Antimony not only improves its manufacturability, but also retards positive grid growth and stress deformation caused by corrosion and active material sulfation. Antimony migrates into, and alters the morphology of, the positive active mass which improves the PbO_2 adhesion and cyclability (179).

Antimony from the grid also diffuses into the electrolyte and migrates to the negative electrode and plates out. This plating or poisoning increases the self-discharge rate of the negative plate. It also reduces the negative plate's hydrogen gassing overpotential, which increases charge inefficiency and water loss resulting from gas generation. These adverse effects have been minimized by using low antimony (1–2%) content alloys (180,181). Such alloys use low percentages of other additives, eg, selenium, tin, sulfur [7704-34-9], and arsenic (182) to improve their strength. The low antimony content alloy reduces the detrimental gassing effects of its high antimony counterpart. It also has improved castability and enhanced corrosion resistance.

Lead alloy containing calcium (0.075–0.1%) as its stiffening agent is used in some maintenance-free batteries. This alloy is more difficult to cast than antimonial–lead alloys. The grids formed from this alloy have somewhat lower self-discharge and water loss than lead–antimony alloy grids, but batteries containing lead–calcium alloy positive grids do not give as much cycle service as those made with lead–antimony alloy positive grids (181). However, this difference is nullified if the battery is in float (constant voltage) or limited cycling applications (183).

Other alloying ingredients in lead, eg, arsenic (0.5–0.7%) and silver [7440-22-4] (0.1–0.15%), inhibit grid growth on overcharge and reduce positive grid corrosion. Tin added to a lead alloy produces well-defined castings that are readily adapted to mass production techniques (184).

The proper selection of the lead alloy depends on the intended use and the economics of the lead–acid battery application. The metallurgical and electrochemical aspects of the lead are discussed in the literature in a comprehensive manner (181,185–187) as are trends of lead alloy use for manufacture of battery grids (188).

Grid Casting.   The grid casting machine consists of a center parting grid, or book mold, trimming mechanism, and melting pots. The grid mold consists of two cast iron parts, each having a grid design for a face. The mold design can also be a single grid or multiple grid configuration. The grid mold is heated to 135–180°C to prevent premature solidification of the lead alloy and holes in the mold release trapped air while the mold is being filled with the molten metal.

The melting pot is heated either electrically or by gas to 427–524°C. The pot capacity is typically over 100 kg of lead alloy, and periodically the top of the molten metal must be skimmed to remove the dross. The pot fumes must be removed by adequate ventilation (forced suction). When the molten metal has reached the proper temperature and flow characteristics, it is transported by pump to the grid mold.

Prior to casting, the grid mold halves are sprayed using a mold-release agent. The releasing agent also serves to control grid thickness and weight during casting. After filling with a slight excess of molten lead alloy, the mold is cooled for several seconds before opening. The cast grid is trimmed to remove rough edges, gates, and minor imperfections before stacking for use in the plate processing operation.

Another common manufacturing technique is expanding cast or wrought lead strip. The strip is typically a calcium alloy of lead, although antimonial alloys have been used. The process consists of casting or rolling a continuous strip and coiling it into large rolls which are subsequently expanded. The expanding process is a multistage process in which the strip is perforated with a cutting tool and progressively stretched apart to form a diamond pattern grid, punched to form lugs, and trimmed to form the singular grids (Fig. 34), before or after pasting.

The advent of higher compression, more powerful engines in compact engine compartments and more electrical loads has precipitated the need for smaller, more powerful batteries. This has driven battery manufacturers to look at thinner plate technology. Grid designs and materials are being reevaluated to establish components to meet future demands. Technologies such as dual material (169), bipolar (189), and folded plate designs (155) are broadening the role of the grid and what it is expected to do.

Plate Pasting.   Two types of machines are commonly used to apply paste to grids. They are the belt type and the fixed orifice type. Each of these types of pasters has a paste hopper that supplies paste to the machine and is fed by the paste mixer. The belt paster presses paste into the grid which is being trans-ported by a belt into the paste stream. A fixed orifice paster squeezes the paste into the grid and then pushes the grid and paste through an orifice to smooth both plate surfaces. In both instances the pasted plate is flash dried and/or papered to remove surface moisture, allowing the plates to be stacked and trans-ported to a curing area. In flash drying the plate passes quickly through a high temperature oven which effectively dries only the surface of the plate. When paper is used, a thin film is applied to both surfaces of the plate directly after pasting. Moisture content of the paste in plates after flash drying is typically 10%.

The tubular positive plate uses rigid, porous fiber glass tubes covered with a perforated plastic foil as the active material retainer (Fig. 31). Dry lead oxide, PbO, and red lead, Pb_3O_4, are typically shaken into the tubes which are threaded over the grid spines. The open end is then sealed by a polyethylene bar. Patents describe a procedure for making a type of tube for the tubular positive plate (190) and a method for filling tubular plates of lead–acid batteries (191). Tubular posi-tive plates are pickled by soaking in a sulfate solution and are then cured. Some proceed directly to formation and do not require the curing procedure.

Plate Curing. Curing strengthens or hardens and dries the paste and establishes a cohesive paste–grid bond. It is a critical process in battery fabrica-tion because it increases the plate's structural integrity and creates the electro-de's active material morphology. A typical curing process consists of placing pasted plates in an elevated temperature, high humidity environment for some extended period of time such as from 16 to 48 hours. There, the plate is subjected to a controlled drying scenario. Unreacted lead particles oxidize in an exothermic reaction, creating heat that slowly drives off plate moisture. The rate of reaction of lead to lead oxide is dependent on the moisture content of the paste and the manner in which this happens affects the formation of the lead sulfate–lead oxide crystal structure (192). A normal curing environment results in small crystals of tribasic lead sulfate, $3PbO \cdot PbSO_4 \cdot H_2O$. If the temperature is elevated to $>57°C$, the result is coarse tetrabasic lead sulfate, $4PbO \cdot PbSO_4$, crys-tals. This is especially critical for the positive electrode where tribasic sulfate converts readily to PbO_2 during battery formation but tetrabasic sulfate does not (193).

After curing, the plates are allowed to finish the drying process in ambient or elevated temperature air. The moisture and metallic lead content of the cured plates should be substantially reduced to less than 2%.

Separators. The separator's purpose is to isolate the positive and nega-tive electrodes electronically while allowing ionic exchange between the two. This is accomplished using a microporous, electronically nonconductive material. Separators are generally designed with a backweb that has ribs on one side. The microporous backweb is kept thin to reduce resistance to ionic diffusion and migration. The ribs, which are normally placed on the positive electrode side, act as standoffs, allowing a reservoir of electrolyte to exist between the plates.

The separator must be structurally sound to withstand the rigors of battery manufacturing, and chemically inert to the lead–acid cell environment. Numer-ous materials have been used for separators ranging from wood, paper, and rub-ber to glass and plastic. Glass fibers are included in this list (194). The majority of separators used are either nonwoven–bound glass or microporous plastic such as PVC or polyethylene.

Separator containing efficiency improving additions to lead–acid batteries have been patented (195). Separator manufacturing procedures and materials are discussed in the literature (196,197) and separator test procedures are also available (198).

Strap. Plates of similar polarity within a cell, whether positive or negative, are connected in parallel by a strap. Most batteries use either a burned-on or a cast-on strap (COS). In the case of COS, a stacked element is aligned and the lugs are placed in a mold where molten lead is cast around them to form a strap and vertical projection. When burning is used, the lugs are placed into a precast comblike strap and fused by melting the lugs and comb together. The strap has a large vertical projection on one end (Fig. 30) that is used to connect the elements in series once inside the container.

Container. The battery container is made up of a cover, vent caps, lead bushings, and case. Cost and application are the two primary factors used to select the materials of construction for container components. The container must be fabricated from materials that can withstand the abusive environment the battery is subjected to in its application. It must also be inert to the corrosive environment of the electrolyte and solid active materials, and weather, vibration, shock, and thermal gradients while maintaining its liquid seal.

The case is the largest portion of the container. The case is divided into compartments which hold the cell elements. The cores normally have a mud-rest area used to collect shed solids from the battery plates and supply support to the element. Typical materials of construction for the battery container are polypropylene, polycarbonate, SAN, ABS, and to a much lesser extent, hard rubber. The material used in fabrication depends on the battery's application. Typical material selections include a polypropylene–ethylene copolymer for SLI batteries; polystyrene for stationary batteries; polycarbonate for large, single cell standby power batteries; and ABS for certain sealed lead–acid batteries.

Covers for the battery designs in Figures 30 and 31 are typically molded from materials identical to that of the respective case, and vent plugs are frequently made of molded polypropylene. Other combinations are possible, eg, containers molded of polyethylene or polypropylene may be mated with covers of high impact rubber for use in industrial batteries. After the cover is fitted over the terminal post, it is sealed onto the case. The cover is heat bonded to the case, if it is plastic; it is sealed with an epoxy resin or other adhesive, if it is vulcanized rubber. Vent caps are usually inserted into the cover's acid fill holes to facilitate water addition and safety vent gasses, except for nonaccessible maintenance-free or recombinant batteries. In nonaccessible batteries, the vent is fabricated as part of the cover.

2.6. Electrolyte. *Sulfuric Acid.* Sulfuric acid is a primary active material of the battery. It must be present to provide sufficient sulfate ions during discharge and to retain suitable conductivity. Lead–acid batteries generally use an aqueous solution of acid in either a free-flowing or in an immobilized state.

Sulfuric acid solutions are quantified by density or specific gravity. Concentrated sulfuric acid has a density ≈ 1.840 g/mL, sp gr = 1.840 at room temperature; water is 1.000 g/mL, at room temperature; mixtures vary between these values. The temperature of the electrolyte affects its specific gravity, thus temperature should be noted and correction factors applied when measurements are made. At room temperature a common range of specific gravities for lead–acid batteries is 1.210 to 1.300. The electrolyte's specific gravity changes as the battery is charged and discharged. This specific gravity change occurs as sulfate ions transfer from liquid H_2SO_4 to solid $PbSO_4$. Thus the specific gravity of the electrolyte is often used as a means of approximating battery condition or state of charge.

When acid is used in the immobilized state, it is either absorbed into a high void-volume separator, or it is gelled. Electrolyte is gelled using one of several procedures (199–201). Materials such as silicon dioxide and sodium silicate [1344-09-8], Na_2SiO_4, are used as gelling agents. The components are mixed and poured into the battery cell where the gel is allowed to set for some preset time. In this case the battery has normally been formed and its liquid electrolyte removed, after which the gelled electrolyte is added.

When a battery is designed with absorbed electrolyte, it is built with dry separators, filled with acid and formed (Fig. 32). Excess electrolyte may then be removed.

2.7. Battery Assembly. The cell element (Fig. 30) is normally constructed from groupings of positive and negative plates. The number and size of plates of each type is determined by the desired performance level for the battery. Positive and negative plates are alternately stacked using separators in between to form the proper plate count. When the element is assembled, this piece moves to the final assembly.

The elements are inserted into containers such as that shown in Figure 30. The voltage of each charged cell is approximately 2 V, thus three elements are connected in series in 6 V batteries, six in 12 V batteries, and so forth. Once in the container, the strap vertical projections are fused together in a series arrangement to produce an unformed battery. Normal fusing techniques include resistance welding, burning, or lead extrusion. When the fusing operation is complete, the cover is sealed to the container. Most batteries use heat sealing, although adhesives, friction welding, and other methods have application.

The assembled unit is then subjected to a pressure test to ensure leak-free seals around the periphery and between the cells. Finally, positive and negative external terminations (located in the two end cells) are fused to the cover bushings. The battery is filled with electrolyte by immersing it in a bath of diluted sulfuric acid or by pouring electrolyte into the cells. After filling, the battery is placed on charge as quickly as possible to retard the conversion of tribasic lead sulfate to normal lead sulfate. These neutralization reactions cause undesired heat buildup and create products that are difficult to form. The reactions also dilute the acid, increasing the solubility of expanders, allowing them to leach out. This is detrimental to finished battery performance.

Formation. During formation the cured plate material is electrochemically converted into electrode active material, ie, spongy lead in the negative, PbO_2 in the positive, utilizing current supplied by power sources. Formation reactions create heat which, when coupled with ohmic heat and heat of neutralization, can harm electrode active materials. Thus formation is often done using evaporative coolers, water baths, or air tables. The efficiency of formation is highly dependent on the precursor cured plate morphology (196), temperature, formation current, and acid concentration (156). These processes are explained in great detail (160,193,202–204) for the positive and negative plate active materials. Normally between 110% and 130% of the battery's capacity $(A \cdot h)$, dependent on the formation current, is required to form it and formation times range from eight hours to beyond 48 hours.

The majority of batteries manufactured are supplied as wet, ie, acid in the formed battery, battery products, and thus utilize in-container formation. Some dry charged product where plates are washed and dried after formation, and moist, dumped and/or centrifuged, batteries are produced for some applications and markets (205).

Testing.   The finished battery may be tested for voltage, capacity, charge rate acceptance, cycle life, accelerated life, storage stability, overcharge, normal temperature discharge operation, low temperature cranking, shock and vibration, and gas generation (206). The detailed test procedures for automotive battery testing are contained in the specifications of the Battery Council International (BCI) (198) and the Society of Automotive Engineers (SAE) (207). Other battery test procedures are indicative of a specialized application or general all-around use. After completion of testing, the test specimens are frequently disassembled and critically examined by chemical, physical, and metallurgical means to manufacturing methods. The SAE (208) has produced a different test procedure for battery powered vehicles. Method for determining the state of lead–acid rechargeable batteries has been described (209).

2.8. Economic Aspects.   Lead–acid batteries including starting-lighting ignition (SLI) and industrial types continue to be the dominant use for lead (87% of total consumption) SLI battery shipments totaled 106×10^6 units in 2001. This total included original equipment and replacement automative-type batteries (100).

In 2001, there was slower growth in demand for industrial-type sealed lead–acid batteries in backup power systems. Industrial-type batteries include stationary batteries, such as those used in uninterruptible power-supply equipment for hospitals, computer and telecommunications networks, and load leveling equipment for commercial electric power systems and traction batteries (ie, forklifts, airline ground equipment, mining vehicles). Telecommunications companies scaled down investment plans and there was also a decline in seasonally related battery failure.

Value regulated lead–acid batteries for use in hybrid-electric vehicles were displayed at the first Automotive Battery Conference in Las Vegas, Nevada in Feb. 2001. They will be marketed soon and will supply sufficient power for automobile idle step and regenerative power recovery and in car luxuries systems features such as electrically conductive oxide coatings for heating windshields, and powere steering. Potential use in mild hybrid vehicles (cars capable of operating in electric assist mode) is also noted (100).

The market for higher voltage batteries is expected to grow significantly in order to handle greater automotive electric demands. The Absorbent Glass Mat lead–acid battery uses a quantity of lead similar to conventional batteries, but is designed to last twice as long. Its sealed design allows use in a variety of positions without danger of electrolyte spillage and it also permits recombination of hydrogen and oxygen gas to form water resulting in a maintenance-free battery.

Lead–acid batteries are expected to dominate demand for lead. In 2002, North American shipment of SLI batteries were expected to decline 3% to 19.8×10^6 units of automotive original equipment and by 5% to 81.4×10^6 units in the automotive replacement sector (210). Demand for industrial type lead-acid batteries were expected to decline by 2% in the motive power sector and by 5% in the stationary power sector.

The U.S. industrial battery market has grown over a 10-year period rate of 6–8% to posting a 15% drop in demand in 2001. However, for the period 2003–2006 the demand is expected to grow at the rate of 6–8% (211).

2.9. Recycling. Battery Council International reported that the U.S. battery industry recycled 93.3% of the available lead scrap from spent lead-acid batteries during the year 1995–1999 (100). The high rate is the result of a successful collaboration among members of the battery industry, retailers, and consumers.

Law are now in place in 42 states that prohibit the disposal of spent lead–acid batteries and require collection through consumer return procedure when a replacement battery is purchased (212).

3. Other Cells

3.1. Introduction. The proliferation of portable electronic devices has fueled rapid market growth for the rechargeable battery industry. Miniaturization of electronics coupled with consumer demand for lightweight batteries providing ever longer run times continues to spur interest in advanced battery systems. Interest also continues to run strong in electric vehicles (EVs) and the large auto manufacturers continue to develop prototype EVs. It is clear that advances in battery technology are required for a widely acceptable EV. Advanced batteries continue to play a strong role in other applications such as load leveling for the electric utility industry and satellite power systems for aerospace.

The goal of advanced battery research is to develop batteries that supply a high number of watt · hours in a small volume (volumetric energy density) and at low weight (gravimetric energy density). Obviously this increase in energy density must be achieved in a manner that is manufacturable, safe, and of acceptable cost. There has been a significant growth in the number of advanced battery systems in development and several systems are either nearing commercial viability or have been introduced as commercial products in specific applications.

3.2. Ambient Temperature Lithium Systems. Traditionally, secondary battery systems have been based on aqueous electrolytes. Whereas these systems have excellent performance, the use of water imposes a fundamental limitation on battery voltage because of the electrolysis of water, either to hydrogen at cathodic potentials or to oxygen at anodic potentials. The application of nonaqueous electrolytes affords a significant advantage in terms of achievable battery voltages. By far the most actively researched field in nonaqueous battery systems has been the development of practical rechargeable lithium batteries

(213). These are systems that are based on the use of lithium [7439-93-2] metal, Li, or a lithium alloy, as the negative electrode (see LITHIUM AND LITHIUM COMPOUNDS).

The use of lithium as a negative electrode for secondary batteries offers a number of advantages (214). Lithium has the lowest equivalent weight of any metal and affords very negative electrode potentials when in equilibrium with solvated lithium ions resulting in very high theoretical energy densities for battery couples. These high theoretical energy densities have prompted a wealth of research activity in a wide variety of experimental battery systems. However, realization of the technology to commercialize these systems has been slow.

A key technical problem in developing practical lithium batteries has been poor cycle life attributable to the lithium electrode (215). The highly reactive nature of freshly plated lithium leads to reactions with electrolyte and impurities to form passivating films that electrically isolate the lithium metal (216). This isolation results in less than 100% current efficiency for cycling of the lithium electrode, hence limiting cycle life (217). A goal of research for practical rechargeable lithium systems has been to improve the cycling efficiency of the lithium electrode in aprotic electrolytes. Improved cathodes are mentioned in the patent literature. See Refs 218–220 for examples.

The choice of battery electrolyte is of paramount importance to achieving acceptable cycle life because of the high reducing power of the metallic lithium. The formation of surface films on the lithium electrode (221) imparts the apparent stability of the electrolyte to the electrode. It is critical to determining lithium cycling efficiency. In addition to providing a stable film in the presence of lithium, the electrolyte must satisfy additional requirements including good conductivity, being in the liquid range over the battery operating temperature, and electrochemical stability over a wide voltage range. Solubility of the electrolyte salt in the solvent system is important in achieving good conductivity (222). In order to satisfy the various electrolyte system requirements, the use of mixed solvent electrolytes has become common in practical cells. Examples are tetrahydrofuran [109-99-9], C_4H_8O, -based electrolytes (223) or ethylene carbonate [96-49-1], $C_2H_4O_3$, –propylene carbonate [108-32-7], $C_4H_6O_3$, mixed solvent systems (224). Typical electrolyte salts include lithium perchlorate [7791-03-9], $LiClO_4$, lithium hexafluoroarsenate [29935-35-1], $LiAsF_6$, or lithium tetrafluoroborate [14283-07-9], $LiBF_4$ (225). The progress in the development of liquid electrolyte systems that provide near 100% current efficiencies for cycling of lithium has played a critical role in the emergence of rechargeable lithium as a commercially viable technology.

A second class of important electrolytes for rechargeable lithium batteries are *solid* electrolytes. Of particular importance is the class known as solid polymer electrolytes (SPEs). SPEs are polymers capable of forming complexes with lithium salts to yield ionic conductivity. The best known of the SPEs are the lithium salt complexes of poly(ethylene oxide) [25322-68-3] (PEO), $-(CH_2CH_2O)_n-$, and poly(propylene oxide) [25322-69-4] (PPO) (226–228). Whereas a number of experimental battery systems have been constructed using PEO and PPO electrolytes, these systems have not exhibited suitable conductivities at or near room temperature. A new heteroatomic polymer for more efficient solid polymer electrolytes for lithium batteries has been patented. These

Table 6. **Theoretical Energy Densities for Rechargeable Lithium Systems**

Battery couple	Operating voltage range, V	Energy density	
		$W \cdot h/kg$	$W \cdot h/L$
Li–CoO$_2$	4.4–2.5	766	2700
Li–MnO$_2$	3.5–2.0	415	553
Li–V$_2$O$_5$	3.5–2.0	541	1304
Li–TiS$_2$	2.6–1.6	473	1187
Li–NbSe$_3$	2.1–1.5	436	1600
Li–MoS$_2$	2.3–1.3	233	882

polyalkyl or polyfluoroalkyl heteroatomic polymers are superior to similar solid polymer electrolytes composed of polyethylene oxide (229).

The lithium or lithium alloy negative electrode systems employing a liquid electrolyte can be categorized as having either a solid positive electrode or a liquid positive electrode. Systems employing a solid electrolyte employ solid positive electrodes to provide a solid-state cell. Another class of lithium batteries are those based on conducting polymer electrodes. Several of these systems have reached advanced stages of development or initial commercialization such as the Seiko Bridgestone lithium polymer coin cell.

The most important rechargeable lithium batteries are those using a solid positive electrode within which the lithium ion is capable of intercalating. These intercalation, or insertion, electrodes function by allowing the interstitial introduction of the Li$^+$ ion into a host lattice (230,231). The general reaction can be represented by the equation:

$$x\,Li^+ + x\,e^- + MY_n \rightleftharpoons Li_xMY_n$$

where MY_n represents a layered compound. A large number of inorganic compounds have been investigated for their ability to function as a reversible positive electrode in a lithium battery. Intercalation electrodes have found wide application in systems employing both liquid or solid electrolytes. The theoretical electrochemical characteristics of the most advanced of these electrode systems are described in Table 6.

Solid Electrolyte Systems.   Whereas there has been considerable research into the development of solid electrolyte batteries (232–235), development of practical batteries has been slow because of problems relating to the low conductivity of the solid electrolyte. The development of an all solid-state battery would offer significant advantages. Such a battery would overcome problems of electrolyte leakage, dendrite formation, and corrosion that can be encountered with liquid electrolytes.

The general configuration of one system that has reached an advanced stage of development (236) is shown in Figure 35. The negative electrode consists of thin lithium foil. The composite cathode is composed of vanadium oxide [12037-42-2], V$_6$O$_{13}$, mixed with polymer electrolyte. Demonstration batteries have been constructed having energy densities of 320 W·h/L. A key to the technology is a unique radiation cross-linked polymer electrolyte which shows good conductivity at ambient temperatures (237).

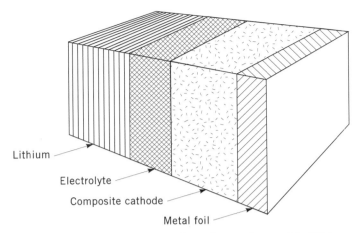

Fig. 35. Configuration for a solid polymer electrolyte rechargeable lithium cell where the total thickness is 100 μm. Courtesy of Mead Corp.

A new all solid-state lithium battery employing a positive electrode comprised of organosulfur polymers, $-(SRS)_n-$, has been reported (238). During discharge of the battery, current is produced by cleavage of the sulfur–sulfur bonds in the polymer, depolymerizing the polymer. On charge, the process is reversed and the disulfides are polymerized back to their original form. This use of a polymerization–depolymerization reaction for a battery electrode is unique and this electrode is expected to offer significantly improved rate capability over intercalation electrodes. The organosulfur electrodes provide excellent stability and reversibility, which should result in long cycle life. Cells constructed to date have employed a PEO electrolyte and hence require operation at elevated temperature (239).

Coin and Button Cell Commercial Systems.   Initial commercialization of rechargeable lithium technology has been through the introduction of coin or button cells. The earliest of these systems was the Li–C system commercialized by Matsushita Electric Industries (MEI) in 1985 (240,241). The negative electrode consists of a lithium alloy and the positive electrode consists of activated carbon [7440-44-0], carbon black, and binder. The discharge curve is not flat, but rather slopes from about 3 V to 1.5 V in a manner similar to a capacitor. Use of lithium alloy circumvents problems with cycle life, dendrite formation, and safety. However, the system suffers from generally low energy density.

A commercial introduction of a rechargeable Li–V_2O_5 coin cell having significantly higher energy density than the previous Li–C cell has been announced (242). This system employs a Li–Al alloy negative electrode coupled with a vanadium pentoxide [1314-62-1], V_2O_5, positive electrode. The cell voltage on discharge is 3 V. This cell is claimed to have twice the energy density of a conventional Ni–Cd cell (243).

A rechargeable coin cell that employs a lithium salt electrolyte but avoids the use of a metallic lithium negative electrode has been commercially introduced by Toshiba. This cell employs a material called linear-graphite hybrid (LGH) in lieu of a lithium electrode to avoid formation of lithium dendrites.

LGH is synthesized by carbonizing organic aromatic polymers. The LGH electrode is coupled with an amorphous V_2O_5 positive electrode to give a battery with a working voltage of 3 V to 1.5 V (244). A similar system employing pyrolytic carbon as a negative electrode has also been reported (245).

A significant development in commercial acceptance of rechargeable lithium was the introduction by Sanyo of rechargeable Li–MnO_2 coin and button batteries (246,247). Manganese dioxide [1313-13-9] is an attractive positive electrode system because of its relatively high and flat discharge voltage and the low cost of MnO_2. The widespread use of Li–MnO_2 primary batteries demonstrates the utility of this system. However, the rechargeability of the MnO_2 conventionally employed in primary cells was not sufficient to yield practical secondary cells. The breakthrough came in the modification of the crystal structure of the MnO_2 to improve rechargeability by a lithiation reaction (248). This modified MnO_2, referred to as composite dimensional manganese oxide (CMDO), is coupled with a Li–Al alloy having special additives to give excellent cycle life and stress resistance to prevent cracking and blistering of the alloy during cycling (249). Lithium batteries with new manganese oxide materials as lithium intercalation hosts have been patented (250). These batteries show significant improvement in cycling performance.

Advanced Systems. Applications for the coin and button secondary lithium cells is limited. However, researchers are working to develop practical "AA"-sized and larger cells. Several systems have reached advanced stages of development.

The first commercially available lithium "AA" cell was introduced in the early 1980s by Moli Energy Ltd. This cell, known as the Molicel, employs a molybdenum disulfide [1317-33-5], MoS_2, positive electrode coupled with a pure Li negative one (251). The cell is constructed in a spirally wound configuration using a microporous separator (Fig. 36). The cells have an open circuit voltage of about 2.3 V when fully charged. The discharge curve slopes, with a midpoint voltage of 1.2 V when discharged. The sloping discharge curve allows indication of state of charge (252). Cell capacity is about 600 mA for "AA" cells. Cycle life was found to be strongly dependent on operating conditions including depth of discharge, recharge and discharge currents, and voltage limits on charge and discharge. However, cycle lives of 300 cycles or more were achieved under normal cycling conditions (253).

One of the most widely studied intercalation electrode materials is titanium disulfide [12039-13-3], TiS_2. A number of factors make TiS_2 attractive for secondary lithium cells including good rate capability, high theoretical energy density, and a highly reversible intercalation reaction (254). Very high cycle lives have been demonstrated for TiS_2 electrodes (255). One of the earliest efforts to develop commercial lithium cells employed TiS_2 (256).

The development of "AA" Li–TiS_2 cells combines the excellent properties of TiS_2 and a proprietary method for preparing polymer bonded electrodes (257) to give a flexible, high capacity bonded electrode. Using this technology, "AA" cells have been constructed giving 1.05 A · h capacity at a voltage of 2.3 V to 1.6 V on discharge. Using excess Li for the negative electrode, cycle lives of 200 cycles are claimed for this cell. Other advantages are a good high rate capability (up to C rate), good gravimetric energy density, and very low self-discharge (258).

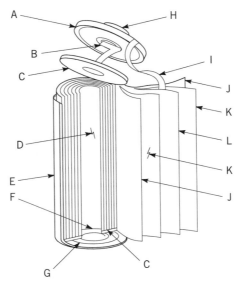

Fig. 36. Configuration for spirally wound rechargeable lithium cell. A, Cap; B, cathode tab; C, insulating disk (2); D, mandrel; E, can; F, ball; G, safety vent; H, glass-to-metal seal (with center pin); I, anode tab; J, cathode; K, separator; L, anode. Courtesy of Moli Energy Ltd.

A rechargeable lithium "AA" cell employing niobium triselenide [12034-78-5], $NbSe_3$, as the positive electrode material has been developed (259,260). The key to this system was a method for thermally growing $NbSe_3$ fibers that can then be pressed onto a current collector to provide a very high energy density electrode. No binder or additional conductive material is required. The excellent properties of this electrode also allow for higher discharge currents than those typically available for rechargeable lithium systems. Discharge rates in excess of C rate have been demonstrated. "AA" cells employing this technology give capacities of 1.1 A·h when cycled between 2.4 and 1.4 V. Cycle lives of over 250 cycles have been demonstrated for this cell (261).

An "AA" Li–MnO_2 cell was announced in the late 1980s, (262,263). This cell is claimed to offer considerable advantages over the conventional Ni–Cd system, offering higher energy density, high operating voltage, high rate capability, and excellent cycle life. A comparison of the parameters of these cells is shown in Table 7 (262).

An advanced Li–MnO_2 battery under the trade name Molicel[2] has been developed by Moli Energy Ltd. (264). The cell has a nominal voltage of 3 V, allowing replacement of two NiCd cells with one Molicel[2], hence significantly reducing battery pack size and volume. Production "AA" cells demonstrate a nominal capacity of 600 mA·h for an energy density of 100 W·h/kg. Cycle life for this cell is reported to be typically 200 cycles (265).

Other solid cathode systems that have been widely investigated include those containing lithium cobalt oxide [12190-79-3], $LiCoO_2$ (266), vanadium pentoxide [1314-62-1], V_2O_5, and higher vanadium oxides, eg, V_6O_{13} (267,268).

Table 7. **Comparison of "AA"-Size Li–MnO$_2$ and NiCd Cells**

| Parameter | Cell system | | Parameter ratio Li–MnO$_2$/Ni–Cd |
	Li–MnO$_2$	Ni–Cd	
cell weight, g	16.0	25.0	0.65
operating voltage, V	2.8[a]	1.2[b]	2.3
cutoff voltage, V	2.0	1.0	
energy at 0.8 W discharge, W·h	2.03	0.85	2.4
self-discharge after 1 mo	1%	25%	0.04
cycles (% DOD[c])			
at 0.85 W·h/cycle	1000 (40%)	500 (100%)	2
at 1.3 W·h/cycle	400 (64%)		
at 1.7 W·h/cycle	200 (84%)		

[a] Average voltage.
[b] Nominal voltage.
[c] DOD = depth of discharge

In addition to cells employing solid positive electrodes, a rechargeable lithium cell employing sulfur dioxide [7446-09-5], SO$_2$, as a liquid electrode has been developed (269). The widespread use of primary Li–SO$_2$ batteries, particularly by the U.S. military, led to a strong interest in developing a rechargeable version. The cell electrolyte for this system consists of liquidsulfur dioxide andlithium tetrachloroaluminate[14024-11-4], LiAlCl$_4$, salt. The overall cell chemistry of the rechargeable system involves the reduction and oxidation of a complex which forms between LiAlCl$_4$, SO$_2$, and the carbon that is employed as the current collector for the positive electrode. Specifications for a prototype Li–SO$_2$ "C"-sized cell are shown in Table 8 (269). Advantages for this system include very high energy densities, a flat running voltage, and the ability to sustain limited overcharge. The principal drawback has been safety concerns over potential cell ruptures on cycling resulting from internal short circuits.

Table 8. **Specifications for a Prototype Li–SO$_2$ "C" Cell**[a]

Parameter	Value
open circuit voltage, V	3.2
operating voltage, at C/6,[b] V	3.0
energy, W·h	5.4
volumetric energy density, W·h/L	220
gravimetric energy density, W·h/kg	134
operating temperature, °C	−30 to 40
cycle life, cycles	50
discharge rate	C/1.8[b,c]
charge rate	C/18[b,d]
charge retention after 9 mo	100%

[a] Where the nominal capacity is1.8 A·h to 2 V.
[b] Where C is the current required to discharge the cell in 1 hour.
[c] C/1.8 = 1 A.
[d] C/18 = 0.1 A.

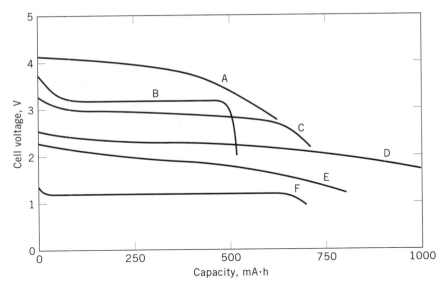

Fig. 37.   Cell voltage profiles as a function of discharge capacity for rechargeable "AA" cells: A, Li ion; B, Li–SO$_2$; C, Li–MnO$_2$; D, Li–TiS$_2$; E, Li–MoS$_2$; F, Ni–Cd.

A related system employs a copper(II) chloride [7447-39-4], CuCl$_2$, electrode in a sulfur dioxide electrolyte (270). "AA" cells of nominal capacity of 500 mA · h at a discharge voltage of 3.4–3.0 V were constructed. Cycle life for this cell is claimed to be 200 cycles. The high voltage of this system offers a significant energy density advantage over conventional Ni–Cd systems.

The discharge performance characteristics of advanced rechargeable lithium "AA" cells and the Ni–Cd cell are illustrated in Figure 37. Whereas it is clear that a number of lithium systems offer energy capacities of significant improvement over conventional Ni–Cd cells, several critical issues remain for cell development before lithium technology gains widespread application. Cycle life continues to be a limitation to most technologies. Unlike Ni–Cd, there is no overcharge or overdischarge protection for the rechargeable lithium cells which generally must be recharged only within closely defined voltage limits. This presents a problem particularly in the design of multicell battery packs. Experiments using electrolyte additives to provide overcharge protection have been carried out (271), but this concept has not been demonstrated in advanced cell designs. Additionally, the high reactivity of lithium metal continues to pose serious safety concerns among battery manufacturers and users. Incidences of lithium cells exploding and/or igniting or venting with flame during use have highlighted the need for extensive safety testing of lithium-containing batteries prior to the widescale commercialization of this technology.

A lithium ion rechargeable cell has apparently been developed in response to safety concerns over the Li–MnO$_2$ cell. This technology is expected to be used for camcorder battery packs. The technical details of this battery have not been reported, but it is understood that the negative electrode consists of a carbon material capable of undergoing an insertion reaction with Li$^+$ ion (272). This electrode is coupled with an insertion positive electrode such as lithium cobalt oxide (273). The resulting cell has a nominal operating voltage of 3.6 V, three times that of NiCd, providing a significant advantage in energy density (see Fig. 37) for multicell battery packs. The cell is claimed to be capable of rapid charge, discharge rates up to 2C, and 1200 cycles (274).

3.3. High Temperature Systems. *Lithium–Aluminum/Metal Sulfide Batteries.* The use of high temperature lithium cells for electric vehicle applications has been under development since the 1970s. Advances in the development of lithium alloy–metal sulfide batteries have led to the Li–Al/FeS system, where the following cell reaction occurs.

$$2 \text{ LiAl}_x + \text{FeS} \rightleftharpoons \text{Fe} + \text{Li}_2\text{S} + 2x \text{ Al}$$

The cell voltage is 1.33 V to give a theoretical energy density of 458 W·h/kg (275). The cell employs a molten salt electrolyte, most commonly a lithium chloride [7447-41-8]/potassium chloride [7447-40-7], LiCl–KCl eutectic mixture. The cell is generally operated at 400–500°C. The negative electrode is composed of lithium–aluminum alloy, which operates at about 300 mV positive of pure lithium. The positive electrode is composed of iron sulfide [1317-37-9] mixed with a conductive agent such as carbon or graphite. Electrodes are constructed by cold pressing powder onto current collectors (276).

Development of practical and low cost separators has been an active area of cell development. Cell separators must be compatible with molten lithium, restricting the choice to ceramic materials. Early work employed boron nitride [10043-11-5], BN, but a more desirable separator has been developed using magnesium oxide [1309-48-4], MgO, or a composite of MgO powder–BN fibers. Corrosion studies have shown that low carbon steel or stainless steel are suitable for the cell housing as well as for internal parts such as current collectors (277).

Li–Al/FeS cells have demonstrated good performance under EV driving profiles and have delivered a specific energy of 115 W·h/kg for advanced cell designs. Cycle life expectancy for these cells is projected to be about 400 deep discharge cycles (278). This system shows considerable promise for use as a practical EV battery.

A similar system under development employs iron disulfide [12068-85-8], FeS_2, as the positive electrode. Whereas this system offers a higher theoretical energy density than does Li–Al/FeS, the FeS_2 cell is at a lower stage of development (279,280).

Sodium–Sulfur. The best known of the high temperature batteries is the sodium [7440-23-5]–sulfur [7704-34-9], Na–S, battery (281). The cell reaction is best represented by the equation:

$$2 \text{ Na} + 3 \text{ S} \rightleftharpoons \text{Na}_2\text{S}_3$$

occurring at a cell voltage of 1.74 V, to give a specific energy of 760 W·h/kg. The cell is constructed using a solid electrolyte typically consisting of β-alumina [1344-28-1], β-Al_2O_3, ceramic, although borate glass fibers have also been used. These materials have high conductivities for the sodium ion. The negative electrode consists of molten sodium metal and the positive electrode of molten sulfur.

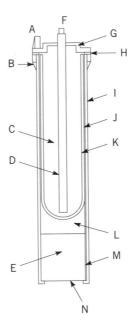

Fig. 38.   Construction of a sodium–sulfur battery. A, Negative terminal; B, springs plus graphite felt; C, sulfur; D, carbon; E, sodium reservoir; F, positive terminal; G, insulator; H, aluminum sealing gaskets; I, steel case; J, film of sodium; K, β-alumina tube; L, carbon felt; M, wick; N, aluminum can (282).

Because sulfur is not conductive, a current collection network of graphite is required. The cell is operated at about 350°C. A typical cell design is shown in Figure 38. The outer cell case is constructed from mild steel. The β-alumina separator tube fits inside the cell case with a small gap in between. The positive electrode of molten sulfur is placed inside the alumina tube with a carbon current collector. The molten sodium is stored below the cell and is wicked into the space between the separator and outer cell casing. The mild steel case acts as the negative electrode current collector (282).

The Na–S battery couple is a strong candidate for applications in both EVs and aerospace. Projected performance for a sodium–sulfur-powered EV van is shown in Table 9 for batteries having three different energies (283). The advantages gained from using a Na–S system rather than the conventional sealed lead–acid batteries are evident.

Table 9. **Electric Vehicle Battery Performance**

	Battery			
Parameter	Lead–acid	Sodium–sulfur		
battery energy, kW · h	40.0	40.0	60.0	85.0
range, km	84.0	113.0	169.0	242.0
max payload, t	0.9	1.7	1.6	1.6
battery weight, kg	1250.0	330.0	424.0	580.0

The Na–S system is expected to provide significant increases in energy density for satellite battery systems (284). In-house testing of Na–S cells designed to simulate midaltitude (MAO) and geosynchronous orbits (GEO) demonstrated over 6450 and over 1400 cycles, respectively.

Difficulties with the Na–S system arise in part from the ceramic nature of the alumina separator: the specific β-alumina is expensive to prepare; and the material is brittle and quite fragile. Separator failure is the leading cause of early cell failure. Cell failure may also be related to performance problems caused by polarization at the sodium/solid electrolyte interface. Lastly, seal leakage can be a determinant of cycle life. In spite of these problems, however, the safety and reliability of the Na–S system has progressed to the point where pilot plant production of these batteries is anticipated for EV and aerospace applications.

A battery system closely related to Na–S is the Na–metal chloride cell (285). The cell design is similar to Na–S; however, in addition to the β-alumina electrolyte, the cell also employs a sodium chloroaluminate [7784-16-9], $NaAlCl_4$, molten salt electrolyte. The positive electrode active material consists of a transition metal chloride such as iron(II) chloride [7758-94-3], $FeCl_2$, or nickel chloride [7791-20-0], $NiCl_2$, (286,287) in lieu of molten sulfur. This technology is in a younger state of development than the Na–S.

Lithium-Polymer Batteries. The lithium-polymer battery differs from other batteries in the type of electrolyte used. The polymer electrolyte replaces the traditional porous separator, which is soaked in electrolytes. The dry polymer design is rugged and has a thin profile. No liquid or gel is used, thus there is no danger of flammability.

The conductivity of the dry lithium-polymer is poor. The internal resistance is too high, and cannot deliver the current bursts needed for modern communication devices. Some dry lithium polymers are used in hot climates as standby batteries. High ambient temperatures do not degrade the service life of the battery. Research continues to develop a battery that performs at room temperature and this should come to pass in 2005 (287).

Most lithium-polymer batteries are a hybrid. The only functioning polymer battery in use today is this hybrid, more correctly know as lithium-ion polymer. The electrolyte in the lithium-ion polymer is solid and replaces the porous separator. Gelled electrolyte is added to enhance ion conductivity.

There have been technical difficulties manufacturing this battery. At present its only advantage is form. Wafer thin batteries are highly desirable in the mobile phone industry.

3.4. Economic Aspects. Almost all major battery manufacturers marketed some type of lithium battery. Research and development continued and innovative rechargeable battery configurations continue to be developed to meet the changing requirements of electronic equipment such as portable phones, computers, and video cameras. Lithium-ion batteries are of interest because they take advantage of the large power capacity available with fewer safety problems. Electric vehicles are considered a large potential market for lithium batteries, but general acceptance of these vehicles is slow. There is also the possibility that fuels cells will be the preferred source of power.

The consumption of lithium compounds in lithium batteries is one of the most rapidly expanding markets for lithium. Demand for lithium metal for batteries is expected to grow. Lithium-ion and lithium polymer batteries appear to posses the greatest potential for growth. First introduced in 1993, with minimal sales, the market for these rechargeable batteries grew to 3×10^9 in 1998 and is expected to top 6×10^9 by 2005 (288).

BIBLIOGRAPHY

The Alkaline Secondary Cells are treated in *ECT* 1st ed., under "Cells, Electric," Vol. 3, pp. 292–342, by W. J. Hamer, National Bureau of Standards, and "Cells, Electric," Suppl. 2, pp. 126–161, by C. K. Morehouse, R. Glicksman, and G. S. Lozier, Radio Corporation of America; "Batteries and Electric Cells, Secondary, Secondary Cells, Alkaline," in *ECT* 2nd ed., pp. 172–249, by W. W. Jakobi, Gould-National Batteries, Inc.; "Batteries and Electric Cells, Secondary, Alkaline Cells," in *ECT* 3rd ed., Vol. 3, pp. 592–639, by A. J. Salkind and E. Pearlman, ESB Technology Company; in *ECT* 4th ed., Vol. 3, pp. 1028–1083, by Alvin J. Salkind and Martin Klein, Rutgers University "Batteries, Electric (Lead-Acid Storage)" in *ECT* 1st ed., Vol. 2, pp. 340–360, by Joseph A. Orsino and Thomas C. Lynes, National Lead Company; "Secondary Cells, Lead–Acid" under "Batteries and Electric Cells, Secondary" in *ECT* 2nd ed., Vol. 3, pp. 249–271, by Joseph A. Orsino, National Lead Company; "Batteries and Electric Cells, Secondary (Lead–Acid)" in *ECT* 3rd ed., Vol. 3, pp. 640–663, by James B. Doe, ESB Technology Company; in *ECT* 4th ed., Vol. 3, pp. 1083–1107, by Kathryn R. Bullouk and John R. Pierson, Johnson Controls, Inc.; "Batteries and Electric Cells, Secondary (Other Cells)" in *ECT* 3rd ed., Vol. 3, pp. 663–670, by J. B. Doe, ESB Technology, Co.; in *ECT* 4th ed., Vol. 3, pp. 1101–1121, by Paul R. Gifford, Ovonic Battery Company.

CITED PUBLICATIONS

1. H. Takeshita, *Presentation at the 20th International Seminar and Exhibit on Primary and Secondary Batteries*, March 17, 2003.
2. Ger. Pat. 491,498 (July 6, 1928), K. Ackermann, P. Gomelin, G. Pfleiderer, and F. Spouna (to I. G. Farbenindustrie, A. G.); Brit. Pat. 331,540 (Jan. 4, 1929), 332,052 (July 8, 1929), 380,242 (1930), 339,645 (1930), to (I. G. Farbenindustrie, A. G.).
3. U.S. Pat. 2,571,927 (Oct. 16, 1951), U. Gottesman and G. Newmann (to Bureau Technique Gautrat Sarl); Brit. Pat. 784,851 (1952), (to Bureau Technique Gaurat).
4. Ger. Pat. 112,351 (1899); Brit. Pat. 15,370 (1899), T. De Michalowski.
5. Swed. Pat. 11,132 (1889), W. Jungner.
6. A. Fleischer, *J. Electrochem. Soc.* **94**, 289 (1948). A. Fleischer, *Proc. 10th Ann. Battery Res. Conf.* **73** (1956). U.S. Pat. 2,771,499 (1951), A. Fleischer.
7. G. W. Briggs, E. Jones, and W. F. K. Wynne-Jones, *Trans. Faraday Soc.* **51**, 1433 (1955); G. W. Briggs and W. F. K. Wynne-Jones, *Trans. Faraday Soc.* **52**, 1972 (1956).
8. A. J. Salkind and P. F. Bruins, *Nickel–Cadmium Sandia Project Reports*, Polytechnic University of Brooklyn, N.Y., 1956–1958; *J. Electrochem. Soc.* **108**, 356–360 (1962).
9. U.S. Pat. 2,819,962 (1958), (to SAFT).
10. J. A. McMillan, *Acta Cryst.* **7**, 640 (1954); *J. Inorg. Nucl. Chem.* **13**, 28 (1960).

11. V. Scatturin, P. Bellon, and A. J. Salkind, *J. Electrochem. Soc.* **108**, 819 (1961).
12. P. L. Bourgault and B. E. Conway, *Can. J. Chem.* **38**, 1557 (1960).
13. S. U. Falk, *J. Electrochem. Soc.* **107**, 661 (1960).
14. U.S. Pat. 3,003,015 and 3,121,029 (1961), J. C. Duddy.
15. A. J. Salkind and J. C. Duddy, *J. Electrochem. Soc.* **109**, 360 (1962).
16. D. Tuomi, *J. Electrochem. Soc.* **112**, 1 (1965).
17. H. Bode, K. Dehmelt, and J. Witte, *Crystal Structures of Nickel Hydroxides*, CITCE Meeting, Strasbourg, France, 1965.
18. Brit. Pat. 917,291 (1963), L. Kandler; U.S. Pat. 3,282,808 (1966), L. Kandler; E. Hausler, *5th Int. Power Sources Symp.* **287** (1966).
19. E. J. McHenry, *Electrochem. Tech.* **5**, 275–279 (1965).
20. U.S. Pat. 3,653,967 (1972), R. L. Beauchamp.
21. U.S. Pat. 3,873,368 (1972), D. F. Pickett.
22. M. A. Gutjahr, H. Buckner, K. D. Beccu, and H. Saufferer, in D. H. Collins, ed., *Power Sources*, Vol. 4, Oriel Press, UK, 1973, p. 79.
23. U.S. Pat. 3,898,099 (1975), B. Baker and M. Klein.
24. J. Edwards, T. Turner, and J. Whittle, *Proc. Int. Power Sources* **7**, (1970). U.S. Pat. 3,785,867 (1974), J. Edwards and J. Whittle.
25. U.S. Pat. 4,215,190 (1980), R. A. Ferrando and R. Satula.
26. U.S. Pat. 4,298,383 (1981); 4,312,670 (1981), J. F. Joyce and S. L. Colucci.
27. B. Bugnet and D. Doniat, *Proceedings of the 31st Power Sources Conference* 1984.
28. *DAUG Co. literature*, Germany, 1985.
29. K. Kordesch, J. Gsellmann, W. Harer, W. Taucher, and K. Tomantschger, *Prog. Batt. Solar Cells* **7**, 194–204 (1988); K. Kordesch and co-workers, *Electrochim. Acta* **26**, 1495 (1981).
30. U.S. Pat. 4,451,543 (1984), M. A. Dzieciuch, N. Gupta, H. Wroblowa, and J. T. Kummer; U.S. Pat. 4,520,005 (1985), Y. Y. Yao.
31. *Commercial literature*, Ovonic Battery Co., Troy, Mich, 1991; Matsushita Battery Co., Seattle, Wash., Oct. 1990; Toshiba Co., Seattle, Wash., Oct. 1990; Gates Battery Co., Gainsville, Fla., 1991; N. Furukawa, *JEC Battery Newsletter #2*, Sanyo Battery Co., JEC Press, Brunswick, Ohio, 1990.
32. Tayata, Honda commercial literature.
33. U.S. Pat. 6,596,436 (July 22, 2003), M. Maruta and co-workers (to Matsushita Electric Industrial Co., Ltd.).
34. U.S. Pat. 6,548,210 (April 15, 2003), K. Shinyama and co-workers (to Sanyo Electric Co., Ltd.).
35. U.S. Pat. 6,562,516 (May 13, 2003), K. Ohta and co-workers (to Matsushita Electric Industrial Co., Ltd.)
36. P. C. Milner and U. B. Thomas, P. Delahay and C. W. Tobias, eds., *Advances in Electrochemistry and Electrochemical Engineering*, Vol. 5, Interscience Publishers, New York, 1967.
37. Swed. Pat. 15,567 (1901), W. Jungner; Ger. Pat. 163,170 (1901), W. Jungner.
38. J. P. Harivel and co-workers, *Proceedings of the 5th International Power Sources Symposium*, Brighton, England, Sept. 1966.
39. B. C. Bradshaw, *Proceedings of the Annual Power Sources Conference* **12**, (1958).
40. F. Kornfeil, in Ref. 39.
41. B. V. Ershler, G. S. Tyvrikov, and A. S. Smirnova, *J. Phys. Chem. (USSR)* **14**, 985 (1940).
42. G. Halpert and L. May, *Extended Abstracts 169th Meeting of the Electrochemical Society*, Washington, D.C., Oct. 1983.
43. J. McBreen, W. E. O'Grady, K. I. Panda, R. W. Hoffman, and D. E. Sayers, *Langmuir* **3**, 428–433 (1987).

44. A. H. Zimmerman and P. K. Effa, *Extended Abstracts, 168th Meeting of the Electrochemical Society*, Oct. 1985.

45. D. T. Fritts, *Extended Abstracts, 164th Electrochemical Society*, Oct. 1983.

46. A. Estelle, *Tekn. Tidskizift* **39**, 105 (1909).

47. G. Halpert, *Proc. 176 Electrochem. Soc.* Oct. 1989.

48. "Cadmium," *Mineral Commodity Summaries*, U.S. Geological Survey, Reston, Va., Jan. 2003; *Minerals Yearbook*, 2001.

49. U.S. Pat. Appl. 20030160590 (Aug. 28, 2003), M. A. Schaefer and R. R. Reinhart (to Linvatec Corp.).

50. L. Ojefors, G. Kramer, and V. A. Olinpunom, *Proceedings of the 9th International Society for Electrochemistry*, Marcoussis, France, 1975; B. Andersson and L. Ojefors, *11th International Power Sources Symposium*, Brighton, England, 1978; *J. Electrochem. Soc.* **123**, 824 (1976).

51. A. Nilsson, private communication, 1991.

52. E. R. Bowerman, "Sintered Iron Electrode," *Proc. 22 Power Sources Conf.* (1968).

53. W. A.Bryant, *J. Electrochem. Soc.* **126**, 1899–1901 (1979).

54. Project reports, Eagle-Picher Co., Joplin, Miss., 1988–1991.

55. A. J. Salkind, J. J. Kelley, and J. B. Ockerman, *Proc. 176 Electrochem. Soc. Meet.* (Oct. 1989).

56. Reports on EPRI Nickel–Iron Electric Vehicle Project, Argonne National Labs, Argonne, Ill., 1988–1991.

57. R. Swaroop, *Reports on Electric Vehicle Batteries*, Electric Power Research Institute (EPRI), Palo Alto, Calif., 1989–1991.

58. T. Iwaki, T. Mitsumata, and H. Ogawa, *Prog. in Batt. Solar Cells* **4**, 255–259 (1982).

59. S. Hills and A. J. Salkind, "Alkaline Iron Electrode," *Proc. 22 Power Sources Conf.* (1968).

60. A. J. Salkind, C. J. Venuto, and S. U. Falk, *J. Electrochem. Soc.* **111**, 493–495 (1964).

61. P. Hersch, *Trans. Faraday Soc.* **51**, 1442–1448 (1955).

62. U.S. Pat. 2,644,022 (1953), J. Moulton and E. F. Schweitzer.

63. U.S. Pat. 3,483,291 (Dec. 16, 1969), M. J. Mackenzie and A. J. Salkind.

64. D. S. Poa, J. F. Miller, and N. P. Yao, *Argonne National Labs Report # ANL/ OEPM-85-2*, Apr. 1985.

65. H. G. André, *Bull. Soc. Franc. Electriciens* **6**, 1, 132 (1941).

66. U.S. Pat. 2,317,711 (Apr. 27, 1943), H. G. André.

67. J. A. McMillan, *J. Inorg. Nucl. Chem.* **13**, 28 (1960).

68. V. Scatturin, P. Bellon, and A. J. Salkind, *J. Electrochem. Soc.* **108**, 819 (1961).

69. A. J. Salkind "Crystal Structures of the Silver Oxides," in A. Fleischer and J. J. Lander, eds., *Zinc–Silver Oxide Batteries*, John Wiley & Sons, Inc., New York, 1971.

70. *The Silver Institute Letter*, Vol. VII, No. 9, Oct. 1977.

71. N. A. Zhulidov and F. I. Yefrennov, *Vestn. Elektropromsti.* **2**, 74 (1963); M. C. Tisgankov, *Electrical Accumulators*, Moscow (1959).

72. V. N. Flerov, *Zh. Prikl. Khim.* **32**, 1306 (1959).

73. P. Goldberg, "Nickel–Zinc Cells Part 1," *Proceedings of the 21st Annual Power Sources Conference*, 1967.

74. P. V. Popat, E. J. Rubin, and R. B. Flanders, "Nickel–Zinc Cells Part 3," *Proceedings of the 21st Annual Power Sources Conference*, 1967.

75. M. J. Sulkes, "Nickel–Zinc Secondary Batteries," *Proceedings of the 23rd Annual Power Sources Conference*, 1969.

76. A. Charkey, "Sealed Nickel–Zinc Cells *Proceedings of the 25th Annual Power Sources Conference*, Atlantic City, N.J., 1972.

77. A. Charkey, "Nickel–Zinc Cells for Sustained High-Rate Discharge," *Proceedings of the 7th Intersociety Energy Confersion Engineering Conference*, 1972.

78. *Design and Cost Study Zinc–Nickel Oxide Battery for Electric Vehicle Propulsion in Final Report, ANL Contract No. 109-38-3543*, Yardney Elect. Corp., Oct. 1976.
79. Final Report, *Design and Cost Study of Nickel–Zinc Batteries for Electric Vehicles, ANL Contract No. 31-109-38-3541*, Energy Research Corp., Oct. 1976.
80. D. P. Boden and E. Pearlman, in D. H. Collins, ed., *Power Sources*, Vol. 4, Academic Press, New York, 1972.
81. J. M. Parry, C. S. Leung, and R. Wells, *Structure and Operating Mechanisms of Inorganic Separators, Report NASA CR-134692*, Mar. 1974.
82. Brit. Pat. 365,125 (1930), J. J. Drumm.
83. Ger. Pat. 2,150,005 (May 25, 1972), O. von Krusenstierna (to AGA Aktiebolaget); U.S. Pat. 3,923,550 (Dec. 2, 1975), O. von Krusenstierna (to AGA Aktiebolaget); O. von Krusenstierna, *SAE Meeting*, Mar. 4, 1977, paper 21.
84. J. Giner and J. D. Dunlop, *Electrochem. Soc.* **122**, 4 (1975).
85. M. Klein and M. George, *Proceedings of the 26th Power Sources Symposium*, Atlantic City, N.J., 1974, p. 18.
86. J. E. Clifford and E. W. Brooman, *Assessment of Nickel–Hydrogen Batteries for Terrestrial Solar Applications*, SAND80-7191, Sandia National Laboratories, 1981.
87. A. J. Salkind, *Samples of Metal Hydrides*, International Battery Materials Association.
88. H. Ogawa, M. Ikoma, H. Kawano, and I. Matsumoto, "Metal Hydride Electrode for High Energy Density Sealed Nickel–Metal Hydride Battery," *Proceedings of the 16th International Power Sources Conference*, UK, 1988.
89. "Nickel," *Mineral Commodity Summaries*, U.S. Geological Survey, Reston, Va., Jan. 2003; *Minerals Yearbook*, 2001.
90. M. Klein, in D. H. Collins, ed., *Power Sources*, Vol. 5, Academic Press, New York, 1974; P. Antoine and P. Fougtere, in J. Thompson, ed., *Power Sources*, Vol. 7, Academic Press, New York, 1979.
91. A. J. Appleby and J. M. Jacquier, *J. Power Sources* **1** (1976).
92. M. Klein and A. Charkey, *Zinc–Oxygen Battery Development*, Electrochemical Society, Atlanta, Ga., Oct. 1977.
93. L. Ojefors and L. Carlson, *J. Power Sources* **2** (1977–1978).
94. B. G. Demezyk, W. A. Bryant, C. T. Liu, and E. S. Buzzelli, "Performance and Structural Characteristics of the Iron–Air Battery System", 15th IECEC Conference, 1980.
95. E. Findl and M. Klein, "Electrolytic Hydrogen–Oxygen Fuel, Cell Battery", *Proceedings of the 20th Power Sources Conference*, Red Bank, N.J., 1966.
96. M. Klein and R. Costa, "Electrolytic Regenerative H2–O2 Secondary Fuel Cell", *Proceedings of Space Technology Conference, ASME*, June 1970.
97. A. L. Almerini and S. J. Bartosh, "Simulated Field Tests on Zinc–Air Batteries", *Proceedings of the 26th Power Sources Symposium*, Atlantic City, N.J., 1974.
98. W. R. Momyer and E. L. Litauer, "Development of a Lithium Water–Air Primary Battery", *Proceedings of the 15th Intersocietal Energy Conversion Engineering Conference*, Seattle, Wash., Aug. 1980.
99. J. F. Cooper, R. V. Homsy, and J. H. Landrum, "The Aluminum–Air Battery for Electric Vehicle Propulsion", in Ref. 98.
100. G. R. Smith, "Lead," *Minerals yearbook*, U.S. Geological Survey, Reston, Va., 2001.
101. *Storage Battery Technical Service Manual*, 10th ed., Battery Council International, Chicago, 1987.
102. K. Eberts and O. Jache, in *Lead 65, Proceedings of the 2nd International Conference on Lead-Arnheim, Belgium*, Pergamon Press, Inc., Elmsford, N.Y., 1967, p. 199.
103. D. H. McClelland, *Prog B&S* **3**, 194 (1980).

104. G. Planté, *Compt. Rend.* **49**, 402 (1859); *Recherches sur l'electricite'*, Gauthier-Villars, Paris, 1883, pp. 35, 36, 40.
105. R. H. Schallenberg, *Bottled Energy: Electrical Engineering and the Evolution of Chemical Energy Storage*, American Philosophical Society, Philadelphia, Pa., 1982.
106. T. A. Edison, *The Beginning of the Incandescent Lamp and Lighting System*, The Edison Institute, Dearborn, Mich., 1976.
107. Brit. Pat. 129 (Jan. 11, 1881), C. Fauré.
108. Brit. Pat. 3926 (Sept. 10, 1881), J. S. Sellon.
109. K. R. Bullock, "The Development and Applications of Storage Batteries—Historical Perspectives and Future Prospects," in *Proceedings, 7th Australian Electrochemistry Conference*, 1988.
110. J. H. Gladstone and A. Tribe, *Nature (London)* **27**, 583 (1883).
111. G. W. Vinal, *Storage Batteries*, 4th ed., John Wiley & Sons, Inc., New York, 1955.
112. W. H. Beck and W. F. K. Wynne-Jones, *Trans. Faraday Soc.* **50**, 136 (1954).
113. A. K. Covington, J. W. Dobson, and W. F. K. Wynne-Jones, *Trans. Faraday Soc.* **61**, 2050 (1965).
114. H. Bode, *Lead–Acid Batteries* (translated by R. J. Brodd and K. V. Kordesch), John Wiley & Sons, Inc., New York, 1977.
115. K. R. Bullock, *J. Power Sources* **35**, 197 (1991).
116. H. E. Wirth, *Electrochim. Acta* **50**, 1345 (1971).
117. W. L. Gardner, R. E. Mitchell, and J. W. Cobble, *J. Phys. Chem.* **73**, 2021 (1969).
118. T. H. Lilley and C. C. Briggs, *Electrochim. Acta* **20**, 257 (1975).
119. R. A. Robinson and R. H. Stokes, *Electrolyte Solutions*, 2nd ed., Butterworths, London, 1959, p. 477.
120. R. H. Gerke, *J. Am. Chem. Soc.* **44**, 1684 (1922); *Chem. Rev.* **1**, 377 (1925).
121. J. A. Duisman and W. F. Giauque, *J. Phys. Chem.* **72**, 562 (1968).
122. W. F. Giauque, E. W. Hornung, J. E. Kunzler, and T. R. Rubin, *J. Am. Chem. Soc.* **82**, 62 (1960).
123. W. H. Beck, K. P. Singh, and W. F. K. Wynne-Jones, *Trans. Faraday Soc.* **55**, 331 (1959).
124. W. H. Beck, J. V. Dobson, and W. F. K. Wynne-Jones, *Trans. Faraday Soc.* **56**, 1172 (1960).
125. K. R. Bullock, *J. Electrochem. Soc.* **127**, 662 (1980).
126. J. Burbank, A. C. Simon, and E. Willihnganz, in C. W. Tobias, ed., *Advances in Electrochemistry and Electrochemical Engineering*, Vol. 8, Wiley-Interscience, New York, 1971, p. 157.
127. J. L. Dawson, in A. T. Kuhn, ed., *The Electrochemistry of Lead*, Academic Press, London, 1979, p. 309.
128. A. T. Kuhn, in Ref. 127, p. 365.
129. D. Pavlov, in B. D. McNicol and D. A. J. Rand, eds., *Power Sources for Electric Vehicles*, Elsevier, Amsterdam, 1984, p. 111.
130. K. R. Bullock, *J. Electroanal. Chem.* **222**, 347 (1987).
131. K. R. Bullock and D. Pavlov, eds., *Proceedings of the Symposium on Advances in Lead Acid Batteries*, 84-14, The Electrochemical Society, Pennington, N.J., 1984.
132. D. A. J. Rand and D. Pavlov, eds., *Proceedings of LABAT-89*, Varna, 1989; *J. Power Sources*, Vols. 30 and 31, 1990.
133. D. A. J. Rand, *J. Power Sources* **15**, B1 (1985); **18**, B31 (1986); **21**, B1 (1987); **27**, B1 (1989).
134. D. Simonsson, *J. Appl. Electrochem.* **3**, 261 (1973); **4**, 109 (1974); *J. Electrochem. Soc.* **120**, 151 (1973).
135. K. Micka and I. Rousar, *Electrochim. Acta.* **18**, 629 (1973); **19**, 499 (1974); **21**, 599 (1976); *Colln. Czech. Chem. Commun.* **40**, 921 (1975).

136. J. Newman and W. Tiedemann, *J.A.I.Ch.E.* **21**, 25 (1975).

137. W. Tiedemann and J. Newman, *J. Electrochem. Soc.* **122**, 70 (1975); S. Gross, ed., *Proceedings of the Symposium on Battery Design and Optimization*, 79-1, The Electrochemical Society, Inc., Princeton, N.J., 1979, pp. 23, 39.

138. W. Tiedemann and J. Newman, in K. R. Bullock and D. Pavlov, eds., *Proceedings of the Symposium on Advances in Lead–Acid Batteries*, 84-14, The Electrochemical Society, Inc., Princeton, N.J., 1984, p. 336.

139. B. L. McKinney, W. Tiedemann, and J. Newman, in Ref. 138, p. 360.

140. A. Eklund, D. Simonsson, R. Karlsson, and F. N. Bark, "Theoretical and Experimental Studies of Free Convection and Stratification of Electrolyte in a Lead–Acid Cell," LABAT-89 Conference, Varna, Bulgaria, 1989.

141. S. Atlung and B. Fastrup, *J. Power Sources* **13**, 39 (1984).

142. D. M. Bernardi, *Extended Abstracts* 88-2, The Electrochemical Society, Inc., Princeton, N.J., 1988, p. 1.

143. N. E. Bagshaw, K. P. Bromelow, and J. Eaton, in D. H. Collins, ed., *Power Sources 6, Proceedings 10th International Power Sources Symposium*, Academic Press, Inc., New York, 1977, p. 1.

144. W. Tiedemann, J. Newman, and E. DeSua, in D. H. Collins, ed., *Power Sources 6*, 1977, p. 15.

145. J. H. Gladstone and A. Tribe, *Nature* **25**, 221 (1882).

146. P. Ruetschi and R. T. Angstadt, *J. Electrochem. Soc.* **105**, 555 (1958).

147. K. R. Bullock and D. H. McClelland, *J. Electrochem. Soc.* **123**, 327 (1976).

148. K. R. Bullock and E. C. Laird, *J. Electrochem. Soc.* **129**, 1393 (1982).

149. J. S. Symanski, B. K. Mahato, and K. R. Bullock, *J. Electrochem. Soc.* **138**, 548 (1988).

150. C. S. C. Bose and N. A. Hampson, *J. Power Sources* **19**, 261 (1987).

151. J. Mrha, K. Micka, J. Jindra, and M. Musilova, *J. Power Sources* **27**, 91 (1989).

152. R. F. Nelson and D. H. McClelland, in S. Gross, ed., *Battery Design and Optimization*, The Electrochemical Society, Princeton, N.J., 1979, p. 13.

153. U.S. Pat. 3,244,562 (Apr. 5, 1966), F. M. Coppersmith and G. J. Vahrenkamp (to National Lead Co.).

154. Ref. 114, pp. 162–176, 196–199, 207–216.

155. U.S. Pat. 4,900,643 (Feb. 13, 1990), M. D. Eskra, W. C. Delaney, G. K. Bowen.

156. V. M. Halsall and J. R. Pierson, "Plate Processing—The Heart of the Lead–Acid Battery," *BCI Proceedings*, June 1972.

157. U.S. Pat. 6,586,136 (July 1, 2003), R. Bkardway and co-workers (to Concorde Battery Corp.).

158. B. K. Mahato, *J. Electrochem. Soc.* **127**(8), 1679–1687 (1980).

159. B. K. Mahato, *J. Electrochem. Soc.* **128**(7), 1416–1422 (1981).

160. J. R. Pierson, P. Gurlusky, A. C. Simon, and S. M. Caulder, *J. Electrochem. Soc.* **117**(12), 1463–1469 (1970).

161. N. A. Hampson, J. B. Lakeman, and K. S. Sodhi, *Surface Technol.* **5**, 377–384 (1980).

162. N. A. Hampson and J. B. Lakeman, *J. Electroanal. Chem.* **119**, 3–15 (1981).

163. J. A. Orsino, "Mixing & Pasting—Part I," *The Battery Man*, Apr. 1982.

164. J. F. Dittmann and H. R. Harner, *Storage Battery Research*, Form No. A-391, Chemical Division, Eagle-Picher Co., Cincinnati, Ohio, 1956.

165. G. W. Vinal, *Storage Batteries*, 4th ed., John Wiley & Sons, Inc., New York, 1955, pp. 33 and 34.

166. S. Hattori and co-workers, *Final Report (ILZRO Project LE-197)*, International Lead–Zinc Research Organization, Inc., New York, 1976.

167. Ref. 114, 216–244.

168. J. Perkins, J. L. Pokorny, and M. T. Coyle, *Naval Postgraduate School (NPS) Report*, NPS-69PS76101, Monterrey, Calif., Oct. 1976, pp. 17–19, 23, 24.
169. J. R. Pierson and R. T. Johnson, *SAE Technical Paper Series 830277*, SAE International Congress & Exposition, Detroit, Mich., 1983.
170. Ref. 144, p. 15.
171. J. E. Puzey and W. M. Orriel, in D. H. Collins, ed., *Power Sources 2, Proceedings 6th International Power Sources Symposium*, Pergamon Press, Inc., Oxford, UK, 1968, p. 121.
172. H. Silverman, *Environmental Protection Agency Final Report*, Contract No. 68-04-0028, EPA, Washington, D.C., 1972.
173. *Prod. Eng.* **43** (1977).
174. U.S. Pat. 3,989,539 (Nov. 2, 1976), N. G. Grabb (to Varta Batteries, Ltd.).
175. U.S. Pat. 3,959,015 (May 25, 1976), J. Brinkmann, G. Trippe, and W. Heissman (to Varta Batterie AG).
176. Fr. Pat. 2,239,020 (Feb. 21, 1975), (to Société Fulmen et Compagnie Europeénne d Accumulateurs).
177. U.S. Pat. 3,956,012 (May 11, 1976), W. R. Scholle (to Scholle Corp.).
178. J. R. Pierson and C. E. Weinlein, "Development of Unique Lightweight, High Performance Lead–Acid Batteries," in J. Thompson, ed., *Power Sources 9*, Academic Press, London, 1983.
179. D. Marshall and W. Tiedemann, *J. Electrochem. Soc.* **123**(12), 1849–1855 (1976).
180. D. L. Douglas and G. W. Mao, in D. H. Collins, ed., *Power Sources 4, Proceedings 8th International Power Sources Symposium*, Academic Press, Inc., New York, 1973, p. 379.
181. R. T. Johnson and J. R. Pierson, "The Impact of Grid Composition on the Performance Attributes of Lead–Acid Batteries," in L. J. Pearce, ed., *Power Sources 11*, International Power Sources Symposium Committee, 1987.
182. D. Berndt and S. C. Nijhawan, *J. Power Sources* **1**, 3 (1976–1977).
183. D. E. Koontz and co-workers, *Bell Syst. Tech. J.* **49**, 1253 (1970).
184. Ref. 114, 251–255.
185. T. J. Hughel and R. H. Hammar, in D. H. Collins, ed., *Power Sources 3, Proceedings 7th International Power Sources Symposium*, Pergamon Press, Inc., Elmsford, N.Y., 1970, p. 209.
186. N. E. Bagshaw, *Lead 68, Proceedings 3rd International Conference on Lead*, Venice, Italy, Pergamon Press, Inc., Elmsford, N.Y., 1969, p. 209.
187. D. L. Douglas and G. W. Mao, in D. H. Collins, ed., *Power Sources 4, Proceedings 8th International Power Sources Symposium*, Academic Press, Inc., New York, 1973, p. 379.
188. M. Torralba, *J. Power Sources* **1**, 301 (1976–1977).
189. U.S. Pat. 4,275,130 (June 23, 1981), W. E. Rippel and B. Edwards.
190. Ger. Pat. 2,346,517 (Apr. 3, 1975), O. Metzler.
191. Ger. Pat. 2,315,984 (Oct. 17, 1974), (to Varta Batterie AG).
192. J. R. Pierson, *Power Sources*, Pergamon Press, New York, 1968, 103–118.
193. J. R. Pierson, *Electrochem. Tech.* **5**, 323–327 (1967).
194. U.S. Pat. 6,495,286 (Dec. 17, 2002), G. C. Zguris and F. C. Harman, Jr., (to Hollingworth and Vise Co.).
195. U.S. Pat. 6,511,775 (June 28, 2003), T. J. Clough (to Ensci Inc.).
196. K. Murata and S. Hattori, "Advantageous Features of Microporous Thin Embossed Waffle Shape Separator," *89th Battery Council International Meeting*, Washington, D.C., Apr. 1977.
197. N. I. Palmer, *Proceedings of the 87th Battery Council International Meeting*, Hollywood-by-the-sea, Fla., April 1975, p. 105.

198. *Battery Technical Manual*, 2nd ed., Battery Council International.
199. U.S. Pat. 3,716,412 (Feb. 13, 1973), K. Peters (to Electric Power Storage Ltd.).
200. U.S. Pat. 3,765,942 (Oct. 16, 1973), O. Jache.
201. U.S. Pat. 3,930,881 (Jan. 6, 1976), J. P. Cestaro and L. J. Crosby (to National Lead Industries, Inc.).
202. D. Pavlov, G. Papazov, and V. Iliev, *J. Electrochem. Soc.* **119**(1), 8–19 (1972).
203. D. Pavlov, V. Iliev, G. Papazov, and E. Bashtavelova, *J. Electrochem. Soc.* **121**(7), 854–860 (1974).
204. D. Pavlov and Papazov, *J. Electrochem. Soc.* **127**(10), (1980).
205. V. M. Halsall and R. R. Wiethaup, "A New Manufacturing Method for Lead–Acid Storage Batteries," SAE 720041, 1972.
206. J. R. Pierson, C. E. Weinlein, and C. E. Wright, "Determination of Acceptable Contaminant Ion Concentration Levels in a Truly Maintenance-Free Lead–Acid Battery," in D. H. Collins, ed., *Power Sources 5-1974*, Pergamon, London and New York, 1975.
207. "SAE Standard Test Procedure for Storage J537—Jun 86," *SAE Recommended Practices*, SAE, New York, June 1986.
208. "SAE J227, Electric Vehicle Test Procedure," *SAE Recommended Practices*, SAE, New York, Mar. 1971.
209. U.S. Pat. 6,392,413 (May 21, 2002), H. L. Huslebrock and co-workers (to VB Autobatteries GmbH).
210. R.L. Armistad, *14th Battery Council International Convention*, Orlando, Fla., April 17, 2002, presentation handout.
211. B. Cullen in Ref. 210.
212. *Advanced Battery Technology* **37**(12), 26 (2001).
213. M. Hughes, N. A. Hampson, and S. A. G. R. Karunathilaka, *J. Power Sources* **12**, 83–144 (1984).
214. E. Yeager, in E. Yeager and co-eds., *Proceedings of the Workshop on Lithium Nonaqueous Battery Electrochemistry*, Vol. 80–7, The Electrochemical Society, Inc., Cleveland, Ohio, 1980, 1–12.
215. V. R. Koch, *J. Power Sources* **6**, 357–370 (1981).
216. M. Garreau, *J. Power Sources* **20**, 9–17 (1987).
217. A. Dey, *Thin Solid Films* **43**, 131–171 (1977).
218. U.S. Pat. 6,569,573 (May 27, 2003), Y. V. Mikhaylik, T. A. Skotheim, and B. A. Trofimov (to Moltech Corp.).
219. U.S. Pat. 6,428,929 (Aug. 6, 2002), J. Kay and co-workers (to NBT GmbH).
220. U.S. Pat. Appl. 20030108793 (June 12, 2003), J. R. Davis and Z. Lu (to 3M Innovative Properties Co.).
221. S. B. Brummer, in Ref. 2, 130–142.
222. Y. Matsuda, *J. Power Sources* **20**, 19–26 (1987).
223. K. M. Abraham, *J. Power Sources* **14**, 179–191 (1985).
224. J. R. Stiles, *New Mater. New Processes* **3**, 89–91 (1985).
225. J. Gabano, *Prog. Batt. Solar Cells* **8**, 149–162 (1989).
226. M. Armand, *Solid State Ionics* **9**, **10**, 745–754 (1983).
227. A. Hooper, in K. M. Abraham and B. B. Owens, eds., *Materials and Processes for Lithium Batteries*, Vol. 89–4, The Electrochemical Society, Inc., Cleveland, Ohio, 1980, 15–32.
228. P. E. Harvey, *J. Power Sources* **26**, 23–32 (1989).
229. U.S. Pat. Appl. 20020192563 (Dec. 19, 2002), R. S. Morris and B. G. Dixon.
230. K. M. Abraham, *J. Power Sources* **7**, 1–43 (1981 and 1982).
231. A. D. Yoffe, *Solid State Ionics* **9**, **10**, 59–70 (1983).
232. B. Scosati, A. Selvaggi, and B. Owens, *Prog. Batt. Solar Cells* **8**, 135–137 (1989).

233. A. Hammou and A. Hammouche, *Electrochimica Acta* **33**, 1719, 1720 (1988).
234. F. Bonino, M. Ottaviani, and B. Scrosati, *J. Electrochem. Soc.* **135**, 12–15 (1988).
235. A. Hooper and J. North, *Solid State Ionics* **9**, **10**, 1161–1166 (1983).
236. D. Shackle, *4th International Seminar on Lithium Battery Technology and Applications*, Deerfield Beach, Fla., Mar. 1989.
237. U.S. Pat. 4,830,939 (May 1989), M. Lee, D. Shackle, and G. Schwab.
238. *Science News* **136**, 342 (1989).
239. M. Liu, S. Visco, and L. DeJonghe, in S. Subbarao and co-workers, eds., *Rechargeable Lithium Batteries*, Vol. 90–5, The Electrochemical Society, Inc., Cleveland, Ohio, 1980, 233–244.
240. N. Koshiba and K. Momose, in Y. Matsuda and C. Schlaikjer, eds., *Practical Lithium Batteries*, JEC Press, Inc., 1988, 79–81.
241. K. Momose, H. Hayakawa, N. Koshiba, and T. Ikehata, *Prog. Batt. Solar Cells* **6**, 56, 57 (1987).
242. *JEC Battery Newsletter*, (1), 7 (Jan./Feb. 1989).
243. *Japan Component News* **7**(2), 4–6 (Feb. 1990).
244. K. Inada and co-workers, in Ref. 26, 96–99.
245. M. Mohiri and co-workers, *J. Power Sources* **26**, 545–551 (1989).
246. U.S. Pat. 4,748,484 (July 19, 1988), N. Furukawa, T. Saito, and T. Nohma (to Sanyo Electric Co., Ltd.).
247. U.S. Pat. 4,904,552 (Feb. 27, 1990), N. Furukawa, T. Saito, and T. Nohma (to Sanyo Electric Co., Ltd.).
248. T. Nohma, T. Saito, and N. Furukawa, *J. Power Sources* **26**, 389–396 (1989).
249. J. Carcone, *3rd International Rechargeable Battery Seminar*, Deerfield Beach, Fla., Mar. 1990.
250. U.S. Pat. 6,465,129 (Oct. 15, 2002), J. Xu, B. B. Owens, and W. H. Smyryl (to the Regents of the University of Minnesota).
251. *Technical Brochure*, Moli Energy Ltd., Burnaby, British Columbia, Canada.
252. J. Stiles, in Ref. 26, 74–78.
253. F. C. Laman and K. Brandt, *J. Power Sources* **24**, 195–206 (1988).
254. E. J. Frazer and S. Phang, *J. Power Sources* **6**, 307–317 (1981).
255. D. H. Shen and co-workers, *Extended Abstract 49*, Vol. 87–2, *172nd Meeting of the Electrochemical Society*, Honolulu, Hawaii, Oct. 1987.
256. U.S. Pat. 4,084,046 (Apr. 1978), M. S. Whittingham (to Exxon Corp.).
257. U.S. Pat. 4,731,310 (Mar. 1988), M. Anderman and J. Lundquist (to W. R. Grace & Co.).
258. M. Anderman and J. Lundquist, *Extended Abstract 49*, Vol. 87–2, 172nd Meeting of the Electrochemical Society, Honolulu, Hawaii, Oct. 1987.
259. U.S. Pat. 3,864,167 (Feb. 4, 1975), J. Broadhead, F. J. DiSalvo, and F. A. Trumbore.
260. B. Vyas, *33rd International Power Sources Symposium*, Cherry Hill, N.J., June 1988.
261. F. Trumbore, *4th International Meeting on Lithium Batteries*, Vancouver, Canada, May 1988.
262. *JEC Battery Newsletter* (2), 26, 27 (1988).
263. U.S. Pat. 4,828,834 (May 1989), T. Nagaura and M. Yokokawa (to Sony Corp.).
264. U.S. Pat. 4,959,282 (Sept. 1990), J. Dahn and B. Way (to Moli Energy, Ltd.).
265. *Product Literature*, Moli Energy Ltd., Burnaby, British Columbia, Canada.
266. E. Plichta and co-workers, *J. Power Sources* **21**, 25–31 (1987).
267. K. Wiesener and co-workers, *J. Power Sources* **20**, 157–164 (1987).
268. K. West, B. Zachua-Christiansen, M. Ostergard, and T. Jacobsen, *J. Power Sources* **20**, 165–172 (1987).
269. A. N. Dey, K. C. Kuo, P. Piliero, and M. Kallianidis, *J. Electrochem. Soc.* **135**, 2115–2120 (1988).

270. C. Cecilio and co-workers, *4th International Seminar on Lithium Battery Technology and Applications*, Deerfield Beach, Fla., Mar. 1989.
271. W. Behl and D. Chin, *J. Electrochem. Soc.* **135**, 16–25 (1988).
272. U.S. Pat. 4,668,595 (May 1987), A. Yoshino, K. Sanechika, and T. Nakajima (to Sony Corp.).
273. Jpn. Pat. 01 122 562, (to Sony Corp.).
274. *JEC Battery Newsletter*, (2), 9–12 (1990).
275. E. J. Cairns, in J. Bockris and co-eds., *Comprehensive Treatise of Electrochemistry*, Vol. 3, Plenum Press, New York, 1981, Chapt. 11.
276. R. W. Glazebrood and M. J. Willars, *J. Power Sources* **8**, 327–339 (1982).
277. P. A. Nelson, *4th International Seminar on Lithium Battery Technology and Applications*, Deerfield Beach, Fla., Mar. 1989.
278. P. A. Nelson and H. Shimotake, *Prog. Batt. Solar Cells* **6**, 150–154 (1987).
279. T. D. Kaun, in Ref. 25, 294–306.
280. T. D. Kaun and co-workers, *Proc. of the 25th IECEC* **3**, 335–340 (1990).
281. J. L. Sudworth and A. R. Tilley, *The Sodium Sulfur Battery*, Chapman and Hall Ltd., 1985, and references therein.
282. D. Pletcher, *Industrial Electrochemistry*, Chapman and Hall Ltd., 1984, 272–274.
283. P. Bindin, *Proc. 20th IECEC Conf.* **2**, 1111–1114 (Aug. 1985).
284. S. Wolanczyk and S. Vukson, *Proc. 25th IECEC Conf.* **3**, 122–124 (Aug. 1990).
285. R. J. Bones, J. Coetzer, R. C. Galloway, and D. A. Teagle, *J. Electrochem. Soc.* **134**, 2379–2382 (1987).
286. R. C. Galloway, *J. Electrochem. Soc.* **134**, 256, 257 (1987).
287. "The Lithium-Polymer Battery, Substance or Hype?," *Powerpulse.net*, Aug. 7, 2003.
288. J. A. Ober, "Lithium," *Minerals Yearbook*, U.S. Geological Survey, Reston, Va., 2002.

ALVIN J. SALKIND
MARTIN KLEIN
Rutgers University

KATHRYN R. BULLOCK
JOHN R. PIERSON
Johnson Controls, Inc.

PAUL R. GIFFORD
Ovonic Battery Company

BEER AND BREWING

1. Introduction

Beer (Latin: bibere, to drink. Old English, beor. Middle English, bere) may be defined as a mildly alcoholic beverage made by the fermentation of an aqueous extract of cereals (grains). Cereals contain carbohydrates, mainly in the form of starch, which brewers' yeasts cannot ferment, and so breakdown of starch to fermentable sugar is a central feature of beer-making processes. This is in contrast to wines in which fermentable sugars are preformed in the raw materials (eg, fruits such as grapes). Thus, sake, for eg, commonly called a rice wine, is in fact a beer. The grain mainly used for beer-making is barley, with rice and corn as adjunct, and some beers are made partly from malted wheat. Others grains such as sorghum can be used, especially in the manufacture of traditional beers, and oats in, eg, a few stouts.

Manufacture of beer has five main stages: (1) malting, (2) brewing, (3) fermentation, (4) finishing, and (5) packaging. This article begins with an overview of the whole process, then reviews malting and malt and other raw materials. It then follows the brewing process and fermentation from beginning to end. It concludes with some brief comments on the history of beer and its healthful value. North American brewing practices are taken as standard in this article.

2. Brief History

Beer has a long history doubtless dating to the first settled communities based on agriculture and growing grain. This was probably in the Tigris and Euphrates valleys some 10,000 or more years ago. Barley was one of the first grains cultivated and was processed in a number of ways, especially, eg, bread-making, to make it edible. From these processes, brewing likely developed; eg, if the ancients (by accident or design) used partially germinated grain to make bread they would have completed the basic processes of malting and mashing. Grain-based beverages arose independently in many cultures, eg, the rice-based drinks of Asia have a history dating to at least 2000 BC. Naturally, the ancients, being unaware of enzymes and microorganisms, ascribed the conversion of water and grain to intoxicating beverages to the intervention of Gods, eg, in Egypt to Osiris and his wife Isis, and to Ceres, a pre-Roman Goddess of agriculture (hence the Spanish word for beer, cerveza). By the first century AD, beer was widely known and the common drink of the Germans, the Britons, and the Irish. Tacitus, the first/second century Roman historian mentions beverages distilled from fermented grains in the British Isles. In these northern European climes barley could be grown where grapes could not and beer remains to this day a preeminent drink in these regions. Beer making has long been associated with the church especially monasteries, and in a few places (eg, the Trappist abbeys of Belgium) remains so. From these origins, and widespread domestic brewing, arose the first commercial enterprises around the twelfth century and, with that, brewers' organizations, regulation, taxation, and licensing as we know it today. Beer came to North America with the first European settlers; indeed the landing at Plymouth Rock was because "—our victuals were much spent especially our beere—". Throughout the early years of history of the United States, beer was actively promoted as the beverage of temperance in preference to spiritous liquors, and our first presidents were ardent home brewers. Beer became embroiled in the turmoil over Prohibition because by the mid-1800s the brewing industry in the USA was almost entirely controlled by Germans, and German was the language of the brewing profession. Anti-German feelings brought on by World War I caused beer to be included in Prohibition. The Volstead Act (Prohibition) passed in 1919 and was repealed in 1933. Brewing was, and is, very much involved with the scientific advances of the day, eg, brewers immediately adopted the discoveries of Louis Pasteur. Brewers enthusiastically embraced new technologies such as the steam engine and refrigeration as they became available. They assiduously followed and contributed to advances in biology, especially enzymology and microbiology, which particularly met the brewers' need to understand their processes in scientific ways. This continues today and brewing laboratories have most modern instruments of analysis, eg, and support and maintain active research programs. Brewers also maintain an interest in contemporary genetics, though the politics of genetically modified organisms (GMOs) is too complex to permit application of these techniques to brewing processes and products.

3. Overview of the Process

Beer is a food product and subject to all the regulations concerning food production and distribution. Malthouses and breweries therefore are impeccably clean and sanitary places not only to meet the provisions of those regulations but also because the brewing process and beer itself are subject to attack by unwanted microorganisms. While these organisms pose no danger from the point of view of transmitting illness to the consumer, they can easily spoil beer flavor, eg, by causing sour tastes and unwanted aromas and by forming hazes.

Malting takes ~8 days. However, the malt spends a good deal more time than that in the malthouse before it is sold to brewers because it needs to be cleaned, matured, analyzed, and then blended to meet brewers' specification before sale. In the malting process, barley is first wetted (steeped) for 2 days and then put to germinate in an appropriate vessel where it is aerated and turned regularly for ~4 or 5 days. Then, heating in a kiln for up to 2 days dries the green malt. This imbues the product with intense malty flavors and color, both of which become part of the character of the beer made from it. *Brewing* (Fig. 1) follows malting. In the brewhouse, five distinct stages occur in a period of ~5–6 h: (*1*) the malt is milled (ground up) and then mixed with suitable water and (*2*) heated through a precise temperature program (mashed). This converts starch to fermentable sugar. The liquid mass is then transferred (*3*) to a device to separate the insoluble spent grains (mainly the husk of the malt and precipitated protein) from the sugary aqueous extract called wort. The wort is then boiled (*4*) with hops, to stabilize it and to impart bitterness. The spent hops are removed (*5*) and the wort is then cooled, which concludes the brewhouse operations. Note that there may be several brews at different stages in the brewhouse at one time, and so one brew might emerge from the brewhouse every 2–3 h or so from each line of brewing vessels. *Fermentation* follows (Fig. 1 continued): A desirable yeast culture is added to the wort and during the course of ~3–9 days (depending on temperature) the yeast converts the sugar present mainly into alcohol and carbon dioxide, but also forms a myriad of flavor compounds that, with the flavors from malt and hops, combine to create the final beer flavor. After this primary fermentation the beer is *finished*, ie matured (aged) and carbonated by further treatments, eg, secondary fermentation. It is then filtered to make it brilliantly clear and stabilized to prevent changes in the market place.

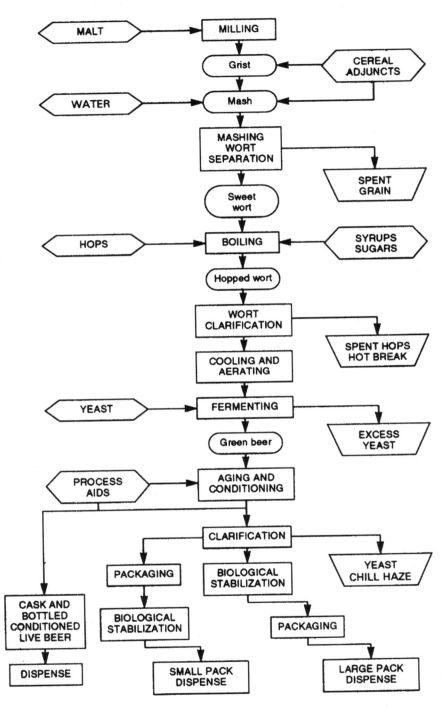

Fig. 1. An outline of brewing and fermentation from raw materials to dispense. (From Ref. 1, with permission).

Finally this beer is *packaged* at high speed into bottles and cans of various convenient sizes and into kegs for sale with strict control of oxygen access.

4. Malting

The objective of malting is to achieve "modification" of the barley grain. Barley lacks suitable enzymes for brewing, lacks friability (ie, it is a hard grain) and so is difficult to mill, it lacks suitable aroma and flavor and color, it lacks simple nitrogenous compounds suitable for yeast growth, and it contains β-glucans, which make aqueous barley extracts viscous and difficult to filter. Modification, which is the sum of changes as barley is converted into malt, and is achieved by partially germinating the grain, solves all of these problems when done well. Barley, unlike wheat, has an adherent husk that protects the grain during malting.

Malt is made from barleys that are approved or selected varieties known as malting barleys and that meet necessary specifications. These barleys generally do not yield as large a crop as feed barleys and so maltsters pay a premium to farmers to plant malting varieties and to meet malting grade. Typically, malting barleys tend to germinate vigorously, modify evenly and completely, have a low nitrogen or protein content, and are relatively plump, ie, tend to have large kernels that contain a lot of starch relative to the amount of husk. In North America, barley for malting is grown in the most northerly United States and Canada and at equivalent latitudes elsewhere; it is harvested only once a year and so it is necessary to store it in silos for up to a year or more before it is used. There are two kinds of malting barleys, six- and two-row, and brewers use blends of both kinds. Generally, malts made from six-row barleys tend to yield rather more enzymes (expressed as diastatic power, DP, or starch-splitting ability) than malts from two-row barleys. On the other hand, malts made from two-row barleys yield a little more extract (more soluble solids to the wort) than six-row barleys.

Barley arrives at the storage silos as a raw agricultural product and the process of converting it to food for humans begins immediately. It is cleaned to remove dust and stones (especially important from the point of view of preventing sparks and potential dust explosions) twigs and straw and separated from foreign seeds and skinned and broken corns. These materials reduce the storage stability of the grain by encouraging insects and molds in the silos. Barley contains living tissue and respiration (breathing) takes place. The grain must therefore be moved from time to time to aerate it and it is recleaned and fumigated if necessary. For stable storage, barley moisture should be ~12–13% and conditions of storage must be dry and cool. The barley variety must be capable of remaining viable (alive) throughout storage. Barley often enters storage in a dormant state depending on the variety and growing conditions and dryness; ie, it is alive but not yet capable of growing (germinating). During storage the grain comes out of dormancy and is then ready for malting.

The cleaned and separated barley is graded before malting, because plump and thin corns germinate at different rates that would affect the evenness of the resulting malt. The plumpest corns are made into malt for use in brewing, but thinner corns can be malted in special ways for use in distilleries or for flavoring malts or baking or confectionery use. There are three stages of malting: steeping, germination, and kilning. In *steeping*, the barley is soaked typically in a conical-bottomed steep tank in potable water at ~10–15°C for up to 48 h. The batch size might be 40,000 lb. The moisture content of the grain rises from ~12% to a level decided in advance by the maltster, which will be in the range of 42–46% moisture. This choice affects the speed and course of germination and so malt characteristics. During steeping the water is changed several times and the steep is vigorously aerated to provide oxygen for the kernels, which are beginning to respire. This also agitates and effectively cleans the grain and 1–2% or more of grain weight is lost by this washing. Toward the end of steeping the "chit" (the tip of the emerging rootlet) appears and at this stage the grain is transferred to the germination vessel where the maltster is better able to meet the needs of the growing grain and to control the process.

One of several designs of *germination* vessels might be used, eg, large horizontal drums that rotate, or large flat boxes that have turners traveling up and down (Fig. 2), or circular vessels with turners rotating about the central axis. There are other choices. In all cases, however, the germination vessel achieves three objectives: (1) to turn the growing grain to assure the rootlets do not entangle to form an impenetrable mass; (2) to pass air evenly through the mass of grain so that the grain can respire; and (3) to maintain control of moisture and temperature (say 18–20°C) through a supply of moisture-saturated cool air. In this way the grain grows at a controlled and steady rate for ~4 or 5 days. During this time there is significant growth of the rootlets, which is obvious but gives little information about the course of the process. The shoot (also called the plumule or acrospire) grows under the husk during germination. Maltsters aim to have the acrospire of all individual kernels grow to $\frac{3}{4}$–1× the length of the kernel

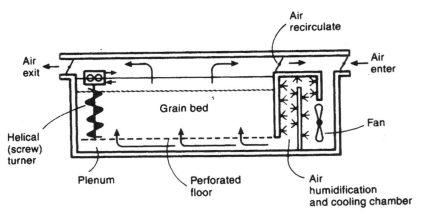

Fig. 2. A barley germination chamber must provide ample air, turning, and temperature–moisture control. Many designs are possible. The chamber might be rectangular or circular in design, or a large revolving drum can be used for germination. This is a schematic of a box or compartment germination chamber showing the typical flow of air through the grain bed. Alternatively, air may be drawn down through the bed. (From Ref. 1, with permission).

because this indicates that sufficient growth and even malt modification has been attained.

The maltster must now arrest further development of the grain and fix the properties imbued in the malt by germination. *Kilning* removes the moisture from the green malt by a flow of warm air (rising over the 2 days from say 50 to 70°C) and halts germination. In some malt houses the germination vessel can also act as a kiln, which is achieved by switching from a flow of wet/cool air to a stream of warm/dry air. These vessels are called GKVs or germination-kilning vessels. For the most part, however, maltsters operate separate facilities for kilning. Initially, kilning provides a relatively high flow of quite cool air to achieve substantial moisture removal. Later, however, more intense heat must be applied to remove moisture deeper within the grain and some associated with large molecules. Thus, there are two stages of kilning, and most kilns therefore have two floors. The hottest air flows through the lower floor, where the partially dried malt resides. Air leaving the lower floor is diluted with fresh air to increase its volume and cool it, and this air then passes through the upper kiln to carry out the initial drying of green malt. Thus, each batch of malt passes through the kiln in two stages, first on the upper floor and then on the lower floor before exiting the kiln. The last stage of kilning is called "curing" and is short period (2 h or so) at high temperature (85–100°C). This serves to reduce the malt moisture to its final value of ~4% and the extra heat increases the flavor and color of the malt by toasting. At the end of kilning the shriveled rootlet can be easily removed. Ordinary pale malt does not look very different from barley, though its texture and flavor are much different. During malting there are significant losses of barley dry weight, including some starch respired to CO_2. This is called malting loss and might be 6–12% of barley dry weight depending on the kind of malt being made and the effectiveness of process control. Moisture also decreases from ~12 to ~4%. Thus from 100 kg of barley only some 80–85 kg or so of malt can be produced.

During the malting process (grain germination) three main changes that are important for brewers take place inside each grain and collectively comprise modification: (*1*) the breakdown of the cell walls (especially the β-glucans they contain) in the barley endosperm (*2*) the accumulation of enzymes not previously available in barley, especially α-amylase, and (*3*) production of low molecular weight nitrogenous compounds, especially amino acids, that help form flavor compounds and will be available for nourishment of yeast during fermentation in the brewery. Note, however, that starch is only partially hydrolyzed under the cool (say 18–20°C) conditions of germination and the bulk of barley starch remains as malt starch.

During steeping water enters each barley kernel through a region of the husk near the embryo called the micropyle and so the embryo is hydrated first (Fig. 3). The embryo also contains a hormone chemical called *gibberellic acid*, which is carried in the water entering the grain into the endosperm and especially to a thin layer of living cells that entirely surround the endosperm, called the *aleurone layer*. This layer reacts to the hormone by synthesizing new enzymes that include amylases (especially α-amylase), β-glucanases, and proteases, which, respectively, are capable of breaking down starch, β-glucans, and proteins. These are the enzymes of modification. They are released into

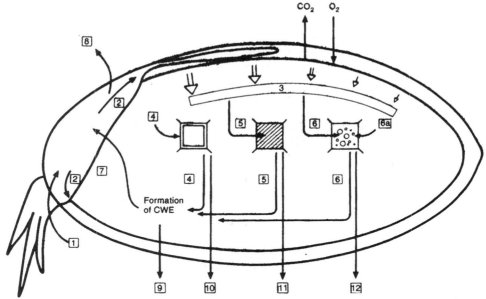

Fig. 3. Summary of events during grain germination (malting). (1) Entrance of water through the micropyle, (2) release of gibberellic acid, and (3) its progressive stimulation of the aleurone, (4) β-glucanases from the aleurone layer attack endosperm cell walls (β-glucans) to increase modification, grain friability, and to lower wort viscosity (10). (5) Proteolytic enzymes attack proteins to form amino acids (FAN) and modification expressed as S/T (soluble nitrogen/total nitrogen %) (11). (6) and (6a) Amylases initiate attack on starch, especially small starch granules, and these enzymes survive into malt (12). Also shown, (7) and (8) nutrition of the embryo causing the rootlet and shoot to grow using materials from actions 4, 5, and 6 with respiration of starch to CO_2 and water both causing malting loss; also (9) formation of low molecular weight compounds from actions 4, 5, and 6, especially amino acids and sugars that also survive into malt and extract into wort. (From Ref. 1, with permission).

the endosperm as they are formed and modify the endosperm material (amylases break down starch only slowly under the cool conditions of malting and so most of it remains in the finished malt). Because some parts of the aleurone layer are closer to the source of the hormone than others, the progress of modification is from the embryo end of the kernel toward the opposite (distal) end; therefore the least modified area of the final malt is found farthest from the embryo. If undermodified, this might be called the "steely" tip; such material is barley-like and extracts poorly in the brewhouse and may cause troublesome hazes and filtration problems downstream.

The quality of malt is gauged by a number of laboratory measures. These include (1) determination of Diastatic Power (DP) that is a measure of the amylase enzymes present (especially β-amylase). The value is usually between 110 and 150° for American malts. There are two amylases, α-amylase and β-amylase, that together are able to break down starch substantially (~65–75% depending on conditions) to fermentable sugar. (2) Ease of milling (using an instrument called the friabilmeter), or direct determination of the β-glucan content, or the ease of extraction (called the coarse/fine difference) is used to gauge the degree

Table 1. **A Typical Malt Analysis/Specification**[a]

Physical properties	
assortment (size) (% on screens)	
on 7/64 screen	60
on 6/64	33
on 5/64	6
through screens	1
growth of shoot (length of kernel) (%)	
0–1/4	0
1/4–1/2	0
1/2–3/4	10
3/4–1	85
overgrown	5
moisture (%)	4
1000-kernel weight (g)	35
Chemical and biochemical analysis	
extract (fine grind) dry basis (%)	81.5
extract (coarse grind) dry basis (%)	80.0
coarse/fine difference (%)	1.5
enzymes	
DP (β-amylase)(dry basis)(°Lintner)	120
DU (α-amylase) (dry basis)	55
total protein (Kjeldahl-N × 6.25)	11
Wort soluble protein (%)	5
soluble protein as % of total protein (S/T)	45
wort pH	5.9
color of wort (°Lovibond)	1.8
Wort viscosity (C)	1.4

[a] From Ref. 1.

of modification. (3) FAN is used to measure the presence of potential yeast nutrients (amino acids, eg). The overall quality of the malt resides in its extract yield; ie, the amount of material that can be dissolved from it in a laboratory scale mashing process; the value is usually ~80%. By these measures and others (Table 1) maltsters and brewers determine the extent to which the necessary changes in barley have been achieved when producing a particular batch of malt. In practice, at the malthouse, many batches of malt are blended together to meet the brewers' specifications. The brewer is concerned with three things: (1) *kernel size* expressed here as assortment (by screening) and as1000-kernel weight; (2) *modification* expressed as growth of shoot (length of kernel), coarse/fine difference % (difference in extract between fine and coarse grind malt), and mash soluble protein expressed as a percentage of soluble over total protein (S/T%), and possibly wort viscosity; (3) *enzyme content* expressed as DP (mainly β-amylase) and as α-Amylase (DU, dexrinizing units).

5. Hops

Hops are a crucial component of beer although only ~4–8 oz/barrel (120–240 g/hL) are used. Their primary role is to give bitterness to beer. Although humans do not usually like bitterness, a sufficient and balanced inclusion of bitter character in beers is necessary for a satisfactory product. How the brewer handles the bitter quality of hops in creating a beer is important in differentiating one beer from another, and meeting the needs of the target consumer population. Hops also can contribute delicate aromas to beers, that, in conjunction with those flavors arising from the yeast and malt, creates the overall impression of a beer. Hops are used in the kettle-boiling process in a brewery. The key chemical reaction of wort boiling is the conversion of relatively insoluble α-acids present in the hops to quite soluble iso-α-acids (Fig. 4) that persist into the beer.

Hops grow in temperate northern climates. Oregon, Washington, and Idaho in the United States but in comparable latitudes in Britain, Germany and

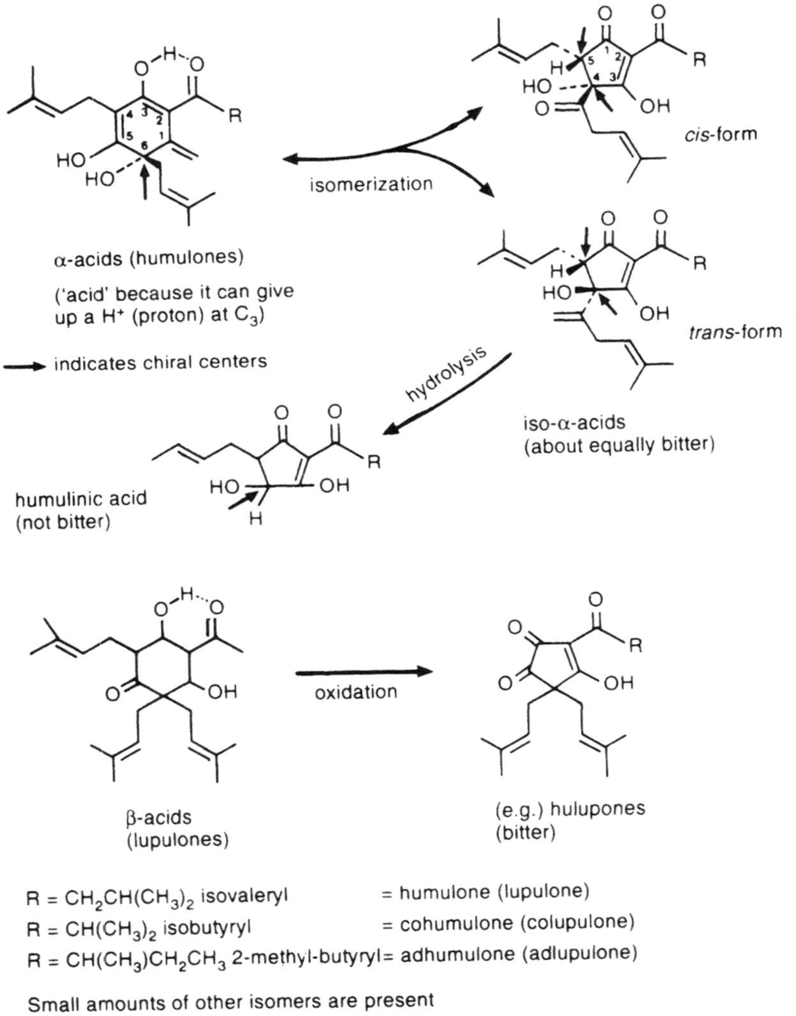

Fig. 4. Hop α-acids (above) and hop β-acids (below) and two chemical reactions of brewing significance, isomerization (above) and oxidation (below). In each case there are three major acids with different side chains, R, as identified. (From Ref. 1, with permission).

Central Europe, China and, in the Southern Hemisphere, in New Zealand and Tasmania. These latitudes are necessary for proper yield because day-length determines flowering and fruit-set and thus adequate commercial yields. Artificial light is used to grow hops at the tip of South Africa that is not quite far enough south for ideal day length. The United States and Germany are the largest producers of hops. Hops plants (*Humulus lupulus*, a member of the family *Cannabinaceae*) are perennial, dioecious (having separate male and female plants) vigorous climbing vines. Normal (ie, non-dwarf) varieties grow up strings to 15–20 ft (4–6 ms) on a strong overhead trellis called a wire-work that are permanently installed in the fields (called yards or gardens). During harvest the vine is cut back to ground level. Only the female plant is planted, although in a few places (Oregon, Britain) a few male plants are permitted because fertilization is thought to protect against some diseases. Generally, however, male plants are ruthlessly removed because brewers do not usually prefer seeded hops.

Hop "cones" are the fruit of the hop vine (though often incorrectly called the "flower"). Each cone is about the size of the top joint of a human thumb and is green and structured like a small artichoke of overlapping bracts. At the base of each bract is a bright yellow powder. These are the lupulin glands that contain the brewing value of the hops (Table 2). The lupulin glands contain the essential oils of hops that have the potential to give powerful aromas to beer and are more prized in some hops (called aroma hops) than in others, and also contain the total resins that comprise (1) α-acids (highly desirable as the prime source of bitterness, and the main measure of hop value and quality), (2) β-acids (of little value, though they can oxidize to yield bitter compounds (Fig. 4), and (3) the so-called hard resins (of no value). Many brewers buy hops on their content of α-acids alone, though others believe that a proper balance of bitterness and aroma is necessary for superior beers. It is necessary to process hops after harvest in such a way that their quality is conserved over perhaps a year or two.

Table 2. **Composition of Whole Dried Hops**[a]

Constituent	Percentage by weight
cellulose and lignin	40–50
soft resins[b]	
α-acids	2–17
β-acids	2–10
proteins (Kjeldahl-N × 6.25)	15
water	10–12
ash	8–10
tannins	3–6
fats and waxes	1–5
pectin	2
simple sugars	2
essential oils	0.5–3.0
amino acids	0.1

[a] The brewing value is in the resins (variable between ~5 and 18%) especially their α-acid content because these contribute to bitterness, and the essential oils (aroma fraction).

[b] Generally those hops most prized for their aroma are lower in content of α-acids.

After harvest and separation from the vine, hops are kilned in a gentle flow of warm air ($\sim$60–80°C) to dry them to $\sim$10% moisture. They are then heavily compacted into bales or pockets, weighing some 80–90 kg, with the intention of minimizing the entrance of air. Air can oxidize the resins (to produce hard resins) and reduce the value of hops considerably. Though cone (or whole) hops in their primary compressed state are used by a few (large) breweries, only $\sim$30% or so of the α-acids they contain reach the beer as *iso*-α-acids; the remainder is lost along the way for various reasons, eg, on yeast. Therefore brewers seek more efficient means of using hops. Most cone hops in commerce today are therefore further processed to conserve them and to increase the ease and lower the cost of handling them, and to increase utilization of α-acids.

Hops can be milled into a powder, some of the extraneous vegetation removed, and then extruded as pellets and packed under inert gas and vacuum. Unlike cone hops, pellets require cool rather than refrigerated storage and are much more compact and easy to transport. Pellets are used in much the same way as cone hops. To produce isomerized hop pellets, some magnesium oxide is mixed into the powdered hop material before pelletizing and the pellets are kept warm ($\sim$50°C) for a few days. These are more easily extracted during kettle boil for increased utilization of the α-acids. Hop pellets can be extracted with solvents to yield syrups that are a very stable and concentrated form of hops. The primary solvent for this is liquid CO_2 though ethanol can also be used. Extracts of hops are used in the wort kettle just as cone hops or pellets might be, but much more conveniently. Extracts, however, can be further modified. By treating the extracts with heat in a mildly alkaline solution the α-acids of the extract can be isomerized to iso-α-acids. Such extracts can be added directly to beer to give bitterness as required, with a high utilization of probably 85% or more. Such syrups can be even further modified. Hop compounds in beer are sensitive to light. This is the reason beer is packaged in dark brown bottles. Light causes the hop compounds to break down to yield a small molecule that, in turn, reacts to yield iso-pentenyl-mercaptan (3-methyl-but-2-enyl-thiol) that has the memorable aroma of skunks. The beer is called "skunky" or "light-struck". By reacting, the iso-α-acids with a powerful reducing agent under special conditions this reaction can be prevented, and beers containing such reduced-iso-α-acids (eg, rho-iso-α-acids or tetra-hydro-iso-α-acids) can be packaged in flint (clear glass) bottles. The reduced iso-α-acids are somewhat more bitter than iso-α-acids and also remarkably improve the foam stability of beer. The products are quite popular. The iso-α-acids and their reduced forms are also somewhat antimicrobial and help to protect beers from contaminating microbes. It is possible that hops originally became popular with brewers because of their preservative power, but have continued in use because of their desirable taste and aroma impacts.

Hop essential oils, the aromatic component of hops, comprises three groups of compounds in low concentration (0.5–3%, Table 2): (*1*) hydrocarbons, (*2*) an oxidized fraction containing alcohols and esters, and (*3*) a sulfur-containing fraction. The composition of the essential oils is complex and is different in each hop variety; exactly which aroma materials (if any) survive boiling and are responsible for the so-called noble aroma of some beers is not known.

6. Water

Water makes up ~95% by volume of most beers, but the quality of water used in brewing, beyond mere potability, can have an impact on its quality and flavor characteristics. Water contains dissolved ions of which the most relevant to brewing are calcium (Ca^{2+}), magnesium (Mg^{2+}), and bicarbonate (HCO_3^-). The first two represent hardness and the last alkalinity. In general, hardness is desirable for brewing purposes (ie, as a beer component) and alkalinity is not. It is a matter of pH or acidity. The hardness ions lower the pH (ie, make the process somewhat more acidic) during mashing by reacting, eg, with phosphate ions of malt. This promotes enzyme action somewhat. Alkalinity raises the pH. High pH tends to extract harsh materials from malt and hops including astringency and increased color and the beers tend to be less crisp and more satiating and dull. Brewers therefore select brewing water with an hardness/alkalinity ratio suitable to the product they intend to brew, or to treat available water to meet their need. Generally, hard water with low alkalinity is best suited for pale ales, but dark beers such as stouts and porters profit from a certain amount of alkalinity because the extractiveness noted above is desirable in such beers and because roasted malts are somewhat acidic. Lagers are made with quite soft water (relatively low in calcium and magnesium ions) of very low alkalinity.

After sand filtration, chlorination, and carbon filtration to clarify and purify raw water, further treatment of brewing water might be as simple as adding acid to neutralize alkalinity (to the level required), or boiling the (hard) water, which tends to break down the bicarbonate ion to give a deposit of calcium carbonate (a reaction familiar to those living in many hard water areas). Reverse osmosis is a popular modern treatment in which water is forced at high pressure through a membrane. Pure water passes through the membrane and the dissolved ions do not.

The bulk of water used in breweries is for cleaning and sanitizing the plant and for raising steam for transporting energy about the brewery. This is at least $4\times$ the volume of water that goes into the beer and might be as high as $10\times$. Soft water (ie, lacking Ca^{2+} and Mg^{2+}) is preferred for these purposes because it does not react with cleaners or deposit "stone" on surfaces being cleaned and sludge in steam boilers. The most common cleaners are based on caustic soda or other strongly alkaline agents. Acid cleaning is used about weekly to remove stone that alkaline cleaning can deposit. Also, acid is a useful cleaner in a CO_2 atmosphere (common in brewers' tanks) because the gas does not dissolve in the solution; this is a problem with alkaline cleaners and can cause tanks to implode. After cleaning, the numbers of bacteria on the beer-contact surfaces are reduced further by either hot water/steam or chemical sanitizers based mostly on halogens such as chlorine or iodine. Water leaving the brewery is effluent, the strength of which is measured as BOD (biological oxygen demand) or COD (chemical oxygen demand), pH (acidity), and suspended solids. Brewers minimize these qualities in out-flowing water because they are charged on the composition of it. Some breweries operate pretreatment plants to minimize these charges. Anaerobic treatment of effluent yields a flow of methane as a useful fuel.

7. Other Products

7.1. Adjuncts. Beers can be made entirely with malt, and many are, but much beer is made with a certain proportion of non-malt material (adjuncts). This material is commonly corn (as yellow corn grits) or rice grits or syrups derived from them. These materials are cheaper forms of extract than malt itself, but they provide only starch or (as syrups) hydrolyzed starch, and none of the complex mix of proteins, polyphenols, enzymes, flavor and color compounds present in malt. They therefore dilute malt character and permit the manufacture of delicately flavored beers that are pale in color, crisp, nonsatiating, and easy to drink. Adjunct beers also tend to be more physically stable because they are low in haze-forming materials derived from malt (especially protein and polyphenols).

7.2. Special Malts and Roasted Materials. They have the opposite effect on beer from adjuncts because they are treated in such a way as to enhance their color and flavor impact. They are used at quite low levels of 5 to perhaps 20% in making dark yellow, brown, reddish, and black beers. There are two ranges of products. The first might be called roasted materials because they are ordinary malt that is heated to a higher temperature than normal, in a roasting drum. The higher the temperature used the darker the color of the malt and the more intense its flavor. Roasting with high heat yields black malts. Barley can also be roasted, rather like coffee, to produce a black product. The second range of products is called crystal malts or caramel malts (the use of these words is not exact). These are made from regular malt that is wetted again and "stewed", ie, heated without drying. This causes the interior of the malt kernel to liquefy (the starch hydrolyses by action of amylases), and when the malt is eventually dried and heated the endosperm crystallizes. Malts with a range of colors and flavors different from roasted materials can be made in this way.

8. Brewing

Suppliers deliver the raw materials necessary for beer manufacture to the brewery as needed. Breweries rarely have more than a few day's or a week's supply of materials on hand because storage is expensive in facilities and capital. Malt is generally delivered in rail cars of 70,000-kg capacity or hopper-bottomed road trailers on a daily basis. Barges on canals are an option for transportation in some parts of the world. Extreme cleanliness around the delivery point and silos is necessary to avoid insect infestations and attracting birds and rodents.

A brewery is divided into three main parts: (1) the brewhouse where the malt is extracted with hot water to make "wort", (2) the cellars where fermentation by yeast takes place and the beer is matured and clarified, and (3) the packaging hall.

8.1. Brewhouse. In the brewhouse, the operations are (1) milling for crushing the malt, (2) mashing for extracting the malt with water, (3) filtration to separate spent grain solids from liquid wort, (4) kettle-boiling for stabilizing the wort and extracting the hops, and (5) wort clarification and cooling. The purpose of the brewhouse is to prepare the malt for extraction by *milling*, produce the extract in *mashing*, recover the extract by filtration (called *lautering/mash filtration*), and stabilize it by *boiling*.

8.2. Milling.   For each brew, malt is moved from the silos over a device that computes the weight delivered and enters the mill in a constant stream. The objective of milling is to crush the malt in such a way that the later processes can operate at maximum efficiency. Thus, finely milled malt will be easily extracted, but it will be difficult to separate the liquid (called "wort") from the insoluble material (called "spent grain"). Depending on the filtration device available therefore, the brewer decides on the most suitable milling strategy. Ideally, the endosperm of the malt is reduced to fine particles and the husk remains intact. Almost all mills in North America are dry mills. That is, the malt enters the mill dry and exits as a dry grist. These are roll mills. They might have three pairs of rolls as in a six-roll mill (Fig. 5) although simpler ones are common, especially in small breweries. After the initial crushing rollers, vibrating screens separate the grist particles so that the subsequent rollers crush only particles that need further reduction. In this way, particle size reduction and particle size control is achieved. Wet mills are also used widely around the world. In

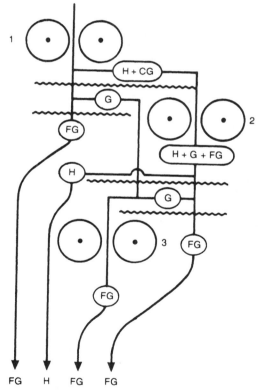

Fig. 5.   Schematic of a six roll mill for crushing malt to a desirable spectrum of malt particles. (1) Break rolls; (2) husk rolls; (3) grit rolls, wavy lines indicate the upper and lower shaker boxes for separation of husks and grits (malt particles), H = husks, G = grits, FG = fine grits and flour. two-Roll and four-roll mill designs are alternatives to this, as well as wet-milling. (From Ref. 1, with permission).

such mills, the grain is wetted before passing through a single pair of rolls and exits the mill as a slurry of malt in water that is pumped directly to the mash vessel. Hammer mills that reduce malt almost to a powder can be used with some kinds of mash filters.

Mashing comprises extracting the milled malt with a predetermined volume of water (which establishes mash thickness, but not <2.5–3 hL/100 kg)

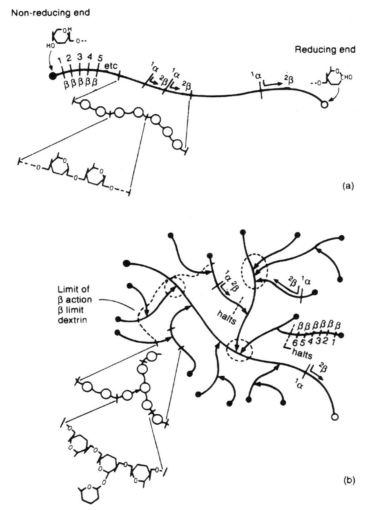

Fig. 6. Attack of α-amylase and β-amylase on (**a**) amylose (25% of total starch) and (**b**) amylopectin (75%), the two components of starch, during mashing, showing the reducing end (open circle) and nonreducing end (closed circle) of the molecules. Small sections are magnified to show the glucose molecules in the straight chain of amylose and at the branch points of amylopectin. Ordered attack by β-amylase (β-1,2,3, etc) from the nonreducing end produces maltose, and, acting alone, would leave a β-limit dextrin of large size from amylopectin. Random attack by α-amylase (^{1}α) permits additional attack by β-amylase (^{2}β). Neither enzyme can attack the branch points of amylopectin (arrowheads) and these survive in beer as glucose polymers called dextrins. (From Ref. 1, with permission).

under closely controlled conditions of temperature, time, and agitation. During mashing the malt enzymes act. Particularly, α- and β-amylase, acting together break down the large amount of starch (a glucose polymer of high molecular weight) present to a mixture of lower molecular weight products (Fig. 6). Roughly 65–70% or more of these products are simple sugars that are fermentable (mostly *maltose*, but also *maltotriose, sucrose, glucose,* and *fructose*) and the remainder is unfermentable *dextrins*. Preformed soluble materials in malt also dissolve in mashing most importantly amino acids, vitamins, and minerals, as well as the color and flavor compounds of malt. The two amylase enzymes in malt, α- and β-amylase, work together but have different functions. Both break the same bond in starch (the α-1-4 bond between adjacent glucose molecules) but α-amylase is random in action and quite heat stable, whereas β-amylase has an ordered action and is rather heat sensitive. The ordered action of β-amylase produces the large quantity of maltose that appears in wort (this sugar is otherwise rare in nature), and the action of α-amylase opens up the interior of the starch molecule to β-amylase attack (Fig. 6). The α-1-6 bonds of starch survive mashing as dextrins (small glucose polymers) and these, comprising ~30% of the original starch, remain in beer because brewers yeast cannot ferment them.

A note on low-calorie beers. Dextrins (unfermentable residues of starch after the action of malt amylases is complete) contribute calories to beers (about one-third of the total) but little else. By adding enzyme(s) (eg, amyloglucosidase from a bacterium, or enzymes from special malt) to the wort, the dextrins break down to fermentable sugar and are converted to alcohol by yeast during fermentation. Brewers dilute this highly alcoholic beer to yield a beer for sale with a normal content of alcohol and yeast-related flavor compounds but no dextrins, and so with about one-third fewer calories than regular beers. Further calorie savings than this can only be achieved by reducing the alcohol content.

The mashing process in American breweries begins with two separate mashes and is therefore called *double mashing*. The first mash is the *cereal or adjunct mash* (corn or rice) that usually comprises some 25–35% of the total extract (though up to 50% is possible in lower cost beers). Some malt is mixed in with the adjunct to prevent setting, and the mass is slowly brought to a boil and boiled for 20 mins or so (Fig. 7). This gelatinizes the starch so that the malt enzymes can attack it most rapidly. Meanwhile the second mash, the *main malt mash*, is started at ~40°C and after a short period of agitation, the boiling adjunct is blended into the main malt mash. This rapidly raises the temperature to ~65–70°C, and there is very rapid starch breakdown by α- and β-amylase acting together to produce fermentable sugars and unfermentable dextrins. After ~30 mins, the temperature of the mash is finally raised to 75°C or even higher, called the mash-off temperature. Mash-off tends for force the last part of the extract into solution, inactivates enzymes, and reduces the viscosity of the mash. Lower viscosity is useful for the next brewhouse stage, wort separation (mash-filtration or lautering). The total period for mashing is between ~2–3 h depending on the particular product being made.

Brewers can use other mashing regimes and do not need to use adjuncts. For example, traditional ale-making practice in Britain, and in many American microbreweries and brewpubs, requires all-malt mashes and *infusion mashing*.

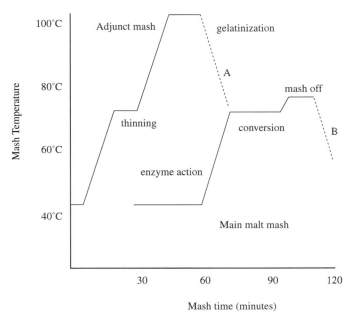

Fig. 7. Temperature profile of a mash using cereal adjunct (eg, rice or corn grits) that must be boiled in the cereal cooker to gelatinize it. The primary event is the hydrolysis of starch to yield fermentable and unfermentable sugars in the main malt mash (during enzyme action/conversion), and the solution of preformed small molecules (such as amino acids), and color and flavor compounds from malt. The dotted lines represent mash transfer: A, boiling adjunct mash into the main malt mash and B, the finished mash to the lauter vessel or mash filter for filtration.

The traditional deep tun (vessel) has a false bottom with slots in it so that mashing and filtration can take place in the same vessel. The cycle time for such a mash is ~4–6 hs. The malt is mixed with hot water to form a thick mash (~2.5 L of water/kg of malt) that is held at a single temperature of ~65°C for an hour or so. Following this, run-off (filtration) of wort begins. *Decoction mashing* is another alternative, traditionally European, mashing regime. This is also usually an all-malt process but (in contrast to infusion mashing) uses a stirred mash with a separate filtration device and a temperature program. In decoction mashing, the temperature program is established by boiling a portion of the malt mash in a small vessel called a mash kettle and returning it to the main malt mash to raise its temperature. In the most traditional forms of this style of mashing, the decoction might be done three times. These days, however, a single decoction is most common.

8.3. Wort Separation. At the end of mashing, the spent grain must be separated from the dense solution (called wort) of sugar and other materials extracted from malt and other grains. This is done by filtration. Two alternative device may be used, a *lauter vessel* or a *mash filter*. The operating principle of both devices is the same. A lauter vessel is a broad flat vessel in which the mash is spread over a false bottom with slots in it. After the mash settles, the wort is drawn slowly through the settled spent grain, where it is clarified, and exits the vessel through the false bottom. In a mash filter, the mash is held in a quite shallow layer against a vertical filter cloth. In either case, the wort, substantially freed of suspended solids, is produced over a period of ~1.5–2 hs and flows to the wort kettle. In both filtration devices, the bed of spent grain is rinsed with fresh hot water to recover as much sugary extract as possible in a process called *sparging*. The spent grain is a brewers' by-product mainly used as animal feed.

8.4. Boiling. Clarified wort from the lauter or mash filter is unstable in several ways: it could possibly contain (*1*) some active enzymes and so be subject to further change, or (*2*) unwanted microorganisms that inevitably find their way to warm moist sugary environments, or (*3*) excessive proteins and polyphenols that could easily cause hazes in beers. By boiling the wort, remnant enzymes are inactivated, bacteria are killed, and much of the protein and polyphenol is precipitated. This precipitate is called "hot trub" or "hot break". In addition, (*4*) the kettle boil concentrates the wort by evaporation of water, (*5*) removes the unwanted volatile components of hops and malt, and, most importantly, (*6*) effects the isomerization of α-acids (which are insoluble in wort and beer) into the bitter and soluble iso-α-acids (Fig. 4). There is also evidence that denaturation (loss of native structure) of proteins during boiling helps form polypeptides that have foam-stabilizing properties.

The wort kettle is a relatively simple device comprising a large (say 500–1500 hL capacity) enclosed insulated vessel with a steam-heated heat-exchange surface called a calandria. The calandria is usually inside the vessel but can be located outside the vessel and the wort pumped through it. Brewers demand a vigorous or "full rolling" boil that is maintained for at least 45–60 min. Hops are added during boiling. Whole (cone) hops, pellets or syrups can be used, or isomerized pellets. The first charge of hops is usually added close to the start of boiling and comprises bittering hops; this is sometimes followed by a middle charge. A third charge is added close to the end of boiling. These are usually aromatic hops that leave some trace of their desirable aroma in the beer.

After boiling, the spent hop material and precipitated trub must be removed. Whole hops, if used, must be removed by a *hop strainer*, but a whirlpool separator best removes the particulate matter from pellets. A whirlpool is a vessel that is about as deep as it is broad; the wort is introduced tangentially so that the wort swirls around the vessel. In this way the particulates quickly settle to the bottom center of the whirlpool where they form a compact sludge pile. Clear hot wort can be run from this vessel to the heat exchangers for *cooling* the wort to a temperature suitable for yeast addition and fermentation. Air or oxygen is gassed into the cool wort stream for yeast nutrition. Upon cooling, a second trub forms called the "cold trub", which can be removed by settlement in a shallow vessel. This concludes the brewhouse processes.

9. Fermentation

Following wort cooling and aeration or oxygenation, yeast is added and the wort–yeast mixture enters the fermentation cellar. Many beer characteristics, especially the alcohol content, are determined by the strength of the wort at the beginning of fermentation. This is expressed as the original specific gravity or O.G., a measure of density. Traditionally the original gravity of wort was 10–12° Plato (°P = % weight/weight (%w/w) or grams of dissolved solids per 100 g of wort. Density is also expressed directly as specific gravity = 1.040–1.048, which is the ratio of the weight of wort to the weight of water). However, these days brewers commonly use *high gravity brewing* throughout the fermentation and finishing processes and then, just before packaging, dilute the beer to sales strength using carbonated water free of oxygen. This practice assures the most efficient use of brewery capacity. In such cases, O.G. might be in the range 15–16° P (common) to 20° P (unusual).

Brewers recover yeast from a completed fermentation to start another, and in this way have nurtured certain yeasts for many centuries. Brewers' yeasts therefore can no longer be found in nature. Brewers have naturally selected yeasts that particularly meet their requirements. Alternatively, one might argue, that certain yeasts behaved in such a way that their recovery was easy. For example, *ale yeast*, when used in small traditional vessels, concentrates at the surface of the fermenting beer where it can be easily recovered by "skimming". Ale yeasts are therefore referred to as "top" yeasts. Because flotation is an unusual behavior, skimming assured the early ale brewers an easy means of recovering a reasonably constant yeast population for reuse. These yeasts are named *Saccharomyces cerevisae*; this designation includes wine yeasts and baker's yeasts, too. *Lager yeast* also has an unusual property that assured the early brewers a reasonably constant yeast supply and a means to recover it: It can grow and ferment at low temperatures and settles readily (bottom yeast). In addition, low temperatures allowed more of the CO_2 evolved in fermentation to remain in solution and so lagers were much more easy to carbonate than ales. Lager yeasts are named *Saccharomyces carlsbergensis* or *Saccharomyces uvarum*. In practice, brewers use lager yeasts at lower temperatures (say 8–14°C) than ale yeasts (say 20°C), and this might well account for the differences in flavor between ales and lagers. Brewers attach great significance to the differences between these two types of yeast, and, indeed, to the nuance of differences among individual strains of these yeasts, because these differences help to define the character of individual beer brands, distinguishing one product from another. Therefore brewers guard their yeast strain(s) jealously, because their yeast strain is a large part of their house flavor character. Nevertheless, the fundamental biochemical differences among ale and lager yeasts are relatively small, and many yeast taxonomists simply call both sorts *S. cerevisae*. Yeasts are fungi (*Saccharomyces* means "sugar fungus") whose growth is primarily unicellular. Brewers' yeasts are spherical to slightly egg-shape and, although the size is quite variable, they are generally large, being some 8–12 µm in diameter (Fig. 8). They increase in cell numbers (grow) by "budding"; in this process a yeast cell grows a small daughter cell attached to it, shares its cell contents and DNA with the daughter, and then splits from it. The split leaves a bud scar on the mother cell. This is asexual reproduction. Brewers recycle yeast from a completed brew to a new one. Some brewers wash this yeast with phosphoric or sulfuric acid at ∼pH 2.3 primarily to help reduce the number of contaminating micro-organisms that might accumulate. Nevertheless, yeast recycling is not done indefinitely in modern practice. A batch of yeast might make 8 or 12

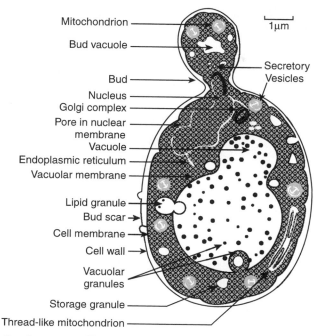

Fig. 8. Schematic of a yeast cell at budding. In fermenting brewers yeast the mitochondria (which are involved in aerobic metabolism) are few and ill-formed. The essential process of fermentation (the Embden-Meyerhof-Parnas pathway or glycolysis) leading to the formation of alcohol and carbon dioxide with the release of some useful biological energy, takes place in the cytoplasm. (From Ref. 2, with permission).

brews and then be discarded (eg, to be used by distillers or for yeast by-products of many kinds) before its performance declines because less vital and even dead cells and bacteria accumulate. New yeast is therefore introduced into the brewery on a regular basis, eg, monthly. This fresh and pure yeast culture is grown in a *propagation plant* (usually located at the headquarters brewery) and distributed to each brewery of a multibrewery company. This helps to keep the product constant at all locations. A modern brewery typically operates with only one yeast strain used as a pure culture. This is considered the best way to assure consistent fermentation.

Yeasts are facultative anaerobes and can grow in the absence of air; ie, they can ferment. When they ferment, however, they are unable to extract much energy from sugar and grow quite poorly compared to aerobic conditions. The main end products of anaerobic yeast metabolism of sugar (2 units) are alcohol (ethanol, 1 unit) and CO_2 (1 unit; much more CO_2 is produced in fermentation than appears in the finished beer), plus some new cell mass and heat ($\sim$140 kcal/ kg of sugar fermented) (Fig. 9). In addition, a host of other compounds, each in relatively low concentration, is produced as by-products of anaerobic metabolism and growth; many of these compounds have flavor, and add greatly and positively to the overall flavor impact of beer when in proper proportions (Table 3). Some flavor compounds are less desirable than others; chief among these is a

Table 3. **Types of Chemical Compounds found in Beers and their Approximate Concentration**[a]

Gross composition	Concentration
water	90–95% volume/volume (v/v)
alcohol (ethanol)	2.5–6% v/v (up to 10% in, eg, some barley wines)
carbon dioxide (CO_2)	1.5–3.0 volumes (~2.5–5 g/L)
carbohydrates	2.0–5% w/v (mainly unfermentable dextrins)[b]
calories (kcal/L)	300–900 (mainly from alcohol and carbohydrate)
flavor compounds	
alcohols (other)	100–400 mg/L (mostly amyl alcohols)
organic acids	200–350 mg/L (mostly lactic and succinic acid)
aldehydes	4–10 mg/L (mostly acetadehyde)
esters	10–60 mg/L (mostly ethyl acetate)
lactones, ketones, hydrocarbons	traces
organic sulfur compounds	traces
inorganic volatile sulfur (SO_2)	5–50 mg/L (below 10 in the United States)
hop compounds	10–50 mg/L (mostly iso-α-acids)
other (nutritional)	
vitamin B complex	4–10 mg/L (mostly niacin)[c]
nitrogenous material	0.2–0.6% w/v (as N × 6.25 = protein)
inorganic salts[d]	200–1000 mg/L

[a] Adapted from Ref. 2.

[b] In calorie-reduced beers dextrins can be ansent and alcohol (and calories) lower.

[c] Beer (1 L) can provide 100% of the daily requirement of folate, vitamin B_{12}, and useful proportion of niacin and biotin as well as some calcium, magnesium, and phosphorus. There is no fat, cholesterol or fat-soluble vitamins in beer.

[d] Mainly the cations potassium, calcium, magnesium and the anions phosphate, sulfate, and chloride.

Note: The composition of beers is extremely variable from the lightest to the heaviest beers, and these values are but rought guides. Further, each category of compounds, though dominated by a few substances as shown, can be made up of dozens if not 100 or more components.

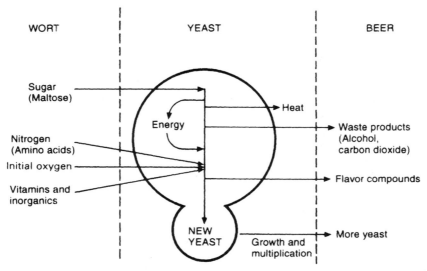

Fig. 9. The main yeast-mediated biochemical events during brewery fermentations. Though maltose (shown) comprises about one-half of the fermentable sugar, glucose, fructose, sucrose, and maltotriose are also present in wort and are fermented. (From Ref. 4, with permission).

compound called diacetyl (2,3-butanedione) that has an aroma like butter, and another is acetaldehyde that suggests a green apple flavor. An important goal of maturation of green beer is the removal or reduction of these compounds.

Of course, yeast cannot grow on sugar alone, and other nutrients derived from malt such as amino acids, vitamins, and minerals, also are necessary for yeast growth. Proper management of (1) wort quality in all its aspects, (2) yeast amount and vitality, and (3) fermentation conditions such as temperature, is necessary to achieve consistent fermentations, and hence consistent beer flavor. Brewers work hard to repeat as exactly as possible wort, yeast, and fermentation conditions from one brew to another to minimize variability. Nevertheless, many individual brews are blended to assure consistent flavor of the final product.

To initiate fermentation, yeast is added to cooled wort in a process called "pitching". Typically some 10–20 million cells per mL of wort are added (often expressed as 1 million cells/° Plato of wort gravity). Detectors monitor yeast addition to assure this addition is done accurately. As it exits the brewhouse, wort is cooled to ~17–20°C for ale fermentations and 8–10°C for lager fermentations. As a result ale fermentations are shorter (~3 days) than lager fermentations (~1 week). Small traditional vessels have now mostly been replaced by large, in some cases, huge vessels. The most imposing are tall, quite narrow *cylindro-conical* vessels (ie, vessels with a cylindrical body and a conical bottom) with a working capacity of up to 6000 hL or even more. These are commonplace in modern breweries as they have many advantages. In these vessels, the release of CO_2 bubbles in a rather coherent column rising up the center of the fermenter acts as a stirring device that accelerates fermentation and (later) yeast settlement. The settled yeast is conveniently recovered from the cone of the vessel. These vertical vessels are essentially self-supporting and, if well insulated, do not need to be entirely enclosed in a building. In most breweries, only the cone is inside a protective building because this is where the vessel is filled, emptied, and monitored.

The progress of fermentation is easily measured by the change in specific gravity of the wort as it becomes beer, or the production of alcohol or carbon dioxide. Fermentation is slow at first then, as the yeast begins to grow, becomes much more rapid and the liquid tends to warm up. Alcohol, carbon dioxide, and most flavor compounds are produced roughly in step with yeast growth. Fermentation then slows as the fermentable sugar is exhausted and the yeast begins to fall out of suspension (flocculate). When all the fermentable sugar is used up and there can be no further change in specific gravity, the brewer cools the beer to encourage further yeast settlement. This brings the primary fermentation to an end. The green beer is now ready for the final stages of processing that are designed to (1) mature the green beer, (2) carbonate it, (3) clarify it, and (4) render it stable. These processes are *secondary fermentation* and *finishing*.

10. Secondary Fermentation and Finishing

There are three general strategies for maturing the green beer. The first is simply to cool the beer to low temperature (0°C or below) after primary fermentation and hold it for some period of time such as a week or two. This is called *aging*; carbon dioxide can be injected at some stage to achieve carbonation. Second, *krausening* is a widely used secondary fermentation strategy that involves mixing freshly fermenting wort containing yeast into the green beer. This mixture is held at ~8°C for ~3 weeks. During this time, the yeast slowly ferments the added sugar and the carbon dioxide formed is entrapped for carbonation of the beer. At the same time, undesirable flavor compounds such as diacetyl and acetaldehyde are reduced by the yeast action to more or less flavorless compounds. The third strategy of secondary fermentation is called *lagering*. In this process the beer, toward the end of primary fermentation, is cooled somewhat to flocculate much, but not all, of the yeast and is then moved to a new vessel before all the fermentable sugar is exhausted. The yeast ferments out the last few degree of gravity slowly at ~8°C, again with entrapment of CO_2 for carbonation, and reduction of diacetyl and acetaldehyde. A modern maturation strategy that considerably shortens the maturation time for many beers is called the *diacetyl rest*. At the end of the primary fermentation, the beer is simply held at fermentation temperature (~15°C at this stage for lagers, 20°C for ales) until the diacetyl is reduced to specification, as determined by measurement. These secondary fermentation and maturation processes are all time consuming and relatively expensive because the brewery must have many large refrigerated vessels to contain the beer; they are, however, necessary for a quality product.

10.1. Finishing.   These processes concern (*1*) *filtration* of the beer at low temperature such as minus 2°C, so that it is brilliantly clear and (*2*) treating the beer with *stabilizing agents* so that it remains brilliantly clear during its sojourn in the market place. Beer is often filtered twice: a rough filtration using diatomaceous earth to remove the vast bulk of the suspended particles and then a polish filtration using, eg, sheet filters to achieve brilliant clarity. Centrifugation sometimes replaces rough filtration, or *finings* (isinglass, ie, specially prepared collagen) can be used to coagulate and settle particles. After rough filtration, stabilizing agents are commonly added. Since the kind of haze most feared by brewers is the protein/polyphenol "chill"-haze (which might arise when the consumer cools the beer before consumption) removing these compounds is a prime strategy for beer stabilization. Modern stabilizing agents are, for the most part, insoluble adsorbents. They include specially prepared silica gels, which remove certain proteins from beer, and/or PVPP (polyvinylpyrrolidone), a compound related in structure to nylon, which removes polyphenols from beer. Beer that has gone through these finishing processes is now mature in flavor, properly carbonated, brilliantly clear and stable against chill-haze formation, and free of oxygen. It is conveyed to a large vessel near the packaging hall called the BBT or *bright beer tank* at which point the brewing process ends. The beer is now ready for packaging and presentation to the consumer.

11. Packaging

Packaging beer into bottles and cans is an immensely expensive process; not only are the packaging materials themselves expensive but the sophisticated machinery, large space required, and numerous skilled employees necessary for packaging add much to the cost. It has been said that if the beer in a bottle or can were replace by water, the price of a six-pack would go down by barely 10%. Two technical factors come into play. First, packaging must be very rapid, ~2000 units/min, eg, so that the large volumes of beer produced by modern breweries can be broken down to consumer units in a reasonably short time. And second, oxygen (air) must be rigorously excluded from the package because it harms beer flavor. With time, especially if warm, the presence of oxygen (even an infinitesimally small amount) in beer causes loss of fresh beer flavor. This is replaced by "oxidized" flavor, often described as a papery or cardboardy flavor, or toasty/bready. For this reason, brewers in North America permit their beers to remain on the supermarket shelf for only ~100–120 days (depending on the market) during which time it retains fresh flavor. Most beers have a "best by" or "born on" date clearly legible to the consumer. Over-age or out-dated beer is withdrawn and destroyed. Imported beers often suffer from these "oxidized" off-flavors because they remain too long in transit to the consumer.

Most beer bottles these days are one-way (nonreturnable) in the United States, though not in other countries. This avoids the environmentally unfriendly practice of sorting and washing returnable bottles, though any overall environmental advantage to one-way bottles does require that they be recycled to the glass plant not merely discarded. New bottles for filling are sanitized, rinsed, and drained and then lifted onto a filling head on a filling machine. Such a machine may have as many as 200 heads (depending on the production required) and is designed as a carousel, ie, it is constantly turning at a speed the eye can barely follow. The filling machine carries a small reservoir of beer that is constantly replenished from the bright beer tank as it is packaged. Each bottle, firmly in contact with each filling head, rides one circuit of the carousel and is then lifted off the machine to be immediately replaced by another bottle. In the short time it is on the machine, the bottle is first evacuated to remove air and flushed with CO_2. This might be repeated. The bottle is then pressurized to the same pressure as is above the beer in the reservoir and then the beer flows down into the bottle slowly at first then more rapidly, by gravity. The fill stops at a preset level and the pressure is slowly released down to atmospheric pressure (the "snift") to avoid excessive overfoaming. The full bottle now moves immediately to the crowner. The bottle is jetted with sterile water to bring up a foam of gas from the beer (CO_2) into the neck of the bottle to displace any air, and the crown is immediately put in place and crimped on. Beer is canned in much the same way except that cans cannot be evacuated to remove air because they would collapse. Beer is packaged in a bewildering variety of bottle and can sizes and shapes to meet consumer expectations and demand. Many special packs are available along with the ubiquitous keg (half-barrel or 15.5 galls United States) for draft dispense, especially in bars and taverns. Plastic (PET = polyethylyethylene terephthalate) packages have been in use in many parts of the world for years but have made little progress in the United States to the present time.

Most beer in bottles or cans is *pasteurized*. That is, the beer is heated briefly to kill any microorganisms that might be present that could spoil beer flavor. As previously noted, no pathogenic (disease causing) organisms can survive in beer. Because brewers assure that the brewing process is extremely sanitary, few microbes enter the beer, and as a result they use a mild heat treatment. A

Table 4. Production Statistics of Beer[a]

Country	Population (mill)	Production (m hL)	Imports (m hL)	Exports (m hL)	Consumption (L per head)	draft (%)	Av. Strength (% ABV)
Argentina	36.1	12.4	0.39	0.18	34.9	1	4.8
Australia	18.5	17.5	0.21	0.43	95.0	24	4.3
Austria	8.1	8.8	0.36	0.51	108.1	32	5.1
Belgium[b]	10.6	14.6	0.88	4.9	99.0	40	5.2
Brazil	165.9	88.0	0.26	0.45	52.9	2	
Bulgaria	8.3	3.8	0.013	0.077	45.2	2	4.8
Canada	30.3	22.8	1.17	3.64	67.0	11	5.0
Chile	14.8	3.67	0.14	0.16	24.6	8	4.5
China	1,255.7	196.4	0.33	0.56	15.6	4	
Colombia	38.3	18.3	0.5	0.04	48.9	1	4.2
Croatia	4.5	3.8	0.175	0.523	75.8	7	5.0
Cuba	11.1	1.25	0.046		11.7		5.0
Czech Republic	10.3	18.3	0.154	1.9	160.8	46	4.5
Denmark	5.3	8.1	0.079	2.4	107.7	10	4.6
Finland	5.2	4.7	0.08	0.32	79.1	23	4.6
France	58.7	19.8	5.3	2.4	38.6	26	5.0
Germany	82.0	111.7	2.8	8.4	127.4	20	
Greece	10.4	4.0	0.19	0.3	42.0	5	4.9
Hungary	10.1	7.0	0.18	0.09	70.0	18	4.7
Ireland	3.6	8.5	0.56	3.45	124.2	80	4.1
Italy	57.5	12.2	3.68	0.37	26.9	16	5.1
Japan	126.4	72.2	0.8	0.71	57.2	16	5.0

Korea (Rep)	46.4	14.1	0.011	0.24	29.8	13	4.0
Mexico	95.8	54.8	0.37	7.79	49.4	1	4.0
New Zealand	3.8	3.21	0.181	0.14	84.7	40	4.0
Netherlands	15.7	24.0	0.95	11.7	84.3	31	5.0
Nigeria	106.4	4.2	0.008	0.006	3.9	0	4.5
Norway	4.4	2.2	0.045	0.011	49.7	27	4.5
Peru	24.8	7.2	0.014	0.031	29.0	1	
Philippines	71.4	12.7	0.004	0.092	17.6	1	4.7
Poland	38.7	20.6	0.17	0.12	53.4	21	5.2
Portugal	9.9	6.8	0.29	0.55	65.3	28	5.2
Romania	22.5	9.9	0.06	0.001	44.2	21	4.5
Russia	147.4	32.5	0.73	0.047	22.5		
Slovak Republic	5.4	4.3	0.5	0.46	84	40	4.5
Slovenia	2.0	2.0	0.101	0.433	83.3	13	4.9
South Africa	42.1	25.3	0.42	0.65	59.5	1	5.0
Spain	39.9	25.0	2.0	0.51	66.4	33	5.2
Sweden	8.9	4.6	0.534	0.041	57.3	12	4.0
Switzerland	7.25	3.6	0.72	0.03	59.9	33	4.9
Ukraine	50.5	6.8	0.096	0.06	13.7	36	
UK	59.2	56.7	5.9	3.9	99.4	64	4.1
USA	270.3	235.5	19.1	6.5	83.7	10	4.6
Venezuela	23.2	17.8	0.018	0.49	74.3	1	

[a] From Ref. 3.
[b] Includes Luxembourg, because of inaccuracies introduced by cross-border trading.

pasteurizer is a large tunnel through which the beer cans or bottles move on an endless belt. The containers are sprayed with increasingly hot water to raise their temperature to 60–62°C. They are held at this temperature as long as required, and then cooled by water sprays. One pasteurization unit (PU) is 1 min at 60°C (or its heat equivalent) and most beers are pasteurized in the range of 5–15 PUs. There are two alternative techniques to tunnel pasteurization for dealing with the few microbes that might enter beer. The first is "flash" pasteurization in which the beer before packaging flows through a heat exchanger and is rapidly heated up and cooled down. This minimizes heat damage to the beer, but aseptic (sterile or microbe-free) packaging must follow and that is a challenging and expensive technology. Second, bacteria present can be filtered out of the beer by extremely tight membrane filtration. Again, aseptic packaging must follow this, but advantageously the beer can be marketed as "draft" beer in a bottle or can, because the definition of draft beer (in the United States) is that it be unpasteurized. The bottle is now ready for labeling. The packages are loading into six-pack holders, cased, and enter the warehouse from whence the product is distributed to wholesalers and eventually to consumers.

12. Economic Aspects

Production statistics of beer are listed in Table 4. The United States is the biggest producer of beer although per capita consumption is not high. China is expected to become the biggest producer eventually.

13. Health Value of Beer

Excessive consumption of alcoholic beverages, including beer, is injurious to health, dangerous, and antisocial. However, possible minor nutritional and significant health benefits of beer and other alcoholic beverages are now well documented. When the relationship between all forms of mortality is related to alcohol intake, there is little doubt that moderate daily consumption of alcohol as beer, wine, or spirits, significantly prolongs life, and especially protects against coronary heart disease and stroke among other ailments. Moderate consumption is defined as 1–3 drinks/day. In addition, beer contains some useful levels of B-vitamins (Table 3), especially folate, and some minerals especially calcium, magnesium, potassium, and selenium. Abstention from alcohol on the basis of health alone might therefore be a poor decision.

BIBLIOGRAPHY

" Beer and Brewing" in *ECT* 1st ed., Vol, 2, pp. 382–413, By E. Krabbe and H. Goob, Blatz Brewing Company; in *ECT* 2nd ed., Vol. 3, pp. 297–338, by H. E. Høyrup, the Brewery Association Copenhagen; "Beer" in *ECT* 3rd ed., Vol. 3, pp. 692–735, by H. E. Høyrup, Bryggeriforeningen, Denmark; in *ECT* 4th ed., Vol. 4, pp. 22–63, by J. F. Nissen, The

Denish Brewers' Association; "Beer" in *ECT* (online), posting date: December 4, 2000, by J. F. Nissen, The Danish Brewers' Association.

CITED PUBLICATIONS

1. M. J. Lewis and T. W. Young, *Brewing*, 2nd ed., Kluwer Academic/Plenum Publishers, New York, 2002.
2. G. G. Stewart and I. Russell, *Brewing Science and Technology*, Series lll, Brewers Yeast, Institute and Guild of Brewing, London.
3. C. W. Bamforth, *Tap into the Art and Science of Brewing*, Oxford University Press, New York, 2003.

GENERAL REFERENCES

American Society of Brewing Chemists, *Methods of Analysis*, 8th ed. ASBC, St. Paul, Minn.

S. Baron, *Brewed in America*, Little, Brown. Boston 1962.

D. R. Berry, I. Russell, and G. G. Stewart, eds., *Yeast Biotechnology*, Allen & Unwin, London, Boston.

D. E. Briggs, J. S. Hough, R. Stevens, and T. W. Young, *Malting and Brewing Science*, Vols. I and II, 2nd ed., Chapman and Hall, London, 1982.

D. E. Briggs, *Barley*, Chapman & Hall, London, 1978.

H. M. Broderick, ed., *Beer Packaging*, Master Brewers Association of the Americas, Madison, Wis. 1982.

H. M. Broderick, ed., *The Practical Brewer*, 2nd ed., Master Brewers Association of the Americas, Madison Wis. 1977.

C. Forget, *The Association of Brewers Dictionary of Beer and Brewing*, Brewers Publications, Boulder. Col. 1988.

W. A. Hardwick, ed., *Handbook of Brewing*, Marcel Dekker, New York, 1995.

M. Jackson, *The New World Guide to Beer*, Running Press, Philadelphia, London, 1988.

M. Jackson, *Beer Companion*, Running Press, Philadelphia, London, 1993.

W. Kunze, *Technology Brewing and Malting*, VLB, Berlin, 1996.

H. F. Linskens and J. F. Jackson, eds., *Beer Analysis*, Springer-Verlag, Berlin, 1988.

J. S. Pierce, ed, *Brewing Science and Technology Series I and II*, Institute and Guild of Brewing, London, 1990.

F. G. Priest and I. Campbell, eds., *Brewing Microbiology*, 3rd ed., Kluwer, New York, 2002.

J. R. A. Pollock, *Brewing Science*, Vols. I, II, and III. Academic Press, London, 1979.

A. H. Rose and J. S. Harrison, eds., *The Yeasts*, (4 Vols.) 2nd ed. Van Nostrand Reinhold, New York, 1987.

L. C. Verhagen, ed., *Hops and Hop Products, Manual of Good Practice*, Getranke-Fachverlag Hans Carl, Nurnburg, 1997.

M. Verzele and D. De Keukeleire, *Chemistry and Analysis of Hop and Beer Bitter Acids*, Elsevier, Amsterdam, New York, 1991.

MICHAEL J. LEWIS
University of California

BENZALDEHYDE

1. Introduction

Benzaldehyde [100-52-71] C_6H_5CHO, is the simplest and quite possibly the most industrially useful member of the family of aromatic aldehydes. Benzaldehyde exists in nature, primarily in combined forms such as a glycoside in almond, apricot, cherry, and peach seeds. The characteristic benzaldehyde odor of oil of bitter almond occurs because of trace amounts of free benzaldehyde formed by hydrolysis of the glycoside amygdalin. Amygdalin was first isolated in 1830 from the seeds of the bitter almond (*Prunus amygdalus*). Sometime later, Liebig and Wöhler found that when amygdalin was hydrolyzed with water and emulsin, benzaldehyde, hydrogen cyanide, and D-glucose were formed (1).

2. Physical Properties

Physical properties of benzaldehyde are listed in Tables 1 and 2; boiling points and concentrations of certain selected binary azeotropes are given in Table 3. For a more complete listing of benzaldehyde azeotropes, see (3).

3. Manufacture

The only industrially important processes for the manufacturing of synthetic benzaldehyde involve the hydrolysis of benzal chloride [98-87-31] and the air oxidation of toluene. The hydrolysis of benzal chloride, which is produced by the side-chain chlorination of toluene, is the older of the two processes. It is

Table 1. **Physical Properties of Benzaldehyde**

Property	Value
molecular formula	C_7H_6O
molecular weight	106.12
boiling point, °C at 101.3 kPa[a]	179
melting point, °C	−26
flash point, closed cup, °C	63
autoignition temperature, °C	192
refractive index, n^{20}	1.5455
viscosity, mPa·s (=cP) at 25°C	1.321
density, g/cm³ at 25°C	1.046
specific heat (liquid) at 25°C, J/g·K[b]	1.615
latent heat of vaporization[c], J/g[b]	362
standard heat of combustion, kJ/g[b]	−31.9
solubility in water at 20°C, wt %	∼0.6
solubility of water in at 20°C, wt %	∼1.5

[a] To convert kPa to atm, divide by 101.3.
[b] To convert J to cal, divide by 4.184.
[c] At the boiling point (179°C).

Table 2. **Vapor Pressure vs Temperature**[a]

Temperature, °C	Pressure, kPa[b]
26.2	0.13
50.1	0.67
62.0	1.33
75.0	2.66
90.1	5.32
99.6	8.0
112.5	13.3
131.7	26.6
154.1	53.3
179.0	101.3

[a] Ref. 2.
[b] To convert kPa to mm Hg, multiply by 7.5.

not utilized in the United States but is used in Europe, India, and China. Other processes, including the oxidation of benzyl alcohol, the reduction of benzoyl chloride, and the reaction of carbon monoxide and benzene, have been utilized in the past, but they no longer have any industrial application. [For an historical article regarding the chlorination of toluene and the subsequent production of benzaldehyde, benzyl alcohol, and benzoic acid, see (4).]

The air oxidation of toluene is the source of the majority of the world's synthetic benzaldehyde. Both vapor- and liquid-phase air oxidation processes have been used. In the vapor-phase process, a mixture of air and toluene vapor is passed over a catalyst consisting of the oxides of uranium, molybdenum, or related metals. High temperatures and short contact times are essential to maximize yields. Small amounts of copper oxide may be added to the catalyst mixture to reduce formation of by-product maleic anhydride. Conversion per pass is reported to be low, 10–20%, with equally low yields, 30–50% (5). The vapor-phase oxidation of toluene was the dominant toluene oxidation process in the 1950s and early 1960s, but is no longer of industrial importance. The liquid-phase process now dominates.

In the liquid-phase process, both benzaldehyde and benzoic acid are recovered. This process was introduced and developed in the late 1950s by the Dow Chemical Company, as a part of their toluene-to-phenol process, and by Snia Viscosa for their toluene-to-caprolactam process. The benzaldehyde recovered from the liquid-phase air oxidation of toluene may be purified by either

Table 3. **Binary Azeotropes of Benzaldehyde**[a]

Component	Azeotrope, Boiling point, °C	Benzaldehyde, %
benzyl chloride	177.9	50
o-cresol	192	23
D-limonene	171.2	43
cineole	172	36
phenol	185.6	49

[a] Ref. 3.

Table 4. **National Formulary and Food Chemicals Codex Specifications**

Item	The National Formulary[a]	Food Chemicals Codex[b]
identification		passes FCC Test
assay (by oximation method)	contains not <98.0% and not >100.5% of C_7H_6O.	98.0%, Minimum of C_7H_6O.
specific gravity (@ 25°C)	between 1.041 and 1.046	1.041–1.046
refractive index (@ 20°C)	between 1.544 and 1.546	1.544–1.547
hydrocyanic acid	passes NF Test	passes FCC Test
nitrobenzene	passes NF Test	
chlorinated compounds	passes NF Test	passes FCC Test
organic volatile impurities	passes NF Test	

[a] NF 21st ed.—Effective 1/1/2003.
[b] FCC 4th ed.—Effective 7/1/1996.

batch or continuous distillation. Liquid-phase air oxidation of toluene is covered more fully (see BENZOIC ACID).

4. Economic Aspects

Benzaldehyde is produced in the United States by Noveon Kalama, Inc., Kalama, Washington (formerly Kalama Chemical). The Noveon plant was originally constructed by The Dow Chemical Company in the early 1960s to produce phenol from benzoic acid and currently recovers benzaldehyde as a by-product of that process (6). Production and sales figures for benzaldehyde are not available.

5. Specifications and Test Methods

Benzaldehyde is sold as technical grade or as meeting the specifications of the National Formulary (NF) (7), the Food Chemicals Codex (FCC) (8), or the British Pharmacopeia (BP) (9) (Tables 4 and 5). The test methods used for the analysis of benzaldehyde are standard methods, with the exception of the assay method.

The assay method involves the reaction of benzaldehyde with hydroxylamine hydrochloride in an alcoholic solution. Benzaldehyde oxime, water, and

Table 5. **British Pharmacopeia and Technical Grade Specifications**

Item	British Pharmacopeia[a]	Technical
assay (by oximation method)	benzaldehyde contains not <98.0% w/w and not >100.5% w/w of C_7H_6O	99.0%, minimum
refractive index @ 20°C		
weight per mL @ 20°C		
free acid	not >1.0% w/v, calculated as benzoic acid	
chlorinated compounds	not >0.05% w/v, calculated as Cl^-	none
color		colorless to pale yellow
toluene		0.1 %, maximum

[a] BP 2002—Effective 1 December, 2002.

hydrochloric acid are the products of the reaction. The hydrochloric acid formed is then titrated with standard caustic solution to determine the benzaldehyde assay. Performing the titration to a potentiometric end point, rather than to a colored end point, has been shown to be the more accurate method. Since other carbonyl containing compounds also react to form the oxime and release hydrochloric acid, this test is not specific for benzaldehyde. The levels of trace impurities in the product benzaldehyde are often more important than the product assay. Gas chromatographic (gc) methods for the determination of those trace impurities are widely used.

Benzaldehyde is not included in the European or the Japanese Pharmacopeias.

6. Health and Safety Factors

The oral LD_{50} for benzaldehyde is reported as 1300 mg/kg in rats and as 1000 mg/kg in guinea pigs. Based upon these values, benzaldehyde is considered a moderately toxic substance when ingested. The subcutaneous lethal dose in rats is ~5 g/kg. The fatal oral dose in humans is estimated to be ~56.7 g (2 oz) (10). Benzaldehyde tested negative for mutagenicity in salmonella assays in the 1988 National Toxicology Program. Studies of the carcinogenic effects of benzaldehyde are currently in progress (11). In the industrial setting, exposure to benzaldehyde through eye and skin contact and inhalation is far more prevalent than ingestion incidence. Overexposure to benzaldehyde vapors is irritating to the upper respiratory tract and produces central nervous system depression with possible respiratory failure. Epileptiform convulsions have been observed in rabbits (10). Contact may cause eye and skin irritation. Some individuals are more sensitive to skin contact than others. See (12) for more toxicological information.

7. Handling

The low autoignition temperature of benzaldehyde (192°C) presents safety problems since benzaldehyde can be ignited by exposure to low pressure steam piping, for example. Benzaldehyde may also spontaneously ignite when soaked into rags or clothing or adsorbed onto activated carbon (13). Bulk storage of benzaldehyde should be made under a nitrogen blanket, since benzaldehyde is easily oxidized to benzoic acid upon exposure to air. All storage tank openings should be easily accessible for cleaning, since they will have a tendency to plug with benzoic acid. Benzaldehyde is stored in noninsulated type 304 stainless steel storage tanks. If storage in very cold climates is contemplated, consideration should be given to insulating and steam-tracing the tank. A baked, phenolic, resin-lined tank is also suitable. Copper or brass are to be avoided since they are readily attacked by benzaldehyde and benzoic acid. Because of the low surface tension of benzaldehyde, the use of screwed piping fittings should be avoided (14).

8. Uses

Benzaldehyde is a synthetic flavoring substance, sanctioned by the U.S. Food and Drug Administration (FDA) to be generally recognized as safe (GRAS) for foods (21 CFR 182.60). Both "pure almond extract" and "imitation almond extract" are offered for sale. Each contains 2.0–2.5 wt% benzaldehyde in an aqueous solution containing approximately one-third ethyl alcohol.

"Natural" benzaldehyde can be produced in a number of ways. The FDA regulations regarding natural products are found in 21 CFR 101.22. At present, there is a controversy over what the term natural really means with regard to benzaldehyde. Whether a particular benzaldehyde product is natural or not becomes an issue only if the final product is said to contain natural flavors. There are at least two routes currently being used to produce natural benzaldehyde. Principal flavor houses are reported to market a product that is derived from cassia oil. The chief constituent of cassia oil is cinnamic aldehyde, which is hydrolyzed into its benzaldehyde and acetaldehyde constituents, which is a fermentative retro-aldol reaction. Whether this hydrolysis allows the final benzaldehyde product to be considered natural is of great concern. The FDA has reportedly issued an opinion letter that benzaldehyde produced from cassia oil is not natural (15).

The other significant production method for natural benzaldehyde involves the steam distillation of bitter almond oil, which has been derived from the kernels of fruit such as apricots, peaches, cherries, plums, or prunes. The benzaldehyde product obtained in this fashion is claimed to have a superior flavor profile. The use of peach and apricot pits to produce the more profitable product laetrile apparently affects the supply available to natural benzaldehyde producers. The subject of natural benzaldehyde came to the forefront in 1984 when it was found that a natural benzaldehyde product, labeled "oil of benzaldehyde", was actually made synthetically by the air oxidation of toluene followed by careful fractionation to remove trace impurities. This finding was accomplished by the Center for Applied Isotopic Studies, University of Georgia, and involved measuring the amounts of ^{13}C and ^{14}C in that material.

Benzaldehyde is widely used in organic synthesis, where it is the raw material for a large number of products. In this regard, a considerable amount of benzaldehyde is utilized to produce various aldehydes, such as cinnamic and methyl, butyl, amyl, and hexyl cinnamic aldehydes. The single largest use for benzaldehyde, however, is the production of benzyl alcohol via hydrogenation.

9. Derivatives

Benzoin [119-53-9], α-hydroxy-α-phenylacetophenone, $C_6H_5CH(OH)COC_6H_5$ (mp, 133–137°C; bp, 343–344°C at 101.3 kPa), is formed by the self-condensation of benzaldehyde in the presence of potassium cyanide. It is used on a small scale as a polymerization catalyst in polyester resin manufacture.

Benzil [134-81-6], diphenyl-α,β-diketone, $C_6H_5COCOC_6H_5$ (mp, 95°C; bp, 346–348°C at 101.3 kPa), formed by oxidizing benzoin is used as an intermediate in chemical synthesis.

Benzyl alcohol [100-51-6], $C_6H_5CH_2OH$ (bp, 205AT at 101.3 kPa), produced by the hydrogenation of benzaldehyde is used in color photography, as a parenteral solution preservative, as a general solvent, and as an intermediate in the manufacture of various benzoate esters for the soap, perfume, and flavor industries (see BENZYL ALCOHOL AND β-PHENETHYL ALCOHOL).

Benzoyl chloride [98-88-4], C_6H_5COCl (mp, $-1°C$; bp, $197.2°C$ at $1-01.3$ kPa; d_4^{25}, 1.2070; n_D^{20}, 1.55369), is a colorless liquid that fumes upon exposure to the atmosphere. It has a sharp odor and, in vapor form, is a strong lachrimator. It is decomposed by water and alcohol, and is miscible with ether, benzene, carbon disulfide, and oils.

Benzylamine [100-46-9], $C_6H_5CH_2NH_2$ (bp, 184°C at 101.3 kPa) produced by reaction of ammonia with benzaldehyde and hydrogenation of the resulting Schiffs base, is used as the raw material for the production of biotin (Vitamin H), as an intermediate for certain photographic materials, and as an intermediate in the manufacture of certain pharmaceutical products.

Benzylideneacetone [122-57-6], $C_6H_5CH=CHCOCH_3$ (bp, 260–262°C at 101.3 kPa; mp, 35–39°C) is produced by condensing acetone and benzaldehyde. It is used as an electroplating additive.

Benzylacetone [2550-26-7], $C_6H_5CH_2CH_2COCH_3$ (bp, 233–234°C at 101.3 kPa) is produced by condensing acetone and benzaldehyde, followed by selective hydrogenation, and is used in soap perfumes.

Dibenzylamine [103-49-1], $C_6H_5CH_2NHCH_2C_6H_5$ (bp, 300°C at 101.3 kPa) is produced by reaction of benzyl amine with benzaldehyde and hydrogenation of the Schiffs base. It is used in rubber and tire compounding, as a corrosion inhibitor, and as an intermediate in the production of rubber compounds and pharmaceutical products.

Cinnamaldehyde [14371-10-9], $C_6H_5CH=CHCHO$ (bp, 253°C at 101.3 kPa), produced by the alkaline condensation of benzaldehyde and acetaldehyde is the main ingredient in cassia oil. It is used in soap perfumes and as an intermediate in the production of other flavor and fragrance compounds.

α-Methylcinnamaldehyde [101-39-3], $C_6H_5CH=C(CH_3)CHO$, is produced by the alkaline condensation of benzaldehyde and propionaldehyde. Its principal use is as the raw material for p-tert-butyl-α-methyl dihydrocinnamic aldehyde, [8054-61], a lily of the valley fragrance intermediate.

α-Amylcinnamaldehyde [122-40-7], $C_6H_5CH=C(C_5H_{11})CHO$ (bp, 140°C at 0.7 kPa), produced by the alkaline condensation of benzaldehyde and n-heptaldehyde, produces a jasminelike floralness and is used extensively as a perfume for soap products.

α-Hexylcinnamaldehyde [101-86-0], $C_6H_5CH=C(C_6H_{13})CHO$ (bp, 174–176°C at 2 kPa), produced by the alkaline condensation of benzaldehyde and n-octaldehyde, also produces a jasminelike floralness and is also used extensively as a perfume for soap products.

BIBLIOGRAPHY

"Benzaldehyde" in ECT 1st ed., Vol. 2, pp. 414–420, by R. L. Clark and C. P. Neidig, Heyden Chemical Corporation; in ECT 2nd ed., Vol. 3, pp. 360–367, by A. J. Deinet and E. P. Dibella, Heyden Newport Chemical Corporation; in ECT 3rd ed., Vol. 3, pp. 736–743, by A. E. Williams, Kalama Chemical Corporation; in *ECT* 4th ed., Vol 4, pp. 64–72, by Jarl Opgrande, and others, Kalama Chemical Corporation; "Benzaldehyde" in *ECT* (online), posting date: December 4, 2000, by Jarl L. Opgrande, C. J. Dobratz, Edward Brown, Jason Liang, Gregory S. Conn, Kalama Chemical, Inc., Frederick J. Shelton, Jan With, Chatterton Petrochemical Corp.

CITED PUBLICATIONS

1. L. F. Fieser and M. Fieser, *Organic Chemistry*, D. C. Heath and Co., Boston, 1944, pp. 368–369.
2. R. H. Perry and co-workers, *Perry's Chemical Engineer's Handbook*, 6th ed., McGraw Hill Book Co., New York, 1984, pp. 3–50.
3. L. H. Horsley and co-workers, *Azeotropic Data*, American Chemical Society, Washington, D.C., 1952.
4. W. H. Shearon, H. E. Hall, and J. E. Stevens, *Ind. Eng. Chem.* **41**, 1812 (1949).
5. W. L. Faith, D. B. Keyes, and R. L. Clark, *Industrial Chemicals*, John Wiley & Sons, Inc., New York, 1965, pp. 120–124.
6. W. W. Kaeding and co-workers, *Indust. Eng. Chem. Proc. Des. Dev.* 4(1), 97 (1965).
7. *The National Formulary*, 20th ed., U.S. Pharmacopeial Convention, Inc., Rockville, Md., 2002, p. 2512.
8. *Food Chemicals Codex*, Fourth ed., National Academy Press, Washington, D.C., 1996, pp. 456–457.
9. *British Pharmacopeia*, Her Majesty's Stationery Office, London, 2002, pp. 197–198.
10. R. H. Gosselin and co-workers, *Clinical Toxicology of Commercial Products*, Williams & Wilkins, Baltimore, Md., 1976, pp. 167.
11. National Toxicology Program, Fiscal Year 1989 Annual Plan, U.S. Dept. of Health and Human Services, Washington, D.C., 1989, pp. 48, 83.
12. N. I. Sax, Dangerous Properties of Industrial Materials Report, Vol. 9, No. 6, Van Nostrand Reinhold Co., New York, 1989, pp. 61–70.
13. Benzaldehyde, Material Safety Data Sheet, Noveon Kalama, Inc., Kalama, Wash., July 1, 2002.
14. Benzaldehyde, Product Information Bulletin, Noveon Kalama, Inc., Kalama, Wash., June 15, 2002.
15. *Chem. Mark. Rep.*, (Mar. 5,1990).

JARL L. OPGRANDE
EDWARD BROWN
MARTHA HESSER
JERRY ANDREWS
Noveon Kalama, Inc.

BENZENE

1. Introduction

Benzene [71-43-2], C_6H_6, is a volatile, colorless, and flammable liquid aromatic hydrocarbon possessing a distinct, characteristic odor. Benzene is used as a chemical intermediate for the production of many important industrial compounds, such as styrene (polystyrene and synthetic rubber), phenol (phenolic resins), cyclohexane (nylon), aniline (dyes), alkylbenzenes (detergents), and chlorobenzenes. These intermediates, in turn, supply numerous sectors of the chemical industry producing pharmaceuticals, specialty chemicals, plastics, resins, dyes, and pesticides. In the past, benzene has been used in the shoe and garment industry as a solvent for natural rubber. Benzene has also found limited application in medicine for the treatment of certain blood disorders, such as polycythemia and malignant lymphoma (1), and further in veterinary medicine as a disinfectant. Benzene, along with other light high octane aromatic hydrocarbons such as toluene and xylene, is used as a component of motor gasoline. Although this use has been largely reduced in the United States, benzene is still used extensively in many countries for the production of commercial gasoline. Benzene is no longer used in appreciable quantity as a solvent because of the health hazards associated with it.

Benzene was first isolated by Michael Faraday in 1825 from the liquid condensed by compressing oil gas. He proposed the name bicarburet of hydrogen for the new compound. In 1833, Eilhard Mitscherlich synthesized bicarburet of hydrogen by distilling benzoic acid, obtained from gum benzoin, with lime and suggested the name benzin for the compound. In 1845, A. W. Hoffman and C. Mansfield found benzene in light oil derived from coal tar. The first practical industrial process for recovery of benzene from coal tar was reported by Mansfield in 1849. Coal tar soon became the largest source of benzene. Soon afterward, benzene was discovered in coal gas and this initiated the recovery of coal gas light oil as a source of benzene.

Until the 1940s, light oil obtained from the destructive distillation of coal was the principal source of benzene. Except for part of the World War II period, the quantity of benzene produced by the coal carbonization industry was sufficient to supply the demand even when a large portion of benzene was used for gasoline blending.

After 1950, benzene in motor fuel was largely replaced by tetraethyllead but the demand for benzene in the chemical industry persisted and soon exceeded the total production by the coal carbonization industry. To meet this growing demand, methods for producing benzene directly from petroleum sources were developed.

Since the 1950s, benzene production from petroleum feedstocks has been very successful and accounts for ~95% of all benzene obtained. Less than 5% of commercial benzene is derived from coke oven light oil.

Benzene is the simplest and most important member of the aromatic hydrocarbons and should not be confused with benzine, a low boiling petroleum fraction composed chiefly of aliphatic hydrocarbons. The term benzole, which denotes commercial products that are largely benzene, is not common in the United States, but is still used in Europe.

2. Physical Properties

The physical and thermodynamic properties of benzene are shown in Table 1 (2). Azeotrope data for benzene with selected compounds are shown in Table 2 (3).

Table 1. **Physical and Thermodynamic Properties of Benzene**[a]

Property	Value
mol wt	78.115
freezing point, °C in air at 101.3 kPa[b]	5.530
boiling point, °C at 101.3 kPa[b]	80.094
density, g/cm^3	
20°C	0.8789
25°C	0.8736
vapor pressure, 25°C, kPa[c]	12.6
refractive index, n_D, 25°C	1.49792
surface tension, 25°C, mN/m (= dyn/cm)	28.20
viscosity, absolute, 25°C in mPa · s(= cP)	0.6010
critical temperature, °C	289.01
critical pressure, kPa[b]	4.898×10^3
critical volume, cm^3/mol	259.0
heat of formation	
g, kJ/mol	82.93
L, kJ/mol	49.08
heat of combustion, kJ/mol[d,e]	3.2676×10^3
heat of fusion, kJ/mol	9.866
heat of vaporization, 25°C, kJ/mol	33.899
solubility in H_2O, 25°C, g/100 g H_2O	0.180

[a] Ref.2. Courtesy of the Thermodynamics Research Center, The Texas A&M
University System.
[b] To convert kPa to atm, divide by 101.3.
[c] To convert kPa to mmHg, multiply by 7.5.
[d] To convert kJ to kcal, divide by 4.184.
[e] At 298.15 K and constant pressure to CO_2 and H_2O.

Benzene forms minimum-boiling azeotropes with many alcohols and hydrocarbons. Benzene also forms ternary azeotropes.

2.1. Structure. The representation of the benzene molecule has evolved from the Kekulé ring formula (1) to the more electronically accurate (2), which indicates all carbon–carbon bonds are identical.

Table 2. **Azeotropes of Benzene**[a]

Component	CAS Registry Number	Bp, °C	Azeotrope Bp, °C	Wt % benzene
cyclohexane	[110-82-7]	80.75	77.56	51.9
cyclohexene	[110-83-8]	82.1	78.9	64.7
methylcyclopentane	[96-37-7]	71.8	71.5	9.4
n-heptane	[142-82-5]	98.4	80.1	99.3
2,2-dimethylpentane	[590-35-2]	79.1	75.85	46.3
2,2,4-trimethylpentane	[540-84-1]	99.2	80.1	97.7
methanol	[67-56-1]	64.72	57.50	60.9
ethanol	[64-17-5]	78.3	68.24	67.6
2-propanol	[67-63-0]	82.45	71.92	66.7
2-butanol	[78-92-2]	99.5	78.5	84.6
tert-butyl alcohol	[75-65-0]	82.9	73.95	63.4
water	[7732-18-5]	100	69.25	91.17

[a] Ref.3.

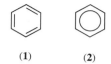

(1)   (2)

The bond angles and distances in benzene are known accurately from X-ray diffraction studies. The six carbon atoms form a regular hexagon in which each carbon atom is 0.139 nm from each of the two adjacent carbon atoms. The carbon–carbon bond lengths in benzene are intermediate in length between single and double carbon–carbon bonds. Each hydrogen atom is 0.108 nm from the carbon atom to which it is bonded. All 12 atoms lie in a single plane. All bond angles in benzene are exactly 120°.

Resonance Stabilization. Benzene has great thermal stability. It has a lower heat of formation from the elements than the corresponding structure (1) possessing three fixed, ethylene-type double bonds. Similarly, when benzene is decomposed into carbon and hydrogen, it absorbs more energy than is predicted by the Kekulé formula. The hydrogenation of benzene is exothermic by ~208 kJ/mol (49.8 kcal/mol), ~151 kJ (36.0 kcal) less than three times the value for cyclohexene. This difference between the energy taken up during the formation of three double bonds and that obtained experimentally for benzene formation is termed the resonance energy for benzene (4,5).

3. Chemical Properties

Benzene undergoes substitution, addition, and cleavage of the ring; substitution reactions are the most important for industrial applications.

3.1. Electrophilic Aromatic Substitution. Benzene undergoes substitution of one or more of its hydrogen atoms by various groups such as halogen, nitro, sulfonic acid, or alkyl. Reactions with chlorine, bromine, or nitric acid are termed electrophilic aromatic substitution because they involve attack of electron-seeking reagents on the delocalized π-electrons of the aromatic ring. Similarly, benzene derivatives substituted with electronegative or electron-with drawing groups undergo nucleophilic substitution reactions with electron-donating reagents. Benzene yields only one monosubstitution product and three possible disubstitution products, classified as ortho, meta, or para.

ortho   meta   para

Table 3 shows the number of structural isomers possible when one, two, three, or four substituents, X, Y, and Z, replace the hydrogens of benzene.

Orientation in Electrophilic Aromatic Substitution. A substituent group that increases the rate of electrophilic substitution relative to benzene itself is

Table 3. **Number of Structural Isomers of the Substitution Products of Benzene**

Substituents	Number of isomers
X	1
X, X	3
X, Y	3
X, X, X	3
X, X, Y	6
X, Y, Z	10
X, X, X, X	3
X, X, X, Y	6
X, X, Y, Y	11
X, X, Y, Z	16

called an activating group. Activating groups often have unshared electron pairs on atoms directly attached to the benzene ring and are characterized by their ability to contribute electron density to the π-orbitals of the aromatic ring, thus stabilizing the electrophile's influence on the ring (5,6). An example of a resonance effect is shown.

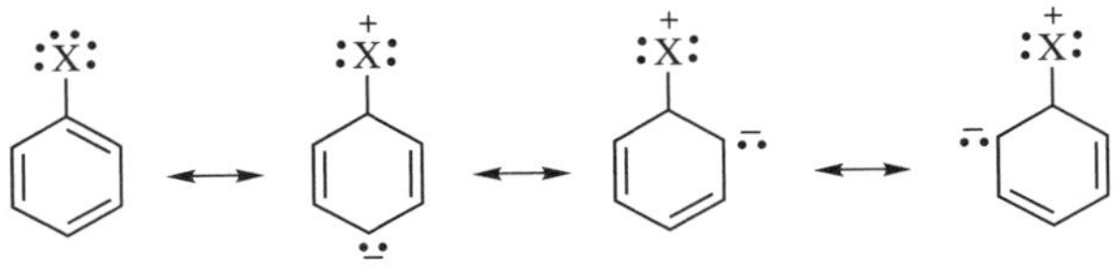

Resonance effects are the primary influence on orientation and reactivity in electrophilic substitution. The common activating groups in electrophilic aromatic substitution, in approximate order of decreasing effectiveness, are $-NR_2$, $-NHR$, $-NH_2$, $-OH$, $-OR$, $-NO$, $-NHCOR$, $-OCOR$, alkyls, $-F$, $-Cl$, $-Br$, $-I$, aryls, $-CH_2COOH$, and $=CH=CH-COOH$. Activating groups are ortho- and para-directing. Mixtures of ortho- and para-isomers are frequently produced; the exact proportions are usually a function of steric effects and reaction conditions.

Deactivating groups decrease the rate at which electrophilic aromatic substitution occurs. They lack an unshared electron pair on the atom directly connected to the aromatic ring and frequently are attached to an electronegative atom by double or triple bonds. The typical deactivating groups, in approximate order of decreasing effectiveness, are $-^+NR_3$, $-NO_2$, $-CN$, $-SO_3H$, $-CHO$, $-COOR$, $-COOH$, $-CONH_2$, and $-CCl_3$. Deactivating groups withdraw electron density from the π-electron cloud making the π-electrons less available for electrophilic reagents. It necessarily follows that, because of resonance effects, deactivating groups direct electrophilic substitution almost exclusively to the meta-position.

The entrance of a third or fourth substituent can be predicted by Beilstein's rule. If a substituent Z-enters into a compound C_6H_4XY, both X and Y exert an influence, but the group with the predominant influence directs Z- to the position it will occupy. Since all meta-directing groups are deactivating, it follows that ortho—para activating groups predominate when one of them is present on the benzene ring.

3.2. Nucleophilic Substitutions of Benzene Derivatives.
Benzene itself does not normally react with nucleophiles such as halide ions, cyanide, hydroxide, or alkoxides (7). However, aromatic rings containing one or more electron-withdrawing groups, usually halogen, react with nucleophiles to give substitution products. An example of this type of reaction is the industrial conversion of chlorobenzene to phenol with sodium hydroxide at 400°C (8).

In nucleophilic aromatic substitutions, required reaction conditions become milder as the number of electron-withdrawing groups on the ring is increased. For example, the conversion of p-nitrochlorobenzene to p-nitrophenol occurs with sodium hydroxide solution at ~160°C. The conversion of 2,4-dinitrochlorobenzene to 2,4-dinitrophenol occurs with sodium carbonate solution at about 130°C. Picric acid (2,4,6-trinitrophenol) is obtained from the chloride by brief warming with water (9). The reaction occurs preferentially at ortho- and para-positions to electron-withdrawing substituents. In contrast to electrophilic aromatic substitution, electron-withdrawing groups such as NO_2 and CN are activating and ortho—para directing in nucleophilic aromatic substitution.

Nucleophilic aromatic substitutions involving loss of hydrogen are known. The reaction usually occurs with oxidation of the intermediate either intramolecularly or by an added oxidizing agent such as air or iodine. A noteworthy example is the formation of 6-methoxy-2-nitrobenzonitrile from reaction of 1,3-dinitrobenzene with a methanol solution of potassium cyanide. In this reaction, it appears that the nitro compound itself functions as the oxidizing agent (10).

3.3. Oxidation.
Benzene can be oxidized to a number of different products. Strong oxidizing agents such as permanganate or dichromate oxidize benzene to carbon dioxide and water under rigorous conditions. Benzene can be selectively oxidized in the vapor phase to maleic anhydride. The reaction occurs in the presence of air with a promoted vanadium pentoxide catalyst (11). Prior to 1986, this process provided most of the world's maleic anhydride [108-31-6], $C_4H_2O_3$. Currently maleic anhydride is manufactured from the air oxidation of n-butane also employing a vanadium pentoxide catalyst. Benzoquinone [106-51-4], $C_6H_4O_2$ (quinone) has been reported as a by-product of benzene oxidation at 410–430°C. Benzene can be oxidized to phenols with hydrogen peroxide and reducing agents such as Fe(II) and Ti(II). Frequently, ferrous sulfate and hydrogen peroxide are used (Fenton's reagent), but yields are generally low (12) and the procedure is of limited utility. Benzene has also been oxidized in the vapor phase to phenol in low yield at 450–800°C in air without a catalyst (13).

3.4. Reduction.
Benzene can be reduced to cyclohexane [110-82-7], C_6H_{12}, or cycloolefins. At room temperature and ordinary pressure, benzene, either alone or in hydrocarbon solvents, is quantitatively reduced to cyclohexane with hydrogen and nickel or cobalt (14) catalysts. Catalytic vapor-phase hydrogenation of benzene is readily accomplished at ~200°C with nickel catalysts. Nickel or platinum catalysts are deactivated by the presence of sulfur-containing impurities in the benzene and these metals should only be used with thiophene-free benzene. Catalysts less active and less sensitive to sulfur, such as molybdenum oxide or sulfide, can be used when benzene is contaminated with sulfur-containing impurities. Benzene is reduced to 1,4-cyclohexadiene

[628-41-1], C_6H_8, with alkali metals in liquid ammonia solution in the presence of alcohols (15).

3.5. Halogenation. Depending on the conditions either substitution or addition products can be obtained by the halogenation of benzene. Chlorine or bromine react with benzene in the presence of carriers, such as ferric halides, aluminum halides, or transition metal halides, to give substitution products such as chlorobenzene or bromobenzene [108-86-1], C_6H_5Br; occasionally para-disubstitution products are formed. Chlorobenzene [108-90-7], C_6H_5Cl, is produced commercially in the liquid phase by passing chlorine gas into benzene in the presence of molybdenum chloride at 30–50°C and atmospheric pressure. This continuous process yields a 14:1 ratio of chlorobenzene to p-dichlorobenzene [106-46-7], $C_6H_4Cl_2$. The reaction of iodine with benzene takes place only in the presence of oxidizing agents such as nitric acid. Iodobenzene [591-50-4], C_6H_5I, is thus produced from reaction of benzene, iodine, and excess nitric acid at 50°C (16). Benzene is fluorinated by direct liquid-phase reaction with fluorine in acetonitrile solution at −35°C. The reaction gives predominantly fluoroben-zene with small amounts of o-, m-, and p-difluorobenzene by-products (17). Direct fluorination of benzene with fluorine has not yet gained commercial importance. Fluorobenzene [462-06-6], C_6H_5F, is most commonly prepared from thermal decomposition of dry benzenediazonium tetrafluoroborate [446-46-8] (18).

Chlorine and bromine add to benzene in the absence of oxygen and presence of light to yield hexachloro- [27154-44-5] and hexabromocyclohexane [30105-41-0], $C_6H_6Br_6$. Technical benzene hexachloride is produced by either batch or continuous methods at 15–25°C in glass reactors. Five stereoisomers are produced in the reaction and these are separated by fractional crystallization. The gamma isomer (BHC), which composes 12–14% of the reaction product, was formerly used as an insecticide. Benzene hexachloride [608-73-1], $C_6H_6Cl_6$, is converted into hexachlorobenzene [118-74-1], C_6Cl_6, upon reaction with ferric chloride in chlorobenzene solution.

3.6. Nitration. The nitration of benzene to nitrobenzene [98-95-3], $C_6H_5NO_2$, occurs in yields often >95% when a mixture of concentrated nitric and sulfuric acids is used at 50–55°C. Because the meta-directing nitro group is deactivating, the extent of nitration is rather easily controlled. To produce m-dinitrobenzene, more vigorous conditions are required, e.g., nitric and sulfuric acids at 100°C. 1,3,5-Trinitrobenzene [99-35-4], $C_6H_3N_3O_6$, is obtained from benzene with a large excess of fuming sulfuric and nitric acids at higher temperatures (19). When benzene reacts with mercuric nitrate and concentrated nitric acid, oxynitration occurs with the formation of either 2,4-dinitrophenol [51-28-5], $C_6H_4N_2O_5$, or 2,4,6-trinitrophenol [88-89-1], $C_6H_3N_3O_7$, depending on the reaction conditions (20).

3.7. Sulfonation. Benzene is converted into benzenesulfonic acid [98-11-3], $C_6H_6SO_3$, upon reaction with fuming sulfuric acid (oleum) or chlorosulfonic acid. m-Benzenedisulfonic acid [98-48-6], $C_6H_6S_2O_6$, is prepared by reaction of benzene-sulfonic acid with oleum for 8 h at 85°C. Often under these conditions, appreciable quantities of p-benzenedisulfonic acid [31375-02-7] are produced. 1,3,5-Benzenetrisulfonic acid [617-99-2], $C_6H_6S_3O_9$, is produced by heating the disulfonic acid with oleum at 230°C (21).

3.8. Alkylation.   Friedel-Crafts alkylation (qv) of benzene with ethylene or propylene to produce ethylbenzene [100-41-4], C_8H_{10}, or isopropylbenzene [98-82-8], C_9H_{12} (cumene) is readily accomplished in the liquid or vapor phase with various catalysts such as BF_3 (22), aluminum chloride, or supported polyphosphoric acid. The oldest method of alkylation employs the liquid-phase reaction of benzene with anhydrous aluminum chloride and ethylene (23). Ethylbenzene is produced commercially almost entirely for styrene manufacture. Cumene [98-82-8] is catalytically oxidized to cumene hydroperoxide, which is used to manufacture phenol and acetone. Benzene is also alkylated with C_{10}–C_{20} linear alkenes to produce linear alkyl aromatics. Sulfonation of these compounds produces linear alkane sulfonates (LAS) that are used as biodegradable detergents.

In recent years, alkylations have been accomplished with acidic zeolite catalysts, most nobably ZSM-5. A ZSM-5 ethylbenzene process was commercialized jointly by Mobil Co. and Badger America in 1976 (24). The vapor-phase reaction occurs at temperatures $>370°C$ over a fixed bed of catalyst at 1.4–2.8 MPa (200–400 psi) with high ethylene space velocities. A typical molar ethylene to benzene ratio is ~1–1.2. The conversion to ethylbenzene is quantitative. The principal advantages of zeolite-based routes are easy recovery of products, elimination of corrosive or environmentally unacceptable by-products, high product yields and selectivities, and high process heat recovery (25,26).

ABB Lummus Crest Inc. and Unocal Corp. have licensed a benzene alkylation process using a proprietary zeolite catalyst. Unlike the Mobil-Badger process, the Unocal-Lummus process is suitable for either ethylbenzene or cumene manufacture (27,28).

3.9. Other Reactions.   Benzene undergoes a number of other useful reactions.

Chloromethylation (Blanc–Quelet Reaction).   Benzene reacts with formaldehyde and hydrochloric acid in the presence of zinc chloride to yield chloromethylbenzene [100-44-7], C_7H_7Cl (benzyl chloride) (29), a chemical intermediate.

Friedel-Crafts Acylation.   The Friedel-Crafts acylation procedure is the most important method for preparing aromatic ketones and their derivatives. Acetyl chloride (acetic anhydride) reacts with benzene in the presence of aluminum chloride or acid catalysts to produce acetophenone [98-86-2], C_8H_8O (1-phenylethanone). Benzene can also be condensed with dicarboxylic acid anhydrides to yield benzoyl derivatives of carboxylic acids. These benzoyl derivatives are often used for constructing polycyclic molecules (Haworth reaction). For example, benzene reacts with succinic anhydride in the presence of aluminum chloride to produce β-benzoylpropionic acid [2051-95-8], which is converted into α-tetralone [529-34-0] (30).

Mercuration–Thallation.   Mercuric acetate and thallium trifluoroacetate react with benzene to yield phenylmercuric acetate [62-38-4] or phenylthallic trifluoroacetate. The arylthallium compounds can be converted into phenols, nitriles, or aryl iodides (31).

Metalation.   Benzene reacts with alkali metal derivatives such as methyl or ethyllithium in hydrocarbon solvents to produce phenyllithium [591-51-5], C_6H_5Li, and methane or ethane. Chloro-, bromo-, or iodobenzene will react with magnesium metal in ethereal solvents to produce phenylmagnesium chloride [100-59-4], C_6H_5MgCl, bromide, or iodide (Grignard reagents) (32).

Pyrolysis.   Benzene undergoes thermal dehydrocondensation at high temperatures to produce small amounts of biphenyls and terphenyls (see BIPHENYL AND TERPHENYLS). Before the 1970s most commercial biphenyl was produced from benzene pyrolysis. In a typical procedure, benzene vapors are passed through a reactor, usually at temperatures > 650°C. The decomposition of benzene into carbon and hydrogen is a competing reaction at temperatures of ∼ 750°C. Biphenyls are also formed when benzene and ethylene are heated to 130–160°C in the presence of alkali metals on activated Al_2O_3 (33).

4. Manufacture

Benzene is a natural component of petroleum, but the amount of benzene present in most crude oils is small, often <1.0% by weight (34). Therefore the recovery of benzene from crude oil is uneconomical and was not attempted on a commercial scale until 1941. To add further complications, benzene cannot be separated from crude oil by simple distillation because of azeotrope formation with various other hydrocarbons. Recovery is more economical if the petroleum fraction is subjected to a thermal or catalytic process that increases the concentration of benzene.

After 1950, the demand for benzene exceeded the output by the coal carbonization industry and to supply the increasing demand, processes were developed for producing and separating benzene directly from petroleum feedstocks. The production of benzene from petroleum increased rapidly thereafter, and by the early 1960s the amount of benzene derived from petroleum was several times greater than that derived from coal. By the late 1970s coal-derived benzene accounted for <10% of total benzene produced. Although coke oven light oil often contains useful quantities of benzene, it is expected to further decrease as a source of aromatics as the number of steel companies that produce metallurgical coke from coal decreases.

Petroleum-derived benzene is commercially produced by reforming and separation, thermal or catalytic dealkylation of toluene, and disproportionation. Benzene is also obtained from pyrolysis gasoline formed in the steam cracking of olefins (35).

4.1. Catalytic Reforming.   Worldwide, ∼30% of commercial benzene is produced by catalytic reforming, a process in which aromatic molecules are produced from the dehydrogenation of cycloparaffins, dehydroisomerization of alkyl cyclopentanes, and the cyclization and subsequent dehydrogenation of paraffins (36). The feed to the catalytic reformer may be a straight-run, hydrocracked, or thermally cracked naphtha fraction in the C_6 to 200°C range. If benzene is the main product desired, a narrow naphtha cut of 71–104°C is fed to the reformer. The reforming catalyst most frequently consists of platinum–rhenium on a high surface area alumina support. The reformer operating conditions and type of feedstock largely determine the amount of benzene that can be produced. The benzene product is most often recovered from the reformate by solvent extraction techniques.

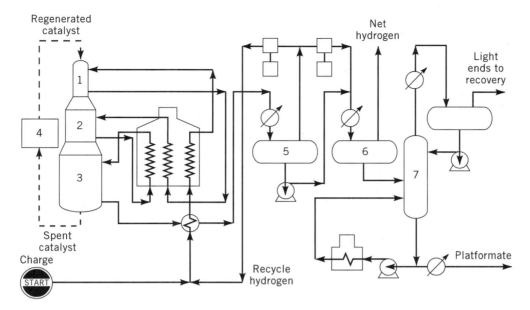

Fig. 1. Universal Oil Products Platforming process. Courtesy Gulf Publishing Co., Houston, Tex. (41).

Several significant reforming processes are in use (37,38). These include Powerforming, Ultraforming Co., Rheniforming (Chevron), Catalytic Reforming, and Platforming. Several other reforming processes are in use but these are limited to a few installations and are not of general interest (39). Platforming is a noteworthy example of current commercial reforming technology.

Platforming. The feedstock is usually a straight-run, thermally cracked, catalytically cracked, or hydrocracked C_6 to 200°C naphtha (40,41) (Fig. 1). The feed is first hydrotreated to remove sulfur, nitrogen, or oxygen compounds that would foul the catalyst, and also to remove olefins present in cracked naphthas. The hydrotreated feed is then mixed with recycled hydrogen and preheated to 495–525°C at pressures of 0.8–5 MPa (116–725 psi) (42). Typical hydrogen charge ratios of 4000 to 8000 standard cubic feet per barrel (scf/bbl) of feed are necessary (40). This is approximately equivalent to 700–1400 m³ hydrogen per m³ of feed.

The feed is then passed through a stacked series of reactors. Usually three or four reactors are used (43,44). All of the reforming catalysts in general use contain platinum chloride or rhenium chloride supported on silica or silica–alumina. The catalyst pellets are generally supported on a bed of ceramic spheres ∼30–40 cm deep. The spheres vary in size from ∼2.5 cm in diameter on the bottom to ∼9 mm in diameter at the top. Two types of Platforming processes are currently in use. The Semiregenerative Platforming process uses three or four reactors in series in which catalyst activity is regenerated at 6–12 month intervals. In the Continuous Platforming process catalyst activity is maintained by continuously withdrawing a small portion of catalyst and passing it through a regeneration tower where coke, a natural by-product of the reforming process, is burned off (45).

The product coming out of the reactor consists of excess hydrogen and a reformate rich in aromatics. Typically the dehydrogenation of naphthenes approaches 100%. From 0 to 70% of the paraffins are dehydrocyclized. The liquid product from the separator goes to a stabilizer where light hydrocarbons are removed and sent to a debutanizer. The debutanized platformate is then sent to a splitter where C_8 and C_9 aromatics are removed. The platformate splitter overhead, consisting of benzene, toluene, and nonaromatics, is then solvent extracted (46).

Aromatics Extraction.   Even when rigorous reforming conditions are employed, the platformate splitter overhead usually contains significant amounts of nonaromatics that must be removed to provide an acceptable commercial benzene product. Numerous solvents are available for extraction of aromatics from an aromatic–aliphatic mixture. These include diethylene glycol (Udex process), *N*-methylpyrrolidinone (Arosolvan process), *N,N*-dimethylformamide (REDEX process), liquid SO_2 extraction (Edelanu process), tetramethylene sulfone [126-33-0] (Sulfolane process), and tetraethylene glycol (Tetra process, Union Carbide). The Udex process was the first solvent extraction process to find widespread usage prior to 1963. Since then, the Sulfolane process has become the most popular. This method, developed by Shell and licensed through UOP, was first reported in 1959 (47). A diagram of the Sulfolane extraction process is shown in Figure 2 (48). Feed is charged to a contactor for the countercurrent extraction of the aromatic components. Solvent from the extractor is charged to an extractive stripper. The stripper vapors are condensed and collected in a separator from which hydrocarbons are returned to the extractor. The stripper bottoms are charged to a recovery column that produces solvent-free aromatics. The aromatics are then fractionated to recover pure benzene.

4.2. Toluene Hydrodealkylation.   Benzene is produced from the hydrodemethylation of toluene under catalytic or thermal conditions. The main

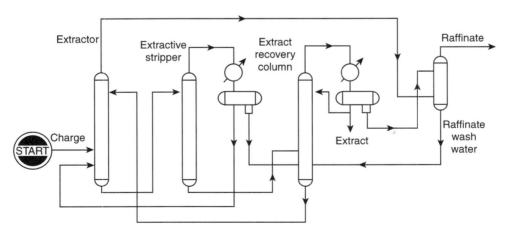

Fig. 2.   Shell-UOP Sulfolane extraction process.Courtesy Gulf Publishing Co., Houston, Tex. (48).

catalytic hydrodealkylation processes are Hydeal and DETOL (49). Two widely used thermal processes are HDA and THD. These processes contribute 25–30% of the world's total benzene supply.

In catalytic toluene hydrodealkylation, toluene is mixed with a hydrogen stream and passed through a vessel packed with a catalyst, usually supported chromium or molybdenum oxides, platinum or platinum oxides, on silica or alumina (50). The operating temperatures range from 500 to 595°C and pressures are usually 4–6 MPa (40–60 atm). The reaction is highly exothermic and the temperature is controlled by injection of quench hydrogen at several places along the reaction. Conversions per pass typically reach 90% and selectivity to benzene is often >95%. The catalytic process occurs at lower temperatures and offers higher selectivities but requires frequent regeneration of the catalyst. Products leaving the reactor pass through a separator where unreacted hydrogen is removed and recycled to the feed. Further fractionation separates methane from the benzene product.

A typical catalytic hydrodealkylation scheme is shown in Figure 3 (49). The most common feedstock is toluene, but xylenes can also be used. Recent studies have demonstrated that C_9 and heavier monoaromatics produce benzene in a conventional hydrodealkylation unit in yields comparable to that of toluene (51). The use of feeds containing up to 100% of C_9–C_{11} aromatics increases the flexibility of the hydrodealkylation procedure that is sensitive to the price differential of benzene and toluene. When toluene is in demand, benzene supplies can be maintained from dealkylation of heavy feedstocks.

4.3. Transalkylation. Two molecules of toluene are converted into one molecule of benzene and one molecule of mixed-xylene isomers in a sequence called transalkylation or disproportionation. Economic feasibility of the process strongly depends on the relative prices of benzene, toluene, and xylene. Operation of a transalkylation unit is practical only when there is an excess of toluene and a strong demand for benzene. In recent years, xylene and benzene prices have generally been higher than toluene prices so transalkylation is presently an attractive alternative to hydrodealkylation (see also BTX PROCESSING).

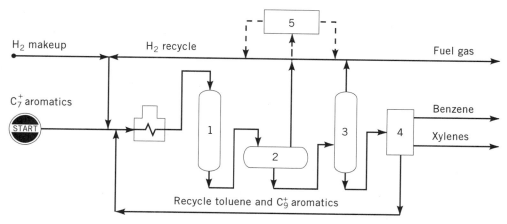

Fig. 3. DETOL hydrodealkylation process. Courtesy Gulf Publishing Co., Houston, Tex. (49).

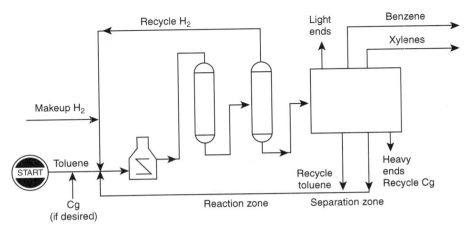

Fig. 4. Tatoray toluene disproportionation process. Courtesy Gulf Publishing Co., Houston, Tex. (52).

An example of toluene disproportionation, the Tatoray process, is shown in Figure 4(52). Toluene and C_9 aromatics are mixed with liquid recycle and recycle hydrogen, heated to 350–530°C at 1–5 MPa (10–50 atm), and charged to a reactor containing a fixed bed of noble metal or rare earth catalyst. Hydrogen to feedstock mole ratios of 5:1 to 12:1 are typically required. Following removal of gases, the separator liquid is freed of light ends and the bottoms are then clay treated and fractionated to produce high purity benzene and xylenes. The yield of benzene and xylene obtained from this procedure is ∼92% of the theoretical. Since the disproportionation is an equilibrium reaction, by varying the feedstock and experimental conditions the ratio of xylene to benzene can be changed.

Lyondell and Sun Oil Co. are the main producers of benzene by disproportionation. Fina Oil Co. of Texas has developed the Fina T2BX process for toluene disproportionation using a proprietary catalyst. The new catalyst is claimed to reduce hydrogen consumption and is suitable for feeds containing small amounts of moisture (53). A process for reacting impure toluene to obtain benzene, toluene, and a para-rich xylene stream that is substantially free of close-boiling nonaromatics by combining aromatization and selective disproportionation has been patented (54).

4.4. Pyrolysis Gasoline. The steam cracking of heavy naphthas or light hydrocarbons such as propane or butane to produce ethylene yields a liquid by-product rich in aromatic content called pyrolysis gasoline, dripolene, or drip oil (55). A typical pyrolysis gasoline contains up to ∼65% aromatics, ∼50% of which is benzene. Approximately 30–35% of benzene produced worldwide is derived from pyrolysis gasoline. The remainder of the product is composed of mono- and diolefins. These olefinic substances are removed by a mild hydrogenation step. Following hydrogenation, the resulting pyrolysis gasoline is used in motor gasoline. Alternatively, pure benzene could be recovered from the pyrolysis gasoline by solvent extraction and subsequent distillation.

4.5. Miscellaneous Sources of Benzene. Benzene has been recovered from coal tar. The lowest boiling fraction is extracted with caustic soda to remove tar acids. The base washed oil is then distilled and further purified by hydrodealkylation.

The synthesis of aromatics from methane, and other light C_2, C_3, and C_4 hydrocarbons has been the subject of investigation since the 1980s. One recent example of this is British Petroleum's Cyclar process, modified with UOPs continuous catalyst regeneration technology (56). The process is claimed to produce high purity benzene, toluene, and xylenes from propane, isobutane, or n-butane feedstocks. The conversion of synthesis gas into aromatic hydrocarbons with metal zeolite catalysts such as ZSM-5 has also been reported (57). ZSM-5 has also been found to convert light paraffins, olefins, and naphthenes to aromatics and light gases (58). The name M2-Forming has been suggested to describe this recently discovered aromatization process. The production of aromatics and gasoline hydrocarbons from methanol has also received attention (59–61). The future may bring practical methods for converting methane into benzene (62,63). A process for generating benzene from reformed gasoline has been reported (64). A purification process for benzene that comprises two permeation stages has also been reported (65).

5. Handling and Shipping

Manufacturers of benzene are required by federal law to publish Material Safety Data Sheets (MSDS) that describe in detail the procedures for its safe handling. Benzene is classified as a flammable liquid and should be stored away from any potential source of ignition. Fire and explosion hazard data for benzene are shown (66).

flash point, closed cup, °C	−11
autoignition temperature, °C	560
flammable limits, vol% in air	1.4 (lower)–8.0 (upper)

Benzene is shipped by rail tank cars, trucks, barges, and tankers. Because of the flammability, toxicity, and volatility of benzene, transfers from one vessel to another are conducted in closed systems. Metal tanks and storage containers should be grounded during transfers. Smaller quantities of benzene are routinely shipped in steel or glass containers. Benzene should be handled only where adequate ventilation is provided; protective clothing and self-contained respirators are recommended. Labeling, packaging, and domestic or international transportation of benzene must comply with regulations described in the *Code of Federal Regulations* (CFR) Title 49. OSHA regulations are described in 29 CFR, Parts 1501, 1502, and 1503. New exposure limits for benzene are published in 29 CFR, Part 1910.1028.

6. Economic Aspects

U.S. producers of benzene and their approximate production capacities are shown in Table 4. These figures are inexact because of the size of the market and instability of benzene prices causes frequent changes in capacity. The data in Table 4 include benzene from naphtha reforming, toluene hydroalkylation, or toluene disproportionation, and extraction from pyrolysis gasoline fractions derived from ethylene plant cracked naphthas or heavier feedstocks.

Table 4. **U.S. Producers of Benzene and Annual Capacities**[a]

Producer	Capacity /10^6 gal[b]
Continental United States	
BP Amoco, Alliance, La.; Lima, Ohio; Texas City, Tex.	270
Chevron, Pascagoula, Miss.; Port Arthur, Tex.; Richmond, Calif.	240
Citgo, Corpus Christi, Tex.; Lake Charles, La.; Lemont, Ill.	170
Coastal, Corpus Christi, Tex.; Westville, N.J.	75
Dow Chemical, Freeport, Tex.; Plaquemine, La.	220
Equilon Enterprises, El Dorado, Kan.	15
Equistar Chemicals, Alvin, Tex.; Channelview, Tex.; Corpus Christi, Tex.	365
Exxon, Baton Rouge, La.; Baytown, Tex.	335
Fina Oil, Port Arthur, Tex.	95
Huntsman, Port Arthur, Tex.	60
Koch Petroleum, Corpus Christi, Tex.	250
Lyondell, Houston, Tex.	50
Marathon Ashland, Catlettsburg, Ky.; Texas City, Tex.	55
Mobile, Beaumont, Tex.; Chalmette, La.	270
Motiva, Delaware City, Del.	15
Nova, Bayport, Tex.	15
Shell Chemical, Deer Park, Tex.	390
Sunoco, Marcus Hook, Pa.; Sunoco, Marcus Hook, Pa.; Philadelphia, Pa.; Toledo Ohio	140
Ultramar Diamond Shamrock, Three Rivers, Tex.	20
Valero Energy, Houston, Tex.	5
Virgin Islands and Puerto Rico	
HOVENSA, St. Croix, V.I.	75
Phillips, Guayama, P.R.; Sweeny, Tex.	145
Total	*3275*

[a] From Ref. 68, as of Dec. 1999.
[b] To convert gal to t, divide by 300.

Estimated demand for 2002 is 2.78 billion gal (68).

Historical (1993–1998) prices were $1.55/gal, barge, spot, works, as a high and $0.82 as a low. In Dec. 1999 the price was $0.75/gal (68).

Benzene production by various foreign countries is shown in Table 5

7. Specifications, Standards, and Test Methods

Several different grades of benzene are commercially available. The most common grades are benzene 535, benzene 485 (nitration grade), benzene 545, and thiophene-free benzene. Specifications and the corresponding American Society

Table 5. **Benzene Production by Various Foreign Countries,**[a] 10^3 t[b]

Producers	Quantity
North America	
Petroleos Mexicanos	400
Western Europe	
Germany	
Deutsche Shell	510
Rühr Ol	380
Redistillationsgemeinschaft	240
Erdölchemie	220
OMV	120
PCK	80
Wintershall	75
DEA	65
Others	60
France	
Elf Atochem	610
Gexaro	200
AAF	110
Total	140
Shell Chimie	95
Italy	
EniChem	700
Netherlands	
Dow Europe	900
Exxon Chemical	480
DSM	300
United Kingdom	
ICI	573
Shell U.K.	240
BP Chemicals	217
Conoco	200
Belgium	
Finaneste	170
Portugal	
Petrogal	60
Spain	
Cepsa	304
Eastern Europe	5,007
Asia	
Japan	
Mitsubishi Chemical	843
Idemitsu	804
Nippon Steel	367
Taiyo Oil	330
Maruzen Petrochemical	350
Sanyo Petrochemical	340
Mitsubishi Oil	345
Koa Oil	204
Mitsui Chemicals	161
Showa Shell Sekiyu	160
Osaka Petrochemical	130
Sumitomo Chemical	122
Tosoh	120

Table 5   (*Continued*)

Producers	Quantity
Kawasaki Steel	116
Nikko Petrochemicals	100
Nippon Petrochemical	90
Others	600
South Korea	
SK	490
Samsung	425
Daelim	190
Hyundai Petrochemical	100
Honam Oil	100
Others	420
Singapore	
Mobil Chemical	250
PCS	220
Taiwan	
Chinese Petroleum	493
China Steel	30
Indonesia	
Pertamina	115
Middle East	
Saudi Arabia	
Sabic	290
Iran	405
South America	
Brazil	
Copene	430
Copesul	263
Petroquimica União	198
Petrobras	38
Others	30
Argentina	
YPF	120
PASA	80
Venezuela	
Pequiven	70
Colombia	
Ecopetrol	65

[a] From Ref. 69. *Sources*: De Witt & Co. (Houston); SRI Consulting (Menlo Park, Calif.); CW estimates.
[b] To convert t to gal, multiply by 300.

for Testing and Material (ASTM) test procedures for these various types are shown in Table 6 (70). American Chemical Society (ACS) specifications for reagent grade benzene are shown in Table 7 (71). Industrial-grade benzene is used primarily in applications that are insensitive to impurities. Nitration-grade benzene is a high quality product used for preparing nitrobenzene and derivatives. Thiophene-free benzene is used as a reagent in ASTM standards and is specially treated to remove thiophene. Thiophene and organic sulfur compounds foul many catalysts used in reactions of benzene.

Table 6. **Specifications for Commercial Grades of Benzene**[a]

ASTM test	Benzene 535[b]	Benzene 485[c]	Industrial-grade[d]
appearance	clear liquid, free from sediment or haze at 18–24°C	clear liquid, free from sediment or haze at 18–24°C	clear liquid, free from sediment at 18–24°C
relative density, 14.56–15.56°C, D3505	0.8820–0.8860	0.8820–0.8860	0.875–0.886
density, 20°C, g/cm³, D4052	0.8780–0.8820	0.8780–0.8820	0.871–0.882
color pt-co scale, D1209	20 max	20 max	20 max
total distillation range, 101.3 kPa,[e] D850	1.0°C max, including the temperature of 80.1°C	1.0°C max, including the temperature of 80.1°C	2.0°C max, including the temperature of 80.1°C
solidification point, D852	5.35°C (anhydrous)	not lower than 4.85°C (anhydrous)	
acid wash color, D848	1 max	2 max	3 max
acidity, D847	none detected	none detected	none detected
H₂S and SO₂, D853	none detected	none detected	none detected
thiophene, D1685	1 mg/kg max		
copper corrosion, D849	pass	pass	copper strip shall-not show iridescence, a gray or black deposit, or discoloration
nonaromatics, D2360	0.15 wt% max		

[a] Ref. 70.
[b] ASTM D2359-85a.
[c] Nitration-grade, ASTM D835-85.
[d] ASTM D836-84.
[e] To convert kPa to mmHg, multiply by 7.5.

The purity of benzene marketed for most laboratory purposes is usually >95.5% with the principal impurities being toluene and other hydrocarbons with boiling points similar to that of benzene. Methods used to assess the quality of benzene include determination of density, boiling point, distillation characteristics, and specific gravity. Benzene of high purity samples is conveniently

Table 7. **Specifications for ACS Reagent-Grade Benzene**[a]

Specification	Value
color (APHA)	not more than 10
boiling range	entirely within 1.0°C range including 80.1 ± 0.1°C
freezing point	not below 5.2°C
residue after evaporation	not more than 0.001%
substances darkened by sulfuric acid	to pass test
thiophene	to pass test (limit ≈1 ppm)
sulfur compounds (as S)	not more than 0.005%
water	not more than 0.05%

[a] Ref. 71.

measured by freezing point, as outlined in ASTM D1016. The acid wash test consists of shaking a mixture of 96% sulfuric acid with benzene and comparing the color of the (lower) acid layer with a set of color standards. Other qualitative tests include those for SO_2 and H_2S determination. The copper strip corrosion test indicates the presence of acidic or corrosive sulfur impurities. The test for thiophene is colorimetric.

7.1. Analysis. The infrared (ir), ultraviolet (uv), and nuclear magnetic resonance (nmr) spectra are distinct and characteristic for benzene and are widely used in analysis (72–74). Benzene also produces diagnostic ions in the mass spectrum (75,76) (see ANALYTICAL METHODS).

The identification of benzene is most easily carried out by gas chromatography (77). Gas chromatographic analysis of benzene is the method of choice for determining benzene concentrations in many diverse media such as petroleum products or reformate, water, soil, air, or blood. Benzene in air can be measured by injection of a sample obtained from a syringe directly into a gas chromatograph (78).

In recent years, gas chromatograph–mass spectrometer (gc–ms) systems have become popular for analyzing trace amounts of benzene (79). The gc–ms method gives higher accuracy and precision than conventional gc methods because components are identified by molecular weight, even when benzene may overlap with other compounds. With multichanneled, double-focusing instruments, detection limits of 0.1 ppb in air or breath samples are claimed by a selective ion monitoring gc–ms procedure (sim–gc–ms) (80).

Rapid, simple, qualitative methods suitable for determining the presence of benzene in the workplace or surroundings have been utilized since the 1930s. Many early tests offered methods for detection of aromatics but were not specific for benzene. A straightforward test allowing selective detection of benzene involves nitration of a sample to *m*-dinitrobenzene and reaction of the resultant ether extract with an ethanolic solution of sodium hydroxide and methyl ethyl ketone (2-butanone), followed by the addition of acetic acid to eliminate interferences from toluene and xylenes. Benzene imparts a persistent red color to the solution (81). The method is claimed to be sensitive to concentrations as low as 0.27 ppm benzene from 10-mL air samples.

Benzene reacts with concentrated sulfuric acid and formaldehyde to produce a brown precipitate. A similar reaction occurs with ferrous sulfate and hydrogen peroxide. The resulting brown solid is dissolved in nitric acid for comparison with color standards.

Colorimetric methods have led to the development of visual devices for measurement of benzene concentration. These visual detection tubes have been popular since the 1960s and have provided a simple and reliable method for evaluating ambient aromatic vapor contamination. These products are available from a number of manufacturers such as Drager (Lubeck, Germany), Gastec (Tokyo, Japan), Kitagawa (Kawasaki, Japan), DuPont (Wilmington, Delaware, USA), and 3M (St. Paul, Minnesota, USA) (79).

Various types of detector tubes have been devised. The NIOSH standard number S-311 employs a tube filled with 420–840 μm (20/40 mesh) activated charcoal. A known volume of air is passed through the tube by either a handheld or vacuum pump. Carbon disulfide is used as the desorbing solvent and the solution is then analyzed by gc using a flame-ionization detector (82). Other adsorbents such as silica gel and desorbents such as acetone have been employed. Passive (diffuse samplers) have also been developed. Passive samplers are useful for determining the time-weighted average (TWA) concentration of benzene vapor (83). Passive dosimeters allow permeation or diffusion-controlled mass transport across a membrane or adsorbent bed, ie, activated charcoal. The activated charcoal is removed, extracted with solvent, and analyzed by gc. Passive dosimeters with instant readout capability have also been devised (79).

Determination of benzene in air samples has been achieved by bubbling contaminated air through various solvents, followed by uv or ir analysis of the solution (84). Methods for identifying benzene in soil, water, and biological media are further described in References 78 and 79.

8. Environmental Considerations

Benzene is classified as a hazardous waste by the Environmental Protection Agency (EPA) under subtitle C of the Resource and Recovery Act (RCRA) (85). Effective Sept. 25, 1990, solid wastes containing more than 0.5-mg/mL benzene must be treated in accordance with applicable RCRA regulations. Benzene is also subject to annual reporting of environmental releases as described in Section 313 of the Emergency and Community Right to Know Act of 1986 (86). Benzene emissions and effluent streams from petroleum refineries or benzene processing plants are also subject to strict federal regulations. Federal waste management procedures must be complied with for any industrial process involving manufacture, transport, treatment, or disposal of benzene. A complete description of the new EPA regulations concerning benzene and other hazardous wastes is found in the *Federal Register* (87). Further information regarding the handling and disposal of toxic or hazardous wastes is in the CFR, Vol. 40.

9. Health and Safety Factors

At room temperature and atmospheric pressure, benzene is sufficiently vaporized to pose an inhalation hazard. Benzene is a toxic substance that can produce both acute and chronic adverse health effects. It is generally recognized that prolonged or repeated exposure to benzene can result in serious damage to the blood-forming elements. The first indications of benzene toxicity during occupational exposure were reported in Sweden in 1897 (88). By the early 1900s, it was clear that humans chronically exposed to benzene suffered bone marrow damage.

Prolonged or repeated exposure to benzene vapor results in blood dyscrasias including lympho-, thrombo-, and pancytopenia, a decrease in all types of circulating blood cells (89–91). The decrease in blood components, caused by the action of benzene on bone marrow, is referred to as aplastic anemia and is the disease most commonly associated with benzene exposure (91). Cases of benzene poisoning resulting in hematological disturbances have been reported from repeated exposures to amounts as low as 60 ppm (92). In the early stages of chronic exposure the blood changes are variable, but as the disease becomes established a decrease in polymorphonuclear leucocytes and a relative lymphocy-tosis are found (93). If benzene exposure is stopped, the blood changes may or may not be reversed, and the blood morphology may require several years to return to normal (94).

A less frequent, but more serious, health complication resulting from chronic benzene exposure is the development of leukemia. The relationship between benzene and leukemia was suggested in the late 1920s. By the late 1930s, 10 cases of leukemia linked to benzene had been documented worldwide. A number of clinical case reports and several epidemiological studies followed and by the late 1970s benzene was clearly recognized as a carcinogen (leukemo-gen). Acute myelogenous leukemia (AML) is the most common form of the dis-ease associated with benzene exposure (88,95,96). It is believed that most, if not all, benzene-induced leukemias are preceded by pancytopenia or aplastic anemia (97) followed by a latency period of at least several years. Currently, the long-term prognosis for AML is poor and the outcome is usually fatal. Benzene has been linked to other, less common forms of leukemia, including lymphoid, myeloblastic, erythroblastic, and the hairy-cell varieties (98,99).

Neither the mechanism by which benzene damages bone marrow nor its role in the leukemia process are well understood. It is generally believed that the toxic factor(s) is a metabolite of benzene (100). Benzene is oxidized in the liver to phenol [108-95-2] as the primary metabolite with hydroquinone [123-31-9], catechol [120-80-9], muconic acid[505-70-4], and1,2,4-trihydroxyben-zene[533-73-3]as significant secondary metabolites (101). Although the identity of the actual toxic metabolite or combination of metabolites responsible for the hematological abnormalities is not known, evidence suggests that benzene oxide, hydroquinone, benzoquinone, or muconic acid derivatives are possibly the ultimate carcinogenic species (89,96,100−105).

Recently, the myelotoxicity has been proposed to occur through initial con-version of benzene to phenol and hydroquinone in the liver, selective accumula-tion of hydroquinone in the bone marrow, followed by conversion of hydroquinone to benzoquinone via bone marrow myeloperoxidase. Benzoquinone is then proposed to react with macromolecules disrupting cellular processes (101).

Benzene is rapidly absorbed from the lungs into the bloodstream. Studies on the inhalation of benzene have given a retention at rest of ~50% (100). The half-life for benzene disappearance in the body is ~0.4−1.6 h (106). Benzene accumulates in fatty tissues and continues to be excreted long after exposure. In one particular study, volunteers were exposed to 200 ppm benzene per hour over an 8-h workday for 5 days. After the fifth day of exposure, the volunteers exhaled twice as much benzene as on the first day of exposure (107). In another study, it was observed that 26% of absorbed benzene was exhaled unmetabolized and was excreted in the urine as 61% phenol, 6.4% catechol, and 2% hydroqui-none (108). Because benzene is oxidized mainly to phenol, the urinary phenol test is a widely used method for detecting benzene exposure. Urinary phenol content of nonoccupationally exposed subjects does not usually exceed 20 mg/L (109). Determining the ratio of inorganic to organic sulfate in urine is no longer recom-mended for evaluating benzene exposure because of its recognized low specificity to phenol (110).

Aplastic anemia and leukemia are not the only health effects ascribed to benzene exposure. A number of recent studies have associated benzene exposure with chromosomal changes (aberrations) (111). Other studies have shown abnormalities in porphyrin metabolism and decrease in leucocyte alkaline phosphatase activity in apparently healthy workers exposed to 10–20 ppm benzene (112,113). Increases in leukoagglutinins, as well as increases in blood fibrinolytic activity, have also been reported and are believed to be responsible for the persistent hemorrhages in chronic benzene poisoning (114,115).

Inhalation of 3000 ppm benzene can be tolerated for 0.5–1 h; 7500 ppm causes toxic effects in 0.5–1 h; and 20,000 ppm is fatal in 5–10 min (116). The lethal oral dose for an adult is ~15 mL (117). Repeated skin contact is reported to cause drying, defatting, dermatitis, and the risk of secondary infection if fissuring occurs.

In chronic benzene intoxication, mild poisoning produces headache, dizziness, nausea, stomach pain, anorexia, and hypothermia. In severe cases, pale skin, weakness, blurred vision, and dyspnea occur on exertion. Hemorrhagic tendencies include petechia, easy bruising, and bleeding gums. Bone marrow depression produces a decrease in circulating peripheral erythrocytes and leucocytes (94). Fatalities from chronic exposure show at autopsy severe bone marrow aplasia, and necrosis or fatty degeneration of the heart, liver, and adrenals (118).

Acute benzene poisoning results in central nervous system (CNS) depression and is characterized by an initial euphoria followed by staggered gait, stupor, coma, and convulsions. Exposure to ~4000 ppm benzene results in complete loss of consciousness. Insomnia, agitation, headache, nausea, and drowsiness may persist for weeks after exposure (119). Continued inhalation of benzene to the point of euphoria has caused irreversible encephalopathy with tremulousness, emotional lability, and diffuse cerebral atrophy (118). In deaths arising from acute exposure, respiratory tract infection, hypo- and hyperplasia of sternal bone marrow, congested kidneys, and cerebral edema have been found at autopsy.

Treatment for acute exposure to benzene vapor involves removing the subject from the affected area, followed by artificial respiration with oxygen; intubation and cardiac monitors may be necessary for severe acute exposures (118,120). Because of its low surface tension, benzene poses a significant aspiration hazard if the liquid enters the lungs. Emesis is indicated in alert patients if more than 1 mL of benzene per kg of body weight has been ingested and less than two hours have passed between ingestion and treatment (120).

Treatment for chronic benzene poisoning is supportive and symptomatic, with chemotherapy and bone marrow transplants as therapeutic agents for leukemia and aplastic anemia (120).

10. Regulations

Because of the potential hazards associated with benzene, exposure to benzene in the workplace has been heavily regulated in the United States. Benzene is considered one of the ~40 known human carcinogens. Benzene is listed as an ACGIH suspected human carcinogen, an NTP human carcinogen, and an

Table 8. National Exposure Limits for Benzene[a]

Country	Concentration		Status
Australia	TWA	5 ppm (16 mg/m^3)	carcinogen
Belgium	TWA	10 ppm (32 mg/m^3)	carcinogen
Czechoslovakia	TWA	10 mg/m^3; STEL 20 mg/m^3	
Denmark	TWA	5 ppm (16 mg/m^3)	skin, carcinogen
Finland	TWA	5 ppm (15 mg/m^3); STEL 10 ppm (30 mg/m^3)	skin
France	TWA	5 ppm (16 mg/m^3)	carcinogen
Germany (DFG MAK)	none		
Hungary	STEL	5 mg/m^3	skin, carcinogen
India	TWA	10 ppm (30 mg/m^3)	carcinogen
Ireland	TWA	5 ppm (16 mg/m^3)	carcinogen
Japan (JSOH)			carcinogen
The Netherlands	TWA	2.3 ppm (7.5 mg/m^3)	skin
The Philippines	TWA	25 ppm (80 mg/m^3)	skin
Poland	TWA	10 mg/m^3; STEL 40 mg/m^3	skin
Russia	TWA	10 ppm (5 mg/m^3); STEL 25 ppm (15 mg/m^3)	skin, carcinogen
Sweden	TWA	1 ppm (3 mg/m^3); STEL 5 ppm (16 mg/m^3),	skin, carcinogen
Switzerland	TWA	5 ppm (16 mg/m^3)	skin, carcinogen
Thailand	TWA	10 ppm (30 mg/m^3); STEL 25 ppm (75 mg/m^3)	skin
Turkey	TWA	20 ppm (64 mg/m^3)	skin
United Kingdom (HSE MEL)	TWA	5 ppm (16 mg/m^3)	

[a] Ref. 122.

IARC human carcinogen. Six foreign countries, including Germany, Italy, Japan, Sweden, and Switzerland, recognize benzene as a carcinogen (88). In the United States, the earliest limit on benzene exposure was recommended in 1927 at 100 ppm (121). Over the decades the upper allowable limits were reduced to 50, 35, then 25 ppm (88). Twenty countries have been reported to limit occupational exposure to benzene by regulation or recommended guideline. These occupational exposure limits are shown in Table 8 (122).

In 1971, the Occupational Safety and Health Administration (OSHA) standard for benzene (20 CFR, Part 1910.0000) adopted a permissible exposure limit (PEL) of 10 ppm benzene measured as an 8-h TWA. In October of 1976, NIOSH updated its earlier criteria document on benzene and recommended that OSHA lower the benzene exposure standard from 10 to 1 ppm. This proposed implementation was blocked by the U.S. Supreme Court in 1980 on the basis of insufficient evidence linking benzene to cancer deaths. OSHA permissible exposure limit (PEL) is 1 ppm (3 mg/m^3) with a short-term exposure limit (STEL) of 5 ppm (15 mg/m^3); the American Conference of Governmental Industrial Hygienists (ACGIH) TLV is 0.5 ppm (1.6 mg/m^3) and an STEL/ceiling level of 2.5 ppm (8 mg/m^3) with an A1 notation. The NIOSH REL is 0.1 ppm with a STEL of 1 ppm (123).

Further, this standard provides for methods of compliance, personal protective equipment, adequate communication of benzene hazards to employees, regulated areas, and medical surveillance of workers who are or may be exposed to benzene. Any employee routinely exposed to benzene should, in addition to wearing protective equipment, receive periodic blood tests.

11. Uses

In the early part of the twentieth century, benzene was used as a universal solvent and degreaser and found widespread use throughout the rubber industry in the manufacture of tires. By the late 1920s, following reports of deaths due to benzene exposure, it was largely replaced by toluene and aliphatic solvents (124).

Before World War II, the largest market for benzene was in gasoline blending to improve octane ratings. After 1950, benzene in gasoline was largely replaced with tetraethyllead. In recent years, with the recognition of the hazards of lead in the environment, the EPA has limited the amount of lead in gasoline to 0.1 g/leaded gallon (125). In addition, the EPA has suggested limits for benzene content of gasoline. The California Air Resources Board (CARB) has proposed limits on benzene in motor gasoline of 0.8 vol% after September, 1990. It is possible in the future that other aromatics in gasoline, especially xylenes, will face similar restrictions. Benzene content of U.S. motor gasoline currently ranges from ~0.1 − 4.4 vol% (126).

Benzene is still used extensively as a gasoline component in Europe and many countries do not limit the benzene content (127). Exceptions are Austria, Norway, Sweden, and Switzerland, which set the maximum at 5.0 vol% (127). Over 90% of European motor gasolines are below the 5.0 vol% limit set by these countries. It is likely that benzene content of European gasoline will be further reduced in the future.

Benzene is now used primarily as an intermediate in the manufacture of industrial chemicals. Approximately 95% of U.S. benzene is consumed by industry for the preparation of polymers, detergents, pesticides, pharmaceuticals, and allied products.

Estimates of benzene consumption for nonfuel uses are shown in Table 9 (68). Benzene consumption worldwide is dominated by the production of three main derivatives, styrene, cumene, and cyclohexane, which account for nearly 86% of the total.

Benzene is alkylated with ethylene to produce ethylbenzene, which is then dehydrogenated to styrene, the most important chemical intermediate derived

Table 9. **United States Use Pattern for Benzene, 1999**[a]

Use	Total consumption, %
ethylbenzene	50
cumene	24
cyclohexane	12
aniline	6
alkylbenzene	2
chlorobenzenes	1
other	5

[a] Ref.68.

from benzene. Styrene is a raw material for the production of polystyrene and styrene copolymers such as acrylonitrile–butadiene–styrene (ABS) and SAN. Ethylbenzene accounted for nearly 50% of benzene consumption in 1999.

Benzene is alkylated with propylene to yield cumene (qv). Cumene is catalytically oxidized in the presence of air to cumene hydroperoxide, which is decomposed into phenol and acetone (qv). Phenol is used to manufacture caprolactam (nylon) and phenolic resins such as bisphenol A. Approximately 24% of benzene produced in 1999 was used to manufacture cumene.

Benzene is hydrogenated to cyclohexane. Cyclohexane is then oxidized to cyclohexanol, cyclohexanone, or adipic acid (qv). Adipic acid is used to produce nylon. Cyclohexane manufacture was responsible for about 12% of benzene consumption in 1999.

Nitration of benzene yields nitrobenzene, which is reduced to aniline, an important intermediate for dyes and pharmaceuticals. Benzene is chlorinated to produce chlorobenzene, which finds use in the preparation of pesticides, solvents, and dyes.

A novel herbicidal and defoliant composition containing substitution benzene compounds has been patented. Control of undesired plants is obtained by low concentrations of these compounds (128).

Some of benzene consumed was used for the manufacture of straight- or branched-chain detergent alkylate. Linear alkane sulfonates (LAS) are widely used as household and laundry detergents.

Prior to 1975, benzene was catalytically oxidized to produce maleic anhydride, an intermediate in synthesis of polyester resins, lubricant additives, and agricultural chemicals.

11.1. Minor Uses.   Small amounts of benzene find use in production of benzene–sulfonic acid. m-Benzenedisulfonic acid is used to produce resorcinol[108-46-3],$C_6H_6O_2$,(1,3-dihydroxybenzene). Benzene is thermally dimerized to yield biphenyl[92-52-4],$C_{12}H_{10}$. Benzene can also be converted into p-diisopropylbenzene[100-18-5],$C_{12}H_{18}$, which is oxidized to hydroquinone(1,4-dihydroxybenzene),a useful antioxidant. Because of its well-recognized toxicity, little benzene is employed for solvent purposes, and then only when no suitable substitutes are available.

Substitute benzene compounds find use as an antiproliferative and cholesterol lowering drug (129), as an immunosuppressive agents, antifungal agent, and hair growth stimulant (130), and in treating ischemic diseases (131).

BIBLIOGRAPHY

"Benzene" in *ECT* 1st ed., Vol. 2, pp. 420–442, by A. H. Cubberley, J. B. Maguire, and C. S. Reeve, The Barrett Division, Allied Chemical and Dye Corporation, and A. E. Remick, Wayne University; in *ECT* 2nd ed., Vol. 3, pp. 367–401, by George W. Ayers, The Pure Oil Company, and Richard E. Muder, Koppers Company, Inc.; in *ECT* 3rd ed., Vol. 3, pp. 744–771, by William P. Purcell, Union Oil Company of California; "Benzene" in *ECT* 4th ed., Vol. 4, pp. 73–103, by William Fruscella, Unocal Corporation; "Benzene" in *ECT* (online), posting date: December 4, 2000, by William Fruscella, Unocal Corporation.

CITED PUBLICATIONS

1. *De Re Medicina*, Eli Lilly and Co., Indianapolis, Ind., 1938, pp. 81, 103.
2. *TRC Thermodynamics Tables-The Hydrocarbons*, Thermodynamics Research Center, Texas A&M University System, College Station, Tex., 1987.
3. L. H. Horsley, *Azeotropic Data—III, Advances in Chemistry Series 116*, American Chemical Society, Washington, D.C., 1973.
4. S. Coffey, ed., *Rodd's Chemistry of Carbon Compounds*, 2nd ed., Vol. III, *Part A, Aromatic Compounds*, Elsevier Publishing Company, New York, 1971, p. 23.
5. G. M. Badger, *The Structure and Reactions of the Aromatic Compounds*, Cambridge University Press, 1957, Chapt. 2, p. 37.
6. J. March, *Advanced Organic Chemistry: Reactions, Mechanisms, and Structure*, 3rd ed., John Wiley & Sons, Inc., New York, 1985, Chapt. 11, pp. 447–571.
7. J. A. Zoltewicz, *Top. Curr. Chem.* **59**, 33 (1975).
8. F. A. Lowenheim and M. K. Moran, *Faith Keyes & Clark's 4th ed. Industrial Chemicals*, Wiley-Interscience, New York, 1975, p. 616.
9. R. T. Morrison and R. N. Boyd, *Organic Chemistry*, 3rd ed., Allyn and Bacon, Inc., Boston, Mass., 1981, p. 827.
10. J. F. Bunnett and R. E. Zahler, *Chem. Rev.* **49**, 273 (1951).
11. M. Malow, *Hydrocarbon Process.* 149 (Nov. 1980).
12. G. Sosnovsky and D. J. Rawlinson in D. Swern, ed., *Organic Peroxides*, Vol. 2, Wiley-Interscience, New York, 1971, pp. 269–336.
13. W. I. Denton, H. G. Doherty, and R. H. Krieble, *Ind. Eng. Chem.* **42**, 777 (1950).
14. E. L. Muetterties and F. J. Hirsekorn, *J. Am. Chem. Soc.* **96**, 4063 (1974).
15. R. L. Augustine, *Reduction: Techniques and Applications in Organic Synthesis*, Marcel-Dekker, New York, 1968, p. 121.
16. A. H. Blatt, ed., *Organic Synthesis Collective Volumes, I*, John Wiley & Sons, Inc., New York, 1941, p. 323.
17. V. Grakauskas, *J. Org. Chem.* **35**, 723 (1970).
18. A. H. Blatt, ed., *Organic Synthesis Collective Volumes, II*, John Wiley & Sons, Inc., New York, 1943, p. 295.
19. L. N. Ferguson, *Chem. Rev.* **50**, 47 (1952).
20. P. B. D. De La Mare and J. H. Ridd, *Aromatic Substitution: Nitration and Halogenation*, Academic Press, New York, 1959, p. 55.
21. E. E. Gilbert, *Sulfonation and Related Reactions*, Wiley-Interscience, New York, 1965, pp. 62–72.
22. A. V. Topchiev, S. V. Zavgorodnii, and V. G. Kryuchkova, *Alkylation with Olefins*, Elsevier Publishing Co., Amsterdam, The Netherlands, 1964, pp. 68–91.
23. A. C. MacFarlane in L. F. Albright, ed., *American Chemical Society Symposium Series 55: Industrial and Laboratory Alkylations*, American Chemical Society, Washington, D.C., 1977, Chapt. 21, pp. 341–359.
24. F. G. Dwyer, P. J. Lewis, and F. H. Schneider, *Chem. Eng.* 90 (Jan. 5, 1976).
25. J. Weitkamp in D. Olson and A. Bisio, eds., *Proceedings of The Sixth International Zeolite Conference*, Butterworths, London, 1984, pp. 271–290.
26. N. Y. Chen and T. F. Degnan, *Chem. Eng. Prog.* 32 (Feb. 1988).
27. U.S. Pat. 4,185,040 (Jan. 22, 1980), J. W. Ward (to Union Oil Co. of Calif.).
28. *Chem. Week*, 12 (Jan. 1990).
29. R. C. Fuson and C. H. McKeever, *Org. React.* **1**, 63 (1942).
30. E. Berliner, *Org. React.* **5**, 229 (1949).
31. A. E. McKillop, E. C. Taylor, W. H. Altland, R. H. Danforth, and G. McGillivray, *J. Am. Chem. Soc.* **92**, 3520 (1970).

32. M. Kharasch and O. Reinmuth, *Grignard Reactions of Non-Metallic Substances*, Prentice Hall, New York, 1954.

33. U.S. Pat. 3,274,277 (Sept. 20, 1966), H. S. Block (to Universal Oil Products Co.).

34. E. G. Hancock, *Benzene and Its Industrial Derivatives*, Halsted Press, a division of John Wiley & Sons, Inc., New York, 1975, p. 55.

35. *Chem. Eng. News*, 16 (Oct. 24, 1983).

36. D. M. Little, *Catalytic Reforming*, Penn Well Books, Tulsa, Okla., 1985. Refs. 38,41,42, and 56 contain excellent descriptions of benzene manufacturing processes.

37. Ref. 36, p. 154.

38. *Hydrocarbon Proc.*, 171 (Sept. 1976).

39. J. H. Gary and G. E. Handwerk, *Petroleum Refining: Technology and Economics*, 2nd ed., Marcel Dekker, New York and Basel, 1984, p. 84.

40. R. A. Meyers, ed., *Handbook of Petroleum Refining Processes*, McGraw-Hill Book Co., New York, 1986.

41. *Hydrocarbon Proc.*, 81 (Sept. 1988).

42. *Hydrocarbon Proc.*, 189 (Sept. 1970).

43. E. A. Sutton, A. R. Greenwood, and F. H. Adams, *Oil Gas J.*, 52 (May 1972).

44. D. P. Thornton, *Petro/Chem. Eng.* **41**(5), 21 (1969).

45. Ref. 39, p. 87.

46. Ref. 40, p. 10-8.

47. C. H. Deal and co-workers, *Pet. Refiner* **38**(9), 185 (1959).

48. *Hydrocarbon Proc.*, 216 (Sept. 1976).

49. *Hydrocarbon Proc.*, 66 (Nov. 1987).

50. D. L. Burdick and W. L. Leffler, *Petrochemicals for the Non-Technical Person*, Penn Well Books, Tulsa, Okla., 1983, p. 29.

51. H. Sardar, A. S. U. Li, and J. L. Gendler, *Oil Gas J. (Technol.)* 91 (Mar. 20, 1989).

52. *Hydrocarbon Proc.*, 83 (Nov. 1983).

53. K. P. Menard, *Oil Gas J. (Technol.)*, 46 (Mar. 16, 1987).

54. U.S. Pat. 6,323,381 (Nov. 7, 2001), G. J. Nacamuli, R. A. Innes, and A. J. Gloyn (to Chevron Corp.).

55. A. M. Brownstein, *Trends in Petrochemical Technology: The Impact of the Energy Crisis*, Petroleum Publishing Company, Tulsa, Okla., 1976, p. 36.

56. *Hydrocarbon Proc.*, 65 (Nov. 1987).

57. R. J. Gormley, V. U. S. Rao, R. R. Anderson, R. R. Schehl, and R. D. H. Chi, *J. Catal.* **113**, 193 (1988).

58. W. E. Garwood, N. Y. Chen, and F. G. Dwyer, *Shape Selective Catalysis in Industrial Applications*, Marcel-Dekker, New York, 1989, pp. 205–218.

59. W. O. Haag, R. M. Lago, and P. G. Rodewald, *J. Mol. Catal.* **17**, 161 (1982).

60. R. Le Van Mao, P. Levesque, B. Sjariel, and D. T. Nguyen, *Can. J. Chem. Eng.* **64**, 462 (1986).

61. G. Pop and co-workers, *Ind. Eng. Chem. Prod. Res. Dev.* **25**, 208–213 (1986).

62. *Chem. Eng. News*, 14 (Nov. 20, 1989).

63. S. T. Ceyer, Q. Y. Yang, A. D. Johnson, and K. J. Maynard, *J. Am. Chem. Soc.* **111**, 8748 (1989).

64. U.S. Pat. 6,124,514 (Sept. 6, 2000), G. Emmerich and co-workers (to Krupp Uhde GmbH, Dortmund)

65. U.S. Pat. 5,905,182 (May 18, 1999), C. Streicher and P. Provost (Institute Françis du Petrole).

66. D. Walsh, ed., *Chemical Safety Data Sheets*, Vol. 1, *Solvents, Benzene*, The Royal Society of Chemistry, Science Park, Cambridge, UK, 1988, p. 5.

67. R. Hoag, "Benzene," *Chemical Economics Handbook*, SRI, Menlo Park, Calif., Oct. 2000.

68. "Benzene," *Chemical News and Data, Chemical Profiles*, http://www.chemexpo.com/new/PROFILE99106.cfm, revised Dec. 6, 1999.

69. *Chemweek Marketplace*, http://www.chemweek.co/marketplace/product-focus/000/ Benzene.html, accesses12/31/01.

70. *1989 Annual Book of ASTM Standards*, Section 6, Vol. 06.03, American Society for Testing and Materials, Philadelphia, Pa., 1989.

71. *Reagent Chemicals*, 7th ed., American Chemical Society, Washington, D.C., 1986.

72. *Sadtler Research Laboratories Standard ^{13}C NMR, Proton NMR, IR and UV Spectral Data*, Philadelphia, Pa., 1980.

73. R. M. Silverstein, G. C. Bassler, and T. C. Morrill, *Spectrometric Identification of Organic Compounds*, 4th ed., John Wiley & Sons, Inc., New York, 1981.

74. C. J. Pouchart, *Aldrich Library of FT–IR Spectra Edition I*, Vol. 1, Aldrich Chemical Co., Inc., Milwaukee, Wis., 1985.

75. F. W. McLafferty, *Interpretation of Mass Spectra*, 3rd ed., University Science Books, Mill Valley, Calif., 1980.

76. *NIST Mass Spectral Library*, Gaithersburg, Md., 1990, data base.

77. H. Hachtenberg, *Industrial Gas Chromatographic Trace Analysis*, Heyden and Son Ltd., London, 1973.

78. M. Aksoy, *Benzene Carcinogenicity*, CRC Press, Boca Raton, Fla., 1988, p. 14.

79. L. Fishbein and I. K. O'Neill, eds., *Environmental Carcinogens Methods of Analysis and Exposure Measurement*, Vol. 10, International Agency for Research on Cancer, Lyon, France, 1988, Chaps. 7–12.

80. L. D. Gruenke, J. C. Craig, R. C. Wester, and H. I. Maibach, *J. Anal. Toxicol.* **10**, 225 (Nov.–Dec. 1986).

81. B. H. Dolin, *Ind. Eng. Chem. Anal. Ed.* **15**, 242 (1943).

82. *NIOSH Manual of Analytical Methods*, Vol. 3, 2nd ed., HEW Publ. No. (NIOSH) 77.157C, National Institute for Occupational Safety and Health, U.S. Department of Health, Education, and Welfare, Washington, D.C., 1977, S311-11.

83. E. D. Palmes and A. F. Gunnison, *Am. Ind. Hyg. Assoc. J.* **32**, 78 (1971).

84. E. Steger and H. Kahl, *Chem. Technol.* **21**(8), 483 (1969).

85. *Chem. Eng. News*, 4 (Mar. 12, 1990).

86. Agency for Toxic Substances and Disease Registry, *Toxicological Profile for Benzene*, U.S. Department of Commerce, Atlanta, Ga., May 1989.

87. *Fed. Regist.* **55**(61), 11,798 (Mar. 29, 1990).

88. J. D. Graham, L. C. Green, and M. J. Roberts, *In Search of Safety: Chemicals and Cancer Risk*, Harvard University Press, Cambridge, Mass., 1988, Chapt. 5, pp. 115–150.

89. G. F. Kalf, *CRC Crit. Rev. Toxicol.* **18**, 141 (1987).

90. B. D. Goldstein, *Adv. Mod. Environ. Toxicol.* **4**, 51 (1983).

91. R. Snyder, *Fundam. Appl. Toxicol.* **4**, 692 (1984).

92. H. H. Cornish, in J. Doull, C. D. Klaassen, and M. O. Amadur, eds., *Casarett and Doull's Toxicology: The Basic Science of Poisons*, MacMillan, New York, 1980, Chapt. 18, pp. 485–488.

93. H. N. MacFarland, *Occup. Med.* **3**(3), 445 (July–Sept. 1988).

94. T. J. Haley and W. O. Berndt, eds., *Handbook of Toxicology*, Hemisphere Publishing Corporation, Washington, D.C., 1987, p. 509.

95. C. Maltoni and co-workers, *Environ. Health Perspect.* **82**, 110 (1989).

96. L. Fishbein, *Sci. Total Environ.* **40**, 189 (1984).

97. K. Bergman, *Scand. J. Work Environ. Health*, **5**(Suppl. 1) 29 (1979).

98. Ref. 78, Chapt. 6.

99. M. Aksoy, *Brit. J. Haematol.* **67**, 203 (1987).

100. M. Berlin and A. Tunek, *Biol. Monit. Surveill. Work Exposed Chem. Proc. Int. Course 1980*, 67 (1984).

101. D. A. Eastmond, M. T. Smith, and R. D. Irons, *Toxicol. Appl. Pharmacol.* **91**, 85 (1987).

102. G. Witz, G. S. Rao, and B. D. Goldstein, *Toxicol. Appl. Pharmacol.* **80**, 511 (1985).

103. L. Latriano, B. D. Goldstein, and G. Witz, *Proc. Natl. Acad. Sci. USA* **83**, 8356 (1986).

104. T. G. Rossman, C. B. Klein, and C. A. Synder, *Environ. Health Perspect.* **81**, 77 (1989).

105. H. Glatt and co-workers, *Environ. Health Perspect.* **82**, 81 (1989).

106. K. P. Pandya, *J. Sci. Ind. Res.* **44**, 615 (1985).

107. M. Berlin, S. Holm, P. Knutsson, and A. Tunek, *Arch. Toxicol.* (Suppl. 2), 305 (1979).

108. J. Teisinger, V. Fiserova-Bergerova, and J. Kudrna, *Prac. Lek.* **4**, 175 (1952) (in Polish).

109. Ref. 79, Chapt. 12, p. 207.

110. National Institute for Occupational Safety and Health, *Occupational Exposure to Benzene; Criteria for a Recommended Standard*, U.S. Department of Health, Education, and Welfare, Washington, D.C., 1974.

111. Ref. 79, Chapt. 2, and references cited therein.

112. H. Kahn and V. Muzyka, *Work. Environ. Health* **110**, 140 (1973).

113. M. I. Mallein, R. Girard, J. Bertholon, P. Coeur, and J. C. Evreux, *Arch. Mal. Prof. Med. Trav. Secur. Soc.* **31**, 3 (1970).

114. M. Aksoy, *Environ. Health Perspect.* **82**, 193 (1989).

115. A. Craveri, *Med. Lav.* **53**, 722 (1962).

116. F. Flury, *Arch. Exp. Pathol. Pharmakol.* **138**, 65 (1928).

117. H. W. Gerarde and W. B. Deichmann, *Toxicology of Drugs and Chemicals*, Academic Press, New York, 1969, p. 142.

118. R. H. Dreisbach, *Handbook of Poisoning*, 8th ed., Lange Medical Publications, Los Altos, Calif., 1974, p. 165.

119. M. J. Ellenhorn and D. G. Barceloux, *Medical Toxicology: Diagnosis and Treatment of Human Poisoning*, Elsevier, New York, 1988, p. 948.

120. Ref. 119, pp. 949–950.

121. P. N. Chereminisoff and A. C. Morresi, *Benzene—Basic and Hazardous Properties*, Marcel Dekker, New York, 1979.

122. E. Bingham, B. Cohrssen, and C. H. Powell, eds., *Patty's Toxicology*, 5th ed., Vol. 8, John Wiley & Sons, Inc., New York, 2001, p. 1177.

123. R. F. Henderson, in E. Bingham, B. Cohrssen, and C. H. Powell, eds., *Patty's Toxicology*, 5th ed., Vol. 4, John Wiley & Sons, Inc., New York, 2001, p. 252.

124. N. K. Weaver, R. L. Gibson, and C. W. Smith, *Adv. Mod. Environ. Toxicol.* **4**, 63 (1983).

125. R. A. Corbett, *Oil Gas J. Ann. Ref./Petrochem. Rep.*, 33 (Mar. 21, 1988).

126. C. L. Dickson and P. W. Woodward, *Motor Gasolines, Winter 1988–1989*, National Institute for Petroleum and Energy Research, Bartlesville, Okla., p. 25.

127. *Consequences of Limiting Benzene Content of Motor Gasoline*, CONCAWE, den Haag, the Netherlands, Dec. 1983, pp. 1–17.

128. U.S. Pat. 6,355,799 (March 12, 2002), S. Gupta and co-workers (to ISK Americas, Inc.).

129. U.S. Pat. 6,284,923 (Sept. 4, 2001), J. C. Medina and co-workers (to Tularik Inc.).

130. U.S. Pat. 6,187,821 (Feb. 13, 2001), T. Fujita and co-workers (to Welfide Corporation).

131. U.S. Pat. 5,998,452 (Dec. 7, 1999), N. Ohi and co-workers (to Chugai Seiyaku Kabushiki Kaisha, Tokyo).

WILLIAM FRUSCELLA
Unocal Corporation

BENZOIC ACID

1. Introduction

Benzoic acid [65-85-0], C_6H_5COOH, the simplest member of the aromatic carboxylic acid family, was first described in 1618 by a French physician, but it was not until 1832 that its structure was determined by Wöhler and Liebig. In the nineteenth century benzoic acid was used extensively as a medicinal substance and was prepared from gum benzoin. Benzoic acid was first produced synthetically by the hydrolysis of benzotrichloride. Various other processes such as the nitric acid oxidation of toluene were used until the 1930s when the decarboxylation of phthalic acid became the dominant commercial process. During World War II in Germany the batchwise liquid-phase air oxidation of toluene became an important process. In the United States, all other processes have been completely phased out and virtually all benzoic acid is manufactured by the continuous liquid-phase air oxidation of toluene. In the late 1950s and the early 1960s both Dow Chemical and Snia Viscosa constructed facilities for liquid-phase toluene oxidation because of large requirements for benzoic acid in the production of phenol and caprolactam. Benzoic acid, its salts, and esters are very useful and find application in medicinals, food and industrial preservatives, cosmetics, resins, plasticizers, dyestuffs, and fibers.

2. Occurrence

Benzoic acid in the free state, or in the form of simple derivatives such as salts, esters, and amides, is widely distributed in nature. Gum benzoin (from *styrax benzoin*) may contain as much as 20% benzoic acid in the free state or in combinations easily broken up by heating. Acaroid resin (from Xanthorrhoca haslilis) contains from 4.5 to 7%. Smaller amounts of the free acid are found in natural products including the scent glands of the beaver, the bark of the black cherry tree, cranberries, prunes, ripe cloves, and oil of anise seed. Peru and Tolu balsams contain benzyl benzoate; the latter contains free benzoic acid as well. The urine of herbivorous animals contains a small proportion of the glycine derivative of benzoic acid, hippuric acid [495-69-2], $(C_6H_5CONHCH_2COOH)$. So-called natural benzoic acid is not known to be available as an item of commerce.

3. Properties

Selected physical properties of benzoic acid are given in Table 1, solubilities in water in Table 2, solubilities in various organic solvents in Table 3, and vapor pressures in Table 4. In its chemical behavior benzoic acid shows few exceptional properties; the reactions of the carboxyl group are normal, and ring substitutions take place as would be predicted.

Table 1. **Physical Properties of Benzoic Acid**

molecular formula	$C_7H_6O_2$
mp, °C	122.4
bp, at 101.3 kPa,[a] °C	249.2
density	
solid, d_4^{24}	1.316
liquid, d_4^{180}	1.029
refractive index, n_D[b], liquid	1.504
viscosity at 130°C, mPa·s (= cP)	1.26
surface tension at 130°C, mN/m (= dyn/cm)	31
specific heat, J/g[c]	
solid	1.1966
liquid	1.774
heat of fusion, J/g[c]	147
heat of combustion, kJ/mol[c,d]	3227
heat of formation at 26.16°C, kJ/mol[c], solid[e]	−385
heat of vaporization,[f] at 140°C, J/g[c]	534
at 249°C, J/g[c]	425
dissociation constant, K_a, at 25°C	6.339×10^{-5}
flash point, °C	121–131
autoignition temperature, °C, in air	573
pH of saturated aqueous solution at 25°C	2.8

[a] To convert kPa to atm, divide by 101.3.
[b] At 131.9°C.
[c] To convert J to cal, divide by 4.184.
[d] Refs. (1,2).
[e] Ref. 3.
[f] Ref. 4.

Table 2. **Solubilities in Water**

Temperature, °C	g/100 g[a]	Temperature, °C	g/100 g[a]
0	0.17	50	0.85
10	0.21	60	1.20
20	0.29	70	1.77
25	0.34	80	2.75
30	0.42	90	4.55
40	0.60	95	6.80

[a] Grams benzoic acid per 100 g water.

Table 3. **Solubilities in Nonaqueous Solvents at 25°C**

Solvent	g/100 g[a]	Solvent	g/100 g[a]
acetone	55.6	ethyl ether	40.8
benzene	12.2	hexane, 17°C	0.9
carbon tetrachloride	4.1	methanol, 23°C	71.5
chloroform	15.0	toluene	10.6
ethanol (abs)	58.4		

[a] Grams benzoic acid per 100 g solvent.

Table 4. **Vapor Pressure of Benzoic Acid**[a]

Temperature, °C	Pressure, kPa[b]	Temperature, °C	Pressure, kPa[b]
96.0	0.13	172.8	8.0
119.5	0.67	186.2	13.3
132.1	1.33	205.8	26.6
146.7	2.66	227.0	53.3
162.6	5.32	249.2	101.3

[a] Ref. 5.

[b] To convert kPa to mm Hg, multiply by 7.5.

4. Manufacture

Benzoic acid is almost exclusively manufactured by the cobalt catalyzed liquid-phase air oxidation of toluene [108-88-3]. Large-scale plants have been built for benzoic acid to be used as an intermediate in the production of phenol (by Dow Chemical) and in the production of caprolactam (by Snia Viscosa) (6–11). The basic process usually consists of a large reaction vessel in which air is bubbled through pressurized hot liquid toluene containing a soluble cobalt catalyst as well as the reaction products, a system to recover hydrocarbons from the reactor vent gases, and a purification system for the benzoic acid product.

4.1. Reaction.   Typical liquid-phase toluene oxidizer reaction conditions may be as follows:

reactor pressure	200–700 kPa (∼2–7 atm)
reactor temperature	136–160°C
cobalt catalyst concentration	25–1000 ppm
reactor benzoic acid concentration	10–60 wt%

A number of different cobalt salts have been used in the oxidation of toluene, the most common being cobalt acetate [71-48-7], cobalt naphthenate, and cobalt octoate [1588-79-0]. Manganese has also been suggested as a cocatalyst. There is some indication that manganese adversely affects the reactor equilibrium such that the coproduction of benzaldehyde [100-52-7] suffers. Those benzoic acid producers who also produce benzaldehyde do not use manganese in their systems.

Catalysts other than the above cobalt salts have been considered. Several patents suggest that cobalt bromide gives improved yields and faster reaction rates (12–16). The bromide salts are, however, very corrosive and require that expensive materials of construction, such as Hastalloy C or titanium, be used in the reaction system.

4.2. Purification.   Small amounts of reaction by-products are produced during the liquid-phase oxidation of toluene. These by-products include acetic and formic acids, benzene, benzaldehyde, benzyl alcohol, aliphatic benzyl esters such as benzyl formate and benzyl acetate, biphenyl, 2-, 3-, and 4-methylbiphenyls, and phthalic acid. Of these only benzaldehyde and benzene [71-43-2] are currently separated commercially.

The recovery and purification of benzoic acid from a liquid-phase toluene oxidizer may involve distillation alone or it may involve a combination of distillation followed by extraction and crystallization.

In either case, the initial distillation involves separating toluene and any material lower boiling than benzoic acid and recycling those low boilers to the toluene oxidizer. The benzoic acid and higher boiling fractions are then distilled and/or subjected to an extraction and crystallization process to produce the desired product.

The toluene-to-phenol production plants have a significant advantage regarding cost in producing benzoic acid. These plants produce an industrial-grade benzoic acid by distillation, and that industrial-grade product then serves as the feedstock to the phenol plant as well as to the technical and USP/FCC production facilities. Utilizing distillation or extraction and crystallization, these plants reject undesirable impurities, such as the methyl biphenyls, into the phenol reactors in dilute concentrations with benzoic acid. These impurities are actually beneficial to the phenol plant, assisting in the removal of reaction tars. Stand-alone benzoic acid producers are forced to spend large amounts of capital and/or suffer significant unit ratio penalties to produce a similar product.

In addition to the presence of organic trace impurities, the color and color stability of the benzoic acid are often important to customers. Various techniques are utilized to improve color and color stability. Most if not all of these are considered trade secrets.

The USP/FCC grade of benzoic acid is usually produced by crystallization from solution or from the melt. Toluene, water, and methanol have all been used as solvents and each is capable of producing a high quality benzoic acid product.

4.3. Hydrocarbon Recovery.   Toluene is typically recovered from the oxidizer vent gases through the use of refrigeration followed by activated carbon adsorption. Thermal oxidation is then employed to remove carbon monoxide and the remaining traces of hydrocarbons.

The vapor-phase oxidation of toluene to produce benzoic acid and benzaldehyde has been tried utilizing several different catalysts, but yields are low and the process cannot compete with the liquid-phase process (see BENZALDEHYDE). Other processes for the production of benzoic acid are presently of little commercial importance.

5. Economic Aspects

The growth of demand for benzoic acid is expected to increase at a rate of between 1 and 2%/year (17). Glycol dibenzoate plasticizers have been growing at close to 10% annually for the past several years, in part due to environmental and product labeling concerns with regard to phthalate plasticizers (qv). The growth of the diet soft drink market has increased the demand for sodium and potassium benzoates (17). All of the benzoic acid producers in the United States employ the liquid-phase toluene air oxidation process. As toluene becomes more important in the gasoline pool as an octane booster, the benzoic acid producers have to compete with gasoline marketers for the available toluene. If the attractiveness of toluene as an octane booster continues, the cost of producing benzoic acid will most likely increase. The North American producers of benzoic acid and their estimated production capacities (17) are as follows:

Producer	Capacity (mt/year)
Noveon–Kalama, Kalama, Wash.	100,000
Velsicol Chemical, Chattanooga, Tenn.	~30,000

The bulk of this benzoic acid production capacity is consumed internally by these producers. Noveon–Kalama converts over one-half of its production to phenol. A large portion of Velsicol's benzoic acid production is utilized in the manufacture of glycol dibenzoate plasticizer esters.

6. Specifications, Analysis, Packaging, and Shipment

Benzoic acid is available as technical grade as well as grades meeting the specifications of the *United States Pharmacopeia (USP)* (18), the *Food Chemicals Codex (FCC)* (19), or the *British Pharmacopeia (BP)* (20). Typical specifications are listed in Table 5. Analytical methods required for testing to meet the specifications listed in regulatory texts are described in those texts.

Trace impurities typically present in technical grade benzoic acid include methyl diphenyls and phthalic acid. Gas chromatography (gc) and high pressure liquid chromatography (hplc) are useful for determining the concentrations of those impurities.

Technical grade benzoic acid is available in molten as well as solid forms (called flakes or chips). USP/FCC grade is available in solid form, either as crystals or powder. The solid forms of technical grade is usually packaged in 25-kg polylined bags and also in a flexible intermediate bulk container (FIBC), each FIBC containing from 500 to 1000 kg of product. USP/FCC grade is usually packaged in polylined fiber drums, each containing 100 lb (45.5 kg).

Molten technical benzoic acid may be transported in type 316 stainless steel tank cars, usually 76 m^3 (20,000 gal) of product, or in ~5000 gal (19 m^3) 316 stainless steel tank trucks.

7. Health and Safety Aspects

Benzoic acid's toxicity is rated as moderate based upon its LD_{50} (oral-rat) of 2530 mg/kg. Healthy individuals may tolerate small doses (<0.5 g of benzoates per day) mixed with food without ill effects. Large doses, up to 4 g of sodium benzoate per day, have mainly digestive effects such as gastric pain, nausea, and vomiting. A 67-kg man reportedly ingested single doses of 50 g without ill effects, although the mean lethal dose in dogs and cats is 2.5 g/kg (21).

In the early 1900s, several food inspection decisions regarding the use of benzoic acid and sodium benzoate were issued, the latter based upon human feeding studies. As a result of these decisions, since 1909 sodium benzoate and

Table 5. **Specifications for Benzoic Acid**

Item	Technical	USP	FCC	BP/EP [a]
identification	passes test	passes test	passes test	passes test
assay	99.5%, min	nlt 99.5% and nmt 100.5% of C_7H_6O, on anhydrous basis	nlt 99.5% and nmt 100.5% of C_7H_6O, on anhydrous basis	contains nlt 99.0% and nmt 100.5% of benzenecarboxylic acid
appearance	white flakes			
chlorinated compounds	none			
odor	characteristic			
halogenated compounds and halides				passes test (300 ppm, max)
heavy metals (as pb)		nmt 10 ppm	nmt 10 ppm	10 ppm, max
readily carbonizable substances		passes test	passes test	
readily oxidizable substances		passes test	passes test	
residue on ignition		nmt 0.05%	nmt 0.05%	
solidification point			between 121 and 123°C	
congealing range		between 121 and 123°C		
melting point				121–124°C
water		nmt 0.7%	nmt 0.7%	
appearance of solution				passes test
carbonisable substances				passes test
oxidisable substances				passes test
sulphated ash				nmt 0.1%

[a] The specifications for benzoic acid contained in the monographs of the British and European Pharmacopeias have been harmonized.

benzoic acid have been allowed to be added to foods at concentrations not to exceed 0.1% (22). A hazard analysis of benzoic acid and a detailed reference on the toxicity of benzoic acid are available (23,24) . Manufacturer's product and information bulletins provide an excellent source for information regarding the safety and handling of benzoic acid.

The principal safety concern in handling molten benzoic acid is its elevated temperature. Thermal burns may result from improper handling of the molten product (25,26).

8. Uses

Although the main uses for benzoic acid are as a chemical raw material, it also has numerous direct uses. Benzoic acid is used in substantial quantities to improve the properties of various alkyd resin coating formulations, where it tends to improve gloss, adhesion, hardness, and chemical resistance. Benzoic acid terminates chain propagation in alkyd resins (qv) and promotes crystallinity in the final product.

Benzoic acid is also used as a down-hole drilling mud additive where it functions as a temporary plugging agent in subterranean formations. Since this is a secondary oil recovery application, this use is heavily dependent on the price of crude oil.

In medicine, the internal uses of benzoic acid are relatively unimportant. Its principal medicinal use is external; it is used in dermatology as an antiseptic stimulant and irritant. Combined with salicylic acid [69-72-7], benzoic acid is employed in the treatment of ringworm of the scalp and other skin diseases (Whitfield's ointment).

The largest use for benzoic acid is as a chemical raw material in the production of phenol, caprolactam, glycol dibenzoate esters, and sodium and/or potassium benzoate.

8.1. Phenol.   In the early 1960s The Dow Chemical Company built three phenol (qv) plants utilizing benzoic acid as the feedstock (6,27). Dow is no longer involved with these plants. Two of the original three are currently operating and another came on line in Japan in 1991. In this process, benzoic acid is air-oxidized to phenol [108-95-2] in a liquid-phase reaction utilizing copper and magnesium catalysts according to the following:

The hydroxyl group of the resulting phenol is situated immediately adjacent to where the carboxyl group was previously located. This same liquid-phase copper oxidation process chemistry has been suggested for the production of cresols by the oxidation of toluic acids. m-Cresol would be formed by the oxidation of either ortho or para toluic acids; a mixture of o- and p-cresols would be produced from m-toluic acid (6). A process involving the vapor-phase catalytic oxidation of benzoic acid to phenol has been proposed, but no plants have ever been built utilizing this technology (27).

8.2. Caprolactam.   At the same time that Dow was constructing toluene to phenol plants, Snia Viscosa (28–30) introduced two processes for the manufacture of caprolactam (qv) from benzoic acid. The earlier process produced ammonium sulfate as a by-product, but the latter process did not. In either process benzoic acid is hydrogenated to cyclohexanecarboxylic acid [98-89-5], which then reacts with nitrosylsulfuric acid to form caprolactam [105-60-2].

8.3. Glycol Dibenzoates.   The benzoate esters of several glycols are another large use for benzoic acid. These high boiling, chemically stable esters find application as plasticizers in the manufacture of floor coverings, vinyl extrusions, plastisols, adhesives, and coatings. The most common types of resins modified with glycol dibenzoates are poly(vinyl acetate) and poly(vinyl chloride). A wide variety of glycol esters have been prepared and evaluated as plasticizers (qv).

The bulk of the commercial production consists of the dibenzoate esters of diethylene and dipropylene glycol. (These products are mixtures of glycol mono- and dibenzoate esters, containing from 5 to 13% monobenzoate.) The largest volume product is a 50:50 wt blend of those two. The glycol dibenzoates are fast-fusing plasticizers that compete favorably with butyl benzyl phthalate. The properties of these two esters are shown in Table 6 (31). Propylene glycol dibenzoate [19224-26-1] and polyethylene glycol 200 dibenzoate also have applications in certain areas. Dipropylene glycol dibenzoate and diethylene glycol dibenzoate both have FDA approval for use in adhesive and food packaging applications (21 CFR 175.105; 21 CFR 176.170/180)

Table 6. **Physical Properties of Selected Glycol Dibenzoates**

Property	Dipropylene glycol dibenzoate	Diethylene glycol dibenzoate
CAS Registry Number	[27138-31-4]	[120-55-8]
molecular formula	$C_{20}H_{22}O_5$	$C_{18}H_{18}O_5$
mol wt	342	314
specific gravity, 25°C	1.129	1.178
freezing point, °C	−30	28
bp at 0.7 kPa[a], °C	232	240
refractive index, 25°C	1.5282	1.5424
flash point, tcc, °C	>149	>149
viscosity, 20°C, mPa·s (= cP)	170	70

[a] To convert kPa to mm Hg, multiply by 7.5.

8.4. Sodium and Potassium Benzoate. These salts are available in grades meeting the specifications of the *National Formulary* (18), the *Food Chemicals Codex* (19), and the *British Pharmacopeia* (20) (Table 7). Sodium benzoate [532-32-1] is produced by the neutralization of benzoic acid with caustic soda and/or soda ash. The resulting solution is then treated to remove trace impurities as well as color bodies and then dried in steam heated double drum dryers. The product removed from the dryers is light and fluffy and in order to reduce

Table 7. **Specifications for Sodium and Potassium Benzoate**

Item	Sodium Benzoate[a] NF/FCC	Sodium Benzoate[a] BP/EP	Potassium Benzoate[a] FCC
identification	passes tests	passes test	passes tests
assay	nlt 99.0% and nmt 100.5% of $C_7H_5NaO_2$, calculated on the anhydrous basis	contains nlt 99.0% and nmt the equivalent of 100.5% of sodium benzenecarboxylate, calculated with reference to the dried substance	nlt 99.0% and nmt 100.5% of $C_7H_5NaO_2$, calculated on the anhydrous basis
alkalinity (as NaOH)	nmt 0.04%	passes test	
alkalinity (as KOH)		passes test	nmt 0.06%
acidity (as benzoic acid)			
heavy metals	nmt 10 mg/kg	10 ppm, max	nmt 10 mg/kg
water	nmt 1.5%		nmt 1.5%
organic volatile impurities (USP)	passes test		
appearance of solution		passes test	
ionized chlorine		200 ppm, max	
total chlorine		300 ppm, max	
loss on drying		nmt 2.0%	

[a] Not more than = nmt.
Not less than = nlt.

shipping and storage space the sodium benzoate is normally compacted. It is then milled and classified into two product forms; dense granular and dense powder. Sodium Benzoate is also available in extruded form. This form has the advantage of being almost totally "dust free". It also dissolves more quickly that the dense granulat form in both water and antifreeze.

Potassium benzoate [582-25-2] is produced by neutralizing benzoic acid with caustic potash. The resulting solution is processed in a fashion nearly identical to that of sodium benzoate. Potassium benzoate is usually available in the dense granular form.

Sodium and potassium benzoate are employed in a wide range of preservative applications because they provide an effective combination of antimicrobial action, low cost, and safety. Although sodium and potassium benzoate are the preservatives offered in the marketplace, the actual active ingredient being sold is free (ie, undissociated) benzoic acid. The benzoate ion has essentially no antimicrobial properties. Since it is the undissociated benzoic acid that provides the antimicrobial action, sodium benzoate and potassium benzoate are recommended for use in application areas where the pH is at 4.5 or lower (Table 8).

Benzoic acid is supplied to this market in the form of salts because the benzoate salts have a high solubility in water and aqueous stock solutions of up to 35% can easily be prepared. In addition, it is easier, and therefore cheaper, to purify sodium and potassium benzoate than to produce the USP/FCC grade of benzoic acid.

Sodium and potassium benzoate are substances that may be added directly to human food and are affirmed as GRAS (33–35). Benzoic acid and sodium and potassium benzoate are now used as preservatives in such foods as sauces, pickles, cider, fruit juices, wine coolers, syrups and concentrates, mincemeat and other acidic pie fillings, margarine, egg powder, fish (as a brine dip component), bottled carbonated beverages, and fruit preserves, jams, and jellies. The popularity of diet soft drinks has led to increased demand for both benzoate salts.

Nonfood preservative applications of sodium and potassium benzoate are found in pharmaceutical and cosmetic preparations, such as toothpastes and powders, tobacco, pastes and glue, as well as starch and latex (36,37).

The use of the potassium salt of benzoic acid as a soft drink preservative originally resulted from concerns regarding sodium intake and its possible relationship to high blood pressure. Later it was determined that in combination with aspertame, potassium benzoate had positive taste attributes.

Sodium benzoate also has application as a corrosion inhibitor. It is incorporated into paper wrapping materials for the prevention of rust or corrosion in the production of such diverse items as razor blades, engine parts, bearings, etc. It is

Table 8. **Undissociated (Free) Benzoic Acid vs pH**

pH	Free Benzoic acid (%)	pH	Free Benzoic acid (%)
2.5	98.0	4.5	32.9
3	93.9	5	13.4
3.5	83.0	5.5	4.7
4	60.8	6	1.5
4.19	50.0	6.5	0.15

also used in the automotive industry as a corrosion inhibitor in engine cooling systems (at ~1.5%). Unlike in its application as a preservative where free benzoic acid is required to provide antimicrobial action, it appears to be the benzoate ion that provides the corrosion protection.

Sodium benzoate is also employed as a nucluating agent for polypropylene plastics, where it imparts strength and reduces processing times.

9. Benzoic Acid Derivatives

Benzoyl chloride, [98-88-4], C_6H_5 COCl, mp, $-1°C$; bp, $197.2°C$ at 101.3 kPa; d^{25}_4, 1.2070; n^{20}_D, 1.55369. Benzoyl chloride is a colorless liquid that fumes upon exposure to the atmosphere, has a sharp odor, and in vapor form is a strong lachrimator. It is decomposed by water and alcohol, and is miscible with ether, benzene, carbon disulfide, and oils. Benzoyl chloride may be prepared in several ways, including the partial hydrolysis of benzotrichloride, the chlorination of benzaldehyde, and from benzoic acid and phosphorus pentachloride. The most common method is the reaction of benzoic acid and benzotrichloride [98-07-7]. Since benzoic acid may be easily obtained from benzotrichloride, the latter is used as the sole raw material for large-scale production of benzoyl chloride.

Benzoyl chloride is an important benzoylating agent. In this use the benzoyl radical is introduced into alcohols, phenols, amines, and other compounds through the Friedel-Crafts reaction and the Schotten-Baumann reaction. Other significant uses are in the production of benzoyl peroxide [94-36-0], benzophenone [119-61-9], and in derivatives employed in the fields of dyes, resins, perfumes, pharmaceuticals, and as polymerization catalysts.

Benzoic anhydride, [93-97-0] $(C_6H_5CO)_2O$, mp, $42°C$; bp, $360°C$ at 101.3 kPa; d^{15}_4, 1.1989; n^{15}_D, 1.157665. Almost insoluble in water, benzoic anhydride is soluble in most common solvents. A number of methods for the preparation of benzoic anhydride are reported (38). Probably the best is the reaction of benzoyl chloride and benzoic acid (39).

Benzoic anhydride is not manufactured on a large scale. Its primary use is as a benzoylating agent in the manufacture of pharmaceuticals and chemical intermediates.

10. Benzoic Acid Salts

Ammonium benzoate [1863-63-41], $C_6H_5COONH_4$, mp, $198°C$. This is a dull white powder which gradually loses ammonia on exposure to air. Its aqueous solution, it is slightly acidic. Ammonium benzoate has been suggested as a component in certain rubber formulations (40) and as a preservative in paints and glues.

Sodium benzoate [532-32-1], C_6H_5COONa, is highly soluble in water (61.2 g dissolve in 100 g of water at $25°C$) and somewhat soluble in ethyl alcohol, glycerol, and methanol. A 25% aqueous solution of sodium benzoate exhibits a pH of 7.5–8.

Lithium Benzoate [553-54-8], $LiC_7H_5O_2$, soluble in water. Used in medicines and to improve the mechanical properties and transparency of polypropylene.

Potassium Benzoate [582-25-2], C_6H_5COOK, is even more soluble in water than sodium benzoate (73.6 g dissolve in 100 g of water at 25°C). A 25% aqueous solution of potassium benzoate exhibits a pH of 8–8.5.

11. Benzoic Acid Esters

Benzyl benzoate [120-51-4], $C_6H_5COOCH_2C_6H_5$, mp, 21°C, d_4^{25}, 1.118; bp, 323–324°C at 101.3 kPa; n_D^{21}, 1.5681. This compound is a colorless, oily liquid with a faint, pleasant aromatic odor and a sharp, burning taste. It occurs naturally in Peru and Tolu balsams, is sparingly volatile with steam, and is insoluble in water. Benzyl benzoate is prepared commercially by the direct esterification of benzoic acid and benzyl alcohol or by reaction of benzyl chloride and sodium benzoate. The pleasant odor of benzyl benzoate, like other benzoic esters, has long been utilized in the perfume industry, where it is employed as a solvent for synthetic musks and as a fixative. It has also been used in confectionery and chewing gum flavors.

Benzyl benzoate has been used as an insect repellent in formulations for repelling mosquitoes, chiggers, ticks, and fleas, and in the control of livestock insects. Benzyl benzoate was used in the Vietnam War to eradicate and repel certain ticks and mites. It has also found some usage in medicine, cosmetics, and as a plasticizer.

Butyl benzoate [136-60-7], $C_6H_5COOC_4H_9$, mp, −22°C; bp, 250°C at 101.3 kPa. This ester is a thick, oily liquid that has found usage as a dye carrier for polyester fibers.

Ethyl benzoate [93-89-0], $C_6H_5COOC_2H_5$, mp, −35°C; bp, 212°C at 101.3 kPa; d_4^{20}, 0.8788. Used in synthetic ylang–ylang oil, ethyl benzoate is similar in odor to methyl benzoate but is reportedly smoother.

n-Hexyl benzoate [6789-88-4], $C_6H_5COOC_6H_{13}$, bp, 272°C at 103.9 kPa. This compound is used in perfumery as a fixitive and has a melonlike odor.

Methyl benzoate [93-58-3], $C_6H_5COOCH_3$, bp, 198–200°C at 101.3 kPa; d_4^{15}, 1.094; n_D^{15}, 1.5205. Insoluble in water, this is a colorless, transparent liquid solidifying at about 15°C. Methyl benzoate is prepared by the direct esterification of benzoic acid and methanol. It is used in the fragrance industry and in the production of other benzoate esters (via transesterification). A technical-grade methyl benzoate is available as a by-product in the manufacture of dimethyl terephthalate [120-61-6].

Phenyl benzoate [93-99-2], $C_6H_5COOC_6H_5$, mp, 70–71°C; bp, 314°C at 101.3 kPa. This compound has been suggested as an antioxidant (qv) for certain high temperature lubricants (41). Phenyl benzoate exists as a nonisolated intermediate in the production of phenol from benzoic acid.

Alkyl (C12-15) Benzoate [68411-27-8], $C_{20}H_{32}O_2$ (av), bp, 300°C at 101.3 kPa. A widely used emmolient for personal care and cosmetic products. Ester of benzoic acid and mixed C12–15 alcohols.

BIBLIOGRAPHY

"Benzoic Acid" in *ECT* 1st ed., Vol. 2, pp. 459–477, by C. Conover, Monsanto Chemical Company, A. W. Dawes, General Aniline Works Division, General Analine Film Corporation (*o*-Aminobenzoic acid), and H. R. Rosenberg, E. I. du Pont de Nemours Co., Inc., (*p*-Aminobenzoic acid); in *ECT* 2nd ed., Vol. 3, pp. 420–439, by C. Drucker, Monsanto Chemical Company; in *ECT* 3rd ed., Vol. 3, pp. 778–791, by A. E. Williams, Kalama Chemical Inc; in *ECT* 4th ed., Vol. 4, pp. 105–115, by Jarl L. Opgrande, and co-workers; "Benzoic Acid" in *ECT* (online), posting date: December 4, 2000, by Jarl L. Opgrande, C. J. Dobratz, Edward E. Brown, Jason C. Liang, Gregory S. Conn, Kalama Chemical, Inc., Jan With, Frederick J. Shelton, Chatterton Petrochemical Corp.

CITED PUBLICATIONS

1. R. S. Jessup, *J. Res. Natl. Bur. Stand.* **29**, 247 (1942).
2. *Ibid.* **36**, 421 (1946).
3. G. T. Furukawa, R. E. McClosky, and G. J. King, *J. Res. Natl. Bur. Stand.* **47**, 256 (1951).
4. S. Klosky, L. P. Woo, and R. J. Flangian, *J. Am. Chem. Soc.* **49**, 1280 (1927).
5. R. H. Perry and co-workers, *Perry's Chemical Engineer's Handbook*, 6th ed., McGraw-Hill, New York, 1984, pp. 3–50.
6. W. W. Kaeding, *Hydrocarbon Process.* **43**, 173 (1964).
7. W. W. Kaeding and co-workers, *I. EC Process Des. Dev.* **4**(1), 97 (Jan. 1965).
8. *Hydrocarbon Process.* **56**(11), 134 (Nov. 1977).
9. *Hydrocarbon Process.* **44**(11), 255 (Nov. 1965).
10. U.S. Pat. 3,816,523 (June 11, 1974), H. Sidi and M. Sidey (to Tenneco Chemicals).
11. Brit. Pat. 1,219,453 (Aug. 16, 1971), Sioli and co-workers (to Snia Viscosa).
12. Brit. Pat. 804,912 (Nov. 19, 1958), E. T. Crisp (to ICI).
13. Brit. Pat. 833,440 (Apr. 27, 1960), W. A. O'Neil (to ICI).
14. Brit. Pat. 841,053 (July 13, 1960), G. H. Whitfield (to ICI).
15. U.S. Pat. 2,963,509 (Dec. 6, 1960), R. S. Borker (to Midcentury).
16. U.S. Pat. 3,163,671 (Dec. 29, 1964), N. Froyen (to Std Oil Co. Indiana).
17. *Chem. Mark. Rep.* (May 3, 1999).
18. *United States Pharmacopeia*, 25th revision, U.S. Pharmacopeial Convention, Rockville, Md.
19. *Food Chemicals, Codex*, 4th ed., National Academy Press, Washington, D.C.
20. *British Pharmacopeia* 2002, Her Majesty's Stationery Office, London.
21. R. H. Gosselin and co-workers, *Clinical Toxicology of Commercial Products*, Williams Wilkins, Baltimore/London, 1976, p. 137.
22. *Food Inspection Decision 104*, United States Department of Agriculture, Washington, D.C., March 3, 1909.
23. N. I. Sax, *Dangerous Properties of Industrial Materials Report*, Vol. 9, No. 6, Van Nostrand Reinhold Company, New York, 1989, pp. 11–29.
24. *GRAS (Generally Recognized as Safe) Food Ingredients: Benzoic Acid and Sodium Benzoate* (PB-221, PB-228), National Technical Information Service, U.S. Department of Commerce, Washington, D.C., September 1972.
25. *Benzoic Acid Product Information Bulletin*, Noveon–Kalama Inc, Kalama, Wash.
26. *Benzoic Acid Material Safety Data Sheet*, Noveon–Kalama Inc., Kalama, Wash.
27. A. P. Gelbein and A. S. Nislick, *Hydrocarbon Process.* **58**(11), 125 (Nov. 1978).
28. G. Messina, *Hydrocarbon Process.* **43**, 191 (1964).

29. *Hydrocarbon Process.* **50**(11), 142 (Nov. 1971).
30. *Hydrocarbon Process.* **59**(11), 145 (Nov. 1979).
31. *Modern Plastics Encyclopedia*, Vol. 53 (10A), McGraw-Hill, New York, 1976–1977, p. 688.
32. *K-FLEX Glycol Dibenzoate Plasticizers—Product Information and Use Bulletin*, Noveon–Kalama Inc., Kalama, Wash.
33. *Direct Food Substances Affirmed as GRAS (Benzoic Acid)*, **21** CFR 184.1021, U.S. Government Printing Office, Washington, D.C., April 1991.
34. *Direct Food Substances Affirmed as GRAS (Sodium Benzoate)*, **21** CFR 184.1733, U.S. Government Printing Office, Washington, D.C., April 1991.
35. *Potassium Benzoate Considered to be GRAS*, Division of Regulatory Guidance, Bureau of Foods, Washington, D.C., Sept. 22, 1982.
36. *Sodium Benzoate Product Information and Use Bulletin*, Noveon–Kalama, Inc., Kalama, Wash.
37. *Potassium Benzoate Product Information and Use Bulletin*, Noveon–Kalama Inc., Kalama, Wash.
38. H. T. Clarke and E. J. Rahrs, *Organic Synthesis Collective*, Vol. 1, John Wiley & Sons, Inc., New York, 1941, pp. 91–94.
39. G. B. Bliss, *Acta Unio Int. Contra Cancrum* **19**, 499 (1963).
40. USSR Pat. 436,834 (July 25, 1974), R. Sh. Frenkel and co-workers.
41. U.S. Pat. 3,151,082 (Sept. 29, 1964), W. G. Archer (to The Dow Chemical Co.).

JARL L. OPGRANDE
EDWARD E. BROWN
MARTHA HESSER
JERRY ANDREWS
Noveon Kalama, Inc.

BERYLLIUM, BERYLLIUM ALLOYS AND COMPOSITES

1. Beryllium

Beryllium [7440-41-7], Be, specific gravity $= 1.848$ g/mL, and mp $= 1287°$C, is the only light metal having a high melting point. The majority of the beryllium commercially produced is used in alloys, principally copper–beryllium alloys (see CAST COPPER ALLOYS). The usage of unalloyed beryllium is based on its nuclear and thermal properties, and its uniquely high specific stiffness, ie, elastic modulus/density values. Beryllium oxide ceramics (qv) are important because of the very high thermal conductivity of the oxide while also serving as an electrical insulator. The only commercial extraction plant operating in the Western world is that of Brush Wellman at Delta, Utah using both beryl and bertrandite ores as input.

1.1. Occurence. The beryllium content of the earth's surface rocks has been estimated at 4–6 ppm (1). Although 45 beryllium-containing minerals have been identified, only beryl [1302-52-9] and bertrandite [12161-82-9] are of commercial significance.

Gemstone beryl (emerald, aquamarine, and beryl) approaches a pure beryllium–aluminum–silicate composition, $3BeO \cdot Al_2O_3 \cdot 6SiO_2$. Beryl is widely distributed in fine-grained, unzoned pegmatite dikes and in pockets in zoned pegmatite dikes. Beryl is usually obtained as a by-product from mining zoned pegmatite deposits to recover feldspar [68476-25-5], spodumene [1302-37-0], or mica [12001-26-2]. The crushed ore is hand sorted to yield the characteristically hexagonal-shaped beryl crystals that are frequently green or blue in color. Beryl is primarily obtained from Brazil but commercial deposits also occur in China, Argentina, Africa, India, and Russia. A BeO content of 10% is considered necessary for the economic extraction of beryllium from beryl ore.

Bertrandite, $4BeO \cdot 2SiO_2 \cdot H_2O$, became of commercial importance in 1969 when the deposits of Spor Mountain in the Topaz district of Utah were opened. These deposits are believed to have been derived from fluorine-rich hydrothermal solutions at shallow depths (2). Whereas economical beneficiation of these ores averaging <1% BeO has not been achieved, these deposits are commercially viable because of the large reserves present, open-pit mining, and the fact that the beryllium may be extracted by leaching with sulfuric acid. Although some beryl is processed, the majority of beryllium is now obtained from bertrandite.

1.2. Properties. A summary of physical and chemical constants for beryllium is compiled in Table 1 (3–7). One of the more important characteristics of beryllium is its pronounced anisotropy resulting from the close-packed hexagonal crystal structure. This factor must be considered for any property that is known or suspected to be structure sensitive. As an example, the thermal expansion coefficient at 273 K of single-crystal beryllium was measured (8) as 10.6×10^{-6} parallel to the a-axis and 7.7×10^{-6} parallel to the c-axis. The actual expansion of polycrystalline metal then becomes a function of the degree of preferred orientation present and the direction of measurement in wrought beryllium.

Beryllium has high X-ray permeability approximately seventeen times greater than that of aluminum. Natural beryllium contains 100% of the ^9Be isotope. The principal isotopes and respective half-life are ^6Be, 0.4 s; Be, 53 d; ^8Be, 10^{-16} s; ^9Be, stable; ^{10}Be, 2.5×10^6 y. Beryllium can serve as a neutron source through either the (α,n) or $(n,2n)$ reactions. Beryllium has a low $(9 \times 10^{-30} m^2)$ absorption cross-section and a high $(6 \times 10^{-28} m^2)$ scatter cross-section for thermal neutrons making it useful as a moderator and reflector in nuclear reactors (qv). Such application has been limited, however, because of gas-producing reactions and the reactivity of beryllium toward high temperature water.

At ambient temperatures beryllium is quite resistant to oxidation; highly polished surfaces retain the brilliance for years. At 700°C, oxidation becomes noticeable in the form of interference films, but is slow enough to permit the working of bare beryllium in air at 780°C. Above 850°C, oxidation is rapid to a loosely adherent white oxide. The oxidation rate at 700°C is parabolic but may become linear at this temperature after 24–48 h of exposure. In the presence of moisture this breakaway oxidation occurs more rapidly and more extensively. Beryllium oxide [1304-56-9], BeO, forms rather than beryllium nitride [1304-54-7], Be_3N_2, but in the absence of oxygen, nitrogen attacks beryllium >900°C.

Table 1. **Physical and Chemical Properties of Beryllium**

Parameter	Value
at no.	4
at wt.	9.0122
electronic structure	$1s^2 2s^2$
at radius, pm	112.50
at vol at 298 K, mL/mol	4.877
crystal lattice constants, pm	$a = 228.56$,
α-Be, hexagonal close-packed (hcp)	$c = 358.32$,
	$c/a = 1.5677$
β-Be, body-centered cubic (bcc) at 1523 K	$a = 255.0$
transformation pt, hcp to bcc, K	1527
mp, °C	1287
bp, °C	2472
density, g/mL	
at 298 K	1.8477
at 1773 K	1.42
heat of fusion, ΔH_{fus}, J/g[a]	1357
heat of sublimation, ΔH_s, kJ/g[a]	35.5–36.6
heat of vaporization, ΔH_v, kJ/g[a]	25.5–34.4
heat of transformation, J/g[a]	837
standard entropy, S°, J/(g·K)[a]	1.054
standard enthalpy, H°, J/g[a]	216
contraction on solidification, %	3
vapor pressure, MPa[b]	
at 500 K	5.7×10^{-29}
at 1000 K	4.73×10^{-12}
at 1560 K	4.84×10^{-6}
specific heat, J/(g·K)[a]	
at 298 K	1.830
at 700 K	2.740
thermal conductivity at 298 K, W/(m·K)	220
linear coefficient of thermal expansion, 278–333 K[c]	11.4×10^{-6}
electrical resistivity at 298 K, Ω·m	4.31×10^{-8}
reflectivity, %	
white light	50–55
infrared (10.6 μm)	98
sound velocity, m/s	12,600

[a] To convert J to cal, divide by 4.184.

[b] To convert MPa to psi, multiply by 145.

[c] Value is for unworked, isostatically pressed powder metallurgy metal.

Beryllium is susceptible to corrosion under aqueous conditions especially when exposed to solutions containing the chloride ion. It is rapidly attacked by seawater. High purity water, containing small amounts of HNO_3 to passivate stainless steel in the system, was quite inert to beryllium over a period of years in a primary U.S. nuclear test reactor. At high temperatures beryllium reduces water, releasing hydrogen and forming BeO. Due to its position in the emf series, beryllium undergoes galvanic corrosion when coupled in a corrosive environment to the common structural metals; manganese, zinc, and magnesium are the only such metals anodic to beryllium. Protective systems used for beryllium include chromic acid passivation, chromate conversion coatings, chromic acid anodizing, electroless plating (qv), and paints.

Beryllium reacts readily with sulfuric, hydrochloric, and hydrofluoric acids. Dilute nitric acid attacks the metal slowly, whereas concentrated nitric acid has little effect. Hot concentrated alkalis give hydrogen and the amphoteric beryllium hydroxide [13327-32-7], Be (OH)$_2$. Unlike the aluminates, the beryllates are hydrolyzed at the boil.

Beryllium reacts with fused alkali halides releasing the alkali metal until equilibrium is established. It does not react with fused halides of the alkaline earth metals to release the alkaline earth metal. Water-insoluble fluoroberyllates, however, are formed in a fused-salt system whenever barium or calcium fluoride is present. Beryllium reduces halides of aluminum and heavier elements. Alkaline earth metals can be used effectively to reduce beryllium from its halides, but the use of alkaline earth's other than magnesium [7439-95-4] is economically unattractive because of the formation of water-insoluble fluoroberyllates. Formation of these fluorides precludes efficient recovery of the unreduced beryllium from the reaction products in subsequent processing operations.

Chemically, beryllium is closely related to aluminum [7429-90-5] from which complete separation is difficult.

1.3. Ore Processing. *Sulfate Extraction of Beryl.* The Kjellgren-Sawyer sulfate process (9) is used commercially for the extraction of beryl. The ore is melted at 1650°C and quenched by pouring into water. The resulting noncrystalline glass is heat-   treated at 900–950°C to further increase the reactivity of the beryllium component. After grinding to <74 μm (200 mesh), a slurry of the powder in concentrated sulfuric acid [7664-93-9] is heated to 250–300°C converting the beryllium and aluminum to soluble sulfates. The silica fraction remains in the dehydrated, water-insoluble form. The nearly dry mass is leached with water using a countercurrent decantation washing procedure and the resulting solution is fed to the same type of solvent extraction process as that used for bertrandite extraction.

Extraction of Bertrandite. Bertrandite-containing tuff from the Spor Mountain deposits is wet milled to provide a thixotropic, pumpable slurry of below 840 μm (−20 mesh) particles. This slurry is leached with sulfuric acid at temperatures near the boiling point. The resulting beryllium sulfate [13510-49-1] solution is separated from unreacted solids by countercurrent decantation thickener operations. The solution contains 0.4–0.7 g/L Be, 4.7 g/L Al, 3–5 g/L Mg, and 1.5 g/L Fe, plus minor impurities including uranium [7440-61-1], rare earths, zirconium [7440-67-7], titanium [7440-32-6], and zinc [7440-66-6]. Water conservation practices are essential in semiarid Utah, so the wash water introduced in the countercurrent decantation separation of beryllium solutions from solids is utilized in the wet milling operation.

A beryllium concentrate is produced from the leach solution by the countercurrent solvent extraction process (10). Kerosene [8008-20-6] containing di(2-ethylhexyl) phosphate [298-07-7] is the water immiscible beryllium extractant. Warming accelerates the slow extraction of beryllium at room temperature. The raffinate from the solvent extraction contains most of the aluminum and all of the magnesium contained in the leach solution.

The loaded organic phase is stripped of beryllium using an aqueous ammonium carbonate [506-87-6] solution, apparently as a highly soluble ammonium beryllium carbonate [65997-36-6] complex, $(NH_4)_4Be(CO_3)_3$. All of the iron [7439-89-6] contained in the leach solution is coextracted with the beryllium. Heating the strip solution to $\sim70°C$ separates the iron and a small amount of coextracted aluminum as hydroxide or basic carbonate precipitates, which are removed by filtration. The stripped organic phase is treated with sulfuric acid to recover the di(2-ethylhexyl)phosphate.

Heating the ammonium beryllium carbonate solution to 95°C causes nearly quantitative precipitation of beryllium basic carbonate [66104-24-3], $Be(OH)_2 \cdot 2BeCO_3$. Evolved carbon dioxide and ammonia are recovered for recycle as the strip solution. Continued heating of the beryllium basic carbonate slurry to 165°C liberates the remaining carbon dioxide and the resulting beryllium hydroxide [13327-32-7] intermediate is recovered by filtration. The hydroxide is the basic raw material for processing into beryllium metal, copper–beryllium and other alloys, and beryllia [1304-56-9] for ceramic products. This process recovers $\sim90\%$ of the beryllium content of bertrandite.

1.4. Production of Beryllium Metal. *Reduction of Beryllium Fluoride with Magnesium.* The Schwenzfeier process (11) is used to prepare a purified, anhydrous beryllium fluoride [7787-49-7], BeF_2, for reduction to the metal. Beryllium hydroxide is dissolved in ammonium bifluoride solution to give a concentration of 20 g/L Be at pH 5.5. The solution is made basic by the addition of solid calcium carbonate and heating to 80°C precipitating residual aluminum. Lead dioxide is added to the solution to convert manganese [7439-96-5] to insoluble MnO_2 and to precipitate chromium [7440-47-3] as insoluble lead chromate. After filtration, ammonium sulfide is added to the filtrate to remove heavy-metal impurities and any solubilized lead from the lead dioxide treatment. After filtration and balancing to the proper stoichiometry, ammonium fluoroberyllate [14874-86-3], $(NH_4)_2BeF_4$, is crystallized by concurrent evaporation under vacuum. The salt is continuously removed by centrifugation and washed lightly, the mother liquor and washings being returned to the evaporator. The $(NH_4)_2 BeF_4$ is charged into inductively heated, graphite-lined furnaces where it is thermally decomposed to beryllium fluoride and ammonium fluoride. The ammonium fluoride is vaporized into fume collectors for recycle to the dissolving process. The molten beryllium fluoride flows continuously from the bottom of the furnace and is solidified as a glassy product on a water-cooled casting wheel (12).

$$BeF_2 + Mg \rightarrow Be + MgF_2$$

The reduction of beryllium fluoride using magnesium has not been forced above an 85% yield. Complications include: volatilization of unreacted magnesium resulting from the exotherm of the reaction; oxidation of the beryllium because, unlike most metals, beryllium floats on the reaction slag and is not protected from oxygen; and the viscous nature of the magnesium fluoride slag at the melting point of beryllium making complete metal collection difficult.

In commercial practice (13), $\sim70\%$ of the stoichiometric magnesium is used, which gives an excess of beryllium fluoride principally to provide a fluid slag under reduction conditions enabling metal collection. Magnesium metal and beryllium fluoride in the solid form are charged into a graphite crucible at a temperature of $\sim900°C$. When the exothermic reaction is completed, the reaction products are heated to $\sim1300°C$ to allow molten beryllium to separate and float on top of the slag. The molten metal and slag are then poured into a graphite-receiving pot where both solidify. The mixed reaction product is then crushed and water-leached in a ball mill. The excess beryllium fluoride quickly dissolves causing disintegration of the reaction mass and liberation of the beryllium as generally spherical pebbles. The leach liquor in this step is continuously passed through the ball mill removing the fine, insoluble magnesium fluoride particles and leaving the beryllium pebble in the mill body. The magnesium fluoride is filtered from the leach water and discarded. The leach water containing the excess beryllium fluoride is recycled to the aqueous portion of the fluoride preparation process. The beryllium pebble contains ∼97% beryllium along with entrapped reduction slag and unreacted magnesium metal.

Electrolytic Processes.   The electrolytic procedures for both electrowinning and electrorefining beryllium have primarily involved electrolysis of the beryllium chloride [7787-47-5], $BeCl_2$, in a variety of fused-salt baths. The chloride readily hydrolyzes making the use of dry methods mandatory for its preparation (see BBERYLLIUM COMPOUNDS). For both ecological and economic reasons there is no electrolytically derived beryllium available in the market place.

Commercial electrorefining of beryllium has been carried out to obtain a purer metal than the magnesium-reduced beryllium. The most notable purification obtained with respect to iron was specified as 300 ppm maximum, and typically between 100 and 200 ppm Fe as contrasted with the 500–1000 ppm, found in the Mg-reduced beryllium metal. There are no metallurgical advantages to having a metal of improved purity. However, high purity foil is used in some instrumentation X-ray windows.

Vacuum Melting and Casting.   A vacuum melting operation is required for beryllium regardless of its origin. The magnesium-reduced pebble contains trapped slag and unreacted magnesium. The electrolytically derived materials contain entrapped electrolyte not removed by aqueous leaching. Vacuum melting is carried out in induction-heated vacuum furnaces using MgO crucibles and graphite ingot molds. The free magnesium and excess beryllium fluoride or electrolyte vaporize during the melting cycle. Nonvolatiles, such as beryllium oxide, magnesium fluoride, and beryllium carbide [506-66-1], separate from the molten metal as a dross that sinks to and adheres to the bottom of the crucible. The purified metal is poured into ingots weighing ∼180 kg. This operation also serves as the recycle point for valuable beryllium scrap such as machining chip and trimmings.

Because beryllium is primarily used as a powder metallurgy product or as an alloying agent, casting technology in the conventional metallurgical sense is not commonly utilized with the pure metal.

1.5. Fabrication.   Beryllium has a close-packed hexagonal crystal structure. At room temperature there are no slip systems operating in a direction outside the basal plane, which sharply restricts ductility. Extensive attempts to increase ductility by purification have not been successful, apparently because of a degree of covalent bonding along the *c* axis, although bonding along the *a* axis is metallic at temperatures above 200°C beryllium exhibits substantial ductility. Alloying has not as yet been found advantageous and the metal is used alone.

Most beryllium hardware is produced by powder metallurgy techniques achieving fine-grained microstructure having a nearly random crystallographic orientation thus providing a strong material with substantial ductility at room temperature. For some specialized applications, sheet and foil have been rolled from cast beryllium ingot. Such material exhibits an average grain size of 50–100 µm as compared to the typical 12 µm or less of the powder metallurgy products.

Beryllium powder is manufactured from vacuum-cast ingot using impact grinding or jet milling. The casting is first reduced to chip by a machining operation such as lathe turning. The chip is then pneumatically directed against a beryllium target using high pressure, dry air producing a powder of <44 µm (−325 mesh) after appropriate screening. Finer powders, eg, less than 20 µm, are prepared by ball milling with tungsten carbide or steel balls followed by air classification.

Beryllium powder is consolidated by a variety of powder metallurgy processes to near-full density bodies (99+% of theoretical density) (see METALLURGY, POWDER). Vacuum hot-pressing of right circular cylinders at 1000–1200°C at 7 MPa (1000 psi) using graphite tooling has been the most commonly used procedure, particularly where large shapes are desired. The pressing sizes that can be produced range from 18 to 183 cm in diameter and 15 to 168 cm in length. The largest pressing made to date weighed ~3000 kg.

Hot-isostatic-pressing (HIP) is replacing the vacuum hot-pressing procedure for all but the largest shapes. This process, in the case of beryllium involves loading the beryllium powder into a mild steel can of the desired configuration, outgassing the powder under vacuum, subsequently sealing the can, and applying argon gas pressure to the can in an isostatic manner in a cold-wall pressure vessel having an internal furnace heated to the appropriate temperature. The usual processing conditions for beryllium are 103 MPa (15,000 psi) at 1000°C. The advantages of the HIP process include the economics of consolidating the powder to near the final desired shape as contrasted to the right circular cylinder limitation of the vacuum hot-pressing procedure, a shorter floor-to-floor time for a given shape, essentially full density (at 100% of theoretical), and higher strengths than the hot-pressed material because there is little or no grain growth during the HIP procedure.

Cold-isostatic-pressing followed by vacuum sintering or HIP is also used to manufacture smaller intricate shapes. In this instance, beryllium powder is loaded into shaped rubber bags and pressed isostatically in a pressure chamber up to 410 MPa (60,000 psi). After the pressing operation the rubber bag is stripped from the part which is then vacuum sintered to ~99% of theoretical density at about 1200°C. If full theoretical density is required, the sintered part may be simply given a HIP cycle because there is no open porosity after vacuum sintering. In a similar manner, conventional axial cold-pressing followed by vacuum sintering is commercially used for small (under 1 kg) parts where appropriate.

Beryllium sheet is produced by rolling powder metallurgy billets clad in steel cans at 750–790°C. Beryllium foil down to 12.5 µm (0.0005 in.) in gauge is commercially available. Extrusion is also carried out in this temperature region, again using steel cans to contain the powder metallurgy billet. Working of beryllium results in the establishment of a high degree of preferred crystallographic orientation, generally enhancing the properties in the direction of working, but impairing properties normal to the working direction. This is

Table 2. Commercial Grades of Vacuum Hot-Pressed Beryllium, Composition by weight

	S-65	S-200F	I-70A	I-220B
Be, min %	99.0	98.5	99.0	98.0
BeO, max %	1.0	1.5	0.7	2.2
Al, max ppm	600	1000	700	1000
C, max ppm	1000	1500	700	1500
Fe, max ppm	800	1300	1000	1500
Mg, max ppm	600	800	700	800
Si, max ppm	600	600	700	800
other, each max ppm	400	400	400	400

particularly true for tensile elongation. Cross-rolling schedules are followed in rolling that ensure good tensile elongation in the plane of the sheet (10% minimum by specification; 20% is not unusual), but the strain capacity in the thickness direction is limited. The preferred orientation problem has limited the use of wrought beryllium; many shapes other than common structurals are usually machined from billets where standard metallurgical practice with other metals would involve rolled, extruded, or forged components.

The chemical composition and guaranteed tensile properties of the available commercial grades of beryllium in the vacuum hot-pressed form are summarized in Tables 2 and 3. Other consolidation procedures have similar specification property levels. The S-65 grade is of particular interest in that a room temperature minimum tensile elongation of 3% is guaranteed when tested in any direction. This strain capacity is achieved through control of the beryllium oxide content to <1%, control and balancing of impurities such as iron and aluminum, and consolidation by techniques which maximize randomization of the crystallographic texture.

1.6. Economic Aspects. The largest consumption of beryllium is in the form of alloys, principally the copper–beryllium series. The consumption of the pure metal has been quite cyclic in nature depending on specific governmental programs in armaments, nuclear energy, and space. The amount of beryllium extracted from bertrandite has ranged between 150 and 270 metric tons per year since 1986 (14). Small quantities of beryl were also processed during this period.

Table 3. Tensile Properties of Beryllium Commercial Grades at Ambient Temperatures[a]

	S-65B	S-200F	I-70A	I-220B
Tensile strength, min MPa[b]				
Ultimate	290	324	241	379
Yield strength[c]	207	241	172	276
Elongation, min %	3	2	2	2
Microyield, min MPa[b]		27	12	34

[a] Young's Modulus is 303 GPa (4.4×10^7 psi) and the Poisson's ratio = 0.07.
[b] To convert MPa to psi, multiply by 145.
[c] 0.2% offset.

The price of beryllium oxide powder was \$154/kg in 2001. The beryllium content of copper–beryllium master alloy was \$352/kg. Pure beryllium powder was priced at \$615/kg whereas simple shapes in vacuum hot-pressed material were priced at about \$685/kg in 2001.

1.7. Analysis. Instrumental methods such as atomic absorption and emission spectrometry, and gamma activation are employed in most beryllium determinations; however, gravimetric and tritrimetric methods remain useful when high accuracy is required.

Beryllium in reference standards is determined by precipitation of beryllium hydroxide using ammonia. The precipitate is ignited to beryllium oxide and weighed. Interfering elements that precipitate must be removed or masked. Excess ammonia minimizes coprecipitation of manganese, cobalt, copper, nickel, and zinc. Ethylenedinitrilo tetraacetate (EDTA) minimizes precipitation of aluminum, chromium, and iron and further reduces coprecipitation of manganese, cobalt, copper, nickel, and zinc. Alternatively, aluminum, iron, titanium, zirconium, cobalt, nickel, copper, cadmium, and zinc can be removed by precipitation with 8-hydroxyquinoline prior to addition of ammonia. Fluoride, citrate, and tartrate prevent complete precipitation of beryllium and must be absent. Fluoride can be removed by strong fuming in sulfuric acid; phosphate can be removed by precipitation with ammonium molybdate; and silica contamination of the ignited beryllium oxide can be eliminated by adding sulfuric and hydrofluoric acids to the crucible, fuming to dryness, and re-igniting to constant weight (15,16).

Assay of beryllium metal and beryllium compounds is usually accomplished by titration. The sample is dissolved in sulfuric acid. Solution pH is adjusted to 8.5 using sodium hydroxide. The beryllium hydroxide precipitate is redissolved by addition of excess sodium fluoride. Liberated hydroxide is titrated with sulfuric acid. The beryllium content of the sample is calculated from the titration volume. Standards containing known beryllium concentrations must be analyzed along with the samples, as complexation of beryllium by fluoride is not quantitative. Titration rate and hold times are critical; therefore use of an automatic titrator is recommended. Other fluoride-complexing elements such as aluminum, silicon, zirconium, hafnium, uranium, thorium, and rare earth elements must be absent, or must be corrected for if present in small amounts. Copper–beryllium and nickel–beryllium alloys can be analyzed by titration if the beryllium is first separated from copper, nickel, and cobalt by ammonium hydroxide precipitation (15,16).

Optical emission or atomic absorption spectrophotometry usually analyzes beryllium alloys. Low voltage spark emission spectrometry is used for the analysis of most copper–beryllium alloys. Spectral interferences, other interelement effects, metallurgical effects, and sample inhomogeneity can degrade accuracy and precision and must be considered when constructing a method (17).

Inductively coupled argon plasma (ICP) and direct current argon plasma (DCP) atomic emission spectrometry are solution techniques that have been applied to copper–beryllium, nickel–beryllium, and aluminum–beryllium alloys, beryllium compounds, and process solutions. The internal reference method, essential in spark source emission spectrometry, is also useful in minimizing drift in plasma emission spectrometry (17). Electrothermal (graphite furnace) atomic absorption spectrophotometry is employed if the beryllium concentration is very low (17–19).

The commercial ores, beryl and bertrandite, are usually decomposed by fusion using sodium carbonate. The melt is dissolved in a mixture of sulfuric and hydrofluoric acids and the solution is evaporated to strong fumes to drive off silicon tetrafluoride, diluted, then analyzed by atomic absorption or plasma emission spectrometry. If sodium or silicon are also to be determined, the ore may be fused with a mixture of lithium metaborate and lithium tetraborate, and the melt dissolved in nitric and hydrofluoric acids (17).

Metallic impurities in beryllium metal were formerly determined by dc arc emission spectrography, following dissolution of the sample in sulfuric acid and calcination to the oxide (16). This technique is still used to determine less common trace elements in nuclear-grade beryllium. However, the common metallic impurities are more conveniently and accurately determined by dc plasma emission spectrometry, following dissolution of the sample in a hydrochloric–nitric–hydrofluoric acid mixture. Thermal neutron activation analysis has been used to complement dc plasma and dc arc emission spectrometry in the analysis of nuclear-grade beryllium.

The methods of choice for beryllium oxide in beryllium metal are inert gas fusion and fast neutron activation. In the inert gas fusion technique, the sample is fused with nickel metal in a graphite crucible under a stream of helium or argon. Beryllium oxide is reduced, and the evolved carbon monoxide is measured by infrared absorption spectrometry. Beryllium nitride decomposes under the same fusion conditions and may be determined by measurement of the evolved nitrogen. Oxygen may also be determined by activation with 14-MeV neutrons (20). The only significant interferents in the neutron activation technique are fluorine and boron, which are seldom encountered in beryllium metal samples.

Total carbon in beryllium is determined by combustion of the sample, along with an accelerator mixture of tin, iron, and copper, in a stream of oxygen (15,16). The evolved carbon dioxide is usually measured by infrared absorption spectrometry. Beryllium carbide can be determined without interference from graphitic carbon by dissolution of the sample in a strong base. Beryllium carbide is converted to methane, which can be determined directly by gas chromatography. Alternatively, the evolved methane can be oxidized to carbon dioxide, which is determined gravimetrically (16).

Chlorine and fluorine in beryllium metal are isolated by pyrohydrolysis or by distillation (21). Fluoride and chloride in the condensate are determined by ion-selective electrode or colorimetrically.

The gamma activation (photoneutron) method is virtually interference-free and applicable to all types of samples. The sample is irradiated with gamma rays from an ^{124}Sb source and detectors arrayed around the samples count emitted neutrons. The neutron flux is proportional to the beryllium content of the sample. A nearly linear response from 0.01 to 100% beryllium is obtained using 25 g of sample and counting times of 200 s. The method is nondestructive, rapid, and requires minimal sample preparation. Minor interferences result from very high concentrations of neutron- or gamma-absorbing elements. Solid and liquid samples can both be analyzed directly; however, sample geometry is critical. Heavy lead shielding is needed to protect the operator. The source, which requires a license in the United States, must be replaced several times per year to maintain a satisfactory counting rate (22,23).

Environmental and biological samples are usually analyzed for beryllium by atomic absorption, using a nitrous oxide—acetylene flame for high concentrations and the graphite furnace for low concentrations. Organic matter in biological samples and air samples collected on filter paper is removed by wet ashing with an oxidizing acid mixture. If refractory beryllium oxide is present, hydrofluoric acid is added to complete its dissolution. Plasma emission spectrometry offers beryllium detection limits that are nearly as good as graphite furnace atomic absorption, as well as reduced interferences and multielement capability. Beryllium in environmental samples has also been converted to volatile complexes for determination by gas chromatography, but this technique has not achieved widespread use (17,24,25).

Spectrophotometric and fluorometric reagents, once used extensively for determination of beryllium (26,27), are seldom employed. Reviews of beryllium analysis are available (15–17,24–30).

1.8. Safe Handling. Beryllium, beryllium-containing alloys and composites, and beryllium oxide ceramic in solid or massive form present no special health risk (31). However, like many industrial materials, beryllium, beryllium-containing alloys and composites, and beryllium oxide ceramic may present a health risk if handled improperly. Care must be taken in the fabrication and processing of beryllium products to avoid inhalation of airborne beryllium-containing particulate such as dust, mist, or fume in excess of the prescribed occupational exposure limits. Inhalation of fine airborne beryllium may cause chronic beryllium disease, a serious lung disorder, in certain sensitive individuals. The biomedical and environmental aspects of beryllium have been summarized (32).

1.9. Occupational Exposure Limits. The U.S. Occupational Safety and Health Administration (OSHA) has set mandatory limits for occupational respiratory exposures. The OSHA Permissible Exposure Limits (PELs) for beryllium are (1) the time-weighted average (TWA) exposure over an 8-h day is not to exceed beryllium concentrations of 2 $\mu g/m^3$ of air; (2) the Ceiling exposure should not exceed beryllium concentrations of 5 $\mu g/m^3$ of air for a 30-min period; and, (3) the Peak exposure should never exceed 25 $\mu g/m^3$ (33). It remains the best practice to maintain levels of all forms of airborne beryllium as low as reasonably achievable. To protect the general public from environmental exposure to airborne beryllium, the U.S. Environmental Protection Agency has established a beryllium standard of 10 g/day as a permissible emission into the air surrounding a plant (34).

Control Measures. Operations capable of generating airborne beryllium-containing particulate, such as melting, machining, welding, grinding, etc, can be effectively controlled using a combination of engineering controls such as local exhaust ventilation and other work practice controls such as wet methods, personal protective equipment (PPE), and housekeeping. The type and capacity of local exhaust ventilation required will depend upon the amount and speed of particle generation. Protective overgarments or work clothing must be worn by persons who may come in contact with dust, mist or fume. Used disposable clothing should be containerized and disposed of in a manner that prevents airborne exposure during subsequent handling activities. Contaminated work clothing and overgarments must be managed in a manner that prevents secondary airborne exposure to family or laundry personnel handling soiled work clothing. When controls are inadequate, or are being developed and potential exposures are above the occupational exposure limits, approved respirators must be used as specified by an Industrial Hygienist or other qualified professional. To determine the effectiveness of engineering and work practice controls and measure the extent of potential exposure, workplace air monitoring should be periodically conducted by prescribed air sampling and analytical methods. Detailed environmental, health and safety guidance is provided by the manufacturer in the Material Safety Data Sheet (MSDS).

Recycling.   Beryllium, beryllium-containing alloys, and beryllium oxide ceramic can be recycled. Because of the high cost of producing beryllium, producers repurchase clean scrap from customers for recycling and reuse.

1.10. Uses.   The applications for beryllium center around its nuclear and thermal properties, uniquely high specific modulus or stiffness considered on a weight basis, and excellent dimensional stability along with good machinability. These properties are all combined with a relatively high melting point for a light metal. Beryllium is used extensively as a radiation window, both in source and detector applications, because of its ability to transmit radiation, particularly low energy X-rays.

Beryllium is used in the space shuttle orbiter as window frames, umbilical doors, and the navigation base assembly. An important application for beryllium is inertial guidance components for missiles and aircraft. Here the lightweight, high elastic modulus, dimensional stability, and the capability of being machined to extremely close tolerances are all important.

Beryllium is important as a sensor support material in advanced fire-control and navigation systems for military helicopters and fighter aircraft utilizing the low weight and high stiffness of the material to isolate instrumentation from vibration. It is also used for scanning mirrors in tank fire-control systems.

Beryllium is used in satellite structures in the form of both sheet and extruded tubing and is a very important material for all types of space optics. Beryllium oxide ceramic applications take advantage of high room temperature thermal conductivity, very low electrical conductivity, and high transparency to microwaves in microelectronic substrate applications.

2. Beryllium Composites

Beryllium's reactivity limits its uses in the formulation of composite materials. There are only two composite formulations available. These are beryllium–aluminum and beryllium–beryllium oxides. The composition and important physical and mechanical properties for these composites are given in Tables 4, 5, and 6.

Beryllium-Aluminum (Be–Al) was reintroduced into the marketplace in 1990. Invented in the 1960s, a 62 wt% beryllium balance aluminum formulation was patented and trademarked as Lockalloy. Production of Lockalloy ceased in the 1970s. In 1990, BeAl was reintroduced into the market as AlBeMet. It's initial application replaced an aluminum alloy in an actuator arm (the device

Table 4. **Composition of Beryllium Composites**

Composite	Trade name	Beryllium	Aluminum	Beryl oxide
beryllium aluminum	AlBeMet			
	AlBeMet 162	62 wt %	38 wt %	
	AlBeMet 140	40 wt %	60 wt %	
beryllium– beryllium oxide	E-Material			
	E-60	40 vol %		60 vol %
	E-40	60 vol %		40 vol %
	E-20	80 vol %		20 vol %

that holds the read–write heads) for a computer hard disk drive assembly. Significant applications for beryllium aluminum currently include aircraft structures, avionics, satellite structures, electronics and automobile racecars. Beryllium–aluminum is available in wrought, cast and semisolid product (35) forms. The beryllium content of these materials varies from a low of 20 wt% to a high of 65%. The attractive properties of this composite are: low density, high stiffness, high thermal conductivity, high heat capacity and a moderate coefficient of thermal expansion. For parts where stiffness is important to their design, aluminum beryllium will save one-half the weight over an equivalent aluminum part.

A beryllium–beryllium oxide (Be–BeO) composite (36,37) was introduced in 1990. This composite is available in 60, 40, and 20 vol% beryllium oxide. The attributes of Be and BeO are a low density, high stiffness, high thermal conductivity, high heat capacity and a low-coefficient of thermal expansion (CTE). The CTE can be tailored to match the CTE of other electronic materials such as gallium arsenide and silicon by varying the volume percent beryllium oxide in the composite. This composite is used primarily in weight sensitive space and aircraft electronics. The Be–BeO stiffness/density ratio translates into thermal management components that are one-half to one-tenth the weight of the same component in competing materials.

2.1. Fabrication. The primary method of manufacturing composites of beryllium is powder processes. The beryllium–beryllium oxide composite (36) is produced by blending S200F beryllium powder with a relatively coarse crystalline beryllium oxide powder (38). The coarse beryllium oxide crystals are necessary to obtain the high thermal performance of this composite. The beryllium/ beryllium oxide powders are blended, loaded into a steel can, the can's gas is evacuated, and the can is sealed. This sealed, evacuated can is consolidated to a 99+% dense block using hot isostatic pressing. Beryllium–beryllium oxide parts are fabricated from this block using conventional sawing and machining practices.

Most of the metal–metal beryllium–aluminum composite is manufactured from a gas atomized beryllium–aluminum powder. This powder is produced by melting beryllium and aluminum in the proper proportion. This molten metal is poured through a round orifice to produce a thin stream of liquid metal. High pressure gas is used to break this steam up into small droplets and cool these droplets to a fine spherical solid powder.

Table 5. Important Physical and Mechanical Properties of Beryllium–Aluminum

Product	Product form	Tensile ultimate strength (MPa)	Tensile yield strength (0.2% offset) MPa	Elong. %	Density g/cm³	StiffnessYoung's modulus GPa	CTE pPM K	Thermal conductivity W/M/K
		Composite—beryllium aluminum						
AlBeMet 162	extruded[a]	427	317	11	2.10	200	13.9	220
AlBeMet 162	hot istotatic pressed block	289	220	5	2.10	200	13.9	220
AlBeMet 140	rolled sheet	270	185	14	2.26	170	16.4	210

[a] Longitudinal direction.

650

Table 6. **Important Physical and Mechanical Properties of Beryllium–Beryllium Oxide**

Product	Product form	Density g/cm^3	Stiffness Young' modulus GPa	CTE PPM/K	Thermal conductivity w/m/k
		Composite—beryllium–beryllium oxide			
E-60	hot isostatic pressed block	2.52	330	6.1	230
E-40	hot isostatic pressed block	2.30	317	7.5	220
E-20	hot isostatic pressed block	2.06	303	8.7	210

The most popular consolidation method for beryllium–aluminum powder is hot isostatic pressing. Beryllium–aluminum powder is loaded into a steel container. The gas in this container is removed using a vacuum pump while the powder-filled can is heated. The sealed, degassed steel can is consolidated using hot isostatic pressure.

An alternative to consolidation by hot isostatic pressure is to extrude the powder filled can through a die using an extrusion press. Here, the sealed, degassed cans are right cylindrically shaped to accommodate the extrusion press. These cans are heated and placed into the extrusion press and pressed or forced through a die significantly smaller than the powder-filled can.

Either the hot isostatic pressing or extruding produces essentially 100% dense metal. Extrusion would be the preferred method of producing right regular shapes such as a rod or tube. Irregular shapes or parts too large for extrusion (extrusion equipment limits one to a 10-in. diameter cylindrical rod) are produced by conventional machining practices from the hot isostatically pressed block. Blocks as large as 150 cm in diameter and 250 cm long can be produced.

There is a small amount of beryllium–aluminum produced by an investment casting (39,40), process. It is very difficult to produce good castings of beryllium aluminum because of the 618°C difference in the melting points of the two metals. In casting, the beryllium solidifies first producing a rigid but porous network of solid surrounded by molten aluminum. As the cast cools, this aluminum solidifies shrinking in volume. This volume reduction leaves behind pores or voids as the aluminum has difficulty feeding this solidifying contracting aluminum pocket through the rigid beryllium network. These pores compromise the mechanical and physical properties of the beryllium aluminum composite such that casting can only currently produce a few parts.

3. Alloys Containing Beryllium

A small beryllium addition produces strong effects in several base metals. In copper and nickel this alloying element promotes strengthening through precipitation hardening. In aluminum alloys a small addition improves oxidation resistance, castability, and workability. Other advantages are produced in magnesium, gold, and zinc. Many other alloying compositions have been researched (41), but no alloy with commercial importance approaching these dilute alloys has emerged.

3.1. Copper Beryllium Alloys.   Wrought copper–beryllium alloys rank high among copper alloys in attainable strength and, at this high strength, useful levels of electrical and thermal conductivity are retained (see COPPER ALLOYS). Applications include uses in electronic components where their strength-formability-elastic modulus combination leads to use as electronic connector contacts (42); electrical equipment where fatigue strength, conductivity, and thermal relaxation resistance leads to use as switch and relay blades; control bearings where antigalling features are important; housings for magnetic sensing devices where low magnetic susceptibility is critical; and resistance welding systems where hot-hardness and conductivity are important in structural components.

Hardness, thermal conductivity, and castability are important in most casting alloy applications. For example, casting alloys are used in molds for plastic component production where fine cast-in detail such as wood or leather texture is desired. These alloys are also used for thermal management in welding equipment, waveguides, and mold components such as core pins. High strength alloys are used in sporting equipment such as investment cast golf club heads. Cast master alloys of beryllium in copper, nickel, and aluminum are used in preparing casting alloys or otherwise treating alloy melts.

Because these alloys are precipitation hardenable, they can be customized for specific requirements across a wide range of property combinations. Advances in composition control, processing, and recycling technology have broadened the capabilities and expanded the range of application. Data sheets published by the manufacturers and others (43) give compositions, properties, and typical applications.

Composition and Properties.   Commercial wrought copper–beryllium alloys contain 0.2–2.0 wt % beryllium, and 0.2–2.7 wt % cobalt [7440-48-4] or up to 2.2 wt % nickel [7440-02-0], in copper [7440-50-8]. Casting alloys are somewhat richer, having up to 2.85 wt % beryllium. Within this composition window, two distinct classes, which are referred to as the "high strength" alloys, and the "high conductivity" alloys, are available. Beryllium in the high strength alloys imparts a gold luster whereas the high conductivity alloys appear reddish, like copper, in color. Compositions and physical properties of these alloys are given in Table 7.

Alloy C17200 is foremost among the wrought high strength alloys in industrial importance. A free-machining version, containing a small lead addition to C17200 and available only as rod and wire, is designated C17300. The wrought high conductivity alloys have traditionally contained 0.2–0.7 wt % Be and up to 2.5 wt % Co or Ni. Alloy developments have focused on leaner wrought high conductivity compositions such as C17410, having up to 0.4 wt % Be and 0.6 wt % Co and, most recently, C17460 with similar Be content and up to 1.4 wt % Ni. The high strength casting alloys contain 1.6–2.85 wt % Be, nominally 0.5 wt % Co, and a small silicon addition. A minor titanium addition or increased cobalt

Table 7. **Physical Properties of Cast and Wrought Copper–Beryllium Alloys**[a]

Product form	Alloy	Major constituents(a)					Density (g/cm³)	Elastic modulus (GPa)	Thermal expansion coefficient (ppm/C) 20–200 C	Thermal conductivity (w/m c)	Melting range (C)
		Be	Co	Ni	Si	Pb					
wrought	C17200	1.80–2.00	0.25				8.36	131	17	105	870–980
	C17300	1.80–2.00	0.25			0.3	8.41	131	17	105	890–1000
wrought	C17510	0.2–0.6		1.7			8.83	138	18	240	1000–1070
	C17410	0.15–0.50	0.5				8.80	138	18	230	1020–1070
	C17460	0.15–0.50		1.25			8.80	138	18	222	1030–1080
cast	C82500	1.90–2.25	0.4		0.25		8.30	130	18	97	870–970
	C82800	2.50–2.85	0.4		0.25		8.14	130	18	90	850–930
cast	C82200	0.35–0.80		1.7			8.83	140	18	195–250	1040–1080

[a] Weight percent (wt %). Balance copper + residual elements. Nominal unless range is shown

content is used for grain refinement. The high conductivity casting alloys contain up to 0.8 wt % Be. In all cases, the third element addition, either cobalt or nickel, is needed to restrict grain growth during annealing by establishing a dispersion of beryllide particles in the matrix. In the high strength alloys this third element addition also retards softening from overaging during precipitation hardening.

Thermal Treatments.   The copper–beryllium alloys are classic precipitation strengthening systems. Hardening occurs because the solubility of beryllium is much less at low temperatures than at higher temperatures. In practice, the hardening process is conducted in two steps: solution treatment, commonly called solution annealing or simply annealing, followed by precipitation hardening, also known as age hardening. Some users of the alloys prefer to do the age hardening themselves after part forming.

Solution annealing consists of heating below the solidus temperature to dissolve beryllium in the copper matrix, then rapidly quenching to room temperature to retain beryllium in supersaturated solid solution. The high strength alloys are typically annealed in the range 760–800°C and the high conductivity alloys in the range of 900–955°C. It is not necessary to hold the metal at the annealing temperature for more than a few minutes to affect solution treatment. As a guide, thin strip or wire are annealed in less than two minutes, heavy section products, once they reach the annealing temperature, are usually held at temperature for 30 min or less. In this state the alloy is soft and highly workable and therefore may be readily rolled or drawn into strip or wire.

Precipitation hardening involves reheating solution annealed, or solution annealed and cold-worked, material for a time sufficient to nucleate and grow the submicroscopic beryllium-rich precipitates responsible for hardening. For the high strength alloys, age hardening is typically performed in the range of 260–400°C for 0.1–4 h. The high conductivity alloys are age hardened in the range of 425–565°C for 0.5–8 h. Cold-work between solution annealing and hardening increases both the magnitude and rate of strengthening response for wrought products. Up to 37% cold work imparted by cold rolling or drawing, can be provided in commercial products.

During the precipitation process, strength increases, passes through a peak, and then decreases more gradually, ultimately reaching a steady-state level. Electrical conductivity is lowest in the solution-annealed condition because of the beryllium dissolved in the copper matrix. During age hardening, electrical conductivity increases steadily as dissolved beryllium precipitates. Characteristic curves describing this behavior at various hardening temperatures are useful in process control.

The mechanical and electrical properties of selected high strength alloys in cast and wrought forms are provided in Table 8. A similar compilation for the high conductivity alloys is given in Table 8. The mechanical properties shown in the tables correspond to standard hardening times and temperatures and therefore are close to peak conditions. Considerable latitude exists for achieving a wide variety of special mechanical and electrical property combinations.

Melting, Casting, and Hot Working.   The first step in the manufacture of copper–beryllium is production of a nominally 4 wt % Be master alloy by carbothermic reduction of beryllium oxide under molten copper in an electric arc furnace. This master alloy is remelted in coreless induction furnaces and diluted

Table 8. **Mechanical Properties and Electrical Conductivity of Selected Cast and Wrought High Conductivity Copper–Beryllium Alloys**

Product form	Alloy	Temper	Precipitation heat treatment	Yield strength 0.2% offset (MPa)	Ultimate tensile strength (MPa)	Elongation (% in 50.8 mm)	Rockwell hardness	Electrical conductivity (%IACS)
wrought	C17510	annealed/age hardened	2–3 h at 480C or Mill	550–890	680–900	10–25	B20–45	45–60
		cold rolled/age hardened	Mill	650–830	750–940	8–20	B78–88	48–60
	C17410	cold rolled/age hardened[a]	Mill	550–825	655–895	7–20	B89–102	45–55
	C17460	cold rolled/age hardened[a]		655–860	795–860	10 min		50 min
cast	C82200	as-cast		150–240	310–410	15–25	B50–65	45–50
		cast/aged	3 h at 340C	170–380	380–520	10–20	B65–90	45–50
		annealed/aged	Annealed + 3 h at 340C	480–550	620–760	3–15	B92–100	45–50

[a] Combined property range for two ascending cold rolled/age hardened (mill hardened) tempers of each of these alloys.

655

with additional copper, cobalt or nickel, and recycled scrap to adjust the final composition. Melts are directly chill-cast into rectangular or round billets for hot-working to wrought product forms or are poured as ingots for remelting into cast products. The semicontinuously cast rectangular billets are hot-rolled to plate or to coils of hot band for conversion to strip. Round billets are hot-extruded to bar, seamless tube, or rod coil. Hot-working temperatures typically coincide with solution annealing temperatures; about 705–815°C for the high strength alloys and ~815–925°C for the high conductivity alloys. Hot-worked products are softened as needed by solution annealing before further processing.

Finishing. Subsequent processing of wrought copper–beryllium alloys typically includes one or more cycles of cold-working and intermediate solution annealing. Chemical cleaning is performed after each anneal. Processing after the final anneal may include cold-rolling, heat treatment to specified strength levels (mill hardening, with subsequent chemical cleaning), and, for strip, slitting to specified width. Wrought products are also treated with corrosion inhibiting films to extend shelf life. During manufacture, mill products are monitored for stringent control of as-cast composition, nonmetallic inclusion content, intermediate and finish annealed grain size, dimensional consistency, as-shipped mechanical properties, age hardening response, and surface condition. Mill hardened products are solderable and platable as-shipped. User-age hardened products typically require chemical cleaning prior to soldering or plating.

Cast Products. The copper–beryllium alloys can be melted in resistance, gas, induction, and electric arc furnaces. Induction furnaces, in particular, allow close control of melt temperature and agitation to minimize gas absorption, beryllium loss, and dross formation. Drossing is minimized by melting under inert gas or in air melting by use of a graphite cover. Furnace refractories suitable for melting copper–beryllium casting alloys include clay graphite, silicon carbide, alumina, magnesia, and zirconia. High silica refractories may react with copper–beryllium melts. Most common casting methods for copper-base alloys are applicable to copper–beryllium. These include pressure casting, investment casting, centrifugal casting, the Shaw process, die casting, and casting in permanent, ceramic, and various types of sand molds. Shrinkage is similar to tin bronze and less than that of aluminum, silicon, or manganese bronze. Metal or graphite chills may be placed in sand molds to promote directional solidification and reduce shrinkage porosity.

Impurities above maximum levels indicated in published specifications (44) can affect the properties of the finished casting. Silicon, eg, is normally added to many of the copper–beryllium casting alloys to promote fluidity, but excess silicon reduces ductility. Excessive zinc, tin, phosphorus, lead, and chromium behave similarly. Aluminum and iron reduce age hardening response, and degrade electrical and thermal properties.

Copper–beryllium alloy castings can be precipitation hardened in the as-cast condition, however, maximum precipitation hardened strength is achieved by solution annealing the castings, followed by age hardening.

Nickel–Beryllium Alloys. Dilute alloys of beryllium in nickel, like their copper–beryllium counterparts, are age hardenable (45). Nickel–beryllium alloys are distinguished by very high strength; good bend formability in strip; and high resistance to fatigue, elevated temperature softening, stress relaxation, and corrosion. Wrought nickel–beryllium is available as strip, rod, and wire and is used in mechanical, electrical, and electronic components that must exhibit good spring properties at elevated temperatures. Examples include thermostats, bellows, pressure sensing diaphragms, other high reliability mechanical springs, plus burn-in connectors and sockets.

A variety of nickel–beryllium casting alloys exhibit strengths nearly as high as the wrought products with castability advantages. Casting alloys are used in molds and cores for glass and plastic molding, and in jewelry and dental applications by virtue of their high replication of detail in the investment casting process. Nickel-beryllium has also been used in golf club heads by reason of its good castability, excellent hardness and ductility combination and corrosion resistance.

Composition and Properties.   A single composition, UNS NO3360, is supplied in wrought form. Nickel–beryllium casting alloys include 6 wt % Be master alloy, a series with 2.2–2.6 wt % Be including one with a minor carbon addition for enhanced machinability, and a series of ternary nickel-base alloys with up to 2.75 wt % Be and 12 wt % Cr. Composition and physical properties of several of these alloys are presented in Table 9; mechanical properties are given in Table 10. Although displaying only a fraction of the conductivity of copper–beryllium, nickel–beryllium exceeds the conductivity of stainless steel and most other nickel-base alloys by a factor of 2–3 because of its relatively low total alloy content.

Thermal Treatments.   Processing of nickel–beryllium alloys is analogous to processing high strength copper–beryllium. The alloys are solution annealed at a temperature high in the alpha nickel region to dissolve a maximum amount of beryllium, then rapidly quenched to room temperature to create a supersaturated solid solution. Precipitation hardening involves heating to a temperature

Table 9. Properties of Cast and Wrought Nickel–Beryllium Alloys[a]

	Constituents wt %[b]						
Alloy[c]	Be	Other	Density g/ml	Elastic modulus GPa[d]	Thermal Expansion coefficient ppm/°C[e]	Thermal conductivity W/(mK)	Melting range °C
NO3360[f]	1.80–2.05	0.5 Ti	8.27	195–210	4.5	28	1195–1325
M220C (NO3220)	2.0	0.5 C	8.08–8.19	179–193	4.8	36.9[g] 51.1[h]	1150 (solidus)
42C	2.75	12.0 Cr	7.8	193		34.6[i]	1165 (solidus)

[a] Tabulated properties apply to age hardened products
[b] The remainder is nickel and residual elements. Nominal unless range is shown.
[c] Alloy is cast unless otherwise indicated.
[d] To convert GPa to psi, multiply by 1.45×10^5
[e] From 20 to 550°C
[f] Wrought alloy.
[g] At 38°C
[h] At 538°C
[i] At 93°C

Table 10. **Mechanical Properties of Cast and Wrought Beryllium Nickel Alloys**

Temper	Heat treatment[a]	Yield strength 0.2% offset MPa[b]	Ultimate tensile strength, MPa[b]	Elongation, % in 50.8 mm[c]	Rockwell hardness	Electrical conductivity IACS %c
Wrought Alloy NO3360						
annealed		275–485	655–895	30	A39–57	4
cold-rolled		1035–1310	1065–1310	1	A55–75	4
annealed/age hardened	2.5	1035[c]	1480[c]	12	15N78–86	6
cold-rolled/age/hardened	1.5	1585[c]	1860[c]	8	15N83–90	6
mill hardened		690–860	1065–1240	14		5
mill hardened		1515–1690	1790–2000	8		5
Cast Alloy M220C (NO3220)						
annealed		345[c]	760[c]	35	B95[c]	
annealed/age hardened	3	1380[c]	1620[c]	4	C54[c]	
Cast Alloy 42C						
annealed/age hardened	3		1034[c]	6	C38[c]	5

[a] Time given is at 510°C.
[b] To convert from MPa to ps, multiply by 145.
[c] Value given is minimum value.

below the equilibrium solvus to nucleate and grow metastable Be -rich precipitates that harden the matrix. Wrought NO3360 is typically solution annealed at ~1000°C. Cold-work up to ~40% may be imparted between solution annealing and aging to increase the rate and magnitude of the age hardening response. Aging to peak strength is performed at 510°C, up to 2.5 h for annealed and 1.5 h for cold-worked material. Under-, peak-, and overaging behavior are displayed in Table 10. To improve the machinability of N03360 rod the solution annealed and optionally cold worked material may be underaged briefly at 412°C to a hardness of about Rockwell C 25-30, which reduces gumminess and benefits machined surface finish. Final peak age hardening at 510°C is performed after machining to impart full hardness. The cast binary alloys are solution annealed at ~1065°C and aged at 510°C for 3 h. Cast ternary alloys are annealed near 1090°C and given the same aging treatment. Castings are typically used in the solution annealed and aged condition for maximum strength. Unlike copper–beryllium alloy castings, nickel–beryllium castings exhibit little or no age hardening response in the as-cast condition.

Production. Manufacturing of nickel–beryllium products commences with induction melting of a charge consisting of 6% Be–Ni master alloy, additive elements, and recycled scrap. The 6% Be–Ni master is produced by induction melting commercial purity beryllium and nickel rather than by carbothermic reduction of beryllium oxide as for copper–beryllium master. Rectangular or round billets are semicontinuously cast for hot-working to strip or round products. Hot-rolling and extrusion are performed in the vicinity of the solution annealing temperature, ~980°C. The hot-worked products are brought to a ready-to-finish size by one or more iterations of solution annealing and cold-working. Mechanical or chemical cleaning is required after each anneal to remove oxide films. A final solution anneal establishes the finished grain size and age hardening response. Cold-working and mill age hardening may follow the final anneal.

Nickel–beryllium casting alloys are readily air melted, in electric or induction furnaces. Melt surface protection is supplied by a blanket of argon gas or an alumina-base slag cover. Furnace linings or crucibles of magnesia are preferred, with zirconium silicate or mullite also adequate. Sand, investment, ceramic, and permanent mold materials are appropriate for these alloys. Beryllium in the composition is an effective deoxidizer and scavenger of sulfur and nitrogen.

Aluminum-Beryllium Alloys. Small additions of beryllium to aluminum systems are known to improve consistency (38). When as little as 0.005–0.05 wt % beryllium is added as a master alloy to an aluminum alloy during melting, a protective surface oxide film is formed. This film reduces drossing, increases cleanliness, and improves fluidity. Preferentially oxidizable alloy additions such as magnesium and sodium are protected from oxidation during melting and casting. Hydrogen absorption is also reduced, as are mold reactions. Castings thus have improved surface finish, consistent strength, and higher ductility. Additional benefits cited include reduced tarnishing, improved buffing and polishing response, and consistency of aging response, particularly in alloys containing magnesium or silicon. Applications include aircraft skin panels and aircraft structural castings in alloy A357.

BIBLIOGRAPHY

"Beryllium and Beryllium Alloys" in *ECT* 1st ed., Vol. 2, pp. 490–505 and "Beryllium" under "Beryllium and Beryllium Oxides" in Suppl. Vol. 2, pp. 81–86, by B. R. F. Kjellgren, The Brush Beryllium Company; "Beryllium and Beryllium Alloys" in *ECT* 2nd ed., Vol. 3, pp. 450–474, by C. W. Schwenzfeier, Jr., The Brush Beryllium Company; in *ECT* 3rd ed., Vol. 3, pp. 803–823, by J. Ballance, A. J. Stonehouse, R. Sweeney, and K. Walsh, Brush Wellman, Inc.; "Beryllium and Beryllium Alloys" in *ECT* 4th ed., Vol. 4, pp. 126–146, by A. James Stonehouse, Raymond K. Hertz, William Spiegelberg, John Harkness, Brush Wellman Inc.; "Beryllium and Beryllium Alloys" in *ECT* (online), posting date: December 4, 2000, by A. James Stonehouse, Raymond K. Hertz, William Spiegelberg, John Harkness, Brush Wellman Inc.

CITED PUBLICATIONS

1. M. Fleischer, *U.S. Geol. Surv. Circ.* **285** 1 (1953).
2. D. A. Lindsey, H. Ganow, and W. Montjoy, *U.S. Geol. Surv. Prof. Pap.* **818-A**, 1 (1973).
3. D. W. White and J. E. Burke, eds., *The Metal Beryllium*, American Society for Metals, Novelty, Ohio, 1955.

4. G. E. Darwin and J. H. Buddery, *Beryllium*, Butterworths Scientific Publications, London, 1960.
5. H. H. Hausner, ed., *Beryllium, Its Metallurgy and Properties*, University of California Press, Berkeley, Calif., 1965.
6. D. Webster and G. J. London, eds., *Beryllium Science and Technology*, Vol. 1, Plenum Press, New York, 1979.
7. D. R. Floyd and J. N. Lowe, eds., *Beryllium Science and Technology*, Vol. 2, Plenum Press, New York, 1979.
8. R. W. Meyerhoff and J. F. Smith, *J. Appl. Phys.* **33**(1), 219 (1962).
9. B. R. F. Kjellgren, *Trans. Electrochem. Soc.* **89**, 247 (1946).
10. U.S. Pat. 3,259,456 (July 5, 1966), R. L. Maddox and R. A. Foos (to The Brush Beryllium Co.).
11. U.S. Pat. 2,708,618 (May 17, 1955), C. W. Schwenzfeier, Jr. (to the U.S. Atomic Energy Commission).
12. C. W. Schwenzfeier, Jr., in Ref. 3, p. 91.
13. U.S. Pat. 2,381,291 (Aug. 7, 1945), B. R. F. Kjellgren (to Brush Beryllium Co.).
14. *Annual Report*, Brush Wellman Inc., Elmore, Ohio, 1990.
15. P. C. Kempchinskey, *Encyclopedia of Industrial Chemical Analysis*, Vol. 7, John Wiley & Sons, Inc., New York, 1968, pp. 103–117 and 129–141.
16. C. L. Rodden and F. A. Vinci, in Ref. 3, pp. 641–691.
17. R. K. Hertz, *Chemical Analysis of Metals, ASTM STP 944*, American Society for Testing and Materials, Philadelphia, Pa., 1987, pp. 74–78.
18. W. J. Price, *Spectrochemical Analysis by Atomic Absorption*, John Wiley & Sons, Inc., New York, 1979, p. 292.
19. *Analytical Methods for Atomic Absorption Spectrophotometry*, The Perkin Elmer Corp., Norwalk, Conn., 1971.
20. *Annual Book of ASTM Standards, ASTM E 385-80*, Vol. 12.02, American Society of Testing and Materials, Philadelphia, Pa., 1989, pp. 113–117.
21. C. J. Rodden, *Analysis of Essential Nuclear Reactor Materials*, U.S. Atomic Energy Commission, Washington, D.C., 1964, pp. 350–405.
22. C. A. Kienbergr, *Determination of Beryllium by Gamma Activation, Y-1733*, Union Carbide Corp., Oak Ridge, Tenn., 1970.
23. Kh. B. Mezhiborskaya, *Photoneutron Method of Determining Beryllium*, Consultants Bureau Enterprises, Inc., New York, 1961.
24. J. S. Drury and co-workers, *Review of the Environmental Effects of Pollutants: VI Beryllium, ORNL/EIS-87, EPA-600/1-78-028*, Oak Ridge National Laboratory, Oak Ridge, Tenn., 1978, pp. 38–76.
25. L. Fishbein, *Int. J. Environ. Anal. Chem.* **17**, 134 (1984).
26. E. B. Sandell, *Colorimetric Determination of Traces of Metals*, 3rd ed., Interscience Publishers, New York, 1959, pp. 304–324.
27. V. Novoselova and L. R. Batsanova, *Analytical Chemistry of Beryllium*, Israel Program for Scientific Translations, Jerusalem, Israel, 1968.
28. Ref. 26, pp. 118–129.
29. R. F. Kjellgren, C. W. Schwenzfeier, Jr., and E. S. Melick, *Treatise on Analytical Chemistry*, Vol. 6, Part 2, John Wiley & Sons, Inc., 1964, pp. 450–474.
30. J. A. Hurlbut, *The History, Uses, Occurrence, Analytical Chemistry, and Biochemistry of Beryllium, RFP-2152*, Dow Chemical U.S. Golden, Colo., 1974, pp. 3–9. (Available from National Technical Information Service, TID-4500-R62).
31. A. J. Breslin, in H. E. Stokinger, ed., *Beryllium, Its Industrial Hygiene Aspects*, Academic Press, New York, 1966.
32. *Occupational Exposure to Beryllium, U.S. PHS Publication No. HSM-72-10268*, Washington, D.C., 1972, p. IV-4.

33. H. E. Stokinger, in Ref. 31, p. 238.

34. M. D. Rossman, O. P. Preuss, and M. B. Powers, eds., *Beryllium—Biomedical and Environmental Aspects*, Williams & Wilkins, Baltimore, Md., 1991, 319 pp.

35. U.W. Pat. 5,124,119 (June 23, 1992) Fritz C. Grensing

36. U.S. Pat. 5,268,334 (December 7, 1993), Mark N. Emly and Donald J. Kaczynski (to Brush Wellman Engineered Materials Inc.)

37. U.S. Pat. 5,304,426 (April 19, 1994), Fritz C. Grensing (to Brush Wellman Inc.)

38. T. H. Sanders, Jr. and E. A. Starke, Jr., eds., *Aluminum Lithium Alloys, Proceedings of the First International Lithium Conference*, The Metallurgical Society of AIME, 1980, p. 344.

39. U.S. Pat. 5,551,997 (September 3, 1996), James M. Marder and Warren J. Haws (to Brush Wellman Inc.)

40. U.S. Pat. 5,642,773 (July 1, 1997), Fritz C. Grensing (to Brush Wellman Inc.)

41. H. Okamoto and L. E. Tanner, eds., *Phase Diagrams of Binary Beryllium Alloys*, ASM International, Materials Park, Ohio, 1987.

42. J. Rose, ed., *Connectors and Interconnections Handbook, Materials*, 2nd ed., Vol. 1, International Institute of Connectors and Interconnection Technology, Inc., Deerfield, Ill., Oct. 1990.

43. *Copper Development Association Standards Handbook Part 2—Alloy Data Wrought Copper and Copper Alloy Mill Products*, 8th ed., Copper Development Association, Inc., Greenwich, Conn., 1985.

44. *1990 Annual Book of ASTM Standards, Section 2, Nonferrous Metals and Alloys*, Vol. 02.01, American Society for Testing and Materials, Philadelphia, Pa., 1990.

45. T. V. Nordstrom and C. R. Hills, *J. Mater. Sci.* **13**(8), 1700 (1978).

DONALD J. KACZYNSKI
Brush Wellman Inc.

BERYLLIUM COMPOUNDS

1. Beryllium Compounds

1.1. Beryllium Carbide.

Beryllium carbide [506-66-1], Be_2C, may be prepared by heating a mixture of beryllium oxide and carbon to 1950–2000°C, or heating a blend of beryllium and carbon powders to 900°C under mechanical pressure of 3.5–6.9 MPa (500–1000 psi). The metal–carbon reaction is easier to carry out and is accompanied by a substantial exotherm. The reaction mass is quite friable and readily converted to a powder for consolidation by hot pressing at temperatures on the order of 1800°C.

Beryllium carbide slowly hydrolyzes to beryllium oxide and methane in the presence of atmospheric moisture although months may be required to complete the reaction. Any carbon contained in beryllium metal is present as the carbide because the solubility of carbon in beryllium is extremely low.

The crystal structure of beryllium carbide is cubic, density = 2.44 g/mL. The melting point is 2250–2400°C and the compound dissociates under vacuum at 2100°C (1). This compound is not used industrially, but Be_2C is a potential first-wall material for fusion reactors, one on the very limited list of possible candidates (see FUSION ENERGY).

1.2. Beryllium Carbonates.

Beryllium carbonate tetrahydrate [60883-64-9], $BeCO_3 \cdot 4H_2O$, has been prepared by passing carbon dioxide through an aqueous suspension of beryllium hydroxide. It is unstable and is obtained only when the solution is under carbon dioxide pressure. Beryllium oxide carbonate [66104-25-4] is precipitated when sodium carbonate is added to a beryllium salt solution. Carbon dioxide is evolved. The precipitate appears to be a mixture of beryllium hydroxide and the normal carbonate, $BeCO_3$, and usually contains two to five molecules of $Be(OH)_2$ for each $BeCO_3$.

Soluble beryllium carbonate complexes are produced by dissolving beryllium oxide carbonate or hydroxide in ammonium carbonate. Iron and aluminum hydroxides are insoluble in this solution; hence, the reaction can be used to separate these two elements from beryllium. The resulting solution appears to approach the stoichiometry of a solution of tetraammonium beryllium tricarbonate [65997-36-6], $(NH_4)_4Be(CO_3)_3$. After removal of insoluble impurities, hydrolysis of $(NH_4)_4Be(CO_3)_3$ just below the boiling point gives a granular precipitate of di(beryllium carbonate) beryllium hydroxide [66104-24-3], $2BeCO_3 \cdot Be(OH)_2$, which can be dried to constant weight at 100°C. Decomposition to BeO is nearly complete after 5 days at 200°C. The continued addition of $2BeCO_3 \cdot Be(OH)_2$ and $(NH_4)_2CO_3$ to a warmed solution of $(NH_4)Be(CO_3)_3$ has produced solutions containing up to 42 g/L of Be in which the empirical composition is $(NH_4)_2Be(CO_3)_2$. The solid beryllium oxide carbonate intermediates are obtained by a laboratory procedure for preparing pure beryllium salt solutions by reaction with aqueous mineral or organic acids.

1.3. Beryllium Carboxylates.

The beryllium salts of organic acids can be divided into normal carboxylates, $Be(RCOO)_2$, and beryllium oxide carboxylates, $Be_4O(RCOO)_6$. The latter are prepared by dissolving beryllium oxide, hydroxide, or the oxide carbonate in an organic acid, followed by evaporation to give either a solid or an oily liquid. The oxide carboxylate is extracted using chloroform or petroleum ether and recrystallized from the solvent. These compounds are nonelectrolytes, soluble in organic solvents, insoluble in cold water, possess sharp melting points, and can usually be sublimed or distilled without decomposition. The oxide acetate is used as a high purity intermediate for the preparation of commercial beryllium reference solution. The oxide formate requires special preparation by heating the normal formate to 250–260°C or by boiling it with a water suspension containing the calculated amount of beryllium oxide carbonate. The normal beryllium carboxylates must be prepared under strictly anhydrous conditions. The normal acetate is made by treating the oxide acetate with glacial acetic acid and acetyl chloride.

1.4. Beryllium Halides.

The properties of the fluoride differ sharply from those of the chloride, bromide, and iodide. Beryllium fluoride is essentially an ionic compound, whereas the other three halides are largely covalent. The fluoroberyllate anion is very stable.

Beryllium fluoride [7787-49-7], BeF_2, is produced commercially by the thermal decomposition of diammonium tetrafluoroberyllate [14874-86-3], $(NH_4)_2BeF_4$. The fluoride and the fluoroberyllates show a strong similarity to silica and the silicates. Like silica, beryllium fluoride readily forms a glass, which on heating >230°C crystallizes spontaneously to give the quartz modification. This quartz modification exists in two forms: The low temperature α-form is converted to the high temperature β-form at 227°C. The melting point of the quartz form of beryllium fluoride appears to be 552°C (2).

Beryllium fluoride is hygroscopic and highly soluble in water, although its dissolution rate is slow. Fluoroberyllates can be readily prepared by crystallization or precipitation from aqueous solution. Compounds containing theBeF^{2-} ion are the most readily obtained, though compounds containing other fluoroberyllate ions can also be obtained, eg, NH$_4$BeF$_3$, depending on conditions.

Beryllium chloride [7787-47-5], BeCl$_2$, is prepared by heating a mixture of beryllium oxide and carbon in chloride at 600–800°C. At pressures of 2.7–6.7 Pa (0.02–0.05 mm Hg) beryllium chloride sublimes at 350–380°C. It is easily hydrolyzed by water vapor or in aqueous solutions. Beryllium chloride hydrate [14871-75-1] has been obtained by concentrating a saturated aqueous solution of the chloride in a stream of hydrogen chloride. Chloroberyllate compounds have not been isolated from aqueous solutions, but they have been isolated from anhydrous fused salt mixtures.

Beryllium bromide [7787-46-4], BeBr$_2$, and beryllium iodide [7787-53-3], BeI$_2$, are prepared by the reaction of bromine or iodine vapors, respectively, with metallic beryllium at 500–700°C. They cannot be prepared by wet methods. Neither compound is of commercial importance and special uses are unknown.

1.5. Beryllium Hydride. Beryllium hydride [7787-52-2], BeH$_2$, is best prepared by the controlled pyrolysis of di-*tert*-butyl beryllium [20841-21-7], C$_8$H$_{18}$Be, at 200°C. Pressure densification of the amorphous pyrolysis product yields 96% pure crystalline BeH$_2$ having a density near 0.6 g/mL. Di-*tert*-butyl beryllium is prepared by the reaction, in ether, of BeCl$_2$ and *tert*-butyl Grignard reagent, *t*-(C$_4$H$_9$)–MgCl (see GRIGNARD REACTION). Metallic beryllium does not react with hydrogen directly to give the hydride (3). Thermally stable to 240°C, crystalline beryllium hydride is resistant to attack by water and common organic solvents. Interest in beryllium hydride has centered on its potential use as a solid propellant rocket fuel. Theoretically, BeH$_2$ has the highest specific impulse of any fuel material except solid hydrogen.

1.6. Beryllium Hydroxide. Beryllium hydroxide [13327-32-7], Be(OH)$_2$, exists in three forms. On addition of alkali to a beryllium salt solution to obtain a slightly basic pH, a slimy, gelatinous beryllium hydroxide is produced. Aging this amorphous product results in a metastable tetragonal crystalline form, which after months of standing transforms into a stable orthorhombic crystalline form. The orthorhombic modification is also precipitated from a sodium beryllate solution containing >5 g/L db Be by hydrolysis near the boil. This granular beryllium hydroxide is the readily filtered product from the sulfate extraction processing of beryl to obtain metallic beryllium. When heated, beryllium hydroxide loses water. Most of the water comes off in the 600–700°C region, but temperatures on the order of 950°C are required for complete dehydration to the oxide. There is evidence that beryllium hydroxide exists in the vapor phase above 1200°C (4). Water vapor reacts with BeO to form beryllium hydroxide vapor, which has a partial pressure of 73 Pa (0.55 mm Hg) at 1500°C.

1.7. Beryllium Intermetallic Compounds. Beryllium forms intermetallic compounds, referred to as beryllides, with most metals. They are usually

Table 1. **High Temperature Oxidation Resistant Beryllides**

Beryllide system	Compound formula	CAS Registry Number	Melting point, °C	X-ray density, g/mL	Be, wt %
hafnium	$HfBe_{13}$		1595	3.93	39.7
	Hf_2Be_{17}		<1750	4.78	30.0
molybdenum	$MoBe_{12}$		~1705	3.03	53.2
niobium	$NbBe_{12}$	[12010-12-7]	1690	2.92	53.8
	Nb_2Be_{17}	[12010-34-3]	1705	3.28	45.2
titanium	$TiBe_{12}$	[12232-67-6]	1595	2.26	69.3
	Ti_2Be_{17}		1630	2.46	61.5
tantalum	$TaBe_{12}$	[12010-13-8]	1850	4.18	37.4
	Ta_2Be_{17}		1990	5.05	29.8
zirconium	$ZrBe_{13}$	[12010-33-2]	1925	2.72	56.2
	Zr_2Be_{17}		1980	3.08	45.7

prepared by a solid-state reaction of the blended powder constituents at ~1260°C. Fabrication of the reacted powders into specific shapes is carried out by standard powder metallurgical techniques such as vacuum hot pressing or hot isostatic pressing (see METALLURGY, POWDER). The properties exhibited by some beryllides include excellent oxidation resistance, high strength at elevated temperature, good thermal conductivity, and low densities as compared with refractory metals and ceramic materials (see CERAMICS; REFRACTORIES). Table 1 lists melting points and densities of some of the more promising oxidation-resistant beryllides (5).

The beryllides, being intermetallic compounds, are hard, strong materials which exhibit little ductility at room temperature. Strength properties increase gradually as a function of temperature up to ~870°C, above which a sharp increase in strength occurs, peaking in the region of 1260°C; the modulus of rupture values exceed 280 MPa (40,000 psi) at this latter temperature.

Similar to some other intermetallic compounds, most notably molybdeum disilicide [12136-78-6], $MoSi_2$, certain beryllides show anomalous oxidation behavior exhibiting excellent oxidation resistance at high temperature, eg, 1260°C, but little or no oxidation resistance in some lower temperature range. Such behavior was observed in the 700–870°C range for Nb_2Be_{17}, $NbBe_{12}$, Zr_2Be_{17}, and $ZrBe_{13}$, but not for other compounds listed in Table 1 (6). Complete disintegration of the vulnerable beryllides into powder occurred within 24 h. The addition of small amounts of aluminum [7429-90-5] metal, or the nickel–aluminum (1:1) [12003-78-0] compound NiAl solved this problem.

The beryllides continue to be of interest for high temperature aerospace applications because of their oxidation resistance, low density, and high strength at elevated temperature (7). The limited strain capacity of the materials, particularly at low temperatures, has thus far prevented actual use.

1.8. Beryllium Nitrate. Beryllium nitrate tetrahydrate [13516-48-0], $Be(NO_3)_2 4H_2O$, is prepared by crystallization from a solution of beryllium hydroxide or beryllium oxide carbonate in a slight excess of dilute nitric acid. After dissolution is complete, the solution is poured into plastic bags and cooled to room temperature. The crystallization is started by seeding. Crystallization from more concentrated acids yields crystals with less water of hydration. On heating >100°C, beryllium nitrate decomposes with simultaneous loss of water and oxides of nitrogen. Decomposition is complete >250°C.

1.9. Beryllium Nitride.   Beryllium nitride [1304-54-7], Be_3N_2, is prepared by the reaction of metallic beryllium and ammonia gas at 1100°C. It is a white crystalline material melting at 2200°C with decomposition. The sublimation rate becomes appreciable in a vacuum at 2000°C. Beryllium nitrate is rapidly oxidized by air at 600°C and like the carbide is hydrolyzed by moisture. The oxide forms on beryllium metal in air at elevated temperatures, but in the absence of oxygen, beryllium reacts with nitrogen to form the nitride. When hot pressing mixtures of beryllium nitride and silicon nitride, Si_3N_4, at 1700°C, beryllium silicon nitride [12265-44-0], $BeSiN_2$, is obtained. $BeSiN_2$ may have application as a ceramic material.

1.10. Beryllium Oxalate.   Beryllium oxalate trihydrate [15771-43-4], $BeC_2O_4 3H_2O$, is obtained by evaporating a solution of beryllium hydroxide or oxide carbonate in a slight excess of oxalic acid. The compound is very soluble in water. Beryllium oxalate is important for the preparation of ultrapure beryllium hydroxide by thermal decomposition >320°C. The latter is frequently used as a standard for spectrographic analysis of beryllium compounds.

1.11. Beryllium Oxide.   Beryllium oxide [1304-56-9], BeO, is the most important high purity commercial beryllium chemical. In the primary industrial process, beryllium hydroxide extracted from ore is dissolved in sulfuric acid. The solution is filtered to remove insoluble oxide and sulfate impurities. The resulting clear filtrate is concentrated by evaporation and upon cooling high purity beryllium sulfate, $BeSO_4·4H_2O$, crystallizes. This salt is calcined at carefully controlled temperatures between 1150 and 1450°C, selected to give tailored properties of the beryllium oxide powders as required by the individual beryllia ceramic fabricators. Commercial beryllium oxide powder calcined at 1150°C consists of crystallites predominately 0.1–0.2 µm in size. Powder particles are made up of clusters or aggregates of the smaller crystallites.

Ceramic-grade beryllium oxide has also been manufactured by a process wherein organic chelating agents (qv) were added to the filtered beryllium sulfate solution. Beryllium hydroxide is then precipitated using ammonium hydroxide, filtered, and carefully calcined to obtain a high purity beryllium oxide powder.

High purity beryllium oxide powder is fabricated by classical ceramic-forming processes such as dry pressing, isostatic pressing, extrusion, tape casting, and slip casting. Additives consisting of the oxides of magnesium, aluminum, or silicon, or various combinations are frequently included in the ceramic mixes to improve the reproducibility of sintering and resultant properties. The green compact of formed beryllia is commonly sintered at 1500–1600°C in dry air or dry hydrogen. Moisture in the sintering atmosphere affects the surface characteristics such as roughness, texture, and microstructure (8). The sintering operation produces beryllia ceramics at 95–97% of the theoretical density with an average grain size between 6 and 30 µm. Higher density may be achieved by hot pressing high purity beryllia powder.

Beryllia ceramics offer the advantages of a unique combination of high thermal conductivity and heat capacity with high electrical resistivity (9). Thermal conductivity equals that of most metals; at room temperature, beryllia has a

Table 2. **Properties of High Purity Beryllium Oxide Ceramics**

Property	Value
specific heat, J/(g K)[a]	1.050
thermal conductivity, W/(mK)	
at 25°C	290–330
at 100°C	190–220
dielectric constant (loss tangent)	
1 MHz at 25°C	6.55–6.72 (0.00005–0.00016)
1 MHz at 100°C	6.55–6.75 (0.00007–0.00019)
1 GHz at 25°C	6.72–6.75 (0.00006–0.00035)
1 GHz at 100°C	6.72–6.81 (0.00014–0.00051)
9.3 GHz at 25°C	6.77 (0.00007–0.00031)
9.3 GHz at 100°C	6.77 (0.00026–0.00047)
volume resistivity, Ωm	
at 25°C	$2.0 \times 10^{14} - 1.3 \times 10^{15}$
at 100°C	$1.4 \times 10^{11} - 5.0 \times 10^{11}$
coefficient of thermal expansion, K^{-1}	
at 100°C	9.7×10^{-6}
at 500°C	13.3×10^{-6}
tensile strength, MPa[b]	150
compressive strength, MPa[b]	1400
modulus of rupture, MPa[b]	250
modulus of elasticity, GPa[b]	345
Poisson's ratio	0.164–0.380

[a] To convert J to cal, divide by 4.184.
[b] To convert MPa to psi, multiply by 145.

thermal conductivity above that of pure aluminum and 75% that of copper. Properties illustrating the utility of beryllia ceramics are shown in Table 2.

Beryllia ceramic parts are frequently used in electronic and microelectronic applications requiring thermal dissipation (see CERAMICS AS ELECTRICAL MATERIALS). Beryllia substrates are commonly metallized using refractory metallizations such as molybdenum–manganese or using evaporated films of chromium, titanium, and nickel–chromium alloys. Semiconductor devices and integrated circuits (qv) can be bonded by such metallization for removal of heat.

Beryllium oxide is used in automotive ignition systems, lasers, electronic circuits for computers heat sinks, and microwave oven components (10).

1.12. Beryllium Sulfate. Beryllium sulfate tetrahydrate [7787-56-6], $BeSO_4 \cdot 4H_2O$, is produced commercially in a highly purified state by fractional crystallization from a beryllium sulfate solution obtained by the reaction of beryllium hydroxide and sulfuric acid. The salt is used primarily for the production of beryllium oxide powder for ceramics. Beryllium sulfate dihydrate [14215-00-0], is obtained by heating the tetrahydrate at 92°C. Anhydrous beryllium sulfate [13510-49-1] results on heating the dihydrate in air to 400°C. Decomposition to BeO starts at ~650°C, the rate is accelerated by heating up to 1450°C. At 750°C the vapor pressure of SO_3 over $BeSO_4$ is 48.7 kPa (365 mm Hg).

2. Economic Aspects

Beryllium is principally consumed in the metallic form, either as an alloy constituent or as the pure metal. Consequently, there is no industry associated with beryllium compounds except for beryllium oxide, BeO, which is commercially important as a ceramic material. Beryllium oxide powder was available at $ 45/kg ($ 100/lb) in 2001 (11).

3. Health and Safety Factors

Beryllium-containing materials can be potentially harmful if mishandled. Care must be taken in the fabrication and processing of beryllium products to avoid inhalation of airborne beryllium particulate matter such as dusts, mists, or fumes in excess of prescribed work place limits. Inhalation of fine airborne beryllium may cause chronic beryllium disease, a serious lung disorder, in certain sensitive individuals. However, most people, perhaps as many as 99%, do not react to beryllium exposure at any level (see BERYLLIUM AND BERYLLIUM ALLOYS). International Agency for Research on Cancer (IARC) classified beryllium and beryllium compounds as carcinogenic to humans (10,12).

BIBLIOGRAPHY

"Beryllium Compounds" in *ECT* 1st ed., Vol. 2, pp. 505–509, by B. R. F. Kjellgren, The Brush Beryllium Company; "Beryllium and Beryllium Oxides" in Suppl. 2, pp. 86–89, by B. R. F. Kjellgren, The Brush Beryllium Company; "Beryllium Compounds" in *ECT* 2nd ed., Vol. 3, pp. 474–480, by C. W. Schwenzfeier, Jr., The Brush Beryllium Company; "Beryllides" in Suppl. 2, pp. 73–80, by A. J. Stonehouse, The Brush Beryllium Company; "Beryllium Compounds" in *ECT* 3rd ed., Vol. 3, pp. 824–829, by K. Walsh and G. H. Rees, Brush Wellman Inc.; in *ECT* 4th ed., Vol. 4, pp. 147–153, by A. James Stonehouse and Mark N. Emly, Brush Wellman, Inc.

CITED PUBLICATIONS

1. W. W. Beaver, in D. W. White and J. E. Burke, eds., *The Metal Beryllium, American Society for Metals*, Novelty, Ohio, 1955, pp. 570–598.
2. A. R. Taylor and T. E. Gardner, *U.S. Bur. Mines Rep. Invest.* 6664 (1964).
3. R. W. Baker and co-workers, *J. Org. Chem.* **159**, 123 (1978).
4. W. A. Young, *J. Phys. Chem.* **64**, 1003 (1960).
5. A. J. Stonehouse and co-workers, in J. T. Weber and co-workers, *Compounds of Interest in Nuclear Reactor Technology*, AIME, New York, 1964, pp. 445–455.
6. R. M. Paine and co-workers, in Ref. 5, pp. 495–509.
7. R. L. Fleischer and R. J. Zabala, *Metall. Trans. A* **20**(7), 1279 (July 1989).
8. W. W. Beaver and co-workers, *J. Nucl. Mater.* **14**, 326 (1964).
9. A. Goldsmith, H. J. Hirschhorn, and T. E. Waterman, *Thermophysical Properties of Solid Materials*, Vol. 3, *Ceramics*, Armour Research Foundation, WADC-TR-58-476, revised, Nov. 1960, pp. 67, 69–70, 81.
10. M. M. Mroz and co-workers, in E. Bingham, B. Cohrseen, and C. H. Powell, eds., *Patty's Toxicology*, 5th ed., Vol. 2, John Wiley & Sons, Inc., New York, 2001, Chapt. 27.
11. L. D. Cunningham, "Beryllium", *Mineral Commodity Summaries*, U.S. Geological Survey, Jan 2002.

12. International Agency for Research on Cancer (IARC), *Monographs on the Evaluation of the Carcinogenic Risk of Chemicals to Humans; Beryllium, cadmium, mercury, and exposures in the glass manufacturing industry.* Vol. 58, IARC, Lyon, France, 1993, pp. 41–117.

GENERAL REFERENCES

D. A. Everest, *The Chemistry of Beryllium*, Elsevier Publishing Company, Amsterdam/London/New York, 1964.

"Argnoberyllium compounds" in *G. Melins Handbook of Inorganic Chemistry*, Part 1, 8th ed., 1987.

DONALD J. KACZYNSKI
Brush Wellman Inc.

BIOCATALYSIS

1. Introduction

Bioorganic catalysis can be defined as the use of biological systems (whole cells or pure enzymes) to produce organic compounds. Bioorganic catalysis has been practiced from the ancient times mainly for the production of foods and beverages. The production of alcohol via fermentation, of vinegar via oxidation of ethanol by acetic acid bacteria, and the production of cheese via enzymatic breakdown of milk proteins are well-known examples. Biocatalysts are ever increasingly being exploited for the production of industrially important materials, in many cases competing with traditional chemical methods and in some instances performing reactions that traditional chemistry methods cannot. As a result, biocatalysis is now being considered as another implement in the chemist's arsenal for tackling chemical transformations.

One of the earliest examples of an industrial application of bioorganic catalysis is the chemoenzymatic synthesis of L-ephedrine and psuedoephedrine (1). In this process exogenous benzaldehyde was condensed with pyruvic acid produced within a yeast fermentation to yield L-phenylacetylcarbinol (L-PAC), which is a key intermediate in the production of L-ephedrine and psuedoephedrine. The synthesis of L-PAC is catalyzed by the pyruvate decarboxylase enzyme and can be produced at a price that competes with traditional chemical synthesis (2). In spite of the prevalent use of biocatalysts in organic synthesis some myths still exist, such as biocatalysts are sensitive, biocatalysts are expensive, biocatalysts have narrow substrate range, and biocatalysts cannot perform all possible chemical reactions.

Debunking the myth of biocatalysts is an ongoing task and there are many examples being presented to elucidate the many advantages of using biocatalysts

Table 1. **Enzymes Sales in Millions**[a]

Category	1980	1990	1999
detergent additives (protease, lipase, amylase, cellulase)	6.2	40	325
starch conversion (amylase, amyloglucosi-dase, fungal amylase, glucose isomerase)	69	135	181
food processing (lipase, pectinase, catalase, glucose oxidase, protease, papain, renin, pepsin)	47	90	146
textile processing (amylase, cellulase, catalase)	60	50	70
diagnostic	20	35	65
research	12	25	50
recombinant DNA	2	30	60
therapeutic (TPA, urokinase, others)	0	500	155
synthesis	0	0	10

[a] Dollar amounts do not incorporate inflation (Dr. Bernard Wolnak, data presented at "Enzymes for the Next Millenium, Chicago, USA, 2000").

for organic reactions. Even though most biocatalysts work in relatively moderate reaction conditions, advances in isolation and expression of extremophilic enzymes and genetic modifications of mesophilic enzymes have created biocatalysts that can tolerate substantially harsh conditions (3,4). Some enzymes isolated from hyperthermophilic organisms are active at temperature as high as 140°C and others isolated from psychrophilic organisms are active at temperatures as low as 4°C. Some enzymes from barophilic organisms isolated near deep-sea vents tolerate pressures as high as 100 bar. Even though enzymes are specific with respect to the type of reaction they catalyze, most of them have activity on a wide range of substrates. In addition, there are many instances where enzymes have been tailored using genetic engineering to suit the conditions in biotransformation processes. There are biologically catalyzed equivalents for almost all of the chemical reactions, even the Diels–Alder reaction and the Claisen rearrangement (5). Advances in genetic engineering and fermentation technology have enabled the production of many enzymes at high concentrations in high cell density fermentations, resulting in a substantial decrease in cost of manufacture of biocatalysts. This is evidenced by the increase in enzyme sales from $130 million in 1980 to $700 million at present (Dr. Bernard Wolnak, data presented at "Enzymes for the Next Millenium, Chicago, USA, 2000"). A breakdown in the enzyme sales during this time period in the many industries is shown in Table 1. However, not all enzymes are amenable to manipulations and cost contribution of the biocatalyst has to be evaluated on a case-by-case basis.

2. Whole Cells Versus Pure Enzymes

There are mainly two biological entities that can compete as bioorganic catalysts, and they are isolated enzymes and whole cells (microbial, plant, or animal) (6).

Several parameters that ultimately affect the cost of a process (product purity, throughput, stability) are important while selecting a particular form of biocatalyst in an industrial biotransformation process. An ideal process would be one where the culturing of the cells would also accomplish the biotransformation at a high product concentration. However, a lot of the substrates and products at any significant level are toxic to cell growth. In addition, separation of the product from the cell broth can be very difficult and there may be undesirable side reactions during cell growth. This necessitates the separation of cell growth and biotransformation. The only time whole cell biocatalysis is advantageous is when the biotransformation involves multiple enzymes, where cofactor regeneration is necessary, or when enzyme isolation is needed and is difficult.

The number and complexity of the steps in the preparation of the biocatalyst is proportional to the cost of the biocatalyst. Progress in genetic engineering has allowed the production of recombinant enzymes in microbial cells. However, production of all recombinant enzymes in microbial cells is not automatic and sometimes impossible. The cost of biocatalyst preparation has to be justified within the context of the biotransformation process involved. Purified enzymes have been used in many cases to demonstrate feasibility of a particular reaction. Separation of soluble enzymes from fermentation broth is cumbersome, and in addition there is the issue of enzyme stability during isolation. The overall cost of a biotransformation process can be reduced if the enzyme can be easily separated and reused. Immobilization of enzymes allows the recycle of enzymes, the ease of which is dependent on the method of immobilization. More than a hundred techniques for immobilization had been published by 1983 (6) and many more since (7). There are advantages and disadvantages to each method of immobilization and the decision of choosing one method over another is made on a case-by-case basis. The properties of an enzyme can change dramatically upon immobilization and this needs to be taken into account when immobilization is contemplated. Some of the problems encountered are enzyme inactivation, lowered enzyme activity, altered allosteric properties, and cost of immobilization. There are some advantages when enzymes are immobilized such as increased stability, high enzyme loading, altered pH, and temperature optima (6,7).

Immobilized cells could be used instead of immobilized enzymes to eliminate the costly isolation step while retaining the ability to recycle the biocatalyst. Typical cell immobilization methods are adsorption, gel entrapment, and compartmentalization in polymer matrices. The advantages for immobilized cells remain the same as for immobilized enzymes, except that improved enzyme stability is rarely observed. The disadvantages may be exacerbated because of the immobilization method and the fact that there is another barrier for the substrate to get to the enzyme, namely the cell membrane.

Crude enzyme probably represents the simplest form of prepared biocatalyst. The advantage is that there is little preparatory cost compared to immobilized enzymes, cells, or purified enzymes. The main disadvantage is that the biocatalyst cannot be recycled. Crude enzyme can be prepared by spray drying or freeze drying the cell culture. Freeze drying which is prevalent at the laboratory scale cannot be practiced economically at the large scale. The critical parameters in spray drying are the operating inlet and outlet temperatures and the residence time in the dryer. Aminotransferases synthesized in recombinant

Escherichia coli have been shown to retain >95% of their original activity after spray drying (Chiragene, Inc., unpublished data). Some of the advantages are that these spray-dried biocatalysts can be stored easily at room temperature, retain activity for an extended period of time, and can be used directly in the reaction without further processing.

3. Biocatalyst Performance

Biocatalyst performance parameters such as activity, selectivity, and stability can be altered by modifying the enzyme or the environment around it. The latter approach uses the principle of changing the solvent environment around the enzyme molecule (commonly referred to as solvent engineering) while the former approach uses the redesigning of enzyme to meet the desired goals. Solvent properties such as dipole, dielectric constant, hydrophobicity, and density have been shown to cause predictable effect on enzyme stability (8–10), activity, and enantioselectivity (11–16). The solvent engineering approach has been demonstrated mainly at small scale but has not been scaled up into commercial processes because enzyme catalytic efficiencies are 2–6 orders of magnitude lower in non-aqueous media than in aqueous solutions (17).

The other approach to improving enzyme performance parameters is by enzyme modification at the molecular level (commonly referred to as enzyme engineering). The enzyme can be altered to tackle the problem of production cost, poor activity, stereoselectivity, and stability. The earliest approach to engineering enzymes is through rational design on the basis of structure–function relationship (18). This requires knowledge regarding the protein structure either by x-ray crystallography or molecular modeling or a combination of both. This can be tedious in an industrial setting where the enzyme has to be improved within a matter of months if not weeks.

Error-prone polymerase chain reaction (PCR) is a powerful tool to introduce errors in the DNA sequence coding for the enzyme of interest (19–21). The advantage of this method is the speed at which mutations can be introduced and the fact that they can be targeted into the gene of interest. The result is a random mutation of the enzyme, resulting in many candidates in a short amount of time. A proper screening or selection method can then be used to pick the enzyme with the required property. With a little more knowledge on the effect of amino acid changes on changes in properties, this mutation strategy can be directed to improving a specific property by building on the initial mutations. Examples are the directed evolution of an esterase and subtilisin to perform in high concentrations of dimethylformamide (20,22,23). An alternative to single amino acid changes is the DNA shuffling method that recombines sequences from various versions of the enzyme to create a new enzyme with new properties (24). There are several published examples where significant improvements in enzyme activity, selectivity, stability, thermostability, or solvent tolerance have been achieved using both these methods (20,22,23,25–32).

In cases where there is no natural enzyme to catalyze the reaction catalytic antibodies have provided a starting point (33,34). The first report on the use of catalytic antibodies involved simple acyl transfer reactions (35,36). The list of transformations has grown considerably since then to include ester and amide hydrolysis, lactonization, group eliminations, reductions, C—C bond formation and cleavage, Claisen rearrangement, and the Diels—Alder reaction (33,34). This technology has not matured to the stage where antibodies can be raised or produced in large quantities to carry out reactions at rates similar to existing enzymes. Another variation is the use of transferred active sites to inert protein scaffolds to create a synthetic enzyme (37,38). There is enough evidence in literature to indicate that any given enzyme can be modified significantly to cater to the needs of the biotransformation process.

4. Bioorganic Catalysis

The enzymes catalyzing oxidation/reduction reactions constitute an important class from an industrial perspective since industrial chemistry involves many oxidation/reduction reactions. The use of oxidoreductases for organic synthesis has been under intense investigation for the past several decades (39) and still continues.

Dehydrogenases, which constitute the largest class within the oxidoreductases, are enzymes that catalyze the reduction and oxidation of carbonyls and alcohols, respectively. Reduction of the carbonyl group can create a chiral center and therefore is very important from an industrial perspective. Oxidation reactions generally destroy the chiral center and can be useful where resolution of a racemic alcohol is needed. Detailed structural, mechanistic, and specificity data are available for this class of oxidoreductases (39). A broad range of substrate alcohols, from simple aliphatic to complex polycyclic, can be oxidized by three alcohol dehydrogenases (yeast, Horse liver, and *Pseudomonas testosteroni*) with overlapping specificities (40). Even though many oxidation/reduction reactions have been accomplished using dehydrogenases, there have been few that have been scaled for large-scale production. The main obstacles to the large-scale use of these enzymes are the cost of enzyme, cost of cofactor, and solubility of substrate.

Reduction of enzyme cost and increasing solubility of substrate are problems that apply to a lot of biotransformations. The solution to the problem of enzyme cost is to improve the performance significantly so as to utilize less of the enzyme and/or to produce the enzyme in recombinant organisms to improve productivity. The problem of solubility is often dealt with by using co-solvents to increase solubility of substrates or by running reactions in pure solvents. The more tenacious problem associated with biological oxidation/reduction reactions is the need for expensive cofactors such as NADH, NADPH, and FADH. These cofactors are usually required in stoichiometric amounts and that typically renders the biological route expensive. The only way the cost contribution from the cofactor can be reduced is via recycling of the cofactor. Several solutions for cofactor recycling have been presented and practiced with varied amount of success. Efficiency of cofactor recycling can be measured as the number of cycles the cofactor undergoes before it is destroyed (turnover number can be expressed as moles of product formed per mole of cofactor). Turnover numbers greater than tens of thousands need to be achieved in order for the recycling method to be

Table 2. **Turnover Numbers Achieved in Various Cofactor Recycling Methods**[a]

Method	Turnover number
chemical (sodium dithionite)	<100
electrochemical and photochemical	<1000
enzymatic (single or coupled)	10^3-10^5

[a] Ref. 50.

considered feasible at a large scale. Table 2 gives turnovers numbers for each of the methods used for cofactor recycling (5).

Chemical methods for cofactor regeneration using sodium dithionite, phenazine methosulfate, and flavin mononucleotide have been fairly successful. However the utility of chemical regeneration method depends on ease of separation of product and number of cofactor turnover in the reaction system (5).

Electrochemical methods have been studied as a means of regenerating cofactors (5,41). Electrochemical methods, although widely used in biosensors, need to demonstrate economic feasibility (high turnover number) before being accepted as method for regenerating cofactors. Another method is to use a second enzyme system to recycle the cofactor, and this has been successfully used in a small-scale process producing multi-kilogram quantities (42). Polyethylene glycol (PEG) derivatized NAD was used as cofactor in an ultrafiltration membrane reactor that allowed separation of cofactor/enzymes and substrates/products. Regeneration was provided by formate dehydrogenase (FDH) that catalyzed the oxidation of formate to carbon dioxide with the concomitant generation of PEG-NADH. FDH cost has been substantially reduced in order to make this recycling process economically feasible in some processes (39). Using the coupled enzyme process, cofactor turnover numbers of greater than 10^5 can be achieved, making it one of the best available methods (42). Unfortunately, the FDH cofactor regenerating system cannot be used with NADPH because the FDH does not accept NADP as a substrate.

A glucose dehydrogenase (GDH) cofactor regenerating system has been used to recycle NADPH and NADH via the oxidation of glucose to gluconolactone. Gluconolactone then spontaneously hydrolyses to form gluconic acid, making the reaction scheme favorable for both NADH and NADPH generation. Since glucose is a cheap substrate, this method can be very inexpensive, provided gluconic acid can be separated from the other products of the reaction. As with FDH the cost of GDH is a factor that has to be addressed. Other regenerating systems have been used such as glucose-6-phosphate dehydrogenase, alcohol dehydrogenases, and hydrogenase. The latter methods are less attractive than the FDH or GDH system for a variety of reasons that are not presented here (43).

The enzyme recycling principle can be applied using whole living cells instead of isolated enzymes (44–49). In this case, the main enzymatic reaction and the cofactor regeneration reaction are carried out during the growth of living cells. Growing cells produce cofactors using their intrinsic metabolic pathways from cheap carbohydrates and regenerate cofactor during their growth cycle. A mixed culture of *E. coli*, one expressing glucose dehydrogenase for cofactor recycling and the other expressing aldehyde reductase for asymmetric reduction of ethyl 4-chloro-3-oxobutanoate, was used in a two-phase chiral alcohol production system successfully at laboratory scale (45). *Trichosporon capitatum, Geotrichum candidum*, Baker's yeast, and other whole cells have been used in a similar fashion to regenerate cofactor *in situ* during reductions of a variety of substrates (44,47–49). Even though some aspects of this method are attractive, there are drawbacks to whole-cell biocatalysis. Nonnatural substrates and products can be toxic to the cells at very low concentrations (0.1–0.3%) (5). Recovery of low concentration products from the reaction mixture may be troublesome with by-products from the growth of cells contaminating the product. Multiple enzymes can act on the substrate and impact the yield as well as stereoselectivity of the product. Side reactions on either substrate or product may reduce yield and purity. There are some solutions to these problems and they need to be evaluated on a case-by-case basis.

Enolate reductases which reduce C–C unsaturated bonds are another class of enzymes that require cofactor recycling. Since the products of these reactions can result in a chirally pure product, they constitute an important class of reactions. Typical biotransformations have utilized whole-cell biocatalysis rather than of isolated enzymes because of the ease of cofactor recycling on the laboratory scale. In addition to the problem of cofactor recycling, enolate reductases are inactivated by traces of oxygen (5).

Biological oxidation reactions achieve heteroatom oxygenation, aromatic hydroxylation, Bayer–Villiger oxidation, double bond epoxidation, and nonactivated carbon atom hydroxylation of substrates, which is difficult via conventional chemistry. The biological oxidation of primary and secondary alcohols to aldehydes is not of practical interest because these reactions are just as easily accomplished using conventional chemical methods, are thermodynamically unfavorable, have unfavorable reaction conditions, and in the case of secondary alcohols destroy an asymmetric center (5). The only context where it is meaningful is during the resolution of racemic alcohols. As discussed earlier, the regeneration of cofactor is a major stumbling block in this scheme.

The regioselective oxidation of polyols is of practical interest because biocatalysts can selectively oxidize one hydroxyl group without requiring any protection of the remaining hydroxyl groups. This is a feat that cannot be achieved by conventional chemical oxidants. Oxidation of glucose to gluconic acid using glucose oxidase is a prominent example (50). Another example is pyranose oxidase that is used in the synthesis of D-fructose from D-glucose and 5-keto-D-fructose from L-sorbopyranose. Selective oxidation of hydroxyl groups in steroids is an important reaction carried by cholesterol oxidase. Since substrate solubility is a problem in aqueous systems, steroid oxidations can be carried out in organic solvents using PEG modification of the enzymes as a method to make the enzymes soluble in organic solvents (5).

Oxygenases are enzymes that incorporate molecular oxygen directly into the substrate. Oxygenase-catalyzed oxidations are important since direct addition of molecular oxygen into unactivated organic substrates is very difficult to accomplish using conventional chemistry. Monooxygensases incorporate one atom of oxygen whereas dioxygensases incorporate two atoms of oxygen into a substrate. Because these enzymes are membrane bound and are difficult to isolate, most oxidations are carried out using whole cells. Main problems in this reaction are further metabolism of products, low yield due to side reactions, and substrate and product toxicity. Many examples exist where these oxygenases were used to synthesize small quantities of material (5). The stereoselective hydroxylation of nonactivated carbons is very important because there is no equivalent traditional chemical method. This field burgeoned in the 1950s when steroid modifications were being investigated, and more recently some of the hydroxylations have been scaled up to produce commercial quantities of steroid (51). By screening for the appropriate organism it is now possible to selectively hydroxylate virtually any center in a steroid (5).

Another class of reactions that is very important is the Baeyer–Villiger reactions where ketones are oxidized into esters and lactones. The fact that there is chiral recognition by the enzymes sets them apart from conventional methods (5). Since flavin- and nicotinamide-dependent monooxygenases are usually involved in these reactions, whole-cell biocatalysis is utilized most of the time. One of the latest examples of a dioxygenase-based process is the production of indanediol, which is an intermediate in the synthesis of the protease inhibitor Crixivan (Merck & Co.) used in the treatment of AIDS (52,53). Growing recombinant *E. coli*, cells carrying the *Pseudomonas putida* toleuene dioxygenase and dihydrodiol dehydrogenase were used for the production of *cis*-(1*S*),(2*R*)-indandiol with an optical purity >99% ee.

Aldolase-catalyzed asymmetric C–C bond formations which are carried out in a neutral pH aqueous environment are a very important class of reactions. Stereocontrolled synthesis of D- or L-threo-phenylserine using L- or D-threonine aldolase has been accomplished on a preparative scale (54). This aldolase technology can generate other β-substituted serines and other derivatives starting from modified substrates. Other commercially important reactions catalyzed by aldolases are for the synthesis of unusual sugars, polyhydroxylated alkaloids, novel C–C polymers, and analogues of *N*-acetylneuraminic acid (5,55,56).

The formation of cyanohydrins is catalyzed by oxynitrilase. These enzymes catalyze the asymmetric additon of hydrogen cyanide to the carbonyl group of an aldehyde or ketone forming a chiral cyanohydrin (57). Chiral cyanohydrins are important intermediates in the synthesis of pharmaceuticals (58–60), agrochemicals (61), or liquid crystals (62,63). Since oxynitrilases are obtained mainly from plant sources, biocatalyst cost is a major issue. This limitation has been overcome in certain cases with the use of enzyme recycling either by immobilization or by using biphasic reaction systems (57).

Hydrolases catalyze the hydrolysis of various bonds such as amides and esters. Among the hydrolases, lipases, esterases, and proteases are most widely used enzymes. Hydrolases are routinely used in organic synthesis since they do not require cofactors and a large number of them possessing relaxed substrate specificities are available from different sources. Lipases are the most widely used hydrolases and they catalyze the hydrolysis of triglycerides into fatty acids and glycerol. On the basis of triglyceride hydrolysis, microbial lipases can be classified into two groups. Lipases of first group have no regiospecificity and release fatty acids from all three positions of glycerol. They completely hydrolyze the triglycerides to fatty acids and glycerol with diacylglycerol and monoacylglycerol as intermediates of the reaction. In contrast, lipases of second group release fatty acids regioselectively from the outer 1- and 3-positions of triglycerides. Lipases have enormous potential in chemical synthesis because of several reasons: (*1*) lipases are stable in organic environment; (*2*) lipases possess broad substrate specificity; (*3*) lipases exhibit high regio- and enantioselectivity. A number of lipases have been isolated from fungi and bacteria and characterized. Lipases and esterases are used to prepare enantiomerically pure esters, acids, and alcohols (5). Two major procedures are used: hydrolysis of racemic ester in water and acylation (transesterification, esterification, transaminolysis) of alcohols in nonaqueous media such as organic solvents and supercritical fluids (64). Lipases and esterases from different sources were used to resolve the racemic mixture to produce S-enantiomers of Ibuprofen and Naproxen in excess of 95% ee (65–67). Another important application of lipases is the interesterification of triglycerides that are useful in the preparation of cocoa butter equivalents (68–70). The steadily growing interest in lipases for the organic synthesis is reflected by an increasing number of review articles covering application of lipases in biocatalysis (71,72).

What has been presented is a small sample of the variety of bioorganic transformations being carried out. In order to better describe the many hurdles faced in biotransformation processes, a case study is presented where an aminotransferase was used to convert a substituted tetralone to a chiral aminotetralin.

5. Case Study

Aminotrasferases are enzymes that transfer an amine group from a donor molecule to an acceptor molecule, resulting in a chiral amine. Chiral amines play an important role in pharmaceutical and fine chemicals (73,74). Chiral amines are also used as resolving agents for the preparation of chiral carboxylic acid. Chiragene, Inc., uses proprietary aminotransferase or transaminase technology to produce chiral amine or chiral amine derived molecules. Figure 1 shows the

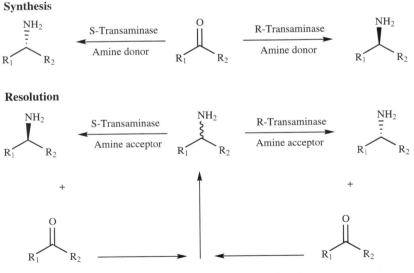

Fig. 1. Aminotransferase route to chiral amine.

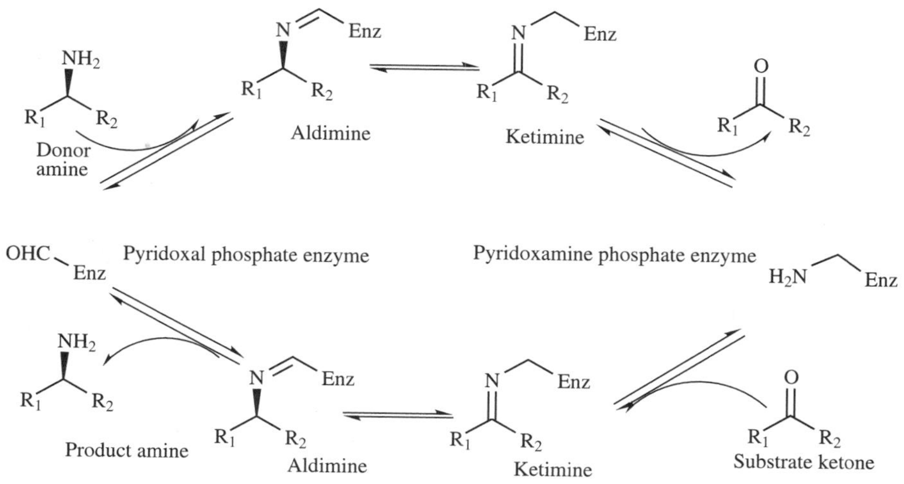

Fig. 2. Mechanism of transamination.

two possible reaction routes involved in a given transamination process. Synthesis involves the transfer of the amine group from a cheap amine donor (isopropylamine or aminobutane) to a prochiral ketone, whereas resolution involves selective transfer of an amine group from one isomer to an acceptor carbonyl group, leaving behind the corresponding ketone. Figure 2 describes the transamination reaction in detail. These enzymes require pyridoxal phosphate (PLP) as a cofactor to act as a shuttle to transfer the amine moiety. This cofactor is tightly bound to the enzyme and is not needed in stoichiometric amounts and hence does not pose the cofactor regeneration problems encountered in oxidation/reduction reactions.

Transamination is a cyclic process involving two steps to complete one cycle (75). As shown in Figure 2, during the first half of the cycle, the amine donor condenses with the aldehyde group of PLP to form a Schiff's base with the release of water. The imine complex thus formed tautomerizes to form a ketamine intermediate that undergoes hydrolysis to release the first product which is the ketone corresponding to the amine donor. At this point the amine group is transferred to the PLP attached to enzyme, converting it to pridoxamine phosphate (PMP). In the second half, the enzyme-PMP complex is attacked by acceptor ketone, followed by a reversal of steps observed in the first half of the reaction. At the end of the second half of the reaction, the acceptor ketone gets converted to the product amine and enzyme–PLP complex is generated. In the resolution mode racemic amine acts as an amine donor while any suitable carbonyl acts as an amine acceptor. Since aminotransferases are selective, only one of the isomers reacts, leaving the desired isomer untouched.

Figure 3 shows a block diagram of aminotranferase process. A typical reaction consists of buffer, PLP, amine donor, enzyme, and amine acceptor in aqueous solution. The pH is maintained using buffer at desired level. After the reaction is complete the biomass is removed by centrifugation. The product amine is isolated from the reaction mixture by solvent extraction. Both

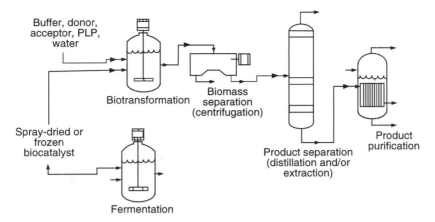

Fig. 3. Process diagram for chiral amine production using aminotransferase.

the synthesis and resolution approaches are tested in a laboratory scale to figure out the most economical route of producing the desired chiral amine molecule in hand. The route that is selected has performance criteria attached, such as the required yield and concentration. Some of the typical hurdles to attain the performance requirements are equilibrium limitation, low chiral purity, enzyme inactivation, and substrate solubility. The enzyme is tailored to meet the performance criteria using error-prone PCR and a colorimetric screen. Once a suitable biocatalyst is identified in the screen it is fermented at 16 L scale and spray dried for testing at small scale (50–100 mL reaction). If the desired performance is not reached, the best mutants are subjected to a second round of tailoring, screening, and testing. This is continued until the desired performance is reached, and this is illustrated in Figure 4. Once the desired improvement for a given process is

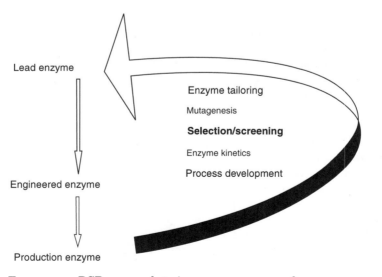

Fig. 4. Error-prone PCR approach to improve enzyme performance parameters.

reached the process is scaled up from 50 mL to 200 gal to 700 gal to 2000 gal reaction volume. Since the biotransformation reactions involve nearly homogenous reaction mixtures there is not much of a deviation from the laboratory-scale results. The decision to use synthesis or resolution depends on relative availability of substrate, cost of substrate, and enzyme activity. The direct synthesis of chirally pure amine is especially valuable where the ketone is expensive. One such example is the production of substituted 2-aminotetralins, starting from prochiral-substituted β-tetralones. Typical costs for the ketone range from $100 to $5000/kg, and in such a situation, an amine yield in excess of 90% is paramount.

After initial feasibility studies with one of the substituted β-tetralones, the concentration achieved in the synthesis mode was low (1–6 g amine in a liter). This throughput in the biotransformation process competes with the traditional chemical route. The performance criteria needed to make the biotransformation feasible required the throughput and the rate to be increased threefold while maintaining the stereoselectivity. Experiments showed that the reaction was equilibrium limited and the enzyme was experiencing inhibition/inactivation by substrate and product amine. Two possible ways of shifting equilibrium in the desired direction are (1) by increasing the concentration of one or both substrates and (2) by removing one or both of the products in the reaction. The latter method is more difficult because the product ketone or amine has to be removed selectively (ie, without affecting the concentration of substrate ketone or amine). The former method is easier because all it needs is additional substrate. Since the ketone is expensive it makes economic sense to increase the concentration of amine donor (isopropylamine or 2-aminobutane) that is relatively inexpensive. However, increasing the concentration of amine donor necessitates that the enzyme be active under these conditions. Since the other hurdle was enzyme inhibition/inactivation by amine engineering an enzyme tolerant to high donor amine concentration could conceivably result in overcoming both hurdles at the same time.

Using the error-prone PCR and a colorimetric screen, several thousand clones were screened for improved performance. The screen was designed to pick up clones that showed activity in the presence of high concentrations of donor amine. An additional indicator that was used to select the clones was the time it required to display activity. After three to four rounds of enzyme modification, an enzyme was identified that was tolerant to high concentrations of donor amine. This enzyme was tested in the biotransformations at small scale (20–50 mL) using the high donor amine concentration, and the product amine concentration was measured as a function of time. The results from this experiment are shown in Figure 5. The modification of the enzyme in conjunction with the appropriate screen resulted in an enzyme that achieved a fourfold increase in product amine concentration within 8–10 h without any change in stereoselectivity (>99% ee). This biotransformation process was scaled up to 2000 gal with reproducible results. Figures 6 and 7 show some of the improvements attained by the enzyme-modification technique in similar transamination reactions. Figure 6 illustrates a case where inactivation/inhibition by product amine was reduced and the enzyme stability was increased, resulting in a sixfold improvement in final product concentration. In addition to improving enzyme

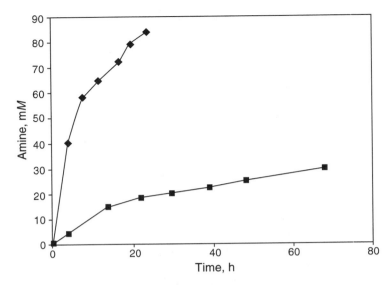

Fig. 5. Comparison of wild-type and modified enzyme. Improvements were made targeting amine tolerance and reaction rate. ◆ Modified; ■ wild type.

activity and stability it is possible to improve selectivity using the same technique. Figure 7 shows the improvement in enantioselectivity that was achieved starting from ~6% ee to >99% ee. These examples demonstrate the power of enzyme modification and its impact on achieving performance and economic targets for an industrial biotransformation process.

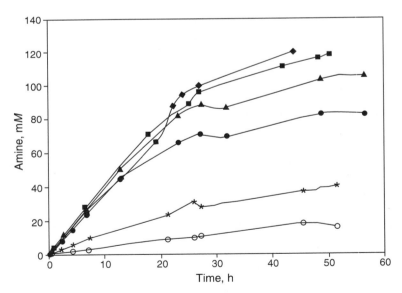

Fig. 6. Improvement of enzyme stability and acitivity using error-prone PCR. ◆ 4th generation; ■ 3rd generation; ▲ 2nd generation; ● 1st generation; ＊ 1st generation; ○ wild type.

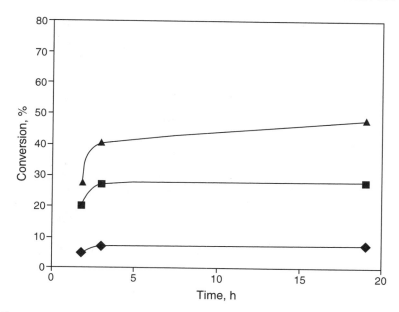

Fig. 7. Improvement in enzyme selectivity using error-prone PCR. ◆ Wild type, 6.5% ee; ■ 1st generation, 37.8% ee; ▲ 2nd generation, 99.4% ee.

6. Conclusions

Biocatalysis is growing very fast because of its unmatched stereoselectivity, regioselectivity, minimal waste, and relatively mild reaction conditions. In addition, recent development in solvent and protein engineering has made redesigning biocatalysts to suit the particular process easier. Major pharmaceutical companies are investing heavily in biocatalysis, indicating the utility and acceptance of biocatalysis as an additional tool in a scientist's hand. Enzymes from all different classes have been used to prepare chiral fine chemicals and pharmaceutical intermediates. Both whole cells and isolated enzymes have been used to catalyze the synthesis of various different chemical reactions. Although there is much to be done in this area, the future holds great promise in this burgeoning field.

7. Acknowledgments

The authors would like to acknowledge the contributions of all their colleagues at the Cambrex Technical Center, especially Dr. Ron Carroll.

BIBLIOGRAPHY

1. C. Neuberg and J. Hirsch, *Biochem. Z. Allg. Microbiol.* **115**, 282 (1921).
2. H. S. Shin and O. P. Ward, *Appl. Microbiol. Biotechnol.* **44**, 7–14 (1995).

3. M. W. W. Adams and R. M. Kelly, *Chem. Eng. News*, 32–42 (Dec. 18, 1995).

4. J. E. Brenchley, *J. Ind. Microb.* **17**, 432–437 (1996).

5. K. Faber, *Biotransformations in Organic Chemistry*, Springer-Verlag, Berlin, 1995, p. 3.

6. A. M. Klibanov, *Science* **219**, 722–727 (1983).

7. K. B. Storey and D. Y. Shafhauser-Smith, *Biotechnol. Gen. Eng. Rev.* **12**, 409–465 (1994).

8. A. Zaks and A. M. Klibanov, *Science* **224**, 1249–1251 (1984).

9. D. B. Volkin and co-workers, *Biotechnol. Bioeng.* **37**, 843–853 (1991).

10. G. Ayala and co-workers, A., *FEBS Lett.* **203**, 41–43 (1986).

11. C. R. Wiscott and A. M. Klibanov, *BBA Biochem. Biophys. Acta.* **1206**, 1–9 (1993).

12. T. Sakurai and co-workers, *J. Am. Chem. Soc.* **110**, 7236–7237 (1988).

13. S. V. Kamat, E. J. Beckman, and A. J. Russell, *J. Am. Chem. Soc.* **115**, 8845–8846 (1993).

14. S. Tawaki and A. M. Klibanov, *J. Am. Chem. Soc.* **114**, 1882–1884 (1992).

15. A. Zaks and A. M. Klibanov, *J. Am. Chem. Soc.* **108**, 2767–2768 (1986).

16. P. A. Fitzpatrick and A. M. Klibanov, *J. Am. Chem. Soc.* **113**, 3166–3171 (1991).

17. J. S. Dordick, *Biotechnol. Prog.* **8**, 259–267 (1992).

18. D. R. Rubingh, *Curr. Opin. Biotech.* **8**, 417–422 (1997).

19. O. Kuchner and F. H. Arnold, *TIBTECH* **15**, 523–530 (1997).

20. J. C. Moore and F. H. Arnold, *Nature Biotech.* **14**, 458–467 (1996).

21. F. H. Arnold, *Chem. Eng. Sci.* **51**, 5091–5102 (1996).

22. K. Chen and F. H. Arnold, *Proc. Natl. Acad. Sci. USA* **90**, 5618–5622 (1993).

23. L. You and F. H. Arnold, *Protein Eng.* **9**, 77–83 (1995).

24. W. P. C. Stemmer, *Nature* **370**, 389–391 (1994).

25. A. Crameri and co-workers, *Nature Biotech.* **14**, 315–319 (1996).

26. P. A. Patten, R. J. Howard, and W. P. C. Stemmer, *Curr. Opin. Biotech.* **8**, 724–733 (1997).

27. W. P. C. Stemmer and co-workers, *Gene* **164**, 49–53 (1995).

28. W. P. C. Stemmer, *Proc. Natl. Acad. Sci. USA* **91**, 10747–10751 (1994).

29. H. Zhao and F. Arnold, *Protein Eng.* **12**, 47–53 (1999).

30. H. Zhao and co-workers, *Nature Biotechnol.* **16**, 258–261 (1998).

31. K. Chen and co-workers, *Biotechnol. Prog.* **7**, 125–129 (1991).

32. V. Schellenberger, *ASM News* **64**, 634–638 (1998).

33. J. W. Jacobs, *Bio/Technology* **9**, 258–262 (1991).

34. G. M. Blackburn and P. Wentworth, *Genet. Eng. Biotechnol.* **14**, 9–24 (1994).

35. S. J. Pollack, J. W. Jacobs, and P. G. Schultz, *Science* **234**, 1570–1573 (1991).

36. A. Tramontano, K. D. Janda, and R. A. Lerner, *Science* **234**, 1566–1570 (1991).

37. C. Vita, *Curr. Opin. Biotechnol.* **8**, 429–434 (1997).

38. K. Hao and co-workers, *J. Am. Chem. Soc.* **118**, 10702–10706 (1996).

39. S. W. May and S. R. Padgette, *Biotechnology* **1**, 677–686 (1983).

40. J. B. Jones, *Tetrahedron* **42**, 3351–3403 (1986).

41. W. A. C. Somers and co-workers, *TIBTECH* **15**, 495–500 (1997).

42. E. Schmidt, in J. M. S. Cabral, D. Best, L. Boross, and J. Tramper, eds., *Applied Biocatalysis*, Harwood Academic Publishers, Chur, Switzerland, 1994, p. 133.

43. K. Drauz and H. Waldmann, eds., *Enzyme Catalysis in Organic Synthesis*, Vol. **VII**, 1995, p. 598.

44. J. Reddy and co-workers, *J. Ferm. Bioeng.* **81**, 304–309 (1996).

45. M. Kataoka and co-workers, *Biosci. Biotechnol. Biochem.* **62**, 167–169 (1998).

46. O. P. Ward and C. S. Young, *Enzyme Microb. Technol.* **12**, 482–493 (1990).

47. R. N. Patel and co-workers, *Enzyme Microb. Technol.* **14**, 731–738 (1992).

48. H. Yamada and S. Shimizu, *Angew. Chem., Int. Ed. Engl.* **27**, 622–642 (1988).

49. O. P. Ward, *Can. J. Bot.* **73**, S1043–S1048 (1995).
50. H. E. Swaisgood, in A. I. Laskin, ed., *Enzymes and Immobilized Cells*, The Benajmin/Cummings Publishing Co., Inc., Menlo Park, Calif., 1985, p. 20.
51. M. D. Lilly, *Chem. Eng. Sci.* **49**, 151–159 (1994).
52. M. Chartrain and co-workers, *Ann. N.Y. Acad. Sci.* **799**, 612–619 (1996).
53. G. Matcham and co-workers, in Chiral '94 USA Proceedings, Reston, Va., 1994.
54. J. A. Littlechild and H. Watson, *BFE* **7**, 158–160 (1990).
55. M. Mahmoudian and co-workers, *Enz. Microb. Tech.* **20**, 393–400 (1997).
56. W. T. Loos and co-workers, *Biocatal. Biotransform.* **12**, 255–266 (1995).
57. N. Matsuo and N. Ohno, *Tetrahedron Lett.* **26**, 5533–5534 (1985).
58. C. G. Kruse, J. Brussee, and A. Van Der Gen, *Spec. Chem.* **12**, 184–192 (1992).
59. T. Sugai and H. Ohta, *Agric. Biol. Chem.* **55**, 293–294 (1991).
60. G. M. R. Tombo and D. Bellus, *Angew. Chem.* **113**, 1219–1241 (1991).
61. I. G. Shenouda and L. C. Chien, *Macromolecules* **26**, 5020–5023 (1993).
62. L. C. Chien and co-workers, *Liq. Cryst.* **15**, 497–511 (1993).
63. A. L. Margolin, *Enzyme Microb. Technol.* **15**, 266–280 (1993).
64. G. P. Stahly and R. M. Starreett, *Chirality in Industry II*, John Wiley & Sons, Inc., New York, 1997.
65. A. L. Margolin, *Enzyme Microb. Technol.* **15**, 266–280 (1993).
66. A. Zaks and D. R. Dodds, *DDT* **2**, 513–531 (1997).
67. A. R. Macrae, *Biocatalyst in Organic Syntheses*, Elsevier Science Publishers, Amsterdam, 1985, p. 195.
68. M. K. Chang and co-workers, *JAOCS* **67**, 833–834 (1990).
69. L. Majovic and co-workers, *Enzyme Microb. Technol.* **15**, 443–448 (1993).
70. K. Jaeger and M. T. Reetz, *TIBTECH* **16**, 396–403 (1997).
71. A. Pandey and co-workers, *Biotechnol. Appl. Biochem.* **29**, 119–131 (1999).
72. D. I. Stirling, *Chirality in Industry*, John Wiley & Sons, Inc., New York, 1992, p. 209.
73. J. Shin and B. Kim, *Biotech. Bioeng.* **55**, 348–358 (1997).
74. J. Shin and B. Kim, *Biotech. Bioeng.* **60**, 534–540 (1998).
75. A. L. Lehninger, *Biochemistry*, Worth Publishers, Inc., New York, 1978, p. 562.

SACHIN PANNURI
ROCCO DISANTO
SANJAY KAMAT
Cambrex Technical Center

BIOMASS ENERGY

1. Introduction

Energy supplies in the United States and around the world are primarily based on the supply of fossil fuels, coal, petroleum, and natural gas. Concerns about the long-term stability of supply of these fossil fuel resources coupled with concerns about the environmental impacts of the use of these fuels has led to the investigation of the use of more renewable resources. Primary among these renewable resources is biomass. Biomass currently provides a ~10–11% of the worlds primary energy, however, the various forms of biomass can potentially supply a quantity of energy several times that of the current global demand (1,2). In the United States, a ~4% of the primary energy supply is provided by biomass. A significant increase to 20% of primary energy is possible within know sustainable supplies of biomass, without having an adverse effect on prices of biomass derived products such as lumber or food (4). Developing countries utilize more biomass, proportionately, than industrialized countries. The United Nations reports that 38% of primary energy is supplied by biomass in developing countries (5). In these developing countries, biomass is used for heating, cooking, and in some instances, power generation. In the more industrialized countries, these energy uses have been replaced by fossil fuels.

Even in the more developed countries, biomass played a key role until the second World War. In the mid-1800s, 85% of the United States' primary energy was derived from biomass. This figure declined to only 2.5% when, in the mid-1970s, the energy crisis resulted in a renewed emphasis by government and the public to expand the use of alternative energy sources, including biomass. Increased environmental pressures including concerns relative to air quality and to the disposal of biomass containing wastes such as construction and demolition wastes, municipal solid wastes, sewage sludges, and urban biomass (yard) wastes further expanded the interest in energy production from biomass materials.

2. What Is Biomass?

Biomass quite simply is any organic material that is or was derived from plants or animals. The most common example of biomass is fuelwood. Other typical forms include crops, animal manures, agricultural residues (examples include sugarcane baggasse, corn stover, rice hulls, nutshells, and straw), sludges from municipal or pulp and paper wastewater treatment, and even portions of municipal solid waste (food residues, yard wastes, and paper). There are also a number of aquatic based biomass forms such as kelp and seaweed. A more legalistic definition has been proposed by researchers and several governments throughout the world. This definition describes biomass as "all nonfossil organic materials that have an intrinsic chemical energy content. This includes all water- and land-based vegetation and trees, or virgin biomass, and organic components of waste materials such as municipal solid waste (MSW), municipal biosolids (sewage) and animal wastes (manures), forestry and agricultural residues, and certain types of industrial wastes" (6).

3. Why Biomass Energy?

Biomass derived energy is considered a "renewable" energy source. That is, it is a source of energy that can be utilized without depleting the reserves. As long as the conditions exist (sunlight, water, and the organic substrate) additional biomass can be produced to replenish that used for energy. Interest in renewable energy is increasing due to dwindling supplies of petroleum and natural gas, two fossil fuels, and to mitigate global climate change. For some time, a worldwide debate continued regarding the case of an accumulation of carbon dioxide in the atmosphere. As more and more data becomes available, there is an increasing concensus that the most significant cause of this accumulation is human activity the affects the natural carbon cycle. The primary causes of the increase are the use of fossil fuels and the deforestation of large land areas (6). During the 1990s, 6.3 Gtonnes of carbon were released to the atmosphere each year from the burning of fossil fuels and an additional 1.6 Gtonnes released from deforestation activities. Of this amount, 4.6 Gtonnes is absorbed annually by the world's oceans and terrestrial vegetation resulting in an annual increase in the atmosphere of 3.3 Gtonnes (7).

An increase in the use of biomass energy provides a means to both reduce the consumption of fossil fuels and to reclaim a portion of land that has been deforested. Because solar energy is an integral part of the production of most biomass, some consider biomass a form of stored solar energy. Biomass based energy, however, provides the advantage of being available at all times (dispatchable) regardless of atmospheric conditions. Further, as biomass is produced from sunlight by the following reaction,

$$CO_2 + H_2O + \text{light} + \text{chlorophyll} \rightarrow (CH_2O) + O_2$$

[The generic carbohydrate formula, (CH_2O), represents the primary building block of the organic product. Woody biomass is richer in lignocellulosic material and has the average formula of $CH_{1.4}O_{0.6}$ due to the higher lignin content.] carbon dioxide, a primary greenhouse gas, is consumed as new biomass is formed in a quantity equal to that produced when biomass is used as an energy source. As a result, biomass provides environmental advantages in relation to greenhouse gases when compared to fossil fuels. When considered as a stand alone process, photosynthesis is a very low efficiency process (<1% or the solar energy is converted into biomass) (8), but due to its simplicity and wide range of products produced, it provide the potential for not only food (energy of a sort for other living organisms) but as a simple to acquire form of thermal energy.

Biomass as an energy source is not free of environmental issues. Biomass is renewable and carbon-neutral only if replacement through new growth occurs at, at least, the same rate at which it is harvested. Furthermore, in many undeveloped countries, cooking and heating using biomass is accomplished in relatively crude stoves or fireplaces. These devices can release carbon monoxide, unburned hydrocarbons, and particulates (as smoke) (9). More advanced, high efficiency biomass-to-energy conversion systems introduce their own unique set of environmental and safety concerns. Intermediate products such as synthesis gas or biomass derived liquids can be hazardous, flammable, or toxic. Most, if not all, of these potential intermediates can be handled safely through proper application of industry accepted safety and health guidelines without any widespread expansion.

Biomass ash, from any conversion system, is generally recyclable to the soil unless the incoming biomass has been contaminated with heavy metals (as is the case with some municipal residues). However, biomass ash, particularly from combustion devices, is a very fine (sometimes <5 μ) material leading to potential emissions of fine particulate matter, PM-2.5. The control of PM-2.5 emissions will be necessary from biomass based sources as newer environmental regulations are implemented.

4. Historical Uses of Biomass Energy

Biomass in the United States historically was used for simple heating by direct combustion in wood stoves or fireplaces. Such technologically simple uses continue today in developing countries throughout the world. Biomass, due to its ready availability, provides a simple and reliable energy supply that can be adapted for home and simple industrial purposes. In the United States, the primary form of biomass used for these simple applications is fuelwood. Fuelwood, in the form of logs or brush is burned in air to provide heat. The combustion of biomass in a fireplace or wood stove is accomplished by reacting the biomass feed material in air according to the following reactions:

$$C + O_2 \rightarrow CO_2 \tag{1}$$

and

$$2\,H + 1/2\,O_2 \rightarrow H_2O \tag{2}$$

Fuelwood continues to be the primary heating fuel in some 3 million homes in the United States (10). Such simple heating with wood is by nature a very low efficiency operation when practiced in an open fireplace ($\sim$7% efficiency) or even in a fireplace with convective tubes ($\sim$15%) providing only a small fraction of the energy available in the incoming biomass to heat, the remainder being lost to the flue or the general environment as unburned fuel. Efficiency improvements have been realized in modern log, chip, and pellet fuel appliances by finer control of the combustion process and by the use of factory fabricated combustion equipment (50–76% overall efficiency) (11). At larger scale, efficiency improvements are realized by larger scale combustion processes in boilers or furnaces, thus improving heat recovery and therefore overall efficiency.

4.1. Power Generation from Biomass. Boiler applications further provide the opportunity to generate electric power from the incoming biomass. Biomass is burned to generate steam. The steam is then used to turn a turbine for the generation of electric power. In the United States, several hundred such biomass based power plants have a combined capacity of >10.3 GW and annually generate 65 Teraw-h of electricity (12,13). The majority of these biomass-based power plants are found in the pulp and paper industry where biomass residues from the pulping operation are used to generate necessary power for plant operations.

Types of Biomass Combustion Devices. To adequately process biomass, a stoker-grate boiler is the usual combustion device of choice. While suspension-type boilers, similar to those used in pulverized coal combustion systems, are also used, these require the biomass to be ground to a fine particle size, thereby consuming additional energy (estimated at >80 kWh tonne^{-1}) and further reducing the efficiency of the power generation process (14). Stoker-grate boilers, on the other hand, can accept much larger particle size material (up to three inches or larger) less reducing overall plant energy consumption. Furthermore, because an inventory of material is continuously present on the grate, significant changes in the feedstock can be tolerated (15,16). Biomass by its very nature is heterogeneous therefore combustion systems that are tolerant of changes in feedstock provide advantages over other types of combustion systems. Furthermore a stoker-grate boiler allows the use of low heating value and high moisture content feedstocks. When viewed on a moisture and ash-free basis, however, much of the apparent variability of biomass fuels disappears. Typically, dry woody biomass will have a heating value in the range of 18.5 MJ kg^{-1}. A similar dry, ash free heating value is obtained for many agricultural residues (17). There is a considerable variation, however, in the moisture content of biomass materials. Most woody biomass will have an as fired moisture content between 40 and 50% thereby lowering its effective heating value by 50%. By comparison, the typical coal feedstock used for power generation will have a heating value in the range of 27.6 MJ kg^{-1} and a moisture content in the range of 10% or less. The moisture in the biomass fuels has the effect of lowering the overall power generation efficiency of stoker-grate boilers to an average of ~20% due to the high excess air levels of air required and the energy required to evaporate the moisture in the incoming biomass fuel. The maximum efficiency for this type of conversion device is ~25% (18).

4.2. Other Commercial Uses of Biomass Combustion.

The pulp and paper industry is a primary user of biomass combustion for its energy needs. The industry as a whole derives ~56% of its total energy from wood and wood residues (19). In addition to direct combustion of wood and woody residues as described above, a significant portion of the energy in a modern pulp mill is derived from the combustion of black liquor, a by-product of the pulping operation. Black liquor contains the spent pulping chemicals and lignin content from the incoming wood pulp. The lignin content provides the energy content in the liquor. When black liquor is burned for energy recovery, the pulping chemicals are recovered and recycled, thus further increasing energy efficiency and reducing costs.

Wood and wood residues (bark, twigs, etc) are collectively known as hog fuel. In a pulp mill, the hog fuel is burned in a power boiler and produces steam for power generation. Typically, the power produced will provide nearly all of the requirements for the mill, and in some cases power for export to the grid.

4.3. District Heating and Combined Heat and Power.

In many European countries, biomass based combined heat and power (CHP) systems provide the primary energy for large segments of the population. These systems typically combust biomass to produce high pressure steam that is used for power generation. Lower pressure steam is then extracted from the turbine system and used for district heating. In some of the Scandinavian countries, biomass enjoys an economic advantage over fossil fuels thus encouraging the development of more advanced CHP systems (20). In Sweden, eg, two-thirds of the power generated from CHP facilities is biofuel based amounting to approximately 18 GWe with an additional 22 GWth supplied through 9600 km of mains as district heating. In Finland, 37% of the country's electrical energy is derived from CHP plants (21).

4.4. Charcoal.

While not used extensively in the United States as an energy source, charcoal plays a major role in many countries as a primary source of fuel for heating, cooking, power production, and metal processing. In some Southeast Asian countires, eg, up to 15% of the primary energy supply of the country is derived from charcoal (22). Likewise, in many African countries, charcoal consumption is higher than gross electricity consumption (23). Charcoal can be made from virtually any organic material. The primary source is fuelwood or coconut shells. The production of charcoal involves heating the incoming biomass under starved air conditions (in a pit, or a kiln) to drive off water, volatiles, and tars. The remaining material, which is primarily carbon, with traces of ash and a few remaining volatiles is charcoal. In a typical charcoal making process, ~20% of the incoming biomass is converted to charcoal. Charcoal has a heating value approximately twice that of fuelwood on an equal weight basis (28–30 MJ kg^{-1}) thereby increasing its ease of transportation. The increased heat content also allows a smaller volume of fuel to be used for domestic heating applications. As many of the biomass materials used in the production of charcoal are residues from agricultural sources, the energy lost in charcoal production is outweighed by the increased flexibility of the charcoal fuel.

Environmental concerns regarding charcoal have recently become important. As the production of charcoal releases volatile organics into the environment and the burning of charcoal releases CO_2, some environmentalists believe that the production of charcoal should be halted completely. World consumption of charcoal, however, is increasing raising concerns about the supply of fuelwood and other biomass resources. If charcoal is produced on a sustainable basis, ie, without deforestation, many of these concerns can be alleviated.

5. Other Energy Recovery Systems

In an attempt to enhance the supply of petroleum in the 1970s a number of methods were used to recover energy from biomass sources much of this work was supported by the United States government through research grants and support for the construction of larger scale facilities. These included wastes to energy facilities utilizing post consumer residues (municipal solid waste) and additional fuelwood fired power facilities and the construction of the number of ethanol producing facilities.

5.1. Waste-to-Energy Facilities. Waste-to-energy facilities typically use the same type of stoker-grate combustor as described above for wood fired power generation. The incoming postconsumer residues from municipalities are generally combusted in a "mass burn" system. In such systems, the incoming residues are dumped in mass into a large holding pit and then transferred by an overhead crane directly into the combustion chamber. No additional preparation of the fuels is performed. Large pieces of oversize material (including large metallic items such a bicycles, containers, etc) are removed along with ash from the base of the combustion chamber. As might be expected, maintenance on these combustors is quite high due to the large variety of materials that is being combusted. An improvement in waste-to-energy facilities is accomplished by performing some preparation of the incoming feed material prior to combustion. This preparation removes much of the metallic materials and glass from the feed material and additionally is used to reduce the size of the incoming feed material so that combustion properties can be enhanced. The resulting prepared municipal waste is referred to as refuse derived fuel (RDF). The RDF plants generally have lower maintenance requirements and higher efficiency and mass burn facilities but are much more expensive than a similar size mass burn facility. Both of these methods require large capital investments thus requiring subsidies in the form of "tipping fees" in excess (sometimes double or triple the fee) of those charged for landfill operations to provide adequate returns on the capital investment. A step-change in technology is required to overcome these financial limitations.

Environmental Considerations. Furthermore, waste-to-energy facilities have potential environmental impacts on the surrounding community. Many of the environmental contaminants present in the exhaust from a waste to energy facility can be controlled by the use of flue gas scrubbers. Contaminants such as nitrogen oxides and particulate are removed in this manner from larger scale coal fired utility boilers. The environmental control systems for waste to energy facilities utilize this same technology. Because waste-to-energy facilities process a fuel that can contain significant quantities of chlorine (from materials such as polyvinyl chloride), much more hazardous contaminants can be produced. These included dioxins and furans. These contaminants, too, can be removed from the exhaust of the facilities but the cost of scrubbing these materials is significantly higher than to remove nitrogen oxides or particulates.

These potential environmental issues have resulted in a reluctance by municipalities to install additional waste to energy facilities thus limiting the amount of energy that can be derived from these wastes. Other forms of urban derived biomass wastes have been included, by association, in the same category as MSW. These other wastes, however, do not contain the same contaminants (such as chlorine) present in MSW, and therefore would not be expected to have the same environmental issues. Included in these other biomass wastes are tree trimmings, yard wastes, and construction and demolition wastes. In large urban areas, hundreds of tons of these materials are produced on a daily basis providing a significant resource for energy recovery (24).

5.2. Ethanol. Ethanol is a form of alcohol found in beverages such as wine and beer. It is readily produced through the natural fermentation of the starches and complex sugars present in many forms of biomass. Grains such as wheat or corn are typically used as the primary sources for ethanol production. The fermentation reactions are accomplished by naturally occurring biological organisms. When used for fuel, ethanol is blended with other liquid fuels (typically gasoline) (25,26).

Ethanol Production. Large-scale production of ethanol for energy is produced in a process similar to brewing beer. Fermentation of the sugars and starches in the biomass is accomplished by blending with yeast and water. The mixture is then heated to ~30°C to initiate the fermentation reactions (27). The yeast breaks down the complex sugars and starches into simple sugars and ultimately into ethanol. The fermentation reactions take place in an anaerobic environment and take ~1–2 days for completion. After fermentation the ethanol is filtered and then distilled to increase its concentration from 10 to 95%. The concentrated ethanol is then blended with petroleum-based fuels to the desired concentration. About 5.7 GL of ethanol are produced in this manner in the United

States annually (28,29). The ethanol industry has shown significant growth in the period 1980 to 2002 by producing over 20 times the product annually that was produced in 1980. This dramatic growth is possible due to advances in both chemical and biological process engineering (30,31). In addition to the primary use as a blend in gasoline, ethanol can also be used as a chemical intermediate in the production of other organic chemicals.

Ethanol is also produced from cellulosic biomass such as agricultural residues, forestry residues, waste paper, yard wastes, portions of municipal residues, and some industrial residues. The cellulose and hemicellulose in these materials are long-chain polymers made up of sugar molecules. These materials are treated with acids or enzymes to break down the molecules into smaller fractions that can be readily fermented. Dilute acid hydrolysis is carried out at high temperatures (~240°C) with dilute sulfuric acid to hydrolyze the cellulose and hemicellulose. In this case, the sugars can be degraded due to the high reaction temperature, thus reducing the yield. Conversely, concentrated acids can be used at lower temperatures thereby increasing the yield but necessitating more complex acid recovery systems. Research continues on acid hydrolysis to reduce acid levels and maintain high conversion levels. Alternatively, a small fraction of the biomass can be used to grow fungi or other organisms to produce enzymes called cellulases. These enzymes then hydrolyze the cellulose in pretreated biomass to glucose that can be fermented into ethanol.

Use of Ethanol Containing Fuels. In 1990, the United States government mandated the use of oxygenated transportation fuels to help improve environmental conditions in parts of the country. These mandates have created a strong demand for ethanol–gasoline blends. Mixtures of up to 10% ethanol referred to as (E10) can be utilized in most gasoline designed engines with no modification. More concentrated blends, namely, the E85 and E95 blends (85 and 95% ethanol, respectively) require specifically designed engines, referred to as "flexible fuel" engines to perform properly. These flexible fuel engines can run on either gasoline or the high concentration ethanol blends. Automobile manufacturers are producing more vehicles that can use these higher ethanol blends each year thus increasing demand.

The spark ignition, Otto-cycle engines used in today's automobiles, even in their early stages of development, were designed to operate with ethanol containing fuels. Henry Ford designed the early Model T to use ethanol as a major fuel source. Such plans were changed subsequently when storage and transportation difficulties combined with high corn prices at the time caused the supply of ethanol containing blends to be reduced. Further efforts to use ethanol containing fuels continued into the 1940s when petroleum became the primary automotive fuel. Today, the petroleum industry is reluctant to accept ethanol on a widespread basis as it is not a fungible fuel due to the storage and transportation issues as well as the potential for phase separation in blends with gasoline.

6. Developing Methods for Conversion of Biomass into Energy

The United States Department of Energy (DOE) along with other similar organizations around the world have been supporting the development of a wide range of biomass to energy conversion technologies. Among these are gasification, pyrolysis, advanced combustion, anaerobic digestion, bio-diesel, and advanced cellulose to ethanol systems. Each of these has unique characteristics and has the potential to provide some energy from biomass. The following sections discuss the characteristics and projected markets for each of these technologies.

6.1. Gasification.

Gasification is quite simply the conversion of a solid or liquid material into a gaseous fuel. The resulting fuel gas has an energy content or "heating value" ranging from 10 to 50% of the heating value of natural gas. The wide variation in heating values is a direct result of both the reactor type used and the reactants chosen for the gasification reactions. Gasification is probably the most flexible conversion system for biomass materials as the fuel gas produced can be used directly as a fuel for heating applications, be used for the production of power in gas turbines or fuel cells, or used as a synthesis gas for the production of liquid fuels, chemicals, or hydrogen. Figure 1 shows the many pathways that can be used to produce energy products by gasification.

Gasification, despite the current development efforts underway, is a technology that has been utilized at large scale for many years. Coal and biomass gasifiers (referred to as gas producers) were in commercial use in Germany in the mid-1800s. Similarly, in the United States, during the early 1900s gasification plants were producing "town gas" in most major cities (32). This town gas was used for residential and industrial heating, lighting, and cooking at that time. It was not until the 1950s that natural gas replaced the town gas as the gaseous fuel of choice. As late as the 1980s some areas of Europe used gasification derived fuel gas rather than natural gas as a major energy supply (33).

The primary fuel for gasification has historically been coal. But many biomass gasifiers utilizing similar reactor schemes have extensive commercial operating experience. Today's gasification development efforts are focused on improving the efficiency of these early designs and improving the environmental performance of the technology.

Gasification Reactions. The primary gasification reactions are as follows:

Reaction	Reaction heat	
$C + O_2 = CO_2$	408 MJ mol^{-1}	(3)
$C + CO_2 = 2\,CO$	-162 MJ mol^{-1}	(4)
$2\,C + O_2 = 2\,CO$	246 MJ mol^{-1}	(5)
$2\,CO + O_2 = 2\,CO_2$	570 MJ mol^{-1}	(6)
$C + H_2O = CO + H_2$	-120 MJ mol^{-1}	(7)
$C + 2\,H_2O = CO_2 + 2\,H_2$	$-78.6 \text{ MJ mol}^{-1}$	(8)
$CO + H_2O = CO_2 + H_2$	41.8 MJ mol^{-1}	(9)

In addition to these reactions, the volatile matter in the incoming feedstock is thermally broken down into a range of aliphatic and aromatic hydrocarbons (34). The simplest of these hydrocarbons, methane (CH_4) and benzene (C_6H_6), can enhance the heating value of the product gas while heavier hydrocarbons (C_{12} and higher) can be problematic when using the fuel gas in downstream equipment. These heavier hydrocarbons are referred to collectively as tars and are a byproduct of all gasification processes.

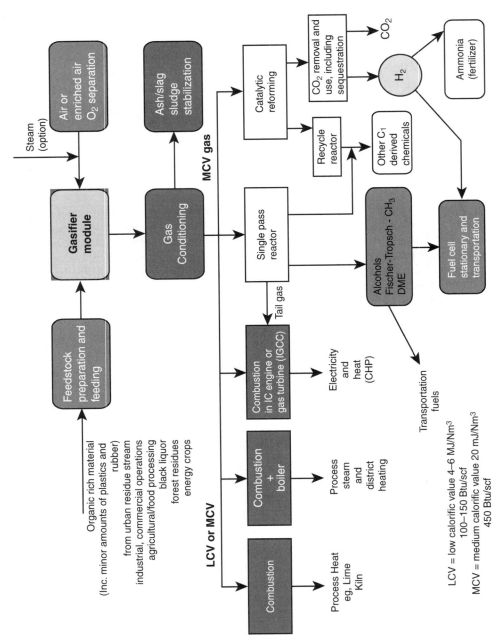

Fig. 1. Gasification technologies provide end-use flexibility.

LCV = low calorific value 4—6 MJ/Nm^3
100–150 Btu/scf
MCV = medium calorific value 20 mJ/Nm^3
450 Btu/scf

Some additional methane can be formed by the reaction:

$$C + 2\,H_2 = CH_4 \tag{10}$$

Reaction 10 is favored by higher pressures in the reactor.

Within a gasifier, all of these reactions take place to one degree or another. The specific proportions of the products of the reactions determine the heating value of the product gas produced. Biomass is a more reactive fuel (35) thereby providing the opportunity to process larger quantities of material in the same reactor and at lower temperatures than would be present in a coal gasifier. Biomass gasifiers typically operate at 800–900°C, while coal gasifiers may operate at temperatures of 1000°C or more. In general the reactivity (the ease with which a material can be converted) of the various fuels that are used in gasification systems can be expressed as:

Higher reactivity – liquids > biomass > coal > coke – lower reactivity

Biomass can exhibit a reactivity as much as a factor of 2 to 3 higher than that of coke, eg.

Gasification Reactions. There are three main types of reactors used for biomass gasification. These are fixed, fluidized, and entrained bed. In addition, there are some hybrid reactor types such as circulating fluidized bed and steam reforming fluidized bed that utilize properties of two reactor types to enhance the conversion reactions. In a fixed bed reactor, the incoming biomass is piled into a refractory lined chamber and reacted with an oxidant (air or oxygen) and/or steam to produce the fuel gas. There are two main types of fixed-bed reactor, updraft, illustrated in Figure 2 and downdraft, illustrated in Figure 3. In

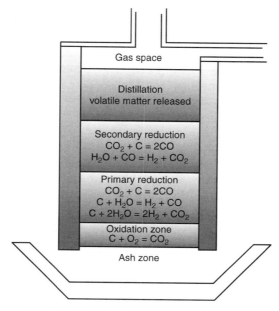

Fig. 2. Fixed-bed gasifier updraft mode.

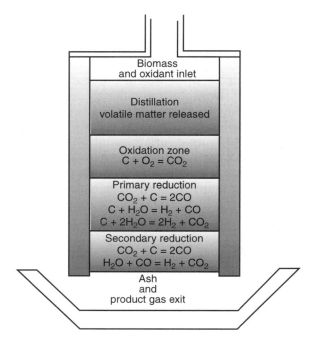

Fig. 3. Fixed-bed gasifier downdraft mode.

the updraft reactor, incoming biomass travels countercurrently to the oxidant source (usually air) while in the downdraft gasifier oxidant and biomass travel cocurrently through the reactor. The volatile matter in the incoming biomass is released at a lower temperature in an updraft reactor and has little opportunity to react with oxidant or steam before exiting the reactor. In contrast, in a downdraft gasifier, the volatiles are released in a higher temperature zone and continue through the reactor bed before exiting the reactor. The product gas from an updraft gasifier therefore will typically have a higher heating value and a higher tar content than the product gas from a downdraft gasifier.

In fixed-bed gasifiers, the inventory within the reactor must be supported by the fuel bed in order to maintain the chemical reactions. As the biomass is converted within the reactor, only the inorganic material (ash) in the biomass remains. Ash strength is therefore an important characteristic of the feedstock. In order to increase the effective strength of the ash in fixed-bed gasifier systems, the incoming biomass is either pelletized to increase its effective density or uniform larger "cubes" ($\sim$10 cm) of biomass are used. The height of the fixed bed within the reactor is also restricted to limit the weight that must be supported by the ash bed. The general limit for bed height is approximately twice the bed diameter. This length to diameter (L/D) ratio is an important design parameter in the other reactor types as well.

The fluidized bed reactor (sometimes referred to as fluid bed reactor) provides more flexibility with regard to feedstock size and shape due to the manner in which the incoming biomass is handled. Within the reactor, incoming biomass particles are suspended in a bed of a finely divided ($\sim$50-particles), inert material

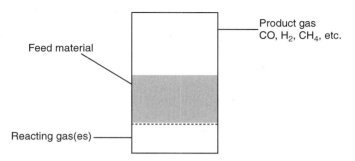

Fig. 4. Fluidized bed gasifier.

(typically sand). This inert bed is "fluidized" by the oxidant and steam reactants entering the reactor. These gaseous reactants are distributed across the reactor by a gas distributor. As they leave the distributor, bubbles are formed in the bed, lifting the particles of the bed material. Because of this bubble formation phenomenon, this reactor type is sometimes referred to as a bubbling fluidized bed. When the incoming velocity of the gas reaches the "minimum fludization velocity," all of the bed material is suspended and begins to behave as a liquid (hence the terminology "fluidized" bed). Fluidized beds are inherently well mixed chemically and thermally thus providing an excellent medium for the gasification reactions (36). A simplified schematic of a fluid bed reactor is shown in Figure 4.

Biomass feed material is transported into the reactor either above or directly into the fluidized bed. It rapidly mixes with the inert bed material and reacts to form the product gas. Because the inert bed material rather than the biomass provides the inertia within the reactor, fluidized beds can accept a wider range of biomass particle sizes. As a general rule, biomass particles from ∼200 μm to 10 cm can be utilized without difficulty in most fluidized bed reactors.

The L/D ratio for a fluidized bed is likewise more flexible, having a typical range from just over 1.0 to ∼5.0. Larger L/D ratios are avoided in fluidized bed design to eliminate an operational problem called slugging. Slugging is caused when bubbles within the fluidized bed grow to the reactor internal diameter and lift the bed material as a mass. The slugging results in larger pressure fluctuations within the bed and reduces the effectiveness of the gas–solids contacting within the reactor. The Renugas gasifier developed by the Institute of Gas Technology and being commercialized by Carbona is an example of a bubbling fluidized bed reactor system.

If gas velocities within the reactor are increased beyond minimum fluidization so that materials are carried through the reactor, the reactor is referred to as a "circulating fluidized bed" or an "entrained bed". In a pure entrained bed no inert medium is used. The incoming biomass is reduced in size to >200 μm and conveyed through the reactor as it reacts with the incoming oxidant and steam. The L/D ratios are less critical in this type of reactor system and may exceed 20 to 1 in some cases. Examples of CFB gasifiers are the Lurgi biomass gasifier and the TPS gasifier, both European developers (37,38). The hybrid "circulating fluidized bed" or CFB uses an inert medium (as in the bubbling fluidized bed, sand is

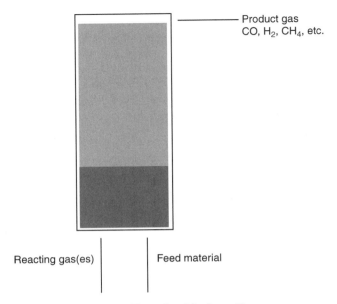

Fig. 5.   Entrained bed gasifier.

the typical material) to allow for a wider range in biomass particle sizes to be utilized. The inert bed material is recycled through the reactor at high rates (sometimes 10 times the biomass feed rate or more) and serves as a flywheel stabilizing temperatures and chemical reactions. By using the inert medium, incoming biomass materials may have a much wider range of particle size ranging from the 200 μm used in pure entrained beds to over 5 cm. This flexibility reduces overall operating costs for fuel preparation and expands the range of biomass materials that can be gasified. A CFB reactor schematic is shown in Figure 5.

The CFB reactor in many applications uses an oxidant as the primary conveying gas thus generating product in the same manner as in the fixed or bubbling fluidized bed reactors. Alternatively, a second CFB can be coupled to the gasfier to burn char produced during gasification and heat the circulating sand phase. Such a process, the SilvaGas process, is being commercialized by FERCO Enterprises, Inc. (39). The heated sand is recycled to the gasifier, providing the energy for gasification as shown in Figure 6. In this configuration, referred to as an indirectly heated gasifier, only steam is used as the conveying medium effectively increasing the heating value to a "medium calorific value" (a gas having a heating value in the range of 17 MJ m^{-3}) of the product gas in much the same way as using pure oxygen as the reactant gas in other reactor configurations. The separation of oxygen from air is an expensive unit operation and, therefore, due to the relatively small scale of biomass gasifiers is considered an uneconomical alternative for the production of a medium calorific value gas. An alternative indirect heating method utilizes heat transfer tubes imbedded in a bubbling fluidized bed to provide the heat necessary for the gasification reactions. This type of gasification reactor is being developed and commercialized by Thermochem, Inc. (36) and is shown in Figure 7. As in the CFB case, a medium calorific value gas is produced, but with a slightly higher hydrogen content

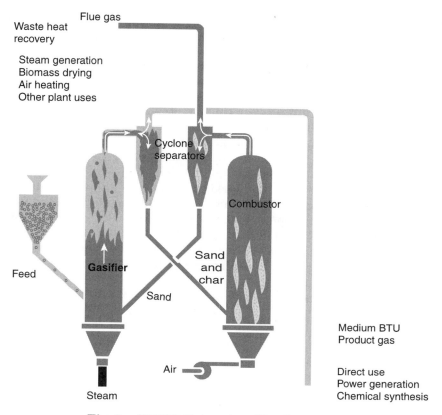

Flue gas

Waste heat
recovery

Steam generation
Biomass drying
Air heating
Other plant uses

Cyclone
separators

Combustor

Feed

Gasifier

Sand
and
char

Sand

Medium BTU
Product gas

Air

Direct use
Power generation
Chemical synthesis

Steam

Fig. 6. FERCO Enterprises SilvaGas gasifier.

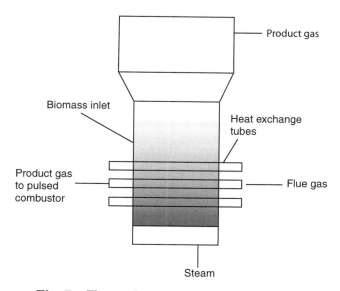

Product gas

Biomass inlet

Heat exchange
tubes

Product gas
to pulsed
combustor

Flue gas

Steam

Fig. 7. Thermochem indirectly heated gasifier.

as a result of reaction 9, the water gas shift reaction, more closely approaching equilibrium due to the increased residence time available in the bubbling bed.

The Products from Biomass Gasification.   The high reactivity of biomass helps determine the products produced during gasification. In all reactor types listed above, a mixture of gas, char, and liquids (tars) will be produced at varying levels. The levels of each component vary due to specific reactor conditions (temperature, pressure), reactant gases, and reactor heating method. The primary product of interest for gasification reactors is the product gas. In pyrolysis systems, discussed below, the primary product of interest is the liquid product. A characteristic product gas analysis from a number of gasification reactors is shown in the table below (37,39,40).

In addition to the product gas listed, coproducts of char and tars are produced to varying degrees. Char chemically consists of devolatilized biomass and has a higher concentration of carbon and a lower concentration of hydrogen than the incoming biomass. The production of char as a coproduct, therefore, results in a significant reduction in the overall efficiency of the gasification process. Operation of the gasification processes is adjusted in an attempt to minimize the formation of char.

Higher molecular weight hydrocarbons (typically above C_{12}) produced during gasification can condense in downstream unit operations and introduce operational difficulties. These tars vary from a few hundreds ppms to a few percent of the product gas depending on the reactor configuration. In order of the uncontrolled tar concentration produced, the reactor types discussed above are

$$\text{Low tar} < \text{downdraft fixed bed} < \text{indirect CFB} < \text{CFB}$$
$$= \text{bubbling bed} < \text{updraft fixed bed}$$

Use of the Gas from Gasification.   As shown in Figure 1, gases derived from the gasification of biomass have a potentially wide range of end uses. In the simplest application, the gas may simply be used as a fuel for heating or in industrial furnaces such as lime kilns in the pulp and paper industry. Either low calorific value gases from air-blown gasifiers or medium calorific value gases from oxygen blown or indirectly heated gasifiers can be used in this manner. Similarly, both low and medium calorific value gases can be used as boiler fuels for the production of steam for direct use or for the production of electric power.

Higher overall efficiencies can be achieved by use of the gases in more demanding applications (those requiring a higher degree of cleanup of the gases). These include, use in an internal combustion engine or a gas combustion turbine for the direct production of power. As much as a 60% increase in power generation efficiency can be realized compared to the less demanding steam cycle approach. Again, both low calorific value and medium calorific value gases are suitable for power generation applications (41).

The diluent effect of nitrogen present in low calorific value gases limits their application for more advanced chemical synthesis applications. Medium calorific value gases, on the other hand, are well suited for synthesis applications. A simple and the most commonly considered synthesis application is the production of methanol or other alcohols from the biomass derived gas. This approach provides a means to introduce biomass derived liquid fuels into the existing petrochemical (gasoline) infrastructure in much the same way as the direct production of ethanol. Like ethanol, however, methanol (and higher alcohols) are not fungible fuels, therefore limiting their acceptance by the petroleum industry. Methanol as a fuel is even more susceptible to phase separation in gasoline blends than ethanol. Synthesis gas from biomass can be utilized to generate essentially any product that would be produced from a petrochemical based synthesis gas. This includes chemical intermediates, polymers, fuel additives, or hydrogen.

6.2. Pyrolysis. Pyrolysis is a similar technology to gasification in that biomass feedstocks are heated in an anaerobic (no air present) environment to break down the biomass into primarily liquid hydrocarbons. In pyrolysis systems currently under development, the production of liquids is enhanced by rapidly heating the biomass in a "fast pyrolysis" mode. Fast pyrolysis is a process that yields a liquid product that is referred to by many names including pyrolysis liquid, pyrolysis oil, bio-crude-oil, bio-oil, bio-fuel-oil, pyroligneous tar, pyroligneous acid, wood liquids, wood oil, wood distillates and liquid wood.

A variety of reactor types are used in fast pyrolysis systems. Because a high heating rate enhances the liquid yield, reactors of the fluidized bed type are favored by most developers. Figures 8 and 9 illustrate fluidized bed and CFB reactor systems for the production of oils from biomass (bio-oil). To further enhance liquid yields, reaction temperatures are controlled in the range of 500–600°C (42).

The bio-oil produced in fast pyrolysis systems is a dark brown liquid with a heating value about one-half that of conventional fuel oil (16–19 MJ/kg^{-1}). In

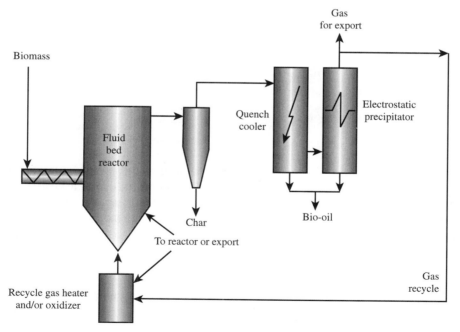

Fig. 8. Fluidized bed fast pyrolysis system.

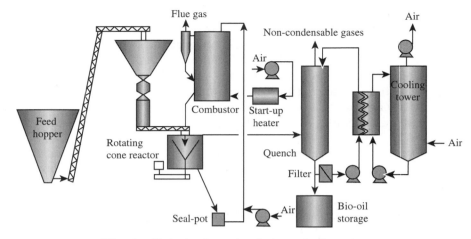

Fig. 9.   Entrained reactor fast pyrolysis system.

order to achieve the high heat up rates necessary for the fast pyrolysis reactions, the incoming biomass must be finely ground (~2 mm, typically). Furthermore, the residence times within the heated zones in the reactor must be minimized to prevent polymerization of the bio-oil into undesired products. As in gasification, gases and char are also produced as coproducts of the pyrolysis reactions. These coproducts are typically used as fuels to provide the energy required for the pyrolysis reactions. The incoming biomass is also typically dried to <10% moisture to minimize the quantity of water in the bio-oil product.

Bio-oil Characteristics.   Typical characteristics of bio-oil are found in the following table.

Physical property	Typical value
moisture content	15–30%
pH	2.5
specific gravity	1.2
viscosity	40–100 cP

Bio-oil has a distinctive odor, much like that of wood smoke, caused primarily by the phenolic materials present in the oil. It is a mixture of a large number (several hundred, typically) of primarily aromatic hydrocarbons all derived from the thermal decomposition of the incoming biomass. The proportion of the specific compounds varies widely from process to process and is influenced by the specific biomass used as feedstock as well as by the reactor conditions used.

Bio-oils cannot be completely vaporized once they have been recovered from the vapor phase. When they are heated, the oil tends to polymerize and form a char like material and heavier liquid hydrocarbons, that are undesirable as fuels. These polymerization reactions also tend to take place at room temperature, but at a much slower rate, thereby allowing reasonable storage periods for bio-oil products.

Uses of Bio-Oil.   Bio-oil can be used in a variety of primarily industrial applications. The primary use is as a substitute or cofired fuel with fuel oil. Such applications include boilers, furnaces, and engines. Because the bio-oils are chemically similar to some organic chemical intermediates, other uses include resins (used in plywood manufacture and as adhesives), agri-chemicals, and other specialty chemicals. Bio-oils can be further upgraded by steam reforming, a process similar to steam gasification, producing a synthesis gas and a hydrogen enriched liquid product (43).

Relatively small quantities of bio-oils are sold as food flavorings and additives. The liquid smoke products found in supermarkets are typical of this type of bio-oil product.

Bio-oils are storable and easier to transport than the gaseous products from gasification systems, however, the end uses are less well defined due to their varying chemical nature.

6.3. Biodiesel.   Biodiesel is a term applied to a fuel derived from the transesterification of used vegetable oils or animal fats. In the production of biodiesel, the triglycerides in the fats and oils are reacted with methanol to make methyl esters and glycerine. The glycerine produced can be sold as a by-product, however, due to large supplies of glycerine produced during soap manufacture, the income from the sale is likely to be small. The process uses a catalyst, typically NaOH or KOH to enhance the reaction rates. In the process, a small amount of the oil is converted to soap that is removed from the biodiesel prior to use. Free fatty acids formed during initial use of the vegetable oils and animal fats can be present in high concentrations. These contaminants can interfere with the biodiesel production and therefore the used oils may have to undergo some "pretreatment" prior to reaction with methanol. The reactions take place at low temperature ($\sim$65°C) and at modest pressure (2 atm). The biodiesel is purified after production by washing and evaporation to remove any remaining methanol. Current dedicated production capacity in the United States is between 60 and 80 million gal/ year (44,45).

Uses of Biodiesel.   Much like ethanol, biodiesel is blended with traditional diesel fuel in varying proportions (typically 20% biodiesel to 80% petroleum diesel fuel). Unlike ethanol, however, much higher concentrations of biodiesel can be blended without requiring a modification to the engine.

Biodiesel has undergone extensive testing and conforms to an industry specification (ASTM D6751). As such it is recognized by environmental agencies (notably the US EPA) and is sold as a motor fuel.

6.4. Anaerobic Digestion.   Anaerobic digestion is the biological degradation of organic material in the absence of oxygen. The product of such digestion is a gas containing primarily methane (CH_4) and carbon dioxide (CO_2). The digestion process occurs naturally in landfills, manure disposal sites, and other such residue disposal sites. It has also been applied in a more industrial setting in controlled digestors or "biogas" reactors (46,47).

The digestion process is carried out by naturally occurring bacteria. In the case of the industrial digestors, these bacteria have been carefully selected so that both the production rate and the methane content of the biogas produced are enhanced. Biogas typically has a heating value between 50 and 70% of that of natural gas. The energy content of biogas is derived entirely from the methane content of the gas. There are four steps that take place as the organic materials are digested (1) hydrolysis—high molecular weight organic molecules are broken down in to smaller molecules like sugars, amino acids, fatty acids, and water; (2) acidogenisis—the smaller molecules are further broken down into organic acids, carbon dioxide, hydrogen sulfide, and ammonia; (3) acetagenisis—the products of acidogenisis are converted into acetates, carbon dioxide, and hydrogen; and (4) methanogenisis—methane is produced from the hydrogen, carbon dioxide, and acetates produced in stages 2 and 3 (48).

The digestion bacteria have temperature ranges in which they are most productive. This results in specific temperature ranges where the digestion can be optimally carried out. These ranges are called mesophilic and thermophilic ranges. The mesophilic range is 25–38°C while the thermophilic range is 50–70°C. In general, the mesophilic digestion is more robust, but requires a longer residence time (typically 15–30 days). Thermophilic digestion, by contrast, reduces the residence time requirements to 12 to 15 days but requires much tighter control of process conditions. Thermophilic digestors produce a higher concentration of methane (up to 70%) thereby increasing the usefulness of the gas produced.

An important benefit of both types of digestion is the reduction of pathogens contained in the feedstock. This characteristic makes digestion processes well suited for the treatment of organic residues such as waste-water sludges and manures. The remaining solids after digestion can be used as a soil conditioner or further processed to produce an organic compost.

Commercial Application of Anaerobic Digestion. The gas produced (landfill gas) from anaerobic digestion in MSW landfills is being collected and used as a fuel for industrial heating and power generation at over 330 MSW landfills in the U.S. More than 500 other MSW landfills flare the gas. Combustion of the landfill gas helps to reduce the environmental impact of landfill gas emissions. Health and environmental concerns are related to the uncontrolled surface emissions of landfill gas into the air. As previously mentioned, landfill gas contains primarily carbon dioxide and methane, but can also contain trace quantities of volatile organic compounds (VOC) or other hazardous materials that can adversely affect public health and the environment. Carbon dioxide and methane are both greenhouse gases that contribute to global climate change. Methane is of particular concern because it is 21 times more effective at trapping heat in the atmosphere than carbon dioxide. Emissions of VOC contribute to ground-level ozone formation (smog). Ozone is capable of reducing or damaging vegetation growth as well as causing respiratory problems in humans.

In the United States, 1.6 GW of electric power are produced from such systems. The gas is cleaned to remove any sulfur compounds and then used as a fuel in gas turbine power generation systems, diesel engines, or other power generation systems (49).

6.5. Advanced Ethanol Production Methods.   Lignocellulosics (wood, straw, and grasses) are in abundant supply and can potentially supply a source for the production of ethanol. Over 1 trillion metric tons of cellulose are estimated to be produced by plants annually. Effective utilization of this resource can greatly enhance the quantity of ethanol that is produced for energy. The primary chemical components of these materials, however, are cellulose (∼40–50%)

and hemicellulose (~25–30%). From a theoretical standpoint, the same quantity of ethanol could be realized from these materials as from high sugar containing biomass such as corn. A more complex conversion process is required to realize this potential as the cellulose and hemicellulose must first be converted into sugars so that fermentation can take place. Cellulose converts via hydrolysis into six-carbon sugars (primarily glucose) that are readily fermented with yeasts to produce ethanol.

Hemicellulose, on the other hand, is converted into mainly five-carbon sugar precursors with xylose as a major product. The C_5 sugars are not readily fermented into ethanol without additional conversion steps and they inhibit the hydrolysis reactions once they are formed. Additional pretreatment stages can be added to enhance the conversion to sugars, including steam, acid, or alkali treatments (50).

Research into the conversion of cellulose and hemicellulose into sugars and the subsequent conversion steps necessary for effective fermentation is under-way in the United States and a number of European countries. Other methods being developed include the simultaneous saccharification and fermentation of cellulose to remove the sugars as they are produced by immediately converting them into ethanol.

7. Environmental Benefits from Biomass Energy

The use of biomass as an energy source, superficially results in a net zero change in carbon dioxide as illustrated in equation 1 above. Carbon dioxide is the predo-minant greenhouse gas found in the atmosphere. However, if forests are cleared for agricultural applications or development, this balance is no longer valid. Other greenhouse gases, namely, methane and nitrous oxide, are produced both by fossil energy conversion systems and during the growing and harvesting of biomass. The balance of these emissions and their ultimate impacts on the environment can only be accurately determined by the use of life cycle assess-ment (LCA) techniques.

The LCA uses rigorous methods to identify the complete production chain of biomass energy systems from seedling to end use. All energy and material inputs are included so that a true picture of the environmental impact can be developed. The results of LCA studies (6,51–54) show that biomass based power generation systems, regardless of the technology employed, have a net negative production in total greenhouse gas emissions (carbon dioxide, methane, and nitrous oxide) due to the elimination of biomass from landfills, thus offsetting emissions of methane that would otherwise be produced. Note that these studies include all of the fossil fuel necessary to transport the biomass to the power station and carry ash away from the station. Fossil fuels, on the other hand, produce 40–250% more greenhouse gases including cases where natural gas is the primary fuel and carbon dioxide sequestration is applied.

Reductions in greenhouse gas emissions are also realized when ethanol is used as a transportation fuel additive. Even with the inefficiencies of internal combustion engines included in the analysis, even a 10% blend of ethanol can

Table 1. **Typical Gas Analyses**

Component	Air blown CFB low press	Air blown bubbling bed high press	Fixed-bed downdraft	Indirect CFB	Indirect bubbling bed
hydrogen	16.7	8.5	19	21.2	46.8
carbon monoxide	21.4	12.3	18	43.2	34.4
carbon dioxide	12.6	15.9	14	13.5	6.6
methane	3.4	7.5	5	15.8	1.6
ethane	0.04	0.02		0.5	0.1
ethylene	1.3	0.7		0.6	0.2
nitrogen	39.6	40.7	44	0	0
water vapor	5.0	14.3	dry	dry	10.3

reduce greenhouse gas emissions by 19–25% depending on the source of the ethanol.

Other environmental benefits beyond the reduction of greenhouse gases can be realized by the use of biomass derived energy products. Cofiring of biomass with fossil fuels in boiler systems results in reductions of both sulfur and nitrogen oxides proportionate to the quantity of biomass being used. If minor modifications to the boiler are made the reduction in nitrogen oxides can be further enhanced. By cofiring the gas produced from a biomass gasifier in either a boiler or a gas turbine, nitrogen oxides can be reduced to levels significantly below those produced with the fossil fuel alone.

In some boiler applications, medium calorific value gas from biomass has been used as a reburn fuel in fossil fuel boilers. The reburn fuel provided ~15% of the energy input of the boiler. By proper placement of the reburn fuel introduction points, nitrogen oxides were reduced by up to 70% when compared to the fossil fuel alone.

8. Conclusion

The steady growth in biomass energy is expected to continue. Biomass is one of the largest renewable resources and provides advantages when compared to other renewable resources. Furthermore, biomass is the only renewable carbon-based resource with minimal greenhouse gas emission potential. This anticipated growth is subject only to the economic climate and not constrained by the supply of biomass feedstocks. Due to the flexibility of biomass energy technologies, a wide range of products can be produced ranging from basic energy products to refined chemicals, pharmaceuticals and fertilizers.

BIBLIOGRAPHY

"Fuels from Biomass" in *ECT* 3rd ed., Vol. 11, pp. 334–392, by D. L. Klass, Institute of Gas Technology; "Fuels from Biomass" in *ECT* 4th ed., Vol. 12, pp. 16–110, by D. L. Klass, Institute of Gas Technology; "Fuels from Biomass" in *ECT* (online), posting date: December 4, 2000, by D. L. Klass, Entech International, Inc.

CITED PUBLICATIONS

1. S. Kyritis, Welcome Address, First World Conference on Biomass for Energy and Industry, Seville, Spain, June 5–9, 2000.
2. IEA *World Energy Outlook*, Paris, France, Organization for Economic Co-operation and Development, 2002.
3. United States Energy Information Administration, Renewable Energy Annual 1999, DOE/EIA 0603 (99), Washington, D.C.
4. American Biomass Association, "Biomass Clean Energy for America", internet site, www.biomass.org.
5. UNEP Division of Technology, Industry and Economics, Energy and OzonAction Unit, "Bioenergy Fact Sheet".
6. Biomass Energy Research Association, internet site, www.beral.org.
7. R. Matthews and K. Robertson, "Answers to ten frequently answered questions about bioenergy, carbon sinks, and their role in global climate change," IEA Bioenergy Task 38, internet site, www.joanneum.at/iea-bioenergy-task38/, 2001.
8. D. O. Hall and K. K. Rao, *Photosynthesis*, Cambridge, Cambridge University Press, 1999.
9. K. R. Smith, "In Praise of Petroleum?", *Science* **298**, 1847 ff (2002).
10. National Renewable Energy Laboratory, Chemistry for Bioenergy Systems Division, "Biomass Energy", 2002.
11. J. E. Houck and P. E. Tiegs, *Residential Wood Combustion Review Volume 1, Technical Report* U. S. Environmental Protection Agency, Office of Research and Development, 1998.
12. United States Department of Energy, Office of Energy Efficiency and Renewable Energy, "Bioenergy, an Overview", 2002.
13. C. Demeter, "Biomass Power in Today's Energy Landscape", World Energy Engineering Conference, Atlanta, Ga., October, 2002.
14. T. Miles, "Biomass Preparation for Thermochemical Conversion," *First European Workshop on Thermochemical Processing of Biomass*, Birmingham, U.K., April 12–13, 1983.
15. K. Taupin, "Modern Wood Fired Boiler Designs—History and Technology Changes", Second Biomass Conference of the Americas, Portland Oregon, August 21–24, 1995.
16. *Steam—It's Generation and Use*—40th ed., Babcock and Wilcox, 1992.
17. E. S. Domalski, T. L. Jobe, Jr., and co-workers, *Thermodynamic Data for Biomass Materials and Waste Components*, ASME, New York, 1987.
18. G. Wiltsee, T. McGowin, and E. Hughes, "Biomass Combustion Technologies for Power Generation", First Biomass Conference of the Americas, Burlington, Vt., 1993.
19. American Forest and Paper Association, internet site www.afandpa.org, 2003.
20. S. Salat, and co-workers "District Heating Biomass for Southern Europe", First World Conference on Biomass for Energy and Industry, Sevilla, Spain, June 5–9, 2000.
21. International Energy Agency, *Energy Policies of IEA Countries, Sweden 2000 Review*, 2000.
22. T. N. Bhattarai, "Charcoal and its Socio-Economic Importance in Asia: Prospects for Promotion", Regional Training on Charcoal Production, Pontianak, Indonesia, February, 1998.
23. R. van der Plas, "Burning Charcoal Issues", FPD Energy Note No. 1, The World Bank Group, April, 1995.
24. G. Wiltsee, "Urban Wood Waste Resource Assessment", Appel Consultants, Inc., National Renewable Energy Laboratory Report NREL/SR-570-25918, November, 1998.

25. General Biomass Corporation, "Bioenergy and Biofuels," corporate internet site www.generalbiomass.com.

26. United States Department of Energy, Office of Transportation Technologies, "History of Biofuels", 2002.

27. Solar Energy Research Institute, "Fuel from Farms", Report SERI/SP-451-519, February, 1980.

28. National Corn Growers Association, "Ethanol—America's Clean Renewable Fuel", July, 2002.

29. Renewable Fuels Association, "How Ethanol is Made", 2000.

30. C. E. Wyman, *Handbook on Bioethanol: Production and Utilization*, Applied Energy Technology Series, Taylor and Francis, Washington, D.C., 1996.

31. C. Wyman, "Commercializing New Bioethanol Technology", First World Conference on Biomass for Energy and Industry, Sevilla, Spain, June 5–9, 2000.

32. H. H. Lowery, *The Chemistry of Coal Utilization*, Vol. II, John Wiley & Sons, Inc., New York, 1945.

33. E. Mangold and co-workers, Coal Liquefaction and Gasification Technologies, Ann Arbor Science, Mich., 1982.

34. R. Overend and M. Paisley, "A First Gasification Course", UBECA Conference, Washington, D.C., November, 1999.

35. N. Rambush, Modern Gas Producers, Benn Brothers, LTD, London, U.K., 1923.

36. F. Zenz and D. Othmer, *Fluidization and Fluid Particle Systems*, Reinhold Publishing Co., New York, 1960.

37. P. Tam and co-workers, "Forest Sector Table: Assessment of Gasification Technologies and Prospects for Their Commercial Application", National Climate Change Program, April, 1999.

38. T. Reed and S. Gaur, *A Survey of Biomass Gasification, 2001*, BEF Press, 2001.

39. M. Paisley and R. Overend, "Verification of the Performance of the FERCO Gasifier", Pittsburgh Coal Conference, September, 2002.

40. Thermochem, Inc., Corporate Brochure, 2002.

41. M. Welch, "Power Generation from Biomass Using a Small Industrial Gas Turbine", Bioenergy 2002, Boise, Ind., September, 2002.

42. T. Bridgewater, *A Guide to Fast Pyrolysis*, Aston University, Birmingham, U.K., 2001.

43. Pyrolysis Network, internet site, www.pyne.co.uk.

44. National Biodiesel Board, "biodiesel Basics", 2002.

45. T. Reed, *Biodiesel*, BEF Press, 2003, www.woodgas.com.

46. H. Kolk, "Anaerobic Digestion", Biomass Technology Group, Amsterdam, The Netherlands, 2002.

47. British Biogen, "Anaerobic Digestion of Farm and Food Processing Residues—Good Practice Guidelines", 2001.

48. California Energy Commission, "Anaerobic Digestion", 2002.

49. U.S. Environmental Protection Agency, Landfill Outreach Program, Office of Atmospheric Programs, Climate Protection Partnersips Division, Methane and Sequestration Branch, internet site, http://www.epa.gov/lmop/products/factsheet.htm.

50. H. Chum and R. Overend, "Biomass and Bioenergy in the United States", *Advances in Solar Energy*, Vol. 15, American Solar Energy Society, 2002.

51. M. Mann and P. Spaeth, Life Cycle Assessment of a Biomass Gasification Combined Cycle Power System, National Renewable Energy Laboratory, TP-430-23076, 1997.

52. M. Mann and P. Spaeth, Life Cycle Assessment of a Direct-Fired Biomass Power Generation System, National Renewable Energy Laboratory, TP-570-26942, 2000.

53. M. Mann and P. Spaeth, Life Cycle Assessment of Biomass Cofiring in a Coal Fired Power Plant, National Renewable Energy Laboratory, TP-430-26963, 2000.

54. R. Overend and E. Chornet, *Biomass a Growth Opportunity in Green Energy and Value Added Products*, Proceedings of the Fourth Biomass Conference of the Americas, Pergamon, New York, 1999.

Mark A. Paisley
FERCO Enterprises, Inc.

BIOMATERIALS, PROSTHETICS, AND BIOMEDICAL DEVICES

1. Introduction

Prosthetics or biomedical devices are objects which serve as body replacement parts for humans and other animals or as tools for implantation of such parts. An implanted prosthetic or biomedical device is fabricated from a biomaterial and surgically inserted into the living body by a physician or other health care provider. Such implants are intended to function in the body for some period of time in order to perform a specific task. Medical devices may replace a damaged part of anatomy, eg, total joint replacement; simulate a missing part, eg, mammary prosthesis; correct a deformity, eg, spinal plates; aid in tissue healing, eg, burn dressings; rectify the mode of operation of a diseased organ, eg, cardiac pacemakers; or aid in diagnosis, eg, insulin electrodes.

Prosthetics and biomedical devices are composed of biocompatible materials, or biomaterials. In the early 1930s the only biomaterials were wood, glass, and metals. These were used mostly in surgical instruments, paracorporeal devices, and disposable products. The advent of synthetic polymers and biocompatible metals in the latter part of the twentieth century has changed the entire character of health care delivery. Polymers, metals, and ceramics originally designed for commercial applications have been adapted for prostheses, opening the way for implantable pacemakers, vascular grafts, diagnostic/therapeutic catheters, and a variety of other orthopedic devices. The term prosthesis encompases both external and internal devices. This article concentrates on implantable prostheses.

2. Biomaterials

A biomaterial is defined as a systemic, pharmacologically inert substance designed for implantation or incorporation within the human body (1). A biomaterial must be mechanically adaptable for its designated function and have the required shear, stress, strain, Young's modulus, compliance, tensile strength, and temperature-related properties for the application. Moreover, biomaterials ideally should be nontoxic, ie, neither teratogenic, carcinogenic, or mutagenic;

nonimmunogenic; biocompatible; biodurable, unless designed as bioresorbable; sterilizable; readily available; and possess characteristics allowing easy fabrication. The traditional areas for biomaterials are plastic and reconstructive surgery, dentistry, and bone and tissue repair. A widening variety of materials are being used in these areas. Artificial organs play an important role in preventive medicine, especially in the early prevention of organ failure.

To be biocompatible is to interact with all tissues and organs of the body in a nontoxic manner, not destroying the cellular constituents of the body fluids with which the material interfaces. In some applications, interaction of an implant with the body is both desirable and necessary, as, for example, when a fibrous capsule forms and prevents implant movement (2).

Polymers, metals, ceramics, and glasses may be utilized as biomaterials. Polymers, an important class of biomaterials, vary greatly in structure and properties. The fundamental structure may be one of a carbon chain, eg, in polyethylene or Teflon, or one having ester, ether, sulfide, or amide bond linkages. Polysilicones, having a $-Si-O-Si-$ backbone, may contain no carbon.

Plastics are found in implants and components for reconstructive surgery, as components in medical instruments, equipment, packaging materials, and in a wide array of medical disposables. Plastics have assumed many of the roles once restricted to metals and ceramics.

Metals are used when mechanical strength or electrical conductivity is required of a device. For example, as of 1995 the femoral component of a hip replacement device was metal, as were the conductors of cardiac pacemaker leads. Titanium and titanium alloys (qv) are well tolerated in the body. This is partly the result of the strongly adhering oxide layer that forms over the metal surface, making the interface between the body and biomaterial effectively a ceramic rather than a metal. Titanium finds wide use as the femoral component of the artificial hip, where it exhibits great strength, comparatively light weight (the density of titanium is 4.5 g/cm^3), and excellent fatigue resistance. Another area in which titanium has replaced all other metals and alloys is as the casing material for cardiac pacemakers, neural stimulators, and implantable defibrillators.

Stainless steel alloys are also useful in orthopedic applications (see STEEL). Stainless steel alloys are used in the manufacture of staples, screws, pins, etc. These alloys are used primarily in applications requiring great tensile strength. Elgiloy, an interesting cobalt-based alloy, was originally developed for the mainspring of mechanical watches. This is used essentially as the conductor of neural stimulator leads, which require excellent flexibility and fatigue resistance. Nitinol, an unusual alloy of nickel and titanium, exhibits shape memory. Its main application has been in dentistry (see DENTAL MATERIALS), where its resilience rather than its shape-memory characteristic is of value.

Ceramics (qv) include a large number of inorganic nonmetallic solids that feature high compressive strength and relative chemical inertness. Low temperature isotropic (LTI) carbon has excellent thromboresistance and has found use in heart valves and percutaneous connectors. LTI carbon, known as LTI, was originally developed for encapsulating nuclear reactor fuel. This material was adapted for biomedical applications in the 1970s. LTI is formed by pyrolysis of hydrocarbons at temperatures between 1000 and 2400°C. Aluminum oxide

[1344-28-1], Al_2O_3, forms the basis of dental implants (see DENTAL MATERIALS). In the polycrystalline form this ceramic is suitable for load-bearing hip prostheses.

Bioglasses are surface-active ceramics that can induce a direct chemical bond between an implant and the surrounding tissue. One example is 45S5 bioglass, which consists of 45% SiO_2, 6% P_2O_5, 24.5% CaO, and 24.5% Na_2O. The various calcium phosphates have excellent compatibility with bone and are remodeled by the body when used for filling osseous defects.

3. Medical Devices

Medical devices are officially classified into one of three classes. Class I devices are general controls that are primarily intended as devices that pose no potential risk to health, and thus can be adequately regulated without imposing standards or the need for premarket review. Manufacturers of these devices must register with the United States Food and Drug Administration (FDA), provide a listing of products, maintain adequate reports, and comply with good manufacturing practices. Examples are stethoscopes, periodontic syringes, nebulizers, vaginal insufflators, etc.

Class II devices have performance standards and are applicable when general controls are not adequate to assure the safety and effectiveness of a device, based on the potential risk to health posed by the device. To classify a device in the Class II category, the FDA must find that enough data are available on which to base adequate performance standards that would control the safety and effectiveness of the device. Examples are diagnostic catheters, electrocardiographs, wound dressings, percutaneous catheters, gastrointestinal irrigation systems, etc.

Class III devices require premarket approval. When a device is critical, ie, life-supporting and/or life-sustaining, unless adequate justification is given for classifying it in another category, it is a Class III device. Class III also contains devices after 1976 that are not sufficiently similar to pre-1976 devices, and devices that were regulated as new drugs before 1976. Examples are bronchial tubes, ventilators, vascular grafts, pacemakers, cardiopulmonary bypass, surgical meshes, etc.

4. Cardiovascular Devices

Treatment of cardiovascular diseases is a vast and growing industry (see CARDIOVASCULAR AGENTS). Cardiovascular disease is a progressive condition which can eventually block the flow of blood through the coronary arteries to the heart muscle, thereby causing heart attacks and other life-threatening situations. The same plaque deposits occur in the peripheral arteries, leading to gangrene, amputations, aneurysms, and strokes. Despite enormous progress in cardiovascular medicine since World War II, challenges and unmet needs abound. Mortality rates have declined significantly, millions of people have been helped to lead normal lives, but the prevalence and incidence of cardiovascular diseases remain high. Open-heart surgery, cardiac pacing, heart transplants, implantable valves, and coronary angioplasty and clot busters, have been developed, but none of the great advances in cardiovascular medicine is preventive. There has been no "Salk vaccine" to preclude the buildup of fatty deposits or plaque in the arteries.

4.1. Cardiovascular Problems.   Despite its durability and resilience, different aspects of the cardiovascular system can malfunction. Some problems are congenital; many are inherited. Diseases can also be caused by infection such as damaged heart valves owing to rheumatic fever. Cardiomyopathy, a diseased heart muscle which may become enlarged, can result from infection or an unknown cause. Other problems may be a function of age. Pacemaker patients often have conduction systems that have simply started to wear out. Lifestyle also plays a role. Although poor diet and smoking cause or contribute to multiple problems such as hypertension and lung disease, when it comes to cardiovascular problems, the main culprit, regardless of its origin, is atherosclerosis. Atherosclerosis is a disease of the arteries resulting from the deposit of fatty plaque on the inner walls.

Plaque.   A heart attack, or myocardial infarction, results from insufficient delivery of oxygen to parts of the heart muscle owing to restricted blood flow in the coronary arteries. If heart muscle tissue is deprived of oxygen long enough, it may infarct or die (Fig. 1). The heart attack is often precipitated by a clot, or thrombus, which forms on a severely narrowed portion of a coronary artery. Silent ischemia is somewhat reduced blood supply from narrowing of the arteries. As the name implies, the disease provides no symptomatic warning of an impending problem. When coronary arteries are blocked to the degree that they cannot meet the heart's temporary demand for more oxygenated blood, angina pectoris, or sharp pain, may result. Further progression of the blockage then brings on the myocardial infarction. Atheroma is the medical term used to describe what plaque, the fatty deposits, does to the walls of the arteries. Plaque

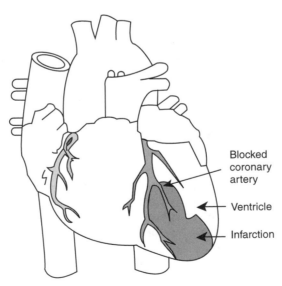

Blocked
coronary
artery

Ventricle

Infarction

Fig. 1.   Myocardial infarction occurs during insufficient delivery of oxygen to a portion of the heart muscle.

also causes other problems such as strokes and aneurysms, as well as complications of peripheral vascular disease.

Lethal Arrhythmias. Arrhythmias are a second significant source of cardiovascular problems. An arrhythmia is an abnormal or irregular heart rhythm. Bradyarrhythmias result in heart rates that are too slow; tachyarrhythmias cause abnormally fast rates. A bradyarrhythmia can be debilitating, causing a person to be short of breath, unable to climb stairs, black out, or even to go into cardiac arrest. Tachyarrhythmias can be unsettling and painful at best, life-threatening at worst.

Arrhythmias are caused by disturbances of the normal electrical conduction patterns synchronizing and controlling heartbeats. The wiring leading to the ventricles might, in effect, break or become frayed, causing a slowdown in the signals getting through, or perhaps result in intermittent electrical impulses. If damage to heart muscle tissue occurs, for example, from a myocardial infarction, this could create new electrical pathways. These in turn set up a separate focus of electrical activity (like another natural pacemaker) generating extra beats which can be highly disruptive. If a tachyrhythmia (tachycardia) occurs in the ventricles, the pumping chambers of the heart, the problem can be severely uncomfortable or even cause death if it deteriorates into ventricular fibrillation. Fibrillation is uncontrolled electrical activity. In this chaotic situation, cells become uncoordinated so that the heart muscle only quivers or twitches and no longer contracts rhythmically. Approximately three-fourths of the more than 500,000 deaths per year in the United States from coronary heart disease are sudden deaths.

There is a close correlation between myocardial infarctions and tachyarrhythmias, illustrated by the presence of complex ventricular arrhythmias among heart attack victims which are estimated to affect one-third of the survivors each year. Frequently, the immediate cause of sudden death is ventricular fibrillation, an extreme arrhythmia that is difficult to detect or treat. In the majority of cases, victims have no prior indication of coronary heart disease.

Valvular Disease. Valve problems severely limit the efficiency of the heart's pumping action bringing forth definitive symptoms. There are two types of conditions, both of which may be present in the same valve. The first is narrowing, or stenosis, of the valve. The second condition is inability of the valve to close completely. Narrowing of the mitral valve, for example, can result in less blood flowing into the left ventricle and subsequently less blood being pumped into the body. If the same valve does not close completely, blood may also back up or regurgitate into the left atrium when the ventricle contracts, preventing even more blood from properly flowing. The backward pressure which results can cause a reduction in the efficiency of the lungs.

Cardiomyopathy. Cardiomyopathy, or diseased heart muscle, may reach a point at which the heart can no longer function. It arises from a combination of factors, including hypertension, arrhythmias, and valve disease. Other problems, such as congestive heart failure, cause the interrelated heart–lung system to break down. Because the heart can no longer adequately pump, fluid builds up in the lungs and other areas.

4.2. Device Solutions. The first big step in cardiovascular devices was the development of a heart–lung machine in 1953. The ability to shut down the

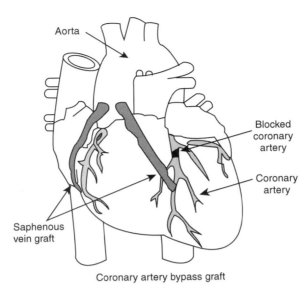

Coronary artery bypass graft

Fig. 2.   In coronary bypass, an autologous saphenous vein is used to provide critical blood to the heart muscle, bypassing a blockage in the coronary artery.

operation of the heart and lungs and still maintain circulation of oxygenated blood throughout the body made open-heart surgery possible. Open heart, really a misnomer, actually refers to opening up the chest to expose the heart, not opening up the heart itself. The principal components of the heart–lung machine, the oxygenator and pump, take over the functions of the lungs and heart.

Atherosclerosis.   The first solution to the problem of atherosclerosis was the coronary artery bypass graft (CABG) procedure, first performed in 1964. In a coronary bypass procedure, a graft is taken from the patient's own saphenous vein. The graft is attached to the aorta (Fig. 2) where the coronary arteries originate and the opposite end is connected to the artery below the blocked segment. Blood can then bypass the obstructed area and reach the surrounding tissue below. Extensions of this useful surgery are CABG procedures which utilize mammary arteries of the patient instead of saphenous veins.

The second step toward solving cardiovascular disease from atherosclerosis, ie, angioplasty, was preceded by the diagnostic tool of angiocardiography by nearly 20 years. Angiocardiography, or angiography, permits x-ray diagnosis using a fluoroscope. A radiopaque contrast medium is introduced into the arteries through a catheter (see RADIOPAQUES), and angiography allows accurate location of the plaque blockage. Percutaneous transluminal coronary angioplasty (PTCA), a nonsurgical procedure, emerged in the 1980s as a viable method for opening up blocked arteries. A PTCA catheter has a balloon at its tip which is inflated after it is positioned across the blocked segment of the artery. Plaque is then compressed against the arterial walls, permitting blood flow to be restored. The same solutions of bypass surgery and angioplasty have been applied to atherosclerosis in the peripheral arteries.

Arrhythmias.   The first solution to cardiovascular problems arising from arrhythmias came about as a result of a complication caused by open-heart surgery. During procedures to correct congenital defects in children's hearts, the electrical conduction system often became impaired, and until it healed, the heart could not contract sufficiently without outside electrical stimulation. A system that plugged into a wall outlet was considered adequate until an electrical storm knocked out power, leading to the development of the first battery-powered external pacemaker.

The first implantable pacemaker, introduced in 1960, provided a permanent solution to a chronic bradyarrhythmia condition. This invention had a profound impact on the future of medical devices. The pacemaker was the first implantable device which became intrinsic to the body, enabling the patient to lead a normal life.

Early pacemakers paced the heart continuously at a fixed rate, were larger than a hockey puck, and had to be replaced frequently owing to power source technology limitations. Advances in electronics, materials, and knowledge have yielded pacemakers about the size of a U.S. 50-cent piece that last five years or more. More importantly, pacemakers restore the heart to functioning in a completely natural way. The pacemaker senses the electrical activity of the heart and kicks in only when something is wrong. If the impulses initiated by the SA node cannot get all the way through to the lower part of the ventricles, the pacemaker takes over completing the electrical process at the same rate indicated by the heart's natural pacemaker. If the SA node is dysfunctional and cannot put out an appropriate signal, sensors (qv) in rate-responsive pacemakers can correlate other data such as sound waves from body activity, body temperature, or the respiratory rate to compute the proper heart rate.

The first automatic implantable cardioverter defibrillator (AICD) was implanted in 1980. As for pacemakers, early generations of AICDs were bulky and cumbersome, did not last very long, and required open-heart surgery. However, these kept people alive by automatically shocking the heart out of its chaotic electric state whenever it went into ventricular fibrillation. Future devices are being designed to provide the full spectrum of arrhythmia control, including pacing, cardioversion, and defibrillation. Techniques are also being developed to map, ie, locate, the source of certain tachyarrhythmias (an ectopic focus or scar tissue) and remove it without open-heart surgery.

External defibrillation was first performed in 1952 and continues as a routine procedure in hospitals and ambulances. The problem of external defibrillation has not been a technological one, but rather a legal one. Only in the 1990s have laws been passed to permit people other than doctors and paramedics to operate semiautomatic defibrillators to provide help when it is needed. New and better defibrillation devices continue to come to market and are easier and safer to use.

Valve Problems.   The primary solution to valve problems has been implantable replacement valves. The introduction of these devices necessitates open-heart surgery. There are two types of valves available: tissue (porcine and bovine) and mechanical. The disadvantage of tissue valves is that these have a limited life of about seven years before they calcify, stiffen, and have to be replaced. The mechanical valves can last a lifetime, but require anticoagulant therapy. In some patients, anticoagulants may not be feasible or may be contraindicated. Of the valves which require replacement, 99% are mitral and aortic valves. The valves on the left side of the heart are under much greater pressure because the left ventricle is pumping blood out to the entire body, instead of only to the lungs. Occasionally, two valves are replaced in the same procedure.

Cardiomyopathy. The best available solution to cardiomyopathy may be one that is less sophisticated than transplant surgery or the artificial heart. The cardiomyoplasty-assist system combines earlier electrical stimulation technology with a new surgical technique of utilizing muscle from another part of the body to assist the heart.

Efforts to develop an artificial heart have resulted in a number of advancements in the assist area. The centrifugal pump for open-heart surgery, the product of such an effort, has frequently been used to support patients after heart surgery (post-cardiotomy), or as a bridge to life prior to transplant. Other efforts have led to the development of ventricular assist devices to support the heart for several months and intra-aortic balloon pumps (IABPs) which are widely used to unload and stabilize the heart.

4.3. Interventional Procedures. The emergence of angioplasty created a specialty called interventional cardiology. Interventional cardiologists not only implant pacemakers and clear arteries using balloon catheters, but they also use balloons to stretch valves (valvuloplasty). In addition, they work with various approaches and technologies to attack plaque, including laser (qv) energy, mechanical cutters and shavers, stents to shore up arterial walls and deliver drugs, and ultrasound to break up plaque or to visualize the inside of the artery.

Typically, procedures have become less invasive as technology evolves. Early pacemaker procedures involved open-heart surgery to attach pacemaker leads (wires) to the outside of the heart. Later, leads could be inserted in veins and pushed through to the interior of the heart, no longer necessitating opening a patient's chest. Using fluoroscopy, the physician can visualize the process, so that the only surgery needed is to create a pocket under the skin for the implantable generator to which the leads are connected.

Clinical evaluation is underway to test transvenous electrodes. Transvenous leads permit pacemakers to be implanted under local anesthesia while the patient is awake, greatly reducing recovery time and risk. As of 1996, the generation of implantable defibrillators requires a thoracotomy, a surgical opening of the chest, in order to attach electrodes to the outside of the heart. Transvenous electrodes would allow cardiologists to perform pacemaker procedures without a hospital or the use of general anesthesia.

Coronary bypass surgery and angioplasty are vastly different procedures, but both procedures seek to revascularize and restore adequate blood flow to coronary arteries. Balloon angioplasty, which looks much like a pacemaker lead except that it has a tiny balloon at the end instead of an electrode, involves positioning a catheter inside a coronary artery under fluoroscopy. The balloon is inflated to compress the offending plaque. Angioplasty is far less invasive than bypass surgery and patients are awake during the procedure. For many patients, angioplasty may not be indicated or appropriate.

Interventional cardiology is but one specialty that has arisen in cardiovascular medicine. Another is interventional radiology for similar procedures in the peripheral arteries, in addition to conventional bypass graft surgery. Competition has been intense among surgeons, cardiologists, and radiologists. Because coronary artery disease is progressive, many patients who are candidates for peripheral and/or coronary angioplasty may be future candidates for bypass surgery.

Cardiologists may be described in terms of three overlapping specialties: interventional, who perform most angioplasty; invasive, who implant about 70% of the pacemakers in the United States; and diagnostic. A subspecialty of diagnostic cardiology, electrophysiology, has grown in importance because it is critical to the treatment of tachrhythmia patients, especially those who are prone to ventricular fibrillation. The further development of implantable devices in this last area depends on close cooperation between companies and electrophysiologists.

Cardiovascular devices are being employed by a wider diversity of specialists and are thus finding applications in other medical areas. This has been particularly true for devices developed to support open-heart surgery. Oxygenators and centrifugal pumps, which take over the functions of the lungs and heart, are used in applications such as support of angioplasty and placing a trauma or heart attack victim on portable bypass in the emergency room. Some devices are finding utility by improving surgical techniques. For example, cardiac surgeons are working with balloon catheters and laser angioplasty systems as an augmentation to regular bypass surgery.

Other cardiovascular devices developed initially for use in open-heart surgery are used extensively in other parts of the hospital and, in many cases, outside the hospital. Patients have been maintained for prolonged periods of time on portable cardiopulmonary support systems while being transported to another hospital or waiting for a donor heart. Blood pumps and oxygenators may take over the functions of the heart and lungs in the catheterization lab during angioplasty, in extracorporeal membrane oxygenation (ECMO) to support a premature baby with severe respiratory problems, or in the emergency room to assist a heart attack victim. It is possible that future patients could be put on portable bypass at the site of the heart attack or accident. The market for cardiac assist devices and oxygenators plus related products such as specialized cannulae and blood monitoring devices is expected to expand rapidly into these areas.

4.4. Biomaterials for Cardiovascular Devices. Perhaps the most advanced field of biomaterials is that for cardiovascular devices. For several decades bodily parts have been replaced or repaired by direct substitution using natural tissue or selected synthetic materials. The development of implantable-grade synthetic polymers, such as silicones and polyurethanes, has made possible the development of advanced cardiac assist devices (see SILICON COMPOUNDS, SILICONES; URETHANE POLYMERS).

Implantable devices to pace, cardiovert, and defibrillate the heart without the need for open-heart surgery should become widely accepted. Dramatic developments and growth are also taking placein other areas such as the use of laser systems intended to ablate significant amounts of plaque. Laser ablation systems hold considerable promise if restenosis (reblocking of the arteries) rates are reduced. Mechanical or atherectomy devices to cut, shave, or pulverize plaque have been tested extensively in coronary arteries. Some of these have also been approved for peripheral use. The future of angioplasty, beyond the tremendous success of conventional balloon catheters, depends on approaches that can reduce restenosis rates. For example, if application of a drug to the lesion site turns out to be the solution to restenosis, balloon catheters would be used for both dilating the vessel and delivering the drug. An understanding of what happens to the arterial walls, at the cellular level, when these walls are subjected to the various types of angioplasty may need to come first.

A primary aspect of cardiovascular devices through the twenty-first century is expected to involve the incorporation of diagnostic and visualization capabilities. A separate ultrasound system has been approved for this purpose. Laser angioplasty systems under development include visualization capabilities to distinguish plaque from the arterial wall. Future pacemakers, which already utilize sensors to determine an appropriate heart rate, are expected to incorporate various other sensors for diagnostic purposes. The biggest challenge in averting sudden death is not so much to perfect a life-sustaining device, but to gain the ability to identify the susceptible patient. Appropriate screening and diagnoses for patients having silent ischemia must be developed. If the presence and extent of coronary artery disease can be identified early, intervention could save thousands of people from an untimely death and help others to live a fuller life. Sensors and specific diagnostic devices are expected to play a large role at about the same time as effective implantable defibrillators.

One of the more intriguing cardiovascular developments is cardiomyoplasty where implantable technologies are blended with another part of the body to take over for a diseased heart. One company, Medtronic, in close collaboration with surgeons, has developed a cardiomyoplasty system to accompany a technique of wrapping back muscle around a diseased heart which can no longer adequately pump. A combination pacemaker and neurological device senses the electrical activity of the heart and correspondingly trains and stimulates the dorsal muscle to cause the defective heart to contract and pump blood. Cardiomyoplasty could greatly reduce the overwhelming need for heart transplants. It might also eliminate the need for immunosuppressive drugs. Development of appropriate materials and manufacturing methods are needed to maintain patency without damaging blood in grafts below 4 mm in diameter.

Pacemakers.   The implantable cardiac pacemaker (Fig. 3) has been a phenomenal technological and marketing success. In the early 1980s, however, many critics were predicting the demise of these devices and the industry was the subject of congressional investigations over sales practices, alleged overuse, and excessive prices. Critics advocated low priced generic pacemakers, and pacemaker unit volume and prices declined about 10% on average. However, costs have been reduced by curtailing the length of time patients need to stay in the hospital following the implantation procedure and by selection of the correct pacemaker for each patient. Significantly lower cost is attached to a single-chamber device having limited longevity than to the far more expensive dual-chamber device which may be indicated for a young and active patient.

As of the mid-1990s, the market for bradyarrhythmia devices is fully penetrated in Western countries. Some growth is expected to result from an aging population but, by and large, the market is mature. The market for tachyrhythmia devices, in contrast, is only beginning.

Implantable tachyrhythmia devices, available for some years, address far less dangerous atrial tachyarrhythmias and fibrillation. The technical barriers

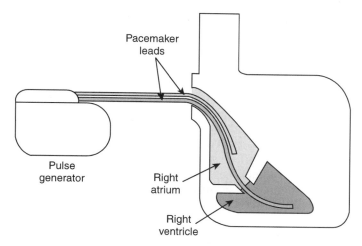

Fig. 3. A pacemaker provides electrical impulses to the heart in an effort to correct potentially fatal arrhythmias.

to counteracting ventricular tachyarrhythmias and fibrillation using massive shocks have been formidable and are compounded by the possibility of causing the very problem the shock is designed to overcome. Newer tachyrhythmia devices are being readied that can safely regulate arrhythmias across the full spectrum.

Surgical Devices.   Surgical devices comprise the equipment and disposables to support surgery and to position implantable valves and a variety of vascular grafts. Central to open-heart surgery is the heart–lung machine and a supporting cast of disposable products. Two devices, the oxygenator and the centrifugal pump, amount to significant market segments in their own right. Other disposables include cardiotomy reservoirs, filters, tubing packs, and cardioplegia products to cool the heart. The oxygenator market has been driven more recently by the conversion from bubbler to membrane devices which account for about 80% of the oxygenators used in the United States.

Centrifugal pumps are increasingly being used as a safer and more effective alternative to the traditional roller pump in open-heart surgery and liver transplants. As of the mid-1990s, about 45% of open-heart procedures use a centrifugal pump. In the latter 1980s, that number was less than 10%. Implantable valves, particularly mechanical valves which continue to encroach on tissue valves, are unique. Methods such as valvuloplasty, mitral valve repair, or use of ultrasound are unlikely to reduce the number of valve replacements into the twenty-first century. Valve selection remains in the hands of the surgeon because of the critical nature of the procedure. If anything goes wrong, the result can be catastrophic to the patient.

Vascular grafts are tubular devices implanted throughout the body to replace blood vessels which have become obstructed by plaque, atherosclerosis, or otherwise weakened by an aneurysm. Grafts are used most often in peripheral bypass surgery to restore arterial blood flow in the legs. Grafts are also frequently employed in the upper part of the body to reconstruct damaged portions of the aorta and carotid arteries. In addition, grafts are used to access the vascular system, such as in hemodialysis to avoid damage of vessels from repeated needle punctures. Most grafts are synthetic and made from materials such as Dacron or Teflon. Less than 5% of grafts utilized are made from biological materials.

Cardiac-Assist Devices. The principal cardiac-assist device, the intra-aortic balloon pump (IABP), is used primarily to support patients before or after open-heart surgery, or patients who go into cardiogenic shock. As of the mid-1990s, the IABP was being used more often to stabilize heart attack victims, especially in community hospitals which do not provide open-heart surgery. The procedure consists of a balloon catheter inserted into the aorta which expands and contracts to assist blood flow into the circulatory system and to reduce the heart's workload by about 20%. The disposable balloon is powered by an external pump console.

Other devices, which can completely take over the heart's pumping function, are the ventricular assist devices (VADs), supporting one or both ventricles. Some patients require this total support for a period of time following surgery (post-cardiotomy); others require the support while being transported from one hospital to another, or while waiting for a donor heart (bridge-to-transplant). Several external and implantable devices are being evaluated for short-term and long-term applications. Considerable interest has emerged in devices providing cardiopulmonary support (CPS), ie, taking over the functions of both the heart and lungs without having to open up the chest. There are several applications for other portable bypass systems or mini-heart–lung machines. Thus far, CPS has been used most frequently in support of anigoplasty prophylactically in difficult cases which could not be otherwise undertaken. The greatest potential is in the emergency room to rest the heart and lungs of heart attack and trauma victims.

Other specialized applications of cardiac arrest devices include extracorporeal membrane oxygenation (ECMO) which occurs when the lungs of a premature infant cannot function properly. The market segments for cardiopulmonary support devices are potentially significant.

Artificial Hearts. Congestive heart failure (CHF) is a common cause of disability and death. It is estimated that three to four million Americans suffer from this condition. Medical therapy in the form of inotropic agents, diuretics (qv), and vasofilators is commonly used to treat this disorder (see CARDIOVASCULAR AGENTS). Cardiac transplantation has become the treatment of choice for medically intractable CHF. Although the results of heart transplantation are impressive, the number of patients who might benefit far exceeds the number of potential donors. Long-term circulatory support systems may become an alternative to transplantation (3).

In 1980, the National Heart, Lung and Blood Institute of NIH established goals and criteria for developing heart devices and support techniques in an effort to improve the treatment of heart disease. This research culminated in the development of both temporary and permanent left ventricular-assist devices that are tether-free, reliable over two years, and electrically powered. The assist devices support the failing heart and systemic circulation to decrease cardiac work, increase blood flow to vital organs, and increase oxygen supply to the myocardium. The newer ventricular assists are required to have no external venting, have a five-year operation with 90% reliability, pump blood at a rate of 3–7 L/min into the aorta at a mean arterial pressure of 90 mm Hg (12 kPa) when assisting the human left ventricle, and have a specific gravity of 1.0 for the implantable ventricular assist device.

In contrast, the total artificial heart (TAH) is designed to overtake the function of the diseased natural heart. While the patient is on heart–lung bypass, the natural ventricles are surgically removed. Polyurethane cuffs are then sutured to the remaining atria and to two other blood vessels that connect with the heart.

One successful total artificial heart is ABIOMED's electric TAH. This artificial heart consists of two seamless blood pumps which assume the roles of the natural heart's two ventricles. The pumps and valves are fabricated from a polyurethane, Angioflex. Small enough to fit the majority of the adult population, the heart's principal components are implanted in the cavity left by the removal of the diseased natural heart. A modest sized battery pack carried by the patient supplies power to the drive system. Miniaturized electronics control the artificial heart which runs as smoothly and quietly as the natural heart. Once implanted, the total artificial heart performs the critical function of pumping blood to the entire body (4).

Heart Valves.   Since the early 1960s nearly 50 different heart valves have been developed. The most commonly used valves as of the mid-1990s include mechanical prostheses and tissue valves. Caged-ball, caged disk, and tilting-disk heart valves are the types most widely used.

Blood Salvage.   In a growing awareness that a patient's own blood is the best to use when blood is needed, newer techniques are reducing the volume of donor blood used in many cardiovascular and orthopedic surgeries. Surgical centers have a device, called the Cell Saver (Haemonetics), that allows blood lost during surgery to be reused within a matter of minutes, instead of being discarded. This device collects blood from the wound, runs it through a filter that catches pieces of tissue and bone, then mixes the blood with a salt solution and an anticoagulant. The device then cleanses the blood of harmful bacteria. Subsequently the blood is reinfused back to the same patient through catheters inserted in a vein in the arm or neck, eliminating the worry of cross-contamination from the HIV or hepatitis viruses (see BLOOD, COAGULANTS AND ANTICOAGULANTS; FRACTIONATION, BLOOD).

Use of intraoperative autotransfusion (IAT) eliminates disease transmission, compatibility testing, and immunosuppression that may result from the use of homologous blood products, reduces net blood loss of the patient, and conserves the blood supply. During vascular surgery, the principal indications for the use of the Cell Saver are ruptured spleen, ruptured liver, aneurysms, and vascular trauma. During orthopedic surgery the principal indications are total hip arthroplasty, spinal fusions, total knee, and any procedure that has wound drains (5).

Blood Access Devices.   An investigational device called the Osteoport system allows repeated access to the vascular system via an intraosseous infusion directly into the bone marrow. The port is implanted subcutaneously and secured into a bone, such as the iliac crest. Medications are administered as in any conventional port, but are taken up by the venous sinusoids in the marrow cavity, and from there enter the peripheral circulation (6).

Blood Oxygenators.   The basic construction of an oxygenator involves any one of several types of units employing a bubble-type, membrane film-type, or hollow-fiber-type design. The most important advance in oxygenator development was the introduction of the membrane-type oxygenator. These employ conditions very close to the normal physiological conditions in which gas contacts occur indirectly via a gas-permeable membrane. Blood trauma is minimized by the use of specialized biomaterials such as PTFE, PVC, and cellophane, although lately silicone rubber and cellulose acetate have predominated. A silicone–polycarbonate copolymer, ethylcellulose perfluorobutyrate, and poly-(alkyl sulfone) were introduced in the mid-1980s, and tend to dominate this field.

4.5. Polyurethanes as Biomaterials.   Much of the progress in cardiovascular devices can be attributed to advances in preparing biostable polyurethanes. (see URETHANE POLYMERS). Biostable polycarbonate-based polyurethane materials such as Corethane (7) and ChronoFlex (8) offer far-reaching capabilities to cardiovascular products. These and other polyurethane materials offer significant advantages for important long-term products, such as implantable ports, hemodialysis, and peripheral catheters; pacemaker interfaces and leads; and vascular grafts.

Implantable Ports.   The safest method of accessing the vascular system is by means of a vascular access device (VAD) or port. Older VAD designs protruded through the skin. The totally implanted ports are designed for convenience, near absence of infection, and ease of implantation. Ports allow drugs and fluids to be delivered directly into the bloodstream without repeated insertion of needles into a vein. The primary recipients of totally implanted ports are patients receiving chemotherapy, bolus infusions of vesicants, parenteral nutrition, antibiotics, analgesics, and acquired immune disease syndrome (AIDS) medications.

Vascular access ports typically consist of a self-sealing silicone septum within a rigid housing which is attached to a radiopaque catheter (see RADIOPAQUES). The catheter must be fabricated from a low modulus elastomeric polymer capable of interfacing with both soft tissue and the cardiovascular environment. A low modulus polyurethane-based elastomer is preferred to ensure minimal trauma to the fragile vein.

Placement of vascular access ports is similar to that of a long-term indwelling arterial catheter. A small incision is made over the selected vein and a second incision is made lower in the anterior chest to create a pocket to house the port. The catheter is tunneled subcutaneously from its entry point into the vein with the tip inside the right atrium. The final position of the catheter is verified by fluoroscopy, secured with sutures, and the subcutaneous pocket is closed. The port septum is easily palpable transcutaneously, and the system may be used immediately. A surgeon typically inserts the vascular access port in an outpatient setting.

To use the port, the overlying skin is prepared using conventional techniques. A local anesthetic is sometimes used to decrease pain of needle insertion, though this is usually not necessary using techniques which utilize small-bore needles. A special point needle is used to puncture the implanted ports as the point of these needles is deflected so it tears the septum rather than coring it, allowing multiple entries. The septum reseals when the needle is removed.

The primary advantages of implantable ports are no maintenance between uses other than periodic flushing with heparinized saline every 28 days to ensure patency, lower incidence of clotting and thrombosis, no dressing changes, insignificant infection incidence, unobtrusive cosmetic appearance, and no restriction on physical activity.

Pacemaker Interfaces and Leads. Problems of existing pacemaker interfaces and pacemaker lead materials made from silicones and standard polyurethanes are environmental stress cracking, rigidity, insulation properties, and size.

Technical advances in programmable pacemakers that assist both the tachycardia and bradycardia have led to the requirement of implanting a two-lead system. Owing to the ridigity and size of silicones, the only material that fulfills this possibility without significantly impeding blood flow to the heart is polyurethane. The primary needs in this medical area are reduction in making frequent changes and in failure rate, and the ability to have multiple conductors to handle advanced pacemaker technology.

Vascular Grafts. Although the use of vascular grafts in cardiovascular bypass surgery is widely accepted and routine, numerous problems exist in these surgeries for the materials available. Biocompatibility is often a problem for vascular grafts which also tend to leak and lead to scarring of the anastomosis. The materials are not useful for small-bore ($\leq$6 mm) grafts. The primary needs that materials can be developed to address are matching compliance to native vessels, having a lesser diameter for small-bore grafts which would serve as a replacement for the saphenous vein in coronary bypass, thinner walls, biostability, controlled porosity, and greater hemocompatibility for reduced thrombosis.

The advent of newer polyurethane materials is expected to lead to a new generation of cardiovascular devices. The characteristics of polyurethanes, combined with newer manufacturing techniques, should translate into direct medical benefits for the physician, the hospital, and the patient. This field offers exciting growth opportunities.

5. Orthopedic Devices

Bone, or osseous tissue, is composed of osteocytes and osteoclasts embedded in a calcified matrix. Hard tissue consists of about 50% water and 50% solids. The solids are composed of cartilaginous material hardened with inorganic salts, such as calcium carbonate and phosphate of lime.

Bone is formed through a highly complex process that begins with the creation of embryonic mesenchymal cells. These cells, found only in the mesoderm of the embryo, migrate throughout the human body to form all the types of skeletal tissues including bone, cartilage, muscle, tendon, and ligament. Mesenchymal cells differentiate into the various types of progenitor cells: osteoblasts, chondroblasts, fibroblasts, and myoblasts. Bone tissue begins to form when osteoblasts and chondroblasts synthesize cartilage-like tissue by secreting at least one potent bone cell growth factor (protein hormone), referred to as IGF-II. Bone growth factor is theorized to stimulate an increased expression of IGF-II receptors on the cell walls of other bone cells. This growth factor helps to initiate a cascade of other subcellular activity, leading to the formation of cartilage. The cartilage then hardens into bone when osteoblasts become lodged within the cartilage matrix, and cease to function. These types of mature cells are then known as osteocytes.

A bone is classified according to shape as flat, long, short, or irregular. A living bone consists of three layers: the periosteum, the hard cortical bone, and the bone marrow or cancellous bone. The periosteum is a thin collagenous layer, filled with nerves and blood vessels, that supplies nutrients and removes cell wastes. Because of the extensive nerve supply, normal periosteum is very sensitive. When a bone is broken, the injured nerves send electrochemical neural messages relaying pain to the brain.

Next is a dense, rigid bone tissue, referred to as the hard compact or cortical bone. It is cylindrical in shape and very hard. This dense layer supports the weight of the body and consists mostly of calcium and minerals. Because it is devoid of nerves, it experiences no pain. The innermost layer, known as cancellous bone or spongy bone marrow, is honeycombed with thousands of tiny holes and passageways. Through these passageways run nerves and blood vessels that supply oxygen and nutrients. This material has a texture similar to gelatin. The marrow produces either red blood cells, white blood cells, or platelets.

Rigid bones are needed for kinetic motion, support of internal organs, and muscle strength. The bones that compose the human thigh are pound for pound stronger than steel. Nature meets these needs by separating the skeleton into several bones and bone systems, creating joints where the bones intersect.

Joints are structurally unique. They permit bodily movement and are bound together by fibrous tissues known as ligaments. Most larger joints are encapsulated in a bursa sac and surrounded by synovial fluid which lubricates the joint continuously to reduce friction. The skeleton is constructed of various types of moveable joints. Some joints allow for no movement, such as those connecting the bones of the skull. Other joints permit only limited movement. For example, the joints of the spine allow limited movement in several directions. Most joints have a greater range of motion than the joints of the skull and spine.

The bearing surface of each joint is cushioned by cartilage. This tissue minimizes friction. The cartilage also reduces force on the bone by absorbing shock. The joint area is a narrow space known as an articular cavity which allows freedom of movement.

Ligaments are composed of bands of strong collagenous fibrous connective tissue. This tissue is originally formed by the mesenchymal cells which differentiate into fibroblast cells. These fibroblast cells then further differentiate into specialized cells known as fibrocytes. When fibrocytes mature, they are inactive and compose the ligaments. Ligaments function to tie two bones together at a joint, maintain joints in position preventing dislocations, and restrain the joint's movements. Ligaments may be reattached to bone by the use of an orthopedic anchor.

Tendons are composed of fibrous connective tissue. Tendon tissue is also formed by the fibroblast cells, similar to the way ligaments are formed. These fibroblast cells then further differentiate into other specialized cells known as fibrocytes. Mature fibrocytes are inactive and compose the cellular portion of tendons. The function of the tendon is to attach muscles to bones and other parts.

The meniscus is skeletal system fibrocartilage-like tissue. It is a type of cartilage found in selected joints, which are subjected to high levels of force. Meniscal tissue originates from mesenchymal cells which differentiate into chondroblast cells. These chondroblast cells then further differentiate into specialized cells known as chondrocytes. When chondrocytes mature, they are inactive and comprise the menisci and other forms of cartilage. The function of the meniscus is to absorb shock by cushioning and distributing forces evenly throughout a joint, and provide a smooth articulating surface for the cartilages of the adjoining bones.

In the knee, the menisci form an interarticular fibrocartilage base for femoral and tibial articulation. The menisci form a crescent shape in the knee. The lateral meniscus is located on the outer side of the knee, and the medial meniscus is located on the inside of the knee. If the knee bends and twists the menisci can overstretch and tear. Menisci tears occur frequently and the knee can sustain more than one tear at a time. If not treated appropriately, however, a menisci tear can roughen the cartilage and lead to arthritis. A meniscus tear acts like grit in the ball bearings of a machine. The longer the torn tissue remains affected, the more irritation it causes.

Meniscus surgery can repair or remove the torn cartilage, depending on the nature of the tear. Arthroscopy, a procedure through small skin incisions to visualize and repair the affected joint, is frequently employed. This procedure is sometimes performed on an out-patient basis. If repair of the menisci is not possible the surgeon removes as little of the meniscus as possible.

The body's frame or skeleton is constructed as a set of levers powered or operated by muscle tissue. A typical muscle consists of a central fibrous tissue portion, and tendons at either end. One end of the muscle, known as the head, is attached to tendon tissue, which is attached to bone that is fixed, and known as the point of origin. The other end of the muscle is attached to a tendon. This tendon is attached to bone that is the moving part of the joint. This end of the muscle is known as the insertion end. An example is the bicep muscle which is connected to the humerus bone of the upper arm at its head or origin. The insertion end of the muscle is connected to the radius bone of the forearm, otherwise known as the moving part of the elbow joint.

Muscle tissue is unique in its ability to shorten or contract. The human body has three basic types of muscle tissue histologically classified into smooth, striated, and cardiac muscle tissues. Only the striated muscle tissue is found in all skeletal muscles. The type of cells which compose the muscle tissue are known as contractile cells. They originate from mesenchymal cells which differentiate into myoblasts. Myoblasts are embryonic cells which later differentiate into contractile fiber cells.

The human body has more than 600 muscles. The body's movement is performed by muscle contractions, which are stimulated by the nervous system. This system links muscle tissue to the spinal cord and brain. The network of nerve cells which carries the brain's signals directs the flow of muscular energy.

Most muscular activity occurs beyond the range of the conscious mind. The body, working through the neuromuscular network, manages its own motion.

Typically, in order for motion to occur, several muscle sets must work together to perform even the simplest movements. The bicep is a two-muscle set; the tricep is a three-muscle set. Each set works in tandem. Within each muscle group, muscle fibers obey the all or none principle, ie, all muscle fibers contract or none contract. Therefore, if the muscle fibers of a muscle group are stimulated enough by nerve impulses to contract, they contract to the maximum.

Bones function as levers; joints function as fulcrums; muscle tissue, attached to the bones via tendons, exert force by converting electrochemical energy (nerve impulses) into tension and contraction, thereby facilitating motion. Muscle tissue works only by becoming shorter. It shortens and then rests, ie, a muscle can only pull, it cannot push. Muscles produce large amounts of heat as they perform work. Involuntary contraction of muscle tissue releases chemical energy. This energy produces heat which warms the body, an action known as shivering.

5.1. Soft Tissue Injuries.   Some of the more common soft tissue injuries are sprains, strains, contusions, tendonitis, bursitis, and stress injuries, caused by damaged tendons, muscles, and ligaments. A sprain is a soft tissue injury to the ligaments. Certain sprains are often associated with small fractures. This type of injury is normally associated with a localized trauma event. The severity of the sprain depends on how much of the ligament is torn and to what extent the ligament is detached from the bone. The areas of the human body that are most vulnerable to sprains are ankles, knees, and wrists. A sprained ankle is the most frequent injury. The recommended treatment for a simple sprain is usually rest, ice, compression, and elevation (RICE). If a ligament is torn, however, surgery may be required to repair the injury.

A strain is the result of an injury to either the muscle or a tendon, usually in the foot or leg. Strain is a soft tissue injury resulting from excessive use, violent contraction, or excessive forcible stretch. The biomechanical description is of material failure resulting from force being applied to an area causing excessive tension, compression, or shear stress loading, leading to structural tissue distortion and the constant release of energy. The strain may be a simple stretch in muscle or tendon tissue, or it may be a partial or complete tear in the muscle and tendon combination. The recommended treatment for a strain is also RICE, usually to be followed by simple exercise to relieve pain and restore mobility. A serious tear may need surgical repair.

A contusion is an injury to soft tissue in which the skin is not penetrated, but swelling of broken blood vessels causes a bruise. The bruise is caused by a blow of excessive force to muscle, tendon, or ligament tissue. A bruise, also known as a hematoma, is caused when blood coagulates around the injury causing swelling and discoloring skin. Most contusions are mild and respond well to rest, ice, compression, and elevation of the injured area.

Tendonitis, an inflammation in the tendon or in the tendon covering, is usually caused by a series of small stresses that repeatedly aggravate the tendon, preventing it from healing properly, rather than from a single injury. Orthopedic surgeons treat tendonitis by prescribing rest to eliminate the biomechanical tissue stress, and possibly by prescribing antiinflammatory medications, such as steroids (qv). Specially chosen exercises correct muscle imbalances and help to restore flexibility. Continuous stress on an inflamed tendon occasionally causes it to rupture. This usually necessitates casting or even surgery to reattach the ruptured tendon.

A bursa, a sac filled with fluid located around a principal joint, is lined with a synovial membrane and contains synovial fluid. This fluid minimizes friction between the tendon and the bone, or between tendon and ligament. Repeated small stresses and overuse can cause the bursa in the shoulder, hip, knee, or ankle to swell. This swelling and irritation is referred to as bursitis. Some patients experience bursitis in association with tendonitis. Bursitis can usually be relieved by rest and in some cases by using antiinflammatory medications. Some orthopedic surgeons also inject the bursa with additional medication to reduce the inflammation.

5.2. Bone Fractures.   A dislocation occurs when sudden pressure or force pulls a bone out of its socket at the joint. This is also known as subluxation. Bone fractures are classified into two categories: simple fractures and compound, complex, or open fractures. In the latter the skin is pierced and the flesh and bone are exposed to infection. A bone fracture begins to heal nearly as soon as it occurs. Therefore, it is important for a bone fracture to be set accurately as soon as possible.

In certain diseases, such as osteomalacia, syphilis, and osteomyelitis, bones break spontaneously and without a trauma. The severity of the fracture usually depends on the force that caused the fracture. If a bone's breaking point was exceeded only slightly, then the bone may crack rather than break all the way through. If the force is extreme, such as in an automobile collision or a gunshot, the bone may shatter. An open or compound fracture is particularly serious because infection is possible in both the wound and the bone. A serious bone infection can result in amputation.

Stress fractures occur when microfractures accumulate because muscle tissue becomes fatigued and no longer protects from shock or impact. These heal, however, if given adequate rest. Without proper rest, unprotected bone beomes fatigued from absorbing the stress which is normally absorbed by muscle. Isolated microfractures become larger and then join together, forming a continuous stress fracture. These are often referred to as fatigue fractures. Stress fractures were first referred to as march fractures.

Stress or fatigue fractures are very painful. Most often symptoms occur after athletic activity or physical exertion. Gradually pain worsens and becomes more constant. Stress fractures do not show up on standard x-rays. A bone scan may be used to confirm the diagnosis. Stress fractures usually occur in the weight-bearing bones of the lower leg and foot. Stress fractures of the tibia account for half of all stress fractures, resulting mostly from athletic activity. These stress fractures are often mistaken for shin splints. In addition to the tibia, the fibula and other small bones of the foot are prone to stress fractures.

5.3. Fracture Treatment.   The movement of a broken bone must be controlled because moving a broken or dislocated bone causes additional damage to the bone, nearby blood vessels, and nerves or other tissues surrounding the bone. Indeed, emergency treatment requires splinting or bracing a fracture injury before further medical treatment is given. Typically, x-rays determine whether there is a fracture, and if so, of what type. If there is a fracture, a doctor reduces it by restoring the parts of the broken bone to their original positions. All treatment forms for fractures follow one basic rule: the broken pieces must be repositioned and prevented from moving out of place until healed. Broken bone ends heal by growing back together, ie, new bone cells form around the edge of the broken pieces. Specific bone fracture treatment depends on the severity of the break and the bone involved, ie, a broken bone in the spine is treated differently from a broken rib or a bone in the arm.

Treatments used for various types of fractures are cast immobilization, traction, and internal fixation. A plaster or fiber glass cast is the most commonly used device for fracture treatment. Most broken bones heal successfully once properly repositioned, ie, fixed in place via a cast. This type of cast or brace is known as an orthosis. It allows limited or controlled movement of nearby joints. This treatment is desirable for certain fractures.

Traction is typically used to align a bone by a gentle, constant pulling action. The pulling force may be transmitted to the bone through skin tapes or a metal pin through a bone. Traction may be used as a preliminary treatment, before other forms of treatment or after cast immobilization.

In internal fixation, an orthopedist performs surgery on the bone. During this procedure, the bone fragments are repositioned (reduced) into their normal alignment and then held together with special screws or by attaching metal plates to the outer surface of the bone. The fragments may also be held together by inserting rods (intramedullary rods) down through the marrow space into the center of the bone. These methods of treatment can reposition the fracture fragments very exactly. A common internal fixation procedure is to surgically fix the femoral neck (broken hip), as shown in Figure 4.

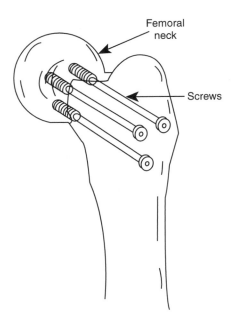

Fig. 4.   Internal femur fixation.

5.4. Joint Replacement. The most frequent reason for performing a total joint replacement is to relieve the pain and disability caused by severe arthritis. The surface of the joint may be damaged by osteoarthritis, ie, a wearing away of the cartilage in a joint. The joint may also be damaged by rheumatoid arthritis, an autoimmune disease, in which the synovium produces chemical substances that attack the joint surface and destroy the cartilage. The swelling, heat, and stiffness that occur in an arthritic joint cause inflammation, the body's natural reaction to disease or injury. Inflammation is usually temporary, but in arthritic joints it is long-lasting and causes disability. When arthritis has caused severe damage to a joint, a total joint replacement may allow the person to return to normal everyday activities.

A total joint replacement is a radical surgical procedure performed under general anesthesia, in which the surgeon replaces the damaged parts of the joint with artificial materials. For example, in the knee joint the damaged ends of the bone that meet at the knee are replaced, along with the underside of the kneecap. In the hip joint, the damaged femoral head is replaced by a metal ball having a stem that fits down into the femur. A new plastic socket is implanted into the pelvis to replace the old damaged socket. This is shown schematically in Figure 5. Whereas hips and knees are the joints most frequently replaced, because the scientific understanding of these is best, total joint

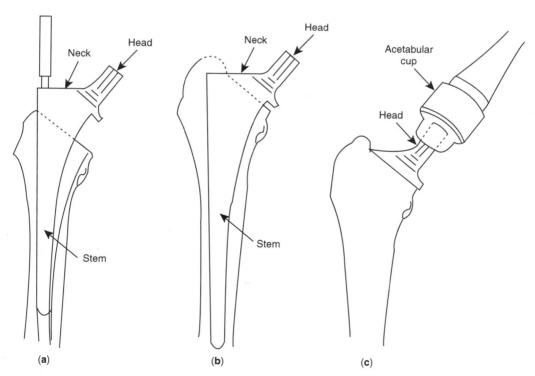

(a)   (b)   (c)

Fig. 5. Three views of a hip implant for joint replacement: (**a**) insertion of the implant into the femur; (**b**) implant in place; and (**c**) femur and implant connected to plastic socket fitted into the pelvis.

replacement can be performed on other joints as well, including the ankle, shoulder, fingers, and elbow.

The materials used in a total joint replacement are designed to enable the joint to function normally. The artificial components are generally composed of a metal piece that fits closely into bone tissue. The metals are varied and include stainless steel or alloys of cobalt, chrome, and titanium. The plastic material used in implants is a polyethylene that is extremely durable and wear-resistant. Also, a bone cement, a methacrylate, is often used to anchor the artificial joint materials into the bone. Cementless joint replacements have more recently been developed. In these replacements, the prosthesis and the bone are made to fit together without the need for bone cement. The implants are press-fit into the bone.

The recovery period following total joint arthropathy depends on both the patient and the affected joint. In general the patient is encouraged to use the joint soon after replacement. In the case of a hip or knee replacement the patient should be standing and beginning to walk within several days. If the shoulder, elbow, or wrist joint is replaced, use of the new joint can begin very soon after surgery, as these are not weight-bearing joints. The patient generally performs appropriate exercises to strengthen and move the joint during the recovery period.

The main benefits to the patient after total joint replacement are pain relief, which often is quite dramatic, and increased muscle power, which was lost because the painful arthritic joint was not used and usually returns with exercise once pain is relieved. Motion of the joint generally improves as well. The extent of movement depends on how stiff the joint was before the joint was replaced. An extremely stiff joint continues to be stiff for some period of time after replacement.

The principal complication for total joint replacement is infection, which may occur just in the area of the incision or more seriously deep around the prosthesis. Infections in the wound area, which may even occur years after the procedure has been performed, are usually treated with antibiotics (qv). Deep infections may require further surgery, prosthesis removal, and replacement.

Loosening of the prosthesis is the most common biomechanical problem occurring after total joint replacement surgery. Loosening causes pain. If loosening is significant, a second or revision total joint replacement may be necessary. Another complication which sometimes occurs after total joint replacement, generally right after the operation, is dislocation, the result of weakened ligaments. In most cases the dislocation can be relocated manually by the orthopedic surgeon. Very rarely is another operation necessary. A brace may be worn after dislocation occurs for a short time. Although some wear can be measured in artificial joints, wear occurs slowly. Whereas wear may contribute to looseness, it is rarely necessary to do corrective surgery because of wear alone.

Breakage of an implanted joint is rare. Breakage occurs when the bone flexes and the metal implant does not flex as much, thereby exceeding its mechanical fatigue point causing the implant to break or crack. A revision joint replacement operation is necessary if breakage occurs.

Nerves are rarely damaged during the total joint replacement surgery. However, nerve damage can occur if considerable joint deformity must be corrected in order to implant the prosthesis. With time these nerves sometimes return to normal function.

Osteoarthritis, the most common arthritic disorder, affects some 30 million Americans each year. Caused by daily wear and tear on joints or injury, osteoarthritis is painful and restricts daily activity. It can affect the basal joint of the thumb, as well as the knee, hip, and other joints.

Hip Joints.   Successful hip joint replacement surgery was introduced in the late 1950s. Since that time design and scientific advances have brought increasingly better clinical results. In excess of 200,000 patients in the United States seek pain relief annually through hip joint replacement. About 18–20% are revision hip systems, ie, second replacement implants. A hip usually becomes painful when the cartilage that lines the hip socket starts to wear out. As total hip systems evolved, designers attempted to eliminate features which led to failure. Modifications were made to each element of the hip system including the femoral stem and acetabular cup. Wear and tear arthritis may be the result of a genetic defect that prevents the body from manufacturing cartilage rugged enough to last a lifetime. Increased life-expectancy, stresses owing to certain occupations, and prior injury that places abnormal stress on cartilage over a long period may also contribute to the development of osteoarthritis. Often arthritic pain can be controlled through the use of antiinflammatory medication. However, if hip pain becomes intolerable, hip replacement surgery may be elected.

In 1974 a prosthesis introduced by Howmedica combined a biomechanically high strength material, Vitallium, with a professionally engineered geometry. This prosthesis marked the first design departure from the diamond-shaped cross-sectional geometry previously used. Sharp corners were eliminated and replaced by broad, rounded medial and lateral borders. The total sectional area was much greater than any of the previous hip joint implant stems. The result of these combined factors was decreased unit stresses on the cement mantle. This system also marked the first time surgeons could choose components from a selection large enough to provide fit for most primary and revision total hip replacement patients.

The next advance in total hip arthroplasty came with the development of various porous surface treatments which allow bone tissue to grow into the metal porous coating on the femoral stem of the hip implant and on the acetabular component of the total joint replacement. These developments arose because of patients who were not able to tolerate cemented implants because of allergies to the cement, methylmethacrylate. More youthful patients are better served by a press-fit implant as well. Figure 6 shows the difference between textured and beaded surface-treated orthopedic prostheses.

Hydroxyapatite (HA) coating on the surface of the hip stem and the acetabular cup is the most recent advancement in artificial hip joint implant technology. This substance is a form of calcium phosphate, which is sprayed onto the hip implant. It is a material found in combination with calcium carbonate in bone tissue, and bones can easily adapt to it. When bone tissue does grow into HA, the tissue then fixes the hip joint implant permanently in position. These HA coatings are only used in press-fit, noncemented implants.

The acetabular component is as integral to successful total hip arthroplasty as is the femoral hip stem component. The life of the acetabular component

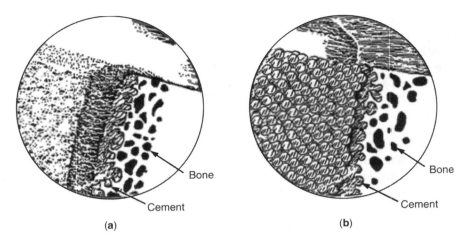

Fig. 6.   Surface treatments: (**a**) textured and (**b**) beaded.

depends on proper placement and bone preparation in the acetabular region of the hip girdle, proper use of bone cement, and superior component design.

The history of the development of the acetabular component parallels that of the femoral component. In 1951 an acetabular prosthesis based on a chromium–cobalt alloy having screw-in sockets was successfully introduced. Beginning in 1955, methylmethacrylate was used as a cementing agent, and work was undertaken to find suitable materials for use as an articulation surface in the acetabular. Teflon appeared to provide a good lubricating, articulating surface. However, over time Teflon exhibited poor wear conditions when it contacted the metal femoral head, and as the body developed systemic reaction to Teflon particles, failures occurred. Since then ultrahigh density polyethylene has proven successful as acetabular cups.

In 1971 a metal-backed polyethylene acetabular cup was introduced. This cup provided an eccentric socket which was replaceable, leaving the metal and replacing only the polyethylene. Because of the success of this component, metal-backed high density polyethylene (HDPE) liner is standard for prosthetic acetabular components. Research confirms that metal-backing reduces the peak stresses in the bone cement, and that HDPE forms a successful articulating surface for the prosthetic joint.

Over time a large variety of materials have been used, including ivory, stainless steel, chromium–cobalt, and ceramics for the acetabular component. None proved sufficient. The implant material composition must provide a smooth surface for joint articulation, withstand hip joint stresses from normal loads, and the substance must disperse stress evenly to the cement and surrounding bone.

The material in use as of the mid-1990s in these components is HDPE, a linear polymer which is tough, resilient, ductile, wear resistant, and has low friction (see OLEFIN POLYMERS, POLYETHYLENE). Polymers are prone to both creep and fatigue (stress) cracking. Moreover, HDPE has a modulus of elasticity that is only one-tenth that of the bone, thus it increases the level of stress transmitted to the cement, thereby increasing the potential for cement mantle failure. When the acetabular HDPE cup is backed by metal, it stiffens the HDPE cup. This results in function similar to that of natural subchondral bone. Metal backing has become standard on acetabular cups.

The femoral component is composed of the head, neck, collar, and stem (see Fig. 5). The head, or ball, is the surface component which articulates with the acetabular cup of the total hip implant. This is an important element in the implant design because this surface absorbs the greatest stress and has the most force applied to it. Consequently the head gets the greatest wear. The diameter of the femoral head affects the distribution of forces in both the femoral and acetabular components. This variable also influences the range of motion that the implant permits, therefore affecting the stability of the ball-in-socket prosthetic joint. The most common head diameters range from 22 to 32 mm. Each size offers advantages and disadvantages. No general consensus exists as to which size is better. As a result, many manufacturers offer more than one head diameter to suit surgeon preference and patient requirements.

The 22-mm diameter is preferred by some doctors who believe that a small-diameter head encourages mechanical fixation of the socket without cement fixation. Other benefits associated with the 22-mm head size include its suitability for use in a patient having a small acetabulum, and the fact that it allows for a thicker acetabular component, which permits more wear and absorbs more energy than do thinner walled components. The larger (32-mm) head diameter range is recommended by other surgeons and has been incorporated into most hip systems. Using this large diameter the surface of the head has a greater area, resulting in decreased stress per unit area. Other advantages include less chance of subluxation (joint dislocation) at the extremes of motion and therefore improved prosthetic joint stability. However, some doctors express concern that the increased articulating surface contact area promotes increased frictional torque.

The stem/neck length and cross-sectional geometry of the neck affects the forces acting on both the neck and the stem. The neck must be large enough to prevent failure, but not so large as to limit range of motion. Further, the neck length should be consistent with the anatomy of the patient.

The most important design consideration in the neck of the femoral component is that it support body weight without breaking. This requires that the head/neck ratio be appropriate. Neck length is measured from the center of the femoral head to the collar of the stem. Variations to neck length, combined with offset neck/stem angle, and head diameter, permit surgeons to adjust leg length of the total hip implant leg to be consistent with that of the opposite leg. It was for this reason that surgeons introduced the concept of providing different neck lengths.

The stem/neck offset, ie, distance from the center of the head to the center of the stem, changes upon a change in the neck length. Increased neck length, and therefore increased offset, raises the bending moment of the stem, thereby increasing the chances for prosthetic failure. The various neck geometries used by designers of prostheses represent attempts to find a satisfactory combination of shape, bulk, and material which can withstand cyclic loading and joint forces without breaking. Most neck/stem angles are neutral (135°) in order to estimate average human anatomy and equalize the moment arms.

An important issue of stem design is length. Increased stem length means more stem area for improved stress distribution. Another benefit to a longer implant stem is engagement of the isthmus, the most narrow portion of the femur. Expanding the stem into this area of contact increases prosthetic stability, helps prevent the stem from shifting position, decreases the amount of micromotion, and achieves better alignment along the neutral axis of the femur. Stems are available in varied lengths to match human anatomy and improve isthmic engagement. The most advanced hip implants on the market are totally modular, so that they are nearly custom made to fit into the femur and the acetabulum of the pelvic girdle.

Starting from the bottom of the hip implant, a modular implant begins with a press-fit, distal, high density, ultrahigh molecular weight, polyethylene (HDPE) plug tip, at the bottom of the femur stem. Then a machined and polished titanium, chromium–cobalt–molybdenum, or vanadium–aluminum metallic alloy diaphyseal–endosteal grooved stem segment is added, followed by a machined, polished, and hydroxyapatite-coated metaphyseal metallic alloy stem segment, the upper portion of the hip implant stem. A custom-fit metallic alloy collar plate, which rests upon the resected (cut-away) head of the femur, comes next, followed by a sized modular metallic alloy neck upon which a chromium–cobalt–molybdenum or zirconium ceramic head, ie, ball that sets upon the hip stem neck, rests. The head then articulates with the acetabular cup liner (see Fig. 5).

The head of the femoral component then articulates with an ion-bombarded, HDPE, high walled, acetabular liner which fits into a screwed in, machined, titanium, chromium–cobalt–molybdenum or vanadium–aluminum metallic alloy hydroxyapatite-coated acetabular shell/cup. Each of the separate parts of the modular system for total hip arthroplasty is manufactured in several different sizes.

Total hip implants of the nature described have hospital list prices in the range of $5000–$8000. Fully custom-made implants cost approximately $10,000. The low end basic total hip implant is forged or cast stainless steel, cemented in place, one size fits all, and costs $1000.

Prosthesis Design.   The challenge in prosthesis design is to create an implant that mimics the material characteristics and the exact anatomical functions of the joint. Stress and loading forces on the hip joint and femur are extraordinary. Stresses on the hip joint exceed 8.3 MPa (1200 psi). Standing on one leg produces a loading force on the hip joint of 250% of the total body weight. Running increases these forces to five times the total body weight. The hip joint is surrounded by the most powerful muscle structure in the body enabling movement while supporting sufficient structural force and loads. Proper surgical technique is as critical to the success of an implant procedure as is the design of the device itself. Therefore, matching surgeon skill, an appropriate implant design for the patient, and the correct tools generally forms the best solution for a successful procedure.

A significant aspect of hip joint biomechanics is that the structural components are not normally subjected to constant loads. Rather, this joint is subject to unique compressive, torsion, tensile, and shear stress, sometimes simultaneously. Maximum loading occurs when the heel strikes down and the toe pushes off in walking. When an implant is in place its ability to withstand this repetitive loading is called its fatigue strength. If an implant is placed properly, its load is shared in an anatomically correct fashion with the bone.

Design variables introduced on various prostheses represent efforts to share the stresses and normal loading characteristics of human locomotion. The size, shape, and tissue structure of bone are most commonly affected in the healing of fractures. Bone remodeling was first described in 1892 by the German physician Julius Wolff in *The Law of Bone Transformation*. In terms of force loading and stresses, Wolff's law states that bone responds to mechanical demends by changing its size, shape, and structure.

Resorption of bone tissue occurs in total hip joint replacement patients if sufficient stresses are not adequately transmitted to the remaining bone in exactly the same way that the bone transmitted those stresses originally. Therefore, the design and proper placement of the neck collar and hip stem must be effective in recreating anatomical structure.

Bone remodeling is the ability of bone to change its size, shape, and structure by adapting to mechanical demands that are placed on it. Bone grows where it is needed, and resorbs where it is not needed. The type of bone tissue that grows depends on the stress level it sustains. Someone who performs strenuous exercise undergoes cortical bone changes resulting in bigger, denser bones. On the other hand, someone who performs minimal physical activity loses bone density through resorption. This is a problem for those who must stay in bed for prolonged periods. The problem of bone density loss owing to minimal physical stress has also been a unique concern for NASA astronauts. Special exercises have been designed for the astronauts to counteract the effects of weightlessness and to slow down bone resorption during long orbital flights.

The process of aging reduces bone size and strength. Thinning and resorption occur in the cancellous bone. Also, cortical bone resorbs and bone shrinks in diameter and thickness. The older the person, the more fragile the bone.

Research based on Wolff's law of bone transformation has resulted in some other important observations. Fluctuating loads, such as those that occur in walking, are better for bone than consistently applied loads, such as weight gain. However, if the effective applied load becomes extreme, pressure necrosis, ie, bone death, occurs. Pressure necrosis is a significant concern in hip arthroplasty. Necrosis means the localized death of living tissue. Undue pressure on living cells causes death. Some total hip replacement failures are the direct result of pressure necrosis.

Some of the early design hip prostheses, created without a complete understanding of stress forces and anatomical loading characteristics of normal activity, had sharp points at the distal end and along the medial and lateral sides of the stem. Improperly seated, or merely subjected to normal forces, these stems directed concentrated stresses into the interfacing cement. This point loading resulted in cement fracturing into fragments and bone tissue suffering pressure necrosis, resulting in implant failure. More rounded prosthetic designs, and the tools and instrumentation to properly seat them, distribute the load over the widest area. This distribution mimics that of natural bone and prevents pressure necrosis.

Biomaterials.   Just as stem designs have evolved in an effort to develop an optimal combination of specifications, so have the types of metals and alloys employed in the construction of total joint implants. Pure metals are usually too soft to be used in prosthesis. Therefore, alloys which exhibit improved characteristics of fatigue strength, tensile strength, ductility, modulus of elasticity, hardness, resistance to corrosion, and biocompatibility are used.

Titanium alloy, composed of titanium, aluminum, and vanadium, is preferred by some orthopedic surgeons primarily for its low modulus of elasticity, which allows for transfer of more stress to the proximal femur. This alloy also exhibits good mechanical strength and biocompatibility (9). The stem flexibility optimizes the transfer of stress directly to the bone, and offers adequate calcar loading to minimize femoral resorption.

Vitallium FHS alloy is a cobalt–chromium–molybdenum alloy having a high modulus of elasticity. This alloy is also a preferred material. When combined with a properly designed stem, the properties of this alloy provide protection for the cement mantle by decreasing proximal cement stress. This alloy also exhibits high yields and tensile strength, is corrosion resistant, and biocompatible. Composites used in orthopedics include carbon–carbon, carbon–epoxy, hydroxyapatite, ceramics, etc.

Tools and Procedures.   Arthroscopy is a surgical procedure used to visualize, diagnose, and treat injuries within joints. The term arthroscopy literally means to look inside the joint. During this procedure the orthopedic surgeon makes an incision into the patients skin and inserts a pencil-shaped arthroscope. An arthroscope is a miniature lens and lighting systems that magnifies and illuminates the structures inside the joint. A television screen which is attached to the arthroscope displays the image of the joint on screen.

This is a minimally invasive procedure (MIP) resulting in a shorter hospital stay, faster recovery, and less evident scar in comparison to other types of surgery. Arthroscopic surgery gives the surgeon a precise, direct view of the affected bones and soft tissues. This procedure allows the surgeon to see areas of the joint that are difficult to see on x-rays and more of the joint than is possible even after making a large incision during open surgery. Arthroscopy can be performed under local, general, or spinal anesthesia. The area surrounding the joint is sterilized, and then the joint is expanded to make room for the arthroscope by injecting a sterile solution into the joint. The surgeon makes a small incision into the skin through which the arthroscope is inserted. A surgical instrument probes various parts of the joints to determine the injury. Surgical repair, if needed, is performed using specially designed surgical instruments which are inserted into the joint through the small incisions. This surgery can be viewed on a television screen. The small incisions that were made during surgery are closed usually using only one or two sutures.

Patients' immediate post-operative pain is lower compared to a standard operation and healing and rehabilitation more rapid. Patients can resume near-normal activities in just days. In some cases athletes, who are in prime physical condition, can return to challenging athletic activities within a few weeks. Complications are rare, but do occur on occasion. Most complications associated with this surgery are infection, phlebitis, excessive swelling or bleeding, blood clots, or damage to blood vessels or nerves.

6. Bioresorbable Polymers

Biomaterials scientists have worked diligently to synthesize polymeric structures which exhibit biocompatibility and long-term biostability. Devices made from these polymers are intended to be implanted in the body for years, and in some cases decades.

The concept of using biodegradable materials for implants which serve a temporary function is a relatively new one. This concept has gained acceptance as it has been realized that an implanted material does not have to be inert, but can be degraded and/or metabolized *in vivo* once its function has been accomplished (10). Resorbable polymers have been utilized successfully in the manufacture of sutures, small bone fixation devices (11), and drug delivery systems (qv) (12).

Several groups have experimented with bioresorbable polymers that have a predictable degree of bioresorbability when exposed to the physiological environment. By the judicious choice of bioresorbability rate it is hoped that as the polymer is resorbed it will leave surface voids where natural tissue would grow, resulting in autologous organ regeneration. The temporary nature of the device will impart initial mechanical functionality to the implant, but after time will be resorbed as the natural tissue regenerates. This concept has been experimentally applied to the regeneration of tissue such as in the liver (13), skeletal tissue (14), cartilage (15), and the vascular wall (16).

One area in which predictable biodegradation is used is the area of degradable surgical sutures. An incision wound, when held together with sutures, heals to about 80% of initial strength within four weeks. Surgical suture is one of the earliest clinical implants in recorded history. Catgut suture, obtained from ovine or bovine intestinal submucosa, was known in 150 AD in the time of Galen, who built his reputation by treating wounded gladiators (17).

Catgut is infection-resistant. The biodegradation of catgut results in elimination of foreign material that otherwise could serve as a nidus for infection or, in the urinary tract, calcification. As a result, chromic catgut, which uses chromic acid as a cross-linking agent, is still preferred in some procedures. Chromic catgut is considered by some to be the most suitable suture material for vaginal hysterectomy owing to its extensibility and rapid absorption. Treatment of natural catgut with synthetic polymers exemplifies the merging of old and new technology. Coating catgut with a polyurethane resin allows catgut to retain its initial tensile strength longer (18).

The first synthetic polyglycolic acid suture was introduced in 1970 with great success (19). This is because synthetic polymers are preferable to natural polymers since greater control over uniformity and mechanical properties are obtainable. The foreign body response to synthetic polymer absorption generally is quite predictable whereas catgut absorption is variable and usually produces a more intense inflammatory reaction (20). This greater tissue compatibility is crucial when the implant must serve as an inert, mechanical device prior to bioresorption.

6.1. Polylactic Acid. Polylactic acid (PLA) was introduced in 1966 for degradable surgical implants. Hydrolysis yields lactic acid, a normal intermediate of carbohydrate metabolism (21). Polyglycolic acid sutures have a predictable degradation rate which coincides with the healing sequence of natural tissues.

Polylactic acid, also known as polylactide, is prepared from the cyclic diester of lactic acid (lactide) by ring-opening addition polymerization, as shown below:

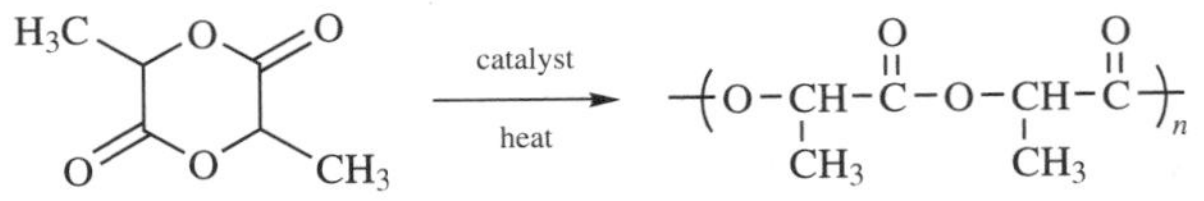

Lactic acid is an asymmetric compound existing as two optical isomers or enantiomers. The L-enantiomer occurs in nature; an optically inactive racemic mixture of D- and L-enantiomers results during synthesis of lactic acid. Using these two types of lactic acid, the corresponding L-lactide (mp 96°C) and DL-lactide (mp 126°C) have been used for polymer synthesis. Fibers spun from poly-L-lactide (mp 170°C) have high crystallinity when drawn, whereas poly-DL-lactide (mp 60°C) fibers display molecular alignment on drawing but remain amorphous. The crystalline poly-L-lactide is more resistant to hydrolytic degradation than the amorphous DL form of the same homopolymer. Therefore, pure DL-lactide displays greater bioresorbability, whereas pure poly-L-lactide is more hydrolytically resistant.

The actual time required for poly-L-lactide implants to be completely absorbed is relatively long, and depends on polymer purity, processing conditions, implant site, and physical dimensions of the implant. For instance, 50–90 mg samples of radiolabeled poly-DL-lactide implanted in the abdominal walls of rats had an absorption time of 1.5 years with metabolism resulting primarily from respiratory excretion (22). In contrast, pure poly-L-lactide bone plates attached to sheep femora showed mechanical deterioration, but little evidence of significant mass loss even after four years (23).

Improved techniques for polylactide synthesis have resulted in preparation of exceptionally high molecular weight polymer. Fiber processing research has resulted in fiber samples having tensile breaking strength approaching 1.2 GPa (174,000 psi). This strength value was obtained by hot-drawing filaments spun from good solvents (24).

6.2. Polyglycolic Acid.   Polyglycolic acid (PGA), also known as polyglycolide, was first reported in 1893, but it wasn't until 1967 that the first commercially successful patent was granted for sutures (25). Like polylactide, polyglycolide is synthesized from the cyclic diester as shown below:

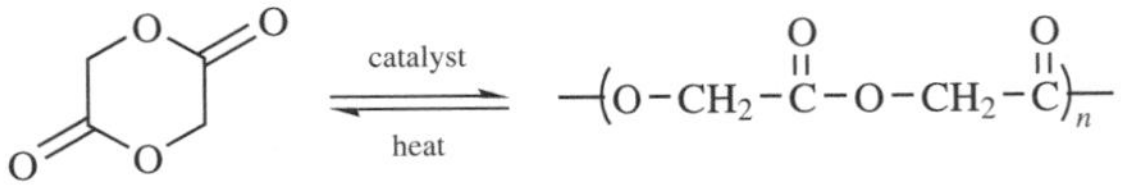

An important difference between polylactide and polyglycolide, is that polyglycolide (mp 220°C) is higher melting than poly-L-lactide (mp 170°C). Although the polymerization reaction in both cases is reversible at high temperature, melt processing of polyglycolide is more difficult because the melting temperature is close to its decomposition temperature.

Unlike poly-L-lactide which is absorbed slowly, polyglycolide is absorbed within a few months post-implantation owing to greater hydrolytic susceptibility. *In vitro* experiments have shown the effect on degradation by enzymes (26), pH, annealing treatments (27), and gamma irradiation (28). Braided polyglycolide sutures undergo surprisingly rapid hydrolysis *in vivo* owing to cellular enzymes released during the acute inflammatory response following implantation (29).

Low humidity ethylene oxide gas sterilization procedures and moisture-proof packaging for polyglycolic acid products are necessary because of the susceptibility to degradation resulting from exposure to moisture and gamma sterilization.

6.3. Poly(lactide-*co*-glycolide).

Mixtures of lactide and glycolide monomers have been copolymerized in an effort to extend the range of polymer properties and rates of *in vivo* absorption. Poly(lactide-*co*-glycolide) polymers undergo a simple hydrolysis degradation mechanism, which is sensitive to both pH and the presence of enzymes (30).

A 90% glycolide, 10% L-lactide copolymer was the first successful clinical material of this type. Braided absorbable suture made from this copolymer is similar to pure polyglycolide suture. Both were absorbed between 90 and 120 days post-implantation but the copolymer retained strength slightly longer and was absorbed sooner than polyglycolide (31). These differences in absorption rate result from differences in polymer morphology. The amorphous regions of poly(lactide-*co*-glycolide) are more susceptible to hydrolytic attack than the crystalline regions (32).

Similar to pure polyglycolic acid and pure polylactic acid, the 90:10 glycolide:lactide copolymer is also weakened by gamma irradiation. The normal *in vivo* absorption time of about 70 days for fibrous material can be decreased to less than about 28 days by simple exposure to gamma radiation in excess of 50 kGy (5 Mrads) (33).

The crystallinity of poly(lactide-*co*-glycolide) samples has been studied (34). These copolymers are amorphous between the compositional range of 25–70 mol % glycolide. Pure polyglycolide was found to be about 50% crystalline whereas pure poly-L-lactide was about 37% crystalline. An amorphous poly(L-lactide-*co*-glycolide) copolymer is used in surgical clips and staples (35). The preferred composition chosen for manufacture of clips and staples is the 70/30 L-lactide/glycolide copolymer.

6.4. Polydioxanone.

Fibers made from polymers containing a high percentage of polyglycolide are considered too stiff for monofilament suture and thus are available only in braided form above the microsuture size range. The first clinically tested monofilament synthetic absorbable suture was made from polydioxanone (36). This polymer is another example of a ring-opening polymerization reaction. The monomer, *p*-dioxanone, is analogous to glycolide but yields a poly(ether–ester) as shown below:

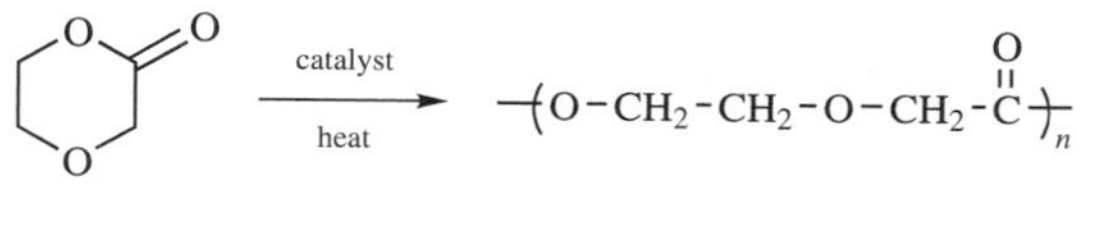

Polydioxanone (PDS) is completely eliminated from the body upon absorption. The mechanism of polydioxanone degradation is similar to that observed for other synthetic bioabsorbable polymers. Polydioxanone degradation *in vitro* was affected by gamma irradiation dosage but not substantially by the presence of enzymes (37). The strength loss and absorption of braided PDS, but not monofilament PDS, implanted in infected wounds, however, was significantly greater than in noninfected wounds.

6.5. Poly(ethylene oxide)–Poly(ethylene terephthalate) Copolymers. The poly(ethylene oxide)–poly(ethylene terephthalate) (PEO/PET) copolymers were first described in 1954 (38). This group of polymers was developed in an attempt to simultaneously reduce the crystallinity of PET, and increase its hydrophilicity to improve dyeability. PEO/PET copolymers with increased PEO contents produce surfaces that approach zero interfacial energy between the implant and the adjacent biological tissue. The collagenous capsule formed around the implant is thinner as the PEO contents increase. The structure of a PEO/PET copolymer is shown below:

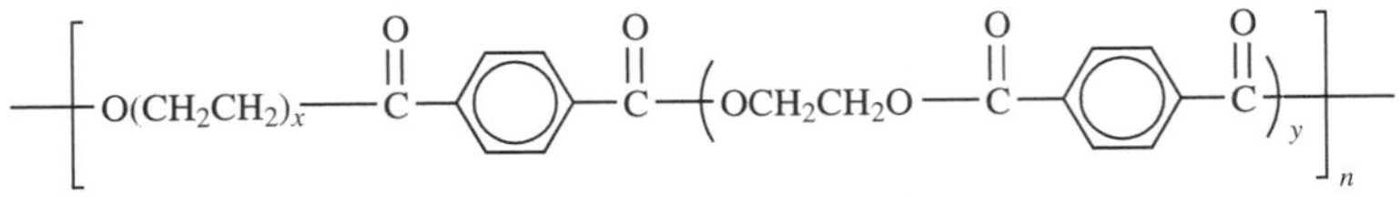

A family of PEO/PET copolymers has been synthesized and the characterized structures found to be close to those expected in theory (39). A wide degradation envelope has been achieved by adjusting the PEO-to-PET ratio. Mechanical properties prove useful for medical applications, and the 60/40 PEO/PET composition is reported as optimal.

6.6. Poly(glycolide-*co*-trimethylene carbonate). Another successful approach to obtaining an absorbable polymer capable of producing flexible monofilaments has involved finding a new type of monomer for copolymerization with glycolide (40). Trimethylene carbonate polymerized with glycolide is shown below:

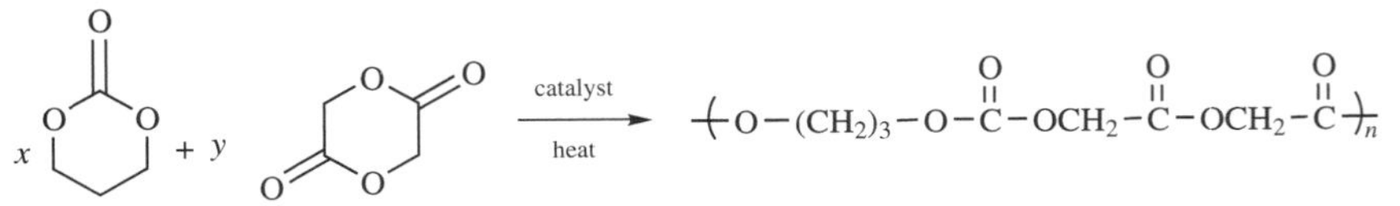

In order to achieve the desired fiber properties, the two monomers were copolymerized so the final product was a block copolymer of the ABA type, where A was pure polyglycolide and B, a random copolymer of mostly poly(trimethylene carbonate). The selected composition was about 30–40% poly(trimethylene carbonate). This suture reportedly has excellent flexibility and superior *in vivo* tensile strength retention compared to polyglycolide. It has been absorbed without adverse reaction in about seven months (41). Metabolism studies show that the route of excretion for the trimethylene carbonate moiety is somewhat different from the glycolate moiety. Most of the glycolate is excreted by urine whereas most of the carbonate is excreted by expired CO_2 and urine.

6.7. Poly(ethylene carbonate). Like polyesters, polycarbonates (qv) are bioabsorbable only if the hydrolyzable linkages are accessible to enzymes and/or water molecules. Thus pellets of poly(ethylene carbonate), $+OCOOCH_2CH_2+_n$ weighing 200 mg implanted in the peritoneal cavity of rats, were bioabsorbed in only two weeks, whereas similar pellets of poly(propylene carbonate), $+OCOOCH(CH_3)CH_2+_n$ showed no evidence of bioabsorption after two months (42). Because poly(ethylene carbonate) hydrolyzes more rapidly *in vivo* than *in vitro*, enzyme-catalyzed hydrolysis is postulated as a contributing factor in polymer absorption. Copolymers of polyethylene and polypropylene carbonate have been developed as an approach to achieving the desired physical and pharmacological properties of microsphere drug delivery systems.

6.8. Polycaprolactone. Polycaprolactone is synthesized from epsilon-caprolactone as shown below:

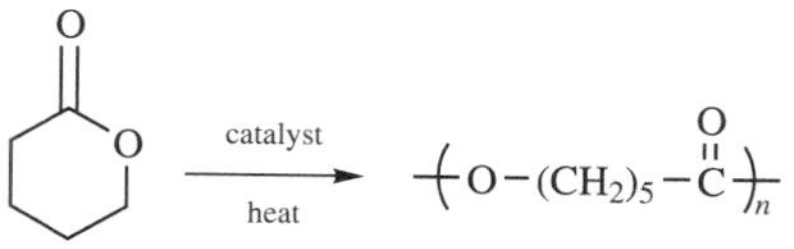

This semicrystalline polymer is absorbed very slowly *in vivo*, releasing ε-hydroxycaproic acid as the sole metabolite. Degradation occurs in two phases: nonenzymatic bulk hydrolysis of ester linkages followed by fragmentation, and release of oligomeric species. Polycaprolactone fragments ultimately are degraded in the phagosomes of macrophages and giant cells, a process that involves lysosome-derived enzymes (43). *In vitro*, polycaprolactone degradation is enhanced by microbial and enzymatic activity. Predictably, amorphous regions of the polymer are degraded prior to breakdown of the crystalline regions (44).

Copolymers of ε-caprolactone and L-lactide are elastomeric when prepared from 25% ε-caprolactone and 75% L-lactide, and rigid when prepared from 10% ε-caprolactone and 90% L-lactide (45). Blends of poly-DL-lactide and polycaprolactone polymers are another way to achieve unique elastomeric properties. Copolymers of ε-caprolactone and glycolide have been evaluated in fiber form as potential absorbable sutures. Strong, flexible monofilaments have been produced which maintain 11–37% of initial tensile strength after two weeks *in vivo* (46).

6.9. Poly(ester–amides). Another approach to obtaining improvements in the properties of synthetic absorbable polymers is the synthesis of polymers containing both ester and amide linkages. The rationale for designing poly(ester–amide) materials is to combine the absorbability of polyesters (qv) with the high performance of polyamides (qv). Two types have been reported. Both involve the polyesterification of diols that contain preformed amide linkages. Poly(ester–amides) obtained from bis-oxamidodiols have been reported to be absorbable only when oxalic acid is used to form the ester linkages (47). Poly(ester–amides) obtained from bis-hydroxyacetamides are absorbable regardless of the diacid employed, although succinic acid is preferred (48).

The absorption rate has been examined *in vivo* for a series of poly(ester–amides) having the following formula:

Polymers, where $x = 6, 8$, and 10, are absorbed within six months; the polymer where $x = 12$ requires over 19 months for complete absorption. Absorption correlates with the water solubility of the starting amidediol monomers. All are at least sparingly soluble except for the $x = 12$ amidediol which is virtually insoluble. *In vivo* strength retention of poly(ester–amides) in fiber form is greatest for the $x = 12$ polymer. This material loses very little strength for four weeks then slowly decreases to 50% strength at 8–10 weeks depending on molecular weight and fiber processing conditions.

The metabolic rate of poly(ester–amide) where $x = 6$ has been studied in rats using carbon-14 labeled polymer. This study indicates that polymer degradation occurs as a result of hydrolysis of the ester linkages whereas the amide linkages remain relatively stable *in vivo*. Most of the radioactivity is excreted by urine in the form of unchanged amidediol monomer, the polymer hydrolysis product (49).

6.10. Poly(orthoesters). The degradation of a bioresorbable polymer occurs in four stages: hydration, loss of strength, loss of integrity, and loss of mass. This typical behavior limits most of the previously mentioned polymers for use as matrices for slow release drug delivery implants because incorporated drugs that are water soluble have been found simply to leach out at a first-order rate. Thus bioabsorbable polymers which are extremely hydrophobic have been developed to prevent hydration yet still possess hydrolytically unstable linkages. This results in degradation of polymer on the exposed surfaces only thereby releasing the drug content at a more uniform rate. Such polymers have been termed bioerodible.

Poly(orthoesters) represent the first class of bioerodible polymers designed specifically for drug delivery applications (50). *In vivo* degradation of the polyorthoester shown, known as the Alzamer degradation, yields 1,4-cyclohexanedimethanol and 4-hydroxybutyric acid as hydrolysis products (51).

6.11. Poly(anhydrides). Poly(anhydrides) are another class of synthetic polymers used for bioerodible matrix, drug delivery implant experiments. An example is poly(bis(*p*-carboxyphenoxy)propane) (PCPP) which has been prepared as a copolymer with various levels of sebacic anhydride (SA). Injection molded samples of poly(anhydride)/drug mixtures display zero-order kinetics in both polymer erosion and drug release. Degradation of these polymers simply releases the dicarboxylic acid monomers (52). Preliminary toxicological evaluations showed that the polymers and degradation products had acceptable biocompatibility and did not exhibit cytotoxicity or mutagenicity (53).

7. Shape Memory Alloys

TiNi shape memory alloy (SMA) has attracted much attention for biomedical applications such as implants (bone plate and marrow needle) and for surgical and dental instruments, devices and fixtures, such as orthodontic fixtures and biopsy forceps (see Shape Memory Alloys). This is due to its excellent biocompatibility and mechanical characteristics. Research on biomedical applications of SMA was started in the 1970s with animal experiments initially, followed by clinical tests. The first example of successful biomedical and dental applications of SMA are available and many new applications are being developed.

SMAs' properties which led to their wide acceptance in biomedical applications include biocompatibility, superelasticity, shape memory effect, hysteresis, and fatigue resistance (54). Studies show that TiNi has superior corrosion resistance, due to the formation of a passive titanium-oxide layer (TiO_2) similar to that found on Ti alloys. This oxide layer increases the stability of the surface layers by protecting the bulk material from corrosion and creates a physical and chemical barrier against Ni oxidation.

7.1. Shape Memory Effect.   Pre-deformed SMAs have the ability to remember their original shape before deformation and are able to recover the shape when heated if the plastic deformation takes place in the martensite phase (55,56). The shape recovery is the result of transformation from the low-temperature martensite phase to the high-temperature austenite phase when it is heated. The shape memory effect makes it easy to deploy an SMA appliance in the body and makes it possible to create a pre-stress after deployment when necessary. That is, the SMA appliances are first packed up in a compact state during deployment and then restored to its expanded shape by means of heating. If the phase transformation temperature of an SMA is below the body temperature, the heat of the body can easily induce shape recovery. In the case where the phase transformation temperature is higher than body temperature, the SMA appliances are usually heated by warm salt water or a high frequency magnetic field.

The shape memory effect has been utilized also for actuator functions in medical applications as a urethral valve and artificial sphincter.

7.2. Superelasticity.   SMAs exhibit superelasticity when they are in the austenite phase (54,56). Figure 7 shows the typical superelastic stress–strain curve (solid line) compared with the stress–strain curve of stainless steel (dashed line). An important feature of superelastic materials is that they exhibit constant loading and unloading stesses over a wide range of strain.

As shown in Figure 7, the effective strain range $\Sigma_{eff}(TN)$ of TiNi corresponding to an optimal force zone is much larger than $\Sigma_{eff}(SS)$ of stainless steel. Hence, a superelastic device can provide a constant pressing force even if the pressed part recedes by a limited amount during the installed period. On the contrary, the pressing force of the appliance made from stainless steel will drop drastically if the pressed part deforms, so that the performance will deteriorate. An orthodontic arch wire was the first product to take advantage of this property. This characteristic is put into use in superelastic eyeglass frames (57). These eyeglass frames have become very popular in the United States, Europe and Japan, and are available in almost every optician's store. These frames can be twisted a full

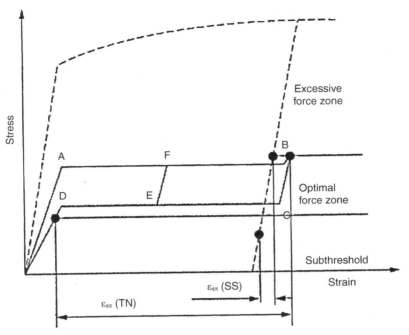

Fig. 7. Typical stress–strain curve of superelastic materials and stainless steel. The superelastic materials exhibit constant unloading stress over a wide range of strain.

180°, but more importantly the frames press against the head with a constant and comfortable stress not only is "fit" less important, but small bends and twists that may develop do not cause discomfort to the wearer.

The superelasticity of SMAs makes it easy to depoly SMA stents. Stents made from stainless steel are expanded against the vessel wall by plastic deformation caused by the inflation of a balloon placed inside the stent. TiNi stents, on the other hand, are self-expanding.

7.3. Hysteresis of SMA. Superelastic SMA demonstrates a hysteretic stress–strain relationship (see Fig. 7), that is, the stress from A to B in the loading phase and the stress from C to D in the unloading are different. Hysteresis is usually regarded as a drawback for traditional engineering application, but it is a useful characteristic for biomedical applications. If the SMA is set at some stress–strain state, for example E, upon unloading during deployment, it should provide a light and constant chronic force against the organ wall even with a certian amount of further strain release (eg, from E to D). On the other hand, it would generate a large resistive force to crushing if it is compressed in the opposite direction, since it takes the loading path from E to F. Hence, the SMA material exhibits a biased stiffness at point E, which is very important in the design of the SMA stent. Since the stress at the loading phase from A to B and the stress in the unloading phase from C to D depends on the material composition of the SMA, the desirable stress–strain curve can be obtained optimizing material composition.

7.4. Anti-kinking Properties. The stress of stainless steel remains nearly constant in the plastic region (see Fig. 7). This means that a small increase of stress in the plastic region could lead to a drasticincrease of strain or the failure of the medical appliance made from stainless steel (54). On the other hand, the stiffness of superelastic TiNi increases drastically after point B at the end of the loading plateau. The increase in stiffness would prevent the local strain in the high strain areas from further increasing and cause the strain to be partitioned in the areas of lower strain. Hence, strain localization is prevented by creating a more uniform strain than could not be realized with a conventional material.

7.5. Applications. *Orthopedic Marrow Needles.* Figures 8 and 9 show two types of marrow needles, which are used in the repair of a broken thighbone (55,57–59). When the Kunster marrow needle of stainless steel is used, the blood flow inside the bone can be blocked and recovery can be delayed. It also has the drawback of low torsional strength. On the other hand, a Kunster marrow needle of SMA can be inserted into the bone in its initial straight shape and turned to curved shape by heating as shown in Figure 8. The Kunster marrow needle shown in Figure 9 has a complicated shape for the purpose of reinforcement, which makes it difficult to insert the needle in the broken bone. Using the shape memory effect, insertion can be greatly improved as shown in the figure, without loosing the reinforcing function, because the needles can be inserted in a

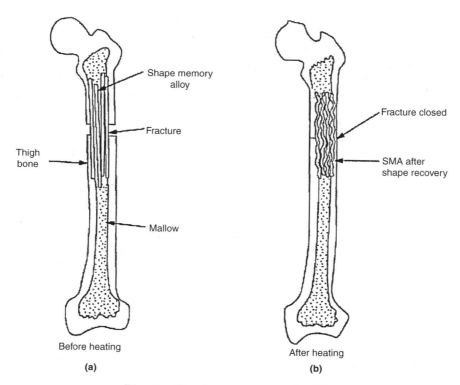

Fig. 8. Kunster marrow needle (57).

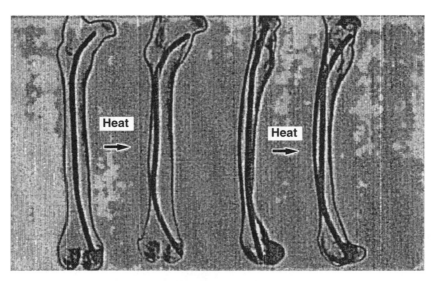

Fig. 9.   Marrow needles before and after heating (56,59).

simpler shape and the necessary size and shape are recovered by heating the needle in the marrow.

Currently available joint prostheses are made of bone cement to be fixed in the bone. Stress acting on the joint prosthesis is quite intense and severe; three to six times the body weight of the patient under nominal action and under such stress being cycled up to 10^6 times. Conventional bone cement causes several inconveniences: gradual loosening after implantation and resultant infection and other complications. The prosthetic joint made of TiNi SMA was developed to avoid such problems. High wear resistance is also another advantage of the TiNi prosthetic joint.

Bone Staple and Bone Plate.   Bone staple and bone plate are used to fix broken bones (55–57). A bone staple (Fig. 10) made of SMA can be inserted at low temperature in holes opened in the bone and then heated by the body temperature to recover its original shape to provide a compressive force in the surfaces of the broken bone. Bone plates are attached with screws for fixing broken bones. Bone plates made of TiNi SMA (Fig. 11) are more effective in connecting the

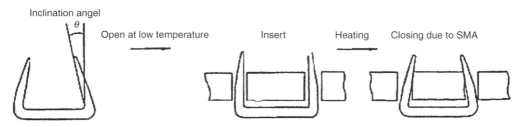

Fig. 10.   Bone staple used to fix broken bones (55,57)

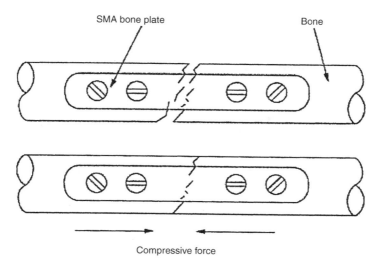

Fig. 11. Bone plate used to fix broken bones (55–57).

broken bones than the bone plates made of conventional material because the SMA bone plates can provide compressive force on the fracture surface of the broken bones as well as a repair as shown in Figure 11. Healing proceeds faster under uniform compressive force.

Dental Applications. Owing to its superelasticity, TiNi has found many applications in dentistry. It is obvious that superelasticity presents the orthodontists with better mechanical characteristics as compared to the conventional elastic materials such as stainless (56,60). When fixtures made of conventional elastic material such as stainless steel are used, the reforming force drops and the fixture loosens with the movement of the teeth. Hence, the fixture must be replaced several times before the treatment is finished. When SMA fixture is used, it can maintain a constant reforming force in a wide range of teeth movement owing its superelasticity so that no future replacement is required after the initial installation. Clinical results show also a faster movement of the teeth and a shorter chair time as compared with stainless steel wire.

Among the methods for restoring the mastication funtion of patients missing more than one tooth, a teeth-root prosthesis is considered to be the method that creates the most natural mastication function. Blade-type implants made of TiNi SMA have been used of Japan (56,57). The open angle of the blade is used to ensure a tight initial fixation and to avoid accidental sinking on mastication. But to make the insertion operation easy, a flat shape teeth-root prosthesis is implanted in the jaw-bone and then the opened shape is changed by heating.

The key to partial denture is the development of an attachment used for connecting the partial denture with the retained teeth for which clasps have been conventionally used. One of the drawbacks of clasps made of conventional elastic materials is loosening during use; this can be improved by replacing the elastic materials with a superelastic. TiNi alloy (56,61). Another drawback of clasps is of esthetic nature, since they are visible with the teeth alignment. In order to overcome this problem the size of the attachment must be smaller than the width of the teeth so that it can be embedded in the teeth completely. A precision attachment using a small screw has recently become available, but they have to be designed and fabricated very precisely so that they lack the flexibility to follow the change in the setting condition due to the shape change of the jawbone during long-term use. Because of its flexibility, using an attachment made of SMA can solve this problem. The SMA attachment consists of two parts: the fixed part, which is made of a conventional dental porcelain-fusible cast alloy and attached to the full cast crown on the anchor teeth, and the movable part, which is made of TiNi SMA and fixed on the side of the partial denture.

Surgical Instruments.   Since superelastic tubing became available in the early to mid 1990s, a variety of catheter products and other endovascular devices using TiNi have appeared on the market. Early applications of TiNi are retrieval baskets with TiNi kink-resistant shafts, as well as a superelastic basket to retrieve stones from kidneys, bladders, bile ducts, etc. An interesting example is the interaortic balloon pump (IABP) used in cardiac assist procedures. The use of NiTi has allowed reduction in the size of the device compared with the polymer tube based designs, and increased the flexibility and kink resistance compared with stainless steel tube designs (54).

Biopsy forceps made from stainless steel are very delicate instruments that can be destroyed by even very slight mishandling. TiNi instruments, on the other hand, can handle considerable bending without buckling, kinking or permanent deformation. For example, a 1.5 mm biopsy forcep that consists of a thin wall TiNi tubing together with a TiNi actuator wire inside are able to be bent around a radius of less than 3 cm without kinking, and still allow for the opening and closing of the distal grasper jaws without increased resistance. The instrument continues to operate smoothly even while bent around tortuous paths.

Stent.   The term stent is used for devices that are used to scaffold or brace the inside circumference of tubular passages or lumens, such as the esophagus bilary duct, and most importantly, a host of blood vessels including coronary, carotid, iliac, arota and femoral arteries (54). Stenting in the cardiovascular system is most often used as a follow-up to balloon angioplasty, a procedure in which a balloon is placed in the diseased vessel and expanded in order to reopen a clogged lumen. Ballooning provides immediate improvement in blood flow, but 30% of the patients have restenosed within a year and need further treatment. The placement of a stent immediately after angioplasty has been shown to significantly decrease the propensity for restenosis. Stents are also used to support grafts, eg, in the treatment of aneurysms.

Most stents today are stainless steel and are expanded against a vessel wall by plastic deformation caused by the inflation of a balloon placed inside the stent. TiNi stents, on the other hand, are self-expanding. They are shape-set to the open configuration, compressed into a catheter, then pushed out of the catheter and allowed to expand against a vessel wall. Typically, the manufactured stent outer diameter is about 10% greater than the vessel in order to assure that the stent anchors firmly in place. The flexibility of TiNi is about 10–20 times greater than the stainless steel and can bear as high as 10% reversible strain. The NiTi stenta are made of knitted or welded wire, laser cut or photoetched sheet, and laser cut tubing. The preferred devices are laser cut tubing avoiding overlaps and welds.

7.6. Applications Under Development. *Artificial Urethral Valve.*
Urinary incontinence is the involuntary discharge of urine caused by the weakness of the urinary canal sphincter muscles due to aging and the expansion of the prostate gland. An artificial urethral valve system driven by an SMA actuator is a potential solution to the problem (62).

The artificial urethral valve should be compact, should have no protrusions so it could be easily implanted in the lower abdominal and attached onto the urethra. A compact cylindrical such valve of stainless steel shells and a circular-arc nitinol plate is currently being considered (Fig. 12). The valve can be opened by the actuation of a SMA element, which is cylindrical at body temperature and goes flat with increased heat. The valve is closed by the force of the bias spring in the normal state and opened to release the choked urethra and allow urinary flow. To heat the SMA, a nicrome wire , insulated with a polyimide membrane, is placed on the surface of the nitinol plate.

The energy to drive an in-dwelled valve is supplied from outside the body by a transcutaneous energy transformer system (63,64)

Artificial Sphincter. Similar to the urethral value, the development of an artificial sphincter is required for medical treatment of patients with fecal incontinence due to a colostomy, congenitally anorectal malformation or surgical operations for anorectal diseases. The lack of anal sphincter is the main reason

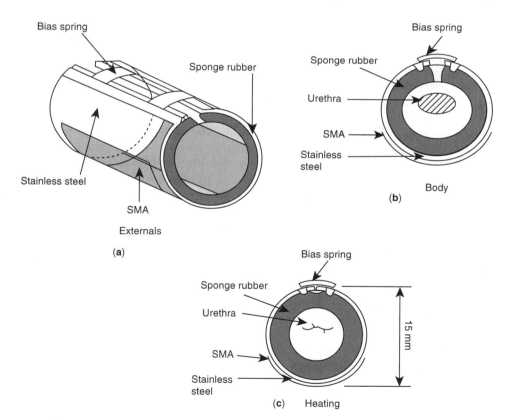

Fig. 12. Geometry of urethral valve: (**a**) externals; (**b**) body; (**c**) heating.

for the problem. An artificial sphincter using SMA actuator is one of the solutions to this condition (65).

A proposed such actuator consists of two SMA plates joined by two hinges and heating coils attached to the SMA plates. The material used for the SMA plates, Ti51at % Ni, is known to exhibit an all-round shape memory effect (ARSME), ie, shows a shape change reverse to the "memorized" one in its martensitic phase. Since the highest temperature for the complete reverse transformation might reach 55°C, thermal insulated materials; cork sheets and sponge rubber sheets are covered at the outer and inner sides of the SMA plates respectively. When electric power is applied to the coils for heating, the reverse transformation occurs in the SMA plates, accompanied by a shape changes from a flat shape to an arc, ie, the restrained shape during annealing. The shape change results in a gap between two SMA plates for opening the intestines. After switching off the electric power, the shape of the SMA plates is recovered by natural cooling, and the intestines will be closed again.

7.7. Piezoelectric Materials and Sample Applications. Piezoelectrical material can convert a mechanical signal to an electric signal (See SMART MATERIALS). Electrical voltage generated by mechinical stress in piezoelectric materials decays very fast due to the charge dissipation. The voltage signal takes the form of a very brief potential wave at the onset of the applied force, and a similar brief wave at termination. It increases with an applied force but drops to zero when the force remains constant. There is no response during the stationary plateau of the applied stimulus. Voltage drops to a negative peak as the pressure is removed and subsequently decays to zero (66). The response is quite similar to the response of the Pacinian corpusclein the human skin (67), one of the sensory receptors in the dermis. In addition PVDF[poly(vinylidence fluoride)] piezofilm is suitable for uses in the biomedical field, since it is very flexible and sensitive to the fast variation of the stress or strain. This section reviews several recent studies on medical applications of PVDF film.

Active Palpation Sensor for Detecting Prostatic Cancer and Hypertrophy. Prostatic carcinoma and hypertrophy are examined in general by the rectal palpation using the doctor's index finger as a probe together with ultrasonic tomography. The morphological features of the two lesions typical to these conditions are key in their diagnosis. The prostatic hypertrophy is a symmetric enlargement of the prostate glands with the stiffness varying from soft to hard. Prostatic cancer, on the other hand, assumes a hard asymmetrical uneven tumor. The palpation depends on the tectile perception of the forefinger, which is said to be ambiguous, subjective, and much affected by the physician's experiences. The development of a palpation sensor for detecting the prostatic cancer and hypertrophy is, hence, important.

The tip probe of the active palpation sensor is mounted on a linear z-translation aluminum bar, which fits into acylindrical outer aluminum shell by a DC micro motor and crank mechanism (Fig. 13). The mechanism for driving is essentially the same as that of an electric toothbrush. The oscillating probe is positioned with its face to the prostate gland.

The probe is an assembly of layered media. The base is a thin aluminum circular plate of 10 mm diameter, on which a cylindrical sponge rubber, a PVDF piezopolymer film of 6 mm across and 28 µm thick as the sensory receptor,

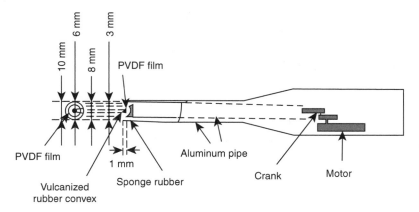

Fig. 13.   Active palpation sensor that has a recessed sensor head.

and a thin acetate film as the protective agent of piezopolymer film are stacked in sequence (68). The sensor head is pressed sinusoidally against the object and the output signal from the piezopolymer film is collected and sent to a digital storageoscilloscope and further forwarded to a personal computer for processing.

The output voltage from the piezopolymer film is proportional to the rate of the straininduced in the film, which means that the maximum amplitude of the signal from the sensor is rather superposed by noises from the measuring system. Data analysis can be then performed by using the absolute output signal of the sensor integrated over the period of data collection.

Haptic Sensor for Monitoring Skin Conditions.   Skin health and its appearance are related to its morphological features such as rashes, chaps or wrinkles. Assessment of the pharmacological action of liniments on skin disease is a matter of importance, which has drawn much attention to the development of objective techniques for measuring morphologic features of skin (69). The evaluation of cosmetic efficacy of toiletries is another use of objective measuring techniques. Noninvasive technology has made great progress in the studies of dermatology during the last decade. Several methods have been developed to measure the mechanical properties of the dermis including the measurement of transepidermal water loss using evaporimeter (70) and the image processing of an egative replica of dermis (71). Those methods, however, fall under the category of indirect measuring techniques of dermis. In this section the development of haptic tribosensors for monitoring skin conditions and distinguishing atophic and normal healthy skins directly are introduced.

Tribo-sensor.   A tactile sensor for the measurement of skin surface conditions is a layered medium, the construction of which is analogous to the human finger (Fig. 14). It is composed of an aluminum shell as the phalanx, a sponge rubber as the digital pulp, a PVDF piezopolymer film of 28 μm thick and 12 mm across as the sensory receptor, an acetate film as the protective agent of piezofilm, and a gauze on the surface as the fingerprint that enhances the tactile sensitivity of the sensor. The sensor is attached onto the tip of an acrylic elastic beam and a strain gauge is mounted on the surface of the beam to monitor the applied force to the skin from the sensor.

Fig. 14.   Schematic of PVDF piezofilm sensor.

The sensor in the tribosensor measurement system is moved by hand over a skin sample, attempting to maintain a constant speed and force (Fig. 15). The voltage signal from the PVDF sensory film is sent to a digital storage oscilloscope and then transmitted to a personal computer for signal processing. It is necessary to reduce the potential difference between the surface of the skin and the sensor to minimize the overlap of noises on the sensor signal during measurement. To achieve this, the subject and the sensor are grounded and a bandpass filter is inserted after the sensor to remove noises from the power sources. During skin identification and signal processing, an area of infected skin of a subject is compared with the skin of healthy subjects. A method of identifying skin samples employing signal processing and neural network-based training is then introduced.

The material in this section on shape memory alloys has been adapted from the longer article that appears in Ref. 72.

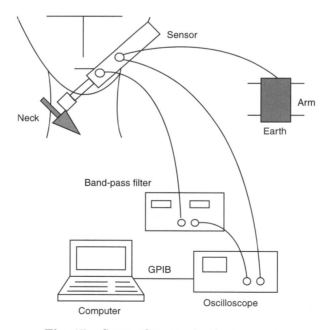

Fig. 15.   Setup of measuring instruments.

BIBLIOGRAPHY

"Prosthetic and Biomedical Devices" in *ECT* 3rd ed., Vol. 19, pp. 275–313, by C. G. Gebelein, Youngstown State University; in *ECT* 4th ed., Vol. 20, pp. 351–395, by Michael Szycher, PolyMedica Industries, Inc.

CITED PUBLICATIONS

1. *Szycher's Dictionary of Biomaterials and Medical Devices*, Technomic Publishing Co., Inc., Lancaster, Pa., 1992.
2. R. L. Whalen, "Connective Tissue Response to Movement at the Prosthesis/Tissue Interface," in *Biocompatible Polymers, Metals and Composites*, Technomic Publishing Co., Lancaster, Pa., 1983.
3. J. R. Hogness and M. VanAntwerp, eds., *The Artificial Heart: Prototypes, Policies, and Patients*, National Academy Press, Washington, D.C., 1991.
4. *ABIOMED 1994 Annual Report*, Danvers, Mass.
5. R. Turner, *Clin. Orth.* **256**, 299–305 (1990).
6. Technical data, Lifequest Medical, Inc., San Antonio, Tex., 1994.
7. U.S. Pat. 5,229,431 (1993), L. Pinchuk (to Corvita Corp.).
8. U.S. Pat. 5,254,662 (1993), M. Szycher (to PolyMedica Industries, Inc.).
9. J. Lemons, K. M. Nieman, and A. B. Wiess, *J. Biomed. Mater. Res.* **7**, 549–553 (1976).
10. S. I. Ertel and J. Kohn, *J. Biomed. Mater. Res.* **28**(8), 919 (1994).
11. *J. Bone Joint Surg.* **73**, 148–153 (1991).
12. R. Langer, *Science*, Sept. 28, 1527–1533 (1990).
13. L. G. Cima, D. E. Ingber, J. P. Vacanti, and R. Langer, *Biotech. Bioeng.* **38**, 145–158 (1991).
14. C. T. Laurencin, M. E. Norman, H. M. Elgendy, and S. F. El-Amin, *J. Biomed. Mat. Res.* **27**, 963–973 (1993).
15. C. A. Vacanti, R. Langer, B. Schloo, and J. P. Vacanti, *Plast. Reconstr. Surg.* **88**, 753–759 (1991).
16. H. P. Greisler, D. Petsikas, T. M. Lam, and co-workers, *J. Biomed. Mat. Res.* **27**, 955–961 (1993).
17. T. H. Barrows, *Clin. Mat.* **1**(4), 233 (1986).
18. D. Borloz, W. Bichon, and A. L. Cassano-Zoppi, *Biomaterials* **5**, 255–268 (1984).
19. W. H. McCarthy, *Aust NZ J. Surg.* **39**, 422–424 (1970).
20. A. Pavan, M. Bosio, and T. Longo, *J. Biomed. Mater. Res.* **13**, 477–496 (1979).
21. R. K. Kulkarni, K. C. Pani, C. Neuman, and F. Leonard, *Arch. Surg.* **93**, 839–843 (1966).
22. J. M. Brady, D. E. Cutwright, R. A. Miller, G. C. Battistone, and E. E. Hunsuck, *J. Biomed. Mater. Res.* **8**, 218 (1973).
23. M. Vert, H. Garreau, M. Audion, F. Chabot, and P. Christel, *Trans. Soc. Biomater.* **8**, 218 (1985).
24. S. Gogolewski and A. J. Pennings, *J. Appl. Polym. Sci.* **28**, 1045–1061 (1983).
25. U.S. Pat. 3,297,033 (1967), E. E. Schmitt (to Ethicon, Inc.).
26. M. Persson, K. Bilgrav, L. Jensen, and F. Gottrup, *Eur. Surg. Res.* **18**, 122–128 (1986).
27. A. Browning and C. C. Chu, *J. Biomed. Mater. Res.* **20**, 613–632 (1986).
28. C. C. Chu and N. D. Campbell, *J. Biomed. Mater. Res.* **16**, 417–430 (1982).
29. D. F. Williams, in Syrett and Acharya, eds., *Corrosion and Degradation of Implant Materials*, ASTM STP684, American Society for Testing and Materials, Philadelphia, Pa., 1979, 61–75.

30. A. M. Reed and D. K. Gilding, *Polymer* **22**, 459–504 (1981).
31. P. H. Craig, J. A. Williams, and K. W. Davide, *Surg. Gynecol. Obstet.* **141**, 1–10 (1975).
32. R. J. Fredericks, A. J. Melveger, and L. J. Dolegiewitz, *J. Poly. Sci.* **22**, 57–66 (1984).
33. E. Pines and T. J. Cunningham, *Eur. Pat. Appl.* **109**, 197A (1984).
34. D. K. Gilding and A. M. Reed, *Polymer* **20**, 1459–1464 (1979).
35. U.S. Pat. 4,523,591.
36. J. A. Ray, N. Doddi, D. Regula, J. A. Williams, and A. Melveger, *Surg. Gynecol. Obstet.* **153**, 497–507 (1981).
37. D. F. Williams, C. C. Chu, and J. Dwyer, *J. Appl. Poly. Sci.* **29**, 1865–1877 (1984).
38. Brit. Pat. 682,866 (1952), D. Coleman (to ICI).
39. D. K. Gilding and A. M. Reed, *Polymer* **20**, 1454–1458 (1979).
40. M. S. Roby, D. J. Casey, and R. D. Cody, *Trans. Soc. Biomater.* **8**, 216 (1985).
41. A. R. Katz, D. P. Mukherjee, A. L. Kaganov, and S. Gordon, *Surg. Gynecol. Obstet.* **161**, 213–222 (1985).
42. T. Kawaguchi, M. Nakano, K. Juni, S. Inoue, and Y. Yoshida, *Chem. Pharm. Bull.* **31**, 1400–1403 (1983).
43. S. C. Woodward, P. S. Brewer, F. Moatamed, A. Schindler, and C. G. Pitt, *J. Biomed. Mater. Res.* **19**, 437–444 (1985).
44. P. Jarrett, C. Benedict, J. P. Bell, J. A. Cameron, and S. J. Huang, *Polym. Prepr.* **24**(1), 32–33 (1983).
45. U.S. Pat. 3,057,537.
46. T. E. Lawler, J. P. English, A. J. Tipton, and R. L. Dunn, *Trans. Soc. Biomater.* **8**, 209 (1985).
47. U.S. Pat. 4,209,607 (June 24, 1980), W. Shalaby (to Ethicon, Inc.).
48. T. H. Barrows, D. M. Grussing, and D. W. Hegdahl, *Trans. Soc. Biomater.* **6**, 109 (1983).
49. T. H. Barrows, S. J. Gibson, and J. D. Johnson, *Trans. Soc. Biomater.* **7**, 210 (1984).
50. U.S. Pat. 4,093,709 (June 6, 1978), G. Heller and N. S. Choi (to Alza Corp.).
51. S. L. Sendelbeck and C. L. Girdin, *Drug Metab. Dispos.* **13**, 29–95 (1985).
52. K. W. Leong, B. C. Brott, and R. Langer, *J. Biomed. Mater. Res.* **19**, 941–964 (1985).
53. K. W. Leong, P. D. Amore, M. Marletta, and R. Langer, *J. Biomed. Mater. Res.* **20**, 51–64 (1986).
54. T. Duerig, A. Pelton, and D. Stockel, *An Overview of Nitinol Medical-Applications, Material Science, and Engineering*, **A293–275**, 194–161 (1999).
55. K. Tanaka, Tobuse, and S. Miyazaki, *Mechanical Properties of Shape Memory Alloys*, Yokendo Ltd., 1993.
56. K. Otsuka and C. M. Wayman, *Shape Memory Materials*, Cambridge University Press, 1998.
57. Y. Suzuki, *Topics on Shape Memory Alloys*, The Nikkan Kogyo Shimbun, Ltd., 1988.
58. Ishikawa, Kinashi, and Miwa, *Collections of SMA Applications*, Kogyo Chosakai Publishing Co., Ltd., Tokyo, 1987.
59. H. Ohnishi, *Artificial Organs* **12**, 862 (1983).
60. R. Sachdeva and S. Miyazaki, *Proc. MRS International Mtg on Advanced Materials*, Vol. 9, 1989, p. 605.
61. S. Miyazaki, S. Fukutsuji, and M. Taira, *Proc. ICOMAT-92*, Monterey, Calif., 1993, p. 1235.
62. M. Tanaka and co-workers, Artificial SMA valve for treatment of urinary incontinence: Upgrading of valve and introduction of transcutaneous transformer, *Bio-Medical Materials Eng.* **9**, 97–112 (1999).

63. J. C. Schuder, H. E. Stephenson, Jr., and J. F. Townsent, High-level-Electromagneticenergy Transfer through a Closed Chest Wall, *IRE Internet. Conv. Rec.* Pt9-9, 1961, pp. 119–126.
64. H. Matsuki, M. Shiiki, K. Murakami, and co-workers, *IEEE Transactions on Magnetics* **26**, 1548–1550 (1990).
65. T. Takagi and co-workers, *J. Soc. Advanced Sci.*, 2000.
66. G. Harsanyi, *Polymer Films in Sensor Applications*, Technomic Publishing Company, Inc., Lancaster, Basel, 1995, p. 97.
67. G. M. Shepherd, "The Somatic Senses," in *Neurobiology*, 3rd ed., Oxford University Press, New York, Oxford, 1994, p. 272.
68. S. Chonan, Z. W. Jiang, M. Tanaka, T. Kato, and M. Kamei, *Int. J. Appl. Elect.* Mech. **9**, 25–38 (1998).
69. H. Tagami, *Fragrance J.* **10**, 11–15 (1993).
70. T. Yamamura, *Fragrance J.* **10**, 35–41 (1993).
71. M. Takahashi, *Fragrance J.* **10**, 16–26 (1993).
72. J. Qui and M. Tanaka, "Biomedical Applications," in M. Schwartz, ed., *Encyclopedia of Smart Materials*, John Wiley & Sons, Inc., New York, 2002.

GENERAL REFERENCES

M. Szycher, *Szycher's Dictionary of Biomaterials and Medical Devices*, Technomic Publishing Co., Inc., Lancaster, Pa., 1992.
C. P. Sharma and M. Szycher, *Blood Compatible Materials and Devices*, Technomic Publishing Co., Inc., Lancaster, Pa., 1991.
M. Szycher, *Biocompatible Polymers, Metals and Composites*, Technomic Publishing Co., Inc., Lancaster, Pa., 1983.
M. Szycher, *High Performance Biomaterials*, Technomic Publishing Co., Inc., Lancaster, Pa., 1991.
M. Szycher, *Introduction to Biomedical Polymers*, ACS Audio Courses, American Chemical Society, Washington, D.C., 1989.
I. Wickelgren, *Science* **272**, 668–670 (1996).

MICHAEL SZYCHER
PolyMedica Industries, Inc.

JINHAO QIU
MAMI TANAKA
Institute of Fluid Science, Tohoku University

BIOREMEDIATION

1. Introduction

Bioremediation is the process of judiciously exploiting biological processes to minimize an unwanted environmental impact; usually it is the removal of a contaminant from the biosphere. Like most definitions in biology, that of bioremediation is the subject of some debate. A narrow definition might focus on the conversion of contaminating organic molecules to carbon dioxide, water, and inorganic ions, and the oxidation or reduction of contaminating inorganic ions. A broader definition would include biological processes for ameliorating extremes of pH, concentrating contaminants so that they can be more easily removed by physical techniques, converting toxic species to less toxic or less bioavailable forms that pose less of a threat to the environment, and restoring functional ecosystems to contaminated or disturbed sites when the contaminants or disturbance cannot be removed.

The concept of "judiciously exploiting biological approaches" is also a subject of debate:

Some would restrict it to providing a nutrient that is otherwise limiting the most effective growth of organisms catalyzing the desired reaction, whether it is the degradation of an organic compound, the reduction or oxidation of an inorganic ion, or the accumulation of a contaminant. This simple approach has been successful with a range of contaminants. The nutrient might be a fertilizer providing nitrogen, phosphorus and other essential minerals, or an electron acceptor such as oxygen.

Others would extend the fertilizer concept to the simultaneous addition of readily biodegradable substrates along with the fertilizer nutrients to stimulate the growth of contaminant-degrading organisms most rapidly, and to aid in the rapid utilization of the fertilizer nutrients before they might be leached from the contaminated area. The specific requirements for the most efficacious substrates is an area of current research.

An alternative use of added readily degradable substrates is to drive the local environment toward anaerobiosis so that reactions such as reductive dechlorinations or reductive removal of nitro-groups are promoted.

A broader view would include the addition of a substrate to stimulate the growth of organisms known to degrade the contaminant of interest only as an incidental part of their metabolism, one might almost say serendipitously. This process is sometimes called co-metabolism, and it too has had success.

Others would include the addition of materials aimed at increasing the bioavailability of the contaminant to the degrading organisms. The most studied compounds are surfactants, but cations have been reported to increase the bioavailability of some organic compounds, and sorbents and clays are also considered. The dispersion of spilled oil on water by the application of dispersants is perhaps the major commercial use of this idea.

Another important option is the addition of remediating organisms. While the addition of contaminant-degrading bacteria has not yet had much documented success with natural products such as hydrocarbons, there is reason to expect that it will be efficacious with pollutants that are more recent additions to the environment. The planting of specific plants, in the process known as phytoremediation, is also a promising approach. Not only do the plants themselves have remediating activities, such as the accumulation of certain metal ions, but they also have extensive bacterial and fungal populations associated with their root systems, and inoculation of this rhizosphere is widely practiced. It is, thus, possible that planting seeds with microbial inoculants will become an option for bioremediation. There is much talk of genetically modifying organisms so that their remediative potential is increased, and in the future this may well become an important option.

The broadest view of "aiding and abetting" includes doing nothing, but merely watching natural processes occur without further intervention. This has been termed "Intrinsic Bioremediation", and it too has met with success. From an environmental point of view, although unfortunately not always from a regulatory viewpoint, it is important that any remediation intervention yield a clear net environmental benefit. Sometimes very mild stimulation of intrinsic processes may be the most environmentally responsible option.

Bioremediation overlaps some older biotechnologies. Municipal and industrial wastewater treatment is a well-established industry, and although it can be distinguished from bioremediation in that the pollutants are under physical control during treatment, the fundamental biological processes have much in common. Similarly composting is a well-established phenomenon, currently gaining popularity in the municipal solid waste treatment industry, and the biofiltration of waste gases is becoming a useful technology. Developments of these technologies, where the contaminant is already under physical control, will undoubtedly aid the development of bioremediation as an accepted tool for dealing with similar wastes when they have escaped control. This article focuses on biological treatments for contaminants when they have escaped into the environment.

Bioremediation is already a commercially viable technology, with estimates of aggregate bioremediation revenues of $2–3 billion for the period 1994–2000 (1). There are significant opportunities to enlarge upon this success. Bioremediation has applications in the gas phase, in water, and in soils and sediments. For water and soils, the process can be carried out *in situ*, or after the contaminated medium has been moved to some sort of contained reactor (*ex situ*). The former is generally rather cheaper, but the latter may result in such a significant increase in rate that the additional cost of manipulating the contaminated material is overshadowed by the time saved. Bioremediation may explicitly exploit bacteria, fungi, algae, or higher plants. Each, in turn, may be part of a complex food-web, and optimizing the local ecosystem may be as important as focusing solely on the primary degraders or accumulators.

Bioremediation usually competes with alternative approaches to achieving an environmental goal. Bioremediation is typically among the least expensive options, but an additional important consideration is that in many cases bioremediation is a permanent solution to the contamination problem, since the contaminant is completely destroyed or collected. Some of the alternatives technologies, such as thermal desorption and destruction of organics, are also permanent solutions, but the simplest, removing the contaminant to a dump site, merely moves the problem, and may well not eliminate the potential liability. Furthermore, by its very nature bioremediation addresses the bioavailable part of any contamination, and when biodegradation or bioaccumulation ceases this probably means that the bioavailable part of the contamination has been addressed. Residual concentrations of contaminants, although perhaps detectable by today's sensitive analytical techniques, may in fact have no residual environmental impact. The same cannot necessarily be said for nonbiological technologies, which may leave bioavailable contaminants at low levels.

Bioremediation also has the advantage that is can be relatively nonintrusive, and can sometimes be used in situations where other approaches would be severely disruptive. For example, bioremediation has been used to clean up hydrocarbon spills under buildings, roads, and airport runways without interfering with the continued use of these facilities.

On the other hand, bioremediation is usually slower than most physical techniques, and may not always be able to meet some very strict clean-up standards. Nevertheless, it is becoming a widely used technology. This article addresses bioremediation in its broadest sense, focusing on the contaminants that can be treated, the underlying biological processes that can mitigate the contamination, and the technologies that have been used, or are being developed, to treat them.

2. General Biological Aspects

The biosphere plays an important role in the great elemental cycles of the earth (2), and bioremediation must be placed in this context if it is to be appreciated in its broadest ramifications. One of the underlying fundamental truths of biological diversity is that if there is free energy available in the metabolism of a substrate, there is probably a guild of organisms that has evolved to make use of it. This is particularly germane to the biodegradation of organic molecules. For example, crude oil seeps to both land and water have occurred for millennia, and as a consequence, aerobic oil-degrading microorganisms are ubiquitous. If biology does not yet take advantage of a source of free energy, then it can be expected that there will be a strong selection pressure in favor of any organism that develops an ability to exploit it. This has been seen with by-products of nylon manufacture, where a *Pseudomonas aeruginosa* has gained the ability to degrade the novel compound 6-aminohexanoate linear dimer, a by-product of nylon-6 manufacture, as the sole source of carbon and nitrogen (3). The successful bioremediation of xenobiotic compounds, such as pesticides and herbicides, may well represent a similar acquisition of traits.

Not all organic molecules provide a source of free energy, however. Some, such as small halogenated solvents, provide no significant source of nutrients or energy, and their aerobic destruction can only occur co-metabolically with the degradation of a more nutritious substrate (4). The white-rot fungi provide another variation on this theme. These organisms seem unique in their ability to degrade lignin, the structural polymer of higher plants. They may not gain any direct energetic benefit from lignin degradation, but it clearly allows access to cellulose which is a substrate for growth. Lignin degradation is catalyzed by a group of extracellular peroxidases that generate nonspecific oxidants, and there have been several proposals to use these systems for destroying contaminating organic compounds (5).

With successful bioremediation, organic compounds can eventually be converted to carbon dioxide, water, and biomass. Similarly, nitrogenous molecules, such as excess ammonia or nitrate in ground water, can be mineralized to gaseous nitrogen. Alternatively they can stimulate the growth of plants, either terrestrial or marine, and the plant biomass can eventually be harvested so that the nitrogen is effectively removed from the local environment. Other nonorganic contaminants provide a different challenge for bioremediation. A few, such as mercury and selenium, are volatilized by some biological processes, but it is not clear that this is always beneficial. In some cases, such as chromium and arsenic, there is a dramatic difference in environmental toxicity depending on the redox state of the contaminant. Bioremediation has sometimes focused on this detoxification, usually by bacterial processes. A more satisfying approach would be to use a biological process to accumulate and concentrate the contaminant so that it can be removed for safe disposal. Fungi, algae, and higher plants have all been used in these efforts.

Table 1 explains a few of the biological terms that are widely used in discussing bioremediation, and which are used in the following text.

Table 1. **Some Biological Definitions Relevant to Bioremediation**

Term	Explanation
aerobic	conditions with free oxygen
anaerobic	conditions scrupulously free of oxygen
anoxic	conditions with very low levels of oxygen
autotrophic	growth using atmospheric CO_2 as sole source of carbon
co-metabolic degradation	biodegradation of a contaminant only fortuitously with degradation of a true substrate
denitrification	the reduction of nitrate to gaseous nitrogen
Eukaryotes	organisms with a membrane-bound nucleus; the protozoa, fungi, plants, and animals
eutrophic	very rich nutrient conditions, especially of nitrogen compounds
heterotrophic	growth at the expense of complex organic substrates
lignolytic	growth of white-rot fungi under conditions where they synthesize lignin-degrading peroxidases
methanogenic	very anaerobic conditions, where carbon dioxide is reduced to methane
methanotrophic	aerobic growth with methane as sole source of carbon and energy
mineralization	conversion of a contaminant to its simplest forms, eg, CO_2, H_2O, CH_4, N_2, Cl^-
nitrate-reducing, denitrifying	anoxic conditions, where nitrate is reduced to nitrogen gas
nitrification	the biological oxidation of ammonia to nitrite and nitrate
oligotrophic	very low nutrient conditions
Prokaryotes	organisms lacking a membrane-bound nucleus; the bacteria and archaea
recalcitrant	very resistant to biodegradation
reductive dehalogenation, reductive dechlorination	the sequential loss of halogen substituents under anaerobic, usually methanogenic, conditions
rhizosphere	the soil around plant roots; this zone has different microbial populations from the bulk soil
sulfate-reducing	very anaerobic conditions, where sulfate is reduced, by sulfate-reducing bacteria, to sulfide
vadose zone	the part of the soil above the water table

3. General Technological Aspects

Successful bioremediation hinges upon the effective application of the biology discussed above. Sometimes the contaminant is on the surface, so access to it is reasonably simple. Indeed the required technology may be as simple as broadcast spreaders or sprayers to apply fertilizers, or tilling the soil to allow good aeration. Of course this is not necessarily as simple as it sounds, since contaminated sites are often very different from agricultural fields, and the technology has to be significantly stronger to "plow" the soil. Frequently the contaminant is below the surface, and applying even simple bioremediation strategies can be very involved. Table 2 lists some of the technologies in use today.

Table 2. **Some Technological Definitions Relevant to Bioremediation**

Technology	Description
air sparging; aquifer sparging; biosparging	injection of air to stimulate aerobic degradation; may also stimulate volatilization
air stripping	injection of air to stimulate volatilization
aquifer bioremediation	*in situ* bioremediation in an aquifer, usually by adding nutrients or co-substrates
aquifer sparging	injection of air into a contaminated aquifer to stimulate aerobic degradation, may also stimulate volatilization
batch reactor	a bioreactor loaded with contaminated material, and run until the contaminant has been consumed, then emptied, and the process is repeated
bioactive barrier; bioactive zone; biowall	a zone, usually subsurface, where biodegrada tion of a contaminant occurs so that no contaminant passes the barrier
bioaugmentation	addition of exogenous bacteria with defined degradation potential (or rarely indigenous bacteria cultivated in a reactor and reapplied)
biofilm reactor	a reactor where bacterial communities are encouraged on a high surface area support, biofilms often have a redox gradient so that the deepest layer is anaerobic while the outside is aerobic
biofiltration	usually an air filter with degrading organisms supported on a high surface area support such as granulated activated carbon
biofluffing	augering soil to increase porosity
bioleaching	extracting metallic contaminants at acid pH
biological fluidized bed; fluidized-bed bioreactor	bioreactor where the fluid phase is moving fast enough to suspend the solid phase as a fluid-like phase
biopile; soil heaping	an engineered pile of excavated contaminated soil, with engineering to optimize air, water, and nutrient control
bioslurping	vacuum extraction of the floating contaminant, water, and vapor from the vadose zone; the air flow stimulates biodegradation
biostimulation	optimizing conditions for the indigenous biota to degrade the contaminant
biotransformation	the biological conversion of a contaminant to some other form, but not to carbon dioxide and water

Table 2   (*Continued*)

Technology	Description
biotrickling filter	a reactor where a contaminated gas stream passes up a reactor with immobilized micro-organisms on a solid support, while nutrient liquor trickles down the reactor
bioventing	vacuum extraction of contaminant vapors from the vadose zone, thereby drawing in air that stimulates the biodegradation of the remainder
borehole bioreactor; in-well bioreactor	the addition of nutrients and electron acceptor to stimulate biodegradation *in situ* in a contaminated aquifer
closed-loop bioremediation	groundwater recovery, a bioreactor, and low-pressure reinjection to maximize nutrient use, and maintain temperature in cold climates
composting	addition of biodegradable bulking agent to stimulate microbial activity; optimal composting generally involves self-heating to 50–60°C
constructed wetland	artificial marsh for bioremediation of contaminated water
continuous stirred tank reactor (CSTR)	a completely mixed bioreactor
digester	usually an anaerobic bioreactor for digestion of solids and sludges that generates methane
ex-situ bioremediation	usually the bioremediation of excavated contaminated soil in a biopile, compost system or bioreactor
fixed-bed bioreactor	bioreactor with immobilized cells on a packed column matrix
land-farming; land treatment	application of a biodegradable sludge as a thin layer to a soil to encourage biodegradation; the soil is typically tilled regularly
natural attenuation; intrinsic bioremediation	unassisted biodegradation of a contaminant
phytoextraction	the use of plants to remove and accumulate contaminants from soil or water to harvestable biomass
phytofiltration	the use of completely immersed plant seedlings, to remove contaminants from water
phytoremediation	the use of plants to effect bioremediation
phytostabilization	the use of plants to stabilize soil against wind and water erosion
pump and treat	pumping groundwater to the surface, treating, and reinjection or disposal
rhizofiltration	the use of roots to immobilize contaminants from a water stream
rotating biological contactor	bioreactor with rotating device that moves a biofilm through the bulk water phase and the air phase to stimulate aerobic degradation
sequencing batch reactor	periodically aerated solid phase or slurry bioreactor operated in batch mode
soil-vapor extraction	vacuum-assisted vapor extraction

4. Organic Contaminants

4.1. Hydrocarbons. *Constituents.* Hydrocarbons get into the environment from biogenic and fossil sources. Methane is produced by anaerobic bacteria in enormous quantities in soils, sediments, ruminants and termites, and it is consumed by methanotrophic bacteria on a similar scale. Submarine methane seeps support substantial oases of marine life, with a variety of invertebrates possessing symbiotic methanotrophic bacteria (6). Thus, methanotrophic bacteria are ubiquitous in aerobic environments. Plants generate large amounts of volatile hydrocarbons, including isoprene and a range of terpenes (7). These compounds provide an abundant substrate for hydrocarbon-degrading organisms.

Crude oil has been part of the biosphere for millennia, leaking from oil seeps on land and in the sea. Crude oils are very complex mixtures, primarily of hydrocarbons although some components do have heteroatoms such as nitrogen (eg, carbazole) or sulfur (eg, dibenzothiophene). Chemically, the principal components of crude oils and refined products can be classified as aliphatics, aromatics, naphthenics, and asphaltic molecules. Representative examples are shown in Figure 1. The ratios of these different classes varies in different oils, but a typical crude oil might contain the four classes in a ratio of approximately 30:30:30:10. Most crude oils contain hydrocarbons ranging in size from methane to molecules with hundreds of carbons, although the lightest molecules are usually absent in oils that have been partially biodegraded in their reservoir. When crude oils reach the surface environment the lighter molecules evaporate, and are either destroyed by atmospheric photooxidation or are washed out of the atmosphere in rain, and are biodegraded. Some molecules, such as the smaller aromatics (benzene, toluene, etc) have significant solubilities, and can be washed out of floating slicks, whether these are at sea, or on terrestrial water tables. Fortunately the majority of molecules in crude oils, and refined products made from them, are biodegradable, at least under aerobic conditions.

Biodegradation. Methane and the volatile plant terpenes are fully biodegradable by aerobic organisms, and most refined petroleum products are essentially completely biodegradable under aerobic conditions. Estimates for crude oil biodegradability range up to 90% (8), and the least biodegradable material, principally polar molecules and asphaltenes, lacks the "oily" feel and properties that are associated with oil. These are essentially impossible to distinguish from more recent organic material in soils and sediments, such as the humic and fulvic acids, and appear to be biologically inert.

Numerous bacterial and fungal genera have species able to degrade hydrocarbons aerobically and the pathways of degradation of representative aliphatic, naphthenic and aromatic molecules have been well characterized in at least some species (8). Other organisms, such as algae and plants, do not seem to play a very important role in the biodegradation of hydrocarbons. It is a truism that the hallmark of an oil-degrading organism is its ability to insert oxygen atoms into the hydrocarbon, and there are many ways in which this is achieved. Figures 2 and 3 show the most well-studied. Once a hydrocarbon possesses a carboxylate or alcohol functionality it is almost invariably a readily degradable compound. A simple example at the human level is the difference between oleic acid, a high calorie

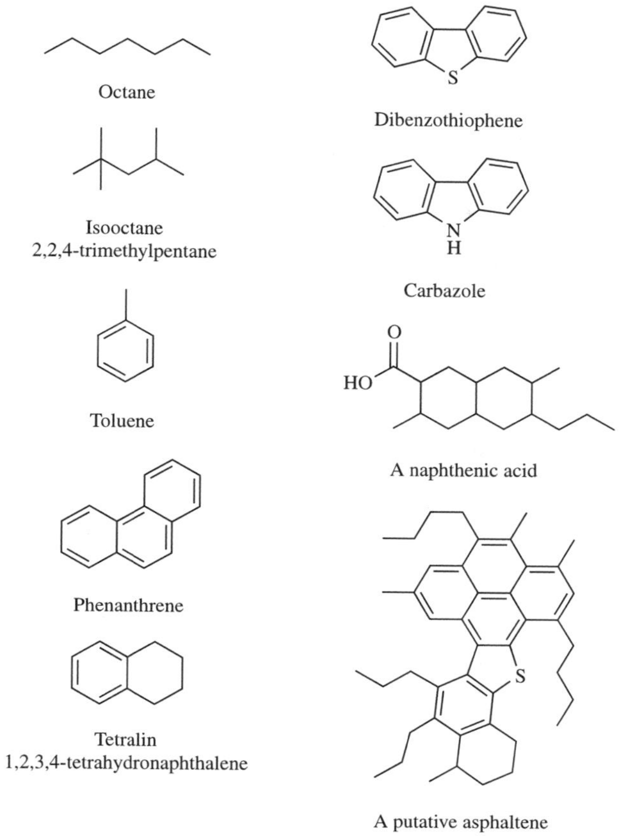

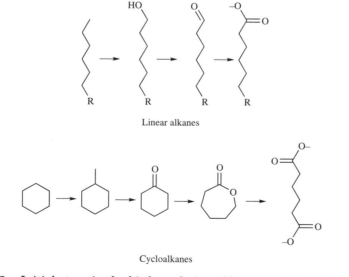

Octane

Isooctane
2,2,4-trimethylpentane

Toluene

Phenanthrene

Tetralin
1,2,3,4-tetrahydronaphthalene

Dibenzothiophene

Carbazole

A naphthenic acid

A putative asphaltene

Fig. 1. Some representative hydrocarbons found in crude oil.

Linear alkanes

Cycloalkanes

Fig. 2. Initial steps in the biodegradation of linear and cyclic alkanes.

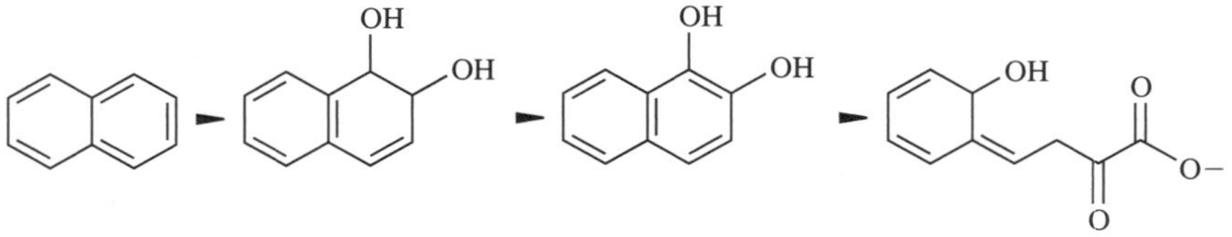

Fig. 3. Initial steps in the aerobic degradation of naphthalene, as a representative multiringed aromatic, and toluene. The different initial steps of toluene degradation are examples of the diversity found in different organisms.

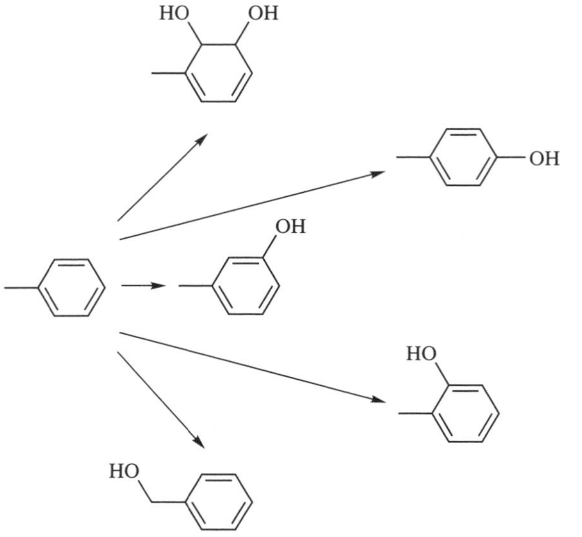

food, and octadecane, present in mineral oil, which is so inert that it serves as an intestinal lubricant!

For many years it was assumed that oil biodegradation was an exclusively aerobic process, since any degradation must involve oxidation. Indeed the very existence of oil reservoirs indicates that anaerobic degradative processes in such environments must be very slow. Nevertheless, in recent years it has become clear that at least some hydrocarbons are oxidized by bacteria under completely anaerobic conditions, where the oxygen is probably coming from water. Limited hydrocarbon biodegradation has now been shown under sulfate-, nitrate-, carbon dioxide- and ferric iron-reducing conditions (Table 3). The phenomenon is still poorly understood, however, and at present the largest molecules demonstrated to undergo biodegradation under these conditions are hexadecane, heptadecene, and phenanthrene. The pathways of degradation are only beginning to be addressed. Figure 4 shows the intermediates identified in anaerobic toluene degradation in different organisms. It is noteworthy that while organisms capable of aerobic oil biodegradation seem to be ubiquitous, organisms capable of the anaerobic degradation of hydrocarbon have to date only been found in a few places.

Table 3. **Hydrocarbons That Have Been Shown to be Biodegraded Under Anaerobic Conditions**

Electron acceptor	Substrate
nitrate (to nitrogen)	heptadecene
	toluene, ethylbenzene, xylene
	naphthalene
	terpenes
iron(III) (to iron(II))	toluene
manganese(IV) (to Mn(II))	toluene
sulfate (to sulfide)	hexadecane, alkylbenzenes
	benzene
	naphthalene, phenanthrene
CO_2 (to methane)	toluene, xylene

Although the majority of molecules in crude oils and refined products are hydrocarbons, the U.S. Clean Air Act amendment of 1990 mandated the addition of oxygenated compounds to gasoline in many parts of the United States. The requirement is usually that 2% (w/w) of the fuel be oxygen, which requires that 5–15% (v/v) of the gasoline be an oxygenated additive (eg, methanol, ethanol, methyl *tert*-butyl ether (MTBE), etc). Although methanol and ethanol are readily degraded under aerobic conditions, the degradability of MTBE remains something of an open question. The compound was previously very rare in the environment, but now it is one of the major chemicals in commerce. At first it seemed that the compound was completely resistant to biodegradation, but complete mineralization has now been reported (9). Whether biodegradation can be optimized for effective bioremediation remains to be seen.

Bioremediation.   Crude oil and refined products are readily biodegradable under aerobic conditions, but they are only incomplete foods since they lack any

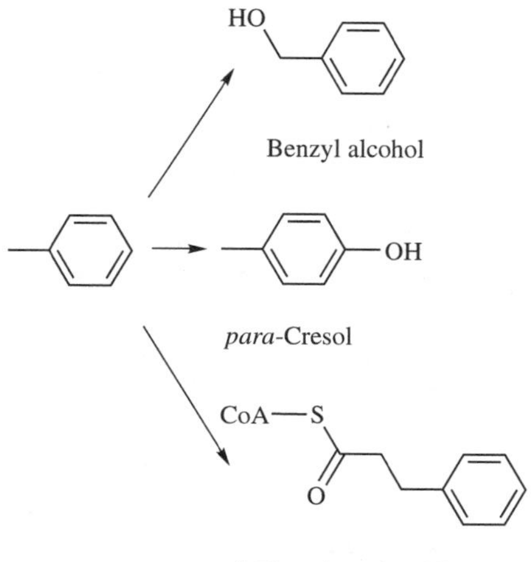

Fig. 4. Proposed initial steps in the anaerobic biodegradation of toluene in different organisms.

significant nitrogen, phosphorus, and essential trace elements. Bioremediation strategies for removing large quantities of hydrocarbon must therefore include the addition of fertilizers to provide these elements in a bioavailable form.

Air.   Hydrocarbon vapors in air are readily treated with biofilters. These are typically rather large devices with a very large surface area provided by bulky material such as a bark or straw compost. The contaminated air, perhaps from a soil vapor-extraction treatment, or from a factory using hydrocarbon solvents, is blown through the filter, and organisms, usually indigenous to the filter material or provided by a soil or commercial inoculum, grow and consume the hydrocarbons. Adequate moisture must be maintained for effective operation. Alternatively, trickling biofilters with recycled water are also in use. Both bacteria and fungi readily colonize such filters, and they can be very effective. Nevertheless, biofilters are usually equipped with a small granulated activated carbon "backup" filter to handle any sudden pulse loads that might overwhelm the biological capacity of the filter. Biofilters compete with granulated activated carbon filters, and are often cheaper because they minimize the cost of the granulated activated carbon, and the energy required to destroy the contaminant and the granulated activated carbon when the latter is saturated (10). Potential problems include plugging and uneven air or water flow, but successful designs work for many years with minimal maintenance except the occasional addition of nutrients and stirring of the bed.

Sea.   Crude oil spills at sea are perhaps the most widely covered environmental incidents in the national and international media. Despite their notoriety, catastrophic tanker spills and well blow-outs are fortunately rather rare, and their total input into the world's oceans is approximately equivalent to that from natural seeps; significantly more oil reaches the world's oceans from municipal sewers (11). Physical collection of the spilled oil is the preferred remediation option, but if skimming is unable to collect the oil, biodegradation and perhaps combustion or photooxidation are the only routes for elimination of the spill. One approach to stimulating biodegradation is to disperse the oil with chemical dispersants. Early dispersants had undesirable toxicity, but modern dispersants and application protocols can stimulate biodegradation by increasing the surface area of the oil available for microbial attachment, and perhaps providing nutrients to stimulate microbial growth (12). Patents have been issued for dispersant formulations that specifically include nitrogen and phosphorus nutrients (13), but the products are not currently commercially available.

Bioremediation by the addition of oil-degrading microbes is often promoted as a treatment option for floating spills, but this approach has not yet met with any documented success (13).

Shorelines.   The successful bioremediation of shorelines affected by the spill from the *Exxon Valdez* in Prince William Sound, Alaska, was perhaps the largest bioremediation project to date (14, 15). More than 73 miles of shoreline were treated in 1989 and similar amounts of fertilizer were used in 1990. Oil had typically penetrated into the surface gravel on these shorelines, occasionally getting as deep as 30 cm into the sediment. Since the gravel was typically very permeable, oxygen availability was unlikely to be the limiting factor for biodegradation, and indeed this was subsequently shown to be correct. Bioremediation thus focused on the addition of nitrogen and phosphorus fertilizers to partially remove the nutrient-limitation on oil degradation. Of course the addition of fertilizers was complicated by the fact that oiled shorelines were washed by tides twice a day. These tides would have rapidly removed any soluble fertilizer, so a strategy was sought that would provide nutrients for a significant length of time. Various approaches to applying fertilizers were tried, including both standard and slow release nutrients, oleophilic nutrients and solutions of liquid fertilizers. Two fertilizers were used in the full-scale applications; one, an oleophilic product known as Inipol EAP22 (trademark of CECA, Paris, France), was a microemulsion of a concentrated solution of urea in an oil phase of oleic acid and trilaurethphosphate, with butoxyethanol as a cosolvent. This product was designed to adhere to oil, and to release its nutrients to bacteria growing at the oil-water interface. The other fertilizer was a slow-release formulation of inorganic nutrients, primarily ammonium nitrate and ammonium phosphate, in a polymerized vegetable oil skin. This product, known as Customblen (trademark of Grace-Sierra, Milpitas, California), released nutrients with every tide, and these were distributed throughout the oiled zone as the tide fell. Fertilizer application rates were carefully monitored so that the nutrients would cause no harm, and the rate of oil biodegradation was stimulated between two- and five-fold (14,15).

A wide range of fertilizers, including agricultural and horticultural fertilizers, and bone and fish meals have been tried at the pilot scale, usually with at least modest success (13). Some current work is aimed at addressing whether providing a readily degradable substrate with the fertilizer nutrients helps immobilize the nutrients in biomass at the oiled site. Of course nutrient-supplementation is only likely to markedly stimulate the rate of biodegradation where nutrient levels are naturally low. It is unlikely that fertilizers will have a dramatic effect in situations where agricultural or municipal run-off maintains elevated levels of nutrients, such as happens in some estuaries and bays. Here aeration is likely to be most effective.

Bioremediation by the addition of oil-degrading microbes has been promoted as a treatment option for oiled shorelines, but this approach has not yet met with any documented success (13).

Areas where there are currently few remediation options include oiled marshes, mangroves, and coral reefs. These environments are generally easily damaged by human intrusion and physical cleaning options may not provide any net environmental benefit. Bioremediation may provide some attractive options, and some success has been claimed, on a small scale, with fertilizer applications (13). Marshes and mangroves offer the additional complication that they are typically anoxic. Perhaps the anaerobic degradation of oil could be stimulated by inoculation with anaerobic hydrocarbon degrading microbes, or perhaps gentle aeration or the addition of slow release oxygen compounds, such as some inorganic peroxides, might stimulate aerobic degradation without significantly changing the redox balance of these environments. This is an area where research is very much in its infancy and there are no well-documented success stories to date.

Bioremediation also offers options for dealing with oiled material, such as seaweed, that gets stranded on shorelines; composting has been shown to be effective.

Groundwater.   Spills of refined petroleum product on land, and leaking underground storage tanks, sometimes contaminate groundwater. Bioremediation is becoming an increasingly popular treatment for such situations.

Hydrocarbons typically have a specific gravity of less than 1, and refined products usually float on the water table if they penetrate soil that deeply. In the parlance of the remediation industry, such floating spills are often called NAPLs (nonaqueous phase liquids). Indeed they are sometimes known as LNAPLs for light nonaqueous phase liquids, to distinguish them from more dense materials, such as halogenated compounds, which are more likely to sink in groundwater. Stand-alone bioremediation is an option for these situations, but "pump and treat" is the more usual treatment. Contaminated water is brought to the surface, free product is removed by flotation, and the cleaned water re-injected into the aquifer or discarded. Adding a bioremediation component to the treatment, typically by adding oxygen and low levels of nutrients, is an appealing and cost-effective way of stimulating the degradation of the residual hydrocarbon not extracted by the pumping. This approach is becoming widely used.

Hydrocarbons are not very soluble in water, but the most soluble components leach out of a spill if there is continual flushing. Typically only small aromatic molecules, the infamous BTEX (benzene, toluene, ethylbenzene, and xylenes, Fig. 5), are soluble enough to contaminate groundwater. Although with the advent of oxygenated gasolines, it is expected that these oxygenates (ethanol, methanol, MTBE (methyl-*tert*-butyl ether) etc) will also be found in groundwater. In the past, remediation of such situations has usually used pump-and-treat methodologies. These methods are slow and may leave reservoirs of contaminants in pockets that are poorly connected to the main water body. Of course the contaminant is biodegradable, and some biodegradation is probably already occurring when the contamination is discovered. The cheapest approach to remediation is, thus, to allow this intrinsic process to continue. Evidence that it is indeed occurring can be found in the selective disappearance of the most biodegradable compounds in the contaminant mixture, and the

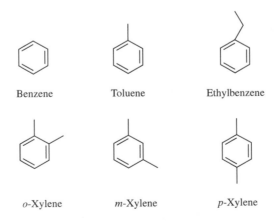

Benzene   Toluene   Ethylbenzene

o-Xylene   *m*-Xylene   *p*-Xylene

Fig. 5.   The components of BTEX.

concomitant disappearance of electron acceptors from the groundwater. Thus oxygen is depleted as the preferential terminal electron acceptor for metabolism, followed by nitrate, ferric iron, sulfate, and finally CO_2 for methanogenesis (16).

Intrinsic bioremediation is becoming an acceptable option in locations where the contaminated groundwater poses little threat to environmental health. Nevertheless, although intrinsic bioremediation is appealingly simple, it may not be the lowest cost option if there are extensive monitoring and documentation costs involved for several years. In such cases it may well be more cost effective to optimize conditions for biodegradation.

One approach is to optimize the levels of electron acceptors. Oxygen can be pumped in as the pure gas, or as air, although this is relatively energy intensive since oxygen is so poorly soluble. Hydrogen peroxide has been used in some situations, but there have been problems with biomass plugging near the injection wells. Slow release formulations of inorganic peroxides, such as magnesium peroxide, have recently been used with success (17). Nitrate may be added, although there are sometimes regulatory limitations on the amount of this material that may be added to groundwater (18). Ferric iron availability may be manipulated by adding ligands (19).

If there are significant amounts of both volatile and nonvolatile contaminants, remediation may be achieved by a combination of liquid and vapor extraction of the former, and bioremediation of the latter. This combination has been termed "bioslurping", where the act of pumping out the liquid contaminant phase draws in air at other wells to stimulate aerobic degradation (20). Such bioremediation requires that there be enough nutrients to allow microbial growth, and fertilizer nutrients are frequently added at the air injection wells. Bioslurping has had a number of well-documented successes.

The majority of remediation operations include stopping the source of the contamination, but in some cases this is impossible, either because of the location of the spill, or because it is over a large area, and not a point source. In these situations it may be possible to intercept the flow of contaminated groundwater off-site, and ensure that no contamination passes. The simplest intervention is a line of wells for pump and treat, but including a biological component may be more cost effective. This can range from the installation of a sparge line for aerating the contaminated plume, to installing some form of semicontained bioreactor where nutrients can be applied with some modicum of control. Often these designs are combined with barriers to ensure that all the contaminated plume passes through the reactive zone. These designs have a variety of names, including biowall, trench biosparge, funnel and gate, bubble curtain, sparge curtain and engineered trenches and gates. Both aerobic and anaerobic designs have been successfully installed.

Where there are large volumes of contaminated water under a small site, it is sometimes most convenient to treat the contaminant in a biological reactor at the surface. Considerable research has gone into reactor optimization for different situations and a variety of stirred reactors, fluidized-bed reactors, and trickling filters have been developed. Such reactors are usually much more efficient than *in situ* treatments, although correspondingly more expensive.

Of course the presence of a liquid phase of hydrocarbon in a soil gives rise to vapor contamination in the vadose zone above the water table. This can be treated by vacuum extraction, and the passage of the exhaust gases through a biofilter (see above) can be a cheap and effective way of destroying the contaminant permanently.

Soil. Hydrocarbon contamination of soils runs the gamut from crude oils at production well and pipeline spills, to the full slate of refined products at refineries, distribution centers, service stations and accident sites. Significant hydrocarbon contamination is also often found at manufactured gas plants, now mainly abandoned, wood treatment facilities, railroad rights of way and terminals, and various military bases. Sometimes the contamination is the result of leaking underground storage tanks and pipelines, leading to subsurface contamination, but surface spills also occur. Physical removal of gross contamination is an obvious first step at all locations, and bioremediation is an appealing option for remediating residual contamination in many of these sites.

Spills from production facilities and pipelines often involve both oil and brine, since most oil reservoirs float on top of concentrated brines, and both are produced in later stages of production. The brine is typically separated from the oil and re-injected into the reservoir, but some is retained in many production pipelines. The environmental impact of spilled brine can be quite deleterious. Not only is salt toxic to most plants, and can inhibit many soil bacteria, but it also can have a major effect on the soil structure by altering the physical properties of clays. Successful bioremediation strategies must therefore include remediating the brine. In wet regions the salt is eventually diluted by rainfall, but in arid regions, and to speed the process in wetter regions, gypsum is often added to restore soil porosity.

Many hydrocarbons bind quite tightly to soil components, and are thereby less available to microbial degradation. The kinetics of binding seem to be complex, and the process of "aging" is only poorly understood. Nevertheless, it seems clear that hydrocarbons that have been in contact with soil for a long time are not as available for biodegradation as fresh spills. Several groups of researchers have suggested the addition of surfactants to overcome this limitation, but this approach is not yet widely used. A significant potential concern is that the surfactant will be degraded in preference to the contaminant of concern.

Intrinsic biodegradation occurs, but it usually only removes the lightest refined products, such as gasoline, diesel and jet fuel. Active intervention is typically required. Usually the least expensive approach is *in situ* remediation, typically with the addition of nutrients, and the attempted optimization of moisture and oxygen by tilling. Various approaches to applying fertilizers have been tried, including both standard and slow-release nutrients, oleophilic nutrients and solutions of liquid fertilizers. Oxygen is a likely limiting nutrient in many cases, and soil tilling is widely practiced. This *in situ* bioremediation of hydrocarbon-contaminated soils is akin to the old practice of "land-farming", wherein sludges and other refinery wastes were deliberately spread onto soil and tilled and fertilized to stimulate biodegradation. Although this practice is now discontinued in the United States, it was quite widely used.

Deeper contamination may be remedied with bioventing, where air is injected through some wells, and extracted through others to both strip volatiles and provide oxygen to indigenous organisms. Fertilizer nutrients may also be added. This is usually only a viable option with lighter refined products.

A recent suggestion has been to use plants to stimulate the microbial degradation of the hydrocarbon (hydrocarbon phytoremediation). This has yet to receive clear experimental verification, but the plants are proposed to help deliver air to the soil microbes, and to stimulate microbial growth in the rhizosphere by the release of nutrients from the roots. The esthetic appeal of an active phytoremediation project can be very great.

When soil contamination extends to some depth it may be preferable to excavate the contaminated soil and put it into "biopiles" where oxygen, nutrient and moisture levels are more easily controlled. Biopiles can also be kept warm during winter months, increasing the amount of time available for biodegradation in colder climates. Since the soil is well mixed during the construction of the pile, there is an opportunity to add selected microbial and fungal strains in an additional attempt to maximize biodegradation.

Composting by the addition of readily degradable bulking agents is also a useful option for relatively small volumes of excavated contaminated soil. Since efficient composting invariably involves self-heating as biodegradation proceeds, this also offers an option for extending the bioremediation season into the winter months in cold climates. A potential drawback of composting is that it usually increases the volume of contaminated material, but if fully successful the finished compost can be returned to the site as a positive contribution to soil quality.

Slurry bioreactors offer the most aggressive approach to maximizing contact between the contaminated soil and the degrading organisms. Both lagoons and reactor vessels have been used, but the former are often not optimally designed for all the soil to be partially suspended by the mixing impellers. Contained reactor designs include mixing tank, airlift, and fluidized-bed aeration. A major advantage of contained slurry bioreactors is the potential ability to optimize nutrients, aeration and degradative inocula as fresh soil is added, and the control of waste materials, including gases continually. Slurry bioreactors are usually the most expensive bioremediation option because of the large power requirements, but under some conditions this cost is offset by the rapid biodegradation that can occur.

In all these cases it is important to bear in mind that although the majority of hydrocarbons are readily biodegraded, some, such as the steranes and hopanes, are very resistant to microbial attack. Estimates of oil biodegradation range from 60–95% for different crude oils, so fresh spills of crude oils are readily treated by bioremediation (21). Refined products, such as gasoline, diesel, jet fuels, and heating oils are usually more biodegradable than typical whole oils, but the various heavy fractions of crude oils, such as the asphalts, are far less biodegradable, and are not such attractive targets for bioremediation. Some crude oils have already been extensively biodegraded in their reservoirs, and these are also poor targets for bioremediation. An example is Orimulsion, a heavy oil in water emulsion (70% bitumen) stabilized by low levels of surfactants, used as a fuel for electricity generation. Similarly, old spills may have already undergone significant biodegradation and the residue may be relatively biologically inert. It is thus important to run laboratory studies to ensure that the contaminant is sufficiently biodegradable that clean-up targets can be met.

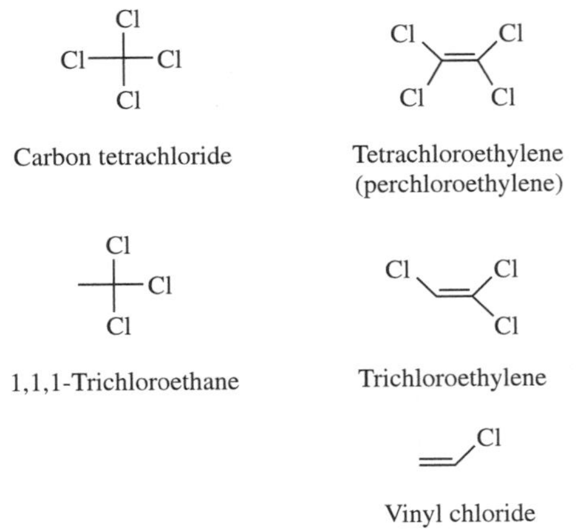

Fig. 6. Some representative halogenated solvents.

4.2. Halogenated Organic Solvents.   *Constituents.*   Halogenated organic solvents are widely used in metal processing, electronics, dry cleaning and paint, paper and textile manufacturing, and some representative examples are shown in Figure 6. These solvents have been used for more than fifty years, and unfortunately they are fairly widespread contaminants. Unlike the hydrocarbons, which usually float on water, the halogenated solvents typically have specific gravities greater than 1, and they generally sink to the bottom of any groundwater, and float on the bedrock. For this reason they are sometimes known as DNAPLs for dense nonaqueous phase liquids.

Biodegradation.   Halogenated solvents are degraded under aerobic and anaerobic conditions. The anaerobic process is typically a reductive dechlorination that progressively removes one halide at a time (Fig. 7). For example, under methanogenic conditions, carbon tetrachloride is sequentially dechlorinated to chloroform, dichloromethane, methyl chloride and methane, while trichloroethylene

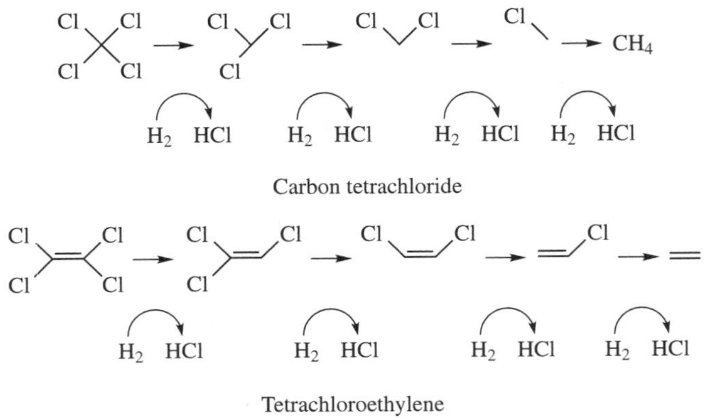

Fig. 7. Reductive dechlorination of carbon tetrachloride and tetrachloroethylene.

is sequentially reduced to ethylene. Some of these compounds are also dehalogenated under sulfate-reducing conditions, and under denitrifying conditions there
are reports that the final product can be CO_2 (22). Chloromethane and dichloromethane have been shown to be the sole carbon source for several anaerobic
organisms (23), and it seems there is much to be learned about the microbial
diversity of anaerobic microorganisms capable of dechlorinating solvents.

The simplest chlorinated alkanes, alkenes, and alcohols (eg, chloromethane,
dichloromethane, chloroethane, 1,2-dichloroethane, vinyl chloride, and 2-chloro-
ethanol) serve as substrates for aerobic growth for some bacteria, but the majority of halogenated solvents cannot support growth (24). Nevertheless these compounds are mineralized under aerobic conditions, albeit with no apparent benefit
to the degrading organism. Indeed, the oxidation appears to be fortuitous and it
occurs during the metabolism of a growth substrate. The phenomenon is therefore known as co-metabolism or co-oxidation. Numerous bacteria are able to catalyze the oxidation of trichloroethylene; some use monooxygenases (for example
methane and ammonia oxidizing species); others contain dioxygenases (eg, some
toluene oxidizing species). The difference between these two classes of enzymes is
the fate of the two atoms of molecular oxygen. Monooxygenases insert one oxygen
atom into their substrate, and reduce the other to water. Dioxygenases insert
both atoms into their substrate. The effect of these two types of enzyme is
illustrated in Figure 8. In either case, biodegradation proceeds to complete
mineralization (25).

The biodegradation of trichloroethylene is the most studied since this is
probably the most widespread halogenated solvent contaminant. Several substrates drive trichlorethylene co-oxidation, including methane, propane, propylene, toluene, isopropylbenzene, and ammonia (25). The enzymes that
metabolize these substrates have subtly different selectivities with regard to
the halogenated solvents, and to date none are capable of co-oxidizing carbon
tetrachloride or tetrachloroethylene. Complete mineralization of these compounds can, however, be achieved by sequential anaerobic and aerobic process.

Bioremediation. Air. Biofilters are an effective way of dealing with air
from industrial processes that use halogenated solvents such chloromethane,
dichloromethane, chloroethane, 1,2-dichloroethane and vinyl chloride, that support aerobic growth (26). Both compost-based dry systems and trickling filter wet
systems are in use. Similar filters could be incorporated into pump-and-treat
operations.

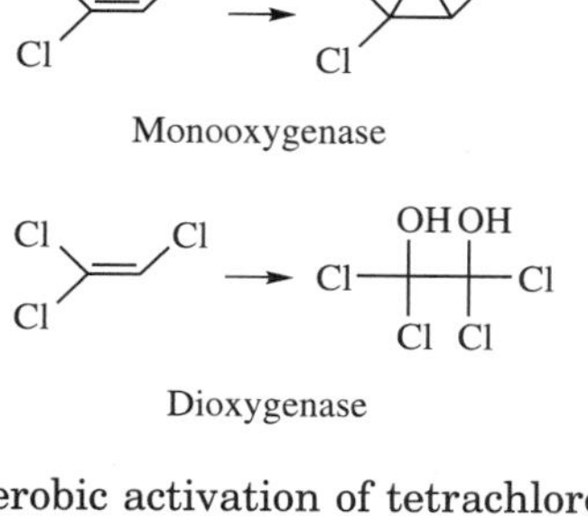

Monooxygenase

Dioxygenase

Fig. 8.   Aerobic activation of tetrachloroethylene.

Groundwater and Soil.   Halogenated solvents have contaminated soils and groundwater throughout the industrialized world, and remediation has a high priority. Since the solvents are so dense, they are typically found on the bedrock underlying aquifers. Pumping out the liquid phase is an obvious first step if the contaminant is likely to be mobile, but *in situ* bioremediation is a promising option. Thus, the U.S. Department of Energy is investigating the use of anaerobic *in situ* degradation of carbon tetrachloride in an aquifer some 76 m below the Hanford, Washington site, with nitrate as electron acceptor, and acetate as electron donor (22).

Trichloroethylene is the most frequent target of remediation, and as discussed above, this is only metabolized cometabolically. Remediation operations thus incorporate the addition of the cometabolized substrate. Methane was used successfully at the U.S. Department of Energy site at Savannah River, near Aiken, South Carolina (27) which had both an air-stripping and a biological component. Horizontal wells were used to pump methane and air below the contaminant, while an upper horizontal well in the vadose zone was used to withdraw these gases through the contaminated zone. Optimum biodegradation performance seemed to come from alternate injection of air and methane in air, and the inclusion of nitrous oxide and triethylphosphate, both gases, to give a C:N:P ratio of 100:10:1.

Plants may have a role to play in enhancing microbial biodegradation of halogenated solvents, for it has recently been shown that mineralization of radiolabelled trichlorothylene is substantially greater in vegetated rather than unvegetated soils (28), indicating that the rhizosphere provides a favorable environment for microbial degradation of organic compounds.

Methane has also been used in aerobic bioreactors that are part of a pump-and-treat operation, and toluene and phenol have also been used as co-substrates at the pilot scale (29). Anaerobic reactors have also been developed for treating trichloroethylene. For example, Wu and co-workers (30) have developed a successful upflow anaerobic methanogenic bioreactor that converts trichloroethylene and several other halogenated compounds to ethylene.

Groundwater contaminated with other halogenated solvents can also be treated in aboveground reactors. Aerobic reactors are useful for those compounds that can support growth. For example, a membrane reactor has been designed for treating 1,2-dichloroethane (31), and bubble columns and packed bed reactors have been developed for the aerobic degradation of 2-chloroethanol (32). As mentioned above, sequential anaerobic and aerobic reactors are capable of mineralizing tetrachloroethylene (33).

4.3. Halogenated Organic Compounds. *Constituents.* Complex halogenated organic compounds have been widely used in commerce in the last fifty years. A few representative examples are shown in Figure 9; pentachlorophenol has been widely used as a wood preservative, and also for termite control. (2,4-Dichlorophenoxy)acetate (2,4-D) is widely used as a broad-leaf herbicide, DDT was widely used as an insecticide, and hexachlorophene has been widely used as a germicide. Polychlorinated biphenyls (PCBs) were sold with varying levels of chlorination for a range of purposes. They ranged from light oily fluids (with two, three, or four chlorines) to viscous oils (five chlorines) to greases and waxes (six or more chlorines), and their names indicated the level of chlorination.

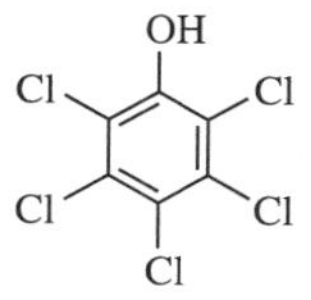

Pentachlorophenol

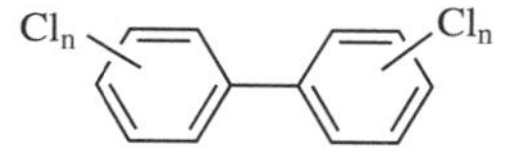

Polychlorinated biphenyl

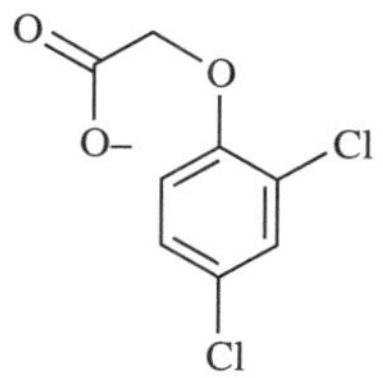

(2,4-Dichlorophenoxy)acetate

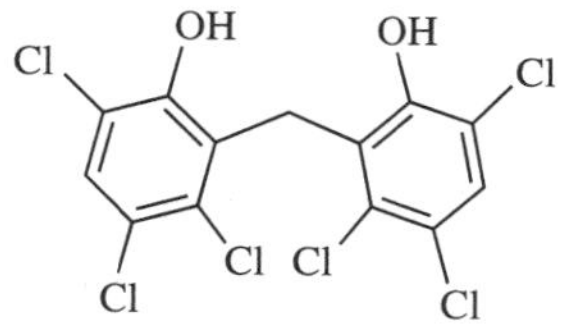

Hexachlorophene

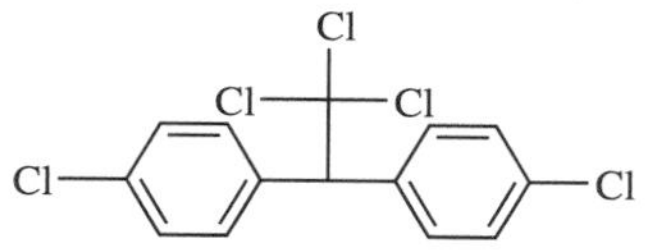

DDT

Fig. 9. Representative halogenated organic contaminants.

Thus Aroclor 1242 (trademark of Monsanto, U.S.), Clophen A30 (trademark of Farbenfabriken Bayer AG, Germany) and Kanechlor 300 (trademark of Kanegafuchi Chemical Industries, Japan) all contained 42% chlorine by weight, and an average of three chlorines per biphenyl. An important property shared by all these compounds is their relative resistance to biodegradation, so at first glance they may not seem a good target for bioremediation. Indeed complex halogenated organic compounds were widely thought to be almost exclusively anthropogenic in origin, so that there would have been little time for biodegradation pathways to evolve. This view is being corrected, for in fact a variety of organisms, particularly marine algae and some fungi, produce significant quantities of these compounds (34). There is, thus, good reason to expect that halogenated-organic degrading organisms will be found in the biosphere, and this has been borne out in practice.

Biodegradation. An important characteristic of degradation is the cleavage of carbon–chlorine bonds, and the enzymes that catalyze these reactions, the dehalogenases, are being characterized (35). The reductive dechlorination seen with carbon tetrachloride and tetrachloroethylene (see Fig. 7) seems to be a general phenomenon, and even compounds as persistent as DDT and the polychlorinated biphenyls are reductively dechlorinated under some conditions, particularly under methanogenic conditions. Some compounds, such as pentachlorophenol, can be completely mineralized under anaerobic conditions, but the more recalcitrant ones require aerobic degradation after reductive dehalogenation.

Pentachlorophenol can be mineralized aerobically and anaerobically, and both processes have been exploited for bioremediation. Under methanogenic conditions a reductive dehalogenation, analogous to that seen with halogenated solvents, eventually generates phenol, although some of the intermediate congeners are quite recalcitrant (36). The phenol is further mineralized to methane and carbon dioxide, and sulfate-reducing bacteria may be involved (37).

Several bacterial isolates are able to grow aerobically using pentachlorophenol as sole source of carbon and energy, but many grow rather better when supplemented with a more nutritions substrate. Unfortunately, in many cases it seems that the more vigorous growth with these substrates does not enhance the biodegradation of pentachlorophenol (38). The initial step in pentachlorophenol biodegradation in one *Flavobacterium* species is an NADPH-dependent oxygenolytic dechlorination, where the *para*-chloro group is replaced by a hydroxyl to generate tetrachlorohydroquinone. Other species seem to produce the same metabolite by a hydrolytic process (38). Fungal degradation of pentachlorophenol, apparently using the lignolytic apparatus, has also been reported (39).

(2,4-Dichlorophenoxy)acetate (2,4-D) has been one of the world's most popular herbicides. Although it is somewhat resistant to biodegradation, it is biodegraded by several bacterial isolates. It is a general truism that the more halogens on a molecule, the slower its biodegradation, and this is borne out with the related herbicide 2,4,5-T ((2,4,5-trichlorophenoxy)acetate). Nevertheless, bacterial degradation has been seen under both aerobic and anaerobic conditions, the latter involving reductive dechlorination via 2,4-D. Aerobic degradation removes acetate from (2,4-dichlorophenoxy)acetate to yield 2,4-dichlorophenol, which is subsequently hydroxylated to 3,5-dichlorocatechol, followed by ring cleavage and complete mineralization (40). Genetic engineering has been used to construct strains that are particularly adept at consuming these compounds, but whether these will overcome the regulatory hurdles to allow their use outside the laboratory remains to be seen.

DDT (1,1,1-trichloro-2,2-bis(*p*-chlorophenyl)ethane) is a remarkably resistant molecule, which explains both its efficacy as an insecticide, and its accumulation at the top of the food-chain. Nevertheless, there are indications that it can be biodegraded, both anaerobically with initial dechlorination and aerobically with initial ring hydroxylation (40). DDT and its partially degraded congeners are very hydrophobic, and biodegradation seems to be stimulated by adding surfactants. White-rot fungi also degrade DDT under lignolytic conditions, although there is little mineralization to CO_2 (5).

The polychlorinated biphenyls are quite recalcitrant. Some lightly chlorinated biphenyls are readily mineralized under aerobic conditions (41), and indeed the structure of an enzyme that catalyzes the key ring-cleavage oxygenation has recently been determined by x-ray crystallography (42). More chlorinated congeners are resistant to aerobic degradation, but they are reductively dechlorinated under anaerobic conditions. Complete degradation of the commercial mixtures, thus, generally requires an anaerobic process followed by an aerobic one. A major issue for the oxidative process is that biphenyl seems to be required for significant expression of the biodegradative system (41); the chlorinated compounds do not induce the enzymes that would degrade them. The biochemistry of the biodegradative process is only beginning to be unraveled, but already there are suggestions that there is considerable diversity in the enzyme systems able to degrade these compounds. There is a lot of effort aimed at engineering organisms to degrade polychlorinated biphenyls, and in finding ways to make these very hydrophobic compounds more bioavailable by the use of chemical oxidants, such as Fenton's reagent (43), or surfactants.

Bioremediation. Soil. Pentachlorophenol has been the target of bioremediation at a number of wood-treatment facilities, and good success has been achieved in several applications. It is rarely the sole contaminant, and is often present with polynuclear hydrocarbons from coal creosote. *In situ* degradation has been stimulated by bioventing, where air is injected through some wells, and extracted through others to both strip volatiles and provide oxygen to indigenous organisms (44). Just as with the halogenated solvents, it seems that plants stimulate microbial degradation of pentachlorophenol in the rhizosphere (45).

The kinetics of such *in situ* degradation are rather slow, however, and more active bioremediation is usually attempted. For example, contaminated soil at the Champion Superfund site in Libby, Montana, was placed into 1-acre land treatment units in 6-in. layers, and irrigated, tilled, and fertilized. Under these conditions, the half-lives of pentachlorophenol, pyrene, and several other polynuclear aromatic hydrocarbons, initially present at around 100–200 ppm, were on the order of 40 days (46). This success relied on the indigenous microbial populations in the soil, but many groups are focusing on the addition of organisms (eg, 47), perhaps immobilized on some sort of carrier. Composting, and bioremediation focusing on the use of white-rot fungi, has also met with success at the pilot scale. Others have used fed-batch or fluidized-bed bioreactors to stimulate the biodegration of pentachlorophenol. This allows significant optimization of the process and increases in rates of degradation by tenfold (48).

A major concern when remediating wood-treatment sites is that pentachlorophenol was often used in combination with metal salts, and these compounds, such as chromated copper–arsenate, are potent inhibitors of at least some pentachlorophenol degrading organisms (49). Sites with significant levels of such inorganics may not be suitable candidates for bioremediation.

The phenoxy-herbicide, 2,4-D, has been successfully bioremediated in a soil contaminated with such a high level of the compound (710 ppm) that it was toxic to microorganisms (50). There were essentially no indigenous bacteria in the soil. Success relied on washing a significant fraction of the contaminant off the soil and adding bacteria enriched from a less contaminated site. Success was achieved in remediating both soil washwater and soil in a bioslurry reactor (50). 2,4-D is also effectively degraded in composting, with about half being completely mineralized, and the other half becoming incorporated in a nonextractable form in the residual soil organic matter (51).

The bioremediation of polychlorinated biphenyls in soils in receiving significant attention because these compounds are quite widely distributed in the environment, either from leaking electrical transformers or sometimes because they were applied as part of road maintenance. In the latter case, the contamination usually includes petroleum hydrocarbons, and unfortunately it seems that the two contaminants inhibit the degradation of each other. Nevertheless, cultures are being found that can degrade both polychlorinated biphenyls and petroleum hydrocarbons. There is also interest in the role of rhizosphere organisms in polychlorinated biphenyl degradation, particularly since some plants exude phenolic compounds into the rhizosphere that can stimulate the aerobic degradation of the less chlorinated biphenyls.

Groundwater.   A successful groundwater bioremediation of pentachlorophenol is being carried out at the Libby Superfund site described above. A shallow aquifer is present at 5.5 to 21 m below the surface, and a contaminant plume is nearly 1.6 km in length. Nutrients and hydrogen peroxide were added at the source area and approximately half way along the plume, and pentachlorophenol concentrations decreased from 420 ppm to 3 ppm where oxygen concentrations were successfully raised. A membrane oxygen dissolution system has now been installed to replace the hydrogen peroxide additions, and costs have been substantially lowered without an apparent decrease in remediation performance (52).

Pentachlorophenol is readily degraded in biofilm reactors (53), so bioremediation is a promising option for the treatment of contaminated groundwater brought to the surface as part of a pump-and-treat operation.

River and Pond Sediments.   Much of the work on polychlorinated biphenyls has focused on the remediation of aquatic sediments, particularly from rivers, estuaries, and ponds. As noted above, a few of the most lightly chlorinated compounds are mineralized under aerobic conditions, but the more chlorinated species seem completely resistant to aerobic degradation, even by white rot fungi (54). On the other hand, there is extensive dechlorination of highly chlorinated forms under anaerobic conditions, particularly methanogenic conditions. Bioremediation thus requires anaerobic and aerobic regimes. Intrinsic biodegradation of polychlorinated biphenyls can be recognized by the changing "fingerprint" of the individual isomers as biodegradation proceeds (41,55). The anaerobic dechlorination of the most recalcitrant congeners can apparently be primed by adding a readily dehalogenated congener, such as 2,5,3',4'-tetrachlorobiphenyl (56), but whether this is a realistic approach for *in situ* bioremediation remains to be seen. Harkness and co-workers (57) have successfully stimulated aerobic biodegradation in large caissons in the Hudson River by adding inorganic nutrients, biphenyl, and hydrogen peroxide, but found that repeated addition of a polychlorinated-biphenyl degrading bacterium (*Alcaligenes eutrophus* H850) had no beneficial effect. Essentially no biodegradation occurred in the stirred control caissons, but losses on the order of 40% were seen in the caissons that received nutrients and peroxide, regardless of whether the stirring was aggressive or rather gentle. Whether this approach can be scaled-up for large-scale use, with a net environmental benefit, remains to be seen.

4.4. Nonchlorinated Pesticides and Herbicides.   *Constituents.*   A vast number of compounds are used as herbicides, pesticides, fungicides, etc, and a few are shown in Figure 10. In order to be effective they must have their effect before they are degraded in the environment, but on the other hand they must not be so resistant to degradation that they accumulate where they are used, or accumulate in food chains. It is unusual for these compounds to become contaminants where they are applied correctly, but manufacturing facilities, storage depots and rural airfields where crop-dusters are based have had spills that can lead to long lasting contamination. Bioremediation is a promising

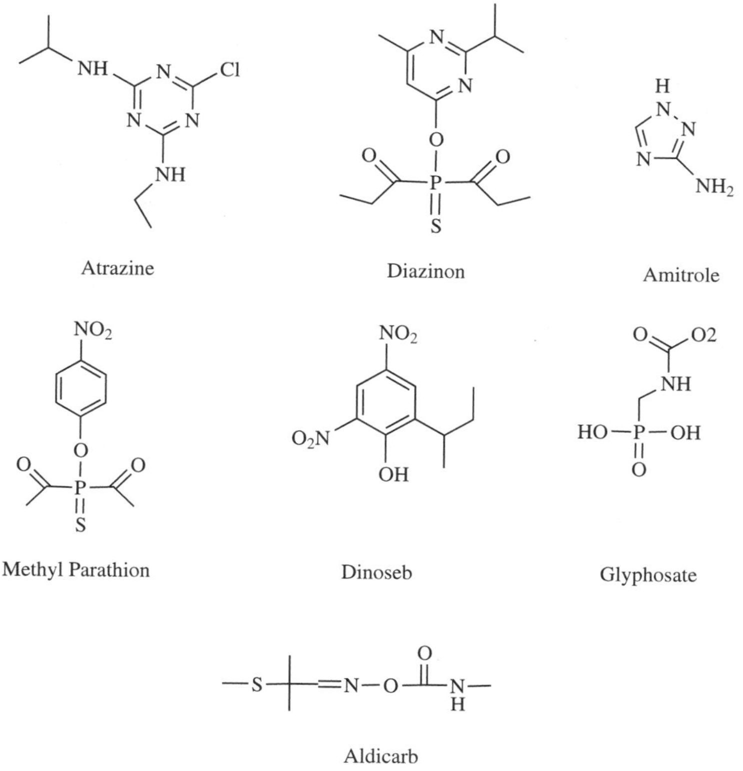

Fig. 10. Examples of herbicides and pesticides amenable to bioremediation.

technology to remediate such sites. There are also some locations were ground-water has become contaminated by these chemicals, and again bioremediation may be a cost-effective remediation strategy.

Biodegradation. The vast majority of pesticides, herbicides, fungicides, and insecticides in use today are biodegradable, although the intrinsic biodegrad-ability of individual compounds is one of the variables used in deciding which compound to use for which task. Some herbicides are acutely toxic to plants that absorb them, but are so readily degraded in soil that seeds can be planted at the same time as the herbicide application. Other herbicides are known to be effective at preventing plant growth for many months, and are used in situations where this is the desired goal. Very few degradation pathways of these com-pounds have been worked out in detail, but some generalizations can be made.

Compounds with organophosphate moieties, such as Diazinon, Methyl Parathion, Coumaphos and Glyphosate are usually hydrolyzed at the phosphorus atom (40,58). Indeed several *Flavobacterium* isolates are able to grow using para-thion and diazinon as sole sources of carbon.

Triazines pose rather more of a problem, probably because the carbons are in an effectively oxidized state so that no metabolic energy is obtained by their metabolism. Very few pure cultures of microorganisms are able to degrade tria-zines such as Atrazine, although some *Pseudomonads* are able to use the compound as sole source of nitrogen in the presence of citrate or other simple carbon substrates. The initial reactions seem to be the removal of the ethyl or isopropyl substituents on the ring (41), followed by complete mineralization of the triazine ring.

Nitroaromatic compounds, such as Dinoseb, are degraded under aerobic and anaerobic conditions (59). The nitro group may be cleaved from the molecule as nitrite, or reduced to an amino group under either aerobic or anaerobic conditions. Alternatively, the ring may be the subject of reductive attack. Thus, while these molecules are sometimes quite long-lived in the environment, they can be completely mineralized under appropriate conditions (59). Recent work has isolated a *Clostridium bifermentans* able to anaerobically degrade dinoseb cometabolically in the presence of a fermentable substrate. The dinoseb was degraded to below detectable levels, although only a small fraction was actually mineralized to CO_2 (60).

Carbamates such as Aldicarb undergo degradation under both aerobic and anaerobic conditions. Indeed the oxidation of the sulfur moiety to the sulfoxide and sulfone is part of the activation of the compound to its most potent form. Subsequent aerobic metabolism can completely mineralize the compound, although this process is usually relatively slow so that it is an effective insecticide, acaricide and nematocide. Anaerobically these compounds are hydrolyzed, and then mineralized by methanogens (61).

Bioremediation. Groundwater. Atrazine dominated the world herbicide market in the 1980s, and contamination of groundwater has been reported in several locations in the U.S., Europe, and South Africa. There are several reports that once in groundwater it is very recalcitrant, suggesting that atrazine-degrading organisms are not widespread. Nevertheless, successful biodegradation has been achieved with indigenous organisms in laboratory mesocosms after a lag phase, and once activity was found, it remained (62). Interestingly the degradation was somewhat slowed by the addition of low concentrations of readily assimilate carbon, such as lactate, and it is not clear how biodegradation might be stimulated in the field. Nevertheless, it is clear that intrinsic remediation is likely to lead to the disappearance of atrazine from groundwaters once atrazine-utilizers have become abundant, and perhaps inoculation with atrazine-metabolizing species will be effective.

If more active treatment is required, such as pump-and-treat, it is possible that biological reactors will be a cost-effective replacement for activated carbon filters (63).

Marsh and Pond Sediments. Herbicides and pesticides are detectable in marsh and pond sediments, but intrinsic biodegradation is usually found to be occurring. Little work has yet been presented where the biodegradation of these compounds has been successfully stimulated by a bioremediation approach.

Soil. Herbicides and pesticides are of course metabolized in the soil to which they are applied, and there are many reports of isolating degrading organisms from such sites. Degradation activities are typically much higher at sites that have seen product application, indicating that natural enrichment processes occur. Much current effort is aimed at assessing the diversity of degradative pathways, and in many cases it seems that several different natural metabolic pathways can degrade individual pollutants. Little work has yet been presented where the biodegradation of these compounds has been successfully stimulated by a bioremediation approach, but inoculation with active organisms may be a promising approach (64,65).

It is a general observation that herbicide degradation occurs more readily in cultivated than fallow soil, suggesting that rhizosphere organisms are effective herbicide degraders. Whether this can be effectively exploited in a phytoremediation strategy remains to be seen.

4.5. Military Chemicals. *Constituents.* The military use a range of chemicals as explosives and propellants, which are sometimes termed "energetic molecules". Generally speaking, modern explosives are cyclic, often heterocyclic, composed of carbon, nitrogen and oxygen. Perhaps the most well known is 2,4,6-trinitrotoluene (Fig. 11), but RDX (Royal Demolition eXplosive; hexahydro-1,3,5-trinitro-1,2,3-triazine) and HMX (High Melting eXplosive; octahydro-1,3,5,7-tetranitro-1,3,5,7-tetrazocine), are even more powerful. *N,N*-Dimethylhydrazine is used as a solid rocket fuel. These compounds are sometimes present at quite high levels in soils and groundwater on military bases and production sites. One quite infamous problem at the latter is "pink water", a relatively undefined mixture of photodegradation products of TNT. Bioremediation is a promising new technology for treating sites contaminated with such compounds.

Bioremediation may also be an appropriate tool for dealing with chemical agents such as the mustards and organophosphate neurotoxins (66), but little work on actual bioremediation has been published.

Biodegradation. Natural nitrosubstituted organic compounds are quite unusual, and it was once thought that their degradation was principally by abiotic processes. As shown above in the case of Dinoseb, however, nitrosubstituted compounds are subject to a variety of degradative processes. The biodegradation of TNT is well established (67). Under anaerobic conditions it is readily reduced to the corresponding aromatic amines and subsequently deaminated to toluene. As shown in the section on hydrocarbons, the latter can be mineralized under

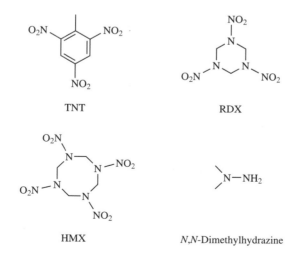

Fig. 11. Military explosives and a rocket propellant.

anaerobic conditions, leading to the potentially complete mineralization of TNT in the absence of oxygen.

Under aerobic conditions TNT can be mineralized by a range of bacteria and fungi, often co-metabolically with the degradation of a more degradable substrate. There is even evidence that some plants are able to deaminate TNT reductively. However, there is also ample evidence that under some conditions, the TNT is converted to insoluble large molecular weight compounds. This probably occurs by addition reactions to soil components, such as humic and fulvic acids, or cellular material in the case of plants. Lignolytic fungi are also yielding promising results for the degradation of both TNT and "pink water", particularly if the latter is pretreated with uv irradiation.

RDX and HMX are rather more recalcitrant, especially under aerobic conditions, but there are promising indications that biodegradation can occur under some conditions, especially composting (67). Several strains of bacteria able to use RDX (and Triazine) as a sole source of nitrogen for growth have recently been isolated, and this is an area where rapid progress is being made.

Little work has been reported on the biodegradation of dimethylhydrazine, but it may become an important target for remediation at some sites (68).

Bioremediation. Groundwater. Nitrotoluenes have been detected in groundwater in some areas, and intrinsic remediation may be occurring at some sites by anaerobic degradation. Research into whether this can be stimulated with a net environmental benefit is in its very earliest stages, and no clear evidence for success has been presented.

A commercial technology (69), the SABRE process, treats contaminated water and soil in a two-stage process by adding a readily degradable carbon and an inoculum of anaerobic bacteria able to degrade the contaminant. An initial aerobic fermentation removes oxygen so that the subsequent reduction of the contaminant is not accompanied by oxidative polymerization.

Soil. Composting of soils contaminated by high explosives is being carried out at the Umatilla Army Depot near Hermiston, Oregon (70). Soil from munitions washout lagoons is being treated indoors in compost rows of 2,000 m^3, and the estimated cost is less than one-third the estimated cost of incineration. If this is successful, there are 30 similar sites on the National Priority List that could be treated in a similar way.

4.6. Other Organic Compounds. The majority of organic compounds in commerce are biodegradable, so bioremediation is a potential option for cleaning up after industrial and transportation accidents. For example, Ref. 71 reports the successful bioremediation of 23,000 liters of vinyl acetate spilled from a railroad tank car in Albany, New York, at a cost approximately half that of excavation and disposal of the 1,100 m^3 of contaminated soil. They also report that *in situ* biological treatment has been used to remediate spills of other organics, including acrylonitrile, styrene, 2-butoxyethanol, and ethacrylate, at other sites where there had been railroad accidents. Bioremediation is thus already an important tool in remediating accidental spills of organic compounds.

Bioremediation is also an option when spills of such compounds contaminate groundwater. For example, bioremediation seems a feasible treatment for aquifers contaminated with alkylpyridines (72) and phenol (73).

5. Inorganic Contaminants

5.1. Nitrogen Compounds. *Constituents.*
Nitrogen-containing compounds are of concern for several reasons. Nitrate levels are regulated in groundwater because of concerns for human and animal health. Ammonia is regulated in streams and effluents as a potential fish toxicant, and any nitrogenous contaminant is a potential problem in water because of its stimulatory effect on the growth of algae. Other nitrogenous contaminants include cyanides in mine waters. Fortunately, all are amenable to biological treatment.

5.2. Biodegradation.
The biological mineralization of fixed nitrogen is well studied; ammonia is oxidized to nitrite, and nitrite to nitrate, by autotrophic bacteria, and nitrate is reduced to nitrogen by anaerobic bacteria. Urea in sewage and industrial wastes is readily hydrolyzed to ammonia and CO_2 by many bacteria, and cyanides and cyanates are used as sole sources of carbon and nitrogen by some organisms. Wastewater treatment facilities utilize these organisms in assuring that municipal and industrial effluents meet strict water quality standards, but this biological process is outside the scope of this article (see WATER, INDUSTRIAL WATER TREATMENT; WATER, MUNICIPAL WATER TREATMENT). On the other hand, ammonia and nitrate are essential nutrients for plant and bacterial growth, so one option is to use these organisms to take up and use the contaminants.

Bioremediation. Surface Water. One example of exploiting biology to handle excess nitrogen in a surface water is the case of the Venice Lagoon in Italy. About a million tons of sea lettuce (*Ulva*) grows in the lagoon annually because of the high levels of nitrogenous nutrients in this relatively landlocked bay. This material is harvested, composted and sold as a low-cost remediation of this problem (74). In areas where water quality is sufficiently high, growing even more valuable crops, such as nori (*Porphyra*) may be an attractive bioremediation option.

A more constrained opportunity for nitrate bioremediation arose at the US-DoE Weldon Spring Site near St. Louis, Missouri. This site had been a uranium and thorium processing facility, and treatment of the metal had involved nitric acid. The wastestream, known as raffinate, was discharged to surface inpoundments and neutralized with lime to precipitate the metals. Two pits had nitrate levels that required treatment before discharge, but heavy rains in 1993 threatened to cause the pits to overflow. Bioremediation by the addition of calcium acetate as a carbon source successfully treated more than 19 million liters of water at a reasonable cost (75).

Groundwater. One approach to minimizing the environmental impact of excess nitrogen in groundwater migrating into rivers and aquifers is to intercept the water with rapidly growing trees, such as poplars, that will use the contaminant as a fertilizer.

An alternative approach is to add a readily degradable substrate to the contaminated aquifer, in the absence of oxygen, to stimulate bacterial denitrification. Soluble substrates such as ethanol have been suggested, but a recent alternative suggestion is to provide vegetable oil, either down the well itself, or on a gravel trickling filter. Since the oil is insoluble in water, it becomes immobilized on the aquifer material, allowing denitrification as the water flows through (76).

5.3. Metals and Metalloids.   A wide range of metallic and nonmetallic elements are present as contaminants at industrial and agricultural sites throughout the world, both in ground- and surface water, and in soils. They pose a quite different problem from that of organic contaminants, since they cannot be degraded so that they disappear. Some metal and metalloid elements have radically different bioavailabilities and toxicities depending on their redox state, so one option is stabilization by converting them to their least toxic form. This can be a very effective way of minimizing the environmental impact of a contaminant, but if the contaminant is not removed from the environment, there is always the possibility that natural processes, biological or abiological, may reverse the process. Removing the contaminants from water phases is relatively straightforward, and the wastewater treatment industry practices this on an enormous scale. Pump-and-treat systems that mimic waste water treatments are already being used for several contaminants, and less complex systems involving biological mats are a promising solution for less demanding situations.

Microbial leaching of metals from ores is a promising adjunct to more aggressive metal recovery technologies (77), but is generally achieved by oxidative processes that generate very acidic waters. It seems unlikely that similar approaches will be of much value in removing contaminant metals and metalloids from soils.

In the past, removing metal and metalloid contaminants from soil has been impossible, and site clean-up has meant excavation and disposal in a secure landfill. An exciting new approach to this problem is phytoextraction, where plants are used to extract contaminants from the soil and harvested.

Immobilization and Toxicity-Minimization.   Adsorption.   Biomass, often agricultural by-products, has been widely used as an adsorbent for metals and other contaminants in water (78), but this is outside the scope of this article.

Microbial Processes.   Most elements display a range of solubilities and biological effects depending on their chemical form. For example, chromium, although it may be an essential micronutrient for many organisms, is known to be toxic at higher levels. Indeed chromium has a range of available redox states that differ significantly in their environmental impact. Cr(VI) is generally more soluble and more toxic than Cr(III), but a wide range of organisms, both bacteria and fungi, are able to reduce the former to the latter under both aerobic and anaerobic conditions. Thereby the environmental impact of the contaminant is reduced. Since Cr(III) precipitates as insoluble hydroxides under neutral and alkaline conditions, microbial reduction is a potential treatment for soils and waters (79).

Similarly selenium is a micronutrient that is toxic at higher levels. It also has quite different bioavailability in its different redox states, with the elemental form being the least biologically available. Many microorganisms, both bacteria and fungi, are able to reduce more oxidized species, especially selenite (Se(IV)) to the red elemental form, providing an appealing remediation option for this element. An alternative approach to remediating selenium contamination is to encourage methylation to volatile dimethylselenide (80). Although the microbiology of this process is not yet very well understood, it seems to be the result of degradation of the selenium-containing amino acid, selenomethionine.

Arsenic is another element with different bioavailability in its different redox states. Arsenic is not known to be an essential nutrient for eukaryotes, but arsenate (As(V)) and arsenite (As(III)) are toxic, with the latter being rather more so, at least to mammals. Nevertheless, some microorganisms grow at the expense of reducing arsenate to arsenite (81), while others are able to reduce these species to more reduced forms. In this case it is known that the element can be immobilized as an insoluble polymetallic sulfide by sulfate reducing bacteria, presumably adventitiously due to the production of hydrogen sulfide (82). Indeed many contaminant metal and metalloid ions can be immobilized as metal sulfides by sulfate reducing bacteria.

A rather more specific mechanism of microbial immobilization of metal ions is represented by the accumulation of uranium as an extracellular precipitate of hydrogen uranyl phosphate by a *Citrobacter* species (83). Staggering amounts of uranium can be precipitated; more than 900% of the bacterial dry weight! Recent work has shown that even elements that do not readily form insoluble phosphates, such as nickel and neptunium, may be incorporated into the uranyl phosphate crystallites (84). The precipitation is driven by the production of phosphate ions at the cell surface by an external phosphatase. Although the process requires the addition of a phosphate donor, such as glycerol-2-phosphate, it may be a valuable tool for cleaning water contaminated with radionuclides. An alternative mode of uranium precipitation is driven by sulfate-reducing bacteria such as *Desulfovibrio desulfuricans*, which reduce U(VI) to insoluble U(IV). When combined with bicarbonate extraction of contaminated soil, this may provide an effective treatment for removing uranium from contaminated soil (85).

Microbial processes can also detoxify mercury ions and organic compounds by reducing the mercury to the elemental form, which is volatile (86). This certainly reduces the environmental impact of compounds such as methylmercury, however, such a bioprocess would have to include a mercury capture system before it could be exploited on a large scale with public support.

Rhizofiltration.   Rhizofiltration is the use of plant root systems to remove contaminating metal ions from water (87). The plants must have substantial root systems, and success has been reported with both hydroponically grown terrestrial plants, such as sunflowers (*Helianthus annuus*), and floating aquatic plants such as water hyacinth (*Eichhornia crassipes*), and water milfoil (*Myriophyllum spicatum*). A filamentous cyanobacterium, *Phormidium* sp., has also proven quite effective at removing metals from water. Understanding rhizofiltration at the molecular level is at an early stage, but it is already clear that several processes are involved. Adsorption of the contaminant to the root surface must occur first, and presumably happens in all plants, probably to polygalacturonic acids for cations, and positively charged polypeptides for anions. Some plants actively translocate the metal ions into their cells and some transport the metal to the leafy shoots. This can be exploited in phytoextraction of metal ions from soils (see below). This phenomenon may actually be undesirable in rhizofiltration, because if the metals are retained in the roots there will be a smaller volume of contaminated material to treat at the end of the water treatment. Perhaps of more importance for rhizofiltration is the process of root assisted precipitation, where metal ions are precipitated on the roots as insoluble inorganic complexes. For example, lead is precipitated as lead phosphates on roots of Indian Mustard

(*Brassica juncea*) to a remarkable level; up to 45% of dry weight (87). This phenomenon seems to require active exudation of precipitants from the roots, for it is dramatically reduced when dead roots are used.

Phytoextraction.   The fact that certain plant species are able to survive on soils with such high levels of metal ions that the growth of most plants is inhibited has been known for centuries. However, it was not until this century that it became clear that some of these plants actually accumulated the potentially toxic metals and somehow resisted their toxic effects. These plants, known as hyperaccumulators, are exemplified by Alpine Pennycress, *Thlaspi caerulescens*, which is able to accumulate a wide range of metals, including cadmium, chromium, copper, lead, nickel, and zinc, more than several thousand fold in its roots, and somewhat less in its leaves (88). Unfortunately for a use in bioremediation, most hyperaccumulators are small, slow growing plants. Researchers have, therefore, turned to more rapidly growing plants which, while they might not be quite as effective as the hyperaccumulators on a weight basis, might be able to extract more metal in a growing season (88). Indian mustard, *Brassica juncea*, was found to be almost as effective as *T. caerulescens* and to grow much faster. This plant is now being used in field trials of phytoextraction.

Phytoextraction of metals from soils requires plants that have substantial root systems to maximize contact with the contaminated soil, and effective transport mechanisms to get the metal from the root to harvestable biomass. Merely accumulating the metal in or on the roots would be less desirable, because harvesting would likely increase the environmental impact of the contaminant by creating dust. It has been known for some time that metals such as cadmium are often stored in plant tissues as phytochelatin-complexes. Phytochelatins are small peptides rich in cysteines that chelate metals via the cysteinyl sulfur (89). Other elements are stored as metalphytate precipitates (88), and in both cases seem to be concentrated in the vacuolar sap. But how the metals are transported in the xylem sap from the root to the shoot is only beginning to be explored. Cadmium seems to be transported as a soluble salt of organic acids (90), at least in *B. juncea*, while nickel is thought to be transported as a soluble histidine complex in *Alyssum* species (91). Optimizing this process will be an important consideration for the commercial success of phytoremediation.

It will also be important to understand the rhizosphere ecology around the roots of metal accumulating plants fully. Maximizing the bioavailability of the contaminant metals in this zone may require the optimization of the microbial communities, or perhaps the addition of soil amendments. There are early indications that such intervention may be beneficial (88), but research in this area is at a very early stage.

Phytoremediation is also being developed for dealing with soils contaminated with high levels of selenium in California; again *B. juncea* seems to be particularly effective in accumulating the contaminant from soil, and all plants tested were more effective at removing selenate than selenite (92). This is an interesting contrast to bacterial systems, where selenite reduction is more commonly found than selenate reduction.

There have also been reports that some plants, including *B. juncea*, stimulate volatilization of dimethylselenide, although it seems that this is an indirect effect where the plant roots stimulate bacterial evolution of the gas (93). Presumably the selenium is eventually oxidized in the atmosphere and returned to the soil as rain. Since much of the world is marginally selenium deficient, such a process might have no deleterious environmental effects. A similar volatilization approach may eventually be used for mercury contamination. Ref. 94 describes the effective transfer of bacterial mercury reduction genes into a plant. In the future, this approach might offer an option for cleaning mercury-contaminated soils, albeit with some form of mercury capture technology.

Phytostabilization.   In some cases it may be important to stabilize a contaminated site to minimize harmful environmental impacts. Barren soils are more prone to erosion, leaching, and dust formation than vegetated soils, so establishing plants as a ground cover may be a very beneficial treatment, even if the contaminant remains in place. In some cases, planting metal tolerant species is enough to establish a plant community, but in other situations, such as mine waste piles, it may be essential to restore fertility to the soil by adjusting the pH and adding organic composts. Where phytoextraction requires plants that actively transport metals to the above-ground biomass, this would be an undesirable phenomenon in phytostabilization programs because it would promote the movement of the contaminating metals into the biosphere.

It may also be possible to stimulate bioreductions in soils, such as Cr(VI) to Cr(III), more effectively by growing plants than by adding bacteria or nutrients to stimulate the microbial process.

Bioremediation.   Water.   Groundwater can be treated in anaerobic bioreactors that encourage the growth of sulfate reducing bacteria, where the metals are reduced to insoluble sulfides, and concentrated in the sludge. For example, such a system is in use to decontaminate a zinc smelter site in the Netherlands (95).

Bacterial remediation of selenium oxyanions in San Joaquin, California, drainage water is under active investigation (96, 97), but has not yet been commercialized. Agricultural drainage rich in selenium is also typically rich in nitrates, so bioremediation must also include conditions that stimulate denitrification (98). Phytoextraction of selenium is also being tested, but is not yet being used on a large scale.

Phytoremediation is also not yet being used commercially, but results at several field trial suggest that commercialization is not far away. Perhaps the biggest success to date is the successful rhizofiltration of radionuclides from a Department of Energy site at Ashtabula, Ohio, where uranium concentrations of 350 ppb were reduced to less than 5 ppb, well below groundwater standards, by Sunflower roots (99). Small field tests also suggest that rhizofiltration will be a remediation option for removing radionuclides, such as ^{137}Cs and ^{90}Sr, from contaminated lakes and ponds near the Chernobyl nuclear reactor in Ukraine.

Mine Drainage.   Natural drainage water that come into contact with active and abandoned metal and coal mines can become seriously contaminated with a range of heavy-metal ions, and/or often become quite acidic, with a pH near 2. While underground, the water is typically anoxic, and any iron is present as soluble ferrous species. When this mixes with aerobic surface water, the iron precipitates as bright orange ferric hydroxide, and this can have a serious environmental impact. In recent years it has become clear that the environmental impact of acid mine drainage can be minimized by the construction of artificial wetlands that combine geochemistry and biological treatments. These systems are being designed for a range of wastewaters, most of which fall outside the scope of this article.

The precipitation of ferric hydroxide is typically biologically mediated by iron-oxidizing bacteria at acid pH, but is usually rather slow. Abiological oxidation becomes more important at pH values above 5, and this is usually much faster than the biological process. Most constructed wetlands for treating acid mine waters thus start with a zone designed to raise the pH. A bed of crushed limestone often suffices to raise the pH significantly, and it is important that this be kept anoxic to prevent rust precipitation on the limestone, which would prevent further production of alkalinity. Once the pH is near neutral, the water is discharged into an aerobic wetland to encourage the precipitation of iron and aluminum oxides, and the co-precipitation of arsenic, if this is present (100,101).

If heavy metals are present in the mine water, the iron-free water can be made to flow into an anaerobic part of the constructed wetland, where organic material, such as compost, manure or sawdust, provides reductants to sulfate-reducing bacteria that become established. These bacteria reduce sulfate to sulfide, which precipitates the heavy metal ions as insoluble sulfides. It is important that this part of the constructed wetland be kept anaerobic, to prevent oxidation and remobilization of the precipitated metals. For this reason, this part of the wetland is typically kept flooded and free of aquatic plants that might introduce oxygen through their roots (100). Finally, an aerobic algal mat can act as a polishing step to complete the removal of contaminants, particularly manganese (101,102). Mine drainage that is not acidic may not need such complicated systems and individual parts of the treatment train described above may suffice in some situations.

Soil.   The first reported field trial of the use of hyperaccumulating plants to remove metals from a soil contaminated by sludge applications has been reported (103). The results were positive, but the rates of metal uptake suggest a time scale of decades for complete cleanup. Trials with higher biomass plants, such as *B. juncea*, are underway at several chromium and lead contaminated sites (88), but data are not yet available.

The bacterial reduction of Cr(VI) to Cr(III) discussed above is also being used to reduce the hazards of chromium in soils and water (104).

6. Conclusions

Bioremediation has been successfully used to treat a wide range of contaminants, including crude oils and refined petroleum products, halogenated solvents, pesticides, herbicides, military chemicals, and mine waters. Much of this success has come by small adjustments of the local environment to encourage the growth of remediating organisms. Fertilizer addition has been a successful treatment for terrestrial and marine oil spills, and the addition of co-substrates, particularly methane, has been a successful treatment for remediating halogenated solvents, such as trichloroethylene. Composting is proving to be a successful treatment for a range of contaminants, and constructed wetlands are successfully treating a range of wastewaters, including those emanating from mines.

One area where there is considerable disagreement between academic scientists and engineering practitioners of bioremediation is the area of bioaugmentation, the addition of selected microorganisms to a site to encourage biodegradation. Dozens of companies sell bacteria for this purpose and claim success in the field, but efforts to demonstrate effectiveness rigorously have met with little success (105,57). Perhaps the most startling test was performed by the U.S. EPA when testing potential inoculants for stimulating oil biodegradation in Alaska following the *Exxon Valdez* oil spill (106). Eight products were tested in small laboratory reactors that allowed substantial degradation of a test oil by the indigenous organisms of Prince William Sound. All eight microbial inocula had a greater stimulatory effect on alkane degradation, at least for the first 11 days, when they were sterilized by autoclaving prior to addition! This suggests that the indigenous organisms readily out-competed the added products, but that autoclaving the products released some trace nutrient that was able to stimulate the growth of the endogenous organisms.

Of course, bioaugmentation may prove more effective with contaminants that have only recently entered the biosphere, such as methyl-*tert*-butyl ether, or with organisms that have a dramatic selective advantage over indigenous organisms. For example, modern molecular biology may offer opportunities for moving effective degradation pathways into organisms native to a contaminated site, improving biodegradation pathways by broadening the substrate range of degradative enzymes, or removing regulatory constraints to maximize degradative activities. There are, however, many technical and regulatory hurdles to be surmounted before this potential can be tested.

Another area where there is controversy is in the role of surfactants. Many bacteria, particularly those that degrade hydrocarbons, produce surfactants as they grow. Some release them into the medium, others incorporate them into their cell exterior, and there have been elegant experiments to show that inhibiting the production of these compounds inhibits the ability of the bacteria to degrade oil. There have thus been many suggestions to add surfactants, either bacterial or synthetic, to stimulate biodegradation. This seems to be beneficial in the case of some oil dispersants at sea (12), but there have not yet been any clear demonstrations of efficacy on a large scale in soils. The role of surfactants in bioremediation is an area that requires further study.

Bioremediation has many advantages over other technologies, both in cost and in effectively destroying or extracting the pollutant. An important issue is thus when to consider it, and a series of questions may lead to the appropriate answer (see Table 4).

If the answers to the questions in Table 4 lead to the selection of bioremediation, it then becomes important to assess the success of the bioremediation strategy in achieving the clean-up criteria. A major disadvantage of bioremediation is that it is typically rather slower than competing technologies such as thermal treatments. How can regulators and responsible parties gain confidence during this time that success will indeed be achieved? The National Research Council (107) has recently addressed this issue, and suggested a three-fold strategy for "proving" bioremediation: (*1*) a documented loss of contaminants from the site; (*2*) laboratory tests showing the potential of endogenous microbes to

Table 4. **Will Bioremediation be a Suitable Treatment for a Site Contaminated with Organic, Nitrogenous, or Organic Contaminants?**

Organic

Is the contaminant biodegradable? If the contaminant is a complex mixture of components, are the individual chemical species biodegradable? If the contaminant has been at the site for some time, biodegradation of the most readily degradable components may have already occurred. Is the residual contamination biodegradable?

Are degrading organisms present at the site?

What is limiting their growth and activity? Can this be added effectively?

Are the levels of contaminant amenable to bioremediation? Are they toxic to microorganisms? Are they so abundant that even substantial microbial activity will take too long to clean the site?

Are the clean-up standards reasonable? Are biological processes known to degrade substrates down to the levels required?

Nitrogenous

Are appropriate microorganisms present at the site?

What is limiting their growth and activity? Can this be added effectively?

Are the levels of contaminant amenable to bioremediation? Are they toxic to microorganisms? Are they so abundant that even substantial microbial activity will take too long to clean the site?

Can the nitrogenous compound be used by plants?

Are the clean-up standards reasonable? Are biological processes known to degrade substrates down to the levels required?

Inorganic

Can the contaminant be made less hazardous by changing its redox state?

Can the contaminant be brought to a reactor or constructed wetland where biological systems, microbial or plant, can extract and immobilize the contaminant?

Can plants extract the contaminant from the soil matrix?

Are the clean-up standards reasonable? Are biological processes known to accumulate contaminants down to the levels required?

catalyze the reactions of interest; and (*3*) some evidence that this potential is achieved in the field.

Although laboratory tests to demonstrate the potential of endogenous organisms are relatively straightforward, the other two are often not as simple as they seem, especially in soils and sediments. Documenting the loss of contaminant is often difficult because of the heterogeneous distribution of the contaminant in the field; large numbers of samples may be needed in order to be able to detect statistically significant decreases in absolute concentration. Providing evidence that biodegradation has indeed been stimulated is also a challenge, but if the contaminant is a complex mixture, such as a crude oil, refined petroleum, or a mixture of polychlorinated biphenyls, the least degradable compounds in the mixture can be used as conserved internal markers for quantifying the degradation of the more degradable components.

For example, hopane (Fig. 12) is a conserved marker in crude oils in at least the early stages of biodegradation (up to 80% degradation) (108); its concentration increases in the residual oil as biodegradation proceeds. Basing estimates of biodegradation on hopane allowed us to quantify the effect of the successful bioremediation strategy following the *Exxon Valdez* oil-spill (14,15). Even in refined products, the least degradable detectable analyte can serve this role. Thus, the

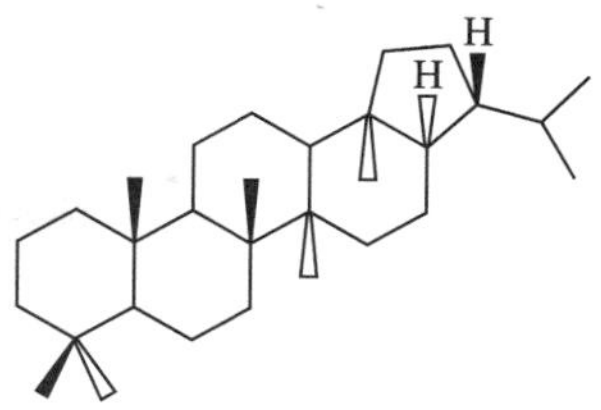

Fig. 12.　17α(*H*),21β(*H*)-hopane.

trimethyl-phenanthrenes can be used to estimate qualitatively the degradation of diesel oils in the environment, even though these molecules are themselves biodegradable (109). Of course, since these molecules are themselves biodegradable, estimates of the rates of biodegradation of more readily biodegradable compounds are systematically underestimated using this approach, but it can still provide very valuable information. The more halogenated polychlorinated biphenyls can serve a similar role in assessing the environmental fate of these products (55), and benzene is typically the last compound in BTEX plumes to be degraded, particularly under anaerobic conditions.

In situations where conserved internal markers cannot be used, such as in spills of essentially pure compounds, the evidence for enhanced biodegradation may have to be more indirect. Oxygen consumption, increases in microbial activity or population, and carbon dioxide evolution have all been used with success.

Finally, a caveat. Despite its documented success in many situations, bioremediation may not always be able to meet current clean-up criteria for a particular site. Some standards are so tight that they are essentially "detection limit" standards, and it is not clear that biological processes will be able to remove contaminants to such low levels. For example, the level of contaminant may be so low that it does not induce the microorganisms to produce the enzymes necessary for biodegradation. Or perhaps the contaminant is bound to soil or sediment particles in such a way that it is not available for biodegradation, although it is still extractable with aggressive solvents in analytical procedures. These are areas that require further research, but bioremediation will be more likely to fulfill its promise as an important tool in contaminated site remediation if there is progress towards standards based on bioavailability and net environmental benefit from the clean up, rather than on arbitrary absolute standards.

BIBLIOGRAPHY

"Bioremediation," in *ECT* 4th ed., Suppl. Vol., pp. 48–89, by Roger C. Prime, Exxon Research and Engineering Co.

CITED PUBLICATIONS

1. K. Devine, in R. E. Hinchee, J. A. Kittel, and H. J. Reisinger, eds., *Applied Bioremediation of Hydrocarbons*, Battelle Press, Columbus, Ohio, 1995, 53–59.

2. S. S. Butcher, R. J. Charlson, G. H. Orians, and G. V. Wolfe, *Global Biogeochemical Cycles*, Academic Press, New York, 1992.

3. I. D. Prijambada, S. Negoro, T. Yomo, and I. Urabe, *Appl. Environ. Microbiol.* **61**, 2020–2022 (1995).

4. P. Adriaens and T. M. Vogel, in L. Y. Young and C. E. Cerniglia, eds., *Microbial Transformation and Degradation of Toxic Organics*, Wiley-Liss, New York, 1995, 435–486.

5. A. Paszczynski and R. L. Crawford, *Biotechnol. Progr.* **11**, 368–379 (1995).

6. J. J. Childress, *Amer. Zool.* **35**, 83–90 (1995).

7. T. D. Sharkey and E. L. Singsaas, *Nature* **374**, 769 (1995).

8. R. C. Prince, Bioremediation of Crude Oil, in R. A. Meyers, ed., *Encyclopedia of Environmental Analysis and Remediation*, John Wiley & Sons, Inc. (in press) (1998).

9. J. P. Salanitro, *Curr. Opini. Biotechnol.* **6**, 337–340 (1995).

10. J. M. Yudelson and P. D. Tinari, in R. E. Hinchee, G. D. Sayles and R. S. Skeen, eds., *Biological Unit Processes for Hazardous Waste Treatment*, Battelle Press, Columbus, Ohio, 1995, 205–209.

11. National Research Council *Oil in the Sea: Inputs, Fates and Effects*, National Academy Press, Washington, D.C., 1985.

12. R. Varadaraj, M. L. Robbins, J. Bock, S. Pace, and D. MacDonald, in *Proceedings of the 1995 International Oil Spill Conference*, American Petroleum Institute, Washington, D.C., 1995, 101–106.

13. R. C. Prince, *Critical Reviews Microbiology* **19**, 217–242 (1993).

14. R. C. Prince, J. R. Clark, J. E. Lindstrom, E. L. Butler, E. J. Brown, G. Winter, M. J. Grossman, R. R. Parrish, R. E. Bare, J. F. Braddock, W. G. Steinhauer, G. S. Douglas, J. M. Kennedy, P. Barter, J. R. Bragg, E. J. Harner, and R. M. Atlas, in R. E. Hinchee, B. C. Alleman, R. E. Hoeppel and R. N. Miller, eds., *Hydrocarbon Remediation*, Lewis Publishers, Boca Raton, Fla., 1994, 107–124.

15. J. R. Bragg, R. C. Prince, E. J. Harner, and R. M. Atlas, *Nature* **368**, 413–418 (1994).

16. R. C. Borden, C. A. Gomez, and M. T. Becker, *Ground Water* **33**, 180–189 (1995).

17. R. D. Norris, in Ref. 1, 483–487 (1995).

18. G. Battermann and M. Meier-Löhr, in Ref. 1, 155–164.

19. D. R. Lovley, J. C. Woodward, and F. H. Chapelle, *Nature* **370**, 128–131 (1994).

20. B. A. Keet, in Ref. 1, 329–334.

21. S. J. McMillen, N. R. Gray, J. M. Kerr, A. G. Requejo, T. J. McDonald, and G. S. Douglas, in R. E. Hinchee, G. S. Douglas, and S. K. Ong, eds., *Monitoring and Verification of Bioremediation*, Battelle Press, Columbus, Ohio, 1995, 1–9.

22. B. M. Peyton, M. J. Truex, R. S. Skeen, and B. S. Hooker, in R. E. Hinchee, A. Leeson, and L. Semprini, eds., *Bioremediation of Chlorinated Solvents*, Battelle Press, Columbus, Ohio, 1995, 111–116.

23. D. Kohler-Staub, S. Frank, and T. Leisinger, *Biodegradation* **6**, 229–236 (1995).

24. D. B. Janssen, F. Pries, and J. R. van der Ploeg, *Ann. Rev. Microbiol.* **48**, 163–191 (1994).

25. D. J. Arp, *Curr. Opin. Biotechnol.* **6**, 352–358 (1995).

26. S. J. Ergas, K. Kinney, M. E. Fuller, and K. M. Scow, *Biotechnol. Bioengineer.* **44**, 1048–1054 (1994).

27. K. H. Lombard, J. W. Borten, and T. C. Hazen, in R. E. Hinchee, ed., *Air Sparging for Site Remediation*, Lewis Publishers, Boca Raton, Fla., 1994, 81–96.

28. T. A. Anderson and B. T. Walton, *Environ. Toxicol. Chem.* **14**, 2041–2047 (1995).

29. J. A. Oleskiewicz and M. Elektorowicz, *J. Soil Contam.* **2**, 205–208 (1993).

30. W. M. Wu, J. Nye, R. F. Hickey, M. K. Jain and J. G. Zeikus, in R. E. Hinchee, A. Leeson and L. Semprini, eds., *Bioremediation of Chlorinated Solvents*, Battelle Press, Columbus, Ohio, 1995, 45–52.

31. L. M. F. Dossantos and A. G. Livingston, *Appl. Microbiol. Biotechnol.* **42**, 421–431 (1994).

32. M. Knippschild and H. J. Rehm, *Appl. Microbiol. Biotechnol.* **44**, 253–258 (1995).

33. J. Gerritse, V. Renard, J. Visser, and J. C. Gottschal, *Appl. Microbiol. Biotechnol.* **43**, 920–928 (1995).

34. G. W. Gribble, *J. Chem. Ed.* **71**, 907–911 (1994).

35. S. Fetzner and F. Lingens, *Microbiol. Rev.* **58**, 641–673 (1994).

36. P. Juteau, R. Beaudet, G. Mcsween, F. Lepine, and J. G. Bisaillon, *Can. J. Microbiol.* **41**, 862–868 (1995).

37. C. Kennes, W. M. Wu, L. Bhatnagar and J. G. Zeikus, *Appl. Microbiol. Biotechnol.* **44**, 801–806 (1996).

38. K. A. McAllister, H. Lee, and J. T. Trevors, *Biodegradation* **7**, 1–40 (1996).

39. B. C. Okeke, A. Paterson, J. E. Smith, and I. A. Watson-Craik, *Letts. Appl. Microbiol.* **19**, 284–287 (1994).

40. J. Aislabie and G. Lloyd-Jones, *Aust. J. Soil. Res.* **33**, 925–942 (1995).

41. D. A. Abramowicz, *Crit. Rev. Biotechnol.* **10**, 241–251 (1990).

42. S. Han, L. D. Eltis, K. N. Timmis, S. W. Muchmore, and J. T. Bolin, *Science* **270**, 976–980 (1995).

43. B. N. Aronstein and L. E. Rice, *J. Chem. Technol. Biotechnol.* **63**, 321–328 (1995).

44. J. L. Gentry and T. J. Simpkin, in R. E. Hinchee, R. M. Miller, and P. C. Johnson, eds., *In situ Aeration: Air Sparging, Bioventing and Related Remediation Processes*, Battelle Press, Columbus, Ohio, 283–289 (1995).

45. A. M. Ferro, R. C. Sims, and B. Bugbee, *J. Environ. Qual.* **23**, 272–279 (1994).

46. S. G. Huling, D. F. Pope, J. E. Matthews, J. L. Sims, R. C. Sims, and D. L. Sorenson, in R. E. Hinchee, R. E. Hoeppel and D. B. Anderson, eds., *Bioremediation of Recalcitrant Organics*, Battelle Press, Columbus, Ohio, 1995, 101–109.

47. G. M. Colores, P. M. Radehaus, and S. K. Schmidt, *Appl. Biochem. Biotechnol.* **54**, 271–275 (1995).

48. M. P. Otte, J. Gagnon, Y. Comeau, N. Maatte, C. W. Greer, and R. Samson, *Appl. Microbiol. Biotechnol.* **40**, 926–932 (1994).

49. A. J. Wall and G. W. Stratton, *Water Air Soil Poll.* **82**, 723–737 (1995).

50. F. Baud-Grasset and T. M. Vogel, in R. E. Hinchee, J. E. Fredrickson and B. C. Alleman, eds., *Bioaugmentation for Site Remediation*, Battelle Press, Columbus, Ohio, 1995, 39–48.

51. F. C. Michel, C. A. Reddy, and L. J. Forney, *Appl. Environ. Microbiol.* **61**, 2566–2571 (1995).

52. C. J. Gantzer and D. Cosgriff, in Ref. 44, 543–549.

53. R. U. Edgehill, *Water Res.* **30**, 357–363 (1996).

54. D. Dietrich, W. J. Hickey, and R. Lamar, *Appl. Environ. Microbiol.* **61**, 3904–3909 (1995).

55. D. L. Bedard and R. J. May, *Environ. Sci. Technol.* **30**, 237–245 (1996).

56. D. L. Bedard, S. C. Bunnell and L. A. Smullen, *Environ. Sci. Technol.* **30**, 687–694 (1996).

57. M. R. Harkness, J. B. McDermott, D. A. Abramowicz, J. J. Salvo, W. P. Flanagan, M. L. Stephens, F. J. Mondello, R. J. May, H. J. Lobos, K. M. Carroll, M. J. Brennan, A. A. Bracco, K. M. Fish, G. L. Warner, P. R. Wilson, D. K. Dietrich, D. T. Lin, C. B. Morgan, C. B. and W. L. Gately, *Science* **259**, 503–507 (1993).

58. R. E. Dick and J. P. Quinn, *Appl. Microbiol. Biotechnol.* **43**, 545–550 (1995).

59. R. H. Kaake, D. L. Crawford, and R. L. Crawford, *Biodegradation* **6**, 329–337 (1995).

60. T. B. Hammill and R. L. Crawford, *Appl. Environ. Microbiol.* **62**, 1842–1846 (1996).

61. J. Kazumi and D. G. Capone, *Appl. Environ. Microbiol.* **62**, 2820–2829 (1995).

62. I. Mirgain, G. Green, and H. Monteil, *Environ. Technol.* **16**, 967–976 (1995).

63. C. J. Hapeman, J. S. Karns, and D. R. Shelton, *J. Agric. Food Chem.* **43**, 1383–1391 (1995).

64. D. R. Shelton, S. Khader, J. S. Karns, and B. M. Pogell, *Biodegradation* **7**, 129–136 (1996).

65. J. A. Entry, P. K. Donnelly, and W. H. Emmingham, *Appl. Soil Ecol.* **3**, 85–90 (1996).

66. D. L. Kaplan, *Curr. Opinion Biotechnol.* **3**, 253–260 (1992).

67. T. Gorontzy, O. Dryzga, M. W. Kahl, D. Bruns-Nagel, J. Breitung, E. von Loew and K.-H. Blotevogel, *Crit. Rev. Microbiol.* **20**, 265–284 (1994).

68. N. S. Kasimov, V. B. Grebenyuk, T. V. Koroleva, and Y. V. Proskuryakov, *Eurasian Soil Science* **28**, 79–95 (1996).

69. U.S. Pat. 5,387,271 (1995), D. L. Crawford, T. O. Stevens and R. L. Crawford.

70. S. Eversmeyer, P. Faessler, and C. Bird, *Soil Groundwater Cleanup* (Dec. 27–29, 1995).

71. E. Flathman, B. J. Krupp, P. Zottola, J. R. Trausch, J. H. Carson, R. Yao, G. J. Laird, P. M. Woodhull, D. E. Jerger, and P. R. Lear, *Remediation* **6**, 57–79 (1996).

72. Z. Ronen, J. M. Bollag, C. H. Hsu, and J. C. Young, *Ground Water* **34**, 194–199 (1996).

73. M. H. Essa, S. Farooq, and G. F. Nakhla, *Water Air Soil Poll.* **87**, 267–281 (1996).

74. V. Cuomo, A. Perretti, I. Palomba, A. Verde, and A. Cuomo, *J. Appl. Phycol.* **7**, 479–485 (1995).

75. G. C. Schmidt and M. B. Ballew, in R. E. Hinchee, J. L. Means, and D. R. Burris, eds., *Bioremediation of Inorganics*, Battelle Press, Columbus, Ohio, 1995, 109–116.

76. W. J. Hunter and R. F. Folett, in *Clean Water—Clean Environment—21st Century*, Vol. II: *Nutrients.* American Society of Agricultural Engineers, St Joseph, Mich., 1995, 79–82.

77. D. K. Ewart and M. N. Hughes, *Adv. Inorg. Chem.* **36**, 103–135 (1991).

78. B. Volesky and Z. R. Holan, *Biotechnol. Prog.* **11**, 235–250 (1995).

79. C. E. Turick, W. A. Apel, and N. S. Carmiol, *Appl. Microbiol. Biotechnol.* **44**, 683–688 (1996).

80. W. T. Frankenberger and U. Karlson, *Geomicrobiol. J.* **12**, 265–278 (1994).

81. D. Ahmann, A. L. Roberts, L. R. Krumholz, and F. M. M. Morel, *Nature* **371**, 750 (1994).

82. K. A. Rittle, J. I. Drever, and P. J. S. Colberg, *Geomicrobiol. J.* **13**, 1–11 (1995).

83. L. E. Macaskie, R. M. Empson, F. Lin, and M. R. Tolley, *J. Chem. Technol. Biotechnol.* **63**, 1–16 (1995).

84. K. M. Bonthrone, G. Basnakova, F. Lin, and L. E. Macaskie, *Nature Biotechnology* **14**, 635–638 (1996).

85. E. J. P. Phillips, E. R. Landa, and D. R. Lovley, *J. Ind. Microbiol.* **14**, 203–207 (1995).

86. E. Saouter, M. Gillman, and T. Barkay, *J. Ind. Microbiol.* **14**, 343–348 (1995).

87. V. Dushenkov, P. B. A. N. Kumar, H. Motto, and I. Raskin, *Environ. Sci. Technol.* **29**, 1239–1245 (1995).

88. D. E. Salt, M. Blaylock, N. P. B. A. Kumar, V. Dushenkov, B. D. Ensley, I. Chet, and I. Raskin, *Bio-Technology* **13**, 468–474 (1995).

89. W. E. Rauser, *Ann. Rev. Biochem.* **59**, 61–86 (1990).

90. D. E. Salt, R. C. Prince, I. J. Pickering, and I. Raskin, *Plant Physiol.* **109**, 1427–1433 (1995).

91. U. Kramer, J. D. Cotterhowells, J. M. Charnock, A. J. M. Baker, and J. A. C. Smith, *Nature* **379**, 635–638 (1996).

92. G. S. Banuelos and D. W. Meek, *J. Environ. Qual.* **19**, 772–777 (1990).

93. A. M. Zayed and N. Terry, *J. Plant. Physiol.* **143**, 8–14 (1994).

94. C. L. Rugh, H. D. Wilde, N. M. Stack, D. M. Thomson, A. O. Summers, and R. B. Meagher, *Proc. Natl. Acad. Sci. USA* **93**, 3182–3187 (1996).

95. L. J. Barnes, P. J. M. Scheeren, and C. J. N. Buisman, in J. L. Means and R. E. Hinchee, eds., *Emerging Technology for Bioremediation of Metals*, Lewis Publishers, Boca Raton, Fla., 1994, 38–49.

96. J. M. Macy, S. Lawson, and H. DeMott-Decker, *Appl. Microbiol. Biotechnol.* **40**, 588–594 (1993).

97. L. P. Owens, K. C. Kovac, J. A. L. Kipps, and D. W. J. Hayes, in Ref. 75, 89–94.

98. J. L. Kipps, in J. L. Means and R. E. Hinchee, eds., *Emerging Technology for Bioremediation of Metals*, Lewis Publishers, Boca Raton, Fla., 1994, 105–109.

99. C. M. Cooney, *Environ. Sci. Technol.* **30**, A194 (1996).

100. G. Robb and J. Robinson, *Mining Environ. Manage.*, 19–21 (Sept. 1995).

101. M. E. Dodds-Smith, C. A. Payne, and J. J. Gusek, *Mining Environ. Manage.*, 22–24 (Sept. 1995).

102. J. Bender and P. Phillips, *Mining Environ. Manage.*, 25–27 (Sept. 1995).

103. A. J. M. Baker, S. P. McGrath, C. M. D. Sidoli, and R. D. Reeves, *Resource Conserv. Recycling* **11**, 41–49 (1994).

104. L. J. DeFilippi, *Environ. Sci. Pollut. Control Ser.* **8**, 437–457 (1994).

105. P. H. Pritchard, *Curr. Opinion Biotechnol.* **3**, 232–243 (1992).

106. A. D. Venosa, J. R. Haines, W. Nisamaneepong, R. Govind, S. Pradhan, and B. Siddique, *J. Ind. Microbiol.* **10**, 13–23 (1992).

107. National Research Council, *In situ Bioremediation. When Does it Work?* National Academy Press, Washington, D.C., 1993.

108. R. C. Prince, D. L. Elmendorf, J. R. Lute, C. S. Hsu, C. E. Haith, J. D. Senius, G. J. Dechert, G. S. Douglas, and E. L. Butler, *Environ. Sci. Technol.* **28**, 142–145 (1994).

109. G. S. Douglas, K. J. McCarthy, D. T. Dahlen, J. A. Seavey, W. G. Steinhauer, R. C. Prince, and D. L. Elmendorf, *J. Soil Contam.* **1**, 197–216 (1992).

GENERAL REFERENCES

M. Alexander, *Biodegradation and Bioremediation*, Academic Press, San Diego, Calif., 1994.

R. E. Hinchee, J. T. Wilson and D. C. Downey, eds., *Intrinsic Bioremediation*, Battelle Press, Columbus, Ohio, 1995.

R. E. Hinchee, R. M. Miller, and P. C. Johnson, eds., *In situ Aeration: Air Sparging, Bioventing and Related Remediation Processes*, Battelle Press, Columbus, Ohio, 1995.

R. E. Hinchee, J. E. Fredrickson, and B. C. Alleman, eds., *Bioaugmentation for Site Remediation*, Battelle Press, Columbus, Ohio, 1995.

R. E. Hinchee, A. Leeson, and L. Semprini, eds., *Bioremediation of Chlorinated Solvents*, Battelle Press, Columbus, Ohio, 1995.

R. E. Hinchee, G. S. Douglas, and S. K. Ong, eds., *Monitoring and Verification of Bioremediation*, Battelle Press, Columbus, Ohio, 1995.

R. E. Hinchee, J. A. Kittel, and H. J. Reisinger, eds., *Applied Bioremediation of Hydrocarbons*, Battelle Press, Columbus, Ohio, 1995.

R. E. Hinchee, R. E. Hoeppel, and D. B. Anderson, eds., *Bioremediation of Recalcitrant Organics*, Battelle Press, Columbus, Ohio, 1995.

R. E. Hinchee, C. M. Vogel, and F. J. Brockman, eds., *Microbial Processes for Bioremediation*, Battelle Press, Columbus, Ohio, 1995.

R. E. Hinchee, G. D. Sayles, and R. S. Keen, eds., *Biological Unit Processes for Hazardous Waste Treatment*, Battelle Press, Columbus, Ohio, 1995.

R. E. Hinchee, J. L. Means, and D. R. Burris, eds., *Bioremediation of Inorganics*, Battelle Press, Columbus, Ohio, 1995.

National Research Council, *In situ Bioremediation. When Does it Work?* National Academy Press, Washington, D.C., 1993.

Organization for Economic Cooperation and Development, *Bioremediation: the Tokyo '94 Workshop*, OECD, Paris, 1995.

L. Y. Young and C. E. Cerniglia, eds., *Microbial Transformation and Degradation of Toxic Organic Chemicals*, Wiley-Liss, New York, 1995.

R. E. Hinchee, A. Leeson, L. Semprini, and S. K. Ong, eds., *Bioremediation of Chlorinated and Polycyclic Aromatic Hydrocarbons*, Lewis Publishers, Ann Arbor, Mich., 1994.

R. E. Hinchee, B. C. Alleman, R. E. Hoeppel, and R. N. Miller, eds., *Hydrocarbon Bioremediation*, Lewis Publishers, Ann Arbor, Mich., 1994.

R. E. Hinchee, D. B. Anderson, F. B. Metting, Jr., and G. D. Sayles, eds., *Applied Biotechnology for Site Remediation*, Lewis Publishers, Ann Arbor, Mich., 1994.

J. L. Means and R. E. Hinchee, eds., *Emerging Technology for Bioremediation of Metals*, Lewis Publishers, Ann Arbor, Mich., 1994.

R. E. Hinchee, ed., *Air Sparging for Site Remediation*, Lewis Publishers, Ann Arbor, Mich., 1994.

R. F. Hickey and G. Smith, eds., *Biotechnology in Industrial Waste Treatment and Bioremediation*, Lewis Publishers, Boca Raton, Fla., 1996.

R. H. Kadlec and R. L. Knight, *Treatment Wetlands*, Lewis Publishers, Boca Raton, Fla., 1996.

ROGER C. PRINCE
Exxon Research and Engineering Company

BIOSENSORS

1. Introduction

A chemical sensor is a device that transforms chemical information, ranging from the concentration of a specific sample component to total composition analysis, into an analytically useful signal. Chemical sensors usually contain two basic components connected in series: a chemical (molecular) recognition system (receptor) and a physicochemical transducer (1). Biosensors are chemical sensors in which the recognition system utilizes a biochemical mechanism (2,3). Historically, the term biosensor has been variously used to a number of devices either used to monitor living system or incorporating biotic elements. However, the current consensus is that this term should be reserved for sensors incorporating a biological element such as an enzyme, antibody, nucleic acid, microorganism, or cell where the biological element is in intimate contact with the transducer. A modern definition of a biosensor is as follows: *"A compact analytical device incorporating a biological or biologically derived sensing element either integrated within or intimately associated with a physicochemical transducer. The usual aim of a biosensor is to produce either discrete or continuous digital electronic signals that are proportional to a single analyte or a related group of analytes."* (4).

Figure 1 presents a generalized scheme of a biosensor.

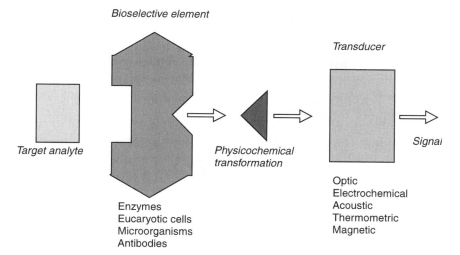

Bioselective element

Transducer

Target analyte

Physicochemical transformation

Signal

Enzymes
Eucaryotic cells
Microorganisms
Antibodies

Optic
Electrochemical
Acoustic
Thermometric
Magnetic

Fig. 1. Scheme of a generic biosensor.

The physicochemical transformation occurring in a biosensor due to the interaction between the biological element and the analyte target is converted into an usable signal by the transducer. The main purpose of the recognition system is to provide specificity to the biosensor, thus creating a device able to detect either a specific molecular target or a related family of compounds. Some biosensors are very selective for a single analyte (eg, glucose using glucose oxidase) and others are class specific, since they use eg, enzymes with broad substrate specificity (ie, phenolic biosensors based on tyrosinase activity) or intact microorganisms (eg, in BOD sensors).

In a biosensor, the biological component and the transducer are in intimate contact; this distinguishes this class of device from an analytical system that incorporates additional processing steps like separation processes or incubating chambers (ie, as in flow injection analysis). Thus, a biosensor should be a reagentless analytical device, although the presence of ambient cosubstrates (eg, oxygen for oxidoreductase) is often required for the detection of the analyte (1). Immobilization of the biological component can be performed using a variety of methods such as chemical or physical adsorption, physical entrapment within a membrane or gel, cross-linking of molecules, or covalent binding. The biological components used in biosensor construction can be divided into two broad categories: catalytic biosensors, where the primary sensing event results from a chemical transformation (catalyzed by an enzyme either isolated or retained in a microorganism or tissue); or affinity biosensors, in which the sensing event is dependent on an essentially irreversible binding of the target molecules (eg, affinity sensors based on antigen–antibody or nucleic acid interactions) (5). Recently, artificial receptors have emerged that may offer viable alternative recognition elements for biosensors, and this has become a fast-growing area for research (6). The transducer, or transducing microsystem, serves to transfer the signal from the recognition system to the electronic device, and it can be electrochemical, optical, thermometric, piezoelectric, or magnetic (7).

Catalytic and affinity-based biosensors have multifarious applications; eg, they can be used to measure blood glucose levels, detect pollutants and pesticides in the environment, monitor food-borne pathogens in the food supply, or to detect biological warfare agents. The future promises high density arrays of biomolecular sensors that rival microelectronics in size and capacity, deliver the recognition and discrimination of complex analytical equipment, and furnish scientists with biologically relevant information (8).

Biosensors may be categorized in different ways; it is possible to group biosensors according to the mode of signal transduction (ie, electrochemical, optical, or acoustic), or according to the biomolecule used (ie, biocatalytic or affinity based). In this article, we classified biosensors primarily on the basis of the biological element used.

2. Catalytic Biosensors

Catalytic biosensors are based on a reaction involving macromolecules that are present in their original biological environment, have been isolated previously, or have been manufactured (1). Biocatalytic elements normally used are enzymes (mono- or multisystems), whole cells from microorganisms such as bacteria or fungi, or tissues directly excised from the original organism. Considering a substrate S and the enzymatic product P, the reaction can be described (eg, 1):

$$S + S' \xrightarrow{\text{cat}} P + P' \tag{1}$$

S′ represents a second substrate involved in the biocatalytic reaction, which can be a reagent added directly during the reaction or something that is present in the environment (ie, oxygen dissolved in the solution). The reaction can be monitored by the transducer by measuring the consumption of S′, or the formation of the product P. Another product P′ can be formed from the reaction, and this can also be quantified. The biological components most commonly used to build biocatalytic biosensors are enzymes. There are a large number of enzymes commercially available, and increasing interest in enzymatic biosensors and assays has stimulated the number of enzymes that can readily be purchased. Advantages of the use of enzymes as biological elements are the possibility to operate in many environments such as aqueous solutions (9), organic phases (10,11) and air (12).

In the simplest type of catalytic biosensor, an enzyme is immobilized onto an electrode surface and the electrode is immersed in the test solution (Fig. 2). If the target analyte (ie, the enzymatic substrate) is present in the solution, a reaction occurs that can be monitored by, eg, amperometry. The signal is related to the substrate concentration.

Such enzyme-modified electrodes can be constructed using enzymes that catalyze oxidation–reduction reactions (where there is a passage of electrons between the enzyme and the electrode), or changes of pH or other ions in the solution tested. Once a substrate molecule has been transformed to product, the active site of the enzyme becomes available to react again with a new molecule.

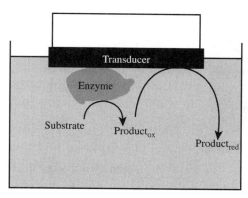

Fig. 2.   Scheme of a biocatalytic biosensor based on enzyme catalysis. The example reported is regarding an electrochemical reduction.

The development of enzyme-based sensing devices, particularly those based on electrochemistry, continues to attract considerable attention. Of the electrochemical transducing systems available (amperometric, potentiometric, conductometric, and impedimetric), amperometric biosensors have dominated both research and commercial activity to date, largely because of their relative simplicity and flexibility (13–15). In general experimental amperometric systems employ a three-electrode design comprising a working electrode (to which a potential is applied in order to drive the electrochemical reaction), a reference electrode, and an auxiliary electrode. When a substance that can be oxidized or reduced at the potential applied is present in the solution, a current is generated. Electroactive species that can be detected using this type of approach include biochemicals such as reduced nicotinamide adenine denucleotide (NADH), inorganic species like molecular oxygen, or products of enzymatic reactions like hydrogen peroxide (produced by oxidases), benzoquinone (from phenol oxidation), or mediators (artificial electron acceptors such as ferrocyanide or ferrocene derivatives) involved in enzymatic reactions. The redox status of the biocatalyst's active center, cofactor, or prosthetic group in the presence of substrate S can also be determined by using an immobilized mediator that reacts sufficiently rapidly with the biological element and for this reason can be easily detected at the transducer.

The most common example of an amperometric biosensor is the mediated amperometric glucose biosensor (16); this device involves the use of the enzyme glucose oxidase (GOD) as the biological recognition element, which catalyzes the following reactions (Fig. 3):

Initially, glucose reacts with the oxidized form of the enzyme to produce gluconolactone, which is then spontaneously hydrolized to gluconic acid. During this transition, two electrons and two protons are produced, with the concomitant reduction of the enzyme GOD. At this stage, a mediator such as a ferrocene derivative or quinone can shuttle electrons from the reduced enzyme to the electrode surface. There, the reduced form of the mediator is reoxidized to yield an amperometric signal.

Biocatalytic elements are also used in the preparation of potentiometric biosensors. These are based on the measurement of the difference in potential

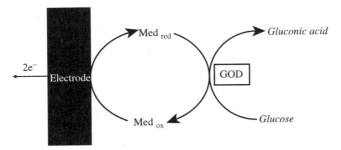

Fig. 3.   Glucose oxidase-mediated electrochemistry.

between the working electrode and a reference electrode at near zero current flow. The potentiometric approach has largely been used to develop biosensors employing ion-selective electrodes modified with membranes. The most common potentiometric devices are pH electrodes, but several other ion (F^-, I^-, CN^-, Na^+, NH_4^+) or gas (CO_2, NH_3) specific electrodes can be used (1). In principle, the potential differences between the working and the reference electrodes varies logarithmically with the ion activity or gas concentration as described by the Nernst equation. This ideal condition is true when the selectivity of the membrane used is complete, or when the concentration of any interferent ions is very low and when the potential due to the contact between the various phases (solution, electrode surface, or membrane surface) is negligible or constant. Many examples of potentiometric catalytic biosensors can be found in the literature (17−19). One of the first applications of potentiometry in biosensors was for the detection of urea (20). In this work, a layer of a jack bean meal was used as a biocatalytic element. This native meal contains the enzyme urease, which catalyzes the following reaction (eg. 2):

$$Urea + H_2O \longrightarrow 2\,NH_3 + CO_2 \tag{2}$$

The CO_2 produced can be detected using a gas-sensing electrode. Similarly, a biosensor for pyruvate detection was constructed (21). In this case, plant tissue (corn kernels) was used, which contains pyruvate decarboxilase catalyzing the following reaction (eg. 3):

$$Pyruvate + H_2O \longrightarrow Acetaldehyde + CO_2 \tag{3}$$

As in equation 3, the CO_2 produced is correlated with the enzymatic reaction to generate a signal that is correlated to the analyte concentration.

The simplest enzyme electrodes employ a single enzyme, but this affords only a limited scope of potential analytes. More frequently now, two or more enzyme reactions are coupled. A recent example is the bienzymatic biosensor developed for oxalate detection (22). This biosensor involves two enzymes, oxalate oxidase (OXO) and horseradish peroxidase (HRP), incorporated into a carbon paste electrode modified with silica gel and coated with titanium oxide containing toluidine blue. The OXO was immobilized on silica gel modified with a titanium oxide surface using glutaraldehyde cross-linking. The HRP

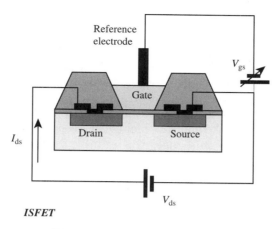

Fig. 4. Diagram of an ISFET.

was immobilized using covalent binding with carbodiimide on graphite powder. The biosensor showed a good performance with a linear response range between 0.1 and 2.0 mM of oxalate. It could be used for 80 determinations when stored in a succinate buffer at pH 3.8 in a refrigerator between measurements. The response time was 0.5 s.

An important transducer in the development of enzymatic biosensors has been the ion-selective field effect transistor (ISFET). A FET is a semiconductor device used for monitoring charge at the surface of an electrode, which has accumulated or been applied to a metal gate between the source and drain electrodes. The integration of an ISE with FET is realized in the ISFET (23) (Fig. 4). In the ISFET, the surface potential varies with the analyte concentration.

ISFETs can be combined with a biocatalytic element to create a biosensor. If the biological element is an enzyme, this is called ENFET and the term IMFET or immunoFET is used when antibodies are used. The invention of the ISFET in the early 1970s represented a major step forward in the design of chemical sensors (24). The pH-dependent potential at the interface between an electrolyte and a dielectric layer (eg, Si_3N_4 or Al_2O_3) is used to control the FET channel conductivity (25) of the ISFET. By immobilizing an enzyme to an ISFET gate, the ENFET (enzyme FET) was created (26). The enzyme catalyzes the conversion of a specific substrate (eg, urea, glucose, or penicillin) into a product (eg, a weak acid) that can be detected by the pH sensitive FET. The literature reports a pH-ISFET urea biosensor based on an enzyme-modified pH-sensitive Si_3N_4 gate ISFET with chemical immobilization of urease on the silicon nitride surface performed with the use of glutaraldehyde. The device obtained gave good results in terms of sensitivity and reproducibility and a linear range covering the physiological as well as pathophysiological ranges (27). Urea ENFETs have now been incorporated into commercial devices for bedside monitoring in critical care situations. ISFET technology has also been applied to develop immunosensors (IMMFET), but application of this technology has been more limited.

3. Affinity Biosensors: Immunosensors

Affinity biosensors are analytical devices comprising a biological affinity element such as an antibody, receptor protein, biomimetic material, or DNA, interfaced to a signal transducer, to convert the concentration of an analyte to a measurable electronic signal (28).

Because of their high affinity, versatility, and commercial availability, antibodies are the most widely reported biological recognition elements used in affinity-type biosensors; in this case, the affinity biosensor is known as an "immunosensor". Antibodies are polymers containing hundreds of individual amino acids arranged in a highly ordered sequence. These polypeptides are produced by immune system cells (B lymphocytes) when exposed to antigenic substances or molecules. Antibodies contain in their structure recognition/binding sites for specific molecular structures of the antigen. According to the "key-lock" model, an antibody interacts in a highly specific way with this unique antigen. The interaction is reversible, as determined by the law of mass action, and it is based on electrostatic forces, hydrogen bonding, and hydrophobic and van der Waals interactions.

Antibody-based biosensors (immunosensors), where the antibody or antigen is immobilized to the transducer, have been constructed in a variety of ways, but generally fall into one of three basic configurations. These formats involve direct noncompetitive assays, competitive (direct or indirect) assays, or sandwich-type assay formats (Fig. 5). In the case of the direct noncompetitive assay format, a unique optical or electrochemical property of the analyte of interest is observed as the target compound binds to the recognition site of the bioaffinity element and accumulates on the sensor surface.

Recent developments in electrochemical transducers have also allowed the noncompetitive detection of small molecular weight molecules. For these sensors, the formation of an antibody–antigen complex at the surface of a conductive thin film induces direct or indirect ionic movements resulting in a current flow through the conductive support. Although these types of format show a great deal of promise, one of their primary limitations involves nonspecific binding of nontarget compounds. More specifically, cocontaminants (any small molecules) that nonspecifically bind to the surface of the sensor cannot be distinguished from the analyte of interest.

The simplest biochemical immunosensor consists of an antibody or antigen immobilized to a transducer and results in a signal generated from the binding of an analyte to an antibody at the sensor surface. If the analyte (itself) can be detected by the signal transducer, this forms the basis of a noncompetitive assay. For example, the fluorescent analyte benzo(a)pyrene can be directly detected when binding to an antibody immobilized to a fiberoptic signal transducer (29). For these assay formats, the signal is directly proportional to the analyte concentration.

Because many sensors cannot detect the binding of an analyte, competitive formats are often used in conjunction with optical or electrochemical tracers. The primary advantages in the use of tracers in a direct immunoassay format are the high sensitivity and low potential for interference from matrix components afforded by these (usually fluorescent) molecules. The primary disadvantage in

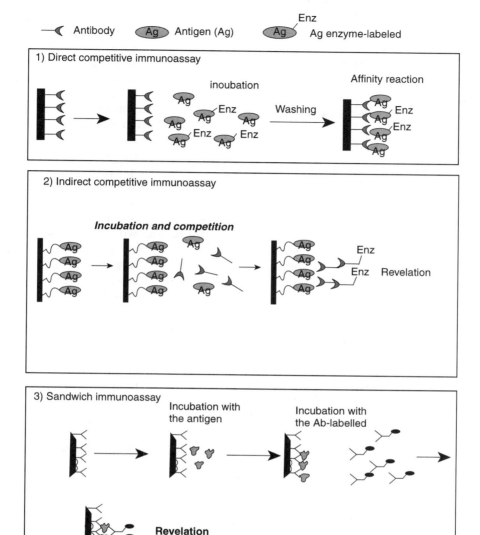

Fig. 5.   Scheme of different enzyme immunoassay formats.

the use of tracers is the need to add a reagent other than the analyte of interest. More specifically, because the signal is inversely proportional to the concentration of analyte, it is often the case that the lowest analyte concentrations are measured as a small change in the maximum signal, which increases the relative signal noise.

Electrochemical immunosensors are highly diffuse. Most electrochemical immunosensors are based on enzyme-linked immunosorbent assay (ELISA) principles, where antibodies or antigens are labeled with, eg, peroxidase, alkaline phosphatase, or GOD, and measured amperometrically by following the respective product iodine, p-aminophenol, or hydrogen peroxide. Glucose oxidase also has been used as an enzyme label in immunosensors development; a GOD labeled immunosensor for human serum albumin has been described using amperometric detection of hydrogen peroxide (30). In this system, nonspecific adsorption of enzyme-labeled antigen to the membrane was only 4.4% of specific binding, and the range of measurement, 0.5–200 mg L^{-1} of human serum albumin, was satisfactory for the diagnosis and for clinical applications.

Whereas the preferred method of measurement for catalytic biosensors is electrochemical, affinity biosensors have generally proved more amenable to optical detection methods (10). The commercialization of real-time bioaffinity monitors based on surface plasmon resonance (SPR) has provided a powerful new tool to the research community and to the pharmaceutical industry in particular. A surface plasmon is an evanescent electromagnetic field generated at the surface of a metal conductor (usually Ag or Au) excited from a light of an appropriate wavelength that impacts on the surface at a particular angle (θ_p). A sharp minimum in light reflectance can be observed when the angle of incidence is varied (31). The value of this "critical angle" is strictly dependent on the properties of the medium that is in contact with the metal layer, but is principally determined by their dielectric constant. When the metal surface interacts in an affinity reaction, the dielectric constant is altered and this can be quantified by measuring the change in the critical angle (Fig. 6).

This principle can be very useful in biosensor technology, because it allows the direct detection of a biological interaction. Surface plasmon resonance has been used to study the binding of immunoglobulin G (IgG) to gold and anti-IgG to immobilized IgG layers (32). In these studies, both a monoclonal mouse and polyclonal sheep IgG were used as receptor layers for anti-IgG. The kinetics of binding were investigated by monitoring the reflectivity of light at an angle close to plasmon resonance. Both the initial rate of change and final reflectivity

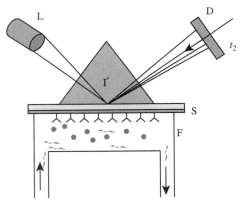

Fig. 6.   Surface plasmon resonance detection unit. L: light source, D: photodiod array, P: prism, S: sensor surface, F: flow cell. The two dark lines in the reflected beam projected on to the detector symbolize the light intensity drop following the resonance phenomenon at time $= t_1$ and t_2. The line projected at t_1 corresponds to the situation before binding of antigens to the antibodies on the surface and t_2 is the position of resonance after binding.

were measured during and after protein binding. The amount of protein bound to the surface was found to be less for the monoclonal mouse IgG compared to the polyclonal sheep IgG, these two IgG nominally being of the same dimensions and molecular weight. Further, anti-IgG binding produced greater changes in reflectivity than the initial IgG layers. SPR was used also to investigate the effect of pesticides such as imazetaphyr, triazines, and parathion. Advances in the research and development of SPR detection of immunointeractions resulted in the introduction of commercially available instruments such as the BIAcore (Uppsala, Sweden). This instrument consists of a sensor chip, a SPR detector unit, and a liquid handling system that has two precision pumps and an integrated microfluidics cartridge (IFC). The autosampler and the liquid handling system together control delivery of sample plugs into a buffer stream that passes continuously across the sensor chip surface. The entire system is computer controlled, including data collection and analysis, resulting in a fully automated analytical system. The sensor chip (signal transducer) is a glass slide with a thin layer (50 nm) of gold deposited on one side. The gold film is in turn covered with a covalently bound matrix on which biomolecules can be immobilized. This commercial apparatus, available for coating with the desired antibody or antigen, was used, eg, to develop an SPR-based immunosensor for atrazine detection (33) together with other environmental applications.

Optical immunosensors based on fluorescence have also been realized. Sepaniak and co-workers (34) described the measurement of benzo(*a*)pyrene, which is itself fluorescencent; the benzopyrene is trapped by an antibody solution placed on the tip of an optical fiber and protected from the external environment by a membrane. Optical immunosensors have been developed for many environmental applications including pesticide analysis for analytes such as parathion (35) and triazine (36). Optical evanescent-wave technology has been used to streamline the design of affinity biosensors that contain a label or marker. For example, many immunoassays use a fluorescent marker to indicate when antibody binds antigen. An impressive recent example is provided by immunosensors designed to detect microbial warfare agents (37, 38). The integration of photochromic dyes into the antibodies of immunoassays has facilitated the production of high affinity sensors that can be optically switched to low affinity so that the devices can be regenerated and used again (39). A recent advance in affinity biosensor immunoassays is the introduction of paramagnetic particles attached to antibodies as the label. Binding of antibody to antigen attached to a solid substrate, can be detected with an electronic device that measures the magnetic field induced by the paramagnetic beads. The detector can be fabricated into a small hand-held device, and the approach offers the added bonus of providing a permanent record because the contents of the assay stick can be remeasured at any time, like a piece of magnetic recording tape.

Piezoelectric biosensors, based on the piezoelectric quartz crystal microbalance (QCM), have been reported for a range of applications in biochemistry and affinity biosensors construction (40). This type of sensor comprises an oscillator circuit and a thin slice of AT-cut (~35° rotation angle) piezoelectric quartz crystal. Metal film (gold or platinum) electrodes are deposited onto both sides of the quartz and crystal. The sensor operates by observation of the propagation of an acoustic wave through the solid-state device. Sensing is usually achieved by

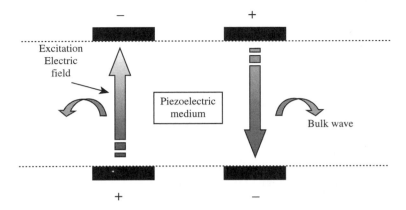

Fig. 7. Diagram of a piezoelectric system.

correlating the acoustic wave-propagation variations to the amount of analyte captured at the surface, and then to the amount or concentration of analyte present in the sample exposed to the sensor. Alternatively, changes in the physical properties of interfacial thin films in response to analyte may be measured (41). A diagram illustrating the principle of the measurement is shown in Figure 7.

The adsorption of a substance increases the rigid mass bound to the quartz crystal surface and, under dry conditions, causes the crystal to change its oscillation frequency according to the Sauerbrey equation (eg. 4) (42):

$$\Delta F = -2.26 \times 10^{-6} F^2 \left(\Delta m / A \right) \tag{4}$$

where ΔF (Hz) is the change in oscillation frequency of the coated quartz crystal, Δm (g), the change in the mass adsorbed onto the crystal, and A (cm^2) is the area of the coated quartz crystal, F (Hz), the resonance frequency of the quartz crystal. This approach can be used to precisely quantify analyte with a detection sensitivity in the nanogram range.

Piezoelectric crystals (Fig. 8) have been used as microbalances and microviscometers owing to their small size, high sensitivity, and stability, simplicity of construction and operation, light weight, and low power requirement. They have been applied in the determination of thin layer thickness, the study of general chemical species adsorption, and the analysis of the microrheology of liquid crystals and electrochemically polymerized thin films. A major application of

Fig. 8. Gold-modified piezoelectric crystals.

piezoelectric crystals has been in biosensor construction, especially immunosensors. By using these piezoelectric biosensor devices, it is possible to measure the immuno-interaction directly; such devices are often known as "microgravimetric immunosensors" (31). Recent developments in microgravimetric immunosensors and their application have been described in a number of reviews (43–45). A flow-through immunoassay system based on piezoelectric detection for label-free determination of antibodies against human immunodeficiency virus (HIV) has been described by Aberl and co-workers (46). This system involves a high level of automation and allows determination of target analytes in the concentration ranges found in HIV diagnostics. The piezoelectric detection method has been used in the determination of small molecules in environmental and clinical analysis. In a typical case, a competitive assay mechanism was used, employing haptens conjugated to a protein. A piezoelectric immunosensor for methamphetamine was realized by immobilizing the target molecule on the surface of the piezoelectric crystal. The assay procedure involved a competitive scheme, where both free antibodies and antigen were contained in the reaction medium (47). Antibodies binding with methamphetamine-modified immunosensor surface were in competition to bind with the target analyte.

4. Affinity Biosensors: DNA Biosensors

DNA biosensors represent a very important class of affinity biosensor. In this developing aspect of the field, DNA strands are used to detect the binding of oligonucleotides (gene probes). Such devices are known as "DNA biosensors" or more generally as "DNA Chips". The DNA biosensors present an enormous potential for early clinical diagnosis of genetically inherited diseases, on site detection of food contamination, forensic studies, and environmental monitoring. DNA biosensors involve the use of nucleic acids as biological recognition elements to explore novel hybridization probes, transduction strategies, and potential practical applications. Transduction strategies that have been reported include electrochemical, acoustic, and optical techniques. For each of these strategies, the biosensor format typically relies on immobilization of single stranded (ss) DNA to the sensor surface followed by hybridization of a complimentary sequence. Detection of the formation of double stranded (ds) DNA has been facilitated through the use of a variety of electrochemical and optical tracers that primarily bind or intercalate into ds DNA.

Electrochemical techniques have been used to differentiate between ss (pre-hybridized) and ds (hybridized) DNA using several approaches. A potentially major application of a DNA biosensor could be for the testing of water, food, soil, and plant samples for the presence of pathogenic microorganisms and for the presence of analytes (carcinogens, drugs, mutagenic pollutants, etc.) with binding affinities for the DNA molecule (Fig. 9). The detection of small molecule binding to DNA and general DNA damage resulting form ionizing radiation, dimethyl sulfate, etc has been described by following variation in the electrochemical signal derived from guanine. These approaches include the use of redox active intercalators that accumulate into ds DNA, transition metal complexes that mediate oxidation of the guanine base, or the direct oxidation of guanine.

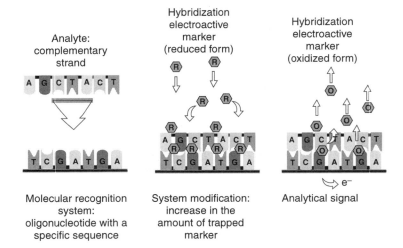

Fig. 9. Schematic of DNA biosensor for hybridization detection.

Marrazza and co-workers (48) reported the construction of disposable electrochemical biosensors for environmental applications based on intercalation and hybridization principles. For the former, a graphite screen-printed electrode was modified by the immobilization (as a probe) of specific synthetic oligonucleotides with a defined sequence for *Chlamydia trachomatis*, a bacterial species. The oligonucleotide immobilized on the surface was hybridized with different concentrations of the complementary sequences and the amount of hybrid was evaluated by chronopotentiometric stripping analysis (PSA) using daunomycin as an indicator of the hybridization reaction. The increase in the area of the daunomycin peak was used to detect the presence and the amount of the complementary sequence. This sensor could be used to detect the presence of bacteria following PCR (polymerase chain reaction) amplification of the sample. The latter DNA biosensor was realized by immobilizing calf thymus DNA at a fixed potential onto an electrode surface. The calf thymus DNA sensor was then immersed in a sample solution containing the analyte. After 2 min of interaction, a PSA was carried out to evaluate the oxidation of guanine residues on the electrode surface. In this case, it was possible to evaluate the electrochemical effects resulting from the presence of genotoxic compounds (49).

The mass change associated with DNA hybridization may also be detected by employing piezoelectric devices. By immobilization of a ss DNA onto quartz crystals, it is possible to detect the mass change after hybridization. Nucleic acid strands were covalently attached to the modified surface of piezoelectric crystal. When these immobilized probe strands were melted and incubated with complementary target strands in solution, association of probe and target to form duplexes resulted in an increase in mass that was detectable as a decrease of several hundred hertz in the crystal's resonance frequency relative to control crystals on which noncomplementary strands were attached.

A further example of piezoelectric DNA-based biosensors has been developed to detect bacteria toxicity in environmental samples (50). In this system,

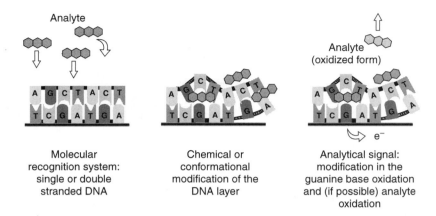

Fig. 10. Schematic of DNA biosensor to detect molecular intercalators.

a biotinylated oligonucleotide probe was immobilized on a streptavidin coated gold disk. Streptavidin was covalently linked to the thiol/carboxylated dextran modified gold surface. The immobilized probe was then reacted with a solution of the target oligonucleotide. The interactions of the immobilized DNA strand with a complementary and a noncomplementary sequence in solution were studied. The hybridization reaction was also performed on real samples of DNA extracted from different *Aeromonas* strains isolated from water, vegetables, or human specimens and amplified by PCR. These experiments showed that it was possible to distinguish between strains that contain the aerolysin gene and those that do not, hence furnishing an assay for the toxicity of the bacterium. The PCR has been widely used in conjunction with various sensor systems to

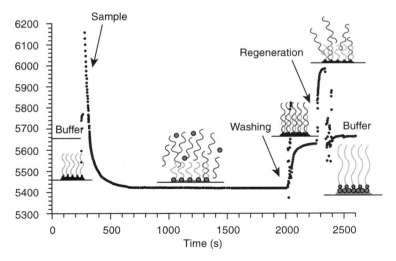

Fig. 11. Frequency profile for piezoelectric crystal modified in order to detect specific DNA sequences.

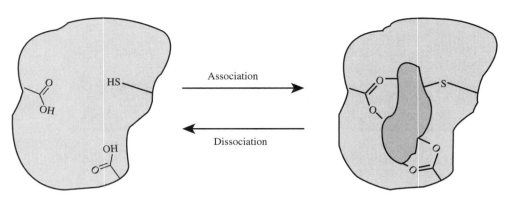

Fig. 12. Schematic representation of the association and dissociation of a molecule with the specific binding cavity generated in a molecularly imprinted polymer.

increase sensitivity by amplifying DNA in a sample, eg, for clinical applications, to detect apolipoprotein E polymorphisms (51).

5. Synthetic Receptors and Biomimetic Sensors

In biosensor technology, the recognition element is conventionally isolated from living systems, measuring interactions such as antigen–antibody or utilizing the substrate specificity obtained with enzymes. However, for a number of potential applications, no biocomponent is available or requirements such as stability cannot be met using biological molecule. For this reason, synthetic bioreceptors have attracted considerable attention as a potential new avenue for biosensor development. Semisynthetic receptors have been created by modifying enzymes and antibodies at both the pre- and posttranslational level. For example, antibody fragments have been produced and libraries of variants generated by random or point mutations at a genetic level. The design of semisynthetic receptors in biosensors may be superseded by totally synthetic ligands produced with the aid of computational chemistry, combinatorial chemistry, molecular imprinting, self-assembly, rational design, or combinations of one or more of these. For example, molecularly imprinted polymers (MIPs) have attracted a lot of attention because they can behave as artificial receptors for molecular recognition. This methodology is based on the prearrangement of print molecules and functional monomers prior to initiation of polymerization. In this way, a rigid cross-linked, macroporous polymer is formed that contains sites complementary to the print molecule both in shape and in the arrangement of functional groups (52). Removal of the print molecule by extraction leaves sites specific for print molecules and free for binding. Many imprinted polymers specific for different compounds have been prepared in several laboratories over the last few years, including specific polymers that have been used as antibody mimics. The striking resemblance of the binding properties (affinity and specificity) of MIPs to those of natural receptors, together with their inherent stability, low cost, and ease of preparation for industrial application have made them an attractive alternative to antibody-based technologies. Conventional imprinted polymers are usually synthesized empirically or semiempirically using a limited number of common monomers in a trial-and-error approach. Three particular features have made MIPs the target of intense investigation: (*1*) their high affinity and selectivity, which are similar to those of natural receptors; (*2*) their unique stability that is superior to that demonstrated by natural biomolecules; and (*3*) the simplicity of their preparation and the ease of adaptation to different practical applications 3–5. A great variety of chemical compounds have been imprinted successfully. The resulting polymers are robust, inexpensive and, in many cases, possess an affinity and specificity that is suitable for industrial application.

Molecularly imprinted polymers are proposed for a number of applications including use as the stationary phase in high performance liquid chromatography (HPLC) and, in thin-layer chromatography (TLC) to separate chiral products and as a replacement for antibodies in biosensor construction. One example is a morphine-sensitive sensor (53). The method involved two steps. In the first step, morphine was bound selectively to the molecularly imprinted polymer in the sensor, a platinum wire electrode. In the second step, an electroinactive competitor (codeine) was added in excess, displacing some of the bound morphine. The released morphine was detected using an amperometric method. The system was able to measure morphine in the concentration range 0.1–10 μg/mL (morphine concentration of 0.5 μg/mL gave a peak current by oxidation of 4 nA) and was stable to autoclaving, demonstrated long-term stability, and was resistant to harsh chemical environments.

Molecularly imprinted polymer (MIP) technology has also been applied for environmental applications. Molecularly imprinted polymer membranes, containing artificial recognition sites for atrazine, have been prepared by photopolymerization using atrazine as a template, methacrylic acid as a functional monomer, and tri(ethylene glycol) dimethacrylate as a cross-linker (54). To obtain thin, flexible, and mechanically stable membranes, oligourethane acrylate was added to the monomer mixture. Reference membranes were prepared with the same monomer mixture but in the absence of the template. Imprinted membranes were tested as a recognition element for an atrazine-sensitive conductometric sensor. The influence of the polymer composition and type of solvent used as a porogen on the magnitude of the sensor response was investigated. The sensor developed demonstrated high selectivity and sensitivity with a detection limit of 5 nM for atrazine. The membranes synthesized exhibited the same recognition characteristics over a period of 6 months.

Transducing the binding event in molecularly imprinted polymers into a detectable signal has proved quite a challenge. However, creative solutions have been achieved with both optical and electrochemical configurations. In one example, a "bite-and-switch" approach has been used to produce sensors that detect creatine and creatinine in blood. In this two-step recognition process, a broadly specific chemical reaction is complemented by a three-dimensional recognition pocket to produce a strong "bite", which is followed by a "switch" to the fluorescent form of the indicator. A thioacetal reaction between the polymer and the amine groups in creatine and creatinine—results in the formation of a fluorescent isoindole complex; this reaction was made more specific for creatine and creatinine by molecular imprinting (55). In a further example of a screen-printed design, an electrochemical sensor was developed that detected the herbicide 2,4-D by the displacement of homogentisic acid from a MIP (56).

6. Mass Production of Biosensors: Thick-Film Technology

Over the past two decades, there has been increasing interest in the application of simple, rapid, inexpensive, and disposable biosensors in fields such as clinical, environmental, or industrial analysis. The most common disposable biosensors are those produced by thick-film technology. A thick-film biosensor configuration is normally considered to be one that comprises layers of special inks or pastes deposited sequentially onto an insulating support or substrate. One of the key factors that distinguishes a thick-film technique is the method of film deposition, namely, screen printing, which is possibly one of the oldest forms of graphic art reproduction. The fabrication process of a screen-printed device requires the application of procedures and materials appropriate to the particular devices. To create a thick-film electrode, a conductive or dielectric film is applied to a substrate (57). The film is applied through a mask contacting the substrate and deposited films are obtained by pattern transfer from the mask. Conventionally, thick-film electrodes were baked at temperature ranging from 300 to 1200°C to drive off solvents from the applied paste and to cure the pattern paste. In commercial biosensors, this step is usually avoided since it would damage incorporated enzymes. Alternatives include cold-cured ink formulations and a photocured process using ultraviolet(uv) light.

More complex structures can be built up by repeating the print process using the materials appropriate to the specific design and a range of mask designs. A variety of screen-printed thick-film devices can thus be fabricated (Fig. 13) as a base for disposable electrochemical immunosensors, DNA sensors, and enzyme electrodes. These planar devices present many advantages including disposability, which is a very important characteristic when working with real samples, and small dimensions, which facilitates the design of portable measuring systems. The working electrode of the system can be fabricated using

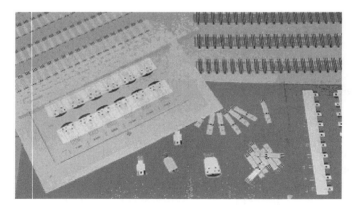

Fig. 13.   Various screen-printed electrodes.

different kind of inks such as graphite, gold, or platinum, and then modified with the appropriate bioselective element.

Screen-printed electrodes have been used extensively in recent years to produce a wide range of sensors and biosensors. They have proved very versatile for work on real samples in environmental and clinical analysis. The most well-kown application of screen-printed electrode technology has been in the clinical analysis of blood glucose levels in people with diabetes. Historically, glucose sensing has dominated the biosensor literature and has delivered huge commercial successes to the field. The deceptively simple combination of a fungal enzyme (GOD) with an electrochemical detector has effectively met the needs of the 1–2% of the world's population that have diabetes. Complications associated with insulin-dependent diabetes such as blindness, kidney and heart failure, and gangrene (resulting in amputation) can be reduced by up to 60% through stringent personal control of blood glucose, including frequent monitoring of glucose levels. People with non-insulin-dependent (Type II) diabetes can also benefit from strict monitoring of glucose levels. Enzyme-based electrode biosensors have been used to test glucose levels in blood samples since 1975, but the emergence of a convenient, hand-held commercial format revolutionized their use (58). The commercial realization of the mediated glucose sensor came with the foundation of Genetics International (later to become MediSense) in Boston, with the launch of the pen-sized Exatech glucose sensors in 1987 (Fig. 14).

The system was invented, designed, and developed at Cranfield University in collaboration with Oxford University in the U.K. and consists of a small, disposable, single-use, glucose-sensitive electrode (based on a screen-printed mixture of GOD and mediator in a conductive carbon-paste binder) and a pen-sized (later designs adopted a credit card or computer mouse shape) meter containing the electronics and LCD display. Patients prick their finger and deposit the blood onto the sensor, and within 20–30 it is possible to quantify the glucose present in blood. The response of an amperometric biosensor is either a steady-state or a transient response, but it is never an equilibrium response, because the

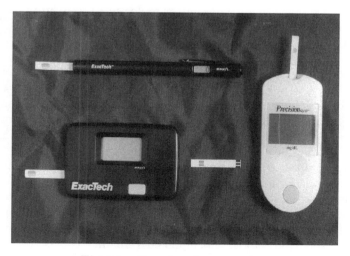

Fig. 14. Pen-size glucose meter.

substrate has to diffuse into the sensor from the bulk solution, pass through the enzymatic membrane and react with the biocatalytic element.

Screen-printed electrodes have also been fabricated to detect environmental pollutants. A disposable amperometric sensor to detect hydrazines has been described (59) in which the working surface of the electrode was modified using cobalt phtalocyanine (CoPC) and a mixed-valent ruthenium cyanide. These mediators are able to catalyze the oxidation of various hydrazines, and then can be detected by amperometric measurement. The configuration of the sensor as a disposable strip facilitates on-site environmental and industrial monitoring.

Screen-printed electrodes can also be designed to incorporate bioaffinity molecules like antibodies, in order to obtain disposable immunosensors for clinical and environmental applications. The electrode constitutes both the solid phase for the immunoassay and the electrochemical transducer. A recent example is a disposable immunosensor (based on enzymatic labeling of one of the reagents) to detect polychlorinated biphenyls (PCBs) (60). In this case, the surface of the electrode was modified using different strategies to produce two different formats of immunochemical tests based on indirect and competitive assays. The graphite surface of the screen-printed electrodes was modified by immobilization of antibodies or antigens in order to obtain a selective surface for the realization of the immunological reaction. In the indirect competitive immunochemical test, a bovine serum albumin (BSA) conjugate (4,4'-dichlorobiphenyls-BSA) was used for the PCB immobilization procedure. The IgG anti-PCB reacts with the sample for a fixed time (30 min). Then a small amount (10 mL) was added to the electrode surface with immobilized PCB–BSA for the competition reaction. The IgG anti-PCB captured on the electrode surface was evaluated using a secondary, labeled HPR antibody. In the direct format, a fixed amount of anti-PCB were immobilized onto the solid phase. In this case, the competition was realized between the antigen in the sample and a limiting amount of the HRP-labeled congener added to it. In both cases, the extent of the affinity reaction was evaluated by chronoamperometric measurement of the products of the enzymatic activity of the label. Then, after molecular recognition, the electrodes were used as electrochemical cells.

7. Applications of Biosensors

The vast majority of commercial activity to date, in the field of biosensors, has been focused on medical applications and a substantial market now exists for such devices especially for home blood glucose measurement. Medically related opportunities for biosensors are expected to increase especially in the areas of pharmaceuticals and drug development and to serve the new genomics and proteomics industries. The second most reported application area for biosensors is for environmental analysis. This activity has mainly been supported by public money to underpin current and impending legislation to protect the environment. In the text above, these two application areas have dominated the examples cited. Less reported, but of enormous current interest, is the application of biosensors for defence (detection of biological and chemical warfare agents) and security (detection of drugs, explosives, identity, etc). Biosensors are also expected to impact on the food and process industries. A steady but limited business has developed for food testing, eg, for sugars, amino acids, and organic acids. Expansion in the future is likely to be in the food safety and labeling area eg, to identify pathogens, toxins, allergens, and genetically manipulated crops.

Enzyme-based biosensors have generally been used to measure batch samples, but a number of attempts have been made to monitor processes on-line (61). In most cases, the analysis must be performed outside the reactor or process line (*ex situ*). Consequently, aseptic on-line sample withdrawal from the bioreactor is necessary, because the biological component of biosensors is not easily sterilizable. Other disadvantages of biosensors for process monitoring are that they are invasive, temperature dependent, subject to fouling, and need to be recalibrated and regenerated frequently. Signal drift causes unreliability in these sensors, and the enzyme stability is also limited. Most biosensors are only able to measure one substrate at a time but, with a combination of different biosensors, various substances (glucose, ammonium, etc) can be measured simultaneously. Applications of on-line biosensors have included process optimization in pilot scale and control of animal cell culture where the product is very valuable. Bioluminescent sensors have also been developed for on-line applications and consist of a bioluminescent enzyme and an optical transducer. Optical biosensors provide a rapid and highly selective detection system. The use of optical biosensors is expanding (62; 63). Improving the reliability of *in situ* biosensors for industrial applications and the integration of biosensors into flow-injection analysis (FIA) is an important objective. Recent developments have been directed toward overcoming some of the common disadvantages of biosensors. An on-line resonant mirror optical biosensor was used to detect DNA–DNA hybridization in real time (64). Biotinylated oligonucleotide probes were immobilized on the sensor surface and hybridization of a complementary target oligonucleotide was monitored. The limit of detection was lower than other optical biosensors using surface-plasmon resonance, fiber-optic fluorescence, or light-activated potentiometric devices. An evanescent field biosensor was used to detect protein A produced by *Staphylococcus aureus* in blood-culture bottles in < 30 min (65). By using laser light at 488 nm, a plastic optical fiber and anti-protein-A antibodies conjugated with fluorescein isothiocyanate in a sandwich immunoassay adsorbed onto the fiber, a detection limit of 1 ng/mL was achieved.

Effective on-line sensing is arguably the most difficult technical hurdle facing biosensor technology. Progress in the related area of real-time *in vivo* has advanced significantly in recent years. Two new commercial devices have been launched for monitoring glucose *in vivo*. One offers a miniaturized, sterilizable, and implantable enzyme electrode produced using microfabrication technology, while the other is a transcutaneous device, which obtains a sample using reverse ionophoresis. It is clear that designs can be developed to deliver the required stability, accuracy, and robustness for this patient-centerd type of device. Future advances, however, are expected to be supported by developments in the biomimetic and imprinting areas, since these offer the hope of far greater stability and resistance to sterilization while maintaining the exquisite specificity and sensitivity associated with traditional biological elements.

BIBLIOGRAPHY

"Biosensors," in *ECT* 4th ed., Vol. 4, pp. 208–221, by William Pietro, York University; "Biosensors" in *ECT* (online), posting date: December 4, 2000, by William Pietro, York University.

CITED PUBLICATIONS

1. D. R. Thévenot, K. Toth, R. A. Durst, and G. S. Wilson, *Biosens Bioelectron.* **16**, 121 (2001).
2. K. Cammann, *Fresenius J. Anal. Chem.* **287**, 1 (1977).
3. A. P. F. Turner, I. Karube, G. S. Wilson, eds., *Biosensors: fundamentals and applications*. Oxford University Press, Oxford, U.K., 1987.
4. J. D. Newmann, P. J. Tigwell, P. J. Warner, and A. P. F. Turner, *Sensor Rev.* **21**, 268 (2001).
5. M. J. Dennison and A. P. F. Turner, *Biotech. Adv.* **13**, 1 (1995).
6. J. D. Newmann and A. P. F. Turner, in K.F. Tipton, eds., Portland Press, London vol. 27, 1993, pp. 147–159.
7. A. P. F. Turner, *McGraw Hill Yearbook of Science & Technology*, McGraw Hill, New York, 1999, pp. 39–42.
8. A. P. F. Turner, *Science* **290**, 1315 (2000).
9. N. Ivanov, G. A. Evtugyn, R. E. Gyurcsányi, K. Tóth, and H. C. Budnikov, *Anal. Chim. Acta*, **40**, 55 (2000).
10. S. Saini, G. F. Hall, M. E. A. Downs, and A. P. F. Turner, *Anal. Chim. Acta* **249**, 1 (1991).
11. J. Wang, N. Naser, and D. Lopez, *Biosens. Bioelectron* **9**, 225 (1994).
12. M. J. Tierney, H. L. Kim, *Anal. Chem.*, **65**, 730 (1993).
13. M. J. Tierney, J. A. Tamada, R. O. Potts, L. Jovanovic, S. Garg, and Cygnus Research Team, *Bios. Bioelectr.* **16**, 621 (2001).
14. S. Cosnier, M. Stoytcheva, A. Senillou, H. Perrot, R. P. Furriel, F. A. Leone, *Anal. Chem.* **71**, 3692 (1999).
15. J. P. Pemberton, J. P. Hart, and J. A. Foulkes, *Electron Acta* **43**, 3567 (1998).
16. A. E. G. Cass, and co-workers *Anal. Chem.* **56**, 667 (1984).
17. T. Yao, and G. A. Rechnitz, *Anal. Chem.* **59**, 2115 (1987).
18. J. de Gracia, M. Poch, D. Martorell, and S. Alegret, *Biosens. Bioelectron.* **11**, 53 (1996).
19. M. Trojanowicz, and M. L. Hitchman, *Biosens. Bioelectron.* **11**, xviii (1996).
20. M. A. Arnold, and S. A. Glazier, *Biotech. Lett.* **6**, 313 (1984).
21. S. Kuriyama, and G. A. Rechnitz, *Anal. Chim. Acta*, **131**, 91 (1981).
22. E. F. Perez, G. de Oliveira Neto, and L. T. Kubota, *Sensors Actuators B: Chem.*, **72**, 80 (2001).
23. A. K. Covington, *Pure Appl. Chem.* **66**, 565 (1994).
24. P. Bergveld, *IEEE Trans. Biomed. Eng.* **17**, 70 (1970).
25. R. E. G. Van Hal, J. C. T. Eijkel, and P. Bergveld, *Sensors Actuators B* **24**, 201 (1995).
26. S. Caras, and J. Janata, *Anal. Chem.* **52**, 1935 (1980).
27. D. G. Pijanowska, and W. Torbicz, *Sensors Actuators B* **44**, 370 (1997).
28. K. R. Rogers, *Mol. Biotechnol.* **14**, 109 (2000).
29. T. Vo-Dinh, B. J. Tromberg, G. D. Griffin, K. R. Ambrose, M. J. Sepaniak, and E. M. Gardenhire, *Appl. Spectrosc.* **41**, 735 (1987).
30. S. Kaku, S. Nakanishi, K. Horiguchi, and M. Sato, *Anal. Chim. Acta* **272**, 213 (1993).
31. A. L. Ghindilis, P. Atanasov, M. Wilkins, and E. Wilkins, *Biosens. Bioelectron.* **13**, 113 (1998).

32. N. J. Geddes, A. S. Martin, F. Caruso, R. S. Urquhart, D. N. Furlong, J. R. Sambles, K. A. Than, and J. A. Edgar, *J. Immunol. Methods* **175**, 149 (1994).
33. M. Minunni, and M. Mascini, *Anal. Lett* **26**, 1441 (1993).
34. M. J. Sepaniak, B. J. Tromberg, J.-P. Alarie, J. R. Bowyer, A. M. Hoyt, and T. Vo-Dinh, *ACS Symp. Ser.* **403**, 318 (1989).
35. F. F. Bier, R. Jockers, and R. D. Schmid, *Analyst* **119**, 437 (1994).
36. P. Orozlan, C. Thommen, M. Wehrli, and M. Gert Ehrat, *Anal. Methods Instrum.* **1**, 43 (1993).
37. S. S. Iqbal, M. W. Mayo, J. G. Bruno, B. V. Bronk, C. A. Batt, J. B. Chambers, *Biosens. Bioelectron.*, **15**, 549 (2000).
38. C. A. Rowe-Taitt, J. W. Hazzard, K. E. Hoffman, J. J. Cras, J. P. Golden, and F. S. Ligler, *Biosens. Bioelectron.* **15**, 579 (2000).
39. D. G. Weston, and D. C. Cullen, in *Biosensors 2000—The 6th World Congress on Biosensors*, 24–26 May 2000, San Diego, Calif., Elsevier, Oxford, U.K., 2000), p. 92.
40. J. Hu, L. Liu, B. Danielsson, X. Zhou, and L. Wang, *Anal. Chim. Acta*, **423**, 215 (2000).
41. Y. H. Chang, T. C. Chang, E.-F. Kao, and C. Chou, *Biosci. Biotechnol. Biochem.* **60**, 1571 (1996).
42. G. Z. Sauerbrey, *Phys.* **155**, 206 (1959).
43. M. Liu, Q. X. Li, and G. A. Rechnitz, *Anal. Chim. Acta* **387**, 29 (1999).
44. J. Horácek and P. Skládal, *Anal. Chim. Acta* **347**, 43 (1997).
45. P. B. Luppa, L. J. Sokoll, and D. W. Chan, *Clin. Chim. Acta* 1 (2001).
46. F. Aberl, C. Kosslinger, S. Drost, P. Woias, and S. Koch, *Sensors Actuators B* **18**, 271 (1994).
47. N. Miura, H. Higobashi, G. Sakai, A. Takeyasu, T. Uda, and N. Yamazioe, *Sensors Actuators B* **13**, 188 (1993).
48. G. Marrazza, I. Chianella, and M. Mascini, *Anal. Chim. Acta* **389**, 297 (1999).
49. F. Lucarelli, I. Palchetti, G. Marrazza, and M. Mascini, *Talanta*, **56**, 949–957 (2002).
50. S. Tombelli, M. Mascini, L. Braccini, M. Anichini, and A. P. F. Turner, *Biosens. Bioelectron.* **15**, 363 (2000).
51. S. Tombelli, M. Mascini, and C. Sacco, A. P. F. Turner, *Anal. Chim. Acta* **418**, 1 (2000).
52. D. Kriz, and K. Mosbach, *Anal. Chim. Acta*, **300**, 71 (1995).
53. D. Kriz, O. Ramström, A. Svennson, and K. Mosbach, *Anal. Chem.* **67**, 2142 (1995).
54. T. A. Sergeyeva, S. A. Piletsky, A. A. Brovko, E. A. Slinchenko, L. M. Sergeeva, and A. V. El'skaya, *Anal. Chim. Acta*, **392**, 105 (1999).
55. S. Subrahmanyam, S. A. Piletsky, E. V. Piletska, B. Chen, K. Karim, and A. P. F. Turner, *Biosens. Bioelectron.* **16**, 631 (2001).
56. S. Kröger, A. P. F. Turner, K. Mosbach, and K. Haupt, *Anal., Chem.* **71**, 3698 (1999).
57. C. A. Galán-Vidal, J. Muñoz, C. Domínguez, and S. Alegret, *TrAC Trends Anal. Chem.* **14**, 225 (1995).
58. M. F. Cardosi, and A. P. F. Turner, in K. G. M. M. Alberti, and L. P. Krall, eds., *Diabetes Annual*, Elsevier, Amsterdam, the Netherlands, 1990), Vol. 5, pp. 254–272.
59. J. Wang, and P. V. A. Pamidi, *Talanta* **42**, 463 (1995).
60. S. Laschi and M. Fránek, *Electroanalysis* **12**, 1293 (2000).
61. A. P. F. Turner and S. F. White, *Biocatal. Biosep.* 2057 (1999).
62. K. Schügerl, B. Hitzmann, H. Jurgens, T. Kullick, R. Ulber, and B. Weigal, *Trends Biotechnol.* **14**, 21 (1996).
63. R. Freitag, *Appl. Biosens.* **4**, 75 (1993).
64. H. J. Watts, D. Yeung, and H. Parkes, *Anal. Chem.* **67**, 4283 (1995).
65. S.-M. Chang, H. Muramatsu, C. Nakamura, and J. Miyake, *Mater. Sci. C* **12**, 111 (2000).

A. P. F. TURNER
Cranfield University at Silsoe

BIOSEPARATIONS

1. Introduction

The large-scale purification of proteins and other bioproducts is the final production step, prior to product packaging, in the manufacture of therapeutic proteins, specialty enzymes, diagnostic products, and value-added products from agriculture. These separation steps, taken to purify biological molecules or compounds obtained from biological sources, are referred to as bioseparations. Large-scale bioseparations combine art and science. Bioseparations often evolve from laboratory-scale techniques, adapted and scaled up to satisfy the need for larger amounts of extremely pure test quantities of the product. Uncompromising standards for product quality, driven by commercial competition, applications, and regulatory oversight, provide many challenges to the scale-up of protein purification. The rigorous quality control embodied in current good manufacturing practices, and the complexity and lability of the macromolecules being processed provide other practical issues to address (1).

Recovery and purification of new biotechnology products is the fastest growing area of bioseparations. Biotechnology was broadly defined in 1991 by the U.S. Office of Technology Assessment as "any technique that uses living organisms (or parts of organisms) to make or modify products, to improve plants or animals, or to develop microorganisms for specific uses." The new biotechnology, introduced in 1970, involves directed manipulation of the cell's genetic machinery through recombinant deoxyribonucleic acid (DNA) techniques and cell fusion. The new biotechnology was first applied on an industrial scale in 1979. Since then it has fundamentally expanded the utility of biological systems, so that biological molecules for which there is no other means of industrial production can be generated. Substantial manufacturing capability is expected to be needed to bring about the full application of this biotechnology (2). The recovery, purification, and packaging of biotechnological products for delivery to the consumer is undergoing unprecedented growth.

Manufacturing approaches for selected bioproducts of the new biotechnology impact product recovery and purification. The most prevalent bioseparations method is chromatography (qv). Thus the practical tools used to initiate scaleup of process liquid chromatographic separations starting from a minimum amount of laboratory data are given.

2. Economic Aspects

The development of biotechnology processes in the biopharmaceutical and bioproduct industries is driven by the precept of being first to market while achieving a defined product purity, and developing a reliable process to meet validation requirements. The economics of bioseparations are important, but are likely to be secondary to the goal of being first to market. The cost of a lost opportunity in a tightly focused market where there is room for only a few manufacturers can be devastating for products which take 5 to 10 years and $100 to 200×10^6 to develop. After process and product are validated, the cost of change in any

Table 1. **Unit Values and Relative Production Quantities for Selected Approved Biopharmaceuticals, 1990–1991**[a]

Product	Year approved	Selling price, $/g[b]	Quantity for 200×10^6 in sales, kg
human insulin	1982	375	530.0
tissue plasminogen activator	1987	23,000	8.7
human growth hormone	1985	35,000	5.7
erythropoetin (Epogen)	1989	840,000	0.24
GM CSF	1991	384,000	0.52
G-CSF	1991	450,000	0.44

[a] Adapted from Ref. 2 with additional data from Ref. 4.
[b] Values are approximate and are likely to decrease.

portion of the procedure can also be great, if only to satisfy regulatory constraints. Hence, once the manufacturing process is in place, changes are likely to be considered only if significant improvements result.

The three main sources of competitive advantage in the manufacture of high value protein products are first to market, high product quality, and low cost (3). The first company to market a new protein biopharmaceutical, and the first to gain patent protection, enjoys a substantial advantage. The second company to enter the market may find itself enjoying only one-tenth of the sales. In the absence of patent protection, product differentiation becomes very important. Differentiation reflects a product that is purer, more active, or has a greater lot-to-lot consistency.

2.1. Biopharmaceuticals and Protein Products.

Purification of proteins is a critical and expensive part of the production process, often accounting for $\geq 50\%$ of total production costs (2). Hence, bioseparation processes have a significant impact on manufacturing costs. For small-volume, very high value biotherapeutics (Table 1), however, these costs may be considered secondary to the first to market principle unless a lower cost competitor surfaces. Annual 1995 sales were $700 million for human insulin (5), $300 million for tissue plasminogen activator, and $220 million for human growth hormone (6). The most successful bioproduct in biotechnology history, recombinant erythropoetin (EPO), had worldwide sales estimated at $1.6 to $2.6 billion in 1995 (5, 7, 8). Epogen is a genetically engineered version of erythropoetin [11096-26-7], which is produced by the kidneys and stimulates blood stem cells to mature into red blood cells. Epogen can reverse the severe anemia often caused by kidney disease. Amgen's sales of this product, together with Neupogen (a recombinant protein that directs blood stem cells to become bacteria-fighting neutrophils), was about $1.8 billion in 1995 (9).

3. Bioproduct Separations

The task of quickly specifying, designing, and scaling-up a bioproduct separation is not simple. These separations are carried out in a liquid phase using macromolecules which are labile, and where conformation and heterogeneous chemical structure undergoing even subtle change during purification may result in an unacceptable product. A typical purification scheme for biopharmaceutical proteins involves the harvesting of protein-containing material or cells, concentration of protein using ultrafiltration (qv), initial chromatographic steps, viral clearance steps, additional chromatographic steps, again concentration of protein using ultrafiltration, and finally formulation (10).

3.1. Biosynthetic Human Insulin from *E. coli.*

Insulin [9004-10-8], a polypeptide hormone, stimulates anabolic reactions for carbohydrates, proteins, and fats thereby producing a lowered blood glucose level. Porcine insulin [12584-58-6] and bovine insulin [11070-73-8] were used to treat diabetes prior to the availability of human insulin [11061-68-0]. All three insulins are similar in amino acid sequence. Eli Lilly's human insulin was approved for testing in humans in 1980 by the U.S. FDA and was placed on the market by 1982 (11, 12).

Human insulin was the first animal protein to be made in bacteria in a sequence identical to the human pancreatic peptide. Expression of separate insulin A and B chains were achieved in *Escherichia coli* K-12 using genes for the insulin A and B chains synthesized and cloned in frame with the β-galactosidase gene of plasmid pBR322 (13, 14). Insulin's small size, 21 amino acids for the A-chain, mol wt = 2300; and 30 for the B-chain, mol wt = 3400, together with the absence of methionine (Met) and tryptophan (Trp) residues, were critical elements both in the decision to undertake cloning of this peptide hormone and in the rapid development of the manufacturing process. The Met and Trp residues, produced as a consequence of engineering and expression in *E. coli*, are hydrolyzed by reagents used during the recovery process. The presence of these amino acids in insulin would have resulted in the hydrolysis and destruction of the product (12).

Recovery and Purification. The production of Eli Lilly's human insulin requires 31 principal processing steps of which 27 are associated with product recovery and purification (13). The production process for human insulin, based on a fermentation which yields proinsulin, provides an instructive case study on the range of unit operations which must be considered in the recovery and purification of a recombinant product from a bacterial fermentation. Whereas the exact sequence has not been published, the principle steps in the purification scheme are outlined in Figure 1a.

The fermentation product is a fusion protein where a portion of the Trp protein is connected to proinsulin through a Met residue (Fig. 1b). The *E. coli* contains a plasmid having the proinsulin gene connected to the Trp promoter. The Trp operon is turned on when the fermentation media is allowed to become depleted of tryptophan and the production of a fusion protein of proinsulin occurs. An inclusion body, ie, a large body of aggregated protein and nucleic acids occupying about half of the cell volume is formed. Because formation of inclusion bodies causes cell growth to stop, premature formation of inclusion bodies results in lower productivity. Hence, the Trp switch is an important practical tool in maximizing productivity. At this point the fermentation is complete, and protein recovery, dissolution, protein refolding, and purification is carried out (12). Following inclusion body recovery, CNBr, a hydrolytic agent which specifically attacks Met and Trp linkages, cleaves away the fusion protein from the proinsulin (see Fig. 1b). No Met or Trp occurs in proinsulin, so the proinsulin

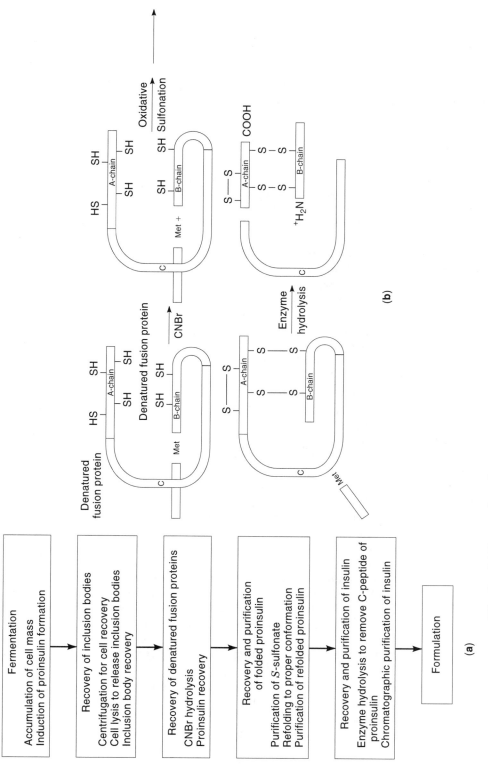

Fig. 1. (**a**) Process flow sheet for human insulin production, recovery, and purification (12); (**b**) corresponding steps in recovery of biosynthetic human insulin.

molecule is left intact. The proinsulin is then subjected to oxidative sulfitolysis, refolding to its proper conformation, purification, and enzyme treatment to remove the peptide connecting the insulin A- and B-chains. The crude insulin consisting of A- and B-chains in their proper conformation is then further purified using a sequence of ion-exchange (qv), reversed-phase, and size-exclusion chromatography (3,12,14).

Desamidation of asparagine or glutamine residues can occur readily in either acidic or neutral solutions; disulfide exchange reactions can occur at alkaline pH and cause formation of isomeric monomers or aggregated forms (multimers) of the protein (13). Deamidation products of insulin, referred to as desamido insulin, can form during processing. These insulin variants require high resolution chromatography techniques to remove. Therefore, a multimodal sequence of chromatographic separations for the crude recombinant insulin is required.

Ion-exchange chromatography removes most of the impurities and is followed by reversed-phase chromatography which separates insulin from structurally similar insulin-like components. Then size-exclusion (gel-permeation) chromatography is introduced to remove salts and other small molecules from the insulin. The best pH range for the acetonitrile mobile phase for reversed-phase chromatography is reported to be 3.0–4.0. This is well below the isoelectric pH of 5.4, gives excellent resolution, and minimizes deamidation of the insulin if the residue time in the reversed-phase column is less than several hours (12,14). This sequence (14) follows the principle of orthogonality of separation sequence, ie, each step is based on a different property, in this case charge, solubility, and size, respectively (1). Near the end of the chromatography sequence, the insulin may be concentrated by precipitation to form insulin zinc crystals.

Process Equipment Volumes.   Product recovery involves cell lysis, centrifugation, refolding, buffer exchange, chromatography, precipitation, and filtration. Some of these steps are repeated. The volumes of the individual chromatography columns are estimated to range from 50 to 1000 L. These volumes are small compared to other types of chemical recovery processes, but are large in the context of biotechnology manufacturing. For example, the reversed-phase step uses an 80 L column volume for an insulin loading of 1.2 kg per run (3). Assuming the total amount of recombinant insulin produced annually in a typical plant is on the order of 1000 to 2000 kg, this size column would enable processing as much as 30% (>300 kg) of the annual output of insulin.

Yield Losses.   The numerous steps incur a built-in yield loss. For example, if only 2% yield loss were to be associated with each step, the overall yield for a purification sequence of 10 steps would be as in equation 1:

$$\eta = 100(1 - L/100)^n = 100(1 - 0.02)^{10} = 81.7\% \tag{1}$$

where η denotes yield, L the percent yield loss, and n the number of steps. If the yield loss at each step were 5%, the overall yield would only be 60%. Maximizing recovery at each step is important.

The purification of human insulin involves five separate alterations in the molecular structure, and hence, changes in physicochemical properties during its recovery and purification. The various forms are fusion protein, denatured aggregate, denatured monomers, properly folded proinsulin, and finally insulin. Whereas various purification procedures are used repeatedly, thus introducing more steps in the process, the change of removing contaminants is maximized because the contaminants are not as likely to change chemically in the same way as the insulin molecule. The final purification steps rely on multiple properties of the insulin, such as size, hydrophobicity, ionic charge, and crystallizability (13). The final purity level is reported to be >99.99% (15).

3.2. Tissue Plasminogen Activator from Mammalian Cell Culture

Tissue-type plasminogen activator or tissue plasminogen activator [105857-23-6] (t-PA) was originally identified in tissue extracts in the late 1940s (15). Other known plasminogen activators include streptokinase from bacteria, urokinase from urine, and prourokinase from plasma (16). In 1981 the Bowes melanoma cell line was found to secrete t-PA (known as mt-PA) at 100× higher concentrations, making possible the isolation and purification of this enzyme in sufficient quantities that antibodies could be generated and assays developed to lead to cloning of the gene for this enzyme and subsequent expression of the enzyme in both *E. coli* and a Chinese hamster ovary (CHO) cell line (15,17).

Comparison of the melanoma and recombinant forms of the enzymes showed the same activity toward dissolution of blood clots. Comparison to urokinase, another thrombolytic agent, served as the basis for introducing recombinant t-PA into clinical trials in 1984 (17). Two pilot studies demonstrated that mt-PA resulted in thrombolysis without significant fibrinogenolysis. Fibrinogen, the precursor to fibrin, is important to the clotting of blood. Because mt-PA was available in limited quantities, recombinant t-PA (rt-PA) was used to carry out the first significant clinical trial. Doses of 0.375–0.75 mg rt-PA/kg body weight was found to be effective in humans for achieving 70% recannalization. Another pilot study confirmed that a dose of 80 mg over three hours gave the same results (17). The comparison of rt-PA (injected intravenously) to streptokinase IV (injected intracoronary) produced sufficiently favorable results to end the trial early, and make the results public in 1985, resulting in the use of t-PA for heart attacks (15). A trial completed in 1996 showed that t-PA administered within three hours of a stroke caused by a clot in the brain facilitated full recovery of 31% of stroke patients. Hence, another use of rt-PA is likely to develop (18).

Approximately 500,000 Americans suffer strokes each year. Many of the 80% that survive suffer paralysis and impaired vision and speech, often needing rehabilitation and/or long-term care. Hence, whereas treatment using rt-PA is likely to be expensive (costs are $2200/dose for treating heat attacks), the benefits of rt-PA could outweigh costs. In the case of heart attacks, the 10 times less expensive microbially derived streptokinase can be used. There is currently no competing pharmaceutical for treatment of strokes (18, 19). Consequently, the cost of manufacture of rt-PA may not be as dominant an issue as would be the case of other types of bioproducts.

Characteristics of t-PA. Tissue plasminogen activator, a proteolytic, hydrophobic enzyme, has a molecular weight of 66,000, 12 disulfide bonds, four possible glycosylation sites, and a bridge of 6 amino acids connecting the principal protein structures (17,20,21). Only three of the sites (Asn-117–118, -448) are actually glycosylated (16). When administered to heart attack victims it dissolves clots consisting of platelets in a fibrin protein matrix and acts by clipping plasminogen, an active precursor protein found in the blood, to form plasmin, a potent protease that degrades fibrin (17,21). Whereas plasminogen activator is found in blood and tissues, concentrations are low (17).

The concentration of t-PA in human blood is 2–5 ng/mL, ie, 2–5 ppb. Plasminogen activation is accelerated in the presence of a clot, but the rate is slow. The dissolution of a clot requires a week or more during normal repair of vascular damage (17). Prevention of irreversible tissue damage during a heart attack requires that a clot, formed by rupture of an atherosclerotic plaque, be dissolved in a matter of hours. This rapid thrombolysis (dissolution of the clot) must be achieved without significant fibrinogenolysis elsewhere in the patient.

rt-PA is derived from a biological source, transformed CHO cells, and by definition is a biologic, not a drug. It is generally not possible to define biologics as discrete chemical entities or demonstrate a unique composition. Other biologics include blood fractionation products such as albumin and Factor VIII, and both live and killed viral vaccines.

The process used to make a biologic is closely monitored and regulated by regulatory agencies, because a significant change in the process may result in a product which is different from that previously reviewed and regulated, and hence may require a new license. Process changes made during the investigational new drug (IND) development stage, and before the license is approved, are more easily incorporated into a new product (from a regulatory point of view) than after the license is generated (15).

t-PA Production.   Recombinant technology provides the only practical means of rt-PA production. The amount of t-PA required per dose is on the order of 100 mg. Cell lines of transformed CHO cells, selected for high levels of rt-PA expression using methotrexate, are grown in large fermenters (21). The purification steps for rt-PA must therefore separate out cells, virus, and DNA. The literature on the industrial practice of recovery and purification of rt-PA generated by suspension culture of chinese hamster ovary cells is limited (15). Recovering a protein derived from mammalian cells involves a number of steps (15). One possible scheme is shown in Figure 2. The culture medium is separated from the cell by sterile filtration (see MICROBIAL AND VIRAL FILTRATION). This is followed by additional removal by cross-flow filtration, ultrafiltration (qv), and chromatography to remove DNA and remaining viruses. The product protein then undergoes purification by chromatography.

The separation of cells from the culture media or fermentation broth is the first step in a bioproduct recovery sequence. Whereas centrifugation is common for recombinant bacterial cells (see CENTRIFUGAL SEPARATION), the final removal of CHO cells utilizes sterile-filtration techniques. Safety concerns with respect to contamination of the product with CHO cells were addressed by confirming the absence of cells in the product, and their relative noninfectivity with respect to immune competent rodents injected with a large number of CHO cells.

The possibility that DNA from recombinant immortal cell lines such as CHO cells could cause oncogenic (gene altering) events resulting in cancer (22) was a concern during development of the rt-PA purification sequence. Data suggest that DNA, by itself, is inactive *in vivo*; removal of the DNA, however, is still a concern. The goal for rt-PA purification is to reduce the DNA to

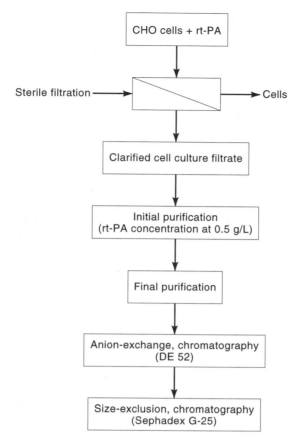

Fig. 2. Outline of possible steps in the recovery and purification sequence for recombinant tissue plasminogen activator derived from recombinant CHO cells.

<10 pg/dose (10^{-11} g/dose). A level of 0.1 pg has been achieved, representing a ~9 log reduction in the DNA, and requiring special assays to detect and quantify these very low levels of DNA in the final product (15).

Sensitive, specific, and if possible, rapid assays for product and potential contaminants are an essential part of separation methods selected for the purification sequence. Ultrafiltration (qv) followed by ion-exchange (qv) chromatography (qv) and then a final round of ultrafiltration concentrate the dilute protein while purifying it (23). Precipitation prior to chromatography could remove unwanted proteins before the sample is injected into the first liquid chromatography column. At the initial purification stage the rt-PA concentration is 0.5 g/L; the DNA is at 0.11 ng/mL. The use of anion-exchange chromatography (DE step) appears to be particularly effective in removing the DNA. Studies using another product, IgM, derived from cell culture showed DNA clearance may be enhanced by predigestion (hydrolysis) of the DNA using nucleases prior to the anionic-exchange chromatography step (24).

Independent Assays for Proving Virus Removal. Retroviruses and viruses can also be present in culture fluids of mammalian cell lines (15,24). Certainly the absence of virus can be difficult to prove. Model viruses, eg, NIH Rauscher leukemia virus and NZB Xenotropic virus, were spiked into fluids being purified, and their removal subsequently validated when subjected to the same purification sequence as used for the product.

Viral clearance can be achieved by use of chaotropes such as urea or guanidine, pH extremes, detergents, heat, formaldehyde, proteases or DNA'ses organic solvents such as formaldehyde, or ion-exchange or size-exclusion chromatography. The protein product must be stable at the conditions used to deactivate or remove the virus. Because only the inactivation or removal which can be measured counts as validation, a sequence of orthoganol removal/fractionation steps must be used (1,15,24). For a fluid spiked with 10^6 virus particles/mL, if the sensitivity of detection after each treatment is 10^2 particles/mL, the analytical technique could only show a removal of 10^4 particles/mL. Hence to achieve evidence of 12 logs of clearance ($10^6 - 10^{-6}$), three different mechanisms of analysis would need to be used assuming each gives 4 logs of clearance (20). It is not valid to use the same approach, ie, the same step repeated three times, to achieve the 12 log reduction. Total clearance, based on the product of three separation steps ($10^4 \times 10^4 \times 10^4 = 10^{12}$), requires that the three steps be totally independent or orthogonal.

Another example of virus clearance is for IgM human antibodies derived from human B lymphocyte cell lines where the steps are precipitation, size exclusion using nucleases, and anion-exchange chromatography (24). A second sequence consists of cation-exchange, hydroxylapatite, and immunoaffinity chromatographies. Each three-step sequence utilizes steps based on different properties. The first sequence employs solubility, size, and anion selectivity; the second sequence is based on cation selectivity, adsorption, and selective recognition based on an anti-u chain IgG (24).

Purification of Human Cell-Line t-PA.   The sequence of steps making up the initial and final purifications of recombinant t-PA from CHO cells is proprietary. A detailed experimental protocol for t-PA derived from normal, nonrecombinant human cultured cells (ATCC CRL-1459, American Type Culture Collection, Rockville, Maryland), however, is available (16). This provides insights into the types of chromatography steps which might be employed for purification of rt-PA. Human cell t-PA also contains urokinase plasminogen activator (u-PA) which, except for a single glycosylation (at Asn-302), is structurally similar to t-PA and tends to co-purify. The sequence in Figure 3 shows the steps for fractionating the two proteins. The yield is only 20 mg from 1400 L, illustrating the critical role of a recombinant cell line in obtaining both high yields and higher selectivity in producing a specific type of protein.

Because the culture media contained both t-PA and u-PA, this separation required several extra affinity chromatography steps, as well as dialysis/buffer exchange between the different chromatography columns (see Fig. 3). The salts and buffers added during the purification sequence must also be removed from the product at various points, adding significant complexity to the purification sequence. Desalting the buffer exchange constitute significant separation steps in the production of almost all biotechnology protein products.

Adsorption of t-PA to process equipment surfaces consisting of either stainless steel or glass was minimized by adding the detergent polyoxyethylene

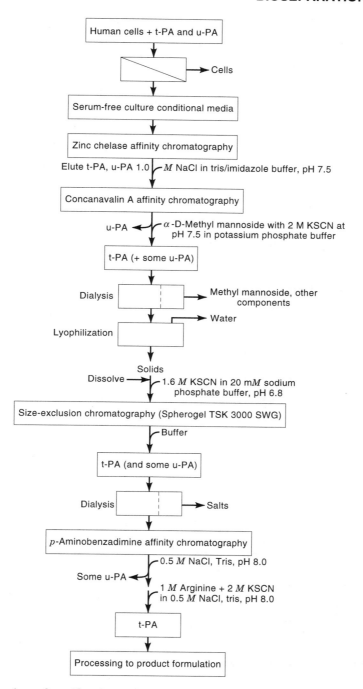

Fig. 3. Overview of purification sequence for the nonrecombinant tissue plasminogen activator (t-PA) which also contains urokinase plasminogen activator (u-PA). Serum-free culture conditional media is from normal human cell line. The temperature for all steps, except for size-exclusion chromatography (22°C), was 4°C. Adapted from Ref. 16.

sorbitan monooleate (Tween 80) to the serum-free culture conditioned media at 0.01% (vol/vol). The equipment was also rinsed, before use, with phosphate buffered saline (PBS) containing 0.01% Tween 80. Hydrophilic, plastic equipment was used whenever possible. All buffers were sterile filtered. Sterile filtration of liquids and gases is usually carried out using 0.2 or 0.45 μm filters.

4. Manufacture of Biologics and Government Regulation

The difference between biologics and drugs is not only a matter of definition, it is also a process design issue. To compensate for the incomplete analytical capability to define biologics, regulatory agencies include parameters of the process used to make biologics in the control and monitoring. Changes in the process may yield a different product from that previously reviewed and approved. A different product requires a new license (15). Thus substantial barriers exist in terms of effort, money, and time to making significant changes in processes used to produce licensed biologies. Process changes are to be expected during the investigational new drugs (IND) phase and before the license is approved, but significant changes are rarely made after licensing. The time which can elapse between conception of an idea for a process change and granting of a new license can be as much as two years and cost several million dollars.

The definition of biologics versus drugs continues to evolve. Assignment is made on a case by case basis (25). Section 351 of the Public Health Service Act defines a biologic product as "any virus, therapeutic serum, toxin, antitoxin, vaccine, blood, blood component or derivative, allergenic product, or analogous product ... applicable to the prevention, treatment, or cure of diseases or injuries in man." Biologics are subject to licensing provisions that require that both the manufacturing facility and the product be approved. All licensed products are subject to specific requirements for lot release by the FDA. In comparison, drugs are approved under section 505 of the FD & C act (21 USC 301–392), where there is not lot release by the FDA except for insulin products. Insulin, growth hormone, and many other hormones have been treated as drugs, whereas erythropoietin (EPO), which also fulfills the criteria of a hormone, was reviewed in the biologic division of the FDA. Insulin is derived from a bacterial fermentation; EPO is obtained from mammalian cell culture. Hormones, for the most part, are expected to be reviewed as drugs.

The design of bioseparation unit operations is influenced by these governmental regulations. The constraints on process development grow as a recovery and purification scheme undergo licensing for commercial manufacture.

5. Protein Chromatography

Proteins and nucleotides are macromolecular biomolecules. Mixtures of biomolecules are fractionated based on differences in charge; molecular weight, shape, and size; solubility in organic solvents; surface hydrophobic character; and types of active sites using ion-exchange, size-exclusion (gel-permeation), and reversed-phase, hydrophobic interaction (surface hydrophobicity), and affinity chromatographies, respectively. The appropriate separation may be selected from these five basic classes of chromatography. More than 30% of the purification steps for laboratory-scale protein purification procedures use ion-exchange and/or gel filtration, and at least 20% use affinity chromatography (23). This pattern is likely to be consistent with industrial practice. The following represent some chromatography column options for biopharmaceutical proteins (10). Sepharose, Sephadex, and Sephacryl resins are supplied by Pharmacia; Spherodex, Spherosil, and Trisacryl resins are supplied by Sepracor, Inc.; Toyopearl resins are supplied by TosoHaas; Fractogel resins are supplied by E. Merck Separations; Bakerbond resins and silicas are supplied by J. T. Baker. For adsorption, silica may be used.

Ion exchange	*Gel permeation*	*Hydrophobicity*
DEAE Sepharose Fast Flow LC	Agarose, 16%	Toyopearl Bulyl-650 M
DEAE Sepharose Fast Flow HC	Sephadex G25 Medium	Bulyl Spherodex M
DEAE Spherodex M	Sephadex G75	Toyopearl Ether-650 M
DEAE Spherosil M	Trisacryl Plus GF 03 M	Octyl Spherodex M
DEAE Trisacryl Plus M	Trisacryl Plus GF 10 M	Phenyl Spherodex M
Toyopearl DEAE-650 (M)	Trisacryl Plus GF 20 M	Toyopearl Phenyl-650 M
Fractogel EMD DMAE-650 (M)	Sephacryl S-100HR	Bakerbond Hi-Propyl
Fractogel EMD DEAE-650 (M)	Sephacryl S-200HR	
Q Sepharose Fast Flow	Toyopearl HW-50F	
QMA Spherodex M	Toyopearl HW-55F	
QMA Spherosil M		
		Affinity
QMA Trisacryl Plus M		
Toyopearl QAE-550 C		Blue Sepharose CL-6B
SP Sepharose Fast Flow		Red Sepharose CL-6B
SP Sepharose High Performance		Blue Spherodex M
SP Sepharose Big Bead		Baseline Blue Trisacryl M
SP Trisacryl Plus M		Heparin Sepharose CL-6B
Toyopearl SP-650 M		Heparin Spherodex M
		Toyopearl AF-Chelate-650 M
Fractogel EMD SO_3-650 M		(Copper)
		Toyopearl AF-Chelate-650 M
		(Zinc)

Reversed-phase chromatography is widely used as an analytical tool for protein chromatography, but it is not as commonly found on a process scale for protein purification because the solvents which make up the mobile phase, ie, acetonitrile, isopropanol, methanol, and ethanol, reversibly or irreversibly denature proteins. Hydrophobic interaction chromatography appears to be the least common process chromatography tool, possibly owing to the relatively high costs of the salts used to make up the mobile phases.

5.1. Liquid Chromatographs. The basic equipment for liquid chromatography is shown in Figure 4. The column is packed with an adsorbent, ie, the stationary phase. The mixture to be separated is pushed through the column by the eluent or mobile phase (26). Isocratic chromatography, carried out at a constant flow rate, buffer composition, and temperature, is usually associated with size-exclusion separations. Gradient chromatography typically uses a constant flow rate and temperature, but the composition of the element is altered by

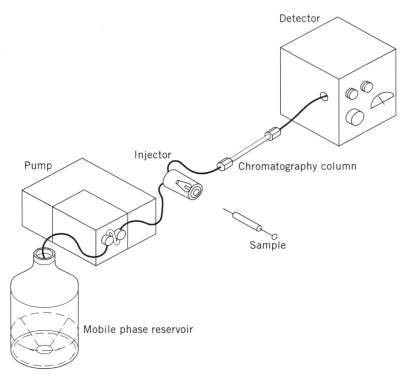

Fig. 4.   Schematic of a process liquid chromatography system (16).

mixing two or more buffer reservoirs to achieve a steadily changing salt concentration or changes in pH. The gradients formed are reported in terms of concentration at the inlet of the chromatography column; protein is detected at the column outlet. A chromatogram of the type illustrated in Figure 5 results.

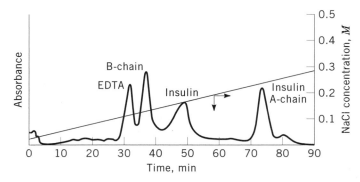

Fig. 5.   Anion-exchange separation of insulin, and insulin A- and B-chains, over diethylaminoethyl (DEAE) in a 10.9×200 mm column having a volume of 18.7 mL. Sample volume is 0.5 mL and protein concentration in 16.7 mM Tris buffer at pH 7.3 is 1 mg/mL for each component in the presence of EDTA. Eluent (also 16.7 mM Tris buffer, pH 7.3) flow rate is 1.27 mL/min, and protein detection is by uv absorbance at 280 nm. The straight line depicts the salt gradient. Courtesy of the American Chemical Society (48).

The concentration profile of the gradient at the outlet of the column can be significantly different from the profile at the inlet when a component (referred to as a modulator) in the eluting buffer adsorbs onto the stationary phase in a non-ideal manner causing the gradient to deform as it passes through the column. What appears to be anomalous peak behavior can result including self-concentration of a peak and the appearance of shoulders or multiple peaks for single sample components known to be homogeneous. This can occur for gradients used in reversed-phase, ion-exchange, or affinity chromatography (27–29).

5.2. Ion-Exchange Chromatography. Ion-exchange chromatography is initiated by eluting an injected sample through a column using a buffer but no NaCl or other displacing salt. The protein, which has charged sites spread over its surface, displaces anions or cations previously equilibrated on the stationary phase, ie, the protein sites exchange with the salt counterions associated with the ion-exchange stationary phase. A protein having a greater number and/or density of charged sites displaces or exchanges more ions and hence binds more strongly than a protein having a lower charge number or charge density.

Proteins deform and change shape in response to the environment. Hence, a protein left on the surface of an ion-exchange resin for a day or longer may slowly start to unfold exposing an increasing number of charged sites to bind with the ion-exchange resin. It is possible that this process can continue until the protein binds so strongly that it is impossible to desorb the protein without dissolving it, in NaOH, for example, and destroying it. To prevent such a situation, ion-exchange chromatography must be completed in a matter of hours or less.

After the column is loaded, proteins of similar size and shape are separated by differential desorption from the ion exchanger by using an increasing salt gradient of the mobile phase. The more weakly bound macromolecules elute first; the most tightly bound elute last, at the highest salt concentration. Figure 5 is an example of an anion-exchange separation (48). Prior to injection of the sample, the column was equilibrated with the 16.7 mM Tris buffer; the EDTA stabilized the solubility of the insulin sample injected onto the column. All of the proteins are initially retained on the anion-exchange stationary phase (DEAE 650 M) during loading of the sample onto the column. Subsequent application of the NaCl gradient, formed by the controlled mixing of buffers from two reservoirs of mobile phase, elutes the proteins. One reservoir contains only the 16.7 mM Tris buffer; the second contains 0.5 M NaCl in the same buffer. Following the elution of the last peak, the column may be flushed using a buffer at a high (2.5 M NaCl) salt concentration to verify that all proteins are desorbed. In some cases, a cleaning procedure is performed by passing methanolic NaOH through the column. The column is then re-equilibrated using the salt-free buffer, by pumping approximately 10 column volumes of the buffer through the stationary phase, or until the pH of the effluent and influent are the same to prepare the column for another injection.

Amphoteric Properties Determine Conditions for Ion-Exchange Chromatography. Proteins, amphoteric polymers of acidic, basic, and neutral amino acid residues, carry both negatively and positively charged groups on the surface, the ratio depending on pH (30). The isoelectric point (pI) is the pH at which a protein has an equal number of positive and negative charges. Proteins in solutions at a pH > pI have a net negative charge. Below the pI,

Table 2. **Ion-Exchange Groups Used in Protein Purification**[a]

Name	Abbreviation	Formula
	Weak anion	
aminoethyl	AE	$-C_2H_4NH_3^+$
diethylaminoethyl	DEAE	$-C_2H_4NH(C_2H_5)_2^+$
	Weak cation	
carboxy	C	$-COO^-$
carboxymethyl	CM	$-CH_2COO^-$
	Strong anion	
trimethylaminoethyl	TAM	$-CH_2N(CH)_3^+$
triethylaminoethyl	TEAE	$-C_2H_4N(C_2H_5)_3^+$
diethyl-2-hydroxypropyl-aminoethyl	QAE	$-C_2H_4N^+(C_2H_5)_2CH_2CH(OH)CH_3$
	Strong cation	
sulfo	S	$-SO_3^-$
sulfomethyl	SM	$-CH_2SO_3^-$
sulfopropyl	SP	$-C_3H_6SO_3^-$

[a] Courtesy of IRL Press (30).

proteins have a net positive charge. Many proteins have a pI < 7 and are processed using buffers having a pH of 7 to 8. Thus anion exchangers (positively charged stationary phases) are popular for protein chromatography. Ion-exchange matrices derivatized having negatively charged groups are cation exchangers. These bind positively charged proteins, ie, cations, when the mobile phase pH is $<$pI.

The selection of the pH of the buffer, as well as the type of ion-exchange (anion or cation) stationary phase is a function of the amphoteric nature of the protein and protein stability as a function of pH. For example, for a protein stable at pH > 6.5, having pI $= 5.5$, an anion exchanger is appropriate when the separation is run in a buffer of pH > 6.5. If this protein were stable at pH $= 5$, and the pI $= 5.5$, a cation exchanger and buffer of pH < 5.0 would be appropriate.

The ion-exchange (qv) groups used to derivatize stationary phases for the purification of proteins are summarized in Table 2. Corresponding buffers are given in Table 3. Strong anion and cation exchangers are almost fully ionized

Table 3. **Buffers for Ion-Exchange Chromatography**[a]

Buffer	pK	Buffering range
	Anion exchange	
L-histadine	6.15	5.5–6.0
imidazole	7.00	6.6–7.1
triethanolamine	7.77	7.3–7.7
Tris	8.16	7.5–8.0
diethanolamine	8.80	8.4–8.8
	Cation exchange	
acetic acid	4.76	4.8–5.2
citric acid	4.76	4.2–5.2
Mes	6.15	5.5–6.7
phosphate	7.20	6.7–7.6
hepes	7.55	7.6–8.2

[a] Courtesy of IRL Press (30).

at pH $= 3$–11 and coincide with the pH range of protein purification. Weak anion and cation exchangers have a narrower pH range over which they are ionized. Anion exchangers are preferred because desorption of the protein is more readily accomplished at lower salt concentrations.

Size Exclusion (Gel-Permeation) Chromatography. Size-exclusion chromatography is often referred to as gel-permeation chromatography because the stationary phases are usually made up of soft spherical particles which resemble gels. Separation occurs by a molecular sieving effect (see MOLECULAR SIEVES; SIZE SEPARATION). The larger molecules, which explore less of the intraparticle void fraction (ie, pores) than smaller molecules, elute first because the former spend less time inside the stationary phase than the latter. Separation can be achieved if the porosity of the stationary phase is properly selected and there is a significant difference in the sizes of the molecules to be separated. This size difference is measured in terms of hydrodynamic ratio. To select the stationary phase having the appropriate pore size requires that the size of the proteins be known.

The apparatus utilized to carry out size-exclusion (gel-permeation) chromatography is analogous to that used for isocratic operating conditions (see Fig. 4). The column is packed with a gel-filtration stationary phase, selected according to the molecular weight of the protein of interest (31). A variety of commercially available gel-filtration matrices facilitates separations ranging from molecular weights of 50 to 10^8 (Fig. 6). However, a single gel having a porosity which is capable of sieving molecules over the entire separation range does not exist.

An example of a size-exclusion chromatogram is given in Figure 7 for both a bench-scale (23.5 mL column) separation and a large-scale (86,000 mL column) run. The stationary phase is Sepharose CL-6B, a cross-linked agarose with a nominal molecular weight range of $\sim$5000–2×10^6 (see Fig. 6) (31).

Buffer Exchange and Desalting. A primary use of size-exclusion chromatography (sec) is for removal of salt or buffer from the protein, ie, desalting and buffer exchange (32). The difference in molecular weights is large; salts generally have a mol wt < 200, whereas mol wts of proteins are between 10,000 and 60,000.

Alternative methods of desalting and buffer exchange include continuous diafiltration, countercurrent dialysis (ccd), a membrane separation technique, and cross-flow filtration, which uses membranes (see MICROBIAL AND VIRAL FILTRATION). Both of those methods rely on filtration at a molecular scale, using membranes having porosities which reject proteins but allow passage of salts. Membrane methods are often preferred for an unspecified protein because these procedures are less costly and enable higher throughput than size-exclusion chromatography. Buffer exchange, used to remove denaturing agents in order to induce refolding of proteins, to remove buffers between purification steps, or to remove buffers and other reagents from the final product, is usually carried out at later steps in a recovery sequence (see Figs. 2 and 3). Equations for calculating separation efficiencies and recovery yields for all three methods for a specific case study using a recombinant protein of 160,000 mol wt are available (32). Size-exclusion chromatography using Sephadex G-25M gel-filtration media in this case had disadvantages compared to the membrane filtration techniques giving 100 $\times$ less complete ion removal, 130–1200% greater dilution, 30% higher cost, 66% higher (eluting) buffer requirement, 50% higher space on the plant

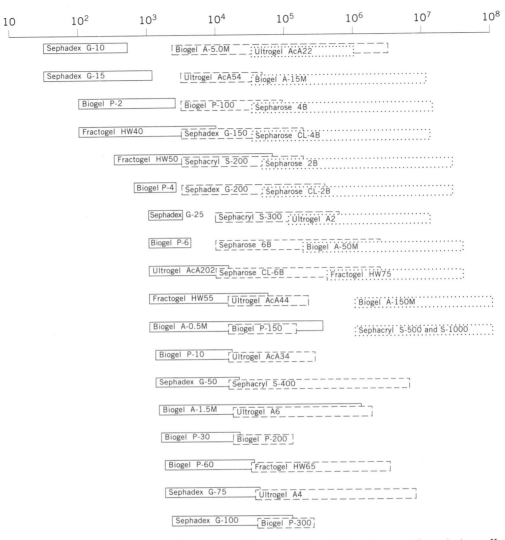

Fig. 6. Fractionation ranges of commercially available gel-filtration matrices: (□) small, (⌐⌐) medium, and (⋮⋮) large (31).

floor, 50% higher operating time, and half the throughput. However, cross- or tangential-flow filtration (tff) required up to 90 passes through a pump whereas sec and ccd were single pass. Other disadvantages of tff include frequent change-out of membranes and relatively large volumes of protein feed being required for laboratory-scale tests, a particular disadvantage for recombinant products. The more recent testing of a novel size-exclusion stationary phase, however, which facilitates rapid preparative (process-scale) separation of salts from protein in less than seven minutes, shows that process size-exclusion chromatography is capable of high throughput while reducing the volume needed to obtain proper refolding of the protein. Salts causing the protein to be denatured were rapidly

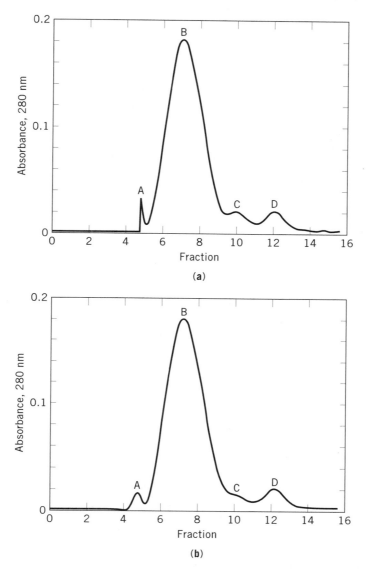

Fig. 7. Chromatograms of size-exclusion separation of IgM (mol wt = 800,000) from albumin (69,000) where A–D correspond to IgM aggregates, IgM, monomer units, and albumin, respectively, using (**a**) FPLC Superose 6 in a 1×30-cm long column, and (**b**) Sepharose CL-6B in a 37-cm column. Courtesy of American Chemical Society (24).

removed and reduced to a level where protein refolding occurred (33). The use of sec is likely to continue to be a widely practiced technology in industry. Rapid size-exclusion columns for the purpose of buffer exchange have been developed which enable desalting to be achieved at linear velocities of 500–600 cm/h (33), significantly increasing throughput and reducing operating time and plant floor space. Further, sec using gel-filtration media on cellulosic-based material has a special niche for the partial and controlled separation of denaturing salts from recombinant proteins for purposes of refolding. The development of such rapid desalting techniques is important because of the larger volumes of proteins needing to be processed in industry.

Column Size. Size-exclusion chromatography columns are generally the largest column on a process scale. Separation is based strictly on diffusion rates of the molecules inside the gel particles. No proteins or other solutes are adsorbed or otherwise retained owing to adsorption, thus, significant dilution of the sample of volume can occur, particularly for small sample volumes. The volumetric capacity of this type of chromatography is determined by the concentration of the proteins for a given volume of the feed placed on the column.

The volume of the solvent between the point of injection and the peak maximum of the eluting protein is defined as the elution volume, V_e (Fig. 8). The fluid volume between the particles of the stationary phase is the extraparticulate void volume or exclusion volume, V_o. The porosity of the stationary phase, determined by the extent of cross-linking of the polymers which make up the particles of a gel-permeation matrix, determines the extent to which a protein or other solute can explore the intraparticulate void volume. The higher the cross-linking, the smaller the effective pore size and the lower the molecular weight (or size) of the molecule which is excluded from the gel. Hence, the apparent porosities of a gel-permeation column are a function of the molecular probes used to measure it. For a large molecule that is completely excluded, the void volume is equivalent to the extraparticulate void volume, V_o. Molecules that are small enough to penetrate the gel have an elution volume $> V_o$. A small molecule, such as a salt, can potentially explore almost all of the bed volume and has the following elution volume:

$$V_e = V_o + K_d V_s \tag{2}$$

where K_d represents the fraction of the volume of the mobile phase inside the particle, V_s, which can be explored by the molecular probe. For a probe small enough to explore all of the intraparticle void volume, K_d is 1 and the elution volume is $V_o + V_s$. Because the combined volume of the fluid between the particles, V_o, and inside the particles, V_s, cannot exceed the total volumes of the column, V_t, V_e must be less than column volume V_t. All components injected into a size-exclusion column thus elute in one column volume.

The total stationary-phase volume required to process a given feed stream is proportional to the inlet concentration and volume of the feed. For example, for a typical inlet concentration of protein of 10 g/L, in a 100 L volume of feed, a column volume of at least 100 L is needed for size-exclusion chromatography. In comparison, an ion-exchange column having an adsorption capacity of 50 g/L would only require 20 L of column volume for the same feed.

Elution and Sample Volumes. The elution volume, V_e, is measured from the beginning of the injection to the center of the peak maximum for a Gaussian peak, if the sample volume is negligible relative to the elution volume (see Fig. 8**a**). A negligible injection volume is defined as being $\leq 2\%$ of the elution volume. The elution volume for samples larger than this are measured from the halfway point of the volume injected to the center of the eluting peak (see Fig. 8**b**). In samples which are so large that a plateau region is obtained having

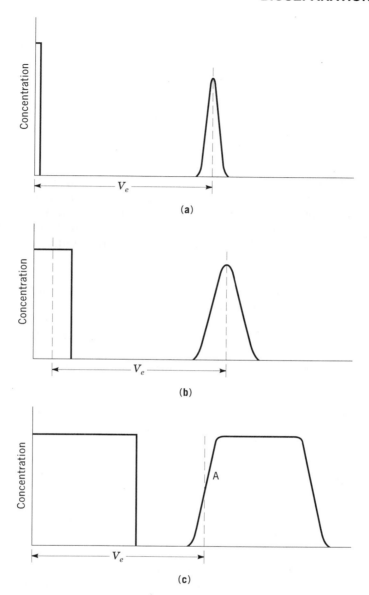

Fig. 8. Representation of measurement of elution volume, V_e, as a function of sample volume: (**a**) <2% of bed volume, (**b**) >2% and (**c**) >2% and giving a plateau region which has the same concentration as the injected sample; A represents the inflection point. See text. Courtesy of Pharmacia (34).

the same concentration as the sample (see Fig. 8**c**), the elution volume is measured from the start of the sample injection to the inflection point of the peak.

Sample dilution in gel permeation can be 10-fold or more for small volumes. Hence, proper representation of the inlet concentration (eg, see Fig. 8**a**), would require that the injection pulse be much higher because the area under the eluting peak and under the injected sample should be the same. Similarly, dilution of

Table 4. **Properties of Standard Proteins**[a]

Protein	Mol wt	pI	Radius, nm	Asymmetry, f/fo[b]
bovine serum albumin	66,000	5.74	3.50	1.29
ovalbumin	45,000	5.08	2.78	1.16
α-lactalbumin	14,200	4.57	2.30	1.18
myoglobin	16,900	7.10	2.40	1.18

[a] Adapted from Ref. 53.
[b] Frictional coefficients of f, solvated protein, and fo, nonsolvated sphere.

the injected sample would occur in the cases represented by Figures 8**b** and 8**c**, although the difference in heights between injected pulse and the maximum of the eluting peak would be smaller, because diffusion of the solute away from the center of mass occurs at the leading (left side) and tailing (right side) edges of the peak. If the peak is broad enough, ie, the sample volume is large enough, the solute at the center of the peak is not diluted owing to diffusion and the peak maximum is therefore equal to that of the injected sample.

Distribution Coefficients. Gel-permeation stationary-phase chromatography normally exhibits symmetrical (Gaussian) peaks because the partitioning of the solute between mobile and stationary phases is linear. Criteria more sophisticated than those represented in Figure 8 are seldom used (34).

The elution volume, V_e, and therefore the partition coefficient, K_D, is a function of the size of solute molecule, ie, hydrodynamic radius, and the porosity characteristics of the size-exclusion media. A protein of higher molecular weight is not necessarily larger than one of lower molecular weight. The hydrodynamic radii can be similar, as shown in Table 4 for ovalbumin and α-lactalbumin. The molecular weights of these proteins differ by 317%; their radii differ by only 121% (53).

Some types of size-exclusion phases, based on silica or macroporous polymeric materials, having rigid pores and defined pore size distributions. The dominant types of gels used in industry are dextran cross-linked epichlorophydran (Sephadex), agarose prepared from agar (Sepharose), or allyl dextran cross-linked with *N,N'*-methylenebisacrylamide (Sephacryl). These materials imbibe significant quantities of water and have bed volumes ranging from 2 to 3 mL/g dry weight of stationary phase for Sephadex G-10 (nominal molecular weight cutoff of 10,000) to 20–25 mL/g dry weight of stationary-phase Sephadex G-200 (nominal molecular weight cutoff of 200,000). Their structures resemble a cross-linked spider web, where the extent of cross-linking or association between hydrated polymer chains, rather than specific pore sizes, determine the apparent pore size distribution. The hydrophilic character of the polymers which make up these gels require cross-linking to prevent dissolution. The hydrophilic character is compatible with the majority of industrially relevant proteins, most of which can be denatured by hydrophobic surfaces but preserved in active confirmation at hydrophilic conditions. This property can be offset by poor flow properties, however, particularly for lightly cross-linked gels, because these gels are soft and have a tendency to compress when flow rates exceed a threshold which decreases with decreasing extents of cross-linking. Hence, Sephadex G-10 has the highest cross-linking and flow stability, and the lowest specific bed volume,

Table 5. **Comparison of Gel-Permeation Stationary Phase**[a]

Sephadex G-X[b]	Specific volume water mL/g dry gel	Permeability, K_o	Operating pressure[c], kPa[d]	Flow rate[c] water, mL/(cm² · h)
10	2–3	19	f	f
15	2.5–3.5	18	f	f
25	4–6	9–290[e]	f	f
50	9–11	13.5–400[e]	f	f
75	12–15		160	77
100	15–20		96	50
150	20–30		36	23
200	30–40		16	12

[a] Adapted from Ref. 34.
[b] Corresponds to the nominal cutoff value for wt $\times$ 10³, eg, G-10 has a mol wt cutoff value of $\sim$10,000.
[c] Value is maximum unless otherwise noted.
[d] To convert kPa to cm H_2O, multiply by 10.2.
[e] Depends on particle size (dp); as dp increases, K_o increases.
[f] May be calculated using Darcy's law: $U = K_o (\Delta P/L)$, where U is linear flow as mL/(cm²·h), L is bed length in cm, ΔP is pressure drop over gel bed in kPad, and the maximum pressure is 30.4 kPad (310 cm H_2O).

but also has the lowest effective pore size or porosity, limiting its sieving capabilities. Sephadex G-200 has the lowest cross-linking and flow stability and the highest specific bed volume and effective pore size. Sephadex 200 is useful for separating high molecular weight proteins, but at relatively low flow rates (Table 5).

The distribution coefficient, K_d, represents the fractional volume of a specific stationary phase explored by a given solute, represented by equation 3:

$$K_d = \frac{V_e - V_o}{V_s} \qquad (3)$$

where V_o is the void volume, V_s is the volume of the solvent (usually acqueous buffer) inside the gel which is available to very small molecules, and V_e is the elution volume of a small volume of injected molecular probe. The measurement of V_s is difficult, requiring use of an ion or small molecule which freely diffuses into all of the fluid volume inside the gel particles and then is readily detected at the outlet of the column. D_2O and radioactive ^{23}Na have been used. The latter is detected by a refractive index detector. An indirect measurement of V_s is more convenient and adequate. The column void volume (Fig. 9) may be measured using a soluble, high molecular weight target molecule which, because it does not explore any of the internal fluid volume of the stationary phase, it only distributed in the mobile phase. Blue dextran, a water-soluble, sulfonated, blue-colored dextran having mol wt $> 669,000$, manufactured by Pharmacia, and DNA (Type III from salmon tests) have been employed (26). The total column volume, V_t, can be calculated from the dimensions of the bed, although the direct measurement of column volume using water displacement before packing is more accurate. (It should be noted that total column volume is also represented by V_c in the recent literature). The difference, $V_t - V_o$, is then taken as an approximation of V_s. On this basis, K_{av}, the fraction of stationary phase volume

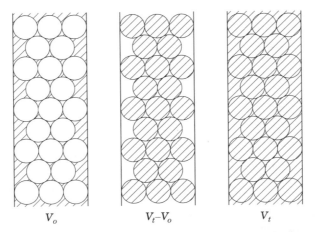

Fig. 9. Diagrammatic representation of V_t and V_o. $V_t - V_o$ includes the volume of the solid material forming the matrix of each bead. Courtesy of Pharmacia (34).

available for a given solute species, is defined as in equation 4:

$$K_{\mathrm{av}} = \frac{V_e - V_o}{V_t - V_o} \qquad (4)$$

The constant K_{av}, is not a true partition coefficient because of difference, $V_t - V_o$, includes the solids and the fluid associated with the gel or stationary phase. By definition, V_s represents only the fluid inside the stationary-phase particles and does not include the volume occupied by the solids which make up the gel. Thus K_{av} is a property of the gel, and like K_d it defines solute behavior independently of the bed dimensions. The ratio of K_{av} to K_d should be a constant for a given gel packed in a specific column (34).

Selectivity curves result from measured values of K_{av} plotted vs log (mol wt) enabling molecular weight determination of globular proteins having similar asymmetry factors. A sphere has an asymmetry factor of 1; an ellipsoid has a factor >1 (51). Such curves are linear (Fig. 10), the intercept increasing with increasing porosity (decreasing cross-linking). Extrapolation of these curves through the x-axis yields the molecular weights of probes which should be completely excluded from the gels because these target molecules are larger than the largest pores. Theoretically, the K_{av} for a given molecular probe should have a value between 0 and 1. A completely excluded molecule has $K_{\mathrm{av}} = 0$; molecules able to completely explore the fluid inside the stationary-phase particle have $K_{\mathrm{av}} = 1$. If K_{av} is less than zero, then channeling owing to a poorly packed bed is a probably cause. If the K_{av} is greater than 1, an interaction (adsorption) of the molecular probe with the stationary phase is a likely explanation.

Gel-permeation media are extremely versatile and may be used for separation of particles such as viruses (Fig. 11) as well as proteins (34). Separations of proteins and other particles having sizes equivalent to a molecular weight of 40×10^6 are possible using the agar-based Sepharose-type gel. This particular

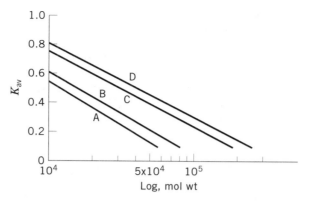

Fig. 10. Selectivity curves A–D for Sephadex G-75, G-100, G-150, and G-200, respectively, for globular proteins. Courtesy of Pharmacia (34).

gel has a limited temperature range for operation, however. It melts upon heating to 40°C (34).

The gels having a larger extent of cross-linking, particularly Sephadex G-10, G-15, and G-25, may retain through weak adsorption some types of aromatic molecules, and consequently impart reversed-phase properties to the gel. This may be the result of the weakly hydrophobic character of the cross-linking agents used in the synthesis of these gels. A recently introduced product, based on cross-linked agarose combined with crosslinked dextran (Superdex), exhibits low non-specific interactions as well as enhanced resolution (34). An excellent overview of gel permeation chromatography is given in Reference 51.

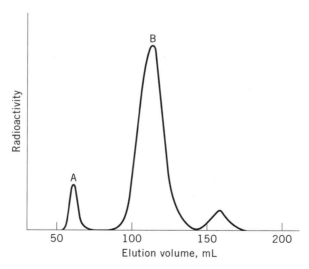

Fig. 11. Separation of ^{32}P-labeled A, adenovirus Type 5, and B, poliovirus Type 1 on Sepharose 2B where the column is 2.1 × 56 cm; eluent, 0.002 M sodium phosphate, pH 7.2, containing 0.15 M NaCl; and flow rate, 2 mL/(cm$^2 \cdot$ h). Courtesy of Pharmacia (34).

Reversed-Phase Chromatography. Stationary phases for reversed-phase chromatography consist principally of silica particles, silica supports having a hydrocarbon bonded phase, or polymeric materials based on either vinyl or styrene–divinylbenzene copolymers (35–41). Mobile phases commonly used in reversed-phase chromatography are aqueous methanol, 2-propanol, and acetonitrile. These solvents are often mixed with acidic buffers containing small amounts of acids such as trifluoroacetic acid or hexafluorobutyric acid. These acids reduce the pH of the mobile phase to 3 and give sharper peaks by suppressing ionization of silanol groups of silica-based reversed-phase supports, and minimizing ionic effects (42).

The most prevalent type of stationary phase is made of silica or another type of inorganic support derivatized and bonded with an octadecyl (C_{18}) or octyl (C_8) coating. Nonderivatized silicas are sometimes utilized for process chromatography. Much like ion-exchange chromatography, the organic component, sometimes referred to as a modifier, is mixed with water or buffer to form an increasing gradient of the modifier. This gradient serves to elute the components, which are initially adsorbed onto the stationary phase in order of increasing hydrophobic character. Methanol is used as the modifier for eluting weakly adsorbed, hydrophilic peptides; 2-propanol is used to elute strongly adsorbed, hydrophobic peptides. Acetonitrile is widely used for separations of proteins and many other types of molecules because this solvent exhibits favorable mass-transfer properties, lower viscosity (and back-pressure) than the other solvents, and good eluting strength. Methanol has a higher aqueous heat of mixing than the other solvents, and this can lead to solvent degassing and bubble formation in the column. Bubbles interfere both with the operation of the column, causing peak dispersion, and detection of the peaks at the column outlet. Bubbles give anomalous peaks in spectrophotometric detectors and refractive index detection (42).

Product Monitoring and Peptide Mapping. Reversed-phase chromatography is widely used for analysis of proteins. Historically, the principal use of reversed-phase chromatography has been in the analysis and process separations of peptides, amino acids, and organic compounds which are characterized by lower molecular weight and solubility in acetonitrile or alcohol gradients. These mobile phases denature proteins and some polypeptides. Hence, reversed-phase chromatography is infrequently used for process-scale purification of proteins. Rather the excellent protein resolving power of this type of chromatography is employed on an analytical scale using columns packed with 2–10 mL of stationary phases having 1–5 µm particle sizes and for sample volumes which typically range from 1 to 10 µL.

One purpose in monitoring a protein product is to detect the presence of a change in which as little as one amino acid has been chemically or biologically altered or replaced during the manufacturing process. Variant amino acid(s) in a protein may not affect protein retention during reversed-phase chromatography if the three-dimensional structure of the polypeptide shields the variant residue from the surface of the reversed-phase support (20). Reversed-phase chromatography discriminates between different molecules on the basis of hydrophobicity. Because large proteins may contain only small patches of hydrophobic residues, these patches may not correlate to the molecular modifications which a reversed-phase analytical method seeks to detect. The reversed-phase method must therefore be completely validated, and preferably combined with controlled chemical and/or proteolytic hydrolysis followed by chromatography or electrophoresis (see ELECTROSEPARATIONS) of the cleared protein to give a map of the resulting peptide fragments (20,43).

A peptide map is generated by cleaving a previously purified protein using chemicals or enzymes. Hydrolytic agents having known specificity are used to perform limited proteolysis followed by resolution and identification of all the peptide fragments formed. Identification of changes, and reconstruction of the protein's primary structure, is then possible. Reagents and enzymes which cleave specific bonds are discussed in the literature (44).

An example of a peptide map generated by trypsin hydrolysis of recombinant tissue plasminogen activator (rt-PA) is shown in Figure 12. The chromatogram shows the resolving power or reversed-phase high performance liquid chromatography in separating peptides obtained from t-PA in which the disulfide bonds had been reduced and alkylated prior to enzyme hydrolysis. The small peptides formed have little or no three-dimensional structure. Hence, measurable shifts in elution profiles occur when there are variant amino acids because a single amino acid change in a peptide has a larger effect on its solubility and retention than the same change has in a protein. The replacement of arginine at position 275 in a normal t-PA molecule with glutamic acid results in a significant peak shift (see Fig. 12) (43), showing how tryptic mapping can be a suitable method for monitoring lot-to-lot consistency of this particular recombinant product (20).

Reversed-phase high performance liquid chromatography has come into use for estimating the purity of proteins and peptides as well. However, before employed, a high performance liquid chromatographic (hplc) profile of a given protein must be completely validated (43).

Insulin Purification. An example of the purification of recombinant product by reversed-phase chromatography is recombinant insulin, a polypeptide hormone. Insulin consists of 51 amino acid residues in two chains and is relatively small. Reversed-phase chromatography is used after most of the other impurities have been removed by a prior ion-exchange step (see Fig. 1) (12). The method utilizes a process-grade C_8 reversed-phase support (Zorbex) having a particle size of 10 μm (14). Partially purified insulin crystals, dissolved in a water-rich mobile phase, are applied to the column and then eluted in a linear gradient generated by mixing 0.25 M aqueous acetic acid to 60% acetonitrile. The acidic mobile phase gives excellent resolution of insulin from structurally similar insulin-like components. The ideal pH is from 3.0 to 4.0, below insulin's isoelectric point, pI = 5.4. Under mildly acidic conditions insulin may deamidate to monodesamido insulin, but if the reversed-phase separation is done within a matter of hours, the deamidation can be minimized.

This reversed-phase chromatography method was successfully used in a production-scale system to purify recombinant insulin. The insulin purified by reversed-phase chromatography has a biological potency equal to that obtained from a conventional system employing ion-exchange and size-exclusion chromatographies (14). The reversed-phase separation was, however, followed by a size-exclusion step to remove the acetonitrile eluent from the final product (12,14).

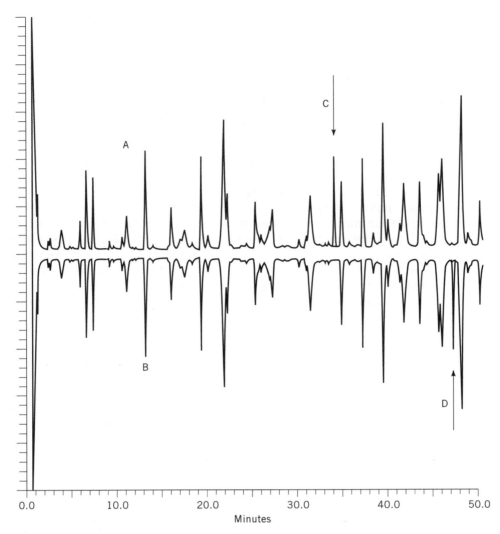

Fig. 12. Tryptic map of rt-PA (mol wt = 66,000) showing peptides formed from hydrolysis of reduced, alkylated rt-PA. Separation by reversed-phase octadecyl (C_{18}) column using aqueous acetonitrile with an added acidic agent to the mobile phase. Arrows show the difference between A, normal, and B, mutant rt-PA where the glutamic acid residue, D, has replaced the normal arginine residue, C, at position 275. Courtesy of Marcel Dekker (43).

Whereas recombinant proteins produced as inclusion bodies in bacterial fermentations may be amenable to reversed-phase chromatography (42), the use of reversed-phase process chromatography does not appear to be widespread for higher molecular weight proteins.

Reversed-Phase Process Chromatography. Polypeptides, peptides, antibiotics, alkaloids, and other low molecular weight compounds are amenable to process chromatography by reversed-phase methods. There are numerous examples of bioproducts which have been purified using reversed-phase chromatography. The manufacture of salmon calcitonin, a 32-residue peptide used for treatment of post-menopausal osteoporosis, hypercalcemia, and Paget's disease of the bone, includes reversed-phase chromatography. This peptide, commercially prepared on a kilogram scale by a solid-phase synthesis, is then purified by a multimodal purification train. Reversed-phase chromatography is the dominant technique used by Rhône-Poulenc Rorer (45).

Another example is the purification of a β-lactam antibiotic, where process-scale reversed-phase separations began to be used around 1983 when suitable, high pressure process-scale equipment became available. A reversed-phase microparticulate (55–105 μm particle size) C_{18} silica column, with a mobile phase of aqueous methanol having 0.1 M ammonium phosphate at pH 5.3, was able to fractionate out impurities not readily removed by liquid–liquid extraction (37). Optimization of the separation resulted in recovery of product at 93% purity and 95% yield. This type of separation differs markedly from protein purification in feed concentration ($\approx$50–200 g/L for cefonicid vs 1 to 10 g/L for protein), molecular weight of impurities (<5000 compared to 10,000–100,000 for proteins), and throughputs ($\approx$1–2 mg/(g stationary phase·min) compared to 0.01–0.1 mg/(g·min) for proteins).

Reversed-phase separation was also found to purify diastereomer precursors used in the chemical synthesis of the insect sex phermone of *Limantria dispar*, a pest which attacks oak trees. The liquid chromatography columns tested had dimensions of up to 15 cm id by 130 cm long, and were able to purify up to 708 g of starting material in 4.1 L sample using a column having 23 L of stationary phase. The throughput is estimated to have been on the order of 0.2–0.4 mg/(g·min), where separation was obtained using a gradient of hexane and diethyl ether.

Small Particle Silica Columns. Process-scale reversed-phase supports can have particle sizes as small as 5–25 μm. Unlike polymeric reversed-phase sorbents, these small-particle silica-based reversed-phase supports require high pressure equipment to be properly packed and operated. The introduction of axial compression columns has helped promote the use of high performance silica supports on a process scale. Resolution approaching that of an analytical-scale separation can be achieved using these columns that can also be quickly packed. These columns consist of a plunger fitted into a stainless steel column. The particles are placed into the column in a slurry. The plunger then squeezes or compacts the bed in an axial direction to give a stable, tightly packed bed. This type of column must be operated at pressures of up to 10 MPa (100 bar), but also gives excellent resolution in run times of an hour or less (36).

Hydrophobic Interaction Chromatography. Hydrophobic interactions of solutes with a stationary phase result in their adsorption on neutral or mildly hydrophobic stationary phases. The solutes are adsorbed at a high salt concentration, and then desorbed in order of increasing surface hydrophobicity, in a decreasing kosmotrope gradient. This characteristic follows the order of the lyotropic series for the anions: phosphates > sulfates > acetates > chlorides > nitrates > thiocyanates. Anions which precipitate proteins less effectively than chloride (nitrates and thiocyanates) are chaotropes or water structure breakers, and have a randomizing effect on water's structure; the anions preceding chlorides, ie, phosphates, sulfates, and acetates, are polar kosmotropes or water structure makers. These promote precipitation of proteins. Kosmotropes also promote adsorption of proteins and other solutes onto a hydrophobic stationary phase (46). These kosmotropes have other beneficial characteristics which include increasing the thermal stability of enzymes, decreasing enzyme inactivation, protecting against proteolysis, increasing the association of protein subunits, and increasing the refolding rate of denatured proteins. Hence, utilization of hydrophobic interaction chromotography is attractive for purification of proteins where recovery of a purified protein in an active and stable conformation is desired (46,47).

   Salt Effects.   The definition of a capacity factor k' in hydrophobic interaction chromatography is analogous to the distribution coefficient, K_{av}, in gel permeation chromatography:

$$k' = \frac{V_e - V_o}{V_o} \tag{5}$$

However, because protein retention owing to adsorption can occur, the value of k' can be greater than one, ie, elution of the most retained peak need not occur after one column volume of mobile phase has passed through the column. The retention behavior of lysozyme on a polymeric hydrophobic interaction support follows the preferential interaction parameter of the lyotropic series of anions (Fig. 13). The preferential interaction parameter is a measure of the net salt inclusion or exclusion of the hydration layer. The higher the value, the larger the disrupting effect of the salt. This analysis led to derivation and experimental validation of the capacity factor for lysozyme with respect the lyotropic number of the anion for a hydrophilic vinyl polymer support having an average particle diameter of 30 μm, and average pore size of 100 nm. This capacity factor has the following

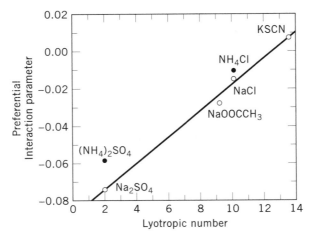

Fig. 13.   Preferential interaction parameter vs lyotropic number for lysozyme on (○) bovine serum albumin and (●) Toyopearl. Courtesy of American Chemical Society (47).

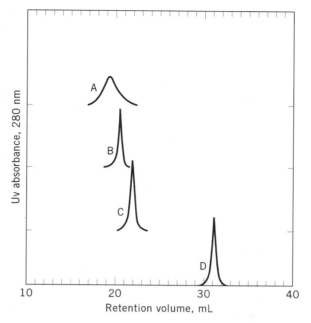

Fig. 14. Chromatographic retention of 20 μL of a 3 mg/mL solution of lysozyme on Toyopearl HW-65S using a 50 cm × 8 min ID column in 1.3 M ammonium salt, 20 mM Tris mobile phase at 1 mL/min for A, NH$_4$I; B, NH$_4$Cl; C, NH$_4$OOCCH$_3$; and D, (NH$_4$)$_2$-SO$_4$. Courtesy of American Chemical Society (47).

form:

$$K' = a[C]^d \, [N_x - b] + h \tag{6}$$

where a, b, d, and h are protein specific parameters, N_x is the lyotropic number, and C is salt concentration in M. Hydrophobic interaction parameters can then be estimated for experimental peak retention data as changes in retention time upon a change in salt or salt concentration. An example of how salt type and concentration affect retention of lysozyme is illustrated in Figure 14. A similar functional relation was found for myoglobin with respect to a hydrophilic vinyl polymer derivatized using butyl groups (47).

Various types of proteins have been purified using hydrophobic interaction chromatography including alkaline phophatase, estrogen receptors, isolectins, strepavidin, calmodulin, epoxide hydrolase, proteoglycans, hemoglobins, and snake venom toxins (46). In the case of cobra venom toxins, the order of elution of the six cardiotoxins supports the hypothesis that the mechanism of action is related to hydrophobic interactions with the phospholipids in the membrane.

The recovery of recombinant chymosin from a yeast fermentation broth showed that large-scale hydrophobic interaction chromatography could produce an acceptable product in one step. Chymosin, which used to be obtained from the lining of the stomachs of calves, is used in cheesemaking, and its cost is an issue. Because the capacity of the hydrophobic interaction stationary-phase is limited, an alternative method has been developed in which the enzyme is extracted into a two-phase polyethylene glycol (PEG) salt system. The partition for the chymosin into PEG coefficient is 100, and hence enables efficient recovery of this in one step. Together with a subsequent ion-exchange step, this method gives a suitably purified chymosin. The use of hydrophobic interaction chromatography may have helped to indicate that two-phase extraction is a viable approach (10).

5.3. Affinity Chromatography. The concept of affinity chromatography, credited to the discovery of biospecific adsorption in 1910, was reintroduced as a means to purify enzymes in 1968 (49). Substrates and substrate inhibitors diffuse into the active sites of enzymes irreversibly or reversibly binding there. Conversely, if the substrate or substrate inhibitor is immobilized through a covalent bond to a solid particle of stationary phase having large pores, the enzyme should be able to diffuse into the stationary phase and bind with the substrate or inhibitor. Because the substrate is small (mol wt < 500) and the enzyme large ($>15,000$), the diffusion of the enzyme to its binding partner at a solid surface can be sterically hindered. The placement of the substrate at the end of an alkyl or glycol chain tethered to the stationary phase's surface reduces hindrance and forms the basis of affinity chromatography. This concept has also been applied to ion-exchange chromatography under the names of tentacle or fimbriated stationary phases.

The realization that enzymes could be selectively retained in a chromatography column packed with particles of immobilized substrates or substrate analogues led to experiments with other pairs of binding partners. Numerous applications of affinity chromatography developed, given the specific and reversible yet strong affinity of biological macromolecules for numerous specific ligands or effectors. These interactions have been exploited for purposes of highly selective, but often expensive protein purifications, recovery of messenger ribonucleic acid (mRNA) in some recombinant DNA applications, and study of mechanisms of protein binding with effector molecules (49).

Minimization of Nonspecific Binding. The purpose of affinity chromatography is the highly selective adsorption and subsequent recovery of the target biomolecule. Loss of specificity occurs when macromolecules, other than the targeted materials, adsorb onto the stationary phase owing to hydrophobic or ionic interactions. For example, a spacer arm, which allows the binding ligand on the column to be located away from the matrix surface, can improve accessibility and reduce steric hindrance to the immobilized ligand, often decreasing selectivity. Hexamethylenediamine, a common spacer arm used initially in affinity chromatography, has been the source of strong, nonspecific binding. This hydrophobic character has been decreased by interposing an ether or secondary amine, such as 3,3-diaminodipropylamine, in the middle of the spacer arm.

The ideal matrices for anchoring binding ligands are nonionic, hydrophilic, chemically stable, and physically robust. The most popular matrices are polysaccharide based, principly owing to their hydrophilic character and history of use as size-exclusion or gel-permeation gels (see Fig. 6), although glass beads, polyacrylamide gels, cross-linked dextrans (Sephadex), and agarose synthesized into a bead form have all been used (49). In particular, agarose such as Biogel A (Bio-Rad), Ultragel A (IBF), and Sepharose (Pharmacia) are popular (49, 50). Cross-linked agaroses (Sepharose CL and Sepharose FF by Pharmacia) are physically more stable than Sepharose and are suitable for attaching affinity ligands. Both forms of the agaroses have an open porosity which allows proteins to readily diffuse inside. Affinity chromatography results are usually reported in terms of specific activity of the final product (activity per mg of material) and the amount of biomolecule recovered (% yield).

Activation of the Stationary-Phase Surface. Activation of polysaccharide, silica, or polyacrylamide stationary phases involve the formation of a reactive intermediate, covalently attached to the surface, to which a difunctional alkyl-, aryl-, or glycol spacer is subsequently joined. The other end of the spacer is subsequently reacted with the ligand. Cyanogen bromide, CNBr, has been widely used to activate agarose and dextran gels (49). The attachment of ligands, and sometimes activation of supports, is generally carried out in the laboratories of the chromatography process developers because fully prepared affinity stationary phases are not as widely available as stationary phases for the other types of chromatography.

Multistep Processes. An excellent synopsis and industrial viewpoint of affinity chromatography and its fit with other bioseparations unit operations is available (49, 50). Ligands range from the low molecular weight components, eg, arginine and benzamidine, which both bind trypsin-like proteases, triazine dyes, and metal chelates; to high molecular weight ligands, eg, protein A, immunoglobins, and monoclonal antibodies. The blood factor VIII, purified by monoclonal affinity techniques, was approved by the U.S. FDA in 1988. Limitations of affinity chromatography as an industrial separation technique can be due to leaching of bound ligands from the column into the product at ppm levels, nonspecific interactions resulting in contamination of the target molecule, and failure of the affinity ligand to differentiate all variant forms of a protein or polypeptide (52). For example, polyclonal antibodies do not distinguish desamido-insulin, which contains a deamidated asparagine, from insulin. Because many antibody preparations cannot differentiate between minor structural changes in proteins, affinity chromatography must be followed by other separation steps, and does not provide a one-step purification.

Receptor Affinity Chromatography. Receptor affinity chromatography is a selective form of immunoaffinity chromatography which is based on antigen-antibody interactions (52). Protein or polypeptide ligands used in preparing receptor affinity supports are themselves products of fermentation of recombinant microorganisms and are subjected to a separate sequence of purification steps, prior to being reacted with a functionalized stationary phase to form the affinity support. The resulting affinity chromatography columns are expensive when viewed on the basis of cost of support/unit volume of stationary phase. The cost/benefit ratio would still be attractive because process-scale columns can be small (volumes on the order of 1–10 L). Moreover, as with other types of affinity chromatography, purification of dilute but highly active protein is possible.

BIBLIOGRAPHY

"Bioseparations," in *ECT* 4th ed., Suppl. Vol., pp. 89–122, by Michael R. Ladisch, Purdue University.

CITED PUBLICATIONS

1. R. C. Willson and M. R. Ladisch, in M. R. Ladisch, R. C. Willson, C-C. Painton, and S. E. Builder, eds., ACS Symp. Ser. 427, 1–13 (1990).
2. Committee on Bioprocess Engineering, National Research Council, Putting Biotechnology to Work: Bioprocess Engineering, National Academy of Sciences, Washington, D.C., 1992, 2–22.
3. S. M. Wheelwright, Protein Purification: Design and Scale-up of Downstream Processing, Hanser Publishers, Munich, Germany, 1991, 1–9, 61, 213–217.
4. C. A. Bisbee, GEN 13(14), 8–9 (1993).
5. A. M. Thayer, C&EN 74(33), 13–21 (1996).
6. A. M. Thayer, C&EN 74(1), 22–23 (1996).
7. R. Rawls, C&EN 74(32), 31 (1996).
8. J. A. Wells, Science 273(5274), 449–450 (1996).
9. W. Roush, Science 273(5273), 300–301 (1996).
10. V. B. Lawlis and H. Heinsohn, LC-GC 11(10), 720–729 (1993).
11. M. Bernon and J. Bodelle, in Y-Y. H. Chien and J. L. Gueriguian, eds., Drug Biotechnol. Reg. 13, xv–xxiii (1991).
12. M. R. Ladisch and K. L. Kohlmann, Biotechnol. Prog. 8(6), 469–478 (1992).
13. W. F. Prouty, in Ref. 11, 221–262.
14. E. P. Kroeff, R. A. Owens, E. L. Campbell, R. D. Johnson, and H. I. Marks, J. Chromatog. 461, 45–61 (1989).
15. S. E. Builder, R. van Reis, N. Paoni, and J. Ogez, "Process Development and Regulatory Approval of Tissue-Type Plasminogen Activator," in Proceedings of the 8th International Biotechnology Symposium, Paris, July 17–22, 1989.
16. N. K. Harakas, J. P. Schaumann, B. T. Connolly, A. J. Wittwer, J. V. Orlander, and J. Feder, Biotechnol. Prog. 4(3), 149–158 (1988).
17. S. E. Builder and E. Grossbard, Transfus. Med. Rec. Technol. Adv., 303–313 (1986).
18. M. Barinaga, Science 272(5262), 664–666 (1996).
19. J. O'C. Hamilton, Business Week, 3478, 118, 122 (1996).
20. S. E. Builder and W. S. Hancock, Chem. Eng. Prog. 84(8), 42–46 (1988).
21. J. D. Watson, M. Gilman, J. Witkowski, and M. Zoller, Recombinant DNA, 2nd ed., W. H. Freeman and Co., New York, 1992, 458–460.
22. B. Alberts, D. Bray, J. Lewis, M. Raff, K. Roberts, and J. D. Watson, Molecular Biology of the Cell, 2nd ed., Garland Publishing, New York, 1989, 1203–1212.
23. S. V. Ho, in Ref. 1, 14–34.
24. G. B. Dove, G. Mitra, G. Roldan, M. A. Shearer, and M-S. Cho, in Ref. 1, 194–209.
25. S. Sobel, in Ref. 11, 499–511.
26. J. K. Lin, B. J. Jacobsen, A. N. Pereira, and M. R. Ladisch, in W. A. Wood and S. T. Kellog, eds., Methods in Enzymology, Vol. 160, Academic Press, Inc., San Diego, Calif., 1988, 145–159.
27. A. Velayudhan and M. R. Ladisch, Anal. Chem. 63(18), 2028–2032 (1991).
28. A. Velayudhan, R. L. Hendrickson, and M. R. Ladisch, AIChE J. 41(5), 1184–1193 (1995).
29. A. Velayudhan and M. R. Ladisch, Ind. Eng. Chem. Res. 34(8), 2805–2810 (1995).
30. S. Roe, in E. L. V. Harris and S. Angal, eds., Protein Purification Methods— A Practical Approach, IRL Press, Oxford, U.K., 1989, 200–216.
31. A. Z. Preneta, in Ref. 30, 293–306.
32. R. T. Kurnik, A. W. Yu, G. S. Blank, A. R. Burton, D. Smith, A. M. Athalye, and R. van Reis, Biotechnol. Bioeng. 45, 149–157 (1995).

33. K. H. Hamaker, J. Liu, R. J. Seely, C. M. Ladisch, and M. R. Ladisch, Biotechnol. Prog. 12, 184–189 (1996).

34. Gel Filtration—Theory and Practice, Pharmacia Fine Chemicals, Uppsala, Sweden, 1979, 1–19, 30–35, 46, 50, and 57; Gel Filtration—Principles and Methods, 6th ed., Pharmacia Biotech, 1997.

35. P. C. Sadek, P. W. Carr, R. M. Doherty, M. J. Kamlet, R. W. Taft, and M. H. Abraham, Anal. Chem. 57, 2971–2978 (1985).

36. H. Colin, P. Hilaireau, and J. De Tournemire, LC-GC 8(4), 302–312 (1990).

37. A. M. Cantwell, R. Calderone, and M. Sienko, J. Chromatog. 316, 133–149 (1984).

38. R. L. Gustafson, R. L. Albright, I. Heisler, J. A. Lirio, and O. T. Reid, Ind. Eng. Chem. Prod. Res. Dev. 7(2), 107–115 (1968).

39. D. J. Pietrzyk and J. D. Stodola, Anal. Chem. 53(12), 1822–1828 (1981).

40. D. J. Pietrzyk and C-H. Chu, Anal. Chem. 49(6), 757–764 (1977).

41. J. Morris and J. S. Fritz, LC-GC 11(7), 513–517 (1993).

42. G. K. Sofer and L. E. Nystrom, Process Chromatography: A Practical Guide, Academic Press, Ltd., London, 1989, 128–129.

43. R. L. Garnick, M. J. Ross, and R. A. Baffi, in Ref. 11, 263–313.

44. E. A. Carrey, in T. E. Creighton, ed., Peptide Mapping in Protein Structure: A Practical Approach, IRL Press, Oxford, U.K., 1990, 117–144.

45. E. Flanigan, High Performance Liquid Chromatography in the Production and Quality Control of Salmon Calcitonin, Purdue University, West Lafayette, Ind., 1991, p. 207.

46. B. F. Roettger and M. R. Ladisch, Biotechnol. Adv. 7, 15–29 (1989).

47. B. R. Roettger, J. A. Myers, M. R. Ladisch, and F. E. Regnier, in Ref. 1, 80–92.

48. M. R. Ladisch, R. L. Hendrickson, and K. L. Kohlmann, in Ref. 1, 93–103.

49. I. Parikh and P. Cuatrecases, CEN 63, 17–32 (1985).

50. J. A. Asenjo and I. Patrick, in E. L. V. Harris and S. Angal, eds., Protein Purification Applications—A Practical Approach, IRL Press, Oxford, U.K., 1990, 1–28.

51. L. Hagel, "Gel Filtration" in J-C. Janson and L. Ryden, ed., Protein Purification: Principles, High Resolution Methods, and Applications, VCH Publishers, N.Y., 1989, 63–106.

52. P. Bailon and D. V. Weber, Nature 335(6193), 839–840 (1988).

53. S. K. Basak and M. R. Ladisch, Anal. Biochem. 226, 51–58 (1995).

MICHAEL R. LADISCH
Purdue University